Electronic Configurations of the Elements

Element	K	L	M	4s	4p	4d	4f	5s	5p	5d	5f	5g	6s	6p	6d	6f	6g	6h	7
55. Cs	2	8	18	2	6	10		2	6				1						
56. Ba	2	8	18	2	6	10		2	6				2						
57. La	2	8	18	2	6	10		2	6	1			2						
58. Ce	2	8	18	2	6	10	2	2	6				2						
59. Pr	2	8	18	2	6	10	3	2	6				2						
60. Nd	2	8	18	2	6	10	4	2	6				2						
61. Pm	2	8	18	2	6	10	5	2	6				2						
62. Sm	2	8	18	2	6	10	6	2	6				2						
63. Eu	2	8	18	2	6	10	7	2	6				2						
64. Gd	2	8	18	2	6	10	7	2	6	1			2						
65. Tb	2	8	18	2	6	10	9	2	6				2						
66. Dy	2	8	18	2	6	10	10	2	6				2						
67. Ho	2	8	18	2	6	10	11	2	6				2						
68. Er	2	8	18	2	6	10	12	2	6				2						
69. Tm	2	8	18	2	6	10	13	2	6				2						
70. Yb	2	8	18	2	6	10	14	2	6				2						
71. Lu	2	8	18	2	6	10	14	2	6	1			2						
72. Hf	2	8	18	2	6	10	14	2	6	2			2						
73. Ta	2	8	18	2	6	10	14	2	6	3			2						
74. W	2	8	18	2	6	10	14	2	6	4			2						
75. Re	2	8	18	2	6	10	14	2	6	5			2						
76. Os	2	8	18	2	6	10	14	2	6	6			2						
77. Ir	2	8	18	2	6	10	14	2	6	7			2						
78. Pt	2	8	18	2	6	10	14	2	6	9			1						
79. Au	2	8	18	2	6	10	14	2	6	10			1						
80. Hg	2	8	18	2	6	10	14	2	6	10			2						
81. Tl	2	8	18	2	6	10	14	2	6	10			2	1					
82. Pb	2	8	18	2	6	10	14	2	6	10			2	2					
83. Bi	2	8	18	2	6	10	14	2	6	10			2	3					
84. Po	2	8	18	2	6	10	14	2	6	10			2	4					
85. At	2	8	18	2	6	10	14	2	6	10			2	5					
86. Rn	2	8	18	2	6	10	14	2	6	10			2	6					
87. Fr	2	8	18	2	6	10	14	2	6	10			2	6					1
88. Ra	2	8	18	2	6	10	14	2	6	10			2	6					2
89. Ac	2	8	18	2	6	10	14	2	6	10			2	6	1				2
90. Th	2	8	18	2	6	10	14	2	6	10			2	6	2				2
91. Pa	2	8	18	2	6	10	14	2	6	10	2		2	6	1				2
92. U	2	8	18	2	6	10	14	2	6	10	3		2	6	1				2
93. Np	2	8	18	2	6	10	14	2	6	10	5		2	6					2
94. Pu	2	8	18	2	6	10	14	2	6	10	6		2	6					2
95. Am	2	8	18	2	6	10	14	2	6	10	7		2	6					2
96. Cm	2	8	18	2	6	10	14	2	6	10	7		2	6	1				2
97. Bk	2	8	18	2	6	10	14	2	6	10	8		2	6	1				2
98. Cf	2	8	18	2	6	10	14	2	6	10	10		2	6					2
99. Es	2	8	18	2	6	10	14	2	6	10	11		2	6					2
100. Fm	2	8	18	2	6	10	14	2	6	10	12		2	6					2
101. Md	2	8	18	2	6	10	14	2	6	10	13		2	6					2
102. No	2	8	18	2	6	10	14	2	6	10	14		2	6					2
103. Lr	2	8	18	2	6	10	14	2	6	10	14		2	6	1				2

To
Mike
with love
cuddles & millions
of kisses & hugs &
Bullwethers
Joy xx

ADVANCED INORGANIC CHEMISTRY

A Comprehensive Text

THIRD EDITION

ADVANCED INORGANIC CHEMISTRY

A Comprehensive Text

F. ALBERT COTTON

ROBERT A. WELCH PROFESSOR OF CHEMISTRY
TEXAS A AND M UNIVERSITY
COLLEGE STATION, TEXAS, USA

and

GEOFFREY WILKINSON, F.R.S.

PROFESSOR OF INORGANIC CHEMISTRY
IMPERIAL COLLEGE OF SCIENCE AND TECHNOLOGY
UNIVERSITY OF LONDON, ENGLAND

Third Edition, completely revised from the original literature

INTERSCIENCE PUBLISHERS
A DIVISION OF JOHN WILEY & SONS
NEW YORK · LONDON · SYDNEY · TORONTO

Library of Congress Cataloging in Publication Data

Cotton, Frank Albert, 1930–
    Advanced inorganic chemistry. Third edition

    1. Chemistry, Inorganic.   I. Wilkinson, Geoffrey,
1921–   joint author.   II. Title.
QD151.2.068   1972   546   70–161693
ISBN 0-471-17560-9

Printed in the United States of America

10 9 8 7 6 5 4 3 2 1

In Memoriam

to

Ronald Sydney Nyholm, F.R.S.

29.1.17–4.12.71

whose teaching and research have contributed much to the Renaissance of Inorganic Chemistry

Preface to the Third Edition

Since the second edition of this text, the literature in inorganic chemistry has continued to grow at an extremely rapid rate. The work has been characterised by increasing sophistication in the use of physical methods as well as in concepts and insights. Although these developments have posed serious problems, we have maintained the same basic approach with the object of providing the student with a background sufficient for the comprehension of current research literature in the field.

We have attempted to include factual material appearing up to around mid-1971 and a new set of references covers progress since the second edition. Since this book is intended primarily as a student textbook, the citations are not exhaustive and do not impute priority or originality being intended solely as a guide to the literature.

In order to accommodate new material, several changes have been made. The first four chapters have been modified so as to eliminate the more elementary aspects of atomic structure and give more coverage of symmetry and molecular structure. Various rearrangements of chapters and of material within sections have been made. One new chapter, on selected aspects of homogeneous catalysis by transition metal organometallic compounds has been added while some information on the biochemistry of iron, copper, cobalt, zinc and molybdenum is now provided.

We thank all those who have offered comments on the previous editions and suggestions for corrections or improvements to this edition would be welcome.

F. A. COTTON
Cambridge, Massachusetts

G. WILKINSON
London, England

Preface to the Second Edition

Although the basic structure of the text is unaltered, we have rearranged several sections and have brought up to date essentially all of the factual material. The vast amount of recent literature has meant an increase in the size of the book, but it is intended to be a *teaching* text and not a reference book and it is our view that it is better to have too much material on hand rather than too little, since sections can always be omitted.

In response to numerous requests, we have improved on the handling of documentation of which there are three levels. First, for the great majority of long known and well established facts and theories, no explicit reference is given since such material can be readily located through standard reference texts and treatises, listed at the end of the text.

Secondly, some material not so available appears in review articles and monographs; a pertinent list is provided at the end of each chapter.

Finally, we have introduced as footnotes in each chapter, some original research references. These cover broadly the period from January 1962 to August 1965 and are intended primarily for teachers and research workers as guide references to recent work.

We take this opportunity to thank all those who gave us their comments on the first edition.

F. A. COTTON
Cambridge, Massachusetts

G. WILKINSON
London, England

Preface to the First Edition

It is now a truism that, in recent years, inorganic chemistry has experienced an impressive renaissance. Academic and industrial research in inorganic chemistry is flourishing, and the output of research papers and reviews is growing exponentially.

In spite of this interest, however, there has been no comprehensive text-book on inorganic chemistry at an advanced level incorporating the many new chemical developments, particularly the more recent theoretical advances in the interpretation of bonding and reactivity in inorganic compounds. It is the aim of this book, which is based on courses given by the authors over the past five to ten years, to fill this need. It is our hope that it will provide a sound basis in contemporary inorganic chemistry for the new generation of students and will stimulate their interest in a field in which trained personnel are still exceedingly scarce in both academic and industrial laboratories.

The content of this book, which encompasses the chemistry of all of the chemical elements and their compounds, including interpretative discussion in the light of the latest advances in structural chemistry, general valence theory, and, particularly, ligand field theory, provides a reasonable achievement for students at the B.Sc. honors level in British universities and at the senior year or first year graduate level in American universities. Our experience is that a course of about eighty lectures is desirable as a guide to the study of this material.

We are indebted to several of our colleagues, who have read sections of the manuscript, for their suggestions and criticism. It is, of course, the authors alone who are responsible for any errors or omissions in the final draft. We also thank the various authors and editors who have so kindly given us permission to reproduce diagrams from their papers: specific acknowledgements are made in the text. We sincerely appreciate the secretarial assistance of Miss C. M. Ross and Mrs. A. B. Blake in the preparation of the manuscript.

F. A. COTTON
Cambridge, Massachusetts

G. WILKINSON
London, England

Contents

PART ONE

General Theory

PART TWO

Chemistry of Nontransition Elements

PART THREE

Chemistry of the Transition Elements

Abbreviations

1. Chemicals, Ligands, Radicals, etc.

acacH	acetylacetone
am	ammonia (or occasionally an amine)
Ar	aryl or arene (ArH)
aq	aquated, H_2O
bcc	body centered cubic
bu	butyl (prefix *n*, *i* or *t*, normal, iso or tertiary butyl)
bz	benzene
ccp	cubic close packed
Cp	cyclopentadienyl, C_5H_5
diars	*o*-phenylenebisdimethylarsine, $o\text{-}C_6H_4(AsMe_2)_2$
diglyme	diethyleneglycoldimethylether, $CH_3O(CH_2CH_2O)_2CH_3$
dipy	2,2′-dipyridine
DMF	*N,N′*-dimethylformamide, $HCONMe_2$
$DMGH_2$	dimethylglyoxime
DMSO	dimethylsulfoxide, Me_2SO
$EDTAH_4$	ethylenediaminetetraacetic acid
$EDTAH_{4-n}^{n-}$	anions of $EDTAH_4$
en	ethylenediamine, $H_2NCH_2CH_2NH_2$
Et	ethyl
glyme	ethyleneglycoldimethylether, $CH_3OCH_2CH_2OCH_3$
hcp	hexagonal close packed
L	ligand
M	central (usually metal) atom in compound
Me	methyl
Me_6 tren	tris-(2-dimethylaminoethyl)amine, $N(CH_2CH_2NMe_2)_3$
$NTAH_3$	nitrilotriacetic acid, $N(CH_2COOH)_3$
ox	oxalate ion, $C_2O_4^{2-}$

Ph	phenyl, C_6H_5
phen	1,10-phenanthroline
PNP	Bis-(2-diphenylphosphinoethyl)amine, $HN(CH_2CH_2PPh_2)_2$
pn	propylenediamine (1,2-diaminopropane)
Pr	propyl (prefix *i* for isopropyl)
QAS	Tris-(2-diphenylarsinophenyl)arsine, $As(o\text{-}C_6H_4AsPh_2)_3$
QP	Tris-(2-diphenylphosphinophenyl)phosphine, $P(o\text{-}C_6H_4PPh_2)_3$
R	alkyl or aryl group
Salen	bis-salicylaldehydeethylenediimine
TAN	Tris-(2-diphenylarsinoethyl)amine, $N(CH_2CH_2AsPh_2)_3$
TAP	Tris-(3-dimethylarsinopropyl)phosphine, $P(CH_2CH_2CH_2AsMe_2)_3$
TAS	Bis-(3-dimethylarsinopropyl)methylarsine, $MeAs(CH_2CH_2CH_2AsMe_2)_2$
THF	tetrahydrofuran
TMED	N,N,N',N'-tetramethylethylenediamine
tn	1,3-diaminopropane (trimethylenediamine)
TPN	Tris-(2-diphenylphosphinoethyl)amine, $N(CH_2CH_2PPh_2)_3$
trien	Triethylenetetraamine, $(CH_2NHCH_2CH_2NH_2)_2$
tren	Tris-(2-aminoethyl)amine, $N(CH_2CH_2NH_2)_3$
TSN	Tris-(2-methylthiomethyl)amine, $N(CH_2CH_2SMe)_3$
TSP	Tris-(2-methylthiophenyl)phosphine, $P(o\text{-}C_6H_4SMe)_3$
TSeP	Tris-(2-methylselenophenyl)phosphine, $P(o\text{-}C_6H_4SeMe)_3$
TTA	thenoyltrifluoroacetone, $C_4H_3SCOCH_2COCF_3$
X	halogen or pseudohalogen

2. Miscellaneous

Å	Angstrom unit, 10^{-10} m
asym	asymmetric or antisymmetric
B.M.	Bohr magneton
b.p.	boiling point
cm^{-1}	wave number
CFSE	crystal field stabilization energy
CFT	crystal field theory
d	decomposes
d-	dextorotatory

esr	electron spin resonance
eV	electron volt
(g)	gaseous state
h	Planck's constant
Hz	herz, sec^{-1}
ICCC	International Coordination Chemistry Conference
ir	infrared
IUPAC	International Union of Pure and Applied Chemistry
(1)	liquid state
l-	levorotatory
LCAO	linear combination of atomic orbitals
LFT	ligand field theory
m.p.	melting point
MO	molecular orbital
nmr	nuclear magnetic resonance
R	gas constant
(s)	solid state
spy	square pyramid(al)
str	vibrational stretching mode
sub	sublimes
sym.	symmetrical
tbp	trigonal bipyramid(al)
U	lattice energy
uv	ultraviolet
VB	valence bond
Z	atomic number
ε	molar extinction coefficient
ν	frequency (cm^{-1} or Hz)
μ	magnetic moment in Bohr magnetons
χ	magnetic susceptibility
θ	Weiss constant

Units and Conversion Factors

1. The SI Units

In recent years many scientific and engineering groups have recommended or adopted a set of units called the Système International d'Unités, the SI units. While these have not yet been universally accepted, even in pure science, there is a growing trend toward general acceptance. In this book some SI units have been adopted, while others have not. We summarize here the SI units and comment on choices made for use in this book.

The SI system is based on the following set of defined units:

Physical Quantity	Name of Unit	Symbol for Unit
length	meter	m
mass	kilogram	kg
time	second	s
electric current	ampere	A
temperature	kelvin	K
luminous intensity	candela	cd

Multiples and fractions of these are specified using the following prefixes:

Multiplier	Prefix	Symbol
10^{-1}	deci	d
10^{-2}	centi	c
10^{-3}	milli	m
10^{-6}	micro	μ
10^{-9}	nano	n
10^{-12}	pico	p
10	deka	da
10^{2}	hecto	h
10^{3}	kilo	k
10^{6}	mega	M
10^{9}	giga	G
10^{12}	tera	T

In addition to the defined units, the system includes a number of derived units, of which the following are the main ones

Physical quantity	SI unit	Unit symbol
Force	newton	$N = kg\, m\, s^{-2}$
Work, energy, quantity of heat	joule	$J = Nm$
Power	watt	$W = J s^{-1}$
Electric charge	coulomb	$C = As$
Electric potential	volt	$V = WA^{-1}$
Electric capacitance	farad	$F = As V^{-1}$
Electric resistance	ohm	$\Omega = VA^{-1}$
Frequency	hertz	$Hz = s^{-1}$
Magnetic flux	weber	$Wb = Vs$
Magnetic flux density	tesla	$T = Wb\, m^{-2}$
Inductance	henry	$H = Vs A^{-1}$

The *energy unit* immediately derived from the basis units is the *joule*, J, $kg\, m^2\, sec^{-2}$. We have adopted this unit, which is equal to $(4.184)^{-1}$ calories, as the unit of energy. In most places it is more convenient to use the kilojoule, kJ. In a few places, e.g., for ionization energies, the *electron volt*, eV, has been used, as before. One $eV = 96.5\, kJ\, mol^{-1}$. The eV is the energy acquired by an electron accelerated by a potential difference of one volt.

In common with many chemists and crystallographers, we have continued to use the Angstrom, Å, defined as 10^{-8} cm, as a unit of length on the atomic and molecular scale, although SI units are the nanometer, $nm(10^{-9}$ m) or picometer, pm $(10^{-12}$ m). The C—C bond length in diamond is 1.54 Å = 0.154 nm = 154 pm.

We have also retained other familiar units, such as dyne and atmosphere (pressure). This is in line with the objective of preparing the student to read the contemporary literature, which is still written almost exclusively in such units.

Signs. When energy is absorbed in a process, the energy of that process will be defined as positive.

For electrochemical processes we shall use the conventions recommended by the International Union of Pure and Applied Chemistry (IUPAC). Because these differ, often confusingly, from the Latimer or American convention still found in many publications in the USA, a detailed explanation of the International system is provided on pages 164–165.

2. Conversion Factors

Some useful relationships between different units are given below. The symbols N and J stand for Newton and Joule, the SI units of force and energy, respectively.

1 atm = 760 mm Hg (= torr) = 1.01325×10^6 dyne cm^{-2} = 101.325 Nm^{-2}

1 bar = 10^6 dyne cm^{-2} = 0.987 atm = 105 Nm^{-2}

1 Bohr Magneton = 9.273×10^{-24} Am^2 molecule^{-1} = 9.273×10^{-21} erg gauss^{-1}

1 calorie (thermochemical) = 4.184 J

1 coulomb = 0.10000 emu = 2.9979×10^9 esu = 1 amp sec

1 dyne = 10^{-5} N

1 erg = 2.3901×10^{-8} cal = 10^{-7} J

1 eV = 8066 cm^{-1} = 23.06 kcal/mole = 1.602×10^{-12} erg = 1.602×10^{-19} J

1 gauss = 10^{-4} T

1 kJ mole^{-1} = 83.54 cm^{-1}

1 mass unit, mu, = 931.5 Mev = 1.660×10^{-24} erg

Molar Magnetic Susceptibility (SI) = Molar Susceptibility (cgs) $\times 4\pi \times 10^{-6}$

$RT(T = 300 \text{K}) = 0.1425$ kJ mol^{-1} = 208.4 cm^{-1}

3. Fundamental Constants, etc.

Avogadro's number ($C^{12} = 12.0000...$), $N_A = 6.02252 \times 10^{23}$ mol^{-1}

Boltzmann's constant $k = 1.3805 \times 10^{-16}$ erg deg^{-1} = 1.3805×10^{-23} JK^{-1}

Bohr radius, $\alpha_0 = 0.52917 \times 10^{-10}$ m

Curie–Weiss Law: $\mu = 2.84 [\chi_M (T-\theta)]^{1/2} = 797.5 [\chi_M (T-\theta)]^{1/2}$ (SI)

Electron charge, $e = (4.8030 \pm 0.0001) \times 10^{-10}$ abs. esu = 1.602×10^{-19} C

Electron mass, $m = 9.1091 \times 10^{-31}$ kg = 0.00054860 mu = 0.5110 Mev

Faraday constant, $F = 96,487$, coulomb. g. equiv^{-1} = 9.6487×10^4 C mol^{-1}

Gas constant, $R = 1.9872$ defined cal deg^{-1} mol^{-1} = 8.3143 JK^{-1} mol^{-1} = 0.082057 liter atm deg^{-1} mol^{-1}

Ice point: 273.150 ± 0.01 K

Molar volume (ideal gas, 0°C, 1 atm) = 22.414×10^3 cm mol^{-1} = 2.241436×10^{-2} m^3 mol^{-1}

Planck's constant, $h = 6.6256 \times 10^{-27}$ erg. sec = 6.6256×10^{-34} Js

Proton mass, $M_p = 1.6725 \times 10^{-24}$ g

Velocity of light in vacuum = 2.99795×10^8 m sec^{-1}

$\pi = 3.14159$; e = 2.7183; ln 10 = 2.3026

S.T.P. = 1.013×10^5 Nm^{-2} and 273.15°K

ADVANCED INORGANIC CHEMISTRY

A Comprehensive Text

THIRD EDITION

PART ONE

General Theory

1
Symmetry and Structure

THE SYMMETRY GROUPS

Molecular symmetry and ways of specifying it with mathematical precision are important for several reasons. The most basic reason is that *all* molecular wave functions—those governing electron distribution as well as those for vibrations, nmr spectra, etc.—must conform, rigorously, to certain requirements based on the symmetry of the equilibrium nuclear framework of the molecule. When the symmetry is high these restrictions can be very severe. Thus, from a knowledge of symmetry alone it is often possible to reach useful qualitative conclusions about molecular electronic structure and to draw inferences from spectra as to molecular structures. The qualitative application of symmetry restrictions is most impressively illustrated by the crystal-field and ligand-field theories of the electronic structures of transition-metal complexes, as described in Chapter 20, and by numerous examples of the use of infrared and Raman spectra to deduce molecular symmetry. Illustrations of the latter occur throughout the book, but particularly with respect to some metal carbonyl compounds in Chapter 22.

A more mundane use for the concept and notation of molecular symmetry is in the precise description of a structure. One symbol, such as D_{4h}, can convey precise, unequivocal structural information which would require long verbal description to duplicate. Thus, if we say that the $Ni(CN)_4^{2-}$ ion has D_{4h} symmetry we imply that: (*a*) it is completely planar; (*b*) the Ni—C—N groups are all linear; (*c*) the C—Ni—C angles are all equal, at 90°; (*d*) the four CN groups are precisely equivalent to one another; and (*e*) the four Ni—C bonds are precisely equivalent to one another. The use of symmetry symbols has become increasingly common in the chemical literature, and it is now necessary to be familiar with the basic concepts and rules of notation in order to read many of the contemporary research papers in inorganic and, indeed, also organic chemistry with full comprehension. It thus seems appropriate to begin this book with a brief survey of molecular symmetry and the basic rules for specifying it.

1-1. Symmetry Operations and Elements

When we say that a molecule has *symmetry*, we mean that *certain parts of it can be interchanged with others without altering either the identity or the orientation of the molecule.* The interchangeable parts are said to be equivalent to one another by symmetry. Consider, for example, a trigonal-bipyramidal molecule such as PF_5 (1-I). The three equatorial P—F bonds, to

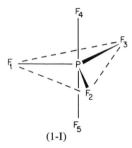

(1-I)

F_1, F_2, and F_3, are equivalent. They have the same length, the same strength, and the same type of spacial relation to the remainder of the molecule. Any permutation of these three bonds among themselves leads to a molecule indistinguishable from the original. Similarly, the axial P—F bonds, to F_4 and F_5, are equivalent. *But*, axial and equatorial bonds are different types (e.g., they have different lengths) and if one of each were to be interchanged the molecule would be noticeably perturbed. These statements are probably self-evident, or at least readily acceptable, on an intuitive basis, but for systematic and detailed consideration of symmetry certain formal tools are needed. The first set of tools is a set of *symmetry operations*.

Symmetry operations are geometrically defined ways of exchanging equivalent parts of a molecule. There are four kinds which are used conventionally and these are sufficient for all our purposes.

1. Simple rotation about an axis passing through the molecule by an angle $2\pi/n$. This operation is called a *proper rotation* and is symbolized $\mathbf{C}_n$. If it is repeated n times, of course the molecule comes all the way back to the original orientation.

2. Reflection of all atoms through a plane which passes through the molecule. This operation is called *reflection* and is symbolized $\mathbf{\sigma}$.

3. Reflection of all atoms through a point in the molecule. This operation is called *inversion* and is symbolized $\mathbf{i}$.

4. The combination, in either order, of rotating the molecule about an axis passing through it by $2\pi/n$ and reflecting all atoms through a plane which is perpendicular to the axis of rotation is called *improper rotation* and is symbolized $\mathbf{S}_n$.

These operations are *symmetry operations if, and only if*, the appearance of the molecule is *exactly* the same after one of them is carried out as it was before. For instance, consider rotation of the molecule H_2S by $2\pi/2$ about an axis passing through S and bisecting the line between the H atoms. As

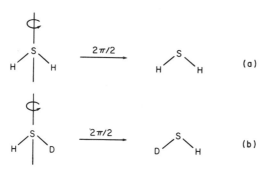

Fig. 1-1. Sketches showing that the operation C_2 carries H_2S into an orientation indistinguishable from the original, but HSD into an observably different orientation.

shown in Fig. 1-1, this operation interchanges the H atoms and interchanges the S—H bonds. Since these atoms and bonds are equivalent, there is no physical (i.e. physically meaningful or detectable) difference after the operation. For HSD, however, the corresponding operation replaces the S—H bond by the S—D bond, and *vice versa*, and one can see that a change has occurred. Therefore, for H_2S, the operation C_2 is a symmetry operation; for HSD it is not.

These types of symmetry operation are graphically explained by the diagrams in Fig. 1-2, where it is shown how an arbitrary point (0) in space is affected in each case. Filled dots represent points above the xy plane and open dots represent points below it. Let us examine first the action of proper rotations, illustrated here by the C_4 rotations, i.e. rotations by $2\pi/4 = 90°$. The operation C_4 is seen to take the point 0 to the point 1. The application of C_4 twice, designated C_4^2, generates point 2. C_4^3 gives point 3 and, of course, C_4^4 which is a rotation by $4 \times 2\pi/4 = 2\pi$ regenerates the original point. The set of four points, 0, 1, 2, 3 are permutable, cyclically, by repeated C_4 proper rotations and are equivalent points. It will be obvious that, in general, repetition of a C_n operation will generate a set of n equivalent points from an arbitrary initial point, provided that point lies off the axis of rotation.

The effect of reflection through symmetry planes perpendicular to the xy plane, specifically, σ_{xz} and σ_{yz} is also illustrated in Fig. 1-2. The point 0 is related to point 1 by the σ_{yz} operation and to the point 3 by the σ_{xz} operation. By reflecting either point 1 or point 3 through the second plane, point 2 is obtained.

The set of points generated by the repeated application of an improper rotation will vary in appearance depending on whether the order of the operation, S_n, is even or odd, order being the number n. A crown of n points, alternately up and down, is produced for n even, as illustrated for S_6. For n odd there is generated a set of $2n$ points which form a right n-sided prism, as shown for S_3.

Finally, the operation i is seen to generate from point 0 a second point, 1, lying on the opposite side of the origin.

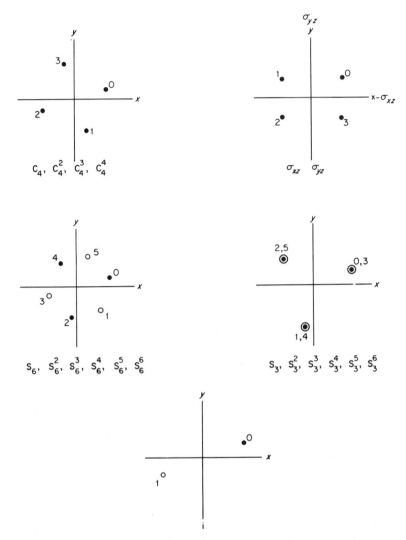

Fig. 1-2.   Diagrams showing the effects of symmetry operations on an arbitrary point, designated 0, thus generating sets of points.

Let us now illustrate the symmetry operations for various familiar molecules as examples. As this is done it will be convenient to employ also the concept of *symmetry elements*. A symmetry element is an *axis* (line), *plane* or *point* about which symmetry operations are performed. The existence of a certain symmetry operation implies the existence of a corresponding symmetry element, while, conversely, the presence of a symmetry element means that a certain symmetry operation or set of operations is possible.

Consider the ammonia molecule, Fig. 1-3. The three equivalent hydrogen atoms may be exchanged amongst themselves in two ways: (1) by proper

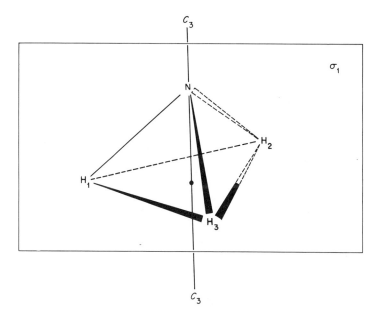

Fig. 1-3. The ammonia molecule, showing its 3-fold symmetry axis, C_3, and one of its three planes of symmetry, σ_1, which passes through H_1 and N and bisects the H_2—H_3 line.

rotations; (2) by reflections. The molecule has an axis of 3-fold proper rotation; this is called a C_3 axis. It passes through the N atom and through the center of the equilateral triangle defined by the H atoms. When the molecule is rotated by $2\pi/3$ in a clockwise direction H_1 replaces H_2, H_2 replaces H_3, and H_3 replaces H_1. Since the three H atoms are physically indistinguishable, the numbering having no physical reality, the molecule after rotation is indistinguishable from the molecule before rotation. This rotation, called a $\mathbf{C}_3$ or 3-fold proper rotation, is a symmetry operation. Rotation by $2 \times 2\pi/3$ also produces a configuration different, but physically indistinguishable, from the original and is likewise a symmetry operation; it is designated $\mathbf{C}_3^2$. Finally, rotation by $3 \times 2\pi/3$ carries each H atom all the way around and returns it to its initial position. This operation, $\mathbf{C}_3^3$, has the same net effect as performing no operation at all, but for mathematical reasons has to be considered as an operation generated by the C_3 axis. This, and other operations which have no net effect, are called *identity* operations and are symbolized by $\mathbf{E}$. Thus, we may write $\mathbf{C}_3^3 = \mathbf{E}$.

The interchange of hydrogen atoms in NH_3 by reflections may be carried out in three ways; that is, there are three planes of symmetry. Each plane passes through the N atom and one of the H atoms, and bisects the line connecting the other two H atoms. Reflection through the symmetry plane containing N and H_1 interchanges H_2 and H_3; the other two reflections interchange H_1 with H_3, and H_1 with H_2.

Inspection of the NH_3 molecule shows that no other symmetry operations besides these six (three rotations, C_3, C_3^2, $C_3^3 \equiv E$, and three reflections, σ_1, σ_2, σ_3) are possible. Put another way, the only symmetry elements the molecule possesses are C_3 and the three planes which we may designate σ_1, σ_2, and σ_3. Specifically, it will be obvious that no sort of improper rotation is possible, nor is there a center of symmetry.

As a more complex example, in which all four types of symmetry operations and elements are represented, let us take the $Re_2Cl_8^{2-}$ ion, which has the shape of a square parallelepiped or right square prism, Fig. 1-4. This ion has altogether six axes of proper rotation, of four different kinds. First, the Re_1–Re_2 line is an axis of 4-fold proper rotation, C_4, and four operations, C_4, C_4^2, C_4^3, $C_4^4 \equiv E$, may be carried out. This same line is also a C_2 axis, generating the operation C_2. It will be noted that the C_4^2 operation means rotation by $2 \times 2\pi/4$, which is equivalent to rotation by $2\pi/2$, i.e. to the C_2 operation. Thus the C_2 axis and the C_2 operation are implied by, and not independent of, the C_4 axis. There are, however, two other types of C_2 axis which exist independently. There are two of the type that passes through the centers of opposite vertical edges of the prism, C_2' axes, and two more which pass through the centers of opposite vertical faces of the prism, C_2'' axes.

The $Re_2Cl_8^{2-}$ ion has three different kinds of symmetry plane [see Fig. 1-4(b)]. There is a unique one which bisects the Re—Re bond and all the vertical edges of the prism. Since it is customary to define the direction of the highest proper axis of symmetry, C_4 in this case, as the vertical direction, this symmetry plane is horizontal and the subscript h is used to identify it, σ_h. There are then two types of vertical symmetry plane, namely, those two that contain opposite vertical edges, and two others that cut the centers of opposite vertical faces. One of these two sets may be designated $\sigma_v^{(1)}$ and $\sigma_v^{(2)}$, the v implying that they are vertical. Since those of the second vertical set bisect the dihedral angles between those of the first set, they are then designated $\sigma_d^{(1)}$ and $\sigma_d^{(2)}$, the d standing for dihedral. Both pairs of planes are vertical and it is actually arbitrary which are labelled σ_v and which σ_d.

Continuing with $Re_2Cl_8^{2-}$, we see that an axis of improper rotation is present. This is coincident with the C_4 axis and is an S_4 axis. The S_4 operation about this axis proceeds as follows. The rotational part, through an angle of $2\pi/4$, in the clockwise direction has the same effect as the C_4 operation. When this is coupled with a reflection in the horizontal plane, σ_h, the following shifts of atoms occur:

$$
\begin{array}{lll}
Re_1 \rightarrow Re_2 & Cl_1 \rightarrow Cl_6 & Cl_5 \rightarrow Cl_2 \\
Re_2 \rightarrow Re_1 & Cl_2 \rightarrow Cl_7 & Cl_6 \rightarrow Cl_3 \\
& Cl_3 \rightarrow Cl_8 & Cl_7 \rightarrow Cl_4 \\
& Cl_4 \rightarrow Cl_5 & Cl_8 \rightarrow Cl_1
\end{array}
$$

Finally, the $Re_2Cl_8^{2-}$ ion has a center of symmetry, i, and the inversion operation, i, can be performed.

In the case of $Re_2Cl_8^{2-}$ the improper axis, S_4, might be considered as merely the inevitable consequence of the existence of the C_4 axis and the σ_h, and, indeed, this is a perfectly correct way to look at it. However, it is impor-

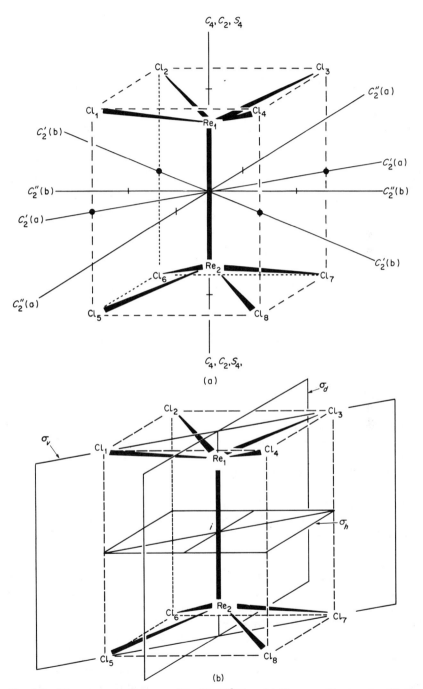

Fig. 1-4. The symmetry elements of the $Re_2Cl_8^{2-}$ ion. (a) The axes of symmetry. (b) One of each type of plane and the center of symmetry.

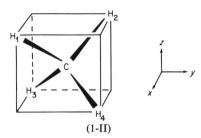

(1-II)

tant to emphasize that there are cases in which an improper axis, S_n, exists without independent existence of either C_n or σ_h. Consider, for example, a tetrahedral molecule as depicted in (1-II), where the methane molecule is shown inscribed in a cube and Cartesian axes, x, y, and z are indicated. Each of these axes is an S_4 axis. For example rotation by $2\pi/4$ about z followed by reflection in the xy plane shifts the H atoms as follows:

$$H_1 \rightarrow H_3 \qquad H_3 \rightarrow H_2$$
$$H_2 \rightarrow H_4 \qquad H_4 \rightarrow H_1$$

Note, however, that the Cartesian axes are not C_4 axes (though they are C_2 axes) and the principle planes, viz., xy, xz, yz, are not symmetry planes. Thus, we have here an example of the existence of the S_n axis without C_n or σ_h having any independent existence. The ethane molecule in its staggered configuration has an S_6 axis and provides another example.

1-2. Symmetry Groups

The complete set of symmetry operations that can be performed on a molecule is called the *symmetry group* for that molecule. The word group is used here, not as a mere synonym for "set" or "collection," but in a technical, mathematical sense, and this meaning must first be explained.

Introduction to Multiplying Symmetry Operations. We have already seen in passing that if a proper rotation, $\mathbf{C}_n$, and a horizontal reflection, $\boldsymbol{\sigma}_h$, can be performed, then there is also an operation that results from the combination of the two which we call the improper rotation, $\mathbf{S}_n$. We may say that $\mathbf{S}_n$ is the product of $\mathbf{C}_n$ and $\boldsymbol{\sigma}_h$. Noting also that the order in which we perform $\boldsymbol{\sigma}_h$ and $\mathbf{C}_n$ is immaterial,* we can write:

$$\mathbf{C}_n \times \boldsymbol{\sigma}_h = \boldsymbol{\sigma}_h \times \mathbf{C}_n = \mathbf{S}_n$$

This is an algebraic way of expressing the fact that successive application of the two operations shown has the same effect as applying the third one. For obvious reasons, it is convenient to speak of the third operation as being the product obtained by multiplication of the other two.

The above example is not unusual. Quite generally, any two symmetry operations can be multiplied to give a third. For example, in Fig. 1-2 the effects of reflections in two mutually perpendicular symmetry planes are illustrated. It will be seen that one of the reflections carries point 0 to

---

* This is, however, a special case; in general, order of multiplication matters (see page 13).

point 1. The other reflection carries point 1 to point 2. Point 0 can also be taken to point 2 by way of point 3 if the two reflection operations are performed in the opposite order. But a moment's thought will show that a direct transfer of point 0 to point 2 can be achieved by a C_2 operation about the axis defined by the line of intersection of the two planes. If we call the two reflections $\sigma(xz)$ and $\sigma(yz)$ and the rotation $C_2(z)$, we can write:

$$\sigma(xz) \times \sigma(yz) = \sigma(yz) \times \sigma(xz) = C_2(z)$$

It is also evident that:

$$\sigma(yz) \times C_2(z) = C_2(z) \times \sigma(yz) = \sigma(xz)$$

and
$$\sigma(xz) \times C_2(z) = C_2(z) \times \sigma(xz) = \sigma(yz)$$

It is also worth noting that if any one of these three operations is applied twice in succession, we get no net result or, in other words, an identity operation, viz.:

$$\sigma(xz) \times \sigma(xz) = E$$
$$\sigma(yz) \times \sigma(yz) = E$$
$$C_2(z) \times C_2(z) = E$$

Introduction to a Group. If we pause here and review what has just been done with the three operations $\sigma(xz)$, $\sigma(yz)$ and $C_2(z)$, we see that we have formed all of the nine possible products. To summarize the results systematically, we can arrange them in the annexed tabular form. Note that we have added seven more multiplications, namely, all those in which the identity operation, E, is a factor. The results of these are trivial since the product of any other, non-trivial operation with E must be just the non-trivial operation itself, as indicated.

	E	$C_2(z)$	$\sigma(xz)$	$\sigma(yz)$
E	E	$C_2(z)$	$\sigma(xz)$	$\sigma(yz)$
$C_2(z)$	$C_2(z)$	E	$\sigma(yz)$	$\sigma(xz)$
$\sigma(xz)$	$\sigma(xz)$	$\sigma(yz)$	E	$C_2(z)$
$\sigma(yz)$	$\sigma(yz)$	$\sigma(xz)$	$C_2(z)$	E

The set of operations, E, $C_2(z)$, $\sigma(xz)$ and $\sigma(yz)$ evidently has the following four interesting properties:

1. There is one operation, E, the identity, that is the trivial one of making no change. Its product with any other operation is simply the other operation.

2. There is a definition of how to multiply operations: we apply them successively. The product of any two is one of the remaining ones. In other words, this collection of operations is self-sufficient, all of its possible products being already within itself. This is sometimes called the property of *closure*.

3. Each of the operations has an *inverse*, that is, an operation by which it may be multiplied to give E as the product. In this case, each operation is its own inverse, as shown by the occurrence of E in all diagonal positions of the table.

4. It can also be shown that if we form a triple product, this may be sub-divided in any way we like without changing the result, thus:

$$\sigma(xz) \times \sigma(yz) \times C_2(z)$$
$$= [\sigma(xz) \times \sigma(yz)] \times C_2(z) = C_2(z) \times C_2(z)$$
$$= \sigma(xz) \times [\sigma(yz) \times C_2(z)] = \sigma(xz) \times \sigma(xz)$$
$$= E$$

Products which have this property are said to obey the *associative law* of multiplication.

The four properties just enumerated are of fundamental importance. They are the properties—and the *only* properties—that any collection of symmetry operations must have in order to constitute a *mathematical group*. Groups consisting of symmetry operations are called *symmetry groups* or sometimes *point groups*. The latter term arises because all of the operations leave the molecule fixed at a certain point in space. This is in contrast to other groups of symmetry operations, such as those that may be applied to crystal structures in which individual molecules move from one location to another.

The symmetry group we have just been examining is one of the simpler ones; but nonetheless, an important one. It is represented by the symbol C_{2v}; the origin of this and other symbols will be discussed below. It is not an entirely representative group in that it has some properties that are *not* necessarily found in other groups. We have already called attention to one, namely, the fact that each operation in this group is its own inverse; this is actually true of only three kinds of operation: reflections, two-fold proper rotations and inversion i. Another special property of the group C_{2v} is that all multiplications in it are *commutative*, that is, every multiplication is equal to the multiplication of the same two operations in the opposite order. It can be seen that the group multiplication table is symmetrical about its main diagonal, which is another way of saying that all possible multiplications commute. In general, multiplication of symmetry operations is *not* commutative, as subsequent discussion will illustrate.

For another simple, but more general, example of a symmetry group, let us recall our earlier examination of the ammonia molecule. We were able to discover six and only six symmetry operations that could be performed on this molecule. If this is indeed a complete list, they should constitute a group. The easiest way to see if they do is to attempt to write a multiplication table. This will contain 36 products, some of which we already know how to write. Thus we know the result of all multiplications involving E, and we know that:

$$C_3 \times C_3 = C_3^2$$
$$C_3 \times C_3^2 = C_3^2 \times C_3 = E$$

It will be noted that the second of these statements means that C_3 is the inverse of C_3^2 and *vice versa*. We also know that E and each of the σ's is its own inverse. So all operations have inverses, thus satisfying requirement 3.

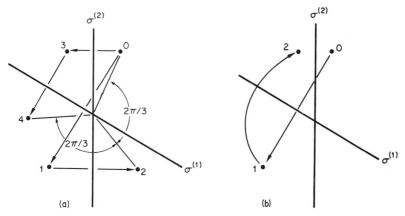

Fig. 1-5. Pictorial representations of the multiplication of symmetry operations. (a) Reflection times reflection. (b) Reflection followed by C_3.

To continue, we may next consider the products when one σ_v is multiplied by another. A typical example is shown in Fig. 1-5(a). When point 0 is reflected first through $\sigma^{(1)}$ and then through $\sigma^{(2)}$, it becomes point 2. But point 2 can obviously also be reached by a clockwise rotation through $2\pi/3$, that is, by the operation C_3. Thus we can write:

$$\sigma^{(1)} \times \sigma^{(2)} = C_3$$

If, however, we reflect first through $\sigma^{(2)}$ and then through $\sigma^{(1)}$, point 0 becomes point 4, which can be reached also by $C_3 \times C_3 = C_3^2$. Thus, we write:

$$\sigma^{(2)} \times \sigma^{(1)} = C_3^2$$

Clearly, the reflections $\sigma^{(1)}$ and $\sigma^{(2)}$ do not commute. The reader should be able to make the obvious extension of the geometrical arguments just used to obtain the following additional products.

$$\sigma^{(1)} \times \sigma^{(3)} = C_3^2$$
$$\sigma^{(3)} \times \sigma^{(1)} = C_3$$
$$\sigma^{(2)} \times \sigma^{(3)} = C_3$$
$$\sigma^{(3)} \times \sigma^{(2)} = C_3^2$$

There remain, now, the products of C_3 and C_3^2 with $\sigma^{(1)}$, $\sigma^{(2)}$, and $\sigma^{(3)}$. Fig. 1-5(b) shows a type of geometric construction that yields these products. For example, we can see that the reflection $\sigma^{(1)}$ followed by the rotation C_3 carries point 0 to point 2, which could have been reached directly by the operation $\sigma^{(2)}$. By similar procedures all the remaining products can be easily determined. The complete multiplication table for this set of operations is given on page 14.

The successful construction of this table demonstrates that the set of six operations does indeed form a group. This group is represented by the symbol C_{3v}. The table shows that its characteristics are more general than those of the group C_{2v}. Thus, it contains some operations that are not, as well as some which are, their own inverse. It also involves a number of multiplications that are not commutative.

	E	C_3	C_3^2	$\sigma^{(1)}$	$\sigma^{(2)}$	$\sigma^{(3)}$
E	E	C_3	C_3^2	$\sigma^{(1)}$	$\sigma^{(2)}$	$\sigma^{(3)}$
C_3	C_3	C_3^2	E	$\sigma^{(3)}$	$\sigma^{(1)}$	$\sigma^{(2)}$
C_3^2	C_3^2	E	C_3	$\sigma^{(2)}$	$\sigma^{(3)}$	$\sigma^{(1)}$
$\sigma^{(1)}$	$\sigma^{(1)}$	$\sigma^{(2)}$	$\sigma^{(3)}$	E	C_3	C_3^2
$\sigma^{(2)}$	$\sigma^{(2)}$	$\sigma^{(3)}$	$\sigma^{(1)}$	C_3^2	E	C_3
$\sigma^{(3)}$	$\sigma^{(3)}$	$\sigma^{(1)}$	$\sigma^{(2)}$	C_3	C_3^2	E

1-3. Some General Rules for Multiplication of Symmetry Operations

In the preceding Section several specific examples of multiplication of symmetry operations have been worked out. On the basis of this experience, the following general rules should not be difficult to accept:

1. The product of two proper rotations must be another proper rotation. Thus, while rotations can be created by combining reflections [recall: $\sigma(xz) \times \sigma(yz) = C_2(z)$] the reverse is not possible.

2. The product of two reflections in planes meeting at an angle θ is a rotation by 2θ about the axis formed by the line of intersection of the planes [recall: $\sigma^{(1)} \times \sigma^{(2)} = C_3$ for the ammonia molecule].

3. When there is a rotation operation C_n and a reflection in a plane containing the axis, there must be altogether n such reflections in a set of n planes separated by angles of $2\pi/2n$, intersecting along the C_n axis [recall: $\sigma^{(1)} \times C_3 = \sigma^{(2)}$ for the ammonia molecule].

4. The product of two C_2 operations about axes that intersect at an angle θ is a rotation by 2θ about an axis perpendicular to the plane containing the two C_2 axes.

5. The following pairs of operations always commute:
   (a) two rotations about the same axis;
   (b) reflections through planes perpendicular to each other;
   (c) the inversion and any other operation;
   (d) two C_2 operations about perpendicular axes;
   (e) C_n and σ_h, where the C_n axis is vertical.

1-4. A Systematic Listing of Symmetry Groups, with Examples

The symmetry groups to which real molecules may belong are very numerous. However, they may be systematically classified by considering how to build them up using increasingly more elaborate combinations of symmetry operations. The outline which follows, while neither unique in its approach nor rigorous in its procedure, affords a practical scheme for use by most chemists.

The simplest non-trivial groups are those of order 2, that is, those containing but one operation in addition to **E**. The additional operation must be one that is its own inverse, and thus the only groups of order 2 are:

$$C_s: \quad \mathbf{E}, \quad \sigma$$
$$C_i: \quad \mathbf{E}, \quad \mathbf{i}$$
$$C_2: \quad \mathbf{E}, \quad \mathbf{C}_2$$

The symbols for these groups are rather arbitrary, except for C_2 which, we shall soon see, forms part of a pattern.

Molecules with C_s symmetry are fairly numerous. Examples are the thionyl halides and sulfoxides (1-III), and secondary amines (1-IV). Molecules having a center of symmetry as their *only* symmetry element are quite rare; two types are shown as (1-V) and (1-VI). The reader should find it very challenging, though not impossible, to think of others. Molecules of C_2 symmetry are fairly common, two examples being (1-VII) and (1-VIII).

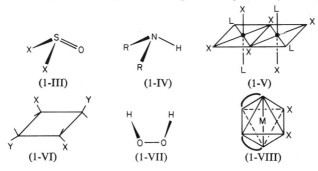

(1-III)	(1-IV)	(1-V)

(1-VI)	(1-VII)	(1-VIII)

The Uniaxial or C_n Groups. These are the groups in which the operations are all due to the presence of a proper axis as the sole symmetry element. The general symbol for such a group, and the operations in it, are:

$$C_n: \mathbf{C}_n, \mathbf{C}_n^2, \mathbf{C}_n^3 \ldots \mathbf{C}_n^{n-1}, \mathbf{C}_n^n \equiv \mathbf{E}$$

A C_n group is thus of order n. We have already mentioned the group C_2. Molecules with pure axial symmetry other than C_2 are rare. Two examples of the group C_3 are shown in (1-IX) and (1-X).

(1-IX)	(1-X)

The C_{nv} Groups. If, in addition to a proper axis of order n, there is also a set of n vertical planes, we have a group of order $2n$, designated C_{nv}. This type of symmetry is found quite frequently and is illustrated in (1-XI) to (1-XV), where the value of n should in each case be obvious.

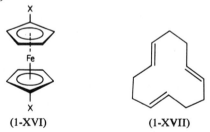

(1-XI)   (1-XII)   (1-XIII)

(1-XIV)   (1-XV)

The C_{nh} Groups. If, in addition to a proper axis of order n, there is also a horizontal plane of symmetry, we have a group of order $2n$ designated C_{nh}. The $2n$ operations include S_n^m operations which are products of C_n^m and σ_h for n odd, to make the total of $2n$. Thus, for C_{3h} the operations are:

$$C_3, C_3^2, C_3^3 \equiv E$$
$$\sigma_h$$
$$\sigma_h \times C_3 = C_3 \times \sigma_h = S_3$$
$$\sigma_h \times C_3^2 = C_3^2 \times \sigma_h = S_3^5$$

Molecules of C_{nh} symmetry are relatively rare; two examples are shown in (1-XVI) and (1-XVII).

(1-XVI)   (1-XVII)

The D_n Groups. When a vertical C_n axis is accompanied by a set of n C_2 axes perpendicular to it, the group is D_n. Molecules of D_n symmetry are, in general, rare, but there is one very important type, namely the trischelates (1-XVIII) of D_3 symmetry.

(1-XVIII)

The D_{nh} Groups. If, to the operations making up a D_n group, we add reflection in a horizontal plane of symmetry, the group D_{nh} is obtained. It should be noted that products of the type $C_2 \times \sigma_h$ will give rise to a set of reflections in vertical planes. These planes *contain* the C_2 axes; this will be an important point in regard to the distinction between D_{nh} and D_{nd} to be mentioned next. D_{nh} symmetry is found in a number of important molecules, a few of which are benzene (D_{6h}), ferrocene in an eclipsed configuration (D_{5h}), $Re_2Cl_8^{2-}$ which we examined above (D_{4h}), $PtCl_4^{2-}$ (D_{4h}) and the boron halides (D_{3h}). All right prisms with regular polygons for bases have some form of D_{nh} symmetry, as illustrated in (1-XIX) and (1-XX).

 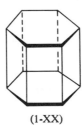

| (1-XIX) | (1-XX) |

The D_{nd} Groups. If to the operations making up a D_n group we add a set of vertical planes that bisect the angles between pairs of C_2 axes (note the distinction from the vertical planes in D_{nh}) we have a group called D_{nd}. The D_{nd} groups have no σ_h. Perhaps the most celebrated examples of D_{nd} symmetry are the D_{3d} and D_{5d} symmetries of ethane and ferrocene in their staggered configurations (1-XXI) and (1-XXII).

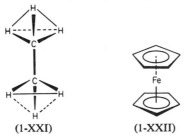

| (1-XXI) | (1-XXII) |

Two comments about the scheme so far outlined may be helpful. The reader may have wondered why we did not consider the result of adding to the operations of C_n *both* a set of $n\sigma_v$'s and a σ_h. The answer is that this is simply another way of getting to D_{nh}, since a set of C_2 axes is formed along the lines of intersection of the σ_h with each of the σ_v's. By convention, and in accord with the symbols used to designate the groups, it is preferable to proceed as we did. Secondly, in dealing with the D_{nh} type groups, if a horizontal plane is found, there must be only the n vertical planes *containing* the C_2 axis. If dihedral planes were also present, there would be, in all $2n$ planes and hence, as shown above, a principal axis of order $2n$, thus vitiating the assumption of a D_n type of group.

The S_n Groups. Our scheme has, so far, overlooked one possibility, namely, that a molecule might contain an S_n axis as its only symmetry element (except for others that are directly subservient to it). It can be shown that, for n odd, the groups of operations arising would actually be those forming the group C_{nh}. For example, take the operations generated by an S_3 axis:

$$S_3$$
$$S_3^2 \equiv C_3^2$$
$$S_3^3 \equiv \sigma_h$$
$$S_3^4 \equiv C_3$$
$$S_3^5$$
$$S_3^6 \equiv E$$

Comparison with the list of operations in the group C_{3h}, given on page 16, shows that the two lists are identical.

It is only when n is an even number that new groups can arise that are not already in the scheme. For instance, consider the set of operations generated by an S_4 axis:

$$S_4$$
$$S_4^2 \equiv C_2$$
$$S_4^3$$
$$S_4^4 \equiv E$$

This set of operations satisfies the four requirements for a group and is not a set that can be obtained by any procedure previously described. Thus S_4, S_6, etc., are new groups. They are distinguished by the fact that they contain no operation that is not an S_n^m operation even though it may be written in another way, as with $S_4^2 \equiv C_2$ above.

Note that the group S_2 is not new. A little thought will show that the operation S_2 is identical with the operation i. Hence the group that could be called S_2 is the one we have already called C_i.

An example of a molecule with S_4 symmetry is shown in (1-XXIII). Molecules with S_n symmetries are not very common.

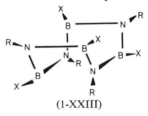

(1-XXIII)

Linear Molecules. There are only two kinds of symmetry for linear molecules. There are those represented by (1-XXIV) that have identical ends. Thus, in addition to an infinite-fold rotation axis, C_∞, coinciding with the molecular axis, and an infinite number of vertical symmetry planes, they have a horizontal plane of symmetry and an infinite number of C_2 axes perpendicular to C_∞. The group of these operations is $D_{\infty h}$. A linear molecule with different ends (1-XXV), has only C_∞ and the σ_v's as symmetry elements. The group of operations generated by these is called $C_{\infty v}$.

A—B—C—B—A   A—B—C—D
(1-XXIV)   (1-XXV)

1-5. The Groups of Very High Symmetry

The scheme followed in the preceding Section has considered only cases in which there is a single axis of order equal to or greater than 3. It is possible to have symmetry groups in which there are several such axes. There are, in fact, seven such groups, and several of them are of paramount importance.

The Tetrahedron.   We consider first a regular tetrahedron. Fig. 1-6 shows some of the symmetry elements of the tetrahedron, including at least one of each kind. From this it can be seen that the tetrahedron has altogether 24 symmetry operations, which are as follows:

There are three S_4 axes, each of which gives rise to the operations S_4, $S_4^2 \equiv C_2$, S_4^3 and $S_4^4 \equiv E$. Neglecting the S_4^4's, this makes $3 \times 3 = 9$.

There are four C_3 axes, each giving rise to C_3, C_3^2 and $C_3^3 \equiv E$. Again omitting the identity operations, this makes $4 \times 2 = 8$.

There are six reflection planes, only one of which is shown in Fig. 1-6, giving rise to six σ_d operations.

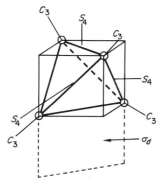

Fig. 1-6.   The tetrahedron, showing some of its essential symmetry elements. All S_4 and C_3 axes are shown but only one of the six dihedral planes σ_d.

Thus there are $9+8+6+$ one identity operation $= 24$ operations. This group is called T_d. It is worth emphasizing that, despite the considerable amount of symmetry, there is no inversion center in T_d symmetry. There are, of course, numerous molecules having full T_d symmetry, such as CH_4, SiF_4, ClO_4^-, $Ni(CO)_4$ and $Ir_4(CO)_{12}$, and many others where the symmetry is less but approximates to it.

If we remove from the T_d group the reflections, it turns out that the S_4 and S_4^3 operations are also lost. The remaining 12 operations (E, four C_3 operations, four C_3^2 operations and three C_2 operations) form a group, designated T. This group is, in itself, of little importance as it is very rarely, if ever, encountered in real molecules. However, if we then add to the operations in the group T a different set of reflections in the three planes defined so that each one contains two of the C_2 axes, and work out all products of operations we get a new group of 24 operations (**E**, four $\mathbf{C_3}$, four $\mathbf{C_3^2}$, three $\mathbf{C_2}$, three $\boldsymbol{\sigma}_h$, **i**, four $\mathbf{S_6}$, four $\mathbf{S_6^5}$) denoted T_h. This, too, is rare, but has some importance since it is the symmetry group for certain complexes such as the hexanitrato complex partly shown in (1-XXVI); for clarity, only three

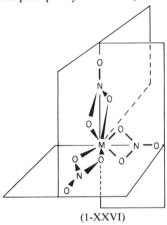

(1-XXVI)

of the six NO_3 groups are indicated; one lies in each of the planes shown. The complete set would be obtained by adding a second one in each plane opposite to the one already there. The three planes shown are, of course, the new symmetry planes.

The Octahedron and the Cube.   These two bodies have the same elements, as shown in Fig. 1-7. In this sketch the octahedron is inscribed in a cube; the centers of the six cube faces form the vertices of the octahedron. Conversely, the centers of the eight faces of the octahedron form the vertices of a cube. Fig. 1-7 shows one of each of the types of symmetry element that

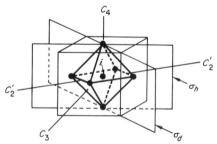

Fig. 1-7.   The octahedron and the cube, showing one of each of their essential types of symmetry element.

these two polyhedra possess. The list of symmetry operations is as follows:

There are three C_4 axes, each generating $\mathbf{C_4}$, $\mathbf{C_4^2} \equiv \mathbf{C_2}$, $\mathbf{C_4^3}$, $\mathbf{C_4^4} \equiv \mathbf{E}$. Thus there are $3 \times 3 = 9$ rotations, excluding $\mathbf{C_4^4}$'s.

There are four C_3 axes giving four $\mathbf{C_3}$'s and four $\mathbf{C_3^2}$'s.

There are six C_2' axes bisecting opposite edges, giving six $\mathbf{C_2'}$'s.

There are three planes of the type σ_h and six of the type σ_d, giving rise to 9 reflection operations.

The C_4 axes are also S_4 axes and each of these generates the operations $\mathbf{S_4}$, $\mathbf{S_4^2} \equiv \mathbf{C_2}$ and $\mathbf{S_4^3}$, the first and last of which are yet not listed, thus adding $3 \times 2 = 6$ more to the list.

The C_3 axes are also S_6 axes and each of these generates the new operations $\mathbf{S_6}$, $\mathbf{S_6^3} \equiv \mathbf{i}$, and $\mathbf{S_6^5}$. The $\mathbf{i}$ counts only once, so there are then $4 \times 2 + 1 = 9$ more new operations.

The entire group thus consists of the identity $+9+8+6+9+6+9 = 48$ operations. This group is denoted O_h. It is, of course, a very important type of symmetry since octahedral molecules (e.g., SF_6), octahedral complexes [$Co(NH_3)_6^{3+}$, $IrCl_6^{3-}$] and octahedral interstices in solid arrays are very common. There is a group O, which consists of only the 24 proper rotations from O_h, but this, like T, is rarely if ever encountered in Nature.

The Pentagonal Dodecahedron and the Icosahedron. These are shown in Fig. 1-8. They are related to each other in the same way as are the octahedron

(a)   (b)

Fig. 1-8. The two regular polyhedra having I_h symmetry. (a) The pentagonal dodecahedron. (b) The icosahedron.

and the cube, the vertices of one defining the face centers of the other and *vice versa*. Both have the same symmetry operations, a total of 120! We shall not list them in detail but merely mention the basic symmetry elements: six C_5 axes; ten C_3 axes, 15 C_2 axes, and 15 planes of symmetry. The group of 120 operations is designated I_h, and is often called the icosahedral group.

There is as yet no known example of a molecule that is a pentagonal dodecahedron, but the icosahedron is a key structural unit in boron chemistry, occurring in all forms of elemental boron as well as in the $B_{12}H_{12}^{2-}$ ion.

If the symmetry planes are omitted, a group called I consisting of only proper rotations remains. This is mentioned purely for the sake of completeness, since no example of its occurrence in Nature is known.

1-6. Molecular Dissymmetry and Optical Activity

Optical activity, that is, rotation of the plane of polarized light coupled with unequal absorption of the right- and left-circularly polarized components, is a property of a molecule (or an entire three-dimensional array of atoms or molecules) that is not superposable on its mirror image. When the number of molecules of one type exceeds the number of those that are their non-superposable mirror images, a net optical activity results. To predict when optical activity will be possible it is necessary to have a criterion to determine when a molecule and its mirror image will not be identical, that is, superposable.

Molecules that are not superposable on their mirror images are called *dissymmetric*. This term is preferable to asymmetric since the latter means, literally, without symmetry, whereas dissymmetric molecules can and often do possess some symmetry, as will be seen.*

A compact statement of the relation between molecular symmetry properties and dissymmetric character is: *A molecule that has no axis of improper rotation is dissymmetric.*

This statement includes and extends the usual one to the effect that optical isomerism exists when a molecule has neither a plane nor a center of symmetry. It has already been noted that the inversion operation, $\mathbf{i}$, is equivalent to the improper rotation $\mathbf{S}_2$. Similarly, $\mathbf{S}_1$ is a correct although unused way of representing σ, since it implies rotation by $2\pi/1$, equivalent to no net rotation, in conjunction with the reflection. Thus σ and $\mathbf{i}$ are both really special cases of improper rotations.

However, even when σ and $\mathbf{i}$ are absent a molecule may still be identical with its mirror image if it possesses an $\mathbf{S}_n$ axis of some higher order. A good example of this is provided by the $(-RNBX-)_4$ molecule shown in (1-XXIII). This molecule has neither a plane nor a center of symmetry, but inspection shows that it can be superposed on its mirror image. As we have noted, it belongs to the symmetry group S_4.

Dissymmetric molecules either have no symmetry at all, or they belong to one of the groups consisting only of proper rotation operations, that is the C_n or D_n groups. (Groups T, O, and I are, in practice, not encountered, though molecules in these groups must also be dissymmetric.) Important examples are the bis-chelate and tris-chelate octahedral complexes (1-VIII), (1-X), and (1-XVIII).

MOLECULAR SYMMETRY

1-7. Coordination Compounds

Historically it has been customary to treat *coordination compounds* as a special class separate from *molecular compounds*. On the basis of actual fact only, i.e. neglecting the purely traditional reasons for such a distinction, there is very little, if indeed any, basis for continuing this dichotomy.

Coordination compounds are conventionally formulated as consisting of

---

* Dissymmetry is sometimes called chirality, and dissymmetric chiral, from the Greek word χειρ for hand, in view of the left-hand/right-hand relation of molecules that are mirror images.

a *central atom* or ion surrounded by a set (usually 2 to 9) of other atoms, ions or small molecules, the latter being called *ligands*. The resulting conglomeration is often called a *complex* or, if it is charged, a *complex ion*. The set of ligands need not consist of several small, independent sets of atoms (or single atoms) but may involve fairly elaborate arrangements of atoms connecting those few that are directly bound to—or *coordinated* to—the central atom. However, there are many molecular compounds of which the same description may be given. Consider, for illustration, the following:

$$SiF_4 \qquad SiF_6^{2-} \qquad Cr(CO)_6 \qquad Co(NH_3)_6^{3+}$$
$$SF_6 \qquad PF_6^- \qquad Cr(NH_3)_6^{3+} \qquad CoCl_4^{2-}$$

Conventionally the first two are called molecules and five of the other six are called complexes; $Cr(CO)_6$ can be found referred to in either way depending on the context. Obviously, one basis for the different designations is the presence or absence of a net charge, only uncharged species being called molecules. Beyond this, which is really quite a superficial characteristic as compared with such basic ones as geometric and electronic structures, there is no logical reason for the division. The essential irrelevancy of the question of overall charge is well demonstrated by the fact that $Pt(NH_3)_2Cl_2$, $Cu(acac)_2$, $CoBr_2(Ph_3P)_2$, and scores of similar compounds are quite normally called complexes. The *molecules* are really only *complexes* which happen to have a charge of 0 instead of $+n$ or $-m$.

Thus SiF_6^{2-}, PF_6^-, and SF_6 are isoelectronic and isostructural. While the character of the bonds from the central atom to fluorine atoms doubtless varies from one to another, there is no basis for believing that SF_6 differs more from PF_6^- than the latter does from SiF_6^{2-}.

It might be argued that the terms "complex" or "coordination compound" should be applied only when the central atom, in some oxidation state, and the ligands, can be considered to exist independently, under reasonably normal chemical conditions. Thus Cr^{3+} and NH_3 would be said to so exist. However, Cr^{3+} actually exists under normal chemical conditions not as such but as Cr^{3+}(aq) which is, in detail, $Cr(H_2O)_6^{3+}$, another species that would, itself, be called a complex. Again, in a similar vein, the argument that PF_6^- and SiF_6^{2-} can be considered to consist respectively, of PF_5+F^- and SiF_4+2F^-, whereas there is no comparable breakdown of SF_6 is a poor one; once the set of six fluorine atoms is completed about the central atom, they become equivalent. The possibility of their having had different origins has no bearing on the nature of the final complex.

The terms "coordination compound" and "complex" may therefore be broadly defined to embrace all species, charged or uncharged, in which a central atom is surrounded by a set of outer or ligand atoms.

Having thus defined coordination compounds in a comprehensive way, we can proceed to discuss their structures in terms of only two properties: (1) *coordination number*, the number of outer, or ligand, atoms bonded to the central one, and (2) *coordination geometry*, the geometric arrangement of these ligand atoms and the consequent symmetry of the complex. We shall consider in detail coordination numbers 2–9, discussing under each the principal ligand arrangements. Higher coordination numbers will be discussed only briefly as they occur much less frequently.

Coordination Number Two. There are two geometric possibilities, linear and bent. If the two ligands are identical the general types and their symmetries are: linear L—M—L, $D_{\infty h}$; bent L—M—L, C_{2v}. This coordination number is, of course, found in numerous molecular compounds of divalent elements, but is relatively uncommon otherwise. In many cases where stoichiometry might imply its occurrence, a higher coordination number actually occurs because some ligands form "bridges" between two central atoms. In terms of the more conventional types of coordination compound——those with a rather metallic element at the center—it is restricted mainly to some complexes of Cu^I, Ag^I, Au^I, and Hg^{II}. Such complexes have linear arrangements of the metal ion and the two ligand atoms, and typical ones are $[ClCuCl]^-$, $[H_3NAgNH_3]^+$, $[ClAuCl]^-$, and $[NCHgCN]$. The metal atoms in cations such as $[UO_2]^{2+}$, $[UO_2]^+$, and $[PuO_2]^{2+}$, which are linear, may also be said to have coordination number 2, but these oxo cations interact fairly strongly with additional ligands and their actual coordination numbers are much higher; it is true, however, that the central atoms have a specially strong affinity for the two oxygen atoms. Linear coordination also occurs in the several trihalide ions, such as I_3^-, $ClBrCl^-$, etc.

Coordination Number Three. The two most symmetrical arrangements are planar (1-XXVII) and pyramidal (1-XXVIII), with D_{3h} and C_{3v} symmetry, respectively. Both these arrangements are found often among molecules formed by trivalent central elements. Among complexes of the metallic elements this is a rare coordination number; nearly all compounds or complexes of metal cations with stoichiometry MX_3 have structures in which sharing of ligands leads to a coordination for M that exceeds 3. There are, however, a few exceptions such as the planar HgI_3^- ion which occurs in $[(CH_3)_3S^+][HgI_3^-]$, the MN_3 groups which occur in $Cr(NR_2)_3$ and $Fe(NR_2)_3$[1] where $R = (CH_3)_3Si$, and the CuS_3 groups found in $Cu[SC(NH_2)_2]_3Cl$[2] and $Cu(SPPh_3)_3ClO_4$.[3]

In a few cases, e.g., ClF_3 and BrF_3, a T-shaped form (1-XXIX) of 3-coordination (symmetry C_{2v}) is found.

(1-XXVII)   (1-XXVIII)   (1-XXIX)

Coordination Number Four. This is a highly important coordination number, occurring in hundreds of thousands of compounds, including, *inter alia*, most of those formed by the element carbon, essentially all those

[1] D. C. Bradley *et al.*, *Chem. Comm.*, **1969**, 14; **1970**, 1715.
[2] W. A. Spofford and E. L. Amma, *Chem. Comm.*, **1968**, 405.
[3] P. G. Eller and P. W. R. Corfield, *Chem. Comm.*, **1971**, 105.

formed by silicon, germanium and tin, and many compounds and complexes of other elements. There are three principal geometries. By far the most prevalent is tetrahedral geometry (1-XXX), which has symmetry T_d when ideal. Tetrahedral complexes or molecules are practically the only kind of four-coordinate ones formed by non-transition elements; whenever the central atom has no electrons in its valence shell orbitals except the four pairs forming the σ bonds to ligands, these bonds are disposed in a tetrahedral fashion. With many transition-metal complexes, square geometry (1-XXXI)

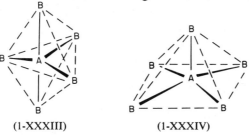

(1-XXX)	(1-XXXI)	(1-XXXII)

occurs because of the presence of additional valence-shell electrons and orbitals (i.e. partially filled d orbitals), although there are also many tetrahedral complexes formed by the transition metals. In some cases, e.g., with Ni^{II}, Co^{II} and Cu^{II} in particular, there may be only a small difference in stability between the tetrahedral and the square arrangement and rapid interconversions may occur (page 40).

Square complexes are also found with non-transitional central atoms when there are two electron pairs present beyond the four used in bonding; these two pairs lie above and below the plane of the molecule. Examples are XeF_4 and $(ICl_3)_2$. Similarly, when there is one "extra" electron pair, as in SF_4, the irregular arrangement, of symmetry C_{2v} (1-XXXII), is adopted. More detailed discussions of these non-transition element structures will be found in Chapter 4.

Coordination Number Five. Though less common than numbers 4 and 6, this is still very important.[4] There are two principal geometries and these may be conveniently designated by stating the polyhedra that are defined by the set of ligand atoms. In one case the ligand atoms lie at the vertices of a trigonal bipyramid (*tbp*) (1-XXXIII), and in the other at the vertices of a square pyramid (*spy*) (1-XXXIV). The *tbp* belongs to the symmetry group D_{3h}; the *spy* belongs to the group D_{4h}. It is interesting and highly important that these two structures are similar enough to be interconverted without

(1-XXXIII)	(1-XXXIV)

[4] A comprehensive, general review is given by E. L. Muetterties and R. A. Schunn, *Quart. Rev.*, 1966, **20**, 245.

great difficulty, and for most real cases they are of similar thermodynamic stability. This is the basis for one of the most prominent types of stereochemical non-rigidity, as will be explained in Section 1-9.

The similar energy of the two structures is rather dramatically illustrated by the occurrence of both forms of $Ni(CN)_5^{3-}$ in the same crystalline compound (cf. Sect. 25-G-2).

A large fraction of the known 5-coordinate species have structures that are appreciably distorted from one or other of the two prototype structures; in some cases the distortion is such that it becomes uncertain which of the two ought to be taken as the idealized geometry. This ready deformability is characteristic of 5-coordination and closely connected with the dynamic stereochemical non-rigidity so prevalent among these structures.

Coordination Number Six. This is perhaps the most common coordination number, and the six ligands almost invariably lie at the vertices of an octahedron or a distorted octahedron. The very high symmetry, group O_h, of the regular octahedron, has been discussed in detail on page 20.

There are three principal forms of distortion of the octahedron. One is *tetragonal*, elongation or contraction along a single C_4 axis; the resultant symmetry is only D_{4h}. Another is *rhombic*, changes in the lengths of two of the C_4 axes so that no two are equal; the symmetry is then only D_{2h}. The third is a *trigonal* distortion, elongation or contraction along one of the C_3 axes so that the symmetry is reduced to D_{3d}. These three distortions are illustrated in Fig. 1-9.

The tetragonal distortion most commonly involves an elongation of one C_4 axis and, in the limit, two *trans* ligands are lost completely, leaving a square, four-coordinated complex. The trigonal distortion transforms the octahedron into a trigonal antiprism.

Another type of 6-coordinate geometry, which is very rare, is that in which the ligands lie at the vertices of a *trigonal prism* (1-XIX); the symmetry (as noted, page 17) is D_{3h}. This was first discovered in MoS_2 and WS_2 many years ago, but only very recently have further examples been encountered. Some examples are the coordination of the Nb and Ta ions in MNb_3S_6 and MTa_3S_6 (M = Mn, Fe, Co, or Ni),[5] several complexes of $M(S_2C_2R_2)_3$ type,[6] the central Co^{II} in $\{[Co(OCH_2CH_2NH_2)_3]Co[Co(OCH_2CH_2NH_2)_3]\}^{2+}$, which is coordinated by six oxygen atoms,[7] and a few cases[8,9] in which there are rigid polydentate ligands especially designed to encapsulate the metal ion in a coordination environment of this geometry. There is at least one case where octahedral geometry is distorted by rotation of one triangular face relative to the opposite one part way from the antiprismatic toward the prismatic configuration, namely, $Fe(S_2CNBu_2^n)_3$.[10]

[5] J. M. van den Berg and P. Cossee, *Inorg. Chim. Acta*, 1968, **2**, 143.
[6] R. Eisenberg and J. A. Ibers, *J. Amer. Chem. Soc.*, 1965, **87**, 3776.
[7] J. A. Bertrand, J. A. Kelly and E. G. Vassian, *J. Amer. Chem. Soc.*, 1969, **91**, 2394.
[8] W. O. Gillum *et al.*, *Chem. Comm.*, **1969**, 843.
[9] J. E. Parks, B. E. Wagner and R. H. Holm, *J. Amer. Chem. Soc.*, 1970, **92**, 3500.
[10] B. F. Hoskins and B. P. Kelly, *Chem. Comm.*, **1968**, 1517.

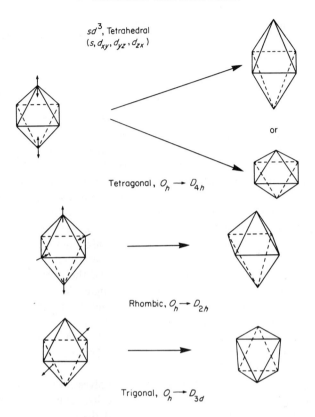

Fig. 1-9. Sketches showing the three principal types of distortion found in real octahedral complexes.

Coordination Number Seven.[11] Three geometrical arrangements are known. The most regular is the pentagonal bipyramid (D_{5h}), which is found in $[UO_2F_5]^{3-}$, as shown in (1-XXXV), and in $[UF_7]^{3-}$, $[ZrF_7]^{3-}$, and $[HfF_7]^{3-}$. The second arrangement, of C_{3v} symmetry, that can be considered to result from addition of a seventh atom at the center of one face of an octahedron, which is distorted mainly by the spreading apart of the three atoms defining this face, has been found in one modification of the lanthanide

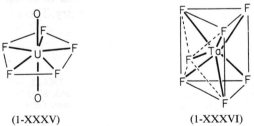

(1-XXXV)   (1-XXXVI)

[11] E. L. Muetterties and C. M. Wright, *Quart. Rev.*, 1967, **21**, 109.

oxides, M_2O_3, and in $[NbOF_6]^{3-}$. The third arrangement, which occurs in $[NbF_7]^{2-}$ and $[TaF_7]^{2-}$ ions, is derived by inserting a seventh atom above the center of one of the rectangular faces of a trigonal prism, as shown in (1-XXXVI).

Coordination Number Eight.[11,12] It is conceptually convenient to begin with the most symmetrical polyhedron having eight vertices, namely, the cube, which has O_h symmetry. Cubic coordination occurs only rarely in discrete complexes (namely, in the octafluoro anions in the compounds Na_3MF_8, M = Pa, U or Np^{13}), although it occurs in various solid compounds where the anions form continuous arrays, as in the CsCl structure. Its occurrence is infrequent presumably because there are several ways in which the cube may be distorted so as to lessen repulsions between the X atoms while maintaining good M—X interactions.

The two principal ways in which the cube may become distorted are shown in Fig. 1-10. The first of these, rotation of one square face by 45° relative to the one opposite to it, clearly lessens repulsions between non-bonded atoms while leaving M—X distances unaltered. The resulting polyhedron is the square antiprism (symmetry D_{4d}). It has square top and bottom and eight isosceles triangles for its vertical faces. The second distortion shown can be best comprehended by recognizing that the cube is composed of two interpenetrating tetrahedra. The distortion occurs when the vertices of one of these tetrahedra are displaced so as to decrease the two vertical angles, that is, to elongate the tetrahedron, while the vertices of the other one are displaced to produce a flattened tetrahedron. The resulting polyhedron is called a dodecahedron, or more specifically, to distinguish it from several

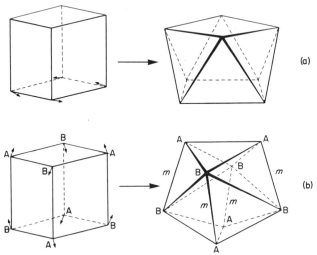

Fig. 1-10.   The two most important ways of distorting the cube: (a) to produce a square antiprism; (b) to produce a dodecahedron.

[12] S. J. Lippard, *Progr. Inorg. Chem.*, 1967, **8**, 109.
[13] D. Brown, J. F. Easey and C. E. F. Rickard, *J. Chem. Soc.*, A, **1969**, 1161.

other kinds of dodecahedron, a triangulated dodecahedron. It has D_{2h} symmetry and it is important to note that its vertices are not all equivalent but are divided into two bisphenoidal sets, those within each set being equivalent.

Detailed analysis of the energetics of M—X and X—X interactions suggests that there will in general be little difference between the energies of the square antiprism and the dodecahedral arrangement, unless other factors, such as the existence of chelate rings, energies of partially filled inner shells, exceptional opportunities for orbital hydridization or the like, come into play. Both arrangements occur quite commonly, and in some cases, e.g., the $M(CN)_8^{n-}$ (M = Mo or W; $n = 3$ or 4) ions, the geometry varies from one kind to the other with changes in the counter-ion in crystalline salts[14] and on changing from crystalline to solution phases.[15]

A form of eight-coordination, which is a variant of the dodecahedral arrangement, is found in several compounds containing bidentate ligands in which the two coordinated atoms are very close together (ligands said to have a small "bite"), such as NO_3^- and O_2^{2-}. In these, the close pairs of ligand atoms lie on the m edges of the dodecahedron [see Fig. 1-10(b)]; these edges are then very short. Examples of this are the $Cr(O_2)_4^{3-}$ and $Co(NO_3)_4^{2-}$ ions and the $Ti(NO_3)_4$ molecule.

Three other forms of octacoordination, which occur much less often and are essentially restricted to actinide and lanthanide compounds, are the hexagonal bipyramid (D_{6h}) (1-XXXVII), the bicapped trigonal prism (D_{3h}) (1-XXXVIII), and the bicapped trigonal antiprism (D_{3d}) (1-XXXIX). The

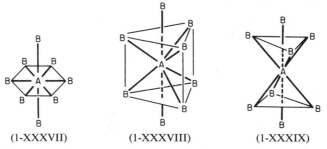

(1-XXXVII)   (1-XXXVIII)   (1-XXXIX)

hexagonal bipyramid is restricted almost entirely to the oxo ions, where an OMO group defines the axis of the bipyramid, though it is occasionally found elsewhere.[16]

Coordination Number Nine. There is only one symmetrical arrangement that occurs with regularity for this coordination number. It is derived from a trigonal prism by placing the three additional atoms outside the centers of the three vertical faces, as shown in Fig. 1-11; it retains the D_{3h} symmetry of the trigonal prism. Among the compounds in which this arrangement

[14] S. S. Basson, L. D. C. Bok, and J. G. Leipoldt, *Acta Cryst.*, 1970, *B*, **26**, 1209.
[15] R. V. Parish *et al.*, *J. Chem. Soc.*, A, **1968**, 2883.
[16] C. D. Garner and S. C. Wallwork, *J. Chem. Soc.*, A, **1970**, 3092.

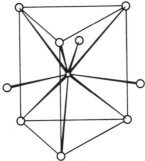

Fig. 1-11.   The structure of many 9-coordinate complexes.

occurs are $La(OH)_3$, UCl_3, $PbCl_2$, hydrated lanthanide compounds such as $Nd(BrO_3)_3 \cdot 9H_2O$, certain hydrated salts of Sr^{2+} and the $[ReH_9]^{2-}$ ion.

Higher Coordination Numbers.[11]   These can occur only in compounds of the largest, heaviest atoms. The geometry is usually irregular, although in only a relatively few cases are diffraction data available.

Coordination number 10 has been demonstrated in $[La(EDTA)\,(H_2O)_4]$ and presumably occurs in various complexes of Th and U, such as the tropolone complexes $[M(O_2C_7H_5)_5]^-$. A bicapped square antiprismatic arrangement has been found in $K_4Th(O_2CCO_2)_4 \cdot 4H_2O$.[17]

Coordination number 11 has been observed for Th in $Th(NO_3)_4 \cdot 5H_2O$, and coordination number 12 is found in several compounds where a large cation is coordinated by a set of oxo anions each supplying several oxygen atoms. This happens, for example, in several hexanitrato species, $M(NO_3)_6$, where each nitrate ion is bidentate and the 12 oxygen atoms then form a distorted icosahedron. The Ba^{2+} ion often has a high coordination number in salts with complex anions. For example, in $BaSiF_6$ and $BaGeF_6$ each barium ion is surrounded by 12 nearly equidistant fluorine atoms. In the perovskite structure (Fig. 2-5), the barium ion is 12-coordinate.

1-8. Cage and Cluster Structures

The formation of polyhedral cages and clusters has only recently been recognized as an important and widespread phenomenon, but discoveries in this field have been frequent in the last decade and examples may now be found in nearly all parts of the Periodic Table. In this Section each of the principal polyhedra will be mentioned and illustrations given. Further details may be found under the chemistry of the particular elements and in the sections on metal-atom clusters (Sect. 19-11) and polynuclear metal carbonyls (Sect. 22-3).

A cage or cluster is in a certain sense the antithesis of a complex and yet there are many similarities due to common symmetry properties. In each

[17] M. N. Akhtar and A. J. Smith, *Chem. Comm.*, **1969**, 705.

type of structure a set of atoms defines the vertices of a polyhedron, but in the one case—the complex—these atoms are each bound to one central atom and not to each other, while in the other—the cage or cluster—there is no central atom and the essential feature is a system of bonds connecting each atom directly to its neighbors in the polyhedron.

The principal polyhedra found in cages and clusters are in many cases the same as those in coordination compounds, e.g., the tetrahedron, trigonal bipyramid, octahedron, and tricapped trigonal prism. However, there are cages, especially those formed by boron and carbon, that have a larger number of vertices than do most coordination polyhedra. We omit here triangular clusters since these are not truly polyhedral although in fact some of these, e.g., those in $Re_3Cl_{12}^{3-}$, $Os_3(CO)_{12}$, are not essentially different from such truly polyhedral species as, say, $(Mo_6Cl_8)^{4+}$ and $Ir_4(CO)_{12}$.

Just as all ligand atoms in a set need not be identical, so the atoms making up a cage or cluster may be different; indeed, to exclude species made up of more than one type of atom would be to exclude the majority of cages and clusters, including some of the most interesting and important ones.

Four Vertices. Tetrahedral cages or clusters have long been known for the P_4, As_4, and Sb_4 molecules and in more recent years have been found in polynuclear metal carbonyls such as $Co_4(CO)_{12}$, $Ir_4(CO)_{12}$, $[h^5\text{-}C_5H_5Fe(CO)]_4$,* $RSiCo_3(CO)_9$, $Fe_4(CO)_{13}^{2-}$, $Ni_4(CO)_6(PR_3)_4$, and a number of others. B_4Cl_4 is another well-known example and doubtless many more will be encountered.

Five Vertices. Polyhedra with five vertices are the trigonal bipyramid (*tbp*) and the square pyramid (*spy*). Both are found among the boranes and carboranes, e.g., the *tbp* in $B_3C_2H_5$ and the *spy* in B_5H_9, as well as among the transition elements. Examples of the latter are the *tbp* cluster, Pt_3Sn_2, in $(C_8H_{12})_3Pt_3(SnCl_3)_2$ and the *spy* clusters in $Fe_5(CO)_{15}C$ (1-XL) and the $Fe_3(CO)_9E_2$ (E = S or Se) species (1-XLI). The central arrangement of two metal atoms and three bridging atoms in the confacial bioctahedron, (1-XLII) provides a fairly common example of a *tbp*.

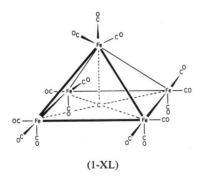

(1-XL)

* h^5 is a symbol denoting the number of atoms (5 C) involved in π-bonding to the metal (Chap. 23).

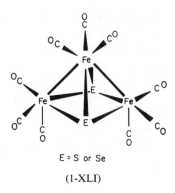

E = S or Se

(1-XLI)

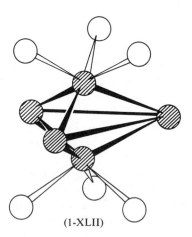

(1-XLII)

Six Vertices. Octahedral clusters and cages are almost the only six-fold type found, and are, indeed, rather numerous. There are several octahedral polynuclear metal carbonyls, such as $Rh_6(CO)_{16}$ and $[Co_6(CO)_{14}]^{4-}$, and

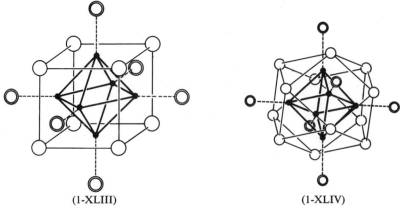

(1-XLIII)   (1-XLIV)

the extensive series of metal-atom cluster compounds formed by Nb, Ta, Mo and W are all based on octahedral sets of metal atoms, the principal types being of M_6X_8 (1-XLIII) and M_6X_{12} (1-XLIV) stoichiometry. The $B_6H_6^{2-}$ and $B_4C_2H_6$ species are also octahedral. The borane B_6H_{10}, however, has a pentagonal pyramid of boron atoms. B_6 octahedra also occur in the class of borides of general formula MB_6.

Seven Vertices. Polyhedra with seven vertices are rare. The $B_7H_7^{2-}$ ion and $B_5C_2H_7$ are examples and presumably have a pentagonal bipyramidal (D_{5h}) structure. The stable form of the compound $Fe_3(CO)_8(PhC)_4$ also contains this polyhedron, with iron atoms at both axial and one of the equatorial positions, as shown in the partial structure (1-XLV). The P_4S_3 molecule has structure (1-XLVI), which is a kind of cage but cannot be described as a simple polyhedron since the sulfur atoms are not bonded to one another.

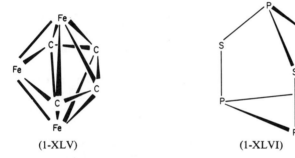

(1-XLV)   (1-XLVI)

Eight Vertices. Eight-atom polyhedral structures are very numerous. By far the most common polyhedron is the *cube*; this is in direct contrast to the situation with 8-fold coordination where a cubic arrangement of ligands is extremely rare because it is disfavored relative to the square antiprism and the triangulated dodecahedron in which ligand–ligand contacts are reduced. In the case of a cage compound, of course, it is the structure in which contacts between atoms are maximized that will tend to be favored (provided good bond angles can be maintained) since bonding rather than repulsive interactions exist between neighboring atoms.

The only known cases with eight *like* atoms in a cubic array are the hydrocarbon *cubane*, C_8H_8, and the $Cu_8(i\text{-MNT})_6^{4-}$ ion[18a] [$i\text{-MNT} = S_2CC(CN)_2^{2-}$]. The other cubic systems all involve two different species of atom which alternate as shown in (1-XLVII). In all cases either the A atoms or the

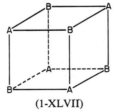

(1-XLVII)

[18a] J. P. Fackler *et al.*, *J. Amer. Chem. Soc.*, 1968, **90**, 7357.

B atoms or both have appended atoms or groups. In the annexed list are collected some of the many[18b, 19] cube species, the elements at the alternate vertices of the cube being given in bold type.

A (and appended groups)	B (and appended groups)
Mn(CO)₃	SEt
Os(CO)₃	O
PtMe₃ or **PtEt₃**	Cl, Br, I, OH
CH₃Zn	OCH₃
Tl	OCH₃
h^5-C₅H₅Fe	S
Me₃AsCu	I
PhAl	NPh
Co(CO)₃	Sb
Ni(EtOH)[C₆H₆O(CH₂O)]	OCH₃

While the polyhedron in cubane, or in any other $(AR)_6$ molecule, has the full O_h symmetry of a cube, the A_4B_4-type structures can have at best tetrahedral, T_d, symmetry since they consist of two interpenetrating tetrahedra.

It must also be noted that only when the two interpenetrating tetrahedra happen to be exactly the same size will all the ABA and BAB angles be equal to 90°. Since the A and the B atoms differ it is not in general to be expected that this will occur. In fact, there is, in principle, a whole range of bonding possibilities. At one extreme, represented by $[(h^5\text{-}C_5H_5)Fe(CO)]_4$ one set of atoms (the Fe atoms) are so close together that they must be considered as directly bonded, while the other set (the C atoms of the CO groups) are not at all bonded among themselves but only to those in the first set. In this extreme, it seems best to classify the system as having a tetrahedral cluster (of Fe atoms) supplemented by bridging CO groups. This is indeed what we have done (see page 31). At the other extreme are those A_4B_4 systems in which all A—A and B—B distances are too long to admit of significant A—A or B—B bonding, and thus the system can be regarded as genuinely cubic (even if the angles differ somewhat from 90°). This is true of all the systems listed above. However, the atoms in the smaller of the two tetrahedra will have some degree of direct interaction with one another and there is some evidence from Raman spectroscopy on species such as $Pb_4(OH)_4^{4+}$ to suggest that direct though very weak metal—metal bonding does occur.[20]

A relatively few species are known in which the polyhedron is, at least approximately, a triangulated dodecahedron [Fig. 1-10(b)]. These are the boron species $B_8H_8^{2-}$, $B_6C_2H_8$, and B_8Cl_8. The S_4N_4 and As_4S_4 molecules (1-XLVIII) are perhaps transitional between simple cyclic compounds and cages since the S—S, N—N, and As—As distances are relatively long though too short to be mere non-bonded contacts.

[18b] R. S. Nyholm, M. R. Truter and C. W. Bradford, *Nature*, 1970, **228**, 648.
[19] A. S. Foust and L. F. Dahl, *J. Amer. Chem. Soc.*, 1970, **92**, 7337.
[20] T. G. Spiro, *Progr. Inorg. Chem.*, 1970, **11**, 1.

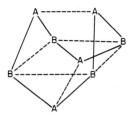

S_4N_4;   A = S,   B = N

As_4S_4;   A = As,   B = S

(1-XLVIII)

Nine Vertices.   Cages and polyhedra with nine vertices are known but are not numerous. Representatives are the Bi_9^{5+} ion which occurs in $Bi_{24}Cl_{26}$, and the $B_9H_9^{2-}$ and $B_7C_2H_9$ species; in all three the structure is that of a trigonal prism capped on its rectangular faces (Fig. 1-11), the most common geometry also for 9-coordinate complexes.

Ten Vertices.   Species with ten vertices are well-known. In $B_{10}H_{10}^{2-}$ and $B_8C_2H_{10}$ the polyhedron (1-XLIX) is a square antiprism capped on the square faces (symmetry D_{4d}). But there is a far commoner structure for

(1-XLIX)   (1- L)

10-atom cages which is commonly called the *adamantane* structure after the hydrocarbon, adamantane, $C_{10}H_{16}$, which has this structure; it is depicted in (1-L); it consists of two subsets of atoms: a set of four (A) which lie at the vertices of a tetrahedron and a set of six (B) which lie at the vertices of an octahedron; the entire assemblage has the T_d symmetry of the tetrahedron. From other points of view it may be regarded as a tetrahedron with a bridging atom over each edge or as an octahedron with a triply bridging atom over an alternating set of four of the eight triangular faces. Two of the most familiar examples of this structure are provided by P_4O_6 and hexamethylenetetr-amine, $(CH_2)_6N_4$. Additional ones are $(CH_3Si)_4S_6$ and $(CH_3N)_6As_4$. Here again, as with the cubic A_4B_4 structure, the symmetry alone does not entirely define the structure. It is possible that the tetrahedral set of atoms may be so compact and closely bonded to each other that the situation is better regarded as a tetrahedral A_4 cluster with a bridging B across each edge as in the case of $Ni_4(CO)_6(PR_3)_4$ mentioned above where strong Ni—Ni bonds exist and the CO groups merely supplement this as bridges over each edge; the PR_3 groups are attached externally, one to each Ni atom.

Eleven Vertices. Perhaps the only known eleven-atom cages are $B_{11}H_{11}^{2-}$ and $B_9C_2H_{11}$.

Twelve Vertices. Twelve-atom cages are not widespread but play a dominant role in boron chemistry. The most highly symmetrical arrangement is the icosahedron [Fig. 1-8(b)], which has 12 equivalent vertices and I_h symmetry. Icosahedra of boron atoms occur in all forms of elemental boron, in $B_{12}H_{12}^{2-}$ and in the numerous carboranes of the $B_{10}C_2H_{12}$ type. A related polyhedron, the cuboctahedron (1-LI) is found in several borides of stoichiometry MB_{12}. As shown on page 249, the icosahedron and cuboctahedron have a rather close relationship.

(1-LI)

1-9. Stereochemical Non-rigidity

Most molecules have a single, well-defined nuclear configuration. The atoms execute more-or-less harmonic vibrations about their equilibrium positions, but in other respects the structures may be considered rigid. There are, however, a significant number of cases in which molecular vibrations or intramolecular rearrangements carry a molecule from one nuclear configuration into another. When such processes occur at a rate such that they can be detected by at least some physical or chemical method, the molecules are designated as *stereochemically non-rigid*. In some cases the two or more configurations are not chemically equivalent and the process of interconversion is called *isomerization* or *tautomerization*. In other cases the two or more configurations are chemically equivalent, and this type of stereochemically non-rigid molecule is called *fluxional*.

The rearrangement processes involved in stereochemically non-rigid molecules are of particular interest when they take place rapidly, although there is a continuous gradation of rates and no uniquely defined line of demarcation can be said to exist between "fast" and "slow" processes. The question of the speed of rearrangement most often derives its significance when considered in relation to the *time scale* of the various physical methods of studying molecular structure. In some of these methods, such as electronic and vibrational spectroscopy and gas-phase electron diffraction, the act of observation of a given molecule is completed in such an extremely short time ($<10^{-11}$ sec) that processes of rearrangement may seldom if ever be fast enough to influence the results. Thus, for a fluxional molecule, where all configurations are equivalent, there will be nothing in the observations to indicate the fluxional character. For interconverting tautomers, the two (or more) tautomers will each be registered independently and there will be nothing in the observations to show that they are interconverting.

It is the technique of nmr spectroscopy that most commonly reveals the occurrence of stereochemical non-rigidity since its time scale is typically in the range of 10^{-2} to 10^{-5} sec. The rearrangements involved in stereochemically non-rigid behavior are rate processes with activation energies. When these activation energies are in the range 25–100 kJ mol^{-1} the rates of the rearrangements can be brought into the range of 10^2–10^5 sec^{-1} at temperatures in the range $+150°$ to $-150°$C. Thus by proper choice of temperature, many such rearrangements can be controlled so that they are slow enough at lower temperatures to allow detection of individual molecules, or environments within the molecules, and rapid enough at higher temperatures for the signals from the different molecules or environments to be averaged into a single line at the mean position. Thus by studying nmr spectra over a suitable temperature range, the rearrangement processes can be examined in much detail.

An example of this type of study of interconverting tautomers is provided by the system[21] of *cis*- and *trans*-isomers of $(h^5\text{-}C_5H_5)_2Fe_2(CO)_4$. The

cis   trans

protons of the cyclopentadienyl rings appear at different positions for the two isomers and the activation energy is about 50 kJ mol^{-1}. Hence, at $-70°$ the rate of interconversion is only about 8×10^{-2} sec^{-1} and each isomer exhibits its own separate proton resonance signal, while at room temperature, where the rate of interconversion is about 4×10^3 sec^{-1}, the resonances from both molecules are found in one sharp signal at the mean position. At intermediate temperatures the separate signals broaden and collapse into the single peak as shown in Fig. 1-12.

An example of a fluxional molecule in which the averaging of two different nuclear environments occurs as the result of intramolecular rearrangement is provided[22] by $(C_5H_5)_4Ti$, which has the structure (1-LII). Even at a temperature as low as $-30°$ the five protons in the b and b′ rings are indistinguishable owing to a rapid internal rotation which is characteristic of rings bonded in this way to metal atoms (see Chap. 23). Thus, at $-30°$ there are, effectively, two types of proton in the molecule, ten of one kind in the a and a′ rings and ten of a second kind in the b and b′ rings, and the

[21] J. G. Bullitt, F. A. Cotton and T. J. Marks, *J. Amer. Chem. Soc.*, 1970, **92**, 2155.
[22] J. L. Calderon *et al.*, *J. Amer. Chem. Soc.*, 1970, **92**, 3801.

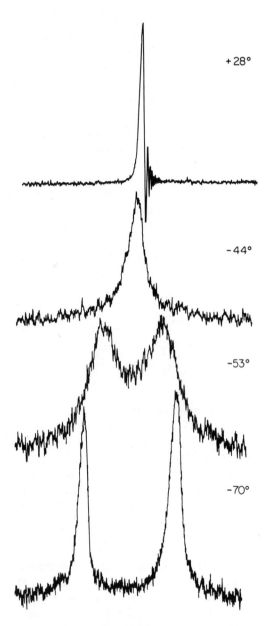

+28°

−44°

−53°

−70°

Fig. 1-12.   The proton nmr spectra of the system *cis*- and *trans*-[h^5-C$_5$H$_5$)Fe(CO)$_2$]$_2$
at several temperatures.

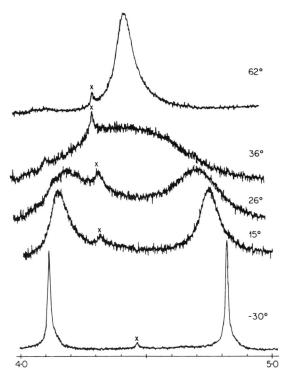

(1-LII)

rate at which these two ring types interchange their roles is slow enough for two separate, sharp signals to be observed. As the temperature is raised, a process in which rings of types a and b exchange roles becomes more and more rapid and eventually rapid enough for the two types of proton environment to be no longer distinguishable in the nmr experiment, as shown in Fig. 1-13. By detailed analysis of the spectra shown in Fig. 1-13 it is possible to show that the fluxional behavior is described by a unimolecular rate constant k given by $k = Ae^{-E_a/RT}$ with $A = 10^{13.5 \pm 0.5}$ and $E_a = 67.4 \pm \pm 1.2$ kJ mol^{-1}.

Fig. 1-13. The proton *nmr* spectra of $(C_5H_5)_4Ti$ at several temperatures. The intermediate spectra were run at higher gain. The peak marked X is due to an impurity.

One of the commonest (Fig. 1-14) types of fluxional behavior is the inversion of pyramidal molecules.[23] In the cases of NH_3 and other simple non-cyclic amines the activation energies, which are equal to the difference between the energies of the pyramidal ground configurations and the planar transition states, are quite low (24–30 kJ mol^{-1}) and the rates of inversion extremely high, e.g., 2.4×10^{10} sec^{-1} for NH_3. Actually, in the case of NH_3 the inversion occurs mainly by quantum-mechanical tunnelling through the barrier rather than by passage over it. In most cases, however, passage over a barrier, that is, a normal activated rate process is operative. With phosphines, arsines, R_3S^+ and R_2SO species the barriers are much higher (>100 kJ mol^{-1}) and inversions are slow enough to allow separation of enantiomers in cases such as RR'R"P and RR'SO.

Fig. 1-14. The inversion of a pyramidal molecule WXYZ. Note that if X, Y and Z are all different the *invertomers* are enantiomorphous.

Among 4-coordinate transition-metal complexes fluxional behavior based on planar–tetrahedral interconversions is of considerable importance. This is especially true of nickel(II) complexes[24] where planar complexes of the type $Ni(R_3P)_2X_2$ have been shown to undergo planar $\rightleftarrows$ tetrahedral rearrangements with activation energies of about 45 kJ mol^{-1} and rates of $\sim 10^5$ sec^{-1} at about room temperature.

Trigonal-bipyramidal Molecules. A class of fluxional molecules of great importance are those with a trigonal-bipyramidal (*tbp*) configuration. Because of their importance and the great amount of information on them, they will be discussed here at some length. When all five appended groups are identical single atoms, as in AB_5, the symmetry of the molecule is D_{3h}. The two apical atoms, B_1 and B_2 [Fig. 1-15(a)] are equivalent but distinct from the three equatorial atoms, B_3, B_4, B_5, which are equivalent among themselves. Experiments, such as measurement of nmr spectra of B nuclei, which can sense directly the kind of environmental difference represented by B_1, B_2 *vs.* B_3, B_4, B_5, should, in general, indicate the presence of two sorts of

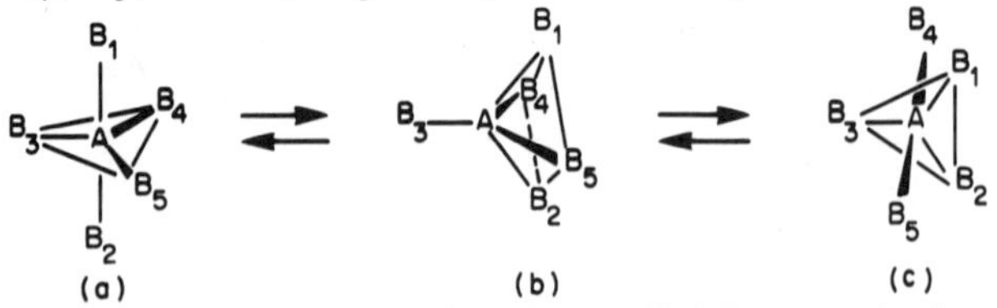

Fig. 1-15. The *tbp–spy–tbp* interconversion, the so-called Berry mechanism or pseudo-rotation for 5-coordinate molecules.

[23] A. Rauk, L. C. Allen and K. Mislow, *Angew. Chem. Internat., Edn.*, 1970, **9**, 400.
[24] L. H. Pignolet, W. DeW. Horrocks, Jr., and R. H. Holm, *J. Amer. Chem. Soc.*, 1970, **92**, 1855.

B nuclei in *tbp* molecules. In many cases, e.g., the ^{13}C spectrum of $Fe(CO)_5$, and the ^{31}P spectrum of PF_5 (to name the two cases where such observations were first made), all five B nuclei appear to be equivalent in the nmr spectrum, even though other experimental data with a shorter time scale, such as diffraction experiments and vibrational spectroscopy, confirm the *tbp* structure.

The accepted explanation for these surprising observations was first suggested by R. S. Berry in 1960 and has been extensively tested by experiment.[25] As noted above (page 26), the *tbp* and *spy* configurations probably differ little in energy in most cases. Berry pointed out that they can also be interconverted by relatively small and simple angle-deformation motions and that in this way axial and equatorial vertices of the *tbp* may be interchanged. Fig. 1-15 shows the "Berry mechanism" for this interchange. The *spy* intermediate (b) is reached by simultaneous closing of the B_1AB_2 angle from $180°$ and opening of the B_4AB_5 angle from $120°$ so that both attain the same intermediate value, thus giving a square set of atoms B_1, B_2, B_4, B_5, all equivalent to each other. This *spy* configuration may then return to a *tbp* configuration in either of two ways, one of which simply recovers the original while the other, as shown, places the erstwhile axial atoms, B_1, B_2, in equatorial positions and the erstwhile equatorial atoms, B_4, B_5, in the axial positions. Note that B_3 remains an equatorial atom and also that the molecule after the interchange is, effectively, rotated by $90°$ about the $A—B_3$ axis. Because of this apparent, but not real, rotation the Berry mechanism is often called a pseudorotation and the atom B_3 is called the pivot atom. Of course, the process can be repeated with B_4 or B_5 as the pivot atom, so that B_3 too will change to an axial position.

If a random series of these pseudorotations occurs rapidly enough, say one every 10^{-4} sec, the B atoms will be moving from one environment to the other so rapidly that they will be "seen" in a conventional nmr experiment as being all in one and the same environment; this is because the characteristic time for the nmr observation, under normal conditions, is many times greater than each individual period of residence of a nucleus at a single vertex.

Because the rearrangement of a trigonal bipyramid is so widely important, it is useful to have a graphical representation of how the various permutations of vertex atoms are interrelated by the relevant pathway for rearrangement, which is assumed to be the Berry mechanism or pseudorotation just discussed. Necessary but not sufficient criteria for defining such a graph are:

1. There must be 20 discrete points, since there are 20 distinct permutations* of atoms 1, 2, 3, 4, 5 over the five vertices of the *tbp*.

[25] G. M. Whitesides and H. L. Mitchell, *J. Amer. Chem. Soc.*, 1969, **91**, 4115; R. R. Holmes, R. M. Deiters and J. A. Golen, *Inorg. Chem.*, 1969, **9**, 2612.

* The number 20 is obtained thus: The *total* number of permutations of the 5 atoms is $5 \times 4 \times 3 \times 2 \times 1 = 120$. However, these may be grouped into sub-sets, each consisting of those that can be interchanged by proper rotations of the *tbp*; those in such a set are not "different" in a physically meaningful sense. The proper rotational symmetry operations for the *tbp* are C_3, C_3^2, C_3^3, $\equiv E$, C_2, C_2', C_2''. Thus each sub-set consists of six permutations. Therefore only $120/6 = 20$ are truly "different" ones. This type of argument a perfectly general one.

2. There must be 30 lines, each representing one pseudorotation and each connecting a pair of points, since each pseudorotation converts any given permutation uniquely into one other one.

3. Three such lines must radiate from each point since there are three different ways to chose the "pivot bond" in executing the pseudorotation.

In addition to these requirements, a correct graph must also meet additional criteria as to the sequence in which one may proceed from one permutation to others not obtainable from it in a single pseudorotation. No simple polyhedron in three-dimensional space can meet *all* requirements;* and thus more complex graphs, that can be regarded as projections of regular polyhedra in higher-dimensional space on three-dimensional space, must be invoked. One that seems particularly useful is shown in Fig. 1-16. It will be

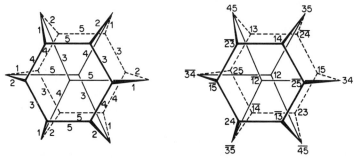

Fig. 1-16. A graph representing the 20 physically different permutations of B nuclei in AB_5 and the 30 ways of interconverting them. See text for explanation of numbers. [Adapted from Figures in K. Mislow, *Accounts Chem. Res.*, 1970, **3**, 321.]

seen to satisfy the three criteria provided we take all 20 vertices to be equivalent, even though, as seen in three-dimensional space projection, they do not appear to be. There are two complementary ways of labelling the graph. One is to identify the permutation represented by each point. This can be done stating the numbers of the atoms at the apical vertices and then distinguishing the two mirror images (1-LIII) and (1-LIV) of 14, for example, by

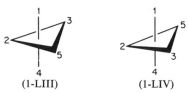

(1-LIII)   (1-LIV)

14 and $\overline{14}$. The unbarred number is used when the equatorial numbers run clockwise viewed from the lower-numbered axial vertex. Alternatively, the graph may be labelled by numbering each edge with the number of the pivot atom.

_____

* The pentagonal dodecahedron, Fig. 1-7(a), meets each of the three necessary but *not sufficient* criteria just listed. The reader may profit by convincing himself that it is, nevertheless, not a correct graph.

Note that this graph states that the fewest number of distinct pseudo-rotations that will return a given permutation to itself is 6, while enantiomorphous permutations can be interconverted in no less than 5 such steps. By direct testing of the Berry pseudorotations it may be demonstrated that these statements are correct and that this particular graph does faithfully express all the relationships involved.

Some important, specific applications of the foregoing analysis to the chemistry of phosphorus will be found in Chapter 13.

Systems with Coordination Numbers of Six or More. The octahedron is usually rather rigid, and fluxional or rapid tautomeric rearrangements generally do not occur in octahedral complexes unless metal–ligand bond-breaking is involved. Among the few exceptions are certain iron and ruthenium complexes of the type $M(PR_3)_4H_2$.[26] The *cis*- and *trans*-isomers of $Fe[PPh(OEt)_2]_4H_2$, for example, have separate, well-resolved signals at $-50°$ which broaden and collapse as the temperature is raised until at $+60°$ there is a single sharp multiplet indicative of rapid interconversion of the two isomeric structures. The preservation of the $^{31}P-^1H$ couplings affords proof that the rearrangement process is non-dissociative. The distortion modes postulated[26] to account for the interconversions are shown in Fig. 1-17.

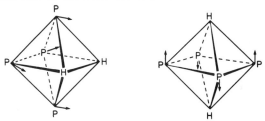

Fig. 1-17.   The types of distortion postulated to lead to interconversion of *cis*- and *trans*-isomers of $Fe[PPh(OEt)_2]_4H_2$.

Stereochemical non-rigidity, especially if of fluxional character, seems likely to be consistently characteristic of complexes with coordination numbers of seven or greater. All seven-coordinate complexes so far investigated by nmr techniques have shown ligand-atom equivalence even though there is no plausible structure for a seven-coordinate complex that would give static or instantaneous equivalence. Thus, ReF_7 and IF_7, for example, are presumed to be fluxional.

There is evidence that eight-coordinate complexes with dodecahedral structures [Fig. 1-10(b)], in which there are two non-equivalent sets of ligands, can interchange ligands rapidly among these sets.[27] A very ready means of doing this can be envisioned,[28] although there is no proof that this mechanism is correct. The dodecahedron is shown in Fig. 1-18(a) in a different orientation from that in Fig. 1-10(b), but the A- and B-type vertices are

[26] E. L. Muetterties, *Accounts Chem. Res.*, 1970, **3**, 266.
[27] E. L. Muetterties, *Inorg. Chem.*, 1965, **4**, 769.
[28] E. L. Muetterties, *J. Amer. Chem. Soc.*, 1969, **91**, 1636.

again labelled. It will be noted that the A and B vertices differ in that each A vertex has only four next-nearest neighbors, while a B vertex has five. This is because a given A vertex is adjacent to three B vertices and one other A vertex, while a given B vertex is adjacent to three A vertices and two other B vertices. Note that, if the B_1—B_2 and B_3—B_4 edges of the dodecahedron are lengthened so that $A_1B_2A_2B_1$ and $A_3B_3A_4B_4$ become square sets, then the dodecahedron is transformed into a square antiprism [Fig. 1-18(b)]. It would be

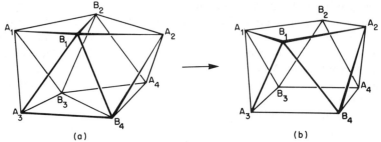

(a)   (b)

Fig. 1-18. A sketch showing how a dodecahedral set (a) of eight ligands may easily rearrange to a square antiprismatic configuration (b).

possible to transform it into a differently labelled square antiprism by lengthening the B_1—B_4 and B_2—B_3 edges appropriately. The square antiprism can then return to the dodecahedron whence it came or, by having the A_1—A_2 and A_3—A_4 pairs approach each other, it may become a dodecahedron in which the former B ligands are A ligands and *vice versa*. Thus, just as the *spy* is a transition state or short-lived intermediate in the interconversion of differently permuted sets of *tbp* ligands, so the square antiprism serves as the necessary connecting link between differently permuted dodecahedral sets.

In the case of 9-coordinate species, where the ligands adopt the D_{3h} capped trigonal prism arrangement shown in Fig. 1-11, there is also an easy pathway for interchanging ligands of the two sets, and for species such as ReH_9^{2-}, $ReH_8PR_3^-$, and $ReH_7(PR_3)_2$ attempts to detect by nmr the presence of hydrogen atoms in two different environments have failed. Fig. 1-19 shows the probable form of the rearrangement which causes the rapid exchange.

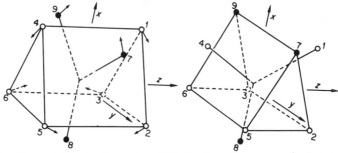

Fig. 1-19. The postulated pathway by which the ligands may pass from one type of vertex to the other in the D_{3h} tricapped trigonal prism.

While coordination numbers higher than nine are not of general importance, polyhedra with 10 and 12 vertices are known. Those that are of greatest importance and undergo intramolecular rearrangements are formed by boron, and discussion of this will be found in Chapter 8.

Organometallic Molecules. Many organometallic molecules involving unsaturated organic groups are fluxional. More detailed discussion and references will be found in Chapter 23, but a few illustrative examples may be cited here.

The compound $C_8H_8Ru(CO)_3$ has the structure (1-LV) in the crystal, and at low temperatures ($< -130°$) in solution its nmr spectrum indicates the same structure since it implies the existence of four different environments for protons. At room temperature, however, the proton resonance spectrum consists of a single sharp line. A detailed study[29] of the way in which the low-temperature spectrum collapses led to the conclusion that the rearrangement involves an infinite, reversible sequence of shifts of the type shown schematically in (1-LVI).

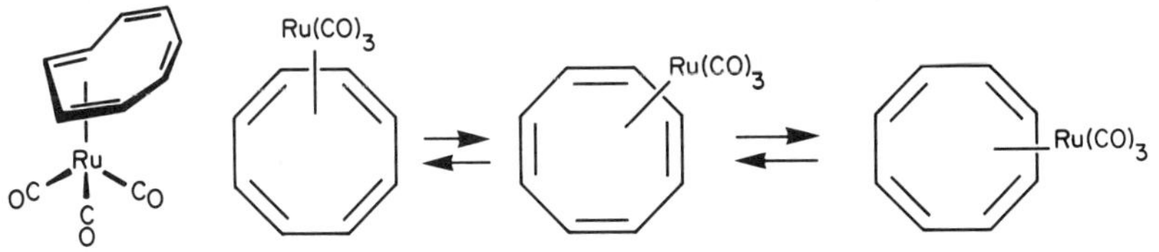

(1-LV)   (1-LVI)

The compound (1-LVIIa) has a spectrum at $-60°$ consistent with the structure shown, which is the one to be expected, but at $+30°C$ it has only a single sharp line. The $Fe(CO)_4$ group must move rapidly among the four positions shown in (1-LVIIb), thereby averaging the environments of the methyl groups. Detailed study[30] indicates that these shifts occur unimolecularly with an activation energy of 38 ± 8 kJ mol^{-1}.

(1-LVIIa)   (1-LVIIb)

[29] F. A. Cotton et al., J. Amer. Chem. Soc., 1969, **91**, 6598.
[30] R. Ben-Shoshan and R. Pettit, J. Amer. Chem. Soc., 1967, **89**, 2231.

Further Reading

Cotton, F. A., Accounts Chem. Res., 1968, **1**, 257 (review of fluxional organometallic molecules).
Cotton, F. A., Chemical Applications of Group Theory, 2nd ed., Wiley-Interscience, **1971**.
Cotton, F. A., Quart. Rev., 1966, **20**, 389 (reviews polyhedra of transition metal atoms).
Emsley, J. W., J. Feeney and L. H. Sutcliffe, High Resolution Nuclear Magnetic Resonance Spectroscopy, Vol. 1, Pergamon Press, **1965** (Chapter 9 covers basic theory of the effect of site exchange on nmr line shapes).

Hawkins, C. J., *Absolute Configuration of Metal Complexes*, Wiley-Interscience, **1971**.

Jaffe, H. H. and M. Orchin, *Symmetry in Chemistry*, Wiley, **1965**.

Kepert, D. L. and K. Vrieze, in *Halogen Chemistry*, Vol. 3, V. Gutmann, ed., Academic Press, **1967**, p. 1 (review of metal-atom clusters containing M_6X_{12}, M_6X_8, M_3X_9 and other groupings).

Mislow, K., *Accounts Chem. Res.*, 1970, **3**, 321 [role of psuedorotation (of PX_5 systems) in displacement reactions].

Mislow, K., *Introduction to Stereochemistry*, W. A. Benjamin, **1965** (the examples are nearly all organic, but the treatment is sound and of value to inorganic chemists).

Muetterties, E. L., *Accounts Chem. Res.*, 1970, **3**, 267 [some aspects of stereochemically non-rigid (fluxional) molecules; references to nearly all previous literature are given here].

Muetterties, E. L. and W. Knoth, *Polyhedral Boranes*, Marcel Dekker, **1968**.

Penfold, B. R., in *Perspectives in Structural Chemistry*, Vol. 2, J. D. Dunitz and J. A. Ibers, eds., Wiley, **1968** (encyclopedic review of metal-atom cluster and cage structures).

Schonland, D. S., *Molecular Symmetry: An Introduction to Group Theory and Its Uses in Chemistry*, Van Nostrand, 1965.

Ugi, I., et al., *Angew. Chem. Internat. Edn.*, 1971, **10**, 687 (an extensive review on non-rigidity).

Wells, A. F., *Structural Inorganic Chemistry*, 3rd ed., Oxford University Press, **1962**.

Wheatley, P. J., *The Determination of Molecular Structure*, Oxford University Press, **1960** (good survey of experimental methods).

Wyckoff, R. W. G., *Crystal Structures*, 2nd ed., Volumes 1–5, Wiley, **1963-1966** (encyclopedic and critical collection of structural data obtained by crystallography).

2

Ionic Solids and Other Extended Arrays

Inorganic chemistry deals not only with substances built of discrete molecules, but also with a great many substances that consist of infinite arrays of atoms. Most of the elements themselves have the latter type of structure. Thus, more than half the elements are metals in which close-packed arrays of atoms are held together by delocalized electrons, while others, such as carbon, silicon, germanium, red and black phosphorus and boron involve infinite networks of more localized bonds. There are also many compounds, such as SiO_2, SiC, etc., in which the array is held together by localized heteropolar bonds. The degree of polarity varies, of course, and this class of substances grades off towards the limiting case of the ionic arrays in which there are well-defined ions held together principally by the coulombic forces between those of opposite charge.

There are also solids that consist neither of small, well-defined molecules nor of well-ordered infinite arrays of atoms; examples are the glasses and polymers, which, for reasons of space, will not be explicitly discussed here. It is, of course, true that most molecular substances form a crystalline solid phase; but, because of the relatively weak intermolecular interactions crystallinity is usually of little chemical importance, though, of course, of enormous practical significance in that it facilitates the investigation of molecular structures, namely, by X-ray crystallography.

2-1. Close Packing of Spheres

The packing of spherical atoms or ions in such a way that the greatest number occupy each unit of volume is one of the most fundamental structural patterns of Nature. It is seen in its simplest form in the solid noble gases, where spherical atoms are concerned, in a variety of ionic oxides and halides where small cations can be considered to occupy interstices in a close-packed array of the larger spherical anions and in metals where close-packed arrays of metal ions are permeated by a cloud of delocalized electrons binding them together.

All close-packed arrangements are built by stacking of close-packed layers

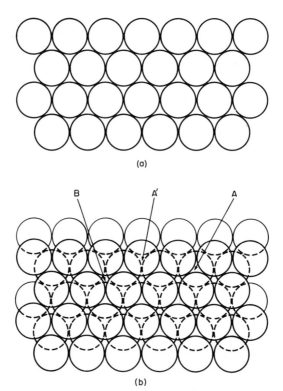

Fig. 2-1. (a) One close-packed layer of spheres. (b) Two close-packed layers showing
how tetrahedral (A, A′) and octahedral (B) interstices are formed.

of the type shown in Fig. 2-1(a); it should be evident that this is the densest
packing arrangement in two dimensions. Two such layers may be brought
together as shown in Fig. 2-1(b), spheres of one layer resting in the declivities
of the other; this is the densest packing arrangement of the two layers; it
will be noted that between the two layers there exist interstices of two types:
tetrahedral and octahedral.

When we come to add a third layer to the two already stacked, two possibilities arise. The third layer can be placed so that its atoms lie directly
over those of the first layer, or with a displacement relative to the first
layer, as in Fig. 2-2(a). These two stacking arrangements may be denoted
ABA and ABC, respectively. Each may be continued in an ordered fashion
so as to obtain

Hexagonal close packing (*hcp*): ABABAB...
Cubic close packing (*ccp*): ABCABCABC...

It is immediately obvious that the *hcp* arrangement does indeed have
hexagonal symmetry, but the cubic symmetry of what has been designated
ccp may be less evident. Fig. 2-2(b) provides another perspective, which
emphasizes the cubic symmetry; it shows that the close-packed layers lie

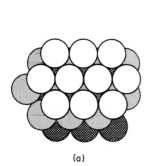

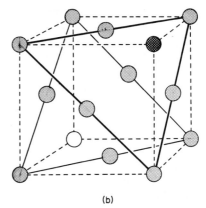

(a)   (b)

Fig. 2-2. (a) The stacking of close-packed layers in the ABC pattern which, on repetition, gives cubic closest packing, *ccp*. (b) Another view of the *ccp* pattern emphasizing its cubic symmetry.

perpendicular to body diagonals of the cube. Moreover, the cubic unit cell is not primitive but face-centered.

There are, of course, an infinite number of stacking sequences possible within the definition of close packing, all of them, naturally, having the same packing density. The *hcp* and *ccp* sequences are those of maximum simplicity and symmetry. Some of the more complex sequences are actually encountered in Nature, though far less often than the two just described.

In *any* close-packed arrangement, each atom has 12 nearest neighbors, six surrounding it in its own close-packed layer, three above and three below this layer. In the *hcp* structure each layer is a plane of symmetry and the set of nearest neighbors of each atom has D_{3h} symmetry. In *ccp* the set of nearest neighbors has D_{3d} symmetry.

2-2. Ionic Radii and Ionic Crystal Structures

Ionic Radii. It is obvious from the nature of wave functions that no ion or atom has a precisely defined radius. The only way radii can be assigned is to determine how closely the centers of two atoms or ions actually approach each other in solid substances and then to assume that such a distance is equal or closely related to the sum of the radii of the two atoms or ions. Even this procedure is potentially full of ambiguities, and further provisions and assumptions are required to get an empirically useful set of radii. The most ambitious attempt to handle this problem is that of Pauling, and some of his arguments and results will be briefly summarized here.

We begin with the four salts NaF, KCl, RbBr and CsI, in each of which the cation and anion are isoelectronic and the radius ratios ($r_{cation}/r_{anion} = r_+/r_-$) should be similar in all four cases. Two assumptions are then made:

(1) The cation and anion are assumed to be in contact so that the internuclear distance can be set equal to the sum of the radii.

(2) For a given noble-gas electron configuration the radius is assumed to be inversely proportional to the effective nuclear charge felt by the outer electrons.

The implementation of these rules may be illustrated by using NaF, in which the internuclear distance is 2.31 Å. Hence,

$$r_{Na^+} + r_{F^-} = 2.31 \text{ Å}$$

Next, using rules developed by Slater to estimate how much the various electrons in the $1s^2 2s^2 2p^6$ configuration shield the outer electrons from the nuclear charge, we obtain 4.15 for the shielding parameter. The effective nuclear charges, Z, felt by the outer electrons are then, for Na^+ with $Z = 11$:

$$11 - 4.15 = 6.85$$

and for F^-, with $Z = 9$:

$$9 - 4.15 = 4.85$$

According to rule (2), the radius ratio r_{Na^+}/r_{F^-} must be inversely proportional to these numbers; hence

$$r_{Na^+}/r_{F^-} = \frac{1}{6.85/4.85} = 0.71$$

Solving this and the previous equation for the sum of the radii simultaneously we obtain

$$r_{Na^+} = 0.95 \text{ Å}$$
$$r_{F^-} = 1.36 \text{ Å}$$

This method, with certain refinements, was used by Pauling to estimate individual ionic radii. Earlier, V. M. Goldschmidt, using a somewhat more empirical method, also estimated ionic radii. The radii for a number of important ions, obtained by the two procedures, are given in Table 2-1. A more recent set of most probable ionic radii is also given in Table 2-1. These are based on a combination of shortest interatomic distances and experimental electron density maps.

Important Ionic Crystal Structures. Fig. 2-3 shows six of the most important structures found among essentially ionic substances. In an ionic structure each ion is surrounded by a certain number of ions of the opposite sign; this number is called the *coordination number* of the ion. In the first three structures shown, namely, the NaCl, CsCl and CaF_2 types, the cations have the coordination numbers 6, 8 and 8, respectively.

The question now arises as to why a particular compound crystallizes with one or another of these structures. To answer this, we first recognize that, ignoring the possibility of metastability, which seldom arises, the compound will adopt the arrangement providing the greatest stability, that is, the lowest energy. The factors that contribute to the energy are the attractive force between oppositely charged ions, which will increase with increasing coordination number, and the forces of repulsion, which will increase very

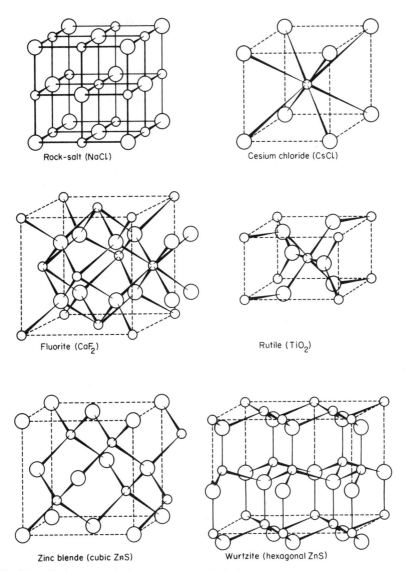

Fig. 2-3. Six important ionic structures. Small circles denote metal cations, large circles denote anions.

rapidly if ions of the same charge are "squeezed" together. Thus the optimum arrangement in any crystal should be the one allowing the greatest number of oppositely charged ions to "touch" without requiring any squeezing together of ions with the same charge. The ability of a given structure to meet these requirements will depend on the relative sizes of the ions.

Let us analyze the situation for the CsCl structure. We place eight negative ions of radius r^- around a positive ion with radius r^+ so that the M^+ to X^-

distance is $r^+ + r^-$ and the adjacent X^- ions are just touching. Then the X^- to X^- distance, a, is given by

$$a = \frac{2}{\sqrt{3}}(r^+ + r^-) = 2r^-$$

or

$$r^-/r^+ = 1.37$$

Now, if the ratio r^-/r^+ is greater than 1.37, the only way we can have all eight X^- ions touching the M^+ ion is to squeeze the X^- ions together. Alternatively, if r^-/r^+ is greater than 1.37, and we do not squeeze the X^- ions, then they cannot touch the M^+ ion and a certain amount of electrostatic stabilization energy will be unattainable. Thus when r^-/r^+ becomes equal

TABLE 2-1

Goldschmidt (G),[a] Pauling (P)[a] and Ladd (L)[a,b] Ionic Radii (in Å)

Ion	G	P	L	Ion	G	P	L
H^-	1.54	2.08	1.39	Pb^{2+}	1.17	1.21	—
F^-	1.33	1.36	1.19				
Cl^-	1.81	1.81	1.70	Mn^{2+}	0.91	0.80	0.93
Br^-	1.96	1.95	1.87	Fe^{2+}	0.83	0.76	0.90
I^-	2.20	2.16	2.12	Co^{2+}	0.82	0.74	0.88
				Ni^{2+}	0.68	0.69	—
O^{2-}	1.32	1.40	1.25	Cu^{2+}	0.72	—	—
S^{2-}	1.74	1.84	1.70				
Se^{2-}	1.91	1.98	1.81	Bi^{3+}	0.2	0.20	—
Te^{2-}	2.11	2.21	1.97	Al^{3+}	0.45	0.50	—
				Sc^{3+}	0.68	0.81	—
Li^+	0.78	0.60	0.86	Y^{3+}	0.90	0.93	—
Na^+	0.98	0.95	1.12	La^{3+}	1.04	1.15	—
K^+	1.33	1.33	1.44	Ga^{3+}	0.60	0.62	—
Rb^+	1.49	1.48	1.58	In^{3+}	0.81	0.81	—
Cs^+	1.65	1.69	1.84	Tl^{3+}	0.91	0.95	—
Cu^+	0.95	0.96	—				
Ag^+	1.13	1.26	1.27	Fe^{3+}	0.53	—	—
Au^+	—	1.37	—	Cr^{3+}	0.53	—	—
Tl^+	1.49	1.40	1.54				
NH_4^+	—	1.48	1.66	C^{4+}	0.15	0.15	—
				Si^{4+}	0.38	0.41	
Be^{2+}	0.34	0.31	—	Ti^{4+}	0.60	0.68	—
Mg^{2+}	0.78	0.65	0.87	Zr^{4+}	0.77	0.80	—
Ca^{2+}	1.06	0.99	1.18	Ce^{4+}	0.87	1.01	—
Sr^{2+}	1.27	1.13	1.32	Ge^{4+}	0.54	0.53	—
Ba^{2+}	1.43	1.35	1.49	Sn^{4+}	0.71	0.71	—
Ra^{2+}	—	1.40	1.57	Pb^{4+}	0.81	0.84	—
Zn^{2+}	0.69	0.74	—				
Cd^{2+}	1.03	0.97	1.14				
Hg^{2+}	0.93	1.10	—				

[a] These radii are obtained by using the *rock-salt type of structure* as standard (i.e. six coordination); small corrections can be made for other structures. For effective ionii radii of M^{3+} in corundum-type oxides of Al, Cr, Ga, V, Fe, Rh, Ti, In and Tl see C. T. Prewitt *et al., Inorg. Chem.*, 1969, **9**, 1985, and for M^{4+} in rutile or closely related oxides of Si, Ge, Mn, Cr, V, Rh, Ti, Ru, Ir, Pt, Re, Os, Tc, Mo, W, Ta, Nb, Sn and Pb see D. B. Rogers *et al., Inorg. Chem.*, 1969, **8**, 841.
[b] M. F. C. Ladd, *Theor. Chim. Acta*, 1968, **12**, 333.

to 1.37 the competition between attractive and repulsive coulomb forces is balanced, and any increase in the ratio may make the CsCl structure unfavorable relative to a structure with a lower coordination number, such as the NaCl structure.

In the NaCl structure, in order to have all ions just touching but not squeezed, with radius r^- for X^- and r^+ for M^+ we have

$$2r^- = \sqrt{2}(r^+ + r^-)$$

which gives for the critical radius ratio:

$$r^-/r^+ = 2.44$$

If the ratio r^-/r^+ exceeds 2.44, then the NaCl structure becomes disfavored, and a structure with cation coordination number 4, for which the critical value of r^-/r^+ is 4.44, may become more favorable. To summarize, in this simple approximation, packing considerations lead us to expect the various structures to have the following ranges of stability in terms of the r^-/r^+ ratio:

CsCl and CaF_2 structures: $1 < r^-/r^+ < 1.37$
NaCl and rutile structures: $1.37 < r^-/r^+ < 2.44$
ZnS structures: $2.44 < r^-/r^+ < 4.44$

Obviously, similar reasoning may be applied to other structures and other types of ionic compound. In view of the fact that the model we are using is a rather crude approximation, we must not expect these calculations to be more than a rough guide. We can certainly expect that in compounds where $r^- \approx r^+$ the CsCl structure will be found, whereas when $r^- \gg r^+$ a structure such as that of ZnS will be preferred.

The more common ionic crystal structures shown in Fig. 2-3 will be mentioned repeatedly throughout the text. The *rutile structure*, named after one mineralogical form of TiO_2, is very common among oxides and fluorides of the MF_2 and MO_2 types (e.g., FeF_2, NiF_2, ZrO_2, RuO_2, etc.) where the radius ratio favors coordination number 6 for the cation. Similarly, the *zinc blende* and *wurtzite structures*, named after two forms of zinc sulfide, are widely encountered when the radius ratio favors four coordination, while the *fluorite structure* is common when eight-coordination of the cation is favored.

When a compound has stoichiometry and ion distribution opposite to that in one of the structures just mentioned, it may be said to have an *anti* structure. Thus, compounds such as Li_2O, Na_2S, K_2S, etc. have the *antifluorite* structure in which the anions occupy the Ca^{2+} positions and the cations the F^- positions of the CaF_2 structure. The antirutile structure is sometimes encountered also.

Structures with Close Packing of Anions.   Many structures of halides and oxides can be regarded as close-packed arrays of anions with cations in the octahedral and/or tetrahedral interstices. Even the NaCl structure can be thought of in this way (*ccp* array of Cl^- ions with all octahedral interstices

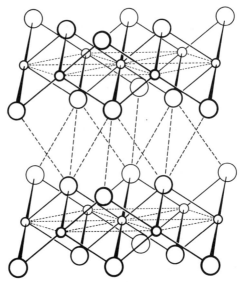

Fig. 2-4. A portion of the CdI_2 structure. Small spheres represent metal cations.

filled) although it is not ordinarily useful to do so. $CdCl_2$ also has *ccp* Cl^- ions with every other octahedral hole occupied by Cd^{2+}, while CdI_2 has *hcp* I^- ions with Cd^{2+} ions in half the octahedral holes. It is noteworthy that the $CdCl_2$ and CdI_2 structures (the latter being shown in Fig. 2-4) are *layer structures*. The particular pattern in which cations occupy half the octahedral holes is such as to leave alternate layers of direct anion–anion contact.

Corundum, the α form of Al_2O_3, has an *hcp* array of oxide ions with two-thirds of the octahedral interstices occupied by cations and is adopted by many other oxides, e.g., Ti_2O_3, V_2O_3, Cr_2O_3, Fe_2O_3, Ga_2O_3 and Rh_2O_3. The BiI_3 structure has an *hcp* array of anions with two-thirds of the octahedral holes in each alternate pair of layers occupied by cations, and it is adopted by $FeCl_3$, $CrBr_3$, $TiCl_3$, VCl_3 and many other AB_3 compounds. As indicated, all the *structures* just mentioned are adopted by numerous *substances*. The structures are usually named in reference to one of these substances. Thus we speak of the NaCl, $CdCl_2$, CdI_2, BI_3 and corundum (or α-Al_2O_3) structures.

Some Mixed Oxide Structures. There are a vast number of oxides (and also some stoichiometrically related halides) having two or more different kinds of cation. Most of them occur in one of a few basic structural types, the names of which are derived from the first or principal compound shown to have that type of structure. Three of the most important such structures will now be described.

1. *The Spinel Structure.* The compound $MgAl_2O_4$, which occurs in Nature as the mineral spinel, has a structure based on a *ccp* array of oxide ions. One-eighth of the tetrahedral holes (of which there are two per anion)

are occupied by Mg^{2+} ions and one-half of the octahedral holes (of which there is one per anion) are occupied by Al^{3+} ions. This structure, or a modification to be discussed below, is adopted by many other mixed metal oxides of the type $M^{II}M_2^{III}O_4$ (e.g., $FeCr_2O_4$, $ZnAl_2O_4$ and $Co^{II}Co_2^{III}O_4$), by some of the type $M^{IV}M_2^{II}O_4$ (e.g., $TiZn_2O_4$ and $SnCo_2O_4$) and by some of the type $M_2^{I}M^{VI}O_4$ (e.g., Na_2MoO_4 and Ag_2MoO_4). This structure is often symbolized as $A[B_2]O_4$, where square brackets enclose the ions in the octahedral interstices. An important variant is the *inverse spinel structure*, $B[AB]O_4$, in which half of the B ions are in tetrahedral interstices and the A ions are in octahedral ones along with the other half of the B ions. This often happens when the A ions have a stronger preference for octahedral coordination than do the B ions. So far as is known, all $M^{IV}M_2^{II}O_4$ spinels are inverse, e.g., $Zn[ZnTi]O_4$, and many of the $M^{II}M_2^{III}O_4$ ones are also, e.g., $Fe^{III}[Co^{II}Fe^{III}]O_4$, $Fe^{III}[Fe^{II}Fe^{III}]O_4$ and $Fe[NiFe]O_4$.

There are also many compounds with *disordered spinel structures* in which only a fraction of the A ions are in tetrahedral sites (and a corresponding fraction in octahedral ones). This occurs when the preferences of both A and B ions for octahedral and tetrahedral sites do not differ markedly.

2. *The Ilmenite Structure.* This is the structure of the mineral ilmenite, $Fe^{II}Ti^{IV}O_3$. It is closely related to the corundum structure except that the cations are of two different types. It is adopted by ABO_3 oxides when the two cations, A and B, are of about the same size, but they need not be of the same charge so long as their total charge is $+6$. Thus in ilmenite itself and in $MgTiO_3$ and $CoTiO_3$ the cations have charges $+2$ and $+4$ while in α-$NaSbO_3$ the cations have charges of $+1$ and $+5$.

3. *The Perovskite Structure.* The mineral perovskite, $CaTiO_3$, has a structure in which the oxide ions and the large cation (Ca^{2+}) form a *ccp* array with the smaller cation (Ti^{4+}) occupying those octahedral holes formed exclusively by oxide ions, as shown in Fig. 2-5. This structure is often slightly distorted (in $CaTiO_3$ itself, for example). It is adopted by a great many ABO_3 oxides in which one cation is comparable in size to O^{2-} and the other much smaller, with the cation charges variable so long as their sum is $+6$. It is found in $Sr^{II}Ti^{IV}O_3$, $Ba^{II}Ti^{IV}O_3$, $La^{III}Ga^{III}O_3$, $Na^{I}Nb^{V}O_3$ and $K^{I}Nb^{V}O_3$, and also in some mixed fluorides, e.g., $KZnF_3$ and $KNiF_3$.

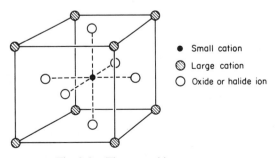

Fig. 2-5. The perovskite structure.

2-3. Ionization Energies and Electron Affinities

The energy necessary to detach an electron from an isolated gaseous atom, ion or molecule is called the ionization energy, measured in kJ mol^{-1}; these same energies are often referred to as ionization potentials, measured in electron volts, eV (per atom). Ionization energies normally have a positive sign according to prevailing convention since energy is normally absorbed. For a given species the energy required to remove the first electron is called the first ionization energy (or potential), and so on.

The first ionization potentials of the elements vary in relation to their positions in the Periodic Table, as shown in Fig. 2-6. With the exception of mercury, the maxima in the curve all occur at noble gases, and the deeper minima all occur with the alkali metals. These facts show that the closed configurations of the noble gases are most difficult to disturb by removal of an electron, whereas the lone electron outside a noble-gas configuration, which is the common feature of each alkali-metal atom, is very easy to remove. Furthermore, although there are irregularities, the potentials rise steeply in going from an alkali metal to the following noble gas. These facts can be accounted for on the basis of the shielding of one electron by others.

As we go from, say, sodium, with a nuclear charge of 11 to argon with a nuclear charge of 18, the eight electrons which are also added are all in the same principal shell. Because the radial wave functions of all electrons in the same shell are either identical or very similar, no one electron spends much time between the nucleus and any other electron. Thus the extent to which these electrons shield one another from the steadily increasing nuclear charge is slight. For example, the effective nuclear charge for the $3s$ electron in sodium is ~ 2, whereas that for the s or p electron in argon is ~ 6.7. It is this substantially greater restraining force which is responsible for the much tighter binding of the electrons.

Close inspection of Fig. 2-6 will show that there is a consistent pattern of deviation from complete linearity in the several series of non-transitional ele-

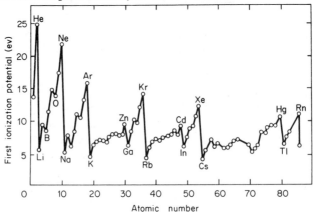

Fig. 2-6.   Variation in first ionization potentials with atomic number.

ments, i.e., the sequences Li→Ne, Na→Ar, Ga→Kr and In→Xe. The Li→Ne series is representative and will suffice for explicit illustration. There are actually small decreases from Be to B and from N to O, which may be explained as follows. At Be the $2s$ shell is filled and the next electron, required in the B atom, enters a $2p$ orbital. The p orbitals are very effectively shielded by the $2s$ electrons and thus the first $2p$ electron is relatively easily removed compared to the second $2s$ electron. The discontinuity between N and O occurs because at N the configuration is $2s^2 2p_x 2p_y 2p_z$, that is, each p orbital is singly occupied. The next p electron to enter must encounter repulsion as it enters an already half-occupied orbital and its binding energy is thus lessened.

For a given atom the ionization potentials always increase, though not always uniformly, in the order $I_1 < I_2 < I_3 < \cdots < I_n$. This is obviously due to the fact that removal of a negative charge from a species of charge $+k$ ($k = 0, 1, 2, \ldots$) must be easier than removal of a negative charge from a similar species of charge $+(k+1)$ or greater.

In general, it is also possible to attach an additional electron to any atom, ion or molecule. The energy which is *released* when this takes place is called the *electron affinity*, E, of the species. It will be noted that this sign convention is *opposite* to the conventional one, i.e.:

$$X(g) + e^- = X^-(g) \qquad \Delta H = -E \, (kJ \, mol^{-1})$$

but this definition of electron affinity is so well established that we must accept it. Negative electron affinities do exist—that is, when the species concerned does not "want" another electron and must be forced to accept it— but the cases of greatest interest are those in which the electron affinity is positive.

Table 2-2 lists some electron affinities.

TABLE 2-2

Electron Affinities for the Process
$X(g) + e \to X^-(g)$ in $kJ \, mol^{-1} [-\Delta H]$ and electron volts (E)

	H							He
	72							−54
	(0.75)							(−0.56)
Li	Be	B	C	N	O	F		Ne
57	−66	15	121	−31	142	333		−99
(0.59)	(−0.68)	(0.16)	(1.25)	(−0.32)	(1.47)	(3.45)		(−1.03)
Na	Mg	Al	Si	P	S	Cl		
21	−67	26	135	60	200	348		
(0.22)	(−0.69)	(0.27)	(1.40)	(0.62)	(3.07)	(3.61)		
						Br		
						324		
						(3.36)		
						I		
						295		
						(3.06)		
						At		
						256		
						(2.69)		

Important examples of negative electron affinities, i.e., those pertaining to energy-absorbing processes, are those for the formation of the oxide and sulfide ions from the oxygen and sulfur atoms. The relevant thermochemical equations are given in Table 2-3.

TABLE 2-3
Thermochemical Equations for Formation
of Oxide and Sulfide Ions

	$-\Delta H$ (kJ mol^{-1})	E (eV)
$O(g)+e \rightarrow O^-(g)$	142	1.47
$O^-(g)+e \rightarrow O^{2-}(g)$	-844	-8.75
$O(g)+2e \rightarrow O^{2-}(g)$	-702	-7.28
$S(g)+e \rightarrow S^-(g)$	200	2.07
$S^-(g)+e \rightarrow S^{2-}(g)$	-532	-5.51
$S(g)+2e \rightarrow S^{2-}(g)$	-332	-3.44

While the addition of one electron to each neutral atom releases energy, as expected, the addition of the second electron in each case requires high energy input, such that the overall 2-electron acquisition is endothermic. This energy can be supplied, as will be seen in the next Section, by the exothermic formation of ionic crystals.

2-4. Lattice Energies

The energy of a set of ions arranged in a regular array, as, for example, in the rock-salt (NaCl) or the rutile (TiO$_2$) structure, is called the *lattice energy* of that array. It is, in general, possible to calculate this energy, by employing the known geometry of the structure and certain properties of the ions. As an example we shall use a compound MX with the NaCl structure (Fig. 2-3).

Let us call the shortest M to X distances r. Using Fig. 2-3 and simple trigonometry it can be shown that an M^+ ion is surrounded by six X^- ions (nearest neighbors) at a distance r, twelve M^+ ions (second nearest neighbors) at a distance $\sqrt{2}r$, eight more distant X^- ions at $\sqrt{3}r$, six M^+ ions at $2r$, twenty-four more X^- ions at $\sqrt{5}r$, etc. The electrostatic interaction energy of the M^+ ion with each of these surrounding ions is equal to the product of the charge on each ion, Z^+ and Z^-, divided by the distance. Hence the total electrostatic energy of this positive ion can be written:

$$E = \frac{6e^2}{r}(Z^+)(Z^-) + \frac{12e^2}{\sqrt{2}r}(Z^+)^2 + \frac{8e^2}{\sqrt{3}r}(Z^+)(Z^-) + \frac{6e^2}{2r}(Z^+)^2 \dots$$

$$= -\frac{e^2|Z|^2}{r}\left(6 - \frac{12}{\sqrt{2}} + \frac{8}{\sqrt{3}} - \frac{6}{2} + \frac{24}{\sqrt{5}} \dots\right) \qquad \dots (2\text{-}1)$$

Actually, it is possible to set up the general formula for the terms in equation 2-1 from geometrical considerations. The sum of all these terms, that is, the sum of an infinite series, is called the *Madelung constant*. It should be clear that the value of the Madelung constant is characteristic of the geometrical arrangement and independent of the particular ions or their charges (i.e., they might both be doubly or triply charged). The above series converges to a value of 1.747558... and can be evaluated to any required degree of accuracy. In many cases the series diverge, and the evaluation of the Madelung constants for such structures requires considerable mathematical manipulation. The Madelung constants for many commonly occurring lattices have been calculated.[1] There are many structures, including the rutile structure, in which the ratios of distances are not uniquely fixed by the symmetry of the structure. For these the Madelung "constant" will vary with the variation of the adjustable parameter or parameters. In such cases, the electrostatic summation (equivalent to a Madelung constant) must be evaluated for an explicit set of ion positions.

If we made a mole (N ion pairs) of the compound MX, then NE would be the energy of the process

$$M^+ (g) + X^- (g) = MX(s) \qquad \ldots (2\text{-}2)$$

This is so because if we wrote out an expression for the electrostatic energy of an X^- ion it would be identical with equation 2-1. If we added the electrostatic energies for a mole of each ion, the result would be twice the true electrostatic energy per mole because we would be counting each pairwise interaction twice. Thus NE is the electrostatic potential energy per mole for an ionic solid with the NaCl structure.

The reason that the ions reside a finite distance, r, from each other and do not move still closer together is because there are short-range repulsive forces due to overlapping of their electron clouds which come into play as they approach. Born proposed the simple assumption that the repulsive force between two ions could be represented by the expression B'/r^n in which B' and n are constants, as yet undetermined, characteristic of the ion pair concerned. We can therefore write, for the repulsive energy of a particular ion in a crystal,

$$E_{rep} = \frac{B}{r^n} \qquad \ldots (2\text{-}3)$$

where B is related to B' by the crystal geometry. The total electrostatic energy of the crystal, that is, the *inherently negative* lattice energy, U, is then given by

$$U = \frac{N(Z^+)(Z^-)Ae^2}{r} + \frac{NB}{r^n} \qquad \ldots (2\text{-}4)$$

where A is the Madelung constant. This energy, U, is exactly the energy of reaction 2-2, if these are the only forces involved. There is a relation between

[1] See T. C. Waddington, *Adv. Inorg. Chem. Radiochem.*, 1959, **1**, 157, for a tabulation.

B and n which may be determined if we recognize that in the equilibrium state of the crystal ($r = r_0$) the energy is a minimum as a function of r. Thus, we have

$$\left(\frac{dU}{dr}\right)_{r=r_0} = 0 = -\frac{A(Z^+)(Z^-)Ne^2}{r_0^2} - \frac{nNB}{r_0^{n+1}} \qquad \text{... (2-5)}$$

which yields

$$B = -\frac{A(Z^+)(Z^-)e^2}{n} r_0^{n-1} \qquad \text{... (2-6)}$$

Substituting equation 2-6 into equation 2-4, we obtain:

$$U = N(Z^+)(Z^-)A\frac{e^2}{r_0}\left(1 - \frac{1}{n}\right) \qquad \text{... (2-7)}$$

The numerical value of n can be derived from measurements of the compressibility of the solid and may also be estimated theoretically. The experimentally derived values and the values calculated by Pauling for noble gas-like ions are given in Table 2-4. It can be seen that the experimental

TABLE 2-4

Sample n Values

Determined by experiment		Estimated from theory	
Compound	n	Noble-gas configuration of ion	n
LiF	5.9	He	5
LiCl	8.0	Ne	7
LiBr	8.7	Ar	9
NaCl	9.1	Kr	10
NaBr	9.5	Xe	12

values are reasonably close to the averages of the appropriate theoretical estimates. It can also be seen that even if the n used is off by 1.0, the lattice energy will be in error only by 1–2%.

In very accurate calculations some correction factors are required because equation 2-4 does not take account of other minor forces. There are three main refinements.

1. *Inclusion of van der Waals Forces.* van der Waals forces operate between all atoms, ions or molecules, but are relatively very weak. They are due to attractions between oscillating dipoles in adjacent atoms and vary approximately as $1/r^6$. They can be calculated from the polarizabilities and ionization potentials of the atoms or ions.

2. *Use of a More Rigorous Expression for the Repulsive Energy.* The simple Born expression (equation 2-3) for the repulsive energy is not strictly correct from quantum-mechanical considerations. More refined expressions do not greatly change the results, however.

3. *Consideration of "Zero Point Energy" of the Crystal.* The "zero point energy" of the crystal is that energy of vibration of the ions which the crystal

TABLE 2-5

Components of Lattice Energy (in electron volts)

Energy	LiF	NaCl	CsI
Coulomb	−12.4	−8.92	−6.4
Repulsion	+1.9	+1.03	+0.63
van der Waals	−0.17	−0.13	−0.48
Zero point	+0.17	+0.08	+0.3

possesses even at the absolute zero. This can be calculated from the lattice vibration frequencies.

The data given in Table 2-5 indicate the relative importance of the various contributions to the lattice energy.

The calculation of lattice energies of ionic compounds is very important since, in general, there is no *direct* way to measure them experimentally, although they can be obtained from certain experimental data using the Born–Haber cycle which is discussed immediately below. For example, the heat of vaporization of NaCl does not give the lattice energy because up to the highest temperatures at which accurate measurements can be made the gas phase consists of NaCl molecules (or ion pairs), and it has so far proved impossible to get an accurate estimate of the heat of dissociation of NaCl(g) into $Na^+(g)$ and $Cl^-(g)$ since NaCl(g) normally dissociates into atoms.

2-5. The Born–Haber Cycle

The heats of formation of various ionic compounds show tremendous variations. In a general way, we know that many factors contribute to the over-all heat of formation, namely, the ionization potentials, electron affinities, heats of vaporization and dissociation of the elements, and the lattice energy of the compound. The Born–Haber cycle is a thermodynamic cycle that shows the interrelation of these quantities and enables us to see how variations in heats of formation can be attributed to the variations in these individual quantities. In order to construct the Born–Haber cycle we consider the following thermochemical equations, using NaCl as an example

$$
\begin{array}{ll}
Na(s) = Na(g) & \Delta H_{subl}(Na) \\
Na(g) = Na^+(g) + e^- & I_{Na} \\
\tfrac{1}{2}Cl_2(g) = Cl(g) & \tfrac{1}{2}\Delta H_{diss}(Cl_2) \\
Cl(g) + e^- = Cl^-(g) & E_{Cl} \\
Na^+(g) + Cl^-(g) = NaCl(s) & U \\
\hline
Na(s) + \tfrac{1}{2}Cl_2(g) = NaCl(s) & \Delta H_f(NaCl)
\end{array}
$$

The net change expressed by the last equation can be achieved by carrying out the preceding five steps successively, as indicated in Fig. 2-7, which is an example of the Born–Haber cycle.

The energies are interrelated by the equation

$$ \Delta H_f = \Delta H_{subl} + I + \tfrac{1}{2}\Delta H_{diss} - E_{Cl} + U \qquad \dots (2\text{-}8) $$

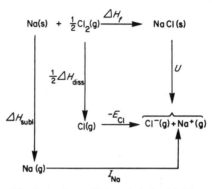

Fig. 2-7. Born–Haber cycle for NaCl.

The Born–Haber cycle is used to calculate any one of the quantities in equation 2-8 when all the others are known, or to provide a check on the internal consistency of a complete set of these quantities. Normally, ΔH_f, ΔH_{subl}, I and ΔH_{diss} are known. Direct measurement of electron affinities is usually rather difficult, and only for the halogens have really accurate values been obtained. In these cases the cycle can then be used as a check on the calculated lattice energies, which, when all refinements are included, are found to be quite accurate. For example, the calculated lattice energy of NaCl is 7.94 eV, and the value obtained from the Born–Haber cycle is 7.86 eV, a difference of $\sim 1\%$. Since we have such checks to give us confidence in the accuracy of computed lattice energies, the cycle is more commonly used to determine electron affinities. For example, the electron affinity of oxygen $(O^{2-}, -7.3$ eV) is a very important quantity which cannot be measured directly, if for no other reason than that it is highly endothermic (see above). It is obtained by applying the Born–Haber cycle to various ionic oxides.

The Born–Haber cycle is also valuable as a means of analyzing and correlating the variations in stability of various ionic compounds. As an example, it enables us to explain why MgO is a stable ionic compound despite the fact that the Mg^{2+} and O^{2-} ions are both formed endothermally, not to mention the considerable energies required to vaporize Mg(s) and to dissociate $O_2(g)$. ΔH_f is highly negative despite these opposing tendencies because the lattice energy of MgO more than balances them out.

The Born–Haber cycle also enables us to understand why most metals fail to form stable ionic compounds in low valence states (e. g., compounds such as CaCl, AlO and ScCl$_2$). Let us consider a metal with 1st and 2nd ionization energies of 600 and 1200 kJ mol^{-1}, which are fairly typical values (of Ca, for example). Let us suppose that this metal forms a $+2$ ion with a radius of 1.00 Å and that its dichloride would have the fluorite structure (as does CaCl$_2$). The M^+ ion would have to be appreciably larger than the M^{2+} ion and a radius of ~ 1.20 Å is a fair estimate. With a radius ratio of ~ 1.5, MCl may be expected to have the NaCl structure. For the two compounds, MCl and MCl$_2$ then, the Born–Haber cycles are as shown in Table 2-6.

TABLE 2-6

Both–Haber cycles for MCl and MCl_2

MCl	Energy $(kJ\ mol^{-1})$
(1) $M(s) = M(g)$	ΔH_{subl}
(2) $M(g) = M^+(g)$	600
(3) $\frac{1}{2}Cl_2(g) = Cl(g)$	121
(4) $Cl(g) + e^- = Cl^-(g)$	-348
(5)[a] $M^+(g) + Cl^-(g) = MCl(s)$	-719
(6) $M(s) + \frac{1}{2}Cl_2(g) = MCl(s)$	$-346 + \Delta H_{subl}$

MCl_2	Energy $(kJ\ mol^{-1})$
(1) $M(s) = M(g)$	ΔH_{subl}
(2) $M(g) = M^{2+}(g)$	1800
(3) $Cl_2(g) = 2Cl(g)$	242
(4) $2Cl(g) + 2e^- = 2Cl^-(g)$	-696
(5)[a] $M^{2+}(g) + 2Cl^-(g) = MCl_2(s)$	-2218
(6) $M(s) + Cl_2(g) = MCl_2(s)$	$-872 + \Delta H_{subl}$

[a] Calculated by using Madelung constants of 1.748 and 5.039 for the NaCl and CaF_2 structures, respectively, with $r_{Cl^-} = 1.81$ Å and n (the Born exponent) equal to 9.

It is evident that, although the energies of reactions (1) to (4) favor MCl over MCl_2 (by 973 kJ mol^{-1}), this is overbalanced by the superior lattice energy of MCl_2. From the above figures we can calculate for the reaction

$$2MCl(s) = MCl_2(s) + M(s)$$

that $\Delta H = -180 - \Delta H_{subl}$. Since ΔH_{subl} will be 130–150 kJ mol^{-1} for a divalent metal such as Ca, the ΔH of the above reaction will be -310 to -330 kJ mol^{-1}. It should be noted that, of course, the true measure of the stability of MCl against disproportionation would be ΔF for the above reaction, not ΔH. However, it may be stated with confidence that the entropy change for such a reaction, in which all reactants and products are solids, must be very close to zero. The enthalpy is thus a reliable guide in *this* type of problem.

2-6. Covalent Solids

Elements. Those elements that form extended covalent (as opposed to metallic) arrays are boron, all the Group IV elements except lead, also phosphorus, arsenic, selenium and tellurium. All other elements form either only metallic phases or only molecular ones. Some of the above elements, of course, have allotropes of metallic or molecular type in addition to the phase or phases that are extended covalent arrays. For example, tin has a metallic allotrope (white tin) in addition to that with the diamond structure (grey tin), and selenium forms two molecular allotropes containing Se_8 rings, isostruc-

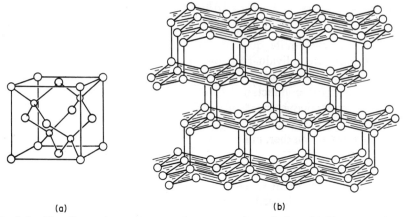

(a)   (b)

Fig. 2-8. The diamond structure seen from two points of view. (a) The conventional cubic unit cell. (b) A view showing how layers are stacked; these layers run perpendicular to the body diagonals of the cube.

tural with the rhombic and the monoclinic form of sulfur. For tellurium we have a situation on the borderline of metallic behavior.

The structures of the principal allotropic forms of all the elements will be discussed in detail as the chemistry of each element is treated. For illustrative purposes, we shall mention here only one such structure, the diamond structure, since this is adopted by several other elements and is a point of reference for various other structures. It is shown from two points of view in Fig. 2-8. The structure has a cubic unit cell with the full symmetry of the group T_d. However, it can, for some purposes, be viewed as a stacking of puckered infinite layers. It will be noted that the zinc blende structure (Fig. 2-3) can be regarded as a diamond structure in which one-half of the sites are occupied by Zn^{2+} (or other cation) while the other half are occupied by S^{2-} (or other anion) in an ordered way. In the diamond structure itself all atoms are equivalent, each being surrounded by a perfect tetrahedron of four others. The electronic structure can be simply and fairly accurately described by saying that each atom forms a localized two-electron bond to each of its neighbors.

Compounds. As soon as one changes from elements, where the adjacent atoms are identical and the bonds are necessarily non-polar, to compounds, there enters the vexatious question of when to describe a substance as ionic and when to describe it as covalent. No attempt will be made here to deal with this question in detail for the practical reason that, very largely, there is no need to have the answer—even granting, for the sake of argument only, that any such thing as "the answer" exists. Suffice it to say that bonds between unlike atoms all have some degree of polarity and (a) when the polarity is relatively small it is practical to describe the bonds as polar covalent ones and (b) when the polarity is very high it makes more sense to consider that the substance consists of an array of ions.

2-7. Metals

It is evident, even on the most casual inspection, that metals have many physical properties quite different from those of other solid substances. Although there are individual exceptions to each, the following may be cited as the characteristic properties of metals as a class: (1) high reflectivity, (2) high electrical conductance, decreasing with increasing temperature, (3) high thermal conductance, and (4) mechanical properties such as strength and ductility. An explanation for these properties, and for their variations from one metal to another, must be derived from the structural and electronic nature of the metal.

Metal Structures. Practically all metal phases have one of three basic structures, or some slight variation thereof, although there are a few exceptional structures which need not concern us here. The three basic structures

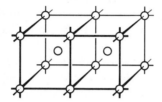

Fig. 2-9. A body-centred cubic, *bcc*, structure.

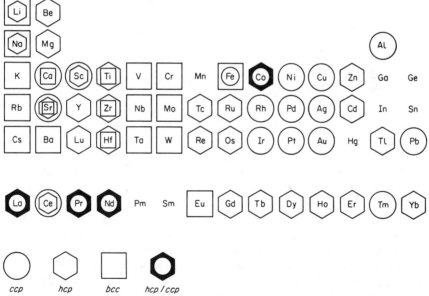

Fig. 2-10. The occurrence of hexagonal close-packed (*hcp*), cubic close-packed (*ccp*) and body-centered (*bcc*) structures among the elements. Where two or more symbols are used, the largest represents the stable form at 25 °C. The symbol labelled *hcp/ccp* signifies a mixed ... ABCABABCAB ... type of close packing, with overall hexagonal symmetry. [Adapted, with permission, from H. Krebs, *Grundzüge der anorganishen Kristallchemie*, F. Enke Verlag, 1968.]

are cubic and hexagonal close-packed, which have already been presented in Section 2.1, and body-centered cubic, *bcc*, illustrated in Fig. 2-9. In the *bcc* type of packing each atom has only eight instead of 12 nearest neighbors, although there are six next nearest neighbors that are only about 15% further away. It is only 92% as dense an arrangement as the *hcp* and *ccp* structures. The distribution of these three structure types, *hcp*, *ccp*, and *bcc*, in the Periodic Table is shown in Fig. 2-10. The majority of the metals deviate slightly from the ideal structures, especially those with *hcp* structures. For the *hcp* structure the ideal value of c/a, where c and a are the hexagonal unit-cell edges, is 1.633, while all metals having this structure have a smaller c/a ratio (usually 1.57–1.62) except zinc and cadmium for which c/a values are 1.86 and 1.89, respectively. While such deviations cannot in general be predicted, their occurrence is not particularly surprising since for a given atom its six in-plane neighbors are not symmetrically equivalent to the set of six lying above and below it, and there is consequently no reason why its bonding to those in the two non-equivalent sets should be precisely the same.

Metallic Bonding. The characteristic physical properties of metals as well as the high coordination numbers (either 12 or 8 nearest neighbors plus 6 more that are not too remote) suggest that the bonding in metals is different from that in other substances. Clearly there is no ionic contribution, and it is also obviously impossible to have a fixed set of ordinary covalent bonds between all adjacent pairs of atoms since there are neither sufficient electrons nor sufficient orbitals. Attempts have been made to treat the problem by invoking an elaborate resonance of electron-pair bonds among all the pairs of nearest neighbor atoms and this approach has had a certain degree of success. However, the main thrust of theoretical work on metals is in terms of the *band theory*, which gives in a very natural way an explanation for the electrical conductance, luster and other characteristically metallic properties. While a detailed explanation of band theory would necessitate a level of mathematical sophistication beyond that appropriate for this book, the qualitative features are sufficient.

Let us imagine a block of metal expanded, without change in the geometric relationships between the atoms, by a factor of, say, 10^6. The interatomic distances would then all be 10^2 greater, that is, about 300–500 Å. Each atom could then be described as a discrete atom with its own set of well-defined atomic orbitals. Now let us suppose the array contracts, so that the orbitals of neighboring atoms begin to overlap and hence interact with each other. Since there are so many atoms involved this gives rise, at the actual internuclear distances in metals, to sets of states so close together as to form essentially continuous energy bands, as illustrated in Fig. 2-11. Spacially these bands are spread through the metal, and the electrons that occupy them are completely delocalized. In the case of sodium, shown in Fig. 2-11, the $3s$ and $3p$ bands overlap.

Another way to depict energy bands is that shown in Fig. 2-12. Here energy is plotted horizontally and the envelope indicates on the vertical the

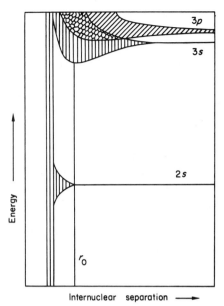

Fig. 2-11. Energy bands of sodium as a function of internuclear distance. r_0 represents the actual equilibrium distance. [Reproduced by permission from J. C. Slater, *Introduction to Chemical Physics*, McGraw-Hill Book Co., 1939.]

number of electrons that can be accomodated at each value of the energy. Shading is used to indicate filling of the bands.

Completely filled or completely empty bands, as shown in Fig. 2-12(a) do not permit net electron flow and the substance is an insulator. Covalent solids can be discussed from this point of view (though it is unnecessary to do so) by saying that all electrons occupy low-lying bands (equivalent to the

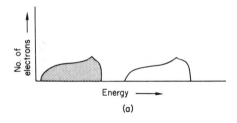

(a)

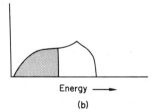

(b)

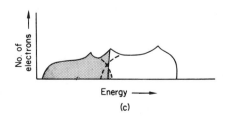

(c)

Fig. 2-12. Envelopes of energy bands, with shading to indicate filling.

bonding orbitals) while the high-lying bands (equivalent to antibonding orbitals) are entirely empty. Metallic conductance occurs when there is a partially filled band, as in Fig. 2-12(b); the transition metals, with their incomplete sets of d electrons, have partially filled d bands and this accounts for their high conductances. The alkali metals have half-filled s bands formed from their s orbitals, as shown in Fig. 2-11; actually these s bands overlap the p bands and it is because this overlap occurs also for the Ca Group metals, where the atoms have filled valence shell s orbitals, that they nevertheless form metallic solids, as indicated in Fig. 2-12(c).

Cohesive Energies of Metals. The strength of binding among the atoms in metals can conveniently be measured by the enthalpies of atomization. Fig. 2-13 shows a plot of the energies of atomization of the metallic elements, lithium to bismuth, from their standard states. It is first notable that cohesive

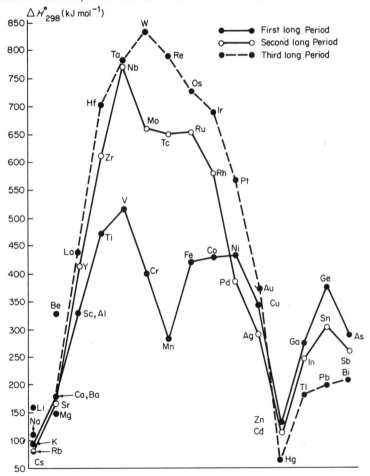

Fig. 2-13. Heats of atomization of metals, ΔH°_{298} for $M(s) \rightarrow M(g)$. [Reproduced by permission from W. E. Dasent, *Inorganic Energetics*, Penguin Books, Ltd., 1970.]

energy tends to maximize with elements having partially filled d shells, i.e., with the transition metals. However, it is particularly with the elements near the middle of the 2nd and 3rd transition series, especially Nb–Ru and Hf–Ir, that the cohesive energies are largest, reaching 837 kJ mol^{-1} for tungsten. It is noteworthy that these large cohesive energies are principally due to the structural nature of the metals whereby high coordination numbers are achieved. For a *hcp* or *ccp* structure, there are six bonds per metal atom (since each of the 12 nearest neighbors has a half share in each of the 12 bonds). Therefore, each bond, even when cohesive energy is 800 kJ mol^{-1}, has an energy of only 133 kJ mol^{-1}, roughly half the C—C bond energy in diamond where each carbon atom has only four near neighbors.

2-8. Defect Structures

All the foregoing discussion of crystalline solids has dealt with their perfect or ideal structures. Such perfect structures are seldom if ever found in real substances and, while low levels of imperfections have only small effects[2] on their chemistry, the physical (i.e., electrical, magnetic, optical and mechanical) properties of many substances are often crucially affected by their imperfections. It is, therefore, appropriate to devote a few paragraphs to describing the main types of imperfection, or defect, in real crystalline solids. We shall not, however, discuss the purely mechanical imperfections such as mosaic structure, stacking faults, and dislocations, all of which are some sort of mismatch between lattice layers.

Stoichiometric Defects. There are some defects that leave the stoichiometry unaffected. One type is the *Schottky defect* in ionic crystals. A Schottky defect consists of vacant cation and anion sites in numbers proportional to the stoichiometry; thus, there are equal numbers of Na^+ and Cl^- vacancies in NaCl and $2Cl^-$ vacancies per Ca^{2+} vacancy in $CaCl_2$. A Schottky defect in NaCl is illustrated in Fig. 2-14(a).

When an ion occupies a normally vacant interstitial site, leaving its proper site vacant, the defect is termed a Frenkel defect. Frenkel defects are most common in crystals where the cation is much smaller than the anion, for instance, in AgBr, as illustrated schematically in Fig. 2-14(b).

Non-stoichiometric Defects. These often occur in transition-metal compounds, especially oxides and sulfides, because of the ability of the metal to exist in more than one oxidation state. A well-known case is "FeO" which consists of a *ccp* array of oxide ions with all octahedral holes filled by Fe^{2+} ions. In reality, however, some of these sites are vacant, while others—sufficient to maintain electroneutrality—contain Fe^{3+} ions. Thus the actual stoichiometry is commonly about $Fe_{0.95}O$. Another good example is "TiO" which can readily be obtained with compositions ranging from $Ti_{0.74}O$ to $Ti_{1.67}O$ depending on the pressure of oxygen gas used in preparing the material.

[2.] J. M. Thomas, *Chem. in Britain*, 1970, 6, 60.

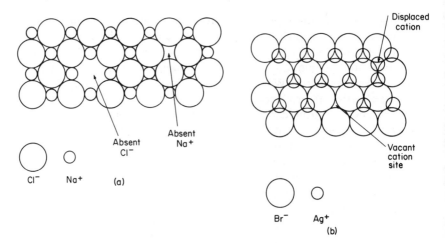

Fig. 2-14.   (a) A Schottky defect in a layer of NaCl. (b) A Frenkel defect in a layer of AgBr. The defects need not be restricted to one layer, as shown here.

Non-stoichiometric defects also occur, however, even when the metal ion has but one oxidation state. Thus, for example, CdO is particularly liable to lose oxygen when heated, to give yellow to black solids of composition $Cd_{1+\gamma}O$. A comparable situation arises when NaCl is treated with sodium vapor, which it absorbs to give a blue solid of composition $Na_{1+\gamma}Cl$. An appropriate number ($n\gamma$) of anion sites are then devoid of anions but occupied by electrons. The electrons in these cavities behave roughly like the simple particle in a box and there are excited states accessible at energies corresponding to the energy range of visible light. Hence these cavities containing electrons are color centers, commonly called F centers, from *Farbe* (German for color).

The existence of defects has a simple thermodynamic basis. The creation of a defect in a perfect structure has an unfavorable effect on the enthalpy; some coulombic or bonding energy must be sacrificed to create it. However, the introduction of some irregularities into an initially perfect array markedly increases the entropy; a $T\Delta S$ term large enough to cancel the unfavorable ΔH term will thus arise up to some limiting concentration of defects. It is possible to write an expression for the concentration of defects in equilibrium with the remainder of the structure just as though a normal chemical equilibrium were involved.

Finally, there are defects resulting from the presence of impurities. Some of these, when deliberately devised and controlled, constitute the basis for solid-state electronic technology.[3] For example, a crystal of germanium (which has the diamond structure) can be "doped" with traces of either gallium or arsenic. A gallium atom can replace one of germanium but an electron vacancy is created. An electron can move into this hole, thus creating

[3]   S. J. Bass, *Chem. in Britain*, 1969, **5**, 100.

a hole elsewhere. In effect, the hole can wander through the crystal and under the influence of an electrical potential difference travel through the crystal in a desired direction. The hole moves in the same direction as would a positive charge and gallium-doped germanium is therefore called a p-type (for positive) semiconductor. Whenever an arsenic atom replaces a germanium atom an electron is introduced into a normally unfilled energy band of germanium. These electrons can also migrate in an electric field and the arsenic-doped material is called an n-type (for negative) semiconductor.

Doped silicon and germanium are technologically the most important types of semiconducting material. In order to have reproducible performance the type and level of impurities must be strictly controlled and thus superpure silicon and germanium must first be prepared (cf. Chapter 11) and the desired doping then carried out. In addition to silicon or germanium as the basis for creating semiconductors, it is also possible to prepare certain isoelectronic III–V or II–VI compounds, such as GaAs or CdSb. Holes or conduction electrons can then be introduced by variation of the stoichiometry or by addition of suitable impurities.

Semiconductor behavior is to be found in many other types of compound, as for example in $Fe_{1-y}O$ and $Fe_{1-y}S$ where electron transfer from Fe^{2+} to Fe^{3+} causes a virtual migration of Fe^{3+} ions and hence p-type conduction.

Further Reading

Addison, W. E., *Allotropy of the Elements*, Oldbourne Press, 1966.

Dasent, W. E., *Inorganic Energetics*, Penguin Books, Ltd., 1970 (excellent, readable coverage of basic principles).

Galasso, F. S., *Structure and Properties of Inorganic Solids*, Pergamon Press, 1970.

Greenwood, N. N., *Ionic Crystals, Lattice Defects and Non-Stoichiometry*, Butterworths, 1968 (excellent short introduction).

Hannay, N. B., *Solid-State Chemistry*, Prentice-Hall, Inc., 1967 (good general survey).

Johnson, D. A., *Some Thermodynamic Aspects of Inorganic Chemistry*, Cambridge University Press, 1968 (a good outline of fundamentals).

Krebs, H., *Fundamentals of Inorganic Crystal Chemistry*, McGraw-Hill, 1968 (excellent discussion of structures and bonding).

Kröger, F. A., *Chemistry of Imperfect Crystals*, North Holland Publ. Co., 1964.

McDowell, C. A., in *Physical Chemistry*, Vol. III, H. Eyring, D. Henderson and W. Jost, ed., Academic Press, 1969, p. 496 (evaluation of electron affinities).

Vedeneyev, V. I., *et al.*, *Bond Energies, Ionization Potentials, and Electron Affinities*, Arnold, 1965 (data tables).

Wells, A. F., *Structural Inorganic Chemistry*, 3rd edn., Oxford University Press, 1962 (a fascinating book with superb illustrations and a wealth of data; 1028 pages of text).

3

The Nature of Chemical Bonding

THE VALENCE BOND APPROXIMATION

3-1. The Two-center Electron-pair Bond

Normally, we use the word bond to describe the linkage between a particular pair of atoms as, for example, the H and Cl atoms in HCl, or the N and one of the H atoms in NH_3. The student will already be familiar with the Lewis electron dot symbols for atoms and molecules whereby we would represent the HCl and NH_3 molecules by (3-I) and (3-II). The basic idea of

$$H : \ddot{C}l :$$

$$\begin{array}{c} H \\ : \ddot{N} : H \\ H \end{array}$$

$$(3\text{-I}) \qquad (3\text{-II})$$

the Lewis theory is that chemical bonds are due to the sharing of one or more pairs of electrons between two atoms. Thus the bond in HCl results from the sharing of one hydrogen electron and one chlorine electron (equation 3-1):

$$\dot{H} \cdot + \cdot \ddot{C}l : \rightarrow H : \ddot{C}l : \qquad \dots (3\text{-}1)$$

In other cases, though less commonly, two or three pairs may be shared (equation 3-2):

$$: \dot{N} \cdot + \cdot \dot{N} : \rightarrow : N : : : N : \qquad \dots (3\text{-}2)$$

Furthermore, it is not necessary that the electrons constituting the bond be contributed equally by the two atoms. For example, there is the *coordinate bond*, illustrated by equation 3-3:

$$\begin{array}{c} F \\ F : \ddot{B} + : O \\ F \end{array} \begin{array}{c} C_2H_5 \\ \diagup \\ \diagdown \\ C_2H_5 \end{array} \rightarrow \begin{array}{c} F \\ F : \ddot{B} : O \\ F \end{array} \begin{array}{c} C_2H_5 \\ \diagup \\ \diagdown \\ C_2H_5 \end{array} \qquad \dots (3\text{-}3)$$

All these various bonds, whether single or multiple, are called *covalent bonds*. However, it is possible even in this simple approach to recognize the difference between non-polar covalent bonding and polar covalent bonding. In a homonuclear diatomic molecule, the shared electron pair or pairs must be shared equally and thus there is no polarity in the system nucleus–electrons–nucleus. In a heteronuclear diatomic molecule, however, one of the atoms will, in general, have a greater affinity for the electrons than the other, and the bond will therefore be polar as in (3-III).

$$\overset{\delta+}{H} \overset{\delta-}{:\ddot{\underset{..}{C}l}:}$$

(3-III)

The intrinsic polarity of the bond should not be confused with the polarity that the molecule as a whole may have for other reasons.

These simple ideas were formulated before the advent of wave mechanics. Quantum theory not only justifies their use but enables us to refine and extend them. In attempting quantitative quantum-mechanical treatment of chemical bonds, approximations must be made. Traditionally, there have been two broad groups of approximations, called the *valence bond* (VB) and the *molecular orbital* (MO) treatments. The former is essentially a direct attempt to invest the qualitative ideas just outlined with quantum-mechanical validity, and it is therefore logical to continue the discussion with a summary of the valence-bond formalism, including such concepts as resonance, valence states and hybridization that arise within this framework. The molecular-orbital formalism will be presented in a following Section.

If we have two atoms—for simplicity, two hydrogen atoms—infinitely far apart, the wave function of this system is

$$\psi = \psi_A \psi_B \qquad \ldots (3\text{-}4)$$

where ψ_A is the wave function for the first hydrogen atom, ψ_B is the wave function for the second hydrogen atom and ψ is the joint wave function for the system. Equation 3-4 implies that neither atom disturbs the other, which is, of course, to be expected if they are far apart. We can also see that the total energy of the system ought to be

$$E^0 = E_A + E_B = 2E_H \qquad \ldots (3\text{-}5)$$

This result can easily be shown to follow by putting equation 3-4 into the wave equation, thus:

$$\mathscr{H}\psi = E\psi$$
$$\mathscr{H}(\psi_A\psi_B) = \psi_A \mathscr{H}\psi_B + \psi_B \mathscr{H}\psi_A$$
$$= \psi_A E_B \psi_B + \psi_B E_A \psi_A$$
$$= (E_A + E_B)(\psi_A\psi_B) = E^0\psi \qquad \ldots (3\text{-}6)$$

What we have done so far is completely rigorous, but it is also rather useless since we are interested in calculating the energy of the system when the two hydrogen atoms approach one another closely and form a bond. The energy of the system is then E', which is less than E^0 by an amount we call the *bond energy*. However, our first attempt to calculate the bond energy makes use of the above considerations regarding the wave function.

We begin by assuming that the wave function in equation 3-4 remains a reasonably good approximation, even when the atoms approach one another closely. We shall rewrite it in the following more explicit form:

$$\psi_1 = \psi_A(1)\,\psi_B(2) \qquad \ldots (3\text{-}7)$$

where we have specifically assigned electron 1 to atom A and electron 2 to atom B. Now in order to calculate the energy of the molecule we must solve the wave equation for small values of the internuclear distance. To do so we first rearrange it in the following way:

$$\mathcal{H}\psi = E\psi$$

$$\psi^* \mathcal{H}\psi = \psi^* E\psi = E\psi^*\psi$$

$$\int \psi^* \mathcal{H}\psi \, d\tau = E \int \psi^*\psi \, dr$$

or

$$E = \frac{\int \psi^* \mathcal{H}\psi \, d\tau}{\int \psi^*\psi \, d\tau} \qquad \ldots \text{(3-8)}$$

Note that in these arrangements we have written $\psi^* E\psi = E\psi^*\psi$ because E is simply a numerical factor, whereas $\psi^* \mathcal{H}\psi$ cannot be so rearranged because $\mathcal{H}$ is not simply a number but instead a symbol for an operation to be performed on whatever function follows it.

At close distances the Hamiltonian contains terms involving the reciprocals of the distance of electron 1 from nucleus B, the distance of electron 2 from nucleus A and r_{12}, the distance between the two electrons. Now when the atoms are far apart these reciprocal distances are so small that these terms are negligible, but as the atoms move together they become progressively more important. Thus the numerator on the right-hand side of equation 3-8 is a function of the internuclear distance, r. If we compute the energy as a function of the internuclear distance, which can be done by laboriously

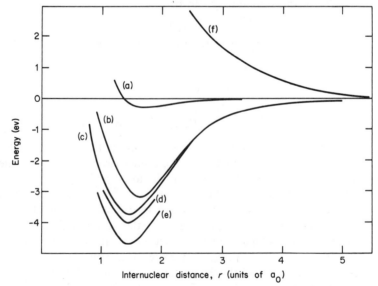

Fig. 3-1.   Potential energy curves for the hydrogen molecule.

solving equation 3-8, we obtain the results shown in curve a of Fig. 3-1. Thus we find that the energy of the system does decrease (by $\sim 1/4$ eV at the minimum) as the atoms approach, giving a minimum at $r = 1.7a_0$.[†] Curve e is the actual potential energy function for H_2, and it can be seen that our calculated results are qualitatively right but quantitatively rather disappointing.

We can make a dramatic improvement in our results by correcting an error in our wave function. Equation 3-7 states that we know that electron 1 is associated with atom A and electron 2 is associated with atom B. But in fact we do not and cannot know this. The electrons are not labeled so we cannot tell them apart to begin with, and, more importantly, we cannot attempt to follow any one electron in a system containing several. This is because wave mechanics does not tell us where any particular electron is but only the probability of finding an electron at a given place. Thus we are no more entitled to use ψ_1 as an approximate wave function than to use ψ_2:

$$\psi_2 = \psi_A(2)\,\psi_B(1) \qquad \ldots (3\text{-}9)$$

Thus if either one is equally likely to be right, we use both, namely:

$$\psi_{cov} = (\psi_1 + \psi_2) \qquad \ldots (3\text{-}10)$$

If now we solve for the energy as a function of r using equation 3-10 we obtain curve b of Fig. 3-1, which is obviously a vast improvement and indicates that, provided we allow for the indistinguishability of electrons, the basic approach we have taken is not unrealistic.

Further improvement in the results can be made by modifying ψ_{cov} in accord with the dictates of physical intuition. ψ_{cov} is only a combination of the hydrogenic wave functions. Although we are correct in assuming that one electron does not affect the other at large distances, we may well expect that when they are close together the forms of ψ_A and ψ_B will differ at least quantitatively from that of the unperturbed hydrogen orbital. Just as in a multielectron atom we use hydrogenic wave functions, but recognize that the electrons mutually shield one another to varying degrees from the nuclear charge, so we can allow for the existence of shielding in the H_2 molecule. We can do this again by using a number Z^*, rather than $Z = 1$, for the effective nuclear charge. When this is done, curve c (Fig. 3-1) is obtained. We now have a binding energy of 3.76 eV, which is a reasonably good approximation to the experimental value of 4.72 eV. There is another refinement that can be made fairly straightforwardly, and a consideration of this introduces the concept of *resonance*.

Any wave function of the type 3-10, however we modify the exact forms of ψ_A and ψ_B, still states that we consider only the possibility that one electron is associated with one atom and the other electron is associated with the other

---

[†] a_0 is the so-called Bohr radius, the radius of the most stable hydrogen-atom orbit in Bohr's semiclassical theory, and also the radius at which the $1s$ wave function maximizes in the wave-mechanical theory. It is thus a "natural" unit for atomic radii, with a magnitude 0.529 Å.

atom. But there is, of course, some finite chance that both electrons will occasionally be in the same atomic orbital. Wave functions describing such a state would be $\psi_A(1)\psi_A(2)$ and $\psi_B(1)\psi_B(2)$, and, because ψ_A and ψ_B are identical in form, we have no reason to prefer one over the other. We therefore use both as follows:

$$\psi_{ion} = \psi_A(1)\ \psi_A(2) + \psi_B(1)\ \psi_B(2) \qquad \text{... (3-11)}$$

We have called the wave function 3-10 ψ_{cov} because it states that the electrons are shared (in this case shared *exactly equally* because the two atoms are identical); similarly, we call the wave function 3-11 ψ_{ion} since it represents a state to which ionic forms contribute. The way in which we combine equations 3-10 and 3-11 is to write:

$$\psi = (1 - 2\lambda + 2\lambda^2)^{-1/2}\ [(1-\lambda)\psi_{cov} + \lambda\psi_{ion}] \qquad \text{... (3-12)}$$

in which λ is a *mixing coefficient*, which tells how much of the ionic wave function is mixed in with the covalent wave function. We have also introduced a normalizing factor. It is necessary to do a considerable amount of tedious calculation to determine what value of λ gives the best (that is, lowest) value of the energy. It turns out to be $\sim 1/5$, and the new value of the minimum energy is 4.10 eV. Curve d (Fig. 3-1) shows part of the potential energy function obtained with a wave function of the type 3-12.

The formulation we have described for the bond in H_2 is, in essence, the following. We have selected an orbital on each atom, ψ_A and ψ_B, each containing one electron, and combined the wave functions of these orbitals into a wave function for the two electrons together, namely, $\psi_A(1)\psi_B(2) + \psi_A(2)\psi_B(1)$. This process is called pairing the electrons, and the valence-bond (VB) method is sometimes called the method of electron pairs. There is another important feature of the process which we have not as yet mentioned explicitly. The wave function for the bonding electrons can, like a single atomic wave function, be described by a set of quantum numbers. Whatever these quantum numbers may be, it is clear that both electrons in the bond have the same set, since by the nature of equation 3-12 their distributions in space are identical. They must therefore differ in their m_s values in order to satisfy the exclusion principle. We shall call this adoption of different m_s values by two electrons *spin pairing*, or just *pairing*. We can in fact state the general rule that only when spin pairing occurs does there result an attractive force and hence an electron-pair bond. Curve f (Fig. 3-1) shows the potential energy curve when two hydrogen atoms approach in such a way that their spins remain unpaired, that is, with the same quantum number or direction. It is seen that the net force of interaction is strongly repulsive, and this state is called a *repulsive state*.

It is obvious that these principles can be extended to describe many chemical bonds and that they represent a rationalization in terms of wave mechanics of G. N. Lewis' idea of the electron-pair bond. Whenever we have two atoms, each with at least one unpaired electron, they may unite to form a bond in which these two electrons are paired. For example, two lithium atoms, each with the configuration $1s^2 2s$, combine to form Li_2 by pairing their $2s$ electrons, and two chlorine atoms, each with the configuration $[Ne]3s^2 3p_x^2 3p_y^2 3p_z$, combine by pairing their odd $3p_z$ electrons.

Multiple bonding may also be treated in this way. For example, two nitrogen atoms, each with the configuration $1s^2 2s^2 2p_x 2p_y 2p_z$, unite and form a triple bond by pairing electrons in corresponding p orbitals. Similarly, we can consider a methylene radical, $H_2C:$, to have two unpaired electrons, so that two of them unite to give ethylene with a double bond. There is one case in which the simple VB method leads to a qualitatively incorrect prediction concerning electronic structure, and that is O_2. An oxygen atom has the ground state configuration $1s^2 2s^2 2p_x^2 2p_y 2p_z$, and we should therefore expect two of them to unite forming two electron-pair bonds. Actually the O_2 molecule has a bond energy indicative of a double bond, but it also has two unpaired electrons, a combination which is awkward to rationalize, much less predict, in simple VB terms. This failure is due, not to any fundamental error in the VB method, but to our use of too rough an approximation. However, the molecular-orbital (MO) method, even in a very crude approximation, gives the correct result for O_2, as will be seen below.

3-2. Resonance

The wave function (equation 3-12) has an interpretation in terms of simple electron dot pictures. The three *canonical forms* (3-IVa, b and c) correspond,

$$\text{H:H} \longleftrightarrow \overset{+}{\text{H}} \text{ :}\bar{\text{H}} \longleftrightarrow \bar{\text{H}}: \overset{+}{\text{H}}$$
$$\text{(3-IVa)} \qquad \text{(3-IVb)} \qquad \text{(3-IVc)}$$

respectively, to the wave functions $\psi_A(1)\psi_B(2) + \psi_A(2)\psi_B(1)$, $\psi_A(1)\psi_A(2)$ and $\psi_B(1)\psi_B(2)$. The double-headed arrows between them indicate that they are in *resonance* with one another, or, to put it another way, that the actual electronic state of the molecule is a *resonance hybrid* of these three structures. This concept of resonance is a useful one provided it is not misinterpreted. H_2 *never has, at any time in its normal ground state life*, any one of the three structures shown. Taken one at a time they are only figments of our imagination but, as shown in the preceding Section, we obtain a satisfactory description of the molecule by supposing it to be in a state that is a hybrid of all three in certain proportions.

Before proceeding to consider further examples of resonance, a few remarks about the mathematical justification for the concept should be made. We have already seen that, for H_2, when we used the normalized function $(1-\lambda)\psi_{cov} + \lambda\psi_{ion}$ instead of just ψ_{cov}, we calculated a lower and more nearly correct energy for the system. If we had used ψ_{ion} alone, we should have obtained an extremely high energy. Yet, the *linear combination* of the two gives us an energy which is lower than that for either one separately. The amount by which the energy of the mixed state lies below the energy of the more stable of the two single states is called the *resonance energy*. It would be simpler to see how it comes naturally out of the solution of the wave equation if we take an example in which the two canonical forms are equivalent. We can write for NO_2^-, the nitrite ion, the two canonical forms (3-Va) and (3-Vb). Each of these forms can be described by a wave function,

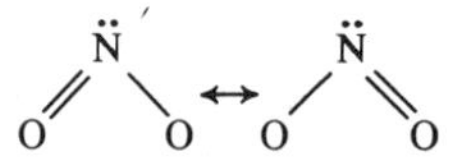

$$(3\text{-Va}) \qquad (3\text{-Vb})$$

namely, ψ_{Va} and ψ_{Vb}, and each of these wave functions would give the same energy, E^0, if used in the wave equation with the appropriate Hamiltonian, $\mathcal{H}$:

$$E^0 = \frac{\int \psi_{Va}^* \mathcal{H} \psi_{Va}\, d\tau}{\int \psi_{Va}^* \psi_{Va}\, d\tau} = \int \psi_{Va}^* \mathcal{H} \psi_{Va}\, d\tau$$

$$= \int \psi_{Vb}^* \mathcal{H} \psi_{Vb}\, d\tau = \frac{\int \psi_{Vb}^* \mathcal{H} \psi_{Vb}\, d\tau}{\int \psi_{Vb}^* \psi_{Vb}\, d\tau} \qquad \ldots (3\text{-}13)$$

where we have assumed that ψ_{Va} and ψ_{Vb} are each already normalized. If we wish to take account, mathematically, of the resonance depicted above, we must solve the wave equation, using the same Hamiltonian but with a linear combination of ψ_{Va} and ψ_{Vb}. From the symmetry it is obvious that this must be the symmetrical combination $(\psi_{Va} + \psi_{Vb})$ where the two are given equal weights. Let us now see what happens when we solve for the energy, E', using this wave function:

$$E' = \frac{\int (\psi_{Va}^* + \psi_{Vb}^*)\, \mathcal{H}\, (\psi_{Va} + \psi_{Vb})\, d\tau}{\int (\psi_{Va}^* + \psi_{Vb}^*)(\psi_{Va} + \psi_{Vb})\, d\tau}$$

$$= \frac{\int \psi_{Va}^* \mathcal{H} \psi_{Va}\, d\tau + \int \psi_{Va}^* \mathcal{H} \psi_{Vb}\, d\tau + \int \psi_{Vb}^* \mathcal{H} \psi_{Va}\, d\tau + \int \psi_{Vb}^* \mathcal{H} \psi_{Vb}\, d\tau}{\int \psi_{Va}^* \psi_{Va}\, d\tau + \int \psi_{Va}^* \psi_{Vb}\, d\tau + \int \psi_{Vb}^* \psi_{Va}\, d\tau + \int \psi_{Vb}^* \psi_{Vb}\, d\tau}$$

$$\ldots (3\text{-}14)$$

Taking account of the normalization, using the following definitions

$$\int \psi_{Va}^* \psi_{Vb}\, d\tau = \int \psi_{Vb}^* \psi_{Va}\, d\tau = S$$

$$\int \psi_{Va}^* \mathcal{H} \psi_{Vb}\, d\tau = \int \psi_{Vb}^* \mathcal{H} \psi_{Va}\, d\tau = E_R$$

and, remembering equation 3-13, we can rewrite equation 3-14 as:

$$E' = \frac{E^0 + E_R}{1 + S} \approx E^0 + E_R \qquad \dots (3\text{-}15)$$

since S, called the overlap integral, is usually small compared to 1. Thus we see that E' is lower than E^0 by E_R, the resonance energy.[†]

The equality of the two integrals called E_R is only true so long as ψ_{Va} and ψ_{Vb} are wave functions giving the same energy. Moreover, the value of such an integral diminishes rapidly the greater the difference in the energies given by ψ_{Va} and ψ_{Vb} separately in case the two wave functions are not equivalent. Thus stabilization of a molecule by resonance is greatest when the canonical forms contributing to the hybrid are close in energy, and best of all, identical. We might, for example, have taken the trouble to make our calculations with a wave function $(\psi_{Va} + \psi_{Vb} + \lambda \psi_{Vc})$ where ψ_{Vc} describes structure (3-Vc).

(3-Vc)

For various reasons (for example, the O—O distance is too great to permit the formation of a strong O—O bond, whereas we are at the same time giving up a rather strong N—O bond), the species (3-Vc) would have a far higher energy than (3-Va) or (3-Vb), and this calculation using $(\psi_{Va} + \psi_{Vb} + \lambda \psi_{Vc})$ will not give us a significantly lower energy than we can obtain by using only $(\psi_{Va} + \psi_{Vb})$.

The hydrogen molecule has provided an example of *covalent–ionic resonance* in a particular bond. Because structures (3-IVb) and (3-IVc) are of importance in an accurate description of the bond from the VB point of view, we say that the bond has some ionic character. However, the polarity that (3-Vb) introduces is exactly balanced by the polarity that (3-Vc) introduces, so that the bond has no net polarity. It is therefore called a *nonpolar covalent* bond. It is important not to confuse polarity and ionic character, although, unfortunately, the literature contains many instances of such confusion. When we turn to a heteronuclear diatomic molecule, we necessarily have bonds that have both ionic and polar character. Even for the pure covalent canonical structure of HCl (3-Ia) there is bond polarity

$$\text{H:}\ddot{\text{C}}\text{l:} \leftrightarrow \text{H} \text{ :}\ddot{\bar{\text{C}}}\text{l:} \leftrightarrow \text{H:} \overset{+}{\ddot{\text{C}}}\text{l:}$$
(3-Ia)   (3-Ib)   (3-Ic)

because the two different atoms necessarily have different affinities for the electron pair. This pair is thus shared, but not equally shared. It is also to be expected that the ionic structures (3-Ib) and (3-Ic) make some contribution.

Owing to the facts that (*a*) hydrogen and chlorine have about equal first ionization potentials, but (*b*) chlorine has a far higher electron affinity than

---

[†] It is *lower* because all three are intrinsically negative.

hydrogen, (3-Ic) is much less favorable energetically and hence contributes much less to the resonance hybrid than (3-Ib). Since the whole scheme we are using here is essentially qualitative, we normally ignore (3-Ic) and consider the ionic–covalent resonance in HCl to be adequately described by (3-Ia $\leftrightarrow$ 3-Ib).

The CO molecule can be considered as a resonance hybrid of the following canonical structures:

$$:C::\overset{..}{O}: \quad \leftrightarrow \quad :\overset{+}{C}:\overset{\bar{..}}{\underset{\cdot}{O}}: \quad \leftrightarrow \quad :\overset{-}{C}::\overset{+}{\underset{\cdot}{O}}:$$

$$\text{(3-VIa)} \qquad\qquad \text{(3-VIb)} \qquad\qquad \text{(3-VIc)}$$

Pauling has estimated that all three are of comparable importance, which is in accord with the low (0.1 D) dipole moment of the molecule, although the very short bond distance (1.13 Å) would suggest that (3-VIc) is predominant. The low dipole moment can be perhaps explained on the basis of smaller contributions from (3-VIa) and (3-VIb) if polarity due to lone-pair moments (page 120) is taken into account.

One final comment should be made here concerning the error of attributing any physical reality to a particular canonical form. The two or more wave functions we combine to make the total wave function must represent only differences in the distribution of electrons about a *fixed and constant* nuclear framework. Now, if such a molecule as is represented by (3-Va) or (3-Vb) for NO_2^- really existed, we should expect that the N=O distance would be appreciably shorter than the N—O distance. Thus (3-Vb) could not be obtained from (3-Va) simply by redistributing electrons. We should also have to shift the nuclear framework, and it is just this that is foreign to the whole concept of resonance. Thus the structures (3-Va) and (3-Vb) cannot represent real molecules, but only hypothetical ones. In our subsequent discussion of molecular-orbital theory we shall examine an alternative method of getting the same results.

3-3. The States Derived from Electronic Configurations

In developing the valence-bond (VB) approach to chemical bonding our next consideration must be promotion energies to so-called valence states. However, in order to speak meaningfully about this, it is first necessary to make a diversion into the concept of *electronic states* that arise from electron configurations. For example, the statement that an atom has a p^2 configuration gives a very incomplete description of its condition because there are three states each having different values for the properties: angular momentum, spin, and, particularly, energy. These three states have energies that differ by as much as 400 kJ mol^{-1}! Not only will an understanding of how these different states arise from a given configuration be essential in the furtherance of the outline of VB theory, but the subject is indispensable in the presentation of crystal and ligand field theories of bonding in transition-metal complexes (Chapter 20).

A *state* may be defined for present purposes by specifying the configuration from which it arises, its energy, its orbital angular momentum and its spin. A state so characterized corresponds to the spectroscopist's *multiplet*. The term multiplet is used because, in fact, there are a number of components of the state which, in general, differ in energy by much less than one such state differs from another. These components of the state or multiplet have different values of the total angular momentum, which is a result of the combination of the orbital and spin angular momenta. Now the three characteristic quantities of a state of the system have magnitudes that are determined by the manner in which the three corresponding quantities for each of the individual electrons combine to produce the resultant quantities for the entire group of electrons. Even for the simplest cases, this is a complicated matter that cannot be dealt with entirely rigorously. However, we are fortunate in finding, from experiment, that for the lighter atoms (up to, approximately, the lanthanides) Nature follows a scheme that can be understood to a fairly accurate level of approximation by using a set of relatively simple rules. This set of rules may be designated the *Russell–Saunders* or *LS* coupling scheme, and it forms the subject of this Section.

Each electron in an atom has a set of quantum numbers, n, l, m_l, m_s; it is the last three that are of concern here. Just as l is used, as $\sqrt{l(l+1)}$, to indicate the orbital angular momentum of a single electron, there is a quantum number L such that $\sqrt{L(L+1)}$ gives the total orbital angular momentum of the atom. The symbol M_L is used to represent a component of L in a reference direction and is analogous to m_l for a single electron. Similarly, we use a quantum number S to represent the total electron spin angular momentum, given by $\sqrt{S(S+1)}$, in analogy to the quantum number s for a single electron. There is the difference here that s is limited to the value $\frac{1}{2}$, whereas S may take any integral or half-integral value beginning with 0. Components of S in a reference direction are designated by M_S, analogous to m_s.

Symbols for the states of atoms are analogous to the symbols for the orbitals of single electrons. Thus the capital letters S, P, D, F, G, H ... are used to designate states with $L = 0, 1, 2, 3, 4, 5$ ... The use of S for both a state and a quantum number is unfortunate, but in practice seldom causes any difficulty. The complete symbol for a state also indicates the total spin, but not directly in terms of the value of S. Rather, the number of different M_S values, which is called the *spin multiplicity*, is used. Thus, for a state with $S = 1$, the spin multiplicity is 3 because there are three M_S values, 1, 0, -1. In general the spin multiplicity is equal to $2S+1$, and is indicated as a left superscript to the symbol for L. The following examples should make the usage clear:

For $M_L = 4$, $S = \frac{1}{2}$, the symbol is ^2G
For $M_L = 2$, $S = \frac{3}{2}$, the symbol is ^4D
For $M_L = 0$, $S = 1$, the symbol is ^3S

In speaking or writing of states with spin multiplicities of 1, 2, 3, 4, 5, 6 ... we call them respectively, *singlets, doublets, triplets, quartets, quintets, sextets* ... Thus, the three states shown above would be called doublet G, quartet D, and triplet S, respectively.

As in the case of a single electron, we may sometimes be interested in the total angular momentum, that is the vector sum of L and S. For the entire atom this is designated J. When required, J values are appended to the symbol as right subscripts. For example, a ^4D state may have any of the following J values, the appropriate symbols being as annexed.

L	M_S	J	Symbol
2	$\frac{3}{2}$	$\frac{7}{2}$	$^4D_{7/2}$
2	$\frac{1}{2}$	$\frac{5}{2}$	$^4D_{5/2}$
2	$-\frac{1}{2}$	$\frac{3}{2}$	$^4D_{3/2}$
2	$-\frac{3}{2}$	$\frac{1}{2}$	$^4D_{1/2}$

In order to determine what states may actually occur for a given atom or ion, we begin with the following definitions, which represent the essence of the approximation we are using:

$$M_L = m_l^{(1)} + m_l^{(2)} + m_l^{(3)} + \dots + m_l^{(n)}$$
$$M_S = m_s^{(1)} + m_s^{(2)} + m_s^{(3)} + \dots + m_s^{(n)}$$

in which $m_l^{(i)}$ and $m_s^{(i)}$ stand for the m_l and m_s values of the ith electron in an atom having a total of n electrons.

In general it is not necessary to pay specific attention to all the electrons in an atom when calculating M_L and M_S since those groups of electrons that completely fill any one set of orbitals (s, p, d, etc.) collectively contribute zero to M_L and to M_S. For instance, a complete set of p electrons includes two with $m_l = 0$, two with $m_l = 1$, and two with $m_l = -1$, the sum, $0 + 0 + 1 + 1 - 1 - 1$, being zero. At the same time, half of the electrons have $m_s = \frac{1}{2}$ and the other half have $m_s = -\frac{1}{2}$, making M_S equal to zero. The generalization to any filled shell should be obvious. Therefore, we need only concern ourselves with partly filled shells.

For a partly filled shell, there is always more than one way of assigning m_l and m_s values to the various electrons. All ways must be considered except those that are either prohibited by the exclusion principle or are physically redundant, as will be explained below. For convenience we shall use symbols in which $+$ and $-$ superscripts represent $m_s = +\frac{1}{2}$ and $m_s = -\frac{1}{2}$, respectively. Thus, when the first electron has $m_l = 1$, $m_s = +\frac{1}{2}$, the second electron has $m_l = 2$, $m_s = -\frac{1}{2}$, the third electron has $m_l = 0$, $m_s = +\frac{1}{2}$, etc., we shall write $(1^+, 2^-, 0^+ \dots)$. Such a specification of m_l and m_s values of all electrons will be called a microstate.

Let us now consider the two configurations $2p3p$ and $2p^2$. In the first case, our freedom to assign quantum numbers m_l and m_s to the two electrons is unrestricted by the exclusion principle since the electrons already differ in their principal quantum numbers. Thus microstates such as $(1^+, 1^+)$ and $(0^-, 0^-)$ are permitted. They are not permitted for the $2p^2$ configuration, however. Secondly, since the two electrons of the $2p3p$ configuration can be distinguished by their n quantum numbers, two microstates such as $(1^+, 0^-)$ and $(0^-, 1^+)$ are physically different. However, for the $2p^2$ configuration, such a pair are actually identical since there is no *physical* distinction between "the first electron" and "the second electron." For the $2p3p$ configuration there are thus $6 \times 6 = 36$ different microstates, while for the $2p^2$ configuration six of these are nullified by the exclusion principle and the remaining 30 consist of pairs that are physically redundant. Hence, there are but 15 micro-states for the $2p^2$ configuration.

TABLE 3-1

Tabulation of Microstates for a p^2 Configuration

(a)

M_L	M_S		
	1	0	-1
2		$(1^+, 1^-)$	
1	$(1^+, 0^+)$	$(1^+, 0^-)(1^-, 0^+)$	$(1^-, 0^-)$
0	$(1^+, -1^+)$	$(1^+, -1^-)(0^+, 0^-)(1^-, -1^+)$	$(1^-, -1^-)$
-1	$(-1^+, 0^+)$	$(-1^+, 0^-)(-1^-, 0^+)$	$(-1^-, 0^-)$
-2		$(-1^+, -1^-)$	

(b)

M_L	M_S		
	1	0	-1
2			
1	$(1^+, 0^+)$	$(1^-, 0^+)$	$(1^-, 0^-)$
0	$(1^+, -1^+)$	$(1^-, -1^+)(0^+, 0^-)$	$(1^-, -1^-)$
-1	$(-1^+, 0^+)$	$(-1^-, 0^+)$	$(-1^-, 0^-)$
-2			

Table 3-1a shows a tabulation of the microstates for the $2p^2$ configuration, in which they are arranged according to their M_L and M_S values. It is now our problem to deduce from this array the possible values for L and S. We first note that the maximum and minimum values of M_L are 2 and -2, each of which is associated with $M_S = 0$. These must be the two extreme M_L values derived from a state with $L = 2$ and $S = 0$, namely, a ^1D state. Also belonging to this ^1D state must be microstates with $M_S = 0$ and $M_L = 1$, 0 and -1. If we now delete a set of five microstates appropriate to the ^1D state, we are left with those shown in Table 3-1b. Note that it is not important which of the two or three microstates we have removed from a box which originally contained several, since the microstates occupying the same box actually mix among themselves to give new ones. However, the *number* of microstates per box is fixed, whatever their exact descriptions may be. Looking now at Table 3-1b, we see that there are microstates with $M_L = 1$, 0, -1 for each of the M_S values, 1, 0, -1. Nine such microstates constitute the components of a ^3P state. When they are removed, there remains only a single microstate with $M_L = 0$ and $M_S = 0$. This must be associated with a ^1S state of the configuration. Thus, the permitted states of the $2p^2$ configuration —or any np^2 configuration—are ^1D, ^3P and ^1S. It is to be noted that the sum of the degeneracies of these states must be equal to the number of micro-states. The ^1S state has neither spin nor orbital degeneracy; its degeneracy number is therefore 1. The ^1D state has no spin degeneracy but is orbitally $2L+1 = 5$-fold degenerate. The ^3P state has 3-fold spin degeneracy and 3-fold orbital degeneracy giving it a total degeneracy number of $3 \times 3 = 9$. The sum of these degeneracy numbers is indeed 15.

For the $2p3p$ configuration the allowed states are again of the types S, P and D, but now there is a singlet and a triplet of each kind. This can be demonstrated by making a table of the microstates and proceeding as before. It can be seen perhaps more easily by noting that for *every* combination of $m_l^{(1)}$ and $m_l^{(2)}$ there are four microstates, with spin assignments $++$, $+-$, $-+$ and $--$. One of these, either $+-$ or $-+$, can be taken as belonging to a singlet state and the other three belong then to a triplet state. It will be noted that the sum of the degeneracy numbers for the six states ^3D, ^1D, ^3P, ^1P, ^3S, ^1S is 36, the number of microstates.

For practice in using the LS coupling scheme, the reader may verify, by the method used for np^2, that an nd^2 configuration gives rise to the states ^3F, ^3P, ^1G, ^1D and ^1S, and that an np^3 configuration gives the states ^4S, ^2P, ^2D.

While the method shown for determining the states of an electron configuration will obviously become very cumbersome as the number of electrons increases beyond perhaps 5, there is, fortunately, a relationship that makes many of the problems with still larger configurations tractable. This relationship is called the *hole formalism*, and with it a partially filled shell of n electrons can be treated either as n electrons or as $N-n$ positrons, where N represents the total capacity of the shell. As far as electrostatic inter-actions of electrons among themselves are concerned, it makes no difference whether they are all positively charged or all negatively charged since the energies of interaction are all proportional to the product of two charges. It is actually rather easy to see that the hole formalism must be true, for, whenever we select a microstate for n electrons in a shell of capacity N, there remains a set of m_l and m_s values that could be used by $N-n$ electrons.

The several states derived from a particular configuration have different energies. However, purely theoretical evaluation of these energy differences is neither easy nor accurate, since they are expressed as certain integrals representing electron–electron repulsions, which cannot be precisely evalu-ated by computation. However, when there are many terms arising from a configuration, it is usually possible to express all the energy differences in terms of only a few integrals. Thus, when the energies of just a few of the states have been measured, the others may be estimated with fair accuracy, though not exactly, because the coupling scheme itself is only an approximation. The magnitudes of the energy differences are generally comparable to energies of chemical bonds and chemical reactions. For example, the energies of the ^1D and ^1S excited states of the carbon atom in the configuration $1s^2 2s^2 2p^2$ are ~ 105 and ~ 135 kJ mol^{-1}, respectively, above the ^3P ground state.

In this book, the major use for the energy differences between states will be in the treatment of transition-metal ions by the crystal and ligand field theories. Thus, further discussion of the energies will be deferred until then (Chapter 20).

As mentioned above, each state of the type ^{2S+1}L actually consists of a group of substates with different values of the quantum number J. The energy differences between these substates are generally an order of magnitude less than the energy differences between the various states themselves, and usually they can be ignored in ordinary problems of chemical interest. However, in certain cases, for example, in understanding the magnetic properties of the lanthanides (see Chapter 27) and in nearly all problems with the very heavy elements (those of the third transition series and the actinides) these energy differences are of great significance and cannot be ignored. Indeed, for the very heavy elements they become comparable in magnitude to the energy differences between the ^{2S+1}L states. When this happens, the LS coupling method becomes inherently unreliable, so that different and more complex treatments must be used.

The cause of the separation between substates with different J values is direct coupling between the spin and the orbital angular momenta of the electrons. In the LS coupling scheme, this is assumed to be negligible. Thus, as stated, we assume that the $m_l^{(i)}$ add only to one another and the $m_s^{(i)}$ add only to one another. When coupling between $m_l^{(i)}$ and $m_s^{(i)}$ for each electron becomes *very* strong, it is possible to utilize another relatively simple method to determine states of the electron configuration. In this method, known as the jj coupling scheme, one assumes that the states arise from the various combinations of j values for each electron. In the jj coupling scheme, the quantum numbers M_L and M_S are no longer meaningful and the states are characterized by other quantum numbers. However, since jj coupling will not find any direct application in the later parts of this book, we shall not discuss it further here.

3-4. Promotion Energies to Valence States

An important factor in understanding both the strengths and the spacial distribution of a set of chemical bonds is the electronic condition of the central atom when it has—hypothetically—been promoted from its ground state to a state in which it is fully ready to form the set of bonds. Actually each atom in the molecule, whether central or peripheral, should be considered to undergo a promotion to a valence state, but it is usually the central atom that is of greatest concern.

Let us consider, for illustration, the elements of Group IV, C, Si, Ge, Sn and Pb, as they exercise their valence of 4:

$$M(g)+4X(g) = MX_4(g)$$
(M = C, Si, Ge, Sn, or Pb;
X = H, F, Cl, etc.) ... (3-16)

In order to be prepared to form four single bonds to the four X groups, the M atom, which has a ground electronic state derived from the s^2p^2 configuration, must be promoted to a state in which the four valence electrons are each in a different orbital and have their spins uncoupled from each other. Fig. 3-2 shows the actual (spectroscopically observed) states of the carbon atom; the patterns for those of Si and Ge are not very different. It is seen that the s^2p^2 configuration gives rise to three states; these are very similar to those of a p^2 configuration, which have been derived in Section 3-3, since the s^2 pair has $L=0$ and $S=0$. The lowest state from the sp^3 configuration is one with $S=2$, i.e., all electron spins parallel. There are various triplet states derived from sp^3 as well as some singlet states (not shown in Fig. 3-2) at very high energies. From the experimental information on the separation of the states with different orbital angular momenta and spin multiplicities it is possible to calculate values for the mean spin–spin interaction energy and

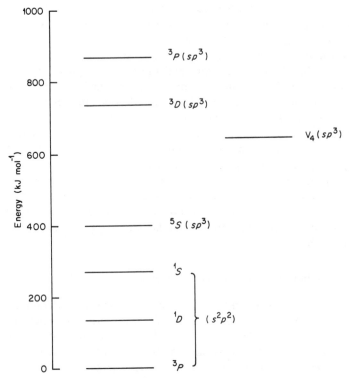

Fig. 3-2. The spectroscopic (real) electronic states of the carbon atom and the valence state (hypothetical) called V_4.

orbital couplings. These may then be subtracted from the energies of the various sp^3 states and the results may then be averaged to give an energy that is the best estimate (but not a rigorous one) of the energy the atom would have if its four valence electrons were all in different, equivalent orbitals with their spins completely random. This is the valence state, called $V_4(sp^3)$, whose energy is also shown in Fig. 3-2.

The final step then is to combine the atom in the valence state, V_4, with the four X atoms (which have also been suitably promoted to a valence state) to give the product $MX_4(g)$. The energy released in this final step is sometimes called the intrinsic bond energy. For CH_4 and SiH_4 the various energies (in kJ mol^{-1}) involved are as follows:

	C	Si
$M(s^2p^2)(g) \rightarrow M(sp^3)(g)$	404	399
$M(sp^3)(g) \rightarrow M(V_4)(g)$	230	85
$M(V_4)(g) + 4H(g) \rightarrow MH_4(g)$	−2290	−1780
$M(s^2p^2)(g) + 4H(g) \rightarrow MH_4(g)$	−1656	−1296
Thermochemical M—H bond energy	414	324
Intrinsic M—H bond energy	572	445

The foregoing discussion implicitly raises two questions. (1) Why do the Group IV elements not form MX_2 compounds, since this would save most of the promotion energy to the V_4 valence state, and (2) how is it that the V_4 valence state leads to the formation of a tetrahedral MX_4 molecule, in which all M—X bonds are equivalent when an sp^3 configuration has electrons in two differently shaped orbitals. We reserve the second of these questions for the next Section. The first can be answered fairly easily.

Even though the promotion energy to reach a valence state V_2 based on a configuration s^2p^2 would be much less than that to reach the state V_4, the intrinsic M—X bond energy is so high that the energy released on forming four instead of two M—X bonds more than compensates for this. However, note that for SiH_4 the intrinsic bond energy (as well as the thermochemical bond energy) is lower than the C—H bond energy. In general, M—X bond energies decrease down a group and the result is a trend toward less stability for the higher valence compounds, although the picture is more complicated than this because many of the lower-valence compounds become ionic rather than molecular.

3-5. Hybridization

In discussing the promotion energy required to transform a Group IV atom from its divalent s^2p^2 ground-state configuration to the tetravalent state necessary to form its normal compounds (e.g., CH_4), we discussed, but did not fully explain, the energy required to promote the atom from the observable, stationary state ^5S based on the configuration sp^3, to what we called its valence state, V_4. The sp^3 configuration referred to is, more specifically, the configuration $2s2p_x2p_y2p_z$. Now, according to the preceding dis-

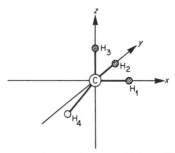

Fig. 3-3.   Geometric structure that methane would have if the carbon atom used pure
hydrogenic orbitals, $2s$, $2p_x$, $2p_y$ and $2p_z$ to form C—H bonds.

cussion, if one atom, say A, with an orbital ψ_A containing one electron,
is to form the strongest bond with another atom, B, then the A—B axis
should be along the direction in which ψ_A has its maximum value. Thus if
atom A is to use its p_x orbital, we should expect the A—B axis to be collinear
with the x axis. For an s orbital, of course, all directions are equivalent.
Thus if the valence state of the carbon atom were really the $2s2p_x2p_y2p_z$
stationary state, we should expect three hydrogen atoms to form bonds to
the p_x, p_y and p_z orbitals, and thus lie along the x, y, and z axes. The fourth
hydrogen atom would bond to the s orbital and probably take up a position
equidistant from the other three.

   This hypothetical state of affairs is depicted in Fig. 3-3. It is easily seen that
the three angles H_1—C—H_2, H_1—C—H_3 and H_2—C—H_3 are 90°, and
the three equivalent angles such as H_1—C—H_4 are $\sim 125°$. Moreover,
since H_1, H_2, and H_3 are bonded to carbon through its p orbitals, all these
bonds should be equivalent and hence of equal length, whereas the C—H_4
bond is formed by using a carbon s orbital and would not be expected to
have the same length. It is known with complete certainty, however, that
methane does not have such a structure. Actually all C—H distances and
H—C—H angles are equal, and the molecule is a regular tetrahedron. From
this we infer that the valence state of the carbon atom is one in which its
four valence electrons are in four equivalent orbitals that have their lobes
directed toward the apices of a tetrahedron. It is this valence state that we
have denoted as V_4. Our next question, then, is how can we express such a
set of equivalent orbitals in terms of the basic set of hydrogenic orbitals $2s$,
$2p_x$, $2p_y$, and $2p_z$?

   If we are to have four completely equivalent orbitals, they must each
have the same fraction of s *character*, namely, one-fourth, and the same
fraction of p *character*, namely, three-fourths, and the resulting orbitals
must be normalized. We can pick the direction for the first one arbitrarily,
so let us put it along the z axis. Neither p_x nor p_y can contribute to this one
since they have precisely the value zero along the z axis, so this first one will
consist entirely of s and p_z. Its general form must therefore be

$$N(s+kp_z)$$

where k is a mixing coefficient yet to be evaluated, and N is a normalizing factor. k can be shown to be $\sqrt{3}$ since the probability of the electron having p_z character must be three times greater than of its having s character, and these probabilities are proportional to the squares of the wave functions, namely:

$$\frac{\int (kp_z)^2 \, d\tau}{\int (s)^2 \, d\tau} = 3$$

and since p_z and s are separately normalized, $k = \sqrt{3}$. The value of N is then easily found by normalizing $(s + \sqrt{3}p_z)$, namely:

$$\int [N(s+\sqrt{3}\,p_z)]^2 \, d\tau = 1$$

$$N^2 = \left[\int s^2 \, d\tau + 3 \int p_z^2 \, d\tau \right]^{-1}$$

$$= [1+3]^{-1}$$

$$N = 1/2$$

where we have also made use of the orthogonality of the s and p_z orbitals.

The expressions for the other three orbitals, equivalent to this one and having their lobes lying along directions 109° from the z axis, can be derived by similar reasoning, but trigonometric complications make it impractical to carry out the derivation here. (A complete derivation of all orbitals in a set is given later for simpler cases.) This set of four equivalently directed orbitals is one example of a set of *hybrid orbitals*; these four are commonly called sp^3 hybrids to indicate their composition in terms of hydrogenic atomic orbitals. In common—though somewhat loose—parlance it is said that the carbon atom "has sp^3 hybridization."

Before proceeding to further discussion of hybridization, we may complete our explanation of the difference between the valence state V_4 and the sp^3 stationary state. In order to promote the carbon atom from the sp^3 stationary state to the valence state, two things are done: (1) the s and the three p orbitals are hybridized to produce four sp^3 hybrids which are each occupied by one electron, and (2) any preferred orientations of the spins of these electrons due to interactions of their spins with one another or with their orbital motions are destroyed, leaving them completely free, random, and ready to be paired with electron spins of other atoms. Both these processes require input of energy, the total in this case being 230 kJ mol^{-1}. Let us emphasize again that the valence state is in general not identical with any observable stationary state of the atom, and the idea of "promotion to the valence state" is only a mental construction which is useful in thinking about the bonding.

We return now to systematic consideration of hybridization, beginning with the simplest common type, namely, sp hybridization. Beryllium in its ground state has the configuration $1s^2 2s^2$, and without promotion to a

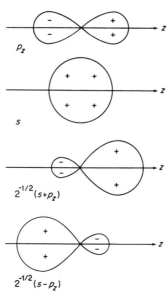

Fig. 3-4. Schematic illustration of the formation of sp digonal hybrid orbitals by combination of an s and a p orbital.

higher state could form no electron pair bonds. Normally, it forms compounds such as $BeCl_2$, $BeBr_2$, and $Be(CH_3)_2$, which, as free gaseous molecules, are linear. To explain this we assume that the atom is first promoted to the state $2s2p$ followed by promotion to a valence state involving sp hybridization. The physical reasoning behind the construction of these hybrids can be understood with reference to Fig. 3-4; the algebraic details in this case are quite simple. Since the two hybrid orbitals are to be equivalent, they must have equal fractions of s and p character, that is, the s orbital must contribute equally to both and the p_z orbital must contribute equally to both. Thus they must be of the forms $N(s+p)$ and $N'(s-p)$, where N and N' are normalizing factors. It is easily shown that these two orbitals are orthogonal:

$$\int N(s+p)N'(s-p)\,d\tau = NN'\int s^2\,d\tau - NN'\int p^2\,d\tau$$

$$= NN'(1-1) = 0$$

and that $N = N' = \sqrt{1/2}$:

$$N^2\int (s+p)^2\,d\tau = N^2\int s^2\,d\tau + N^2\int p^2\,d\tau = 2N^2 = 1$$

An atom, X, using a pair of such sp or digonal hybrid orbitals, will form two equivalent bonds to two other atoms or univalent groups, Y, giving a linear YXY molecule. The gaseous halides (though not the solid compounds) of most of the Group II elements have linear structures which may be attributed to the sp hybridization of the metal atoms. Of course, a completely ionic molecule $Y^-X^{2+}Y^-$ would tend to be linear for electrostatic reasons, so linearity does not, by itself, demonstrate that the bonds are covalent; in many cases, however, there are various kinds of evidence suggesting appreciable covalent character in the bonds. Mercuric halides and pseudohalides, e.g., $Hg(CN)_2$, are doubtless predominantly covalent. In these linear molecules we postulate that mercury with a configuration $[Xe]5d^{10}6s^2$ uses $6s6p$ hybrid orbitals.

The elements of Group III which have ground state configurations ns^2np form many molecules of the type MX_3 where M is B, Al, Ga, In or Tl, and X is a halogen or an organic radical, CH_3, C_6H_5, etc. In all these the monomeric molecules (some dimerize; e.g., Al_2Cl_6) are known or presumed to have the shape of a planar equilateral triangle. Thus we assume that the metal atoms, B, for example, must first be excited to a stationary state based on the configuration sp_xp_y and then further promoted to a valence state in which there is a set of equivalent sp^2 hybrid orbitals. The correct algebraic expressions and schematic "balloon" picture of a set of sp^2 or trigonal hybrids are shown in Fig. 3-5. Each one has its maximum along one of a set of three axes, lying in the same plane making angles of 120° with one another. Moreover, the orbitals are equivalent in shape. Each one, if rotated 120°, would be exactly superposed on the orbital already on that axis.

We can easily work out the expressions for these orbitals by invoking the requirements that they be equivalent, meaning that the proportion of s character to p character be the same in all of them, and that they be normalized. We shall see that in this particular case the orthogonality takes care of itself, but not all cases are as overdetermined as this one.

Let us choose to direct the first hybrid along the x axis (see Fig. 3-5), and make the coefficient of p_x $\sqrt{2}$ times that of s since the orbital must have twice as much p character as s character. This gives us

$$\psi_1 = N(s + \sqrt{2}\,p_x)$$

and the normalization factor is easily shown to have the value of $1/\sqrt{3}$. In order to work out the expressions for the other two orbitals to which s, p_x,

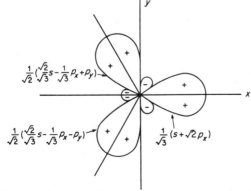

Fig. 3-5. Schematic illustration of sp^2 trigonal hybrid orbitals.

and p_y all contribute, we make use of the fact that orbitals behave like vectors in the sense that if an orbital makes a contribution χ in a certain direction it makes a contribution $\chi \cos \theta$ in a direction θ from the first. Thus, considering ψ_2 to be the orbital lying in the second quadrant (Fig. 3-5) and ψ_3 to be the one in the third quadrant, we proceed as follows. The contribution of the s orbital is isotropic and the three hybrid orbitals are all to have the same amount of s character. Moreover, by virtue of the vectorial properties of the atomic and hybrid orbitals, p_y and p_x must contribute to ψ_2 in the proportions $\cos 30°(\sqrt{3}/2)$ and $\cos 120°(-1/2)$ and to ψ_3 in the proportions $-\cos 30°(-\sqrt{3}/2)$ and $\cos 240°(-1/2)$. Thus we can write

$$\psi_2 = (1/\sqrt{3})s + \alpha[(-1/2)p_x + (\sqrt{3}/2)p_y]$$
$$\psi_3 = (1/\sqrt{3})s + \beta[(-1/2)p_x - (\sqrt{3}/2)p_y]$$

It is now necessary to adjust α and β so that each of these orbitals is normalized. For α:

$$\int \psi_2\psi_2\, d\tau = \tfrac{1}{3}\int ss\,d\tau + \tfrac{1}{4}\alpha^2 \int p_x p_x\, d\tau + \tfrac{3}{4}\alpha^2 \int p_z p_z\, d\tau = \tfrac{1}{3} + \alpha^2 = 1$$

whence α equals $\sqrt{2}/\sqrt{3}$. The same value is obtained for β and the final expressions for the three trigonal hybrid orbitals are therefore:

$$\psi_1 = (1/\sqrt{3})s + (\sqrt{2}/\sqrt{3})p_x$$
$$\psi_2 = (1/\sqrt{3})s - (1/\sqrt{6})p_x + (1/\sqrt{2})p_y$$
$$\psi_3 = (1/\sqrt{3})s - (1/\sqrt{6})p_x - (1/\sqrt{2})p_y$$

It is now easily shown that these orbitals are mutually orthogonal. For example:

$$\int \psi_1\psi_2\, d\tau = \int [(1/\sqrt{3})s]^2\, d\tau - \int [(\sqrt{2}/\sqrt{3})p_x][(1/\sqrt{6})p_x]\, d\tau$$
$$= 1/3 - 1/3 = 0$$

It is also easy to show that the ratio of p to s character is 2:1 in ψ_2 and ψ_3 as well as in ψ_1:

$$\frac{(-1/\sqrt{6})^2 + (\pm 1/\sqrt{2})^2}{(1/\sqrt{3})^2} = \frac{1/6 + 1/2}{1/3} = \frac{2}{1}$$

It can be seen from examination of Figs. 3-4 and 3-5 that hybrid orbitals provide much greater concentrations of the electron cloud in particular directions than do the simple hydrogenic orbitals of which they are constructed. Thus hybrid orbitals can provide better overlap with orbitals of other atoms along these preferred directions, and they consequently make certain configurations of the molecule preferred. In general, the increased overlap means that the bonds are stronger, and this more than compensates for the promotion energy required to attain the hybridized valence state.

Hybridization is not limited to s and p orbitals, but may, in general, involve the mixing of all types of atomic orbitals. Hybrids involving d orbitals

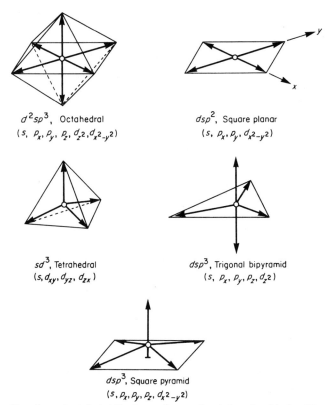

Fig. 3-6.   Five important hybridization schemes involving d orbitals. Heavy arrows show the directions in which the lobes point.

occur quite commonly among the heavier elements and are particularly important in complexes of the transition elements. We shall mention five of the most important hybridizations involving one or more d orbitals, each of which is illustrated in Fig. 3-6.

1. d^2sp^3, *Octahedral hybridization.*   When the $d_{x^2-y^2}$ and d_{z^2} orbitals are combined with an s orbital and a set of p_x, p_y, and p_z orbitals, a set of equivalent orbitals with lobes directed to the vertices of an octahedron can be formed.

2. dsp^2, *Square-planar hybridization.*   A $d_{x^2-y^2}$ orbital, an s orbital and p_x and p_y orbitals can be combined to give a set of equivalent hybrid orbitals with lobes directed to the corners of a square in the xy plane.

3. sd^3, *Tetrahedral hybridization.*   An s orbital and the set d_{xy}, d_{xz}, d_{yz} may be combined to give a tetrahedrally directed set of orbitals.

4. dsp^3, *Trigonal-bipyramidal hybridization.*   The orbitals s, p_x, p_y, p_z, and d_{z^2} may be combined to give a non-equivalent set of five hybrid orbitals directed to the vertices of a trigonal bipyramid.

5. dsp^3, *Square-pyramidal hybridization.*   The orbitals s, p_x, p_y, p_z, and

$d_{x^2-y^2}$ may be combined to give a non-equivalent set of five hybrid orbitals directed to the vertices of a square pyramid.

We have now examined some hybridization schemes from the point of view of the kinds of atomic orbital required to construct them. So long as the required orbitals are available, the existence of a particular set of hybrids is possible from this point of view. However, there are some energy considerations that are also important. If one or more of the orbitals required in the hybridization lie at a much higher energy than the others, it may not be energetically possible for the atom actually to achieve full hybridization. For example, with methane, if the energy of promotion from the ground state to V_4 were somewhat higher, say about 700 kJ mol^{-1} instead of $\sim 630 \text{ kJ mol}^{-1}$, CX_2 molecules might often be more stable than CX_4 molecules. It is possible to have a mixture of hybrid states for energetic reasons. The two hybridization schemes giving a set of tetrahedrally directed orbitals, namely, sp^3 and sd^3, are only extremes, and it is possible to have a set of tetrahedral hybrids using one s orbital and portions of each of the two sets $d_{xy}d_{xz}d_{yz}$ and $p_xp_yp_z$. For carbon, the amount of d character is doubtless negligible since the lowest available d orbitals, the $3d$'s, are so far above the $2p$'s that their use could only be a great energetic disadvantage. With silicon, germanium, tin, and lead, this will not be so clear-cut. Outer d orbitals may well play at least some part in the bonding in these MX_4 compounds. In the tetrahedral ions MnO_4^-, CrO_4^{2-}, etc., the $3d$ orbitals are of about the same energy as the $4s$ orbitals, and the $4p$ orbitals are somewhat higher. The hybridization of the Mn and Cr atoms in these cases is thus probably a mixture of sd^3 and sp^3, with d character greater than p character.

3-6. The Overlap Criterion of Bond Strength

The best theoretical criterion we have for the strength of a given electron-pair bond is, of course, the energy we calculate to be released when the bond is formed. We have seen, however, that even in the simple, prototype case of H_2 lengthy and tedious calculations are required to obtain even an approximate value for this energy. Obviously a simpler, if less fundamentally correct, criterion is desirable, and such a criterion does exist. Pauling and Mulliken have pointed out that there is a qualitative and, under some well-defined conditions, even a semiquantitative relation between bond energies and the overlap of the atomic orbitals used in forming the bonds. It is not difficult to see qualitatively why good overlap makes for strong bonding. The more the two bonding orbitals overlap, the more the bonding electrons are concentrated between the nuclei where they can minimize the nuclear repulsion and maximize the attractive forces between themselves and both nuclei jointly. The overlap, S, between a pair of atomic orbitals is given by the equation

$$S = \int \psi_A \psi_B \, d\tau \qquad \ldots (3-17)$$

and S is called the overlap integral. This integral may have a value which is positive, negative or exactly zero. When its value is positive, there is a build-up of electronic charge between the nuclei and a bond can be formed. When the overlap integral is negative, there is reduction in the electron density between the nuclei. This increases the repulsion between them and they tend to move apart. When the overlap is zero, there is no net interaction, attractive or repulsive. Thus, in the use of the overlap criterion there are two stages. First, there is the qualitative question of whether the overlap will be positive, negative, or zero. Secondly, when a positive overlap occurs, we may wish to know its magnitude in order to estimate the strength of the bond.

The qualitative nature of overlaps can generally be determined by examination of simple pictures of the orbitals involved.

Examples are presented in Fig. 3-7. The overlaps shown in (a)–(d) are all positive, while those in (e)–(h) are negative. It should be noted that in each of these examples the magnitude and sign of the overlap depend on the internuclear distance and on the relative sizes of the orbitals. The situations depicted are those in which the orbitals are of comparable size with inter-

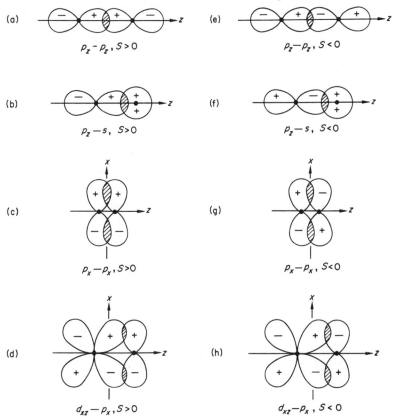

Fig. 3-7. Some representative positive and negative overlaps.

nuclear distances chosen relative to the sizes so as to correspond to conditions typical of an actual molecule.

There are many cases in which the prediction of precisely zero overlap can be made, irrespective of the relative sizes of the orbitals or the internuclear distance. These predictions follow from the *symmetry properties* of the orbitals and are thus independent of scale factors, because there must always be *exactly* equal areas (volumes) of positive and negative overlap. It should now become obvious why it is important to remember not only the shapes but also the *signs* of the lobes of orbitals.

Fig. 3-8 shows four examples in which the net overlap must be zero on symmetry grounds alone.

When non-zero overlaps are expected, it is often desirable to estimate their magnitudes. This can be done by expressing ψ_A and ψ_B in equation 3-17 as functions of the spacial coordinates about the nuclei A and B and then carrying out the integration. Algebraically, this is tedious, but there are now extensive tables of numerical values of overlaps over the useful ranges of parameters. The parameters are the sizes of the orbitals (which depend on the effective nuclear charges) and the internuclear distance. These tables have been computed by using simple exponential expressions (3-18) for the radial factors, $R(r)$, of the orbitals which were proposed many years ago by Slater.

$$R(r) = N r^{n-1} e^{-\mu r/a_0} \qquad \ldots (3\text{-}18)$$

Orbitals that use the usual angular factors obtained by solving the wave equation for hydrogen together with these simple radial functions are called *Slater-type orbitals* (STO's).

In the expression for the radial part of an STO (equation 3-18), N is a normalizing factor, n is the principal quantum number, a_H is the Bohr radius (0.529 Å) and μ is a function of the effective nuclear charge for an electron in the orbital concerned. Slater proposed rules for adjusting the value of

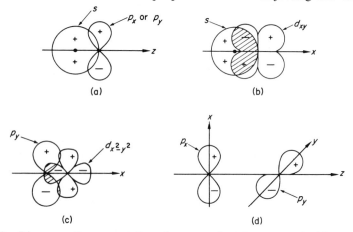

Fig. 3-8. Diagrammatic representation of some overlaps that are required by symmetry to be zero.

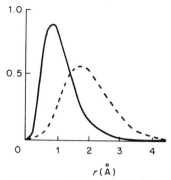

Fig. 3-9. A comparison of a Hartree–Fock $3d$ wave function for the Cr^{3+} ion (———) with a Slater-type wave function (– – –).

μ so as to make the energy of the orbitals agree well with experiment. For overlap calculations, however, it is more important that the shape of the orbital in its outer region (where overlaps occur) match as closely as possible the true shape of the orbital. In general, STO's are only a fair approximation in the latter respect, as shown in Fig. 3-9 where a Hartree–Fock $3d$ orbital for the Cr^{3+} ion is compared with the STO. Thus overlaps computed by using single STO's are rough approximations. However, it is possible to take linear combinations of a few STO's so as to approximate very closely the true shape of an orbital. The overlap between two orbitals each expressed as a combination of several (say, p and q) STO's can be evaluated by looking up the necessary number ($p \times q$ in general) of overlaps in the tables mentioned above. Thus, in summary, with the aid of existing tables, overlaps of accurate (i.e., Hartree–Fock) orbitals for atoms may be computed without excessive labor.

In order to estimate actual bond energies by using the overlap criterion one usually uses the molecular-orbital method. Hence, we shall return to this matter again after the molecular-orbital method has been introduced.

THE MOLECULAR ORBITAL (MO) APPROXIMATION

3-7. Introduction

In the preceding Section we discussed the valence bond (VB) or electron-pair theory of bonding. The basic qualitative idea in this theory is essentially Lewis' idea that each bonded atom pair in a molecule is held together by an electron pair or perhaps several electron pairs. These electron pairs are *localized* between particular pairs of nuclei. Moreover, it is assumed that the wave functions for these electrons are just the products of atomic wave functions. The MO theory starts with a qualitatively different assumption.

In building up a multielectron atom we start with a nucleus and a set of *one-center orbitals* about that nucleus and feed the required number of electrons into these orbitals in increasing order of orbital energy. We discovered the forms of these orbitals by solving exactly the problem with only one electron and then assumed that when many electrons are present the orbitals have the same form but that their relative energies are affected by the shielding of one electron by another. In other words, we treat each electron as if it moves in the effective field produced by the nucleus and all the other electrons. In its essentials, the molecular-orbital theory treats a molecule in the same way. We start with several nuclei, arranged as they are in the complete molecule. We then determine the various orbitals that *one* electron would have in the field of this set of nuclei. These *multicenter orbitals* are taken as the set to be filled with as many electrons as are required in the molecule under consideration. Again it is understood that the mutual shielding of the electrons and other interactions between them will have an important effect on the relative energies of the various molecular orbitals.

Although this scheme is just as good in principle for molecules as for atoms, it has a severe limitation in practice. As we have seen, we can get our basic set of atomic orbitals readily by exact solution of the wave equation of the hydrogen atom. In general, the problem of an electron moving in the field of several nuclei cannot be solved exactly. Therefore, we must begin by using only an approximate form for our one-electron MO's.

3-8. Linear Combination of Atomic Orbitals (LCAO) Approximation

The LCAO method is a simple and qualitatively useful approximation. It is based on the very reasonable idea that as the electron moves around in the nuclear framework it will at any given time be close to one nucleus and relatively far from others, and that when near a given nucleus it will behave more or less as though it were in an atomic orbital belonging to that nucleus. To develop this idea more concretely we shall use the hydrogen molecule ion H_2^+. This is a prototype for homonuclear diatomic molecules just as the hydrogen atom is for atoms in general.

If the electron belonged to either of the hydrogen nuclei A or B alone, its behavior in the ground state would be described by ϕ_A or ϕ_B alone. When it is in some general position with reference to the nuclear framework it can be described approximately by a superposition of both, that is, by $\phi_A \pm \phi_B$. Such an algebraic sum of functions is called a linear combination. The two normalized LCAO wave functions written out fully, are:

$$\psi_b = N_b(\phi_A + \phi_B) \quad \left.\begin{array}{r}\\\\\end{array}\right\} \dots (3\text{-}19)$$
$$\psi_a = N_a(\phi_A - \phi_B)$$

The two normalization constants are readily shown to have the following values:

$$N_b = \frac{1}{\sqrt{2+2S}} = 0.56 \qquad N_a = \frac{1}{\sqrt{2-2S}} = 1.11$$

The numerical values were obtained by inserting the correct numerical value (0.59) of the overlap integral, S. Generally, the overlap integrals are not so large, but rather around 0.25. For $S = 0.25$ the two normalization constants would have been 0.63 and 0.82, whereas, if S is neglected altogether, both normalization constants would have been 0.71. The small errors thus introduced by neglect of S in the normalization constants are usually tolerable in simple LCAO–MO theory, and overlap is neglected.

In order to understand the physical meaning of the two LCAO–MO's, let us first examine Fig. 3-10. The solid lines show ψ_a^2 and ψ_b^2, the squares

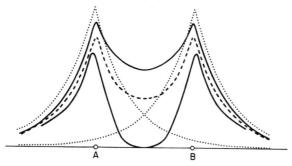

Fig. 3-10. A plot of the atomic orbital electron probabilities, ϕ_A^2 and ϕ_B^2 (.....), the molecular orbital electron probabilities, ψ_a^2 and ψ_b^2 (———), and the probability distribution of a single electron equally distributed over ϕ_A and ϕ_B (– – – –) along the internuclear line for H_2^+, according to LCAO–MO theory, including overlap.

being used because we are at the moment chiefly interested in how the electron density is distributed along the internuclear line. The dotted lines show the electron density in the individual atomic orbitals, that is, they show ϕ_A^2 and ϕ_B^2. It is evident that an electron in ψ_b has a high distribution between the nuclei, while one in ψ_a has a very low distribution in this region. ψ_a^2 actually goes to zero at the midpoint. As a further indication of the significance of ψ_a^2 and ψ_b^2, they may also be compared with the broken line, which is a plot of $\sqrt{1/2}\,\phi_A^2 + \sqrt{1/2}\,\phi_B^2$; this function gives the distribution of one electron spending its time equally in ϕ_A and in ϕ_B, these remaining, however, as separate atomic orbitals. The factors of $\sqrt{1/2}$ normalize the total electron density in each ϕ to 1/2. It is clear that ψ_b and ψ_a put more and less electron density, respectively, between the nuclei than does the simple sum of non-interacting atomic orbitals. This explains why one MO is given the subscript b, meaning *bonding*, and the other the subscript a, meaning *antibonding*.

In order to estimate the energies of the LCAO–MO's, ψ_a and ψ_b, we insert the expressions (3-19) into the wave equation. For ψ_b we have

$$E_b = \frac{\displaystyle\int \psi_b \mathscr{H} \psi_b \, d\tau}{\displaystyle\int \psi_b \psi_b \, d\tau} = \frac{N_b^2 \displaystyle\int (\phi_A + \phi_B)\, \mathscr{H}\, (\phi_A + \phi_B)\, d\tau}{1}$$

$$= N_b^2 \left[\int \phi_A \mathscr{H} \phi_A \, d\tau + \int \phi_A \mathscr{H} \phi_B \, d\tau + \int \phi_B \mathscr{H} \phi_A \, d\tau \right.$$

$$\left. + \int \phi_B \mathscr{H} \phi_B \, d\tau \right] \qquad \qquad \ldots \text{(3-20)}$$

We do not immediately attempt to evaluate the four integrals in equation 3-20. Instead, we give them the following symbols:

$$Q_A = \int \phi_A \mathscr{H} \phi_A \, d\tau$$

$$Q_B = \int \phi_B \mathscr{H} \phi_B \, d\tau$$

$$\beta = \int \phi_A \mathscr{H} \phi_B \, d\tau = \int \phi_B \mathscr{H} \phi_A \, d\tau$$

Since atoms A and B here are identical, $Q_A = Q_B = Q$ and equation 3-20 takes the following simple form:

$$E_b = 2 N_b^2 (Q + \beta) \qquad \qquad \ldots \text{(3-21)}$$

It follows, similarly, that
$$E_a = 2 N_a^2 (Q - \beta) \qquad \qquad \ldots \text{(3-22)}$$

Now the integral Q is obviously just the energy of an electron in the orbital ϕ_A or ϕ_B, that is, it is equal to the ground-state energy of the hydrogen atom. The integral β represents the energy of interaction between the orbitals ϕ_A and ϕ_B. It is called an exchange integral or resonance integral, and it can be shown that it is inherently negative. Thus an electron occupying the bonding MO, ψ_b, is more stable and an electron in ψ_a is less stable than one in a pure atomic orbital, ϕ_A or ϕ_B. The actual energies of stabilization and destabilization can be obtained in units of β by inserting the normalization coefficients into (3-21) and (3-22). If we make the approximation that overlap can be ignored, so that $N_a^2 = N_b^2 = 1/2$, we find that

$$E_a = Q + \beta$$
$$E_b = Q - \beta$$

These results are expressed in the energy level diagram, Fig. 3-11.

Its qualitative features are common to all energy level diagrams showing the energies of MO's formed between two atoms, each supplying one orbital.

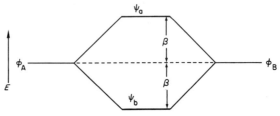

Fig. 3-11.   Energy level diagram showing the formation of bonding and antibonding MO's from two equivalent atomic orbitals in a homonuclear diatomic molecule.

If the two atomic orbitals do not have the same energy, then ψ_a and ψ_b will be placed at approximately equal distances above and below the mean energy of the two atomic orbitals. Fig. 3-11 is also a typical energy level diagram in its arrangement. The atomic orbitals of the constituent atoms are placed on each side, and the resulting MO's are placed in the center. It is to be noted that when the overlap is included in the normalization coefficients the problem becomes more complicated algebraically and the energy level diagram no longer possesses the simple, symmetrical form of Fig. 3-11. We shall not, however, go into this matter further here since all applications of LCAO–MO theory in this book will utilize the approximation of neglecting overlap.

In discussing the valence-bond theory the overlap integral, S, was introduced and discussed. It was noted that, in general, the larger the value of S for two orbitals ϕ_A and ϕ_B on two atoms, the stronger would be the bond which is formed on using these orbitals. In terms of simple LCAO–MO theory, this idea may be given a more quantitative form, by following a suggestion originally made by Mulliken. In essence, Mulliken proposed that integrals of the type β should be roughly proportional to the corresponding overlap integrals. It is also necessary to build into the relation expressing this proportionality a dependence on the average value of the energies of the orbitals that are forming the bond. Thus the following general equation is written:

$$\int \phi_A \mathscr{H} \phi_B \, d\tau = \beta = CQ_{av}S = CQ_{av}\int \phi_A \phi_B \, d\tau$$

in which C is a numerical constant, and Q_{av} is sometimes taken to be the arithmetic average,

$$Q_{av} = (Q_A + Q_B)/2$$

and by other workers as the geometrical average, viz.,

$$Q_{av} = (Q_A Q_B)^{1/2}$$

The proportionality constant, C, is expected to be approximately 2.0 and is often assigned exactly this value. Molecular-orbital calculations made with this approximation are now quite common, especially for transition-metal complexes and will be discussed further in Chapter 20.

We can now proceed to build the electronic structures of some other molecules just as we used the aufbau principle to build the electron configurations of atoms. If we add an additional electron to H_2^+ we have, of course, the hydrogen molecule H_2. This second electron will enter ψ_b along with the first, since ψ_b is the lowest energy orbital with a vacancy in it. In order to satisfy the exclusion principle, the spins of the two electrons must be paired. Thus we can write the electron configuration of H_2, from the MO viewpoint, as $(\psi_b)^2$. The binding energy will be approximately -2β with some correction for the mutual shielding effect. Thus in this simple case MO theory gives us a physical description of the bond in H_2 which is rather similar to what we obtain from VB theory, namely, that there are two electrons with spins paired which are concentrated between the nuclei. In more complicated cases we shall presently see more clearly how the two theories differ.

Let us continue applying the aufbau principle in the present situation. Suppose we bring together a helium atom and a hydrogen atom, so that again we have a bonding orbital, $\psi_b = \phi_{He} + \phi_H$, and an antibonding orbital, $\psi_a = \phi_{He} - \phi_H$. Again ψ_b will be more stable than the mean energy of the He $1s$ and H $1s$ orbitals, and ψ_a will be less stable, by roughly equal amounts. Because the exclusion principle prevents us from placing more than two electrons in one MO, the three electrons must occupy the two MO's in the following way: $(\psi_b)^2(\psi_a)^1$. Thus the molecule HeH should be stable by about the energy β. Such a molecule is known in the vapor phase.

Finally, let us suppose we bring together two helium atoms. Then we have four electrons to be housed in two MO's of the sort shown in Fig. 3-11. Clearly, two must be placed in the bonding orbital and two in the antibonding orbital so that, according to the LCAO approximation, the binding energy of He_2 is precisely zero. A similar argument may be framed for all the noble gases, and this explains why they are all monoatomic.

Bond Orders. The bond order is defined in MO theory as the number of electron pairs occupying bonding MO's minus the number of electron pairs occupying antibonding MO's. Thus the bond orders in H_2^+, H_2, HHe, and He_2 would be 1/2, 1, 1/2, and 0, respectively.

3-9. Homonuclear Diatomic Molecules

So far we have considered only cases in which the only important MO's are formed by overlap of an s orbital on each of the two atoms. MO's of this type are called σ (*sigma*) MO's and the property that so classifies them is their cylindrical symmetry about the internuclear axis. That such symmetry must exist should be clear from the fact that each of the two s orbitals composing them is symmetrical about this axis.

We must next consider how p orbitals may combine, in terms of the LCAO approximation, to form MO's in a homonuclear diatomic molecule. Suppose we define the internuclear axis as the z axis. If each atom has available a p_z orbital, they may be combined into a bonding MO, $p_z(1) + p_z(2)$ (Fig. 3-12a), and an antibonding MO, $p_z(1) - p_z(2)$. These MO's are also σ MO's. Note that a p_z orbital on one atom may also combine with an s orbital on the other to produce bonding (Fig. 3-12b) and antibonding MO's. In general, any kind of $s-p_z$ hybrid orbitals on the two atoms may combine to give σ MO's. Because of its ability to contribute to a σ MO, a p orbital (in this case, the p_z orbital, since we have chosen to identify the internuclear axis with the z axis of our coordinate system) lying along the molecular axis is called a $p\sigma$ orbital. An s orbital is simply understood to be of σ character and is not so denoted.

If the p_z orbital is a $p\sigma$ orbital, then the p_x and p_y orbitals are not. They

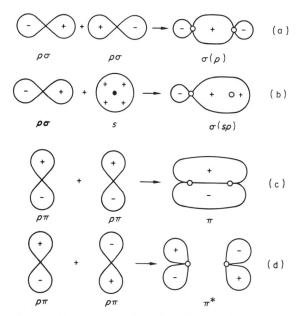

Fig. 3-12.   Diagrams illustrating the formation of some simple two-center MO's from atomic orbitals.

are $p\pi$ orbitals. A π-type orbital is one whose nodal plane includes the molecular axis. It is not therefore cylindrically symmetric about this axis, but has equal electron density on either side of a plane containing the axis, while the wave function itself is of opposite sign on the two sides. Two such $p\pi$ orbitals can be combined into a bonding πMO, $p\pi(1)+p\pi(2)$, and an antibonding πMO, $p\pi(1)-p\pi(2)$. These two are simply denoted π and π^*, respectively. Their formation is illustrated in Figs. 3-12c and 3-12d.

Let us consider an element in the first short Period having $2s$ and $2p$ orbitals in its valence shell. When two such atoms are combined into a homonuclear diatomic molecule, the two sets of atomic orbitals may combine into various MO's. Before we can specify the electronic structures of the diatomic molecules of these elements, we must know the relative energies of these MO's.

We may begin by treating the three types of orbitals, s, $p\sigma$, and $p\pi$ entirely separately. Thus the s orbitals give rise to σ_s and σ_s^*, bonding and antibonding MO's, respectively, just as in the case of H_2^+ and H_2. Similarly, the two $p\sigma$ orbitals combine to give σ_p and σ_p^* MO's. The two p_x orbitals combine to give bonding and antibonding MO's and so also do the two p_y orbitals. We note, however, that since the p_x and p_y orbitals on each atom are of the same energy, and entirely equivalent to one another except that one is rotated 90° from the other about the internuclear axis, the two π-type bonding MO's are equivalent to one another and of equal energy and so are the two π-type antibonding MO's. The two MO's of each pair are said to be *degenerate*.

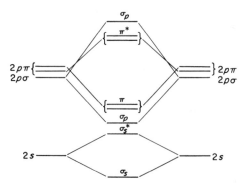

Fig. 3-13. The energy level diagram obtained in a molecular orbital treatment of a homonuclear diatomic molecule in which only s and p orbitals are used and $s\sigma$–$p\sigma$ interactions are neglected.

It is found that the magnitudes of the three sorts of overlaps decrease in the order s–$s \approx p\sigma$–$p\sigma > p\pi$–$p\pi$, and therefore, according to the overlap criterion of bond strength, the corresponding values of β should also decrease in the same order. On the basis of these considerations, together with the knowledge that for the elements of the first short Period the p orbitals have higher energies than the s orbitals, the energy level diagram shown in Fig. 3-13 can be drawn.

However, this diagram ignores one factor in the problem and is therefore not entirely correct. We have assumed that the s orbitals interact only with one another and the $p\sigma$ orbitals only with one another. While it is true that the greatest interactions should occur between the orbitals closest in energy (in this instance, those identical in energy) it is wrong to neglect entirely the other interactions permitted by symmetry. Actually, there are interactions between s and $p\sigma$ orbitals. One way to take account of this is to retain the diagram of Fig. 3-13 as a first approximation and then introduce the effects of further interactions. These will be essentially repulsions between pairs of orbitals of the same type, i.e. the two σ orbitals and the two σ^* orbitals. Fig. 3-14 shows the effect of introducing these interactions; part (a) is the ordering previously obtained (Fig. 3-13) while part (b) shows the new ordering. Now referring to Fig. 3-14 we see that if these shifts are sufficiently large, there may be a change in the qualitative order of the levels, namely, the third lowest orbital may be a π orbital instead of a σ orbital, as shown in Fig. 3-14b. The extent of the energy shifts will be greatest when the energy difference between the s and p orbitals is least, because then they can interact most. The s–p energy difference is relatively small for Li (~ 2 eV) and increases steadily until it is about 27 eV for Ne. Thus, while we might definitely expect the pattern shown in Fig. 3-14b at the beginning of the Period, the pattern in Fig. 3-14a *might* be correct at the end (i.e. for F_2). Whether this is so, we do not know for certain, but on the basis of experimental data, the pattern in Fig. 3-14b does appear to persist as far, at least, as N_2.

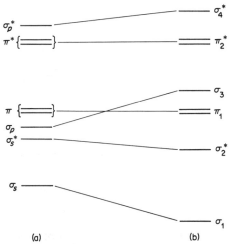

Fig. 3-14. The effect of $s\sigma$–$p\sigma$ interactions on the level order in a homonuclear diatomic molecule: (a) is the level order transcribed from Fig. 3-13; (b) is the level order after appreciable $s\sigma$–$p\sigma$ interaction is introduced.

We shall now consider the electron configurations of each of the homonuclear diatomic molecules of the elements in the first short Period on the basis of the LCAO–MO theory outlined above.

Li$_2$. The two valence electrons should enter the MO σ_1. The configuration is thus $(\sigma_1)^2$, the bond order is 1 and the molecule should be diamagnetic. The situation is, of course, quite analogous to that in H_2, but there are quantitative differences. The bond in Li_2 is much longer (2.67 Å vs 0.74 Å) and weaker (105 kJ mol^{-1} vs 432 kJ mol^{-1}) than in H_2. Two factors account for these differences. First, the $2s$ and $2p$ orbitals of Li are much larger and more diffuse than the $1s$ orbital of hydrogen. Thus they tend to overlap less effectively and with maximum overlap at a greater internuclear distance than in the case of the hydrogen $1s$ orbitals. Secondly, the electrons occupying the $1s$ shells of the Li atoms cause a repulsion between the atoms, which lessens the net energy of interaction and prevents closer approach of the atoms.

Be$_2$. For this molecule, the four electrons would be expected to give the MO configuration $(\sigma_1)^2(\sigma_2^*)^2$. The bond order would thus be 0. Consistent with this, Be_2 is not stable.

B$_2$. The six electrons should give the electron configuration $(\sigma_1)^2(\sigma_2^*)^2(\pi_1)^2$ on the basis of Fig. 3-14b. The two electrons in the π orbital should have their spins unpaired in accord with Hund's first rule, which applies to the filling of MO's as well as to atomic orbitals. On the other hand, according to Fig. 3-14a, the configuration would be $(\sigma_1)^2(\sigma_2^*)^2(\sigma_3)^2$, with no unpaired electrons. Experiment shows that B_2 has two unpaired electrons in a π orbital; thus, Fig. 3-14b is still correct at B_2. The bond order is 1 (one net π bond, since the σ interactions cancel). This is consistent with the bond energy of 289 kJ mol^{-1}.

C_2. For this molecule, the level order in Fig. 3-14b predicts a configuration $(\sigma_1)^2(\sigma_2^*)^2(\pi_1)^4$, while that in Fig. 3-14a predicts $(\sigma_1)^2(\sigma_2^*)^2(\sigma_3)^2(\pi_1)^2$. In the event that the σ_3 and π levels are very close, a $(\sigma_1)^2(\sigma_2^*)^2(\sigma_3)^1(\pi_1)^3$ configuration is also possible. Experimental results indicate that the ground state is derived from the $(\sigma_1)^2(\sigma_2^*)^2(\pi_1)^4$ configuration although a state derived from the $(\sigma_1)^2(\sigma_2^*)^2(\sigma_3)^1(\pi_1)^3$ configuration lies only 7.3 kJ mol^{-1} higher. A bond order of 2 is expected, and this is consistent with the observed bond energy of 630 kJ mol^{-1}.

N_2. For N_2, with a very high bond energy (946 kJ mol^{-1}) and no unpaired electron, either of the configurations $(\sigma_1)^2(\sigma_2^*)^2(\pi_1)^4(\sigma_3)^2$ or $(\sigma_1)^2(\sigma_2^*)^2(\sigma_3)^2(\pi_1)^4$ would be acceptable, each giving a bond order of 3. The former is presumably correct since the odd electron in N_2^+ is in a sigma orbital according to spectroscopic evidence.

O_2. The electron configuration might be either $(\sigma_1)^2(\sigma_2^*)^2(\sigma_3)^2(\pi_1)^4(\pi_2^*)^2$ or $(\sigma_1)^2(\sigma_2^*)^2(\pi_1)^4(\sigma_3)^2(\pi_2^*)^2$. Each provides a bond order of 2 and two unpaired electrons. Both these predictions are in good accord with experiment, the experimental bond energy being 493 kJ mol^{-1}. It is also interesting that the addition of electrons to O_2 causes the bond length to *increase* ($d_{O_2} = 1.21$ Å; $d_{O_2^-} = 1.26$ Å; $d_{O_2^{2-}} = 1.49$ Å) while the loss of an electron causes the bond to shorten ($d_{O_2^+} = 1.12$ Å). Since bond length varies inversely with bond strength, these results are in excellent accord with the MO electron configuration. The orbital to or from which electrons are added or removed is antibonding; thus removal of an electron strengthens the bond (the bond order in O_2^+ is 2.5), while addition of electrons weakens the bond (the bond orders in O_2^- and O_2^{2-} are 1.5 and 1.0).

F_2. All orbitals are filled except σ_4^*. The molecule should have no unpaired electron and a low bond energy corresponding to the bond order of 1. F_2 is diamagnetic with a bond energy of 158 kJ mol^{-1}.

Ne_2. This molecule has not been observed. Since the electron configuration would involve filling of all the MO's, the predicted bond order is 0, in accord with non-existence of Ne_2.

3-10. Heteronuclear Diatomic Molecules

The treatment of heteronuclear diatomic molecules by LCAO–MO theory is not fundamentally different from the treatment of homonuclear diatomics, except that the MO's are not symmetric with respect to a plane perpendicular to and bisecting the internuclear axis. The MO's are still constructed by forming linear combinations of atomic orbitals on the two atoms, but since the atoms are now different we must write them $\phi_A + \lambda\phi_B$, where λ is not in general equal to ± 1. Thus these MO's will not in general represent non-polar bonding. As examples let us consider HCl, CO, and NO.

In treating HCl we find it necessary to mention explicitly another factor influencing the stability of a bonding MO. Even if two atomic orbitals are capable of combining from the point of view of symmetry, the extent to which they will actually mix—that is, lose their individuality and merge to form a bonding and an antibonding MO—will depend upon whether their energies are comparable to begin with. If their energies are vastly different, they will scarcely mix at all. Mathematically, the two unnormalized MO's would be

$$\psi_b = \phi_A + \lambda\phi_B$$
$$\psi_a = \phi_B - \lambda\phi_A$$

where λ would be very small so that $\psi_b \approx \phi_A$ and $\psi_a \approx \phi_B$. In other words, when the energies are not similar, we can treat ϕ_A and ϕ_B as though they do not mix at all. This is more or less what occurs in HCl. The H $1s$ orbital and the Cl $3p\sigma$ orbital mix fairly effectively to form a bonding and an antibonding MO, but the Cl $3p\pi$ and $3s$ orbitals are so much lower in energy than any other hydrogen orbitals such as $2s$ or $2p\pi$ that no significant mixing occurs. Thus we call the $3p\pi$ and $3s$ orbitals of Cl in this case non-bonding because they neither help nor hinder the bonding to a significant extent.

The heteronuclear molecule CO may be regarded as a perturbed nitrogen molecule. C and O, differing in atomic number by only two, have atomic orbitals that are quite similar; the formation of MO's will therefore be almost the same as shown in Figs. 3-13 and 3-14 for a homonuclear diatomic, although the energies of the two sets of atomic orbitals will not now match exactly. In fact, the oxygen orbitals will be somewhat more stable, so that they will contribute more to the bonding MO's than will the carbon orbitals, whereas the carbon orbitals will contribute more to the antibonding MO's. Thus, although the ten electrons are comprised of six from oxygen and four from carbon, we can explain the low polarity of the molecule because eight of them are in bonding orbitals where they are held closer to O than to C, thus tending to neutralize the greater nuclear charge of the oxygen core. As for N_2, a bond order of 3 is predicted and is in accord with the experimental bond energy of 1073 kJ mol^{-1}.

The electron configuration of NO might be derived by either removing one electron from O_2 or adding one to N_2. Either procedure would predict that there should be one π^* electron and this is actually so. NO readily loses this electron to form the NO$^+$ ion, which is found to have a stronger bond than does NO (cf. the earlier discussion of the ions derived from O_2).

3-11. Polyatomic Molecules

MO theory is extensively used for those polyatomic molecules in which multiple bonding occurs. We have already considered how NO$_2^-$ is formulated in VB theory as a resonance hybrid (3-V). In MO theory it can be treated in the following way. We first assume that a set of σ bonds is formed using four electrons and that several other electron pairs are non-bonding. Thus we write a framework of nuclei and σ electrons (3-VII). There are still four electrons to be assigned and each of the three atoms has an empty $p\pi$ orbital (one whose nodal plane coincides with the molecular plane). If we start with

$$\ddot{N}$$
$$:\underset{}{O} \qquad \underset{}{O}:$$

(3-VII)

three atomic orbitals it must be possible to combine them into three molecular orbitals. A discussion of the methods by which the correct combinations are derived would be beyond the scope of this text, but their approximate forms are illustrated for the present case in Fig. 3-15. ψ_b is bonding, having the lowest energy, ψ_n is non-bonding and ψ_a is antibonding. If we place our four remaining electrons in this set of π MO's, the configuration will be $(\psi_b)^2(\psi_n)^2(\psi_a)^0$. The net energy is thus favorable to bonding. Note that in general these MO's are linear combinations of all three atomic orbitals, and this description of the bonding does not therefore refer to *localized* electron pair bonds, but to *delocalized* electrons moving in MO's that extend over the entire molecule. Moreover, from the nature of the occupied MO's, ψ_b and ψ_n, it is obvious that the distribution of the four π electrons is symmetric in the two NO links. This description is sometimes symbolized by (3-VIII), where the broken line indicates bonding due to delocalized electrons in molecular orbitals. This set of π MO's is formally similar to one of the types of three-center σ bonding described in the next Section.

Benzene is described in VB theory as a resonance hybrid of the Kekulé, Dewar, and other canonical structures, is well known. In MO theory, it is treated by assuming the formation of extensive π MO's. Each carbon atom is assumed to use its s, p_x and p_y orbitals and three of its electrons to form the σ-bonded framework (3-IX). Each carbon atom still has a p_z orbital (which is

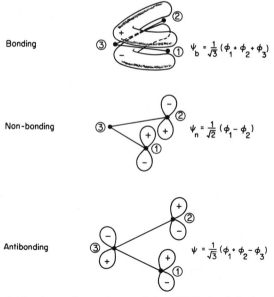

Bonding $\psi_b = \dfrac{1}{\sqrt{3}}(\phi_1 + \phi_2 + \phi_3)$

Non-bonding $\psi_n = \dfrac{1}{\sqrt{2}}(\phi_1 - \phi_2)$

Antibonding $\psi = \dfrac{1}{\sqrt{3}}(\phi_1 + \phi_2 - \phi_3)$

Fig. 3-15.   Approximate shapes of the π MO's in nitrite ion, NO_2^-.

H
H—C—H
C   C
C
C   C
H   H
C
H

(3-VIII)   (13-IX)   (3-X)

a $p\pi$-type orbital) containing an electron. These $p\pi$ orbitals will merge into various MO's, and the electrons in them will thus not be localized in definite double bonds but will be free to wander around the circular π MO's. This view can be rendered symbolically with a broken line (3-X).

3-12. Multicenter Bonding

Although the bonding in the great majority of chemical compounds presently known can be described satisfactorily by using two electrons to form a bond between each adjacent pair of atoms, that is, in terms of so-called two-center, two-electron ($2c$–$2e$) bonds, there are also many cases where this type of bonding is an inadequate, or at least a very inconvenient, method of describing the electronic structures. Particularly cogent examples are the boranes (Section 8-8) and many polyhedral species such as borane anions (Sections 8-10 and 8-11) and metal-atom clusters (Section 19-11). For these, various forms of multicenter bonding are invoked, of which the most basic is some type of three-center bonding.

Three-center Bonding. We begin by dividing three-center bonds into two geometric types: open and closed. In the former, which are the more common, the three atoms are so arranged that two terminal atoms are each close to a central one but not close to each other, as in (3-XI); the angle ϕ may be 180° or less, so long as it does not approach 60° in which case

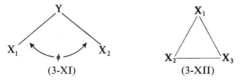

(3-XI)   (3-XII)

direct X—X bonding would occur. In idealized closed three-center bonding three identical atoms form an equilateral triangle, as in (3-XII), so that all three atom pairs, X_1—X_2, X_2—X_3, X_3—X_1, are equally strongly bonded. We shall first discuss open three-center bonding.

The details of open three-center bonding vary according to the type of orbital used by the central atom and the number of electrons available. Two principal cases, (a) and (b) in Fig. 3-16, may be distinguished. In case (a) the central atom employs an orbital, ϕ_3, which is symmetric with respect to the two-fold axis or plane of symmetry that interchanges the end atoms. As shown in Fig. 3-16a, this symmetrical orbital may be an s orbital or some sort of s–p hybrid, including the limiting case of a pure p orbital appro-

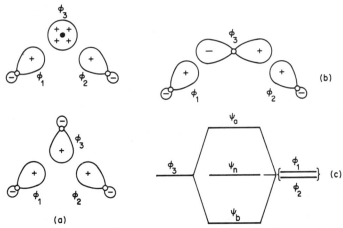

Fig. 3-16. Open 3-center bonding. (a) Two situations in which the central orbital, ϕ_3, is symmetric. (b) The case where ϕ_3 is antisymmetric. (c) The type of energy level diagram that arises in either case (a) or case (b).

priately oriented. It is not difficult to see that with the atomic orbitals, ϕ_1, ϕ_2, and ϕ_3 defined as they are, the following three linear combinations may be constructed:

$$\psi_b \approx \phi_1 + \phi_2 + \phi_3 \qquad \text{bonding}$$
$$\psi_n \approx \phi_1 - \phi_2 \qquad \text{non-bonding}$$
$$\psi_a = \phi_1 + \phi_2 - \phi_3 \qquad \text{antibonding}$$

Similarly, in case (b), where ϕ_3 is a p orbital oriented so as to be antisymmetric to the two-fold axis or plane of symmetry, the three linear combinations are:

$$\psi_b \approx -\phi_1 + \phi_2 + \phi_3 \qquad \text{bonding}$$
$$\psi_n \approx \phi_1 + \phi_2 \qquad \text{non-bonding}$$
$$\psi_a \approx \phi_1 - \phi_2 + \phi_3 \qquad \text{antibonding}$$

The important point to note here is that despite the difference in the type of atomic orbital constituting ϕ_3, which leads to algebraically different expressions for the three linear combinations, the final result is the same: three 3-center orbitals, one bonding, one non-bonding, and one antibonding, are formed. Thus, both case (a) and case (b) can be represented with the type of energy level diagram shown in Fig. 3-16c. Let it be noted again that either case (a) or case (b) can exist with a completely linear chain of atoms.

Given the orbital diagram in Fig. 3-16c, there are two important electron distributions, namely, 2-electron and 4-electron populations. If the three atoms, between them, supply only two electrons, these will occupy the bonding orbital, ψ_b. Thus two electrons will serve to unite three atoms. Since the orbital ψ_b is made up of ϕ_1, ϕ_2, and ϕ_3, electrons occupying it are spread fairly uniformly over the three atomic centers, thus leading to a basically non-polar distribution. This situation, in which three centers are united by two paired electrons is called 3-center, 2-electron bonding, abbreviated to $3c$–$2e$ bonding. Its principal application is in the case of the bridging hydrogen atoms in diborane and other boron hydrides (Section 8-9).

There are also cases of open 3-center bonding in which a total of four electrons are employed. This situation is called 3-center, 4-electron ($3c$–$4e$) bonding. In addition to the rather uniformly distributed electron pair in ψ_b, there is also a pair of electrons in ψ_n; these electrons are entirely concentrated on the end atoms, as may be seen from the expressions given above for ψ_n. Thus $3c$–$4e$ bonding is relatively polar, with the end atoms negative relative to the center one. The $3c$–$4e$ bonding finds application in the hypervalent compounds of the heavier non-transition elements (page 133) and in cases of strong hydrogen bonding (page 153).

In terms of the nomenclature introduced here for three-center bonds, the conventional electron-pair bond between two atoms may be called a $2c$–$2e$ bond.

SOME EMPIRICAL CONSIDERATIONS

3-13. Bond Energies

We have already frequently used the term bond energy, assuming that the reader's previous knowledge and/or the context would make the meaning sufficiently clear. The subject is here examined more closely. For a diatomic molecule the bond energy, D, is equal to the enthalpy of the reaction

$$XY(g) = X(g) + Y(g) \qquad \Delta H^\circ = D \qquad \qquad \text{... (3-23)}$$

where the molecule and the atoms are all in their ground states. To be exact, this bond energy is a function of temperature, and the best value to use would be that for the hypothetical reaction at $0\,^\circ K$. The differences between the values at $0\,^\circ K$ (properly denoted D_0^0, to indicate that the temperature is $0\,^\circ K$ and that the diatomic molecule is in its lowest, that is, 0, vibrational state) and those at room temperature, denoted D_{300} are always small. They cannot exceed $\sim 10\ \text{kJ mol}^{-1}$ and must always be such that $D_{300} > D_0^0$. For example, D_0^0 for H_2 is 432, while D_{300} is 436 kJ mol^{-1}. For our purposes here, the differences between D_{300} and D_0^0 values are not important, and we shall simply speak of bond dissociation energies as though they were independent of temperature and use the symbol D.

Although the bond energy of a diatomic molecule is actually an experimental datum, bond energies in polyatomic molecules must be carefully defined to be meaningful. Let us consider the simplest case, that of an AB_n-type molecule where all B's are equivalently bonded to A and not to one another. BF_3 is an example. Since all the B—F bonds are equivalent, all B—F bond energies, D_{B-F}, must be equal and the relation (3-24)

$$BF_3(g) = B(g) + 3F(g) \qquad \Delta H = 3D_{B-F} \qquad \text{... (3-24)}$$

is obvious. Thus, if we know the heats of formation of $BF_3(g)$, $B(g)$ and $F(g)$, we can readily calculate D_{B-F}. This value is called the *mean thermochemical bond energy*. However, it is *not* the energy, ΔH_1, of the process (3-25)

$$BF_3(g) = BF_2(g) + F(g) \qquad \Delta H_1 \qquad \qquad \text{... (3-25)}$$

since when the first bond is broken, the nature of the remaining two will necessarily be altered to some extent. Also ΔH_2 and ΔH_3 (3-26 and 3-27)

$$BF_2(g) = BF(g) + F(g) \qquad \Delta H_2 \qquad \qquad \text{... (3-26)}$$
$$BF(g) = B(g) + F(g) \qquad \Delta H_3 \qquad \qquad \text{... (3-27)}$$

are not likely to be equal to one another, or to ΔH_1 (equation 3-25), or to D_{B-F}. It is, of course, true that $\Delta H_1 + \Delta H_2 + \Delta H_3 = 3D_{B-F}$, since the sum of equations 3-25, 3-26 and 3-27 is equal to equation 3-24 and Hess' law can be applied. The question of how much ΔH_1, ΔH_2, ΔH_3 and D_{B-F} will differ cannot be answered since there are not sufficient data available to calculate them all.

It is probable that the differences between successive dissociation energies, and between any one of them and the mean, will be fairly small so long as no one step involves a unique change in electron configuration of the central atom. For example, the following data are available for H_2O:

$$H_2O(g) = 2H(g) + O(g) \qquad \tfrac{1}{2}\Delta H = D_{O-H} = 460 \text{ kJ mol}^{-1}$$
$$H_2O(g) = H(g) + OH(g) \qquad \Delta H_1 \qquad = \sim 489 \text{ kJ mol}^{-1}$$
$$HO(g) = H(g) + O(g) \qquad \Delta H_2 \qquad = \sim 431 \text{ kJ mol}^{-1}$$

Since the oxygen atom in its ground state has the two unpaired electrons required to form the two O—H bonds in H_2O, there will be only relatively small changes in valence states in the different processes and no one of them will have any particularly large promotion energy peculiar to itself. In the case of the mercuric halides, however, we have a quite different situation, as the following data show:

$$HgCl_2(g) = Hg(g) + 2Cl(g) \qquad \tfrac{1}{2}\Delta H = D_{Hg-Cl} = 222 \text{ kJ mol}^{-1}$$
$$HgCl_2(g) = HgCl(g) + Cl(g) \qquad \Delta H_1 = 339 \text{ kJ mol}^{-1}$$
$$HgCl(g) = Hg(g) + Cl(g) \qquad \Delta H_2 = 105 \text{ kJ mol}^{-1}$$

Whereas breaking the first ClHg—Cl bond results in only a small change in the state of the Hg atom, when the second bond, Hg—Cl, is broken the mercury atom drops from some sort of *sp* configuration into its s^2 ground state, releasing considerable energy which partially offsets the energy required to break the second bond. Hence, $\Delta H_2 \ll \Delta H_1$.

The type of molecule in which the concept of bond energies becomes decidedly empirical and not uniquely defined is the commonest type, namely, that in which there are two or more different kinds of bond. Consider, for example, a molecule such as H_3CGeCl_3, in which there are three equivalent C—H bonds, three equivalent Ge—Cl bonds and one Ge—C bond. The only entirely straightforward and rigorous thermochemical equation involving bond energies that we can write is:

$$H_3CGeCl_3(g) = 3H(g) + 3Cl(g) + C(g) + Ge(g) \qquad \text{... (3-28a)}$$
$$\Delta H = 3\,D_{C-H} + 3\,D_{Ge-Cl} + D_{Ge-C} \qquad \text{... (3-28b)}$$

The apportionment of ΔH among the three terms on the right-hand side of (3-28b) is settled, in practice, by a trial and error scheme such that values of

TABLE 3-2

Some Average Thermochemical Bond Energies at 25° in kJ mol^{-1}

A. Single Bond Energies

	H	C	Si	Ge	N	P	As	O	S	Se	F	Cl	Br	I
H	436	416	323	289	391	322	247	467	347	276	566	431	366	299
C		356	301	255	285	264	201	336	272	243	485	327	285	213
Si			226	—	335	—	—	368	226	—	582	391	310	234
Ge				188	256	—	—	—	—	—	—	342	276	213
N					160	~200	—	201	—	—	272	193	—	—
P						209	—	~340	—	—	490	319	264	184
As							180	331	—	—	464	317	243	180
O								146	—	—	190	205	—	201
S									226	—	326	255	213	—
Se										172	285	243	—	—
F											158	255	238	—
Cl												242	217	180
Br													193	151
I														151

B. Multiple Bond Energies

C=C 598	C=N 616	C=O 695	N=N 418
C≡C 813	C≡N 866	C≡O 1073	N≡N 946

D_{C-H}, D_{Ge-Cl}, D_{Ge-C}, and various other bond energies are adjusted so that a given set of such energies will fit the atomization energies of a large number of molecules with the least mean deviation. A list of such average thermochemical bond energies is given in Table 3-2.

3-14. Electronegativities

The qualitative concept of electronegativity is one to which most chemists subscribe, but precise definition is elusive. The original qualitative definition of Pauling is: "Electronegativity is the power of an atom *in a molecule* to attract electrons to itself." It is not, however, the same as electron affinity, as will be seen.

The first attempt to assign numerical estimates of electronegativities was the thermochemical method of Pauling. The rationale was that the energy of an X—Y bond almost invariably exceeds the mean of the X—X and Y—Y bond energies and that the excess, Δ, can be attributed to ionic–covalent resonance X—Y↔X$^+$ Y$^-$ if the electronegativity of Y is greater than that of X. It was then proposed that electronegativity values, x_X, x_Y, etc., be assigned so as to satisfy the relationship (3-29) for as wide a range of bond energy data as possible.

$$|x_X - x_Y|^2 = \frac{\Delta}{96} \qquad \dots (3\text{-}29)$$

Since Pauling worked out the first electronegativity scale in this way, bond energy values have been revised and electronegativities recalculated. The values listed in Table 3-3 as "Pauling electronegativities" were calculated by Pauling's method but are not, in general, his original values.

Among the dozens of other methods proposed for estimating experimentally or calculating electronegativity values, we shall mention only two. Mulliken showed by theoretical arguments that the tendency of an atom *in a molecule* to compete with another atom to which it is bound in attracting the shared electrons should be proportional to $(I+A)/2$, that is, to the average of its ionization potential and its electron affinity. Physically this is quite reasonable since we should expect that the overall ability of an atom to attract shared electrons might be the average of quantities related to the tendencies of the free atom to hold its own electrons and to attract additional electrons. It is found that when the I and A values are expressed in electron volts the Mulliken electronegativity values can be adjusted to give the least mean deviation from Pauling's by dividing them by 3.15. Representative Mulliken electronegativities are listed in Table 3-3.

An empirical method due to Allred and Rochow seems to have certain advantages arising precisely from its complete and strict empiricism. Pauling's qualitative definition of electronegativity may be interpreted as

$$\text{Force} = \frac{Z^* e^2}{r^2}$$

TABLE 3-3
Electronegativities of the Elements

(Values in bold type are calculated by using the Allred–Rochow formula; those in italics are estimated by Pauling's method and those in Roman type are calculated by Mulliken's method.)[a]

I	II	III	IV	II	II	II	II	II	II	II	II	III	IV	III	II	I[b]	
H **2.20**																	He
Li **0.97** *0.98* 1.46	Be **1.47** *1.57*											B **2.01** *2.04* 2.01	C **2.50** *2.55* 2.63	N **3.07** *3.04* 2.33	O **3.50** *3.44* 3.17	F **4.10** *3.98* 3.91	Ne
Na **1.01** *0.93* 0.93	Mg **1.23** *1.31* 1.32											Al **1.47** *1.61* 1.81	Si **1.74** *1.90* 2.44	P **2.06** *2.19* 1.81	S **2.44** *2.58* 2.41	Cl **2.83** *3.16* 3.00	Ar
K **0.91** *0.82* 0.80	Ca **1.04** *1.00*	Sc **1.20** *1.36*	Ti **1.32** *1.54*	V **1.45** *1.63*	Cr **1.56** *1.66*	Mn **1.60** *1.55*	Fe **1.64** *1.83*	Co **1.70** *1.88*	Ni **1.75** *1.91*	Cu **1.75** *1.90* 1.36	Zn **1.66** *1.65* 1.49	Ga **1.82** *1.81* 1.95	Ge **2.02** *2.01*	As **2.20** *2.18* 1.75	Se **2.48** *2.55* 2.23	Br **2.74** *2.96* 2.76	Kr
Rb **0.89** *0.82*	Sr **0.99** *0.95*	Y **1.11** *1.22*	Zr **1.22** *1.33*	Nb **1.23**	Mo **1.30** *2.16*	Tc **1.36**	Ru **1.42**	Rh **1.45** *2.28*	Pd **1.35** *2.20*	Ag **1.42** *1.93* 1.36	Cd **1.46** *1.69* 1.4	In **1.49** *1.78* 1.80	Sn **1.72** *1.96*	Sb **1.82** *2.05* 1.65	Te **2.01**	I **2.21** *2.66* 2.56	Xe
Cs **0.86** *0.79* 0.82	Ba **0.97** *0.89*	*La **1.08** *1.10*	Hf **1.23**	Ta **1.33**	W **1.40** *2.36*	Re **1.46**	Os **1.52**	Ir **1.55** *2.20*	Pt **1.44** *2.28*	Au **1.42** *2.54*	Hg **1.44** *2.00*	Tl **1.44** *2.04*	Pb **1.55** *2.33*	Bi **1.67** *2.02*	Po **1.76**	At **1.96**	Rn
Fr **0.86**	Ra **0.97**	**															

*Lanthanides:

La **1.08** *1.10*	Ce **1.06** *1.12*	Pr **1.07** *1.13*	Nd **1.07** *1.14*	Pm **1.07**	Sm **1.07** *1.17*	Eu **1.01**	Gd **1.11** *1.20*	Tb **1.10**	Dy **1.10** *1.22*	Ho **1.10** *1.23*	Er **1.11** *1.24*	Tm **1.11** *1.25*	Yb **1.06**	Lu **1.14** *1.27*

**Actinides:

Ac **1.00**	Th **1.11**	Pa **1.14**	U **1.22** *1.38*	Np **1.22** *1.36*	Pu **1.22** *1.28*	Am	Cm	Bk	Cf	Es	Fm	Md

Am ————— ~1.2 (estimated) ————→ Md

[a] Allred–Rochow values from *J. Inorg. Nuclear Chem.*, 1958, **5**, 264; Pauling-type values from A. L. Allred, *J. Inorg. Nuclear Chem.*, 1961, **17**, 215; Mulliken-type values from H. O. Pritchard and H. A. Skinner, *Chem. Rev.*, 1955, **55**, 745.

[b] Roman numerals at the top give the oxidation states used for the Pauling-type values.

where Force is the attraction felt by an electron toward one of the two nuclei in a bond, Z^* is the effective nuclear charge that the electron feels and r is its mean distance from the nucleus. Z^* is estimated by using a set of shielding parameters derived many years ago by Slater, and r is taken as the covalent radius of the atom, which for a homonuclear diatomic molecule is half the internuclear distance. (Covalent radii are discussed below.) The following equation places Allred and Rochow's electronegativities (Table 3-3) on the same numerical scale as those of Pauling:

$$x_{AR} = 0.359 \frac{Z^*}{r^2} + 0.744$$

It is important to note that there must be a variation of the electronegativity of an element from one compound or one bonding situation to another depending on its valence state. This may be best appreciated by considering Mulliken's definition. For rigorous application of Mulliken's definition, one should take, not the I and A values applying to the ground state atom, but those applying to the *valence state* (see page 85) of the atom in a particular compound, and these will vary with the nature of the valence state. For example, if nitrogen is in the valence state s^2p^3, its Mulliken electronegativity is 2.33, whereas if it is in the state sp^4 the x_M value is 2.55. The former is given in Table 3-3 since the valence state for N in its common trivalent compounds is probably nearest to s^2p^3. It is clearly absurd to assume that the electronegativity of an element is independent of its valence, for S in SCl_2 must surely have a different electronegativity from S in SF_6. Clearly, an atom will have a greater attraction for electrons when it is in a high oxidation state than when it is in a low one. Thus, the numbers in Table 3-3 should not be taken as exact measures of electronegativities but only as rough guides, perhaps as the median numbers in a range for each element.

3-15. Bond Lengths and Covalent Radii

The lengths of bonds, that is, the internuclear distances in molecules, can be measured in many ways, and a considerable body of such data is available. If we consider a homonuclear diatomic molecule with a single bond, such as F_2 or Cl_2, we can assign to the atoms F and Cl *covalent single bond* radii equal to one-half of the internuclear distances in the respective molecules. It is then gratifying to find that very often the sums of these covalent radii are equal to the internuclear distances in the interhalogens such as Cl—Br (calculated 2.13, found 2.14). For elements that do not form diatomic molecules with single bonds, other methods of estimating the radii are used. For example, the C—C distance in diamond and a host of organic molecules is found to be 1.54 ± 0.01 Å, so the covalent radius of C is taken as 0.77. To obtain the covalent radius of nitrogen, 0.77 is subtracted from the C—N distance in H_3C—NH_2, yielding 0.70. In this fashion a table of single-bond covalent radii can be compiled (Table 3-4).

TABLE 3-4

Some Single-bond Covalent Radii (in Å)

H[a]	0.28	O	0.66
C	0.77	S	1.04
Si	1.17	Se	1.17
Ge	1.22	Te	1.37
Sn	1.40	F	0.64
N	0.70	Cl	0.99
P	1.10	Br	1.14
As	1.21	I	1.33
Sb	1.41		

[a] One-half the bond length of H_2 is 0.375, but this value does not apply when H is bonded to other atoms. 0.28 was obtained by subtracting the radius of X from various H—X bond lengths.

Multiple-bond radii can also be obtained. For example, the triple-bond radii of carbon and nitrogen can be calculated from the bond lengths in $HC\equiv CH$ and $N\equiv N$ as 0.60 and 0.55, giving 1.15 for $C\equiv N$ as compared with experimental values of ~ 1.16. It may be stated, as a general rule, that the higher the order of a bond between two atoms, the shorter it is. Thus, for carbon—carbon bonds the following are typical lengths: C—C, 1.54; C=C, 1.33; C≡C, 1.21.

It is also true, however, that hybridization effects are important. Thus, strictly, the C—C distance of 1.54 refers to a bond between two sp^3 hybridized carbon atoms, and the C=C distance usually refers to a bond between two sp^2 hybridized carbon atoms, whereas the C≡C bond necessarily occurs between two sp hybridized carbon atoms. Since the $2s$ orbital of carbon has a smaller mean radius than the $2p$ orbitals, it would be expected that the greater the s character in the hybrid orbitals used, the shorter would be the internuclear distance at which the best balance of overlap and repulsion would occur in the σ bonds. Hence, at least part of the decreases in the bond lengths cited is attributable to this effect rather than to the π bonding. It has been estimated that the single bond radii for the carbon atom in the several states of hybridization are: sp^3, 0.77; sp^2, 0.74; sp, 0.70. When this effect is taken into account, certain previously accepted conclusions about the importance of single bond–multiple bond resonance become questionable. For example, the C—C bond in cyanogen is only 1.37 Å. If a carbon—carbon single bond distance is taken to be $2 \times 0.77 = 1.54$ Å, then there is evidently a large shortening and one assumes that in addition to (3-XIIIa), the canonical forms (3-XIIIb) and (3-XIIIc) are of major importance in a VB description of the electronic structure. However, if one notes that the carbon atoms have sp hybridization, the "expected" length of the C—C bond in (3-XIIIa) is only $2 \times 0.70 = 1.40$ Å, and the importance of (3-XIIIb) and (3-XIIIc) appears to be slight.

$$:N\equiv C-C\equiv N: \quad \leftrightarrow \quad :\overset{+}{N}=C=C=\overset{..}{\overset{-}{N}}: \quad \leftrightarrow \quad :\overset{..}{\overset{-}{N}}=C=C=\overset{+}{N}:$$

$$\text{(3-XIIIa)} \qquad\qquad \text{(3-XIIIb)} \qquad\qquad \text{(3-XIIIc)}$$

This subject is unfortunately not as tidy as might be imagined from the above examples, which were especially chosen to illustrate how the system works *when it works*. There are many cases in which it leaves much to be desired. The reader might have wondered why the radius of nitrogen (0.70) was obtained from the C—N bond in methylamine instead of from the N—N bond in H_2N—NH_2, which would be more analogous to the procedure used for the halogens. The answer is simply that one-half of the N—N distance in hydrazine is 0.73 and this does not fit as consistently with the bulk of data on X—N bond lengths as does 0.70. (The long bond in N_2H_4 is best attributed to repulsion between lone pairs of electrons on the N atoms.) Even more striking, however, are cases such as SiF_4, for which an Si—F distance of 1.81 would be calculated while the actual distance is ~ 1.54. Again, for BF_3 the calculated distance would be ~ 1.5 (the covalent radius of B is not easy to evaluate unambiguously), whereas the measured value is 1.30. Schomaker and Stevenson proposed that since these "shrinkages" generally occur in bonds between atoms of disparate electronegativities they may be due to bond-strengthening and hence shortening due to ionic-covalent resonance (3-XIVa)$\leftrightarrow$(3-XIVb) (three similar ionic forms) and

$$\begin{array}{ccc}
\underset{\underset{F}{|}}{\overset{\overset{F}{|}}{F-Si-F}} \leftrightarrow \underset{\underset{F}{|}}{\overset{\overset{F}{|}}{F-Si^{+}}} \ F^{-} & \qquad & F-B \ \ \leftrightarrow \ \ B_{+} \\
(3\text{-XIVa}) \qquad\qquad (3\text{-XIVb}) & & (3\text{-XVa}) \qquad (3\text{-XVb})
\end{array}$$

(3-XVa$\leftrightarrow$b) (two similar ionic forms). They therefore proposed an equation known as the Schomaker–Stevenson relationship which takes account of this by making the bond distance depend on the electronegativity difference, which is an index of the ionic character of the bond. The equation they suggested is

$$r_{A-B} = r_A + r_B - 0.09|x_A - x_B|$$

where r_A is the covalent radius of atom A. This relationship is not really very satisfactory except qualitatively, since it predicts too little shortening for some bonds, for example, Si—O and Si—F, and too much for others, for example, C—Cl. In the case of BF_3 the bond shortening can also be attributed to B—F double bonding since boron has a vacant p_z orbital and the fluorine atoms have filled p_z orbitals (3-XVIa$\leftrightarrow$b) and two other similar forms.

$$\begin{array}{cc}
\ddot{F}-B \overset{\ddot{F}}{\underset{\ddot{F}}{}} & \leftrightarrow & F\overset{-}{=}B \overset{\ddot{F}}{\underset{\ddot{F}}{}} \\
(3\text{-XVIa}) & & (3\text{-XVIb})
\end{array}$$

In SiF_4 π bonding, using F 2p orbitals and Si 3d orbitals, could contribute to the shortening. It is probable that *both* ionic–covalent resonance and multiple bonding contribute significantly.

In conclusion, we may say that the concept of atomic covalent radii is useful, but we cannot expect close correspondence between experimental interatomic distances and sums of these radii when the environment of either or both of the corresponding atoms differs appreciably from that of the atoms in the classes of substances used to derive the radii. Whether or not the ionicity of a bond can be related quantitatively to deviations from covalent radius sums, it does seem logical that covalent radii cannot be expected to describe appreciably ionic bonds very exactly.

3-16. van der Waals Radii

In addition to ionic and covalent radii, there remain to be considered the distances between atoms in liquids and solids when these atoms are not bonded to one another either ionically or covalently.

Let us consider, for example, the noble gases. The fact that they can be liquefied and solidified at all proves that there are *some* forces of attraction between the atoms; at the same time the exceedingly low temperatures required to condense them proves that these forces are extremely weak. These forces are usually called van der Waals forces, after the Dutch physicist who first emphasized their importance by taking account of them in his equation of state for imperfect gases, although sometimes they are also called London forces, since their nature was first explained by Fritz London using wave mechanics. We have mentioned them before as minor contributors to the total attractive force in ionic crystals. In crystals of the noble gases, however, there are no electrostatic forces, and these van der Waals forces are the only attractive forces. Again, as in ionic crystals, the equilibrium separation of neighboring atoms is that distance at which the attractive force is balanced by the repulsive force due to overlap of the outer portions of the electron clouds. Since this repulsive force rises very steeply and becomes important only at very short distances, the ionic radius of Br^- and one-half the distance of closest approach of two krypton atoms in solid krypton do not differ very much despite the differences in the nature of the attractive forces. The latter quantity—one half the Kr–Kr separation in solid krypton— is called the van der Waals radius of krypton. The van der Waals radii are much larger than covalent radii, however. Thus, the ionic radius of Br^- is 1.95 Å, the covalent radius of Br is 1.15 Å, and the van der Waals radius of Kr is 2.00 Å.

van der Waals radii for all elements may be estimated if the distances of closest approach of their atoms to other atoms when no chemical bond exists between them are known from structural studies. For instance, in solid bromine the closest approach of non-bonded bromine atoms is 3.90 Å, giving a van der Waals radius of 1.95 Å. If we consider crystals consisting of molecules with permanent dipole moments, then the dipole–dipole attractions will contribute to the stability of the crystals, but the closest distance of approach of two non-bonded atoms can still be taken as the sum of their van der Waals radii. As with ionic and covalent radii, deviations from additivity occur since the basic idea is something of an oversimplification, but a set of radii has been assigned to commonly occurring non-metallic atoms that gives the best overall agreement with a large number of experimental data. A set of van der Waals radii computed by Pauling is given in Table 3-5.

TABLE 3-5
van der Waals Radii of Non-metallic Atoms (in Å)

H	1.1–1.3					He	1.79
N	1.5	O	1.40	F	1.35	Ne	1.6
P	1.9	S	1.85	Cl	1.80	Ar	1.92
As	2.0	Se	2.00	Br	1.95	Kr	1.98
Sb	2.2	Te	2.20	I	2.15	Xe	2.18

Radius of a methyl group, 2.0 Å
Half-thickness of an aromatic ring, 1.85 Å

3-17. Polarities of Bonds and Molecules

The electrical polarity of a molecule is expressed as its *dipole moment*. A system consisting of a positive charge, $+x$, and a negative charge, $-x$, separated by a distance, d, possesses a dipole moment of magnitude xd. If x is equal to the electronic charge (4.80×10^{-10} esu) and d is 1 Å, $xd = 4.80$ D (Debye) units. A dipole moment is a vector quantity since it has a definite direction as well as magnitude. We adopt here the convention of having the vector point in the direction of the negative end of the dipole. The dipole moment of a molecule can be thought of as the vector sum of several interdependent moments due to various parts of the molecule.

Any heteronuclear diatomic molecule, such as HCl, must have a dipole moment since its two ends are different. However, an explanation of the magnitude of such a dipole moment must deal explicitly with two contributing factors.

1. *Polar nature of a heteronuclear bond.* If we place a proton and a Cl^+ ion—the latter being assumed, for the moment, to have spherical symmetry (hence no dipole within itself)—1.27 Å apart and then introduce the two bonding electrons between them, the system as a whole is neutral. If we place the centroid of charge of these two bonding electrons exactly half way between the nuclei, then the system will have no dipole moment. However, since the hydrogen orbital is much smaller than the chlorine orbital, the centroid of charge for the bonding electrons must be closer to H than to Cl and a dipole will result, namely,

$$\overleftarrow{\text{H—Cl}}$$

This moment has been estimated to be ~ 1.0 D.

2. *Orbital moments of unshared electrons.* Let us consider the internuclear axis to be the z axis of a coordinate system with the chlorine atom at the origin. If the chlorine atom uses a pure p_z orbital to form the bond, then the configuration of the remaining, unshared or non-bonding, electrons will be $s^2 p_x^2 p_y^2$, and they will contribute nothing to the polarity of the system.

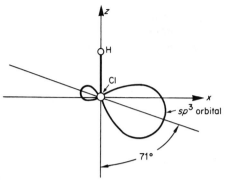

Fig. 3-17. Diagram illustrating the electrical asymmetry of electrons in an sp^3 hybrid orbital of chlorine in HCl.

The s electrons are spherically distributed about the chorine nucleus, and the p_x and p_y electrons lie in a disc perpendicular to the z axis with the Cl nucleus at its center. Now let us suppose instead that the chlorine atom has full sp^3 hybridization and uses one of these hybrid orbitals to form the bond. The remaining three pairs of electrons will lie in the three equivalent hybrid orbitals, which have approximately the shape shown in Fig. 3-17. It is easily seen that electrons in such an orbital are much more concentrated below the xy plane than above it, and hence there will be an *orbital dipole moment* which can be represented by a vector of magnitude **v** pointing along the axis of the orbital. There is, then, a dipole moment contribution in the bond direction of $-\mathbf{v} \cos 71°$, or a total from the three such orbitals of $-3\mathbf{v}\cos 71°$, namely:

$$\overset{-3\mathbf{V}\cos 71°}{\overline{\text{H}-\text{Cl}}}$$

(The vector sum in the xy plane is, of course, zero.) Now, we do not expect the chlorine atom to have full sp^3 hybridization and the unshared pairs to make this maximal contribution; rather, we expect some lesser degree of hybridization and hence some smaller contribution from the non-bonding electrons. There are, however, various reasons for believing that there must be *some* hybridization.

The important point illustrated by this discussion of HCl is that the net molecular dipole moment is a vector sum of the bond moment and the lone-pair moments. There is clearly *no necessary relationship between molecular dipole moments and bond polarity alone*. Naturally, an entirely analogous situation prevails with polar polyatomic molecules, as illustrated by NH_3 and NF_3.

The two pyramidal molecules NH_3 and NF_3 have dipole moments of 1.5 and 0.2 D, respectively. Knowing the bond angles ($\angle HNH = 106.75°$; $\angle FNF = 102.5°$), we would easily calculate the N—H bond moment to be ~ 1.33 D and the N—F bond moment to be ~ 0.15 D. However, these results do not appear credible. In the first place, we should expect an $\overrightarrow{\text{N—F}}$ bond moment much larger than 0.15 D. Secondly, if we assume that only the electronegativity difference is responsible for the bond dipole, the $\overrightarrow{\text{N—H}}$ dipole

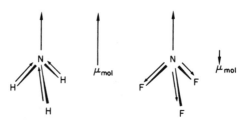

Fig. 3-18.   Diagram illustrating how the molecular dipole moment, μ_{mol}, can be regarded as a vector sum of bond moments and the lone pair moment in NH_3 and NF_3.

should be still smaller than the $\overrightarrow{N—F}$ dipole. If we treat the N—H bond as we did the H—Cl bond and consider that, despite the tendency of electronegativity to produce a moment, $\overleftarrow{N—H}$, the difference in size of the overlapping orbitals will tend to produce the opposite polarity, $\overrightarrow{N—H}$, it is still difficult to believe that the net N—H moment is ~ 9 times the N—F moment. A satisfactory explanation of the molecular dipole moments can be developed, however, if we take account of the hybridization of the nitrogen atom and of the consequent orbital moment of the unshared electron pair on the nitrogen atom in each molecule. The observed bond angles are much larger than 90°, though not so large as 109°, and may be taken to mean that the nitrogen atom in each case is bonding the three atoms by s–p hybrid orbitals with slightly less than the amount of s character that occurs in regular sp^3 hybrids. Consequently, the unshared electrons do not occupy a pure s orbital, in which case they would be distributed spherically around the nitrogen atom and contribute nothing to the polarity of the molecules. Instead, they are in an s–p_z hybrid, so that they are more concentrated above than below the nitrogen atom. Thus, assuming the direction of the N—H bond moments to be as shown, the net polarity of each molecule may be accounted for as the vector sum of bond moments and the lone pair moment as shown in Fig. 3-18. In general, bond moments are not directly relatable to the molecular moment by geometry alone, but hybridization and consequent orbital moments of unshared pairs must also be considered. Once again, however, caution is necessary, for the assumption that the hybridization can be inferred from the bond angles lacks rigorous justification. It is possible that the orbitals are not directed *exactly* along the internuclear lines. That is, the bonds may be a little *bent*. There is evidence that bent bonds actually occur, though they are only slightly bent. However, even slight bending can introduce appreciable error into a calculation of the above kind.

3-18. Some Relations between Bond Properties

Bond lengths and bond energies have now been discussed and it has been observed that both of them vary—the first inversely and the second directly—with the bond order. Another measurable bond property whose magnitude is directly related to the bond order is the *force constant* of the bond. The force constant, f, of a bond between atoms A and B is given by the following equation:

$$v \ (cm^{-1}) = \frac{1}{c} \sqrt{\frac{f}{4\pi^2 \mu}}$$

where v is the vibration frequency of the bond expressed in wave numbers, cm^{-1} (see Units and Conversion Factors, page xix), c is the velocity of light and μ is the reduced mass of the A–B oscillator in grams. When values of the constants are put into this equation, along with a numerical factor to permit inserting the atomic masses in the usual chemical atomic weight units (i.e. for C, a mass of 12.00), we obtain

$$v^2 = 17.0 \times \frac{f}{\mu}$$

Thus, when the vibrational frequency, v, of a bond has been measured, most commonly by observation of the infrared or Raman spectrum, the force constant can be calculated. In polyatomic molecules there are often strong interactions between the vibrations of different bonds, so that the calculation of the force constant for a particular bond cannot be made directly from any one of the observed frequencies. However, there are well-established methods for dealing with this problem.

The qualitative relationships between the three bond properties, length, dissociation energy and force constant are often useful in predicting or interpreting physical and chemical properties. Since we know how each of them depends on the bond order, we can infer the dependence of any one of them on another. Specifically, we have the following:

(1) As bond order increases:
    length decreases
    energy increases
    force constant increases
(2) As bond length increases:
    energy decreases
    force constant decreases

(3) As energy increases:
    length decreases
    force constant increases
(4) As force constant increases:
    length decreases
    energy increases

Generally, over any appreciable range, none of these relations is linear. Figs. 3-19 to 3-21 show some representative plots of one bond parameter against another. The shapes of the curves are easily understood in terms of simple physical ideas. Consider, for example, Fig. 3-21 which is a plot of bond lengths against bond force constants for molybdenum—oxygen bonds. The curve approaches the horizontal axis asymptotically, because as bond length tends to infinity, the force of attraction between the atoms tends to zero. The curve also approaches a vertical line asymptotically at some small value of the bond length. As the force between the atoms increases the bond length decreases, but eventually the repulsive forces, due mainly to repulsion between the inner-shell electrons of the two atoms, must become so great that further increases in attractive force can do little to shorten the bond any further. These repulsive forces have an inverse

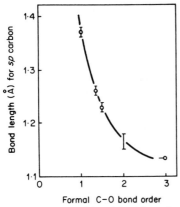

Fig. 3-19. Plot of bond length against bond order for CO bonds (reproduced by permission from F. A. Cotton and R. M. Wing, *Inorg. Chem.*, 1965, **4**, 314).

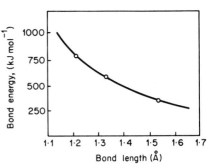

Fig. 3-20. Plot of bond length against bond energy for CC bonds.

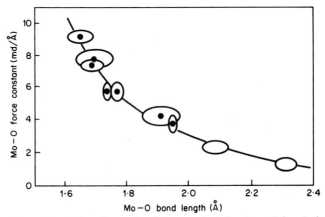

Fig. 3-21. Plot of bond lengths against force constant for Mo—O bonds (reproduced by permission from F. A. Cotton and R. M. Wing, *Inorg. Chem.*, 1965, **4**, 867).

dependence on the bond length, l, with a very high exponent in the neighborhood of 10–12, viz.:

$$\text{Repulsive force is proportional to } 1/l^{10-12}$$

These relationships will be utilized in several places subsequently.

Further Reading

Allred, A. L., *J. Inorg. Nuclear Chem.*, 1961, **17**, 215 (discussion of Allred–Rochow and other electronegativity scales).

Ballhausen, C. J. and H. B. Gray, *Molecular Orbital Theory*, Benjamin, 1964 (qualitative introduction to MO theory).

Coulson, C. A., *Valence*, 2nd ed., Oxford University Press, 1961 (essentially non-mathematical coverage).

Dasent, W. E., *Inorganic Energetics*, Penguin Books, Ltd., 1970 (good sections on several topics covered in this chapter).

Eyring, H., J. Walter and G. E. Kimball, *Quantum Chemistry*, Wiley, 1960 (introduction to mathematically rigorous application of quantum mechanics to chemistry).

Gray, H. B., *Electrons and Chemical Bonding*, Benjamin, 1964 (qualitative introduction to MO theory).

Kauzman, W., *Quantum Chemistry*, Academic Press, 1957 (advanced but lucid mathematical treatment).

LaPaglia, S. R., *Introductory Quantum Chemistry*, Harper and Row, 1971 (moderately mathematical).

Linnett, J. W., *The Electronic Structure of Molecules*, Methuen, 1964 (introductory, semi-mathematical).

Murrell, J. N., S. F. A. Kettle and J. M. Tedder, *Valence Theory*, 2nd ed., Wiley, 1970 (introduction to mathematical quantum treatment).

Pauling, L., *The Nature of the Chemical Bond*, 3rd ed., Cornell University Press, 1960 (covers valence-bond theory comprehensively, but omits MO approach entirely).

Pritchard, H. O. and H. A. Skinner, *Chem. Rev.*, 1955, **55**, 745 (a review of electronegativity scales).

Royer, D. J., *Bonding Theory*, McGraw-Hill, 1968 (moderately mathematical).

Stals, J., *Rev. Pure Appl. Chem.*, 1970, **20**, 1 (extensive discussion of bond length, bond order, stretching force constants, and bond energy correlations).

4

Stereochemistry and Bonding in Compounds of Non-transition Elements

4-1. Introduction

The problem to be considered in this Chapter is that of finding relations between the structures of molecules and the nature of the chemical bonds that they contain. This problem has two aspects. The one to which we shall here devote most attention concerns the shapes of the molecules, that is, the angles between the bonds formed by a given atom. The second and shorter part of the discussion will deal with certain aspects of multiple bonding, that is, with certain questions of bond lengths.

We shall assume implicitly that in two-center bonds the electron density is distributed symmetrically about the internuclear axis. In other words, we shall not explicitly consider the possibility of bent bonds. While this is unquestionably a good and useful approximation—doubtless a much better one than many others usually made in simple valence theory—it is well to remember that, except in special cases where symmetry considerations prohibit it, chemical bonds may well be, *at least a little*, bent.

The difficulty in presenting this subject is that there is still great diversity in the viewpoints taken by different workers in attempting to correlate electronic and molecular structures. Each approach has certain virtues and yet they have very little common ground. It is not that their assumptions are particularly in conflict but rather that they seem unrelated, which is disconcerting since they all purport to be answering the same questions. Because of this situation, the discussion in this Chapter will necessarily have some of the same fragmented character. Each approach to the problem will be presented along the lines generally used by its advocates. Following this, however, an attempt will be made to compare, interrelate and criticize the several models.

4-2. The Valence Shell Electron Pair Repulsion, VSEPR, Model[1]

In this model the arrangement of bonds around the central atom is considered to depend upon how many valence-shell electron pairs, each occupying a localized one- or two-center orbital, are present, and on the relative sizes and shapes of these orbitals. The first rule is: (1) *The preferred arrangement of a given number of electron pairs in the valence shell of an atom is that which maximizes their distance apart.*

Each electron pair is assumed to occupy a reasonably well-defined region of space and other electrons are effectively excluded from this space. Hence, localized electron-pair orbitals behave as if they repel each other and they adopt that arrangement in space that keeps them, on the average, as far apart as possible. For electron pairs in the same valence shell, the arrangements that maximize the least distance apart of any two pairs are listed in Table 4-1.

TABLE 4-1

Predicted Arrangements of Electron Pairs in One Valence Shell

Number of pairs	Polyhedron defined
Two	Linear
Three	Equilateral triangle
Four	Tetrahedron
Five	Trigonal bipyramid, *tbp*
Six	Octahedron
Seven	Monocapped octahedron
Eight	Square antiprism
Nine	Tricapped trigonal prism

In order to apply rule (1) for the qualitative prediction of molecular shapes where only single bonds and unshared pairs are concerned the total number of electron pairs, bonding and non-bonding, is computed, the appropriate arrangement in Table 4-1 is selected and the electron pairs are assigned to it. In the cases of two, three or four pairs, the results are immediately obvious as shown in Fig. 4-1. For example, an AB_4 molecule must be tetrahedral, an AB_3E molecule (E represents an unshared pair) must be pyramidal, and an AB_2E_2 molecule must be bent. There are no known exceptions to these predictions. Table 4-2 lists a number of molecules and ions of the types AB_2E_2 and AB_3E, giving the angles.

In most other cases, additional rules are needed. The next one, rule (2), is: *A non-bonding pair of electrons takes up more room on the surface of an atom than a bonding pair.* This is a plausible idea inasmuch as a non-bonding pair is under the influence of only one nucleus while a bonding pair is constrained by two nuclei. With this rule, the structures of molecules having five and six electron pairs can be explained. The question which needs to be answered first for the cases of AB_4E, AB_3E_2, AB_2E_3 is: Will non-bonding

R. J. Gillespie, *J. Chem. Educ.*, 1970, **47**, 18 (this article should be consulted for more detailed discussion and references to earlier literature).

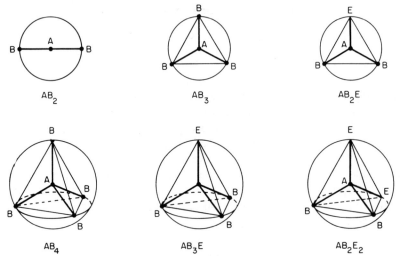

Fig. 4-1.   Diagrams showing how the shapes of molecules are predicted when the valence shell of the central atom contains 2, 3 or 4 electron pairs.

TABLE 4-2

The Angles in Some AB_3E and AB_2E_2 Molecules

AB_2E_2		AB_3E	
Molecule	BAB Angle (°)	Molecule	BAB Angle (°)
OH_2	105	NH_3	107
OF_2	102	NF_3	102
OCl_2	~111	PH_3	94
SH_2	92	PF_3	98[b]
SCl_2	102	PCl_3	100
Cl_2F^+, Cl_3^+	Bent[a]	PBr_3	~102
		PI_3	~102
		AsH_3	92
		AsF_3	96[c]
		$AsCl_3$	98
		$AsBr_3$	100
		AsI_3	100

[a] R. J. Gillespie and M. J. Morton, *Inorg. Chem.*, 1970, **9**, 811, angles not known.
[b] Y. Moreno, K. Kuchitsu and T. Moritani, *Inorg. Chem.*, 1969, **8**, 867; data for all Group V trihalides are summarized here.
[c] F. B. Clippard, Jr., and L. S. Bartell, *Inorg. Chem.*, 1970, **9**, 805.

pairs occupy axial or equatorial orbitals of the *tbp*? It has not proved possible to answer this question unambiguously by purely theoretical means but by making the empirically justified assumption that lone pairs prefer equatorial orbitals consistent results can be obtained within the framework of the basic postulates. Tables 4-3 and 4-4 show the structures of some representative AB_5, AB_4E, AB_3E_2, and AB_2E_3 molecules. All AB_5 molecules, including those with mixed ligands, should be *tbp*, and nearly all

TABLE 4-3

Some Trigonal Bipyramidal Molecules and Their Bond Lengths

Molecule	Axial ligands and bond lengths (Å)		Equatorial ligands and bond lengths (Å)	
PF_5[a]	2F	1.577	3F	1.534
PF_4CH_3[a]	2F	1.612	2F	1.543
			CH_3	1.780
$PF_3(CH_3)_2$[a]	2F	1.643	F	1.553
			$2CH_3$	1.798
AsF_5[b]	2F	1.711	3F	1.656
PF_3Cl_2	2F	—[e]	F	—[e]
			2Cl	—[e]
PCl_5 (gas)	2Cl	2.19	3Cl	2.04
$SbPh_2Cl_3$	2Cl	2.52	Cl	2.43
			2Ph	—[d]
$SbPh_4OH$[c]	Ph	2.22	3Ph	2.11–2.14
	OH	2.05		
$BiPh_3Cl_2$	2Cl	2.57	3Ph	2.12

[a] L. S. Bartell and K. W. Hansen, *Inorg. Chem.*, 1965, **4**, 1777.
[b] F. B. Clippard, Jr. and L. S. Bartell, *Inorg. Chem.*, 1970, **9**, 805.
[c] A. L. Beauchamp, M. J. Bennett and F. A. Cotton, *J. Amer. Chem. Soc.*, 1969, **91**, 297.
[d] Sb–C distances not accurately determined.
[e] Structure known from spectroscopic evidence but bond lengths unknown.

TABLE 4-4

Structures of Some AB_4E and AB_3E_2 Molecules[a]

Molecule	B and A–B distance	BAB angle (°)	B' and A–B' distance	B'AB' angle (°)
SF_4	F, 1.545	102	F, 1.646	173
$Se(CH_3C_6H_4)_2Cl_2$	C, 1.93	107	Cl, 2.38	178
$Se(CH_3C_6H_4)_2Br_2$	C, 1.95	108	Br, 2.55	177
$Te(CH_3)_2Cl_2$	C, 2.09	98	Cl, 2.48	172
$S(CH_2)_4TeI_2$[b]	C, 2.16	100	I, 2.85–2.99	176
ClF_3	F, 1.598	—	F, 1.698	175
BrF_3	F, 1.721	—	F, 1.810	172

[a] See Fig. 4-2 for sketches defining B and B' ligands, and the B'AB' angles.
[b] C. Knobler, J. D. McCullough and H. Hope, *Inorg. Chem.*, 1970, **9**, 797.

known ones are, although there are several definite exceptions (cf. Section 13-1) such as $SbPh_5$, whose structure can best be considered as distorted *spy* (*spy* = square pyramid).[2] It is not impossible that the *tbp* structure of $SbPh_5$ might be more stable in solutions. All AB_4E and AB_3E_2 molecules have the structures shown in Fig. 4-2. All AB_2E_3 molecules are linear; examples include XeF_2, KrF_2 and the many trihalide ions.

With species involving six pairs of valence-shell electrons, rules (1) and (2) lead unambiguously to correct structures. Table 4-5 lists some AB_5E and AB_4E_2 molecules; the characteristic structures are shown in Fig. 4-3.

[2] A. L. Beauchamp, M. J. Bennett, and F. A. Cotton, *J. Amer. Chem. Soc.*, 1968, **90**, 6675.

Fig. 4-2.   The shapes of molecules of types (a) AB_4E and (b) AB_3E_2, showing how the
B′AB′ angles are defined.

TABLE 4-5

Structures of Some AB_5E and AB_4E_2 Molecules[a]

Molecule	B and A–B distance (Å)	B′ and A–B′ distance (Å)	BAB′ angle (°)
ClF_5	4F, 1.72	F, 1.62	—[b]
BrF_5	4F, 1.78	F, 1.68	85
$XeOF_4$	4F, 1.95	O, 1.70	91 ± 2
XeF_4	4F, 1.953		
ICl_4^-	4Cl, 2.46		
BrF_4^-	4F, 1.88		

[a] See Fig. 4-3a for sketch defining B and B′ atoms in AB_5E.
[b] Not determined.

Rule (2) is invoked to account for the square structures of AB_4E_2 molecules.
It is clear that a lone pair encounters less repulsion from a *cis*-bonding pair
than from a *cis*-lone pair; hence the lone pairs are *trans* to each other.

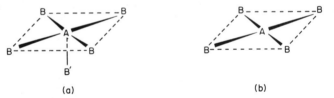

(a)   (b)

Fig. 4-3.   The shapes of molecules of types (a) AB_5E and (b) AB_4E_2.

For the higher numbers of valence-shell pairs, 7, 8 and 9, there are only a
few non-transition element complexes known. In the case of IF_7 the structure appears to be a pentagonal bipyramid, contrary to the entry in Table 4-1.
However, with these higher numbers, the predictions of preferred arrangements necessarily become less certain because the repulsive energy of the set
of electron pairs does not have a pronounced minimum for any one configuration and atom–atom interactions assume greater importance.

The problem becomes acute for the AB_6E type molecules and ions.
Discussion of this case is deferred to Section 4-7, where all the theoretical
models will be considered together *vis à vis* the rather ambiguous experimental data.

In addition to predicting, or at least correlating, gross geometrical features,
as just outlined, the VSEPR model can give a consistent account of certain
finer details of the structures. Thus, according to rule (2) the greater size of

lone pairs might be expected to result in the angles between their axes and bond axes being greater than the ideal values for the polyhedron concerned, with the observable results that the bond angles would be smaller. It will be seen from Tables 4-2 to 4-5 that this is precisely the case for AB_3E, AB_2E_2, AB_4E, AB_3E_2, and AB_5E molecules.

To account for other details two more rules, both natural extensions of the scheme, are required:

Rule (3): *The size of a bonding electron pair decreases with increasing electronegativity of the ligand.*

Rule (4): *The two electron pairs of a double bond (or the three electron pairs of a triple bond) take up more room than does the one electron pair of a single bond.*

Using rule (3) one can rationalize some of the trends in Table 4-2. For instance, the angles in NF_3 and F_2O are less than those in NH_3 and H_2O. Similarly, in a set of halo molecules AB_2E_2 or AB_3E, the BAB angles increase in the order $F < Cl < Br \approx I$. There are, however, often exceptions when hydrides are considered since the PH_3, AsH_3, and SH_2 angles are less than those in any trihalide of the same element.

Rule (4) accounts for the fact that angles in which multiple bonds are involved are generally larger than those involving only single bonds. A few representative examples are shown in Table 4-6. It should be noted that when the double bond is to an atom less electronegative than those to which the single bonds are directed, the operation of rule (3) reinforces the effect of rule (4).

TABLE 4-6

Bond Angles in Some Molecules Containing a Double Bond

Molecule	Angles (°)		
	XCX	XCO	XCC
F_2CO	108	126	
Cl_2CO	111	124	
$(NH_2)_2CO$	118	121	
F_2SO	93	107	
Br_2SO	96	108	
$H_2C = CF_2$	110		125
$H_2C = CCl_2$	114		123
OPF_3	103		
$OPCl_3$	104		

Predictions concerning relative lengths of bonds are also possible by using these rules. Thus for AB_5, AB_4E, AB_3E_2, and AB_5E molecules the A—B bonds and A—B' bonds differ by about 0.1 Å in length. In the first three cases, in which there is a trigonal-bipyramidal distribution of electron pairs, the axial bonds are longer. In AB_5E it is the four basal bonds that are longer. In the cases of five electron pairs with a *tbp* arrangement, the axial pairs have three neighboring pairs on lines at only 90° away while the equatorial ones have only two such closely neighboring pairs; equilibrium is thus attained when the axial pairs move to a somewhat greater distance from the central atom, thus lengthening the axial bonds relative to the equatorial ones. In the case of AB_5E, the greater size of the lone pair will act more strongly on the bonding pairs *cis* to it, thus lengthening the set of basal bonds.

4-3. The Hybridization or Directed Valence Theory

According to this treatment bond directions are determined by a set of hybrid orbitals on the central atom which are used to form bonds to the ligand atoms and to hold unshared pairs. Thus AB_2 molecules are linear owing to the use of linear sp hybrid orbitals. AB_3 and AB_2E molecules should be equilateral triangular and angular, respectively, owing to use of trigonal sp^2 hybrids. AB_4, AB_3E, and AB_2E_2 molecules should be tetrahedral, pyramidal, and angular, respectively, because sp^3 hybrid orbitals are used. These cases are, of course, very familiar and involve no more than an octet of electrons.

For the AB_5, AB_4E, AB_3E_2, and AB_2E_3 molecules, the hybrids must now include d orbitals in their formation. The hybrid orbitals used must obviously be of the sp^3d type, but an ambiguity arises since there are two such sets, viz., $sp^3d_{z^2}$ leading to *tbp* geometry and $sp^3d_{x^2-y^2}$ leading to *spy* geometry. It is impossible to predict with certainty which set should give the more stable molecules and so a decision has to be made empirically. As already noted, all AB_5 molecules whose structures are known, with at most three exceptions, are *tbp*. It is therefore assumed that the $sp^3d_{z^2}$ hybrids and *tbp* geometry will generally be appropriate. Once this assumption is made a consistent correlation of structures follows, exactly as in the VSEPR model where the basic *tbp* arrangement was adopted for a different reason. Again, however, the preferential allocation of lone pairs to equatorial orbitals is essentially arbitrary. It does then follow that AB_4E molecules have the shape shown in Fig. 4-2a, AB_3E_2 molecules have the drooping T shape of Fig. 4-2b and AB_2F_3 molecules are linear.

For AB_6 molecules octahedral sp^3d^2 hybrids are used. AB_5E molecules must, naturally, be *spy*. For AB_4E_2 molecules there is nothing in the directed valence theory itself to show whether the lone pairs should be *cis* or *trans*. The assumption that they must be *trans* leads to consistent results.

It may be seen that, if we assume formation of hybrid orbitals to determine the basic geometry, it is possible then to borrow a portion of the VSEPR dogma, namely, rule (2), that a lone pair takes up more room than a bonding pair in order to rationalize the finer features of certain structures, e.g., the bond angles in AB_4E, AB_3E_2, AB_5E, and AB_4E_2 molecules. In short, one may reject the view that electron-pair repulsion is the primary factor in stereochemistry and assume instead that directed hybrid orbitals have a basic role but still allow that electron-pair repulsions enter into the problem at a secondary level.

4-4. The Three-center Bond Model

This approach is a limited MO treatment, predicated on two main ideas: (1) that the use of outer d orbitals of the central atom is so slight that they may be neglected altogether, and (2) that the persistent recurrence of bond angles close to 90° and 180° in AB_n molecules suggests that orbitals perpendicular to one another, namely, p orbitals, are being used.

Two types of chemical bond are considered. First there is the ordinary two-center, two-electron ($2c–2e$) bond, formed by the overlap of a p orbital of the central atom with a σ orbital of an outer atom. Secondly, there is the linear three-center, four-electron ($3c–4e$) bond (page 109), formed from a p orbital of the central atom and the σ orbitals of two outer atoms.

For molecules with an octet, or less, of electrons in the valence shell of the central atom, the hybridization theory, employing sp, sp^2 and sp^3 orbitals remains valid. It is for molecules in which five or more electron pairs on the central atom must be accounted for that this model was proposed.[3] Since it is, as we shall note below, undoubtedly too great a simplification, it will be applied only to a few illustrative cases. It is worth mentioning, however, since it has a certain heuristic value.

In molecules of the AB_4E type, the central atom is considered to use p_x and p_y orbitals to bind the B atoms (cf. Fig. 4-2a) while the B' atoms are bound using the p_z orbital to form a $3c–4e$ bond. Again some supplementary assumption, such as rule (2) of the VSEPR model must then be invoked to explain the non-linearity of the B'AB' set. Clearly the remaining electron pair is postulated as occupying a pure s orbital and as having no stereochemical role. In an analogous way AB_3E_2 molecules are postulated as involving one $2c–2e$ bond, formed from a single p orbital of A to the B atom (Fig. 4-2b) and a $3c–4e$ bonding system for the B'AB' set, with the two other electron pairs in the s and remaining p orbital. For an octahedral molecule, three mutually perpendicular $3c–4e$ bonding systems are postulated, while for the AB_5E case, two such $3c–4e$ systems and one $2c–2e$ bond are employed.

It is clear that this model does, at least qualitatively, accord with the variation in bond lengths in these molecules. It allots a full electron pair to all the shorter A—B bonds and makes the longer ones members of a $3c–4e$ bond system.

4-5. The Correlation Diagram Approach

This approach was first applied to relatively simple cases many years ago by Walsh[4] and led to certain generalizations, called Walsh's rules, relating the shapes of triatomic molecules to their electronic structures. The basic approach is to calculate, or estimate, the energies of molecular orbitals for two limiting structures, say, linear and bent (to 90°) for an AB_2 molecule,

[3] R. E. Rundle, *Records Chem. Progr.*, 1962, **23**, 195.
[4] A. D. Walsh, *J. Chem. Soc.*, **1953**, 2260, 2268, 2288.

and then draw a diagram showing how the orbitals of one configuration correlate with those of the other. Then, depending on which orbitals are occupied, one or the other structure can be seen to be preferred. By means of approximate MO theory, implemented by digital computers, this approach has been extended and generalized in recent years.[5,6]

Triatomic Molecules. This case will be elaborated in some detail to expound the method. Other cases will then be treated more summarily. The coordinate system for the AB_2 molecule is shown in Fig. 4-4.

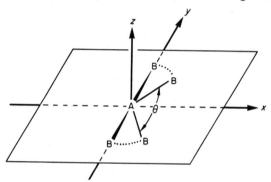

Fig. 4-4.   The coordinate axes used for linear and bent AB_2 molecules.

When it is bent, the AB_2 molecule has C_{2v} symmetry and when linear $D_{\infty h}$ symmetry. However, it will simplify notation if the linear configuration is considered as simply an extremum of the C_{2v} symmetry and the labels given to the orbitals through the range $90° \leqslant \theta < 180°$ are therefore retained even when $\theta = 180°$. The symbols used to label the orbitals are derived from the orbital symmetry properties in a systematic way, but a detailed explanation will not be given here.[7] For present purposes, these designations may be treated simply as labels.

The A atom of an AB_2 molecule will be assumed to have only s, p_x, p_y and p_z orbitals in its valence shell, while the B atoms will each be allowed only a single orbital oriented to form a σ bond to A. In the linear configuration p_x^A and p_z^A are equivalent non-bonding orbitals labelled $2a_1$ and b_1, respectively. The orbitals s^A and p_y^A interact with σ_1^B and σ_2^B, the σ orbitals on the B atoms, to form one very strongly bonding orbital, $1a_1$, one less strongly bonding orbital, $1b_2$, and two antibonding σ orbitals, $3a_1$ and $2b_2$. The ordering of these orbitals and, in more detail, the approximate values of their energies can be estimated by an MO calculation. Similarly, for the bent molecule the MO energies may be estimated. Here only p_z^A will be

[5] R. M. Gavin, Jr., *J. Chem. Educ.*, 1969, **46**, 413.

[6] B. M. Gimarc, *J. Amer. Chem. Soc.*, 1970, **92**, 266; 1971, **93**, 593, 815.

[7] For an explanation of the symbols consult any introductory book on chemical or physical applications of group theory, e.g., F. A. Cotton, *Chemical Applications of Group Theory*, 2nd ed., Wiley-Interscience, 1971, p. 86, or H. H. Jaffé and M. Orchin, *Symmetry in Chemistry*, Wiley, 1965, p. 61.

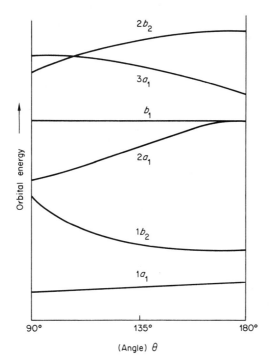

Fig. 4-5. Orbital correlation diagram for AB_2 triatomic molecules where A uses only s and p orbitals.

non-bonding; spacings and even the order of the other orbitals is a function of the angle of bending, θ. The complete pattern of orbital energies, over a range of θ, as obtained with typical input parameters is shown in Fig. 4-5. Calculations in the Hückel approximation are simple to perform and give the correct general features of the diagram[5] but for certain cases (e.g., AB_2E_2, as noted below) very exact computations are needed for an unambiguous prediction of structure.

From the approximate diagram it is seen that an AB_2 molecule (one with no lone pairs) is more stable when linear than when bent. The $1b_2$ orbital drops steadily in energy from $\theta = 90°$ to $\theta = 180°$, while the energy of the $1a_1$ orbital is fairly insensitive to angle. For an AB_2E molecule the results are ambiguous, because the trend in the energy of the $2a_1$ orbital approximately offsets that of the $1b_2$ orbital. The correct result may be a sensitive function of the nature of A and B. Important examples of AB_2E molecules are carbenes (page 284) CR_2, whose structures do indeed seem to vary with changes in R. For AB_2E_2 molecules, the result should be the same as for AB_2E, since the energy of the b_1 orbital is independent (in this rough approximation) of the angle. Thus, it is not clear, in this approach, that AB_2E_2 molecules should *necessarily* be bent, but all known ones are. For AB_2E_3

molecules the behavior of the $3a_1$ orbital would clearly favor a linear structure and this is in accord with all known facts.

Tetratomic Molecules. For AB_3E molecules with a C_3 axis of symmetry, very accurate calculations of energy as a function of angle have been made because of the relevance of this to the problem of barriers to inversion in pyramidal molecules. In the best of these calculations it is found that the relative energies of the pyramidal and the planar configurations are such that the pyramidal configuration is always the more stable, but the energy difference varies considerably (from ~ 25 kJ mol^{-1} for NH_3 to over 150 kJ mol^{-1} for some AsX_3 and SbX_3 molecules). Also, interelectronic and internuclear repulsions as well as bond energies play an important role in determining the configurations.[8]

For molecules with more than an octet of valence-shell electrons on the central atom, the employment of correlation diagrams has been less systematic. Instead, some individual cases have been treated to see what distortions from assumed idealized geometries might be expected. For example, the T-shaped ClF_3 molecule has been treated as indicated in Fig. 4-6, where the results suggest that the angle ϕ should be $\sim 10°$ less than $90°$, in semiquantitative agreement with observation.

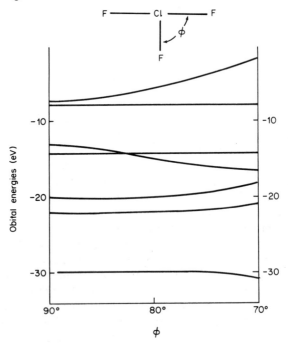

Fig. 4-6.   Orbital energies (eV) as a function of angle for ClF_3 as calculated in the extended Hückel approximation. [Adapted from Gavin.[5]]

[8] Cf. A. Rauk, L. C. Allen, and K. Mislow, *Angew. Chem. Internat. Edn.*, 1970, **9**, 400, for a summary and references.

4-6. More General and Exact Molecular-orbital Calculations; The Problem of Outer d Orbitals

Both the three-center bond model and the correlation diagram treatment, as just outlined, omit all central-atom orbitals except the s and p orbitals of the valence shell. Indeed the three-center bond model neglects even the s orbital except as a storage place for one electron pair. They can be described as very restricted or incomplete MO treatments. They are also inexact, even within their self-imposed limits, since numerical accuracy is neither sought nor obtained in their usual applications. It would not, of course, be sensible to strive for numerical precision after such sweeping assumptions have been made at the outset. On the other hand, the hybridization or directed valence treatment assumes very full involvement of outer d orbitals whenever more than four pairs of electrons must be accommodated. This extreme assumption is also unlikely to be accurate. Finally, the VSEPR model resorts to a simple electrostatic model, which, however successful it may be, can scarcely be taken literally.

Rigorous MO Calculations. It would seem obvious that, in principle, the best approach—one that would supersede all the others if it could be carried through systematically—should be a comprehensive and numerically accurate, *ab initio* MO treatment or, rather, a whole set of such treatments, one at least for each important type of molecule. It must be stressed that, in order to *predict* molecular geometry reliably, the calculation would have to be far more rigorous than the familiar type based on one-electron orbitals. This is because interelectronic repulsion energies, which are of the same order of magnitude as the electron-core attractive energies, play a major role in determining the net energy of a molecule. The importance of repulsive energies in determining preferred structures has been clearly demonstrated in calculations of inversion barriers for some AB_3 molecules.[8] Thus, when we speak of a rigorous, numerically accurate, *ab initio* MO calculation, we mean something an order of magnitude (if not several orders of magnitude) more complex than the familiar forms of one-electron orbital calculations.

Moreover, such calculations would have to be carried out over a whole range of configurations in order to search for the configuration of lowest energy. For a few AB_2 and AB_3 molecules, which contain only first-row atoms (so that only s and p orbitals need be considered), such calculations have been attempted and have had some success.[8] For the AB_2 case there are only two structure parameters, the A—B distance and the BAB bond angle, to be varied; for the AB_3 cases the maintenance of at least C_{3v} symmetry has been required, thus again restricting to two the variable structure parameters, viz., one distance and one angle. Even under these relatively simple conditions the calculations are extremely complex and expensive. The evaluation of all the multicenter attractive and repulsive integrals is an enormous task, and, yet, to omit some would be immediately to introduce uncertainty as to how far the results could be trusted. This is because two configurations may differ in energy by only about 100 kJ mol^{-1} whereas the total energies of each will be around 10^5 kJ mol^{-1}. Hence the percentage error in each total energy must be very small in order to get accurate results for the differences between such energies.

Less Ambitious MO Calculations. As just indicated, it is questionable whether MO calculations can provide a sound guide to completely *a priori* prediction of structure unless carried out at a level of rigor which is at present impracticable for any but the simpler cases. There are, however, reasons for carrying out MO calculations of a less rigorous type, although more rigorous than the very simple ones so far used in the construction of correlation diagrams. Such calculations can be made by using the known geometry of the molecule instead of trying to predict the geometry. The results are useful in understanding ground-state electron distributions and the relative strengths and polarities of bonds, and they help to elucidate the reasons why the observed geometry may be the preferred one.[9,10] These intermediate-level MO calculations are relatively straightforward when only *s* and *p* orbitals need be considered. Criteria for choice of parameters in semi-empirical calculations have been established by careful comparisons between these and more rigorous calculations, especially for hydrogen compounds of boron and carbon.[11] In cases where *d* orbitals play a role in bonding, the calculations are not so straightforward because of uncertainty about the nature of the *d* orbitals in the context of the specific molecule.

The Role of *d* Orbitals. Fig. 4.7 shows a schematic energy-level diagram for an octahedral molecule formed by a central atom from the second or higher row in the non-transition part of the Periodic Table and ligand atoms in which only valence-shell *s* and *p* orbitals are considered. The central atom has not only *ns* and *np* orbitals but also *nd* orbitals. As shown, the interactions of the central and ligand atom *s* and *p* orbitals probably give the principal contributions to bonding, but the *d* orbitals will also play a role. Two of them, d_{z^2} and $d_{x^2-y^2}$, which are jointly denoted e_g orbitals can participate in σ bonding by interacting with a suitable set of ligand *s* and, particularly, *p* orbitals which are also denoted e_g. In addition, the other three *d* orbitals, d_{xy}, d_{xz}, and d_{yz} form a set denoted t_{2g} and these can interact with a combination of ligand *p* orbitals which is of t_{2g} character. In principle, such interactions must occur, but whether they are of sufficient magnitude to matter depends on whether the *d* orbitals are low enough in energy and compact enough in shape.

Spectroscopic data for free atoms such as Si, P, S and Cl, as well as their heavier congeners, show that the valence-shell *d* orbitals are far higher in energy and much more diffuse than the *s* and *p* orbitals, and thus imply that they cannot contribute much to the bonding. It was on this basis that bonding

[9] R. S. Berry *et al.*, *Acta Chem. Scand.*, 1968, **22**, 231.
[10] P. C. van der Voorn and R. S. Drago, *J. Amer. Chem. Soc.*, 1966, **88**, 3255.
[11] M. D. Newton, F. P. Boer, and W. N. Lipscomb, *J. Amer. Chem. Soc.*, 1966, **88**, 2353, 2361, 2367.

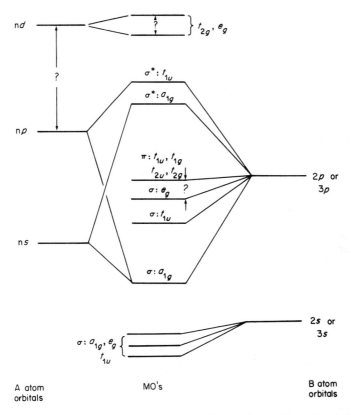

Fig. 4-7. A schematic molecular orbital energy diagram for an AB_6 molecule in which the center atom has ns, np, and nd valence-shell orbitals.

models such as the three-center bonding model, which entirely neglect d orbitals, were proposed.

However, it can be shown that the size and energy of the d orbitals are sensitive functions of both the oxidation state and the electron configuration of the atom. Some detailed conclusions, which will now be summarized, are based on calculations[12-15] for the sulfur atom, but it seems safe to assume that essentially similar ones should apply to other atoms in the later non-transitional groups.

(1) There is a remarkable contraction of the $3d$ orbitals when *two* electrons are promoted to give an sp^3d^2 configuration. The mean radius of the d orbitals in states derived from the s^2p^4, sp^4d and s^2p^3d configurations is 3.0–3.5 Å, but for an sp^3d^2-hybridized valence state this is reduced to ~ 1.6 Å, which may be compared with radii of ~ 0.9 Å for the $3s$ and $3p$ orbitals.

[12] C. A. Coulson and F. A. Gianturco, *J. Chem. Soc.*, A, **1968**, 1618.
[13] G. S. Chandler and T. Thirunamachandran, *J. Chem. Phys.*, 1968, **49**, 3640.
[14] B. C. Webster, *J. Chem. Soc.*, A, **1968**, 2909.
[15] R. G. A. R. Maclagan, *J. Chem. Soc.*, A, **1970**, 2992.

One study[13] explicitly suggests that it is only when two (or more) electrons are promoted to d orbitals that they contract sufficiently to be useful.

(2) Oxidation or partial oxidation of the central atom leads to further reduction of the $3d$ orbital radius. For the (hypothetical) state of the sulfur atom in SF_6 in which a net charge of $+0.6$ might arise by loss of 0.1 of each electron in the sp^3d^2 valence state, the $3d$ orbital radius decreases to ~ 1.4 Å while the $3s$ and $3p$ radii remain at ~ 0.9 Å.

(3) However, the promotion energy to states derived from the sp^3d^2 configuration are very large, namely, in the range of 30–35 eV. Such energies can only be compensated, if at all, by very strong bonds.

In SF_6 and similar compounds the use of outer d orbitals is plausible on the above considerations. In SF_6 there is clearly the opportunity to make good use of a valence state derived from the sp^3d^2 configuration to form six strong bonds, and the electronegativity of fluorine would assist by creating a substantial positive charge ($+1$ or more) on the sulfur atom, thereby improving further (see point 2 above) the participation of the $3d$ orbitals. In short, for SF_6 factors are very favorable for extensive d orbital contributions to the bonding.

It is not clear how far from this optimum situation one may go and still expect major contributions to bonding from the outer d orbitals. More extensive and detailed studies will be required to refine our understanding of this problem. The foregoing considerations do clearly suggest that compounds in which d orbitals might be used are most likely to be formed with the more electronegative ligand atoms (F, O, Cl) and this is in accord with observation.

4-7. The Curious Case of the AB_6E Molecules

According to the VSEPR model, an AB_6E molecule should not have a regular octahedral structure; the six B atoms should occupy six of the seven vertices of a capped octahedron or, perhaps, a pentagonal bipyramid. The directed valence or hybridization approach would normally begin with the postulation of some set of sp^3d^3 hybrid orbitals. One of these orbitals would be occupied by the lone pair leaving a configuration of B atoms which could not be a regular octahedron.

On the other hand, the three-center bond model leads straightforwardly to a regular octahedral configuration. The six pairs of bonding electrons are assigned to the three mutually perpendicular $3c$–$4e$ bonds formed from the p orbitals of atom A; the remaining electron pair is then assigned to the spherically symmetrical s orbital.

The experimental facts on AB_6E systems are not simple. The $SeCl_6^{2-}$, $SeBr_6^{2-}$, $TeCl_6^{2-}$ and $TeBr_6^{2-}$ ions are all octahedral in crystals.[16] However, the vibrational spectra of the TeX_6^{2-} ions are anomalous[17] in several respects

[16] I. D. Brown, *Canad. J. Chem.*, 1964, **42**, 2758.
[17] C. J. Adams and A. J. Downs, *Chem. Comm.*, **1970**, 1699.

as compared with the spectra of such species as GeX_6^{2-} or SnX_6^{2-} where there are no valence-shell electrons other than the six bonding pairs. The abnormalities in the spectra seem to imply that, while the TeX_6^{2-} ions have a regular octahedral ground state, their vibrations are perturbed by the presence of less symmetrical, low-lying excited electronic states. Thus there is a partial breakdown in vibrational selection rules and the deformation modes have abnormal amplitudes and frequencies.

The SbX_6^{3-} ions may have regular octahedral structures in solution,[18] but they appear to be distorted, in an undetermined way, in crystals,[17,18] possibly by spreading of one triangular face to give C_{3v} symmetry as shown in (4-I). Their spectra also show anomalies.

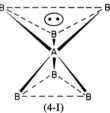

(4-I)

The XeF_6 molecule definitely deviates from the behavior expected for a simple, rigid octahedron.[19,20] Electron diffraction and infrared data suggest that XeF_6 is a very flexible molecule with a non-octahedral configuration that is most probably of C_{3v} symmetry. This might be imagined to result from the influence of the lone pair, which will tend to become localized on a triangular face, thus splaying it out, as in (4-I). Interconversion of the eight equivalent distortions of this type probably occurs rapidly *via* highly anharmonic vibrational deformation modes with the regular octahedron or the C_{2v} structure (4-II) as intermediates with only slightly higher energy than the C_{3v} structure.

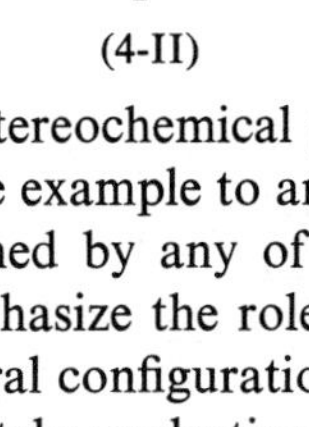

(4-II)

The complex and dynamic stereochemical properties of AB_6E molecules, which vary somewhat from one example to another, are clearly not straightforwardly predicted or explained by any of the simpler models discussed above. Some of these overemphasize the role of the non-bonding electrons in distorting a regular octahedral configuration, while the three-center bond theory goes too far in completely neglecting the role of these electrons.

[18] E. Martineau and J. B. Milne, *J. Chem. Soc., A*, **1970**, 2971.
[19] L. S. Bartell and R. M. Gavin, Jr., *J. Chem. Phys.*, 1966, **48**, 2466.
[20] R. D. Burbank and N. Bartlett, *Chem. Comm.*, **1969**, 545.

4-8. Comparison of Models

Each of the simple models has its limitations and cannot be trusted blindly. On the other hand, all of them produce *mostly* correct predictions as to structure, and they are therefore useful for correlating information even if their rather extreme premises and hence their implications with regard to the actual electron distributions are not to be taken literally.

The VSEPR model is the least sophisticated; it makes practically no use of the mathematical machinery of quantum mechanics. Its emphasis on the influence of interelectronic repulsions is in sharp contrast to the other models, in which repulsions are not explicitly mentioned except perhaps at a secondary stage to refine details of the structural prediction. Nevertheless, this model achieves considerable success and should be regarded as a useful qualitative tool in handling structural problems. It is interesting that the repulsive forces invoked in the VSEPR model may not, in fact, be simple electrostatic ones. Instead, they may be attributable to the combined effects of the orthogonality of orbitals and the Pauli exclusion principle.[9]

The hybridization or directed valence model also achieves considerable success but is subject to certain obvious criticisms. For all cases in which there are more than four electron pairs in the valence shell of the central atom it is necessary to postulate that at least one d orbital becomes fully involved in the bonding. There are both experimental and theoretical reasons for believing that this is too drastic an assumption. The most recent MO calculations[9] and other theoretical considerations suggest that, while the valence-shell d orbitals make a significant contribution to the bonding in many cases, they never play as full a part as do the valence-shell p orbitals. There is fairly direct experimental evidence in the form of nuclear quadrupole resonance studies[21] of the ICl_2^- and ICl_4^- ions which show that, in these species, d orbital participation is very small; this participation is probably greater in species with more electronegative ligand atoms, such as PF_5, SF_6 and $Te(OH)_6$, but not of equal importance with the s and p orbital contributions.

The hybridization theory fails for the SeX_6^{2-} and TeX_6^{2-} ions and there is no simple way of adapting it for these cases. It is also unable to cope with finer details of various structures in any simple way. Thus it provides no obvious explanation either for the different lengths of the bonds in molecules of the AB_3F_2 and AB_5E types, or for the deviations of the angles in such molecules from 90°, but some VSEPR considerations can be grafted on to it to deal with these matters, as already noted.

The use of correlation diagrams tends to be limited by the number of structure parameters that have to be varied in order to cover a sufficient range of structural possibilities. The reliability of the conclusions that can be drawn depends on the thoroughness of this exploration of the possibilities and on the inherent reliability of the method used to compute the energies.

[21] C. D. Cornwell and R. S. Yamasaki, *J. Chem. Phys.*, 1957, **27**, 1060.

The limitations can be no less than those of the computational methods employed.

The three-center bond approach is clearly only an aborted MO treatment. In its *total* neglect of d orbitals it is in error, but in many cases less so than the hybridization approach with its *total* inclusion of d orbitals. In its neglect of the central-atom valence-shell s orbital, it is rather seriously unrealistic. As Fig. 4-7 shows, and as specific calculations such as those on PF_5 and BrF_5 show,[9] the s orbital is heavily involved in the A—B σ bonds and cannot safely be neglected.

Clearly, the most flexible approach to an understanding, not only of structures, but also of the character of the bonds in them is an MO approach. As already noted, it may be impossible to get completely reliable *ab initio* predictions of structure by MO calculations unless they are of a very rigorous nature, but less rigorous calculations performed with an assumed or known structure are very useful. In general, the MO approach, even in a quantitatively rough form, as exemplified by Fig. 4-7, provides a general, flexible conceptual framework within which the models in which "ruthless approximations" are made can be examined and evaluated.

It would be unfair to conclude this discussion without mentioning that there are a few other cases besides the AB_6E molecules that none of the models can accommodate. These are certain non-linear AB_2 molecules containing no unshared valence-shell electrons. In general, when both the A atom and the B atoms are small, as in, e.g., $BeCl_2$ and $CaCl_2$, these molecules are linear, as expected. However, with a large central atom and small ligands, e.g., with BaF_2, the molecules are bent.[22] It may be that in these cases an assumption common to all the models, namely, that the inner shells of the central atom may be entirely neglected, is invalid.

4-9. $d\pi$–$p\pi$ Bonds

While the role of central atom d orbitals in the formation of σ bonds to outer atoms has been a controversial subject, there is another role for d orbitals where their actual participation has been more generally accepted for some time, although here, too, the exact *extent* of that participation is subject to some differences of opinion.

While the heavier non-transition elements show little tendency to engage their p orbitals in π-bond formation, they do form at least partial π bonds to lighter elements, especially to oxygen and nitrogen, by using their outer d orbitals. The experimental indications of this are chiefly the high bond-stretching force constants and the shortness of bonds compared to the force constants and bond lengths to be expected for single bonds. More recently, photoelectron spectroscopy has provided evidence for such bonding even in H_3SiCl.[23]

[22] W. A. Klemperer *et al.*, *J. Chem. Phys.*, 1963, **39**, 2023, 2299; 1964, **40**, 3471; *J. Amer. Chem. Soc.*, 1964, **86**, 4544.
[23] S. Cradock and E. A. V. Ebsworth, *Chem. Comm.*, **1971**, 57.

Tetrahedral Molecules. We first consider what possibilities exist in principle, that is on the basis of compatibility of orbitals, for forming $d\pi$–$p\pi$ bonds. For a tetrahedral AB_4 molecule such as SiF_4 or PO_4^{3-}, each of the B atoms has two filled $p\pi$ orbitals perpendicular to the A—B bond axis and perpendicular to each other. The central atom, A, is assumed to use its s and p orbitals for σ bonding. A detailed examination of the suitability of the d orbitals of A for overlapping with the $p\pi$ orbitals on the B atoms shows that all of them are able to do so, but two, namely d_{z^2} and $d_{x^2-y^2}$, are particularly well suited for this.[24] Each of these two would be expected to have about $\sqrt{3}$ times as much overlap with the $p\pi$ orbitals of the four B atoms as one of the other three d orbitals. Fig. 4-8 shows in a rough schematic way the principal $d\pi$–$p\pi$ overlap possibilities.

Bond length data in the series of ions SiO_4^{4-}, PO_4^{3-}, SO_4^{2-}, ClO_4^{-} indicate that such $p\pi$–$d\pi$ bonding actually does occur. As shown in Table 4-7, the X—O bonds are all short relative to values reasonably expected for

TABLE 4-7

Bond Lengths and $d\pi$–$p\pi$ Overlaps in XO_4^{n-} Ions

Ion	Obs. X—O dist.,[a] in Å	Est. X—O single[a] bond dist., in Å	Shortening[a]	$p\pi$–$d\pi$ Overlap[b]
SiO_4^{4-}	1.63	1.76	0.13	0.33
PO_4^{3-}	1.54	1.71	0.17	0.46
SO_4^{2-}	1.49	1.69	0.20	0.52
ClO_4^{-}	1.46	1.68	0.22	0.57

[a] From ref. 24.
[b] From H. H. Jaffé, *J. Phys. Chem.*, 1954, **58**, 185.

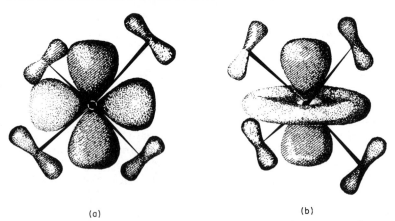

(a)　　　　　　　　　(b)

Fig. 4-8. Quasi-perspective view of the overlap of $p\pi$ orbitals of the B atoms in a tetrahedral AB_4 molecule with (a) the $d_{x^2-y^2}$ and (b) the d_{z^2} orbitals of the A atom. [Adapted from ref. 24 by permission.]

[24] D. W. J. Cruickshank, *J. Chem. Soc.*, **1961**, 5486.

single bonds, and, moreover, the trend in shortening is closely parallel to the calculated trend in $d\pi$–$p\pi$ overlaps.

Similarly, in SiF_4, even after due allowance for the effect of ionic–covalent resonance in strengthening and hence shortening the Si—F bonds, they appear to be around 0.13 Å shorter than the length expected for single bonds, and the implication, therefore, is that π bonding is present as well.

Other Molecules. In less symmetrical molecules the detailed analysis of π bonding is more difficult because the d orbitals of the central atom can interact with different types of outer atoms to different degrees. However, by utilizing the idea (page 123) that bond order and bond length are inversely related, it is possible to deduce approximate, relative degrees of π bonding in various compounds containing SiO, PO, SO and ClO groups, as well as, to a more limited extent, in other cases. For PO and SO bonds the data are most extensive. The various types of P—O bond vary in length from ~ 1.68 to ~ 1.40 Å, and it has been suggested[24] that this last value, about the shortest observed distance for any SO, PO, SN or PN bond, corresponds to about a double bond, i.e. to a π bond order of about 1. However, it is possible that this is an underestimate, since in SF_3N, where the S—N bond must be more nearly triple than double the length is 1.42 Å.

For the molecules and ions SF_2O_2, PF_3O, ClO_3F, ClO_2^-, ClO_3^- and ClO_4^-, *ab initio* MO calculations indicate significant d orbital participation in both σ and π bonds.[25]

In certain instances the existence of $d\pi$–$p\pi$ bonding is indicated by the overall molecular geometry. Thus, the Si_3N and Ge_3N skeletons are planar[26] in $(SiH_3)_3N$ and $(GeH_3)_3N$, and in $(Me_3Si)_2NBeN(SiMe_3)_2$ the $Si_2NBeNSi_2$ group has an allene-like configuration (Si_2NBe groups in two perpendicular planes[27]). In each case the Si—N or Ge—N bonds are 0.05–0.15 Å shorter than expected for single bonds. These structural features can be accounted for by assuming that in the planar configuration about the N atom there is enough $N(2p_z) \rightarrow Si(3d)$ or $Ge(3d)$ π-bonding to stabilize this configuration relative to the pyramidal one found in most other R_3N molecules.

It is worth noting that the mere presence of orbitals suitable for such bonding does not necessarily lead to such pronounced structural effects. Thus, $P(SiH_3)_3$ and $As(SiH_3)_3$ are both pyramidal[28] (Si—X—Si angles of 96.5° and 93.8°, respectively), as is $P(GeH_3)_3$,[29] and $S(SiH_3)_2$ is bent[30] (Si—S—Si = 98°). Apparently the $3p$ orbitals of the P, As and S atoms do not overlap with the $3d$ orbitals of Si and Ge as well as does the $2p$ orbital of a central N atom. In the case of $(SiH_3)_2O$ an intermediate situation occurs: the molecule is bent, but the angle (144°), is quite large.[31] More-

[25] I. H. Hillier and V. R. Saunders, *Chem. Comm.*, **1970**, 1183.
[26] C. Glidewell, D. W. H. Rankin, and A. G. Robiette, *J. Chem. Soc., A*, **1970**, 2935.
[27] A. H. Clark and A. Haaland, *Chem. Comm.*, **1969**, 912.
[28] B. Beagley, A. G. Robiette, and G. M. Sheldrick, *J. Chem. Soc., A*, **1968**, 3002, 3006.
[29] S. Cradock *et al.*, *Chem. Comm.*, **1965**, 515.
[30] A. Almenningen *et al.*, *Acta Chem. Scand.*, 1963, **17**, 2264.
[31] A. Almenningen, *Acta Chem. Scand.*, 1963, **17**, 2455.

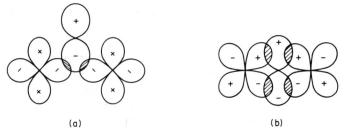

(a)   (b)

Fig. 4-9. Overlaps of a central $p\pi$ orbital with $d\pi$ orbitals on outer atoms for (a) bent and (b) linear configurations.

over the Si—O bonds are about 0.14 Å shorter than the value expected for single bonds, and the bond order may be estimated from this shortening to lie in the range 1.3–1.6.

The $(SiH_3)_3O$ case well illustrates the fact that some $p\pi \rightarrow d\pi$ bonding can occur without there being a completely linear or flat configuration of the molecule, as indicated in Fig. 4-9a. It is probable that the π bonding is strongest in the linear and flat configurations (Fig. 4.9b), but unless the magnitude of the bonding is fairly large it may not be able to overcome other factors favoring the bent configuration, or may, as in $(SiH_3)_2O$, only be able to reach a compromise with these other factors at some intermediate angle.

Further Reading

Coulson, C. A., *J. Chem. Soc.*, **1964**, 1442 (specifically concerned with noble-gas compounds, but the bonding theory pertains equally to many other systems, including all halogen compounds with more than four pairs of valence-shell electrons).

Malm, J. G., H. Selig, J. Jortner, and S. Rice, *Chem. Rev.*, 1965, **65**, 199 (theory of bonding as discussed specifically for Xe compounds has wide applicability).

Mitchell, K. A., *Chem. Rev.*, 1969, **69**, 137 (use of outer d orbitals in bonding).

Pettit, L. D., *Quart. Rev.*, 1971, **25**, 1 (multiple bonding in inorganic compounds).

Wiebenga, E. H., E. E. Havinga, and K. H. Boswijk, *Adv. Inorg. Chem. Radiochem.*, 1961, **3**, 133.

PART TWO

Chemistry of Nontransition Elements

5
Hydrogen

GENERAL REMARKS

5-1. Introduction

Three isotopes of hydrogen are known: ^{1}H, ^{2}H (deuterium or D) and ^{3}H (tritium or T). Although isotope effects are greatest for hydrogen, justifying the use of distinctive names for the two heavier isotopes, the chemical properties of H, D and T are essentially identical except in matters such as rates and equilibrium constants of reactions. The normal form of the element is the diatomic molecule; the various possibilities are H_2, D_2, T_2, HD, HT, DT.*

Naturally occurring hydrogen contains 0.0156% deuterium, while tritium occurs naturally in only minute amounts believed to be of the order of 1 in 10^{17}.

Tritium is formed continuously in the upper atmosphere in nuclear reactions induced by cosmic rays. For example, fast neutrons arising from cosmic-ray reactions can produce tritium by the reaction ^{14}N$(n, {}^3$H$)^{12}$C. Tritium is radioactive (β^-, 12.4 years) and is believed to be the main source of the minute traces of ^{3}He found in the atmosphere. It can be made artificially in nuclear reactors, for example, by the thermal neutron reaction, ^{6}Li$(n,\alpha)^3$H, and is available for use as a tracer in studies of reaction mechanism.

Deuterium as D_2O is separated from water by fractional distillation or electrolysis and by utilization of very small differences in the free energies of the H and D forms of different compounds, the H_2O–H_2S system being particularly favorable in large-scale use:

$$HOH(l) + HSD(g) = HOD(l) + HSH(g) \qquad K \approx 1.01$$

Deuterium oxide is available in ton quantities and is used as a moderator in nuclear reactors, both because it is effective in reducing the energies of fast fission neutrons to thermal energies and because deuterium has a much lower capture cross-section for neutrons than has hydrogen and hence does not

---

* Molecular H_2 (and D_2) have *ortho* and *para* forms in which the nuclear spins are aligned or opposed, respectively. This leads to very slight differences in bulk physical properties and the forms can be separated by gas chromatography.

reduce appreciably the neutron flux. Deuterium is widely used in the study of reaction mechanisms and in spectroscopic studies.

Although the abundance on earth of molecular hydrogen is trivial, hydrogen in its compounds has one of the highest of abundances. Hydrogen compounds of all the elements other than the noble gases are known, and many of these are of transcendental importance. Water is the most important hydrogen compound; others of great significance are hydrocarbons, carbohydrates and other organic compounds, ammonia and its derivatives, sulfuric acid, sodium hydroxide, etc. Hydrogen forms more compounds than any other element.

Molecular hydrogen is a colorless, odorless gas, f.p. 20.28°K, virtually insoluble in water. It is most easily prepared by the action of dilute acids on metals such as Zn or Fe and by electrolysis of water; industrially hydrogen may be obtained by thermal cracking or steam re-forming of hydrocarbons, by the reduction of water by carbon (water-gas reaction) and in other ways.

Hydrogen is not exceptionally reactive. It burns in air to form water and will react with oxygen and the halogens explosively under certain conditions. At high temperatures the gas will reduce many oxides either to lower oxides or to the metal. In the presence of suitable catalysts and above room temperature it reacts with N_2 to form NH_3. With electropositive metals and most non-metals it forms hydrides.

In the presence of suitable catalysts, usually Group VIII metals or their compounds, a great variety of both inorganic and organic substances can be reduced. Heterogeneous hydrogenation may be carried out in the gas phase or in solution, while a number of transition-metal ions and complexes can react with hydrogen, transferring it to a substrate homogeneously in solution (see Chapter 24).

The dissociation of hydrogen is highly endothermic, and this accounts in part for its rather low reactivity at low temperatures:

$$H_2 = 2H \qquad \Delta H_0^0 = 434.1 \text{ kJ mol}^{-1}$$

In its low-temperature reactions with transition-metal species heterolytic splitting to give H^-, bound to the metal, and H^+ may occur; the energy involved here is much lower, probably ~ 125 kJ mol^{-1}. At high temperature, in arcs at high current density, in discharge tubes at low hydrogen pressure, or by ultraviolet irradiation of hydrogen, atomic hydrogen can be produced. It has a short half-life (~ 0.3 sec). The heat of recombination is sufficient to produce exceedingly high temperatures, and atomic hydrogen has been used for welding metals. Atomic hydrogen is exceedingly reactive chemically, being a strong reducing agent.

5-2. The Bonding of Hydrogen

The chemistry of hydrogen depends mainly on three electronic processes:

1. *Loss of the valence electron.* The 1s valence electron may be lost to give the hydrogen ion, H^+, which is merely the proton. Its small size

$(r \sim 1.5 \times 10^{-13}$ cm) relative to atomic sizes $(r \sim 10^{-8}$ cm) and its small charge result in a unique ability to distort the electron cloud surrounding other atoms; the proton accordingly never exists as such, except in gaseous ion beams; in condensed phases it is invariably associated with other atoms or molecules.

2. *Acquisition of an electron.* The hydrogen atom can acquire an electron, attaining the $1s^2$ structure of He, to form the hydride ion, H^-. This ion exists as such essentially only in the saline hydrides formed by the most electropositive metals (Section 5-14).

3. *Formation of an electron-pair bond.* The majority of hydrogen compounds contain an electron-pair bond. The number of carbon compounds of hydrogen is legion and most of the less metallic elements form numerous hydrogen derivatives. Many of these are gases or liquids. Although most metallic elements do not form simple covalent hydrides, a wide variety of complex compounds which contain M—H bonds are known, e.g., $HCo(CO)_4$.

The chemistry of many of these compounds is highly dependent upon the nature of the element (or the element plus its other ligands) to which hydrogen is bound. Particularly dependent is the degree to which compounds undergo dissociation in polar solvents and act as acids:

$$HX \rightleftharpoons H^+ + X^-$$

Also important for chemical behavior is the electronic structure and coordination number of the molecule as a whole. This is readily appreciated by considering the covalent hydrides BH_3, CH_4, NH_3, OH_2, and FH. The first not only dimerizes (see below) but also manifests its coordinative unsaturation in acting as a Lewis acid; methane is chemically unreactive and neutral, ammonia has a lone pair and is a base, and water with two lone pairs can act as a base or as a very weak acid, while FH is a much stronger, though still relatively weak, acid in water.

Except in H_2 itself, where the bond is homopolar, all other H—X bonds possess polar character to some extent. The orientation of the dipole may be either $\overrightarrow{H—X}$ or $\overleftarrow{H—X}$, and important chemical differences arise accordingly. Although the term "hydride" might be considered appropriate only for compounds with the polarization $\overleftarrow{H—X}$, many compounds that act as acids in polar solvents are properly termed covalent hydrides. Thus, although HCl and $HCo(CO)_4$ behave as strong acids in aqueous solution, they are gases at room temperature and are undissociated in non-polar solvents.

4. *Unique bonding features.* The nature of the proton and the complete absence of any shielding of the nuclear charge by electron shells allow other forms of chemical activity which are either unique to hydrogen or particularly characteristic of it. Some of these are the following, which are discussed in some detail subsequently:

(*a*) The formation of numerous compounds, often non-stoichiometric, with metallic elements. They are generally called *hydrides* but cannot be regarded as simple saline hydrides (Section 5-16).

(b) Formation of *hydrogen bridge bonds* in electron-deficient compounds such as in (5-I) or transition-metal complexes as in (5-II).

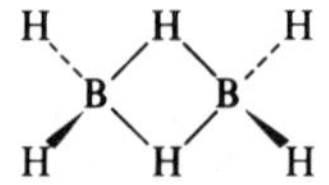

(5-I) 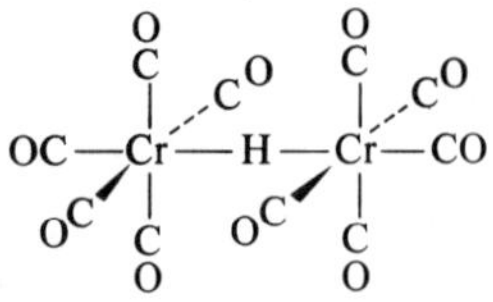   (5-II)

The best-studied examples of bridge bonds are those in boranes and related compounds (Chapter 8) and in certain hydride complexes (pages 183 and 250). Bridge bonds in transition-metal complexes are discussed along with transition-metal hydride complexes in Chapter 22.

(c) *The hydrogen bond.* This bond is important not only because it is essential to an understanding of much other hydrogen chemistry but also because it is one of the most intensively studied examples of intermolecular attraction. Hydrogen bonds dominate the chemistry of water, aqueous solutions, hydroxylic solvents and OH-containing species generally, and they are of crucial importance in biological systems, being responsible *inter alia* for the linking of polypeptide chains in proteins and the base pairs in nucleic acids.

THE HYDROGEN BOND, HYDRATES, HYDROGEN ION, AND ACIDS

5-3. The Hydrogen Bond[1]

The hydrogen bond is the term given to the relatively weak secondary interaction between a hydrogen atom bound to an electronegative atom and another atom which is also generally electronegative and which has one or more lone pairs and can thus act as a base. We can thus refer to proton donors, XH, and proton acceptors, Y, and can give the following generalized representation of a hydrogen bond:

$$\overset{\delta-}{X}-\overset{\delta+}{H}\cdots Y$$

Such interaction is strongest when both X and Y are first-row elements; the main proton donors are N—H, O—H and F—H, and the most commonly encountered hydrogen bonds are O—H$\cdots$O and N—H$\cdots$O. The groups

[1] (a) W. C. Hamilton and J. A. Ibers, *Hydrogen Bonding in Solids*, Benjamin, 1968 (an extensive review of structural methods and results).

(b) G. C. Pimentel and A. L. McClellan, *The Hydrogen Bond*, Freeman, 1960 (a thorough book with extensive tables and bibliography).

(c) A. Rich and N. Davidson, eds., *Structural Chemistry and Molecular Biology*, Freeman, 1968 (contains chapters on hydrogen bonding).

(d) A. K. Covington and P. Jones, eds., *Hydrogen-Bonded Solvent Systems*, Taylor and Francis, 1968 (Symposium Report on various topics).

(e) G. Zundel, *Hydration and Intermolecular Interaction*, Academic Press, 1970 (Infrared spectra of water and hydrates).

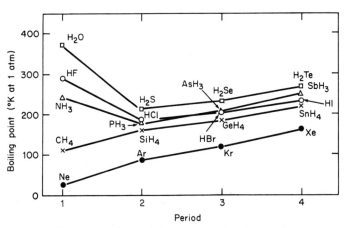

Fig. 5-1. Boiling points of some molecular hydrides.

P—H, S—H, Cl—H, and Br—H can also act as proton donors, and so even can C—H provided that the C—H bond is relatively polar as it is when the carbon is bound to electronegative groups as in $CHCl_3$[2] or the carbon atom is in an *sp*-hybridized state as in HCN or $RC\equiv CH$. The acceptor atoms can be N, O, F, Cl, Br, I, S or P, but carbon never acts as an acceptor other than in certain π-systems noted below.

Much of the earlier experimental evidence for hydrogen bonding came from comparisons of the physical properties of hydrogen compounds. Classic examples are the apparently abnormally high boiling points of NH_3, H_2O, and HF (Fig. 5-1) which imply association of these molecules in the liquid phase. Other properties such as heats of vaporization provided further evidence for association. While physical properties reflecting association are still a useful tool in detecting hydrogen bonding, the most satisfactory evidence for solids comes from X-ray and neutron-diffraction crystallographic studies, and for solids, liquids, and solutions from infrared and nuclear magnetic resonance spectra.

Although H atoms are often observable in X-ray studies, their positions can seldom be ascertained with any accuracy. However, neutron-diffraction data[3] can usually give quite precise locations, because the scattering of neutrons of thermal energies (approximately 0.1 eV, 9.6 kJ mol^{-1}) is roughly similar for all nuclei, regardless of atomic number, whereas the scattering of X-rays depends on electron density and is lowest for hydrogen. Even if accurate location of hydrogen atoms is not possible, the overall X—Y distance is significant. If X—Y distances are significantly shorter than normal van der Waals contacts for non-bonded atoms by, say, 0.2 Å, then we can be fairly certain of the presence of a hydrogen bond, although it is hazardous to

[2] See, e.g., R. D. Green and J. S. Martin, *J. Amer. Chem. Soc.*, 1969, **91**, 3659; R. M. Dieters, W. G. Evans, and D. H. McDaniel, *Inorg. Chem.*, 1968, 7, 1615.
[3] G. Will, *Angew. Chem. Internat. Edn.*, 1969, **8**, 356 (review with examples of hydrogen bonding).

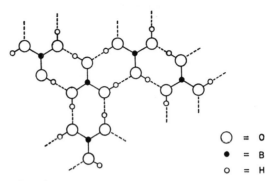

Fig. 5-2. The structure of boric acid showing hydrogen bonds.

use distance criteria alone unless the hydrogen atom is detected. For O—H···O, distances below approximately 3 Å may indicate hydrogen bonding. Thus in crystalline $NaHCO_3$ there are four kinds of O—O distances (between O's of different HCO_3^- ions) having values of 2.55, 3.12, 3.15 and 3.19 Å. The last three are approximately equal to twice the van der Waals radius of O, but the first, 2.55 Å, corresponds to the H-bonded pair O—H···O. A classic example of H bonding is that in the loosely bonded parallel sheets of orthoboric acid (Fig. 5-2), while Fig. 5-3(a) shows the structure typical of carboxylic acid dimers. Many other compounds associate in the latter type of elongated hexagonal structure, and Fig. 5-3(b) shows the type of arrangement in the biologically important purine and pyrimidine base pairs.

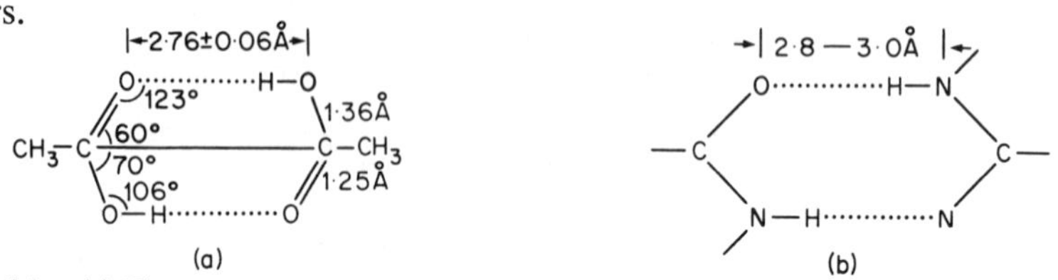

Fig. 5-3. (a) The structure of acetic acid dimer which has an enthalpy of dissociation of 57.9 kJ mol⁻¹ (A. D. H. Clague and H. J. Bernstein, *Spectrochim. Acta*, 1959, **25***A*, 593). (b) Hydrogen bonding between base pairs such as adenine–thymine or cytosine–guanine.

Hydrogen bonding has been much studied by means of infrared[4] and Raman spectra, and the former is an especially convenient experimental tool. When the X—H group enters into H bonding there are three main changes in infrared spectra:

(*a*) The X—H stretching frequency is lowered relative to that in the free molecule.

(*b*) The band due to the X—H stretch is substantially broadened, occasionally up to several hundred cm⁻¹, and the intensity also increases markedly.

(*c*) The X—H bending frequency is usually raised.

[4] L. J. Bellamy, *Advances in Infrared Group Frequencies*, Methuen, 1968 (Chapter 8 is an excellent review).

TABLE 5-1

Some Parameters of Hydrogen Bonds

Bond	Compound	Bond energy (kJ mol^{-1})	Depression of str. freq. (cm^{-1})	Bond length^a (Å)	X—H distance (Å)
F—H—F	KHF$_2$	~113	~2700	2.26	1.13
F—H...F	HF(g)	~ 28.6	700	2.55	
O—H...O	(HCOOH)$_2$	29.8	~ 460	2.67	
O—H...O	H$_2$O(s)	~ 21	~ 430	2.76	0.97
O—H...O	B(OH)$_3$			2.74	1.03
N—H...N	Melamine	~ 25	~120	3.00	
N—H...Cl	N$_2$H$_5$Cl		~460	3.12	
C—H...N	(HCN)$_n$		180	3.2	1.0

^{a} The distance between the hydrogen-bonded atoms X and Y.

High-resolution nuclear magnetic resonance spectra are also potentially a rapid means of studying H bonding. The proton resonance of X—H$\cdots$Y now occurs at lower applied fields owing to deshielding effects that can be attributed in part to the withdrawal of electron density from H by the electronegative atom Y, and partly to the inhibition of electronic circulation about the H atom by the field of Y. On increase in temperature or on dilution, hydrogen bonds may be broken and the resonance line then moves to higher fields, as in the free molecules. Thus it is possible by nuclear magnetic resonance methods to study equilibria such as

$$HF(aq) + F^-(aq) \rightleftharpoons F—H \cdots F^-(aq)$$

The crystallographic, spectroscopic and other studies have shown that there are two main classes of H bond, although, of course, there is no sharp break between them:

(a) Weak hydrogen bonds. Here there is little change in the bond length of X—H and rather small shifts in infrared stretching frequencies. There appear to be fairly smooth relationships between the X—Y distance and the X—H stretching frequency, but the relation between frequency shifts and energies of H bonds is only approximate and of little predictive value.

The energies of these weaker bonds lie in the range 4–40 kJ mol^{-1}, considerably less than energies of normal bonds as in H—X.

(b) Strong hydrogen bonds. Such bonds are found essentially only in the ion F—H—F$^-$, for O—H—O in a number of anions of organic acids,[5a] for ClHCl$^-$,[5b] and possibly for N—H—N in complex cyano acids. The extreme example is the ion F—H—F$^-$. In these cases there are very substantial shifts in the infrared spectra and, where known, very substantial changes in the H—X bond length.

In some cases, the symmetrical location of the hydrogen atom can be proved, as in KHF$_2$, by broad-line nuclear magnetic resonance studies and by the absence of residual entropy (see below). However, it is not always clear

[5a] J. C. Speakman et al., J. Chem. Soc. A, **1971**, 1994, 1997.
[5b] J. S. Swanson and J. M. Williams, Inorg. Nuclear Chem. Letters, 1970, **6**, 271.

whether the potential energy function for the H atom has a single minimum at the center, or whether there is a symmetric double-minimum potential with both minima near the center, thus giving a dynamic mixture of X—H···X and X···H—X and an apparent statistical centering of the H atom.

Parameters for some representative H-bonded substances are given in Table 5-1.

Some other features of hydrogen bonding can now be noted.

(1) Although most H bonds are intermolecular, many cases of *intramolec-ular* bonding are known, examples being *o*-nitrophenol (5-III) and the chloromaleate ion (5-IV).

(5-III)	(5-IV)	(5-V)

(2) Bifurcated H bonds (5-V) are rare but several well-authenticated cases, both symmetric and unsymmetric, are known.[4,6]

(3) X—H···Y bonds need not be, and usually are not, strictly linear.

(4) There is evidence that H bonds can be formed between very polar X—H groups,[7] such as O—H, and polarizable double bonds or aromatic ring systems (Fig. 5-4).

Fig. 5-4. Examples of intramolecular hydrogen bonding of OH groups to polarizable electron clouds.

(5) In some crystals, reorientation of hydrogen bonds may occur at a certain temperature. There is usually a crystallographic change at this point and there may also be important changes in other physical properties, notably ferroelectricity. One of the best studied examples is KH_2PO_4.[8] Below 123°K this has an ordered structure which accounts for the ferroelectricity: the H atom nearer to one oxygen atom of the tetrahedral PO_4 changes its position to lie closer to the other oxygen of the O—H···O bond as the direction of the dielectric polarization is reversed. Above 123°K the structure is disordered and non-ferroelectric. A number of other H-bonded crystals with ordered lattices show similar behaviour.

[6] J. Donohue in A. Rich and N. Davidson, eds., *Structural Chemistry and Molecular Biology*, Freeman, 1958, p. 443.

[7] See, e.g., H bonding of $CHCl_3$ to benzene and other arenes; N. C. Perrins and J. P. Simons, *Trans. Faraday Soc.*, 1969, **65**, 390.

[8] R. M. Hill and S. K. Ichiki, *J. Chem. Phys.*, 1968, **48**, 838.

Theory of Hydrogen Bonding[9]

In order to account for the existence of hydrogen bonds we must consider the possible contributions of covalent bonding, resonance and electrostatic attraction and attempt to assess the importance of each. In short, we must ask how much each of the possible canonical structures (5-VI), (5-VII) and

$$\overset{\delta-}{X}—\overset{\delta+}{H}—\overset{\delta-}{Y} \qquad X—H \; : \; Y \; \leftrightarrow \; \bar{X} \; : \; H—\overset{+}{Y} \qquad \overset{\delta-}{X}—\overset{\delta+}{H}\cdots\overset{\delta-}{Y}$$

(5-VI)	(5-VIIa) (5-VIIb)	(5-VIII)
Covalence	Resonance	Electrostatic attraction

(5-VIII) contributes to the bond energy. It is certain that structure (5-VI), involving the coexistence of two covalent bonds to hydrogen is completely negligible, since it would require the use of the $2s$ or $2p$ orbitals of hydrogen, and these are of such high energy as to be essentially useless for bonding.

Theoretical work is concerned with the relative contributions of (5-VII) and (5-VIII) and leads to the conclusion that the resonance represented by (5-VII) is of importance *only* for the *strongest, shortest bonds*. It has been estimated, for example, that in an O—H$\cdots$O bond with the O—O distance 2.78 Å, and the O—H distance 1.0 Å (fairly typical parameters), structure (5-VIIb) appears in the over-all wave function to the extent of only about 4%. Thus, it is believed that most hydrogen bonds are *basically electrostatic*; but this then raises another question. If unshared electron pairs are concentrated along the direction of hybrid orbitals, will the proton approach the atom Y preferentially along these directions? In other words, does the proton see the atom Y as a structureless concentration of negative charge or as an atomic dipole? The answer to this question is not entirely clear cut, because in most cases where the angle θ in (5-IX) is in accord with the latter idea it is possible to

$$\overset{Y-}{\underset{Z}{\diagup}}\overset{}{\underset{\theta}{}}\text{---H—X}$$

(5-IX)

attribute this to steric requirements, as in carboxylic acid dimers or *o*-nitrophenol, or it can be equally well explained on the simpler theory as in the case of HCN polymers which are linear. However, the case of the $(HF)_n$ polymer, Fig. 5-5 (and a few others) seems to lend strong support to the hypothesis of preferred directions, since there appears to be no other reason why the structure should not be linear.

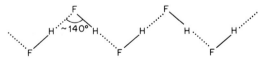

Fig. 5-5.   The structure of crystalline hydrogen fluoride.

The foregoing analysis is basically in valence bond (VB) terminology, in which the relative contributions of canonical structures are considered. It is possible to carry out essentially the same enquiry employing a molecular

[9] (a) J. N. Murrell, *Chem. in Britain*, **1969**, 107; (b) S. H. Linn in *Physical Chemistry*, Vol. 5, *Valency*, H. Eyring, ed., Academic Press, 1970.

orbital (MO) approach and using somewhat different terminology. Such a study has been made and leads to very similar conclusions. Again, for most hydrogen bonds, Coulomb energy provides the largest contribution and delocalization (i.e. resonance) effects are small except for the strong hydrogen bonds. However, simple dipole–dipole interaction does not adequately describe the dominant Coulombic or electrostatic contribution.

Finally, for the extreme case of exceedingly strong hydrogen bonds, as exemplified by the symmetrical FHF^- ion, a formulation in terms of standard $3c$–$4e$ MO theory (page 110) is applicable. Here, in the extreme, highly polar covalent bonding can describe the situation.

In conclusion, let it be emphasized that hydrogen bonds vary considerably in nature and involve many contributing factors. To the question "Where does the energy of the hydrogen bond come from?" ... "there is still no concise answer that seems to be generally acceptable and there probably never will be."[9a]

5-4. Ice and Water[10]

The structural natures of ice and, *a fortiori*, of water are very complex matters which can be but briefly treated here.

There are nine known modifications of ice, each stable over a certain range of temperature and pressure. Ordinary ice, ice I, which forms from liquid water at $0°C$ and one atm, has a rather open structure built of puckered, six-membered rings, as shown in Fig. 5-6. Each H_2O is tetrahedrally sur-

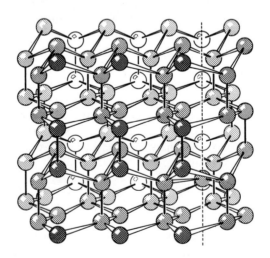

Fig. 5-6.   The structure of ice I. Only the oxygen atoms are shown.

[10] R. A. Horne, *Survey of Progress in Chemistry*, Vol. 4, Academic Press, 1968 (a good review); D. Eisenberg and W. Kautzman, *The Structure and Properties of Water*, Oxford University Press, 1968; A. H. Narten and H. A. Levy, *Science*, 1969, **165**, 447; N. H. Fletcher, *The Chemical Physics of Ice*, Cambridge Univ. Press, 1970.

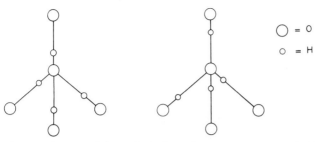

Fig. 5-7. Two possible configurations about an oxygen atom in ice.

rounded by the oxygen atoms of four neighbouring molecules, and the whole array is linked by unsymmetrical hydrogen bonds. The O—H⋯O distance is 2.75 Å (at 100°K) and the H atoms lie 1.01 Å from one oxygen and 1.74 Å from the other. Each oxygen atom has two near and two far hydrogen atoms, but there are six distinct arrangements, two being illustrated in Fig. 5-7, all equally probable. However, the existing arrangement at any one oxygen eliminates certain of those at its neighbors. A rigorous analysis of the probability of any given arrangement in an entire crystal leads to the conclusion that at the absolute zero ice I should have a disordered structure with a zero-point entropy of 0.81 eu, in excellent agreement with experiment. This result in itself constitutes a good proof that the hydrogen bonds are unsymmetrical; if they were symmetrical there would be a unique, ordered structure and hence no zero-point entropy. Entirely similar considerations confirm the presence of a network of unsymmetrical H bonds in KH_2PO_4 and $Ag_2H_3IO_6$, whereas the absence of zero-point entropy in K[FHF] supports the idea that FHF^- has a symmetrical structure.

The structural nature of liquid water is still controversial. The structure is not random, as in liquids consisting of more-or-less spherical non-polar molecules; instead, it is highly structured owing to the persistence of hydrogen bonds; even at 90°C only a few percent of the water molecules appear not to be hydrogen-bonded. Still, there is considerable disorder, or randomness, as befits a liquid.

In an attractive, though not universally accepted, model of liquid water the liquid consists at any instant of an imperfect network, very similar to the network of ice I, but differing in that (a) some interstices contain water molecules that do not belong to the network but instead disturb it, (b) the network is patchy and does not extend over long distances without breaks, (c) the short-range ordered regions are constantly disintegrating and re-forming (they are "flickering clusters"), and (d) the network is slightly expanded compared to ice I. The fact that water has a slightly higher density than ice I may be attributed to the presence of enough interstitial water molecules to more than offset the expansion and disordering of the ice I network. This model of water receives strong support from X-ray scattering studies.

It may be noted that alcohols, which are similar to water in many respects,

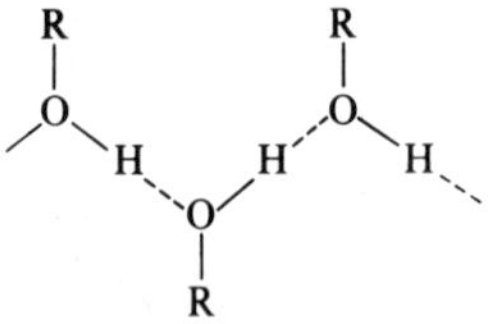

(5-X)

cannot form three-dimensional arrays and hence only linear or cyclic polymers (5-X) exist.

5-5. Hydrates[1a]

Crystalline hydrates, especially of metal ions, are of enormous importance in inorganic chemistry, but hydrates of organic substances, especially those with N—H and O—H bonds are also common. For metal ions, the negative end of the water dipole, i.e. oxygen, is always bound to the metal and the lone pairs on it can be directed towards the metal and involved in the bonding; they can, however, also act as acceptors for H bonds, and H atoms of coordinated water are normally hydrogen-bonded. Thus there is an extreme flexibility in hydrogen bonding and this allows stabilization in lattices of many different types of hydrate structure.

Some of the main structures are given in Fig. 5-8. It must be particularly noted that *gross distortions* from the angles of the ideal models shown are the *rule* and also that large deviations from linearity of the H bonds also occur.

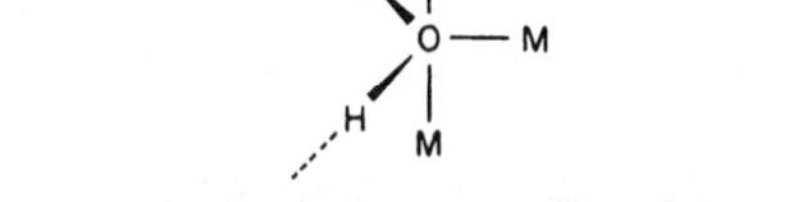

Fig. 5-8.   Idealized structures for H-bonded coordinated water in hydrates. The metal M can also be replaced by HY in hydrates of organic compounds.

5-6. Clathrates, Gas Hydrates and Other Enclosure Compounds

There are certain substances formed by combining one stable compound with another or with an atomic or molecular element without the existence of any chemical bonds between the two components. This occurs when one of the compounds can crystallize in a very open structure containing cavities, holes or channels in which atoms or molecules of the other can be trapped. Those compounds in which the *host* lattice contains cavities like cages are the most important type; they are called *clathrate* compounds, from the Latin *clathratus*, meaning "enclosed or protected by crossbars or grating." Most of these, and the important ones certainly, involve hydrogen bonding.

One of the first clathrate systems to be investigated in detail, and still one of the best understood, comprises the substances in which the host is β-quinol. Quinol clathrates have been prepared enclosing O_2, N_2, CO, NO, CH_4, SO_2, HCl, HBr, Ar, Kr, Xe, HCOOH, HCN, H_2S, CH_3OH, or CH_3CN.

Crystallization of solutions of quinol [*p*-dihydroxybenzene, *p*-$C_6H_4(OH)_2$] in water or alcohol under pressure of 10–40 atm of, say, krypton, produces crystals, often up to 1 cm in length, which are readily distinguishable from the crystals of ordinary quinol (α-quinol), even visually. These crystals contain the noble gas trapped in the lattice of β-quinol. When the crystals are dissolved in water, or heated, the gas is released. The crystals are stable at room temperature and can be kept for years.

Such trapping of the gases is made possible by the occurrence of cavities in the crystal lattices of certain compounds. X-ray analysis indicates that, in β-quinol, three quinol molecules form an approximately spherical cage of free diameter ~4 Å with the quinol molecules bound together by hydrogen bonds. The free volumes are in the form of isolated cavities, and the apertures leading from one cage to another through the crystal are very small in diameter. Molecules trapped within these cavities during formation of the crystal are unable to escape. As a molecule approaches the cage walls, it experiences repulsive forces. Since three quinol molecules are required to form each cavity, the limiting ratio of quinol to trapped atom or molecule for the composition of clathrates is 3:1. This ratio is reached for acetonitrile, but for the noble gases various composition ranges may be obtained depending on conditions—for example, $C_6H_4(OH)_2$/Kr, 3:0.74; $C_6H_4(OH)_2$/Xe, 3:0.88— and normally the cages are incompletely filled.

Since the free diameter of the quinol cage in a clathrate compound is ~4 Å, only molecules of appropriate size may be expected to be trapped. Thus, although CH_3OH forms quinol clathrates, C_2H_5OH is too large and does not. On the other hand, not all small molecules may form clathrates. Helium does not, the explanation being that the He atom is too small and can escape between the atoms of the quinol molecules which form the cage. Similarly, neon has not been obtained in a quinol clathrate as yet. Water, although of a suitable size, also does not form a clathrate; in this case the explanation cannot be a size factor, but may lie in the ability of water molecules to form hydrogen bonds which enables them to approach the cage walls and thus escape through gaps in the walls.

A second important class of clathrates are the *gas hydrates*.[11] When water is solidified in the presence of certain atomic or small-molecular gases, as well as some substances such as $CHCl_3$ that are volatile liquids at room temperature, it forms one of several types of very open structure in which there are

[11] G. A. Jeffrey and R. K. McMullen, *Progr. Inorg. Chem.*, 1967, **8**, 43.

cages occupied by the gas or other guest molecules. These structures are far less dense than the normal form of ice and are unstable with respect to the latter in the absence of the guest molecules. There are two common gas hydrate structures, both cubic. In one the unit cell contains 46 molecules of H_2O connected to form six medium-size and two small cages. This structure is adopted when atoms (Ar, Kr, Xe) or relatively small molecules (e.g., Cl_2, SO_2, CH_3Cl) are used, generally at pressures greater than one atmosphere for the gases. Complete filling of only the medium cages by atoms or molecules, X, would give a composition $X \cdot 7.67H_2O$, while complete filling of all eight cages would lead to $X \cdot 5.76H_2O$. In practice, complete filling of all cages of one or both types is seldom attained and these formulas therefore represent limiting rather than observed compositions; for instance, the usual formula for chlorine hydrate (see Section 16.3) is $Cl_2 \cdot 7.3H_2O$. The second structure, often formed in the presence of larger molecules of liquid substances (and thus sometimes called the liquid hydrate structure) such as chloroform and ethyl chloride, has a unit cell containing 136 water molecules with eight large cages and sixteen smaller ones. The anesthetic effect of substances such as chloroform is due to the formation of liquid hydrate crystals in brain tissue.[12]

A third notable class of clathrate compounds, salt hydrates, is formed when tetraalkylammonium or sulfonium salts crystallize from aqueous solution with high water content, for example $[(n\text{-}C_4H_9)_4N]C_6H_5CO_2 \cdot 39.5H_2O$ or $[(n\text{-}C_4H_9)_3S]F \cdot 20H_2O$. The structures of these substances are very similar to the gas and liquid hydrate structures in a general way though different in detail. These structures consist of frameworks constructed mainly of hydrogen-bonded water molecules but apparently including also the anions (e.g., F^-) or parts of the anions (e.g., the O atoms of the benzoate ion). The cations and parts of the anions (e.g., the C_6H_5C part of the benzoate ion) occupy cavities in an incomplete and random way.

An additional relationship between the gas hydrate and the salt hydrate structures has been revealed by the discovery that bromine hydrate, $Br_2 \cdot \sim 8.5H_2O$, crystallizes in neither of the cubic gas hydrate structures but rather is nearly isostructural with the tetragonal tetra-n-butylammonium salt hydrates. Its ideal limiting composition would be $Br_2 \cdot 8.6H_2O$.

Although not classifiable as clathrate compounds, many other crystalline substances have holes, channels, or honeycomb structures which allow inclusion of foreign molecules, and many studies have been made in this field. Urea is an example of an organic compound which in the crystal has parallel continuous uniform capillaries; it may be utilized to separate straight-chain hydrocarbons from branched-chain ones, the latter being unable to fit into the capillaries.

Among inorganic lattices that can trap molecules, the best known are the so-called "molecular sieves" which are discussed in Section 11-6.

---

[12] J. F. Catchpool in A. Rich and N. Davidson, eds., *Structural Chemistry and Molecular Biology*, Freeman, 1968, p. 325.

5-7. The Hydrogen Ion

For the reaction
$$H(g) = H^+(g) + e$$

the ionization potential, 13.59 eV ($\Delta H = 569$ kJ mol^{-1}), is higher than the first ionization potential of Xe and is high by comparison with Li or Cs and indeed many other elements. Hence, with the possible exception of HF, bonds from hydrogen to other elements must be mainly covalent. For HF the bond energy is 5.9 eV. For a purely ionic bond the energy can be estimated as the sum of (1) 13.6 eV to ionize H, (2) -3.5 ev to place the electron on F and (3) -15.6 eV as an upper limit on the electrostatic energy of the ion pair H^+F^- at the observed internuclear distance in HF. The sum of these terms is -5.5 eV as an upper limit, which is not too far below the actual bond energy. For HCl, on the other hand, the experimental bond energy is 4.5 eV, whereas for a purely ionic situation we would have the sum $+13.6 - 3.6 - 11.3 = -1.5$ eV as an upper limit. Thus purely electrostatic bonding cannot nearly explain the stability of HCl.

Hydrogen can form the hydrogen ion *only* when its compounds are dissolved in media which *solvate* protons. The solvation process thus provides the energy required for bond rupture; a necessary corollary of this process is that the proton, H^+, never exists in condensed phases, but occurs always as solvates—H_3O^+, R_2OH^+, etc. The order of magnitude of these solvation energies can be appreciated by considering the solvation reaction in water (estimated from thermodynamic cycles):

$$H^+(g) + xH_2O = H^+(aq) \qquad \Delta H = -1091 \text{ kJ mol}^{-1}$$

Compounds that furnish solvated hydrogen ions in suitable polar solvents, such as water, are *protonic acids*.

The nature of the hydrogen ion in water, which should more correctly be called the hydroxonium ion, H_3O^+, is discussed below. The hydrogen ion in water is customarily referred to as "the hydrogen ion," implying H_3O^+. The use of other terms, such as hydroxonium, is somewhat pedantic except in special cases. We shall usually write H^+ for the hydrogen ion and assume it to be understood that the ion is aquated, since in a similar manner many other cations, Na^+, Fe^{2+}, Zn^{2+}, etc., are customarily written as such, although there also it is understood that the actual species present in water are aquated species, for example $[Fe(H_2O)_6]^{2+}$.

Water itself is weakly ionized:

$$2H_2O = H_3O^+ + OH^- \qquad \text{or} \qquad H_2O = H^+ + OH^-$$

Other cases of such *self-ionization* of a compound where one molecule solvates a proton originating from another are known; for example, in pure sulfuric acid

$$2H_2SO_4 = H_3SO_4^+ + HSO_4^-$$

and in liquid ammonia

$$2NH_3 = NH_4^+ + NH_2^-$$

In aqueous solutions, the hydrogen ion concentration is often given in terms of pH, defined as $-\log_{10}[H^+]$, where $[H^+]$ is the hydrogen ion activity, which may be considered to approximate to the molar concentration of H^+ ions in very dilute solutions.

At 25° the ionic product of water is

$$K_w = [H^+][OH^-] = 1 \times 10^{-14} M^2$$

This value is significantly temperature-dependent. When $[H^+] = [OH^-]$, the solution is neutral and $[H^+] = 1 \times 10^{-7} M$; that is, pH = 7.0. Solutions of lower pH are acidic; those of higher pH are alkaline.

The standard hydrogen electrode provides the reference for all other oxidation–reduction systems. The hydrogen half-cell or hydrogen electrode is

$$H^+(aq) + e = \tfrac{1}{2}H_2(g)$$

By definition, the potential of this system is zero ($E^0 = 0.000$ V) at all temperatures when an inert metallic electrode dips into a solution of hydrogen ions of unit activity (i.e., pH = 0) in equilibrium with H_2 gas at 1 atm pressure. The potentials of all other electrodes are then referred to this defined zero. However, the absolute potentials of other electrodes may be either greater or smaller, and thus some must have positive and others negative potentials relative to the standard hydrogen electrode. While this subject is not properly an aspect of the chemistry of hydrogen, it will be briefly discussed here as a matter of convenience.

The difficulties that are sometimes caused by the so-called electrochemical sign conventions have arisen largely because the term "electrode potential" has been used to mean two distinct things:

(1) *The potential of an actual electrode.* For example, a zinc rod in an aqueous solution of zinc ions at unit activity ($a = 1$) at 25° has a potential of -0.7627 V relative to the standard hydrogen electrode. There is no ambiguity about the sign because, if this electrode and a hydrogen electrode were connected with a salt bridge, it would be necessary to connect the zinc rod to the negative terminal of a potentiometer and the hydrogen electrode to the positive terminal in order to measure the potential between them. Physically, the zinc electrode is richer in electrons than the hydrogen electrode.

(2) *The potential of a half reaction.* Using the same chemical system as an example, and remembering also that the Gibbs free energy of the standard hydrogen electrode is also defined as zero, we can write:

$$
\begin{array}{ll}
\text{Zn} + 2H^+(a = 1) \rightarrow \text{Zn}^{2+}(a = 1) + H_2(g) \left.\right\} & \Delta G^0 = -147.5 \text{ kJ} \\
\text{Zn} \qquad\qquad\quad \rightarrow \text{Zn}^{2+}(a = 1) + 2e^- \left.\right\} & E^0 = -\Delta G^0/nF = +0.7627 \text{ V} \\
\text{Zn}^{2+}(a = 1) + H_2(g) \rightarrow \text{Zn} + 2H^+(a = 1) \left.\right\} & \Delta G^0 = +147.5 \text{ kJ} \\
\text{Zn}^{2+}(a = 1) + 2e^- \quad \rightarrow \text{Zn} \left.\right\} & E^0 = -\Delta G^0/nF = -0.7627 \text{ V}
\end{array}
$$

Since metallic zinc does actually dissolve in acid solutions, under conditions specified in the definition of a standard electrode, the standard change in Gibbs free energy must be negative for the first pair of reactions and positive for the second pair. The potential of the zinc couple, defined by $\Delta G^0 = -nFE^0$ (n = number of electrons = 2, F = the Faraday), has to change sign accordingly. The half-reaction

$$\text{Zn} \rightarrow \text{Zn}^{2+} + 2e^-$$

involves oxidation and its potential is an *oxidation potential* whose sign is that of the so-called American sign convention. The half-reaction

$$Zn^{2+} + 2e^- \rightarrow Zn$$

involves reduction and its potential is a *reduction potential* associated with the European sign convention. There is no doubt about which potential is relevant provided the half-reaction to which it refers is written out in full.

Inspection shows that the reduction potential has the same sign as the potential of the actual electrode. For this reason we adopt the IUPAC recommendation that *only reduction potentials should be called electrode potentials*. Every half-reaction is therefore written in the form

$$Ox + ne^- \rightleftharpoons Red$$

and the Nernst equation for the electrode potential, E, is

$$E = E^0 + \frac{2.3026\,RT}{nF} \log_{10} \frac{\text{activity of oxidant}}{\text{activity of reductant}} \qquad (5\text{-}1)$$

where E^0 is the standard electrode potential, R the gas constant and T the absolute temperature. Alternatively, we may sometimes speak of the electrode potential of a couple, e.g., Fe^{3+}/Fe^{2+}, giving it the sign appropriate to the half-reaction written as a reduction.

For pure water, in which the H^+ activity is only 10^{-7} mol l^{-1} the electrode potential, according to equation (5-1) is more negative than the standard potential, that is, hydrogen becomes a better reductant:

$$H^+(aq)\,(10^{-7}M) + e = \tfrac{1}{2}H_2 \qquad E_{298} = -0.414\,V$$

In a basic solution, where the OH^- activity is $1M$, the potential is -0.83 V. In the absence of overvoltage (a certain lack of reversibility at certain metal surfaces), hydrogen is liberated from pure water by reagents whose electrode potentials are more negative than -0.414 V. Similarly, certain ions, for example the U^{3+} ion, for which the U^{4+}/U^{3+} standard potential is -0.61 V, will be oxidized by water, liberating hydrogen.

Many electropositive metals or ions, even if they do not liberate hydrogen from water, will be oxidized by a greater concentration of hydrogen ions—thus the reactions of Zn or Fe are normally used to prepare hydrogen from dilute acids.

Finally a word on rates of acid–base reactions. All the protons in water are undergoing rapid migration from one oxygen atom to another, and the lifetime of an individual H_3O^+ ion in water is only approximately 10^{-13} sec. The rate of reaction of H_3O^+ with a base such as OH^- in water is very fast but also diffusion-controlled.[13] Reaction occurs when the solvated ions diffuse to within a critical separation, whereupon the proton is transferred by concerted shifts across one or more solvent molecules hydrogen-bonded to the base.

---

[13] M. C. Rose and J. Stuehr, *J. Amer. Chem. Soc.*, 1968, **90**, 7205.

5-8. Oxonium Ions

There is evidence for the existence, in crystalline hydrates of strong acids, not only of H_3O^+ but also of $H_5O_2^+$ and more heavily hydrated ions. Some examples are given in Table 5-2.

In acids generally, the proton must be present either as (a) an oxonium ion or (b) H-bonded to some other atom in the acid molecule—if there is one suitable. Thus, while it is possible to have hydrates of certain acids, e.g., $H_2PtCl_6 \cdot 2H_2O$, which contain H_3O^+ ions, the anhydrous acid cannot be made. On the other hand, both hydrated and anhydrous forms of other acids such as $H_4[Fe(CN)_6]$ are known and in the anhydrous form there are H-bonds, M—CN—H—NC—M.[14] There are also cases where no oxonium ion is present, such as hydrated oxalic acid, $(COOH)_2 \cdot 2H_2O$, which has a three-dimensional H-bonded structure[15a] and phosphoric acid hemihydrate[15b] which is $2PO(OH)_3 \cdot H_2O$. It is worth noting that adducts of acids are sometimes not all that they might seem; thus $CH_3CN \cdot 2HCl$ could have contained the HCl_2^- ion, but is actually[16] $[CH_3(Cl)C = NH_2]^+Cl^-$. The nitric acid adducts of certain metal complex nitrate salts also do not contain oxonium ions, but rather the ions $[H(NO_3)_2]^-$ and $[H(NO_3)_4]^{3-}$ which have O—H$\cdots$O bonds; other H-bonded anions HX_2^- or MXY^- are known where X may be F, Cl, CO_3, RCOO, etc.[17]

The structural role of H_3O^+ in a crystal often closely resembles that of NH_4^+; thus H_3O^+ ClO_4^- and NH_4^+ ClO_4^- are isomorphous. The important difference is that compounds of H_3O^+ and other oxonium ions generally have much lower melting points than have NH_4^+ salts. The structure of the H_3O^+ ion is that of a rather flat pyramid with an HOH angle of about 115°.

The $H_5O_2^+$ ion has a short oxygen—oxygen distance, 2.42–2.57 Å. In some cases it is not certain whether the central H atom is actually centered or whether it is disordered; in $HAuCl_4 \cdot 4H_2O$, it is definitely disordered. The ion is known in cis, trans, and gauche rotational conformations. In $H_5O_2^+ClO_4^-$ and trans-[Co en$_2$ Cl$_2$]$^+$ $H_5O_2^+$ 2Cl$^-$ the ion is trans as in (5-XI), but in trans-[Co(l-pn)$_2$Cl$_2$]$^+$ $H_5O_2^+$ 2Cl$^-$ it is cis and in $H_5O_2^+$ Cl$^-$ and some other species, gauche.

(5-XI)   (5-XII)

[14] A. N. Garg and P. S. Goel, J. Inorg. Nuclear Chem., 1969, 31, 697, and references therein.

[15a] F. F. Iswasaki and M. I. Y. Saito, Acta Cryst., 1967, 23, 64; R. Telgren and I. Olovsson, J. Chem. Phys., 1971, 54, 127.

[15b] A. D. Mighell, J. P. Smith and W. E. Brown, Acta Cryst., 1969, B, 25, 776.

[16] S. W. Petersen and J. M. Williams, J. Amer. Chem. Soc., 1966, 88, 2866.

[17] D. G. Tuck, Progr. Inorg. Chem., 1968, 9, 161 (an extensive review).

TABLE 5.2

Constitution of Crystalline Acid Hydrates

H_3O^+		$H_5O_2^{+\,a}$		Other	
Formula	Species	Formula	Species	Formula	Species
$HF \cdot H_2O$	$H_3O^+F^-$	$HCl \cdot 2H_2O^e$	$H_5O_2^+, Cl^-$	$HBr \cdot 4H_2O^c$	$H_7O_3^+, H_9O_4^+$
$HCl \cdot H_2O$	$H_3O^+Cl^-$	$HCl \cdot 3H_2O^e$	$H_5O_2^+, H_2O\ Cl^-$		$2Br^-, H_2O^h$
$HClO_4 \cdot H_2O^f$	$H_3O^+ClO_4^-$	$HClO_4 \cdot 2H_2O^b$	$H_5O_2^+, ClO_4^-$		
$HNO_3 \cdot H_2O^g$	$H_3O^+NO_3^-$	$HAuCl_4 \cdot 4H_2O^d$	$H_5O_2^+, AuCl_4^-, 2H_2O$		
$H_2SO_4 \cdot H_2O^g$	$H_3O^+HSO_4^-$				
$H_2SO_4 \cdot 2H_2O$	$2H_3O^+SO_4^{2-}$				
$H_2PtCl_6 \cdot 2H_2O$	$H_3O^+PtCl_6^{2-}$				

[a] For references to $H_5O_2^+$ and $H_9O_4^+$ see J. M. Williams, Special Publ. No. 301, Nat. Bureau Stand., Washington, D.C. The $H_5O_2^+$ ion is also found in a number of acid adducts of complex salts such as trans-[Co en₂Cl₂]⁺[H₅O₂]⁺2Cl⁻; cf. H. E, Le May, Jr., *Inorg. Chem.*, 1968, **7**, 2531.

[b] A. C. Pavia and P. A. Giguère, *J. Chem. Phys.*, 1970, **52**, 3551.

[c] J. O. Lundgren and I. Olovsson, *J. Chem. Phys.*, 1968, **49**, 1068; *Acta Cryst.*, 1970, *B*, **26**, 1893.

[d] J. M. Williams and S. W. Petersen, *J. Amer. Chem. Soc.*, 1969, **91**, 776.

[e] J. O. Lundgren and I. Olovsson, *Acta Cryst.*, 1967, **23**, 966, 971; A. S. Gilbert and N. Sheppard, *Chem. Comm.*, 1971, 337.

[f] D. E. O'Reilly, E. M. Peterson, and J. M. Williams, *J. Chem. Phys.*, 1971, **54**, 96.

[g] These liquids are only slightly ionized and, if frozen quickly, give glasses containing few oxonium ions.

[h] The $H_7O_3^+$ ion also occurs in bipyridinium salts (J. E. Derry and T. A. Hamor, *Chem. Comm.*, 1970, 1284).

There is good evidence that oxonium ions other than H_3O^+ also can exist in solution. Thus the presence of species such as $H_9O_4^+$ (5-XII) has been used to explain many properties of aqueous acid solutions such as the extraction of metal ions into organic solvents. An example is the extraction of the ion $AuCl_4^-$ from hydrochloric acid solutions into benzene containing tributyl phosphate.

STRENGTHS OF PROTONIC ACIDS[18]

One of the most important characteristics of hydrogen compounds, HX, is the extent to which they ionize in water or other solvents, i.e., they act as acids. The strength of an acid depends not only on the nature of the acid itself but very much on the medium in which it is dissolved. Thus CF_3COOH and $HClO_4$ are strong acids in water whereas in 100% H_2SO_4 the former is non-acidic and the latter only a very weak acid. Similarly, H_3PO_4 is a base in 100% H_2SO_4. Although acidity can be measured in a wide variety of solvents, the most important is water for which the pH scale has been discussed above.

5-9. Binary Acids

Although the intrinsic strength of H—X bonds is one factor, other factors are involved, as will be seen by considering the appropriate thermodynamic cycles for a solvent system, as discussed below. The intrinsic strength of H—X bonds and the thermal stability of covalent hydrides seem to depend on the electronegativities and size of the element X. The variation in bond strength in some binary hydrides is shown in Fig. 5-9. There is a fairly smooth *decrease* in bond strength with *increasing Z* in a Periodic Group and a general *increase across* any Period. Thermal stability is only a crude guide to bond strength, but it is useful where precise bond energies are unknown. Thermal stability, in the sense of resistance to the reaction

$$H_nX = \frac{n}{2}H_2 + \frac{1}{x}X_x$$

invariably decreases with increasing Z. Generally, for two elements of about equal electronegativity, the *heavier* element forms the less stable hydride. Thus we have the stability orders $CH_4 > H_2S$ and $PH_3 > TeH_2$.

If we now consider dissolving HX in water, we have dissociation according to the equation:

$$HX(aq) = H^+(aq) + X^-(aq)$$

The dissociation constant, K, is related to the change in Gibbs free energy by the relation

$$\Delta G^0 = -RT \ln K \tag{5-2}$$

[18] D. D. Perrin, *Dissociation of Inorganic Acids and Bases in Aqueous Solution*, Butterworths, 1970; E. J. King, *Acid–Base Equilibria*, Vol. 4, Topic 15, *Internat. Encyclop. Phys. Chem.*, Pergamon, 1965.

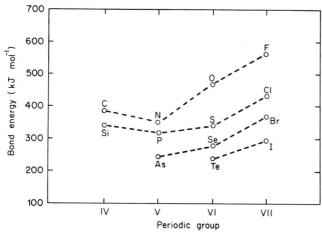

Fig. 5-9. Variation in mean H—X bond energies.

and the free energy change is in turn related to the changes in enthalpy and entropy via the relation

$$\Delta G = \Delta H - T\Delta S \tag{5-3}$$

in which R is the gas constant and T is the absolute temperature, which we shall take to be 298° in the following discussion. The dissociation process may be considered as the sum of several other reactions (that is, as one step in a thermodynamic cycle). Table 5-3 summarizes the Gibbs free energy changes

TABLE 5-3

Free Energy Changes (kJ mol^{-1}) for Dissociation of HX Molecules in Water at 298°

Process	HF	HCl	HBr	HI
$HX(aq) = HX(g)$	23.9	−4.2	−4.2	−4.2
$HX(g) = H(g) + X(g)$	535.1	404.5	339.1	272.2
$H(g) = H^+(g) + e$	1,320.2	1,320.2	1,320.2	1,320.2
$X(g) + e = X^-(g)$	−347.5	−366.8	−345.4	−315.3
$H^+(g) + X^-(g) = H^+(aq) + X^-(aq)$	−1,513.6	−1,393.4	−1,363.7	−1,330.2
$HX(aq) = H^+(aq) + X^-(aq)$	18.1	−39.7	−54.0	−57.3
$pK_a (= \Delta G°/5.71)$	3.2	−7.0	−9.5	−10

for these several steps. It can be seen that HF is out of line with the other three HX acids principally in two respects. (1) It has an exceptionally high free energy of bond breaking, and (2) the F$^-$ ion has an exceptionally high hydration energy. The first of these disfavors acid dissociation while the second favors it. However, the first factor is larger and predominates.

It is further to be noted that the entropies of bond breaking are nearly the same for all the HX molecules (88.8 J mol^{-1} deg^{-1} for HI, increasing smoothly to 99.7 for HF), so that one can say, on the basis of equation 5-3, that the main cause of the weakness of HF as an acid in aqueous solution is the strength of the HF bond. It is also worth noting that relatively small changes in enthalpies or free energies make large changes in equilibrium constants, as shown by equation 5-2. A change of 5.71 kJ mol^{-1} at 298° in the bond energy would change K by a factor of 10. Thus a decrease of only about 25 kJ mol^{-1} in the H—F bond energy would make K_a about 10 and HF, therefore, a strong acid.

The foregoing discussion of the hydrogen halides exemplifies the principles necessary in understanding the strengths of binary acids generally.

5-10. Oxo Acids

The second main class of acid behavior is shown by compounds with X—OH groups; these are called oxo acids, and generally have a formula of the type H_nXO_m, for example H_3PO_4.

For oxo acids certain useful generalizations may be made concerning (a) the magnitude of K_1 and (b) the ratios of successive constants, K_1/K_2, K_2/K_3, etc. The value of K_1 seems to depend upon the charge on the central atom. Qualitatively it is very reasonable to suppose that the greater the positive charge, the more will the process of proton loss be favored on electrostatic grounds. It has been found that if this positive charge is taken to be the so-called formal charge semiquantitative correlations are possible. The formal charge in an oxo acid, H_nXO_m, is computed in the following way, assuming the structure of the acid to be $O_{m-n}X(OH)_n$. Each X—(OH) bond is formed by sharing one X electron and one OH electron and is thus *formally* non-polar. Each X—O bond is formed by using two X electrons and thus represents a

TABLE 5-4

Strengths of Oxo Acids, H_nXO_m, in Water

$(m-n)$	Examples	$-\log K_1$ (pK_1)	$-\log K_2$ (pK_2)	$-\log K_3$ (pK_3)
0	HClO	7.50	—	—
	HBrO	8.68	—	—
	H_3AsO_3	9.22	?	?
	H_4GeO_4	8.59	13	?
	H_6TeO_6	8.80	?	?
	[H_3PO_3	1.8	6.15	—]
	H_3BO_3	9.22	?	?
1	H_3PO_4	2.12	7.2	12
	H_3AsO_4	3.5	7.2	12.5
	H_5IO_6	3.29	6.7	~15
	H_2SO_3	1.90	7.25	—
	H_2SeO_3	2.57	6.60	—
	$HClO_2$	1.94	—	—
	HNO_2	3.3	—	—
	[H_2CO_3	6.38 (3.58)	10.32	—]
2	HNO_3	Large neg. value	—	—
	H_2SO_4	Large neg. value	1.92	—
	H_2SeO_4	Large neg. value	2.05	—
3	$HClO_4$	Very large neg. value	—	—
	$HMnO_4$	Very large neg. value	—	—
[-1(?)]	H_3PO_2	2	?	?]

net loss of one electron by X. Therefore, the formal positive charge on X is equal to the number of X—O bonds, hence equal to $(m-n)$. It may be seen from the data in Table 5-4 that, with the exception of the acids listed in brackets, which are special cases to be discussed presently, the following relations between $(m-n)$ (or formal positive charge on X) and the values of K_1 hold:

For $m-n=0$,
$$pK_1 \sim 8.5 \pm 1.0 \qquad (K \sim 10^{-8} \text{ to } 10^{-9})$$

For $m-n=1$,
$$pK_1 \sim 2.8 \pm 0.9 \qquad (K \sim 10^{-2} \text{ to } 10^{-4})$$

For $m-n \gtrsim 2$,
$$pK_1 \ll 0 \text{ (the acid is very strong)}$$

It will also be noted that the difference between successive pK's is 4–5 with very few exceptions.

H_3PO_3 obviously is out of line with the other acids having $m-n=0$ and seems to fit fairly well in the group with $m-n=1$. This is, in fact, where it belongs, since there is independent evidence (Section 13-14) that its structure is OPH*(OH)$_2$ with H* bonded directly to P. Similarly, H_3PO_2 has a pK_1 that would class it with the $m-n=1$ acids where it, too, belongs since its structure is OP(H*)$_2$(OH), with the two H* hydrogen atoms directly bound to P.

Carbonic acid is exceptional in that the directly measured pK_1, 6.38, does not refer to the process

$$H_2CO_3 = H^+ + HCO_3^-$$

since carbon dioxide in solution is only partly in the form of H_2CO_3, but largely present as more loosely hydrated species, CO_2(aq). When a correction is made for the equilibrium

$$CO_2(aq) + H_2O = H_2CO_3(aq)$$

the pK_1 value of 3.58 is obtained which falls in the range for other $m-n=1$ acids (see also Section 10-4).

We may note finally that many metal ions whose solutions are acidic may be regarded as oxo acids. Thus, although the hydrolysis of metal ions is often written as shown here for Fe^{3+}:

$$Fe^{3+} + H_2O = Fe(OH)^{2+} + H^+$$

it is just as valid thermodynamically and much nearer to physical reality to recognize that the ferric ion is coordinated by water molecules and to write:

$$[Fe(H_2O)_6]^{3+} = [Fe(H_2O)_5(OH)]^{2+} + H^+ \qquad K_{Fe^{3+}} \approx 10^{-3}$$

From this formulation it becomes clear why the ferrous ion, with a lower positive charge, is less acidic or, in alternative terms, less hydrolyzed than the ferric ion:

$$[Fe(H_2O)_6]^{2+} = [Fe(H_2O)_5(OH)]^+ + H^+ \qquad K_{Fe^{2+}} \ll K_{Fe^{3+}}$$

It should be noted that one cannot necessarily compare the acidity of the bivalent ion of one metal with that of the trivalent ion of *another* metal in this way, however. There appears to be no good general rule concerning the acidities of hydrated metal ions at the present time, although some attempts have been made at correlations.

5-11. General Theory of Ratios of Successive Constants

It was shown many years ago by Niels Bjerrum that the ratios of successive acid dissociation constants could be accounted for in a nearly quantitative way by electrostatic considerations. Consider any bifunctional acid HXH:

$$HXH = HX^- + H^+ \qquad K_1$$
$$HX^- = X^{2-} + H^+ \qquad K_2$$

There is a purely statistical effect which can be considered in the following way. For the first process, dissociation can occur in two ways (i.e. there are two protons, either of which may dissociate), but recombination in only one; whereas in the second process, dissociation can occur in only one way, but recombination in two (i.e. the proton has two sites to which it may return and hence twice the probability of recombining). Thus, on purely statistical grounds one would expect $K_1 = 4K_2$. Bjerrum observed that for the dicarboxylic acids, $HOOC(CH_2)_nCOOH$, the ratio K_1/K_2 was always greater than four, but decreased rapidly as n increased (see Table 5-5). He suggested the following explanation. When the two points of attachment of protons are

TABLE 5-5

K_1/K_2 Ratio for Dicarboxylic Acids,
$HOOC(CH_2)_nCOOH$

n	1	2	3	4	5	6	7	8
K_1/K_2	1120	29.5	17.4	12.3	11.2	10.0	9.5	9.3

close together in the molecule, the negative charge left at one site when the first proton leaves strongly restrains the second one from leaving by electrostatic attraction. As the separation between the sites increases, this interaction should diminish.

By making calculations using the Coulomb law,* Bjerrum was able to obtain rough agreement with experimental data. The principal difficulty in obtaining quantitative agreement lies in a choice of dielectric constant since some of the lines of electrostatic force run through the molecule ($D \sim 1$–10), others through neighboring water molecules (D uncertain), and still others through water having the dielectric constant (~ 82) of pure bulk water. More recently, Kirkwood and Westheimer were able to get nearly quantitative agreement with the data by making very elaborate calculations which take into account the variability of the dielectric constant. The important point here for our purposes is to recognize the physical principles involved without necessarily trying to obtain quantitative results.

Thus, the large separations in successive pK's for the oxo acids are attributable to the electrostatic effects of the negative charge left by the dissociation of one proton upon the remaining ones. In bifunctional binary acids, where the negative charge due to the removal of one proton is concentrated on the

---

* $F \propto q_1q_2/Dr$, where F is the force; q_1 and q_2 the charges separated by r; and D the dielectric constant of the medium between them.

very atom to which the second proton is bound, the separation of the constants is extraordinarily great. K_1 and K_2 for H_2S are $\sim 10^{-7}$ and $\sim 10^{-14}$, respectively, whereas for water we have

$$H_2O = H^+ + OH^- \qquad K_1 = 10^{-14}$$
$$OH^- = H^+ + O^{2-} \qquad K_2 < 10^{-36} \text{ (est.)}$$

5-12. Pure Acids and Relative Acidities[19]

The concepts of hydrogen ion concentration and pH discussed above are meaningful only for dilute aqueous solutions of acids. In water-like solvents such as methanol, similar concepts may be developed, but in most organic solvents, in concentrated aqueous solutions and in the anhydrous state the concept is meaningless and some other scale of acidity is required. There are a number of different acidity scales but the one most commonly used is the Hammett acidity function, which allows comparison of the same acid in different media as well as intercomparisons of acids. The function H_0 pertains to the equilibrium between a base, B, its conjugate acid, BH^+, and the proton, H^+:

$$B + H^+ \rightleftharpoons BH^+$$

and is defined as

$$H_0 = pK_{BH^+} - \log \frac{[BH^+]}{[B]}$$

In very dilute solutions

$$K_{BH^+} = \frac{[B][H^+]}{[BH^+]}$$

so that in water, H_0 becomes synonymous with pH. By using suitable organic bases, e.g., p-nitroaniline, and suitable indicators over various ranges of concentration and acidities or by nmr methods, it is possible to interrelate values of H_0 for strong acids extending from dilute solutions to the pure acid. Other acidity scales utilize organic amides or hydrocarbons but a more useful scale for diverse compounds of low basicity depends on heats of protonation in strong acids such as FSO_3H or H_2SO_4.[20]

For a number of strong acids in aqueous solution up to concentrations about $8M$, the values of H_0 are very similar. This suggests that the acidity is independent of the anion. The rise in acidity with increasing concentration can be fairly well predicted by assuming that the hydrogen ion is present as $H_9O_4^+$, so that protonation can be represented as

$$H_9O_4^+ + B \rightleftharpoons BH^+ + 4H_2O$$

Values of H_0 for some pure liquid acids are given in Table 5-6. It is to be noted particularly that for HF the acidity can be very substantially increased by the addition of a Lewis acid or fluoride ion acceptor, for example:

$$2HF + SbF_5 \rightleftharpoons H_2F^+ + SbF_6^-$$

[19] R. E. Bates in J. J. Lagowski, ed., *The Chemistry of Non-Aqueous Solvents*, Vol. 1, Academic Press, 1966; E. A. Arnett, *Progr. Phys. Org. Chem.*, 1963, **1**, 223.
[20] E. M. Arnett, R. P. Quirk and J. J. Burke, *J. Amer. Chem. Soc.*, 1970, **92**, 1260; G. C. Levy, J. D. Cargioli and W. Racela, *J. Amer. Chem. Soc.*, 1970, **92**, 6238.

TABLE 5-6

The Hammett Acidity Function, H_0, for Several Acids

Acid	H_0	Acid	H_0
$HSO_3F + SbF_5 + SO_3$	> 16	HF	10.2
$HF + SbF_5$ (3M)	15.2	$HF + NaF$ (1M)	8.4
HSO_3F	12.6	H_3PO_4	5.0
$H_2S_2O_7$	12.2	H_2SO_4 (63% in H_2O)	4.9
H_2SO_4	11.0	HCOOH	2.2

but its acidity is decreased by addition of NaF owing to formation of the HF_2^- ion. Antimony pentafluoride (page 376) is commonly used as the Lewis acid since it is comparatively easy to handle, being a liquid, and is commercially available. However, other fluorides such as BF_3, NbF_5, and TaF_5 behave in a similar way.

The enhancement of the acidity of fluorosulfuric acid HSO_3F by SbF_5 gives one of the strongest of known acids, but here, though SbF_5 is acting as a Lewis base, it is not a fluoride ion acceptor. The SbF_5—FSO_3H system, which is very complicated, has been thoroughly investigated by nuclear magnetic resonance and Raman spectra.[21,22] The acidity is due to the formation of the $H_2SO_3F^+$ ion. The equilibria depend on the ratios of the components; with low ratios of SbF_5 to FSO_3H the main ones are the following:

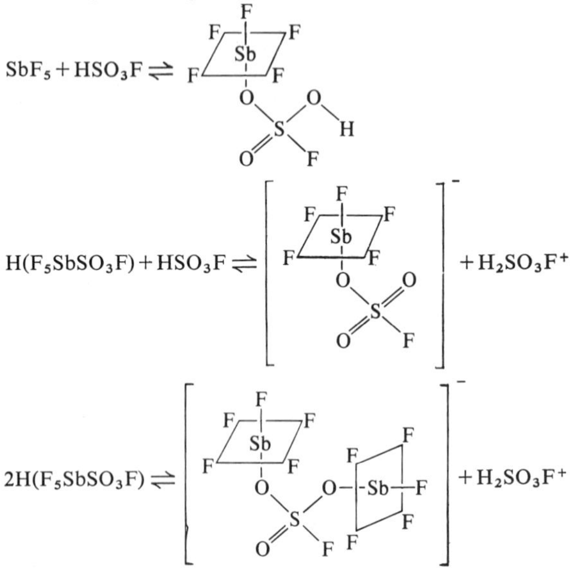

At higher ratios the solutions appear to contain also the ions SbF_6^- and $[F_5Sb-F-SbF_5]^-$, which occur in solutions of SbF_5 in liquid HF, together

[21] A. Commeyras and G. A. Olah, J. Amer. Chem. Soc., 1969, 91, 2929.
[22] R. J. Gillespie, K. Ouchi and G. P. Pez, Inorg. Chem., 1969, 8, 63; R. J. Gillespie and G. P. Pez, Inorg. Chem., 1969, 8, 1233.

with HS_2O_6F and HS_3O_9F, which occur in SO_3—FSO_3H solutions. These species are generated by the additional reactions

$$HSO_3F \rightleftharpoons HF + SO_3$$
$$3SbF_5 + 2HF \rightleftharpoons HSbF_6 + HSb_2F_{11}$$
$$2HSO_3F + 3SO_3 \rightleftharpoons HS_2O_6F + HS_3O_9F$$

The SbF_5—HSO_3F solutions are very viscous and are normally diluted with liquid sulfur dioxide so that better resolution of nuclear magnetic resonance spectra is obtained. Although the equilibria appear not to be appreciably altered for molecular ratios $SbF_5 : HSO_3F < 0.4$, in more concentrated SbF_5 solutions the additional equilibria noted above are shifted to the left by removal of SbF_5 as the stable complex $SbF_5 \cdot SO_2$, which can be obtained crystalline (see Section 15-11). This results in a lowering of the acidity of the system; on the contrary, if SO_3 is added the acidity is increased[22] by raising the concentration of $H_2SO_3F^+$ and the strongest known acid is SbF_5—HSO_3F $\cdot nSO_3$ ($n \gtrsim 3$).

There has been extensive study of very strong acids especially FSO_3H—SbF_5—SO_2, HF—SbF_5, and HCl—Al_2Cl_6 for the protonation of weak bases.[23] It can generally be stated that virtually all organic compounds can be protonated and the resultant species characterized by nuclear magnetic resonance. Thus formic acid at $-60°$ gives equal amounts of the protonated species (5-XIII) and (5-XIV) and protonated formaldehyde[24] (5-XV), while fluorobenzene gives the ion (5-XVI).

(5-XIII)   (5-XIV)   (5-XV)   (5-XVI)

The superacid media can induce hydride abstraction, H–D exchange and other reactions even with saturated hydrocarbons.[25] Carbonium ions are formed, some of which, notably the trimethylcarbonium ion, are quite stable:

$$Me_3CH \rightarrow Me_3C^+ \leftarrow CH_3CH_2CH_2CH_3$$

It is postulated that the attack by H^+ occurs on the electron density of the C—H and C—C single bonds and not on the C and H atoms themselves. The order of reactivity, qualitatively, is: tertiary $CH > C$—$C >$ secondary $CH \gg$ primary CH. Even molecular H_2 may be protonated since H_2–D_2 exchange is observed in the superacids, probably through a planar H_3^+ transition state. The reactions possibly involve "pentacoordinate" carbonium ions, e.g.,

$$R_3CH + H^+ \rightleftharpoons \left[R_3C \begin{smallmatrix} \cdots H \\ \vdots \\ \cdots H \end{smallmatrix} \right]^+ \rightleftharpoons CR_3^+ + H_2$$

[23] R. J. Gillespie, *Accounts Chem. Res.*, 1968, **1**, 202; G. A. Olah *et al.*, *Chem. Rev.*, 1970, **70**, 561, and papers in *J. Amer. Chem. Soc.*
[24] A. M. White and G. A. Olah, *J. Amer. Chem. Soc.*, 1969, **91**, 2943.
[25] G. A. Olah *et al.*, *J. Amer. Chem. Soc.*, 1971, **93**, 1251, 1256, and references therein.

where the two hydrogen atoms are bound to carbon by closed three-center bonds.

The carbonium ions can undergo complex reactions. Thus methane can give carbonium ions with C—C bonds by condensation reactions of the type:

$$CH_4 \underset{}{\overset{H^+}{\rightleftharpoons}} H_2 + CH_3^+ \xrightarrow{CH_4} C_2H_7^+ \rightleftharpoons H_2 + C_2H_5^+, \text{ etc.}$$

CH_3^+ ions have been detected by trapping[26] with CO and subsequent hydrolysis of the acylium ion with water to give acetic acid, e.g.,

$$CH_3^+ + CO \rightarrow CH_3CO^+ \xrightarrow{H_2O} CH_3COOH + H^+$$

It may also be noted the aqueous acid solutions of K_2PtCl_4 also homogeneously catalyze H–D exchange in alkanes with relative rates opposite to the ones above, namely primary CH > secondary CH > tertiary CH.[27] The mechanism of exchange is not clear but it may involve an oxidative addition of C—H to Pt^{II} (Chapter 24).

The use of $HCl-Al_2Cl_6$ or $HF-SbF_5$ to isomerize straight-chain to branched-chain alkanes or vice versa has potential industrial value, and indeed such acidic media are already important in many organic reactions of hydrocarbons such as isomerization, acetylation, alkylation, etc.[28]

Other acids that have been used for protonation studies are fluoroboric acid in propionic anhydride (to remove excess of water) and also liquid hydrogen chloride, although the latter is inconvenient to use.

Considerably more study has been given to protonation of organic compounds, but a number of inorganic compounds have also been studied. Thus many metal carbonyl and organometallic complexes may be protonated[29] on the metal or on the ligand (Section 24-A-2), e.g.:

$$Fe(CO)_5 + H^+ \rightleftharpoons HFe(CO)_5^+$$
$$(h^5\text{-}C_5H_5)_2Fe + H^+ \rightleftharpoons (h^5\text{-}C_5H_5)_2FeH^+$$
$$C_8H_8Fe(CO)_3 + H^+ \rightleftharpoons C_8H_9Fe(CO)_3^+$$

Even protonated carbonic acid, or more properly, the trihydroxycarbonium ion, $C(OH)_3^+$ [cf. (5-XIII, XIV, XV)] has been observed[30] in solutions of carbonates or bicarbonates in $FSO_3H-SbF_5-SO_2$ solutions at $-78°$; the ion is stable to $0°$ in absence of SO_2. It was suggested that $C(OH)_3^+$ might be involved even in biological systems at very acid sites in enzymes such as carbonic anhydrase.

5-13. Properties of Some Common Strong Acids

In Table 5-7 are collected some properties of the more common and useful strong acids in their pure states.

[26] H. Hogeveen et al., Rec. Trav. Chim., 1969, **88**, 703, 719; Chem. Comm., **1969**, 921.
[27] R. J. Hodges, D. E. Webster and P. B. Wells, J. Chem. Soc., **1971**, 3230.
[28] See G. A. Olah, ed., Friedel–Crafts and Related Reactions, Vols. 1 and 2, Interscience-Wiley, 1963; D. M. Brouwer, Rec. Trav. Chim., 1968, **87**, 1435.
[29] M. A. Haas, Organometallic Chem. Rev., A, 1969, **4**, 307; J. J. Kotz and D. G. Pedrotty, Organometallic Chem. Rev., A, 1969, **4**, 479.
[30] G. A. Olah and A. M. White, J. Amer. Chem. Soc., 1968, **90**, 1884.

TABLE 5-7

Properties of Some Strong Acids in the Pure State

Acid	M.p. (°C)	B.p. (°C)	κ^a	ε^b
HF	−83.36	19.51	1.6×10^{-6} (0°)	84 (0°)
HCl	−114.25	−85.09	3.5×10^{-9} (−85°)	14.3 (−114°)
HBr	−86.92	−66.78	1.4×10^{-10} (−84°)	7.33 (−86°)
HI	−50.85	−35.41	8.5×10^{-10} (−45°)	3.57 (−45°)
HNO₃	−41.59	82.6	3.72×10^{-2} (25°)	
HClO₄	−112	(109° extrap).		
HSO₃F	−88.98	162.7	1.085×10^{-4} (25°)	∼120 (25°)
H₂SO₄	10.3771	∼270 d^c	1.044×10^{-2} (25°)	110 (20°)

^{a} Specific conductance in ohm^{-1} cm^{-1}. Values are often very sensitive to impurities.
^{b} Dielectric constant.
^{c} Constant-boiling mixture (338°) contains 98.33% of H₂SO₄, d = with decomposition.

Hydrogen Fluoride.[31]   The acid HF is made by the action of concentrated H_2SO_4 on CaF_2 and is the principal source of fluorine compounds (Chapter 16). It is commercially available in steel cylinders, with purity approximately 99.5%; it can be purified further by distillation. Although liquid HF attacks glass rapidly it can be handled conveniently in apparatus constructed either of copper or Monel metal or of materials such as polytetrafluoroethylene (Teflon or PTFE), Kel-F (a chlorofluoro polymer), etc.

The high dielectric constant is characteristic of hydrogen-bonded liquids. Since HF forms only a two-dimensional polymer it is less viscous than water. In the vapor, HF is monomeric above 80°, but at lower temperatures the physical properties are best accounted for by an equilibrium between HF and a hexamer, $(HF)_6$, which has a puckered ring structure.[32] Crystalline $(HF)_n$ has zigzag chains (Fig. 5-5).

After water, liquid HF is one of the most generally useful of solvents. Indeed in some respects it surpasses water as a solvent for both inorganic and organic compounds, which often give conducting solutions as noted above; it can also be used for cryoscopic measurements.[33a]

The self-ionization equilibria in liquid HF are:

$$2HF \rightleftharpoons H_2F^+ + F^- \qquad K \sim 10^{-10}$$
$$F^- + HF \rightleftharpoons HF_2^- \xrightarrow{HF} H_2F_3^-, \text{ etc.}$$

The formation of the stable hydrogen-bonded anions accounts in part for the extreme acidity. In the liquid acid the fluoride ion is the conjugate base, and ionic fluorides behave as bases. Fluorides of M^+ and M^{2+} are often appreciably soluble in HF, and some such as TlF are very soluble.

The only substances that function as "acids" in liquid HF are those such

[31] (a) M. Kilpatrick and J. G. Jones in *The Chemistry of Non-Aqueous Solvents*, J. J. Lagowski, ed., Vol. 2, Academic Press, 1967.
   (b) H. H. Hyman and J. J. Katz in *Non-Aqueous Solvent Systems*, T. C. Waddington, ed., Academic Press, 1965.
[32] J. Janzen and L. S. Bartell, *J. Chem. Phys.*, 1969, **50**, 3611.
[33a] R. J. Gillespie and D. A. Humphreys, *J. Chem. Soc., A*, **1970**, 2311.

as SbF_5 noted above, which increase the concentration of H_2F^+. The latter ion appears to have an abnormally high mobility in such solutions.

Reactions in liquid HF are known that illustrate also amphoteric behavior, solvolysis, or complex formation. Although HF is water-like, it is not easy, because of the reactivity, to establish an emf series, but a partial one is known.[33b]

In addition to its utility as a solvent system, HF as either liquid or gas is a useful fluorinating agent, converting many oxides and other halides into fluorides.

In *aqueous solution*, HF differs from the other halogen acids in that it is a weak acid[34] in dilute solution (page 169) where the equilibria are:

$$F^- + H^+ \rightleftharpoons HF \qquad \log K_1 = 3.16$$
$$F^- + HF \rightleftharpoons HF_2^- \qquad \log K_2 = 0.7$$

In $5\text{--}15M$ aqueous solution the acidity increases owing to ionization to H_3O^+, HF_2^-, and more complex $(H_nF_{n+1})^-$ species.

Hydrogen Chloride, Bromide and Iodide.[35] These three hydrogen halides are very similar to each other and differ notably from hydrogen fluoride. They are normally pungent gases; in the solid state they have hydrogen-bonded zigzag chains and there is probably some hydrogen-bonding in the liquid. Hydrogen chloride is made by the action of concentrated H_2SO_4 on concentrated aqueous HCl or NaCl; HBr and HI may be made by catalytic reaction of $H_2 + X_2$ over platinized silica gel or, for HI, by interaction of iodine and boiling tetrahydronaphthalene. The gases are soluble in a variety of solvents, especially polar ones. The solubility in water is not exceptional;[36] in moles of HX per mole of solvent at $0°$ and 1 atm the solubilities in water, 1-octanol and benzene, respectively, are: HCl 0.409, 0.48, 0.39; HBr 1.00, 1.30, 1.39; HI 0.065, 0.173, 0.42.

The self-ionization is very small:

$$3HX \rightleftharpoons H_2X^+ + HX_2^-$$

Liquid HCl has been fairly extensively studied as a solvent, and many organic and some inorganic compounds dissolve giving conducting solutions:

$$B + 2HCl \rightleftharpoons BH^+ + HCl_2^-$$

The low temperatures required and the short liquid range are limitations, but conductimetric titrations are readily made.

Salts of the ion H_2Cl^+ have not been isolated, but salts of the HX_2^- ions, $[X-H-X]^-$, are known, especially for HCl_2^-. The HCl_2^- and HBr_2^- ions, as they occur in the compounds $3CsX \cdot H_3O^+ \cdot HX_2^-$ have $X-H-X$ distances of 3.14 and 3.35 Å for $X = Cl$ and Br, respectively.[37] These distances, like

[33b] A. F. Clifford, W. D. Pardieck and M. W. Wadley, *J. Phys. Chem.*, 1966, **70**, 3241.
[34] E. W. Bauman, *J. Inorg. Nuclear Chem.*, 1969, **31**, 3155.
[35] F. Klanberg in *The Chemistry of Non-Aqueous Solvents*, J. J. Lagowski, ed., Vol. 2, Academic Press, 1967; M. E. Peach and T. C. Waddington in *Non-Aqueous Solvent Systems*, T. C. Waddington, ed., Academic Press, 1965; and papers by T. C. Waddington, mainly in *J. Chem. Soc.*
[36] W. Gerrard, *Chem. and Ind. (London)*, **1969**, 295.
[37] L. W. Schroeder and J. A. Ibers, *Inorg. Chem.*, 1968, **7**, 594.

that in HF_2^- (2.26 Å), are ~ 0.5 Å shorter than the sum of van der Waals radii and suggest that there are strong hydrogen-bonds.

Nitric Acid.[38] Nitric acid is made industrially by oxidation of ammonia with air over platinum catalysts.[39] The resulting nitric oxide (Section 12-6) is absorbed in water in the presence of air to form NO_2, which is then hydrated. The normal concentrated aqueous acid (approximately 70% by weight) is colorless but often becomes yellow as a result of photochemical decomposition, which gives NO_2:

$$2HNO_3 \xrightarrow{h\nu} 2NO_2 + H_2O + \tfrac{1}{2}O_2$$

The so-called red "fuming" nitric acid contains dissolved NO_2 in excess of the amount which can be hydrated to $HNO_3 + NO$.

Pure nitric acid can be obtained by treating KNO_3 with 100% H_2SO_4 at 0° and removing the HNO_3 by vacuum-distillation. The pure acid is a colorless liquid or white crystalline solid; the latter decomposes above its melting point according to the equation given above for the photochemical decomposition and hence must be stored below 0°.

The pure acid has the highest self-ionization of the pure liquid acids. The initial protolysis

$$2HNO_3 \rightleftharpoons H_2NO_3^+ + NO_3^-$$

is followed by rapid loss of water:

$$H_2NO_3^+ = H_2O + NO_2^+$$

so that the overall self-dissociation is

$$2HNO_3 \rightleftharpoons NO_2^+ + NO_3^- + H_2O$$

Pure nitric acid is a good ionizing solvent for electrolytes but, unless they produce the NO_2^+ or NO_3^- ions (Section 12-7), salts are sparingly soluble.

In dilute aqueous solution, nitric acid is approximately 93% dissociated at $0.1M$ concentration. Nitric acid of concentration below $2M$ has little oxidizing power. The concentrated acid is a powerful oxidizing agent and, of the metals, only Au, Pt, Rh and Ir are unattacked, although a few others such as Al, Fe, and Cu are rendered "passive," probably owing to formation of an oxide film; magnesium alone can liberate hydrogen and then only initially from dilute acid. The attack on metals generally involves reduction of nitrate. Aqua regia (approximately 3 vols. of conc. HCl + 1 vol. of conc. HNO_3) contains free chlorine and ClNO, and it attacks Au and Pt metals, its action being more effective than that of HNO_3 mainly because of the complexing function of chloride ion. A similar effect is shown by the fact that some metals, notably tantalum, are quite resistant to HNO_3 but dissolve with extreme vigor if HF is added, to give TaF_6^- or similar ions. Non-metals are usually oxidized by HNO_3 to oxo acids or oxides. The ability of nitric acid, especially in the presence of concentrated sulfuric acid, to nitrate many

[38] W. H. Lee in *The Chemistry of Non-Aqueous Solvents*, J. J. Lagowski, ed., Vol. 2, Academic Press, 1967; S. A. Stern, J. T. Mullhaupt and W. B. Kay, *Chem. Rev.*, 1960, **60**, 195 (an exhaustive review of the physical properties).
[39] H. Connor, *Platinum Metals Rev.*, 1967, **11**, 2.

organic compounds is attributable to the formation of the nitronium ion, NO_2^+ (see discussion in Section 12-7).

Gaseous nitric acid has a planar structure (Fig. 5–10) although hindered rotation of OH relative to NO_2 probably occurs.

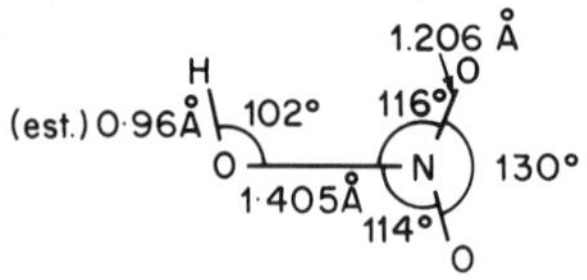

Fig. 5-10.   The structure of nitric acid in the vapor.

Perchloric Acid.[40]   Perchloric acid, $HClO_4$, is commercially available in concentrations 70–72% by weight. The water azeotrope with 72.5% of $HClO_4$ boils at 203° and although some chlorine is produced, which can be swept out by air, there is no hazard involved. The anhydrous acid is best prepared by vacuum-distillation of the concentrated acid in presence of the dehydrating agent $Mg(ClO_4)_2$; it reacts explosively with organic material. The pure acid is stable at room temperature for only 3–4 days, decomposing to give $HClO_4 \cdot H_2O$ (84.6% acid) and Cl_2O_7.

The most important applications of aqueous perchloric acid involve its use as an oxidant. However, at concentrations below 50% and temperatures not exceeding 50–60°, there is no release of oxygen. The hot concentrated acid oxidises organic materials vigorously or even explosively; it is a useful reagent for the destruction of organic matter,[41] especially after pre-treatment with, or in the presence of, sulfuric or nitric acid. The addition of concentrated $HClO_4$ to organic solvents such as ethanol should be avoided where possible, even if the solutions are chilled.

Sulfuric Acid.[42]   Sulfuric acid is prepared on an enormous scale by the lead chamber and contact processes.[43] In the former, SO_2 oxidation is catalyzed by oxides of nitrogen (by intermediate formation of nitrosylsulfuric acid, $HOSO_2ONO$); in the latter, heterogeneous catalysts such as platinum are used for the oxidation. Pure sulfuric acid, H_2SO_4, is a colorless liquid which is obtained from the commercial 98% acid by addition first of sulfur trioxide or oleum and then titration with water until the correct specific conductance or melting point is achieved.

The phase diagram of the H_2SO_4–H_2O system is complicated, and eutectic hydrates such as $H_2SO_4 \cdot H_2O$ (m.p. 8.5°) and $H_2SO_4 \cdot 2H_2O$ (m.p. −38°) occur.

In pure crystalline H_2SO_4 there are SO_4 tetrahedra with S–O distances 1.42, 1.43, 1.52 and 1.55 Å, linked by strong hydrogen bonds. There is also extensive hydrogen bonding in the concentrated acid.

[40] G. S. Pearson, *Adv. Inorg. Chem. Radiochem.*, 1966, **8**, 177 (an exhaustive review).
[41] G. F. Smith, *Talanta*, 1968, **15**, 489.
[42] R. J. Gillespie and E. A. Robinson, in *Non-Aqueous Solvents*, T. C. Waddington, ed., Academic Press, 1965; W. M. Lee in *The Chemistry of Non-Aqueous Solvents*, J. J. Lagowski, ed., Vol. 2, Academic Press, 1967.
[43] T. J. Pearce in *Inorganic Sulphur Chemistry*, G. Nickless, ed., Elsevier, 1968.

Pure H_2SO_4 shows extensive self-ionization resulting in high conductivity. The equilibrium

$$2H_2SO_4 \rightleftharpoons H_3SO_4^+ + HSO_4^- \qquad K_{10°} = 1.7 \times 10^{-4} \text{ mol}^2 \text{ kg}^{-2}$$

is only one factor, since there are additional equilibria due to dehydration:

$$2H_2SO_4 \rightleftharpoons H_3O^+ + HS_2O_7^- \qquad K_{10°} = 3.5 \times 10^{-5} \text{ mol}^2 \text{ kg}^{-2}$$
$$H_2O + H_2SO_4 \rightleftharpoons H_3O^+ + HSO_4^- \qquad K_{10°} = 1 \text{ mol kg}^{-1}$$
$$H_2S_2O_7 + H_2SO_4 \rightleftharpoons H_3SO_4^+ + HS_2O_7^- \qquad K_{10°} = 7 \times 10^{-2} \text{ mol kg}^{-1}$$

Estimates of the concentrations in 100% H_2SO_4 of the other species present, namely, H_3O^+, HSO_4^-, $H_3SO_4^+$, $HS_2O_7^-$, and $H_2S_2O_7$ can be made; for example, at 25°, HSO_4^- is 0.023 molar.[44]

Pure H_2SO_4 and dilute oleums have been much studied as solvent systems,[45] but interpretation of the cryoscopic and other data is often complicated. Sulfuric acid is not a very strong oxidizing agent although the 98% acid has some oxidizing ability when hot. The concentrated acid reacts with many organic materials, removing the elements of water and sometimes causing charring, for example, of carbohydrates. Many substances dissolve in the 100% acid, often undergoing protonation. Alkali-metal sulfates and water also act as bases. Organic compounds may also undergo further dehydration reactions, for example:

$$C_2H_5OH \xrightarrow{H_2SO_4} C_2H_5OH_2^+ + HSO_4^- \xrightarrow{H_2SO_4} C_2H_5HSO_4 + H_3O^+ + HSO_4^-$$

Because of the strength of H_2SO_4, salts of other acids may undergo solvolysis, for example:

$$NH_4ClO_4 + H_2SO_4 \rightleftharpoons NH_4^+ + HSO_4^- + HClO_4$$

There are also examples of acid behavior. Thus H_3BO_3, which behaves initially as a base, gives quite a strong acid:

$$H_3BO_3 + 6H_2SO_4 \rightleftharpoons B(HSO_4)_3 + 3H_3O^+ + 3HSO_4^-$$
$$B(HSO_4)_3 + HSO_4^- \rightleftharpoons B(HSO_4)_4^-$$

The addition of SO_3 to H_2SO_4 gives what is known as *oleum* or fuming sulfuric acid $(SO_3)_n \cdot H_2O$; the constitution of concentrated oleums is controversial, but with equimolar ratios the major constituent is pyrosulfuric (disulfuric) acid, $H_2S_2O_7$. At higher concentrations of SO_3, Raman spectra indicate the formation of $H_2S_3O_{10}$ and $H_2S_4O_{13}$. Pyrosulfuric acid[46] has higher acidity than H_2SO_4 and ionizes thus:

$$2H_2S_2O_7 \rightleftharpoons H_2S_3O_{10} + H_2SO_4 \rightleftharpoons H_3SO_4^+ + HS_3O_{10}^-$$

The acid protonates many materials; $HClO_4$ behaves as a weak base, and CF_3COOH is a non-electrolyte in oleum.

Fluorosulfuric Acid.[47] Fluorosulfuric acid is made by the reaction:

$$SO_3 + HF = FSO_3H$$

or by treating KHF_2 or CaF_2 with oleum at $\sim 250°$. When freed from HF by

[44] P. A. H. Wyatt, *Trans. Faraday Soc.*, 1969, **65**, 585.
[45] R. J. Gillespie in *Inorganic Sulphur Chemistry*, G. Nickless, ed., Elsevier, 1968; A. Vincent and R. F. M. White, *J. Chem. Soc., A*, 1970, **2179**.
[46] R. J. Gillespie and K. C. Malhotra, *J. Chem. Soc., A*, **1968**, 1933.
[47] R. J. Gillespie, *Accounts Chem. Res.*, 1968, **1**, 202; R. C. Thompson in *Inorganic Sulphur Chemistry*, G. Nickless, ed., Elsevier, 1968.

sweeping with an inert gas, it can be distilled in glass apparatus. Unlike $ClSO_3H$, which is explosively hydrolyzed by water, FSO_3H is relatively slowly hydrolyzed.

Fluorosulfuric is one of the strongest of pure liquid acids. It is commonly used in presence of SbF_5 as a protonating system, as noted above (page 174). An advantage over other acids is its ease of removal by distillation in vacuum. The self-ionization

$$2FSO_3H \rightleftharpoons FSO_3H_2^+ + FSO_3^-$$

is much lower than for H_2SO_4 and consequently interpretation of cryoscopic and conductometric measurements is fairly straightforward.

In addition to its solvent properties, FSO_3H is a convenient laboratory fluorinating agent. It reacts readily with oxides and salts of oxo acids at room temperature. For example, K_2CrO_4 and $KClO_4$ give CrO_2F_2 and ClO_3F, respectively.

Another similar acid, (trifluoromethyl)sulfuric acid,[48] CF_3SO_3H (b.p. 162°), is also a very useful strong acid, comparable in strength to $HClO_4$. It is very hygroscopic and forms a hydrate, $CF_3SO_3H \cdot H_2O$ (m.p. 34°). Its main advantage lies in the fact that its salts are similar to perchlorates (Section 16-13), but they are non-explosive, an important feature where the cation contains much organic material.

BINARY METALLIC HYDRIDES

A rough attempt to classify the various types of hydrides is shown in Fig. 5-11. In this Section we deal with binary metallic hydrides.[49]

```
                                                                        | He
  H                                            | B   C   N   O   F  | Ne
  Li  Be                                       | Al  Si  P   S   Cl | Ar
  Na  Mg
  K   Ca | Sc    Ti* V*  Cr* Mn* Fe* Co* Ni*| Cu Zn | Ga  Ge  As  Se  Br | Kr
  Rb  Sr | Y     Zr  Nb  Mo* Tc* Ru* Rh* Pd*| Ag Cd | In  Sn  Sb  Te  I  | Xe
  Cs  Ba | La–Lu Hf  Ta* W*  Re* Os* Ir* Pt*| Au Hg Tl | Pb  Bi  Po  At | Rn
  Fr  Ra | Ac              U, Pu
  Saline hydrides|   Transition metal hydrides    | Borderline | Covalent hydrides
                                                  |  hydrides  |
```

Fig. 5-11. A classification of the hydrides. The starred elements are the transition elements for which complex molecules or ions containing M—H bonds are known.

5-14. The Hydride Ion, H^-; Saline Hydrides

The formation of the unipositive ion H^+ (or H_3O^+, etc.) suggests that hydrogen should be classed with the alkali metals in the Periodic Table. On the other hand, the formation of the hydride ion might suggest an analogy with the halogens. Such attempts at classification of hydrogen with other

[48] Minnesota Mining and Manufacturing Co. Technical Information Bulletin.
[49] G. G. Libowitz, The Solid State Chemistry of Binary Metal Hydrides, Benjamin, 1965; K. M. Mackay, Hydrogen Compounds of the Metallic Elements, Spon, 1966; W. M. Mueller, J. P. Blackledge and G. G. Libowitz, eds., Metal Hydrides, Academic Press, 1969.

elements can be misleading. Hydrogen has a very low electron affinity, and the tendency to form the negative ion is much lower than for the more electronegative halogen elements. This may be seen by comparing the energetics of the formation reactions:

$$\frac{1}{2}H_2(g) \rightarrow H(g) \quad \Delta H = 218 \text{ kJ mol}^{-1} \qquad \frac{1}{2}Br_2(g) \rightarrow Br(g) \quad \Delta H = 113 \text{ kJ mol}^{-1}$$
$$H(g)+e \rightarrow H^-(g) \quad \Delta H = -67 \text{ kJ mol}^{-1} \qquad Br(g)+e \rightarrow Br^-(g) \quad \Delta H = -327 \text{ kJ mol}^{-1}$$
$$\frac{1}{2}H_2(g)+e \rightarrow H^-(g) \quad \Delta H = +151 \text{ kJ mol}^{-1} \qquad \frac{1}{2}Br_2(g)+e \rightarrow Br^-(g) \quad \Delta H = -214 \text{ kJ mol}^{-1}$$

Thus, owing to the endothermic character of the H^- ion, only the most electropositive metals—the alkalis and alkaline earths—form saline or salt-like hydrides, such as NaH and CaH_2. The ionic nature of the compounds is shown by their high conductivities just below or at the melting point and by the fact that on electrolysis of solutions in molten alkali halides hydrogen is liberated at the *anode*.

X-ray and neutron diffraction studies show that in these hydrides the H^- ion has a crystallographic radius between those of F^- and Cl^-. Thus the electrostatic lattice energies of the hydride and the fluoride and chloride of a given metal will be similar. These facts and a consideration of the Born–Haber cycles lead us to conclude that *only* the most electropositive metals *can* form ionic hydrides, since in these cases relatively little energy is required to form the metal ion.

The known saline hydrides and some of their physical properties are given in Table 5-8. The heats of formation of the saline hydrides, compared with

TABLE 5-8

The Saline Hydrides and Some of Their Properties

Salt	Structure	Heat of formation $\Delta H_f(298°)$ (kJ mol^{-1})	M—H distance (Å)	Apparent radius of H^- (Å)[a]
LiH	NaCl type	91.0	2.04	1.36
NaH	NaCl type	56.6	2.44	1.47
KH	NaCl type	57.9	2.85	1.52
RbH	NaCl type	47.4	3.02	1.54
CsH	NaCl type	49.9	3.19	1.52
CaH$_2$	Slightly distorted *hcp*	174.5	2.33[b]	1.35
SrH$_2$	Slightly distorted *hcp*	177.5	2.50	1.36
BaH$_2$	Slightly distorted *hcp*	171.5	2.67	1.34
MgH$_2$	Rutile type	74.5	—	1.30

[a] See text.
[b] Although half the H^- ions are surrounded by four Ca^{2+} and half by three Ca^{2+} the Ca—H distances are the same.

those of the alkali halides, which are about 420 kJ mol^{-1}, reflect the inherently small stability of the hydride ion.

For the relatively simple two-electron system in the H^- ion, it is possible to calculate an effective radius for the free ion, the value 2.08 Å having been obtained. It is of interest to compare this with some other values, specifically, 0.93 Å for the He atom, ~0.5 Å for the H atom, 1.81 Å for the crystallographic radius of Cl^- and 0.30 Å for the covalent radius of hydrogen, as well as with the values of the "apparent" crystallographic radius of H^- given in Table 5-8. The values in the Table are obtained by subtracting the Goldschmidt radii of the metal ions from the experimental M—H distances. The value 2.08 Å for the radius of free H^- is at first sight surprisingly large, being more than twice that for He. This results from the facts that the H^- nuclear charge is only half that in He and that the electrons repel each other and screen each other ($\sim 30\%$) from the pull of the nucleus. It will be seen in Table 5-8 that the apparent radius of H^- in the alkali hydrides never attains the value 2.08 Å and also that it decreases markedly with decreasing electropositive character of the metal. The generally small size is probably attributable in part to the easy compressibility of the rather diffuse H^- ion and partly to a certain degree of covalence in the bonds.

Preparation and Chemical Properties. The saline hydrides are prepared by direct interaction at 300–700°. For complete reaction of lithium, the temperature must be approximately 725°. Sodium normally reacts with H_2 only above 200° and the reaction is slow owing to formation of a coating of an inert hydride. However, studies on continuously clean surfaces[50] show that the reaction obeys first-order kinetics with an activation energy of ca. 70 kJ mol^{-1}. Dispersion of sodium in mineral oil increases the reactivity, but NaH in a very reactive form can be prepared at room temperature and pressure by interaction of H_2 with sodium and naphthalene ($Na^+C_{10}H_8^-$) in tetrahydrofuran, with titanium isopropoxide as catalyst.[51]

The saline hydrides are crystalline solids, white when pure but usually grey owing to traces of metal. They can be dissolved in molten alkali halides and on electrolysis of such a solution, for example, CaH_2 in $LiCl+KCl$ at 360°, hydrogen is released at the anode.

A key to the reactivity of these hydrides lies in the formalism of regarding H—H as the exceedingly weak parent acid (an extrapolation back from HCl, strong, and HF, weak) of the MH salts. Thus H^- and its salts react instantly and completely with any substance affording even the minutest traces of H^+, such as water, according to the reaction:

$$NaH + H^+ = Na^+ + H_2$$

The standard potential of the H_2/H^- couple has been estimated to be -2.25 v, making H^- one of the most powerful reducing agents known.

The hydrides are reactive towards air and water, and those of Rb, Cs and Ba may ignite spontaneously in moist air. Thermal decomposition at high temperatures gives the metal and hydrogen; LiH alone can be melted (m.p. 688°) and it is unaffected by oxygen below red heat or by chlorine or dry HCl.

LiH is seldom used except for the preparation of the more useful complex hydride $LiAlH_4$ discussed later (page 273). However, NaH and CaH_2 are used. NaH is available as a dispersion in mineral oil; although the solid reacts

[50] R. J. Pulham, *J. Chem. Soc., A*, **1971**, 1389.
[51] E. E. van Tamelen and P. B. Fechter, *J. Amer. Chem. Soc.*, 1968, **90**, 6854; cf S. Bank and M. C. Prislopski, *Chem. Comm.*, **1970**, 1624.

violently with water, the reaction of the dispersion is less violent. It is used extensively in organic synthesis[52] and for the preparation of the dimethyl sulfoxide-derived reagent $Na^+[MeS(O)CH_2]^-$, and also for the preparation of $NaBH_4$ (see Section 8-11).

CaH_2 reacts smoothly with water and is a useful source of hydrogen (one liter per g); it is a convenient drying agent for organic solvents and for gases.[52]

5-15. Hydrides of More Covalent Nature

The hydrides BeH_2 and MgH_2 may be obtained by thermal decomposition of the alkyls, $[Me_3C]_2Be$ and Et_2Mg, respectively, but MgH_2 can be made by direct interaction of Mg under pressure or from the alloy Mg_2Cu at 1 atm and 300°, or in a reactive form, by interaction of NaH and $MgBr_2$ in diethyl ether.[53]

BeH_2 is difficult to obtain pure but it is believed to have a polymeric structure with bridged hydrogen atoms as in the boranes (Chapter 8). MgH_2 has a rutile-type structure (page 51), like MgF_2; it also has a low heat of formation and is less stable thermally than the true saline hydrides.

A more important compound is aluminum hydride. There is some evidence for the existence of gaseous AlH_3 and Al_2H_6 at low pressures, but AlH_3 is best obtained as a white powder by action of 100% H_2SO_4 on $LiAlH_4$ in tetrahydrofuran:[54]

$$2LiAlH_4 + H_2SO_4 = 2AlH_3 + 2H_2 + Li_2SO_4$$

The structure[55] and thermodynamic properties[56] have been reported for the hydride prepared by an undisclosed method. The hydride is highly unstable to thermal decomposition; $\Delta H_f^{298} = -11.45$ kJ mol^{-1}. It forms a three-dimensional lattice isostructural with AlF_3 where the Al atoms are octahedrally surrounded by hydrogen atoms and linked together by non-linear hydrogen bridges, Al—H—Al. Aluminum hydride is a useful reducing agent in organic chemistry,[54] and the products formed are significantly different from those of reduction by $LiAlH_4$. Nitriles can be reduced to amines via complex intermediates,[57] and the reduction of alkyl halides is slower than with $LiAlH_4$ so that carboxyl or ester groups in compounds RCOOH and RCOOR′ can be reduced preferentially in presence of R″X.

There are a number of complex hydrides of aluminium; they are discussed in Section 9-9 where also the Lewis acid behavior of AlH_3 is considered.

[52] See L. F. Fieser and M. Fieser, *Reagents for Organic Synthesis*, Wiley, 1967; J. Plesek and S. Hermanek, *Sodium Hydride: Its Use in the Laboratory and Technology*, Butterworths, 1968.
[53] J. J. Reilly and R. M. Wiswall, *Inorg. Chem.*, 1967, **6**, 2220; E. C. Ashby and R. D. Schwartz, *Inorg. Chem.*, 1971, **10**, 355.
[54] N. M. Yoon and H. C. Brown, *J. Amer. Chem. Soc.*, 1968, **90**, 2926.
[55] J. W. Turley and H. W. Rinn, *Inorg. Chem.*, 1969, **8**, 18.
[56] G. C. Sinke, L. C. Walker, F. L. Oetting and D. R. Stull, *J. Chem. Phys.*, 1967, **47**, 2759.
[57] R. Ehrlich and A. R. Young, *J. Inorg. Nuclear Chem.*, 1968, **30**, 53.

There is no evidence for a comparable gallium hydride, but an unstable viscous oil which shows bands in the infrared spectrum due to Ga—H has been obtained by the reaction:

$$Me_3N \cdot GaH_3(s) + BF_3(g) \xrightarrow{-15°} GaH_3(l) + Me_3NBF_3(g)$$

There is also little evidence for hydrides of In and Tl; the hydrides of the metallic elements of Groups IV and V are covalent volatile compounds and are considered under their respective elements in later Chapters.

5-16. Transition-metal Hydrides

Hydrogen reacts with many transition metals or their alloys on heating, to give compounds commonly called hydrides even though in some cases they clearly do not contain hydride ions. Many of these metal hydride systems are exceedingly complicated, showing the existence of more than one phase, often with wide divergences from stoichiometry. The most extensive studies have been made on the most electropositive elements, the lanthanides and actinides, and the Ti and V groups of the d-block elements.

For many of these hydrides there is still discussion whether properties such as magnetic susceptibility, electrical conductivity, etc., are best accounted for by a hydridic model with M^{n+} and H^- ions, a protonic model where the hydrogen electrons are lost to conduction bands in the metal, or an alloy-like model without appreciable charge separation.

Lanthanide Hydrides.   The metals such as La or Nd react with H_2 at 1 atm and at or slightly above room temperature, to give black solids of graphite-like appearance. These products are pyrophoric in air and react vigorously with water. There are the phases MH_2 and MH_3 which are non-stoichiometric, e.g., $LaH_{2.87}$. Normally Eu and Yb give only the dihydride phase, but higher ratios, e.g., $YbH_{2.55}$, can be obtained at 350° under pressure.[58]

The hydrides appear to be predominantly ionic in nature[59] and to contain M^{3+} ions even in the MH_2 phase where the odd valence electron is probably located in a metallic conduction band as in the so-called dihalides such as LaI_2 (Chapter 27); in $YbH_{2.55}$ there is some evidence for both Yb^{2+} and Yb^{3+}.

Actinide Hydrides.   Thorium and other actinides form complex systems with non-stoichiometric and stoichiometric phases. Uranium hydride is of some importance chemically as it is often more suitable for the preparation of uranium compounds than is the massive metal. Uranium reacts rapidly and exothermically with hydrogen at 250–300° to give a pyrophoric black powder. The reaction is reversible:

$$U + \tfrac{3}{2}H_2 = UH_3 \qquad \Delta H_f^0 = -129 \text{ kJ mol}^{-1}$$

The hydride decomposes at somewhat higher temperatures to give extremely reactive, finely divided metal. A study of the isostructural deuteride by X-ray

[58] J. C. Warf et al., Inorg. Chem., 1966, **6**, 1719, 1726, 1728, 1736.
[59] W. G. Bos and H. S. Gutowsky, Inorg. Chem., 1967, **7**, 552.

and neutron diffraction shows that the deuterium atoms lie in a distorted tetrahedron equidistant from four uranium atoms; no U—U bonds appear to be present and the U—D distance is 2.32 Å. The stoichiometric hydride, UH_3, can be obtained, but the stability of the product with a slight deficiency of hydrogen is greater.

Some typical useful reactions are the following:

$$UH_3 \begin{cases} \xrightarrow{H_2O,\ 350°} & UO_2 \\ \xrightarrow{Cl_2,\ 200°} & UCl_4 \\ \xrightarrow{H_2S,\ 450°} & US_2 \\ \xrightarrow{HF,\ 400°} & UF_4 \\ \xrightarrow{HCl,\ 250-300°} & UCl_3 \end{cases}$$

Hydrides of d-Block Transition Metals. Titanium, zirconium, and hafnium absorb hydrogen exothermically to give non-stoichiometric materials such as $TiH_{1.7}$ and $ZrH_{1.9}$. These and the similar hydrides of V, Nb and Ta are greyish-black solids similar in appearance and reactivity to the finely divided metal. They are fairly stable in air but react when heated with air or acid reagents. The Ti and Zr hydrides are used as reducing agents in metallurgical and other processes.

The affinity of many of the other d-block elements for hydrogen is small or zero with the exception of the following two special cases.

Palladium.[60]   One of the unique characteristics of metallic palladium and Pd–Ag or Pd–Au alloys is the high rate of diffusion of hydrogen gas through a metal membrane compared to the rates for other metals such as nickel or iridium. There is no doubt that pressure–temperature–composition curves indicate the presence of palladium hydride phases.

Copper.   There has been much discussion on whether a true hydride exists or not. It appears that an insoluble CuH with a wurtzite structure (page 51) can be obtained by reduction of Cu^{2+} solutions by hypophosphorous acid. An amorphous hydride soluble in organic solvents such as pyridine or alkylphosphines can be obtained from the reaction of CuI and $LiAlH_4$ in pyridine.[61] A bright red crystalline phosphine complex, $[CuH(PPh_3)]_6$, has been isolated,[62] and its structure is shown in Fig. 25-H-2. Although the hydride ligand cannot be detected spectroscopically or by X-ray study, it reacts with C_6H_5COOD to give HD and H_2.

[60] F. A. Lewis, *The Palladium–Hydrogen System*, Academic Press, 1967 (a very detailed review.
[61] J. A. Dilts and D. F. Shriver, *J. Amer. Chem. Soc.*, 1968, **90**, 5769; 1969, **91**, 4088.
[62] S. A. Bezman *et al.*, *J. Amer. Chem. Soc.*, 1971, **93**, 2063.

Further Reading

Abel, H. F., *Die Acidität der C–H Säuren*, G. Thiele Verlag, 1969.
Augustine, R. L., *Catalytic Hydrogenation*, Arnold–Dekker, London, 1965.
Bell, R. P., *Acids and Bases; Their Quantitative Behaviour*, 2nd edn, Methuen, 1969.
Emmett, P. H., ed., *Catalysis*, Vols. 3–5, Reinhold, New York, 1955–57 (these volumse cover various aspects of hydrogenation and other reactions involving hydrogen).

Evans, E. A., *Tritium and its Compounds*, Butterworths, London, 1966.
Frankenberg, W. G., V. I. Komarewsky and E. K. Rideal, eds., *Advances in Catalysis*, Academic Press, annually from 1948 (these volumes discuss various aspects of hydrogenation and other reactions involving hydrogen).
Freifelder, M., *Practical Catalytic Hydrogenation*; *Technique and Applications*, Wiley, 1971.
Gibb, T. R. P., *Progr. Inorg. Chem.*, 1962, **3**, 315 (an extensive discussion of the nature of metallic hydrides).
Gillespie, R. J., in G. A. Olah, ed., *Friedel–Crafts and Related Reactions*, Vol. 1, Interscience–Wiley, 1963 (discussion of protonic acids).
Hadzi, D., ed., *Hydrogen Bonding*, Pergamon Press, 1959.
Hagen, Sister M., *Clathrate Inclusion Compounds*, Reinhold, 1962.
Hinton, J. F., and E. S. Amis, *Chem. Rev.*, 1967, **7**, 367 (ionization equilibria in acids as determined by nmr methods).
Janz, G. J. and S. S. Danyluk, *Chem. Rev.*, 1960, **60**, 209 (conductivities of hydrogen halides in anhydrous polar organic solvents).
Jones, J. R., *Quart. Rev.*, 1971, **25**, 365 (audites of carbon anals).
Katz, J. J., in *Chemical and Biological Studies with Deuterium*, Pennsylvania State Univ., 1965 (physical properties and isotope effects including biological effects).
Krindel, P., and I. Eliezer, *Coord. Chem. Rev.*, 1971, **6**, 217 (water structure models).
Liler, M., *Reaction Mechanisms in Sulphuric Acid and Other Strong Acid Solutions*, Academic Press, 1971.
Mandelcorn, L., ed., *Non-Stoichiometric Compounds*, Academic Press, 1964 (clathrates and inclusion compounds).
Mikheeva, V. I., *Hydrides of the Transition Elements*, U.S. Atomic Energy Commission, A. E. C.-tr-5224, 1962. Office of Technical Service, Dept. of Commerce, Washington, D.C. (a review with 678 references).
Murphy, G. M., ed., *Production of Heavy Water*, National Nuclear Energy Series, Vol. VIII-4F, McGraw-Hill, 1955.
Pourbaix, M., *Atlas of Electrochemical Equilibria in Aqueous Solution*, Pergamon Press, 1966.
Rochester, C. H., *Acidity Functions*, Academic Press, 1970.
Rylander, P. N., *Catalytic Hydrogenation over Platinum Metals*, Academic Press, 1967.
Satchell, D. P. N., and R. S. Satchell, *Quart. Rev.*, 1971, **25**, 171 (quantitative aspects of Lewis acidity).
Siegel, B., *J. Chem. Educ.*, 1961, **38**, 484 (a review of the reactions of atomic hydrogen).
Simon, H., and D. Palm, *Angew. Chem. Internat. Edn.*, 1966, **5**, 920 (isotope effects).
Sokolskii, D. V., *Hydrogenation in Solutions*, Oldbourne Press, 1965 (comprehensive Russian source book).
Tanabe, K., *Solid Acids and Bases*; *Their Catalytic Properties*, Academic Press, 1971 (properties of oxides, sulfates, clay minerals, etc).
Westheimer, F. H., *Chem. Rev.*, 1961, **61**, 265 (hydrogen and deuterium isotope effects).
Wiberg, E., and A. Amberger, *Hydrides of Elements of Main Groups I–IV*, Elsevier, 1971.

6

The Group I Elements: Li, Na, K, Rb, Cs

GENERAL REMARKS

6-1. Introduction

The closely related elements Li, Na, K, Rb and Cs, often termed the alkali metals, have a single s electron outside a noble-gas core. Some relevant data are listed in Table 6-1.

TABLE 6-1

Some Properties of Group I Metals

Element	Electronic configuration	Metal radius (Å)	Ionization potentials, eV		M.p. (°C)	B.p. (°C)	$E^{\circ a}$ (V)	E_{diss}^{b} (kJ mol^{-1})
			1st	2nd				
Li	[He]2s	1.52	5.390	75.62	180.5	1326	−3.02	108.0
Na	[Ne]3s	1.86	5.138	47.29	97.8	883	−2.71	73.3
K	[Ar]4s	2.27	4.339	31.81	63.7	756	−2.92	49.9
Rb	[Kr]5s	2.48	4.176	27.36	38.98	688	−2.99	47.3
Cs	[Xe]6s	2.65	3.893	23.4	28.59	690	−3.02	43.6
Fr	[Rn]7s							

[a] For $M^+(aq) + e = M(s)$.
[b] Energy of dissociation of the diatomic molecule M_2.

The low ionization potentials for the outer electrons and the fact that the resulting M^+ ions are spherical and of low polarizability have as a result that the chemistry of these elements is essentially that of their +1 ions. No other oxidation state is known, nor is any to be expected in view of the magnitudes of the second ionization potentials.

Although the chemistry of the elements is predominantly ionic, some degree of covalent bonding occurs in certain cases. The gaseous diatomic molecules—Na_2, Cs_2 etc.—are covalently bonded, and the bonds to oxygen, nitrogen and carbon in various chelate and organometallic compounds doubtless have some slight covalent character. The tendency to covalence

189

is greatest with lithium and least with cesium, as would be expected from the charge:radius ratios.

The element *francium* is formed in the natural radioactive decay series and in nuclear reactions. All its isotopes are radioactive with short half-lives. Precipitation reactions and solubility and ion-exchange studies have shown that the ion behaves as would be expected from its position in the group.

Of all the groups in the Periodic Table, the Group I metals show most clearly and with least complication the effect of increasing size and mass on chemical and physical properties. Thus all of the following *decrease* through the series: (a) melting points and heats of sublimation of the metals; (b) lattice energies of all salts except those with the very smallest anions (because of irregular radius ratio effects); (c) the effective hydrated radii and the hydration energies (see Table 6-2); (d) the ease of thermal decomposition of nitrates and carbonates; (e) strength of the covalent bonds in the M_2 molecules; (f) heats of formation of fluorides, hydrides, oxides and carbides (because of higher lattice energies with the smaller cations). Other trends also can readily be found.

Lithium has some chemical behavior that resembles the chemistry of Mg. Anomalous properties of Li result mainly from the small size of the atom and the ion; the polarization power of Li^+ is the greatest of all the alkali metal ions and leads to a singularly great tendency toward solvation and covalent bond formation. There is also evidence to suggest that "lithium bonds" comparable with hydrogen bonds exist in, e.g., H—F$\cdots$Li—F and $(LiF)_2$.[1]

The reactivity of the Group I metals toward all chemical reagents except nitrogen increases with increasing electropositive nature (Li $\rightarrow$ Cs). Li is usually the least reactive. Li is only rather slowly attacked by water at 25°, whereas Na reacts vigorously, K inflames, and Rb and Cs react explosively. With liquid Br_2, Li and Na barely react, whereas the others do so violently. Lithium does not replace the weakly acidic hydrogen in $C_6H_5C\equiv CH$, whereas the other alkali metals do so, yielding hydrogen gas. However, with N_2, Li is uniquely reactive to give a ruby-red crystalline nitride, Li_3N (Mg also reacts to give Mg_3N_2); at 25° this reaction is slow, but it is quite rapid at 400° and has been studied in detail.[2] Both Li and Mg can be used to remove nitrogen from other gases. When heated with carbon, both Li and Na react to form the *acetylides* Li_2C_2 and Na_2C_2. The heavier alkali metals also react with carbon, but give non-stoichiometric interstitial compounds where the metal atoms enter between the planes of carbon atoms in the lamellar graphite structure. This difference may be attributed to size requirements for the metal, both in the ionic acetylides ($M_2^+ C_2^{2-}$) and in the penetration of the graphite.

A particularly fundamental chemical difference between lithium and its congeners, attributable to cation size, is the reaction with oxygen. When the

[1] P. A. Kollman, J. F. Liebman and L. C. Allen, *J. Amer. Chem. Soc.*, 1970, **92**, 1142.
[2] C. C. Addison and B. M. Davies, *J. Chem. Soc.*, A, **1969**, 1822, 1827, 1831.

metals are burnt in air or oxygen at 1 atm, lithium forms the oxide, Li_2O, with only a trace of Li_2O_2, whereas the other alkali oxides, M_2O, react further giving as principal products the peroxides, M_2O_2, and (from K, Rb, and Cs) the superoxides, MO_2.

Although the molten metals Na to Cs are miscible in all proportions, Li is immiscible with K, Rb or Cs and miscible with Na only above 380°.[3]

The Li^+ ion is exceptionally small and has, therefore, an exceptionally high charge:radius ratio, comparable to that of Mg^{2+}. The properties of a number of *lithium compounds* are therefore anomalous (in relation to the other Group I elements) but resemble those of magnesium compounds. Many of the anomalous properties arise from the fact that the salts of Li^+ with small anions are exceptionally stable owing to their very high lattice energies, while salts with large anions are relatively unstable owing to poor packing of very large with very small ions. LiH is stable to approximately 900° while NaH decomposes at 350°. Li_3N is stable whereas Na_3N does not exist at 25°. Lithium hydroxide decomposes at red heat to Li_2O, whereas the other hydroxides MOH sublime unchanged; LiOH is also considerably less soluble than the other hydroxides. The carbonate, Li_2CO_3, is thermally much less stable relative to Li_2O and CO_2 than are other alkali-metal carbonates M_2CO_3. The solubilities of Li^+ salts resemble those of Mg^{2+}. Thus LiF is sparingly soluble (0.27 g/100 g H_2O at 18°) and can be precipitated from ammoniacal NH_4F solutions; LiCl, LiBr, LiI and, especially, $LiClO_4$ are soluble in solvents such as ethanol, acetone and ethyl acetate, and LiCl is soluble in pyridine.

$NaClO_4$ is less soluble than $LiClO_4$ in various solvents by factors of 3–12, whereas $KClO_4$, $RbClO_4$ and $CsClO_4$ have solubilities only 10^{-3} of that of $LiClO_4$. Since the spherical ClO_4^- ion is virtually non-polarizable and the alkali-metal perchlorates form ionic crystals, the high solubility of $LiClO_4$ is mainly attributable to strong solvation of the Li^+ ion. LiBr in hot concentrated solution has the unusual property of dissolving cellulose. Lithium sulfate, in contrast to the other M_2SO_4 salts, does not form alums; it is also not isomorphous with the other sulfates.

The elements copper, silver and gold, the so-called coinage metals, are sometimes treated with the alkalis. The only justification for this procedure is that the atom of each of these elements has a single *s* electron outside a closed shell. In the coinage metals, however, the closed shell is a *d* shell of the penultimate principal level. Although these elements do have +1 oxidation states, their over-all chemical resemblance to the alkalis is very slight. They are best considered as close relatives of the transition metals, which they resemble in much of their chemistry, such as formation of complexes, variable oxidation state, etc.

It is pertinent that there are other ions that have chemical behavior closely resembling that of the Group I ions:

[3] J. R. Christman, *Phys. Rev.*, 1967, **159**, 108.

1. The most important of these are the ammonium ions, NH_4^+, RNH_3^+, ...R_4N^+. Salts of NH_4^+ generally resemble those of potassium quite closely in their solubilities and crystal structures.

2. The thallium(I) ion, Tl^+, behaves in certain respects as an alkali-metal ion (although in others more like Ag^+). Its ionic radius (1.54 Å) is comparable to that of Rb^+, although it is more polarizable. Thus thallous hydroxide is a water-soluble, strong base, which absorbs carbon dioxide from the air to form the carbonate. The sulfate and some other salts are isomorphous with the alkali-metal salts.

3. A variety of other types of monopositive, essentially spherical cations often behave like alkali-metal ions of comparable size. For example, the very stable di-(h^5-cyclopentadienyl)cobalt(III) ion and its analogs with similar "sandwich" structures (Chapter 23) have precipitation reactions similar to those of Cs^+, and $[(h^5\text{-}C_5H_5)_2Co]OH$ is a strong base which absorbs carbon dioxide from the air and forms insoluble salts with large anions.

THE ELEMENTS

6-2. Preparation and Properties

Sodium and potassium are in high abundance (2.6% and 2.4%) in the lithosphere and occur in large deposits of sodium chloride and carnallite, $KCl \cdot MgCl_2 \cdot 6H_2O$. Lithium, rubidium and cesium have much lower abundances and occur mainly in a few silicate minerals.

Lithium and sodium are obtained by electrolysis of fused salts or of low-melting eutectics such as $CaCl_2 + NaCl$. Potassium cannot readily be made by electrolysis, owing to its low melting point and ready vaporization, and it is made by treating molten KCl with Na vapor in a counter-current fractionating tower. Rb and Cs are made similarly. All the metals are best purified by distillation. Because there is only one valence electron per metal atom, the binding energies in the close-packed metal lattices are relatively weak and the metals are consequently very soft and have low melting points. Liquid alloys of the metals are known, the most important being the Na–K alloys. The eutectic mixture in this system contains 77.2% of K and melts at $-12.3°$. This alloy, which has a wide liquid range and high specific heat, has been considered as a coolant for nuclear reactors, but sodium is used for this purpose. Lithium is relatively light (density 0.53 g/cm³) and has the highest melting and boiling point and also the longest liquid range of all the alkali metals; it has also an extraordinarily high specific heat. These properties should make it an excellent coolant in heat exchangers, but it is also very corrosive—more so than other liquid metals—which is a great practical disadvantage; it is used to deoxidize, desulfurize and generally degas copper and copper alloys.

Sodium metal may be dispersed by melting on various supporting solids such as sodium carbonate, kieselguhr, etc., or by high-speed stirring of a suspension of the metal in various hydrocarbon solvents held just above the melting point of the metal. Dispersions of the latter type are commercially available; they may be poured in air, and they react with water only with effervescence. They are often used synthetically where sodium shot or lumps would react too slowly. Sodium and potassium, when dispersed on supports such as carbon or K_2CO_3, are used as catalysts for various reactions of alkenes, notably the dimerization of propene to 4-methyl-1-pentene[4] (cf. the use of Li alkyls discussed below).

Studies of the spectra of Group I metal vapors at about the boiling points of the metals show the presence of ~1% of diatomic molecules whose dissociation energies decrease with increasing atomic number (Table 6.1). These molecules provide the most unambiguous cases of covalent bonding of the alkalis; some s–p hybridization is considered to be involved.

All the metals are highly electropositive and react directly with most other elements. The reactions of substances in liquid sodium have been studied in some detail in view of the use of the metal as a reactor coolant. As noted above, the reactivities toward air and water increase down the Group. In air Li, Na, and K tarnish rapidly, and the other metals must be handled in an inert atmosphere, as must Na–K alloys. Although Li, Na, K and Rb are silvery in appearance, Cs has a distinct golden-yellow cast. The metals dissolve in mercury with considerable vigor, to give amalgams. Sodium amalgam (Na/Hg) is a liquid when containing little sodium but a solid when rich in sodium; it is a useful reducing agent and can be employed with aqueous solutions.

The metals also dissolve with reaction in alcohols, to give the alkoxides. Sodium or potassium in ethanol or *tert*-butyl alcohol is commonly used in organic chemistry as reducing agent and also provides a source of the nucleophilic alkoxide ions.

6-3. Solutions of Metals in Liquid Ammonia and Other Solvents[5]

The Group I metals, and to a lesser extent Ca, Sr, Ba, Eu and Yb, are soluble in liquid ammonia and certain other solvents, giving solutions that are blue when dilute. These solutions conduct electricity *electrolytically* and measurements of transport numbers suggest that the main current carrier, which has an extraordinarily high mobility, is the solvated electron. Solvated electrons are now known to be formed in aqueous[6] or other polar media by photolysis, radiolysis with ionizing radiations such as X-rays, electrolysis and probably some chemical reactions. The high reactivity of the electron

[4] J. K. Hambling, *Chem. in Britain*, 1969, **5**, 354.
[5] J. C. Thompson in *The Chemistry of Non-Aqueous Solvents*, J. J. Lagowski, ed., Vol. II, Academic Press, 1967; R. F. Gould, ed., *Adv. Chem. Series No. 50*, American Chemical Society, Washington, D.C., 1965; U. Schindewolf, *Angew. Chem. Internat. Edn.*, 1968, **7**, 190.
[6] D. C. Walker, *Quart. Rev.*, 1967, **21**, 75 (hydrated electron, but the discussion includes other solvents); M. Anbar, *Quart. Rev.*, 1968, **22**, 578 (reactions of hydrated electrons with inorganic compounds).

and its short lifetime[7] (in $0.75M$ $HClO_4$, 6×10^{-11} sec; in neutral water, $t_{1/2}$ approximately 10^{-4} sec) make detection of such low concentrations difficult. Electrons can also be trapped in ionic lattices or in frozen water or alcohol when irradiated and again blue colors are observed.

In very pure liquid ammonia the lifetime of the solvated electron may be quite long (less than 1% decomposition per day), but under ordinary conditions initial rapid decomposition occurs with water present, and with glass[8] this is followed by a slower decomposition.

Solutions in ammonia and other solvents have been extensively studied and it is agreed that in *dilute solutions* the metal is dissociated into solvated metal ions M^+ and electrons. The broad absorption around 15,000 Å accounts for the common blue color; since the metal ions are colorless this absorption must be associated with the solvated electrons. Magnetic and electron spin resonance studies show the presence of "free" electrons, but the decrease in paramagnetism with increasing concentration suggests that the ammoniated electrons can associate to form diamagnetic species containing electron pairs. Although there may be other equilibria, the data can be accommodated by equilibria such as

$$Na(s) \text{ (dispersed)} \rightleftharpoons Na \text{ (in solution)} \rightleftharpoons Na^+ + e$$
$$2e \rightleftharpoons e_2$$

Just how the electrons are associated with the ammonia molecules or the solvated metal ions is still a matter of discussion. However, the most satisfactory models assume that the electron is not localized but is "smeared out" over a large volume so that the surrounding solvent molecules experience electronic and orientational polarization. The electron is trapped in the resultant polarization field, and repulsion between the electron and the electrons of the solvent molecules leads to the formation of a cavity within which the electron has the highest probability of being found. In ammonia this is estimated to be approximately 3–3.4 Å in diameter; this cavity concept is based on the fact that solutions are of much lower density than the pure solvent, i.e. they occupy far greater volume than that expected from the sum of the volumes of metal and solvent.

There is evidence for formation of metal ion clusters as the concentration of metal increases. At concentrations $3M$ or above, the solutions are copper-colored and have a metallic luster, and in various physical properties such as their exceedingly high electrical conductivities they resemble liquid metals. When 20% solutions of Li in liquid ammonia are cooled, a golden-yellow conducting solid, $Li(NH_3)_4$, is obtained.[9] Similar behavior is also shown by the Group II elements, which also give solids, usually non-stoichiometric but approximating to $M(NH_3)_6$.

The metals are also soluble to varying degrees in other amines, and Na

[7] M. J. Bronskill, R. K. Wolff and J. W. Hunt, *J. Phys. Chem.*, 1969, **73**, 1173; U. Schindewolf, H. Kohrmann and G. Lang, *Angew. Chem. Internat. Edn.*, 1969, **8**, 512.
[8] D. C. Jackman and C. W. Keenan, *J. Inorg. Nuclear Chem.*, 1968, **30**, 2047.
[9] N. Mammano and M. J. Sienko, *J. Amer. Chem. Soc.*, 1968, **90**, 6322.

and K are soluble in hexamethylphosphoramide, $P(NMe_2)_3$. Fairly stable solutions of K, Rb and Cs have been obtained in tetrahydrofuran, ethylene glycol dimethyl ether, and even in diethyl ether containing cyclic polyethers that form complexes with the alkali-metal ions (page 200).[9a] The general properties of these solutions, insofar as they have been determined in view of the attack on the solvents, appear to be similar to those of the amine and liquid ammonia solutions; the alkali-metal concentrations in saturated ether solutions are, however, only $\sim 10^{-4}$ mol l^{-1}.

The ammonia and amine solutions of alkali metals are widely used preparatively in both organic and inorganic chemistry. Thus lithium in methylamine shows great selectivity in its reducing properties, but both this reagent and lithium in ethylenediamine are quite powerful and can reduce aromatic rings to cyclic monoolefins. Sodium in liquid ammonia is probably the most widely used system for preparative purposes. The ammonia solution is moderately stable, but the decomposition reaction

$$Na + NH_3(l) = NaNH_2 + \tfrac{1}{2}H_2$$

can occur photochemically and is catalyzed by transition-metal salts. Sodium amide can be conveniently prepared by treatment of sodium with liquid ammonia in the presence of a trace of ferric chloride. Amines react similarly:

$$Li(s) + CH_3NH_2(l) \xrightarrow{50-60°} LiNHCH_3(s) + \tfrac{1}{2}H_2$$

and the lithium dialkylamides especially have been used extensively as preparative reagents for making compounds with M—NR$_2$ bonds.

For the amides of K, Rb and Cs it has been shown[10] that the reaction

$$e^- + NH_3 \rightleftarrows NH_2^- + \tfrac{1}{2}H_2 \qquad K = 5 \times 10^4$$

is reversible, but for LiNH$_2$ and NaNH$_2$ which are insoluble in liquid ammonia we have, for example

$$Na^+(am) + e^-(am) + NH_3(l) = NaNH_2(s) + \tfrac{1}{2}H_2 \qquad K = 3 \times 10^9$$

The physical and chemical properties required of the solvent to make possible the formation of such metal solutions are not fully understood. The dielectric constant of the solvent is important in the same way as in the solution of an ionic solid, namely, to diminish the forces of attraction between the oppositely charged particles—in this case, M$^+$ ions and electrons. Furthermore, if the solvent molecules immediately surrounding these particles interact strongly with them, the energy of the system is further lowered. While the detailed nature of the interaction of the electrons with the surrounding solvent molecules is still debatable, it is fairly clear that the metal ions are solvated in the same way as they would be in a solution of a metal salt in the same solvent (see discussion below).

[9a] J. L. Dye, M. G. De Backer and V. A. Nicely, *J. Amer. Chem. Soc.*, 1970, **92**, 5227.
[10] E. R. Kirschke and W. L. Jolly, *Inorg. Chem.*, 1967, **6**, 855; U. Schindewolf, R. Vogelsgesang and K. W. Böddeker, *Angew. Chem. Internat. Edn.*, 1967, **6**, 1076.

COMPOUNDS OF THE GROUP I ELEMENTS

6-4. Binary Compounds

The Group I metals react directly with most non-metals to give one or more binary compounds; they also form numerous alloys and compounds with other metals such as Pb and Sn. Many of these compounds are described under the appropriate element.

The most important are the *oxides*, obtained by combustion as noted in Section 6-1. Although sodium normally gives Na_2O_2 it will take up further oxygen at elevated pressures and temperatures to form NaO_2. The per- and super-oxides of the heavier alkalis can also be prepared by passing stoichiometric amounts of oxygen into their solutions in liquid ammonia, and ozonides, MO_3, are also known. The structures of the ions O_2^{2-}, O_2^- and O_3^- and of their alkali salts are discussed in Section 14-7. The increasing stability of the per- and super-oxides as the size of the alkali ions increases is noteworthy and is a typical example of the stabilization of larger anions by larger cations through lattice-energy effects.

Owing to the highly electropositive character of the metals, the various oxides (and also sulfides and similar compounds) are readily hydrolyzed by water according to the following equations:

$$M_2O + H_2O = 2M^+ + 2OH^-$$
$$M_2O_2 + 2H_2O = 2M^+ + 2OH^- + H_2O_2$$
$$2MO_2 + 2H_2O = O_2 + 2M^+ + 2OH^- + H_2O_2$$

The oxide Cs_2O has the anti-$CdCl_2$ structure and is the only known oxide with this type of lattice. An abnormally long Cs—Cs distance and short Cs—O distance imply considerable polarization of the Cs^+ ion.

Rb and Cs both form non-stoichiometric suboxides that are metallic in nature.

The *hydroxides*, MOH, are white crystalline solids soluble in water and in alcohols. They can be sublimed unchanged at 350–400°, and the vapors consist mainly of dimers, $(MOH)_2$. KOH at ordinary temperatures is monoclinic with each K surrounded by a distorted octahedron of O atoms while the OH groups form a zigzag hydrogen-bonded chain with $O—H\cdots O = 3.35 Å$. The breaking of these bonds results in the formation of the cubic high-temperature form.

Measurements of the proton affinities[11] of MOH in the gas phase show that the base strength increases from Li to Cs, but this order need not be observed in aqueous or alcoholic solutions where the base strength of the hydroxide is reduced by solvent effects and hydrogen bonding. In suspension in non-hydroxylic solvents such as 1,2-dimethoxyethane, the hydroxides are exceedingly strong bases and can conveniently be used[12] to deprotonate a wide variety of weak bases such as PH_3 ($pK \approx 27$) or C_5H_6 ($pK \approx 16$). The

[11] S. K. Searle, I. Džidić and P. Kebarle, *J. Amer. Chem. Soc.*, 1969, **91**, 2810.
[12] W. L. Jolly, *Inorg. Chem.*, 1967, **6**, 1435; *Inorg. Synth.*, 1968, **11**, 113.

driving force for the reaction is provided by the formation of the stable hydrate:

$$2KOH(s) + HA = K^+A^- + KOH \cdot H_2O(s)$$

6-5. Ionic Salts

Salts of the bases, MOH, with virtually all acids are known. For the most part they are colorless, crystalline, ionic solids. Those that are colored owe this to the anions, except in special cases. The colors of metal ions are due to absorption of light of proper energy to excite electrons to higher energy levels; for the alkali-metal ions with their very stable noble-gas configurations, the energies required to excite electrons to the lowest available empty orbitals could be supplied only by quanta far out in the vacuum ultraviolet (the transition $5p^6 \rightarrow 5p^56s$ in Cs^+ occurs at about 1000 Å). However, colored crystals of compounds such as NaCl are sometimes encountered. This is due to the presence of holes and free electrons, called color centers, in the lattice, and such chromophoric disturbances can be produced by irradiation of the crystals with X-rays and nuclear radiations. The color results from transitions of the electrons between energy levels in the holes in which they are trapped. These electrons behave in principle similarly to those in solvent cages in the liquid ammonia solutions, but the energy levels are differently spaced and consequently the colors are different and variable. Small excesses of metal atoms produce similar effects, since these atoms form M^+ ions and electrons that occupy holes where anions would be in a perfect crystal (see page 69).

The structures and stabilities of the ionic salts are determined in part by the lattice energies and by radius ratio effects, which have been discussed in Chapter 2. Thus the Li^+ ion is usually tetrahedrally surrounded by water molecules or negative ions. On the other hand, the large Cs^+ ion can accommodate eight near-neighbor Cl^- ions, and its structure is different from that of NaCl where the smaller cation Na^+ can accommodate only six near neighbors.

The salts are generally characterized by high melting points, by electrical conductivity of the melts, and by ready solubility in water. They are seldom hydrated when the anions are small, as in the halides,[13] because the hydration energies of the ions are insufficient to compensate for the energy required to expand the lattice. Owing to its small size, the Li^+ ion has a large hydration energy, and it is often hydrated in its solid salts when the same salts of other alkalis are unhydrated, viz., $LiClO_4 \cdot 3H_2O$. For salts of *strong* acids, the lithium salt is usually the *most* soluble in water of the alkali-metal salts, whereas for *weak* acids the lithium salts are usually *less* soluble than those of the other alkalis.

Since there are few salts that are not appreciably water-soluble, there are few important *precipitation reactions* of the aqueous ions. A unique case is the precipitation by methanolic solutions of 4,4'-diaminodiphenylmethane

[13] For detailed discussion of the factors involved in the solubilities of alkali halides in H_2O, see J. Elson, *J. Chem. Educ.*, 1969, **46**, 86.

(L) of Li and Na salts, for example, $NaLCl$.[14] Generally the larger the M^+ ion the more numerous are its insoluble salts. Thus sodium has very few insoluble salts; the mixed sodium zinc and sodium magnesium uranyl acetates (e.g., $NaZn(UO_2)_3(CH_3COO)_9 \cdot 6H_2O$), which may be precipitated almost quantitatively under carefully controlled conditions from dilute acetic acid solutions, are useful for analysis. The perchlorates and hexachloroplatinates of K, Rb and Cs are rather insoluble in water and virtually insoluble in 90% ethanol. These heavier ions may also be precipitated by cobaltinitrite ion, $[Co(NO_2)_6]^{3-}$, and various other large anions. Sodium tetraphenylborate, $NaB(C_6H_5)_4$, which is moderately soluble in water, is a useful reagent for precipitating the tetraphenylborates of K, Rb and Cs from neutral or faintly acid aqueous solutions, and quantitative gravimetric determinations of these ions may be made.

The large size of the Cs^+ and Rb^+ ions frequently allows them to form ionic salts with rather unstable anions, such as various polyhalide anions (Chapter 16) and the superoxides already mentioned.

Although alkali-metal nitrates decompose to nitrites when heated, they can be distilled from their melts at 350–500° in vacuum.[15]

6-6. The M^+ Ions in Solutions

The M^+ ions are hydrated or generally solvated to rather indeterminate degrees and are best written as $M^+(aq)$ or $M^+(solv)$. Lithium salt hydrates seldom contain more than four water molecules and exceptions are probably attributable to hydration of the anions. Lithium ions in solution have the largest hydrated radius (Table 6-2). Li salts generally deviate from ideal

TABLE 6-2

Data on Hydration of Aqueous Group I Ions

	Li^+	Na^+	K^+	Rb^+	Cs^+
Crystal radii,[a] Å	0.86	1.12	1.44	1.58	1.84
Hydrated radii (approx.), Å	3.40	2.76	2.32	2.28	2.28
Approximate hydration numbers[b]	25.3	16.6	10.5	—	9.9
Hydration energies, kJ mol^{-1}	519	406	322	293	264
Ionic mobilities (at ∞ dil., 18°)	33.5	43.5	64.6	67.5	68

[a] Ladd radii; for Pauling radii see Table 2-1.
[b] From transference data.

solution behavior, showing abnormal colligative properties such as very low vapor pressure, freezing point, etc.

X-ray scattering studies have indicated that the primary hydration shell of K^+ contains four water molecules. Since Na^+ forms the very stable $[Na(NH_3)_4]^+$ ion in liquid ammonia (see below), it is probable that it too

[14] T. C. Shields, *Chem. Comm.*, **1968**, 832.
[15] For kinetics of $NaNO_3$ decomposition, see B. D. Bond and P. W. M. Jacobs, *J. Chem. Soc.*, A, **1966**, 1265.

has a primary hydration sphere of four water molecules. Nothing definite in this connection is known about the Rb^+ and Cs^+ ions. It is quite possible that they, especially Cs^+, might have six water molecules in the first hydration shell. However, electrostatic forces are still operative beyond the first hydration sphere, and additional water molecules will be bound in layers of decreasing definiteness and strength of attachment. Apparently, the larger the cation itself, the less it binds additional outer layers, so that, although the crystallographic radii increase down the Group, the hydrated radii decrease as shown in Table 6-2. Also, hydration energies of the gaseous ions decrease. The decrease in size of the hydrated ions is manifested in various ways. The mobility of the ions in electrolytic conduction increases, and so generally does the strength of binding to ion-exchange resins.

In a cation-exchange resin, two cations compete for attachment at anionic sites in the resin, as in the following equilibrium:

$$A^+(aq) + [B^+R^-](s) = B^+(aq) + [A^+R^-](s)$$

where R represents the resin and A^+ and B^+ the cations. Such equilibria have been measured quite accurately, and the order of preference of the alkali cations is usually $Li^+ < Na^+ < K^+ < Rb^+ < Cs^+$, although irregular behavior does occur in some cases. The usual order may be explained if we assume that the binding force is essentially electrostatic and that under ordinary conditions the ions within the water-logged resin are hydrated approximately as they are outside it. Then the ion with the smallest hydrated radius (which is the one with the largest "naked" radius) will be able to approach most closely to the negative site of attachment and will hence be held most strongly according to the Coulomb law.

The reasons for the deviations from this simple pattern as well as selective passage of certain ions through cell walls are not properly understood, and factors other than mere size are doubtless important.

Of importance in connection with the solubility of the metals in liquid ammonia are ammonia solvates such as the $[Na(NH_3)_4]^+$ ion which is formed on treatment of NaI with liquid ammonia. $[Na(NH_3)_4]I$ is a liquid of fair thermal stability. It freezes at $3°$ and at $25°$ has an equilibrium pressure of NH_3 of 420 mm; thus it must be kept in an atmosphere of ammonia with at least this pressure at $25°$. The infrared and Raman spectra indicate the complex ion $[Na(NH_3)_4]^+$ to be tetrahedral with Na—N bonds about as strong as the Zn—N bonds in $[Zn(NH_3)_4]^{2+}$ or the Pb—C bonds in $Pb(CH_3)_4$. Bending and rocking frequencies, however, are quite low, suggesting that the Na—N bonding is mainly due to ion–dipole forces. Thus it may be assumed that Na^+ and other metal ions in the dilute liquid ammonia, amine and ether solutions are strongly solvated in the same way.

The effectiveness of tetrahydrofuran and the dimethyl ethers of ethylene and diethylene glycols ("glyme" and "diglyme," respectively) as media for reactions involving sodium may be due in part to the slight solubility of the metal, but the solvation of ions by ether molecules undoubtedly provides the most important contribution. Indeed there are numerous very stable and

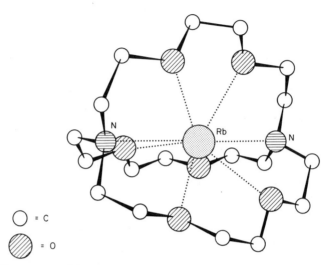

Fig. 6-1.   The structure of the cation in the salt $[RbC_{18}H_{36}N_2O_6]SCN \cdot H_2O$. [Reproduced by permission from M. R. Truter, *Chem. in Britain*, **1971**, 203.]

often crystalline solvates for M^+ (and also M^{2+}) ions with a wide variety of macrocyclic polyethers. Ethers with 3–20 oxygen atoms have been synthesized.[16] One of these, whose formal name is 2,5,8,15,18,21-hexaoxatricyclo [20.4.0.0^{9,14}] hexacosane–abbreviated to cyclohexyl-18-crown-6 is shown in formula (21-XXVI). Equilibrium constants[17] for this ether are in the order $K^+ > Rb^+ > Cs^+$, $Na^+ > Li^+$. Similar macrocyclic polythioethers (which have greater affinity for transition–metal ions) have been made, but more important are the so-called "cryptates," which are tricyclic nitrogen–oxygen macrocycles,[18] e.g., $N[CH_2CH_2OCH_2CH_2OCH_2CH_2]_3N$ (21-XXVII). In complexes of the latter, the metal ion is completely enclosed as in the cation of the salt $[RbC_{18}H_{36}N_2O_6]SCN \cdot H_2O$ shown in Fig. 6-1.[18a] The cryptates have remarkable complexing ability for M^{2+} ions and will, for example, render even $BaSO_4$ soluble.

Salts of more simple ethers such as $[K \text{ diglyme}_3][Mo(CO_5)I]$ are also known, while solvates with other oxygen donors, e.g., $(CH_3)_2SO$ and amides, exist although these are normally unstable in water.

The polyether complexes have been used as models for naturally occurring compounds that are involved in the transport of alkali and alkaline-earth ions across membranes and for the very high selectivities towards Na^+ and K^+ or Ca^{2+} and Mg^{2+} shown by natural systems.[18b]

[16] C. J. Peterson, *J. Amer. Chem. Soc.*, 1970, **92**, 386, 391.
[17] R. M. Izzatt *et al.*, *J. Amer. Chem. Soc.*, 1971, **93**, 1619.
[18] J. M. Lehn and J. P. Sauvage, *Chem. Comm.*, **1971**, 440; B. Metz, D. Moras and R. Weiss, *Chem. Comm.*, **1971**, 444; *J. Amer. Chem. Soc.*, 1971, **93**, 1807.
[18a] M. R. Truter, *Chem. in Britain*, **1971**, 203.
[18b] R. J. P. Williams, *Quart. Rev.*, 1970, **24**, 331 (the biochemistry of Na, K, Ca and Mg).

6-7. Complexes

Apart from solvates of the type just discussed, there are comparatively few complexes of the M^+ ions, and these are mostly chelates of β-diketones, nitrophenols, 1-nitroso-2-naphthol, etc.[19] Certain of these derivatives, especially those containing fluorinated ligands such as hexafluoroacetylacetone, are sublimable[20] at 200°. The bonding in such complexes is nevertheless essentially electrostatic. The anhydrous β-diketonates are usually insoluble in organic solvents, indicating an ionic nature, but in presence of additional coordinating ligands, including water, they may become soluble even in hydrocarbons; for example, sodium benzoylacetylacetonate·$2H_2O$ is soluble in toluene, while tetramethylethylenediaminelithium hexafluoroacetylacetonate is monomeric in benzene.[21]

This behavior has allowed the development of solvent-extraction procedures for alkali-metal ions. Thus, not only can the trioctylphosphine oxide adduct, $Li(PhCOCHCOPh)[OP(octyl)_3]_2$, be extracted from aqueous solutions into p-xylene, but also this process can be used to separate lithium from other alkali-metal ions.[22] Further, Cs^+ can be extracted from aqueous solutions into hydrocarbons by 1,1,1-trifluoro-3-(2'-thenoyl)acetone (TTA) in presence of nitromethane.[23]

It may be noted finally that ^7Li (92.7%) gives nuclear magnetic resonance signals comparable to those given by ^1H, so that complex formation in aqueous solutions can be studied; in this way it was shown[24] that nitrilotriacetic acid, $N(CH_2COOH)_3$ (NTA), probably forms a complex ion, $[Li(NTA)_2]^{5-}$, in solution.

6-8. Organometallic Compounds[25]

One of the most important areas of the chemistry of Group I elements is that of their organic compounds. The most important are those of lithium; organosodium compounds, and to a lesser extent organopotassium ones, are of limited use.

Lithium Alkyls and Aryls.   One of the largest uses of metallic lithium, industrially and in the laboratory, is for the preparation of organolithium compounds. These are of great importance and utility; in their reactions they generally resemble Grignard reagents, although they are usually more reactive. Their preparation is best accomplished by using an alkyl or aryl

[19] A. J. Layton, et al., J. Chem. Soc., A, **1970**, 1894.
[20] R. Belcher, A. W. L. Dudeney and W. I. Stephen, J. Inorg. Nuclear Chem., 1969, **31**, 625.
[21] K. Shobatake and K. Nakamoto, J. Chem. Phys., 1968, **49**, 4792.
[22] D. A. Lee, W. L. Naylor, W. J. McDowell and J. S. Drury, J. Inorg. Nuclear Chem., 1969, **30**, 2807.
[23] P. Crowther and A. Jurriaanse, J. Inorg. Nuclear Chem., 1968, **30**, 3365.
[24] See, e.g., J. W. Akitt and M. Parekh, J. Chem. Soc., A, **1968**, 2195.
[25] G. E. Coates, M. L. H. Green and K. Wade, Organometallic Compounds, 3rd edn., Vol. 1, Methuen, 1967 (a comprehensive account); see also Organometallic Chem. Rev., B (annual surveys).

chloride (eq. 6-1) in benzene or petroleum; ether solutions can be used, but these solvents are attacked slowly by the lithium compounds. Metal–hydrogen exchange (eq. 6-2), metal–halogen exchange (eq. 6-3) and metal–metal exchange (eq. 6-4) may also be used.

$$C_2H_5Cl + 2Li = C_2H_5Li + LiCl \tag{6-1}$$

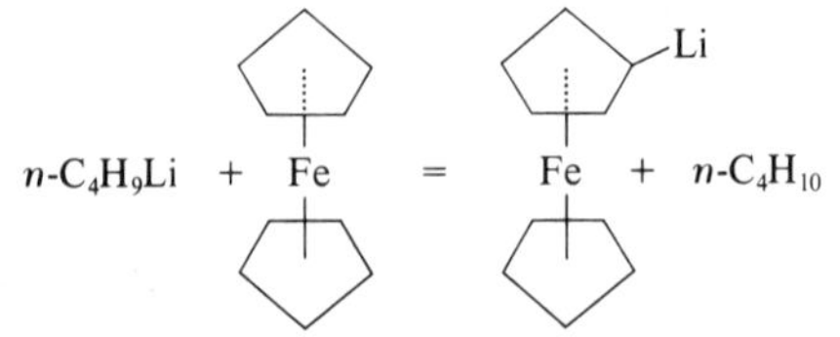

$$n\text{-}C_4H_9Li \ + \ Fe \ = \ Fe \ + \ n\text{-}C_4H_{10} \tag{6-2}$$

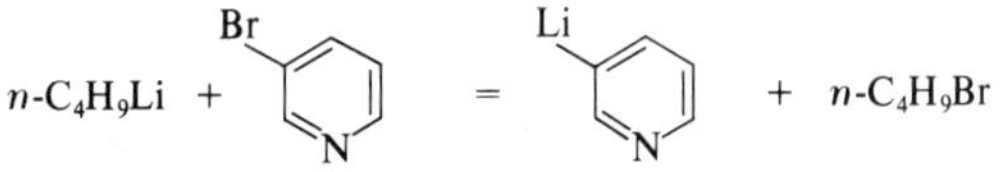

$$n\text{-}C_4H_9Li \ + \ \underset{N}{\overset{Br}{\bigcirc}} \ = \ \underset{N}{\overset{Li}{\bigcirc}} \ + \ n\text{-}C_4H_9Br \tag{6-3}$$

$$2Li + R_2Hg = 2RLi + Hg \tag{6-4}$$

n-Butyllithium in hexane, benzene or ethers is commonly used for such reactions. Methyllithium is also prepared by exchange through the interaction of $n\text{-}C_4H_9Li$ and CH_3I in hexane at low temperatures, whence it precipitates as insoluble white crystals.

Organolithium compounds all react rapidly with oxygen, being usually spontaneously flammable in air, with liquid water and with water vapor. However, lithium bromide and iodide form solid complexes of stoichiometry $RLi(LiX)_{1-6}$ with the alkyls, and these solids are stable in air.

Organolithium compounds are among the very few alkali-metal compounds that have properties—solubility in hydrocarbons or other non-polar liquids and high volatility—typical of covalent substances. They are generally liquids or low-melting solids, and molecular association is an important structural feature.[26]

In the crystals of methyl- (Fig. 6-2) and ethyl-lithium (EtLi, m.p. 90°) the lithium atoms are at the corners of a tetrahedron with the alkyl groups centered over the facial planes. Although the CH_3 group is symmetrically bound to three Li atoms, the α-carbon of the C_2H_5 group is closer to one Li atom than the other two.

The alkyl bridge bonding is of the electron-deficient multi-center type (page 110) found in Be and Al alkyls and in boranes. Aggregate formation is due to Li—C—Li rather than to Li—Li bonding interactions. The electronic structure can be accommodated by MO theory. The question whether any significant direct Li—Li bonding occurs in these tetrahedra remains contro-

[26] T. L. Brown, *Adv. Organometallic Chem.*, 1965, **3**, 365 (structures and mechanisms of lithium alkyls); M. Szwarc, *Carbanions, Living Polymers and Electron Transfer Processes*, Interscience-Wiley, 1968.

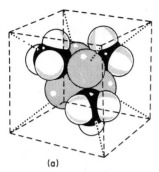
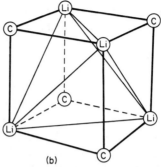

(a)   (b)

Fig. 6-2. The structure of $(CH_3Li)_4$: (a) showing the tetrahedral Li_4 unit with the CH_3 groups located symmetrically above each face of the tetrahedron. [Adapted from E. Weiss and E. A. C. Lucken, *J. Organometallic Chem.*, 1964, **2**, 197.] The structure can also be regarded as derived from a cube (b).

versial. MO calculations[26a] have suggested that there is bonding, but Raman and nmr spectra suggest that there is none.[26b]

Lithium alkyls and aryls are associated in solutions, but the nature of the species depends on the nature of the solvent, the steric nature of the organic radical and temperature.[27] Cryoscopy and nmr study with ^{13}C, ^{7}Li and ^{1}H resonances[28,29] show that in hydrocarbon solvents MeLi, EtLi, n-PrLi and some others are hexamers but *tert*-butyllithium, which presumably is too bulky, is only tetrameric. On addition of ethers or amines, or in these as solvent, solvated tetramers are formed. The formation of dimers, or aggregates less than tetramers, seems not to occur.

However, when chelating ditertiary amines, notably tetramethylethylenediamine (TMED), $Me_2NCH_2CH_2NMe_2$, are used, comparatively stable monomeric alkyllithium complexes are obtained. The alkyls and aryls also form complexes with other metal alkyls such as those of Mg, Cd and Zn, e.g.:

$$2LiC_6H_5 + Mg(C_6H_5)_2 = Li_2[Mg(C_6H_5)_4]$$

This type of complexing, as well as the rates and mechanisms of alkyl-exchange reactions in solution, have been studied by nmr methods.[30]

It is not surprising that there are wide variations in the comparative reactivities of Li alkyls depending upon the differences in aggregation and ion-pair interactions. An example is benzyllithium, which is monomeric in tetrahydrofuran and reacts with a given substrate more than 10^4 times as fast as the tetrameric methyllithium.[31] The monomeric TMED complexes

[26a] A. H. Cowley and W. D. White, *J. Amer. Chem. Soc.*, 1969, **91**, 34; G. R. Peyton and W. H. Glaze, *Theor. Chim. Acta*, 1969, **13**, 259.
[26b] T. L. Brown, L. M. Seitz and B. Kimura, *J. Amer. Chem. Soc.*, 1968, **90**, 3245; W. M. Scovell, B. Kimura and T. G. Spiro, *J. Coordination Chem.*, 1971, **1**, 107.
[27] J. F. Garst in *Solute–Solvent Interactions*, J. F. Coetzee and C. D. Ritchie, eds., Dekker, 1969.
[28] M. Y. Darensbourg *et al.*, *J. Amer. Chem. Soc.*, 1970, **92**, 1236.
[29] H. L. Lewis and T. L. Brown, *J. Amer. Chem. Soc.*, 1970, **92**, 4664.
[30] L. M. Seitz and B. F. Little, *J. Organometallic Chem.*, 1969, **18**, 227.
[31] R. Waack and M. A. Doran, *J. Amer. Chem. Soc.*, 1969, **91**, 2456; S. P. Patterman, I. L. Karle and G. D. Stucky, *J. Amer. Chem. Soc.*, 1970, **92**, 1150.

noted above are also very much more reactive than the corresponding aggregated alkyls. Alkyllithiums can polylithiate acetylenes, acetonitrile, and other compounds;[31a] thus $CH_3C\equiv CH$ gives Li_4C_3 which can be regarded as a derivative of C_3^{4-}.

Reactions of lithium alkyls are generally considered to be carbanionic in nature, but in reactions with alkyl halides free radicals have been detected by electron spin resonance.[32] Lithium alkyls are widely employed as stereospecific catalysts for the polymerization of alkenes, notably isoprene, which gives up to 90% of 1,4-cis-polyisoprene; numerous other reactions with alkenes have been studied.[33] The TMED complexes again are especially active: not only will they polymerize ethylene but they will even metallate benzene and aromatic compounds, as well as reacting with hydrogen at 1 atm to give LiH and alkane.

6-9. Organo-sodium and -potassium Compounds[34]

These compounds are all essentially ionic and are not soluble to any appreciable extent in hydrocarbons; they are exceedingly reactive, being sensitive to air and hydrolyzed vigorously by water. Although alkyl- and particularly aryl-sodium derivatives can be prepared for use as reaction intermediates *in situ*, they are seldom isolated. However, methylpotassium, which is a highly pyrophoric substance, has been obtained by the reaction:

$$(CH_3)_2Hg + 2K-Na \rightarrow 2CH_3K + Na-Hg$$

It has an NiAs-type structure (Fig. 15-5), and the isolated methyl groups are presumably in the lattice as the pyramidal CH_3^- ion.[35]

Sodium and potassium alkyls can be used for metallation reactions as, for example, in eq. 6-2. They can also be prepared from Na or K dispersed on an inert support material, and such solids act as carbanionic catalysts for the cyclization, isomerization or polymerization of alkenes. The so-called "alfin" catalysts for copolymerization of butadiene with styrene or isoprene to give rubbers consist of sodium alkyl (usually allyl) and alkoxide (usually isopropoxide) and NaCl, which are made simultaneously in hydrocarbons.[33]

More important are the compounds formed by acidic hydrocarbons such as cyclopentadiene, indene, acetylenes, etc. These are obtained by reaction with sodium in liquid ammonia or, more conveniently, sodium dispersed in tetrahydrofuran, glyme, diglyme or dimethylformamide.

$$3C_5H_6 + 2Na \rightarrow 2C_5H_5^-Na^+ + C_5H_8$$
$$RC\equiv CH + Na \rightarrow RC\equiv C^-Na^+ + \tfrac{1}{2}H_2$$

[31a] G. A. Gornowicz and R. West, *J. Amer. Chem. Soc.*, 1971, **93**, 1714, 1720.

[32] G. A. Russell and D. W. Lamson, *J. Amer. Chem. Soc.*, 1969, **91**, 3967.

[33] H. Sinn and F. Patat, *Angew. Chem. Internat. Edn.*, 1964, **3**, 93; C. W. Kamienski, *Ind. Eng. Chem. Internat. Edn.*, 1965, **57**, 38; G. G. Eberhardt, *Organometallic Chem. Rev.*, 1966, **1**, 491 (organoalkali compounds in catalysis); L. Reich and A. Schindler, *Polymerisation by Organometallic Compounds*, Interscience-Wiley, 1966.

[34] M. Schlosser, *Angew. Chem. Internat. Edn.*, 1964, **3**, 287, 362.

[35] E. Weiss and G. Sauerman, *Chem. Ber.*, 1970, **103**, 265; *J. Organometallic Chem.*, 1970, **21**, 1.

Many aromatic hydrocarbons, as well as aromatic ketones, triphenyl-phosphine oxide, triphenylarsine, azobenzene, etc., can form highly colored *radical anions*[36] when treated at low temperatures with Na or K in solvents such as tetrahydrofuran. For the formation of such anions it is necessary that the negative charge can be delocalized over an aromatic system. Species such as the benzenide, $C_6H_6^-$, naphthalenide or anthracenide ions can be detected and characterized spectroscopically and by electron spin resonance.[37] The sodium–naphthalene system, $Na^+[C_{10}H_8]^-$, in an ether is widely used as a powerful reducing agent, e.g., in nitrogen-fixing systems employing titanium catalysts (page 345) and for the production of complexes in low oxidation states. The blue solution of sodium and benzophenone in tetrahydrofuran, which contains the "ketyl" or radical ion, is a useful and rapid reagent for the removal of traces of oxygen from nitrogen.

[36] E. de Boer, *Adv. Organometallic Chem.*, 1965, **2**, 115; E. T. Kaiser and J. L. Kevan, *Radical Ions*, Wiley, 1968.
[37] N. Hirota, *J. Amer. Chem. Soc.*, 1968, **90**, 3603; K. Höfelmann, J. Jagur-Grodzinski and M. Szwarc, *J. Amer. Chem. Soc.*, 1969, **91**, 4645.

Further Reading

Foote Mineral Co., Philadelphia 4, Pa. (various bulletins containing physical and chemical data on lithium and its compounds).

Lithium Corp. of America, Minneapolis 2, Minn. (various bulletins containing physical and chemical data on lithium and its compounds).

The Alkali Metals, Special Publ. No. 22, The Chemical Society, London, 1967 (Symposium Reports on wide variety of topics varying from binary compounds to reactions in the liquid metals).

Advances in Chemistry Series no. 19, "*Handling and Uses of Alkali Metals*," American Chemical Society, Washington, D.C., 1957 (recovery, handling, and manufacture of metals, hydrides and oxides of Li, Na, and K).

Fatt, I., and M. Tashima, *Alkali Dispersions*, Van Nostrand, 1962 (extensive review with preparative details).

Liquid Metals Handbook, 3rd ed. (Sodium, NaK Supp.), Jackson, C. B., ed., Atomic Energy Commission and Bureau of Ships, U.S. Dept. of Navy, 1955.

Juza, R., *Angew. Chem. Internat. Edn.*, **3**, 471 (1964) (amides of alkali and alkaline-earth metals; review).

Kaufmann, D. W., *Sodium Chloride* (American Chemical Society Monograph, No. 145), Reinhold, 1960 (an encyclopedic account of salt).

Mellor's Comprehensive Treatise on Inorganic and Theoretical Chemistry, Vol. II, Supplement 2, Li, Na (1961); Supplement 3, K, Rb, Cs, Fr (1963). Longmans Green.

Perel'man, F. M., *Rubidium and Caesium*, Pergamon Press, 1965 (comprehensive reference book).

Stern, K. H., and E. S. Amis, *Chem. Rev.*, 1959, **59**, 1 (ionic size; a comprehensive review on radii in crystals and solutions).

Symons, M. C. R., *Quart. Rev.*, 1959, **13**, 99 (the nature of the solutions of alkali metals in liquid ammonia).

Symons, M. C. R., and W. T. Doyle, *Quart. Rev.*, 1960, **14**, 62 (color centers in alkali halides).

7

Beryllium and the Group II Elements: Mg, Ca, Sr, Ba, Ra

GENERAL REMARKS

7-1. Group Relationships

Some pertinent data for Group II elements are given in Table 7-1. Briefly, Be has unique chemical behavior with a predominantly covalent chemistry, although forming a cation $[Be(H_2O)_4]^{2+}$. Magnesium, the second-row element, has a chemistry intermediate between that of Be and the heavier elements, but it does not stand in as close relationship with the predominantly ionic heavier members as might have been expected from the similarity of Na, K, Rb and Cs. It has considerable tendency to covalent bond formation, consistent with the high charge : radius ratio. For instance, like beryllium, its hydroxide can be precipitated from aqueous solutions, whereas hydroxides of the other elements are all moderately soluble; and it readily forms bonds to carbon.

The metal atomic radii are smaller than those of the Group I metals owing to the increased nuclear charge; the number of bonding electrons in the

TABLE 7-1

Some Physical Parameters for the Group II Elements

Element	Electronic configuration	M.p. (°C)	Ionization potentials, eV		E^0 for $M^{2+}(aq) + 2e = M(s)$ (V)	Ionic radii (Å)[a]	$\dfrac{\text{Charge}}{\text{radius}}$
			1st	2nd			
Be	[He]$2s^2$	1278	9.32	18.21	-1.85^b	0.34	6.5
Mg	[Ne]$3s^2$	651	7.64	15.03	-2.37	0.78	3.1
Ca	[Ar]$4s^2$	843	6.11	11.87	-2.87	1.06	2.0
Sr	[Kr]$5s^2$	769	5.69	10.98	-2.89	1.27	1.8
Ba	[Xe]$6s^2$	725	5.21	9.95	-2.90	1.43	1.5
Ra	[Rn]$7s^2$	700	5.28	10.10	-2.92	1.57	1.3

[a] Ladd radii.   [b] Estimated.

metals is twice as great, so that the metals have higher melting and boiling points and greater densities.

All are highly electropositive metals, however, as is shown by their high chemical reactivities, ionization potentials, standard electrode potentials and, for the heavier ones, the ionic nature of their compounds. Although the energies required to vaporize and ionize these atoms to the M^{2+} ions are considerably greater than those required to produce the M^+ ions of the Group I elements, the high lattice energies in the solid salts and the high hydration energies of the M^{2+}(aq) ions compensate for this, with the result that the standard potentials are similar to those of the Li–Cs group.

The potential E^0 of beryllium is considerably lower than those of the other elements, indicating a greater divergence in compensation by the hydration energy, the high heat of sublimation and the ionization potential. As in Group I, the smallest ion crystallographically, i.e. Be^{2+}, has the largest hydrated ionic radius.

All the M^{2+} ions are smaller and considerably less polarizable than the isoelectronic M^+ ions. Thus deviations from complete ionicity in their salts due to polarization of the cations are even less important. However, for Mg^{2+} and, to an exceptional degree for Be^{2+}, polarization of anions by the cations does produce a degree of covalence for compounds of Mg and makes covalence characteristic for Be. Accordingly only an estimated ionic radius can be given for Be; the charge : radius ratio is greater than for any other cation except H^+ and B^{3+}, which again do not occur as such in crystals. The closest ratio is that for Al^{3+} and some similarities between the chemistries of Be and Al exist. Examples are the resistance of the metal to attack by acids owing to formation of an impervious oxide film on the surface, the amphoteric nature of the oxide and hydroxide, and Lewis acid behavior of the chlorides. However, Be shows just as many similarities to zinc, especially in the structures of its binary compounds (see Sections 18-6 to 18-8) and in the chemistry of its organic compounds. Thus BeS (zinc blende structure) is insoluble in water although Al_2S_3, CaS, etc., are rapidly hydrolyzed.

Calcium, strontium, barium, and radium form a closely allied series in which the chemical and physical properties of the elements and their compounds vary systematically with increasing size in much the same manner as in Group I, the ionic and electropositive nature being greatest for Ra. Again the larger ions can stabilize certain large anions: the peroxide and superoxide ions, polyhalide ions, etc. Some examples of systematic group trends in the series Ca–Ra are: (a) hydration tendencies of the crystalline salts increase; (b) solubilities of sulfates, nitrates, chlorides, etc. (fluorides are an exception) decrease; (c) solubilities of halides in ethanol decrease; (d) thermal stabilities of carbonates, nitrates and peroxides increase; (e) rates of reaction of the metals with hydrogen increase. Other similar trends can be found.

All isotopes of *radium* are radioactive, the longest-lived isotope being ^{226}Ra (α; ~ 1600 yr). This isotope is formed in the natural decay series of

^{238}U and was first isolated by Pierre and Marie Curie from pitchblende. Once widely used in radiotherapy, it is now being supplanted by radioisotopes made in nuclear reactors.

The elements Zn, Cd, and Hg, which have two electrons outside filled penultimate d shells, are also classed in Group II. Although the difference between the calcium and zinc sub-groups is marked, Zn, and to a lesser extent Cd, show some resemblance to Be or Mg in their chemistry. We shall discuss these elements separately (Chapter 18), but it may be noted here that Zn, which has the lowest second ionization potential in the Zn, Cd, Hg group, still has a value (17.89 eV) similar to that of Be (18.21 eV), and its standard potential (-0.76 V) is considerably less negative than that of Mg.

There are a few ions with ionic radii and chemical properties similar to those of Sr^{2+} or Ba^{2+}, notably those of the $+2$ lanthanides (Section 27-17) and especially the europous ion, Eu^{2+}, and its more readily oxidized analogs Sm^{2+} and Yb^{2+}. Because of this fortuitous chemical similarity, europium is frequently found in Nature in Group II minerals, and this is a good example of the geochemical importance of such chemical similarity.

7-2. Lower Oxidation States

Although the differences between the first and second ionization potentials, especially for Be, might suggest the possibility of a stable $+1$ state, there is no evidence to support this. Calculations using Born–Haber cycles show that, owing to the much greater lattice energies of MX_2 compounds, MX compounds would be unstable and disproportionate:

$$2MX \rightarrow M + MX_2$$

Detailed studies of the Ca, Sr and Ba systems confirm the absence of M^+ ions.[1]

There is some evidence for Be^I in fused chloride melts, for example:

$$Be + Be^{II} = 2Be^I$$

but no Be^I compound has been isolated. Some studies of the dissolution of Be from anodes suggested Be^+ as an intermediate, but subsequent work showed that disintegration of the metal occurs during dissolution so that the apparent effect is one of the metal going into solution in the $+1$ state—too much metal is lost for the amount of current passed. The anode sludge, a mixture of Be and $Be(OH)_2$, had been considered to be due to disproportionation of Be^+, but photomicrography indicates that the beryllium in the sludge is due merely to spallation of the anode.

On the other hand, similar studies of anodic dissolution of Mg in pyridine and aqueous salt solutions do provide some evidence for transitory Mg^+ ions, which would account for evolution of H_2 at or near the anode. Electrically generated Mg^+ ions have been used to reduce organic compounds.[2]

[1] H. H. Emons, *Z. anorg. Chem.*, 1963, **323**, 114.
[2] M. D. Rausch, W. E. McEwen and J. Kleinberg, *Chem. Rev.*, 1957, **57**, 417.

BERYLLIUM[3]

7-3. Covalency and Stereochemistry

As a result of the small size, high ionization potential and high sublimation energy of beryllium, its lattice and hydration energies are insufficient to provide complete charge separation and the formation of simple Be^{2+} ions. In fact, in all compounds whose structures have been determined, even those of the most electronegative elements, i.e. BeO and BeF_2, there appears to be substantial covalent character in the bonding. On the other hand, to allow the formation of two covalent bonds, —Be—, it is clear that unpairing of the two $2s$ electrons is required. Where free BeX_2 molecules occur, the Be atom is promoted to a state in which the two valence electrons occupy two equivalent sp hybrid orbitals and the X—Be—X system is linear. However, in such a linear molecule the Be atom has a coordination number of only two and there is a strong tendency for Be to achieve maximum (four-fold) coordination, or at least three-fold coordination. Maximum coordination is achieved in several ways:

1. Polymerization may occur through bridging, as in solid $BeCl_2$ (Fig. 7-1).

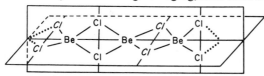

Fig. 7-1.   The structure of polymeric $BeCl_2$ in the crystal. The structure of $Be(CH_3)_2$ is similar.

The coordination of Be is not exactly tetrahedral since the Cl—Be—Cl angles are only 98°, which means that the $BeCl_2Be$ units are somewhat elongated in the direction of the chain axis. In such a situation the exact sizes of the angles are determined by competing factors. If the ClBeCl angles were opened to 109°, the BeClBe angles would decrease to 71°, which would probably weaken the Be—Cl bonds; Be···Be repulsions would also be increased. The observed value is a compromise giving the lowest free energy.

Beryllium chloride readily sublimes; at high temperatures ($\sim750°$) it consists of essentially all monomeric, linear $BeCl_2$ molecules, but at lower temperatures there are appreciable amounts ($\sim20\%$ at 560°) of the dimer, in which Be is three-coordinate.

2. By functioning as Lewis acids, many beryllium compounds attain maximum coordination of the metal atom. Thus the chloride forms etherates, $Cl_2Be(OR_2)_2$, and complex ions such as BeF_4^{2-} and $[Be(H_2O)_4]^{2+}$ exist. In chelate compounds such as the acetylacetonate, $Be(acac)_2$, four approximately tetrahedral bonds are formed with the C—O and Be—O bond lengths equivalent.

[3]   D. A. Everest, *The Chemistry of Beryllium*, Elsevier, 1964; H. E. Stockinger, *Beryllium, its Industrial Hygiene Aspects*, Academic Press, 1966; L. B. Tepper, H. L. Hardy and R. I. Chamberlain, *Toxicity of Beryllium Compounds*, Elsevier, 1961.

The packing in crystals is almost invariably such as to give Be a coordination number of 4, with a tetrahedral configuration. In binary compounds, the structures are often those of the corresponding zinc compounds. Thus the low-temperature form of BeO has the wurtzite structure (Fig. 2-3); the most stable $Be(OH)_2$ polymorph has the $Zn(OH)_2$ structure; and BeS has the zinc blende structure (Fig. 2-3). Be_2SiO_4 is exceptional among the orthosilicates of the alkaline earths, the rest of which have structures giving the metal ion octahedral coordination, in having the Be atoms tetrahedrally surrounded by oxygen atoms. It may be noted that Be with F gives compounds often isomorphous with oxygen compounds of silicon; thus BeF_2 is isomorphous with cristobalite (SiO_2), $BaBeF_4$ with $BaSiO_4$ and $NaBeF_3$ with $CaSiO_3$, and there are five different corresponding forms of Na_2BeF_4 and Ca_2SiO_4.

Three-coordinate Be occurs in some cases, for example, the gaseous dimers Be_2Cl_4 and Be_2Br_4. In Be phthalocyanine, the metal is perforce surrounded by four nitrogen atoms in a plane. However, no more than three electron pairs from the N atoms can be truly coordinated to the Be atom (i.e. occupy bonding MO's), since there are only three atomic orbitals available (the s and two p orbitals) in a given plane. This compound constitutes an example of a *forced configuration* since the Be atom is held strongly in a rigid environment.

There are very few examples of Be compounds existing at room temperature in which Be is 2-coordinate with sp linear bonds; but the monomeric compounds di-*tert*-butylberyllium, $Be(CMe_3)_2$ and the silazane,[4] $Be[N(SiMe_3)_2]_2$, are evidently of this class, presumably because of steric factors.

It is to be noted especially that beryllium compounds are exceedingly poisonous, particularly if inhaled, and great precautions must be taken in handling them.[3]

7-4. Elemental Beryllium

The most important mineral is *beryl*, $Be_3Al_2(SiO_3)_6$, which often occurs as large hexagonal prisms. The extraction from ores is complicated.[5] The metal is obtained by electrolysis of $BeCl_2$ but, since the melt has very low electrical conductivity (about 10^{-3} that of NaCl), sodium chloride is also added.

The grey metal is rather light (1.86 g/cm^3) and quite hard and brittle. Since the absorption of electromagnetic radiation depends on the electron density in matter, beryllium has the lowest stopping power per unit mass thickness of all suitable construction materials and is used for "windows" in X-ray apparatus. It is also added as an antioxidant to copper and phosphor bronzes and as a hardener to copper.

Metallic Be, like Al, is rather resistant to acids unless finely divided or amalgamated, owing to the formation of an inert and impervious oxide film

[4]   A. H. Clark and A. Haaland, *Acta Chem. Scand.*, 1970, 24, 3024.
[5]   *Chem. Eng. News*, **1965**, April 19th, p. 70.

on the surface. Thus, although the standard potential (-1.85 V) would indicate rapid reaction with dilute acids (and even H_2O), the rate of attack depends greatly on the source and fabrication of the metal. For very pure metal the relative dissolution rates are $HF > H_2SO_4 \sim HCl > HNO_3$. The metal dissolves rapidly in $3M$ H_2SO_4 and in $5M$ NH_4F, but very slowly in HNO_3. Like Al, it dissolves also in strong bases, forming what is called the beryllate ion.

7-5. Binary Compounds

The white crystalline *oxide*, BeO, is obtained on ignition of Be or its compounds in air. It resembles Al_2O_3 in being highly refractory (m.p. 2570°) and in having polymorphs; the high-temperature form ($>800°$) is exceedingly inert and dissolves readily only in a hot syrup of concentrated H_2SO_4 and $(NH_4)_2SO_4$. The more reactive forms dissolve in hot alkali hydroxide solutions or fused $KHSO_4$.

Addition of OH^- ion to $BeCl_2$ or other beryllium solutions gives the *hydroxide*. This is amphoteric and in alkali solution the "beryllate" ion, probably $[Be(OH)_4]^{2-}$, is obtained. When these solutions are boiled, the most stable of several polymorphs of the hydroxide can be crystallized.[6]

Beryllium fluoride is obtained as a glassy hygroscopic mass by heating $(NH_4)_2BeF_4$; it has randomly oriented chains of $\cdots F_2BeF_2Be\cdots$ with F bridges. Thus the structure is similar to those of $BeCl_2$ and $BeBr_2$ except that the packing of the chains is disordered. BeF_2 melts (803°) to a viscous liquid that has low electrical conductivity. The polymerization in the liquid may be lowered by addition of LiF which forms the BeF_4^{2-} ion.[7]

Beryllium chloride is prepared by passing CCl_4 over BeO at 800°. On a small scale the chloride and bromide are best prepared pure by direct interaction in a hot tube.[8] The white crystalline chloride (m.p. 405°) dissolves exothermally in water; from HCl solutions the salt $[Be(H_2O)_4]Cl_2$ can be obtained. $BeCl_2$ is readily soluble in oxygenated solvents such as ethers. In melts with alkali halides, chloroberyllate ions, $[BeCl_4]^{2-}$, may be formed but this ion does not exist in aqueous solution.

On interaction of Be with NH_3 or N_2 at 900–1000° the *nitride*, Be_3N_2, is obtained as colorless crystals, readily hydrolyzed by water. The metal reacts with ethylene at 450° to give BeC_2.

7-6. Complex Chemistry

Oxygen Ligands. In strongly acid solutions the *aquo ion* $[Be(H_2O)_4]^{2+}$, occurs, and crystalline salts with various anions can be readily obtained. The water in such salts is more firmly retained than is usual for aquates, indicating

[6] R. A. Mercer and R. P. Miller, *J. Inorg. Nuclear Chem.*, 1966, **28**, 61.
[7] A. L. Matthews and C. F. Baes, Jr., *Inorg. Chem.*, 1968, **9**, 373.
[8] E. C. Ashby and R. C. Arnott, *J. Organometallic Chem.*, 1968, **14**, 1.

strong binding. Thus the sulfate is dehydrated to $BeSO_4$ only on strong heating, while $[Be(H_2O)_4]Cl_2$ loses no water over P_2O_5. Solutions of beryllium salts are acidic; this may be ascribed to the acidity of the aquo ion, the initial dissociation being

$$[Be(H_2O)_4]^{2+} \rightleftharpoons [Be(H_2O)_3(OH)]^+ + H^+$$

The addition of soluble carbonates to beryllium salt solutions gives only basic carbonates. Beryllium salt solutions also have the property of dissolving additional amounts of the oxide or hydroxide. This behavior is attributable to the formation of complex species with Be—OH—Be or Be—O—Be bridges. The rapidly established equilibria[9] involved in the hydrolysis of the $[Be(H_2O)_4]^{2+}$ ion are very complicated and depend on the nature of the anion, the concentration, the temperature, and the pH. The main species, which will achieve 4-coordination by additional water molecules, are considered to be $Be_2(OH)^{3+}$, $Be_3(OH)_3^{3+}$ (probably cyclic) and possibly $Be_5(OH)_7^{3+}$. Various crystalline hydroxo complexes have been isolated.[6] In concentrated alkaline solution the main species is $[Be(OH)_4]^{2-}$.

Other complexes of oxygen ligands are mainly adducts of beryllium halides or alkyls with ethers, ketones, etc., e.g., $BeCl_2(OEt_2)_2$. There are also neutral complexes of β-diketones and similar compounds, of which the acetylacetonate is the simplest, and solvated cationic species such as $[Be(DMF)_4]^{2+}$.[10] The most unusual complexes have the formula $Be_4O(OOCR)_6$ and are formed by refluxing the hydroxide with carboxylic acids. These white crystalline compounds are soluble in organic solvents, even alkanes, but are insoluble in water and lower alcohols; they are inert to water but are hydrolyzed by dilute acids; in solution they are un-ionized and monomeric; X-ray study has shown that they have the structures illustrated in Fig. 7-2. The central

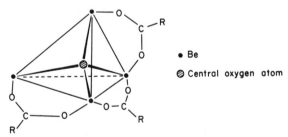

Fig. 7-2. The structure of the basic carboxylate complexes $Be_4O(OOCR)_6$. Only three RCOO groups are shown.

oxygen atom is tetrahedrally surrounded by the four beryllium atoms (this being one of the few cases, excepting solid oxides, in which oxygen is four-coordinate), and each beryllium atom is tetrahedrally surrounded by four oxygen atoms. Zinc also forms such complexes, as does the ZrO^{2+} ion, with

[9] R. E. Mesmer and C. F. Baes, Jr., *Inorg. Chem.*, 1967, **6**, 1951; G. Schwartzenbach and H. Wenger, *Helv. Chim. Acta*, 1969, **52**, 644.
[10] W. G. Movius and N. A. Matwiyoff, *J. Amer. Chem. Soc.*, 1968, **90**, 2542.

benzoic acid. The zinc complexes are rapidly hydrolyzed by water, in contrast to those of beryllium. The acetate complex has been utilized as a means of purifying beryllium by solvent extraction from an aqueous solution into an organic layer. When $BeCl_2$ is dissolved in N_2O_4 in ethyl acetate, crystalline $Be(NO_3)_2 \cdot 2N_2O_4$ is obtained. When heated at 50° this gives $Be(NO_3)_2$, which at 125° decomposes to N_2O_4 and volatile $Be_4O(NO_3)_6$. The structure of the latter appears to be similar to that of the acetate but with bridging nitrate groups. The basic nitrate is insoluble in non-polar solvents.

The only halogeno complexes are the *tetrafluoroberyllates*, which are obtained by dissolving BeO or $Be(OH)_2$ in concentrated solutions or melts of acid fluorides such as NH_4HF_2. The tetrahedral ion has a crystal chemistry similar to that of SO_4^{2-}, and corresponding salts, e.g., $PbBeF_4$ and $PbSO_4$, usually have similar structures and solubility properties. BeF_2 readily dissolves in water to give mainly $BeF_2(H_2O)_2$ according to ^9Be nmr spectra.[11] In $1M$ solutions of $(NH_4)_2BeF_4$ the ion BeF_3^- occurs to 15–20%.[12]

The interaction between Cl^- and $[Be(H_2O)_4]^{2+}$ is very small and may be outer sphere in nature.

Other Complexes. The stability of complexes with ligands containing nitrogen or other atoms is lower than those of oxygen ligands. Thus $[Be(NH_3)_4]Cl_2$ is thermally stable but is rapidly hydrolyzed in water. When $BeCl_2$ is treated with the Li salt of 2,2'-bipyridine, a green paramagnetic complex is formed which is best regarded as a complex of Be^{2+} with the bipyridinyl radical anion.[13]

Most of the other nitrogen complexes are derived from the hydride (page 214) or organoberylliums, although compounds are known such as $[Be(NMe_2)_2]_3$, which have a central four-coordinate Be and terminal three-coordinate Be atoms with both bridge and terminal NMe_2 groups.[14]

7-7. Organoberyllium Compounds[15]

Although beryllium alkyls can be obtained by the interaction of $BeCl_2$ with lithium alkyls or Grignard reagents, they are best made in a pure state[8, 16] by heating the metal and a mercury dialkyl, for example:

$$HgMe_2 + Be \xrightarrow{\ 110°\ } BeMe_2 + Hg$$

[11] R. A. Kovar and G. L. Morgan, *J. Amer. Chem. Soc.*, 1970, **92**, 5067.
[12] J. C. Kotz, R. Schaeffer and A. Clouse, *Inorg. Chem.*, 1967, **6**, 620.
[13] G. E. Coates and S. I. E. Green, *J. Chem. Soc.*, **1962**, 3340.
[14] J. L. Atwood and G. D. Stucky, *J. Amer. Chem. Soc.*, 1969, **91**, 4426.
[15] N. R. Felter, *Organometallic Chem. Rev.* 1968, **3**, 1; B. J. Wakefield, *Adv. Inorg. Chem. Radiochem.*, 1968, **11**, 341; G. E. Coates, M. L. H. Green and K. Wade, *Organometallic Chemistry*, 3rd ed., Vol. 1, Methuen, 1967; E. C. Ashby, *Quart. Rev.*, 1967, **21**, 259 (Grignard reagents); S. T. Yoffe and A. N. Nesmeyanov, *The Organic Compounds of Be, Mg, Ca, Sr and Ba: Methods of Elemento Organic Chem. Series*, Vol. 2, North Holland, 1967; S. T. Yoffe and A. N. Nesmeyanov, *Handbook of Magnesium Organic Compounds*, Vols. I–III, Pergamon Press; G. E. Coates and G. L. Morgan, *Adv. Organometallic Chem.*, 1970, **9**, 195 (beryllium).
[16] J. R. Sanders, Jr., E. C. Ashby and J. H. Clark, *J. Amer. Chem. Soc.*, 1968, **90**, 6385.

The alkyl can be collected by sublimation or distillation in a vacuum. On the other hand, the aryls are made by reaction of a lithium aryl in a hydrocarbon with $BeCl_2$ in diethyl ether in which the LiCl formed is insoluble, for example:

$$2 LiC_6H_5 + BeCl_2 \rightarrow 2 LiCl\downarrow + Be(C_6H_5)_2$$

The beryllium alkyls are liquids or solids of high reactivity, being spontaneously flammable in air and violently hydrolyzed by water. Dimethylberyllium is a chain polymer (cf. $BeCl_2$, Fig. 7-1) with bridging CH_3 groups; for the bonding see page 306. In the vapour,[17] $BeMe_2$ is monomeric and linear (sp); it is also monomeric in ether, presumably as the complex $Me_2Be(OEt_2)_2$. The alkyls readily undergo exchange reactions in solution and, as in Grignard reagents, the equilibrium

$$BeR_2 + BeX_2 \rightleftharpoons 2RBeX$$

lies to the right.

The higher alkyls are progressively less highly polymerized; diethyl- and diisopropyl-beryllium are dimeric in benzene but the t-butyl compound is monomeric; the same feature is found in aluminum alkyls.

As with several other elements, notably Mg and Al, there are close similarities between the alkyls and hydrides, especially in the complexes with donor ligands. For the polymeric alkyls, especially $BeMe_2$, strong donors such as Et_2O, Me_3N or Me_2S are required to break down the polymeric structure.[18] Mixed hydrido alkyls are known; thus pyrolysis of diisopropyl-beryllium gives a colorless, non-volatile polymer:

$$x(iso\text{-}C_3H_7)_2Be \xrightarrow{200°} [(iso\text{-}C_3H_7)BeH]_x + xC_3H_6$$

however, above 100° the $tert$-butyl analog gives pure BeH_2 (page 185). With tertiary amines, reactions of the following types may occur:

$$BeMe_2 + Me_3N \rightarrow Me_3N \cdot BeMe_2$$
$$2BeH_2 + 2R_3N \rightarrow [R_3NBeH_2]_2$$

The trimethylamine hydrido complex[18a] appears to have the structure (7-I).

(7-I)

Beryllium alkyls give colored complexes with 2,2'-bipyridine, for example, $bipyBe(C_2H_5)_2$, which is bright red; the colors of these and similar complexes[13] with aromatic amines given by Be, Zn, Cd, Al and Ga alkyls are believed to be due to electron transfer from the M—C bond to the lowest unoccupied orbital of the amine.

[17] R. A. Kovar and G. L. Morgan, *Inorg. Chem.*, 1969, **8**, 1099.
[18] R. A. Kovar and G. L. Morgan, *J. Amer. Chem. Soc.*, 1969, **91**, 7269.
[18a] L. H. Shepherd, G. L. Ter Haar and E. M. Marlett, *Inorg. Chem.*, 1969, **8**, 976.

MAGNESIUM, CALCIUM, STRONTIUM, BARIUM AND RADIUM

7-8. Occurrence; the Elements

These elements, except radium, are widely distributed in minerals and in the sea. They occur in substantial deposits such as *dolomite*, $CaCO_3 \cdot MgCO_3$; *carnallite*, $MgCl_2 \cdot KCl \cdot 6H_2O$; *barytes*, $BaSO_4$; etc. Calcium is the third most abundant metal terrestrially.

Magnesium is produced in several ways. An important source is dolomite from which, after calcination, the calcium is removed by ion exchange using sea water. The equilibrium is favorable because the solubility of $Mg(OH)_2$ is lower than that of $Ca(OH)_2$:

$$Ca(OH)_2 \cdot Mg(OH)_2 + Mg^{2+} \rightarrow 2Mg(OH)_2 + Ca^{2+}$$

The most important processes for preparation of magnesium are (*a*) the electrolysis of fused halide mixtures (e.g., $MgCl_2 + CaCl_2 + NaCl$) from which the least electropositive metal, Mg, is deposited, and (*b*) the reduction of MgO or of calcined dolomite ($MgO \cdot CaO$). The latter is heated with ferrosilicon:

$$CaO \cdot MgO + FeSi = Mg + \text{silicates of Ca and Fe}$$

and the magnesium is distilled out. MgO can be heated with coke at 2000° and the metal deposited by rapid quenching of the high-temperature equilibrium which lies well to the right:

$$MgO + C \rightleftarrows Mg + CO$$

Calcium and the other metals are made only on a relatively small scale, by electrolysis of fused salts or reduction of the halides with sodium.

Radium is isolated in the processing of uranium ores; after coprecipitation with barium sulfate, it can be obtained by fractional crystallization of a soluble salt.

Magnesium is a greyish-white metal with a surface oxide film which protects it to some extent chemically—thus it is not attacked by water, despite the favorable potential, unless amalgamated. It is readily soluble in dilute acids and is attacked by most alkyl and aryl halides in ether solution to give Grignard reagents. Calcium and the other metals are soft and silvery, resembling sodium in their chemical reactivities although somewhat less reactive. These metals are also soluble, though less readily and to a lesser extent than sodium, in liquid ammonia, giving blue solutions of a similar nature to those of the Group I metals (page 193). These blue solutions are also susceptible to decomposition (with the formation of the amides) and have other chemical reactions similar to those of the Group I metal solutions. They differ, however, in that moderately stable metal ammines such as $Ca(NH_3)_6$ can be isolated on removal of solvent at the boiling point.

7-9. Binary Compounds

Oxides. The oxides, MO, are obtained most readily by calcination of the carbonates. They are white crystalline solids with ionic, NaCl-type lattices.

Magnesium oxide is relatively inert, especially after ignition at high temperatures, but the other oxides react with water, evolving heat, to form the hydroxides. They also absorb carbon dioxide from the air. Magnesium hydroxide is insoluble in water ($\sim 1 \times 10^{-4}$ g/l at $20°$) and can be precipitated from Mg^{2+} solutions; it is a much weaker base than the Ca–Ra hydroxides, although it has no acidic properties and unlike $Be(OH)_2$ is insoluble in excess of hydroxide. The Ca–Ra hydroxides are all soluble in water, increasingly so with increasing atomic number [$Ca(OH)_2$, ~ 2 g/l; $Ba(OH)_2$, ~ 60 g/l at $\sim 20°$], and all are strong bases.

There is no optical transition in the electronic spectra of the M^{2+} ions and they are all colorless. Colors of salts are thus due only to colors of the anions or to lattice defects. The oxides may also be obtained with defects, and BaO crystals with $\sim 0.1\%$ excess of metal in the lattice are deep red.

Halides. The anhydrous halides can be made by dehydration (Sect. 16-8) of the hydrated salts. For rigorous studies, however, magnesium halides are best made[8] by the reaction

$$Mg + HgX_2 \xrightarrow{\text{Boiling ether}} MgX_2(solv) + Hg$$

Magnesium and calcium halides readily absorb water. The tendency to form hydrates, as well as the solubilities in water, decrease with increasing size, and Sr, Ba and Ra halides are normally anhydrous. This is attributed to the fact that the hydration energies decrease more rapidly than the lattice energies with increasing size of M^{2+}.

The fluorides vary in solubility in the reverse order, i.e. Mg < Ca < Sr < Ba, because of the small size of the F^- relative to the M^{2+} ion. The lattice energies decrease unusually rapidly because the large cations make contact with one another without at the same time making contact with the F^- ions.

All the halides appear to be essentially ionic. On account of its dispersion and transparency properties, CaF_2 is used for prisms in spectrometers and for cell windows (especially for aqueous solutions). It is also used to provide a stabilizing lattice for trapping lanthanide +2 ions[19] (cf. Chapter 27).

Carbides. All the metals in the Ca–Ba series or their oxides react directly with carbon[20] in an electric furnace to give the carbides, MC_2. These are ionic acetylides whose general properties [hydrolysis to $M(OH)_2$ and C_2H_2, structures, etc.] are discussed in Chapter 10. Magnesium at $\sim 500°$ gives MgC_2 but, at $500–700°$ with an excess of carbon, Mg_2C_3 is formed, which on hydrolysis gives $Mg(OH)_2$ and propyne and is presumably ionic, that is, $(Mg^{2+})_2(C_3^{4-})$.

Other Compounds. Direct reaction of the metals with other elements can lead to binary compounds such as borides, silicides, arsenides, sulfides, etc. Many of these are ionic and are rapidly hydrolyzed by water or dilute acids. At $\sim 300°$, magnesium reacts with nitrogen[21] to give colorless, crystalline

[19] See, e.g., P. F. Walker, *Inorg. Chem.*, 1966, **5**, 736, 739.
[20] R. L. Faircloth, R. H. Flowers and F. C. W. Pummery, *J. Inorg. Nuclear Chem.*, 1967, **29**, 311.
[21] R. I. Bickley and S. I. Gregg, *J. Chem. Soc., A*, **1966**, 1349.

Mg_3N_2 (resembling Li and Be in this respect). The other metals also react normally to form M_3N_2, but other stoichiometries are known. X-ray study[22] of M_2N indicates a larger lattice of the *anti*-$CdCl_2$ type (page 54) found also for Cs_2O, Ag_2F, etc. Here the metal ions are to be regarded as M^{2+} with the excess electrons essentially free in the lattice. This can account for the graphitic luster and the semiconductor behavior.

The *hydrides* are discussed on page 180; a complex salt $KMgH_3$ has been prepared.[23]

7-10. Oxo Salts, Ions and Complexes

All the elements of this Group form *oxo salts*, those of Mg and Ca often being hydrated. The carbonates are all rather insoluble in water, and the solubility products decrease with increasing size of M^{2+}. The same applies to the sulfates; magnesium sulfate is readily soluble in water, and calcium sulfate has a hemihydrate $2CaSO_4 \cdot H_2O$ (plaster of Paris) which readily absorbs more water to form the very sparingly soluble $CaSO_4 \cdot 2H_2O$ (gypsum), while Sr, Ba, and Ra sulfates are insoluble and anhydrous. The nitrates of Sr, Ba, and Ra are also anhydrous and the last two can be precipitated from cold aqueous solution by addition of fuming nitric acid. Magnesium perchlorate is used as a drying agent.

For water, acetone and methanol solutions, nuclear magnetic resonance studies[24] have shown that the coordination number of Mg^{2+} is 6, although in liquid ammonia[25] it appears to be 5. The $[Mg(H_2O)_6]^{2+}$ ion is not acidic and in contrast to $[Be(H_2O)_4]^{2+}$ can be dehydrated fairly readily; it occurs in a number of crystalline salts.

Complexes. Only Mg and Ca show any appreciable tendency to form complexes and in solution, with a few exceptions, these are with oxygen ligands. $MgBr_2$, MgI_2 and $CaCl_2$ are soluble in alcohols and some other organic solvents, as is $Mg(ClO_4)_2$; cationic solvated ions (see above) may be formed in these solvents. Adducts of ethers are known, e.g., $MgBr_2(OEt_2)_2$ and $MgBr_2(THF)_4$.[26]

Oxygen chelate compounds, among the most important being those of the ethylenediaminetetraacetate (EDTA) type, readily form complexes in alkaline aqueous solution, e.g.:

$$Ca^{2+} + EDTA^{4-} = [Ca(EDTA)]^{2-}$$

The complexing of calcium by EDTA and also by polyphosphates is of some importance, not only for removal of calcium ions from water, but also for the volumetric estimation of calcium.

The only known halo complex $[Et_4N]_2MgCl_4$ has been made by inter-

[22] E. T. Keve and A. C. Skapski, *Inorg. Chem.*, 1968, **7**, 1757.
[23] E. C. Ashby, R. Kovar and R. Arnott, *J. Amer. Chem. Soc.*, 1970, **92**, 2182.
[24] N. A. Matwiyoff and H. Taube, *J. Amer. Chem. Soc.*, 1968, **90**, 2796; S. Nakamura and S. Meiboom, *J. Amer. Chem. Soc.*, 1967, **89**, 1765.
[25] T. J. Swift and H. H. Lo, *J. Amer. Chem. Soc.*, 1967, **89**, 3988.
[26] L. J. Guggenberger and R. E. Rundle, *J. Amer. Chem. Soc.*, 1968, **90**, 5375.

action of $MgCl_2$ and Et_4NCl in $SOCl_2$. Nitrogen ligands generally form weak complexes that exist in the solid state and dissociate in aqueous solution. Both Mg and Ca halides absorb NH_3 or amines to give e.g., $[Mg(NH_3)_6]Cl_2$. Ca, Sr and Ba perchlorates give 9-coordinate ions, $[M \, dien_3] \, (ClO_4)_2$, but again these exist only in the solid state.[27]

An important exception to this rule is provided by the magnesium complexes of tetrapyrrole systems, the parent compound of which is porphine (7-II). These conjugated heterocycles provide a rigid planar environment for Mg^{2+} (and similar) ions. The most important of such derivatives are the *chlorophylls* and related compounds,[28] which are of transcendental importance in photosynthesis in plants.[29] The structure of chlorophyll-*a*, one of the many chlorophylls, is (7-III).

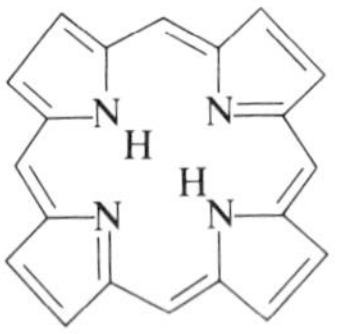

(7-II) 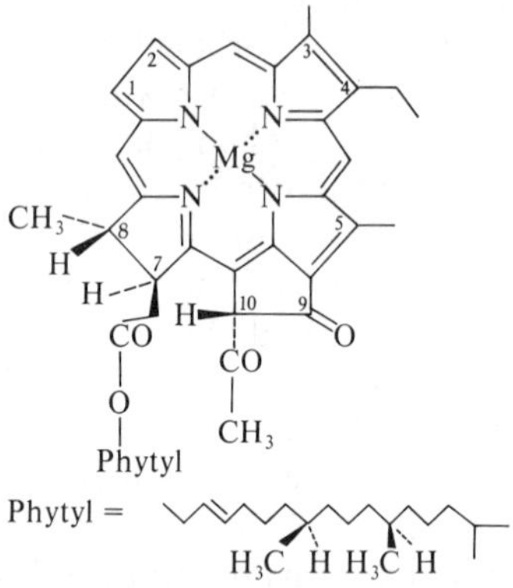   (7-III)

In such porphine compounds, the Mg atom is formally 4-coordinate but further interaction with either water or other solvent molecules is a common, if not universal, occurrence; further, in chlorophyll, interaction with the keto group at position 9 in *another* molecule is also established. It also appears that 5-coordination is preferred over 6-coordination as in the structure of magnesium tetraphenylporphyrin hydrate,[30] where the Mg atom is out of the plane of the N atoms and is approximately square pyramidal. Although Mg and other metalloporphyrins can undergo oxidation[31] by one-electron changes, for Mg it is the macrocycle and not the metal that is involved.

In chlorophylls, hydrogen-bonding interactions lead to polymerization (Fig. 7-3); the hydrates may be monomeric or dimeric in benzene, but ordered aggregates of colloidal dimensions are formed in dodecane. Where a polar solvent is not present, association via coordination of the keto group at position 9 occurs as in solutions of anhydrous chlorophyll in alkanes.[32]

[27] P. S. Gentile, J. Carlotto and T. A. Shankoff, *J. Inorg. Nuclear Chem.*, 1967, **29**, 1427.
[28] L. P. Vernon and G. R. Seely, eds., *The Chlorophylls*, Academic Press, 1966.
[29] E. Rabinowitch and Govindjee, *Photosynthesis*, Wiley, 1969.
[30] R. Timkovitch and A. Tulinsky, *J. Amer. Chem. Soc.*, 1969, **91**, 4430.
[31] Cf. J. H. Fuhrhop and D. Mauzerall, *J. Amer. Chem. Soc.*, 1969, **91**, 4174.
[32] K. Ballschmiter and J. J. Katz, *J. Amer. Chem. Soc.*, 1969, **91**, 2661; *Biochim. Biophys. Acta*, 1969, **180**, 347.

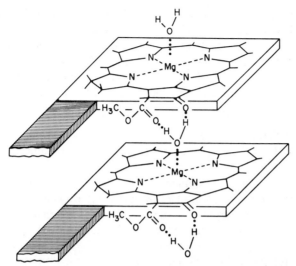

Fig. 7-3. Structure illustrating the chlorophyll-*a*–water–chlorophyll-*a* interaction. The dimensions of the ring and the phytyl chain are not to scale. [Reproduced by permission from K. Ballschmiter and J. J. Katz, *J. Amer. Chem. Soc.*, 1969, **91**, 2661.]

The role of chlorophyll in the photosynthetic reduction of CO_2 by water in plants is to provide a source of electrons that may continue to be supplied for a time in the dark. Electron spin resonance studies of light-irradiated chlorophyll show that radicals are formed.[33] These are probably of the type (7-IV). The electrons are transmitted through chlorophyll micelles to other intermediates involved in the reduction of CO_2.

(7-IV)

7-11. Organomagnesium Compounds[15]

Where they are known, the organic compounds of Ca, Sr and Ba are highly ionic, reactive, and of little utility or importance. Magnesium compounds, of which the Grignard reagents are the best known, are probably the most widely used of all organometallic compounds. They are employed for the synthesis of alkyl and aryl compounds of other elements as well as for a host of organic syntheses.

[33] M. Garcia-Morin, R. A. Uphaus, J. R. Norris and J. J. Katz, *J. Phys. Chem.*, 1969, **73**, 1066.

Magnesium compounds are of the types RMgX—the Grignard reagents—and MgR_2. The former are made by direct interaction of the metal with an organic halide RX in a suitable solvent, usually an ether such as diethyl ether or tetrahydrofuran. The reaction is normally most rapid with iodides RI, and iodine may be used as an initiator. For most purposes, RMgX reagents are used *in situ*. The species MgR_2 are best made[8] by the dry reaction

$$HgR_2 + Mg(\text{excess}) \rightarrow Hg + MgR_2$$

The dialkyl or diaryl is then extracted with an organic solvent. Both RMgX, as solvates, and R_2Mg, are reactive, being sensitive to oxidation by air and to hydrolysis by water.

There has been prolonged controversy concerning the nature of Grignard reagents in solutions. Discordant results have often been obtained because of failure to eliminate impurities, such as traces of water or oxygen, which can aid or inhibit the attainment of equilibrium and the occurrence of exchange reactions. Although recent work has given a reasonable understanding, the following discussion probably applies only to Grignard reagents prepared under strict conditions and not to those normally prepared without special precautions in the laboratory.

X-ray diffraction studies on certain crystalline Grignard reagents have been made.[26,34] In the structure of $C_6H_5MgBr \cdot 2(Et_2O)$ and $C_2H_5MgBr \cdot 2(Et_2O)$ the Mg atom is, essentially, tetrahedrally surrounded by C, Br and two oxygen atoms of the ether as in (7-V). For less sterically demanding ethers such as

$$
\begin{array}{c}
Br \\
| \\
R\text{---}Mg\text{---}OEt_2 \\
| \\
OEt_2
\end{array}
$$

(7-V)

tetrahydrofuran, higher coordination numbers may occur, as in $CH_3MgBr \cdot$ 3THF which is *tbp* (cf. also $MgBr_2 \cdot 2Et_2O$ but $MgBr_2 \cdot 4THF$). Thus it is now clear that in crystals the basic Grignard structure is $RMgX \cdot n(\text{solvent})$.

For diethylmagnesium, $Mg(C_2H_5)_2$, the structure is that of a chain polymer, similar to that of $Be(CH_3)_2$, with bridging methylene groups but again tetrahedral Mg.[35] A special case is that of magnesium cyclopentadienide, $Mg(C_5H_5)_2$ which has a "sandwich" structure similar to that of ferrocene (Section 23-2), but with $C_5H_5^-$ and Mg^{2+}. This compound is readily made by direct action of cyclopentadiene vapor on hot Mg, or by thermal decomposition of C_5H_5MgBr, which in turn is made by action of cyclopentadiene, C_5H_6, on C_2H_5MgBr in solution.

The nature of Grignard reagents *in solution* is complex and depends critically on the nature of the alkyl and halide groups and on the solvent, concentration and temperature. Quite generally, the equilibria involved are

[34] M. Vallino, *J. Organometallic Chem.*, 1969, **20**, 1.
[35] E. Weiss, *J. Organometallic Chem.*, 1965, **4**, 101.

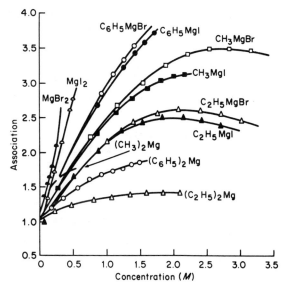

Fig. 7-4. Association of several alkyl- and aryl-magnesium bromides and iodides and related magnesium compounds in diethyl ether. [Reproduced by permission from ref. 36.]

of the type:

$$RMg\overset{\displaystyle X}{\underset{\displaystyle X}{\diagdown\diagup}}MgR \; \rightleftarrows \; 2RMgX \; \rightleftarrows \; R_2Mg + MgX_2 \; \rightleftarrows \; \overset{\displaystyle R}{\underset{\displaystyle R}{\diagup\diagdown}}Mg\overset{\displaystyle X}{\underset{\displaystyle X}{\diagdown\diagup}}Mg$$

Solvation (not shown) occurs and association is predominantly by halide rather than by carbon bridges, except for methyl compounds where bridging by CH_3 groups may occur.

In dilute solutions and in more strongly donor solvents the monomeric species normally predominate; but in diethyl ether at concentrations $> 0.1M$ association occurs[36] and linear or cyclic polymers may be present. The behavior of several compounds is shown in Fig. 7-4, which includes also the halides $MgBr_2$ and MgI_2.

Nmr spectra normally do not distinguish between $RMgX$ and R_2Mg in solution because of rapid exchange of alkyl or aryl groups via a transition state such as (7-VI). However, for C_6F_5MgBr and $(C_6F_5)_2Mg$ the distinc-

$$R\!-\!Mg\overset{\displaystyle R}{\underset{\displaystyle X}{\diagup\diagdown}}Mg\!-\!R$$
$$(7\text{-VI})$$

tion can be made at room temperature, although for normal aryls and alkyls lower temperatures are required. At temperatures below $-70°$ the exchange rates are slow and nmr spectra of $(CH_3)_2Mg$ and CH_3MgBr can be resolved;

[36] F. W. Walker and E. C. Ashby, *J. Amer. Chem. Soc.*, 1969, **91**, 3845.

the distinction can also be made in presence of hexamethylphosphoramide at 25°.[37]

More complicated Mg alkyls can be prepared. Thus $MgMe_2$ dissolves in Al_2Me_6 to give species with CH_3 bridges.[38] Stable complexes with certain

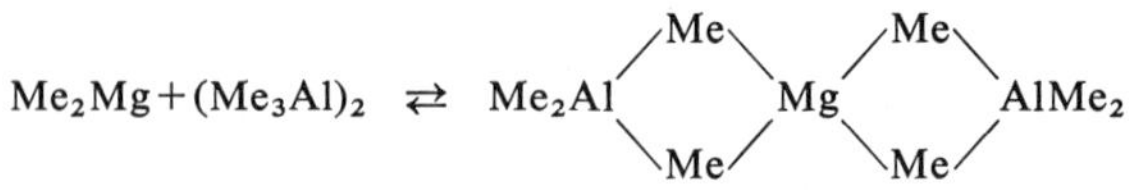

$$Me_2Mg+(Me_3Al)_2 \rightleftharpoons Me_2Al \begin{array}{c} Me \\ \diagdown \\ Me \end{array} Mg \begin{array}{c} Me \\ \diagdown \\ Me \end{array} AlMe_2$$

amines can also be obtained, for example with N,N,N',N'-tetramethylethylenediamine (TMED) which gives $C_6H_5MgBr(TMED)$ and $R_2Mg(TMED)$. With other amines, elimination of alkane may occur to give complexes similar to those of Be (page 214), with bridging alkylamido groups and 3-coordinate Mg.[39]

[37] E. C. Ashby, G. Parris and F. W. Walker, *Chem. Comm.*, **1969**, 1464; D. F. Evans and G. V. Fazakerley, *J. Chem. Soc.*, *A*, **1971**, 184.
[38] J. L. Attwood and G. D. Stucky, *J. Amer. Chem. Soc.*, 1969, **91**, 2538.
[39] G. E. Coates and D. Ridley, *J. Chem. Soc.*, *A*, **1967**, 56.

Further Reading

Bellamy, R. G. and N. A. Hall, *Extraction and Metallurgy of Uranium, Thorium and Beryllium*, Pergamon Press, 1965.
Darwin, F. E. and J. H. Buddery, *Beryllium* (No. 7 of *Metallurgy of the Rarer Metals*), Butterworths, 1960.
Kharasch, M. S. and O. Reinmuth, *Grignard Reactions of Non-Metallic Substances*, Constable and Co., and Prentice-Hall, 1954.
Pannell, E. V., *Magnesium: Its Production and Use*, Pitman, 1948.
Williams, R. J. P., *Quart. Rev.*, 1970, **24**, 331 (biochemistry of Ca, Mg, Na and K).

8
Boron

8-1. Electronic Structure and Bonding

The first ionization potential of boron, 8.296 eV, is rather high, and the next two are much higher. Thus the total energy required to produce B^{3+} ions is far more than would be compensated by lattice energies of ionic compounds or by hydration of such ions in solution. Consequently, simple electron loss to form a cation plays no part in boron chemistry. Instead, covalent bond formation is of major importance, and boron compounds usually resemble those of other non-metals, notably silicon, in their properties and reactions.

Despite the $2s^2 2p$ electronic structure, boron is always trivalent and never monovalent. This is because the total energy released in formation of three bonds in a BX_3 compound exceeds the energy of formation of one bond in a BX compound by more than enough to provide for promotion of boron to a hybridized valence state of the sp^2 type, wherein the three sp^2 hybrid orbitals lie in one plane at angles of 120°. It would therefore be expected, and is indeed found, without exception, that all monomeric, three-covalent boron compounds (trihalides, trialkyls, etc.) are planar with X—B—X bond angles of 120°. The covalent radius for trigonally hybridized boron is not well defined but probably lies between 0.85 and 0.90 Å. There are apparently substantial shortenings of many B—X bonds, and this has occasioned much discussion. For example, the estimated B—F, B—Cl and B—Br distances would be ~ 1.52, ~ 1.87 and ~ 1.99 Å, whereas the actual distances in the respective trihalides are 1.30, 1.75 and 1.87 Å.

Three factors appear to be responsible for the shortness of bonds to boron:

1. Formation of $p\pi$–$p\pi$ bonds using filled $p\pi$ orbitals of the halogens and the vacant $p\pi$ orbital of boron. This is probably most important in BF_3, but of some significance in BCl_3 and BBr_3 as well.

2. Strengthening and hence shortening of the B—X bonds by ionic-covalent resonance, especially for B—F and B—O bonds because of the large electronegativity differences. Evidence that this is important, in addition

to the dative $p\pi–p\pi$ bonding, is afforded by the fact that even in BF_3 complexes such as $(CH_3)_3\overset{+}{N}\overset{-}{B}F_3$ and BF_4^-, where the $p\pi–p\pi$ bond must be largely or totally absent, the B—F bonds are still apparently shortened.[1]

3. Because of the incomplete octet in boron, repulsions between nonbonding electrons may be somewhat less than normal, permitting closer approach of the bonded atoms.

Elemental boron has properties that place it on the borderline between metals and non-metals. It is a semiconductor, not a metallic conductor, and chemically must be classed as a non-metal. In general, boron chemistry resembles that of Si more closely than that of Al, Ga, In and Tl. The main resemblances to Si and differences from Al are the following:

1. The similarity and complexity of the boric and silicic acids is notable. Boric acid, $B(OH)_3$, is weakly but definitely acidic, and not amphoteric, whereas $Al(OH)_3$ is mainly basic with some amphoteric behavior.

2. The hydrides of B and Si are volatile, spontaneously flammable, and readily hydrolyzed, whereas the only binary hydride of Al is a solid, polymeric material. However, structurally, the boron hydrides are unique, having unusual stoichiometries and configurations and unusual bonding because of their *electron-deficient* nature.

3. The boron halides (not BF_3), like the silicon halides, are readily hydrolyzed, whereas the aluminum halides are only partially hydrolyzed in water.

4. B_2O_3 and SiO_2 are similar in their acidic nature, as shown by the ease with which they dissolve metallic oxides on fusion, to form borates and silicates, and both readily form glasses that are difficult to crystallize. Certain oxo compounds of B and Si are structurally similar, specifically the linear $(BO_2)_x$ and $(SiO_3)_x$ ions in metaborates and pyroxene silicates, respectively.

5. However, despite dimerization of the halides of Al and Ga and of the alkyls of Al, they behave as acceptors and form adducts similar to those given by boron halides and alkyls, for example, $Cl_3\overset{-}{Al}\overset{+}{N}(CH_3)_3$. Aluminum, like boron, also forms volatile alkoxides such as $Al(OC_2H_5)_3$, which are similar to borate esters, $B(OR)_3$.

8-2. Acceptor Behavior

In BX_3 compounds the boron octet is incomplete; boron has a low-lying orbital which it does not use in bonding owing to a shortage of electrons, although partial use is made of it in the boron halides through B—X multiple bonding. The alkyls and halides of aluminum make up this insufficiency of electrons by forming dimers with alkyl or halogen bridges, but the boron compounds do not. The reason or reasons for this difference are not known with certainty. The size factor may be important for BCl_3 and BBr_3, since the small boron atom may be unable to coordinate strongly to four atoms as large as Cl and Br. The fact that BCl_4^- and BBr_4^- ions are stable only in crystalline salts of large cations such as Cs^+ or $(CH_3)_4N^+$ might suggest this.

[1] G. Brunton, *Acta Cryst.*, 1968, *B*, **24**, 1703.

The fact that a certain amount of B—X $p\pi$–$p\pi$ bond energy would have to be sacrificed would also detract from the stability of dimers relative to monomers. The size factor cannot be controlling for BF_3, however, since BF_4^- is quite stable. Here, the donor power of the fluorine already bonded to another boron atom may be so low that the energy of the bridge bonds would not be sufficient to counterbalance the energy required to break the B—F π bonding in the monomer. Phenomena of this nature are often difficult to explain with certainty.

An important consequence of the incomplete octet in BX_3 compounds is their ability to behave as acceptors (Lewis acids), in which boron achieves its maximum coordination with approximately sp^3 hybridization. Thus, various Lewis bases, such as amines, phosphines, ethers and sulfides, form 1:1 complexes with BX_3 compounds. The following are representative of the addition compounds formed: $(CH_3)_3NBCl_3$, $(CH_3)_3PBH_3$, $(C_2H_5)_2OBF_3$.

There is good evidence that the relative strengths of the boron halides as Lewis acids are in the order $BBr_3 \geqslant BCl_3 > BF_3$. This order is the opposite of what would be expected both on steric grounds and from electronegativity considerations. It can be explained, at least partially, in terms of the boron–halogen π bonding. In an addition compound this π bonding is largely or completely lost, so that addition compounds of the trihalide with the strongest π bonding will be the most destabilized by loss of the energy of π bonding. Calculations indicate that the π bonding energies of the trihalides are in the order $BF_3 \geqslant BCl_3 > BBr_3$. However, certain properties of the BX_3 adducts with donor molecules suggest that the donor-to-boron bonds may themselves increase in strength in the order $BF_3 < BCl_3 < BBr_3$. No satisfactory explanation has been given for this.

Boron also completes its octet by forming both anionic and cationic complexes. The former type has long been known and includes such important species as BF_4^-, BH_4^-, $B(C_6H_5)_4$ and $BH(OR)_3^-$, as well as chelates such as $[B(o\text{-}C_6H_4O_2)_2]^-$ and the salicylato complex (8-I) which has been partially

(8-I)   (8-II)

resolved by fractional crystallization of its strychnine salts. The cationic species[2] are of three main types: $[(base)_2BH_2]^+$, $[(base)_2BHX]^+$ and $[(base)_3BH]^{2+}$. Representative preparative procedures are:

$$py\,BH_3 + py + [Ph_3C]^+[BF_4]^- \rightarrow Ph_3CH + [(py)_2BH_2]BF_4$$

$$py\,BH_3 \xrightarrow{\;I_2\;} py\,BHI_2 \xrightarrow{\;bipy\;} [(py)(bipy)BH]^{2+}$$

$$(Me_3N)BH_2I \xrightarrow{\;py\;} [(Me_3N)(py)BH_2]^+ \xrightarrow{\;Br_2\;} [(Me_3N)(py)BHBr]^+$$

[2] G. E. Ryschkewitsch and T. E. Sullivan, *Inorg. Chem.*, 1970, 9, 899; L. E. Benjamin *et al.*, *Inorg. Chem.*, 1970, 9, 1844.

These cations have considerable hydrolytic stability, though they are attacked by base; an analog, (8-II), of the last one, above, has been resolved and is optically stable in acid at 25°.

THE ELEMENT

8-3. Occurrence, Isolation and Properties

The most abundant boron mineral is *tourmaline*, a complex aluminosilicate containing about 10% of boron. The principal boron ores are borates, such as borax, $Na_2B_4O_5(OH)_4 \cdot 8H_2O$, which occur in large beds in arid parts of California and elsewhere.

Natural boron consists of two isotopes, ^{10}B (19.6%) and ^{11}B (80.4%). Isotopically enriched boron compounds can be made and are useful in spectroscopic and reaction-mechanism studies. The boron nuclear spins, (^{10}B, $S = 3$; ^{11}B, $S = 3/2$) are also highly useful in structural elucidation. For an example see page 244.

It is exceedingly difficult to prepare elemental boron in a high state of purity because of its high melting point and the corrosive nature of the liquid. It can be prepared in quantity but low purity (95–98%) in an amorphous form by reduction of B_2O_3 with Mg, followed by vigorous washing of the material so obtained with alkali, hydrochloric acid, and hydrofluoric acid. This amorphous boron is a dark powder that may contain some microcrystalline boron but also contains oxides and borides.

The preparation of pure boron in crystalline form is a matter of considerable complexity and difficulty even when only small research-scale quantities are required. There are three allotropic forms whose structures are known in detail, but other allotropes not as yet structurally characterized certainly exist.[3]

α-Rhombohedral boron has been obtained by pyrolysis of BI_3 on tantalum, tungsten and boron nitride surfaces at 800–1000°, by pyrolysis of boron hydrides, and by crystallization from boron–platinum melts at 800–1200°. It is the most dense allotrope, and its structure consists entirely of B_{12} icosahedra (cf. Fig. 1–8b, page 21), which are packed together in a manner similar to cubic closest packing of spheres; there are bonds between the icosahedra which are, however, weaker than those within the icosahedra.

A tetragonal form of boron, which can be obtained by reduction of BBr_3 with H_2 on a tantalum or tungsten filament at 1200–1400°, is the form longest known; it consists of layers of B_{12} icosahedra connected by single boron atoms.

β-Rhombohedral boron is invariably obtained by crystallization of fused boron. It is built entirely of B_{12} icosahedra packed together, with B—B bonds between them, in a more complicated way than in the case of α-rhombohedral

---

[3] For the definitive review of this complex subject see J. L. Hoard and R. E. Hughes, in ref. 4.

boron. It appears that the β-rhombohedral form is the thermodynamically stable one over a considerable range of temperature, though sluggish attainment of equilibria makes this a difficult point to establish with certainty. The melting point is $2250 \pm 50\,°C$.

Crystalline boron is extremely inert chemically. It is unaffected by boiling HCl or HF, only slowly oxidized by hot, concentrated nitric acid when finely powdered, and either not attacked or only very slowly attacked by many other hot concentrated oxidizing agents.

BORON COMPOUNDS[4]

8-4. Borides[5]

Compounds of boron with elements less electronegative than itself (i.e. metals) are called borides. Often compounds of boron with rather less metallic or metalloidal elements (e.g., P, As) are also termed borides. Borides of most but not all elements are known. They are generally hard, refractory substances and fairly inert chemically, and they often possess very unusual physical and chemical properties. For example, the electrical and thermal conductivities of ZrB_2 and TiB_2 are about ten times greater than those of the metals themselves, and the melting points are more than $1000°$ higher. Some of the lanthanide hexaborides are among the best thermionic emitters known. The monoborides of phosphorus and arsenic are promising high-temperature semiconductors, and higher borides of some metalloids, e.g., AsB_6, are remarkably inert to chemical attack.

Industrially, borides are prepared in various ways, including reduction of metal oxides by mixtures of carbon and boron carbide, electrolysis in fused salts, and direct combination of the elements. For research purposes, the last method is usually used. Cobalt boride, made in aqueous solution by reduction of Co^{2+} salts with $NaBH_4$ is an active catalyst for reduction of various substrates.[6]

The borides do not conform to the ordinary concepts of valence either in stoichiometry or in structure. With only a few exceptions, borides are of one of the following main types:

1. Borides with isolated boron atoms. These include most of those with low B to M ratios such as M_4B, M_3B, M_2B, M_5B_2 and M_7B_3. In the M_4B and M_2B structures, boron atoms lie in triangular-prismatic or square-antiprismatic holes between multiple layers of metal atoms. In the others, the metal atoms are arranged in approximately close-packed arrays, with the boron atoms in triangular-prismatic interstices.

2. Borides with single and double chains of boron atoms. As the propor-

[4] E. L. Muetterties, ed., *The Chemistry of Boron and its Compounds*, Wiley, 1967.
[5] B. Aronsson, T. Lundstrom and S. Rundqvist, *Borides, Silicides, and Phosphides*, Methuen, 1965.
[6] J. M. Pratt and G. Swinden, *Chem. Comm.*, **1969**, 1321.

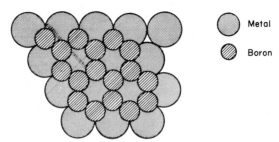

Fig. 8-1.   Parallel layers of metal atoms and boron atoms in MB_2 compounds.

tion of boron atoms increases, so do the possibilities for boron–boron linkages. In V_3B_2 there are pairs of boron atoms. In one modification of Ni_4B_3, two-thirds of the boron atoms form infinite, zigzag chains, while one-third are isolated from other boron atoms; in another modification all the boron atoms are members of chains. MB compounds all have structures with single chains, while in many M_3B_4 compounds there are double chains.

3. Borides with two-dimensional nets. These are represented by MB_2 and M_2B_5 compounds and include some of the best electrically conducting, hardest, and highest-melting of all borides. The crystal structures of the MB_2 compounds are unusually simple, consisting of alternating layers of close-packed metal atoms and "chicken wire" sheets of boron atoms, as shown in Fig. 8-1.

4. Borides with three-dimensional boron networks. The major types have formulae MB_4, MB_6 and MB_{12}. MB_4 compounds may be of several types insofar as structural details are concerned. ThB_4 and CeB_4 contain rather open networks of boron atoms interpenetrating a network of metal atoms. Perhaps as many as twenty other MB_4 compounds have the same structure. The MB_6 structure is fairly easy to visualize with the help of Fig. 8-2: it can be thought of as a CsCl structure, with B_6 octahedra in place of the Cl^- ions; however, the B_6 octahedra are closely linked along the cube edges, so that the boron atoms constitute an infinite three-dimensional network. The MB_{12} compounds also have cubic structures consisting of M atoms and B_{12} cubo-

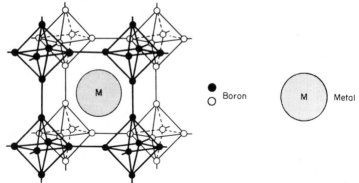

Fig. 8-2.   Atomic arrangement in many MB_6 compounds.

octahedra (8-III) packed in the manner of NaCl; again, the B_{12} polyhedra are closely linked to one another. It has recently been reported[7] that heating NaB_6 at 1000° in argon converts it into NaB_{15}, in which there are icosahedra linked both directly and through B_3 chains.

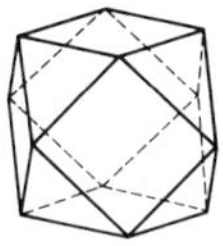

(8-III)

Boron nitride, which can be obtained by interaction of boron with ammonia at white heat, is a slippery white solid with a layer structure very similar to that of graphite (page 288). The units, instead of being hexagonal carbon rings, have alternate B and N atoms 1.45 Å apart with angles of 120° (sp^2 at B). The distance between the sheets is 3.34 Å. The analogy of C—C and B—N further discussed below is heightened by the conversion of graphite-like BN under high temperature and pressure into a cubic form with a diamond-like structure and by the formation of alkali-metal intercalation compounds. The cubic form is extremely hard and will scratch diamond. The nitride is stable in air but slowly hydrolyzed by water.

8-5. Oxygen Compounds of Boron

These are among the most important compounds of boron, comprising nearly all the naturally occurring forms of the element. The structures of such compounds consist mainly of trigonal BO_3 units with the occasional occurrence of tetrahedral BO_4 units. B—O bond energies are 560–790 kJ, rivalled only by the B—F bond in BF_3 (640 kJ) in strength.

Boron Oxides. The principal oxide, B_2O_3, is obtained by fusing boric acid. It usually forms a glass and can be crystallized only with the greatest difficulty. It is acidic, reacting with water to give boric acid, $B(OH)_3$, and, when fused, dissolves many metal oxides to give borate glasses. Both the glassy and the crystalline substances[8] contain infinite chains of triangular BO_3 units, interconnected by weaker B—O bonds.

The lower oxide, BO, is well established though of unknown structure. It apparently contains both B—B and B—O—B bonds and vaporizes to B_2O_2 molecules at 1300–1500°. It is obtained on heating $B_2(OH)_4$ at 250° and <1 mm pressure. $B_2(OH)_4$ itself is obtained in fair yield by the following reaction sequence:

$$2(Me_2N)_2BCl \xrightarrow{\text{Na dispersion}} (Me_2N)_2BB(NMe_2)_2$$
$$\xrightarrow{H_3O^+} (HO)_2BB(OH)_2 + 4Me_2NH_2^+$$

Hydrolysis of the esters $B_2(OR)_4$ at pH 7 affords $B_2(OH)_4$ quantitatively. $B_2(OH)_4$ is believed to have a $(HO)_2B$—$B(OH)_2$ structure.

[7] R. Naslain and J. S. Kasper, *J. Solid State Chem.*, 1970, **1**, 150.
[8] G. E. Gurr *et al.*, *Acta Cryst.*, 1970, B, **26**, 907.

Boric Acid.[9]   Hydrolysis of boron halides, hydrides, etc., affords the acid, $B(OH)_3$, or its salts. The acid forms white, needle-like crystals in which $B(OH)_3$ units are linked together by hydrogen bonds to form infinite layers of nearly hexagonal symmetry; the layers are 3.18 Å apart, which accounts for the pronounced basal cleavage.

Some reactions of boric acid are given in Fig. 8-3.

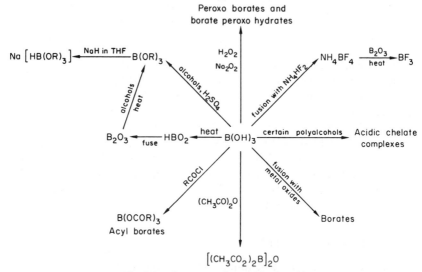

Fig. 8-3.   Some reactions of boric acid.

Boric acid is moderately soluble in water with a large negative heat of solution so that the solubility increases markedly with temperature. It is a very weak and exclusively monobasic acid that is believed to act, not as a proton donor, but as a Lewis acid, accepting OH^-:

$$B(OH)_3 + H_2O \rightleftharpoons B(OH)_4^- + H^+   \qquad pK = 9.00$$

The $B(OH)_4^-$ ion actually occurs in several minerals. At concentrations $\leqslant 0.025M$, essentially only mononuclear species $B(OH)_3$ and $B(OH)_4^-$ are present; but at higher concentrations the acidity increases, and pH measurements are consistent with the formation of polymeric species such as

$$3B(OH)_3 \rightleftharpoons B_3O_3(OH)_4^- + H^+ + 2H_2O   \qquad pK = 6.84$$

There is also good evidence that polymers are present in mixed solutions of boric acid and borates, e.g.,

$$2B(OH)_3 + B(OH)_4^- \rightleftharpoons B_3O_3(OH)_4^- + 3H_2O   \qquad K = 110$$

The predominant polymeric species appears to be the ring polymer (8-IV). Rings of this sort have been characterized in crystalline borates such as $Cs_2O \cdot 2B_2O_3$. The equilibria in solution are rapidly established, as shown, for example, by rapid exchange between boric acid labeled with ^{18}O and borates.

[9] V. F. Ross and J. O. Edwards, in ref. 4, p. 155; R. P. Bell, J. O. Edwards and R. B. Jones, in ref. 4, p. 209.

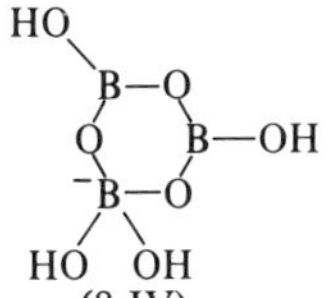

(8-IV)

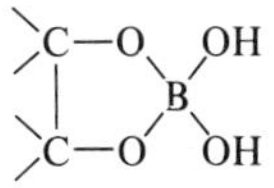

(8-V)

Boric acid and borates form very stable complexes exceedingly rapidly with polyols[10] and α-hydroxy carboxylic acids.[11] These are mainly 1:1 and of type (8-V). The acidity of boric acid is thereby increased; glycerol is commonly used analytically as the acid can then be titrated by aqueous NaOH. Steric considerations are very critical in the formation of these complexes. Thus 1,2- and 1,3-diols in the *cis*-form only, such as *cis*-1,2-cyclopentanediol, are active, and only *o*-quinols react. Indeed, the ability of a diol to affect the acidity of boric acid is a useful criterion of the configuration where *cis-trans*-isomers are possible.

Borates. Many borates occur naturally, usually in hydrated form. Anhydrous borates can be made by fusion of boric acid and metal oxides, and hydrated borates can be crystallized from aqueous solutions. The stoichiometry of borates, for example, $KB_5O_8 \cdot 4H_2O$, $Na_2B_4O_7 \cdot 10H_2O$, CaB_2O_4 and $Mg_7Cl_2B_{16}O_{30}$, gives little idea of the structures of the anions, which are cyclic or linear polymers formed by linking together of BO_3 and/or BO_4 units by shared oxygen atoms. The main principles for determining these structures are similar to those for silicates, to which the borates are structurally and often physically similar in forming glasses. It is interesting to note that in contrast to the borates, the carbonate ion of superficially similar structure forms no polymeric species; this is probably attributable to the formation of strong C—O π bonds.

Examples of complex anhydrous borate anions are the ring anion (8-VI) in $K_3B_3O_6$ and the infinite chain anion (8-VII) in CaB_2O_4.

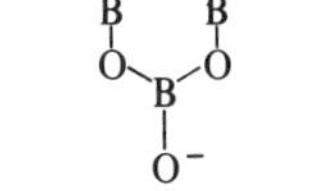

(8-VI)

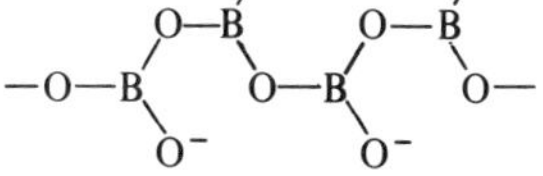

(8-VII)

Hydrated borates also contain polyanions in the crystal, but not all the known polyanions exist as such in solution; only those containing one or more BO_4 groups appear to be stable. Important features of the structures are the following:

1. Both trigonal BO_3 and tetrahedral BO_4 groups are present, the ratio of BO_4 to total B being equivalent to the ratio of the charge on the anion to that on total boron. Thus $KB_5O_8 \cdot 4H_2O$ has one BO_4 and four BO_3, whereas $Ca_2B_6O_{11} \cdot 7H_2O$ has four BO_4 and two BO_3 groups.

[10] See R. F. Nickerson, *J. Inorg. Nuclear Chem.*, 1970, **32**, 1401.
[11] K. Kustin and R. Pizer, *J. Amer. Chem. Soc.*, 1969, **91**, 317.

2. The basic structure is a six-atom ring whose stability depends on the presence of one or two BO_4 groups. Anions that do not have BO_4 groups, such as metaborate, $B_3O_6^{3-}$, or metaboric acid, $B_3O_3(OH)_3$, hydrate rapidly and lose their original structures. The fact that certain complex borates can be precipitated or crystallized from solution does not constitute evidence for the existence of such anions in solution, since other less complex anions can readily recombine during the crystallization process.

3. Other discrete and chain-polymer anions can be formed by linking of two or more rings by shared tetrahedral boron atoms, in some cases with dehydration (cf. metaborate below).

Some known structures are those of $KB_5O_8 \cdot 4H_2O$ (8-VIII) and borax, $Na_2B_4O_7 \cdot 10H_2O$ (8-IX).

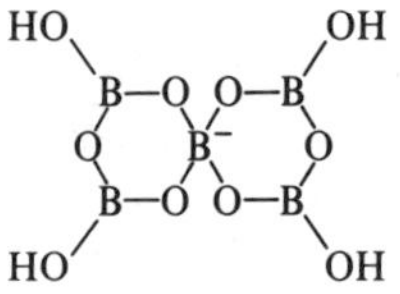

(8-VIII) 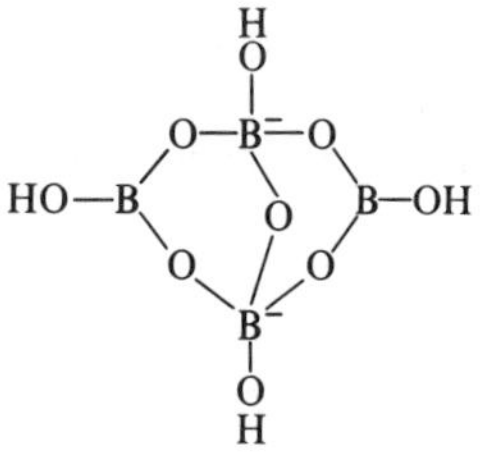   (8-IX)

Simple sharing of one oxygen atom by two BO_3 units would give $[O_2BOBO_2]^{4-}$; this so-called pyroborate anion has been shown to exist in $Co_2B_2O_5$. Also, the compound referred to as boron acetate, prepared by the reaction

$$2B(OH)_3 + 5(CH_3CO)_2O \rightarrow (CH_3COO)_2BOB(OOCCH_3)_2 + 6CH_3COOH$$

has a pyroborate-like structure. Boron phosphate, BPO_4, obtained by the reaction of boric and phosphoric acids, has tetrahedrally coordinated boron with B—O—P bonds.

Treatment of borates with hydrogen peroxide or of boric acid with sodium peroxide leads to products variously formulated as $NaBO_3 \cdot 4H_2O$ or $NaBO_2 \cdot H_2O_2 \cdot 3H_2O$, which are extensively used in washing powders because they afford H_2O_2 in solution. The crystal structure has been found to contain $[B_2(O_2)_2(OH)_4]^{2-}$ units with two peroxo groups bridging the tetrahedral boron atoms. When this salt is heated, paramagnetic solids containing O_2^-, O_3^- and a peroxoborate radical are formed.[12]

Metaborates. When heated, boric acid loses water stepwise:

$$B(OH)_3 \underset{H_2O}{\overset{Heat}{\rightleftharpoons}} HBO_2 \underset{H_2O}{\overset{Heat}{\rightleftharpoons}} B_2O_3$$

The intermediate substance, HBO_2, metaboric acid, exists in three modifications. If the $B(OH)_3$ is heated below 130° HBO_2-III is formed. This has a layer structure in which B_3O_3 six-rings are joined by hydrogen bonding between OH groups on the boron atoms. On continued heating of HBO_2-III at 130–150°, HBO_2-II is formed; this has a more complex structure contain-

[12] J. O. Edwards et al., J. Amer. Chem. Soc., 1969, 91, 1095.

ing both BO_4 tetrahedra and B_2O_5 groups in chains linked by hydrogen bonds. Finally, on heating of HBO_2-II above 150°, cubic HBO_2-I is formed in which all boron atoms are 4-coordinate.

8-6. Trihalides of Boron[13]

Compounds of the type BX_3 exist for all the halogens. When any two of the halides, BF_3, BCl_3, BBr_3, are mixed at room temperature, redistribution of halogen atoms occurs rather rapidly to produce a mixture of the original pure halides with the mixed halides in about statistical proportions. Thus, for example, we have the equilibrium:

$$BF_3 + BCl_3 \rightleftharpoons BFCl_2 + BF_2Cl$$

Nuclear resonance studies of mixtures of three halides have established the existence of BFClBr and BClBrI. These redistribution reactions presumably involve transitory formation of dimers such as $F_2BFClBCl_2$ which may dissociate to $BF_2Cl + BCl_2F$. However, no mixed halide has actually been isolated.

Some reactions of the halides are summarized in Fig. 8-4. (See page 234.)

Boron Trifluoride.   This compound, a pungent, colorless gas (b.p. $-101°$), prepared by heating B_2O_3 with NH_4BF_4 or with CaF_2 and concentrated H_2SO_4, reacts with water to form two "hydrates," which may be written $BF_3 \cdot H_2O$ and $BF_3 \cdot 2H_2O$ and melt at 10.18° and 6.36°, respectively. They are un-ionized in the solid state, and the structure of $BF_3 \cdot H_2O$ is probably that of a normal adduct. This species presumably also exists in the solid dihydrate, but the manner in which the second H_2O is held is at present unknown. Both these hydrates are partially dissociated into ions in their liquid phases, presumably as follows:

$$2(BF_3 \cdot H_2O) \rightleftharpoons [H_3O-BF_3]^+ + [BF_3OH]^-$$
$$BF_3 \cdot 2H_2O \rightleftharpoons H_3O^+ + [BF_3OH]^-$$

Above about 20° they decompose extensively, giving off BF_3. When relatively small amounts of BF_3 are passed into water, a solution of fluoroboric acid (not isolable as a pure substance) is obtained:

$$4BF_3 + 6H_2O \rightarrow 3H_3O^+ + 3BF_4^- + B(OH)_3$$

There is also some hydrolysis of the fluoroborate ion to produce HF and hydroxofluoroborate ion:

$$BF_4^- + H_2O \rightleftharpoons [BF_3OH]^- + HF \qquad K = 2.3 \times 10^{-3}$$

Boron trifluoride is one of the most avid acceptors—that is, strongest Lewis acids—known, and readily unites with water, ethers, alcohols, amines, phosphines, etc., to form adducts. BF_3 is commonly available as its diethyl etherate, $(C_2H_5)_2\overset{+}{O}\overset{-}{B}F_3$. Because of its potency as a Lewis acid and its greater resistance to hydrolysis compared with BCl_3 and BBr_3, BF_3 is widely used to

[13] A. G. Massey, *Adv. Inorg. Chem. Radiochem.*, 1967, **10**, 1; G. A. Olah, ed., *Friedel–Crafts and Related Reactions*, Vols. I, II, Interscience-Wiley, 1963 (contains much chemistry of boron halides).

promote various organic reactions, such as (a) ethers or alcohols + acids →
esters + H_2O or ROH; (b) alcohols + benzene → alkylbenzenes + H_2O;
(c) polymerization of olefins and olefin oxides; and (d) Friedel–Crafts-like
acylations and alkylations. In the first two cases, the effectiveness of BF_3 must
depend on its ability to form an adduct with one or both of the reactants,
thus lowering the activation energy of the rate-determining step in which

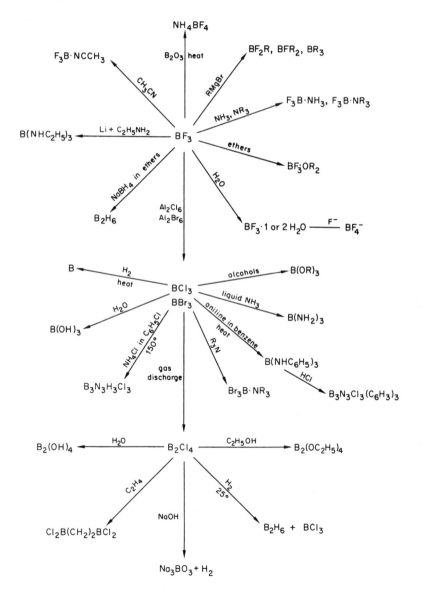

Fig. 8-4.   Some reactions of the halides of boron.

H_2O or ROH is eliminated by breaking of C—O bonds. However, the exact mechanisms of these reactions are not at present known, nor are those of the olefin and olefin oxide polymerizations.

In the case of the Friedel–Crafts-like reactions, isolation of certain intermediates at low temperatures has provided a fairly definite idea of the function of the BF_3. Thus, the ethylation of benzene by ethyl fluoride proceeds as in equation (8-1). With benzene, HF and BF_3, compound (8-X) can be

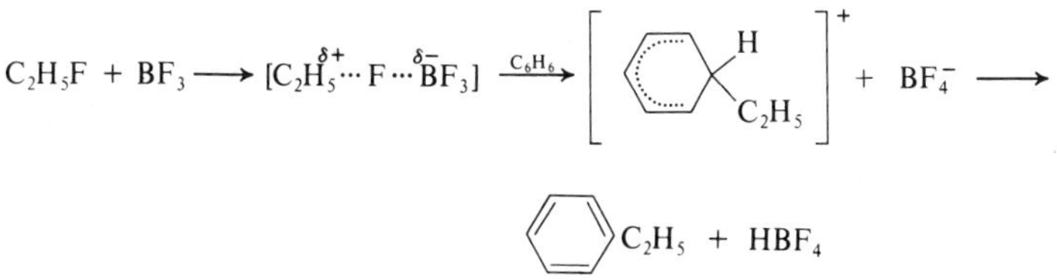

$$C_2H_5F + BF_3 \longrightarrow [C_2H_5^{\delta+} \cdots F \cdots \overset{\delta-}{B}F_3] \xrightarrow{C_6H_6} \left[\begin{array}{c} \overset{H}{\underset{C_2H_5}{}} \end{array} \right]^+ + BF_4^- \longrightarrow$$

$$\bigcirc C_2H_5 + HBF_4$$

$$(8\text{-}1)$$

isolated at low temperatures. It will be seen that the BF_3 is not really "catalytic," but must be present in the stoichiometric amount since it is consumed in the process of tying up the HF as HBF_4.

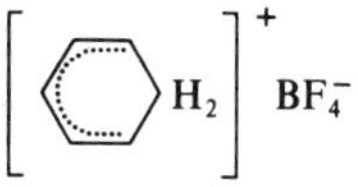

$$\left[\bigcirc H_2 \right]^+ BF_4^-$$

$$(8\text{-}X)$$

Solid salts of tetrafluoroboric acid are readily isolated, those of NH_4^+ and many metals such as the alkalis and alkaline earths being commercially available. Ammonium fluoroborate may be prepared in a dry way by fusing NH_4HF_2 with B_2O_3. $B(OH)_3$ also readily dissolves in HF to form solutions of fluoroboric acid:

$$B(OH)_3 + 4HF = H_3O^+ + BF_4^- + 2H_2O$$

Fluoroboric acid is a strong acid. The fluoroborate ion has a tetrahedral structure, and fluoroborates closely resemble the corresponding perchlorates in their crystal structures and solubilities.

Other Trihalides. *Boron trichloride* is a liquid at room temperature under slight pressure (b.p. 12.5°), and the *bromide* boils at 90°. Both fume in moist air and are completely hydrolyzed by water, e.g.

$$BCl_3 + 3H_2O \rightarrow B(OH)_3 + 3HCl$$

The compounds are prepared by direct interaction of the elements at elevated temperatures.

The rapid hydrolysis by water could indicate that these halides are stronger Lewis acids than BF_3. In fact, the molar heats of solution of the trihalides in nitrobenzene and the heats of reaction with pyridine in nitrobenzene show that under these conditions the electron-acceptor strength decreases in the order $BBr_3 > BCl_3 > BF_3$.

The *triiodide*, a white solid (m.p. 43°), explosively hydrolyzed by water, is prepared by the action of iodine on $NaBH_4$ or of HI on BCl_3 at red heat.

Tetrachloroborates are obtained by addition of BCl_3 to alkali chlorides at high pressures, by cold milling at room temperatures, or by the reaction

$$[(C_2H_5)_4N]^+Cl^- + BCl_3 \xrightarrow{CHCl_3} [(C_2H_5)_4N]^+BCl_4^-$$

The stability of these salts and the corresponding tetrabromoborates and tetraiodoborates is greatest with the largest cations. With a given cation, the stability order is $MBCl_4 > MBBr_4 > MBI_4$, tetraiodoborates occurring only with the largest cations. Mixed ions such as BF_3Cl^- also exist.

The reaction of BCl_3 with anhydrous $HClO_4$ at $-78°$ has been stated to yield $BCl_2(ClO_4)$, $BCl(ClO_4)_2$ and $B(ClO_4)_3$, the product depending on the mole ratio of the reactants. There is as yet no structural information or other indication of how the perchlorate groups are bound to the boron atoms in these highly reactive substances.

8-7. Lower Halides of Boron

All four compounds of stoichiometry B_2X_4 have been reported but only the fluoride and, especially, the chloride have been much studied. In the crystalline state both B_2F_4 and B_2Cl_4 consist of BX_2 groups joined by B—B bonds, and the molecules are planar (D_{2h}), whereas spectroscopic and electron-diffraction evidence suggests that in the liquid and the gaseous state they are non-planar, with mutually perpendicular BX_2 groups (D_{2d}). The following appear to be the most convenient preparative reactions:

$$2(BO)_x + 2xSF_4 \rightarrow xB_2F_4 + 2xSOF_2$$
$$6(BO)_x + 4xBCl_3 \rightarrow 3xB_2Cl_4 + 2xB_2O_3$$
$$2BCl_3 + 2Hg \xrightarrow[\text{arc}]{\text{Low pressure}} B_2Cl_4 + HgCl_2$$

B_2F_4 is a colorless gas and B_2Cl_4 a colorless liquid at room temperature. B_2Cl_4 is a very reactive substance and among its most important reactions are protolytic ones and addition to olefins:

$$B_2Cl_4 + 4ROH \rightarrow B_2(OR)_4 + 4HCl \quad (R = \text{alkyl, aryl or H})$$
$$B_2Cl_4 + RCH{=}CHR \rightarrow Cl_2BCHR{-}CHRBCl_2$$

Both B_2F_4 and B_2Cl_4 react avidly with oxygen, the latter burning with a beautiful green flame: $6B_2Cl_4 + 3O_2 \rightarrow 2B_2O_3 + 8BCl_3$

Tetraboron tetrachloride, B_4Cl_4, first obtained in traces in the arc synthesis of B_2Cl_4, may be more efficiently prepared (10 mg/h) by passing B_2Cl_4 through a mercury arc discharge. It is a pale yellow, volatile solid (m.p. 95°), thermally stable to $\sim200°$. The molecule consists of a B_4 tetrahedron with B—Cl bonds arranged to preserve precise T_d symmetry. B_4Cl_4 is reactive but its chemistry has, as yet, been little explored.

The thermal disproportionation of B_2Cl_4 affords BCl_3 and a wealth of subhalides, most of which are not characterized. Evidently B_2Br_4 and B_2I_4 similarly generate a multitude of unidentified subhalides. Among the chloro species, besides B_4Cl_4, just mentioned, are an intensely red paramagnetic solid, $B_{12}Cl_{11}$ (which is the major product), orange B_9Cl_9 and purple

B_8Cl_8.[14] The last has a distorted B_8 trigonal dodecahedron (D_{2d}) with one Cl on each B atom.

8-8. The Boron Hydrides (Boranes) and Related Compounds[15]

In a remarkable series of papers from 1912 to 1936, Alfred Stock and his co-workers prepared and chemically characterized the following hydrides of boron (boranes): B_2H_6, B_4H_{10}, B_5H_9, B_5H_{11}, B_6H_{10} and $B_{10}H_{14}$. With the exception of diborane, B_2H_6, which was prepared by thermal decomposition of higher boranes, Stock prepared these hydrides by the action of acid on magnesium boride, MgB_2, obtaining in this way a mixture of volatile, reactive and air-sensitive (some spontaneously flammable) compounds. In order to handle compounds with these properties, Stock developed the glass vacuum line and techniques for using it.

Subsequently a number of other boranes have been discovered, so that today there are no less than 14 that are well-characterized chemically, and for most the structures are well-defined. A convenient nomenclature for the boranes specifies the number of hydrogen atoms with an arabic number in parentheses. For example: B_5H_9 pentaborane(9), and B_5H_{11} pentaborane(11). The hydrogen number may be omitted where only one compound with the number of boron atoms in question is known, viz., B_4H_{10} tetraborane. A list of the better characterized boranes is given in Table 8-1.

Preparative methods for the boranes are numerous and highly varied; Stock's original method is now used only for B_6H_{10}. Most preparations begin with B_2H_6 and involve a pyrolysis under a variety of conditions and often in the presence of H_2 or other reagents.[16] The very important $B_{10}H_{14}$, for example, is obtained by pyrolysis of B_2H_6 at about 100°, while B_5H_9 is formed on pyrolysis of B_2H_6 in presence of hydrogen at 250°. An electric discharge through B_2H_6 or other boranes is sometimes used, but this method gives only small yields. The best method for any given borane may be peculiar to it, e.g., reaction of B_5H_{11} with the surface of crystalline hexamethylenetetramine to give B_9H_{15}, and the carefully controlled hydrolysis of the hydroxonium ion salt of $B_{20}H_{18}$ to give a mixture of the isomers of $B_{18}H_{22}$.

Diborane. B_2H_6 is of special interest because it is the starting material for preparation of various other boron hydrides and because of its synthetic uses. It can be prepared in essentially quantitative yield by reaction of metal hydrides with boron trifluoride, a convenient method being to drop boron trifluoride etherate into a solution of sodium borohydride in diglyme (diethylene glycol dimethyl ether):

$$3NaBH_4 + 4BF_3 \rightarrow 3NaBF_4 + 2B_2H_6$$

[14] J. Kane and A. G. Massey, *J. Inorg. Nuclear Chem.*, 1971, **33**, 1195; G. F. Lanthier, J. Kane and A. G. Massey, *J. Inorg. Nuclear Chem.*, 1971, **33**, 1569.

[15] (a) R. L. Hughes, I. C. Smith and E. W. Lawless, *Production of Boranes and Related Research*, Academic Press, 1967; (b) W. N. Lipscomb, *Boron Hydrides*, Benjamin, 1963.

[16] L. H. Long, *J. Inorg. Nuclear Chem.*, 1970, **32**, 1097.

TABLE 8-1

Important Properties of Boranes[a]

Formula	Name	Melting point (°C)	Boiling point (°C)	Reaction with air, at 25°C	Thermal stability	Reaction with water
B_2H_6	Diborane(6)	−164.85	−92.59	Spontaneously flammable	Fairly stable at 25°	Instant hydrolysis
B_4H_{10}	Tetraborane(10)	−120	18	Not spontaneously flammable if pure	Decomposes fairly rapidly at 25°	Hydrolysis in 24 hours
B_5H_9	Pentaborane(9)	−46.6	48	Spontaneously flammable	Stable at 25°; slow decomposition 150°	Hydrolyzed only on heating
B_5H_{11}	Pentaborane(11)	−123	63	Spontaneously flammable	Decomposes very rapidly at 25°	Rapid hydrolysis
B_6H_{10}	Hexaborane(10)	−62.3	108	Stable	Slow decomposition at 25°	Hydrolyzed only on heating
B_6H_{12}	Hexaborane(12)	−82.3	80–90	—	Liquid stable few hours at 25°	Quantitative, to give B_4H_{10}, $B(OH)_3$, H_2
B_8H_{12}	Octaborane(12)	−20	—	—	Decomposes above −20°	—
B_8H_{18}	Octaborane(18)	—	—	—	Unstable	—
B_9H_{15}	Enneaborane(15)	2.6	—	Stable	—	—
$B_{10}H_{14}$	Decaborane(14)	99.7	213 (extrap.)	Very stable	Stable at 150°	Slow hydrolysis
$B_{10}H_{16}$	Decaborane(16)	—	—	Stable	Stable at 25°	—
$B_{18}H_{22}$	Octadecaborane(22)	—	—	—	—	Monoprotic acid
iso-$B_{18}H_{22}$	Isooctadecaborane(22)	—	—	—	—	—
$B_{20}H_{16}$	Icosaborane(16)	196–199	—	Stable	Stable at 25°	Irreversibly gives $B_{20}H_{16}(OH)_2^-$ and $2H^+$

[a] Also B_8H_{14}, B_8H_{16}, iso-B_9H_{15}, $B_{10}H_{18}$, $B_{10}H_{16}$ (?), B_{6-7} (?), B_8 (?); see, e.g., J. Dibson, R. Maruca and R. Schaeffer, *Inorg. Chem.*, 1970, **9**, 2161.

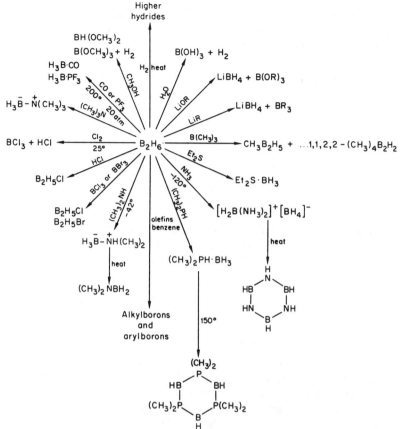

Fig. 8-5.   Some reactions of diborane.

It can also be prepared from B_2O_3 and H_2 at high temperature and pressure with $Al + AlCl_3$ as catalyst, but this is not a laboratory method.

Borane(3), BH_3, appears to have a transitory existence in the thermal decomposition of diborane:

$$2B_2H_6 \rightleftharpoons BH_3 + B_3H_9$$

but it reacts further so that only exceedingly low concentrations are obtained.[16] Appreciable amounts can, however be made by thermal decomposition of borane carbonyl, H_3BCO, diluted in an inert gas in a very fast flow reactor.[17] However, BH_3 can be trapped by interaction of B_2H_6 with donors, and adducts such as H_3BCO (b.p. $-64°$), H_3BPF_3, and H_3BNMe_3 are well known (Fig. 8-5). The borohydride anions discussed below, e.g., BH_4^-, BH_3SH^- and BH_3CN^-, can be regarded as adducts with H^-, SH^- or CN^-.

Diborane is an extremely versatile reagent for the preparation of organo-

[17] G. W. Mappes and T. P. Fehlner, *J. Amer. Chem. Soc.*, 1970, **92**, 1562.

boranes, which in turn are most useful intermediates in organic synthesis.[18] It is also[19] a powerful reducing agent for organic functional groups, e.g., $RCHO \rightarrow RCH_2OH$, and $RCN \rightarrow RCH_2NH_2$. The reaction of B_2H_6 in ethers with unsaturated hydrocarbons, commonly called *hydroboration*, gives predominantly anti-Markownikoff, *cis*-hydrogenation or hydration:

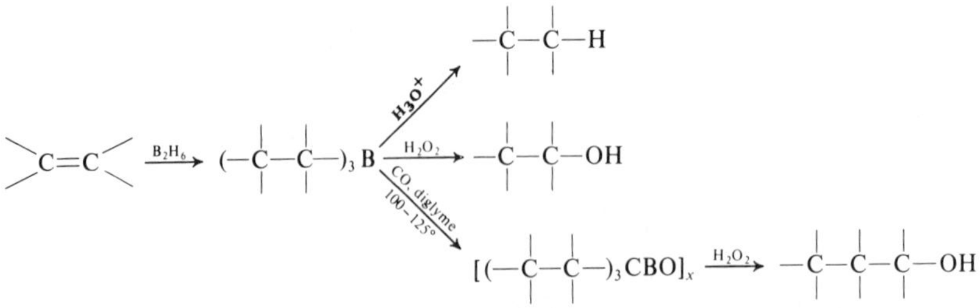

Carbonylations[18b] using CO result in the formation of species in which carbon is "inserted" between the B and C of the alkyl group.

By using dialkylboranes, R_2BH, stereospecificities up to 99% can be obtained, viz.:

$$RCH{=}CH_2 + R_2'BH \rightarrow RCH_2{-}CH_2BR_2'$$

R = C_4H_9, R' = isopentyl: 99%
R = C_6H_5, R' = ethyl: 92%

8-9. Structures and Bonding in the Boranes

The stoichiometries of the boranes, from the simplest, B_2H_6, to the most complex, $B_{20}H_{16}$, together with the number of electrons available, do not permit one to design structures or bonding schemes like those for hydrocarbons or other "normal" compounds of the lighter non-metals. X-ray crystallography and other studies have in fact shown that the structures of the boranes are quite unlike hydrocarbon structures. Fig. 8-6 shows a few of them. Not only are the structures unique but in all of the boranes there is the problem of *electron deficiency*, i.e. there are not enough electrons to permit the formation of conventional 2-electron bonds between all adjacent pairs of atoms (2c-2e bonds). In order to rationalize the structures in terms of acceptable bonding prescriptions *multicenter bonding* of various sorts must be widely employed.

For diborane itself 3c-2e bonds (page 110) are required to explain the B—H—B bridges. The terminal B—H bonds may be regarded as conventional 2c-2e bonds. Thus, each boron atom uses two electrons and two roughly sp^3 orbitals to form 2c-2e bonds to two hydrogen atoms. The boron atom in each BH_2 group still has one electron and two hybrid orbitals for use in further bonding. The plane of the two remaining orbitals is perpendicular to the BH_2 plane. When two such BH_2 groups approach each other as shown in Fig. 8-7, with hydrogen atoms also lying, as shown, in the plane

[18] H. C. Brown, (a) *Hydroboration*, Benjamin, 1962; (b) *Accounts Chem. Res.*, 1969, **2**, 65; (c) R. Köster, *Chimia (Switz.)*, 1969, **23**, 196.
[19] H. C. Brown *et al.*, *J. Amer. Chem. Soc.*, 1970, **92**, 1637.

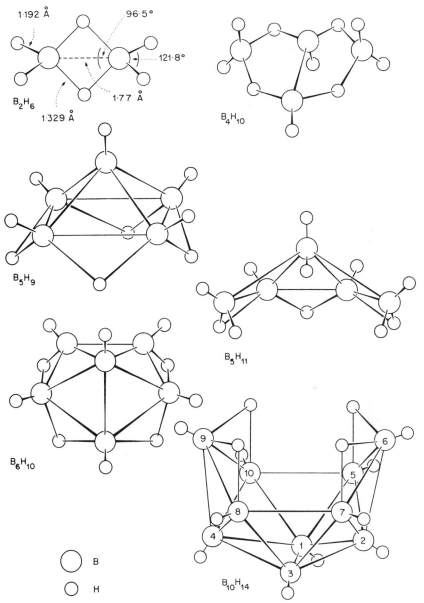

Fig. 8-6.   Structures of some boranes.

of the four empty orbitals, two B—H—B 3c-2e bonds are formed. The total of four electrons required for these bonds is provided by the one electron carried by each H atom and by each BH_2 group.

We have just seen that two structure/bonding elements are used in B_2H_6, viz., 2c-2e BH groups and 3c-2e BHB groups. In order to account for the

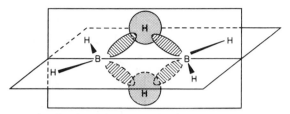

Fig. 8-7. Diagram showing how the approach of two properly oriented BH_2 radicals and two H atoms leads to the formation of two 3c-2e B—H—B bonds.

structures and bonding of the higher boranes, these elements as well as three others are required. The three others are: 2c-2e BB groups; 3c-2e open BBB groups; and 3c-2e closed BBB groups. These five structure/bonding elements be conveniently represented in the following way:

Terminal 2c-2e boron—hydrogen bond   B—H

3c-2e Hydrogen bridge bond   B $\overset{H}{\frown}$ B

2c-2e Boron—boron bond   B—B

Open 3c-2e Boron bridge bond   B $\overset{B}{\frown}$ B

Closed 3c-2e boron bond   B $\overset{B}{\wedge}$ B

By using these five elements, "semitopological" descriptions of the structures and bonding in all of the boranes may be given, as shown by Lipscomb. The scheme is capable of elaboration into a comprehensive, semi-predictive tool for correlating all the structural data. We give here, in Fig. 8-8, only a few examples of its use to depict known structures; further details will be found elsewhere.[15b]

The semitopological scheme does not always provide the best description of bonding in the boranes and related species such as the polyhedral borane anions and carboranes to be discussed below. Where there is symmetry of a high order it is often more convenient and conceptually simpler to think in terms of a highly delocalized molecular-orbital description of the bonding. For instance, in B_5H_9 where the four basal boron atoms are equivalently related to the apical boron atom, it is *possible* to depict a resonance hybrid involving the localized B $\overset{B}{\frown}$ B and B—B elements, viz.:

$$ \begin{matrix} B & & B \\ & B & \\ B & & B \end{matrix} \longleftrightarrow \begin{matrix} B & & B \\ & B & \\ B & & B \end{matrix} $$

but it is ultimately neater and simpler to formulate a set of seven 5-center molecular orbitals with the lowest three occupied by electron pairs. When one approaches the hypersymmetrical species such as $B_{12}H_{12}^{2-}$, use of the full molecular symmetry in an MO treatment becomes the only practical course.

Full MO treatments can, of course, be carried out even for the boranes in which the localized structure/bonding elements used in the semitopological

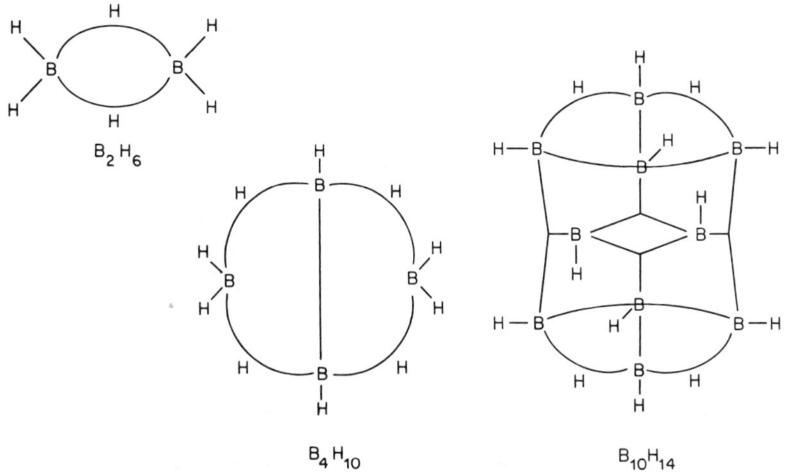

Fig. 8-8. Valence descriptions of boron hydrides in terms of Lipscomb's "semitopological" scheme.

theory appear to be well-defined. It has been shown[20] that in such cases the completely general MO's are equivalent to, and can be readily transformed into, more localized components corresponding to the structure/bonding elements just mentioned.

In addition to the desire to develop a consistent set of principles to explain the structures of boranes and to predict new ones, one of the main motivations for theoretical study of the electronic structures of these molecules is the desire to understand their chemical reactivity. One of the most important types of reaction that the boranes (and also the borane anions and carboranes) undergo is electrophilic substitution. Although there is no *a priori* basis for presuming it to be so, it is an empirical fact that those boron atoms to which bonding theory assigns the greatest negative charge are those preferentially attacked in electrophilic substitution. For example, in $B_{10}H_{14}$, charge-distributions calculated from MO treatments assign considerable (~ 0.25 e) excess negative charge to boron atoms 2 and 4, approximate neutrality to boron atoms 1 and 3 and positive charge to all others. Experiments show consistently that only positions 1, 2, 3 and 4 can be substituted electrophilically and that positions 2 and 4 are perhaps slightly preferred. Similar agreement between experimental results and calculated charge distributions has been obtained for $B_{10}C_2H_{12}$. Thus, it is believed that, at least in a qualitative sense, computations of electronic structures are worth making in order to gain clues as to preferred positions of reactivity.

Structural Study by Nmr.[21]   Though X-ray crystallography is the most precise source of structural information on boranes, nuclear magnetic

[20] G. W. Adamson and J. W. Linnett, *J. Chem. Soc., A,* **1969,** 1697.
[21] G. R. Eaton and W. N. Lipscomb, *Nmr Studies of Boron Hydrides and Related Compounds,* W. A. Benjamin, 1969.

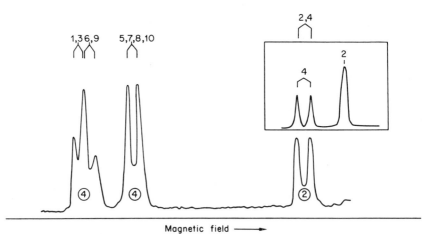

Fig. 8-9. The ^{11}B nuclear magnetic resonance spectrum of $B_{10}H_{14}$. The assignment is shown at the top; circled numbers give intensities of the multiplets. The inset shows the appearance of a portion of the spectrum of 2-iodo-$B_{10}H_{13}$.

resonance is an extremely important adjunct, especially in elucidating the course of substitution reactions. This is particularly true with respect to the polyhedral borane anions and carboranes discussed below. Primarily, it is the abundant ^{11}B isotope that is studied; proton-resonance spectra are rarely useful because the signals are broad, complex multiplets as a result of splitting by the ^{11}B and ^{10}B nuclei.

As an illustration, Fig. 8-9 shows the ^{11}B spectrum of decaborane(14) at 64 mHz together with the assignment in terms of the conventional numbering scheme (Fig. 8-6). Each type of B atom is represented by a signal with intensity proportional to the number of such nuclei; the signals are all doublets because of splitting by the proton attached by a 2c-2e bond to each boron atom. Splitting by bridging protons is not resolved. The insert in Fig. 8-9 shows the principal difference (minor changes occur elsewhere in the spectrum) observed in the spectrum of one of the monoiododecaboranes. It is quite clear that, provided the assignment of the $B_{10}H_{14}$ spectrum itself is correct (it is), the molecule in question has the iodine substituent in the 2(or 4)-position. Note the lack of splitting when no H is bonded to the boron atom.

BORANE ANIONS, CARBORANES AND
TRANSITION-METAL COMPLEXES[22]

The neutral borane molecules, interesting and complex as they are, represent only the beginning of the chemistry of boron–hydrogen com-

[22] (a) E. L. Muetterties and W. M. Knoth, *Polyhedral Boranes*, Dekker, 1968; (b) M. F. Hawthorne, *Endeavour*, **1970**, 146; (c) R. N. Grimes, *Carboranes*, Academic Press, 1971; (d) L. J. Todd, *Adv. Organometallic Chem.*, 1970, **8**, 87.

pounds. We now turn to several other classes of compound, some salt-like and containing $B_nH_m^{x-}$ anions, some neutral or anionic in which elements other than boron, especially carbon, form part of the framework, and still others in which transition-metal ions are complexed in a manner reminiscent of the metal sandwich compounds, such as ferrocene, $(C_5H_5)_2Fe$ (Section 23-2).

Owing to the complexity of these compounds, problems of nomenclature are often acute, and comprehensive systems of notation become quite elaborate.[23] A few very basic points may be noted here. Complex frameworks of boron atoms which include also other atoms *in* the framework—not just appended to it—are generically given compound names, such as carboranes (carbon), phosphaboranes (phosphorus), thiaboranes (sulfur), carbaphosphaboranes, etc. Geometrically, all these molecules fall into two broad classes. First, there are those in which the framework closes on itself forming a polyhedron; the prefix *clovo* (from the Greek for cage) or, alternatively, *closo*, is used to designate these. In contrast, those frameworks which are "open," that is, incomplete polyhedra, are given the prefix *nido*, meaning nest. Several other points of nomenclature will be introduced as needed in connection with the metal complexes.

8-10. Polyhedral (Closo) Borane Anions and Carboranes

The polyhedral borane anions are species of the type $B_nH_n^{2-}$. Theoretical arguments indicate that there should be stable species with all values of n from 5 to 12, and, in fact, those with n from 6 to 12 have all been prepared. In most cases the reactions used are not mechanistically understood. For example, $B_9H_9^{2-}$ salts are obtained, together with BH_4^-, $B_{10}H_{10}^{2-}$ and $B_{12}H_{12}^{2-}$ salts, by pyrolysis under a vacuum at 200–230° of *pure* alkali-metal salts of $B_3H_8^-$. When ether is present the yield of $B_{12}H_{12}^{2-}$ increases and may become dominant. Solutions of $B_9H_9^{2-}$ in ethers such as THF are oxidized by air to afford the $B_8H_8^{2-}$ and $B_7H_7^{2-}$ ions, but $B_3H_8^-$, $B_6H_6^{2-}$ and $B_{12}H_{12}^{2-}$ are also formed in low yields. Many of the salts of these $B_nH_n^{2-}$ ions have considerable thermal stability, especially for $n = 9$, 10, 11, or 12. Hydrolytic stability varies, $B_{10}H_{10}^{2-}$ and $B_{12}H_{12}^{2-}$ being very stable, while $B_6H_6^{2-}$, $B_8H_8^{2-}$ and $B_9H_9^{2-}$ are less so. $B_7H_7^{2-}$ is least stable, slow evolution of hydrogen commencing immediately on dissolution in water.

The polyhedral carboranes may be considered as *formally* (not preparatively) derived from the $B_nH_n^{2-}$ ions on the basis that the CH group is isoelectronic and isostructural with, and may thus replace, the BH^- group. Thus, two such replacements lead to neutral molecules of general formula $B_{n-2}C_2H_n$. All such carboranes from $n = 5$ to $n = 12$ are known, either as such or, for $n = 8–10$, as *C,C*-dimethyl derivatives. In all cases, isomeric forms of these carboranes are possible and in several instances two or more isomers have actually been isolated.

[23] *Inorg. Chem.*, 1968, **7**, 1945.

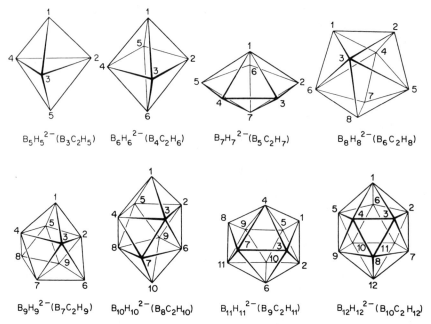

$B_5H_5^{2-}(B_3C_2H_5)$ $B_6H_6^{2-}(B_4C_2H_6)$ $B_7H_7^{2-}(B_5C_2H_7)$ $B_8H_8^{2-}(B_6C_2H_8)$

$B_9H_9^{2-}(B_7C_2H_9)$ $B_{10}H_{10}^{2-}(B_8C_2H_{10})$ $B_{11}H_{11}^{2-}(B_9C_2H_{11})$ $B_{12}H_{12}^{2-}(B_{10}C_2H_{12})$

Fig. 8-10.   The triangulated-polyhedral structures of $B_nH_n^{2-}$ and $B_{n-2}C_2H_n$ species with the conventional numbering schemes also indicated.

Present information indicates that all the *closo*borane anions and carboranes are fully triangulated polyhedra, as shown in Fig. 8-10.

$B_nH_n^{2-}$ **Ions.**   Because of their greater stability the $B_{10}H_{10}^{2-}$ and $B_{12}H_{12}^{2-}$ ions have been studied far more thoroughly than the others. Salts of $B_{12}H_{12}^{2-}$ can be prepared in many ways, the most useful ones, which give nearly quantitative yields when carried out at $\sim 150°$, employ the reactions

$$6B_2H_6+2R_3N \rightarrow 2(R_3NH^+)+B_{12}H_{12}^{2-}+11H_2$$
$$5B_2H_6+2NaBH_4 \rightarrow 2Na^+ + B_{12}H_{12}^{2-}+13H_2$$

With certain changes in conditions, $B_{10}H_{10}^{2-}$ salts can be obtained by analogous reactions. Decaborane(14) can also be used to obtain $B_{10}H_{10}^{2-}$ quantitatively:

$$B_{10}H_{14}+2R_3N \rightarrow 2(R_3NH)^+ + B_{10}H_{10}^{2-}+H_2$$

The reactions of $B_{10}H_{10}^{2-}$ and $B_{12}H_{12}^{2-}$ have been extensively studied and an enormous body of substitution chemistry, reminiscent of aromatic hydrocarbon chemistry, has been discovered. Attack by electrophilic reagents is the most important general reaction type. Representative attacking species are RCO^+, CO^+, $C_6H_5N_2^+$ and Br^+. These reactions, whose mechanisms are not yet well understood, proceed most readily in strongly acid media, and, in general, $B_{10}H_{10}^{2-}$ is much more susceptible to substitution than is $B_{12}H_{12}^{2-}$. Less reactive nucleophiles, such as $C_6H_5N_2^+$ fail to attack $B_{12}H_{12}^{2-}$ and attack $B_{10}H_{10}^{2-}$ selectively at the apical (1, 10) positions.

Both $B_{10}H_{10}^{2-}$ and $B_{12}H_{12}^{2-}$ can be partially and completely halogenated. The perhalogeno ions have extremely high thermal stabilities and are also highly resistant to hydrolysis.

The $B_{10}H_{10}^{2-}$ ion can be oxidized by Fe^{3+} in aqueous solution to the $B_{20}H_{18}^{2-}$ ion (8-XI), which can then be reduced with sodium in liquid ammonia to a $B_{20}H_{18}^{4-}$ ion having the structure (8-XII). This, in turn, can be isomerized to (8-XIII) and (8-XIV).

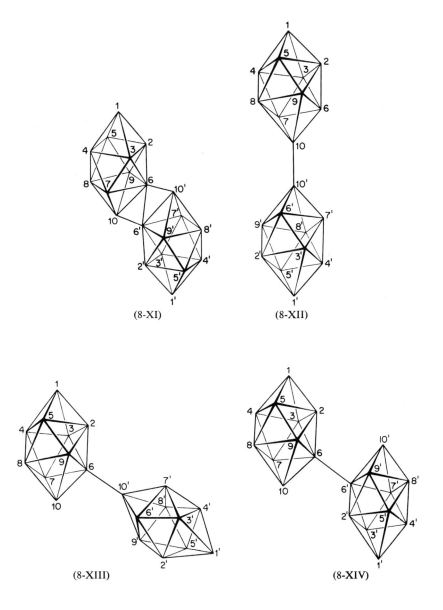

(8-XI)   (8-XII)

(8-XIII)   (8-XIV)

The $B_{2-n}C_2H_n$ Carboranes. These carboranes may be prepared in a variety of ways. The most extensively studied carboranes are $B_{10}C_2H_{12}$ and its C-substituted derivatives, obtained by the reactions

$$B_{10}H_{14} + 2R_2S \rightarrow B_{10}H_{12}(R_2S)_2 + H_2$$
$$B_{10}H_{12}(R_2S)_2 + RC\equiv CR' \rightarrow B_{10}H_{10}C_2RR' + 2R_2S + H_2$$

In this way the 1,2-isomer is obtained, but when this is heated to 450° a smooth rearrangement to the 1,7-isomer occurs. More will be said of this below.

The 1,2- and 1,7-dicarba*closo*dodecaboranes, and their C-substituted derivatives are degraded quantitatively by base to give isomeric *nido*carborane anions, $B_9C_2H_{12}^-$, the structures of which will be discussed below, viz.:

$$B_{10}C_2H_{12}^- + C_2H_5O^- + 2C_2H_5OH \rightarrow B_9C_2H_{12}^- + B(OC_2H_5)_3 + H_2$$

Both isomeric $B_9C_2H_{12}^-$ ions, upon treatment with anhydrous acids followed by heating, are converted in high yield into the *closo*carborane $B_9C_2H_{11}$:

$$B_9C_2H_{12}^- \xrightarrow{H^+} B_9C_2H_{13} \xrightarrow{150°} B_9C_2H_{11} + H_2$$

Several of the lower *closo*carboranes can then be obtained by the reaction sequence:

$$B_9C_2H_{11} \xrightarrow{H_2CrO_4} B_7C_2H_{13} \xrightarrow{200°} \left. \begin{cases} 1,7\text{-}B_6C_2H_8 \\ 1,6\text{-}B_8C_2H_{10} \end{cases} \right\} \text{ good yield}$$
$$B_7C_2H_9 \qquad \text{low yield}$$

The $1,6\text{-}B_8C_2H_{10}$ rearranges quantitatively to the 1,10-isomer at 350°.

Several of the lower *closo*carboranes are best obtained by the following reactions,[24] proceeding through the *nido*carborane $B_4C_2H_8$:

$$B_5H_9 + C_2H_2 \rightarrow 4,5\text{-}B_4C_2H_8 + \tfrac{1}{2}B_2H_6$$

$$4,5\text{-}B_4C_2H_8 \xrightarrow[\text{1--3 sec}]{450°} \begin{cases} B_3C_2H_5 & 40\% \\ B_5C_2H_7 & 40\% \\ B_4C_2H_6 & 20\% \end{cases}$$

Only the 1,2- and $1,7\text{-}B_{10}C_2H_{12}$ carboranes have been studied in much detail, owing in part to the relative inaccessibility of the others. As already noted, C-substituted derivatives can be obtained by using substituted acetylenes in the carborane synthesis. Derivatives may also be prepared beginning with ready replacement of the hydrogen atoms attached to the carbon atoms by lithium. The dilithio derivative reacts with a variety of other reagents, as indicated in the scheme below, where a conventional and self-explanatory abbreviation is used for the $B_{10}H_{10}C_2$ group.

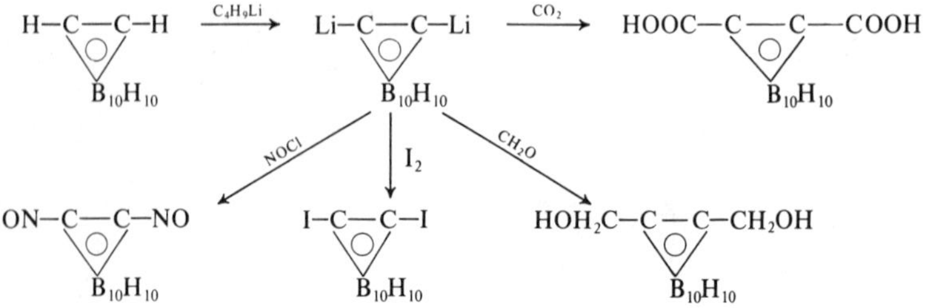

[24] J. F. Ditter, *Inorg. Chem.*, 1968, **7**, 1748.

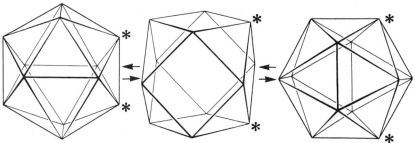

Fig. 8-11.   The interconversion of 1,2- and 1,7-disubstituted icosahedral species *via* a cuboctahedral intermediate.

Reactions of this type have been used to prepare an enormous number of derivatives. One of the principal motivations in developing this chemistry has been to incorporate the thermally stable carborane residue into high polymers, and this has been accomplished, notably in some very stable ones incorporating $1,7\text{-}B_{10}C_2H_{10}$ residues with siloxane linkages.

The 1,2- and $1,7\text{-}B_{10}C_2H_{12}$ molecules are readily chlorinated to yield a number of $B_{10}C_2H_{12-x}Cl_x$ molecules. Chlorination appears to occur preferentially at the most negative boron atoms, in accord with similar results noted earlier for $B_{10}H_{14}$ and other boranes.

One of the intriguing properties of $1,2\text{-}B_{10}C_2H_{12}$ and its derivatives is their ability to isomerize thermally. As mentioned above, $1,2\text{-}B_{10}C_2H_{12}$ rearranges to $1,7\text{-}B_{10}C_2H_{12}$ at 450°. The latter isomer is expected to be the more stable since the relatively positive carbon atoms are more separated from each other. The 1,12-isomer, which, by the same token, ought to be the most stable one, can only be obtained by much more severe heating whereby considerable pyrolysis also occurs. To rationalize these limited results it was suggested that the icosahedron might be able, relatively easily, to open up to a cuboctahedron, as shown in Fig. 8-11; the cuboctahedron would then have a 50/50 chance of regenerating an icosahedron different from the original one. The difference would be precisely that necessary to account for the 1,2- to 1,7- (or *ortho* to *meta*) isomerization. As further scrutiny will show, this particular pathway can never lead to the 1,12- (or *para*) isomer. Other studies,[25] employing substituted $B_{10}C_2H_{12}$ derivatives have shown that, while the simple traverse of cuboctahedral intermediates can account for the gross features of the relatively rapid rearrangements, additional pathways must be invoked to account for all the rearrangements observed. It has been suggested that rotation of triangular sets of atoms by 120° relative to the rest of the cuboctahedron may be the most important of these higher-energy processes.

Other carboranes are also known to isomerize thermally, but detailed studies are lacking.

[25] H. D. Kaesz *et al., J. Amer. Chem. Soc.*, 1967, **89**, 4218; H. V. Hart and W. N. Lipscomb, *J. Amer. Chem. Soc.*, 1969, **91**, 771.

It may be noted that monocarborane anions, $B_{n-1}CH_n^-$, the logical middle members in the sequence $B_nH_n^{2-}$, $B_{n-1}CH_n^-$, $B_{n-2}C_2H_n$, have been prepared in two cases, namely, $B_9CH_{10}^-$ and $B_{11}CH_{12}^-$.

8-11. Other Borane Anions and Carboranes

Besides the polyhedral borane anions and carboranes, the *closo* species, there are others of importance that lack the closure. There are several nest-like or *nido*carboranes, such as $4,5\text{-}B_4C_2H_8$ which has been mentioned and there are two important small borane anions, BH_4^- and $B_3H_8^-$.

The *borohydride* ion, BH_4^-, itself is of great importance and substituted species, e.g., $[BH(OCH_3)_3]^-$, are also very useful synthetically. Borohydrides of a great many metallic elements—the alkali metals, Be, Mg, Al, Ti, Zr, Th, U, etc.—have been made.[26] Typical preparative reactions are:

$$4NaH + B(OCH_3)_3 \xrightarrow{\sim 250°} NaBH_4 + 3NaOCH_3$$
$$NaH + B(OCH_3)_3 \xrightarrow{THF} NaBH(OCH_3)_3$$
$$2LiH + B_2H_6 \xrightarrow{Ether} 2LiBH_4$$
$$AlCl_3 + 3NaBH_4 \xrightarrow{Heat} Al(BH_4)_3 + 3NaCl$$
$$UF_4 + 2Al(BH_4)_3 \longrightarrow U(BH_4)_4 + 2AlF_2BH_4$$

$NaBH_4$ is representative of the alkali borohydrides and it is the most common one. It is a white crystalline substance, stable in dry air and non-volatile. While insoluble in diethyl ether, it dissolves in water, tetrahydro-furan, and glymes (ethylene glycol ethers) to give solutions widely useful in synthetic chemistry as reducing agents and sources of hydride ions.[27] Treatment of $NaBH_4$, with protonic acids (e.g., HCl) or Lewis acids (e.g., BCl_3, $AlCl_3$) generates diborane. With smaller metal ions the BH_4^- ion can react covalently, using bridging hydrogen atoms; it thus serves as an unusual sort of ligand. Examples of this are provided by $Al(BH_4)_3$, where the number of bridging H atoms per BH_4^- is believed (though not proved) to be two, and by $(PPh_3)_2CuBH_4$, where there are two bridging H atoms.[28]

Ionic borohydrides (e.g., $NaBH_4$) contain discrete BH_4^- ions, which are tetrahedral, and structures of some MBH_4 compounds are the same as that of NH_4Cl since NH_4^+ and BH_4^- are isoelectronic and isosteric. One of the most puzzling borohydrides from a structural point of view has been $Be(BH_4)_2$. The seemingly obvious linear structure (8-XV) appears to be inconsistent with much physical evidence, but there is, as yet, no unanimity as to which of several non-linear structures is preferable.[28a]

(8-XV)

[26] B. D. James and M. G. M. Wallbridge, *Progr. Inorg. Chem.*, 1970, **11**, 99 (an extensive review).
[27] See, e.g., P. M. Treichel, J. P. Stenson and J. J. Benedict, *Inorg. Chem.*, 1971, **10**, 1183.
[28] S. J. Lippard and K. M. Melmed, *Inorg. Chem.*, 1967, **6**, 2223.
[28a] T. H. Cook and G. L. Morgan, *J. Amer. Chem. Soc.*, 1970, **92**, 6493.

Of substituted anions, the cyanohydride, $NaBH_3CN$ which is made[29] by action of HCN on $NaBH_4$ in THF, is the most useful since it persists and can be used for reductions in acid solution even at pH 3. Also it will exchange H for D in acid D_2O solutions,[30] possibly by the reactions

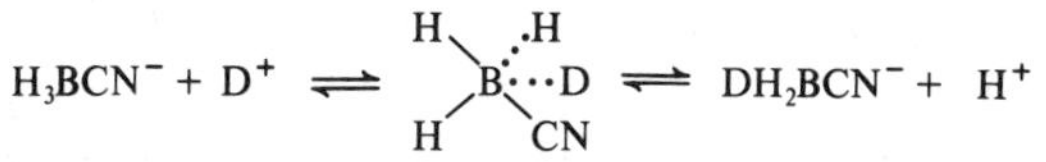

$$H_3BCN^- + D^+ \; \rightleftharpoons \; \underset{H}{\overset{H}{\underset{\diagdown}{}}}\; B\overset{\diagup H}{\underset{\diagdown CN}{\cdots D}} \; \rightleftharpoons \; DH_2BCN^- + H^+$$

In presence of ammonia or amines, aldehydes and ketones can be reduced and aminated to give amines. The isomeric H_3BNC^- is known; it isomerizes to H_3BCN^- in acid solution.

Another important non-polyhedral borane anion is $B_3H_8^-$, most easily prepared by the reaction

$$NaBH_4 + B_2H_6 \; \xrightarrow[\text{diglyme}]{100^\circ} \; NaB_3H_8 + H_2$$

Its structure (8-XVI) has been proved by X-ray diffraction. It is, however, a fluxional species, all three boron atoms appearing as a single signal having

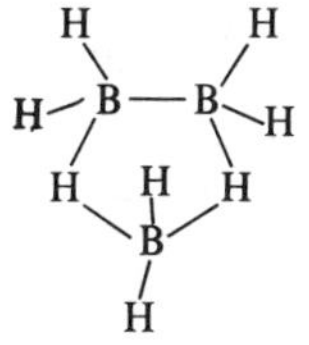

(8-XVI)

the nonet pattern that would arise by splitting due to eight equivalent protons: apparently all the hydrogen atoms undergo rapid site exchange. Like BH_4^-, this anion can serve as a ligand, and the structures of several complexes are known;[31] one, which is representative, is shown in (8-XVII).

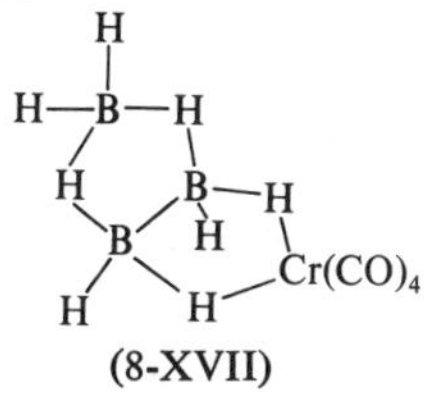

(8-XVII)

$B_9C_2H_{13-n}^{n-}$ **Species.** It was mentioned earlier that both 1,2- and 1,7-$B_{10}C_2H_{12}$ could be degraded by strong bases, e.g., $C_2H_5O^-$, to give isomeric $B_9C_2H_{12}^-$ ions. This removal of a BH^{2+} unit from the parent carborane (see below) may be interpreted as a nucleophilic attack at the most electron-deficient boron atoms of the carborane. Molecular-orbital calculations show

[29] R. C. Wade et al., Inorg. Chem., 1970, 9, 2146; J. R. Berschied and K. F. Purcell, Inorg. Chem., 1970, 2, 624.
[30] R. F. Borch and H. D. Durst, J. Amer. Chem. Soc., 1969, 91, 3906; M.M. Kreevoy and J. E. C. Hutchins, J. Amer. Chem. Soc., 1969, 91, 4330.
[31] L. J. Guggenberger, Inorg. Chem., 1970, 9, 367; S. J. Lippard and K. M. Melmud, Inorg. Chem., 1969, 8, 2755; E. L. Muetterties and C. W. Aligranti, J. Amer. Chem. Soc., 1970, 92, 4114.

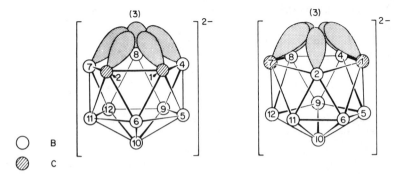

○ B

▨ C

(3) Open position left by the removal of B atom 3

Fig. 8-12.   Structures of the isomeric $B_9C_2H_{11}$ ions.

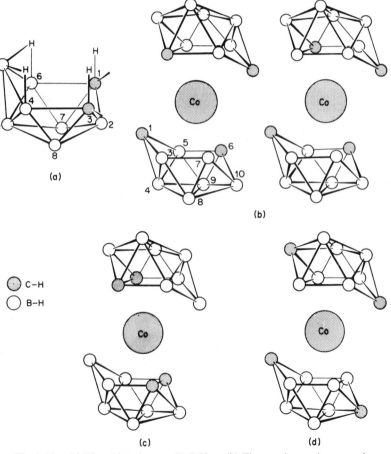

Fig. 8-13.   (a) The *nido*carborane $B_7C_2H_{13}$. (b) The two brown isomers of $[(B_7C_2H_9)_2Co]^-$. (c) The red isomer of $[(B_7C_2H_9)_2Co]^-$. (d) The thermally most stable isomer of $[(B_7C_2H_9)_2Co]^-$.

that the carbon atoms in the $B_{10}C_2H_{12}$ molecules (and in carboranes in general) have considerable electron-withdrawing power, and thus the most electron-deficient boron atoms are those adjacent to C, namely, those at positions 3 and 6 in $1,2-B_{10}C_2H_{12}$ and those at positions 2 and 3 in $1,7-B_{10}C_2H_{12}$ (cf. Fig. 8-10 for the numbering scheme). Therefore, the direct products of the reactions of the isomeric $B_{10}C_2H_{12}$ molecules with strong base should be the isomeric $B_9C_2H_{11}^{2-}$ ions shown in Fig. 8-12. These ions themselves are strong bases and acquire protons to form the $B_9C_2H_{12}^{-}$ ions. It is not known precisely how the twelfth hydrogen atom is bound in these species, but it presumably resides somewhere on the open face of the now incomplete icosahedron and is probably very labile. The $B_9C_2H_{12}^{-}$ ions can be protonated to form the neutral *nido*carboranes, $B_9C_2H_{13}$, which are strong acids.

Another example of a *nido*carborane of importance because of the anion that it forms is $B_7C_2H_{13}$. This has already been mentioned (page 243) as an intermediate, obtained by oxidation of $B_9C_2H_{11}$, in the preparation of several lower *closo*carboranes. It has the structure shown in Fig. 8-13a. The compound $B_7C_2H_{13}$ has several acidic protons, presumably those on the carbon atoms, and the anion $B_7C_2H_{11}^{2-}$ can be prepared by treatment with NaH; further reactions of this anion with transition metals will be discussed in Section 8-12.

8-12. Metal Complexes of Carborane Anions

It was Hawthorne who, in 1964, recognized that the open pentagonal faces of the $B_9C_2H_{11}^{2-}$ ions (Fig. 8-12) bear a strong resemblance, structurally and electronically, to the cyclopentadienyl anion, $C_5H_5^{-}$, which forms strong bonds to the transition metals having the geometry (8-XVIII). He investi-

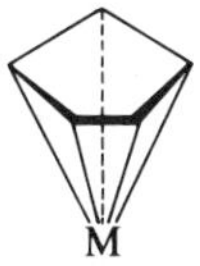

M
(8-XVIII)

gated the possibility of forming comparable bonds with the carborane anions and soon reported positive results. In order to generate the $B_9C_2H_{11}^{2-}$ ions from the $B_9C_2H_{12}^{-}$ ions, which have been discussed in Section 8-11, the very strong base NaH was used:

$$B_9C_2H_{12}^{-} + NaH \rightarrow Na^{+} + B_9C_2H_{11}^{2-} + H_2$$

The $B_9C_2H_{11}^{2-}$ ions were found to combine readily with transition-metal ions such as Fe^{2+} and Co^{3+} which have d^6 configurations, thus leading to species isoelectronic with $(C_5H_5)_2Fe$ and $(C_5H_5)_2Co^{+}$, that is, $(B_9C_2H_{11})_2Fe^{2-}$ and $(B_9C_2H_{11})_2Co^{-}$, respectively. The iron complex undergoes reversible oxidation analogous to that of $(C_5H_5)_2Fe$. The structures of these $(B_9C_2H_{11})_2M^{n-}$ systems are mostly as shown in Fig. 8-14a and have essen-

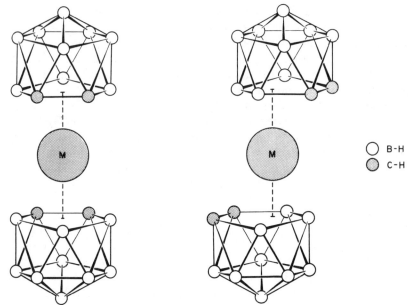

Fig. 8-14.   The structures of some bis(dicarbollide) metal complexes. (a) The symmetrical
structure. (b) The "slipped" structure.

tially D_{5d} symmetry, if the difference between B and C atoms is ignored. In the cases of the $(B_9C_2H_{11})_2Cu^{2-}$ and $(B_9C_2H_{11})_2Cu^-$ species, and probably some others, there is a "slippage" of the ligands as shown in Fig. 8-14b, caused by the presence of too many electrons to be accommodated in the bonding MO's of the symmetrical structure.

Besides the $(B_9C_2H_{11})_2M^{n-}$ species, many mixed complexes in which there is one $B_9C_2H_{11}^{2-}$ ligand and some other ligand or set of ligands, e.g., $C_5H_5^-$, Ph_4C_4, or 3CO, can be prepared. Three are shown in Fig. 8-15.

As suggested by Fig. 8-14, the majority of the compounds made from $B_9C_2H_{11}^{2-}$ ions contain the isomer with adjacent carbon atoms, since the $1,2-B_{10}C_2H_{12}$ carborane is the most accessible starting material; however, isomers obtainable from the 1,7-carborane, with non-adjacent carbon atoms, are also known.

Since systematic nomenclature for the $B_9C_2H_{11}^{2-}$ ion and its complexes is rather unwieldy, the trivial name "dicarbollide" ion has been proposed. This is derived from the Spanish word *olla* meaning *pot*, in view of the shape of the 11-particle icosahedral fragment. The unknown parent "ollide" ion would be $B_{11}H_{11}^{4-}$.

Another series of carborane anion complexes of transition metals is formed from the $B_7C_2H_9^{2-}$ ion.[32] For example, when the $B_7C_2H_{11}^{2-}$ ion mentioned in Section 8-11 is treated with cobalt(II) chloride, the following reaction occurs:

$$2B_7C_2H_{11}^{2-} + 1.5Co^{2+} \rightarrow [(B_7C_2H_9)_2Co]^- + 0.5Co + 2H_2$$

[32] T. A. George and M. F. Hawthorne, *J. Amer. Chem. Soc.*, 1969, **91**, 5475.

Similar reactions have been successful with other transition metals. When the above reaction is carried out at room temperature the anionic complex obtained is red, whereas when the reaction mixture, in 1,2-dimethoxyethane, is refluxed the product is brown. Figure 8-13 shows the structures of these isomeric species; it will be seen that "the" brown isomer is, in fact, a mixture of geometric isomers. Comparison of Fig. 8-13a with 8-13c reveals that the red isomer is derived directly from the skeleton of the $B_7C_2H_{13}$ ancestor, whereas the brown isomer contains ligands with a rearranged skeleton. When heated at 315° for 24 hr, both the red and the brown forms rearrange quantitatively to an orange isomer that has a structure shown in Fig. 8-13d. In view of the shoe-like shape of the $B_7C_2H_9^{2-}$ ions a trivial name based on the Spanish word for shoe, *zapato*, has been assigned to them. Thus, the hypo-

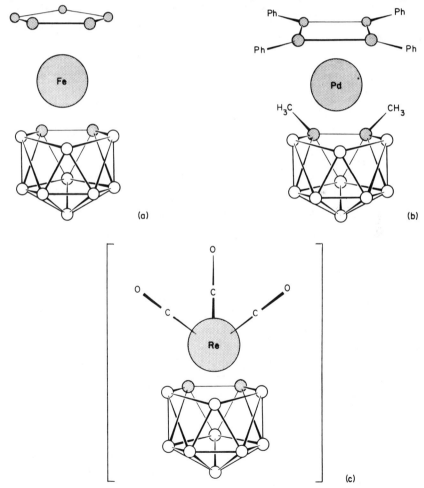

Fig. 8-15. Some complexes containing one dicarbollide ion and ligands of other types, viz., (a) $C_5H_5^-$, (b) C_4Ph_4 and (c) three CO groups.

thetical $B_9H_9^{4-}$ ion would be called the "zapide" ion and $B_7C_2H_9^{2-}$ is then called the "dicarbazapide" ion. The $[(B_7C_2H_9)_2Co]^-$ ions are called bis(dicarbazapyl)cobalt(III) ions.

8-13. Boranes with Hetero Atoms Other Than Carbon

Molecules and ions in which atoms other than B and C make up the framework are in principle possible and some have already been made. However, the study of such species is still only beginning and we merely mention here a few illustrative examples. The formalism for relating isoelectronic species is that P is equivalent to BH^-, and S is equivalent to BH^{2-}.

The thiaborane anions, $B_{10}SH_{11}^-$ and $B_{10}SH_{10}^{2-}$ are known; the latter is isoelectronic with $B_9C_2H_{11}^{2-}$ and is presumably isostructural with it. It reacts with transition-metal ions to form complexes presumably analogous to the bis(dicarbollyl)metal species, e.g.:

$$2B_{10}SH_{10}^{2-} + Fe^{2+} \rightarrow [(B_{10}SH_{10})_2Fe]^{2-}$$

The known carbaphosphaboranes[33] are so far limited to the three isomers of $B_{10}CPH_{11}$. The 1,2-isomer is obtained in the reaction

$$Na_3B_{10}H_{10}CH + PCl_3 \rightarrow 1,2\text{-}B_{10}CPH_{11} + 3NaCl$$

Like $1,2\text{-}B_{10}C_2H_{12}$, $1,2\text{-}B_{10}CPH_{11}$ can be thermally converted into its 1,7- and 1,12-isomers. These react with piperidine to generate isomeric $B_9CPH_{11}^-$ anions that are probably structural analogs of the $B_9C_2H_{12}^-$ anions, and some metal complexes of a related anion have been prepared.[34]

8-14. Compounds of Boron with Other Elements

Aside from the heteroborane compounds just treated, there are a host of more conventional compounds containing bonds between boron and other elements such as N, P, As, S and C. Only a few of the more interesting topics can be covered here.

Boron–Nitrogen Compounds.[35] Compounds of the type $R'_3N \rightarrow BR_3$ have already been mentioned amongst the many donor–acceptor complexes formed by BR_3 compounds. In this Section we are concerned with compounds containing the —NR'—BR— unit. This unit is similar to a —CR'=CR— group and can replace it in many compounds. This analogy is often justified by the assumption that the actual electron distribution in the N to B bond can be described by the resonance (8-XIX), whereby appre-

$$\overset{\backslash}{\underset{/}{N}} \!\!:\!\! - B \!\! \overset{/}{\underset{\backslash}{}} \quad \longleftrightarrow \quad \overset{\backslash}{\underset{/}{\overset{+}{N}}} \!\! = \!\! \bar{B} \overset{/}{\underset{\backslash}{}}$$

(8-XIXa)   (8-XIXb)

[33] L. J. Todd, J. L. Little and H. T. Silverstein, *Inorg. Chem.*, 1968, **8**, 1698.

[34] P. S. Welcker and L. J. Todd, *Inorg. Chem.*, 1970, **9**, 286.

[35] K. Niedenzu and J. W. Dawson in ref. 4; A. Finch, J. B. Leach and J. M. Morris, *Organometallic Chem. Rev.*, (A), 1969, **4**, 1; Advances in Chem. Series, No. 42, Amer. Chem. Soc., Washington.

ciable π bonding is introduced. There is fairly good evidence that such N to B bonds do indeed have appreciable π character, while at the same time, they lack the polarity which would be engendered by (8-XIXb). This apparent paradox is explained by the existence of considerable polarity in the σ bond, in a direction opposite to that in the π bond. Thus the net, or actual, polarity is only the difference between the two.

There are a number of carbon heterocycles, such as (8-XX) and (8-XXI) containing one or two —BR′—NR— units, but perhaps the most interesting

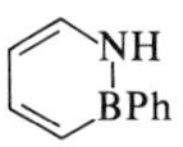
(8-XX)

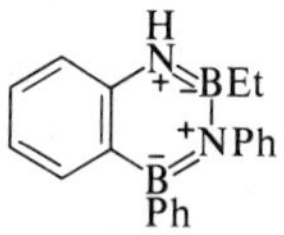
(8-XXI)

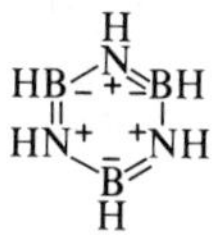
(8-XXII)

compounds are borazine (8-XXII) and its derivatives. Borazine has an obvious formal resemblence to benzene and the physical properties of the two compounds are very similar. However, the analogy should not be too strongly stressed since the chemical reactivity of borazine is generally quite different from that of benzene. Borazine is the more reactive and it readily undergoes addition reactions (which is not true of benzene), such as

$$B_3N_3H_6 + 3HX \rightarrow (\text{—}H_2N\text{—}BHX\text{—})_3 \qquad (X = Cl, OH, OR, \text{etc.})$$

Borazine decomposes slowly on storage and is hydrolyzed at elevated temperatures to NH_3 and $B(OH)_3$. It is of interest that borazine resembles benzene in forming arene–metal complexes (Section 23-3); thus the hexamethylborazine complex, $B_3N_3(CH_3)_6Cr(CO)_3$ has been reported and closely resembles $C_6(CH_3)_6Cr(CO)_3$ but is thermally less stable.[36]

Reaction sequences by which borazine and substituted borazines may be synthesized are the following:

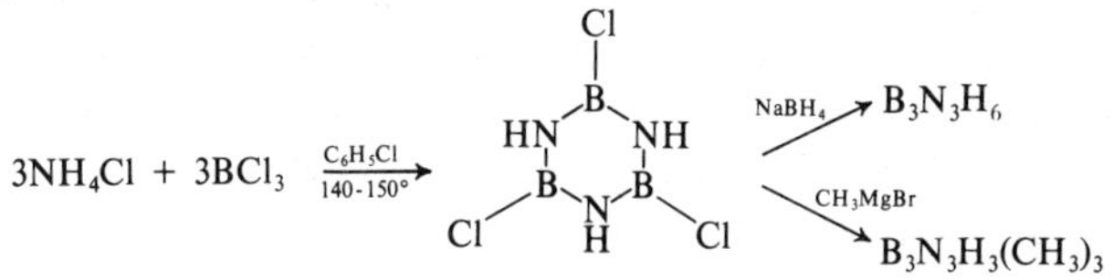

$$CH_3NH_2 + BCl_3 \xrightarrow[C_6H_5Cl]{\text{Boiling}} Cl_3B\cdot NH_2CH_3 \quad (\text{m.p. } 126\text{–}128°)$$

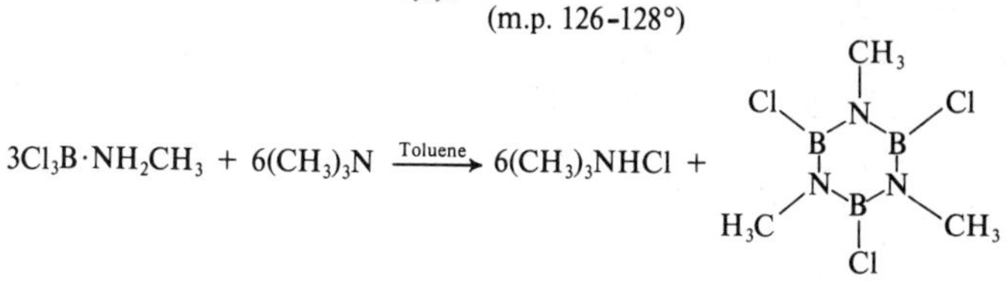
(m.p. 153-156°)

[36] H. Werner, R. Prinz and E. Deckelmann, *Chem. Ber.*, 1969, **102**, 95.

Both MO calculations[37] and experimental results for the transmission of substituent effects through the B_3N_3 ring[38] indicate that the π electrons are partially delocalized. Complete delocalization is not to be expected since the $N\pi$ orbitals should be of appreciably lower energy than the $B\pi$ orbitals. The planarity of the borazine molecule is shown by MO calculations to be stabilized by the π bonding, and the calculations also suggest that the N to B π electron drift is actually outweighed by B to N σ drift, so that the nitrogen atoms are relatively negative.

In view of the extensive aminoboron chemistry, studies have been made of boron–phosphorus and boron–arsenic chemistry. One of the notable B—P compounds is $[(CH_3)_2PBH_2]_3$, which has the cyclic structure (8-XXIII). The

$$\begin{array}{c} Me_2 \\ P \\ H_2B \quad\quad BH_2 \\ | \quad\quad\quad | \\ Me_2P \quad\quad PMe_2 \\ B \\ H_2 \end{array}$$

(8-XXIII)

arsenic analog is also known. This compound and its arsenic analog are extraordinarily stable and inert, a fact that has been attributed to a drift of electron density from the BH_2 groups into the d orbitals of P or As. This has the effect of reducing the hydridic nature of the hydrogen atoms, making them less susceptible to reaction with protonic reagents, and also of off-setting the $\bar{B}$—$\overset{+}{P}$, $\bar{B}$—$\overset{+}{As}$ polar character which the σ bonding alone tends to produce.

Organoboron Compounds.[39]   Thousands of boron compounds with B—C, B—O—C, B—S—C, B—N—C, etc. bonds, whose chemistry is essentially organic in nature, are known.

The *alkyl-* and *aryl-borons* can be made from the halides by lithium or Grignard reagents. However, as noted above (page 240), the alkylborons, which have great importance as intermediates in organic synthesis, are often most conveniently made by hydroboration. The lower alkyls are reactive substances and inflame in air, but the aryls are stable.

When boron halides are treated with four equivalents of alkylating agent, the trialkyl or triaryl reacts further to form an anion of the type BR_4^-. The most important such compound is *sodium tetraphenylborate*, $Na[B(C_6H_5)_4]$; this is soluble in water and is stable in weakly acid solution; it gives insoluble precipitates with larger cations such as K^+, Rb^+ or Me_4N^+, that are suitable for gravimetric analysis. The ion can also act as a ligand wherein one phenyl ring is bound in an arene complex as in (8-XXIV).[40]

[37] P. M. Kuznesof and D. F. Shriver, *J. Amer. Chem. Soc.*, 1968, **90**, 1683; D. R. Armstrong and D. T. Clark, *Chem. Comm.*, **1970**, 99.

[38] O. T. Beachley, Jr., *J. Amer. Chem. Soc.*, 1970, **92**, 5372.

[39] M. F. Lappert in ref. 4; H. Steinberg and R. J. Brotherton, *Organoboron Chemistry*, Vols. 1 and 2, Interscience-Wiley, 1964, 1967 (comprehensive reference texts); *Organometallic Chem. Rev.*, Section B (annual surveys).

[40] R. R. Schrock and J. A. Osborn, *Inorg. Chem.*, 1970, **9**, 2339; R. J. Haines and A. L. du Preez, *J. Amer. Chem. Soc.*, 1971, **93**, 2820.

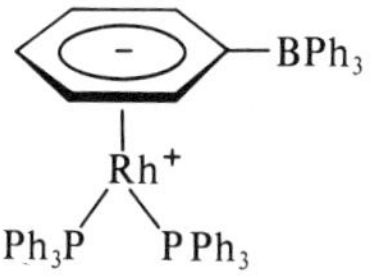

(8-XXIV)

Unlike Na[BPh$_4$], the compound Li[BMe$_4$], m.p. 189°, is readily soluble in benzene and can be sublimed.[41] Although apparently only an ion pair, Li$^+$ BMe$_4^-$, in solution, it is polymeric in the crystal. An unusual type of methyl bridge occurs in addition to the conventional type found in (BeMe$_2$)$_n$ and (AlMe$_3$)$_2$; there are linear Li—C—B chains with Li—C = 2.12 Å and C—B = 1.51 Å. The full implications of this structural feature and its possible occurrence elsewhere are still being assessed.

Other important series of compounds are the *alkyl* and *aryl orthoborates*, B(OR)$_3$, and their complexes such as Na[HB(OR)$_3$], the trialkoxoborohydrides; these may be thought of as derived from B(OH)$_3$.

There are also the so-called *boronic* and *boronous acids*, RB(OH)$_2$ and R$_2$B(OH), and their esters and anhydrides. The boronic acids can be made in various ways, for example:

$$BF_3 \cdot O(C_2H_5)_2 + C_6H_5MgBr \rightarrow C_6H_5BF_2 \xrightarrow{H_2O} C_6H_5B(OH)_2$$

they are quite stable and water-soluble. Their acidities depend on the nature of the alkyl or aryl group. Dehydration of a boronic acid by heat yields a boronic anhydride or *boroxine*:

$$3RB(OH)_2 \rightarrow (RBO)_3 + 3H_2O$$

Boroxines have trimeric, cyclic structures with planar rings of alternating boron and oxygen atoms. The alkyl groups are also in the plane of the ring.

Boron can also be incorporated into numerous carbon ring systems.

[41] D. Groves, W. Rhine and C. D. Stucky, *J. Amer. Chem. Soc.*, 1971, **93**, 1553.

Further Reading

Progress in Boron Chemistry, Vol. 1, 1964 (reviews).
Fortschritte der Chemischen Forschung, F. Boscke, ed., *New Results in Boron Chemistry*, Springer Verlag, 1970 (borates, B—N ring compounds, lower hydrides).
Boron, Metallo-Boron Compounds and Boranes, R. M. Adams, ed., Interscience-Wiley, 1964.
Lappert, M. F., and H. Pyszora, *Adv. Inorg. Chem. Radiochem.*, 1966, **9**, 133 (pseudohalides of boron and other Group III and Group IV elements).
Nemodruk, A. A., and Z. K. Karalova, *Analytical Chemistry of Boron*, Oldbourne Press, 1965.
Schmid, G., *Angew. Chem. Internat. Edn.*, 1970, **9**, 819 (survey of compounds with boron bound to metals).
Thompson, R., *Chem. in Britain*, **1970**, 140 (uses of boron compounds).

9

The Group III Elements: Al, Ga, In, Tl

GENERAL REMARKS

9-1. Electronic Structures and Valences

The electronic structures and some other important fundamental properties of the elements are listed in Table 9-1.

TABLE 9-1

Some Properties of the Group III Elements

Element	Electronic structure	Ionization potentials (eV)				$E°$ (V)^a	M.p. (°C)
		1st	2nd	3rd	4th		
B	[He]$2s^2 2p$	8.30	25.15	37.92	259.30	Not measurable	~2200
Al	[Ne]$3s^2 3p$	5.98	18.82	28.44	119.96	−1.66	660
Ga	[Ar]$3d^{10} 4s^2 4p$	6.00	20.43	30.6	63.8	−0.53	29.8
In	[Kr]$4d^{10} 5s^2 5p$	5.79	18.79	27.9	57.8	−0.338^b	157
Tl	[Xe]$4f^{14} 5d^{10} 6s^2 6p$	6.11	20.32	29.7	50.5	+0.72^c	303

^{a} For $M^{3+}(aq) + 3e = M(s)$.
^{b} $In^+ + e = In(s)$ $E° = -0.147$.
^{c} $Tl^+ + e = Tl(s)$ $E° = -0.3363$.

Aluminum and its congeners, Ga, In and Tl, are considerably larger than boron and hence they are much more metallic and ionic in their character. Elemental aluminum itself is clearly metallic, but it is nevertheless still on the borderline (like beryllium) between ionic and covalent character in its compounds. So also are Ga, In and Tl.

While the trivalent state is important for all four elements, the univalent state becomes progressively more stable as the Group is descended and for thallium the Tl^I–Tl^{III} relationship is a dominant feature of the chemistry. This occurrence of an oxidation state two below the group valence is sometimes attributed to the so-called *inert pair* effect which first makes itself evident here, although it is adumbrated in the low reactivity of mercury in Group II, and it is much more pronounced in Groups IV and V. The term refers to the resistance of a pair of s electrons to be lost or to participate in covalent-bond formation. Thus mercury is difficult to oxidize, allegedly because it contains only an inert pair ($6s^2$), Tl readily forms Tl^I rather than Tl^{III} because of the inert pair in its valence shell ($6s^2 6p$), etc. The concept of the inert pair does not actually tell us anything about the ultimate reasons for the stability of certain lower valence states, but it is useful as a label and is often encountered in the literature. The true cause of the phenomenon is not intrinsic inertness, that is, unusually high ionization potential of the pair of s electrons, but rather the decreasing strengths of bonds as a Group is descended. Thus, for example, the sum of the 2nd and 3rd ionization potentials is lower for In (46.7 eV) than for Ga (51.0 eV) with Tl (50.0 eV) intermediate. There is, however, a steady decrease in the mean thermochemical bond energies, for example, among the trichlorides: Ga, 242; In, 206; Tl, 153 kJ mol^{-1}. The relative stabilities of oxidation states differing in the presence or absence of the inert pair are further discussed in connection with the Group IV elements.

In the trihalide, trialkyl, and trihydride compounds there are some resemblances to the corresponding boron chemistry. Thus MX_3 compounds behave as Lewis acids and can accept either neutral donor molecules or anions to give tetrahedral species; the acceptor ability generally decreases in the order Al > Ga > In, with the position of Tl uncertain. There are, however, notable distinctions from boron. These are in part due to the reduced ability to form multiple bonds and to the ability of the heavier elements to have coordination numbers exceeding four. Thus, while boron gives $Me_2\overset{-}{B}=\overset{+}{N}Me_2$, Al, Ga and In give dimeric species, e.g., $[Me_2AlNMe_2]_2$, in which there is an NMe_2 bridging group and both the metal and nitrogen atoms are 4-coordinate. Similarly, the boron halides are all monomeric, while those of Al, Ga and In are all dimeric. The polymerization of trivalent Al, Ga, In and Tl compounds to achieve coordination saturation is general and four-membered rings appear to be a common way despite the valence-angle strain implied. Secondly, compounds such as $(Me_3N)_2AlH_3$ have trigonal-bipyramidal structures which of course are impossible for boron adducts. Finally, in contrast to boron, there is a well-defined aqueous cationic chemistry; aquo ions, e.g., $[In(H_2O_6]^{3+}$, salts of oxo anions and complexes, all with octahedral stereochemistry, exist; for Al^{3+}(aq) the coordination number six has been proved by ^{17}O nuclear resonance studies.

THE ELEMENTS[1]

9-2. Occurrence, Isolation and Properties

Aluminum, the commonest metallic element in the earth's crust, occurs widely in Nature in silicates such as micas and feldspars, as the hydroxo

[1] I. A. Sheka, I. S. Chaus and T. T. Mityureva, *The Chemistry of Gallium*, Elsevier, 1966; N. N. Greenwood, *Adv. Inorg. Chem. Radiochem.*, 1963, 5, 91 (the chemistry of gallium); A. G. Lee, *The Chemistry of Thallium*, Elsevier, 1971.

oxide (*bauxite*) and as *cryolite*(Na_3AlF_6). The other three elements are found only in trace quantities. Gallium and indium are found in aluminum and zinc ores, but the richest sources contain less than 1% of gallium and still less indium. Thallium is widely distributed; the element is usually recovered from flue dusts from the roasting of certain sulfide ores, mainly pyrites.

Aluminum is prepared on a vast scale from bauxite. This is purified by dissolution in sodium hydroxide and reprecipitation using carbon dioxide. It is then dissolved in molten cryolite at 800–1000° and the melt is electrolyzed. Aluminum is a hard, strong, white metal. Although highly electropositive, it is nevertheless resistant to corrosion because a hard, tough film of oxide is formed on the surface.[2] Thick oxide films, some with the proper porosity when fresh to trap particles of pigment, are often electrolytically applied to aluminum. Aluminum is soluble in dilute mineral acids, but is passivated by concentrated nitric acid. If the protective effect of the oxide film is overcome, for example, by scratching or by amalgamation, rapid attack even by water can occur. The metal is attacked under ordinary conditions by hot alkali hydroxides, halogens, and various non-metals. Highly purified aluminum is quite resistant to acids and is best attacked by hydrochloric acid containing a little cupric chloride or in contact with platinum, some H_2O_2 also being added during the dissolution.

Gallium, indium and *thallium* are usually obtained by electrolysis of aqueous solutions of their salts; for Ga and In this possibility arises because of large overvoltages for hydrogen evolution on these metals. They are soft, white, comparatively reactive metals, dissolving readily in acids; however, thallium dissolves only slowly in sulfuric or hydrochloric acid since the Tl^I salts formed are only sparingly soluble. Gallium, like aluminum, is soluble in sodium hydroxide. The elements react rapidly at room temperature, or on warming, with the halogens and with non-metals such as sulfur.

The exceptionally low melting point of gallium has no simple explanation. Since its boiling point (2070°) is not abnormal, gallium has the longest liquid range of any known substance and finds use as a thermometer liquid.

CHEMISTRY OF THE TRIVALENT STATE

BINARY COMPOUNDS

9-3. Oxygen Compounds

Stoichiometrically there is only one oxide of aluminum, namely, alumina, Al_2O_3. However, this simplicity is compensated by the occurrence of various polymorphs, hydrated species, etc., the formation of which depends on the conditions of preparation. There are two forms of anhydrous Al_2O_3, namely, α-Al_2O_3 and γ-Al_2O_3.* Various other trivalent metals (Ga, Fe) form oxides

---

[2] J. W. Diggle, T. C. Downie and C. W. Golding, *Chem. Rev.*, 1969, **69**, 365.
* "β-Al_2O_3" is actually $Na_2O \cdot 6Al_2O_3$; see R. Scholder and M. Mansmann, *Z. anorg. Chem.*, 1963, **321**, 246.

that crystallize in these same two structures. In α-Al_2O_3 the oxide ions form a hexagonally close-packed array and the aluminum ions are distributed symmetrically among the octahedral interstices. The γ-Al_2O_3 structure is sometimes regarded as a "defect" spinel structure, that is as having the structure of spinel with a deficit of cations (see below).

α-Al_2O_3 is stable at high temperatures and also indefinitely metastable at low temperatures. It occurs in Nature as the mineral corundum and may be prepared by heating γ-Al_2O_3 or any hydrous oxide above 1000°. γ-Al_2O_3 is obtained by dehydration of hydrous oxides at low temperatures ($\sim$450°). α-Al_2O_3 is very hard and resistant to hydration and attack by acids, whereas γ-Al_2O_3 readily takes up water and dissolves in acids. The Al_2O_3 that is formed on the surface of the metal has still another structure, namely, a defect rock-salt structure; there is an arrangement of Al and O ions in the rock-salt ordering with every third Al ion missing.

There are several important hydrated forms of alumina corresponding to the stoichiometries $AlO \cdot OH$ and $Al(OH)_3$. Addition of ammonia to a boiling solution of an aluminum salt produces a form of $AlO \cdot OH$ known as böhmite, which may be prepared in other ways also. A second form of $AlO \cdot OH$ occurs in Nature as the mineral diaspore. The true hydroxide, $Al(OH)_3$, is obtained as a crystalline white precipitate when carbon dioxide is passed into alkaline "aluminate" solutions.

The gallium oxide system quite closely resembles the aluminum oxide system, affording a high-temperature α- and a low-temperature γ-Ga_2O_3, $GaO \cdot OH$ and $Ga(OH)_3$. The trioxide is formed by heating the nitrate, the sulfate or the hydrous oxides that are precipitated from Ga^{III} solutions by the action of ammonia. β-Ga_2O_3 contains both tetrahedrally and octahedrally coordinated gallium with Ga—O distances of 1.83 and 2.00 Å, respectively. Indium gives yellow In_2O_3, which is known in only one form, and a hydrated oxide, $In(OH)_3$. Thallium has only the brown-black Tl_2O_3, which begins to lose oxygen at about 100° to give Tl_2O. The action of NaOH on Tl^{III} salts gives what appears to be the oxide, whereas with Al, Ga and In the initial products are basic salts.

Aluminum, gallium and thallium form mixed oxides with other metals. There are, first, aluminum oxides containing only traces of other metal ions. These include ruby (Cr^{3+}) and blue sapphire (Fe^{2+}, Fe^{3+} and Ti^{4+}). Synthetic ruby, blue sapphire, and white sapphire (gem-quality corundum) are now produced synthetically in large quantities. Second are mixed oxides containing macroscopic proportions of other elements, such as the minerals *spinel*, $MgAl_2O_4$, and *crysoberyl*, $BeAl_2O_4$. The spinel structure has been described and its importance as a prototype for many other $M^{II}M_2^{III}O_4$ compounds noted (page 54). Alkali-metal compounds such as $NaAlO_2$,[3] which can be made by heating Al_2O_3 with sodium oxalate at 1000°, are also ionic mixed oxides.

³ A. F. Reid and A. E. Ringwood, *Inorg. Chem.*, 1968, **7**, 443.

9-4. Halides[4]

All four halides of each element are known, with one exception. The compound TlI_3, obtained by adding iodine to thallous iodide, is not thallium(III) iodide, but rather thallium(I) triiodide, $Tl^I(I_3)$. This situation may be compared with the non-existence of iodides of other oxidizing cations such as Cu^{2+} and Fe^{3+}, except that here a lower-valent compound fortuitously has the same stoichiometry as the higher-valent one. The coordination numbers of the halides are shown in Table 9-2. The fluorides of Al, Ga and In are all high-melting [1290°, 950° (subl.), 1170°, respectively], whereas the chlorides, bromides and iodides have lower melting points. There is, in general, a good correlation between melting points and coordination number. Thus, the three chlorides have the following melting points: $AlCl_3$, 193° (at 1700 mm); $GaCl_3$, 78°; $InCl_3$, 586°.

TABLE 9-2

Coordination Numbers of Metal Atoms
in Group III Halides[a]

	F	Cl	Br	I
Al	6	6	4	4
Ga	6	4	4	4
In	6	6	6	4
Tl	6	6	4	

[a] N. N. Greenwood, D. J. Prince and B. P. Straughan, *J. Chem. Soc., A,* **1968**, 1694.

The halides with coordination numbers 4 can be considered to consist of discrete dinuclear molecules (Fig. 9-1) and since there are no strong lattice forces the melting points are low. In the vapor, aluminum chloride is also dimeric so that there is a radical change of coordination number on vaporization, and these covalent structures persist in the vapor phase at temperatures not too far above the boiling points. At sufficiently high temperatures, however, dissociation occurs into planar, triangular monomers, analogous to the boron halides. There is some evidence that for gallium iodide this dissociation is very extensive even at the boiling point. The Group III halides dissolve readily in many non-polar solvents such as benzene, in which they are dimeric. The enthalpies of dissociation, $Al_2X_6(g) = 2AlX_3(g)$, have been measured and are 46–63 kJ mol^{-1}. As Fig. 9-1 shows, the configuration of halogen atoms about each metal atom is roughly, though far from exactly, tetrahedral. The formation of such dimers is attributable to the tendency of the metal atoms to complete their octets. The dimers may be split by reaction with donor molecules, giving complexes such as R_3NAlCl_3.

[4] (a) *Friedel-Crafts and Related Reactions*, G. A. Olah, ed., Vol. 1, Wiley, 1963 (much information on Al and Ga halides and their complexes). (b) J. Carty, *Coordination Chem. Rev.*, 1969, **4**, 29 (halides and complexes of Ga, In and Tl). (c) R. A. Walton, *Coordination Chem, Rev.*, 1971, **6**, 1 (halides and complexes of thallium).

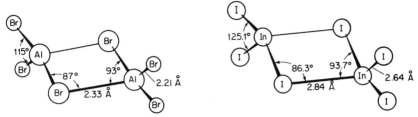

Fig. 9-1.   The structures of Al_2Br_6 and In_2I_6.

The halides dissolve in water, giving acidic solutions from which hydrates may be obtained.

The thallium(III) halides, other than the iodide noted above, are genuine but unstable. The chloride, which is most commonly used, can be prepared conveniently[5] by the following sequence:

$$Tl,\ or\ TlCl,\ or\ Tl_2CO_3 \xrightarrow{ClNO} TlCl_3 \cdot NOCl \xrightarrow{Heat} TlCl_3$$

Solutions of $TlCl_3$ and $TlBr_3$ in CH_3CN which are useful for preparative work are conveniently obtained by treating solutions of the monohalides with Cl_2 or Br_2.[6] Solid $TlCl_3$ loses chlorine at about 40° and above, to give the monochloride, and the tribromide loses bromine at even lower temperatures to give first "$TlBr_2$" which is actually $Tl^I[Tl^{III}Br_4]$. The fluoride is stable to about 500°. These facts provide a very good illustration of the way in which the stability of the lower valence state dominates thallium chemistry.

9-5. Other Binary Compounds

The Group III elements form various compounds such as carbides, nitrides, phosphides and sulfides.

Aluminum carbide, Al_4C_3, is formed from the elements at temperatures of 1000–2000°. It reacts instantly with water to produce methane, and X-ray studies have shown it to contain discrete carbon atoms (C—C = 3.16 Å); for these reasons it is sometimes considered to be a "methanide," that is, a salt containing C^{4-}, but this is probably an oversimplification.

The *nitrides* AlN, GaN and InN are known. Only aluminum reacts directly with nitrogen. GaN is obtained on reaction of Ga or Ga_2O_3 at 600–1000° with NH_3 and InN by pyrolysis of $(NH_4)_3InF_6$. All have a wurtzite structure (Fig. 2-3). They are fairly hard and stable, as might be expected from their close structural relationship to diamond and the diamond-like BN.

Aluminum and especially Ga and In form 1:1 compounds with Group V elements, the so-called III–V compounds, such as GaAs.[7] These compounds

[5] W. O. Groenveld and A. P. Zuur, *Inorg. Nuclear Chem. Letters*, 1967, **3**, 229.
[6] B. F. G. Johnson and R. A. Walton, *Inorg. Chem.*, 1966, **5**, 49.
[7] *Compound Semi Conductors*, R. K. Willardson and H. L. Goering, eds., Vol. 1, Reinhold, 1962; R. K. Willardson and A. C. Beer, *Semi-Conductors and Semi-metals*, Academic Press, 1966, *et seq.* (physics and optical properties of III—V compounds; several volumes).

have semiconductor properties similar to those of elemental Si and Ge, to which they are electronically and structurally similar. They can be obtained by direct interaction or in other ways. Thus GaP can be obtained as pale orange single crystals by the reaction of phosphorus and Ga_2O vapor at 900–1000°.

9-6. The Aquo Ions, Oxo Salts, Aqueous Chemistry

The elements form well-defined octahedral aquo ions, $[M(H_2O)_6]^{3+}$, and a wide variety of salts are formed including hydrated halides, oxo acid salts such as sulfate, nitrate and perchlorate, as well as sparingly soluble salts such as phosphates.

In aqueous solution the octahedral $[M(H_2O)_6]^{3+}$ ions are quite acidic. There is some evidence to suggest that the Tl^{III} aquo ion has two *trans* water molecules which are more strongly bound than the others (cf. stability of $TlCl_2^+$ below). For the reaction

$$[M(H_2O)_6]^{3+} \rightleftharpoons [M(H_2O)_5(OH)]^{2+} + H^+$$

the following constants have been determined: $K_a(Al)$, 1.12×10^{-5}; $K_a(Ga)$, 2.5×10^{-3}; and $K_a(In)$, 2×10^{-4}; $K_a(Tl)$, $\sim 7 \times 10^{-2}$. Although little emphasis can be placed on the exact numbers, the orders of magnitude are important, for they show that aqueous solutions of the M^{III} salts are subject to extensive hydrolysis. Indeed, salts of weak acids—sulfides, carbonates, cyanides, acetates, etc.—cannot exist in contact with water.

Studies of the hydrolysis of perchlorate solutions of Al^{III} have found widely divergent interpretations, in part owing to the slowness in reaching equilibria; chloride solutions are even more complex. Recent studies show that the above hydrolysis equation is too simple in the case of aluminum. Thus, a dimerization[8,9] certainly occurs:

$$2AlOH^{2+}(aq) = Al_2(OH)_2^{4+}(aq) \qquad K = 600 \text{ mol}^{-1} \text{ (30°)}$$

The presence of still more complex species of general formula $Al[Al_3(OH)_8]_m^{m+3}$ has also been postulated[10] to fit potentiometric data. Hydroxo-bridged polymers are known to be present in various crystalline basic salts.

Nmr methods[9,11] using ^{27}Al have been used to study the replacement of H_2O in the aquo ion by other ligands such as THF, Me_2SO, SO_4^{2-}, etc., to give species such as $[Al(H_2O)_5HSO_4]^{2+}$. ^{17}O-nmr allows measurement of exchange rates,[12] as well as determination of coordination number referred to above.

[8] E. Grunwald and D. W. Fong, *J. Phys. Chem.*, 1969, **73**, 650.
[9] J. W. Akitt, N. N. Greenwood and G. D. Lester, *J. Chem. Soc.*, A, **1969**, 803; *Chem. Comm.*, **1969**, 988.
[10] J. Aveston, *J. Chem. Soc.*, **1965**, 4438; F. H. Van Cauwelaert and H. J. Bosmans, *Rev. Chim. minérale*, 1969, **6**, 611; N. Deželic, H. Bilinski and R. H. H. Wolf, *J. Inorg. Nuclear Chem.*, 1971, **33**, 791.
[11] A. Fratiello *et al.*, *Inorg. Chem.*, 1969, **8**, 69.
[12] D. Fiat and R. E. Connick, *J. Amer. Chem. Soc.*, 1968, **90**, 608.

For Ga^{3+} in perchlorate solution, the aquo ion appears to be the main species and the hydrolysis, which is exceedingly slow at 25°, gives only white crystalline GaOOH.[13]

The indium and thallium aquo ions are known in ClO_4^- solution, but in presence of halide and other complexing anions complex species such as $InSO_4^+(aq)$ or the very stable linear $TlCl_2^+(aq)$ are formed.

A particularly important class of aluminum salts, the *alums* are structural prototypes and give their name to a large number of analogous salts formed by other elements. They have the general formula $MAl(SO_4)_2 \cdot 12H_2O$ in which M is practically any common univalent, monatomic cation except for Li^+, which is too small to be accommodated without loss of stability of the structure. The crystals are made up of $[M(H_2O)_6]^+$, $[Al(H_2O)_6]^{3+}$ and two SO_4^{2-} ions. There are actually three structures, all cubic, consisting of the above ions, but differing slightly in details depending on the size of the univalent ion. Salts of the same type, $M^IM^{III}(SO_4)_2 \cdot 12H_2O$, having the same structures are formed by many other trivalent metal ions, including those of Ti, V, Cr, Mn, Fe, Co, Ga, In, Rh and Ir, and all such compounds are referred to as alums. The term is used so generally that those alums containing aluminum are designated, in a seeming redundancy, as aluminum alums.

Thallium carboxylates, particularly the acetate and trifluoroacetate, which can be obtained by dissolution of the oxide in the acid, are extensively used in organic chemistry.[14] Both Tl metal and Tl^I salts such as the acetylacetonate also have specific uses. One example is the use of thallium(III) acetate in controlled bromination of organic substances such as anisole. The trifluoroacetate will directly thallate (cf. aromatic mercuration, Section 18-9) aromatic compounds to give aryl thallium ditrifluoroacetates, e.g., $C_6H_5Tl(OOCCF_3)_2$. It also acts as an oxidant, *inter alia* converting *para*-substituted phenols into *p*-quinones.

Aluminates and gallates. The hydroxides are amphoteric:

$$
\begin{array}{ll}
Al(OH)_3(s) = Al^{3+} + 3OH^- & K \approx 5 \times 10^{-33} \\
Al(OH)_3(s) = AlO_2^- + H^+ + H_2O & K \approx 4 \times 10^{-13} \\
Ga(OH)_3(s) = Ga^{3+} + 3OH^- & K \approx 5 \times 10^{-37} \\
Ga(OH)_3(s) = GaO_2^- + H^+ + H_2O & K \approx 10^{-15}
\end{array}
$$

and not only the hydroxides and oxides,[15] but also the metals, dissolve in alkali bases as well as in acids. The oxides and hydroxides of In and Tl are, by contrast, purely basic; hydrated Tl_2O_3 is precipitated from solution even at pH 1–2.5. The nature of the so-called "aluminate" and "gallate" solutions has been a problem. For the aluminum system from pH 8 to pH 12, according to Raman spectra,[16] the main species appears to be a polymer with octahedral Al and OH bridges, but, at pH > 13 and concentrations below

[13] H. R. Craig and S. Y. Tyree, Jr., *Inorg. Chem.*, 1969, **8**, 591.
[14] E. C. Taylor and A. McKillop, *Accounts Chem. Res.*, 1970, **10**, 338.
[15] See, e.g., A. Packter and H. S. Dhillon, *J. Chem. Soc.*, A, **1970**, 1266.
[16] L. A. Cameron *et al.*, *J. Chem. Phys.*, 1966, **45**, 2216.

ca. 1.5 *M*, [27]Al nmr, infrared and Raman spectra,[17] as well as ion-exchange studies,[18] indicate a tetrahedral $Al(OH)_4^-$ ion. Above 1.5 *M* there is condensation to give the ion $[(HO)_3AlOAl(OH)_3]^{2-}$, which occurs in the crystalline salt $K_2[Al_2O(OH)_6]$ and has an angular Al—O—Al bridge.

COMPLEX COMPOUNDS

The trivalent elements form 4-, 5- and 6-coordinate complexes, which may be cationic, like $[Al(H_2O)_6]^{3+}$ or $[Al(OSMe_2)_6]^{3+}$,[11] neutral, like the adducts of the halides, e.g., $AlCl_3(NMe_3)_2$, or anionic, like AlF_6^{3-} and $In(SO_4)_2^-$(aq).

9-7. Halide Complexes and Adducts [4b, c; 19]

The hydrated fluorides $AlF_3 \cdot nH_2O$ ($n = 3$ or 9) can be obtained by dissolving Al in aqueous HF. The nonahydrate is very soluble in water, and ^{19}F-nmr spectra[20] show the presence of $(H_2O)_3AlF_3$ as well as the ions AlF_4^-, $AlF_2(H_2O)_4^+$ and $AlF(H_2O)_5^{2+}$. At high fluoride concentrations and in crystalline solids the AlF_6^{3-} ion is also formed. The gallium system is similar.

One of the most important salts is *cryolite* whose structure (Fig. 9-2) is important since it is adopted by many other salts containing small cations and large octahedral anions and, in its *anti*-form, by many salts of the same type as $[Co(NH_3)_6]I_3$. It is closely related to the structures adopted by many compounds of the types $M_2^+[AB_6]^{2-}$ and $[XY_6]^{2+}Z_2^-$. The last two structures are essentially the fluorite (or antifluorite) structures (see Fig. 2-3, page 51), except that the anions (or cations) are octahedra whose axes are oriented parallel to the cube edges. The unit cell contains four formula units.

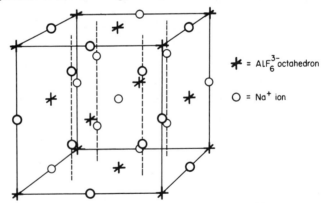

$\bigstar$ = AlF_6^{3-} octahedron

$\bigcirc$ = Na^+ ion

Fig. 9-2.   The cubic structure of cryolite, Na_3AlF_6.

[17] R. J. Moolenaar, J. C. Evans and L. D. McKeever, *J. Phys. Chem.*, 1970, **74**, 3629.
[18] M. Yoshio, H. Waki and N. Ishibashi, *J. Inorg. Nuclear Chem.*, 1970, **32**, 1365.
[19] D. G. Tuck, *Coordination Chem. Rev.*, 1966, **1**, 286 (coordination chemistry of In[III]).
[20] N. A. Matwiyoff and W. E. Wageman, *Inorg. Chem.*, 1970, **9**, 1031.

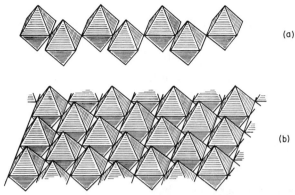

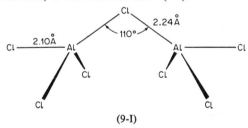

Fig. 9-3. Diagrams showing how AlF_6^{3-} octahedra combine by sharing corners in the compounds (a) Tl_2AlF_5 and (b) NH_4AlF_4.

Addition of four cations per unit cell, one at the center of the cube and one at the midpoint of each edge, gives the cryolite structure. Other complex fluorides of aluminum, such as Tl_2AlF_5 and NH_4AlF_4, also contain octahedrally coordinated aluminum; in the former the octahedra are joined at corners to form chains (Fig. 9-3a), and in the latter they share corners (but not edges on account of the high charge on Al^{3+}) to form sheets (Fig. 9-3b). Aluminum and gallium normally form the *tetrahedral ions*, MX_4^-, by direct interaction of the halide and, say, $Et_4N^+Cl^-$ in an organic solvent; they can be obtained also with mixed halogens.[21]

The tetrahaloaluminates are hydrolyzed by water, but gallium can be extracted from $8M$ HCl solutions into ethers, where the ether phase contains $GaCl_4^-$ ions. Oxonium salts of the type $[(Et_2O)_nH]^+MCl_4^-$ of Al, Ga, In and also Fe, have been isolated as viscous oils by reaction of the chloride in ether with hydrogen chloride.[22] The ion $Al_2Cl_7^-$ has been isolated as a salt of CH_3CO^+ by interaction of acetyl chloride and Al_2Cl_6 in liquid HCl,[23] and with large cations; it has the structure (9-I).[24]

$$Cl$$

Cl —2.10Å— Al   110°   Al ——— Cl
                    Cl   Cl
                Cl   Cl

(9-I)

2.24Å

The ion also occurs in melts[25] and at 170–240° there is the equilibrium

$$2Al_2Cl_7^- \rightleftharpoons Al_2Cl_6 + 2AlCl_4^-$$

[21] R. G. Kidd and D. R. Truax, *J. Amer. Chem. Soc.*, 1968, **90**, 6687.
[22] R. J. H. Clark, B. Crociani and A. Wasserman, *J. Chem. Soc., A*, **1970**, 2450.
[23] M. E. Peach, V. L. Tracy and T. C. Waddington, *J. Chem. Soc., A*, **1969**, 366.
[24] D. A. Lokken, T. W. Couch and J. D. Corbett, Abs., Amer. Chem. Soc. Meeting, Chicago, September, 1970.
[25] H. A. Øye et al., *Acta Chem. Scand.*, 1971, **25**, 559.

Also, aluminum chloride forms complexes with NaCl, $CaCl_2$, $NdCl_3$, UCl_4 and $CrCl_3$, which are volatile at high temperatures,[26] e.g.,

$$CrCl_3(s) + 1.5Al_2Cl_6(g) \rightleftharpoons [CrCl_3 \cdot 3AlCl_3](g)$$

The $Ga_2Cl_7^-$ is also formed in melts of $GaCl_2$ and Ga_2Cl_6, which contain $Ga^+[Ga_2Cl_7]^-$.[27]

Finally, we note that the formation of $AlCl_4^-$ and $AlBr_4^-$ ions is essential to the functioning of Al_2Cl_6 and Al_2Br_6 as Friedel–Crafts catalysts, since in this way the necessary carbonium ions are simultaneously formed:

$$RCOCl + AlCl_3 \rightarrow RCO^+ + [AlCl_4]^- \text{ (ion pair)}$$
$$RCO^+ + C_6H_6 \rightarrow [RCOC_6H_6]^+ \rightarrow RCOC_6H_5 + H^+$$

Gallium trichloride acts similarly.[28]

Indium halogeno complexes are similar to those of Al and Ga but, although in non-aqueous media and in crystalline salts the $InCl_4^-$ ion is tetrahedral,[29] it is probably aquated in aqueous solution; the species $InCl^{2+}$ and $InCl_2^+$ are also formed.[30]

In $[Et_4N]_2[InCl_5]$, made in non-aqueous media, the anion has *spy* geometry,[31] which is unusual for a non-transition element that does not have an additional electron pair in the valence shell (see page 127). Among the few other examples are Ph_5Sb and matrix-isolated SbF_5 (see page 392). The compound $[Et_4N]_2[TlCl_5]$ is isomorphous and presumably isostructural.[32] In nitromethane solution these MCl_5^{2-} ions dissociate to MCl_4^- ions. The complex $(Ph_3P)_2InCl_3$, on the other hand, has *tbp* geometry with unusually long axial In—P bonds.[33] Salts of the ions $InCl_6^{3-}$, $TlCl_6^{3-}$ and $[TlCl_5(H_2O)]^{2-}$ are also known, while thallium forms chloro complexes readily in aqueous solution, and species up to $TlCl_6^{3-}$ occur.[34]

The dichloro ion $TlCl_2^+$ appears to be particularly stable, presumably because it is linear and analogous to the linear, very stable and isoelectronic $HgCl_2$ (cf. also Me_2Tl^+ below). The thermodynamic quantities are also consistent with a change of coordination number on formation of the tetrahedral $TlCl_4^-$ ion, e.g.,

$$trans\text{-}[TlCl_3(H_2O)_3] + Cl^- \rightarrow TlCl_4^- + 3H_2O$$

For TlI_4^- the stability of iodide in contact with Tl^{III} is a result of the stability of the ion, since thallium(III) iodide is itself unstable relative to $Tl^I(I_3)$. Thallium alone forms the ion $Tl_2Cl_9^{3-}$, which has the confacial bioctahedron structure shown in Fig. 19-5a.

[26] K. Lascelles and H. Schäfer, *Angew. Chem. Internat. Edn.*, 1971, **10**, 128.
[27] M. J. Taylor, *J. Chem. Soc., A*, **1970**, 2812.
[28] See H. C. Brown *et al.*, *J. Amer. Chem. Soc.*, 1969, **91**, 4844, 4854.
[29] J. Trotter, F. W. B. Einstein and D. G. Tuck, *Acta Cryst.*, B, 1969, **25**, 603.
[30] M. P. Hanson and R. A. Plane, *Inorg. Chem.*, 1969, **8**, 746.
[31] D. S. Brown, F. W. B. Einstein, and D. G. Tuck, *Inorg. Chem.*, 1969, **8**, 14; D. M. Adams and R. R. Smardzewski, *J. Chem. Soc., A*, **1971**, 714.
[32] D. F. Shriver and I. Wharf, *Inorg. Chem.*, 1969, **8**, 2167.
[33] M. V. Veidis and G. J. Palenik, *Chem. Comm.*, **1969**, 586.
[34] T. G. Spiro, *Inorg. Chem.*, 1965, **4**, 1290; 1967, **6**, 569; J. Gislason, M. H. Lloyd and D. G. Tuck, *Inorg. Chem.*, 1971, **10**, 1907.

Adducts: Lewis Acid Behavior. The trihalides (except the fluorides), and other R_3M compounds such as the trialkyls, triaryls, mixed R_2MX compounds and AlH_3, all function as Lewis acids, forming 1:1 adducts with a great variety of Lewis bases. This is one of the most important aspects of the chemistry of the Group III elements. The Lewis acidity of the AlX_3 groups, where $X = Cl$, CH_3, etc., has been extensively studied thermodynamically,[35] and basicity sequences for a variety of donors have been established.[36]

As already indicated, the MX_3 molecules (X = halide) react with themselves to form dimeric molecules in which each metal atom has distorted tetrahedral coordination. Even in mixed organo halo compounds such as $(CH_3)_2AlCl$, this type of dimerization with *halogen atom bridges* occurs. In addition to the tetrahedral adducts, neutral or ionic as in MX_4^- there are both 5-coordinate and 6-coordinate species. For the series of $MX_3 \cdot 2NMe_3$ compounds ($MX_3 = AlCl_3$, $GaCl_3$, $InCl_3$, $InBr_3$ or InI_3), vibrational spectra indicate *tbp* structures with axial nitrogen atoms.[37] Cationic complexes[38] are octahedral of the type

$$[GaCl_2(bipy)_2]^+ GaCl_4^- \quad \text{or} \quad [InCl_2(bipy)_2]^+ [InCl_4(bipy)]^-$$

9-8. Oxo Complexes

The aquo and aluminate ions have already been noted. Aluminum acetate has a structure similar to that of the basic acetates of Cr^{3+}, Fe^{3+} and other M^{3+} ions and probably contains the basic unit $[Al_3O(OOCMe)_6(H_2O)_3]^+$, where the H_2O molecules can dissociate to form OH or can be replaced by other ligands.

The most important octahedral complexes of the Group III elements are those containing chelate rings. Typical are those of β-diketones, pyrocatechol (9-II), dicarboxylic acids (9-III), and 8-quinolinol (9-IV). The neutral complexes dissolve readily in organic solvents, but are insoluble in water. The acetylacetonates have low melting points ($<200°$) and vaporize without decomposition. The anionic complexes are isolated as the salts of large univalent cations. The 8-quinolinolates are used for analytical purposes. Tropone (T) gives an 8-coordinate anion of indium in $Na[InT_4]$.

(9-II)   (9-III)   (9-IV)

[35] D. P. Eyman *et al.*, *Inorg. Chem.*, 1968, **7**, 1028, 1047; J. K. Gilbert and J. D. Smith, *J. Chem. Soc., A,* **1968**, 233; J. W. Wilson and I. S. Worrall, *J. Chem. Soc., A.,* **1968**, 316, 2389.

[36] B. M. Cohen, A. R. Cullingworth and S. D. Smith, *J. Chem. Soc., A,* **1969**, 2193.

[37] I. R. Beattie, T. Gilson and G. A. Ozin, *J. Chem. Soc., A,* **1968**, 1092; I. R. Beattie, G. A. Ozin and H. E. Blayden, *J. Chem. Soc., A,* **1969**, 2535.

[38] See, e.g., R. A. Walton, *Inorg. Chem.*, 1968, **7**, 640; *J. Chem. Soc., A,* **1969**, 61; G. Beran *et al.*, *Chem. Comm.*, **1970**, 222; R. Restivo and G. J. Palenik, *Chem. Comm.*, **1969**, 807.

The four elements form *alkoxides*, which we can regard as complexes since they are all polymeric even in solution in inert solvents. Only those of aluminum, particularly the isopropoxide, which is widely used in organic chemistry as a reducing agent for aldehydes and ketones,[39] are of importance. They can be made by the reactions

$$Al + 3ROH \xrightarrow[\text{catalyst, warm}]{1\% \text{ HgCl}_2 \text{ as}} (RO)_3Al + \tfrac{3}{2}H_2$$

$$AlCl_3 + 3RONa \longrightarrow (RO)_3Al + 3NaCl$$

The alkoxides hydrolyze vigorously in water. The *tert*-butoxide is a cyclic dimer (9-V) in solvents, whereas the isopropoxide is tetrameric (9-VI) at ordinary temperatures but trimeric at elevated temperatures. Terminal and bridging alkoxyl groups can be distinguished by nmr spectra. Other alkoxides can exist also as dimers and trimers.

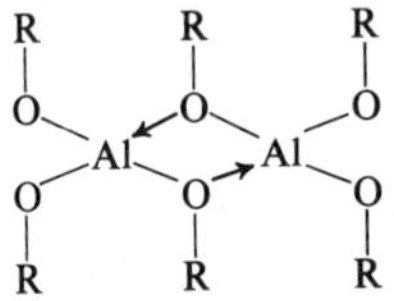

(9-V) 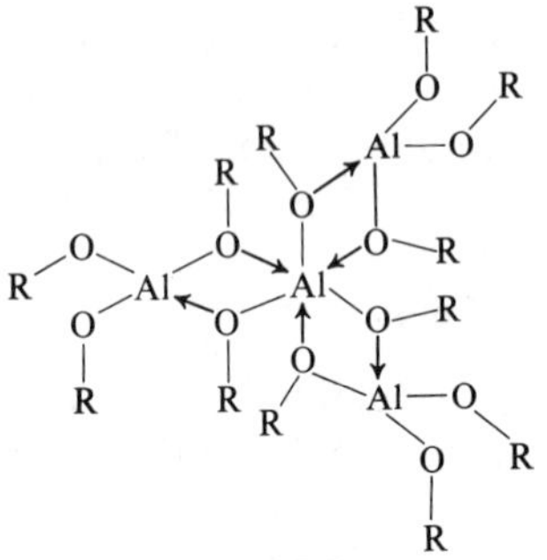   (9-VI)

Gallium[40] and thallium[41] form *nitrato* complexes by the reaction of N_2O_5 with $NO_2^+[GaCl_4]^-$ or $TlNO_3$, respectively. The gallium ion in $NO_2^+[Ga(NO_3)_4]^-$ appears to have unidentate NO_3 groups, by contrast with $[Fe(NO_3)_4]^-$ which has bidentate groups and is 8-coordinate, although the radius of Ga^{III} (0.62) is only slightly smaller than that of Fe^{III} (0.64 Å).

9-9. Complex Hydrides[42]

The binary hydrides of Al and Ga have been discussed on page 183. There is, in addition, an extensive complex chemistry that can be regarded as arising from the Lewis acid behavior of MH_3 (even if the simple molecules are too unstable to be isolated) by formation of adducts either with donor molecules such as NR_3, PR_3, SR_2 or with anions, viz.:

$$GaH_3 + NMe_3 \rightarrow Me_3NGaH_3$$
$$AlH_3 + H^- \rightarrow [AlH_4]^-$$

[39] For mechanism of the Meerwein–Pondorff–Verley reaction see V. J. Shiner, Jr., and D. Whittaker, *J. Amer. Chem. Soc.*, 1969, **90**, 394.

[40] D. Bowler and N. Logan, *Chem. Comm.*, **1971**, 582.

[41] D. W. Amos, *Chem. Comm.*, **1970**, 19.

[42] H. Nöth and E. Wiberg, *Fortsch. Chem. Forsch.*, 1967, **8**, 321; S. Cuccinella, A. Mazzei and W. Marconi, *Inorg. Chim. Acta Rev.*, 1970, **4**, 51 (comprehensive reviews of hydrides and hydride complexes).

In such compounds there are close analogies to borane derivatives (page 237). The reaction of AlH_3 with alcohols,

$$AlH_3 + nROH \rightarrow (RO)_nAlH_{3-n} + nH_2$$

is a useful preparative method for mixed alkoxo hydrido species.[43]

Hydride Anions. The alkali-metal salts of AlH_4^- and GaH_4^- are similar to those of BH_4^-. The thermal and chemical stabilities vary according to the ability of the MH_3 group to act as an acceptor as in the above equation, the order being $B > Al > Ga$. Thus $LiGaH_4$ decomposes slowly even at $25°$ to LiH, Ga and H_2 and is a milder reducing agent than $LiAlH_4$. Similarly, although BH_4^- is stable in water, the Al and Ga salts are rapidly and often explosively hydrolyzed by water:

$$4H_2O + MH_4^- \rightarrow 4H_2 + M(OH)_3 + OH^-$$

Lithium aluminum hydride is an important reducing agent in both organic and inorganic chemistry. It is a non-volatile, crystalline solid, white when pure but usually gray. It is stable below $\sim 120°$ and is soluble in diethyl ether and other ethers such as THF and various glymes. It accomplishes many otherwise tedious or difficult reductions, e.g., —COOH to —CH_2OH; some reactions are shown in Fig. 9-4.

The crystal structure[44] of $LiAlH_4$ shows the presence of tetrahedral AlH_4^- ions, with the average Al—H distance equal to 1.55 Å. The Li^+ ions each have four near (1.88–2.00 Å) and a fifth more remote (2.16 Å) hydrogen neighbor.

Both Al and Ga salts can be prepared by the reaction

$$4LiH + MCl_3 \xrightarrow{\text{Et}_2O} LiMH_4 + 3LiCl$$

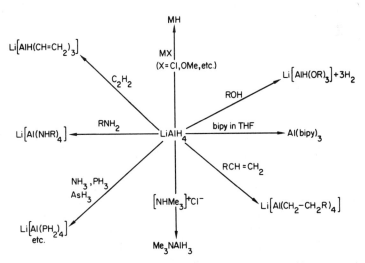

Fig. 9-4.   Some reactions of lithium aluminum hydride.

[43] H. Nöth and H. Suchy, *Z. anorg. Chem.*, 1968, **358**, 44.
[44] N. Sklar and B. Post, *Inorg. Chem.*, 1967, **6**, 669.

However, for AlH_4^- the Li, Na and K salts are more conveniently obtained by direct interaction, e.g.,

$$Na + Al + 2H_2 \xrightarrow[\text{150°/2000 p.s.i./24h}]{\text{THF}} NaAlH_4$$

The salt is obtained by precipitation with toluene and can be efficiently converted into the lithium salt:

$$NaAlH_4 + LiCl \xrightarrow{Et_2O} NaCl(s) + LiAlH_4$$

Although $LiAlH_4$ is stable in ethers, amine solvents can "abstract" AlH_3 to give the amine hydrides discussed below, e.g.,

$$3LiAlH_4 + 4Me_3N \rightarrow Li_3AlH_6 + 2AlH_3(NMe_3)_2$$

The suggestion has been made that $LiAlH_4$ is a covalently bonded species in solution and can undergo reactions of this sort by attack on the Li atom. $NaAlH_4$, which is clearly ionic, does not react in the same way but will give amine solvates, e.g., with tetramethylethylenediamine $(NaTMED)AlH_4$.[45]

In addition to AlH_4^-, the octahedral *hexahydrido ion*, AlH_6^{3-}, is known. Although salts were first made by the reactions

$$2NaH + NaAlH_4 \xrightarrow{\text{Heat in heptane}} Na_3AlH_6$$
$$2LiH + LiAlH_4 \xrightarrow{\text{Et}_3\text{Al as catalyst}} Li_3AlH_6$$

they are best prepared by direct interaction of the metals and hydrogen under pressure.[46] The commercial benzene-soluble reducing agent "Red-Al" is $Na[AlH(OMe)_2OEt]$.

Donor Adducts. These are again similar to borane adducts, the stability order being $B > Al > Ga$ and also similar to adducts of the halides and alkyls where the stability order is halides $>$ alkyls $>$ hydrides. The most studied adducts are the trialkylamine alanes (alane $= AlH_3$). Trimethylamine gives both 1:1 and 1:2 adducts but the latter are stable only in the presence of an excess of amine:

$$Me_3NAlCl_3 + 3LiH \xrightarrow{Et_2O} Me_3NAlH_3 + 3LiCl$$
$$Me_3HN^+Cl^- + LiAlH_4 \xrightarrow[-60°]{Et_2O} Me_3NAlH_3 + LiCl + H_2$$
$$3LiAlH_4 + AlCl_3 + 4NMe_3 \longrightarrow 4Me_3NAlH_3 + 3LiCl$$
$$Me_3NAlH_3 + Me_3N \rightleftharpoons (Me_3N)_2AlH_3$$

The monoamine is a white, volatile, crystalline solid (m.p. 75°), readily hydrolyzed by water, which slowly decomposes to $(AlH_3)_n$. It is monomeric and tetrahedral. The bisamine is trigonal-bipyramidal with axial N atoms and linear N—Al—N. Tetrahydrofuran also gives 1:1 and 1:2 adducts, but diethyl ether, presumably for steric reasons, gives only the 1:1 compound, though a mixed THF–Et$_2$O adduct exists.

Aminoalanes, particularly those containing secondary amines, can be obtained by direct interaction of Al, the amine and hydrogen at 120° under pressure.[47] These compounds are used as reducing agents in organic syntheses and as polymerization catalysts. They have the general formula

[45] J. A. Dilts and E. C. Ashby, *Inorg. Chem.*, 1970, **9**, 855.
[46] E. C. Ashby and B. D. James, *Inorg. Chem.*, 1969, **8**, 2468.
[47] R. A. Kovar and E. C. Ashby, *Inorg. Chem.*, 1971, **10**, 893.

$[HAl(NR_2)_2]_n$, where n may be 1, 2 or 3. The dimers and trimers have bridging N atoms.

A number of polymers with Al—N backbones have been made, e.g., by the reaction of R_3NAlH_3 with ethylamine or acetonitrile.

There are similar monoamine gallanes where the Ga—H stretch at ~ 1850 cm^{-1}, compared to ~ 1770 for the alane, suggests a stronger M—H bond, and indeed the gallanes are less sensitive to hydrolysis. The $(Me_3N)_2GaH_3$ compound is unstable above $-60°$. Preparation of the gallanes illustrates a useful principle regarding the use of a weak donor as solvent:

$$Me_3NGaH_3 + BF_3 \xrightarrow[\text{weak}]{Me_2S} Me_2SGaH_3 + Me_3NBF_3$$
$$\text{strong–weak} \quad \text{strong} \qquad\qquad \text{weak–weak} \qquad \text{strong–strong}$$

Because the weak–weak, strong–strong combination is favored over two weak–strong adducts, the net effect is to displace the strong donor, Me_3N, by the weaker one, Me_2S.

Note that the important complex $Al(BH_4)_3$ has been discussed previously (page 250).

9-10. Organometallic Compounds[48]

Those of aluminum are by far the most important and best known. They may be prepared by the classical reaction of aluminum with the appropriate organomercury compound:

$$2Al + 3R_2Hg \rightarrow 2R_3Al \text{ (or } [R_3Al]_2) + 3Hg$$

or by reaction of Grignard reagents with $AlCl_3$:

$$RMgCl + AlCl_3 \rightarrow RAlCl_2, R_2AlCl, R_3Al$$

More direct methods suitable for large-scale use are now available. These procedures stemmed from studies which showed that aluminum hydride or $LiAlH_4$ reacts with olefins to give alkyls or alkyl anions—a reaction specific for B and Al hydrides:

$$AlH_3 + 3C_nH_{2n} \rightarrow Al(C_nH_{2n+1})_3$$
$$LiAlH_4 + 4C_nH_{2n} \rightarrow Li[Al(C_nH_{2n+1})_4]$$

Although $(AlH_3)_n$ cannot be made by direct interaction of Al and H_2, nevertheless in the presence of aluminum alkyl the following reaction to give the dialkyl hydride can occur:

$$Al + \tfrac{3}{2}H_2 + 2AlR_3 \rightarrow 3AlR_2H$$

This hydride will then react with olefins:

$$AlR_2H + C_nH_{2n} \rightarrow AlR_2(C_nH_{2n+1})$$

[48] Houben–Weyl, *Methoden der Organischen Chemie*, Vol. XIII/4, *Metalloorganische Verbindung der* Al, Ga, In, Tl; G. Thiele Verlag, Stuttgart, 1970; R. Köster and P. Benger, *Adv. Inorg. Chem. Radiochem.*, 1965, 7, 263 (organoaluminum compounds); A. N. Nesmeyanov and R. A. Sokolik, *The Organocompounds of* B, Al, Ga, In *and* Tl, North Holland, 1967; J. J. Eisch, *Organometallic Chem. Rev.*, 1968, 4, 331, 336 (surveys of Ga, In and Tl); A. G. Lee, *Quart. Rev.*, 1970, 24, 310 (thallium compounds); H. Reinheckel, K. Haage and D. Jahnke, *Organometallic Chem. Rev.*, A, 1969, 4, 47 (organoaluminum compounds in organic reactions).

Thus the direct interaction of Al, H_2 and olefin can be used to give either the dialkyl hydrides or the trialkyls.

It may be noted in connection with the direct synthesis of $NaAlH_4$ mentioned above that, if ethylene (or other olefin) is present and $AlEt_3$ is used as catalyst, the direct interaction gives $Na[AlH_{4-n}Et_n]$.

Other technically important compounds are the "sesquichlorides" such as $Me_3Al_2Cl_3$ or $Et_3Al_2Cl_3$. These compounds can be made by direct interaction of Al or Mg–Al alloy with the alkyl chloride. This reaction fails for propyl and higher alkyls since the alkyl halides decompose in presence of the alkyl-aluminum halides to give HCl, alkenes, etc.

The lower aluminum alkyls are reactive liquids, inflaming in air and explosively sensitive to water. All other derivatives are similarly sensitive to air and moisture though not all are spontaneously flammable. The lower alkyls are dimerized (Section 10-7), and the alkyl halides are also dimeric but with *halogen* bridges. The fluorides are somewhat different in that for $[Me_2AlF]_4$ and $[Et_2AlF]_4$ (and for Me_2GaF) ring polymers with single Al—F—Al bridges are formed.[49] Further, the reaction of Al_2Et_6 with KF in toluene gives $K[Et_3AlFAlEt_3]$ where the Al—F—Al bond is linear, possibly on account of $p\pi \rightarrow d\pi$ bonding. It may be noted that aluminum fluorides of the type $AlF_nX_{3-n}(X = Cl, Br$ or $I; n = 1$ or $2)$ are also aggregated, presumably via Al—F—Al bridges, in tetrahydrofuran.[50]

At $-75°$ the proton nuclear resonance spectrum of $Al_2(CH_3)_6$ exhibits separate resonances for the terminal and bridging methyl groups but, on warming, these begin to coalesce and at room temperature only one sharp peak is observed. This indicates that the bridging and terminal methyl groups can exchange places across a relatively low energy barrier. This may occur by partial or complete dissociation of the dimer.[50a]

The alkyls are Lewis acids, combining with donors such as amines, phosphines, ethers, and thioethers to give tetrahedral, four-coordinate species. Thus Me_3NAlMe_3 in the gas phase has C_{3v} symmetry with staggered methyl groups.[51] With tetramethylhydrazine and $(CH_3)_2NCH_2N(CH_3)_2$, 5-coordinate species that appear to be of the kind shown in (9-VII) are

(9-VII)

obtained, although at room temperature exchange processes cause all methyl groups and all ethyl groups to appear equivalent in the proton nuclear resonance spectrum. With $(CH_3)_2NCH_2CH_2N(CH_3)_2$ a complex is formed

[49] J. Weidlein and V. Krieg, *J. Organometallic Chem.*, 1968, **11**, 9; H. Schmidbauer *et al.*, *Chem. Ber.*, 1968, **101**, 2268.

[50] E. E. Clagg and D. L. Schmidt, *Inorg. Nuclear Chem. Letters*, 1969, **31**, 2329.

[50a] D. S. Matteson, *Inorg. Chem.*, 1971, **10**, 1555.

[51] G. A. Anderson, F. R. Forgaard and A. Haaland, *Chem. Comm.*, **1971**, 480.

which has an AlR_3 group bound to each nitrogen atom. Aluminum alkyls also combine with lithium alkyls:

$$(C_2H_5)_3Al + LiC_2H_5 \xrightarrow{\text{In benzene}} LiAl(C_2H_5)_4$$

X-ray study has shown that $LiAl(C_2H_5)_4$ is built up of chains of alternating tetrahedral $Al(C_2H_5)_4^-$ and Li^+ in such a way that each lithium atom is tetrahedrally surrounded by four α-carbon atoms, close enough to indicate weak Li—C bonds, and vibrational spectra suggest that $LiAlMe_4$ is similar.[52]

When primary or secondary amines or phosphines are used, the 1:1 complexes can eliminate one or two hydrocarbon molecules to give bridged species such as (9-VIII) or (9-IX), the latter being comparable to the hydrocarbon cubane.

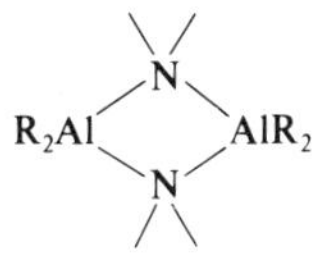

(9-VIII) 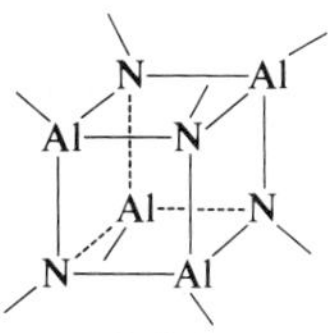   (9-IX)

Triethylaluminum, the sesquichloride $(C_2H_5)_3Al_2Cl_3$ and alkyl hydrides are used together with transition-metal halides or alkoxides or organometallic complexes as catalysts (e.g., Ziegler catalysts) for the polymerization of ethylene, propene and a variety of other unsaturated compounds, as discussed in Chapter 24. They are also used widely as reducing and alkylating agents for transition-metal complexes.

The trialkyls of Ga, In and Tl resemble those of aluminum but they have been less extensively investigated and are increasingly less stable. One signal point of difference is the lack of dimerization of the alkyls of B, Ga, In and Tl at ordinary temperatures, with the exception of the unusual polymerization of crystalline Me_3In and Me_3Tl (see page 306).

The In and Tl phenyls also exhibit weak intermolecular interaction probably involving donation of electrons from a phenyl ring to a vacant p orbital of the metal in another molecule.[53]

A number of dialkyl compounds of Ga, In, and Tl are well characterized and are stable even in aqueous solutions. Thus interaction of Me_2GaCl and ammonia gives $[Me_2Ga(NH_3)_2]^+Cl^-$, while partial hydrolysis of $Me_3Ga \cdot Et_2O$ gives crystals of the OH-bridged tetramer $[Me_2GaOH]_4$, which is soluble in acids to give the cation $[Me_2Ga(H_2O)_2]^+$ and in bases to give $[Me_2Ga(OH)_2]^-$.[54] Thallium gives very stable ionic derivatives of the type R_2TlX (X = halogens, SO_4^{2-}, CN^-, NO_3^-, etc.), which resemble compounds R_2Hg in being unaffected by air and water. The ion $(CH_3)_2Tl^+$ in aqueous solution and in salts is linear, like R_2Hg and the ions Me_2Pb^{2+} and

[52] J. Yamamoto and C. A. Wilkie, *Inorg. Chem.*, 1971, **10**, 1129.
[53] J. F. Malone and W. S. McDonald, *J. Chem. Soc., A*, 1970, 3362.
[54] R. S. Tobias, M. J. Sprague and G. E. Glass, *Inorg. Chem.*, 1968, **7**, 1714.

Me_2Sn^{2+};[55] the reason for the difference in structure of Me_2Ga^+ and Me_2Tl^+ is not obvious.

In crystals of thallium salts, as in dimethyltin salts (Section 11-11), the anions may act as bridges. Thus the sulfinate $(Me_2TlSO_2Me)_2$ is dimeric[56] with sulfinate bridges. Additional coordination can certainly occur and a complex $[Me_2Tl$ py]ClO_4 has been isolated. The $[Me_2Tl$ py]$^+$ ion appears to be T-shaped on the basis of spectroscopic measurements. Bis(pentafluorophenyl)thallium halides give what appear to be 5-coordinate adducts such as $(C_6F_5)_2TlCl(bipy)$.

LOWER VALENT COMPOUNDS

Since the elements have the outer electron configurations ns^2np, it is natural to consider whether monovalent ions might be capable of existence. It may be recalled that there is no evidence for B^I under chemically important conditions.

9-11. Aluminum and Gallium

There is no evidence that compounds containing Al^I exist at ordinary temperatures. Anodic oxidation of aluminum at high current densities evidently produces lower-valent aluminum ions, either Al^I or Al^{II}, or both, but they are ephemeral. There is no doubt that *gaseous* Al^I halide molecules exist at high temperatures, and their spectroscopic properties are well known. In the chloride system the equilibrium

$$AlCl_3(g) + 2Al(s) \rightleftharpoons 3AlCl(g)$$

has been thoroughly studied and its use in purifying aluminum has been proposed. The reaction proceeds to the right at high temperatures, but reverses readily at low temperatures. Similarly, it has been shown that gaseous Al_2O and AlO molecules exist above 1000°, but no solid oxide containing lower-valent aluminum has been shown definitely to exist under ordinary conditions although Al_2O and other M_2O species can be trapped in inert-gas matrices at low temperatures.[57] A red solution obtained by photolyzing $Al(C_6H_5)_3$ in toluene may contain $Al^IC_6H_5$.[57a]

A zerovalent complex, $Al(bipy)_3$, is formed by reduction of $AlCl_3$ with Li bipyridyl in THF; it is exceedingly air-sensitive, green, and paramagnetic ($\mu = 2.32$ B.M.). See, however, Section 22-14.

Gallium(I) *compounds* have been prepared in the gas phase at high temperatures by reactions such as

$$Ga_2O_3(s) + 4Ga(l) \xrightarrow{700°} 3Ga_2O(g)$$
$$Ga(l) + SiO_2(s) \rightleftharpoons Si\ (in\ Ga) + 2Ga_2O(g)$$
$$GaCl_3(g) \rightleftharpoons^{1100°} GaCl(g) + Cl_2$$

[55] R. S. Tobias, *Organometallic Chem. Rev.*, 1966, **1**, 93.
[56] A. G. Lee, *J. Chem. Soc.*, A, **1970**, 467.
[57] D. M. Makowiechi, D. A. Lynch, Jr. and K. D. Carlson, *J. Phys. Chem.*, 1971, **75**, 1963.
[57a] J. J. Eisch and J. L. Considine, *J. Amer. Chem. Soc.*, 1968, **90**, 6257.

While GaCl has not been isolated pure, Ga_2O and Ga_2S can be, although the latter solid is non-stoichiometric.

"Divalent" chalconides GaS, GaSe, and GaTe can be made by direct interaction; they do not contain Ga^{2+}, however, which would lead to paramagnetism, but have Ga–Ga units in a layer lattice, each gallium atom being tetrahedrally surrounded by three S and one Ga. The best-known compounds are the "dihalides" GaX_2; these are known to have the salt-like structure $Ga^I[Ga^{III}X_4]$; the Ga^I ion can also be obtained in other salts such as $Ga[AlCl_4]$. Fused $GaCl_2$ is a typical conducting molten salt. These halides are prepared by the reaction

$$2Ga + 4GaX_3 \rightarrow 3Ga[GaX_4].$$

Salts of the type $[GaL_4][GaCl_4]$ have been obtained with S, Se and As donors.

Anodic dissolution of gallium in $6M$ HCl or HBr at $0°$ followed by addition of Me_4NX precipitates white crystalline compounds that are stable and diamagnetic. These compounds have the composition $[Me_4N]_2[Ga_2X_6]$. Their vibrational spectra indicate that the anions have an ethane-like structure; the Ga—Ga stretching frequencies appear as intense, polarized Raman bands at 162 and 233 cm^{-1} for $Ga_2Br_6^{2-}$ and $Ga_2Cl_6^{2-}$, respectively.[58]

The presence of low-valent gallium, probably Ga^I, species in aqueous solution has been proposed frequently. For example, some delayed reducing ability of gallium dissolved in HCl has been observed, and gallium is one of the few reductants for perchloric acid, which may be due to the reactions

$$Ga + H^+ \rightarrow Ga^+ + \tfrac{1}{2}H_2$$
$$4Ga^+ + 8H^+ + ClO_4^- \rightarrow 4Ga^{3+} + Cl^- + 4H_2O$$

The potential for Ga^I–Ga^{III} in basic solution has been estimated:

$$Ga(OH)_4^- + 2e = Ga^I + 4OH^- \qquad E_B^0 < -1.24 \text{ V}$$

9-12. Indium

Indium(I) can be obtained in low concentration in aqueous solution by using an indium metal anode in $0.01M$ perchloric acid.[59] It is rapidly oxidized by both H^+ ion and by air, as can be seen from the potential

$$In^{3+} + 2e = In^+ \qquad E° = -0.425 \text{ V}$$

and is also unstable to disproportionation:

$$In^{III} + 2In^0 \rightleftharpoons 3In^I \qquad \log K = -8.4.$$

Solutions of In^I in acetonitrile are considerably more stable;[60] thus the situation is comparable to that for the copper(I) ion in the same two solvents (Section 25-H-2). There is no direct evidence for In^{II} but it is believed to be a transitory intermediate in reductions, e.g., of Fe^{3+} by In^+.

There are a number of solid indium halides, including monohalides and

[58] C. A. Evans and M. J. Taylor, *Chem. Comm.*, **1969**, 1201.
[59] R. S. Taylor and A. G. Sykes, *J. Chem. Soc.*, A, **1969**, 2419; **1971**, 1628.
[60] J. B. Headridge and D. Pletcher, *Inorg. Nuclear Chem. Letters*, 1967, **3**, 475.

InF_2. The best known is the chloride system obtained by dissolving indium in fused $InCl_3$.[61] $InCl_2$ appears to be like $GaCl_2$ and to be $In^+InCl_4^-$, whereas In_2Cl_3 is $In_3[In^{III}Cl_6]$; others are In_4Cl_5 and In_4Cl_7. Molten In_2Cl_3 is red and some sort of metal cluster is likely (cf. Bi_5^{3+}, Chapter 13).

9-13. Thallium(I)

With thallium, the unipositive state is quite stable. In aqueous solution it is distinctly more stable than Tl^{III}:

$$Tl^{3+} + 2e = Tl^+ \qquad E° = +1.25 \text{ V } [E_f = +0.77, 1M \text{ HCl}; +1.26, 1M \text{ HClO}_4]$$

The thallous ion is not very sensitive to pH, although the thallic ion is extensively hydrolyzed to $TlOH^{2+}$ and the colloidal oxide even at pH 1–2.5; the redox potential is hence very dependent on pH as well as on presence of complexing anions. Thus, as indicated by the above potentials, the presence of Cl^- stabilizes Tl^{3+} more (by formation of complexes) than Tl^+ and the potential is thereby lowered.

The colorless thallous ion has a radius of 1.54 Å, which can be compared with those of K^+, Rb^+ and Ag^+ (1.44, 1.58 and 1.27 Å). In its chemistry this ion resembles either the alkali or argentous ions; thus it may replace K^+ in certain enzymes and has potential use as a probe for potassium.[62] In crystalline salts, the Tl^+ ion is usually 6- or 8-coordinate. The yellow hydroxide is thermally unstable, giving the black oxide, Tl_2O, at about 100°. The latter and the hydroxide are readily soluble in water to give strongly basic solutions that absorb carbon dioxide from the air; TlOH is a weaker base than KOH, however. Many thallous salts have solubilities somewhat lower than those of the corresponding alkali salts, but otherwise are similar to and quite often isomorphous with them. Examples of such salts are the cyanide, nitrate, carbonate, sulfate, phosphates, perchlorate and alums. Thallous solutions are exceedingly poisonous[63] and in traces cause loss of hair.

Thallous sulfate, nitrate and acetate are moderately soluble in water, but—except for the very soluble TlF—the halides are sparingly soluble.[64] The chromate and the black sulfide, Tl_2S, which can be precipitated by hydrogen sulfide from weakly acid solutions, are also insoluble. Thallous chloride also resembles silver chloride in being photosensitive, darkening on exposure to light. Incorporation of Tl^I halides into alkali halides gives rise to new absorption and emission bands due to formation of complexes of the type that exist also in solutions, most notably TlX_2^- and TlX_4^{3-}; such thallium-activated alkali halide crystals are used as phosphors, e.g., for

[61] J. H. R. Clarke and R. E. Hester, *Inorg. Chem.*, 1969, **8**, 1113.
[62] J. P. Manners, K. G. Morallee and R. J. P. Williams, *J. Inorg. Nuclear Chem.*, 1971, **33**, 2085.
[63] A. Christie, *The Pale Horse*, Collins, London, 1961.
[64] See A. D'Aprano and R. M. Fuoss, *J. Amer. Chem. Soc.*, 1969, **90**, 279; M. F. C. Ladd and W. H. Lee, *Trans. Faraday Soc.*, 1970, **66**, 2767.

scintillation radiation detectors. Thallous chloride is insoluble in ammonia, unlike silver chloride.

With the exception of those with halide, oxygen and sulfur ligands, Tl^I gives rather few complexes. The dithiocarbamates, $Tl(S_2CNR_2)$, made from aqueous Tl_2SO_4 and the sodium dithiocarbamates, are useful reagents for the synthesis of other metal dithiocarbamates from the chlorides in organic solvents, since the Tl dithiocarbamate is soluble while TlCl is precipitated on reaction. The structure of the *n*-propyl complex shows that it is polymeric with $[TlS_2CNPr_2]_2$ dimeric units linked by Tl—S bonds.[65]

The two isotopes, ^{203}Tl and ^{205}Tl (70.48%), both have nuclear spin, and nmr signals are readily detected for thallium solutions or for solids. In solution both Tl^I and Tl^{III} resonances are markedly dependent on concentration and on the nature of anions present; such data have shown that thallous perchlorate is highly dissociated, but salts of weaker acids and TlOH have been shown to form ion pairs in solution.

Electron-exchange reactions in the Tl^I–Tl^{III} system have been intensively studied and appear to be two-electron transfer processes; various Tl^{III} complexes participate under appropriate conditions.

The only known Tl^I organo compound is the polymeric TlC_5H_5 precipitated on addition of aqueous TlOH to cyclopentadiene; InC_5H_5 is similar. In the gas phase these compounds consist of discrete molecules having five-fold symmetry. The metal atoms lie over the centers of the rings and are apparently bound by forces mainly of covalent nature. TlC_5H_5 is a very useful reagent for the synthesis of other metal cyclopentadienyl compounds (Chapter 23).

When oxygen, nitrogen and sulfur bound to organic groups is also bound to Tl^I, the Tl—X bond appears to be more covalent than the bond to

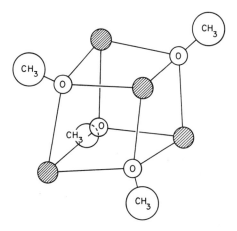

Fig. 9-5.   Structure of tetrameric thallium methoxide.

[65] D. Coucouvanis, *Progr. Inorg. Chem.*, 1970, **11**, 234

alkali-metal ions in similar compounds.[66] Thallium compounds tend to be polymeric rather than ionic. Thus the acetylacetonate is a linear polymer with 4-coordinate Tl. The alkoxides, which are obtained by the reaction

$$4Tl + 4C_2H_5OH \rightarrow (TlOC_2H_5)_4 + 2H_2$$

are liquids with the exception of the crystalline methoxide. All are tetramers and the methoxide has a distorted cube structure (Fig. 9-5) with the Tl and O atoms at corners of regular tetrahedra of different size, so that oxygen is 4-coordinate. Vibrational studies indicate that there are weak direct Tl$\cdots$Tl interactions in these molecules.[67]

[66] A. G. Lee, J. Chem. Soc., A., 1971, 2007.
[67] V. A. Maroni and T. G. Spiro, Inorg. Chem., 1968, 7, 193.

10

Carbon

GENERAL REMARKS

There are more known compounds of carbon than of any other element except hydrogen, most of them best regarded as organic chemicals.

The electronic structure of the carbon atom in its ground state is $1s^2 2s^2 2p^2$, with the two $2p$ electrons unpaired, following Hund's rule. In order to account for the normal four-covalence of carbon, we must consider that it is promoted to a valence state based on the configuration $2s2p_x2p_y2p_z$. The ion C^{4+} does not arise in any normal chemical process; the C^{4-} ion may possibly exist in some carbides. In general, however, carbon forms covalent bonds.

Some cations, anions and radicals of moderate stability can occur, and there is abundant evidence from the study of organic reaction mechanisms for transient species of these types.

Carbonium ions[1] are of the type $R^1R^2R^3C^+$. The triphenylmethyl cation, one of the earliest known, owes its stability primarily to the fact that the positive charge is highly delocalized, as indicated by canonical structures of the type 10-I(a–d). It behaves in some respects like other large univalent cations (Cs^+, R_4N^+, R_4As^+, etc.) and forms insoluble salts with large anions

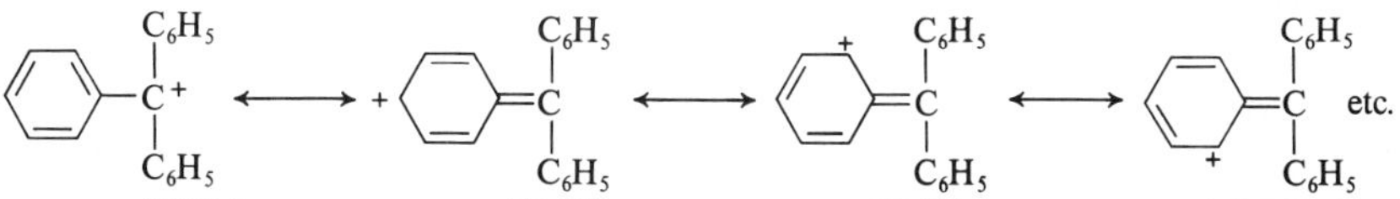

(10-Ia)	(10-Ib)	(10-Ic)	(10-Id)

such as BF_4^-, $GaCl_4^-$. There is good evidence that the cation has a propeller-like arrangement for the phenyl groups which are bound to the central atom by coplanar sp^2 trigonal bonds.

Carbanions[2] are of the type $R^1R^2R^3C^-$ and generally have no permanent

---

[1] G. A. Olah and P. von R. Schleyer, *eds.*, *Carbonium Ions*, Vols. I and II, Wiley, 1968, 1969; D. Bethnell and V. Gold, *Carbonium Ions*, Academic Press, 1967.

[2] D. J. Cram, *Fundamentals of Carbanion Chemistry*, Academic Press, 1965; M. Szwarc, *Carbanions, Living Polymers and Electron Transfer Processes*, Wiley-Interscience, 1968; U. Schöllkopf, *Angew. Chem. Internat. Edn.*, 1970, **9** 763.

existence, except in cases where the negative charge can be effectively delo-calized. The triphenylmethyl carbanion (10-II) is a good example, as is also the cyclopentadienyl anion (10-III). In fact, since the negative charge in the latter case is equally delocalized on all the carbon atoms, the anion is a regular planar pentagon and the π-electron density distribution can be well represented by (10-IV).

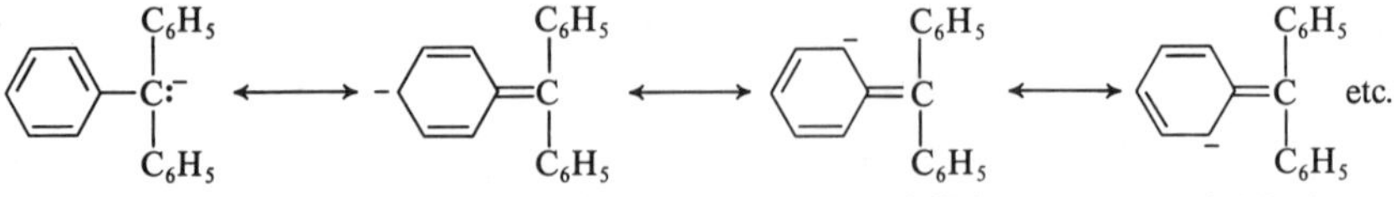

(10-IIa) (10-IIb) (10-IIc) (10-IId)

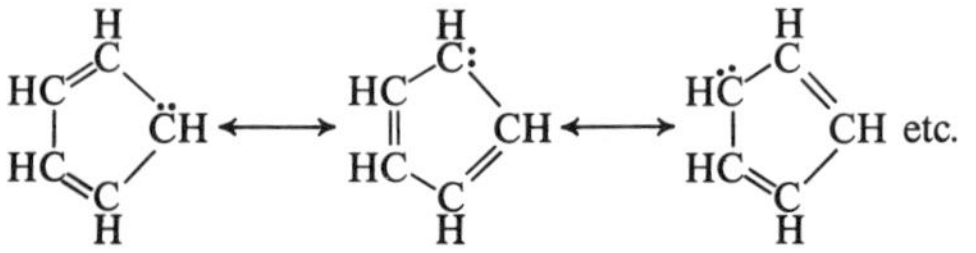

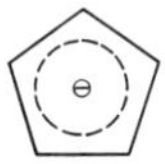

(10III-a) (10-IIIb) (10-IIIc) (10-IV)

Some stable carbanions,[3] isolable in crystalline salts, are (10-Va, b, c); these all have a planar set of bonds to the central carbon atom.

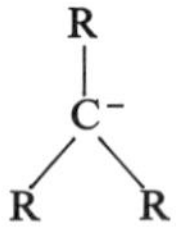

(a) R = −CN
(b) R = −C(CN)$_2$
(c) R = −NO$_2$

(10-V)

There are also a number of *radicals* that are fairly long-lived, such as the triphenylmethyl radical. Here again the stability is due mainly to delocali-zation—in this case of the odd electron—in a set of structures like those of $(C_6H_5)_3C^-$ with the odd electron in place of the electron pair.

Methyl radicals and substituted methyl radicals are often important reac-tion intermediates. Present indications are that CH_3 is planar, while CF_3, CCl_3 and CBr_3 are pyramidal.[4]

Divalent Carbon Compounds. There are a number of :CRR′ species, gen-erally called *carbenes*, which play a role in many reactions even though they are short-lived.[5] A general means of generating carbenes is by photolysis of diazoalkanes; this is done in the presence of the substrate with which the carbene is intended to react, such as an olefin:

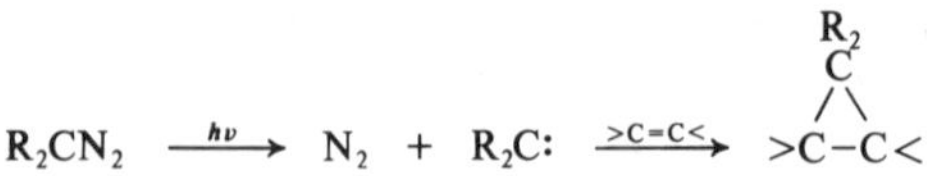

Carbenes so generated are, however, very energetic and their reactions are

[3] P. Anderson, B. Klewe and E. Thom, *Acta Chem. Scand.*, 1967, **21**, 1530; D. A. Bekoe, P. K. Gantzel and K. N. Trueblood, *Acta Cryst.*, 1967, **22**, 657.
[4] J. H. Current and J. K. Burdett, *J. Phys. Chem.*, 1969, **73**, 3505.
[5] T. L. Gilchrist and C. W. Rees, *Carbenes, Nitrenes and Arynes*, Thomas Nelson and Sons, 1969; W. Kirmse, *Carbene Chemistry*, 2nd ed., Academic Press, 1971.

often indiscriminate. Other methods of generating carbenes are therefore used for practical synthetic purposes but, in many if not all of these, truly free carbenes may never be formed. Thus carbenes, especially halocarbenes, may be conveniently generated from organomercury compounds,[6] thermally or by action of sodium iodide, e.g.,

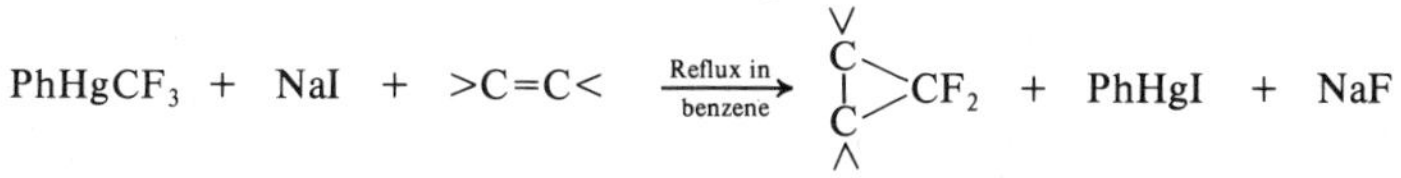

$$\text{PhHgCF}_3 + \text{NaI} + \; >\!\!C\!=\!\!C\!\!< \; \xrightarrow[\text{benzene}]{\text{Reflux in}} \; \overset{\vee}{\underset{\wedge}{\overset{C}{\underset{C}{\Big|}}}}\!\!>\!\!CF_2 + \text{PhHgI} + \text{NaF}$$

The structures and ground states of carbenes are difficult to establish with certainty. Experimental data and calculations indicate that most carbenes have bent structures with two unpaired electrons, with the exceptions of dihalocarbenes and those with O, N or S attached to the divalent carbon. These exceptional compounds probably have no unpaired electrons.

It may be noted that SiF_2 and other Group IV MX_2 compounds (Chapter 11) can be considered to have carbene-like behavior.

Carbene complexes of transition metals are discussed in Chapter 23.

Catenation. A key feature of carbon chemistry is the formation of chains or rings of C atoms, not only with single but also with multiple bonds (10-VI, VII, VIII). Clearly an element must have a valence at least two and must form strong bonds with itself to do this. Sulfur and silicon are the elements next most inclined to catenation but are far inferior to carbon in this respect.

$$R-(C\!=\!C-)_n R \qquad\qquad R-(C\!\equiv\!C-)_n R$$
$$\text{(10-VI)} \qquad\qquad\qquad \text{(10-VII)} \qquad\qquad \text{(10-VIII)}$$

The unusual stability of catenated carbon compounds, compared with those of Si and S, can be appreciated by considering the bond-energy data shown in Table 10-1. Thus the simple *thermal* stability of C—C—C··· chains

TABLE 10-1

Some Bond Energies Involving Carbon, Silicon and Sulfur

Bond	Energy (kJ mol^{-1})	Bond	Energy (kJ mol^{-1})
C—C	356	C—O	336
Si—Si	226	Si—O	368
S—S	226	S—O	~330

is high because of the intrinsic strength of C—C bonds. The relative stabilities toward oxidation follow from the fact that C—C and C—O bonds are of comparable stability, whereas for Si, and probably also for S, the bond to oxygen is considerably stronger. Thus, given the necessary activation energy, compounds with a number of Si—Si links are converted very exothermically into compounds with Si—O bonds.

[6] D. Seyferth *et al.*, *J. Amer. Chem. Soc.*, 1969, **91**, 6536; *J. Organometal. Chem.*, 1971, **33**, C1.

THE ELEMENTS

Naturally occurring carbon has the isotopic composition ^{12}C 98.89%, ^{13}C 1.11%. Only ^{13}C has nuclear spin ($S = \frac{1}{2}$), which provides a useful means of probing the structure and bonding in carbon compounds. The nmr measurements are considerably more difficult than are those for ^{1}H, partly because ^{13}C generally relaxes slowly so that only low power levels may be used and partly because the abundance is low unless ^{13}C-enriched samples are used.

The radioisotope ^{14}C (β^-, 5570 y), which is widely used as a tracer, is made by thermal neutron irradiation of lithium or aluminum nitride, $^{14}N(n, p)^{14}C$. It is available not only as CO_2 or carbonates but also in numerous labeled organic compounds. Its formation in the atmosphere and absorption of CO_2 by living organisms provide the basis of radiocarbon dating.

10-1. Allotropy of Carbon: Diamond; Graphite

The two best-known forms of carbon, diamond and graphite, differ in their physical and chemical properties because of differences in the arrangement and bonding of the atoms. Diamond is denser than graphite (diamond 3.51 g cm^{-3}; graphite 2.22 g cm^{-3}), but graphite is the more stable, by 2.9 kJ mol^{-1} at 300°K and 1 atm pressure. From the densities it follows that in order to transform graphite into diamond, pressure must be applied, and from the known thermodynamic properties of the two allotropes it can be estimated that they would be in equilibrium at 300°K under a pressure of $\sim 15,000$ atm. Of course, equilibrium is attained extremely slowly at this temperature, and it is this which allows the diamond structure to persist under ordinary conditions.

The energy required to vaporize graphite to a monoatomic gas is an important quantity, since it enters into the estimation of the energies of all bonds involving carbon. It is not easy to measure directly because, even at very high temperatures, the vapor contains appreciable fractions of C_2, C_3, etc. Spectroscopic studies established that the value had to be either ~ 520, ~ 574, or 716.9 kJ mol^{-1}, depending on the actual nature of the process measured spectroscopically. The composition of vapors has been determined mass-spectrographically with sufficient accuracy to show that the low values are unacceptable, hence it is now certain that the exact value is 716.9 kJ mol^{-1} at 300°K. In using older tables of bond energies, attention should be paid to what value was used for the heat of sublimation of graphite.

Diamond.[7] The diamond is one of the hardest solids known. This and the higher density are explicable in terms of the structure, which has been discussed in Section 2-4 and is shown in Fig. 2-8 (page 64).

Diamonds can be produced from graphite only by the action of high pressure. Furthermore, in order to get an appreciable rate of conversion,

[7] S. Tolansky, *History and Uses of Diamond*, Methuen, 1962.

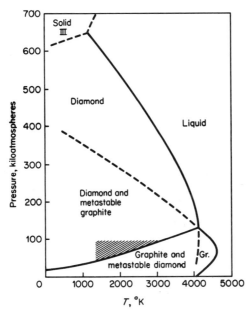

Fig. 10-1. Carbon phase diagram. Shaded area is the most favorable for catalyzed graphite–diamond conversion. [Adapted from F. P. Bundy, *J. Chem. Phys.*, 1963, **38**, 618, 631.]

high temperatures are necessary. Naturally occurring diamonds must have been formed when such conditions were provided by geological processes. Since at least 1880 recognition of these requirements has led many workers to attempt the production of synthetic diamonds. Until 1955 all such attempts ended in failure, inadequately proved claims, and even in a bogus report of success. Modern knowledge of the thermodynamics of the process indicates that none of the conditions of temperature and pressure reported could have been sufficient for success.

The present knowledge, some of it tentative, of the phase diagram for carbon is summarized in Fig. 10-1. Although graphite can be directly converted into diamond at temperatures of ca. 3000°K and pressures above 125 kbar, in order to obtain useful rates of conversion, a transition-metal catalyst such as Cr, Fe or Pt is used. It appears that a thin film of molten metal forms on the graphite, dissolving some and reprecipitating it as diamond, which is less soluble. Diamonds up to 0.1 carat of high industrial quality can be routinely produced at competitive prices. Some gem quality diamonds have also been made but the cost, so far, has been prohibitive of commercial development. Under certain conditions of temperature and pressure a hexagonal form of diamond (the normal is cubic) with a density of 3.33 can be formed;[8] this form evidently occurs in some meteorites, but its thermodynamic stability is not known.

[8] F. P. Bundy and J. S. Kasper, *J. Chem. Phys.*, 1967, **46**, 3437.

The chemical reactivity of diamond is much lower than that of carbon in the form of macrocrystalline graphite or the various amorphous forms. Diamond can be made to burn in air by heating it to 600–800 °C.

Graphite. Graphite has a layer structure as indicated in Fig. 10-2. The separation of the layers is 3.35 Å, which is about equal to the sum of van der Waals radii and indicates that the forces between layers should be relatively slight. Thus the observed softness and particularly the lubricity of graphite can be attributed to the easy slippage of these layers over one another. It will be noted that within each layer each carbon atom is surrounded by only three others. After forming one σ bond with each neighbor, each carbon atom would still have one electron and these are paired up into a system of π bonds (10-IX). Resonance with other structures having different but equivalent arrangements of the double bonds makes all C—C distances equal at 1.415 Å. This is a little longer than the C—C distance in benzene, where the bond order is 1.5, and agrees with the assumption that the bond order in graphite is ~ 1.33.

(10-IX)

Actually two modifications of graphite exist, differing in the ordering of the layers. In no case do all the carbon atoms of one layer lie directly over those in the next layer, but, in the structure shown in Fig. 10-2, carbon atoms in every other layer are superposed. This type of stacking, which may be designated (ABAB...), is apparently the most stable and exists in the commonly occurring hexagonal form of graphite. There is also a rhombic form,

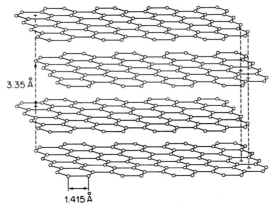

Fig. 10-2. The normal structure of graphite.

frequently present in naturally occurring graphite, in which the stacking order is (ABCABC...); that is, every third layer is superposed. It seems that local areas of rhombic structure can be formed by mechanical deformation of hexagonal crystals and can be removed by heat treatment.

The many forms of so-called amorphous carbon, such as charcoals, soot, and lampblack, are all actually microcrystalline forms of graphite. In some soots the microcrystals are so small that they contain only a few unit cells of the graphite structure. The physical properties of such materials are mainly determined by the nature and magnitude of their surface areas. The finely divided forms, which present relatively vast surfaces with only partially saturated attractive forces, readily absorb large amounts of gases and solutes from solution.[9] Active carbons impregnated with palladium, platinum, or other metal salts are widely used as industrial catalysts.

An important aspect of graphite technology is the production of very strong fibers by pyrolysis, at 1500 °C or above, of orientated organic polymer fibers, e.g., those of polyacrylonitrile, polyacrylate esters, or cellulose. When incorporated into plastics the reinforced materials are light and of great strength. Other forms of graphite such as foams, foils, or whiskers can also be made.[10]

10-2. Lamellar Compounds of Graphite[11]

The very loose, layered structure of graphite makes it possible for many molecules and ions to penetrate between the layers, forming interstitial or lamellar compounds. There are two basic types: those in which the graphite, which has good electrical conductivity, becomes non-conducting and those in which high electrical conductivity remains and is enhanced. Only two substances of the first type are known, namely, graphite oxide and graphite fluoride. Graphite oxide is obtained by treating graphite with strong aqueous oxidizing agents such as fuming nitric acid or potassium permanganate. Its composition is not entirely fixed and reproducible, but approximates to C_2O with a little hydrogen always present; the layer separation increases to 6–7 Å and it is believed that the oxygen atoms are present in C—O—C bridges across *meta*-positions, and in keto and enol groups, the latter being fairly acidic; the graphite layers thus lose their unsaturated character and buckle.

Graphite fluoride, also called poly(carbon monofluoride), is obtained by direct fluorination[12] of graphite at a temperature of ~600°. When lower temperatures are used, grey or black materials deficient in fluorine are

[9] M. Smisek and S. Cerny, *Active Carbon*, Elsevier, 1970.
[10] L. R. Creight, H. W. Rauch and W. H. Sutton, *Ceramic and Graphite Fibers and Whisker*, Academic Press, 1965; O. Vohler *et al.*, *Angew. Chem. Internat. Edn.*, 1970, 9, 414; J.E. Bailey and A.J. Clarke, *Chem. in Britain*, 1970, 484.
[11] R. C. Croft, *Quart. Rev.*, 1960, 14, 1; W. Rudorff, *Angew. Chem. Intern. Edn.*, 1963, 2, 67; *Adv. Inorg. Chem. Radiochem.*, 1959, 2, 224; G. R. Hennig, *Progr. Inorg. Chem.*, 1959, 1, 125; Y. N. Novitov and M. E. Vol'pin, *Uspekhi Khim.*, 1971, 40, 1568.
[12] J. L. Wood *et al.*, *J. Phys. Chem.*, 1969, 73, 3139.

obtained, but under proper conditions white, stoichiometric $(CF)_x$ can be reproducibly obtained. Actually, with the usual small particles of graphite $(10^2 - 10^3$ Å$)$, the formation of CF_2 groups at the edges of layers leads to a stoichiometry around $CF_{1.12}$. The layer spacing is ~ 8 Å and the layers are most likely buckled. $(CF)_x$ has lubricating properties like those of graphite, but it is superior in resisting oxidation by air up to at least 700°.

In the *electrically conducting lamellar compounds*, various atoms, molecules, and ions are inserted or intercalated between the carbon sheets. A large number of the compounds are formed spontaneously when graphite and the reactant are brought into contact. Thus the heavier alkali-metals K, Rb and Cs, the halogens Cl_2 and Br_2, and a great variety of halides, oxides and sulfides of metals, for example, $FeCl_3$, UCl_4, FeS_2 and MoO_3, form lamellar compounds spontaneously. A smaller group of compounds is formed by electrolysis of the reactant using a graphite anode, for example, with sulfuric acid. The manner in which the invading reactant species increase the conductivity of the graphite is not definitely settled, but apparently they do so by either adding electrons to, or removing electrons from, the conduction levels of graphite itself.

CARBIDES

The term carbide[13] is applied to those compounds in which carbon is combined with elements of lower or about equal electronegativity. Thus compounds with oxygen, sulfur, phosphorus, nitrogen, halogens, etc., are not considered in this category, and by convention neither are those with hydrogen. The reasonableness of this division will become apparent below. There are usually considered to be three types of carbide: (1) the salt-like carbides, formed chiefly by the elements of Groups I, II and III; (2) the interstitial carbides formed by most transition metals, especially those in Groups IV, V and VI, and (2a) a borderline type formed by a few of the transition metals with small atomic radii; and (3) covalent carbides, SiC and B_4C.

The general preparative methods for carbides of all three types include: (*a*) direct union of the elements at high temperature (2200° and above); (*b*) heating a compound of the metal, particularly the oxide, with carbon; and (*c*) heating the metal in the vapor of a suitable hydrocarbon. Carbides of Cu, Ag, Au, Zn and Cd, also commonly called acetylides, are prepared by passing acetylene into solutions of the metal salts; with Cu, Ag and Au, ammoniacal solutions of salts of the unipositive ions are used to obtain Cu_2C_2, Ag_2C_2 and Au_2C_2 (uncertain), whereas, for Zn and Cd, the acetylides ZnC_2 and CdC_2 are obtained by passing acetylene into petroleum solutions of dialkyl compounds. The Cu and Ag acetylides are explosive, being sensitive to both heat and mechanical shock.

1. *Salt-like carbides.* The most electropositive metals form carbides having physical and chemical properties indicating that they are essentially

[13] W. A. Frad, *Adv. Inorg. Chem. Radiochem.*, 1968, **11**, 153 (a general review).

ionic. They form colorless, transparent crystals and are decomposed by water and/or dilute acids at ordinary temperatures. The liberated anions are also immediately hydrolyzed, and hydrocarbons are thus evolved. There are ionic carbides containing C^{4-} and C_2^{2-} ions and one which, it has been inferred, contains C_3^{4-} ions.

Carbides containing C^{4-} ions evolve methane on hydrolysis and can be called methanides. Be_2C and Al_4C_3 are of this type. Thus the hydrolysis of the latter may be written

$$Al_4C_3 + 12H_2O \rightarrow 4Al(OH)_3 + 3CH_4$$

The structure of Be_2C is rather simple, being the antifluorite structure (see Fig. 2-3, page 51). The structure of Al_4C_3 is complicated; the details need not concern us except insofar as it is found that the carbon atoms occur singly.

There are a great many carbides that contain C_2^{2-} ions, or anions that can be so written to a first approximation. For the $M_2^IC_2$ compounds, where M^I may be one of the alkali metals or one of the coinage metals, and for the $M^{II}C_2$ compounds where M^{II} may be an alkaline-earth metal, Zn or Cd, and for $M_2^{III}(C_2)_3$ compounds in which M^{III} is Al, La, Pr or Tb, this description is probably a very good approximation. In these cases, the postulation of C_2^{2-} ions requires that the metal ions be in their normal oxidation states. In those instances where accurate structural parameters are known, the C—C distances lie in the range 1.19–1.24 Å. The compounds react with water and the C_2^{2-} ions are hydrolyzed to give acetylene only, e.g.,

$$Ca^{2+}C_2^{2-} + 2H_2O \rightarrow HCCH + Ca(OH)_2$$

There are, however, a number of carbides that have the same structures as those discussed above, meaning that the carbon atoms occur in discrete pairs, but that cannot be satisfactorily described as C_2^{2-} compounds. These include YC_2, TbC_2, YbC_2, LuC_2, UC_2, Ce_2C_3, Pr_2C_3 and Tb_2C_3. For all the MC_2 compounds in this list, neutron-scattering experiments show that (a) the metal atoms are essentially trivalent and (b) the C—C distances are 1.28–1.30 Å for the lanthanide compounds and 1.34 Å for UC_2. These facts and other details of the structures are consistent with the view that the metal atoms lose not only the electrons necessary to produce C_2^{2-} ions (which would make them M^{2+} ions) but also a third electron, mainly to the antibonding orbitals of the C_2^{2-} groups, thus lengthening the C—C bonds (cf. C—C = 1.19 Å in CaC_2). There are actually other, more delocalized, interactions among the cations and anions in these compounds since they have metallic properties. The M_2C_3 compounds have the metals in their trivalent states, C—C distances of 1.24–1.28 Å and also direct metal–metal interactions. These carbides, which cannot be represented simply as aggregates of C_2^{2-} ions and metal atoms in their normal oxidation states, are hydrolyzed by water to give only 50–70% of HCCH, while C_2H_4, CH_4 and H_2 are also produced. There is no detailed understanding of these hydrolytic processes.

Most of the MC_2 acetylides have the CaC_2 structure in which the $[C—C]^{2-}$

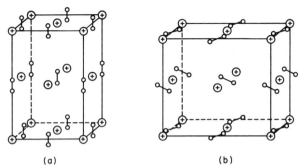

(a)   (b)

Fig. 10-3.   The CaC$_2$ (a) and ThC$_2$ (b) structures (the latter is somewhat simplified here).

ions lie lengthwise in the same direction along the cell axes, thus causing a distortion from cubic symmetry to tetragonal symmetry with one axis longer than the other two. In thorium carbide, the C$_2^{2-}$ ions are lying flat in parallel planes in such a way that two axes are equally lengthened with respect to the third. These structures are shown in Fig. 10-3. Li$_2$C$_2$ has a structure similar to that of CaC$_2$.

2. *Interstitial carbides.*[14]   In interstitial carbides, carbon atoms occupy the octahedral holes in close-packed arrays of metal atoms. The characteristics of interstitial carbides, namely, very high melting points (3000–4800°), great hardness (7–10, mostly 9–10, on Mohs' scale), and metallic electrical conductivity, are thus easily explainable. The electron band structure and other characteristic properties of the pure metal are not fundamentally altered by insertion of the carbon atoms into some of the interstices of the metal lattice; at the same time, the carbon atoms further stabilize the lattice, thus increasing hardness and raising the melting point. The ability of the carbon atoms to enter the interstices without appreciably distorting the metal structure requires that the interstices, and hence the metal atoms, be relatively large, and it can be estimated that a metal atom radius of ~ 1.3 Å or greater is required. An example of a carbon atom in an isolated octahedron of metal atoms is afforded by Ru$_6$C(CO)$_{14}$(mesitylene).[15] The carbon-to-metal bonding must be formulated in terms of MO's.

The metals Cr, Mn, Fe, Co and Ni have radii somewhat smaller than 1.3 Å and therefore they do not form typical interstitial carbides. Instead, the metal lattices are appreciably distorted and the carbon atoms interact directly with one another. The structures can be roughly described as having carbon chains (with C—C distances ~ 1.65 Å) running through very distorted metal lattices. The carbides Cr$_3$C$_2$ and M$_3$C (M = Mn, Fe, Co or Ni) are rather easily hydrolyzed by water and dilute acids, to give a variety of hydrocarbons (even liquid and solid ones and, in the case of Fe$_3$C, free carbon) and hydrogen. They are thus transitional between the typical ionic and interstitial carbides.

[14] L. K. Storms, *The Refractory Carbides*, Academic Press, 1967.
[15] R. Mason and W. R. Robinson, *Chem. Comm.*, **1968**, 468.

3. *Covalent carbides.* Although other carbides, for example, Be_2C, are at least partially covalent, the two elements which approach carbon closely in size and electronegativity, namely, Si and B, give completely covalent compounds. Silicon carbide, SiC, technically known as carborundum, is an extremely hard, infusible, and chemically stable material made by reducing SiO_2 with carbon in an electric furnace. It occurs in three structural modifications, in all of which there are infinite three-dimensional arrays of Si and C atoms, each tetrahedrally surrounded by four of the other kind. Interestingly, no evidence has ever been obtained for a germanium carbide of this or any other type.

Boron carbide, B_4C, is also an extremely hard, infusible, and inert substance, made by reduction of B_2O_3 with carbon in an electric furnace, and has a very unusual structure. The carbon atoms occur in linear chains of three, and the boron atoms in icosahedral groups of twelve (as in crystalline boron itself). These two units are then packed together in a sodium chloride-like array. There are, of course, covalent bonds between carbon atoms and boron atoms as well as between boron atoms in different icosahedra.

SIMPLE MOLECULAR COMPOUNDS

Some of the more important inorganic carbon compounds and their properties are listed in Table 10-2.

10-3. Carbon Halides

Carbon tetrafluoride is an extraordinarily stable compound. It is the end product in the fluorination of any carbon-containing compound. A useful laboratory preparation, for example, involves the fluorination of silicon carbide. The SiF_4 also formed is removed easily by passing the mixture through 20% NaOH solution. The CF_4 is unaffected, whereas the SiF_4 is immediately hydrolyzed; the difference is due to the fact that, in CF_4, carbon is coordinately saturated whereas silicon in SiF_4 has $3d$ orbitals available for coordination of OH^- ions in the first step of the hydrolysis reaction.

Carbon tetrachloride is a common solvent; it is fairly readily photochemically decomposed and also quite often readily transfers chloride ion to various substrates, CCl_3 radicals often being formed simultaneously at high temperatures (300–500°). It is often used to convert oxides into chlorides. Although it is thermodynamically unstable with respect to hydrolysis, the absence of acceptor orbitals on carbon makes attack very difficult.

Carbon tetrabromide is a pale yellow solid at room temperature. It is insoluble in water and other polar solvents but soluble in some non-polar solvents such as benzene.

Carbon tetraiodide is a bright-red crystalline material possessing an odor like that of iodine. Both heat and light cause decomposition to iodine and tetraiodoethylene. The tetraiodide is insoluble in water and alcohol, though

TABLE 10-2

Some Simple Compounds of Carbon

Compound	M.p. (°C)	B.p. (°C)	Remarks
CF_4	-185	-128	Very stable
CCl_4	-23	76	Moderately stable
CBr_4	93	190	Decomposes slightly on boiling
CI_4	171	—	Decomposes before boiling; can be sublimed under low pressure
COF_2	-114	-83	Easily decomposed by H_2O
$COCl_2$	-118	8	"Phosgene"; highly toxic
$COBr_2$	—	65	Fumes in air; $COBr_2 + H_2O \rightarrow CO_2 + 2HBr$
$CO(NH_2)_2$	132	—	Isomerized by heat to $NH_4^+NCO^-$
CO	-205	-190	Odorless and toxic
CO_2	-57 (5.2 atm)	-79	
C_3O_2	—	7	Evil-smelling gas
COS	-138	-50	Flammable; slowly decomposed by H_2O
CS_2	-109	46	Flammable and toxic
$(CN)_2$	-28	-21	Very toxic; colorless; water-soluble
HCN	-13.4	25.6	Very toxic; high dielectric constant (116 at 20°) for the associated liquid

attacked by both at elevated temperatures, and soluble in benzene. It may be prepared by the reaction:

$$CCl_4 + 4C_2H_5I \xrightarrow{AlCl_3} CI_4 + 4C_2H_5Cl$$

The increasing instability, both thermal and photochemical, of the carbon tetrahalides with increasing weight of the halogen correlates with a steady decrease in the C—X bond energies:

C—F 485; C—Cl 327; C—Br 285; C—I 213 kJ mol^{-1}

The *carbonyl halides*, COX_2 (X = F, Cl and Br), are all hydrolytically unstable substances. Mixed carbonyl halides such as $COClBr$ are also known. The compound $CO(NH_2)_2$ is urea. In the molecular structures of both urea and $COCl_2$ the C—O bond length is somewhat longer than the expected value for a C=O double bond, whereas the N—C and Cl—C distances are somewhat shorter than expected for single bonds. These facts lead to the conclusion that these molecules may be viewed as resonance hybrids (10-X).

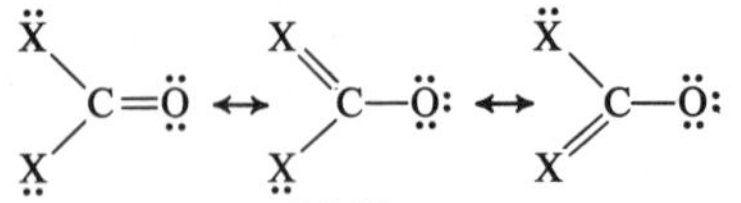

(10-X)

10-4. Carbon Oxides

There are five stable oxides of carbon: CO, CO_2, C_3O_2, C_5O_2 and $C_{12}O_9$. The last is the anhydride of mellitic acid (10-XI) and will not be discussed, nor will unstable oxides such as C_2O, C_2O_3[16] and CO_3.[17]

[16] R. F. Peterson and R. L. Wolfgang, *Chem. Comm.*, **1968**, 1201.
[17] K. V. Krishnamurthy, *J. Chem. Educ.*, 1967, **44**, 594.

O—C≶O   O
‖
O≶C   C   C
   ‖   ‖   ‖
O≶C   C   C   O
   ‖   ‖   ‖
O   C   C   C
‖   ‖
O—C≶O   O

(10-XI)

Carbon monoxide[18] is formed when carbon is burned with a deficiency of oxygen. At all temperatures, the following equilibrium exists, but it is not rapidly attained at ordinary temperatures:

$$2CO(g) \rightleftharpoons C(s) + CO_2(g)$$

The reaction

$$C + H_2O \rightleftharpoons CO + H_2$$

is important commercially, the equimolar mixture of CO and H_2 being called water gas. A convenient laboratory preparation of CO is by the action of concentrated sulfuric acid on formic acid:

$$HCOOH \xrightarrow{-H_2O} CO$$

Although CO is an exceedingly weak Lewis base, one of its most important properties is the ability to act as a donor ligand toward transition metals, giving *metal carbonyls*. For example, nickel metal reacts with CO to form $Ni(CO)_4$, and iron reacts under more forcing conditions to give $Fe(CO)_5$; many other carbonyl complexes are also known. The ability of CO to form bonds to transition-metal atoms is discussed in detail in Chapter 22, and catalytic reactions whereby CO is incorporated into organic compounds are discussed in Chapters 23 and 24.

CO reacts with alkali metals in liquid ammonia to give the so-called alkali-metal "carbonyls"; these white solids contain the $[OCCO]^{2-}$ ion.

Carbon dioxide is obtained by combustion of carbon in the presence of an excess of oxygen or by treating carbonates with dilute acids. Its important properties should already be familiar. It undergoes a number of "insertion" reactions (Section 24-A-3) similar to those of CS_2 noted below, and a few complexes with transition metals are known.

Carbon suboxide, C_3O_2, an evil-smelling gas, is a relatively uncommon substance that is formed by dehydrating malonic acid with P_2O_5 in a vacuum at 140–150°. The C_3O_2 molecule is probably linear with bond distances C—C 1.30 Å (theoretical for C=C 1.33) and C—O, 1.20 Å (theoretical for C=O 1.22). Linearity of the molecule and the observed bond lengths may be attributed to resonance among the canonical structures (10-XII). Although indefinitely stable at $-78°$, C_3O_2 polymerizes at room temperature and

$$O{=}C{=}C{=}C{=}O \leftrightarrow \overset{+}{O}{\equiv}C{-}C{\equiv}C{-}\overset{-}{O} \leftrightarrow \overset{-}{O}{-}C{\equiv}C{-}C{\equiv}\overset{+}{O}$$

(10-XII)

[18] *Carbon Monoxide: A Bibliography with Abstracts*, PHS No. 1503, U.S. Govt. Printing Office, Washington D.C., 1966.

above to give yellow to violet materials. Photolysis of C_3O_2 gives C_2O, which will react with various molecules, e.g.,

$$C_2O + CH_2{=}CH_2 \rightarrow CH_2{=}C{=}CH_2 + CO$$

The existence of a higher oxide, C_5O_2, is uncertain. Note that if resonance of the above type is an important stabilizing factor, then oxides of the type C_nO_2 would be expected only for n odd, so that C_5O_2 would be theoretically possible, whereas C_2O_2 and C_4O_2, which are unknown, would not.

There also exist aromatic anions such as $C_4O_4^{2-}$, $C_5O_5^{2-}$ and $C_6O_6^{2-}$, all having planar carbon rings.[19]

Carbonic Acids.   Carbon monoxide is formally the anhydride of formic acid, but its solubility in water and bases is slight. When heated with alkalis, however, it reacts to give the corresponding formate. C_3O_2 is the anhydride of malonic acid. It combines very vigorously with water to produce the acid and with ammonia and amines to produce malondiamides.

$$C_3O_2 + 2H_2O \rightarrow HOOCCH_2COOH$$
$$C_3O_2 + 2NHR_2 \rightarrow R_2NCOCH_2CONR_2$$

Carbon dioxide is the anhydride of the most important simple acid of carbon, "carbonic acid."[20] For many purposes, the following acid dissociation constants are given for aqueous "carbonic acid":

$$\frac{[H^+][HCO_3^-]}{[H_2CO_3]} = 4.16 \times 10^{-7}$$

$$\frac{[H^+][CO_3^{2-}]}{[HCO_3^-]} = 4.84 \times 10^{-11}$$

The equilibrium quotient in the first equation above is not really correct. It assumes that all CO_2 dissolved and undissociated is present as H_2CO_3, which is not true. In actual fact, the greater part of the dissolved CO_2 is only loosely hydrated, so that the correct first dissociation constant, using the "true" activity of H_2CO_3, has a value of about 2×10^{-4}, a value more nearly in agreement with expectation for an acid with the structure $(HO)_2CO$.

The rate at which CO_2 comes into equilibrium with H_2CO_3 and its dissociation products when passed into water is measurably slow, and this indeed is what has made possible an analytical distinction between H_2CO_3 and the loosely hydrated $CO_2(aq)$. This slowness is of great importance physiologically and in biological, analytical and industrial chemistry.

The slow reaction can easily be demonstrated by addition of a saturated aqueous solution of CO_2 on the one hand and of dilute acetic acid on the other to solutions of dilute NaOH containing phenolphthalein indicator. The acetic acid neutralization is instantaneous whereas with the CO_2 neutralization it takes several seconds for the color to fade.

[19] G. Maahs and P. Hegenberg, *Angew. Chem. Internat. Edn.*, 1966, **5**, 888; R. T. Bailey, *Chem. Comm.*, **1970**, 332.
[20] D. M. Kern, *J. Chem. Educ.*, 1960, **37**, 14; M. J. Welch *et al.*, *J. Phys. Chem.*, 1969, **73**, 3551.

The neutralization of CO_2 occurs by two paths. For $pH < 8$ the principal mechanism is direct hydration of CO_2:

$$CO_2 + H_2O = H_2CO_3 \qquad \text{(slow)} \qquad\qquad (10\text{-}1)$$
$$H_2CO_3 + OH^- = HCO_3^- + H_2O \qquad \text{(instantaneous)}$$

The rate law is pseudo-first order,

$$-d(CO_2)/dt = k_{CO_2}(CO_2); \qquad k_{CO_2} = 0.03 \text{ sec}^{-1}$$

At $pH > 10$, the predominant reaction is direct reaction of CO_2 and OH^-:

$$CO_2 + OH^- = HCO_3^- \qquad \text{(slow)} \qquad\qquad (10\text{-}2)$$
$$HCO_3^- + OH^- = CO_3^{2-} + H_2O \qquad \text{(instantaneous)}$$

where the rate law is

$$-d(CO_2)/dt = k_{OH^-}(OH^-)(CO_2); \qquad k_{OH^-} = 8500 \text{ sec}^{-1} \text{ (mol/l)}^{-1}$$

This can be interpreted, of course, merely as the base catalysis of (10-1). In the pH range 8–10 both mechanisms are important. For each hydration reaction (10-1, 10-2) there is a corresponding dehydration reaction:

$$H_2CO_3 \rightarrow H_2O + CO_2 \qquad k_{H2CO3} = k_{CO_2} \times K = 20 \text{ sec}^{-1}$$
$$HCO_3^- \rightarrow CO_2 + OH^- \qquad k_{HCO3} = k_{OH^-} \times KK_w/K_a = 2 \times 10^{-4} \text{ sec}^{-1}$$

Hence for the equilibrium

$$H_2CO_3 \rightleftharpoons CO_2 + H_2O \qquad\qquad (10\text{-}3)$$
$$K = (CO_2)/(H_2CO_3) = k_{H2CO3}/k_{CO_2} = \text{ca. } 600$$

It follows from (10-3) that the true ionization constant of H_2CO_3, K_a, is greater than the apparent constant as noted above.

An etherate of H_2CO_3 is obtained by interaction of HCl with Na_2CO_3 at low temperatures in dimethyl ether. The resultant white crystalline solid, m.p. $-47°$, which decomposes at about $5°$, is probably $OC(OH)_2 \cdot O(CH_3)_2$.

Carbamic acid, $O = C(OH)NH_2$, can be regarded as derived from carbonic acid by substitution of $-NH_2$ for $-OH$. This is only one example of the existence of compounds that are related in this way; $-NH_2$ and $-OH$ are isoelectronic and virtually isosteric and frequently give rise to isostructural compounds. If the second OH in carbonic acid is replaced by NH_2, we have urea. Carbamic acid is not known in the free state, but many salts are known, all of which, however, are unstable to water, because of hydrolysis:

$$H_2NCO_2^- + H_2O \rightarrow NH_4^+ + CO_3^{2-}$$

10-5. Compounds with C—N Bonds; Cyanides and Related Compounds

An important area of "inorganic" carbon chemistry is that of compounds with C—N bonds. The most important species are the cyanide, cyanate, and thiocyanate ions and their derivatives. We can regard many of these compounds as being pseudo-halogens or pseudo-halides, but the analogies, although reasonably apt for cyanogen, $(CN)_2$, are not especially valid in other cases.

1. *Cyanogen.*[21] This flammable gas (Table 10-2) is stable despite the

[21] T. K. Brotherton and J. W. Lynn, *Chem. Rev.*, 1959, **59**, 841; H. E. Williams, *Cyanogen Compounds*, 2nd edn., Arnold, 1948 (describes most CN compounds including cyanides); G. J. Jantz, *Cyanogen and Cyanogen-like Compounds as Dienophiles in 1,4-Cyclo-addition Reactions*, J. Hamer, ed., Academic Press, 1967.

fact that it is unusually endothermic ($\Delta Hf^\circ_{298} = 297$ kJ mol^{-1}). It can be obtained by direct oxidation of HCN in the gas phase by air over a silver catalyst, by Cl_2 over activated carbon or silica, or by NO_2 over calcium oxide–glass; the last reaction allows the NO produced to be recycled:

$$2HCN + NO_2 \rightarrow (CN)_2 + NO + H_2O$$

Cyanogen can also be obtained from the cyanide ion by aqueous oxidation using Cu^{2+} (cf. the $Cu^{2+} - I^-$ reaction):

$$Cu^{2+} + 2CN^- \rightarrow CuCN + \tfrac{1}{2}(CN)_2$$

or acidified peroxodisulfate. A better procedure for dry $(CN)_2$ employs the reaction:

$$Hg(CN)_2 + HgCl_2 \rightarrow Hg_2Cl_2 + (CN)_2.$$

This reaction also gives some paracyanogen, $(CN)_n$. Although pure $(CN)_2$ is stable, the impure gas may polymerize at 300–500°. The solid polymer reverts to $(CN)_2$ at 800–850° but decomposes above this temperature. The structure of $(CN)_n$ has been inferred from infrared spectroscopy to be (10-XIII).

Cyanogen is slowly hydrolyzed by water, in part giving (11-XIV), which indicates that the order of atoms must be NCCN. This is fully confirmed by structural studies, which show the molecule to be symmetrical and linear

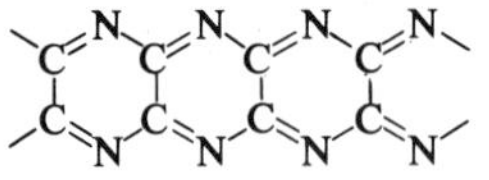

(10-XIII) 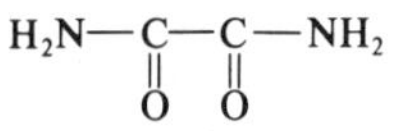   (10-XIV)

with C—C $= 1.37$ Å and C—N $= 1.13$ Å. The C—C distance is only slightly shorter than that expected (1.40) for a single bond between two carbon atoms using sp hybrid orbitals. The single canonical form $:N{\equiv}C{-}C{\equiv}N:$ therefore provides a reasonably good description of the electronic structure. Cyanogen dissociates into CN radicals, and, like RX and X_2 compounds, it can oxidatively (Section 24-A-2) add to lower-valent metal atoms giving dicyano complexes[22], e.g.,

$$(Ph_3P)_4Pd + (CN)_2 \rightarrow (Ph_3P)_2Pd(CN)_2 + 2Ph_3P$$

A further resemblance to the halogens is the disproportionation in basic solution:

$$(CN)_2 + 2OH^- \rightarrow CN^- + OCN^- + H_2O$$

Thermodynamically this reaction can occur in acid solution but is rapid only in base. Cyanogen has a large number of reactions, some of which are shown in Fig. 10-4. A stoichiometric mixture of O_2 and $(CN)_2$ burns producing one of the hottest flames (ca. 5050°K) known from a chemical reaction.

2. *Hydrogen cyanide.* HCN, like the hydrogen halides, is a covalent, molecular substance, but capable of dissociation in aqueous solution. It is an extremely poisonous (though less so than H_2S), colorless gas and is evolved when cyanides are treated with acids. It condenses at 25.6° to a liquid with a very high dielectric constant (107 at 25°). Here, as in similar cases, such as

[22] B. J. Argento *et al.*, *Chem. Comm.*, **1969**, 1427.

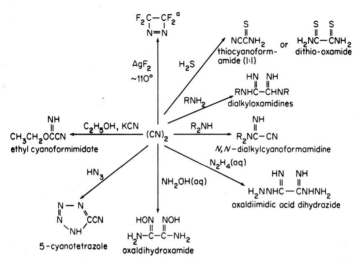

Fig. 10-4. Some reactions of cyanogen. [a]Other products may also be obtained by fluorination, e.g., $CF_3N{=}NCF_3$.

water, the high dielectric constant is due to association of intrinsically very polar molecules by hydrogen bonding. Liquid HCN is unstable[23] and can polymerize violently in the absence of stabilizers: in aqueous solutions polymerization is induced by ultraviolet light.

Hydrogen cyanide is thought to have been one of the small molecules in the earth's primeval atmosphere and to have been an important source or intermediate in the formation of biologically important chemicals.[24] Among the many polymerized products of HCN are: the trimer, aminomalononitrile, $HC(NH_2)(CN)_2$; the tetramer, diaminomalononitrile; and polymers of high molecular weight. Further, under pressure with traces of water and ammonia, HCN pentamerizes to adenine, while HCN can also act as a condensing agent for amino acids to give polypeptides.

In aqueous solution, HCN is a very weak acid, $pK_{25°} = 9.21$, and solutions of soluble cyanides are extensively hydrolyzed.

Hydrogen cyanide is made industrially from CH_4 and NH_3 by catalytic oxidation in a highly exothermic (475 kJ mol^{-1}) reaction (10-4):

$$2CH_4 + 3O_2 + 2NH_3 \xrightarrow[>800°]{Catalyst} 2HCN + 6H_2O \qquad (10\text{-}4)$$

It was once widely used to make acrylonitrile by reaction with acetylene in presence of an aqueous cuprous chloride–ammonium chloride catalyst, but this process is obsolescent and is being displaced by ammoxidation of propene with bismuth molybdate catalysts:

$$CH_3CH{=}CH_2 + NH_3 + 3/2 O_2 \xrightarrow{450°} CH_2{=}CHCN + 3H_2O$$
$$(70\% + HCN + CH_3CN)$$

[23] For thermodynamic properties see J. J. Christensen, H. D. Johnson and R. M. Izzatt, J. Chem. Soc., A, 1970, 454.
[24] S. W. Fox et al., Chem. Eng. News, 1970, June 22, p. 80; J. H. Boyer and H. Dabek, Chem. Comm., 1970, 1204; M. Calvin, Chemical Evolution, Clarendon Press, 1969.

Hydrogen cyanide may, however, be added directly to alkenes; e.g., butadiene gives adiponitrile (for Nylon) in the presence of palladium or zerovalent nickel phosphite catalysts[25] which operate by oxidative-addition and transfer reactions (Chapter 24).

3. *Cyanides.* Sodium cyanide is manufactured by fusion of calcium cyanamide with carbon and sodium carbonate:

$$CaCN_2 + C + Na_2CO_3 \rightarrow CaCO_3 + 2NaCN$$

The cyanide is leached with water. The calcium cyanamide is made in a rather impure form contamined with CaO, CaC_2, C, etc., by the interaction:

$$CaC_2 + N_2 \xrightarrow{\text{ca. } 1100°} CaNCN + C$$

The linear cyanamide ion is isostructural and isoelectronic with CO_2. Cyanamide itself, H_2NCN, can be made by dehydrosulfurization of thiourea with HgO or by acidification of CaNCN. The commercial product is the dimer, $H_2NC(=NH)NHCN$, which also contains much of the tautomer containing the substituted carbodiimide group, $H_2N—C(=NH)—N=C=NH$. Organo-carbodiimides are important synthetic reagents and $CH_3N=C=NCH_3$ is stable enough to be isolated.[26]

Sodium cyanide can also be obtained by the reaction

$$NaNH_2 + C \xrightarrow{500-600°} NaCN + H_2$$

In crystalline alkali cyanides at normal temperatures, the CN^- ion is freely rotating and is thus effectively spherical with a radius of 1.92 Å. Hence, for example, NaCN has the same structure as NaCl.

There are numerous salts of CN^- and those of Ag^I, Hg^I and Pb^{II} are very insoluble. Mercuric cyanide, $Hg(CN)_2$, is made by treating $HgSO_4$ with NaCN and extracting the $Hg(CN)_2$ with ethanol; it is moderately soluble in water and is un-ionized. The cyanide ion is of great importance as a ligand and large numbers of cyano complexes[27] are known of transition metals, Zn, Cd, Hg, etc.; some, like $Ag(CN)_2^-$ and $Au(CN)_2^-$, are of technical importance and cyano complexes are also employed analytically. The complexes sometimes resemble halogeno complexes, e.g., $Hg(CN)_4^{2-}$ and $HgCl_4^{2-}$, but other types exist; they are discussed further in Section 22-13. One respect in which CN shows no close resemblance to halogen is in the relative scarcity of covalent compounds of non-metals; this may be due to lack of efforts to prepare them rather than to inherent lack of stability.

The electrochemical oxidation of CN^- in aqueous solution gives $(CN)_2$, but in CH_3CN as solvent cyanation of aromatic compounds may be achieved.[28]

4. *Cyanogen halides.*[29] Compounds of CN with the halogens (called halogen cyanides or cyanogen halides) are well known. FCN, prepared by

[25] E. S. Brown and E. A. Rick, *Chem. Comm.*, **1969,** 112.

[26] G. Rapi and G. Sbrana, *Chem. Comm.*, **1968,** 128.

[27] M. H. Ford-Smith, *The Chemistry of Complex Cyanides*, H. M. Stationery Office, London, 1964; B. M. Chadwick and A. G. Sharpe, *Adv. Inorg. Chem. Radiochem.*, 1966, **8,** 84.

[28] S. Andreades and E. W. Zachnow, *J. Amer. Chem. Soc.*, 1969, **91,** 4181.

cracking cyanuric fluoride (from fluorination of cyanuric chloride), is stable for weeks as a gas (b.p. $-46°$) but polymerizes as a liquid at $25°$ and can be exploded. ClCN and BrCN are prepared by treating aqueous solutions of CN^- with chlorine or bromine, while ICN is obtained by treating a dry cyanide, usually $Hg(CN)_2$, with iodine. ClCN, BrCN and ICN are all rather volatile compounds that behave like the halogens and other halogenoids. The molecules are linear. Like HCN, the halogen cyanides tend to polymerize forming the cyanuric halides (10-XV). The amide (10-XVI), melamine, can be obtained by polymerization of cyanamide.

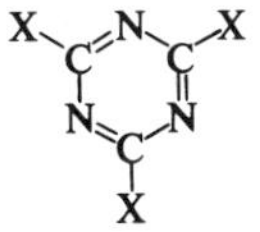

$$(10\text{-}XV) \qquad\qquad\qquad (10\text{-}XVI)$$

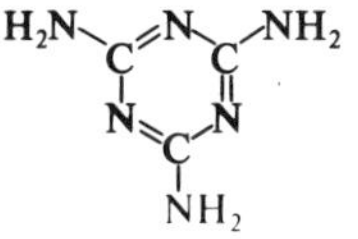

Compounds between CN and other halogenoid radicals are known, such as NCN_3 formed by the reaction

$$BrCN + NaN_3 = NaBr + NCN_3$$

5. *Cyanate and its analogous* S, Se *and* Te *ions.* The linear cyanate ion, OCN^-, is obtained by mild oxidation of aqueous CN^-, e.g.:

$$PbO(s) + KCN(aq) \rightarrow Pb(s) + KOCN(aq)$$

The free acid, $K = 1.2 \times 10^{-4}$, decomposes in solution to NH_3, H_2O and CO_2. There is little evidence for $(OCN)_2$ but covalent compounds such as $P(NCO)_3$ and some metal complexes are known. The compounds are usually prepared from halides by interaction with AgNCO in benzene or NH_4OCN in acetonitrile or liquid SO_2. In such compounds or complexes, either the O or N atoms of OCN can be bound to other atoms and this possibility exists also for SCN. In general most non-metallic elements seem to be N-bonded.

Thiocyanates are obtained by fusing alkali cyanides with sulfur; the reaction of S with KCN is rapid and quantitative, and S in benzene or acetone can be titrated with KCN in 2-propanol with Bromothymol Blue as indicator. Thiocyanogen is obtained by oxidation of aqueous SCN^- with MnO_2:

$$(SCN)_2 + 2e = 2SCN^- \qquad E^0 = +0.77 \text{ V}$$

but since it is rapidly decomposed by water it is best made by action of Br_2 on AgSCN in an inert solvent. In the free state, $(SCN)_2$ rapidly and irreversibly polymerizes to brick-red polythiocyanogen, but it is most stable in CCl_4 or CH_3COOH solution, where it exists as NCSSCN.[30]

The SCN^- ion is a good ligand and the numerous thiocyanate complexes, which may be either S- or N-bonded, are usually stoichiometrically analogous to halide complexes. Similar complexes of $SeCN^-$ are known. Non-metallic thiocyanates are usually S-bonded.

[29] B. S. Thyagarajan, *The Chemistry of Cyanogen Halides*, Intrascience Research Foundation, Santa Monica, Calif., 1968.
[30] C. E. Vanderzee and A. S. Quist, *Inorg. Chem.*, 1966, **5**, 1238.

10-6. Compounds with C—S bonds

Carbon disulfide. This well-known solvent is prepared on a large scale by direct interaction of C and S at high temperatures. A similar yellow liquid CSe_2 is made by action of CH_2Cl_2 on molten Se; it has a worse smell than CS_2 but, unlike it, is non-flammable. The selenide slowly polymerizes spontaneously, but CS_2 does so only under high pressures, to give then a black solid of structure (10-XVII).[31]

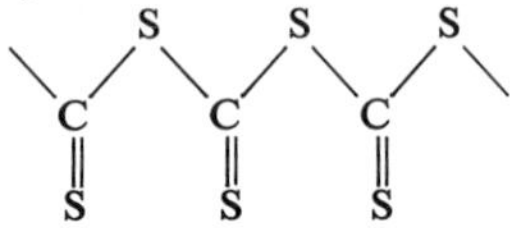

(10-XVII)

In addition to its high flammability in air, CS_2 is a very reactive molecule and has an extensive chemistry, much of it organic in nature. It is used to prepare carbon tetrachloride industrially:

$$CS_2 + 3Cl_2 \rightarrow CCl_4 + S_2Cl_2$$

Carbon disulfide is one of the small molecules that readily undergo the "insertion reaction" (Section 24-A-3) where the —S—C— group is inserted

between Sn—N, Co—Co, or other bonds. Thus with titanium dialkylamides, dithiocarbamates are obtained:

$$Ti(NR_2)_4 + 4CS_2 \rightarrow Ti(S_2CNR_2)_4$$

The CS_2 molecule can also serve as a ligand,[32] being either bound as a donor through sulfur or added oxidatively (Section 24-A-2) to give a three-membered ring as in (10-XVIII).[33] Such ring complexes are believed to be intermediates in the formation of thiocarbonyls (Section 22-9), e.g., *trans*-$RhCl(CS) (PPh_3)_2$.[34]

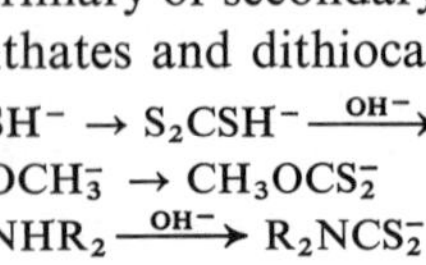

(10-XVIII)

Important reactions of CS_2 involve nucleophilic attacks on carbon by the ions SH^- and OR^- and by primary or secondary amines, which lead respectively to thiocarbonates, xanthates and dithiocarbamates, e.g.:

$$SCS + :SH^- \rightarrow S_2CSH^- \xrightarrow{OH^-} CS_3^{2-}$$
$$SCS + :OCH_3^- \rightarrow CH_3OCS_2^-$$
$$SCS + :NHR_2 \xrightarrow{OH^-} R_2NCS_2^-$$

Thiocarbonates.[35]   Thiocarbonates are readily formed by the action of

[31] A. J. Brown and E. Whalley, *Inorg. Chem.*, 1968, **7**, 1254.
[32] D. Commeruc, I. Douek and G. Wilkinson, *J. Chem. Soc.*, A, 1970, 1771.
[33] R. Mason and A. I. M. Rae, *J. Chem. Soc.*, A, 1970, 1767.
[34] M. Yagupsky and G. Wilkinson, *J. Chem. Soc.*, A, 1968, 2813.
[35] M. Dräger and G. Gattow, *Angew. Chem. Internat. Edn.*, 1968, **7**, 868 (chalcogeno carbonic acids and their salts).

SH$^-$ on CS$_2$ in alkaline solution (cf. reaction above), and numerous yellow salts containing the planar ion are known. Heating CS$_3^{2-}$ with S affords orange tetrathiocarbonates, which have the structure [S$_3$C—S—S]$^{2-}$. The free acids can be obtained from both these ions as red oils, stable at low temperatures.

There is an extensive chemistry of transition-metal complexes[36] with CS$_3^{2-}$, COS$_2^{2-}$, CS$_2$(SR)$^-$, CS$_2$(OR)$^-$, as well as with the dithiocarbamates discussed below.

Dithiocarbamates; thiuram disulfides.[36b, 37] Dithiocarbamates are normally prepared as alkali-metal salts by action of primary or secondary amines on CS$_2$ in presence of, say, NaOH. The Zn, Mn and Fe dithiocarbamates are extensively used as agricultural fungicides, and Zn salts as accelerators in the vulcanization of rubber.[38] Alkali-metal dithiocarbamates are usually hydrated and are dissociated in aqueous solution. When anhydrous, they are soluble in organic solvents in which they are associated.[38a]

Dithiocarbamates form a wide range of complexes with transition metals, and compounds of other metals such as Sn and of non-metals such as Te are also known. The CS$_2^-$ group in dithiocarbamates as well as in xanthates, thioxanthates, and thiocarbonates is usually chelated as in (10-XIX), but a few cases of monodentate dithiocarbamates are known.[39]

$$M \overset{S}{\underset{S}{\diagdown\diagup}} C - X \qquad \begin{array}{l} X = NHR \text{ or } NR_2, \\ OR \text{ or } SR, \\ O, S \text{ or } S-S \end{array}$$

(10-XIX)

On oxidation of aqueous solutions by H$_2$O$_2$, Cl$_2$ or S$_2$O$_8^{2-}$, thiuram disulfides, of which the tetramethyl is the commonest, are obtained:

$$I_2 + 2Me_2NCS_2^- \rightarrow Me_2N\underset{\underset{S}{\|}}{C}-S-S-\underset{\underset{S}{\|}}{C}NMe_2 + 2I^-$$

Thiuram disulfides, which are strong oxidants, are also used as polymerization initiators (for, when heated, they give radicals) and as vulcanization accelerators. Also tetraethylthiuram disulfide is "Antabuse," the agent for rendering the body allergic to ethanol.

The dithiocarbamates of certain metals, notably Zn^{II}, Cd^{II}, Cu^I and Ag^I, may be polymeric. The silver species are hexameric and provide a good exam-

[36] (a) See, e.g., J. P. Fackler and W. Seidel, *Inorg. Chem.*, 1969, **8**, 1631; D. Coucouvanis, S. J. Lippard and J. A. Zubieta, *J. Amer. Chem. Soc.*, 1970, **92**, 334; (b) D. Coucouvanis, *Progr. Inorg. Chem.*, 1970, **11**, 234 (complexes of dithioacids and dithiolates; extensive review); R. Eisenberg, *Progr. Inorg. Chem.*, 1970, **12**, 329 (structures of dithiocarbamate, xanthate and related complexes).

[37] G. D. Thorn and R. A. Ludwig, *The Dithiocarbamates and Related Compounds*, Elsevier, 1962.

[38] See, e.g., Monsanto Rubber Chemicals Handbook, 1968.

[38a] A. Uhlin, *Acta Chem. Scand.*, 1971, **25**, 393.

[39] C. O'Connor, J. D. Gilbert and G. Wilkinson, *J. Chem. Soc.*, A, **1969**, 84; R. Davis *et al.*, *J. Chem. Soc.*, A, **1971**, 994.

ple of metal polyhedra where the metal atoms are bridged by ligands without any metal—metal bonding.[39a]

ORGANOMETALLIC COMPOUNDS

10-7. General Survey of Types

Organometallic compounds are those in which the *carbon* atoms of organic groups are bound to metal atoms. Thus we do not include in this category compounds in which carbon-containing components are bound to a metal through some other atom such as oxygen, nitrogen, or sulfur. For example, $(C_3H_7O)_4Ti$ is not considered to be an organometallic compound, whereas $C_6H_5Ti(OC_3H_7)_3$ is, because in the latter there is one direct linkage of the metal to carbon. Although organic groups can be bound through carbon, in one way or another, to virtually all the elements in the Periodic Table, excluding the noble gases, the term organometallic is usually rather loosely defined and organo compounds of decidedly non-metallic elements such as B, P and Si are often included in the category. Specific compounds are discussed in the Sections on the chemistry of the individual elements since the organo derivatives are usually just as characteristic of any element as are, say, its halides or oxides. However, it is pertinent to make a few general comments here on the various types of compound.

1. *Ionic compounds of electropositive metals.* The organometallic compounds of highly electropositive metals are usually ionic in nature. Thus the alkali-metal derivatives (page 201), with the exception of those of lithium which are fairly covalent in nature, are insoluble in hydrocarbon solvents and are very reactive toward air, water, etc. The alkaline-earth metals Ca, Sr and Ba give poorly characterized compounds that are even more reactive and unstable than the alkali salts. The stability and reactivity of ionic compounds are determined in part by the stability of the carbanion. Compounds containing unstable anions (e.g., $C_nH_{2n+1}^-$) are generally highly reactive and often unstable and difficult to isolate; however, where reasonably stable carbanions exist, the metal derivatives are more stable though still quite reactive [e.g., $(C_6H_5)_3C^-Na^+$ and $(C_5H_5^-)_2Ca^{2+}$].

2. *σ-Bonded compounds.* Organo compounds in which the organic residue is bound to a metal by a normal two-electron covalent bond (albeit in some cases with appreciable ionic character) are formed by most metals of lower electropositivity and, of course, by non-metallic elements. The normal valence rules apply in these cases, and partial substitution of halides, hydroxides, etc., by organic groups is possible, as in $(CH_3)_3SnCl$, $(CH_3)SnCl_3$, etc. In most of these compounds, bonding is predominantly covalent and the chemistry is organic-like, although there are many differences in detail due to factors such as use of higher d orbitals or donor behav-

[39a] P. Jennische and R. Hesse, *Acta Chem. Scand.*, 1971, **25**, 423.

ior as in R_4Si, R_3P, R_2S, etc., incomplete valence shells, or coordinative unsaturation as in R_3B or R_2Zn, and effects of electronegativity differences between M—C and C—C bonds.

While the existence of stable M—C bonds has long been regarded as a normal part of the chemistry of the non-transition metals and metalloids, compounds containing transition metal-to-carbon σ bonds have only in recent years been made in substantial numbers. The reasons for the relative rarity of such compounds are still a subject for investigation, but several points seem clear. First, an important pathway for decomposition of M—R bonds is by a shift of a β-hydrogen atom, followed by olefin elimination, viz.:

$$M—CH_2—CHRR' \rightarrow M \leftarrow \overset{H}{\underset{CCR'}{\overset{|}{\underset{}{CH_2}}}} \rightarrow MH + CH_2=CRR'$$

Thus, molecules designed to impede or render impossible such a process may be relatively stable. The most obvious approach is to use R groups in which there are no β hydrogen atoms. When R is $(CH_3)_3SiCH_2$, thermally stable compounds, e.g., VOR_3, CrR_4, Mo_2R_6, can be made,[40] whereas similar compounds with ordinary alkyl groups are stable only at very low temperatures. On the basis of the above mechanism of decomposition it would also be expected that M—CH$_3$ and M—aryl bonds would be relatively favorable and it is, indeed, true that methyl and aryl compounds are generally more stable than those with, e.g., M—C$_2$H$_5$ bonds (see Sect. 23-6).

The intrinsic strength of M—C bonds is doubtless subject to some variations with the nature of the R group and the nature of the other ligands on the metal atom. It was observed about 1956 that metal atoms bearing CO, R_3P or h^5-C_5H_5, and related strongly interacting ligands (see Chapters 22 and 23) usually form compounds that have M—C bonds of superior stability. It is entirely possible, however, that the stability is due, not to any enhancement of M—C bond strength, but to the effect of such ligands in blocking the initial hydrogen-transfer. In the case of perfluoroalkyl groups, comparison of $CF_3Mn(CO)_5$ and CF_3I C—F force constants, does seem to indicate a genuine enhancement of M—C bond strength,[41] attributable to a π interaction superimposed on the σ bond, although the absence of β-hydrogen atoms may also contribute to stability.

3. *Non-classically bonded compounds.* There are many compounds in which metal-to-carbon bonding cannot be explained as either ionic or covalent in the simple sense of a $2c$-$2e$ M—C bond or bonds. The largest and most important class of these "non-classical" molecules comprises those formed primarily by the transition elements in which unsaturated groups are attached to metal atoms by interaction of the π electrons with metal orbitals. These include simple metal—alkene complexes of the types (10-XX) and (10-XXI) as well as those in which cyclopentadienyl and benzene rings are

[40] G. Yagupsky, W. Mowat, A. Shortland and G. Wilkinson, *Chem. Comm.*, **1970**, 1369.
[41] F. A. Cotton and R. M. Wing, *J. Organometallic Chem.*, 1967, **9**, 511.

bound to metals, viz., (10-XXII) and (10-XXIII). The chemistry and bonding in such compounds form the subject of Chapter 23.

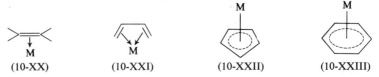

Another, smaller class of non-classical compounds is made up of those with bridging alkyl groups. The elements B, Al, Ga, In and Tl all form fairly stable but reactive trialkyls and triaryls, those of B, Ga, In and Tl being monomeric in the vapor and in solution. $(CH_3)_3In$ and $(CH_3)_3Tl$ form tetramers in their crystals,[42] but the association is weak and of uncertain nature and does not persist in other conditions. The aluminum compounds are unique in Group III in forming several reasonably stable dimers. Thus trimethylaluminum is a dimer in benzene solution and, partly, even in the vapor phase. $AlEt_3$ and $AlPr_3^n$ are also dimeric in benzene solution, but are almost completely dissociated in the vapor phase. $AlPr_3^i$ is a monomer in benzene. The molecules $Al_2(CH_3)_4(C_6H_5)_2$ and $Al_2(C_6H_5)_6$ are also known, as well as the polymeric $[Be(CH_3)_2]_x$. The structures[43,44] of four of these

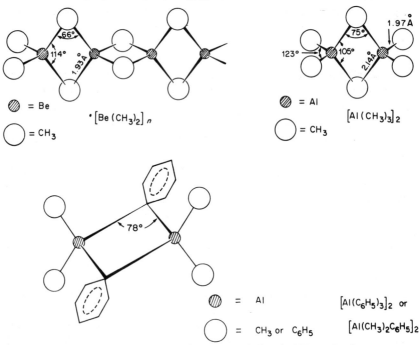

Fig. 10-5.   The structures of $[Be(CH_3)_2]_x$ and several dimeric AlR_3 molecules.

[42] G. M. Sheldrick and W. S. Sheldrick, *J. Chem. Soc., A,* **1970,** 28.
[43] B. G. Vranka and E. L. Amma, *J. Amer. Chem. Soc.,* 1967, **89,** 3121.
[44] J. F. Malone and W. S. McDonald, *Chem. Comm.,* **1967,** 444; **1970,** 280.

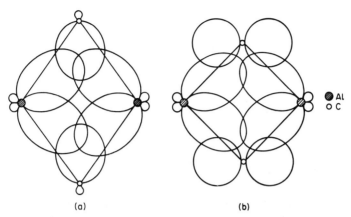

Fig. 10-6.   Schematic indication of orbital overlaps in Al—C—Al bridge bonding. (a) For
a methyl bridge. (b) Possible π component in phenyl bridging.

substances are shown in Fig. 10-5. In all cases the bridging carbon atoms are
equidistant from the metal atoms; in short, the Al—C—Al—C groups have
D_{2h} symmetry.

The bridging is accomplished by means of Al—C—Al 3c-2e bonds
(page 109) where each Al atom supplies an s–p hybrid orbital and so also
does the carbon atom. The situation then is as depicted in Fig. 10-6a. In
the case of $Al_2(CH_3)_6$ this view of the bonding has been strongly supported
by a low-temperature structure determination.[45] In the case of the bridging
phenyl groups, which lie perpendicular to the AlCAlC planes, the larger
Al—C—Al angles and slight inequalities in the C—C distances about the
rings have been taken[44] to mean that the $p\pi$ orbital of the bridging carbon
atom may also play some role in the bridge bonding, as indicated by the
overlap depicted in Fig. 10-6b.

The curious fact that none of the alkyls of the Group III elements except
those of aluminum are dimerized (except perhaps $GaEt_3$ and trivinylgallium
which appear to be dimers in solution) has not yet been satisfactorily explained. It has been proposed that for the larger metals, the small M—C—M
angles required to secure good overlap would introduce large repulsions
between the bulky metal atoms, but this cannot explain why $B(CH_3)_3$ does
not dimerize, especially since hydrogen bridging is quite important in the
boranes.

An important feature of coordinatively unsaturated alkyls, such as those
just noted or those of Mg, Zn, etc., is the moderately rapid exchange of alkyl
groups.[46] The exchanges can be readily studied by nmr methods and it appears
that bridged transition states, or intermediates of the type (10-XXIV) and
the like, provide the means for exchange. The rates of exchange reactions are

[45] J. C. Huffman and W. E. Streib, *Chem. Comm.*, **1971**, 911.
[46] J. P. Oliver, *Adv. Organometallic Chem.*, 1970, **8**, 167 (extensive review).

$$
\begin{array}{c}
\text{Me} \diagdown \overset{H_3}{C} \diagup \text{Me} \\
\overset{|}{\text{M}} \overset{|}{\text{M}} \\
\text{Me} \diagup \underset{H_3}{C} \diagdown \text{Me}
\end{array}
$$

(10-XXIV)

usually slowed by the presence of donor ligands and, if the donor is sufficiently strong to block the coordination sites, the exchange is stopped.

Further Reading

CARBON

Carbon, Pergamon Press (a review journal on physical properties).
Chemistry and Physics of Carbon, Vols. I and II, P. L. Walker, ed., Dekker, 1966.
Modern Aspects of Graphite Technology, L. C. F. Blackman, ed., Academic Press, 1970.
Davidson, H. W. *et al.*, *Manufactured Carbon*, Pergamon Press, 1968.
Reynolds, W. N., *Physical Properties of Graphite*, Elsevier, 1968.

ORGANOMETALLIC COMPOUNDS

See also under individual elements and Chapter 23

Handbook of Organometallic Compounds, N. Hagihara, M. Kumada and R. Okawara, eds., Benjamin, 1969 (data tables).
Organometallic Chem. Rev., Section B, Annual Surveys, D. Seyferth and R. B. King, eds., Vol. I, 1964, Elsevier (comprehensive and critical surveys).
Organometallic Compounds, Vols. 1 and 2, M. Dub and R. Weiss, eds. Springer (methods of synthesis, physical constants and chemical reactions). Vol. 1, 2nd edn., 1966, transition metals; Vol. II, 2nd edn., 1967, Ge-Pb; Vol. III, 2nd edn., 1968, As-Bi. (Exhaustive literature coverage.)
Becker, E. I. and M. Tsutsui, *Organometallic Reactions*, Wiley, Vol. 1, 1971 (a series on synthesis, reactions and mechanism).
Bokranz, A. and H. Plum, *Topics in Current Chem.*, Vol. 16, Issue 3/4, *Organometallic Compounds in Industry*, Springer Verlag.
Coates, G. E., M. L. H. Green and K. Wade, *Organometallic Compounds*, 3rd edn., Methuen, Vol. 1, *Main Group Elements*, 1967; Vol. 2, *Transition Elements*, 1968 (the best short introduction).
Cox, J. D. and G. Pilcher, eds., *Thermochemistry of Organic and Organometallic Compounds*, Academic Press, 1970.
Eisch, J. J., *Chemistry of Organometallic Compounds: the Main Group Elements*, McMillan, 1967 (an organic-chemical approach).
Nesmeyanov, A. N. and K. A. Kocheskov, *Methods of Elemento Organic Chemistry*, North Holland (an authoritative series of volumes covering various elements).
Ramsey, B. G., *Electronic Transitions in Metalloids*, Academic Press, 1969 (data for organo compounds of Groups I–V and Te).
Reutov, O. A. and I. P. Beletskaya, *Reaction Mechanisms of Organometallic Compounds*, North Holland, 1968 (emphasizes non-transition metal and especially organomercury reactions).
Skinner, H. A., *Adv. Organometallic Chem.*, 1965, **2**, 49 (many data on M—C bond strengths).
Stone, F. G. A. and R. West, *Adv. Organometallic Chem.*, Vol. 1, 1964 (review articles on all aspects).
Tsutsui, M., *Characterization of Organometallic Compounds*, Parts 1, 2, Wiley, 1971.

11

The Group IV Elements: Si, Ge, Sn, Pb

GENERAL REMARKS

11-1. Group Trends

There is no more striking example of an enormous discontinuity in general properties between the first and the second row elements followed by a relatively smooth change toward more metallic character thereafter than in this group. Little of the chemistry of silicon can be inferred from that of carbon. Carbon is strictly non-metallic; silicon is essentially non-metallic; germanium is a metalloid; tin and especially lead are metallic. Some properties of the elements are given in Table 11-1.

TABLE 11-1

Some Properties of the Group IV Elements*

Element	Electronic structure	M.p. (°C)	B.p. (°C)	Ionizations potentials (eV) 1st	2nd	3rd	4th	Electro-nega-tivities[a]	Covalent radius[b] (Å)
C	[He]$2s^2 2p^2$	>3550[c]	4827	11.264	24.376	47.864	64.476	2.5–2.6	0.77
Si	[Ne]$3s^2 3p^2$	1410	2355	8.149	16.34	33.46	45.13	1.8–1.9	1.17
Ge	[Ar]$3d^{10} 4s^2 4p^2$	937	2830	7.809	15.86	34.07	45.5	1.8–1.9	1.22
Sn	[Kr]$4d^{10} 5s^2 5p^2$	231.9	2260	7.332	14.63	30.6	39.6	1.8–1.9	1.40[d]
Pb	[Xe]$4f^{14} 5d^{10} 6s^2 6p^2$	327.5	1744	7.415	15.03	32.0	42.3	1.8	1.44[e]

* For further detail see E. A. V. Ebsworth.[1]
[a] H. O. Pritchard and H. A. Skinner, *Chem. Rev.*, 1955, **55**, 745; see also A. J. Smith, W. Adcock and W. Kitching, *J. Amer. Chem. Soc.*, 1970, **92**, 6140.
[b] Tetrahedral, i.e. sp^3 radii.
[c] Diamond.
[d] Covalent radius of Sn^{II}, 1.63 Å.
[e] Ionic radius of Pb^{2+}, 1.21; of Pb^{4+}, 0.775 Å.

Catenation. While not as extensive as in carbon chemistry, catenation is an important feature of Group IV chemistry in certain types of compound. Extensive chains occur in Si and Ge hydrides (up to Si_6H_{14} and Ge_9H_{20}),

[1] E. A. V. Ebsworth in *The Organometallic Compounds of Group IV Elements*, A. G. MacDiarmid, ed., Vol. 1, Part 1, Dekker, 1968.

in Si halides (only Ge_2Cl_6 is known), and in certain organo compounds. For Sn and Pb, catenation occurs only in organo compounds. In lead alloys such as Na_4Pb_4 and Na_4Pb_9 there appear to be cluster anions, and the Pb_9^{4-} ion has been suggested as having a structure similar to Bi_9^{5+} in "bismuth monochloride" (Section 13-5).

On the whole, however, there is a decrease in the tendency to catenation in the order $C \gg Si > Ge \approx Sn \gg Pb$. This general, if not entirely smooth, decrease in the tendency to catenation may be ascribed partly to diminishing strength of the C—C, Si—Si, Ge—Ge, Sn—Sn and Pb—Pb bonds (Table 11-2).

TABLE 11-2

Approximate Average Bond Energies

Group IV[a] element	Energy of bond (kJ mol^{-1}) with:							
	Self	H	C	F	Cl	Br	I	O
C	356	416		485	327	285	213	336
Si	210–250	323	250–335	582	391	310	234	368
Ge	190–210	290	255	465	356	276	213	
Sn	105–145	252	193		344	272	187	

[a] Data derived mainly from MX_4-type compounds which are unstable or non-existent when M = Pb. Pb—C in $PbEt_4$ = 128.8 kJ mol^{-1}; see C. F. Shaw and A. L. Allred, *Organometallic Chem. Rev.*, *A*, 1970, **5**, 96.

Bond Strengths. The strengths of single covalent bonds between Group IV atoms and other atoms (Table 11-2) generally decrease in going down the Group. In some cases there is an initial rise from C to Si followed by a decrease. These energies do not, of course, reflect the ease of heterolysis of bonds which is the usual way in chemical reactions; thus, for example, in spite of the high Si—Cl or Si—F bond energies, compounds containing these bonds are highly reactive. Since the charge separation in a bond is a critical factor, the bond ionicities must also be considered when interpreting the reactivities toward nucleophilic reagents. Thus Si—Cl bonds are much more reactive than Si—C bonds because, though stronger, they are more polar, $Si^{\delta+}$—$Cl^{\delta-}$, rendering the silicon more susceptible to attack by a nucleophile such as OH^-.

Two other points may be noted: (*a*) there is a steady decrease in M—C and M—H bond energies; (*b*) M—H bonds are stronger than M—C bonds.

Electronegativities. The electronegativities of the Group IV elements have been a contentious matter. Although C is generally agreed to be the most electronegative element, certain evidence, some of it suspect, has been interpreted as indicating that Ge is more electronegative than Si or Sn. It is to be remembered that electronegativity is a very qualitative matter and it seems most reasonable to accept a slight progressive decrease Si → Pb.

It can be noted that Zn and hydrochloric acid reduce only germanium halides to the hydrides, which suggests a higher electronegativity for Ge than for Si or Sn. Also, dilute aqueous NaOH does not affect GeH_4 or SnH_4, while SiH_4 is rapidly hydrolyzed by water containing a trace of OH^-, which is consistent with, though not necessarily indicative of, the Ge—H or Sn—H bonds either being non-polar or having the positive charge on hydrogen. Finally, germanium halides are hydrolyzed in water only slowly and reversibly.

The Divalent State. The term "lower valence" indicates the use of fewer than four electrons in bonding. Thus, although the *oxidation state* of carbon in CO is usually formally taken to be 2, this is only a formalism and carbon uses more than two valence electrons in bonding. True divalence is found in carbenes (page 284) and in a few SiX_2 compounds discussed below; the high reactivity of carbenes may result from the greater accessibility of the sp^2 hybridized lone pair in the smaller carbon atom. The stable divalent compounds of the other elements can be regarded as carbene-like in the sense that they are bent with a lone pair and undergo the general type of carbene reactions to give two new bonds to the element, i.e.,

$$\begin{array}{c} R \\ \diagdown \\ R \diagup \end{array} C: \rightarrow \begin{array}{c} R \\ \diagdown \\ R \diagup \end{array} C \begin{array}{c} \diagup \\ \diagdown \end{array}$$

However the divalent state becomes increasingly stable down the Group and is dominant for Pb.

Inspection of Table 11-1 clearly shows that this trend cannot be explained exclusively in terms of ionization potentials, since these are essentially the same for all of the elements Si–Pb; the "inert pair" concept is not particularly instructive either, especially since the non-bonding electrons are known not to be inert in a stereochemical sense (see below).

Other factors which undoubtedly govern the relative stabilities of the oxidation states are promotion energies and bond strengths for covalent compounds and lattice energies for ionic compounds. Taking first the former, it is rather easy to see why the divalent state becomes stable if we remember that the M—X bond energies generally decrease in the order Si—X, Ge—X, Sn—X, Pb—X(?). For methane the factor that stabilizes CH_4 relative to $CH_2 + H_2$, despite the much higher promotional energy required in forming CH_4, is the great strength of the C—H bonds and the fact that two more of these are formed in CH_4 than in CH_2. Thus if we have a series of reactions $MX_2 + X_2 = MX_4$ in which the M—X bond energies are decreasing, it is obviously possible that this energy may eventually become too small to compensate for the $M^{II} \rightarrow M^{IV}$ promotion energy and the MX_2 compound becomes the more stable. The progression is illustrated by ease of addition of chlorine to the dichlorides:

$$GeCl_2 + Cl_2 \rightarrow GeCl_4 \text{ (very rapid at } 25°)$$
$$SnCl_2 + Cl_2 \rightarrow SnCl_4 \text{ (slow at } 25°)$$
$$PbCl_2 + Cl_2 \rightarrow PbCl_4 \text{ (only under forcing conditions)}$$

Note that even $PbCl_4$ decomposes except at low temperatures, while $PbBr_4$ and PbI_4 do not exist, probably owing to the reducing power of Br^- and I^-.

For ionic compounds matters are not so simple but, since the sizes of the (real or hypothetical) ions, M^{2+} and M^{4+}, increase down the group, it is possible that lattice energy differences no longer favor the M^{4+} compound relative to the M^{2+} compound in view of the considerable energy expenditure required for the process

$$M^{2+} \rightarrow M^{4+} + 2e$$

Of course, there are few compounds of the types MX_2 or MX_4 that are entirely covalent or ionic (almost certainly no ionic MX_4 compounds), so that the above arguments are oversimplifications, but they indicate roughly the factors involved. For solutions no simple argument can be given, since Sn^{4+} and Pb^{4+} probably have no real existence.

Multiple Bonding.   Si, Ge, Sn and Pb do not form $p\pi$ multiple bonds under any circumstances. Thus numerous types of carbon compound, such as alkenes, alkynes, ketones, nitriles, etc., have no analogs. Although stoichiometric similarities may occur, e.g., CO_2, SiO_2, $(CH_3)_2CO$, $(CH_3)_2SiO$, there is no structural or chemical relationship[2] and reactions that might have been expected to yield a carbon-like product do not; e.g., dehydration of silanols $R_2Si(OH)_2$ produces $R_2Si(OH)—O—SiR_2(OH)$ and $(R_2SiO)_n$.

However, there is considerable evidence that in certain bonds to Si, notably from O and N, there is some double-bond character involving d orbital overlap, i.e. $d\pi$-$p\pi$ bonding. It is important to note, however, that this does not necessarily lead to conjugation in the sense usual for carbon multiple bond systems.[2] Thus nmr contact shifts for the tetrahedral tropone iminates of Si, Ge and Sn (11-I) show negligible conjugation with the π-system of the ring.[3] Observations of the following types provide evidence for $d\pi$-$p\pi$ bonding.

(*a*) Trisilylamine, $(H_3Si)_3N$, differs from $(H_3C)_3N$ in being planar[3a] rather than pyramidal and in being a very weak Lewis base. Other compounds such as $(H_3Si)_2NH$[4] and (11-II) are also planar; H_3SiNCS has a linear

(11-I)   (11-II)

Si—N—C—S group whereas in H_3CNCS the C—N—C group is bent as expected. These observations can be explained by supposing that nitrogen forms dative π bonds to the silicon atoms. In the planar state of $N(SiH_3)_3$, the non-bonding electrons of nitrogen would occupy the $2p_z$ orbital, if we assume that the N—Si bonds are formed using sp_xp_y trigonal hybrid orbitals of nitrogen. Silicon has empty $3d$ orbitals, which are of low enough energy to

[2]   H. Bock, H. Alt and H. Seidl, *J. Amer. Chem. Soc.*, 1969, **91**, 355; 1970, **92**, 1569.
[3]   D. R. Eaton and W. R. McClellan, *Inorg. Chem.*, 1967, **6**, 2134.
[3a] B. Beagley and A. R. Conrad, *Trans. Faraday Soc.*, 1970, **66**, 2740.
[4]   D. W. H. Rankin *et al.*, *J. Chem. Soc., A*, **1969**, 1224.

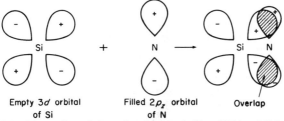

Empty $3d$ orbital of Si   Filled $2p_z$ orbital of N   Overlap

Fig. 11-1. Formation of $d\pi$-$p\pi$ bond between Si and N in trisilylamine.

be able to interact appreciably with the nitrogen $2p_z$ orbital. Thus, the N—Si π bonding is due to the kind of overlap indicated in Fig. 11-1. It is the additional bond strength to be gained by this $p\pi$-$d\pi$ bonding that causes the NSi_3 skeleton to take up a planar configuration, whereas with $N(CH_3)_3$, where the carbon has no low-energy d orbitals, the σ bonding alone determines the configuration, which is pyramidal as expected.

It is of interest that, contrary to earlier reports, $(H_3Si)_3P$ is pyramidal,[5] which may be taken as indicating that $d\pi$-$p\pi$ bonding between Si and at least this second-row element is small.

(b) Siloxane, $(H_3Si)_2O$ is a much weaker base than $(H_3C)_2O$ and here, as in many other compounds with Si—O—Si groups, the bond angle is about 150°.

(c) Silanols, such as $(CH_3)_3SiOH$, are stronger acids than the carbon analogs and form stronger hydrogen bonds; for Ph_3MOH the acidities are in the order $C \approx Si \gg Sn$. The hydrogen bonding can be ascribed to Si—O π bonding involving one of the two unshared pairs of the silanol oxygen and the $3d$ orbital of Si to give a situation somewhat similar electronically to that of the nitrogen atom in an imine $R_2C=NH$. The fact that one unshared pair still remains on the oxygen is consistent with the fact that the *base* character of the silanol is not much lowered in spite of its stronger acidity, compared with the analogous alcohol; the base order is $C \approx Si < Ge < Sn$.

A similar situation arises with the acid strength of R_3MCOOH, where the order is $Si \geqslant Ge > C$; in this case the $d\pi$-$p\pi$ bonding probably acts to stabilize the anion.[6a] The order of π-bonding, $C < Si > Ge \geqslant Sn > Pb$ is obtained from hydrogen-bonding and nmr studies on amines.[6b] Thus $N[Si(CH_3)_3]_3$ is virtually non-basic, the germanium compound is about as basic as a tertiary amine, and the tin compound is more basic than any organic amine. The same order[7] is found in RMX_3 when X = alkyl, but when X = halogen it is $C < Si < Ge < Sn$.

Stereochemistry[8]

The stereochemistries of Group IV compounds are given in Table 11-3.

[5] B. Beagley, A. G. Robiette and G. M. Sheldrick, *J. Chem. Soc., A,* **1968**, 3002, 3006.
[6a] O. W. Steward *et al., J. Chem. Soc. A.,* **1968**, 3119.
[6b] J. Mack and C. H. Yoder, *Inorg. Chem.,* 1969, **8**, 3119.
[7] G. M. Whitesides *et al., J. Organometallic Chem.,* 1970, **22**, 365.
[8] B. J. Aylett, *Progr. Stereochem.,* 1969, **4**, 213; see also *Organometallic Compounds of Group IV Elements,* A. G. McDiarmid, ed., Vol. 1, Part 1, Dekker, 1969.

TABLE 11-3

Valence and Stereochemistry of Group IV Elements

Valence[a]	Coordination number	Geometry[b]	Examples
Si^0	6	Octahedral	$Si(bipy)_3$
Si^{II}, Ge^{II}, Sn^{II}, Pb^{II}	2	ψ-Trigonal (angular)	$SiF_2(g)$, $SnCl_2(g)$, $Pb(C_5H_5)_2$, GeF_2
	3	ψ-Tetrahedral (pyramidal)	$SnCl_2 \cdot 2H_2O$, $SnCl_3^-$, $SnCl_2(s)$
	4	ψ-tbp	Pb^{II} in Pb_3O_4, GeF_2
	5	ψ-Octahedral	SnO (blue-black form)
	6	Octahedral	PbS(NaCl type), $GeI_2(CaI_2$ type)
	7	Complex	$[SC(NH_2)_2]_2PbCl$
	6, 7	ψ-Pentagonal bipyramid + complex ψ-8-coord.[c]	$Sn^{II}[Sn(edta)H_2O] \cdot H_2O$
Si^{IV}, Ge^{IV}, Sn^{IV}, Pb^{IV}	4	Tetrahedral	SiO_4 (silicates, SiO_2), SiS_2, $SiCl_4$, $PbMe_4$, GeH_4
	5	tbp	$Me_3SnClpy$, $SnCl_5^-$, SiF_5^-, $RSiF_4^-$, Me_3SnF (copolymer)
	5	?	$[SiPh_3bipy]^+$
	6	Octahedral	SiF_6^{2-}, $[Siacac_3]^+$, $SnCl_6^{2-}$, GeO_2, SnO_2 (rutile str.), $PbCl_6^{2-}$, cis-$SnCl_4(OPCl_3)_2$, SnF_4, $trans$-$GeCl_4py_2$, Pb^{IV} in Pb_3O_4, $Sn(S_2CNEt_2)_4$ $[M(C_2O_4)_3]^{2-}$; M = Si, Ge or Sn
	8	Dodecahedral ?	$Sn(NO_3)_4$; $Pb(OOCMe)_4$

[a] Sn^{III} species may be intermediates in the reduction of Cu^{II} by Sn^{II} in HCl solution (T. L. Nunes, *Inorg. Chem.*, 1970, **9**, 1325).
[b] ψ indicates that a coordination position is occupied by a lone pair.
[c] F. P. van Remoortere *et al.*, *Inorg. Chem.*, 1971, **10**, 1511.

IV-Oxidation State. Silicon is normally, though not exclusively, tetrahedral and the other elements are commonly so; the expected optical isomers such as those of $SiMePhEt(C_6H_4COOH)$ or $GeHMePh(\alpha$-naphthyl) can be resolved. In view of the possibility of valence-shell expansion by utilization of the outer d orbitals, coordination numbers of 5 and 6 are quite common. Pentacoordination is mainly confined to: (*a*) the ions MX_5^- stabilized in lattices by large cations; (*b*) compounds, especially of Si, with various oxygen and nitrogen chelates;[9a] (*c*) adducts with organic bases such as MX_4L, R_3MXL, etc.; and (*d*) for tin, polymeric compounds of the type R_3SnX where X acts as a bridge group. Although octahedral coordination is well known for all four elements, comparatively few silicon compounds have had their structure determined by X-ray methods.[9b]

[9a] C. L. Frye, *J. Amer. Chem. Soc.*, 1970, **92**, 1205.
[9b] For references see J. J. Flynn and F. P. Boer, *J. Amer. Chem. Soc.*, 1969, **91**, 5756.

For the ions and adducts, it cannot be predicted whether a complex will involve 5- or 6-coordination since that depends upon delicate energy balances.

II-Oxidation State. For Sn^{II}, and to a lesser extent for Ge^{II} and Pb^{II}, it has been shown that the pair of electrons that is unused in bonding has important effects on the stereochemistry.[10]

Thus in the blue-black form of SnO, each Sn atom is surrounded by five oxygen atoms at approximately the vertices of an octahedron, the sixth vertex being presumably occupied by the lone pair. This is called a ψ-octahedral arrangement. In $SnCl_2$, SnS, SnSe (orthorhombic form), $SnCl_2 \cdot 2H_2O$, $K_2SnCl_4 \cdot H_2O$ and $SnSO_4$, there are ψ-tetrahedral groupings, that is, atoms at 3 corners of a tetrahedron and a lone-pair of electrons at the fourth. Thus $SnCl_2 \cdot 2H_2O$ has a pyramidal $SnCl_2OH_2$ molecule, the second H_2O not being coordinated (it is readily lost at 80°), while $K_2SnCl_4 \cdot H_2O$ consists of ψ-tetrahedral $SnCl_3^-$ ions and Cl^- ions. The ψ-tetrahedral SnF_3^- ion is also known, and the $Sn_2F_5^-$ ion consists of two SnF_3^- ions sharing a fluorine atom. Other Sn^{II} compounds such as $SnCl_2$ or SnS similarly involve 3-coordination but with a bridge group between the metal atoms.

An important consequence is that solvated $SnCl_2$ or the $SnCl_3^-$ ion can act as a donor ligand towards transition metals (see page 330); although other Group IV compounds should behave similarly, little proof has so far been obtained except for $GeCl_3^-$ and $(C_5H_5)_2Sn$.

For lead, there is evidence for lone pairs in the structure of PbS and a polymeric dithioate complex (see below).

It may be noted finally that tin has an isotope, ^{119}Sn, suitable for Mössbauer spectral studies (Section 25-E) and that Sn^{II} and Sn^{IV} compounds can be readily distinguished.[11]

THE ELEMENTS

11-2. Occurrence, Isolation and Properties

Silicon is second only to oxygen in weight percentage of the earth's crust ($\sim 28\%$) and is found in an enormous diversity of silicate minerals. Germanium, tin, and lead are relatively rare elements ($\sim 10^{-3}$ weight percent), but they are well known because of their technical importance and the relative ease with which tin and lead are obtained from natural sources.

Silicon is obtained in the ordinary commercial form by reduction of SiO_2 with carbon or CaC_2 in an electric furnace. Similarly, germanium is prepared by reduction of the dioxide with carbon or hydrogen. Silicon and germanium are used as semiconductors, especially in transistors. For this purpose exceedingly high purity is essential, and special methods are required to obtain usable materials. For silicon, methods vary in detail but the following general procedure is followed.

[10] R. E. Rundle and D. H. Olsen, *Inorg. Chem.*, 1964, **3**, 596.
[11] J. J. Zuckerman, *Adv. Organometallic Chem.*, 1970, **9**, 21.

1. Ordinary, "chemically" pure Si is converted, by direct reaction, into a silicon halide or into $SiHCl_3$. This is then purified (of B, As, etc.) by fractional distillation in quartz vessels.

2. The SiX_4 or $SiHCl_3$ is then reconverted into elemental silicon by reduction with hydrogen in a hot tube or on a hot wire, when X is Cl or Br,

$$SiX_4 + 2H_2 \rightarrow Si + 4HX$$

or by direct thermal decomposition on a hot wire when X is I. Very pure Si can also be obtained by thermal decomposition of silane, SiH_4.

3. Pure silicon is then made "super-pure" (impurities $< 10^{-9}$ atom percent) by zone refining. In this process a rod of metal is heated near one end so that a cross-sectional wafer of molten silicon is produced. Since impurities are more soluble in the melt than they are in the solid they concentrate in the melt, and the melted zone is then caused to move slowly along the rod by moving the heat source. This carries impurities to the end. The process may be repeated. The impure end is then removed.

Super-pure germanium is made in a similar way. Germanium chloride is fractionally distilled and then hydrolyzed to GeO_2, when is then reduced with hydrogen. The resulting metal is zone melted.

Tin and lead are obtained from their ores in various ways, commonly by reduction of their oxides with carbon. Further purification is usually effected by dissolving the metals in acid and depositing the pure metals electrolytically.

Silicon is ordinarily rather unreactive. It is attacked by halogens giving tetrahalides, and by alkalis giving solutions of silicates. It is not attacked by acids except hydrofluoric; presumably the stability of SiF_6^{2-} provides the driving force here. A highly reactive form of silicon has been prepared by the reaction

$$3CaSi_2 + 2SbCl_3 \rightarrow 6Si + 2Sb + 3CaCl_2$$

Silicon so prepared reacts with water to give SiO_2 and hydrogen. It has been suggested that it is a graphite-like allotrope, but proof is as yet lacking, and its reactivity may be due to a state of extreme subdivision as in certain reactive forms of amorphous carbon.

Germanium is somewhat more reactive than silicon and dissolves in concentrated sulfuric and nitric acids. Tin and lead dissolve in several acids and are rapidly attacked by halogens. They are slowly attacked by cold alkali, rapidly by hot, to form stannates and plumbites. Lead often appears to be much more noble and unreactive than would be indicated by its standard potential of -0.13 V. This low reactivity can be attributed to a high overvoltage for hydrogen and also in some cases to insoluble surface coatings. Thus lead is not dissolved by dilute sulfuric and concentrated hydrochloric acids.

11-3. Allotropic Forms

Silicon and germanium are normally isostructural with diamond. By use of very high pressures, denser forms with distorted tetrahedra have been produced. The graphite structure is peculiar to carbon, which is understandable since such a structure requires the formation of $p\pi$-$p\pi$ bonds.

Tin has two crystalline modifications, with the equilibria

$$\alpha\text{-Sn} \underset{\text{"grey"}}{\overset{18°}{\rightleftharpoons}} \beta\text{-Sn} \underset{\text{"white"}}{\overset{232°}{\rightleftharpoons}} \text{Sn(l)}$$

α-Tin, or grey tin (density at $20° = 5.75$), has the diamond structure. The metallic form, β or white Sn (density at $20° = 7.31$) has a distorted close-packed lattice. The approach to ideal close packing accounts for the considerably greater density of the β-metal compared with the diamond form.

The most metallic of the Group IV elements, lead, exists only in a *ccp*, metallic form. This is a reflection both of its preference for divalence rather than tetravalence and of the relatively low stability of the Pb—Pb bond.

COMPOUNDS OF GROUP IV ELEMENTS

11-4. Hydrides[12]

These hydrides are listed in Table 11-4; all are colorless.

TABLE 11-4

Hydrides and Halides of Group IV Elements

Hydrides		Fluorides and Chlorides[a]		
MH_4	Other[b]	MF_4	MCl_4	Other
SiH_4 b.p. $-112°$	$Si_2H_6 \rightarrow Si_6H_{14}$[c] b.p. $-145°$	SiF_4 b.p. $-86°$	$SiCl_4$ b.p. $-57.6°$	$Si_2Cl_6 \rightarrow Si_6Cl_{14}$ b.p. $-145°$ $Si_2F_6 \rightarrow Si_{16}F_{34}$ b.p. $-18.5°$
GeH_4 b.p. $-88°$	$Ge_2H_6 \rightarrow Ge_9H_{20}$[c] b.p. $-29°$	GeF_4 m.p. $-37°$	$GeCl_4$ b.p. $83°$	Ge_2Cl_6 m.p. $40°$
SnH_4 b.p. $-52.5°$	Sn_2H_6	SnF_4 subl. $704°$	$SnCl_4$ b.p. $114.1°$	
PbH_4		PbF_4	$PbCl_4$ d. $105°$	

[a] All MX_4 compounds except $PbBr_4$ and PbI_4 are known, as well as mixed halides of Si, e.g., SiF_3I, $SiFCl_2Br$ but not $SiFClBrI$. The halogenodisilanes Si_2Br_6 and Si_2I_6 are also known.

[b] For mixed Si–Ge hydrides see K. M. Mackay, S. T. Hosfield and S. R. Stobert, *J. Chem. Soc.*, A, **1969**, 2937.

[c] Species and their isomers are separable by g.l.c., see, e.g., K. M. Mackay and K. J. Sutton, *J. Chem. Soc.*, A, **1968**, 231.

Silanes. Monosilane, SiH_4, is best prepared[13] on a small scale by heating SiO_2 and $LiAlH_4$ at 150–170°. On a larger scale the reduction of SiO_2 or

[12] B. J. Aylett, *Adv. Inorg. Chem. Radiochem.*, 1968, **11**, 249; J. E. Drake and C. Riddle, *Quart. Rev.*, 1970, **24**, 263 (extensive reviews of hydrides and their derivatives).
[13] J. M. Bellama and A. G. McDiarmid, *Inorg. Chem.*, 1968, **7**, 2070.

alkali silicates is possible by means of an $NaCl$–$AlCl_3$ eutectic (m.p. 120°) containing Al metal with hydrogen at 400 atm and 175°. The original Stock procedure (cf. boranes, page 237) of acid hydrolysis of magnesium silicide (prepared by direct interaction of Mg and Si or SiO_2) gives a mixture of silanes. Chlorosilanes may also be reduced by $LiAlH_4$.

Only SiH_4 and Si_2H_6 are indefinitely stable at 25°; the higher silanes decompose giving hydrogen and mono- and di-silane, possibly indicating SiH_2 as an intermediate.[13a]

The hydridic reactivity of the Si—H bond in silanes and substituted silanes is similar and may be attributed to charge separation $Si^{\delta+}$—$H^{\delta-}$ which results from the greater electronegativity of H than of Si. Silanes are spontaneously flammable in air, e.g.,

$$Si_4H_{10} + \tfrac{13}{2}O_2 \rightarrow 4SiO_2 + 5H_2O$$

Although silanes are stable to water and dilute mineral acids, rapid hydrolysis occurs with bases:

$$Si_2H_6 + (4+2n)H_2O \rightarrow 2SiO_2 \cdot nH_2O + 7H_2$$

The silanes are strong reducing agents. With halogens they react explosively at 25°, but controlled replacement of H by Cl or Br may be effected in presence of AlX_3 to give halogenosilanes such as SiH_3Cl.

Monogermane together with Ge_2H_6 and Ge_3H_8 can be made by heating GeO_2 and $LiAlH_4$[13] or by addition of $NaBH_4$ to GeO_2 in acid solution. Higher germanes are made by electric discharge in GeH_4. Germanes are less flammable than silanes, although still rapidly oxidized in air, and the higher germanes increasingly so. The germanes are resistant to hydrolysis, and GeH_4 is unaffected by even 30% NaOH.

Stannane, SnH_4, is best obtained by interaction of $SnCl_4$ and $LiAlH_4$ in ether at $-30°$. It decomposes rapidly when heated and even at 0° where it gives β-tin.[14] Whilst it is stable to dilute acids and bases, $2.5M$ NaOH causes decomposition to Sn and a little stannate. SnH_4 is easily oxidized and can be used to reduce organic compounds,[15] e.g., C_6H_5CHO to $C_6H_5CH_2OH$, and $C_6H_5NO_2$ to $C_6H_5NH_2$. With concentrated acids at low temperatures, the solvated stannonium ion[15a] is formed by the reaction

$$SnH_4 + H^+ \rightarrow SnH_3^+ + H_2$$

Plumbane, PbH_4, is formed in traces only when Mg–Pb alloys are hydrolyzed by acid or when lead salts are reduced cathodically.

All the elements form organo hydrides R_nMH_{4-n}, and even Pb derivatives are stable; they are readily made by reduction of the corresponding chlorides with $LiAlH_4$. There are also a number of compounds of transition metals with silyl groups, e.g., $H_3SiCo(CO)_4$.

Perhaps the most important reaction of compounds with an Si—H bond,

[13a] M. Bowrey and J. H. Purnell, *J. Amer. Chem. Soc.*, 1970, **92**, 2595.

[14] R. H. Herber and G. I. Paris, *Inorg. Chem.*, 1966, **5**, 769.

[15] G. H. Reifenberg and W. J. Considine, *J. Amer. Chem. Soc.*, 1969, **91**, 2401.

[15a] J. R. Webster and W. L. Jolly, *Inorg. Chem.*, 1971, **10**, 877.

such as Cl_3SiH or Me_3SiH, and one that is of commercial importance, is the Speier, or hydrosilation, reaction of alkenes, e.g.:

$$RCH=CH_2 + HSiCl_3 \rightarrow RCH_2CH_2SiCl_3$$

Normally chloroplatinic acid is used as a catalyst. The mechanism of reaction is discussed in Section 24-B-4.

The unusual reducing properties of $SiHCl_3$ are discussed below.

11-5. Halides

The more important halides are given in Table 11-4.

Fluorides. These compounds, which are of limited utility, are obtained by fluorination of the other halides or by direct interaction; GeF_4 is best made by heating $BaGeF_6$. Si and Ge tetrafluorides are hydrolyzed by an excess of water to the hydrous oxides; the main product from SiF_4 and H_2O in the gas phase is $F_3SiOSiF_3$.[16] In an excess of aqueous HF, the hexafluoro anions, MF_6^{2-}, are formed. SnF_4 is polymeric, with Sn octahedrally coordinated by four bridging and two non-bridging F atoms. PbF_4 is made by action of F_2 on PbF_2; a supposed preparation by the action of HF on $Pb(OOCMe)_4$ in $CHCl_3$ at $0°$ gives the much more reactive $Pb(OOCMe)_2F_2$.[17]

Silicon chlorides. $SiCl_4$ is made by chlorination of Si at red heat. Si_2Cl_6 can be obtained by interaction of $SiCl_4$ and Si at high temperatures or, along with $SiCl_4$ and higher chlorides, by chlorination of a silicide such as that of calcium. The higher members, which have highly branched structures, can also be obtained by amine-catalyzed reactions[18] such as

$$5Si_2Cl_6 \rightarrow Si_6Cl_{14} + 4SiCl_4$$
$$3Si_3Cl_8 \rightarrow Si_5Cl_{12} + 2Si_2Cl_6$$

The products are separated by fractional distillation.

All the chlorides are immediately and completely hydrolyzed by water, but careful hydrolysis of $SiCl_4$ gives $Cl_3SiOSiCl_3$ and $(Cl_3SiO)_2SiCl_2$.

Hexachlorodisilane, Si_2Cl_6, is a useful reducing agent[19] for compounds with oxygen bound to S, N or P; under mild conditions, at $25°$ in $CHCl_3$ chlorooxosilanes are produced. It is particularly useful for converting optically active phosphine oxides, $R^1R^2R^3PO$ into the corresponding phosphine. Since the reduction is accompanied by configurational inversion, the intermediacy of a highly nucleophilic $SiCl_3^-$ ion (cf. PCl_3) has been proposed:

$$Si_2Cl_6 + O=P\lessdot \rightarrow Cl_3SiOP^{+}\lessdot + Si^-Cl_3$$

$$\gtrdot P + Cl_3SiOSiCl_3 \leftarrow \gtrdot P^+SiCl_3 + {}^-OSiCl_3$$

The postulation of $SiCl_3^-$ can also accommodate the equally useful, clean, selective reductions by trichlorosilane (b.p. $33°$) and also the formation of

[16] J. L. Margrave, K. G. Sharp and P. W. Wilson, *J. Amer. Chem. Soc.*, 1970, **92**, 1530.

[17] F. Bornstein and L. Skarlos, *J. Amer. Chem. Soc.*, 1968, **90**, 5044.

[18] G. Urry *et al.*, *J. Inorg. Nuclear Chem.*, 1964, **26**, 409; *Accounts Chem. Res.*, 1970, **9**, 33; D. Kummer and H. Köster, *Angew. Chem. Internat. Edn.*, 1969, **8**, 879.

[19] K. Nauman *et al.*, *J. Amer. Chem. Soc.*, 1969, **91**, 7012, 7023, 7027.

$C{=}C$ and $Si-C$ bonds by reaction of $SiHCl_3$ with CCl_4, RX, RCOCl, and other halogen compounds in presence of amines.[20] In these cases the hypothetical $SiCl_3^-$ could be generated by the reaction

$$HSiCl_3 + R_3N \rightleftarrows R_3NH^+ + SiCl_3^-$$

followed by

$$SiCl_3^- + Cl_3C-CCl_3 \rightarrow SiCl_4 + Cl^- + Cl_2C{=}CCl_2$$
$$SiCl_3^- + RX \rightarrow [R^- + XSiCl_3] \rightarrow RSiCl_3 + X^-$$

There is some precedent for the postulation of the $SiCl_3^-$ ion since trisubstituted organosilanes, R_3SiH, react with bases to give silyl ions R_3Si^-.

Chloride Oxides. A variety of chlorooxosilanes, both linear and cyclic, is known. Thus controlled hydrolysis of $SiCl_4$ with moist ether, or interaction of Cl_2 and O_2 on hot silicon, gives $Cl_3SiO(SiOCl_2)_nSiCl_3$, where $n = 1-4$.

Germanium Tetrachloride. This differs from $SiCl_4$ in that only partial hydrolysis occurs in aqueous $6-9M$ HCl and there are equilibria involving species of the type $[Ge(OH)_nCl_{6-n}]^{2-}$; from concentrated HCl solutions of GeO_2 the tetrachloride can be distilled and separated.

Tin and Lead Tetrachlorides. These are also hydrolyzed completely only in water and in presence of an excess of acid form chloroanions, as discussed below.

11-6. Oxygen Compounds of Silicon[21]

Silica. Pure SiO_2 occurs in only two forms, *quartz* and *cristobalite*. The silicon is always tetrahedrally bound to four oxygen atoms but the bonds have considerable ionic character. In cristobalite the Si atoms are placed as are the C atoms in diamond, with the O atoms midway between each pair. In quartz, there are helices so that enantiomorphic crystals occur and these may be easily recognized and separated mechanically.

The interconversion of quartz and cristobalite on heating requires breaking and re-forming of bonds, and the activation energy is high. However, the rates of conversion are profoundly affected by the presence of impurities, or by the introduction of alkali-metal oxides or other "mineralizers." Studies of the system have shown that what was believed to be another form of quartz, tridymite, is a solid solution of mineralizer and silica.[22]

Slow cooling of molten silica or heating any form of solid silica to the softening temperature gives an amorphous material which is glassy in appearance and is indeed a glass in the general sense, that is, a material with no long-range order but rather a disordered array of polymeric chains, sheets or three-dimensional units.

Dense forms of SiO_2, called coesite and stishovite, were first made under

[20] R. A. Benkeser, *Accounts Chem. Res.*, 1971, **4**, 94.
[21] W. Eitel, ed., *Silicate Science*, Vols. I–V, Academic Press; W. A. Deer, R. A. Howie and J. Zussman, *Rock Forming Minerals*, Vols. I–V, Longmans; N. V. Belov, *Crystal Chemistry of Large Cation Silicates*, Consultants Bureau, 1963; A. A. Hodgson, *Fibrous Silicates*, Royal Inst. Chem. (London), Lecture Series No. 4, 1964.
[22] See R. Wollast, Proc. 8th Conf. Silicate Industry, Akadémiai Kiadô, Budapest, 1966.

drastic conditions (250–1300° at 35–120 k atm), but they were subsequently identified in meteor craters where the impact conditions were presumably similar; stishovite has the rutile structure. Both are chemically more inert than normal SiO_2 to which they revert on heating.

Silica is relatively unreactive towards Cl_2, H_2, acids and most metals at ordinary or slightly elevated temperatures, but it is attacked by fluorine, aqueous HF, alkali hydroxides, fused carbonates, etc.

Silicates. When alkali metal carbonates are fused with silica ($\sim 1300°$), CO_2 is driven off and a complex mixture of alkali silicates is obtained. When the mixtures are rich in alkali, the products are soluble in water, but with low alkali contents they become quite insoluble. Presumably the latter contain very large, polymeric anions. Aqueous sodium silicate solutions appear to contain the ion $[SiO_2(OH)_2]^{2-}$ according to Raman spectra[23] but, depending on the pH and concentration, tetrameric and other polymerized species are also present.[24]

Most of our understanding of silicate structures comes from studies of the many naturally occurring (and some synthetic) silicates of heavier metals. The basic unit of structure is the SiO_4 tetrahedron. These tetrahedra occur singly, or by sharing oxygen atoms, in small groups, in small cyclic groups, in infinite chains or in infinite sheets.

Simple Orthosilicates. A few silicates are known in which there are simple, discrete SiO_4^{4-}, orthosilicate, anions. In such compounds the associated cations are coordinated by the oxygen atoms, and various structures are found depending on the coordination number of the cation. In phenacite, Be_2SiO_4, and willemite, Zn_2SiO_4, the cations are surrounded by a tetrahedrally arranged set of four oxygen atoms. There are a number of compounds of the type M_2SiO_4, where M^{2+} is Mg^{2+}, Fe^{2+}, Mn^{2+}, or some other cation with a preferred coordination number of six, in which the SiO_4^{4-} anions are so arranged as to provide interstices with six oxygen atoms at the apices of an octahedron in which the cations are found. In zircon, $ZrSiO_4$, the Zr^{4+} ion is eight-coordinate although not all Zr—O distances are equal. It may be noted that, although the M—O bonds are probably more ionic than the Si—O bonds, there is doubtless some covalent character to them, and these substances should not be regarded as literally ionic in the sense $[M^{2+}]_2[SiO_4^{4-}]$ but rather as somewhere between this extreme and the opposite one of giant molecules. There are also other silicates containing discrete SiO_4 tetrahedra.

Other Discrete, Non-cyclic Silicate Anions. The simplest of the condensed silicate anions—that is, those formed by combining two or more SiO_4 tetrahedra by sharing of oxygen atoms—is the pyrosilicate ion, $Si_2O_7^{6-}$. This ion occurs in thortveitite ($Sc_2Si_2O_7$), hemimorphite ($Zn_4(OH)_2Si_2O_7$) and in at least three other minerals. It is interesting that the Si—O—Si angle varies from 131 to 180° in these substances.

[23] W. P. Griffith, *J. Chem. Soc., A*, **1969**, 1372.
[24] J. Aveston, *J. Chem. Soc.*, **1965**, 4444.

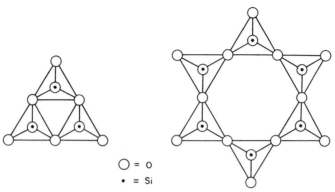

$$\bigcirc = O$$
$$\bullet = Si$$

Fig. 11-2. Examples of cyclic silicate anions.

Cyclic Silicate Anions. Only two such cyclic ions are known, namely, $Si_3O_9^{6-}$ and $Si_6O_{18}^{12-}$, the structures of which are shown schematically in Fig. 11-2. It should be clear that the general formula for any such ion must be $Si_nO_{3n}^{2n-}$. The ion $Si_3O_9^{6-}$ occurs in benitoite, $BaTiSi_3O_9$, and probably in $Ca_2BaSi_3O_9$. The ion $Si_6O_{18}^{12-}$ occurs in beryl, $Be_3Al_2Si_6O_{18}$.

Infinite Chain Anions. These are of two main types, the *pyroxenes*, which contain single-strand chains of composition $(SiO_3^{2-})_n$ (Fig. 11-3) and the *amphiboles* which contain double-strand, cross-linked chains or bands of composition $(Si_4O_{11}^{6-})_n$. Note that the general formula of the anion in a pyroxene is the same as in a silicate with a cyclic anion. Silicates with this general stoichiometry are often, especially in older literature, called "meta-silicates." There is actually neither metasilicic acid nor any discrete meta-silicate anion. With the exception of the few "metasilicates" with cyclic anions, such compounds contain infinite chain anions.

Fig. 11-3. A linear chain silicate anion.

Examples of pyroxenes are enstatite, $MgSiO_3$, diopside, $CaMg(SiO_3)_2$, and spodumene, $LiAl(SiO_3)_2$, the last being an important lithium ore. In the lithium compound there is one unipositive and one tripositive cation instead of two dipositive cations. Indeed the three compounds cited illustrate very well the important principle that, within rather wide limits, *the specific cations or even their charges are unimportant so long as the total positive charge is sufficient to produce electroneutrality.* This may be easily understood in terms of the structure of the pyroxenes in which the $(SiO_3)_n$ chains lie parallel and are held together by the cations which lie between them. Obviously the exact identity of the individual cations is of minor importance in such a structure.

A typical amphibole is tremolite, $Ca_2Mg_5(Si_4O_{11})_2(OH)_2$. Although it would not seem to be absolutely necessary, amphiboles apparently always contain some hydroxyl groups attached to the cations. Aside from this, however, they are structurally similar to the pyroxenes, in that the $(Si_4O_{11}^{6-})_n$ bands lie parallel and are held together by the metal ions lying between them. Like the pyroxenes and for the same reason, they are subject to some variability in the particular cations incorporated.

Because of the strength of the $(SiO_3)_n$ and $(Si_4O_{11})_n$ chains in the pyroxenes and amphiboles, and also because of the relative weakness and lack of strong directional properties in the essentially electrostatic forces between them via the metal ions, we might expect such substances to cleave most readily in directions parallel to the chains. This is in fact the case, dramatically so in the various asbestos minerals, which are all amphiboles.

Infinite Sheet Anions. When SiO_4 tetrahedra are linked into infinite two-dimensional networks as shown in Fig. 11-4, the empirical formula for the anion is $(Si_2O_5^{2-})_n$. Many silicates have such sheet structures with the sheets bound together by the cations which lie between them. Such substances might thus be expected to cleave readily into thin sheets, and this expectation is confirmed in the micas which are silicates of this type.

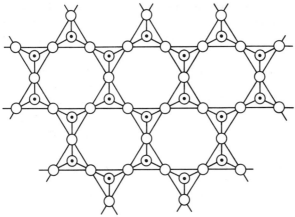

Fig. 11-4.   Sheet silicate anion structure idealized. [For a real example see A. K. Pant *Acta Cryst.*, *B*, 1968, **24**, 1077.]

Framework Minerals.   The next logical extension in the above progression from simple SiO_4^{4-} ions to larger and more complex structures would be to three-dimensional structures in which *every* oxygen is shared between two tetrahedra. The empirical formula for such a substance would be simply $(SiO_2)_n$; that is, we should have silica. However, if, in such a three-dimensional framework structure, some silicon atoms are replaced by aluminum atoms, the framework must be negatively charged and there must be other cations uniformly distributed through it. Aluminosilicates of this type are the feldspars, zeolites and ultramarines which (excepting the last) are among the most widespread, diverse, and useful silicate minerals in nature. Moreover, many synthetic zeolites have been made in the laboratory, and several are

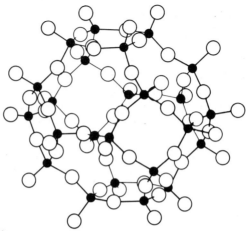

Fig. 11-5. The arrangement of AlO_4 and SiO_4 tetrahedra which gives the cubo-octahedral cavity in some zeolites and felspathoids. ● represents Si or Al.

manufactured industrially for use as ion-exchangers and "molecular sieves." The feldspars are the major constituents of igneous rocks and include such minerals as orthoclase, $KAlSi_3O_8$, which may be written $K[(AlO_2)(SiO_2)_3]$ to indicate that one-fourth of the oxygen tetrahedra are occupied by Al atoms, and anorthite, $CaAl_2Si_2O_8$ or $Ca[(AlO_2)_2(SiO_2)_2]$, in which half of the tetrahedra are AlO_4 and half SiO_4. The ultramarines are silicates that are blue; in addition to cations sufficient to balance the negative charge of the $[(SiAl)O_2]$ framework these substances contain additional cations and anions such as Cl^-, SO_4^{2-}, and sulfide ion, probably S_2^-, which is believed to account for the color. The framework of the ultramarines is rather open, permitting fairly ready exchange of both cations and anions.

Among the most important framework silicates are the *zeolites*.[21,25] Their chief characteristic is the openness of the $[(Al,Si)O_2]_n$ framework (Figs. 11-5 and 11-6), which makes possible their uses as ion-exchangers and selective

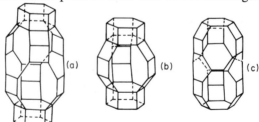

Fig. 11-6. Cavities of different dimensions in
(a) chabazite, $Ca_6Al_{12}Si_{24}O_{72} \cdot 40H_2O$,
(b) gmelinite, $(Na_2Ca)_4Al_8Si_{16}O_{48} \cdot 24H_2O$, and
(c) erionite, $Ca_{4.5}Al_9Si_{27}O_{72} \cdot 27H_2O$.

[25] R. M. Barrer in *Non-Stoichiometric Compounds*, L. Mandelcorn, ed., Academic Press, 1964; D. W. Breck, *J. Chem. Educ.*, 1964, **41**, 678 (molecular sieves).

adsorbants. Some of the many natural zeolites have been synthesized, and numerous synthetic ones are known. The composition is always of the type $M_{x/n}[(AlO_2)_x(SiO_2)_y] \cdot zH_2O$ where n is the charge of the metal cation, M^{n+}, which is usually Na^+, K^+ or Ca^{2+}, and the z is the number of moles of water of hydration, which is highly variable.

Zeolites as ion-exchange materials have been largely displaced by synthetic cationic and anionic exchange resins, but their use as selective absorbants for gases or liquids—"molecular sieves"—is extensive. They are also widely used as catalysts or catalyst-support materials for a variety of heterogeneous reactions.[26] The zeolites used for all these purposes are mainly synthetic. Thus slow crystallization under precisely controlled conditions of a sodium aluminosilicate gel of proper composition, gives the crystalline compound $Na_{12}[(AlO_2)_{12}(SiO_2)_{12}]27 \cdot H_2O$. In this hydrated form it can be used as a cation-exchanger in basic solution. To obtain the "molecular sieve," the water of hydration is removed by heating in a vacuum to 350°. The crystalline substance is of cubic symmetry. The AlO_4 and SiO_4 tetrahedra are linked together so as to form a ring of eight oxygen atoms on each face of the unit cube and an irregular ring of six oxygen atoms across each corner. In the center of the unit cell is a large cavity about 11.4 Å in diameter which is connected to six identical cavities in adjacent unit cells by the eight-membered rings which have inner diameters of about 4.2 Å. In addition, the large cavity is connected to eight smaller cavities, about 6.6 Å in diameter, by the six-membered rings, which provide openings about 2.0 Å in diameter. In the hydrated form all the cavities contain water molecules. In the anhydrous state the same cavities may be occupied by other molecules brought into contact with the zeolite, providing such molecules are able to squeeze through the apertures connecting cavities. Molecules within the cavities then tend to be held there by attractive forces of electrostatic and van der Waals types. Thus the zeolite will be able to absorb and strongly retain molecules just small enough to enter the cavities. It will not absorb at all those too big to enter, and it will absorb weakly very small molecules or atoms which can enter but also leave easily. For example, the zeolite under discussion will absorb straight-chain hydrocarbons but not branched-chain or aromatic ones.

Just as high pressures convert SiO_2 in one of its forms containing SiO_4 tetrahedra into stishovite which contains SiO_6 octahedra, so some silicates, e.g., the feldspar $KAlSi_3O_8$, at 120 kbar and 900° can be transformed into a form containing SiO_6 octahedra, and these are then metastable at normal temperature and pressure.

A hydrated calcium silicate sulfate carbonate has a discrete $Si(OH)_6^{2-}$ ion as part of its structure.[27]

---

[26] J. Turkevitch in *Catalyst Reviews*, Vol. 1, Dekker, 1968; P. B. Venuto and P. S. Landis, *Adv. in Catalysis*, Academic Press, 1968; Vol. 18, J. A. Rabo and M. L. Poutsma, Adv. Chem. Series, No. 102, 1971, pp. 284 et seq. (catalytic cracking on zeolites).
[27] R. A. Edge and H. F. W. Taylor, *Nature*, 1969, **224**, 364.

11-7. Oxygen Compounds of Germanium, Tin and Lead

Oxides and Hydroxides. The oxides GeO_2, SnO_2 and PbO_2 are all well-characterized compounds. GeO_2 resembles SiO_2 in that it exists in two forms; the stable high-temperature form has a cristobalite lattice and the other a rutile lattice; the radius ratio is close to that at which the change from tetrahedral to octahedral coordination should occur theoretically. SnO_2 exists in three different modifications of which the rutile form (in the mineral cassiterite) is most common; PbO_2 shows only the rutile structure. The basicity of the dioxides appears to increase from Si to Pb. SiO_2 is purely acidic, GeO_2 is less so and in concentrated HCl gives $GeCl_4$, while SnO_2 is amphoteric, though when made at high temperatures or by dissolving Sn in hot concentrated HNO_3 it is, like PbO_2, remarkably inert to chemical attack.

There is little evidence that there are true hydroxides, $M(OH)_4$, and the products obtained by hydrolysis of hydrides, halides, alkoxides, etc. are best regarded as hydrous oxides. Thus the addition of OH^- to Sn^{IV} solutions gives a white gelatinous precipitate which when heated is dehydrated through various intermediates and at 600° gives SnO_2.[28]

Oxo Anions. The germanates, stannates and plumbates have been less well studied than silicates. Both metagermanates, $M_2^IGeO_3$, and orthogermanates (for example, Mg_2GeO_4) have been obtained in crystalline form and have been shown to have structures analogous to the corresponding meta- and ortho-silicates. Thus $SrGeO_3$ contains a cyclic $Ge_3O_9^{6-}$ ion. Germanates containing the $Ge(OH)_6^{2-}$ ion are also known. In dilute aqueous solutions the major germanate ions appear to be $[GeO(OH)_3]^-$, $[GeO_2(OH)_2]^{2-}$ and $\{[Ge(OH)_4]_8(OH)_3\}^{3-}$. Fusion of SnO_2 or PbO_2 with K_2O gives K_2MO_3, which has chains of edge-shared MO_5 square pyramids.[29] Crystalline alkali-metal stannates and plumbates can be obtained as trihydrates, for instance, $K_2SnO_3 \cdot 3H_2O$. Such materials contain the octahedral anions $Sn(OH)_6^{2-}$ and $Pb(OH)_6^{2-}$.

11-8. Complexes of Group IV Elements

Most of the complexes of these elements in the IV oxidation state contain halide ions or donor ligands that are O, N, S or P compounds.

Anionic Species. Silicon forms only fluoro anions, normally, SiF_6^{2-}, whose high stability constant accounts for the incomplete hydrolysis of SiF_4 in water:

$$2SiF_4 + 2H_2O \rightarrow SiO_2 + SiF_6^{2-} + 2H^+ + 2HF$$

The ion is usually made by attack of HF on hydrous silica and is stable even in basic solution. While the salts that crystallize are normally those of the SiF_6^{2-} ion, the pentafluorosilicate ion can be stabilized under selected conditions and with cations of the correct size.

This ion was first noted in a salt obtained in a remarkable reaction where

---

[28] E. W. Giesekke, H. S. Gutowsky, P. Kirkov and H. A. Laitenen, *Inorg. Chem.*, 1967, 6, 1294.

[29] B. M. Gatehouse and D. J. Lloyd, *Chem. Comm.*, **1969**, 727.

tetrafluoroethylene, C_2F_4, and $PtHCl(PEt_3)_2$ (Section 26-H-2) were heated in a silica vessel in the expectation of forming $Pt(C_2F_4H)Cl(PEt_3)_2$. The actual product[30] was $[(Et_3P)_2PtCl(CO)]^+SiF_5^-$. The SiF_5^- salts are more conveniently obtained by direct reactions, e.g.:[31,32]

$$SiO_2 + HF(aq) + R_4N^+Cl \xrightarrow{CH_3OH} [R_4N]SiF_5$$
$$SiF_4 + [R_4N]F \xrightarrow{CH_3OH} [R_4N]SiF_5$$

Nmr data[32] for the ion and also for similar species $RSiF_4^-$ and $R_2SiF_3^-$ indicate *tbp* structures, but above $-60°$ exchange processes are occurring.

Germanium, tin and lead also form hexafluoro anions; for example, dissolution of GeO_2 in aqueous HF followed by the addition of KF at $0°$ gives crystals of K_2GeF_6. The Ge and Sn anions are hydrolyzed by bases, but the Pb salts are hydrolyzed even by water. A great variety of tin species, $SnF_{6-n}X_n^{2-}$ have been studied by nmr spectroscopy, and the equilibrium constants have been estimated,[33] e.g.,

$$SnF_6^{2-} + H_2O \rightleftharpoons SnF_5(OH_2)^- + F^- \qquad K = 2.3 \times 10^{-6}$$
$$SnF_6^{2-} + OH^- \rightleftharpoons SnF_5(OH)^{2-} + F^- \qquad K = 7.7 \times 10^6$$

Anhydrous hexafluorostannates can be made by dry fluorination of the stannates, $M_2^ISnO_3 \cdot 3H_2O$.[34]

The hexachloro ions of Ge and Sn are normally made by the action of HCl or M^ICl on MCl_4. The thermally unstable yellow salts of $PbCl_6^{2-}$ are obtained by action of HCl and Cl_2 on $PbCl_2$.[35] Under certain conditions, pentachloro complexes of Ge and Sn may be stabilized, e.g., by the use of $(C_6H_5)_3C^+$ as the cation or by the interaction of MCl_4 and $(C_4H_9)_4N^+Cl^-$ in $SOCl_2$ solution.[36]

Other anionic species include the ions $[Sn(NO_3)_6]^{2-}$ and also thiostannate, written SnS_3^{2-} but of uncertain structure, obtained by dissolving SnS_2 in alkali or ammonium sulfide solution. The most extensive series are the oxalates $[Mox_3]^{2-}$ (where $M = Si$, Ge or Sn) and other carboxylates.[37]

Cationic Species. There are comparatively few cationic complexes, the most important being the octahedral β-diketonates and tropolonates (T) of Si and Ge such as $[Ge \; acac_3]^+$ and SiT_3^+. So-called "siliconium" ions[38] can also be formed by reactions such as

$$Ph_3SiX + bipy \xrightarrow{CH_2Cl_2} \left[\begin{array}{c} Ph \\ \diagdown \\ Ph \diagup \\ Ph \end{array} \begin{array}{c} N \\ | \\ Si-N \\ \end{array} \right]^+ + X^-$$

[30] H. C. Clark, P. W. R. Corfield, K. R. Dixon and J. A. Ibers, *J. Amer. Chem. Soc.*, 1967, **89**, 3360.
[31] H. C. Clark, K. R. Dixon and J. G. Nicolson, *Inorg. Chem.*, 1969, **8**, 450; J. J. Harris and B. Rudnor, *J. Amer. Chem. Soc.*, 1968, **7**, 515.
[32] F. Klanberg and E. L. Muetterties, *Inorg. Chem.*, 1968, **7**, 155.
[33] P. A. W. Dean and D. F. Evans, *J. Chem. Soc.*, A, **1968**, 1154.
[34] P. J. Moehs and H. M. Haendler, *Inorg. Chem.*, 1968, **7**, 2115.
[35] R. D. Whealy and D. R. Lee, *Inorg. Chim. Acta*, 1967, **1**, 397.
[36] I. R. Beattie *et al.*, *J. Chem. Soc.*, A, **1967**, 712; K. M. Harmon *et al.*, *Inorg. Chem.*, 1969, **8**, 1054; J. A. Creighton and J. H. S. Green, *J. Chem. Soc.*, A, **1968**, 808.
[37] P. A. W. Dean, D. F. Evans and R. F. Phillips, *J. Chem. Soc.*, A, **1969**, 363.
[38] See, e.g., T. Tanaka *et al.*, *Inorg. Chim. Acta*, 1969, **3**, 187.

Neutral Species. There is a variety of these, of which the tin complexes[39] $SnCl_2(\beta\text{-diket})_2$ are good examples. Tin dithiocarbamates such as $SnCl_2(S_2CNEt_2)_2$ and $Sn(S_2CNEt_2)_4$ are also known;[40] in the latter the tin is presumably 8-coordinate.

Adducts. The tetrahalides all show Lewis acid behavior, and $SnCl_4$ is a good Friedel–Crafts catalyst. The acid strengths for SnX_4 are known to be $SnCl_4 \gg SnBr_4 > SnI_4$.

The adducts may be 1:1 or 1:2, but unless an X-ray study has been made or other evidence is unequivocal it is not always certain whether the adducts are neutral and 5- or 6-coordinate, i.e., $MX_4 \cdot L_2$, or whether they are salts, e.g., $[MX_2L_2]X_2$. Among the best defined adducts are the pyridine adducts MCl_4py_2 of Si, Ge and Sn, which are molecular and have *trans*-pyridine groups.[41]

11-9. Other Compounds

Alkoxides, Carboxylates and Oxo Salts. All four elements of this Group form alkoxides, but those of silicon, e.g., $Si(OC_2H_5)_4$, are the most important; the surface of glass or silica can also be alkoxylated. Alkoxides are normally obtained by the standard method:

$$MCl_4 + 4ROH + 4 \text{ amine} \rightarrow M(OR)_4 + 4 \text{ amine} \cdot HCl$$

Silicon alkoxides are hydrolyzed by water, eventually to hydrous silica, but polymeric hydroxo alkoxo intermediates occur.[42] Of the carboxylates, lead tetraacetate is the most important as it is used in organic chemistry as a strong but selective oxidizing agent. It is made by dissolving Pb_3O_4 in hot glacial acetic acid or by electrolytic oxidation of Pb^{II} in carboxylic acids. In oxidations the attacking species is generally considered to be $Pb(OOCMe)_3^+$, which is isoelectronic with the similar oxidant, $Tl(OOCMe)_3$, but this is not always so and some oxidations are known to be free radical in nature.[43] The trifluoroacetate is a white solid, which will oxidize even heptane to give $ROOCCF_3$ species whence the alcohol ROH is obtained by hydrolysis; benzene similarly gives phenol.[44]

Oxo Salts. These are few. Tin(IV) sulfate, $Sn(SO_4)_2 \cdot 2H_2O$, can be crystallized from solutions obtained by oxidation of Sn^{II} sulfate; it is extensively hydrolyzed in water.

Tin(IV) nitrate is obtained as a colorless volatile solid by interaction of N_2O_5 and $SnCl_4$; it contains bidentate NO_3^- groups giving dodecahedral coordination.[45] The compound reacts with organic matter.

[39] J. W. Faller and A. Davison, *Inorg. Chem.*, 1967, **6**, 182.
[40] F. Bonati, G. Minghetti and S. Cenini, *Inorg. Chim. Acta*, 1968, **2**, 375.
[41] I. R. Beattie *et al.*, *J. Chem. Soc.*, A, **1969**, 482; **1968**, 2772.
[42] M. F. Bechtold, R. D. Vest and L. Plambeck, Jr., *J. Amer. Chem. Soc.*, 1968, **90**, 4590.
[43] See, e.g., R. J. Ouellette *et al.*, *J. Amer. Chem. Soc.*, 1969, **91**, 97.
[44] R. E. Partch, *J. Amer. Chem. Soc.*, 1967, **89**, 3662.
[45] C. D. Garner, D. Sutton and S. C. Wallwork, *J. Chem. Soc.*, A, **1967**, 1949.

Sulfides. Lead disulfide is not known, but for the other elements direct interaction of the elements gives MS_2. The silicon and germanium compounds are colorless crystals hydrolyzed by water. The structures of SiS_2 and GeS_2 are chains of tetrahedral MS_4 linked by the sulfur atoms (Section 15-9), SnS_2 has a CaI_2 lattice, each Sn atom having six sulfur neighbors.

Silicon—Nitrogen Compounds.[46] There is a very extensive chemistry of compounds with nitrogen bound to silicon. The amide, $Si(NH_2)_4$ is made by action of NH_3 on $SiCl_4$; on being heated, it gives an imide and finally the nitride Si_3N_4.

Most of the chemistry of Si, Ge and Sn with nitrogen is in the area of organo compounds.

11-10. The Divalent State

Silicon.[47,48] Divalent Si species are thermodynamically unstable under normal conditions. However, SiX_2 species have been identified in high-temperature reactions and have been trapped by rapid chilling to liquid-nitrogen temperature. The best studied compound is SiF_2, but SiO, SiS, SiH_2, $SiCl_2$ and some other species are known.

At ca. 1100° and low pressures SiF_4 and Si react to give SiF_2 in ca. 99.5% yield:

$$SiF_4 + Si \rightleftarrows 2SiF_2$$

The compound is stable for a few minutes at 10^{-4} cm pressure. It is diamagnetic and the molecule is angular with a bond angle of 101° both in the vapor and in the condensed phase. The reddish-brown solid gives an esr spectrum and presumably contains also $\cdot SiF_2(SiF_2)_nSiF_2\cdot$ radicals. When warmed it becomes white, cracking to give fluorosilanes up to $Si_{16}F_{34}$.

In the gas phase, SiF_2 reacts with oxygen but is otherwise not very reactive, but allowing the *solid* to warm in presence of various compounds, e.g., CF_3I, H_2S, GeH_4, affords compounds such as SiF_2HSH.[49] The corresponding chloride exists for only milliseconds at 10^{-4} cm pressure since it readily reacts with an excess of $SiCl_4$. By reaction with other halides, e.g., BCl_3, mixed compounds such as Cl_3SiBCl_2 can be prepared.[50]

Germanium. The germanium dihalides are quite stable entities. GeF_2, a white crystalline solid (m.p. 111°) is formed by action of anhydrous HF on Ge in a bomb at 200° or by reaction of Ge and GeF_4 above 100°. It is a fluorine-bridged polymer, the Ge atom having a distorted trigonal-bipyramid

[46] U. Wannagut, *Adv. Inorg. Chem. Radiochem.*, 1964, **6**, 225.

[47] J. L. Margrave and P. W. Wilson, *Accounts Chem. Res.*, 1971, **4**, 145; W. H. Atwell and D. R. Weyenberg, *Angew. Chem. Internat. Edn.*, 1969, **8**, 469 (extensive review of so-called "silylenes" as analogs of carbenes).

[48] J. L. Margrave *et al.*, *J. Amer. Chem. Soc.*, 1970, **92**, 1530; *Inorg. Chim. Acta*, 1969, **3**, 601; *Inorg. Nuclear Chem. Lett.*, 1971, **7**, 103.

[49] K. G. Sharp, J. L. Margrave and P. W. Wilson, *Inorg. Chem.*, 1969, **8**, 2655; *J. Inorg. Nuclear Chem.*, 1970, **32**, 1817.

[50] P. L. Timms, *Inorg. Chem.*, 1968, **7**, 387; R. W. Kirk and P. L. Timms, *J. Amer. Chem. Soc.*, 1969, **91**, 6315.

arrangement of four atoms and an equatorial lone pair.[51] The compound reacts exothermically with solutions of alkali-metal fluorides to give the hydrolytically stable ion GeF_3^-; in fluoride solutions the ion is oxidized by air, and in strong acid solutions by H^+, to give GeF_6^{2-}. GeF_2 vapors[52] contain oligomers $(GeF_2)_n$ where $n = 1$–3.

The other germanium dihalides are less stable and less well studied, but salts of the ion $GeCl_3^-$ are well known.[53] The sulfide, GeS, and a yellow amphoteric hydroxide, "$Ge(OH)_2$," are known; the latter is precipitated by base from Ge^{IV} solutions in H_2SO_4 that have been reduced by zinc.

Tin.[54]   The most important compounds are SnF_2 and $SnCl_2$, which are obtained by heating Sn with gaseous HF or HCl. The fluoride, which is sparingly soluble in water, is commonly used as the source of F^- ion in protective toothpastes.[55] Water hydrolyzes $SnCl_2$ to a basic chloride, but from dilute acid solutions $SnCl_2 \cdot 2H_2O$ can be crystallized. Both halides dissolve in solutions containing an excess of halide ion, thus:

$$SnF_2 + F^- = SnF_3^- \qquad pK \approx 1$$
$$SnCl_2 + Cl^- = SnCl_3^- \qquad pK \approx 2$$

In aqueous fluoride solutions SnF_3^- is the major species, but the ions SnF^+ and $Sn_2F_5^-$ can be detected. However, salts of the $Sn_2F_5^-$ ion can be isolated from melts or from solution.[56] Adducts of SnF_2 with fluoride acceptors such as SbF_5 appear to be ionic, with the cation, SnF^+, joined to the anion, SbF_6^-, by a fluorine bridge.[56a]

The halides readily dissolve in donor solvents such as acetone, pyridine, or DMSO, and pyramidal adducts, SnX_2L, are formed.[57] The lone pair in $SnCl_2L$ or $SnCl_3^-$ can be utilized and numerous transition-metal complexes with tin(II) chloride as ligand are known;[58] Mössbauer studies suggest that such species are best regarded as Sn^{IV} complexes, however.[59] The oxidation on complex formation is more clearly illustrated by the "carbene-like" reactions of $SnCl_2$ with metal—metal bonds, where an "insertion reaction" (Section 24-A-3) occurs, e.g.,

$$[h^5\text{-}C_5H_5(CO)_2Fe]_2 + Sn^{II}Cl_2 \rightarrow h^5\text{-}C_5H_5(CO)_2Fe\text{—}Sn^{IV}Cl_2\text{—}Fe(CO)_2h^5\text{-}C_5H_5$$

The very air-sensitive stannous ion, Sn^{2+}, occurs in acid perchlorate solutions, which may be obtained by the reaction

$$Cu(ClO_4)_2 + Sn/Hg = Cu + Sn^{2+} + 2ClO_4^-$$

[51] J. Trotter, M. Akhtar and N. Bartlett, *J. Chem. Soc., A*, **1966**, 30.
[52] J. L. Margrave, *Inorg. Chem.*, 1968, **7**, 608.
[53] P. S. Poskozim and A. L. Stone, *J. Inorg. Nuclear Chem.*, 1970, **32**, 1391.
[54] J. D. Donaldson, *Progr. Inorg. Chem.*, 1967, **8**, 287 (an extensive review).
[55] W. E. Cooley, *J. Chem. Educ.*, 1970, **47**, 177.
[56] J. D. Donaldson *et al.*, *J. Chem. Soc. A*, **1967**, 1821; **1969**, 2696.
[56a] T. Birchall, P. A. W. Dean and R. J. Gillespie, *J. Chem. Soc., A*, **1971**, 1777.
[57] J. D. Donaldson, D. G. Nicholson and B. J. Senior, *J. Chem. Soc., A*, **1968**, 2928;
    R. J. H. Clark, L. Maresca and P. J. Smith, *J. Chem. Soc., A*, **1970**, 2687.
[58] J. F. Young, *Adv. Inorg. Chem. Radiochem.*, 1968, **11**, 92.
[59] D. E. Fenton and J. J. Zuckerman, *Inorg. Chem.*, 1969, **8**, 1771.

Hydrolysis gives $[Sn_3(OH)_4]^{2+}$ with $SnOH^+$ and $[Sn_2(OH)_2]^{2+}$ as minor contributors:

$$3Sn^{2+} + 4H_2O \rightleftharpoons [Sn_3(OH)_4]^{2+} + 4H^+ \qquad \log K = -6.77$$

The trimeric, probably cyclic, ion appears to provide the nucleus of several basic tin(II) salts obtained from aqueous solutions at fairly low pH. Thus the nitrate appears to be $Sn_3(OH)_4(NO_3)_2$ and the sulfate, $Sn_3(OH)_2OSO_4$.[60] All Sn^{II} solutions are readily oxidized by oxygen and, unless stringently protected from air, normally contain some Sn^{IV}. The chloride solutions are often used as mild reducing agents:

$$SnCl_6^{2-} + 2e = SnCl_3^- + 3Cl^- \qquad E^0 = \text{ca. } 0.0 \text{ V } (1\,M\,HCl, 4\,M\,Cl^-)$$

The addition of aqueous ammonia to Sn^{II} solutions gives the white hydrous oxide which, when heated in suspension at 60–70° in $2\,M\,NH_4OH$ is dehydrated to black SnO, or at 90–100° in presence of hypophosphite to a red modification.[61] The hydroxide is amphoteric and dissolves in alkali hydroxide to give solutions of stannites, which may contain the ion $[Sn(OH)_6]^{4-}$. These solutions are quite strong reducing agents; on storage they deposit SnO and at 70–100° disproportionate slowly to β-tin and Sn^{IV}.

Various other Sn^{II} compounds are known, including carboxylates and complex carboxylate anions,[62] phosphites,[63] thiocyanates[64] and $Sn(ClO_4)_2 \cdot 3H_2O$.[65]

Lead.   Lead has a well-defined cationic chemistry. There are several crystalline salts but with the exception of $Pb(NO_3)_2$ and $Pb(OCOCH_3)_2 \cdot 2H_2O$ (which ionizes incompletely in water) most lead salts are sparingly soluble (PbF_2, $PbCl_2$) or insoluble ($PbSO_4$, $PbCrO_4$, etc.) in water. The halides, unlike the tin(II) halides, are always anhydrous and have complex crystal structures with distorted close-packed halogen lattices. In water, species such as PbX^+ are formed and, on addition of an excess of halogen acid, $PbX_n^{(n-2)-}$; with the fluoride, only PbF^+ occurs, but with Cl^- several complex ions are formed.

The plumbous ion is partially hydrolyzed in water. In perchlorate solutions the first equilibrium appears to be

$$Pb^{2+} + H_2O = PbOH^+ + H^+ \qquad \log K \approx -7.9$$

but at higher degrees of hydrolysis and at high concentrations polymerized species with 3, 4 or 6 Pb atoms are formed. Crystalline salts containing polymeric anions can be obtained by dissolving PbO in perchloric acid and adding an appropriate quantity of base. X-ray investigation of the structure of the Pb_6 species,[66] $[Pb_6O(OH)_6]^{4+}(ClO_4^-)_4 \cdot H_2O$, reveals three tetrahedra of Pb

[60] C. G. Davies and J. D. Donaldson, *J. Chem. Soc.*, A, **1967**, 1790; **1968**, 946.
[61] W. Kwestro and P. H. G. M. Vroman, *J. Inorg. Nuclear Chem.*, 1967, **29**, 2187.
[62] J. D. Donaldson and A. Jelen, *J. Chem. Soc.*, A, **1968**, 1448, 2244.
[63] C. G. Davies, J. D. Donaldson and W. B. Simpson, *J. Chem. Soc.*, A, **1969**, 417.
[64] B. R. Chamberlaine and W. Moser, *J. Chem. Soc.*, A, **1969**, 354.
[65] C. G. Davies and J. D. Donaldson, *J. Inorg. Nuclear Chem.*, 1968, **30**, 2635.
[66] T. G. Spiro, D. H. Templeton and A. Zalkin, *Inorg. Chem.*, 1969, **8**, 856; T. G. Spiro *et al.*, *Inorg. Chem.*, 1969, **8**, 2524.

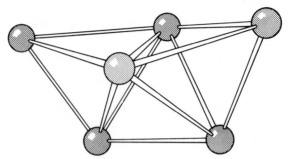

Fig. 11-7.   The three face-sharing tetrahedra of Pb atoms in the $Pb_6O(OH)_6^{4+}$ cluster.

atoms which share faces. The middle tetrahedron has an O atom near the center (Fig. 11-7) while the OH groups lie on the faces of the two end tetrahedra. Addition of more base gives the hydrous oxide, which dissolves in an excess of base to give the plumbate ion. If aqueous ammonia is added to a lead acetate solution, a white precipitate of a basic acetate is formed which on suspension in warm aqueous ammonia and subsequent drying gives very pure lead oxide, the most stable form of which is the tetragonal red form.[67,68]

The oxide Pb_3O_4 (red lead), which is made by heating PbO or PbO_2 in air behaves chemically as a mixture of PbO and PbO_2, but the crystal contains $Pb^{IV}O_6$ octahedra linked in chains by sharing of opposite edges, the chains being linked by Pb^{II} atoms each bound to three oxygen atoms.[68]

Lead(II) also forms numerous complexes which are mostly octahedral, although a phosphorodithioate, $Pb[S_2P(iso-C_3H_7O)_2]_2$, is polymeric with six Pb—S bonds and a stereochemically active lone pair.[69]

11-11. Organo Compounds of Group IV Elements[70]

There is an exceedingly extensive chemistry of the Group IV elements bound to carbon, and some of the compounds, notably silicon—oxygen polymers and alkyl-tin and -lead compounds, are of commercial importance; germanium compounds appear to be of little utility.

Essentially all the compounds are of the M^{IV} type. In the divalent state the only well established compounds are the cyclopentadienyls $(h^5-C_5H_5)_2Sn$ and $(h^5-C_5H_5)_2Pb$; alkyls or aryls of formula R_2Sn are either transitory or non-existent, and the stable substances of this stoichiometry are linear or cyclic

[67] W. Kwestro, J. de Jonge and P. H. G. M. Vroman, *J. Inorg. Nuclear Chem.*, 1967, **29**, 39; M. Neuberger: "Lead oxide" Data sheet D5-155 May 1967, Electronic Properties Information Center, Hughes Aircraft Co., Culver City, California.

[68] B. Dickens, *J. Inorg. Nuclear Chem.*, 1965, **27**, 1495.

[69] S. L. Lawton and G. T. Kokotailo, *Nature*, 1969, **221**, 550.

[70] G. E. Coates, M. L. H. Green and K. Wade, *Organometallic Compounds*, Vol. I, Methuen, 1967 (an excellent account of Ge and Pb chemistry); A. G. MacDiarmid, ed., *Organometallic Compounds of Group IV Elements*, Dekker, Vol. 1, Part I (nature of bond to C; Si—C); Part II (Ge, Sn, Pb to C bonds) (additional volumes are in preparation); R. Weiss, *Organometallic Compounds*, Vol. II, 2nd edn., Ge, Sn, Pb, Springer, 1967; see also Further Reading, page 337.

polymers of tetravalent tin. The cyclopentadienyls are made by action of C_5H_5Na on the chlorides; they are air-sensitive and have a structure in the gas phase similar to that of ferrocene (Section 23-2) but with the metal-to-ring axes at an angle consistent with the presence of a lone pair. The donor properties are shown by the formation of $(C_5H_5)_2SnBF_3$.[71]

For all four elements the compounds can generally be designated $R_{4-n}MX_n$, where R is alkyl or aryl and X can vary widely, being H, Cl, O, COR', OR', NR'_2, SR', $Mn(CO)_5$, $W(CO)_3(h^5\text{-}C_5H_5)$, etc. The elements can also be incorporated into heterocyclic rings of various types.

For a given class of compound those with C—Si and C—Ge bonds have higher thermal stability and lower reactivity than those with bonds to Sn and Pb. In catenated compounds similarly, Si—Si and Ge—Ge bonds are more stable and less reactive than Sn—Sn and Pb—Pb bonds; for example Si_2Me_6 is very stable, but Pb_2Me_6 blackens in air and decomposes rapidly in CCl_4 although it is fairly stable in benzene.[72]

The bonds to carbon are usually made via interaction of Li, Hg, or Al alkyls or RMgX and the Group IV halide, but there are many special synthetic methods, some of which will be noted below.

Silicon.[73] The organo compounds of Si and Ge are very similar in their properties, although Ge compounds[74] have been less extensively studied. We discuss only Si compounds.

Silicon—carbon bond dissociation energies are less than those of C—C bonds but are still quite high, in the region 250–335 kJ mol^{-1}. The tetra-alkyls and -aryls are hence thermally quite stable; $Si(C_6H_5)_4$, for example, boils unchanged at 530°.

The chemical reactivity of Si—C bonds is generally greater than that of C—C bonds because (*a*) the greater polarity of the bond, $Si^{\delta+}$—$C^{\delta-}$, allows easier nucleophilic attack on Si and electrophilic attack on C than for C—C compounds, and (*b*) displacement reactions at silicon are facilitated by its ability to form 5-coordinate transition states by utilization of *d* orbitals.

The mechanism of reactions of silicon compounds has been extensively studied and is complicated.[75] Detailed discussion is beyond the scope of this book; a few illustrations will suffice. The displacement reactions of optically active methyl-1-naphthylphenylsilane usually proceed with inversion of configuration. For R_3SiX in general, if the incoming nucleophile Y is more basic than X and X is a leaving group for which the conjugate acid has a pK $< \sim 6$, then the displacement is by an associative pathway with inversion

[71] P. G. Harrison and J. J. Zuckerman, *J. Amer. Chem. Soc.*, 1970, **92**, 2577.

[72] R. J. H. Clark *et al.*, *J. Amer. Chem. Soc.*, 1969, **91**, 1334.

[73] C. Eaborn, *Organosilicon Compounds*, Butterworth, 1960 (an excellent text); V. Bazant, V. Chvalovsky and J. Rathowsky, *Organosilicon Compounds*, Academic Press, 1965 (comprehensive reference volumes).

[74] M. Lesbre, P. Mazerolles and J. Satgé, *The Organic Compounds of Germanium*, Wiley, 1971.

[75] See L. H. Sommer, *Stereochemistry, Mechanism and Silicon*, McGraw Hill, 1965, and subsequent papers mainly in *J. Amer. Chem. Soc.*, e.g., 1967, **89**, 868; 1969, **91**, 4729, 7040, 7045, 7067.

regardless of the nature of the solvent. Such a path may lead through a 5-coordinate transition state with long axial bonds to X and Y, which would then allow inversion:

$$
\begin{array}{c}
R^1 \\
R^2 \!-\!\!\!\!\text{Si}\!-\! X \\
R^3
\end{array}
\xrightarrow{+Y}
\left[
\begin{array}{c}
X \\
R^1 \\
R^2 \diagup\!\!\!\diagdown R^3 \\
Y
\end{array}
\right]
\xrightarrow{-X}
\begin{array}{c}
R^3 \\
R^2 \!-\!\!\!\!\text{Si}\!-\! Y \\
R^1
\end{array}
$$

Free radicals are less important in silicon than in carbon chemistry,[76] and it is only recently that esr detection of a silicon radical in solution has been made (although radicals have been isolated in matrixes) by the hydrogen abstraction reaction with t-butoxy and other radicals generated photochemically, e.g.,

$$R_3SiH + Me_3CO\cdot \rightarrow R_3Si\cdot + Me_3COH$$

A comparison of the rates of reactions such as

$$p\text{-}XC_6H_4\text{-}MR_3 + H_2O \rightarrow R_3MOH + C_6H_5X$$

in aqueous-methanolic $HClO_4$ gives the order $Si(1) < Ge(36) \ll Sn(3 \times 10^5) \ll \ll Pb(2 \times 10^8)$, which suggests that, with increasing size, there is increased availability of outer orbitals, which allows more rapid initial solvent coordination to give the 5-coordinate transition state.

Alkyl- and Aryl-silicon Halides.[77] These compounds are of special importance because of their hydrolytic reactions. They may be obtained by normal Grignard procedures from $SiCl_4$, or, in the case of the methyl derivatives, by the Rochow process in which methyl chloride is passed over a heated, copper-activated silicon:

$$CH_3Cl + Si(Cu) \rightarrow (CH_3)_nSiCl_{4-n}$$

The halides are liquids that are readily hydrolyzed by water, usually in an inert solvent. In certain cases, the silanol intermediates R_3SiOH, $R_2Si(OH)_2$, and $RSi(OH)_3$ can be isolated, but the diols and triols usually condense under the hydrolysis conditions to siloxanes which have Si—O—Si bonds. The exact nature of the products depends on the hydrolysis conditions and linear, cyclic, and complex cross-linked polymers of varying molecular weights can be obtained. They are often referred to as silicones; the commercial polymers usually have $R = CH_3$, but other groups may be incorporated for special purposes.[78]

Controlled hydrolysis of the alkyl halides in suitable ratios can give products of particular physical characteristics. The polymers may be liquids, rubbers, or solids, which have in general high thermal stability, high dielectric strength and resistance to oxidation and chemical attack.

Examples of simple siloxanes are $Ph_3SiOSiPh_3$ and the cyclic trimer or

[76] P. J. Krusic and J. K. Kochi, *J. Amer. Chem. Soc.*, 1969, **91**, 6161; I. M. T. Davidson, *Quart. Rev.*, 1971, **25**, 111.
[77] R. J. H. Voorhoeve, *Organohalosilanes: Precursors to Silicones*, Elsevier, 1967.
[78] W. Noll *et al.*, *Chemistry and Technology of Silicones*, Academic Press, 1968.

tetramer $(Et_2SiO)_{3 (or 4)}$; linear polymers contain $—SiR_2—O—SiR_2—O—$ chains, whereas the crosslinked sheets have the basic unit

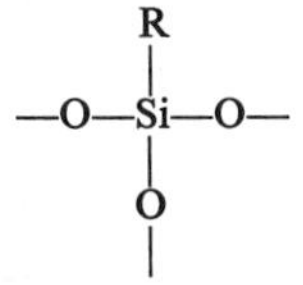

$$
\begin{array}{c}
R \\
| \\
—O—Si—O— \\
| \\
O \\
|
\end{array}
$$

Tin.[11,79]   Where the compounds of tin differ from those of Si and Ge they do so mainly because of a greater tendency of Sn^{IV} to show coordination numbers higher than four and because of ionization to give cationic species.

Trialkyltin compounds of the type R_3SnX of which the best studied are the CH_3 compounds with $X = ClO_4$, F, NO_3, etc., are of interest in that they are always associated in the solid by anion bridging (11-III and IV); the coordination of the tin atom is close to *tbp* with planar $Sn(Me)_3$ groups. When X is RCOO the compounds may in addition be monomeric with unidentate or bidentate carboxylate groups.[80]

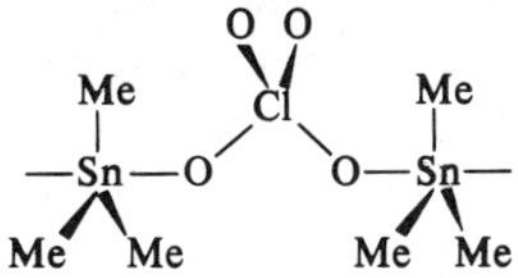

(11-III) 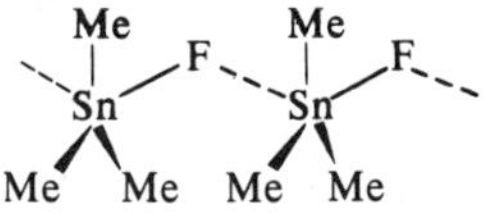   (11-IV)

The R_3SnX (and also R_3PbX) compounds also form 1:1 and 1:2 adducts with Lewis bases and these also generally appear to contain 5-coordinate Sn, with the alkyl groups in axial positions. In water the perchlorate and some other compounds ionize to give cationic species, e.g., $[Me_3Sn(H_2O)_2]^+$.

Dialkyltin compounds, R_2SnX_2, have behavior similar to that of the trialkyl compounds. Thus the fluoride Me_3SnF_2 is again polymeric, with bridging F atoms, but Sn is octahedral and the Me—Sn—Me group is linear. However, the chloride and bromide have low melting points (90° and 74°) and are essentially molecular compounds, but there is weak interaction between neighboring molecules *via* halogen bridges.[81]

The halides also give conducting solutions in water and the aquo ion has the linear C—Sn—C group characteristic of the dialkyl species (cf. the linear species Me_2Hg, Me_2Tl^+, Me_2Cd, Me_2Pb^{2+}), probably with four water molecules completing octahedral coordination. The linearity in these species appears to result from maximizing of *s* character in the bonding orbitals of

[79] R. C. Poller, *Organotin Chemistry*, Logos Press, 1970; W. P: Neumann, *The Organic Chemistry of Tin* (transl. R. Moser), Wiley, 1970; R. Okawara and M. Wada, *Adv. Organometallic Chem.*, 1967, **5**, 137 (structural aspects of organotin compounds); A. G. Davies, *Chem. in Britain*, **1969**, 403 (a short review).

[80] R. J. H. Clark, A. G. Davies and R. J. Puddephatt, *J. Amer. Chem. Soc.*, 1968, **90** 6923; N. W. Alcock and R. E. Timms, *J. Chem. Soc., A*, **1968**, 1873, 1876.

[81] A. G. Davies *et al.*, *J. Chem. Soc., A.*, **1970**, 2862; C. W. Hobbs and R. S. Tobias, *Inorg. Chem.*, 1970, **9**, 1037.

the metal atoms. The ions Me_2SnCl^+ and Me_2SnOH^+ also exist, and, in alkaline solution $trans$-$[Me_2Sn(OH)_4]^{2-}$.

Catenated linear and cyclic organotin compounds are relatively numerous and stable. For example, the reaction of Na in liquid ammonia with $Sn(CH_3)_2Cl_2$ gives "$[Sn(CH_3)_2]_n$," which consists mainly of linear molecules with chain lengths of 12–20 (and perhaps more), as well as at least one cyclic compound, $[Sn(CH_3)_2]_6$. There is no evidence for branching of chains. Similar results have been obtained with other alkyl and aryl groups; e.g., the cyclic hexamer and nonamer of $(C_2H_5)_2Sn$, the cyclic pentamer and hexamer of $(C_6H_5)_2Sn$ and the cyclic tetramer of $(t\text{-}C_4H_9)_2Sn$ have been isolated, as well as linear species. It has been reported that in some cases the terminal groups of the linear species are SnR_2H. The structure of $[Ph_2Sn]_6$ is known; it contains an Sn_6 ring in a chair configuration, with the Sn—Sn bonds of about the same length as those in grey tin.

Finally, organotin hydrides,[82] R_3SnH, which can be made by $LiAlH_4$ reduction of the halide, or in other ways, are useful reducing agents in organic chemistry; some of the reactions are known to proceed by free-radical pathways. The hydrides undergo additional reactions with alkenes or alkynes similar to the hydrosilation reaction, which provides a useful synthetic method for organotin compounds containing functional groups. In contrast to the addition of Si—H (page 319) the hydrostannation reaction is free-radical in nature.[82, 83]

Lead.[84]   There is an extensive organolead chemistry, but the most important compounds are $(CH_3)_4Pb$ and $(C_2H_5)_4Pb$ which are made in vast quantities for use as antiknock agents in gasoline. Although lead alkyls and aryls can be made by alkylation of Pb^{II} compounds, the two tetraalkyls are made otherwise.

The major commercial synthesis is by the interaction of a sodium–lead alloy with CH_3Cl or C_2H_5Cl in an autoclave at 80–100°, without solvent for C_2H_5Cl but in toluene at a higher temperature for CH_3Cl. The reaction is complicated and not fully understood and only a quarter of the lead appears in the desired product:

$$4NaPb + 4RCl \rightarrow R_4Pb + 3Pb + 4NaCl$$

The required re-cycling of the lead is disadvantageous and electrolytic procedures have been developed. One process involves electrolysis of $NaAlEt_4$ with a lead anode and mercury cathode; the sodium formed can be converted into NaH and the electrolyte regenerated:

$$4NaAlEt_4 + Pb \rightarrow 4Na + PbEt_4 + 4AlEt_3$$
$$4Na + 2H_2 \rightarrow 4NaH$$
$$4NaH + 4AlEt_3 + 4C_2H_4 \rightarrow 4NaAlEt_4$$

[82] H. G. Kuivala, *Accounts Chem. Res.*, 1968, **1**, 299; *Synthesis*, **1970**, 499.
[83] R. J. Strunk *et al.*, *J. Amer. Chem. Soc.*, 1970, **92**, 2849.
[84] H. Shapiro and F. W. Frey, *The organic Compounds of Lead*, Wiley, 1968; W. P. Neumann and K. Kühlein, *Adv. Organomet. Chem.*, 1968, **7**, 242 (compounds with Pb bonds to C, H, N and O).

Another process involves electrolysis of solutions of Grignard reagents in ethers at lead anodes; RMgCl is regenerated by adding RCl:

$$4RMgCl \rightleftarrows 4R^- + 4MgCl^+$$
$$4R^- + Pb \xrightarrow{-4e} 4PbR_4$$
$$4MgCl^+ \xrightarrow{+4e} 2Mg + 2MgCl_2$$

The alkyls are non-polar, highly toxic liquids;[85] the methyl member begins to decompose around 200° and the ethyl member around 110°, by free-radical mechanisms.

[85] D. Bryce-Smith, *Chem. in Britain*, **1970**, 54 (toxicology of lead).

Further Reading

Abrikosov, N. Kh., *et al.*, *Semiconducting II—VI, IV—VI and V—VI Compounds*, Plenum Press, 1969.

Abel, E. W., and D. A. Armitage, *Adv. Organometallic Chem.*, 1967, **5**, 2 (organosulfur compounds of Group IV).

Andrianov, K. A., and L. M. Khananashvili, *Organometallic Chem. Rev.*, 1967, **2**, 141 (cyclic organosilicon compounds).

Andrianov, K. A., and A. I. Petrashko, *Organometallic Chem. Rev.*, 1967, **2**, 38 (halogenated derivatives of alkylhalosilanes).

Aylett, B. J., *Organometallic Chem. Rev.*, 1968, **3**, 151 (Si—N polymers).

Bazant, V., J. Joklik and J. Rathousky, *Angew. Chem. Internat. Edn.*, 1968, **7**, 112 (direct synthesis of R_nSiX_{4-n}).

Borisov, S. N., M. G. Voronkov and E. Ya Lukevits, *Organosilicon Heteropolymers and Heterocompounds*, Heyden Press, 1969.

Brook, A. G., *Adv. Organometallic Chem.*, 1968, **7**, 96 (keto derivatives of Group IV).

Bürger, H., *Organometal. Chem. Rev.*, 1968, **3**, 425 (infrared spectra of Me_3Si compounds).

Chernyshev, E. A., and E. F. Bugerenko, *Organometallic Chem. Rev.*, 1968, **3**, 469 (organo Si—P compounds).

Creemers, H. M. J. C., *Hydrostannolysis*, Schotanus and Utrecht, Utrecht, 1967.

Davidov, V. I., *Germanium*, Gordon and Beach, 1966.

Davidsohn, W. E., and M. C. Henry, *Chem. Rev.*, 1967, **67**, 73 (alkynyl compounds).

Ebsworth, E. A. V., *Volatile Silicon Compounds*, Pergamon Press, 1962.

Fritz, G., J. Grobe and D. Kummer, *Adv. Inorg. Chem. Radiochem.*, 1965, **7**, 349 (carbosilanes).

Gielen, M., and N. Sprecher, *Organometallic Chem. Rev.*, 1966, **1**, 455 (adducts of organo compounds of Group IV).

Gilman, H., W. H. Atwell and F. K. Cartledge, *Adv. Organometal. Chem.*, 1966, **4**, 1 (catenated Group IV organo compounds).

Glocking, F., *The Chemistry of Germanium*, Academic Press, 1969.

Goryunova, N. A., *The Chemistry of Diamond-like Semi-conductors*, Methuen, 1965.

Haas, A., *Angew. Chem. Internat. Edn., Engl.*, 1965, **4**, 1014 (silicon—sulfur compounds).

Hofmann, U., *Angew. Chem. Internat. Edn., Engl.*, 1968, **7**, 681 (chemistry of clay).

Ingram, R. K., S. D. Rosenberg and H. Gilman, *Chem. Rev.*, 1960, **60**, 459 (organotin compounds).

Jones, K., and M. F. Lappert, *Organometallic Chem. Rev.*, 1966, **1**, 67 (organo Sn—N compounds).

Kumada, M., and K. Tamao, *Adv. Organometallic Chem.*, 1968, **6**, 19 (aliphatic organopolysilanes).

Lappert, M. F., and H. Pyszora, *Adv. Inorg. Chem. Radiochem.*, 1966, **9**, 133 (pseudohalides of Group IV).

Lindquist, I., *Inorganic Adducts of Molecules of Oxo Compounds*, Springer Verlag, 1963.

Luijten, J. G. A., F. Rijkens and G. J. M. Van der Kerk, *Adv. Organometallic Chem.*, 1965, **3**, 397 (organo compounds containing N).

Lukevitz, E. Ya., and M. G. Voronkov, *Organic Insertion Reactions of Group IV Elements*, Consultants Bureau, 1960.

Mackay, K. M., and R. Watt, *Organometallic Chem. Rev.*, 1969, **4**, 137 (chain compounds of Group IV elements).

Mironov, V. F., and T. K. Gar, *Organometallic Chem. Rev.*, 1968, **3**, 311 (trichlorogermane chemistry).

Moedritzer, K., *Organometallic Chem. Rev.*, 1966, **1**, 179 (redistribution reactions of organocompounds).

Müller, R., *Organometallic Chem. Rev.*, 1966, **1**, 359 (organo fluorosilicates).

Petrov, A. D., B. F. Mironov, V. A. Ponomarenko and E. A. Chernyshev, *Synthesis of Organosilicon Monomers*, Consultants Bureau, 1964.

Samsonov, G. V., and V. N. Bonderev, *Germanides*, Plenum Press, 1969.

Scherer, O. J., *Organometal. Chem. Rev.*, 1968, **3**, 281 (cleavage of Si, Ge and Sn nitrogen compounds; silylamines).

Scherer, O. J., *Angew. Chem. Internat. Edn.*, 1969, **8**, 861 (Group IV nitrogen compounds).

Stone, F. G. A., *Hydrogen Compounds of Group IV Elements*, Prentice Hall, 1962.

12

Nitrogen

GENERAL REMARKS

12-1. Introduction

The electronic configuration of the nitrogen atom in its ground state (^4S) is $1s^2 2s^2 2p^3$, with the three $2p$ electrons distributed among the p_x, p_y and p_z orbitals with spins parallel. Nitrogen forms an exceedingly large number of compounds, most of which are to be considered organic rather than inorganic. It is one of the most electronegative elements, only oxygen and fluorine exceeding it in this respect.

The nitrogen atom may complete its octet in several ways:

1. *Electron gain to form the nitride ion,* N^{3-}. This ion occurs only in the salt-like nitrides of the most electropositive elements, for example, Li_3N. Many non-ionic nitrides exist and are discussed later in this Chapter.

2. *Formation of electron-pair bonds.* The octet can be completed either by the formation of three single bonds, as in NH_3 or NF_3, or by multiple-bond formation, as in nitrogen itself, $:N{\equiv}N:$, azo compounds, $-\ddot{N}{=}\ddot{N}-$, nitro compounds, RNO_2, etc.

3. *Formation of electron-pair bonds with electron gain.* The completed octet is achieved in this way in ions such as the amide ion, NH_2^-, and the imide ion, NH^{2-}.

4. *Formation of electron-pair bonds with electron loss.* Nitrogen can form four bonds, provided an electron is lost, to give positively charged ions R_4N^+ such as NH_4^+, $N_2H_5^+$ and $(C_2H_5)_4N^+$. The ions may sometimes be regarded as being formed by protonation of the lone pair:

$$H_3N: + H^+ \rightarrow [NH_4]^+$$

or generally

$$R_3N: + RX \rightarrow R_4N^+ + X^-$$

Failure to Complete the Octet. There are a few relatively stable species in which, *formally*, the octet of nitrogen is incomplete. In actual fact this incompleteness is shared with the other atoms to which nitrogen is bound. The classic examples are NO and NO_2. There is also an extensive class of paramagnetic compounds called *nitroxides*. The simple dialkyl nitroxide[1] (12-I), is a red liquid, stable to oxygen, water and aqueous alkali at 25°. Many others are known, the chief interest being in their use as "spin labels."

$$
\begin{array}{c}
(CH_3)_3C \\
 \diagdown \\
\quad \ddot{N}\!=\!\ddot{O}\!: \\
 \diagup \\
(CH_3)_3C
\end{array}
$$
$$(12\text{-}I)$$

Nitroxide radicals can be attached to various points in proteins, membranes, etc., and inferences about their environment drawn from the characteristics (*g*-value, hyperfine structure) of the observed esr signals.[2]

There is also the classic case of Fremy's salt (potassium nitrosodisulfonate), prepared as follows:

$$2HSO_3^- + HNO_2 \rightarrow HON(SO_3)_2^{2-} \xrightarrow{-e} ON(SO_3)_2^{2-} \xrightarrow{+K^+} K_2[ON(SO_3)_2]$$

or better by electrolysis of alkaline potassium hydroxylamine-*N,N*-disulfonate.[2a] The violet ion $ON(SO_3)_2^{2-}$ in solution is paramagnetic, having one unpaired electron. The similarity to the nitroxides is evident. Two crystalline forms of Fremy's salt are known. The monoclinic form is diamagnetic and spectral evidence suggests dimerization of the anion through a peroxo bridge. The triclinic form is paramagnetic, having very loose coupling between pairs of anions *via* a relative orientation of N—O groups similar to that in solid NO (see below).[2b]

Formal Oxidation Numbers. Classically, formal oxidation numbers ranging from -3 (e.g., in NH_3) to $+5$ (e.g., in HNO_3) have been assigned to nitrogen. Though they are occasionally useful in balancing redox equations these numbers have no physical significance.

12-2. Types of Covalence in Nitrogen; Stereochemistry

In common with other first-row elements, nitrogen has only four orbitals available for bond formation, and a maximum of four $2c\text{-}2e$ bonds may be formed. However, since formation of three electron-pair bonds completes the octet, $:N(:R)_3$, and the nitrogen atom then possesses a lone pair of electrons, four $2c\text{-}2e$ bonds can only be formed either (*a*) by coordination, as in donor–acceptor complexes, e.g., $F_3\bar{B}\!-\!\overset{+}{N}(CH_3)_3$, or in amine oxides, e.g. $(CH_3)_3\overset{+}{N}\!-\!\bar{O}$, or (*b*) by loss of an electron, as in ammonium ions NH_4^+, NR_4^+. This loss of an electron gives a valence–state configuration for nitrogen (as N^+) with four unpaired electrons in sp^3 hybrid orbitals analogous to that of neutral carbon, while, as noted above, gain of an electron (as in

[1] A. K. Hoffmann and A. T. Henderson, *J. Amer. Chem. Soc.*, 1961, **83**, 4671.
[2] C. L. Hamilton and H. M. McConnell in *Structural Chemistry and Molecular Biology*, A. Rich and N. Davidson, eds., W. H. Freeman and Co., 1968, p. 115; O. H. Griffith and A. S. Waggoner, *Accounts Chem. Res.*, 1969, **2**, 17.
[2a] W. R. T. Cottrell and J. Farrer, *J. Chem. Soc.*, A, **1970**, 1418.
[2b] R. A. Howie, L. S. D. Glasser and W. Moser, *J. Chem. Soc.*, A, **1968**, 3043; D. L. Fillmore and B. J. Wilson, *Inorg. Chem.*, 1968, **7**, 1592.

NH_2^-) leaves only two electrons for bond formation. In this case, the nitrogen atom (as N^-) is isoelectronic with the neutral oxygen atom, and angular bonds are formed. We can thus compare, sterically, the following isoelectronic species:

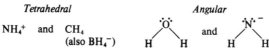

It may be noted that the ions NH_2^-, OH^- and F^- are isoelectronic and have comparable sizes. The amide, imide and nitride ions, which can be considered as members of the isoelectronic series NH_4^+, NH_3, NH_2^-, NH^{2-}, N^{3-} occur as discrete ions only in salts of highly electropositive elements.

In all nitrogen compounds where the atom forms two or three bonds, there remain, respectively, two or one pair of non-bonding electrons, also called lone pairs. As already discussed in Chapter 4, non-bonding electron pairs have a profound effect on stereochemistry. Furthermore, the lone pairs are responsible for the donor properties of the atom possessing them. To illustrate the important chemical consequences of non-bonding electron pairs in nitrogen chemistry, we shall consider one of the most important types of molecule, namely, NR_3, as exemplified by NH_3 and amines.

Three-covalent Nitrogen. Molecules of this type are invariably pyramidal, except in special cases such as the planar N-centered triangular iridium complexes,[3] e.g., $[Ir_3N(SO_4)_6(H_2O)_3]^{4-}$, and trisilylamine noted below, or when multiple bonding is involved. The bond angles vary according to the groups attached to the nitrogen atom.

It is to be noted that pyramidal molecules of the kind $NRR'R''$ should be chiral. No optical isomers have ever been isolated, however, because molecules of this type execute a motion known as inversion, in which the nitrogen atom oscillates through the plane of the three R groups much as an umbrella can turn inside out. As the nitrogen atom crosses from one side of the plane to the other (from one equilibrium position, say $+r_0$, to the other, $-r_0$, in Fig. 12-1a), the molecule goes through a state of higher potential energy, as shown in the potential energy curve (Fig. 12-1b). However, this "potential energy barrier" to inversion is only about 24 kJ mol^{-1}, and the

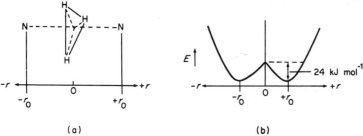

Fig. 12-1. Diagrams illustrating the inversion of NH_3 (see text).

[3] M. Ciechanowicz *et al.*, *Chem. Comm.*, **1971**, 876.

frequency of the oscillation is 2.387013×10^{10} cps (cycles per second) in NH_3. In simple alkylamines generally, barriers are in the range 16–40 kJ mol^{-1} and thus isolation of optical isomers can never be expected. However, hetero atom substitution and incorporation of N into strained rings raises the barrier and there are cases where invertomers have been separated.[4]

Multiple Bonding in Nitrogen and its Compounds. Like its neighbors carbon and oxygen, nitrogen readily forms multiple bonds, differing in this respect from its heavier congeners, P, As, Sb and Bi. Nitrogen thus forms many compounds for which there are no analogs among the heavier elements. Thus, whereas phosphorus, arsenic and antimony form tetrahedral molecules, P_4, As_4 and Sb_4, nitrogen forms the multiple-bonded diatomic molecule $:N\equiv N:$, with an extremely short internuclear distance (1.094 Å) and very high bond strength. The multiple bonding in nitrogen has already been discussed in both the VB and MO approximations in Chapter 3. Nitrogen also forms triple bonds to carbon (in $-C\equiv N:$) and to oxygen (in $:N\equiv O:$, the odd electron being in an antibonding orbital).

In compounds where nitrogen forms one single and one double bond, the grouping $X-\ddot{N}=Y$ is non-linear. This can be explained by assuming that nitrogen uses a set of sp^2 orbitals, two of which form σ bonds to X and Y, while the third houses the lone pair. A π bond to Y is then formed using the nitrogen p_z orbital. In certain cases, stereoisomers result from the non-linearity, for example, cis- and trans-azobenzenes (12-IIa and 12-IIb) and the oximes (12-IIIa and 12-IIIb). These are not readily interconverted, although more easily than are cis- and trans-olefins.

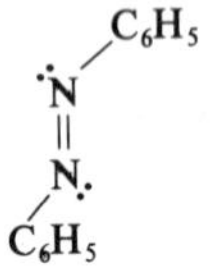

(12-IIa) 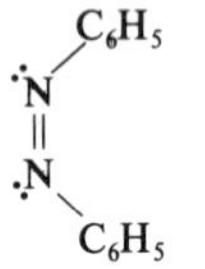   (12-IIb) 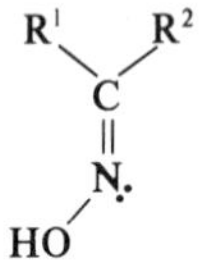   (12-IIIa) 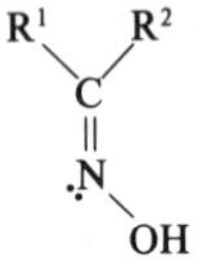   (12-IIIb)

Multiple bonding occurs also in oxo compounds. For example, NO_2^- (12-IV) and NO_3^- (12-V) can be regarded as resonance hybrids in the VB approach. From the MO viewpoint, one considers the existence of a π MO extending symmetrically over the entire ion and containing the two π electrons.

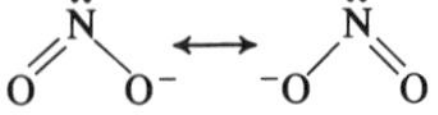

(12-IVa)   (12-IVb)

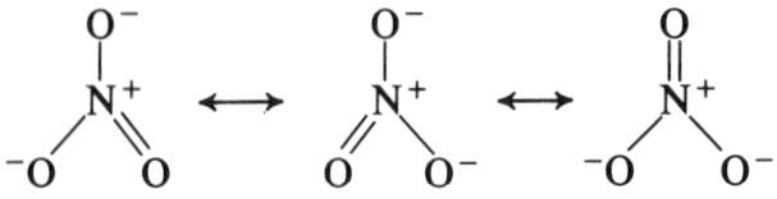

(12-Va)   (12-Vb)   (12-Vc)

[4] A. Rauk, L. C. Allen and K. Mislow, *Angew. Chem. Internat. Edn.*, 1970, **9**, 400; K. Mislow *et al.*, *Tetrahedron Lett.*, **1971**, 3437, 3441.

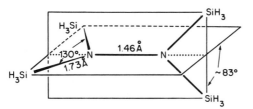

Fig. 12-2. The molecular structure of tetrasilylhydrazine as determined by electron diffraction. The dihedral angle could be 90°.

An unusual but significant case of multiple bonding occurs in trisilylamine, $N(SiH_3)_3$, and other compounds with Si—N bonds, as discussed on page 313, where $d\pi$-$p\pi$ bonds are involved. Another example is tetrasilylhydrazine which appears to have the structure shown in Fig. 12-2, in which the $NNSi_2$ groups are each planar.[5] This should be compared with the structure of hydrazine, N_2H_4 (page 352). In F_2PNH_2 the P, N and H atoms again are planar, providing support for the P—N multiple bonding.[5a]

Donor Properties of Three-covalent Nitrogen; Four-covalent Nitrogen. As noted above, the formation of approximately tetrahedral bonds to nitrogen occurs principally in ammonium cations, R_4N^+, amine oxides, R_3N^+—O^- and in Lewis acid–Lewis base adducts, e.g., $R_3\overset{+}{N}$—$\overset{-}{B}X_3$. In the amine oxides and these adducts the bonds must have considerable polarity; in the amine oxides, for instance, N→O donation cannot be effectively counterbalanced by any back-donation to N. In accord with this is the fact that the stability of amine oxides decreases as the R_3N basicity decreases, since the ability of N to denote to O is the major factor. Similarly, R_3N→BX_3 complexes have stabilities which roughly parallel the R_3N basicity for given BX_3. When R is fluorine, basicity is minimal and $F_3N \rightarrow BX_3$ compounds are unknown. It is therefore curious that F_3NO is an isolable compound (see page 366). Evidently the extreme electronegativity of fluorine coupled with the availability of $p\pi$ electrons on oxygen allows structures (12-VI) to contribute to stability.

$$F—N^+=O \quad \longleftrightarrow \quad F^-—N^+=O \quad \longleftrightarrow \quad F—N^+=O$$

(12-VI)

Catenation and N—N Single-bond Energies. Unlike carbon and a few other elements, nitrogen has little tendency for catenation, primarily owing to the weakness of the N—N single bond. If we compare the single-bond energies in H_3C—CH_3, H_2N—NH_2, H—O—O—H and F—F which are, respectively, about 350, 160, 140 and 150 kJ mol^{-1} it is clear that there is a profound drop between C and N. This difference is most probably attributable to the effects of repulsion between non-bonding lone-pair electrons. The strength

[5] C. Glidewell et al., J. Chem. Soc., A, **1970**, 318.
[5a] K. Cohn, personal communication.

of the N—N bond, and also of the O—O bond, decreases with increasing electronegativity of the attached groups; while increasing electronegativity would perhaps have been expected to reduce repulsion between lone pairs, it obviously will also weaken any homonuclear σ bond.

There are a few types of compound containing chains of three or more nitrogen atoms with some multiple bonds such as $R_2N—N=NR_2$, $R_2N—N=N—NR_2$, $RN=N—NR—NR_2$, $RN=N—NR—N=NR$, and even $RN=N—NR—N=N—NR—N=NR$, where R represents an organic radical (some R's may be H, but known compounds contain only a few H's). There are also cyclic compounds containing rings with up to five consecutive nitrogen atoms. Many of these compounds are not particularly stable, and all are traditionally in the realm of organic chemistry.

Hydrogen Bonding. Since it is one of the most electronegative elements, nitrogen, along with oxygen, fluorine and, to a lesser extent, chlorine, enter extensively into hydrogen-bond formation in its hydrogen compounds, both as a proton donor, N—H$\cdots$X, and as a proton receptor, $\geqq$N$\cdots$H—X.

THE ELEMENT

12-3. Occurrence and Properties

Nitrogen occurs in Nature mainly as the inert diatomic gas N_2 (m.p. 63.1°K, b.p. 77.3°K), which comprises 78% by volume of the earth's atmosphere. Naturally occurring nitrogen consists of ^{14}N and ^{15}N with an absolute ratio $^{14}N/^{15}N = 272.0$. ^{15}N is often useful as an isotopic tracer, and it has been found possible to prepare nitric acid containing up to 99.8% ^{15}N by efficient fractionation of the system.[6]

$$^{15}NO(g) + H^{14}NO_3(aq) = {}^{14}NO(g) + H^{15}NO_3(aq) \qquad K = 1.055$$

The $H^{15}NO_3$ produced can be used to prepare any desired ^{15}N-labeled nitrogen compound.

The heat of dissociation of nitrogen is extremely large:

$$N_2(g) = 2N(g) \qquad \Delta H = 944.7 \text{ kJ mol}^{-1} \qquad K_{25°} = 10^{-120}$$

Because the reaction is endothermic, the equilibrium constant increases with increasing temperature, but still, even at 3000° and ordinary pressures, there is no appreciable dissociation. The great strength of the N≡N bond is principally responsible for the chemical inertness of N_2 and for the fact that most simple nitrogen compounds are endothermic even though they may contain strong bonds. Thus $E(N≡N) \approx 6E(N—N)$ whereas $E(C≡C) \approx 2.5E(C—C)$.

Nitrogen is notably unreactive in comparison with isoelectronic, triply bonded systems such as X—C≡C—X, :C≡O:, X—C≡N: and X—N≡C:. Both —C≡C— and —C≡N groups are known to serve as donors by using their π electrons. The inability of N_2 to form stable linkages in this way may be

[6] Cf. T. Taylor and M. Spindel, *Proc. Internat. Symp. on Isotope Separation*, P. Kistemacher, J. Biegeleisen and O. Nier, eds., North Holland Publ. Co., 1958.

attributed to the fact that its electron configuration (page 106) is $\cdots(\pi)^4(\sigma_2)^2$, that is, the π bonding electrons are even more tightly bound than the σ bonding electrons, and the latter are themselves tightly bound (IP $= 15.5$ eV). In acetylene, on the other hand, the electron configuration is $\cdots(\sigma_g)^2(\pi_u)^4$, and the IP of the π_u electrons is only 11.4 eV. The N_2 molecule can, however, form complexes similar to those formed by CO, though to a much more limited extent, in which there are M—N≡N and M—C≡O configurations. The very different abilities of the two molecules to function in this way are attributable to several quantitative differences in their qualitatively similar electronic structures.[7]

Nitrogen is obtained commercially by liquefaction and fractionation of air; when so prepared it usually contains some argon and, depending on the quality, upwards of ~ 30 ppm of oxygen. The oxygen may be removed by admixture with a little hydrogen and treatment with a platinum catalyst, by passing the gas over hot copper or other metal, or by bubbling it through aqueous solutions of Cr^{2+} or V^{2+} ions. Spectroscopically pure nitrogen is conveniently prepared by thermal decomposition of sodium azide or barium azide, for example:

$$2NaN_3 \rightarrow 2Na + 3N_2$$

The only reactions of N_2 at room temperature are with metallic lithium to give Li_3N, with certain transition-metal complexes, and with nitrogen-fixing bacteria, either free-living or symbiotic on root nodules of clover, peas, beans, etc. The mechanism by which these bacteria fix nitrogen is still unknown and remains one of the most conspicuously challenging, unsolved problems in chemistry.

At elevated temperatures nitrogen becomes more reactive, especially when catalyzed, typical reactions being:

$$N_2(g) + 3H_2(g) = 2NH_3(g) \qquad K_{25°} = 10^3 \text{ atm}^{-2}$$
$$N_2(g) + O_2(g) = 2NO(g) \qquad K_{25°} = 5 \times 10^{-31} \cdot$$
$$N_2(g) + 3Mg(s) = Mg_3N_2(s)$$
$$N_2(g) + CaC_2(s) = C(s) + CaNCN(s)$$

In recent years many efforts have been made to discover simple chemical systems that would mimic the reductive fixation of N_2 by living systems. The common point of departure for these attempts has been the established fact that the living systems employ a metalloenzyme, nitrogenase, which appears to contain molybdenum, iron, and labile thiol (sulfhydryl) groups together with a coenzyme that serves as a reducing agent.[8] Aqueous solutions containing MoO_4^-, various thiols and a reducing agent (e.g., BH_4^-) have shown some activity.[9] A variety of systems in which the catalytic species is some form of lower-valent titanium and the reducing agent is potassium

[7] Cf. J. Chatt, D. P. Melville and R. L. Richards, J. Chem. Soc., A, 1969, 2841; K. G. Caulton, R. L. De Kock and R. F. Fenske, J. Amer. Chem. Soc., 1970, 92, 515.
[8] R. W. F. Hardy and R. C. Burns, Ann. Rev. Biochem., 1968, 37, 311; K. Kuchynka, Catalysis Rev., 1970, 3, 111; L. E. Mortenson, Survey Progr. Chem., 1968, 4, 127; R. W. F. Hardy, R. C. Burns and G. Parshall, Adv. in Chem. Series, No. 100, p. 219, Amer. Chem. Soc., 1971.
[9] R. E. E. Hill and R. L. Richards, Nature, 1971, 233, 114.

naphthalenide, potassium metal or aluminum metal have likewise been successful in varying degrees.[10] Other transition-metal compounds such as $(C_5H_5)_2NbCl_2$, $(C_5H_5)_2ZrCl_2$ and $W(OCH_3)_2Cl_3$ coupled with sodium dihydronaphthalenide have also given finite quantities of NH_3.[11] In several cases low-valent metal complexes of N_2 are formed and can even be isolated, but *catalytic* processes by which the bound nitrogen becomes reduced (or oxidized) have not yet been devised. The N_2 complexes as such are discussed elsewhere (Section 22-10).

Active Nitrogen. When gaseous molecular nitrogen is subjected to an electrical discharge, under suitable conditions, a very reactive[12] form of nitrogen is generated, accompanied by a yellow afterglow which may persist for several seconds after the discharge has been terminated. The high reactivity is largely due to the presence of ground state (^4S) nitrogen atoms. These have a relatively long lifetime in a vessel suitably "poisoned" to minimize wall recombination. The ternary collision process

$$2N(^4S) + X \rightarrow N_2 + X$$

where X is molecular or atomic nitrogen is at least partially responsible for populating the excited molecular states which in turn lead to the afterglow. The afterglow is due mainly to emission of the first positive band system, $N_2(B^3\Pi_g) \rightarrow N_2(A^3\Sigma_u^+)$ of the molecular nitrogen spectrum, though some other band systems also contribute.[13]

NITROGEN COMPOUNDS

12-4. Nitrides

As with carbides, there are three general classes. Ionic nitrides are formed by Mg, Ca, Ba, Sr, Zn, Cd, Li and Th. Their formulas correspond to what would result from combination of the normal metal ions with N^{3-} ions. They are all essentially ionic compounds and are properly written as $(Ca^{2+})_3(N^{3-})_2$ $(Li^+)_3N^{3-}$, etc. Nitrides of the M_3N_2 type are often anti-isomorphous with oxides of M_2O_3 type. This does not in itself mean that, like the oxides, they are ionic. However, their ready hydrolysis to ammonia and the metal hydroxides makes this seem likely. The ionic nitrides are prepared by direct union of the elements or by loss of ammonia from amides on heating, for example:

$$3Ba(NH_2)_2 \rightarrow Ba_3N_2 + 4NH_3$$

There are various covalent "nitrides"—BN, S_4N_4, P_3N_5, etc.—the properties of which vary greatly depending upon the element with which nitrogen is

[10] M. E. Vol'pin *et al.*, *Chem. Comm.*, **1968**, 1704; E. E. van Tamelen *et al.*, *J. Amer. Chem. Soc.*, 1971, **93**, 3526. For reviews of non-enzymic activation of N_2, see G. Henrici-Olivé and S. Olivé, *Angew. Chem. Intern. Ed.*, 1969, **8**, 650, and R. Murray and D. C. Smith, *Coordination Chem. Rev.*, 1968, **3**, 429; K. Kuchynka, ref. 8.

[11] D. R. Gray and C. H. Brubaker, *Chem. Comm.*, **1968**, 1239.

[12] R. Brown and C. A. Winkler, *Angew. Chem. Internat. Edn.*, 1970, **9**, 181.

[13] A. N. Wright and C. A. Winkler, *Active Nitrogen*, Academic Press, 1968.

combined. Such substances are therefore discussed under the appropriate element.

The transition metals form nitrides[14] that are analogous to the transition-metal borides and carbides in their constitution and properties. The nitrogen atoms often occupy interstices in the close-packed metal lattices. These nitrides are often not exactly stoichiometric (being nitrogen-deficient), and they are metallic in appearance, hardness, and electrical conductivity, since the electronic band structure of the metal persists. Like the borides and carbides, they are chemically very inert, extremely hard, and very high-melting. They are usually prepared by heating the metal in ammonia at 1100–1200°. A representative compound, VN, melts at 2570° and has a hardness between 9 and 10.

12-5. Nitrogen Hydrides

Ammonia. On a laboratory scale, ammonia, NH_3, may be generated by treatment of an ammonium salt with a base:

$$NH_4X + OH^- \rightarrow NH_3 + H_2O + X^-$$

Hydrolysis of an ionic nitride is a convenient way of preparing ND_3 (or NH_3):

$$Mg_3N_2 + 6D_2O \rightarrow 3Mg(OD)_2 + 2ND_3$$

On the industrial scale ammonia is obtained by the Haber process in which the reaction

$$N_2(g) + 3H_2(g) = 2NH_3(g) \qquad \Delta H = -46 \text{ kJ mol}^{-1}; \qquad K_{25°} = 10^3 \text{ atm}^{-2}$$

is carried out in the presence of a catalyst at pressures of 10^2–10^3 atm and temperatures of 400–550°. Although the equilibrium is most favorable at low temperature, even with the best available catalysts elevated temperatures are required to obtain a satisfactory rate of conversion. The best catalyst is α-iron containing some oxide in order to widen the lattice and enlarge the active interface.

Ammonia is a colorless pungent gas with a normal boiling point of $-33.35\,°C$ and a freezing point of $-77.8\,°C$. The liquid has a large heat of evaporation (1.37 kJ/g at the boiling point) and is therefore fairly easily handled in ordinary laboratory equipment. Liquid ammonia resembles water in its physical behavior, being highly associated because of the polar nature of the molecules and strong hydrogen bonding. Its dielectric constant (~ 22 at $-34°$; cf. 81 for H_2O at 25°) is sufficiently high to make it a fair ionizing solvent.[15] A system of nitrogen chemistry with many analogies to

[14] R. Juza, *Adv. Inorg. Chem. Radiochem.*, 1967, **9**, 81 (1st transition series).
[15] J. J. Lagowski, ed., *The Chemistry of Non-Aqueous Solvents*, Academic Press, 1967; G. Jander, M. Spandau and C. C. Addison, eds., *Chemistry in Anhydrous Ammonia* (Vol. 1, Part I, Inorganic and General Chemistry, 1966; Part II, Organic Reactions, 1963), Interscience-Wiley; W. L. Jolly and C. J. Hallada, in *Non-Aqueous Solvent Systems*, T. C. Waddington, ed., Academic Press, 1965; G. W. A. Fowles, in *Developments in Inorganic Nitrogen Chemistry*, C. B. Colburn, ed., Vol. 1, Elsevier, 1966.

the oxygen system based on water has been built up. Thus, we have the comparable self-ionization equilibria:

$$2NH_3 = NH_4^+ + NH_2^- \qquad K_{-50°} = [NH_4^+][NH_2^-] = \sim 10^{-30}$$
$$2H_2O = H_3O^+ + OH^- \qquad K_{25°} = [H_3O^+][OH^-] = 10^{-14}$$

Table 12-1 presents a comparison of the ammonia and the water systems. Liquid ammonia has lower reactivity than H_2O towards electropositive metals, such metals reacting immediately with water to evolve hydrogen.

TABLE 12-1

	Ammonia system		Water system	
	Class of compound	Example	Class of compound	Example
Acids	$NH_4^+ X^-$	$NH_4^+ Cl^-$	$H_3O^+X^-$	$H_3O^+Cl^-$
Bases	Amides	$Na^+NH_2^-$	Hydroxides	Na^+OH^-
	Imides	$(Li^+)_2 NH^{2-}$	Oxides	$(Li^+)_2O^{2-}$, $Mg^{2+}O^{2-}$
	Nitrides	$(Mg^{2+})_3(N^{3-})_2$		

Liquid ammonia, on the other hand, dissolves many electropositive metals to give blue solutions containing metal ions and solvated electrons (see also page 193).

Because $NH_3(l)$ has a much lower dielectric constant than water, it is a better solvent for organic compounds but generally a poorer one for ionic inorganic compounds. Exceptions occur when complexing by NH_3 is superior to that by water. Thus AgI is exceedingly insoluble in water but $NH_3(l)$ at 25° dissolves 207 g/100 ml. Primary solvation numbers of cations in $NH_3(l)$ appear similar to those in H_2O, e.g., 5.0 ± 0.2 and 6.0 ± 0.5 for Mg^{2+} and Al^{3+}, respectively.[16]

Reactions of Ammonia.   Ammonia reacts with both oxygen and water. Normal combustion in air follows reaction 12-1. However, ammonia can be

$$4NH_3(g) + 3O_2(g) = 2N_2(g) + 6H_2O(g) \qquad K_{25°} = 10^{228} \qquad (12\text{-}1)$$

made to react with oxygen as shown in equation 12-2, despite the fact that the process of equation 12-1 is thermodynamically much more favorable,

$$4NH_3 + 5O_2 = 4NO + 6H_2O \qquad K_{25°} = 10^{168} \qquad (12\text{-}2)$$

by carrying out the reaction at 750–900° in the presence of a platinum or platinum–rhodium catalyst. This can easily be demonstrated in the laboratory by introducing a piece of glowing platinum foil into a jar containing gaseous NH_3 and O_2; the foil will continue to glow because of the heat of reaction 12-2, which occurs only on the surface of the metal, and brown fumes will appear owing to the reaction of NO with the excess of oxygen to produce NO_2. Industrially, the mixed oxides of nitrogen are then absorbed in water to form nitric acid:

$$2NO + O_2 \rightarrow 2NO_2$$
$$3NO_2 + H_2O \rightarrow 2HNO_3 + NO, \text{ etc.}$$

[16] T. J. Swift and H. H. Lo, *J. Amer. Chem. Soc.*, 1967, **89**, 3988.

Thus, the sequence in industrial utilization of atmospheric nitrogen is as follows:

$$N_2 \xrightarrow[\substack{\text{Haber} \\ \text{process}}]{H_2} NH_3 \xrightarrow[\substack{\text{Ostwald} \\ \text{process}}]{O_2} NO \xrightarrow{O_2+H_2O} HNO_3(aq)$$

Ammonia is extremely soluble in water. Two stable crystalline hydrates are formed at low temperatures, $NH_3 \cdot H_2O$ (m.p. $194.15\,°K$) and $2NH_3 \cdot H_2O$ (m.p. $194.32\,°K$), in which the NH_3 and H_2O molecules are linked by hydrogen bonds. The substances contain neither NH_4^+ and OH^- ions nor discrete NH_4OH molecules. Thus, $NH_3 \cdot H_2O$ has chains of H_2O molecules linked by hydrogen bonds ($2.76\,Å$). These chains are cross-linked by NH_3 into a three-dimensional lattice by $O—H\cdots N$ ($2.78\,Å$) and $O\cdots H—N$ bonds ($3.21–3.29\,Å$). In aqueous solution ammonia is probably hydrated in a similar manner. Although aqueous solutions are commonly referred to as solutions of the weak base NH_4OH, called "ammonium hydroxide," this is to be discouraged, since there is no evidence that undissociated NH_4OH exists and there is reason to believe that it probably does not. Solutions of ammonia are best described as $NH_3(aq)$, with the equilibrium written as

$$NH_3(aq) + H_2O = NH_4^+ + OH^- \qquad K_{25°} = \frac{[NH_4^+][OH^-]}{[NH_3]} = 1.81 \times 10^{-5} \ (pK_b = 4.75)$$

In an odd sense NH_4OH might be considered a *strong* base since it is completely dissociated in water. A $1N$ solution of NH_3 is only $0.0042N$ in NH_4^+ and OH^-.

Nuclear magnetic resonance measurements show that the hydrogen atoms of NH_3 rapidly exchange with those of water by the process

$$H_2O + NH_3 = OH^- + NH_4^+$$

but there is only slow exchange between NH_3 molecules in the vapor phase or in the liquid if water is completely removed.

Ammonium salts. There are many rather stable crystalline salts of the tetrahedral NH_4^+ ion; most of them are water-soluble, like alkali-metal salts. Salts of strong acids are fully ionized, and the solutions are slightly acidic:

$$NH_4Cl = NH_4^+ + Cl^- \qquad K \approx \infty$$
$$NH_4^+ + H_2O = NH_3 + H_3O^+ \qquad K_{25°} = 5.5 \times 10^{-10}$$

Thus, a $1M$ solution will have a pH of ~ 4.7. The constant for the second reaction is sometimes called the hydrolysis constant; however, it may equally well be considered as the acidity constant of the cationic acid NH_4^+, and the system regarded as an acid–base system in the following sense:

$$\underset{\text{Acid}}{NH_4^+} + \underset{\text{Base}}{H_2O} = \underset{\text{Acid}}{H_3O^+} + \underset{\text{Base}}{NH_3(aq)}$$

Ammonium salts generally resemble those of potassium and rubidium in solubility and, except where hydrogen bonding effects are important, in structure, since the three ions are of comparable (Pauling) radii: $NH_4^+ = 1.48\,Å$, $K^+ = 1.33\,Å$, $Rb^+ = 1.48\,Å$.

Many ammonium salts volatilize with dissociation around $300°$, for example:

$$NH_4Cl(s) = NH_3(g) + HCl(g) \qquad \Delta H = 177 \text{ kJ mol}^{-1}; \ K_{25°} = 10^{-16}$$
$$NH_4NO_3(s) = NH_3(g) + HNO_3(g) \qquad \Delta H = 171 \text{ kJ mol}^{-1}$$

Some salts which contain oxidizing anions decompose when heated, with oxidation of the ammonia to N_2O or N_2 or both. For example:

$$(NH_4)_2Cr_2O_7(s) = N_2(g) + 4H_2O(g) + Cr_2O_3(s) \qquad \Delta H = -315 \text{ kJ mol}^{-1}$$
$$NH_4NO_3(l) = N_2O(g) + 2H_2O(g) \qquad \Delta H = -23 \text{ kJ mol}^{-1}$$

Ammonium nitrate volatilizes reversibly at moderate temperatures; at higher temperatures, irreversible decomposition occurs exothermically, giving mainly N_2O. This is the reaction by which N_2O is prepared commercially. At still higher temperatures, the N_2O itself decomposes into nitrogen and oxygen. Ammonium nitrate can be caused to detonate when initiated by another high explosive, and mixtures of ammonium nitrate with TNT or other high explosives are used for bombs. The decomposition of liquid ammonium nitrate can also become explosively rapid, particularly when catalyzed by traces of acid and chloride; there are a number of instances of disastrous explosions following after fires of ammonium nitrate in bulk. Moreover, ammonium perchlorate is important as an oxidizer in solid propellants for rocket fuels, and its thermal decomposition has been studied in detail.[17]

Tetraalkylammonium ions, R_4N^+, prepared generally by the reaction

$$R_3N + RI = R_4N^+I^-$$

are often of use in inorganic chemistry when large univalent cations are required. Various R_4N radicals in the form of apparently crystalline amalgams (~ 12 Hg/R_4N) can be obtained either electrolytically or by reduction of R_4NX with Hg/Na in media where the resulting NaX is insoluble.[18]

Hydrazine. Hydrazine, N_2H_4, may be thought of as derived from ammonia by replacement of a hydrogen atom by the —NH_2 group. It might therefore be expected to be a base, but somewhat weaker than NH_3, which is the case. It is a bifunctional base:

$$N_2H_4(aq) + H_2O = N_2H_5^+ + OH^- \qquad K_{25^\circ} = 8.5 \times 10^{-7}$$
$$N_2H_5^+(aq) + H_2O = N_2H_6^{2+} + OH^- \qquad K_{25^\circ} = 8.9 \times 10^{-16}$$

and two series of hydrazinium salts are obtainable. Those of $N_2H_5^+$ are stable in water while those of $N_2H_6^{2+}$ are, as expected from the above equilibrium constant, extensively hydrolyzed. Salts of $N_2H_6^{2+}$ can be obtained by crystallization from aqueous solution containing a large excess of the acid, since they are usually less soluble than the monoacid salts.

As another consequence of its basicity, hydrazine, like NH_3, can form coordination complexes with both Lewis acids and metal ions. Just as with respect to the proton, electrostatic considerations (and, in these cases, also steric considerations) militate against bifunctional behavior. Although some polymeric complexes having hydrazine bridges (12-VII) have been demonstrated, generally only one nitrogen atom is coordinated,[19] as in, e.g., $[Zn(N_2H_4)_2Cl_2]$.

[17] A. G. Keenan and R. I. Siegmund, *Quart. Rev.*, 1969, **23**, 430; P. W. M. Jacobs and H. M. Whitehead, *Chem. Rev.*, 1969, **69**, 551.
[18] J. D. Littlehailes and B. J. Woodhall, *Chem. Comm.*, **1967**, 665.
[19] C. H. Stapfer and R. W. D'Andrea, *Inorg. Chem.*, 1970, **10**, 1224; F. Bottomly, *Quart. Rev.*, 1970, **24**, 617.

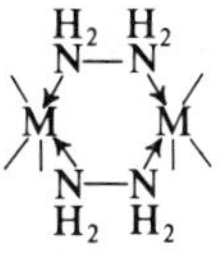

(12-VII)

Anhydrous N_2H_4 (m.p. 2°, b.p. 114°), a fuming colorless liquid with a high dielectric constant ($\varepsilon = 52$ at 25°), is surprisingly stable in view of its endothermic nature ($\Delta H_f^0 = 50$ kJ mol^{-1}). It will burn in air, however, with considerable evolution of heat which accounts for interest in it and certain of its alkylated derivatives as potential rocket fuels.

$$N_2H_4(l)+O_2(g) = N_2(g)+2H_2O(l) \qquad \Delta H^\circ = 622 \text{ kJ mol}^{-1}$$

Aqueous hydrazine is a powerful reducing agent in basic solution; in many of such reactions, diimine (see below) is an intermediate. One reaction, which is quantitative with some oxidants (e.g., I_2), is

$$N_2+4H_2O+4e = 4OH^-+N_2H_4(aq) \qquad E^\circ = -1.16 \text{ V}$$

However, NH_3 and HN_3 are also obtained under various conditions. Air and oxygen, especially when catalyzed by multivalent metal ions in basic solution, produce hydrogen peroxide:

$$2O_2+N_2H_4(aq) \rightarrow 2H_2O_2(aq)+N_2$$

but further reaction occurs in presence of metal ions:

$$N_2H_4+2H_2O_2 \rightarrow N_2+4H_2O$$

In acid solution, hydrazine can reduce halogens:

$$N_2H_4(aq)+2X_2 \rightarrow 4HX+N_2$$

The preparation of hydrazine has been the subject of much study. Many reactions produce it in small amounts under certain conditions, for example:

$$\begin{aligned} N_2+2H_2 &\rightarrow N_2H_4 \\ N_2O+2NH_3 &\rightarrow N_2H_4+H_2O+N_2 \\ 2NH_3(g)+\tfrac{1}{2}O_2 &\rightarrow N_2H_4+H_2O \\ N_2O+3H_2 &\rightarrow N_2H_4+H_2O \end{aligned}$$

However, none of these has ever been developed into a practical method because there are competing, and thermodynamically more favorable, reactions, such as

$$\begin{aligned} 2NH_3+\tfrac{3}{2}O_2 &= N_2+3H_2O \\ 3N_2O+2NH_3 &= 4N_2+3H_2O \\ N_2O+H_2 &= N_2+H_2O \end{aligned}$$

The last three reactions are good illustrations of the effect of the great stability of N_2 on nitrogen chemistry.

The only practical methods for preparing hydrazine in quantity are the Raschig synthesis, discovered in the first decade of this century, and a variant thereof. The overall reaction, carried out in aqueous solution, is

$$2NH_3+NaOCl \rightarrow N_2H_4+NaCl+H_2O$$

Actually, the reaction proceeds in two steps:

$$\begin{aligned} NH_3+NaOCl &\rightarrow NaOH+NH_2Cl \qquad \text{(fast)} \\ NH_3+NH_2Cl+NaOH &\rightarrow N_2H_4+NaCl+H_2O \end{aligned}$$

However, there is a competing and parasitic reaction which is rather fast once some hydrazine has been formed:

$$2NH_2Cl + N_2H_4 \rightarrow 2NH_4Cl + N_2$$

To obtain appreciable yields, it is necessary to add some gelatinous material which serves two essential purposes. First, it sequesters heavy-metal ions that catalyze the parasitic reaction: even the part per million or so of Cu^{2+} in ordinary water will almost completely prevent the formation of hydrazine if no catalyst is used. Since simple sequestering agents such as EDTA are not as beneficial as gelatin, the latter is assumed to have a positive catalytic effect as well. Yields of 60–70% are obtained under optimum conditions. Anhydrous hydrazine may be obtained by distillation over NaOH or by precipitating $N_2H_6SO_4$, which is then treated with liquid NH_3 to precipitate $(NH_4)_2SO_4$. A more recent variant of the Raschig process involves the use of a ketone to catalyze the reaction of Cl_2 with NH_3.

At 25°C N_2H_4 is 100% in the *gauche* form (12-VIII) (cf. N_2F_4, below).

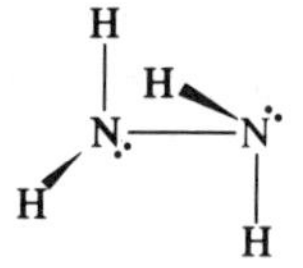

(12-VIII)

Diimine. Although the parent of azo compounds, diimine, HN=NH, cannot be isolated, there is good evidence for its transient existence in the gaseous phase and in solutions. Diimine is obtained commonly in oxidations of hydrazine by two-electron oxidants, e.g., molecular oxygen, peroxides, chloramine-T, etc.:

$$N_2H_4 \xrightarrow{-2e, -2H} N_2H_2$$

It is also formed in alkaline cleavage of chloramine:

$$H_2NCl \xrightarrow{OH^-} HNCl^-$$

$$HNCl^- + H_2NCl \xrightarrow{-Cl^-} \underset{\underset{Cl}{|}}{HN-NH_2} \xrightarrow{-HCl} HN=NH$$

The existence of N_2H_2 has been shown by, *inter alia*, the stereospecific *cis*-hydrogenation of C=C bonds by hydrazine and an oxidant. Diimine can either decompose to N_2 and H_2 or it can disproportionate to N_2 and N_2H_4. These two reactions appear to be competitive.

Hydrazoic Acid and Azides.[20] Although hydrazoic acid, HN_3, is a hydride of nitrogen in a formal sense, it has no essential relationship to NH_3 and N_2H_4. The sodium salt is prepared by the reactions

$$3NaNH_2 + NaNO_3 \xrightarrow{175°} NaN_3 + 3NaOH + NH_3$$
$$2NaNH_2 + N_2O(g) \xrightarrow{190°} NaN_3 + NaOH + NH_3$$

[20] A. D. Yoffe in *Developments in Inorganic Nitrogen Chemistry*, C. B. Colburn, ed., Elsevier, 1966; P. A. S. Smith, *Open Chain Nitrogen Compounds*, Vol. II, p. 211, Benjamin, 1966.

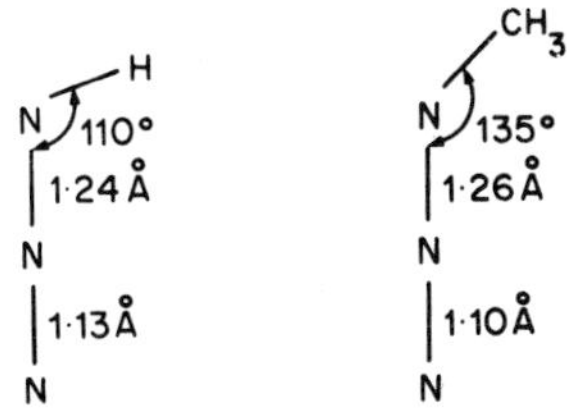

Fig. 12-3. Structures of HN_3 and CH_3N_3.

while the free acid can be obtained in solution by the reaction

$$N_2H_5^+ + HNO_2 \xrightarrow{\text{Aq. soln.}} HN_3 + H^+ + 2H_2O$$

Many other oxidizing agents attack hydrazine to form small amounts of HN_3 or azides. Hydrazoic acid ($pK_a^{25} = 4.75$), obtainable pure by distillation from aqueous solutions, is a colorless liquid (b.p. 37°) and dangerously explosive. Azides of many metals are known: those of heavy metals are generally explosive; Pb, Hg and Ba azide explode on being struck sharply and are used in detonation caps.

Azides of electropositive metals are not explosive and, in fact, decompose smoothly and quantitatively when heated to 300° or higher, for example

$$2NaN_3(s) = 2Na(l) + 3N_2(g)$$

Azide ion also functions as a ligand in complexes of transition metals. In general, N_3^- behaves rather like a halide ion and is commonly considered as a pseudohalide, although the corresponding pseudohalogen $(N_3)_2$ is not known.

The azide ion itself is symmetrical and linear (N—N, 1.16 Å), and its electronic structure may be represented in valence-bond theory as

$$:\ddot{N} = \overset{+}{N} = \ddot{N}: \quad \longleftrightarrow \quad :N \equiv \overset{+}{N} : \overset{2-}{\ddot{N}}: \quad \longleftrightarrow \quad :\overset{2-}{\ddot{N}} : \overset{+}{N} \equiv N:$$

In covalent azides, on the other hand, the symmetry is lost as is evident in HN_3 and CH_3N_3 (Fig. 12-3). In such covalent azides the electronic structure is a resonance hybrid:

$$R : \ddot{N} = \overset{+}{N} = \ddot{N}: \quad \longleftrightarrow \quad R : \overset{-}{\ddot{N}} : \overset{+}{N} \equiv N:$$

Hydroxylamine. As hydrazine may be thought of as derived from ammonia by replacement of one hydrogen by NH_2, so hydroxylamine, NH_2OH, is obtained by replacement of H by OH. Like hydrazine, hydroxylamine is a weaker base than NH_3:

$$NH_2OH(aq) + H_2O = NH_3OH^+ + OH^- \qquad K_{25°} = 6.6 \times 10^{-9}$$

Hydroxylamine is prepared by reduction of nitrates or nitrites either electrolytically or with SO_2, under very closely controlled conditions. It is also made, in 70% yield, by H_2 reduction of NO_2 in HCl solution with platinized active charcoal as catalyst. Free hydroxylamine is a white solid (m.p. 33°) which must be kept at 0 °C to avoid decomposition. It is normally encountered as an aqueous solution and as salts, e.g., $[NH_3OH]Cl$, $[NH_3OH]NO_3$ and $[NH_3OH]_2SO_4$, which are stable, water-soluble, white solids. Although hydroxylamine can serve as either an oxidizing or a reducing agent, it is usually used as the latter.[20a]

12-6. Oxides of Nitrogen

The known oxides of nitrogen are listed in Table 12-2, and their structures are shown in Fig. 12-4.

TABLE 12-2

Oxides of Nitrogen

Formula	Name	Color	Remarks
N_2O	Nitrous oxide	Colorless	Rather unreactive
NO	Nitric oxide	Gas, colorless; liquid and solid, blue	Moderately reactive
N_2O_3	Dinitrogen trioxide	Blue solid	Extensively dissociated as gas
NO_2	Nitrogen dioxide	Brown	Rather reactive
N_2O_4	Dinitrogen tetroxide	Colorless	Extensively dissociated to NO_2 as gas and partly as liquid
N_2O_5	Dinitrogen pentoxide	Colorless	Unstable as gas; ionic solid
NO_3 ; N_2O_6	—	—	Not well characterized and quite unstable

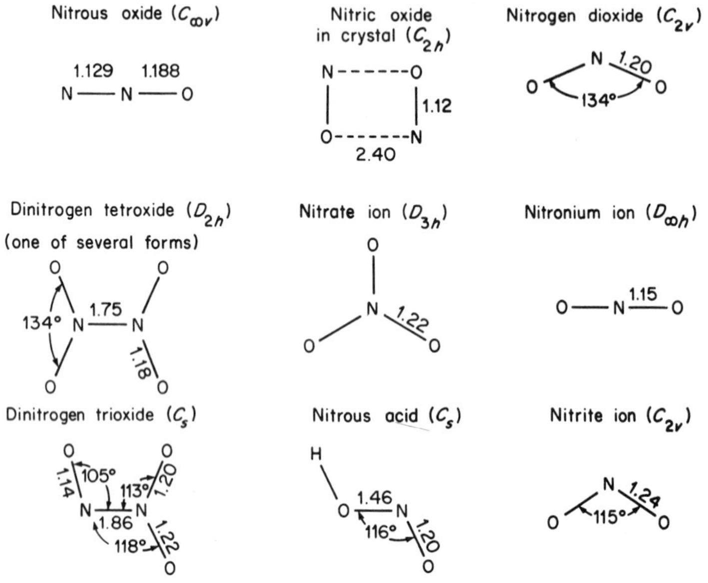

Fig. 12-4. The structures and point group symmetries of some nitrogen oxides and anions (angle in degrees; bond lengths in Å).

[20a] M. N. Hughes and H. G. Nicklin, *J. Chem. Soc., A,* **1971,** 164.

Nitrous Oxide. Nitrous oxide, N_2O, is obtained by thermal decomposition of ammonium nitrate in the melt at 250–260°:

$$NH_4NO_3 \rightarrow N_2O + 2H_2O$$

The contaminants are NO, which can be removed by passage through ferrous sulfate solution and 1–2% of nitrogen. The NH_4NO_3 must be free from Cl^- as this catalytically causes decomposition to N_2. However, heating HNO_3 or H_2SO_4 solutions of NH_4NO_3 with small amounts of Cl^- gives almost pure N_2O. The gas is also produced in the reduction of nitrites and nitrates under certain conditions and by decomposition of hyponitrites.

Nitrous oxide is relatively unreactive, being inert to the halogens, alkali metals and ozone at room temperature. It will oxidize some low-valent transition-metal complexes[21] and forms the complex $[Ru(NH_3)_5N_2O]^{2+}$ (Section 26-F-4). At elevated temperatures it decomposes to nitrogen and oxygen, reacts with alkali metals and many organic compounds, and supports combustion. It has a moderate solubility in cream, and, apart from its anesthetic role, its chief commercial use is as the propellant gas in "whipped" cream bombs.

Nitric Oxide. Nitric oxide, NO, is formed in many reactions involving reduction of nitric acid and solutions of nitrates and nitrites. For example, with $8N$ nitric acid:

$$8HNO_3 + 3Cu \rightarrow 3Cu(NO_3)_2 + 4H_2O + 2NO$$

Reasonably pure NO is obtained by the aqueous reactions:

$$2NaNO_2 + 2NaI + 4H_2SO_4 \rightarrow I_2 + 4NaHSO_4 + 2H_2O + 2NO$$
$$2NaNO_2 + 2FeSO_4 + 3H_2SO_4 \rightarrow Fe_2(SO_4)_3 + 2NaHSO_4 + 2H_2O + 2NO$$

or, dry,
$$3KNO_2(l) + KNO_3(l) + Cr_2O_3(s) \rightarrow 2K_2CrO_4(s, l) + 4NO$$

Commercially it is obtained by catalytic oxidation of ammonia as already noted. Direct combination of the elements occurs only at very high temperatures, and to isolate the small amounts so formed (a few volume per cent at 3000°) the equilibrium mixture must be rapidly chilled. Though much studied, this reaction has not been developed into a practical commercial synthesis.

Nitric oxide reacts instantly with O_2:

$$2NO + O_2 \rightarrow 2NO_2$$

It also reacts with F_2, Cl_2 and Br_2 to form the nitrosyl halides, XNO (see page 365) and with CF_3I to give CF_3NO and I_2. It is oxidized to nitric acid by several strong oxidizing agents; the reaction with permanganate is quantitative and provides a method of analysis. It is reduced to N_2O by SO_2 and to NH_2OH by chromous ion, in acid solution in both cases.

Nitric oxide is thermodynamically unstable at 25° and 1 atm and at high pressures it readily decomposes in the range 30–50°:

$$3NO \rightarrow N_2O + NO_2$$

The NO molecule has the electron configuration $(\sigma_1)^2(\sigma_1^*)^2(\sigma_2, \pi)^6(\pi^*)$ (see page 107). The unpaired π^* electron renders the molecule paramagnetic

[21] R. G. S. Banks, R. J. Henderson and J. M. Pratt, *J. Chem. Soc., A*, **1968**, 2286.

and partly cancels the effect of the π bonding electrons. Thus the bond order is 2.5, consistent with an interatomic distance of 1.15 Å, which is intermediate between the triple bond distance in NO^+ (see below) of 1.06 Å and representative double-bond distances of ~ 1.20 Å.

The NO molecule has only a slight tendency to dimerize. It forms centrosymmetric dimers in the solid state (cf. Fig. 12-4). There is also evidence for weak association in the liquid, and dimers have been matrix-isolated in N_2 at 15°K. In the N_2 matrix *cis*-ONNO species appear to predominate though two other species, of uncertain structure, are also observed.[22] Dimers with a heat of dissociation of ~ 8 kJ mol^{-1} persist in the vapor at the boiling point.[23] Like CO, the molecule has a very small electric dipole moment (0.17 D). The magnetism of NO varies strongly with temperature owing to an unusual form of spin–orbit coupling.

The electron in the π^* orbital is relatively easily lost (IP = 9.23 eV), to give the *nitrosonium ion*, NO^+, which has an extensive and important chemistry. Because the electron removed comes out of an antibonding orbital, the bond is stronger in NO^+ than in NO: the bond length decreases by 0.09 Å and the vibration frequency rises from 1840 cm^{-1} in NO to 2150–2400 cm^{-1} (depending on environment) in NO^+. Numerous ionic compounds of NO^+ are known.

When N_2O_3 or N_2O_4 is dissolved in concentrated sulfuric acid, the ion is formed:

$$N_2O_3 + 3H_2SO_4 = 2NO^+ + 3HSO_4^- + H_3O^+$$
$$N_2O_4 + 3H_2SO_4 = NO^+ + NO_2^+ + 3HSO_4^- + H_3O^+$$

The isolable compound $NO^+HSO_4^-$, nitrosonium hydrogen sulfate, is an important intermediate in the lead-chamber process for manufacture of sulfuric acid. Its salt-like constitution has been shown by electrolysis, conductivity studies and cryoscopic measurements. The compounds $NO^+ClO_4^-$, and $NO^+BF_4^-$, both isostructural with the corresponding ammonium and H_3O^+ compounds, are known; many others such as $(NO)_2PtCl_6$, $NOFeCl_4$, $NOAsF_6$, $NOSbF_6$ and $NOSbCl_6$ may be made in the following general ways:

$$NO + MoF_6 \rightarrow NO^+MoF_6^-$$
$$ClNO + SbCl_5 \rightarrow NO^+SbCl_6^-$$

All such salts are readily hydrolyzed:

$$NO^+ + H_2O \rightarrow H^+ + HNO_2$$

and they must be prepared and handled under anhydrous conditions.

In alkaline solution at 0°, SO_3^{2-} reacts with NO to give a white crystalline solid, potassium *N*-nitrosohydroxylamine-*N*-sulfonate, $K_2SO_3N_2O_2$:[24]

$$O\dot{N} + :SO_3^{2-} = [O\dot{N} \leftarrow SO_3]^{2-}$$

$$[O\dot{N}SO_3]^{2-} + \dot{N}O = [O{=}N \diagdown \diagup SO_3]^{2-}$$
$$\underset{|}{N}$$
$$O$$

[22] W. A. Guillory and C. E. Hunter, *J. Chem. Phys.*, 1969, **50**, 3516.

[23] *Chem. Eng. News*, Sept. 22, 1969, page 42.

[24] R. Longhi, R. O. Ragsdale and R. S. Drago, *Inorg. Chem.*, 1962, **1**, 768; T. L. Nunes and R. E. Powell, *Inorg. Chem.*, 1970, **9**, 1916.

Other species with N_2O_2 groups are obtained by interaction of amines with NO; alcohol in base also gives $[O_2N_2CH_2N_2O_2]^{2-}$.

The NO^+ ion is isoelectronic with CO, and, like CO, will form bonds to metals. Thus, for example, analogous to nickel carbonyl, $Ni(CO)_4$, there is the isoelectronic $Co(CO)_3NO$. These transition-metal nitrosyl complexes are discussed in Section 22-12, but we may note here that the compound responsible for the brown ring in the test for nitrates is a nitrosyl complex of iron with the formula $[Fe(H_2O)_5NO]^{2+}$.

Dinitrogen Trioxide.[25] This oxide, N_2O_3, exists pure only in the solid state at low temperatures; in the vapor it is largely dissociated:

$$N_2O_3(g) = NO(g) + NO_2(g) \qquad \Delta H = 39.7 \text{ kJ mol}^{-1}$$

but dissociation is less extensive at low temperatures in solution or in the liquid.

It is best obtained by interaction of stoichiometric quantities of NO and O_2 or NO and N_2O_4. It is an intensely blue liquid (f.p. ca. $-100°$) and a very pale blue solid. The oxide is formally the anhydride of nitrous acid, and dissolution of an equimolar mixture of NO and NO_2 in alkalis gives virtually pure nitrite. Nitrogen tracer studies have shown rapid exchange between NO and NO_2 consistent with the above equilibrium. The solid is believed to have two forms, an unstable one of structure ONONO and the other with a long N—N bond similar to that in N_2O_4 discussed below.

The gas contains some $ONNO_2$ molecules,[26] whose detailed structure is shown in Fig. 12-4. The $ONNO_2$ structure appears to persist in the liquid state.

Nitrogen Dioxide and Dinitrogen Tetroxide. These two oxides, NO_2 and N_2O_4, exist in a strongly temperature-dependent equilibrium

$$\underset{\substack{\text{Brown} \\ \text{Paramagnetic}}}{2NO_2} \quad \rightleftharpoons \quad \underset{\substack{\text{Colorless} \\ \text{Diamagnetic}}}{N_2O_4}$$

both in solution[27] and in the gas phase,[28] where ΔH°_{298} for dissociation is 57 kJ mol^{-1}. In the solid state, the oxide is wholly N_2O_4. In the liquid, partial dissociation occurs; it is pale yellow at the freezing point ($-11.2°$) and contains 0.01% of NO_2 which increases to 0.1% in the deep red-brown liquid at the boiling point, 21.15°. In the vapor at 100°, the composition is NO_2 90%, N_2O_4 10%, and dissociation is complete above 140°. The monomer, NO_2, has an unpaired electron which appears to be located mainly on the N atom; its properties, red-brown color and ready dimerization to colorless and diamagnetic N_2O_4, are not unexpected for such a molecule. NO_2 can also lose its odd electron fairly readily (IP = 9.91 eV) to give NO_2^+, the *nitronium ion* discussed below.

[25] I. R. Beattie, *Progr. Inorg. Chem.*, 1963, **5**, 1; A. W. Shaw and A. J. Vosper, *J. Chem. Soc., A.*, **1971**, 1592.

[26] A. H. Brattain, A. P. Cox and R. L. Kuczkowski, *Trans. Faraday Soc.*, 1969, **65**, 1963.

[27] T. F. Redmond and B. B. Wayland, *J. Phys. Chem.*, 1968, **72**, 1626; D. W. James and R. C. Marshall, *J. Phys. Chem.*, 1968, **72**, 2923; A. J. Vosper, *J. Chem. Soc., A.*, **1970**, 2192.

[28] A. J. Vosper, *J. Chem. Soc., A*, **1970**, 625.

The dimer of NO_2 is known to exist in three isomeric forms.[29] The most stable by far is the *planar* O_2N—NO_2 molecule (Fig. 12-4). At liquid nitrogen temperature a twisted or non-planar form can be trapped in an inert matrix. At $\sim 4\,°K$ still a third species can be trapped which has an infrared spectrum suggesting that it is $ONONO_2$. While there is no doubt that the O_2NNO_2 molecule constitutes almost the entirety of the $(NO_2)_2$ molecules in the gas and the liquid phase under the usual conditions of chemical reactions, it may well be that small amounts of the evanescent $ONONO_2$ play a key role in the reactions of $(NO_2)_2$. Thus, many reactions of the gas can be most reasonably explained by assuming the presence or ready formation of $NO\bullet$ and NO_3^- and these radicals would be plausible products of homolytic bond scission in ON—ONO_2. For liquid $(NO_2)_2$ most chemical evidence is consistent with the idea that NO^+ and NO_3^- are present or readily formed. These are plausible products of heterolytic dissociation of ON—ONO_2 in the moderately ionizing solvents $(NO_2)_2$ or $(NO_2)_2$ mixed with liquids such as ethyl acetate. Thus, although its equilibrium concentration may be quite small, $ONONO_2$ may provide the pathway between NO_2, O_2NNO_2 and the actual reactive entities, $NO\bullet$, NO^+, NO_3^-, and NO_3^-. This hypothesis is *not* proved, but it is attractive and consistent with available evidence.

Several features of the planar O_2N—NO_2 molecule are unusual: (1) the greater stability of the planar form relative to a twisted one, since the former maximizes $O\cdots O$ repulsions; (2) the unusually long N—N bond, $1.75\,\text{Å}$, compared to 1.47 in H_2NNH_2. Many attempts[29] have been made to explain these features but none is entirely adequate.

The mixed oxides are obtained by heating metal nitrates, by oxidation of nitric oxide in air and by reduction of nitric acid and nitrates by metals and other reducing agents. The gases are highly toxic and attack metals rapidly. They react with water:

$$2NO_2 + H_2O = HNO_3 + HNO_2$$

the nitrous acid decomposing, particularly when warmed:

$$3HNO_2 = HNO_3 + 2NO + H_2O$$

The thermal decomposition

$$2NO_2 \rightleftarrows 2NO + O_2$$

begins at 150° and is complete at 600°.

The oxides are fairly strong oxidizing agents in aqueous solution, comparable in strength to bromine:

$$N_2O_4 + 2H^+ + 2e = 2HNO_2 \qquad E^0 = +1.07\ \text{V}$$

The mixed oxides, "nitrous fumes," are used in organic chemistry as selective oxidizing agents; the first step is hydrogen abstraction:

$$RH + NO_2 = R\bullet + HONO$$

and the strength of the C—H bond generally determines the nature of the reaction.

[29] W. G. Fateley, H. A. Bent and B. Crawford, Jr., *J. Chem. Phys.*, 1959, **31**, 204; H. A. Bent, *Inorg. Chem.*, 1963, **2**, 747.

Dinitrogen tetroxide has been extensively studied as a non-aqueous solvent.[30] The electrical conductivity of the liquid is quite low. It forms molecular addition compounds with a great variety of nitrogen, oxygen, and aromatic donor compounds. Systems involving liquid N_2O_4 mixed with an organic solvent are often very reactive; for example, they dissolve relatively noble metals to form nitrates, often solvated with N_2O_4. Thus copper reacts vigorously with N_2O_4 in ethyl acetate to give crystalline $Cu(NO_3)_2 \cdot N_2O_4$, from which anhydrous, volatile (at 150–200°) cupric nitrate is obtained (Section 25-H-3). Some of the compounds obtained in this way may be formulated as nitrosonium salts, for example, $Zn(NO_3)_2 \cdot 2N_2O_4$ as $(NO^+)_2[Zn(NO_3)_4]^{2-}$.

In anhydrous acids, N_2O_4 dissociates ionically, as in H_2SO_4 above, and in anhydrous HNO_3 almost completely:

$$N_2O_4 = NO^+ + NO_3^-$$

The dissociation in H_2SO_4 is complete in dilute solution; at higher concentrations undissociated N_2O_4 is present, and at very high concentrations nitric acid is formed:

$$N_2O_4 + 3H_2SO_4 = NO^+HSO_4^- + HNO_3 + HSO_4^- + SO_3 + H_3O^+$$

The $NOHSO_4$ actually crystallizes out. The detailed mechanism and intermediates are undoubtedly complex, and the system is not yet completely unraveled.

Dinitrogen Pentoxide. This oxide, N_2O_5, which forms colorless crystals, is usually obtained by dehydration of nitric acid with P_2O_5; the oxide is not too stable (sometimes exploding) and is distilled in a current of ozonized oxygen.

$$2HNO_3 + P_2O_5 \rightarrow 2HPO_3 + N_2O_5$$

It is, conversely, the anhydride of nitric acid:

$$N_2O_5 + H_2O = 2HNO_3$$

It is deliquescent, readily producing nitric acid by the above reaction.

The gaseous compound appears to have a structure of the type (12-IX)

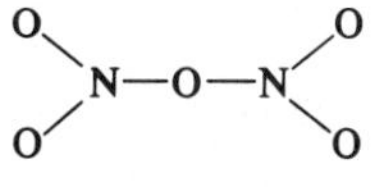

(12-IX)

with a bent N—O—N group, although this angle may be near 180°. Solid N_2O_5 in its stable form is nitronium nitrate, $NO_2^+NO_3^-$, but when the gas is condensed on a surface at $\sim 90°K$, the molecular form is obtained and persists for several hours. On warming to $\sim 200°K$, however, the latter rapidly changes to $NO_2^+NO_3^-$. The structures of the two ions are shown in Fig. 12-4.

---

[30] C. C. Addison in *Chemistry in Non-aqueous Ionizing Solvents*, Vol. 3, Part 1, Pergamon Press, 1967.

As with N_2O_4, ionic dissociation occurs in anhydrous H_2SO_4, HNO_3 or H_3PO_4 to produce NO_2^+, for instance

$$N_2O_5 + 3H_2SO_4 \rightleftharpoons 2NO_2^+ + 3HSO_4^- + H_3O^+$$

Many gas-phase reactions of N_2O_5 depend on dissociation to NO_2 and NO_3, with the latter then reacting further as an oxidizing agent. These reactions are among the better understood complex inorganic reactions.[31]

In the N_2O_5-catalyzed decomposition of ozone, the steady-state concentration of NO_3 can be high enough to allow its absorption spectrum to be recorded.

12-7. Oxo Acids of Nitrogen

Hyponitrous Acid.[32] Salts of hyponitrous acid, $H_2N_2O_2$, are formed by treating NH_2OH with pentyl nitrite in ethanol containing $NaOC_2H_5$. Reduction of nitrites with sodium amalgam also gives hyponitrites. The silver salt is insoluble in water, and the ion is commonly isolated as the silver salt. The sodium salt and also the free acid can be obtained as white crystals, the latter from the silver salt on treatment with HCl in ether. The acid is weak, $pK = 7$, and is moderately stable in solution. Hyponitrites of the alkali metals react with CO_2 to give N_2O.

The hyponitrite ion has the *trans*-configuration (12-X).

A "compound" made by the action of NO on Na in liquid NH_3 and long formulated as NaNO is unlikely to contain NO^- ions nor does it contain the hyponitrite ion. It has been proposed that it may contain a *cis*-hyponitrite ion, but there is no positive evidence for this.[33]

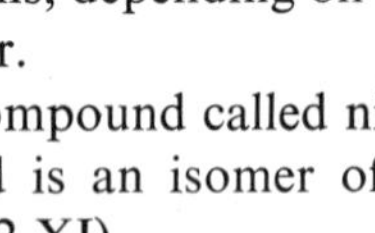

(12-X) 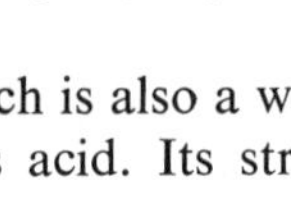   (12-XI)

Hyponitrites undergo various oxidation–reduction reactions in acid and alkaline solutions, depending on conditions; they usually behave as reducing agents, however.

There is a compound called nitramide, which is also a weak acid ($K_{25°} = 2.6 \times 10^{-7}$) and is an isomer of hyponitrous acid. Its structure has been shown to be (12-XI).

Nitrous Acid. Solutions of the weak acid, HNO_2 ($pK_a^{25} = 5.22$), are easily made by acidifying solutions of nitrites. The aqueous solution can be obtained free of salts by the reaction

$$Ba(NO_2)_2 + H_2SO_4 \rightarrow 2HNO_2 + BaSO_4(s)$$

The acid is unknown in the liquid state, but it can be obtained in the vapor

[31] G. Schott and N. Davidson, *J. Amer. Chem. Soc.*, 1958, **80**, 1841.
[32] M. N. Hughes, *Quart. Rev.*, 1968, **22**, 1.
[33] J. Goubeau and K. Laitenberger, *Z. anorg. Chem.*, 1963, **320**, 78.

phase;[34] the *trans*-form (Fig. 12-4) has been shown to be more stable than the *cis*-form by about 2.1 kJ mol^{-1}. In the gas phase the following equilibrium is rapidly established:

$$NO + NO_2 + H_2O = 2HNO_2 \qquad K_{20^\circ} = 1.56 \text{ atm}^{-1}$$

Aqueous solutions of nitrous acid are unstable and decompose rapidly when heated, according to the equation

$$3HNO_2 = HNO_3 + H_2O + 2NO$$

This reaction is reversible. Nitrous acid can behave both as an oxidant, e.g., towards I$^-$, Fe^{2+} or C$_2$O$_4^{2-}$:

$$HNO_2 + H^+ + e = NO + H_2O \qquad E^0 = 1.0 \text{ V}$$

and as a reducing agent:[34a]

$$NO_3^- + 3H^+ + 2e = HNO_2 + H_2O \qquad E^0 = 0.94 \text{ V}$$

Nitrites of the alkali metals are best prepared by heating the nitrates with a reducing agent such as carbon, lead, iron, etc.

Nitrous acid is used in the well-known preparation of diazonium compounds in organic chemistry. Numerous organic derivatives of the NO$_2$ group are known. They are of two types: nitrites, R—ONO, and nitro compounds, R—NO$_2$. Similar tautomerism occurs in inorganic complexes, in which either oxygen or nitrogen is the actual donor atom when NO$_2^-$ is a ligand.

Peroxonitrous acid, HOONO, is formed as an intermediate in the oxidation of HNO$_2$ to HNO$_3$ by H$_2$O$_2$. Although the acid is unstable, the anion is stable in alkaline solution, imparting a yellow color.[34b]

The nitrite ion is bent (Fig. 12-4), as expected.

Nitric Acid and Nitrates. The acid has already been discussed (Chapter 5, page 177).

Nitrates of practically all metallic elements are known. They are frequently hydrated and most are soluble in water. Many metal nitrates can be obtained anhydrous and a number of these, e.g., Cu(NO$_3$)$_2$, sublime without decomposition. Even alkali-metal nitrates sublime in a vacuum, at 350–500°, but decomposition occurs at higher temperatures to yield nitrites or at very high temperatures to yield oxides or peroxides. NH$_4$NO$_3$ gives N$_2$O and H$_2$O (cf. above). In neutral solution, nitrates can be reduced only with difficulty, and the mechanism of the reduction is still obscure. Al or Zn in alkaline solution produces NH$_3$. Nitrate complexes are discussed in Chapter 21.

The Nitronium Ion. This ion, NO$_2^+$, is directly involved, not only in the dissociation of nitric acid itself, but also in nitration reactions and in solutions of nitrogen oxides in nitric and other strong acids. Various early physical measurements by Hantzsch gave evidence for dissociation of HNO$_3$ in sulfuric acid. Thus "HNO$_3$" in H$_2$SO$_4$ shows no vapor pressure, and

[34] P. L. Asquith and B. J. Tyler, *Chem. Comm.*, 1970, 744.
[34a] M. Green and A. G. Sykes, *J. Chem. Soc., A*, 1970, 3209.
[34b] D. J. Benton and P. Moore, *J. Chem. Soc., A.*, 1970, 3179.

cryoscopic studies gave a van't Hoff i factor of 3. Hantzsch proposed therefore

$$HNO_3 + 2H_2SO_4 = H_3NO_3^{2+} + 2HSO_4^-$$

More recent work has shown that Hantzsch's suggestion is not correct in detail, but that ionic dissociation does occur. This work was undertaken by Hughes and Ingold and others to find an explanation for the enormous increases in the rate of nitration of aromatic compounds by HNO_3–H_2SO_4 mixtures as the concentration of the sulfuric acid is increased and to account for variations in rate in other media. For example, the rate of nitration of benzene increases by 10^3 on going from 80% H_2SO_4 to 90% H_2SO_4. Detailed kinetic data, in sulfuric acid, nitromethane and glacial acetic acid solutions, were explicable only by the postulate that the NO_2^+ ion was the attacking species. The origin of the NO_2^+ ion can be explained, for example, by ionizations of the following types:

$$2HNO_3 = NO_2^+ + NO_3^- + H_2O$$
$$HNO_3 + H_2SO_4 = NO_2^+ + HSO_4^- + H_2O$$

The importance of the first type is reflected in the fact that addition of ionized nitrate salts to the reaction mixture will retard the reaction. The actual nitration process can then be formulated as

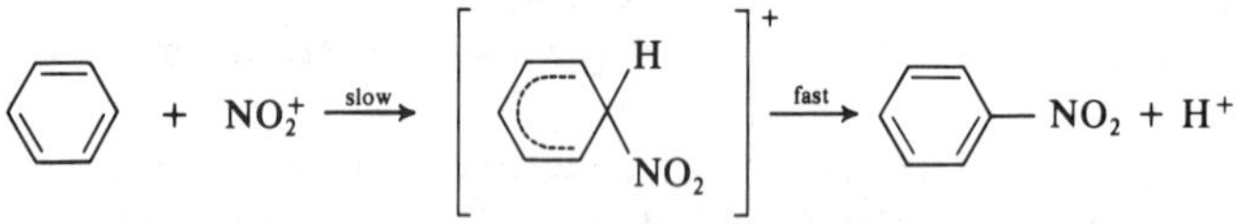

The dissociation of nitric acid in various media has been confirmed by cryoscopic studies, and nitrogen oxides have also been found to dissociate to produce nitronium ions as noted above. Spectroscopic studies have confirmed the presence of the various ions in such solutions. For example, the NO_2^+ ion can be identified by a Raman line at about $1400 \, cm^{-1}$.

Final confirmation of the existence of nitronium ions has been obtained by isolation of nitronium salts which have a symmetrical N—O stretching frequency at $\sim 1400 \, cm^{-1}$ and a bond length of 1.15 Å. Thus, from HNO_3 and $HClO_4$ in nitromethane a mixture of the perchlorates $NO_2^+ ClO_4^-$ and $H_3O^+ ClO_4^-$ can be obtained by crystallization. Other reactions leading to crystalline nitronium salts are:

$$N_2O_5 + HClO_4 \rightarrow [NO_2^+ ClO_4^-] + HNO_3$$
$$N_2O_5 + FSO_3H \rightarrow [NO_2^+ FSO_3^-] + HNO_3$$
$$HNO_3 + 2SO_3 \rightarrow [NO_2^+ HS_2O_7^-]$$

The first two of these reactions are really just metatheses, since N_2O_5 in the solid and in anhydrous acid solution is $NO_2^+ NO_3^-$. The other reaction is one between an acid anhydride, SO_3, and a base(!), $NO_2^+ OH^-$.

Nitronium salts are crystalline and thermodynamically stable, but very reactive chemically. They are rapidly hydrolyzed by moisture; in addition $NO_2^+ ClO_4^-$, for example, reacts violently with organic matter, but it can actually be used to carry out nitrations in nitrobenzene solution.

12-8. Halogen Compounds of Nitrogen

Binary Halides. In addition to NF_3, NF_2Cl, $NFCl_2$ and NCl_3, we have N_2F_2, N_2F_4 and the halogen azides XN_3 ($X = F$, Cl, Br or I). With the exception of NF_3 the halides are reactive, potentially hazardous substances, some of them like $NFCl_2$ explosive, others not. Only the fluorides are important.[35]

Nitrogen trifluoride, NF_3, plus small amounts of dinitrogen difluoride, N_2F_2, is obtained by electrolysis of NH_4F in anhydrous HF, whereas electrolysis of molten NH_4F constitutes a preferred preparative method for N_2F_2. The following reactions have all been proposed as good synthetic ones for the several nitrogen fluorides:

$$2NF_2H \xrightarrow[\text{pH } 1-2]{\text{FeCl}_3 \text{ (aq)}} N_2F_4 \ (\sim 100\%)$$

$$NH_3 + F_2 \text{ (diluted by } N_2) \xrightarrow[\text{reactor}]{\text{Copper-packed}} \begin{cases} NF_3 \\ N_2F_4 \\ N_2F_2 \\ NHF_2 \end{cases}$$

(Predominant product depends on conditions, esp. F_2/NH^3 ratio.)

$$2NF_2H + 2KF \rightarrow 2KHF_2 + N_2F_2$$

N_2F_2 may also be prepared by photolysis of N_2F_4 in presence of Br_2.[36]

Nitrogen trifluoride (b.p. $-129°$) is a very stable gas which normally is reactive only at about $250-300°$ but which reacts readily with $AlCl_3$ at $70°$:

$$2NF_3 + 2AlCl_3 \rightarrow N_2 + 3Cl_2 + 2AlF_3$$

It is not affected by water or most other reagents at room temperature and does not decompose when heated in the absence of reducing metals; when heated in presence of fluorine acceptors such as copper, the metal is fluorinated and N_2F_4 is obtained. The NF_3 molecule has a pyramidal structure but a very low dipole moment, and it appears to be totally devoid of donor properties.[37] However, the following reaction has been reported:[37a]

$$NF_3 + F_2 + Sb(As)F_5 \xrightarrow[\text{85 atm}]{200°} NF_4Sb(As)F_6$$

Tetrafluorohydrazine, N_2F_4, also a gas (b.p. $-73°$), is best prepared by the reaction of NF_3 with copper mentioned above. Its structure is similar to that of hydrazine, but differs in consisting of comparable fractions of *gauche* and *trans* forms,[38] the latter being slightly more stable, by ca. 2 kJ mol^{-1}. It is interesting that N_2F_4 dissociates readily in the gas and the liquid phase according to the equation

$$N_2F_4 = 2NF_2 \qquad \Delta H_{298°} = 84 \text{ kJ mol}^{-1}$$

which accounts for its high reactivity. The esr and electronic spectra of the difluoroamino radical, $\cdot NF_2$, indicate that it is bent (cf. OF_2, O_3^-, SO_2^-, ClO_2) with the odd electron in a relatively pure π molecular orbital.

[35] J. K. Ruff, *Chem. Rev.*, 1967, **67**, 665.
[36] L. M. Zaborski, K. E. Pullen and J. M. Shreeve, *Inorg. Chem.*, 1969, **8**, 2005.
[37] P. Holte *et al.*, *Inorg. Chem.*, 1971, **10**, 201.
[37a] W. E. Tolberg *et al.*, *Inorg. Chem.*, 1967, **6**, 1156.
[38] M. J. Caudillo and S. H. Bauer, *Inorg. Chem.*, 1969, **8**, 2086.

Since N_2F_4 dissociates so readily, it can be expected to show reactions typical of free radicals.[39] Thus it abstracts hydrogen from thiols:

$$2NF_2 + 2RSH \rightarrow 2HNF_2 + RSSR$$

and undergoes other reactions such as:

$$N_2F_4 + Cl_2 \xrightarrow{h\nu} 2NF_2Cl \quad K_{25°} = 1 \times 10^{-3}$$
$$RI + NF_2 \xrightarrow{UV} RNF_2 + \tfrac{1}{2}I_2$$
$$RCHO + N_2F_4 \longrightarrow RCONF_2 + NHF_2$$
$$R_FSF_5 + N_2F_4 \longrightarrow R_FNF_2$$

It reacts explosively with H_2 in a radical chain reaction.[40] It also reacts at 300° with NO and rapid chilling in liquid nitrogen gives the purple nitroso-difluoroamine $ONNF_2$; this is unstable.

Difluorodiazene (dinitrogen difluoride), N_2F_2, is a gas consisting of two isomers (12-XII) and (12-XIII). The *cis*-isomer predominates ($\sim 90\%$) at

b.p. $-105.7°$ b.p. $-111.4°$

(12-XII)   (12-XIII)

$25°C$ and is the more reactive. Isomerization to the equilibrium mixture is catalyzed by stainless steel. The pure *trans*-form can be obtained in about 45% yield by the reaction[41]

$$2N_2F_4 + 2AlCl_3 \rightarrow N_2F_2 + 3Cl_2 + 2AlF_3 + N_2$$

The other nitrogen trihalides are known, but are of less importance. NCl_3 is formed in the chlorination of slightly acid solutions of NH_4Cl and may be continuously extracted into CCl_4. When pure, it is a pale yellow oil (b.p. $\sim 71°$). It is endothermic ($\Delta H_f^0 = 232$ kJ mol^{-1}), explosive, photosensitive, and, generally, very reactive. The vapor has been employed in bleaching flour. The molecule is pyramidal.[42] Nitrogen tribromide is similar to NCl_3.

Concentrated aqueous ammonia reacts with I_2 at $25°$ to give black, explosive crystals of $NI_3 \cdot NH_3$. With liquid ammonia, more complex reactions occur. The structure of $NI_3 \cdot NH_3$ contains zig-zag chains of NI_4 tetrahedra sharing corners, with NH_3 molecules lying between the chains and linking them together.[43]

The mixed halide $NClF_2$ (b.p. $-67°$) is relatively stable.

Haloamines. These are compounds of the type H_2NX and HNX_2, where also H may be replaced by an alkyl radical. Only H_2NCl (chloramine), HNF_2, and H_2NF have been isolated; $HNCl_2$, H_2NBr, and $HNBr_2$ probably exist but are quite unstable. It is believed that, on chlorination of

[39] J. P. Freeman, *Inorg. Chim. Acta, Rev.*, 1967, **1**, 65 (chemistry of NF_2 radical).
[40] L. P. Kuhn and C. Wellman, *Inorg. Chem.*, 1970, **9**, 602.
[41] G. L. Hurst and S. I. Khayat, *J. Amer. Chem. Soc.*, 1965, **87**, 1620.
[42] I. F. Carter, R. F. Bratton and J. F. Jackowitz, *J. Chem. Phys.*, 1968, **49**, 3751.
[43] H. Hartl, H. Bärnighausen and J. Jander, *Z. anorg. Chem.*, 1968, **357**, 225.

aqueous ammonia, NH_2Cl is formed at pH > 8.5, $NHCl_2$ at pH 4.5–5.0, and NCl_3 at pH < 4.4. Difluoroamine, a colorless, explosive liquid (b.p. 23.6°), can be obtained as above or by H_2SO_4 acidification of fluorinated aqueous solutions of urea; the first product, H_2NCONF_2, gives HNF_2 on hydrolysis.[44] It can be converted into chlorodifluoroamine, $ClNF_2$, by action of Cl_2 and KF.[45]

While NF_3 is devoid of donor properties, NHF_2 is a weak donor:

$$NHF_2(g) + BF_3(g) = HF_2NBF_3(s) \qquad \Delta H = -88 \text{ kJ mol}^{-1}$$

Oxo Halides.[46]   There are two series of these oxo halides that might formally be considered as salts of the nitronium and nitrosonium ions, respectively, but since they are, in fact, quite covalent compounds, they were not discussed under the chemistry of these ions. The known compounds and some of their properties are listed in Table 12-3.

TABLE 12-3
Physical Properties of Nitrosyl Halides, HalNO, and Nitryl Halides, HalNO$_2$

	FNO^a	ClNO	BrNO	$FNO_2{}^a$	$ClNO_2$
Color of gas	Colorless	Orange-yellow	Red	Colorless	Colorless
Melting point, °C	−133	−62	−56	−166	−145
Boiling point, °C	−60	−6	~0	−72	−15
Structure	Bent	Bent	Bent	Planar^b	Planar^b
X—N distance, Å	1.52	1.95 ± 0.01	2.14 ± 0.02	1.35	1.840 ± 0.002
N—O distance, Å	1.13	1.14 ± 0.02	1.15 ± 0.04	1.23	1.202 ± 0.001
X—N—O angle, deg	110	116 ± 2	114		
O—N—O angle, deg				125 (assumed)	130.6 ± 0.2

^{a} Uncertainties in structure parameters not known. ^{b} Molecular symmetry, C_{2v}.

The nitrosyl halides can all be obtained by direct union of the halogens with nitric oxide and also in other ways. They are increasingly unstable in the series FNO, ClNO, BrNO. ClNO is always slightly impure, decomposing (to Cl_2 and NO) to the extent of about 0.5% at room temperature, and BrNO is decomposed to ~7% at room temperature and 1 atm.

All three are reactive and are powerful oxidizing agents, able to attack many metals. All decompose on treatment with water producing HNO_3, HNO_2, NO, and HX.

The only known nitryl halides, which may be regarded as derivatives of nitric acid where a halogen atom replaces OH, are FNO_2 and $ClNO_2$.[47] The former is conveniently prepared by the reaction:

$$N_2O_4 + 2CoF_3(s) \xrightarrow{300°} 2FNO_2 + 2CoF_2(s)$$

$ClNO_2$ is not obtainable by direct reaction of NO_2 and Cl_2, but is easily made in excellent yield by the reaction

$$ClSO_3H + HNO_3 \text{ (anhydrous)} \xrightarrow{0°} ClNO_2 + H_2SO_4$$

[44] E. A. Lawton, D. Pilipovitch and R. D. Wilson, *Inorg. Chem.*, 1965, **4**, 118.
[45] W. C. Firth, *Inorg. Chem.*, 1965, **4**, 254.
[46] R. Schmutzler, *Angew. Chem. Internat. Edn.*, 1968, **7**, 440 (review of oxo fluorides).
[47] R. C. Paul, D. Singh and K. C. Malhotra, *J. Chem. Soc.*, A, **1968**, 1396.

Both compounds are quite reactive; both are decomposed by water:

$$XNO_2 + H_2O = HNO_3 + HX$$

Halogen Nitrates. Two of these highly reactive substances, $ClONO_2$ (b.p. 22.3°) and $FONO_2$ (b.p. $-46°$) are known. $ClONO_2$ is not known to be intrinsically explosive, but it reacts explosively with organic matter; $FONO_2$ is liable to explode. The best preparative reactions appear to be[48]

$$ClF + HNO_3 \rightarrow ClONO_2 + HF$$
$$F_2 + HNO_3 \rightarrow FONO_2 + HF$$

Trifluoramine Oxide. This stable, toxic and oxidizing gaseous (b.p. $-87.6°$), which is resistant to hydrolysis, can be prepared[49] by the reactions

$$2NF_3 + O_2 \xrightarrow[\text{discharge}]{\text{Electric}} 2NF_3O$$
$$3NOF + 2IrF_6 \rightarrow 2NOIrF_6 + NF_3O$$

or by fluorination of NO provided that the gas mixture is quenched. Spectroscopic data indicate that it is a "tetrahedral" molecule (C_{3v} symmetry). Since NF_3 is inactive as a simple donor and lacks additional orbitals for a multiple interaction with an additional atom or group, the stability of NF_3O is somewhat surprising. Its electronic structure is perhaps best represented by the following resonance:

The only chemical reactions so far reported are with strong F^- acceptors (e.g., AsF_5, SbF_5), giving salts of NF_2O^+.

[48] C. J. Schack, *Inorg. Chem.*, 1967, **6**, 1938.
[49] W. B. Fox *et al.*, *J. Amer. Chem. Soc.*, 1970, **92**, 9240; P. J. Bassett and D. R. Lloyd, *J. Chem. Soc., A*, **1971**, 3377.

Further Reading

Advanced Propellant Chemistry, Adv. in Chem. Series, No. 54, Amer. Chem. Soc. (chemistry of N_2O_5, N—F compounds, NO_2^+ and $N_2H_5^+$ perchlorates).

Developments in Inorganic Nitrogen Chemistry, C. B. Colburn, ed., Elsevier, 1966 (bonding, azides, S–N compounds, N ligands, P—N compounds, N compounds of B, Al, Ga, In and Tl; reactions in liquid NH_3).

Mellor's Comprehensive Treatise on Inorganic and Theoretical Chemistry, Vol. VIII, Suppl. I and II, Longmans Green, 1967 (encyclopedic coverage of the inorganic chemistry of nitrogen).

Addison, C. C., *et al.*, *Quart. Rev.*, 1971, **25**, 289 (structural aspects of coordinated nitrate).

Kosower, E. M., *Accounts Chem. Res.*, 1971, **4**, 193 (reactions of monosubstituted diazenes including metal complexes).

Smith, P. A. S., *The Chemistry of Open-Chain Organic Nitrogen Compounds*, Vols. I and II, Benjamin, 1966 (contains much of inorganic interest).

13

The Group V Elements: P, As, Sb, Bi

GENERAL REMARKS

13-1. Group Trends and Stereochemistry

The electronic structures and some other properties of the elements are listed in Table 13-1. The valence shells have a structure formally similar to that of nitrogen, but, beyond the stoichiometries of some of the simpler compounds—NH_3, PH_3, NCl_3, $BiCl_3$, for example—there is little resemblance between the characteristics of these elements and those of nitrogen.

TABLE 13-1

Some Properties of P, As, Sb and Bi

	P	As	Sb	Bi
Electronic structure	$[Ne]3s^23p^3$	$[Ar]3d^{10}4s^24p^3$	$[Kr]4d^{10}5s^25p^3$	$[Xe]4f^{14}5d^{10}6s^26p^3$
Sum of 1st three ionization potentials, eV	60.4	58.0	52.3	52.0
Electronegativity[a]	2.06	2.20	1.82	1.67
Radii (Å)				
ionic	2.12 (P^{3-})		0.92 (Sb^{3+})	1.08 (Bi^{3+})
covalent[b]	1.10	1.21	1.41	1.52
Melting point (°C)	44.1 (α-form)	814 (36 atm)	603.5	271.3

[a] Allred–Rochow type.
[b] For trivalent state.

The elements P, As, Sb and Bi show a considerable range in chemical behavior. There are fairly continuous variations in certain properties and characteristics, although in several instances there is no regular trend, for example, in the ability of the pentoxides to act as oxidizing agents. Phosphorus, like nitrogen, is essentially covalent in all of its chemistry, whereas arsenic, antimony, and bismuth show increasing tendencies to cationic behavior. Although the electronic structure of the next noble gas could be achieved by electron gain, considerable energies are involved, e.g.,

367

~ 1450 kJ mol^{-1} to form P^{3-} from P, and thus significantly ionic compounds such as Na$_3$P are few. The loss of valence electrons is similarly difficult to achieve because of the high ionization potentials. The 5+ ions do not exist, but for trivalent antimony and bismuth cationic behavior does occur. BiF$_3$ seems predominantly ionic, and salts such as Sb$_2$(SO$_4$)$_3$ and Bi(NO$_3$)$_3 \cdot$5H$_2$O, as well as salts of the oxo ions SbO$^+$ and BiO$^+$, exist.

Some of the more important trends are shown by the oxides, which change from acidic for phosphorus to basic for bismuth, and by the halides, which have increasingly ionic character, PCl$_3$ being instantly hydrolyzed by water to HPO(OH)$_2$, while the other trihalides give initially clear solutions which hydrolyze to As$_2$O$_3$, SbOCl and BiOCl, respectively. There is also an increase in the stability of the lower oxidation state with increasing atomic number; thus Bi$_2$O$_5$ is the most difficult to prepare and the least stable pentoxide.

Although oxidation states or oxidation numbers can be, and often are, assigned to these elements in their compounds, they are of rather limited utility except in the formalities of balancing equations. The important valence features concern the number of covalent bonds formed and the stereochemistries. The general types of compound and stereochemical possibilities are given in Table 13-2.

TABLE 13-2

No. of bonds to other atoms	Geometry[a]	Examples
3	Pyramidal	PH$_3$, AsCl$_3$, Ph$_3$Sb
4	Tetrahedral	PH$_4^+$, PO(OH)$_3$, Cl$_3$PO
	ψ-Trigonal bipyramidal	KSb$_2$F$_7$, SbCl$_3$(PhNH$_2$),
		K$_2$[Sb$_2$((+)-tart)$_2$]$\cdot$3H$_2$O,
		β-Sb$_2$O$_4$, SbOCl
5	Trigonal bipyramidal	PF$_5$, AsF$_5$, SbCl$_5$, Ph$_5$P, Ph$_5$As
	Square pyramidal	Ph$_5$Sb
	ψ-Octahedral	K$_2$SbF$_5$, [Sb$_4$F$_{16}$]$^{4-}$, Sb$_2$S$_3$, SbCl$_3$(PhNH$_2$)$_2$
6	Octahedral	PF$_6^-$, AsF$_6^-$, [Sb(OH)$_6$]$^-$, SbBr$_6^-$, (SbF$_5$)$_n$,
		[Bi$_6$O$_6$(OH)$_3$]$^{3+}$, [As(DMF)$_6$]$^{3+}$

[a] The designation ψ- indicates a central atom with bonds plus lone pair(s).

The differences between N and P in their chemistries, which are due to the same factors as are responsible for the C—Si and O—S differences, can be summarized as follows:

Nitrogen	Phosphorus
(a) Very strong $p\pi$–$p\pi$ bonds	No known $p\pi$–$p\pi$ bonds
(b) $p\pi$–$d\pi$ bonding is rare	Weak to moderate but important $d\pi$–$p\pi$ bonding
(c) No valence expansion	Valency expansion

The point (a) leads to facts such as the existence of P(OR)$_3$ but not of

$N(OR)_3$, nitrogen giving instead $O=N(OR)$, and the structural differences between nitrogen oxides and oxo acids and oxides such as P_4O_6 or P_4O_{10} and the polyphosphates. Point (b) is associated with rearrangements such as

$$\text{\Large \raisebox{0.5em}{$\diagdown$}}\!\!\!\!\underset{\diagup}{P}\!-\!OH \;\rightleftharpoons\; H\!-\!\underset{\diagup}{P}\!=\!O$$

and with the existence of phosphonitrilic compounds, $(PNCl_2)_n$. Further, while PX_3, AsX_3 and SbX_3 (X = halogen, alkyl or aryl), like NR_3 compounds, behave as donors owing to the presence of lone pairs, there is one major difference: the nitrogen atom can have no function other than simple donation because no other orbital is accessible, but P, As and Sb have empty d orbitals of fairly low energy. Thus, when the atom to which the P, As or Sb donates has electrons in orbitals of the same symmetry as the empty d orbitals, back-donation resulting in overall multiple-bond character may result. This factor is especially important for the stability of complexes with transition elements where $d\pi–p\pi$ bonding contributes substantially to the bonding (see Chapter 22). The consequences of vacant d orbitals are also evident on comparing the amine oxides, R_3NO, on the one hand, with R_3PO or R_3AsO on the other. In the N-oxide the electronic structure can be represented by the single canonical structure $R_3\overset{+}{N}—\bar{O}$, whereas for the others the bonds to oxygen have multiple character and are represented as resonance hybrids:

$$R_3\overset{+}{P}—\bar{O} \leftrightarrow R_3P=O \overset{2}{\leftrightarrow} R_3\overset{+}{P}\!\equiv\!\overset{-}{O}$$

These views are substantiated by the shortness of the P—O bonds ($\sim$1.45 as compared with $\sim$1.6 Å for the sum of the single-bond radii) and by the normal bond lengths and high polarities of N—O bonds. The amine oxides are also more chemically reactive, the P—O bonds being very stable indeed, as would be expected from their strength, $\sim$500 kJ mol^{-1}.

Point (c) is responsible for phenomena such as the Wittig reaction (page 390) and for the existence of compounds such as $(C_6H_5)_5P$, $P(OR)_5$, $[P(OR)_6]^-$, and $[PR_4]^+[PR_6]^-$ in which the coordination number is 5 or 6. The extent to which hybridization employing $3d$ orbitals is involved is somewhat uncertain since the d levels are rather high for full utilization and the higher states may be stabilized in part by electrostatic forces; it is significant that the higher coordination numbers for P^V are most readily obtained with more electronegative groups such as halogens, OR or phenyl.

The stereochemistry of compounds of P^V, As^V and Sb^V gains complexity because the *tbp* and *spy* configurations differ little in energy. While most of the molecular species MX_5 have *tbp* configurations, there are a few [$SbPh_5$, SbF_5 and perhaps $Sb(C_3H_5)_5$] for which the *spy* structure is more stable. In any event, the similar energies of the two configurations provide a pathway for stereochemical non-rigidity (cf. page 40) which is an essential factor in the chemistry of the pentavalent compounds.

It is also noteworthy that the only naturally occurring isotope of phosphorus, ^{31}P, has a nuclear spin of 1/2 and a large magnetic moment.

Nuclear magnetic resonance spectroscopy has, accordingly, played an extremely important role in the study of phosphorus chemistry.[1]

Both phosphorus and arsenic show a significant tendency to catenation, forming a series of cyclic compounds $(RP)_n$ and $(RAs)_n$ where $n = 3–6$, as well as some R_2PPR_2 and R_2AsAsR_2 compounds.

It will be noted in Table 13-2 that the five-coordinate compounds have different stereochemistries depending upon the formal valence state of the element. This is in accord with the principles discussed in Chapter 4. In the pentavalent compounds the central atom has only the five bonding pairs in its valence orbitals and the usual trigonal-bipyramidal arrangement is adopted. In the trivalent species (e.g., SbF_5^{2-}) there are six electron pairs; this anion is isoelectronic with BrF_5 and has the same ψ-octahedral structure (see page 130) where the five bonding and one non-bonding pair of electrons are approximately at the vertices of an octahedron and the larger charge cloud of the non-bonding pair makes the F—Sb—F angles somewhat less than 90°.

13-2. The Elements

Occurrence. Phosphorus occurs in various orthophosphate minerals, notably *fluorapatite*, $3Ca_3(PO_4)_2 \cdot Ca(F, Cl)_2$. Arsenic and antimony occur more widely, though in lower total abundance, and are often associated with sulfide minerals, particularly those of Cu, Pb and Ag. Bismuth ores are rather uncommon, the sulfide being the most important; bismuth also occurs in other sulfide minerals.

Phosphorus. The element is obtained by reduction of phosphate rock with coke and silica in an electric furnace. Phosphorus volatilizes as P_4 molecules (partly dissociated above 800° into P_2) and is condensed under water as white phosphorus.

$$2Ca_3(PO_4)_2 + 6SiO_2 + 10C \rightarrow P_4 + 6CaSiO_3 + 10CO$$

There are three main allotropic forms—white, red and black; each of these is polymorphic and there are at least eleven known modifications,[2] some amorphous, others of some indefinite identity, and all but three of unknown structure. In the liquid and in solid white phosphorus, there are tetrahedral P_4 molecules; in the vapor below 800°, where measurable dissociation to P_2 occurs, the element is also as P_4 molecules. The P—P distances are 2.21 Å; the P—P—P angles, of course, are 60°. The low angle indicates considerable strain, and the strain energy has been estimated to be about 96 kJ mol^{-1}. This means that the total energy of the six P—P bonds in the molecule is that much smaller than would be the total energy of six P—P bonds of the same length formed by phosphorus atoms with normal bond angles. Thus the structure of the molecule is consistent with its high reactivity. It is most likely that pure $3p$ orbitals are involved even though the bonds are

[1] M. M. Crutchfield *et al.*, *Topics in Phosphorus Chem.*, 1968, **5**, 1.
[2] C. C. Stephenson *et al.*, *J. Chem. Thermodynamics*, 1969, **1**, 59.

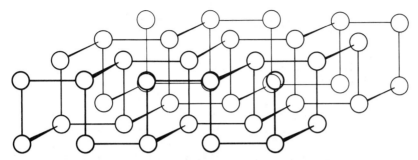

Fig. 13-1. The arrangement of atoms in the double layers found in crystalline black phosphorus.

bent, since hybridization such as pd^2, which would give 60° angles, would require a rather large promotion energy. MO calculations also indicate that d orbital participation is negligible.[3]

Black P is obtained in crystalline form by heating white P either under very high pressure, or at 220–370° for 8 days in the presence of mercury as a catalyst and with a seed of black P. The structure consists of double layers, each P atom being bound to three neighbors (Fig. 13-1). The closest P—P distances within each double layer are 2.17–2.20 Å, i.e., normal single-bond distances. The entire structure consists of a stacking of these double layers with the shortest P—P distance between layers at 3.87 Å. The crystals are therefore flaky, like graphite.

Red phosphorus is made by heating the white form at 400° for several hours. As many as six modifications may exist, none of which is structurally characterized, with one possible exception. Red, crystalline, so-called Hittorf's phosphorus has been structurally characterized,[4] but there is no evidence that this occurs in ordinary red phosphorus, though it may. It is made by a complex process involving slow heating in molten lead, followed by slow cooling. The structure is extremely complex, involving a criss-cross packing of infinite tubular chains of P atoms.

The various main forms of phosphorus show considerable difference in chemical reactivity; the white is by far the most reactive form, and the black the least. White P is stored under water to protect it from the air, whereas the red and black are stable in air; indeed, black P can be ignited only with difficulty. White P inflames in air and is soluble in organic solvents such as CS_2 and benzene. Some reactions of both red and white P are shown in Fig. 13-2.

In the P_4 molecule there should be a lone pair directed outwards from each P atom, and the molecule might thus be expected to have donor ability. Somewhat unstable complexes, of the type $RhCl(PR_3)_2(P_4)$ have been reported, but the authors postulate that Rh is bonded to a tetrahedral face of P_4.[5]

[3] I. H. Hillier and V. R. Saunders, *Chem. Comm.*, **1970**, 1233.
[4] H. Thurn and H. Krebs, *Acta Cryst.*, **1969**, *B*, **25**, 125.
[5] A. P. Ginsberg and W. E. Lindsell, *J. Amer. Chem. Soc.*, 1971, **93**, 2082.

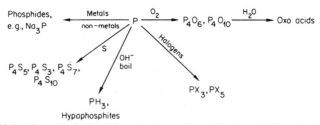

Fig. 13-2.   Some typical and important reactions of red and white phosphorus.

Finally, the techniques used to make S_n^{2+} ions (Section 15-6) have been utilized to obtain P_4^{2+} and P_8^{2+} ions, e.g.,

$$P_4 + H_2S_2O_7 \rightarrow P_4^{2+} + 2HS_3O_{10}^- + 5H_2SO_4 + SO_2$$

The As_4^{2+} and Sb_4^{2+} ions are obtained similarly.[5a]

Arsenic, Antimony and Bismuth.   These elements are obtained by reduction of their oxides with hydrogen or carbon. For As and Sb unstable yellow allotropes, presumably containing tetrahedral As_4 and Sb_4 molecules, can be obtained by rapid condensation of vapors. They are easily transformed into the stable forms, and yellow Sb is stable only at very low temperatures. Bismuth does not occur in a yellow form. The normal forms of As, Sb and Bi are bright and metallic in appearance and have crystal structures similar to that of black P. When heated, the metals burn in air to form the oxides, and they react directly and readily with halogens and some other non-metals. They form alloys with various other metals. Dilute non-oxidizing acids are without effect on them. With nitric acid, As gives arsenic acid, Sb gives the trioxide and Bi dissolves to form the nitrate.

BINARY COMPOUNDS

13-3. Phosphides, Arsenides and Antimonides

Direct interaction of phosphorus with many metals and metalloids gives binary compounds of four major types. (1) Volatile, molecular compounds (mostly with S, Se and Te). (2) A few more or less ionic phosphides, e.g., Na_3P, Ca_3P_2, Sr_3P_2. These, as well as the phosphides of the alkaline-earth, lanthanide and other electropositive metals, are generally rapidly hydrolyzed by water to PH_3. In phosphides such as K_2P_5 it is possible that discrete phosphide ions exist. In general, the structures, which are little known, may be more complex. In α-CdP_2, for example, there are spiral chains of P atoms (alternating P—P distances of 2.05 and 2.39 Å) with Cd atoms in between.[6] (3) Covalently bound, complex polymers. (4) Metal-like compounds ranging from hard, essentially metallic solids to amorphous polymeric powders. Phosphides of transition metals, for example, Fe_2P, are

[5a] R. C. Paul, J. K. Puri and K. C. Malhotra, *Chem. Comm.*, **1971**, 1031.
[6] J. Goodyear and G. A. Steigmann, *Acta Cryst.*, 1969, *B*, **25**, 2371.

commonly grey-black metallic substances, insoluble in water and conductors of electricity. They may also be ferromagnetic. The compositions and structures of these compounds are often very complex.

Arsenic and antimony give similar compounds, but the tendency to form volatile molecular compounds is much less, decreasing in the order $P > As > Sb \gg Bi$.

13-4. Hydrides

All the elements form gaseous hydrides of general formula MH_3, which can be obtained by treating phosphides or arsenides of electropositive metals with acids or by reduction of sulfuric acid solutions of arsenic, antimony or bismuth with an electropositive metal or electrolytically. The stability of these hydrides falls rapidly along the series, so that SbH_3 and BiH_3 are very unstable thermally, the latter having been obtained only in traces. The average bond energies are in accord with this trend in stabilities: E_{N-H}, 391; E_{P-H}, 322; E_{As-H}, 247; and E_{Sb-H}, 255 kJ mol^{-1}.

Phosphine, PH_3, has been most thoroughly studied.[7] The molecule is pyramidal with an HPH angle of 93.7°. Phosphine, when pure, is not spontaneously flammable, but often inflames owing to traces of P_2H_4 or P_4 vapor. However, it is readily oxidized by air when ignited, and explosive mixtures may be formed. It is also exceedingly poisonous. These properties account for its commercial unavailability. Unlike NH_3, it is not associated in the liquid state and it is only sparingly soluble in water; pH measurements show that the solutions are neither basic nor acidic—the acid constant is $\sim 10^{-29}$ and the base constant $\sim 10^{-26}$. It does, however, react with some acids to give phosphonium salts (page 386). It has been shown that the proton affinities of PH_3 and NH_3 (eq. 13-1) differ considerably.[8]

$$EH_3(g) + H^+(g) = EH_4^+(g) \qquad (13\text{-}1)$$
$$\Delta H° = 770 \text{ kJ mol}^{-1} \text{ for } E = P$$
$$\Delta H° = 866 \text{ kJ mol}^{-1} \text{ for } E = N$$

Also, the barrier to inversion[9] for PH_3 is 155 kJ mol^{-1} as compared with only 24 kJ mol^{-1} for NH_3. Quite generally the barriers for R_3P and R_3N compounds differ by about this much. Like other PX_3 compounds, PH_3 (and also AsH_3) forms complexes with transition metals, e.g., *cis*-$Cr(CO)_3(PH_3)_3$.[10]

Arsine, AsH_3, is extremely poisonous. Its ready thermal decomposition to arsenic, which is deposited on hot surfaces as a mirror, is utilized in tests for arsenic, for example, the well known Marsh test. *Stibine* is very similar to arsine but even less stable.

All these hydrides are strong reducing agents and react with solutions of

---

[7] E. Fluck and V. Novobilsky, *Fortschr. Chem. Forsch.*, 1969, **13**, 125.
[8] D. Holtz and J. L. Beauchamp, *J. Amer. Chem. Soc.*, 1969, **91**, 5913; M. A. Haney and J. L. Franklin, *J. Chem. Phys.*, 1969, **50**, 2028.
[9] J. M. Lehn and B. Munsch, *Chem. Comm.*, **1969**, 1327.
[10] E. O. Fischer, E. Lonis and C. G. Kreiter, *Angew. Chem. Internat. Edn.*, 1969, **8**, 377.

many metal ions, such as Ag^I and Cu^{II}, to give the phosphides, arsenides or stibnides or a mixture of these with the metals. In basic solution there is a reaction

$$\tfrac{1}{4}P_4 + 3H_2O + 3e = PH_3 + 3OH^- \qquad E^\circ = -0.89 \text{ V}$$

Phosphorus alone forms other hydrides. Diphosphine, P_2H_4, is generally formed along with phosphine and can be condensed as a yellow liquid. It is spontaneously flammable and decomposes on storage to form polymeric, amorphous yellow solids, insoluble in common solvents and of stoichiometry approximating to, but varying around, P_2H. It differs from N_2H_4 in having no basic properties. On photolysis it gives P_3H_5.[10a]

13-5. Halides[11]

The binary halides are of two main types, MX_3 and MX_5. All the trihalides except PF_3 are best obtained by direct halogenation, keeping the element in excess, whereas all the pentahalides may be prepared by treating the elements with an excess of the appropriate halogen.

All four of the Group V elements give all four trihalides. Besides the sixteen binary trihalides, a few of the many possible mixed trihalides are also known, namely, PF_2Cl, $PFCl_2$, PF_2Br, $PFBr_2$, PF_2I and $SbBrI_2$. Others that apparently cannot be isolated have been identified in mixtures by spectroscopic techniques. Redistribution reactions such as

$$PCl_3 + PBr_3 \rightleftharpoons PCl_2Br + PClBr_2$$

seem to reach equilibrium in a few minutes, though when fluorine is involved the rates are slower. All the trihalides, simple or mixed, are rapidly hydrolyzed by water and are rather volatile; the gaseous molecules have the expected pyramidal structures. Most appear to have molecular lattices, but the iodides AsI_3, SbI_3 and BiI_3 crystallize in layer lattices with no discrete molecules. BiF_3 has an ionic lattice, while SbF_3 has an intermediate structure in which SbF_3 molecules (Sb—F = 1.92 Å) are linked through F bridges (Sb$\cdots$F, 2.61 Å) to give each Sb^{III} a very distorted octahedral environment.[12] There is also some evidence from Raman spectra that $BiCl_3$ and $BiBr_3$ may have structures not entirely describable as molecular,[13] but detailed information is lacking.

Phosphorus trifluoride. This is a colorless gas, best made by fluorination of PCl_3. One of its most interesting properties is its ability to form complexes with transition metals similar to those formed by carbon monoxide. Like CO, it is highly poisonous because of the formation of a hemoglobin complex. PF_3 complexes (and other PX_3 complexes) are described in Section 22-13. Unlike the other trihalides, PF_3 is hydrolyzed only slowly by water,

[10a] T. P. Fehner, *J. Amer. Chem. Soc.*, 1968, **90**, 6062.
[11] D. S. Payne, *Topics in Phosphorus Chem.*, 1967, **4**, 85 (a comprehensive review); L. Kolditz, *Adv. Inorg. Chem. Radiochem.*, 1965, **7**, 1; J. F. Nixon, *Adv. Inorg. Chem. Radiochem.*, 1970, **13**, 364 (fluorophosphines).
[12] A. J. Edwards, *J. Chem. Soc., A*, **1970**, 2751.
[13] R. P. Oertel and R. A. Plane, *Inorg. Chem.*, 1969, **8**, 1188.

but it is attacked rapidly by alkalis. It has not been observed to have any
acceptor properties or to form complexes with F^-.

Phosphorus trichloride. The most common of the phosphorus halides is
a low-boiling liquid that is violently hydrolyzed by water to give phosphorous acid or, under special conditions, other acids of lower-valent
phosphorus. It also reacts readily with oxygen to give $OPCl_3$. The hydrolysis
of PCl_3 may be contrasted with that of NCl_3 which gives $HOCl$ and NH_3.
Fig. 13-3 illustrates some of the important reactions of PCl_3. Many of
these reactions are typical of other MX_3 compounds and also, with obvious
changes in formulas, of $OPCl_3$ and other oxo halides.

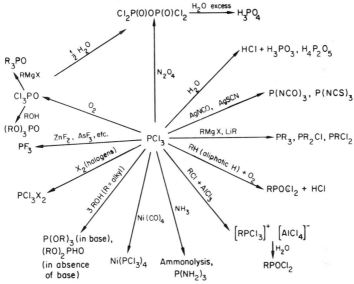

Fig. 13-3.   Some important reactions of PCl_3. Many of these are typical for other MX_3
compounds as well as for MOX_3 compounds.

Arsenic, Antimony and Bismuth Trihalides. Arsenic trihalides are similar
to those of phosphorus in both physical and chemical properties. However,
they have appreciable electrical conductances and chemical evidence suggests
that this may be due to auto-ionization to, for example, SbF_2^+ and SbF_4^-.
Thus, addition of KF or SbF_5 to liquid AsF_3 increases the conductance and
the compounds $KAsF_4$ and $SbF_5 \cdot AsF_3$ (AsF_2^+ SbF_6^-) can be isolated.[14]
$SbCl_3$ (m.p. 73.17°) has a high dielectric constant and is a strong Cl^-
acceptor. Many chlorides dissolve in the melt to give conducting solutions:

$$MCl + SbCl_3 \rightleftharpoons M^+ + SbCl_4^-$$

Certain aromatic polycyclic hydrocarbons also dissolve and give conducting solutions[15] (see also page 380). SbF_3, a white, readily hydrolyzed solid,
finds considerable use as a moderately active fluorinating agent. Both AsF_3

[14] A. J. Edwards and R. J. C. Sills, *J. Chem. Soc.*, A, **1971**, 942.
[15] P. V. Johnson, *J. Chem. Soc.*, A, **1971**, 2856.

and SbF_3 can function as F^- acceptors, although the product is seldom the simple MF_4^- ion. $SbCl_3$ differs from its P and As analogs in that it dissolves in a limited amount of water to give a clear solution that, on dilution, gives insoluble oxo chlorides such as $SbOCl$ and $Sb_4O_5Cl_2$. There is, however, no evidence to suggest that any simple Sb^{3+} ions exist in the solutions. $BiCl_3$, a white, crystalline solid, is hydrolyzed by water to $BiOCl$ but may be obtained from aqueous solution containing concentrated HCl since this reaction is reversible; $BiOCl$ redissolves in concentrated HCl, and $BiCl_3$ is obtained on evaporating such solutions.

The PF_2 group, which has a considerable derivative chemistry, can be introduced[16] by using PF_2I:

$$2PF_2I + Hg \rightarrow F_2PPF_2 + HgI_2$$
$$PF_2I + CuCN \rightarrow PF_2CN + CuI$$
$$PF_2I + RCOOAg \rightarrow RCOOPF_2 + AgI$$

Pentahalides. Seven binary pentahalides are known, namely, the four fluorides, PCl_5, PBr_5, and $SbCl_5$. $AsCl_5$ has never been isolated and phase studies show that it does not exist in stable equilibrium.

Fluorides. PF_5 is easily prepared by the interaction of PCl_5 with CaF_2 at 300–400°. It is a very strong Lewis acid and forms complexes with amines, ethers and other bases as well as with F^- in which phosphorus becomes 6-coordinate. However, these organic complexes are less stable than those of BF_3 and are rapidly decomposed by water and alcohols. Like BF_3, PF_5 is a good catalyst, especially for ionic polymerization.

The only arsenic pentahalide definitely known is the fluoride, which is similar to PF_5. The action of chlorine on AsF_3 at 0° gives a compound whose conductivity in an excess of AsF_3 suggests that it may be $[AsCl_4]^+ [AsF_6]^-$.

Antimony pentafluoride is a viscous liquid (b.p. 150°) which is associated even in the vapor state at the boiling point. Nmr studies of the liquid lead to the conclusion that each antimony atom is surrounded octahedrally by six fluorine atoms, two *cis* fluorines being shared with adjacent octahedra. It is not certain whether SbF_5 is dimeric or whether there are higher polymers but polymers are favored. In the crystal (m.p., 7°) it is tetrameric with a structure similar to that of $(NbF_4)_4$, Fig. 26-B-2, but with alternating fluorine bridge angles of 170° and 141°.[16a] Individual SbF_5 molecules have been possibly observed at 400° in the gas phase but dimers are certainly present.[17] PF_5 and AsF_5 molecules are *tbp*.

Bismuth pentafluoride, made by direct fluorination of liquid bismuth at 600° with fluorine at low pressure, is a white crystalline solid and an extremely powerful fluorinating agent.

AsF_5 and SbF_5, and, to a lesser extent PF_5, are potent fluoride ion acceptors, forming MF_6^- ions or more complex species. The PF_6^- ion is a common

[16] G. G. Flaskerud, K. E. Pullen and J. M. Shreeve, *Inorg. Chem.*, 1969, **8**, 728.
[16a] A. J. Edwards and P. Taylor, *Chem. Comm.*, **1971**, 1376.
[17] E. W. Lawless, *Inorg. Chem.*, 1971, **10**, 2084; L. E. Alexander, *Inorg. Nuclear Chem. Letters*, 1971, **7**, 1053.

and convenient "non-complexing" anion, which has even less coordinating ability than ClO_4^- or BF_4^-.[18]

In liquid HF, PF_5 is a non-electrolyte but AsF_5 and SbF_5 give conducting solutions presumably[18a] because of the reactions

$$2HF + 2AsF_5 \rightarrow H_2F^+ + As_2F_{11}^- \text{ (can be isolated as } Et_4N^+ \text{ salt)}$$
$$2HF + SbF_5 \rightarrow H_2F^+ + SbF_6^-$$

The use of SbF_5 to enhance the acidities of HF and, especially, HSO_3F has been discussed in detail (page 171).

Phosphorus(v) chloride has a *tbp* molecular structure in the gaseous state but the solid is $[PCl_4]^+[PCl_6]^-$. The tetrahedral PCl_4^+ ion (P—Cl$=1.91$ Å) can be considered to arise here by transfer of Cl^- to the Cl^- acceptor, PCl_5. It is not therefore surprising that many salts of the PCl_4^+ ion are obtained when PCl_5 reacts with other Cl^- acceptors,[19] viz.,

$$PCl_5 + TiCl_4 \rightarrow [PCl_4]_2^+[Ti_2Cl_{10}]^{2-} \text{ and } [PCl_4]^+[Ti_2Cl_9]^-$$
$$PCl_5 + NbCl_5 \rightarrow [PCl_4]^+[NbCl_6]^-$$

There is some evidence to suggest that PCl_5 is dimeric in CCl_4 solution, the structure being one with two octahedra sharing edges. In benzene or 1,2-dichloroethane it is monomeric and apparently trigonal-bipyramidal.

Solid phosphorus(v) bromide is also ionic, $[PBr_4]^+Br^-$. There is even an $AsCl_4^+$ ion, which occurs with such large anions as PCl_6^-, $AlCl_4^-$, and $AuCl_4^-$ even though $AsCl_5$ does not exist.

Antimony pentachloride, a powerful chlorinating agent, is normally a fuming yellow liquid (b.p., 79°; m.p., 6°), but colorless when highly pure.

A considerable number of mixed halides, such as PF_4Cl, PF_3Cl_2, PF_3Br_2, $SbCl_3F_2$, $SbCl_2F_3$ and $SbCl_4F$ are known as well as some mixed organo-halides (e.g., PF_3Me_2, PF_2Me_3, PCl_3Me_2 and $BiPh_3Cl_2$) and the two hydrido fluorides, PHF_4 and PH_2F_3. All of these molecules appear to have *tbp* structures[20] and to follow strictly the rule that equatorial positions are preferred by the less electronegative substituents. The compound PCl_4Me contains *tbp* molecules in benzene or CS_2 but appears to be $[PCl_3Me]^+Cl^-$ in the solid or in acetonitrile.[21]

The PHF_4 and PH_2F_3 molecules, which have only recently been characterized, can be prepared in two ways:[22,23]

$$H_2P(O)OH + 3HF \rightarrow PH_2F_3 + 2H_2O$$
$$HP(O)(OH)_2 + 4HF \rightarrow PHF_4 + 3H_2O$$
$$PF_5(g) \xrightarrow{Me_3SnH(g)} PHF_4 + PH_2F_3$$

[18] H. G. Mayfield, Jr., and W. E. Bull, *J. Chem. Soc., A.,* **1971**, 2279.

[18a] P. A. W. Dean *et al., J. Chem. Soc., A,* **1971**, 341.

[19] T. J. Kistenmacher and G. D. Stucky, *Inorg. Chem.,* 1968, 7, 2150; 1971, **10**, 122; P. Reich and H. Preiss, *Z. Chem.,* 1967, 7, 115; W. Wieker and A.-R. Grimmes, *Z. Naturforsch.,* 1967, **22b**, 257, 1220.

[20] J. Goubeau, R. Baumgartner and H. Weiss, *Z. anorg. Chem.,* 1966, **348**, 286; R. R. Holmes, *J. Chem. Phys.,* 1967, **46**, 3718; J. A. Salthouse and T. C. Waddington, *Spectrochim. Acta, A,* 1967, **23**, 1069; A. J. Downs and R. Schmutzler, *Spectrochim. Acta, A,* 1967, **23**, 681.

[21] I. R. Beattie, K. Livingston and T. Gilson, *J. Chem. Soc., A,* **1968**, 1.

[22] R. R. Holmes and R. N. Storey, *Inorg. Chem.,* 1966, **5**, 2146.

[23] P. M. Treichel, R. A. Goodrich and S. B. Pierce, *J. Amer. Chem. Soc.,* 1967, **89**, 2017.

The equatorial positions of the hydrogen atoms are indicated by vibrational[22] and microwave[24] spectra.

Lower Halides. Phosphorus and As form the so-called tetrahalides, P_2Cl_4, P_2I_4 and As_2I_4, which decompose on storage to the trihalide and non-volatile yellow solids and are readily decomposed by air and water. P_2I_4 has been shown to have an I_2P—PI_2 molecular structure with a *trans* rotomeric orientation in the solid[25] but probably staggered (like N_2H_4, page 352) in CS_2 solution.[26]

It has long been known that when metallic bismuth is dissolved in molten $BiCl_3$ a black solid of approximate composition $BiCl$ can be obtained. This solid is actually $Bi_{24}Cl_{28}$ and it has an elaborate constitution, consisting of four $BiCl_5^{2-}$, one $Bi_2Cl_8^{2-}$ and two Bi_9^{5+} ions, the structures of which are depicted in Fig. 13-4. The electronic structure of the Bi_9^{5+} ion, a metal atom cluster, has been successfully treated in terms of delocalized molecular orbitals.[27]

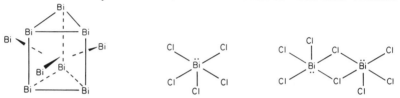

Fig. 13-4.　The structures of the species present in "BiCl," which is, in fact, $Bi_{24}Cl_{28}$.

More recent work suggests that other low-valent bismuth species are present in various molten salt solutions.[28] Among these are Bi^+ and Bi_5^{3+} and possibly others such as Bi_8^{2+}, Bi_4^{4+} and Bi_3^+. The presence of Bi^+ in a crystalline compound has been demonstrated by an X-ray crystallographic study[29] of $Bi_{10}Hf_3Cl_{18}$, which consists of Bi^+, Bi_9^{5+} and $3HfCl_6^{2-}$ ions. The Bi_9^{5+} has the same structure as that found in $Bi_{24}Cl_{28}$.

13-6. Halo Complexes

The trivalent elements form complexes formally derived from the trihalides by addition of further halide ions or other donors. The stereochemistries can be understood by considering the presence of the lone pair in the valence shell and employing the models outlined in Chapter 4.

For phosphorus(III) halides Lewis acidity is extremely limited; the only example appears to be

$$PBr_3 + Br^- \xrightarrow[\text{(C}_3\text{H}_7)_4\text{N}^+]{\overset{\text{in } PhNO_2, \quad ClH_2C-CH_2Cl}{}} \begin{array}{l} PBr_4^- \text{ (yellow-green soln.)} \\ [(C_3H_7)_4N]PBr_4 \text{ (yellow)} \end{array}$$

[24] S. B. Pierce and C. D. Cornwell, *J. Chem. Phys.*, 1968, **48**, 2118.

[25] Y. C. Leung and J. Waser, *J. Chem. Phys.*, 1965, **60**, 539.

[26] M. Baudler and G. Fricke, *Z. Anorg. Chem.*, 1963, **320**, 11.

[27] J. D. Corbett and R. E. Rundle, *Inorg. Chem.*, 1964, **3**, 1408.

[28] N. J. Bjerrum, C. R. Boston and G. P. Smith, *Inorg. Chem.*, 1967, **6**, 1162, 1968; C. R. Boston, *Inorg. Chem.*, 1970, **9**, 389; R. A. Lynde and J. D. Corbett, *Inorg. Chem.*, 1971, **10**, 1746.

[29] R. M. Friedman and J. D. Corbett, *Chem. Comm.*, **1971**, 422.

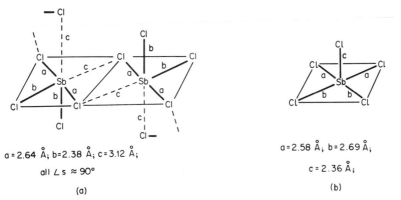

$a = 2.64$ Å; $b = 2.38$ Å; $c = 3.12$ Å;

all ∠s ≈ 90°

(a)

$a = 2.58$ Å; $b = 2.69$ Å;

$c = 2.36$ Å;

(b)

Fig. 13-5. (a) A portion of the anion chain in $(pyH)SbCl_4$. (b) The $SbCl_5^{2-}$ ion found in $(NH_4)_2SbCl_5$.

Infrared and Raman spectra suggest the presence of a discrete PBr_4^- ion with the expected SF_4-like (ψ-tbp) structure.[30]

There has been very little reported concerning the As^{III} halides except that AsF_3, and also SbF_3, have at best only slight Lewis acidity, but some complex ions, e.g., SbF_4^-, $Sb_2F_7^-$ and SbF_5^{2-}, are known.[30a]

Complex formation by Sb^{III} and, to a lesser extent Bi^{III}, has been extensively investigated. A great many compounds with stoichiometry M^ISbX_4 or M^ISbX_3Y have been prepared and spectroscopic evidence for ψ-tbp SbX_4^- or SbX_3Y^- ions has been reported.[31] In one instance, however, where the structure has been examined crystallographically,[32] the $SbCl_4^-$ ions are associated to form infinite, zigzag chains, as shown in Fig. 13-5a. There are also $SbCl_5^{2-}$ and $SbCl_6^{3-}$ complexes. The former, in $(NH_4)_2SbCl_5$ is found to have a discrete anion with a square-pyramidal (ψ-octahedral) structure[33] (Fig. 13-5b), as would be expected. The $SbCl_6^{3-}$ ion appears to have O_h symmetry in aqueous solution but to be distorted in $M_3^ISbCl_6$ crystals.[34] It is an example of the stereochemically anomalous AB_6E species discussed in Section 4.4.

$SbCl_3$ also forms complexes with neutral donors; those with $PhNH_2$ have structures showing that the N—Sb bonds are relatively weak and that the lone pair is stereochemically active, as illustrated in Fig. 13-6. The variations in Sb—Cl distances are in very good accord with the 3-center bond model discussed in Section 4-4.

A number of fluoro complexes of Sb^{III} are also known. The SbF_5^{2-} ion has a spy (ψ-octahedral) structure, like $SbCl_5^{2-}$. KSb_2F_7 has a complex

[30] K. B. Dillon and T. C. Waddington, *Chem. Comm.*, **1969**, 1317.
[30a] S. H. Mastin and R. R. Ryan, *Inorg. Chem.*, 1971, **10**, 1757; S. L. Lawton, R. A. Jacobsen and R. S. Frye, *Inorg. Chem.*, 1971, **10**, 701, 709.
[31] G. Y. Ahlijah and M. Goldstein, *J. Chem. Soc., A*, **1970**, 2590.
[32] S. K. Porter and R. A. Jacobson, *J. Chem. Soc., A*, **1970**, 1356.
[33] M. Webster and S. Keats, *J. Chem. Soc., A*, **1971**, 298.
[34] E. Martineau and J. B. Milne, *J. Chem. Soc., A*, **1970**, 2971.

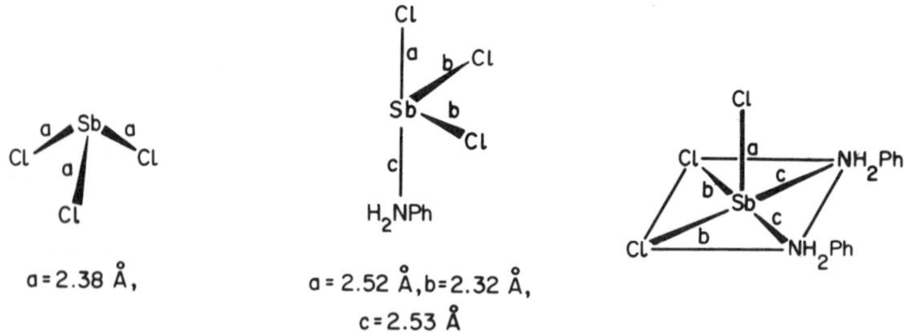

a = 2.38 Å,

a = 2.52 Å, b = 2.32 Å,

c = 2.53 Å

Fig. 13-6. The structures of $SbCl_3$ and two of its aniline complexes. [Reproduced by permission from R. Hulme, D. Mullen and J. C. Scrutton, *Acta Cryst.*, 1969, *A*, **25**, S. 171.]

structure[35] consisting of infinite chains of alternate *tbp* SbF_4^- ions and pyramidal SbF_3 molecules. In $KSbF_4$ there is an $Sb_4F_{16}^{4-}$ ion built of ψ-octahedral SbF_5 units (Fig. 13-7).

Bismuth(III) forms halo complexes that are in general, if not in detail, similar to those of Sb^{III}. Thus various solids with stoichiometries such as M^IBiX_4, where M^I is a large quaternary cation and X = Br or I, have been isolated but not definitively characterized as to structure. Raman studies have shown that Bi^{III} in aqueous HCl forms the ions $BiCl_4^-$, $BiCl_5^{2-}$ and $BiCl_6^{3-}$. Similar bromo ions and the iodo ions BiI_4^-, BiI_6^{3-}, BiI_7^{4-} (but not BiI_5^{2-}) also are formed in aqueous media.[36]

Antimony(III) chloride has long been known to form complexes with aromatic hydrocarbons; recently the structure of the napthalene complex $C_{10}H_8 \cdot 2SbCl_3$ was determined.[37] This can be schematically represented as in (13-I); each Sb atom has roughly ψ-*tbp* coordination with the hydrocarbon in an axial position, but the Sb—C distances are quite long (~ 3.37 Å), implying that the interaction is very weak.

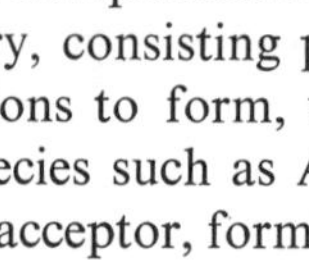

(13-I)

The pentahalides and related compounds have rather limited and straightforward coordination chemistry, consisting primarily of the acquisition by PF_5, AsF_5 and SbF_5 of F^- ions to form, usually, the MF_6^- ions though occasionally more complex species such as $As_2F_{11}^-$ (page 377) are formed. $SbCl_5$ is also a fairly good Cl^- acceptor, forming $SbCl_6^-$, and PCl_6^- occurs in solid PCl_5, as noted above.

[35] S. H. Mastin and R. R. Ryan, Abstracts Amer. Chem. Soc. Meeting, Chicago, September, 1970.
[36] R. P. Oertel and R. A. Plane, *Inorg. Chem.*, 1967, **6**, 1960; A. J. Eve and D. N. Hume, *Inorg. Chem.*, 1967, **6**, 331; A. Nystrom and O. Lindquist, *Acta Chem. Scand.*, 1967, **21**, 2570.
[37] R. Hulme and J. T. Szymanski, *Acta Cryst.*, 1969, *B* **25**, 753.

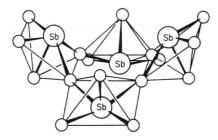

Fig. 13-7.   The structure of the $Sb_4F_{16}^{4-}$ ion in $KSbF_4$.

13-7. Oxides

The following oxides of the Group V elements are known, those in parentheses not having been structurally characterized:

P_4O_6	As_4O_6	Sb_4O_6	(B_2O_3)
(P_4O_7)			
P_4O_8		Sb_2O_4	
P_4O_9			
P_4O_{10}	(As_2O_5)	(Sb_2O_5)	

Phosphorus Oxides.   The phosphorus oxide most thoroughly studied and best understood is phosphorus pentoxide (named according to its empirical formula, P_2O_5, for historical reasons), the correct molecular formula of which is P_4O_{10}, as written above. It is usually the main product of burning phosphorus and, under proper conditions with an excess of oxygen, is the only product. It is a white, crystalline material that sublimes at 360° and 1 atm, and this constitutes an excellent method of purification since the products of incipient hydrolysis, which are the commonest impurities, are comparatively non-volatile. Phosphorus pentoxide exists in three crystalline polymorphs, in an amorphous form, and in a glassy form. A hexagonal crystal form, known as the H-form, is obtained on sublimation. It consists of P_4O_{10} molecules with the structure shown in Fig. 13-8. In this structure,

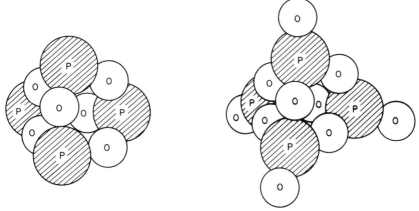

Fig. 13-8.   The structures of P_4O_6 and P_4O_{10}.

the P atoms are at the corners of a tetrahedron with six oxygen atoms along the edges and the remaining four lying along extended three-fold axes of the tetrahedron. The twelve P—O distances between phosphorus atoms and shared oxygen atoms are 1.62 Å, which is about the P—O single-bond distance, but the other four apical P—O distances are only 1.39 Å and indicate considerable $p\pi$–$d\pi$ double bonding. The same molecular units persist in the gas phase.

When the H-form is heated in a closed system for 24 hours at 450°, an orthorhombic form known as the O'-form is obtained. By heating the H-form for only 2 hours at 400°, a metastable orthorhombic form known as the O-form is obtained. In both of these, there are infinite sheets in which each phosphorus atom has essentially the same environment as in the P_4O_{10} molecules, namely, a tetrahedral set of bonds to three shared oxygen atoms and one unshared. When the H-form is melted, a volatile liquid of low viscosity is initially obtained, but this eventually changes into a viscous, non-volatile liquid that is the same as or similar to that obtained directly on melting the O- or the O'-form. It thus appears that the molecular H-form melts to a molecular liquid which requires time to reorganize into a polymeric structure. The glassy solid is obtained on chilling the viscous liquid.

The most important chemical property of P_4O_{10} is its avidity for water. It is one of the most effective drying agents known at temperatures below 100°. It reacts with water to form a mixture of phosphoric acids (see below) whose composition depends on the quantity of water and other conditions. It will even extract the elements of water from many other substances themselves considered good dehydrating agents; for example, it converts pure HNO_3 into N_2O_5 and H_2SO_4 into SO_3. It also dehydrates many organic compounds, e.g., converting amides into nitriles. With alcohols it gives esters of simple and polymeric phosphoric acids depending on reaction conditions. The breakdown of P_4O_{10} with various reagents (alcohols, water, phenols, ethers, alkyl phosphates, etc.) is a very general one and is illustrative also of the general reaction schemes for the breakdown of P_4S_{10} and for the reaction of P_4 with alkali to give PH_3, hypophosphite, etc. Thus the first reaction of an alcohol with P_4O_{10} can be written as equation 13-2, followed by further reaction at the next most anhydride-like linkage, until eventually products containing only one P atom are produced (eq. 13-3).

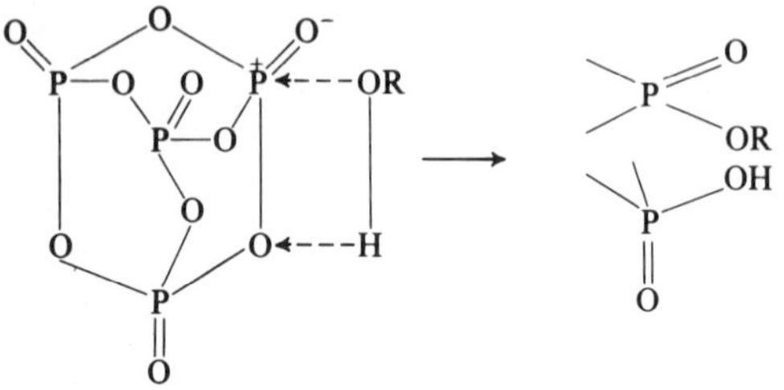

$$P_4O_{10} + 6ROH \longrightarrow 2(RO)_2PO\cdot OH + 2RO\cdot PO(OH)_2 \qquad (13\text{-}3)$$

The fusion of P_4O_{10} with basic oxides gives solid phosphates of various types, the nature depending on experimental conditions.

The other well-characterized oxide of phosphorus is the so-called trioxide, whose true molecular formula is P_4O_6. The structure of this molecule is very similar to that of P_4O_{10} except that the four non-bridging apical oxygens present in the latter are missing. P_4O_6 is a colorless, volatile compound (m.p. 23.8°, b.p. 175°) that is formed in about 50% yield when white phosphorus is burned in a deficit of oxygen. It is difficult to separate by distillation from traces of unchanged phosphorus, but irradiation with ultraviolet light changes the white phosphorus into red, from which the P_4O_6 can be separated by dissolution in organic solvents. The chemistry of P_4O_6 is not well known but it appears to be complex.[38]

When heated above 210°, P_4O_6 decomposes into red P and other oxides, PO_x. It reacts vigorously with chlorine and bromine to give the oxo halides and with iodine in a sealed tube to give P_2I_4. It is stable to oxygen at room temperature. When it is shaken vigorously with an excess of *cold* water, it is hydrated exclusively to phosphorous acid, H_3PO_3, of which it is formally the anhydride; P_4O_6 apparently cannot be obtained by dehydration of phosphorous acid. The reaction of phosphorus trioxide with *hot* water is very complicated, producing among other products PH_3, phosphoric acid and elemental P; it may be noted in partial explanation that phosphorous acid itself, and all trivalent phosphorus acids generally, are thermally unstable, for example,

$$4H_3PO_3 \rightarrow 3H_3PO_4 + PH_3$$

It has been found that materials with compositions PO_x, $1.5 \leqslant x \leqslant 2.5$, are definite substances.[39] PO_2 and $PO_{2.25}$ have been shown to consist of molecules P_4O_8 and P_4O_9 which have structures intermediate between those of P_4O_6 and P_4O_{10}. P_4O_8 and P_4O_9 have two and three outer oxygen atoms, respectively, instead of four as in P_4O_{10}. The oxide P_4O_7 is also known to exist and its gross structure presumably fits into the pattern.

Arsenic Oxides. Arsenic trioxide is formed on burning the metal in air. The gaseous molecules have the formula As_4O_6 and the same structure as P_4O_6. Three crystalline forms are known[40] plus a glass. The ordinary form contains the same tetrahedral As_4O_6 molecules as the gas, while in the other crystalline forms there are AsO_3 pyramids joined through the oxygen atoms to form layers. The As_4O_6 molecule can be thought of as consisting of four such AsO_3 units forming a closed group rather than an infinite sheet. The ordinary form is soluble in various organic solvents as As_4O_6 molecules and in water to give solutions of "arsenious acid." Arsenic pentoxide, whose true molecular formula and structure are unknown, cannot be obtained by direct reaction of arsenic with oxygen. It can be prepared by oxidation of As with nitric acid followed by dehydration of the arsenic acid hydrates so

[38] J. G. Riess and J. R. Van Wazer, *Inorg. Chem.*, 1966, **5**, 178.
[39] B. Beagley *et al.*, *Trans. Faraday Soc.*, 1969, **65**, 1219.
[40] K. A. Becker, K. Plieth and I. N. Stranski, *Progr. Inorg. Chem.*, 1962, **4**, 1.

obtained. It readily loses oxygen when heated to give the trioxide. It is very soluble in water, giving solutions of arsenic acid.

Antimony Oxides.   Antimony trioxide is also obtained by direct reaction of the metal with oxygen. In the vapor there are Sb_4O_6 molecules of the same tetrahedral structure as their P and As analogs. The solid form which is stable up to 570° has a molecular lattice of such molecules; above this temperature there is another solid form with a polymeric structure. The trioxide is insoluble in water and dilute nitric and sulfuric acids, but soluble in hydrochloric and certain organic acids. It dissolves in bases to give solutions of antimonates. Antimony pentoxide is prepared by the action of nitric acid on the metal. It loses oxygen on mild heating, to give the trioxide.

When either oxide of Sb is heated in air at about 900° there is formed a white insoluble powder of stoichiometry SbO_2. It has been found that SbO_2 exists in two structurally different but related forms.[41] Both forms contain a 1:1 mixture of Sb^{III} and Sb^V and thus the formula is usually written Sb_2O_4. The α-form is isostructural with $Sb^{III}Nb^VO_4$ and $Sb^{III}Ta^VO_4$.

Bismuth Oxides.   The only well-established oxide of bismuth is Bi_2O_3, a yellow powder soluble in acids to give bismuth salts but with no acidic character, being insoluble in alkalis. From solutions of bismuth salts, alkali or ammonium hydroxide precipitates a hydroxide, $Bi(OH)_3$, which is a definite compound. Like the oxide, this is completely basic in nature. It appears that a bismuth(v) oxide does exist, but that it is extremely unstable and has never been obtained in a completely pure state. It is obtained by the action of extremely powerful oxidizing agents on Bi_2O_3 and is a red-brown powder which rapidly loses oxygen at 100°.

The oxides of the Group V elements clearly exemplify two important trends that are manifest to some extent in all Groups of the Periodic Table: (1) the stability of the higher oxidation state decreases with increasing atomic number, and (2) in a given oxidation state the metallic character of the elements, and therefore the basicity of the oxides, increase with increasing atomic number. Thus, P^{III} and As^{III} oxides are acidic, Sb^{III} oxide is amphoteric and Bi^{III} oxide is strictly basic.

13-8. Sulfides

Phosphorus and sulfur combine directly above 100° to give a number of sulfides, the most important being P_4S_3, P_4S_5, P_4S_7 and P_4S_{10}. It is possible to obtain any one of these compounds in high yield by heating stoichiometric quantities of red phosphorus and sulfur for a suitable length of time at the proper temperature. A melting-point diagram for the P—S system has shown that these four are probably the only binary compounds that exist under equilibrium conditions, although there is some evidence for a phase

---

[41] D. Rogers and A. C. Skapski, *Proc. Chem. Soc.*, **1964**, 400; A. C. Skapski and D. Rogers, *Chem. Comm.*, **1965**, 611; G. G. Long, J. G. Stevens and L. H. Bowen, *Inorg. Nuclear Chem. Letters*, 1969, **5**, 799.

between P_4S_5 and $P_4S_{6.9}$. P_4S_3 is used commercially in matches and is soluble in organic solvents such as carbon disulfide and benzene. The structures of all four compounds are known, and in terms of the structures the above formulae do not appear as irrational as they otherwise might. P_4S_{10} has the same structure as P_4O_{10}. The others also have structures based on a tetrahedral group of phosphorus atoms with sulfur atoms bonded to

individual P atoms or bridging along the edges of the tetrahedron; their symmetries are respectively C_{3v}, C_i, and C_{2v}. Like P_4O_{10}, P_4S_{10} breaks down with alcohols, but with a different stoichiometry:

$$P_4S_{10} + 8ROH \rightarrow 4(RO)_2P(S)SH + 2H_2S$$

The difference is due to the fact that acids of the type $ROP(S)(SH)_2$ are more reactive than their oxygen analogs and react thus:

$$ROP(S)(SH)_2 + ROH \rightarrow (RO)_2P(S)SH + H_2S$$

These reactions of P_4S_{10} are important in that dialkyl and diaryl dithio-phosphates form the basis of many extreme-pressure lubricants, of oil additives and of flotation agents. Phosphorus gives compounds with Se and Te; P_4Se_3 has been shown to have the same structure as P_4S_3.

Arsenic forms the sulfides As_4S_3, As_4S_4, As_2S_3 and As_2S_5 by direct interaction; the last two can also be precipitated from hydrochloric acid solutions of As^{III} and As^V by H_2S. As_2S_3 is insoluble in water and acids but shows its acidic nature by dissolving in alkali sulfide solutions to give thio anions. As_2S_5 behaves similarly. As_4S_4 has a structure containing an As_4 tetrahedron (see page 435); As_2S_3 has the same structure as As_2O_3. As_4S_3 has the same molecular structure as P_4S_3.[42]

Antimony forms Sb_2S_3 either by direct combination of the elements or by precipitation with H_2S from Sb^{III} solutions; it dissolves in an excess of sulfide to give anionic thio complexes, probably mainly SbS_3^{3-}. Sb_2S_3, as well as Sb_2Se_3 and Bi_2S_3, have a ribbon-like polymeric structure in which each Sb atom and each S atom is bound to three atoms of the opposite kind, forming interlocking SbS_3 and SSb_3 pyramids (see page 434). So-called antimony(v) sulfide, Sb_2S_5, is not a stoichiometric substance and according to Mössbauer spectroscopy[43] contains only Sb^{III}.

Bismuth gives dark brown Bi_2S_3 on precipitation of Bi^{III} solutions by H_2S; it is not acidic. A sulfide, BiS_2, is obtained[44] as grey needles by direct

[42] H. J. Whitfield, *J. Chem. Soc., A*, **1970**, 1800.
[43] G. G. Long *et al.*, *Inorg. Nuclear Chem. Letters*, 1969, **5**, 21; T. Birchall and B. Della Valle, *Chem. Comm.*, **1970**, 675.
[44] M. S. Silverman, *Inorg. Chem.*, **1965**, **4**, 587.

interaction at 1250° and 50 kbar; its structure is unknown but may be $Bi^{3+}[BiS_4]^{3-}$.

Selenides and tellurides of As, Sb and Bi can be made. Some of these, for example, bismuth telluride, have been studied intensively as semi-conductors.

OTHER COMPOUNDS

13-9. Oxo Halides

These compounds are of various stoichiometric types. Among the most important are the phosphoryl halides, X_3PO, in which X may be F, Cl or Br. The most important one is Cl_3PO, obtainable by the reactions

$$2PCl_3 + O_2 \rightarrow 2Cl_3PO$$
$$P_4O_{10} + 6PCl_5 \rightarrow 10Cl_3PO$$

The reactions of Cl_3PO are much like those of PCl_3. The halogens can be replaced by alkyl or aryl groups by means of Grignard reagents, and by alkoxo groups by means of alcohols; hydrolysis by water yields phosphoric acid. Cl_3PO also has donor properties toward metal ions, and many complexes are known. Distillation of the Cl_3PO complexes of $ZrCl_4$ and $HfCl_4$ can be used to separate Zr and Hf, and the very strong Cl_3PO—Al_2Cl_6 complex has been utilized to remove Al_2Cl_6 from adducts with Friedel–Crafts reaction products.

The structure of all X_3PO molecules consists of a pyramidal PX_3 group with the oxygen atom occupying the fourth position to complete a distorted tetrahedron. Corresponding compounds, X_3PS and X_3PSe, exist.

More complex oxo halides containing P—O—P bonds are known; some of these have linear structures, while some form rings. The linear compound $Cl_2(O)P$—O—$P(O)Cl_2$ is obtained either by oxidation of PCl_3 with N_2O_4 or by partial hydrolysis of Cl_3PO; the fluorine analogs exist.

Antimony and bismuth form the important oxo halides SbOCl and BiOCl which are insoluble in water. They are precipitated when solutions of Sb^{III} and Bi^{III} in concentrated HCl are diluted. They have quite different structures; both are complicated.

The only oxo halide of As, F_3AsO (b.p. 26°), is made by fluorination of an equimolar mixture of $AsCl_3$ and As_2O_3.

13-10. Phosphonium Compounds

Although organic derivatives of the type $[MR_4]^+X^-$ are well known for M = P, As and Sb (see page 390), only phosphorus gives the hydrogen-containing prototype, PH_4^+, and this does not form any very stable compounds. As noted in Section 13-4, the proton affinity of PH_3 is substantially less than that of NH_3. The PH_4^+ ion is tetrahedral[45] with P—H = 1.414 Å as

[45] A. Sequeira and W. C. Hamilton, *J. Chem. Phys.*, 1967, **47**, 1818.

compared to P—H $= 1.44$ Å in PH_3. The best known phosphonium salt is the iodide, which is formed as colorless crystals on mixing of gaseous HI and PH_3. The chloride and bromide are even less stable; the dissociation pressure of PH_4Cl into PH_3 and HCl reaches 1 atm below 0°. The estimated basicity constant of PH_3 in water is about 10^{-26}, and phosphonium salts are completely hydrolyzed by water, releasing the rather insoluble gas PH_3:

$$PH_4I(s) + H_2O \rightarrow H_3O^+ + I^- + PH_3(g)$$

PH_3 dissolves in very strong acids such as $BF_3 \cdot H_2O$ and $BF_3 \cdot CH_3OH$ where it is protonated to PH_4^+.[46]

Perhaps the most readily produced organic phosphonium compound is obtained by the interaction of phosphine with formaldehyde in hydrochloric acid solution:

$$PH_3 + 4CH_2O + HCl \rightarrow [P(CH_2OH)_4]^+Cl^-$$

It is a white crystalline solid soluble in water and it is available commercially. On addition of base it forms $P(CH_2OH)_3$.

13-11. Phosphorus–Nitrogen Compounds

A great many compounds are known with P—N and P=N bonds.[47] R_2N—P bonds are particularly stable and occur widely in combination with bonds to other univalent groups, such as P—R, P—Ar and P—halogen. The entire subject of P—N compounds cannot be properly outlined in the space available. Hence, only one part of it, the chemistry of phosphazenes (also called phosphonitrilic compounds), will be described in detail.

The phosphazenes are cyclic or chain compounds that contain alternating phosphorus and nitrogen atoms with two substituents on each phosphorus atom. The three main structural types are the cyclic trimer (13-II), cyclic tetramer (13-III) and the oligomer or high polymer (13-IV). A few cyclic pentamers and hexamers are also known. The alternating sets of single and double bonds in (13-II) to (13-IV) are written for convenience but (see below) should not be taken literally.

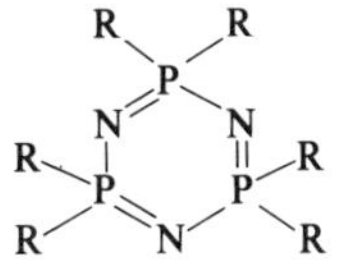

(13-II) 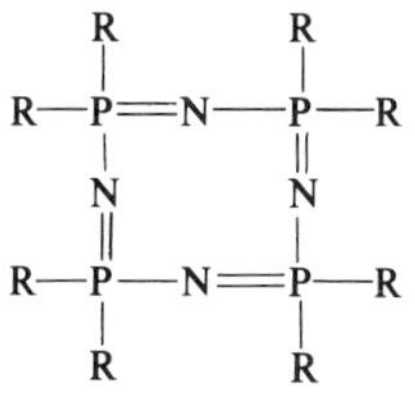   (13-III) 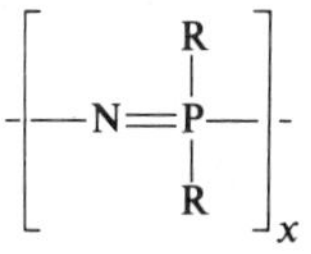   (13-IV)

Hexachlorocyclotriphosphazene, $(NPCl_2)_3$, is a key intermediate in the synthesis of many other phosphazenes and is manufactured commercially. On either a large or a small scale, it is readily prepared by the reaction

$$nPCl_5 + nNH_4Cl \xrightarrow{\text{in } C_2H_2Cl_4 \text{ or } C_6H_5Cl} (NPCl_2)_n + 4nHCl$$

[46] G. M. Sheldrick, *Trans. Faraday Soc.*, 1967, **63**, 1077.
[47] (a) E. Fluck, *Topics in Phosphorus Chem.*, 1967, **4**, 291; (b) H. R. Alcock, *Chem. Eng. News*, **1968**, April 22, p. 68.

This reaction produces a mixture of $[NPCl_2]_n$ species with $n = 3, 4, 5, \ldots$ and low-polymeric linear species. Favorable conditions give 90% yields of the $n = 3$ or 4 species, which can be separated without great difficulty.

The majority of reported phosphazene reactions involve replacement of halogen atoms by other groups (OH, OR, NR_2, NHR, or R) to give partially or fully substituted derivatives:

$$(NPCl_2)_3 + 6NaOR \rightarrow [NP(OR)_2]_3 + 6NaCl$$
$$(NPCl_2)_3 + 6NaSCN \rightarrow [NP(NCS)_2]_3 + 6NaCl$$
$$(NPF_2)_3 + PhLi \rightarrow (NPPh_2)_3 + 6LiF$$

The mechanisms of these reactions, especially those with organometallic reagents, are not fully understood, but in general they appear to involve S_N2 attack on P by an anion. In partly substituted molecules a great many isomers are possible.

There are also P—N ring systems not requiring formal assignment of P=N bonds, although bond lengths suggest that some P—N π bonding does in fact occur. An interesting example[48] is (13-VI), which is more stable than (13-V) from which it is obtained by a thermal rearrangement.

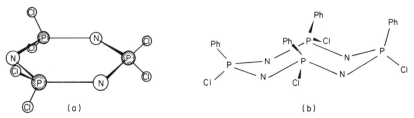

(13-V)   (13-VI)

The rings in $(NPF_2)_x$ where $x = 3$ or 4 are planar and those when $x = 5$ or 6 approach planarity. For other $(NPX_2)_n$ compounds the six-rings are planar or nearly so, but larger rings are generally non-planar with NPN angles of $\sim 120°$ and PNP angles of $\sim 132°$. Fig. 13-9 shows the structures of $(NPCl_2)_3$ and $(NPClPh)_4$.[49] The P—N distances, which are generally equal or very nearly so in these ring systems, lie in the range 1.56–1.61 Å; they are thus shorter than the expected single-bond length of ~ 1.75–1.80 Å.

Considerable attention has been paid to the nature of the P—N π bonding,

(a)   (b)

Fig. 13-9. The structures of two representative cyclic phosphazenes.
(a) $(NPCl_2)_3$. (b) All-*cis*-$(NPClPh)_4$.

[48] G. B. Ansell and G. J. Bullen, *J. Chem. Soc., A,* **1968,** 3026.
[49] G. J. Bullen and P. A. Tucker, *Chem. Comm.,* **1970,** 1185.

which the P—N distances indicate is appreciable, but the matter is still subject to controversy. The main question concerns the extent of delocalization, i.e. whether there is complete delocalization all around the rings to give them a kind of aromatic character, or whether there are more localized "islands" within the NPN segments. Of course, there may be considerable differences between the essentially planar rings and those which are puckered.[50]. The problem is a complicated one owing to the large number of orbitals potentially involved and to the general lack of ring planarity which means that rigorous assignment of σ and π character to individual orbitals is impossible.

The high-polymeric linear phosphazenes are potentially useful materials as far as physical and mechanical properties are concerned, but they have been generally useless because of chemical (especially hydrolytic) instability. Recently, use of perfluoroalkoxy and other side groups has given promise that useful polymers may yet be developed.[47b, 51]

13-12. Organic Derivatives

There is a vast chemistry of organophosphorus compounds, and even for arsenic, antimony and bismuth the literature is voluminous.[52] Consequently only a few topics can be mentioned in this book, a systematic approach being beyond its scope.

With a few exceptions to be mentioned at the end of this Section, the organo derivatives are compounds with only three or four bonds to the central atom. They may be prepared in a great variety of ways, the simplest being by treatment of halides or oxo halides with Grignard reagents:

$$(O)MX_3 + 3RMgX \rightarrow (O)MR_3 + 3MgX_2$$

Trimethylphosphine is spontaneously flammable in air, but the higher trialkyls are oxidized more slowly. The R_3MO compounds, which may be obtained from the oxo halides as shown above or by oxidation of the corresponding R_3M compounds, are all very stable.

There are good methods[53] for preparing optically pure dissymetric phosphine oxides, abcPO, e.g., $(CH_3)(C_3H_7)(C_6H_5)PO$. It is then possible to reduce these to optically pure phosphines with either retention or inversion. The reductant $HSiCl_3$ (page 320) accomplishes this with either retention or inversion depending upon the base used in conjunction with it. Hexachloro-

[50] G. R. Branton et al., J. Chem. Soc., A, 1970, 151.
[51] Chem. Eng. News, 1969, January 13, p. 34.
[52] G. O. Doak and L. D. Freedman, Organometallic Compounds of Arsenic, Antimony and Bismuth, Wiley-Interscience, 1970; Organophosphorus Chemistry, Specialist Periodical Reports, Chemical Society, London; A. J. Kirby and S. G. Warren, The Organic Chemistry of Phosphorus, Elsevier, 1967 (reaction mechanisms); P. G. Harrison, Organometallic Chem. Rev., 1970, 5, 183 (Bi); M. Dub, Organometallic Compounds, Vol. 3, 2nd edn., Springer Verlag (As, Sb, Bi).
[53] K. Mislow et al., J. Amer. Chem. Soc., 1967, 89, 4784; 1968, 90, 4842.

disilane reduces with inversion,[54] to account for which the following mechanism has been proposed:

$$abcPO + Si_2Cl_6 \rightarrow abcP^+OSiCl_3 + SiCl_3^-$$
$$SiCl_3^- + abcP^+OSiCl_3 \rightarrow Cl_3SiP^+abc + SiCl_3O^- \text{ (inversion)}$$
$$Cl_3SiP^+abc + SiCl_3O^- \rightarrow Cl_3SiOSiCl_3 + Pabc \text{ (attack of } Cl_3SiO^- \text{ on SiCl}_3)$$

Interestingly, the same reagent removes S from abcPS with retention; it is presumed that the first step is similar, but that $SiCl_3^-$ then attacks sulfur, rather than phosphorus:

$$abcPS + Si_2Cl_6 \rightarrow abcP^+SSiCl_3 + SiCl_3^-$$
$$abcP^+SSiCl_3 + SiCl_3^- \rightarrow abcP + Cl_3SiSSiCl_3$$

Trialkyl- and triaryl-phosphines, -arsines and -stibines are all good donors toward d-group transition metals and chelating di- and tri-phosphines and -arsines have been especially widely used as π-acid ligands (Section 22-13). The oxides, R_3MO, also form many complexes, but they function simply as donors. Trialkyl- and triaryl-phosphines, -arsines and -stibines generally react with alkyl and aryl halides to form quaternary salts:

$$R_3M + R'X \rightarrow [R_3R'M]^+X^-$$

The stibonium compounds are the most difficult to prepare and are the least common. These quaternary salts, excepting the hydroxides, which are obtained as sirupy masses, are white crystalline compounds. The tetraphenyl-phosphonium and -arsonium ions are useful for precipitating large anions such as ReO_4^-, ClO_4^- and complex anions of metals. Tetraphenylstibonium hydroxide has been shown[55] to contain tbp molecules, with the OH group in an axial position.

Triphenylphosphine, a white crystalline solid, is utilized widely in the Wittig reaction for olefin synthesis. This reaction involves the formation of alkylidenetriphenylphosphoranes from the action of butyllithium or other base on the quaternary halide, for example,

$$[(C_6H_5)_3PCH_3]^+Br^- \xrightarrow{n-\text{Butyllithium}} (C_6H_5)_3P{=}CH_2$$

This intermediate reacts very rapidly with aldehydes and ketones to give zwitterionic compounds (13-VII) which eliminate triphenylphosphine oxide under mild conditions to give olefins (13-VIII):

$$(C_6H_5)_3P{=}CH_2 \xrightarrow{\text{Cyclohexanone}} \text{(13-VII)} \longrightarrow \text{(13-VIII)} + (C_6H_5)_3PO$$

Although these alkylidenephosphoranes were long considered to be too unstable to isolate unless the alkylidene group contained stabilizing substituents, recent work[56] has accomplished the isolation of such substances as $Me_3P{=}CH_2$, $Et_3P{=}CH_2$, $Me_2EtP{=}CH_2$ and $Et_3P{=}CHMe$, all colorless liquids, stable for long periods in an inert atmosphere, and

[54] K. Mislow et al., J. Amer. Chem. Soc., 1969, 91, 7012, 7023.
[55] A. L. Beauchamp, M. J. Bennett and F. A. Cotton, J. Amer. Chem. Soc., 1969, 91, 297.
[56] H. Schmidbauer and W. Tronich, Chem. Ber., 1968, 101, 595, 604; A. Schmidt, Chem. Ber., 1968, 101, 4015.

Me_3As=CH_2, a colorless crystalline solid. The structures of a number of Ph_3P=CXY ylides have been determined, including that of the "parent" Ph_3PCH_2.[57] The P—C distances range from 1.66 Å in Ph_3PCH_2, clearly indicative of double-bond character, to 1.74 Å. A P—C single bond would have a length of 1.80–1.85 Å.

Cyclopoly-phosphines and -arsines.[58] These are compounds of general formula $(RP)_n$ or $(RAs)_n$, where $n = 4, 5, 6$, or more, and there are many other types. Typical preparative reactions are:

$$\tfrac{4}{n}RPH_2 + \tfrac{4}{n}RPCl_2 \rightarrow (RP)_n + nHCl$$
$$nRAsCl_2 + nHg \rightarrow (RAs)_n + nHgCl_2$$

Many, though not all, have good thermal stability, but they are reactive.

The $(RP)_4$ and $(RP)_5$ compounds, with $R = CF_3$, have the structures shown in Fig. 13-10. $(MeAs)_5$ has a very similar structure to that of $(CF_3P)_5$, and $(CF_3As)_4$ is isostructural[59] with $(CF_3P)_4$. In one form of $(C_6H_5P)_n$, six-rings with a chair conformation are found; note that all rings are puckered. Nmr studies of $(CF_3P)_5$ and $(CF_3As)_5$ have shown that they are fluxional.[60] It appears that two processes, one a relatively fast flapping of the out-of-plane ring atom and the other a slower rotation of the position of pucker around the ring are involved.

There are a few linear triphosphines that all contain CF_3 and they are obtained by the reactions[61]

$$2(CF_3)_2PI + CF_3PH_2 + 2(CH_3)_3N \rightarrow (CF_3)_2P—P(CF_3)—P(CF_3)_2 + 2(CH_3)_3NHI$$
$$2(CF_3)_2PCl + CH_3PH_2 + 2(CH_3)_3N \rightarrow CH_3P[P(CF_3)_2]_2 + 2(CH_3)_3NHCl$$

The extensive series of dimethylarsenic compounds, often called "cacodyl" compounds, e.g., Me_2AsCl, cacodyl chloride, is worthy of mention. There are also diarsenic tetraalkyls such as dicacodyl, $Me_2AsAsMe_2$, and $Et_2AsAsEt_2$.

A phosphorus-to-carbon triple bond evidently occurs in HC≡P, which is a pyrophoric substance that polymerizes slowly even at −130°. The C≡P bond length is only 1.54 Å.[62]

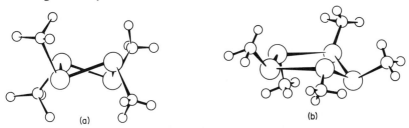

Fig. 13-10. The structures of (a) $(CF_3P)_4$ and (b) $(CF_3P)_5$. Large, medium and small circles represent P, C and F atoms, respectively.

[57] J. C. J. Bart, *J. Chem. Soc., B*, **1969**, 350.
[58] L. Maier, *Fortschr. Chem. Forsch.*, 1967, **8**, 1; A. H. Cowley and R. P. Pinnell, *Topics in Phosphorus Chem.*, 1967, **4**, 1.
[59] N. Mandel and J. Donohue, *Acta Cryst.*, 1971, B **27**, 476.
[60] E. J. Wells *et al.*, *Chem. Comm.*, **1967**, 895; *Canad. J. Chem.*, 1968, **46**, 2733.
[61] A. B. Burg and J. F. Nixon, *J. Amer. Chem. Soc.*, 1964, **86**, 356; A. B. Burg and K. K. Joshi, *J. Amer. Chem. Soc.*, 1964, **86**, 353.
[62] J. K. Taylor, *J. Chem. Phys.*, 1964, **40**, 1170.

Finally, the existence of organo derivatives of the pentavalent elements must be mentioned. These tend to be more stable with the heavier elements, while the corresponding R_3M compounds become less stable. Pentaaryl compounds are much more stable than pentaalkyl ones; none of the latter can be isolated for P or As, though several of antimony are known. The pentaphenyl compounds are perhaps the best characterized pure organic derivatives, but a host of mixed compounds, especially of antimony, of the types Ph_4SbX and Ph_3SbX_2 (X = OH, OR or halogen) are known. In all but two cases these molecules have *tbp* structures, with the more electro-negative ligands at axial positions. A configuration closer to *spy* has been found for crystalline Ph_5Sb, and an *spy* configuration has been postulated on spectroscopic evidence for penta(cyclopropyl)antimony.[63]

The antimony compounds $Me_3Sb(NO_3)_2$ and $Me_3Sb(ClO_4)_2$, which appear to be molecular, with *tbp* structures in the solid, dissolve in water and ionize,[64] apparently to give the planar cation $(CH_3)_3Sb^{2+}$.

13-13. Aqueous Cationic Chemistry

Apart from the quaternary salts mentioned in the preceding Section, there is no cationic chemistry of P and As. Although the reactions

$$H_2O + OH^- + AsO^+ \leftarrow As(OH)_3 \rightarrow As^{3+} + 3OH^-$$

may occur to some slight extent, there is little direct evidence for the existence of significant concentrations of either of the cations even in strong acid solutions.

Antimony has some definite cationic chemistry, but only in the trivalent state, the basic character of Sb_2O_5 being negligible. Cationic compounds of Sb^{III} are mostly of the so-called "antimonyl" ion, SbO^+, although some of the "Sb^{3+}" ion, such as $Sb_2(SO_4)_3$, are known. Antimony salts readily form complexes with various acids in which the antimony forms the nucleus of an anion.

In sulfuric acid, from $0.5M$ to $12M$, Sb^V appears to be present as the $[Sb_3O_9]^{3-}$ ion. For Sb^{III} in sulfuric acid, the species present vary markedly with the acid concentration,[65] namely,

$$SbO^+ \text{ and/or } Sb(OH)_2^+ \text{ in } < 1.5M \ H_2SO_4$$
$$SbOSO_4^-, Sb(SO_4)_2^- \text{ in } 1.0-18M \ H_2SO_4$$

The $Sb(C_2O_4)_3^{3-}$ ion forms isolable salts and has been shown to have the ψ-pentagonal-bipyramid structure (Fig. 13-11a) with a lone pair at one axial position.[66] The tartrate complexes of antimony(III) have been much studied as "tartar emetic" has been used medicinally for over three hundred years. This potassium salt and salts of NH_4^+ and $[Fe(phen)_3]^{2+}$ are known to have a binuclear structure with tartrate bridges as shown in Fig. 13-11b.[67]

[63] A. H. Cowley *et al. J. Amer. Chem. Soc.*, 1971, **93**, 2150.
[64] A. J. Downs and I. A. Steer, *J. Organometallic Chem.*, 1967, **8**, P21.
[65] J. L. Dawson, J. Wilkinson and M. I. Gillibrand, *J. Inorg. Nuclear Chem.*, 1970, **32**, 501.
[66] M. C. Poore and D. R. Russell, *J. Chem. Soc., A*, **1971**, 18.
[67] R. E. Tapscott, R. L. Belford and I. C. Paul, *Coordination Chem. Rev.*, 1969, **4**, 323.

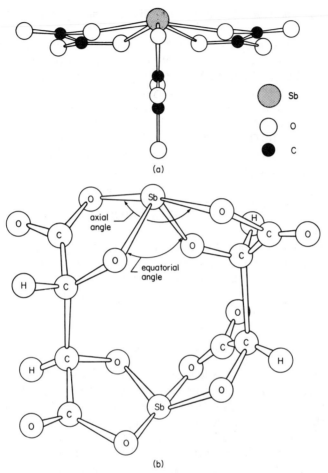

Fig. 13-11. The structures of two oxo acid complexes of Sb^{III}. (a) The Sb$(C_2O_4)_3^{3-}$ ion projected on a plane approximately perpendicular to the basal plane of the pentagonal pyramid. (b) Geometry of the Sb^{III} tartrato-bridged binuclear complex ion of "tartar emetic." [Reproduced by permission from reference 67.]

The Sb atom has ψ-*tbp* geometry as in a number of other Sb^{III} species. This coordination is also found[68] in the complex K[As$(C_6H_4O_2)_2$] derived from pyrocatechol (*o*-dihydroxybenzene).

Only for bismuth can it be said that there is an extensive true cationic chemistry. Aqueous solutions contain well defined hydrated cations but there is no evidence for the simple aquo ion [Bi$(H_2O)_n$]$^{3+}$. In neutral perchlorate solutions the main species is [Bi$_6O_6$]$^{6+}$ or its hydrated form, [Bi$_6$(OH)$_{12}$]$^{6+}$, and at higher pH [Bi$_6O_6$(OH)$_3$]$^{3+}$ is formed. The [Bi$_6$(OH)$_{12}$]$^{6+}$ species contains an octahedron of Bi^{3+} ions with an OH$^-$ bridging each edge.

[68] A. C. Skapski, *Chem. Comm.*, **1966**, 10.

Vibrational analysis suggests some weak bonding directly between the Bi atoms.[69]

There is considerable evidence for the association of Bi^{3+} ion with nitrate ions in aqueous solution. The nitrate ions appear to be mainly bidentate, and all members of the set $Bi(NO_3)(H_2O)_n^{2+} \dots Bi(NO_3)_4^-$ appear to occur.[70] From acid solution various hydrated crystalline salts such as $Bi(NO_3)_3 \cdot 5H_2O$, $Bi_2(SO_4)_3$ and double nitrates of the type $M_3^{II}[Bi(NO_3)_6]_2 \cdot 24H_2O$ can be obtained. Treatment of Bi_2O_3 with nitric acid gives bismuthyl salts such as $BiO(NO_3)$ and $Bi_2O_2(OH)(NO_3)$. Similar bismuthyl salts are precipitated on dilution of strongly acid solutions of various bismuth compounds. Bismuthyl salts are generally insoluble in water.

THE OXO ANIONS

The oxo anions in both lower and higher states are a very important part of the chemistry of phosphorus and arsenic and comprise the only real aqueous chemistry of these elements. For the more metallic antimony and bismuth, oxo anion formation is less pronounced, and for bismuth only ill-defined "bismuthates" exist.

13-14. Oxo Acids and Anions of Phosphorus

All phosphorus oxo acids have POH groups in which the hydrogen atom is ionizable; hydrogen atoms in P—H groups are not ionized. There are a vast number of oxo acids or ions, some of them of great technical importance; but, with the exception of the simpler species, they have not been well understood structurally until quite recently. We can attempt to deal only with some structural principles and some of the more important individual compounds. The *oxo anions* are of main importance, since in many cases the free acid cannot be isolated, though its salts are stable. Both lower (P^{III}) and higher (P^V) acids are known.

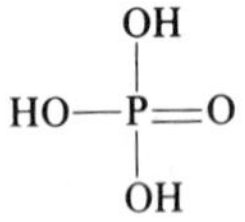

(13-IXa) 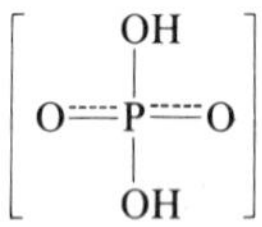   (13-IXb) 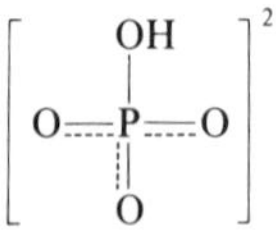   (13-IXc) 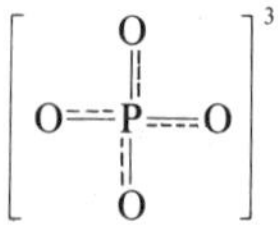   (13-IXd)

The principal higher acid is orthophosphoric acid, (13-IXa), and its various anions, (13-IXb, c, d), all of which are tetrahedral. The phosphorus(III) acid, which might naïvely have been considered to be $P(OH)_3$, has in fact the four-connected, tetrahedral structure (13-Xa); it is only difunctional and its anions are (13-Xb) and (13-Xc). Only in the triesters of

[69] V. A. Maroni and T. G. Spiro, *Inorg. Chem.*, 1968, **7**, 183.
[70] R. P. Oertel and R. A. Plane, *Inorg. Chem.*, 1968, **7**, 1192.

$$\begin{array}{ccc}
\begin{array}{c}
H \\
| \\
HO-\!\!-P-\!\!-OH \\
\| \\
O
\end{array}
&
\left[\begin{array}{c}
H \\
| \\
O\!=\!\!=\!P\!=\!\!=\!O \\
| \\
OH
\end{array}\right]^{-}
&
\left[\begin{array}{c}
H \\
| \\
O\!=\!\!=\!P\!=\!\!=\!O \\
\| \\
O
\end{array}\right]^{2-} \\
(13\text{-Xa}) & (13\text{-Xb}) & (13\text{-Xc})
\end{array}$$

phosphorous acid, $P(OR)_3$, do we encounter three-connected phosphorus, and even these, as will be seen below, have a tendency to rearrange to four-connected species.

Similarly, the acid of formula H_3PO_2, hypophosphorous acid, also has a four-connected structure, (13-XIa), as does its anion (13-XIb).

$$\begin{array}{cc}
\begin{array}{c}
H \\
| \\
H-\!\!-P-\!\!-OH \\
\| \\
O
\end{array}
&
\left[\begin{array}{c}
H \\
| \\
H-\!\!-P\!=\!\!=\!O \\
\| \\
O
\end{array}\right]^{-} \\
(13\text{-XIa}) & (13\text{-XIb})
\end{array}$$

Lower Acids. *Hypophosphorous acid*, $H[H_2PO_2]$. The salts are usually prepared by boiling white phosphorus with alkali or alkaline-earth hydroxide. The main reactions appear to be

$$P_4 + 4OH^- + 4H_2O \rightarrow 4H_2PO_2^- + 2H_2$$
$$P_4 + 4OH^- + 2H_2O \rightarrow 2HPO_3^{2-} + 2PH_3$$

The calcium salt is soluble in water, unlike that of phosphite or phosphate; the free acid can be made from it or obtained by oxidation of phosphine with iodine in water. Both the acid and its salts are powerful reducing agents, being oxidized to orthophosphate. The pure white crystalline solid is a monobasic acid, ($pK = 1.2$); other physical studies, such as nmr, confirm the presence of a PH_2 group, and the anion has been characterized crystallographically.[71] Either or both of the hydrogen atoms can be replaced, by indirect methods, with alkyl groups to give mono- or di-alkyl *phosphonous* compounds.

Phosphorous acid, $H_2[HPO_3]$. As noted above, this acid and its mono- and di-esters have a P—H bond. The free acid is obtained by treating PCl_3 or P_4O_6 with water; when pure, it is a deliquescent colorless solid (m.p. 70.1°, $pK = 1.8$). The presence of the P—H bond has been demonstrated by a variety of structural studies as well as by the formation of only mono and di series of salts. It is oxidized to orthophosphate by halogen, sulfur dioxide and other agents, but the reactions are slow and complex. The mono-, di- and tri-esters can be obtained from reactions of alcohols with PCl_3 alone or in the presence of an organic base as hydrogen chloride acceptor. RPO_3^{2-} ions are called *phosphonate* ions.

The *phosphite triesters*, $P(OR)_3$ (cf. above), are notable for forming donor complexes with transition metals and other acceptors. They are readily oxidized to the respective phosphates:

$$2(RO)_3P + O_2 \rightarrow 2(RO)_3PO$$

[71] T. Matsuzaki and Y. Itaka, *Acta Cryst.*, 1969, *B*, **25**, 1933.

They also undergo the Michaelis–Arbusov reaction with alkyl halides, forming dialkyl phosphonates:

$$P(OR)_3 + R'X \rightarrow [(RO)_3PR'X] \rightarrow RO\overset{\overset{\displaystyle O}{\|}}{\underset{\underset{\displaystyle OR}{|}}{P}}-R' + RX$$

Phosphonium
intermediate

The methyl ester easily undergoes spontaneous isomerization to the dimethyl ester of methylphosphonic acid:

$$P(OCH_3)_3 \rightarrow CH_3PO(OCH_3)_2$$

Higher Acids. *Orthophosphoric acid*, H_3PO_4, commonly called phosphoric acid, is one of the oldest known and most important phosphorus compounds.[72] It is made in vast quantities, usually as 85% sirupy acid, by the direct reaction of ground phosphate rock with sulfuric acid and also by the direct burning of phosphorus and subsequent hydration of the oxide P_4O_{10}. The pure acid is a colorless crystalline solid (m.p. 42.35°). It is very stable and has essentially no oxidizing properties below 350–400°. At elevated temperatures it is fairly reactive toward metals and is reduced; it will also then attack quartz. Fresh molten H_3PO_4 has appreciable ionic conductivity[73] suggesting autoprotolysis:

$$2H_3PO_4 \rightleftharpoons H_4PO_4^+ + H_2PO_4^-$$

Pyrophosphoric acid is also produced:

$$2H_3PO_4 \rightarrow H_2O + H_4P_2O_7$$

but this conversion is temperature-dependent and is slow at room temperature.

The acid is tribasic: at 25°, $pK_1 = 2.15$, $pK_2 = 7.1$, $pK_3 \approx 12.4$. The pure acid and its crystalline hydrates have tetrahedral PO_4 groups connected by hydrogen bonds (Fig. 13-12). These persist in the concentrated solutions and are responsible for the sirupy nature. For solutions of concentration less than $\sim 50\%$, the phosphate anions are hydrogen-bonded to the liquid water rather than to other phosphate anions.

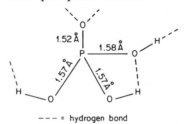

−−− = hydrogen bond

Fig. 13-12.   Structure of anhydrous orthophosphoric acid.

Phosphates of most metal ions and other cations are known. Some of these are of enormous commercial and practical importance, for example, ammonium phosphate fertilizers, alkali phosphate buffers, etc. Natural phosphate minerals are *all* orthophosphates, the major one being fluorapatite; hydroxo-

[72] A. V. Slack, *ed., Phosphoric Acid*, Dekker, 1968.
[73] R. A. Munsen, *J. Phys. Chem.*, 1964, **68**, 3374.

apatites partly carbonated, make up the mineral part of teeth. The role of traces of F^- in strengthening dental enamel is presumably connected with these structural relationships, but a detailed explanation of the phenomenon is still lacking.

Orthophosphoric acid and phosphates form complexes with many transition-metal ions. The precipitation of insoluble phosphates from fairly concentrated acid solution ($3-6N$ HNO_3) is characteristic of $4+$ cations such as those of Ce, Th, Zr, U, Pu, etc.

Large numbers of phosphate *esters* are known. Some of these are important technically, particularly for solvent extraction of metal ions from aqueous solutions.

Condensed phosphates. Condensed phosphates are those containing more than one P atom and having P—O—P bonds. We may note that the *lower* acids can also give condensed species, although we shall deal here only with a few examples of phosphates.

There are three main building units in condensed phosphates: the end unit (13-XII), middle unit (13-XIII) and branching unit (13-XIV).

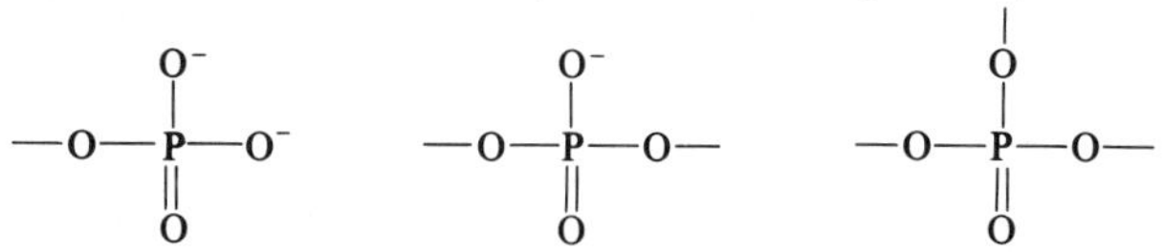

(13-XII)   (13-XIII)   (13-XIV)
$PO_{3.5}^{2-}$ PO_3^- $PO_{2.5}$

These units can be distinguished not only chemically—for example, the branching points are rapidly attacked by water—but also by the ^{31}P nmr spectra. These units can be incorporated into either (*a*) chain or *polyphosphates*, containing 2–10 P atoms, (*b*) cyclic *metaphosphates*, containing 3–7 or more P atoms or (*c*) infinite chain *metaphosphates*. Not all possible combinations of the basic units are known. Some of the most important are:

Linear polyphosphates, which are salts of anions of general formula $[P_nO_{3n+1}]^{(n+2)-}$. Examples are $M_4^IP_2O_7$ (13-XV), a pyrophosphate or dipolyphosphate, and $M_5^IP_3O_{10}$ (13-XVI), a tripolyphosphate.

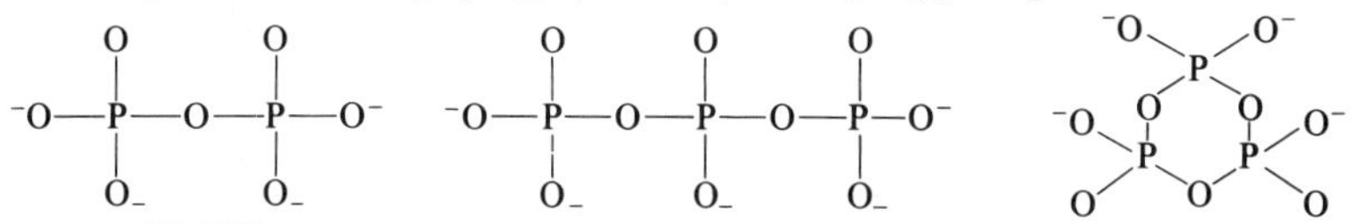

(13-XV)   (13-XVI)   (13-XVII)

Cyclic polyphosphates which are salts of anions of general formula $[P_nO_{3n}]^{n-}$. Examples are $M_3P_3O_9$, a trimetaphosphate (13-XVII), and $M_4P_4O_{12}$, a tetrametaphosphate. The eight-membered ring of the $P_4O_{12}^{4-}$ ion is puckered with equal P—O bond lengths.[74]

An example of an infinite chain metaphosphate is provided by one of the several crystal forms of KPO_3.[75]

[74] D. A. Koster and A. J. Wagner, *J. Chem. Soc., A.*, **1970**, 435.
[75] K. H. Jost, *Acta Cryst.*, 1963, **16**, 623.

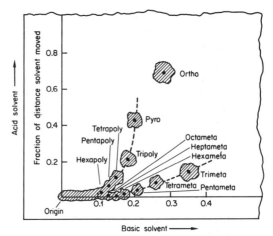

Fig. 13-13.   Chromatographic separation of complex phosphate mixtures. Corner of a two-dimensional paper chromatogram, showing the positions of the pentameta- through octameta-phosphate rings in relation to the positions of the well known ring and chain phosphates. The basic solvent traveled 23 cm in 24 h, whereas the acid solvent traveled 11.5 cm in 5.5 h.

Condensed phosphates are usually prepared by dehydration of ortho-phosphates under various conditions of temperature (300–1200°) and also by appropriate hydration of dehydrated species, as, for example,

$$(n-2)NaH_2PO_4 + 2Na_2HPO_4 \xrightarrow{\text{Heat}} \underset{\text{Polyphosphate}}{Na_{n+2}P_nO_{3n+1}} + (n-1)H_2O$$

$$nNaH_2PO_4 \xrightarrow{\text{Heat}} \underset{\text{Metaphosphate}}{(NaPO_3)_n} + nH_2O$$

They can also be prepared by controlled addition of water or other reagents to P_4O_{10}, by treating chlorophosphates with silver phosphates, etc. The complex mixtures of anions that can be obtained are separated by using ion-exchange or chromatographic procedures as illustrated in Fig. 13-13.

Condensed phosphates form soluble complexes with many metals, and chain phosphates are used industrially for this purpose, for example, as water softeners.

The most important *cyclic* phosphate is *tetrametaphosphate*, which can be prepared by heating copper nitrate with slightly more than an equimolar amount of phosphoric acid (75%) slowly to 400°. The sodium salt can be obtained by treating a solution of the copper salt with Na_2S. Slow addition of P_4O_{10} to ice water gives ~75% of the P as tetrametaphosphate.

Fluorophosphates.   As with many other oxo anions, fluorine can replace OH in phosphate to give mono- and di-fluorophosphate salts and esters. The dialkyl monofluorophosphate esters have been found to inhibit cholinesterase in the body and to be exceedingly toxic. The *hexafluorophosphate* ion, PF_6^-, has been discussed (page 376).

Phosphate Esters.   Esters of orthophosphoric acid can be prepared by reactions such as

$$O{=}PCl_3 + 3ROH \rightarrow O{=}P(OR)_3 + 3HCl$$
$$3RCOCl + Ag_3PO_4 \rightarrow (RCOO)_3P{=}O + 3AgCl$$

Many of the most essential chemicals in life processes are phosphate esters.[76] These include the genetic substances DNA and RNA (representative fragments of the chains of which are shown as (13-XVIII) and (13-XIX), respec-

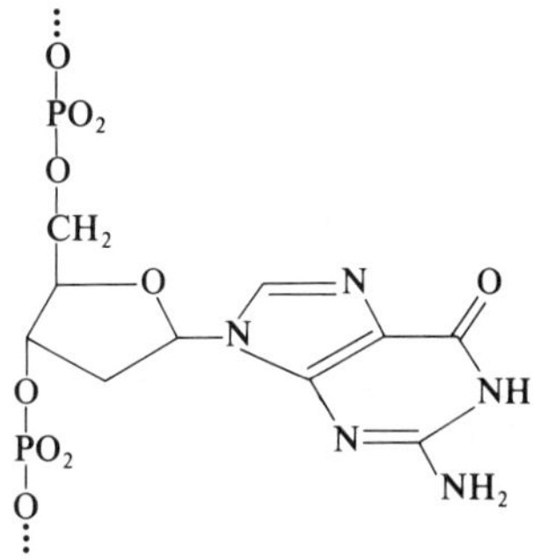

(13-XVIII)

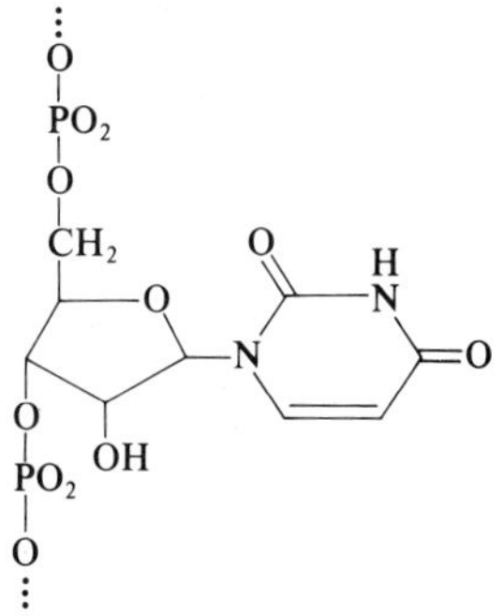

(13-XIX)

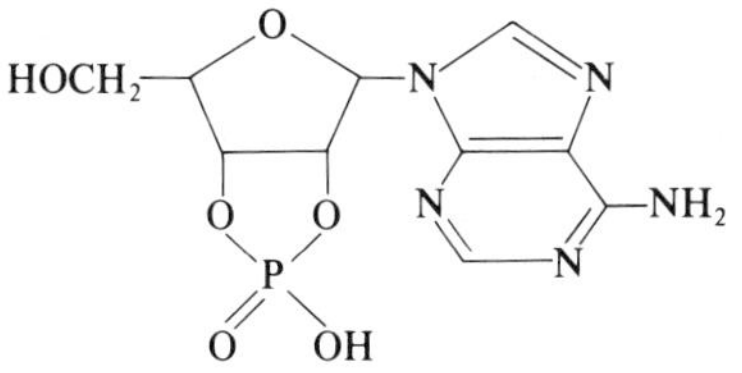

(13-XX)

tively), as well as cyclic AMP (adenosine monophosphate) (13-XX). In addition, the transfer of phosphate groups between ATP and ADP, eq. 13-4, is of fundamental importance in the energetics of biological systems. All the

$$\underset{\text{ATP}}{\text{Adenosine—O—}\overset{\text{O}}{\underset{\text{O}}{\text{P}}}\text{—O—}\overset{\text{O}}{\underset{\text{O}}{\text{P}}}\text{—O—}\overset{\text{O}}{\underset{\text{O}}{\text{P}}}\text{—OH}} \underset{-H_2O}{\overset{+H_2O}{\rightleftharpoons}} \underset{\text{ADP}}{\text{Adenosine—O—}\overset{\text{O}}{\underset{\text{O}}{\text{P}}}\text{—O—}\overset{\text{O}}{\underset{\text{O}}{\text{P}}}\text{—OH}} + H_2PO_4^- \quad (13\text{-}4)$$

biological reactions involving formation and hydrolysis of these and other phosphate esters and polyphosphates are effected by enzyme catalysts, many of which contain as parts of their structure, or require as coenzymes, metal ions.

In no small measure because of the importance of such substances and processes as those just mentioned, the hydrolysis of phosphate esters has received much fundamental study[77] by inorganic and physical organic chemists. It is known that triesters are attacked by OH^- at phosphorus and by H_2O at carbon, as shown in eq. 13-5.

[76] See, for example, A. L. Lehninger, *Biochemistry*, Worth Publishers, 1970, and H. M. Kalkar, *Biological Phosphorylations*, Prentice-Hall, 1969; K. Decker *et al.*, *Angew. Chem. Internat. Edn.*, 1970, **9**, 138.

[77] J. R. Cox and O. B. Ramsey, *Chem. Rev.*, 1964, **64**, 317; T. C. Bruice and S. J. Benkovic, *Bioorganic Mechanisms*, W. A. Benjamin, 1966, Chapters 5–7; C. A. Bunton, *Accounts Chem. Res.*, 1970, **3**, 257; F. H. Westheimer, *Accounts Chem. Res.*, 1968, **1**, 70.

$$\text{OP(OR)}_3 \begin{cases} \xrightarrow{\text{}^{18}\text{OH}^-} \text{OP(OR)}_2(\text{}^{18}\text{OH}) + \text{RO}^- \\ \xrightarrow{\text{H}_2\text{}^{18}\text{O}} \text{OP(OR)}_2(\text{OH}) + \text{R}^{18}\text{OH} \end{cases} \tag{13-5}$$

Diesters, which are strongly acidic (eq. 13-6), are completely in the anionic form at normal (and physiological) pH's:

$$\begin{array}{c} \text{O} \\ \parallel \\ \text{RO} - \text{P} - \text{OR}' \rightleftharpoons \text{ROPO}_2\text{OR}^- + \text{H}^+ \\ | \\ \text{OH} \end{array} \qquad K \approx 10^{-1.5} \tag{13-6}$$

They are thus relatively resistant to nucleophilic attack by either OH^- or H_2O and this is why enzymic catalysis is indispensible to achieve useful rates of reaction.

Relatively little has been firmly established as yet concerning the mechanisms of most phosphate ester hydrolyses, especially the many enzymic ones. However, the following three basic mechanistic types can be envisioned for cases where nucleophilic attack on phosphorus is the rate-limiting step:

1. One-step nucleophilic displacement (S_N2) with inversion:

2. Release of a short-lived "metaphosphate" group, which rapidly recovers the 4-connected orthophosphate structure:

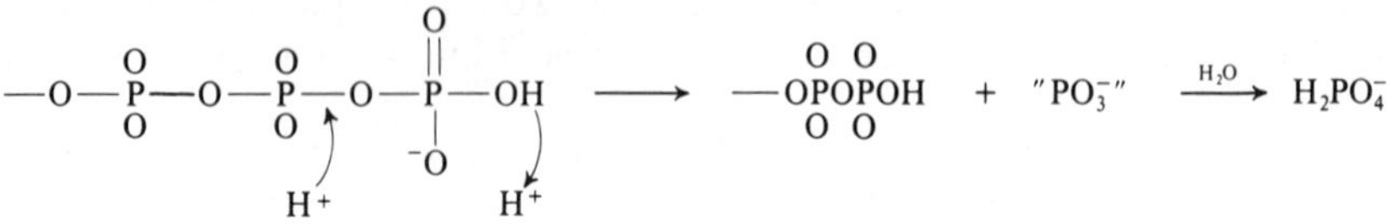

3. Nucleophilic attack in which a cyclic 5-coordinate intermediate is formed which then pseudorotates (cf. page 40):

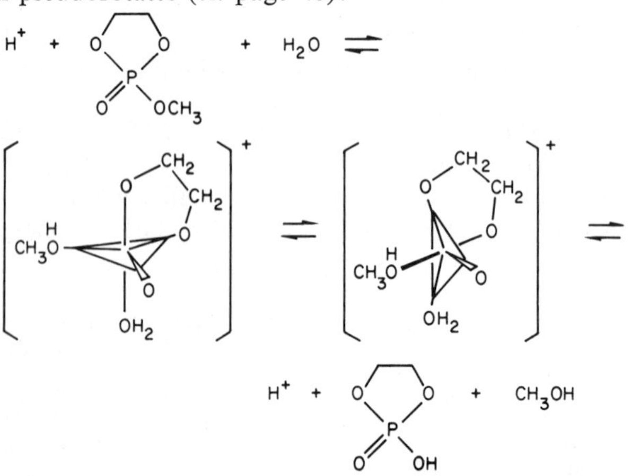

13-15. Oxo Acids and Anions of Arsenic, Antimony and Bismuth

Arsenic. Raman spectra[78] show that in acid solutions of As_4O_6 the only detectable species is $As(OH)_3$. In basic solutions ([OH^-]/[As^{III}] ratios of 3.5–15) the four pyramidal species $As(OH)_3$, $As(OH)_2O^-$, $As(OH)O_2^{2-}$ and AsO_3^{3-} appear to be present. In solid salts the arsenite ion is known in the AsO_3^{3-} form, as well as in more complex ones.

Arsenic acid, H_3AsO_4, is obtained by treating arsenic with concentrated nitric acid to give white crystals, $H_3AsO_4 \cdot \frac{1}{2}H_2O$. Unlike phosphoric acid, it is a moderately strong oxidizing agent in acid solution, the potentials being

$$H_3AsO_4 + 2H^+ + 2e = HAsO_2 + 2H_2O \qquad E° = 0.559 \text{ V}$$
$$H_3PO_4 + 2H^+ + 2e = H_3PO_3 + H_2O \qquad E° = -0.276 \text{ V}$$

Arsenic acid is tribasic but somewhat weaker ($pK_1 = 2.3$) than phosphoric acid. The arsenates generally resemble orthophosphates and are often isomorphous with them.

Condensed arsenic anions are much less stable than the condensed phosphates and, owing to rapid hydrolysis, do not exist in aqueous solution. Dehydration of KH_2AsO_4 gives three forms, stable at different temperatures, of metaarsenate; one form is known to contain an infinite chain polyanion, like that in one form of KPO_3.

There are also fluoroarsenates, such as the $M_2^I[As_2F_8O_2]$ compounds which contain arsenic atoms octahedrally coordinated by four fluoride ions and two bridging oxygen atoms.[79]

Antimony. No lower acid is known but only the hydrated oxide, Sb_2O_3(aq); the antimonites are well defined salts, however. The higher acid is known only in solution, but it gives crystalline antimonates of the type $K[Sb(OH)_6]$. There do not appear to be finite SbO_4^{3-} ions under any circumstances. Some "antimonates" obtained by heating oxides, for example, M^ISbO_3, $M^{III}SbO_4$ and $M_2^{II}Sb_2O_7$, contain SbO_6 octahedra and differ only in the manner of linking in the lattice. They are best regarded as mixed oxides.

Bismuth. When $Bi(OH)_3$ in strongly alkaline solution is treated with chlorine or other strong oxidizing agents, "bismuthates" are obtained, but never in a state of high purity. They can also be made, for example, by heating Na_2O_2 and Bi_2O_3 which gives $NaBi^VO_3$. Bismuthates are powerful oxidizing agents in acid solution.

It has been observed that aqueous Bi^V oxidizes Cl^-, Br^-, and I^- at the same rate, independent of halide ion concentration. The proposed mechanism[80] is

$$Bi^V + H_2O \rightarrow Bi^{IV} + HO\cdot + H^+ \text{ (rate-determining)}$$
$$\left. \begin{array}{l} Bi^{IV} + X^- \rightarrow Bi^{III} + X\cdot \\ HO\cdot + X^- \rightarrow OH^- + X\cdot \end{array} \right\} \text{ fast}$$
$$2X\cdot \rightarrow X_2$$

[78] T. M. Loehr and R. A. Plane, *Inorg. Chem.*, 1968, **7**, 1708.
[79] H. Dunken and W. Haase, *Z. Chem.*, 1963, **3**, 433.
[80] M. H. Ford-Smith and J. J. Habeeb, *Chem. Comm.*, **1969**, 1445.

Further Reading

Phosphorus, Vol. 1, 1971, M. Grayson and L. Horner, eds., Gordon and Breach (new journal devoted to P and Group V).

Topics in Phosphorus Chemistry, Vol. 1, 1964, M. Grayson and E. J. Griffiths, eds., Wiley (a continuing series of detailed reviews).

Mellor's Comprehensive Treatise on Inorganic and Theoretical Chemistry, Supplement III to Vol VIII, *Phosphorus*, Longmans, 1971.

Bent, H. A., *J. Inorg. Nuclear Chem.*, 1961, **19**, 43 (hybridization, bond angles and bond lengths in PX_3, X_3PO and X_3PS compounds).

Berlin, K. D., and G. B. Butler, *Chem. Rev.*, 1960, **60**, 243 (preparation and properties of phosphine oxides).

Booth, G., *Adv. Inorg. Chem. Radiochem.*, 1964, **6**, 1 (complexes of transition metals with phosphines, arsines and stibines).

Burg, A. B., *Accounts Chem. Res.*, 1969, **2**, 353 (fluorophosphine chemistry).

Cadogan, J. G., *Quart. Rev.*, 1962, **16**, 208 (oxidation of PR_3 organo compounds).

Clark, V. M., *Proc. Chem. Soc.*, **1964**, 129; *Angew. Chem. Internat. Edn.*, 1964, **3**, 678 (phosphorylation in organic and biochemistry).

Decker, K., *et al.*, *Angew. Chem. Internat. Edn.*, 1970, **9**, 138 (biological phosphorylations).

Doak, G. O., and L. D. Freedman, *Chem. Rev.*, 1961, **61**, 31 (structure and properties of dialkyl phosphonates).

Frank, A. W., *Chem. Rev.*, 1961, **61**, 389 (phosphonous acids and their derivatives).

Haiduc, I., *The Chemistry of Inorganic Ring Systems*, Wiley, 1970 (Chapter V covers Group V exhaustively).

Harrison, P. G., *Organometallic Chem. Rev.*, *A*, 1970, **5**, 183 (organobismuth chemistry).

Hartley, S. B., *et al.*, *Quart. Rev.*, 1963, **17**, 204 (thermochemistry of phosphorus compounds).

Heath, D. F., *Organophosphorus Poisons*, Pergamon Press, 1961.

Holmes, R. R., *J. Chem. Educ.*, 1963, **40**, 125 (review of phosphorus halides).

Hudson, R. F., *Pure Appl. Chem.*, 1964, **9**, 371; *Adv. Inorg. Chem. Radiochem.*, 1964, **5**, 347 (bonding, structure and reactivity of organophosphorus compounds; the latter reference also contains discussion of $d\pi$–$p\pi$ bonding).

Hudson, R. F., *Structure and Mechanism in Organophosphorus Chemistry*, Academic Press, 1965.

Hudson, R. F., and M. Green, *Angew. Chem. Internat. Edn.*, 1963, **2**, 11 (stereochemistry of displacement reactions at phosphorus atoms).

Huheey, J. E., *J. Chem. Educ.*, 1963, **40**, 153 (review on compounds with P—P bonds).

Kirby, A. J., and S. G. Warren, *The Organic Chemistry of Phosphorus*, Elsevier, 1967.

Maier, L., *Progr. Inorg. Chem.*, 1963, **5**, 27 (chemistry of primary, secondary and tertiary phosphines).

Markl, G., *Angew. Chem. Internat. Edn.*, 1965, **4**, 1023 (heterocycles containing phosphorus).

Mann, F. G., *The Heterocyclic Derivatives of Phosphorus, Arsenic, Antimony, and Bismuth*, 2nd edn., Wiley-Interscience, 1970 (a comprehensive treatise).

Mooney, R. W., and M. A. Ais, *Chem. Rev.*, 1961, **61**, 433 (alkaline-earth phosphates).

Nixon, J. F., *Adv. Inorg. Chem. Radiochem.*, 1970, **13**, 364 (fluorophosphine chemistry).

Paddock, N. L., R. Inst. Chem. Lectures, 1962, No. 2 (structure and reactions of phosphorus compounds); *Quart. Rev.*, 1964, **18**, 168 (phosphonitrilic derivatives).

Payne, D. S., *Quart. Rev.*, 1961, **15**, 173 (the halides of P, As, Sb and Bi).

Payne, D. S., in *Non-Aqueous Solvent Systems*, Academic Press, 1965 (Group V halides and oxo halides as solvents).

Schmutzler, R, *Angew. Chem. Internat. Edn.*, 1965, **4**, 496 (chemistry and nmr spectra of 5-coordinate fluorophosphoranes, R_nPF_{5-n}).

Thilo, E., *Angew. Chem. Internat. Edn.*, 1965, **4**, 1061 (structures of condensed phosphates).

Van Wazer, J. R., *Phosphorus and Its Compounds*, Vol. I, Interscience, 1958 (a comprehensive account of all phases of phosphorus chemistry); Vol. II, *Technology, Biological Functions, and Applications*, 1961.

Van Wazer, J. R., and C. F. Callis, *Chem. Rev.*, 1958, **58**, 1011 (complexing of metals by phosphate).

14

Oxygen

GENERAL REMARKS

14-1. Types of Oxides

The oxygen atom has the electronic structure $1s^2 2s^2 2p^4$. Oxygen forms compounds with all the elements except He, Ne and possibly Ar, and it combines directly with all the other elements except the halogens, a few noble metals and the noble gases, either at room or at elevated temperatures. The earth's crust contains about 50% by weight of oxygen. Most inorganic chemistry is concerned with its compounds, if only in the sense that so much chemistry involves the most important oxygen compound—water.

As a first-row element, oxygen follows the octet rule, and the closed-shell configuration can be achieved in ways that are similar to those for nitrogen, namely, by (a) electron gain to form O^{2-}, (b) formation of two single covalent bonds (e.g., R—O—R) or a double bond (e.g., O=C=O), (c) gain of one electron and formation of one single bond (e.g., in OH^-) and (d) formation of three or four covalent bond (e.g., R_2OH^+, etc.).

There is a variety of binary oxygen compounds of disparate natures. The range of physical properties is attributable to the range of bond types from essentially ionic to essentially covalent. Some representative oxides and their properties are listed in Table 14-1.

The formation of the oxide ion, O^{2-}, from molecular oxygen requires the expenditure of a considerable energy, ~ 1000 kJ mol^{-1}:

$$\begin{aligned}
\tfrac{1}{2}O_2(g) &= O(g) & \Delta H &= 248 \text{ kJ mol}^{-1} \\
O(g) + 2e &= O^{2-}(g) & \Delta H &= 752 \text{ kJ mol}^{-1}
\end{aligned}$$

Moreover, in the formation of an ionic oxide, energy must be expended in vaporizing and ionizing the metal atoms. Nevertheless, many essentially ionic oxides exist and are very stable because the energies of lattices containing the relatively small (1.40 Å), doubly charged oxide ion are quite high. In fact, the lattice energies are often sufficiently high to allow the ionization of metal atoms to unusually high oxidation states. Many metals form oxides in oxidation states not encountered in their other compounds, except perhaps in fluorides or some complexes. Examples of such higher oxides are

TABLE 14-1

Some Representative Oxides

Compound	Nature	Properties
	Crystalline Oxides	
CaO	White solid; m.p. 2580°	Ionic lattice; basic
SiO_2	Colorless crystals; m.p. 1710°[a]	Infinite three-dimensional lattice; acidic
BeO	White solid; m.p. 2570°	Semi-ionic; amphoteric
$Th_{0.7}Y_{0.3}O_{1.85}$	White crystalline solid	Fluorite lattice with some O^{2-} missing; typical mixed metal oxide
$FeO_{0.95}$	Black solid	NaCl lattice with some Fe^{3+} ions and some cation vacancies
NbO	Glistening dark solid	High metallic-type electrical conductance
	Molecular Oxides	
CO	Colorless gas	Inert; no acid or basic properties
SO_2	Colorless gas	Acid anhydride
OsO_4	Pale yellow, volatile solid; m.p. 41°	Readily reduced to Os
Cl_2O_7	Explosive, colorless oil	Anhydride of $HClO_4$

[a] Cristobalite; see page 320.

MnO_2, AgO and PrO_2. Many of these higher ionic oxides are non-stoichiometric.

In some cases the lattice energy is still insufficient to permit complete ionization, and oxides having substantial covalent character, such as BeO or B_2O_3, are formed. Finally, at the other extreme there are numerous oxides, such as CO_2, the nitrogen and phosphorus oxides, SO_2, SO_3, etc., that are essentially covalent molecular compounds. Such compounds are gases or volatile solids or liquids. Even in "covalent" oxides, unusually high *formal* oxidation states are often found, as in OsO_4, CrO_3, SO_3, etc.

In some oxides containing transition metals in very low oxidation states, metal "*d* electrons" enter delocalized conduction bands and the materials have metallic properties.[1] An example is NbO.[2]

In terms of chemical behavior, it is convenient to classify oxides according to their acid or base character in the aqueous system.

Basic Oxides. Although X-ray studies show the existence of discrete oxide ions, O^{2-} (and also peroxide, O_2^{2-}, and superoxide, O_2^-, ions, to be discussed below), these ions cannot exist in any appreciable concentration in aqueous solution owing to the hydrolytic reaction

$$O^{2-}(s) + H_2O = 2OH^-(aq) \qquad K > 10^{22}$$

We have also for the per- and super-oxide ions:

$$O_2^{2-} + H_2O = HO_2^- + OH^-$$
$$2O_2^- + H_2O = O_2 + HO_2^- + OH^-$$

[1] J. M. Honing, *IBM J. Res. Dev.*, 1970, **14**, 232.
[2] G. V. Chandrashekhar, J. Moyo and J. M. Honig, *J. Solid State Chem.*, 1970, **2**, 528.

Thus only those ionic oxides that are insoluble in water are inert to it. Ionic oxides function, therefore, as *basic anhydrides*. When insoluble in water, they usually dissolve in dilute acids, for example

$$MgO(s) + 2H^+(aq) \rightarrow Mg^{2+}(aq) + H_2O$$

although in some cases, MgO being one, high-temperature ignition produces a very inert material, quite resistant to acid attack.

Acidic Oxides. The covalent oxides of the non-metals are usually acidic, dissolving in water to produce solutions of acids. They are termed *acid anhydrides*. Insoluble oxides of some less electropositive metals of this class generally dissolve in bases. Thus:

$$N_2O_5(s) + H_2O \rightarrow 2H^+(aq) + 2NO_3^-(aq)$$
$$Sb_2O_5(s) + 2OH^- + 5H_2O \rightarrow 2Sb(OH)_6^-$$

Basic and acidic oxides will often combine directly to produce salts, such as:

$$Na_2O + SiO_2 \xrightarrow{\text{Fusion}} Na_2SiO_3$$

Amphoteric Oxides. These oxides behave acidically toward strong bases and as bases toward strong acids:

$$ZnO + 2H^+(aq) \rightarrow Zn^{2+} + H_2O$$
$$ZnO + 2OH^- + H_2O \rightarrow Zn(OH)_4^{2-}$$

Other Oxides. There are various other oxides, some of which are relatively inert, dissolving in neither acids nor bases, for instance, N_2O, CO, and MnO_2; when MnO_2 (or PbO_2) does react with acids, e.g., concentrated HCl, it is a redox, not an acid–base, reaction.

There are also many oxides that are non-stoichiometric. These commonly consist of arrays of close-packed oxide ions with some of the interstices filled by metal ions. However, if there is variability in the oxidation state of the metal, non-stoichiometric materials result. Thus ferrous oxide generally has a composition in the range $FeO_{0.90}$–$FeO_{0.95}$, depending on the manner of preparation. There is an extensive chemistry of mixed metal oxides (see also page 54).

It may be noted further that, when a given element forms several oxides, the oxide with the element in the highest formal oxidation state (usually meaning more covalent) is more acidic. Thus, for chromium we have: CrO, basic; Cr_2O_3, amphoteric; and CrO_3, fully acidic.

The Hydroxide Ion.[3] Discrete hydroxide ions, OH^-, exist only in the hydroxides of the more electropositive elements such as the alkali metals and alkaline earths. For such an ionic material, dissolution in water results in formation of aquated metal ions and aquated hydroxide ions:

$$M^+OH^-(s) + nH_2O \rightarrow M^+(aq) + OH^-(aq)$$

and the substance is a strong base. In the limit of an extremely covalent M—O bond, dissociation will occur to varying degrees as follows:

$$MOH + nH_2O \rightleftharpoons MO^-(aq) + H_3O^+(aq)$$

[3] (a) R. F. W. Bader in *The Chemistry of the Hydroxyl Group*, S. Patai, ed., Interscience-Wiley, 1971 (theoretical and physical properties of the hydroxyl ion and group); (b) W. L. Jolly, *J. Chem. Educ.*, 1967, **44**, 304 (intrinsic basicity of OH^-); (c) J. R. Jones, *Chem. in Britain*, **1971**, 336 (highly basic media and applications).

and the substance must be considered an acid. Amphoteric hydroxides are those in which there is the possibility of either kind of dissociation, the one being favored by the presence of a strong acid:

$$M—O—H + H^+ = M^+ + H_2O$$

the other by strong base:

$$M—O—H + OH^- = MO^- + H_2O$$

because the formation of water is so highly favored, i.e.

$$H^+ + OH^- = H_2O \qquad K_{25°} = 10^{14}$$

Similarly, hydrolytic reactions of many metal ions can be written as

$$M^{n+} + H_2O = (MOH)^{(n-1)+} + H^+$$

However, in view of the fact that such ions are coordinated by water molecules, a more realistic equation is

$$M(H_2O)_x^{n+} = [M(H_2O)_{x-1}(OH)]^{(n-1)+} + H^+$$

Thus we may consider that, the more covalent the M—O bond tends to be, the more acidic are the hydrogen atoms in the aquated ion, but at present there is no extensive correlation of the acidities of aquo ions with properties of the metal.

The hydroxide ion has the ability to form bridges between metal ions. Thus there are various compounds of the transition and other metals containing OH bridges between pairs of metal atoms, as in (14-I).

(14-I) (14-II)

The formation of hydroxo bridges occurs at an early stage in the precipitation of hydrous metal oxides. In the case of Fe^{3+}, precipitation of $Fe_2O_3 \cdot xH_2O$—commonly, but incorrectly, written $Fe(OH)_3$—proceeds through the stages

$$[Fe(H_2O)_6]^{3+} \xrightarrow{-H^+} [Fe(H_2O)_5OH]^{2+} \rightarrow [(H_2O)_4Fe(OH)_2Fe(H_2O)_4]^{4+} \xrightarrow{-xH^+}$$
$$\text{pH} < 0 \qquad\qquad 0 < \text{pH} < 2 \qquad\qquad \sim 2 < \text{pH} < \sim 3$$

$$\text{colloidal } Fe_2O_3 \cdot xH_2O \xrightarrow{-yH^+} Fe_2O_3 \cdot zH_2O \text{ ppt.}$$
$$\sim 3 < \text{pH} < \sim 5 \qquad\qquad \text{pH} \sim 5$$

While bridges of the type (14-I) are most common, triply bridging hydroxo groups (14-II) are also known.[4]

Analogous to the OH^- ion[3c] are the alkoxide ions, OR^-. These are even stronger bases as a rule, being immediately hydrolyzed by water:

$$OR^- + H_2O = OH^- + ROH$$

[4] V. Albano et al., Chem. Comm., 1969, 1242; F. A. Cotton and B. H. C. Winquist, Inorg. Chem., 1969, 8, 1304; T. G. Spiro et al., Inorg. Chem., 1968, 7, 2165; H. S. Preston et al., J. Organometallic Chem., 1968, 14, 447.

A considerable number of metal alkoxides are known, many stoichiometrically analogous to the metal hydroxides, for example, $Ti(OH)_4$ and $Ti(OR)_4$ (Section 25-A). These compounds are quite reactive, and as R becomes large they become essentially organic in their physical properties; such alkoxides are usually polymeric, with coordination numbers greater than those indicated by the simple stoichiometry, owing to the existence of OR bridge groups.

14-2. Covalent Compounds; Stereochemistry of Oxygen

Two-coordinate Oxygen. The majority of oxygen compounds contain two-coordinate oxygen, in which the oxygen atom forms two single bonds to other atoms and has two unshared pairs of electrons in its valence shell. Such compounds include water, alcohols, ethers, and a variety of other covalent oxides. In the simple two-coordinate situations, where the σ bonds are not supplemented by π bonds to any significant extent, the X—O—X group is always bent; the angles range from $104.5°$ in H_2O to $111°$ in $(CH_3)_2O$.

In many cases, where the X atoms of the X—O—X group have orbitals (usually d orbitals) capable of interacting with the lone-pair orbitals of the oxygen atom, the X—O bonds acquire some π character. Such interaction causes shortening of the X—O bonds and generally widens the X—O—X angle. The former effect is not easy to document since an unambiguous standard of reference for the pure single bond is generally lacking. However, the increases in angle are self-evident, for example $(C_6H_5)_2O$ ($124°$) and Si—O—Si angle in quartz ($142°$). In the case of H_3Si—O—SiH_3 the angle is apparently $>150°$.

The limiting case of π interaction in X—O—X systems occurs when the σ bonds are formed by two digonal sp hybrid orbitals on oxygen, thus leaving two pairs of π electrons in pure p orbitals; these can then interact with empty $d\pi$ orbitals on the X atoms so as to stabilize the linear arrangement. Many examples of this are known, representative ones being $[Cl_5Ru$—O—$RuCl_5]^{4-}$ and $(C_5H_5)Cl_2Ti$—O—$TiCl_2(C_5H_5)$.

Three-coordinate Oxygen. Both pyramidal and planar geometries are found. The pyramidal type are represented mainly by *oxonium ions*, e.g., H_3O^+, R_2OH^+, ROH_2^+, R_3O^+ and by donor–acceptor complexes such as $(C_2H_5)_2OBF_3$. The formation of oxonium ions is analogous to the formation of ammonium ions such as NH_4^+, $RNH_3^+ \cdots R_4N^+$, except that oxygen is less basic and the oxonium ions are therefore less stable. Water, alcohol and ether molecules serving as ligands to metal ions presumably also have pyramidal structures, at least for the most part.. Like NR_3 compounds (page 341), OR_3^+ species undergo rapid inversion.[5]

---

[5] J. B. Lambert and D. M. Johnson *J. Amer. Chem. Soc.*, 1968, **90**, 1349.

The occurrence of planar geometry is rare and may in general be attributed to interaction of the oxygen lone pair with suitable π orbitals. For example, the three bonds to oxygen are coplanar, or nearly so, in (14-III). The basic acetates of certain trivalent metals contain the cation

$$(CH_3)_3Si-O-Al(CH_3)_2$$
$$(CH_3)_2Al-O-Si(CH_3)_3$$

(14-III)

$[M_3O(O_2CCH_3)_6L_3]^+$ (M = Mn, Cr, Ru or Fe), in which a triangle of metal atoms surrounds a central oxygen atom. It is also reported that a planar or nearly planar Hg_3O group occurs in the $[(CH_3Hg)_3O]^+$ ion.[6]

Four-coordinate Oxygen. While attainment of this coordination number is not common, there are a number of well-documented examples. Certain ionic or partly ionic oxides, e.g., PbO, have this coordination number and it sometimes forms the center of polynuclear complexes.[7] Examples are $Mg_4OBr_6 \cdot 4C_4H_{10}O$, $Cu_4OCl_6(Ph_3PO)_4$ and the long-known $M_4O(OOCR)_6$ compounds, where M = Be or Zn.

Unicoordinate, Multiply-bonded Oxygen. There are, of course, innumerable examples of XO groups, where the order of the XO bond may vary from essentially unity, as in amine oxides, $\geqslant \overset{+}{N} : \overset{..}{O} : ^-$, through varying degrees of π bonding up to a total bond order of 2, or a little more. The simplest π bonding situation occurs in ketones where one well-defined π bond occurs perpendicular to the molecular plane. In most inorganic situations, such as R_3PO or R_3AsO compounds, tetrahedral oxo ions such as PO_4^{3-}, ClO_4^-, MnO_4^-, OsO_4 or species such as $OsCl_4O_2^{2-}$, the opportunity exists for two π interactions between X and O, in mutually perpendicular planes that intersect along the X—O line. Indeed, the symmetry of the molecule or ion is such that the two π interactions *must* be of equal extent. Thus, in principle the extreme limiting structure (14-IVb) must be considered to be mixed with (14-IVa). In general, available evidence suggests that a partially polar bond

$$\overset{+}{X} : \overset{..}{\underset{..}{O}} : \quad \longleftrightarrow \quad \bar{X} ::: \overset{+}{O} :$$

(14-IVa)   (14-IVb)

of order approaching two results; it is important, however, to note the distinction from the situation in a ketone, since a π bond order of 1 in this context does not mean one full π interaction but rather two mutually perpendicular π interactions of order 0.5.

Catenation. As with nitrogen, catenation occurs only to a very limited extent. In peroxides and superoxides there are two consecutive oxygen atoms. Only in O_3, O_3^- and the few $R_FO_3R_F$ molecules[8] are there well-established chains of three oxygen atoms. There is no confirmed report of longer chains.

[6] J. H. R. Clarke and L. A. Woodward, *Spectrochim. Acta, A*, 1967, **23**, 2077.

[7] See, e.g., T. G. Spiro, D. H. Templeton and A. Zalkin, *Inorg. Chem.*, 1969, **8**, 856.

[8] L. R. Anderson and W. B. Fox, *J. Amer. Chem. Soc.*, 1967, **89**, 4313; D. G. Thompson, *J. Amer. Chem. Soc.*, 1967, **89**, 4316.

THE ELEMENT

14-3. Occurrence, Properties and Allotropes

Oxygen occurs in Nature in three isotopic species: ^{16}O (99.759%), ^{17}O (0.0374%) and ^{18}O (0.2039%). The rare isotopes, particularly ^{18}O, can be concentrated by fractional distillation of water, and concentrates containing up to 97 atom % ^{18}O or up to 4 atom % ^{17}O and other labeled compounds are commercially available. ^{18}O has been widely used as a tracer in studying reaction mechanisms of oxygen compounds. Although ^{17}O has a nuclear spin (5/2), its low abundance means that, even when enriched samples are used, a very sensitive nmr spectrometer is required. Nevertheless, a variety of useful studies has been carried out[9] and newer techniques of spectrum accumulation and/or the Fourier transform method should enhance the usefulness of ^{17}O spectra markedly. An important use of ^{17}O resonance studies has been to distinguish between H_2O in a complex, e.g., $[Co(NH_3)_5H_2O]^{3+}$, and solvent water.

Elemental oxygen occurs in two allotropic forms; the common, stable O_2 and ozone, O_3. O_2 is paramagnetic in the gaseous, liquid, and solid states and has the rather high dissociation energy of 496 kJ mol^{-1}. The valence-bond theory in its usual form would predict the electronic structure $:\ddot{O}{=}\ddot{O}:$ which, while accounting for the strong bond, fails to account for the para-magnetism. However, as already shown (page 106) the MO approach, even in first approximation, correctly accounts for the triplet ground state ($^3\Sigma_g^-$) having a double bond. There are several low-lying singlet states which figure importantly in photochemical oxidations; these will be discussed shortly (page 411). Like NO, which has one unpaired electron in an anti-bonding (π^*) MO, oxygen molecules associate only weakly, and true electron pairing to form a symmetrical O_4 species apparently does not occur even in the solid. Both liquid and solid O_2 are pale blue.

Ozone.[10]   Ozone is usually prepared by the action of a silent electric discharge upon O_2; concentrations up to 10% of O_3 can be obtained in this way. Ozone gas is perceptibly blue and is diamagnetic. Pure ozone can be obtained by fractional liquefaction of O_2–O_3 mixtures. There is a two-phase liquid system; one with 25% of ozone is stable, but a deep purple phase with 70% of ozone is explosive, as is the deep blue pure liquid (b.p. $-112°$). The solid (m.p. $-193°$) is black-violet. Small quantities of ozone are formed in electrolysis of dilute sulfuric acid, in some chemical reactions producing elemental oxygen, and by the action of ultraviolet light on O_2. Ozone occurs in traces in the upper atmosphere, the maximum concentrations being at an altitude of ~ 25 km. It is of vital importance in protecting the earth's surface from excessive exposure to ultraviolet light radiation. Ozone is very endothermic:

$$O_3 = \tfrac{3}{2}O_2 \qquad \Delta H = -142 \text{ kJ mol}^{-1}$$

[9]   B. L. Silver and S. Lux, *Quart. Rev.*, 1967, **21**, 458.
[10]   *Ozone Chemistry and Technology*, Advances in Chemistry Series No. 21, Amer. Chem. Soc., 1959.

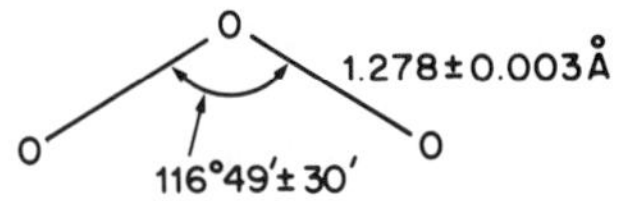

Fig. 14-1. The structure of ozone, O_3.

but it decomposes only slowly at 250° in absence of catalysts and ultraviolet light.

The structure of O_3 is shown in Fig. 14-1. Since the O—O bond distances are 1.49 Å in HOOH (single bond) and 1.21 Å in O_2 ($\sim$ double bond), it is apparent that the O—O bonds in O_3 must have considerable double-bond character.

Chemical properties of O_2 and O_3.   The chemical reactivities of oxygen[11] and ozone differ vastly. Although O_2 combines directly with practically all other elements, in most cases it does so only at elevated temperatures, but ozone is a powerful oxidizing agent and reacts with many substances under conditions where O_2 will not. The reaction

$$O_3 + 2KI + H_2O \rightarrow I_2 + 2KOH + O_2$$

is quantitative and can be used for the determination of ozone. Ozone is commonly used for oxidations of organic compounds;[12] the mechanisms probably involve free radical-chain processes as well as intermediates with —OOH groups. Ozone adducts are noted below.

The following potentials indicate the oxidizing strengths of O_2 and O_3 in aqueous solution:

$$O_2 + 4H^+ + 4e = 2H_2O \qquad\qquad E^0 = +1.229 \text{ V}$$
$$O_2 + 2H_2O + 4e = 4OH^- \qquad\qquad E^0 = +0.401 \text{ V}$$
$$O_2 + 4H^+(10^{-7}M) + 4e = 2H_2O \qquad E^0 = +0.815 \text{ V}$$
$$O_3 + 2H^+ + 2e = O_2 + H_2O \qquad\qquad E^0 = +2.07 \text{ V}$$
$$O_3 + H_2O + 2e = O_2 + 2OH^- \qquad\qquad E^0 = +1.24 \text{ V}$$
$$O_3 + 2H^+(10^{-7}M) + 2e = O_2 + H_2O \qquad E^0 = +1.65 \text{ V}$$

The first step in the reduction of O_2 in aprotic solvents such as dimethyl sulfoxide and pyridine appears to be a one-electron step to give the superoxide anion:

$$O_2 + e = O_2^-$$

whereas in aqueous solution a two-electron step occurs to give HO_2^-:

$$O_2 + 2e + H_2O = HO_2^- + OH^-$$

In acid solution, O_3 is exceeded in oxidizing power only by fluorine, the perxenate ion, atomic oxygen, OH radicals, and a few other such species. The rate of decomposition of ozone drops sharply in alkaline solutions, the half-life being ca. 2 minutes in $1N$ NaOH at 25°, 40 minutes at $5N$, and 83 hours at $20N$; the ozonide ion (see below) is also more stable in alkaline solution.

[11] S. Fallab, *Angew. Chem. Internat. Edn.*, 1967, **6**, 496 (reactions of O_2); H. Taube, *J. Gen. Physiol.*, 1965, **49**, (1) Part 2, p. 29 (mechanisms of oxidation by O_2).
[12] See L. F. Fieser and M. Fieser, *Reagents for Organic Synthesis*, Wiley, 1967, p. 773; D. B. Denny, ed., *Techniques and Methods of Organic and Organometallic Chemistry*, Vol. 9, Dekker, 1969.

It can also be seen that neutral water saturated with O_2 is a fairly good oxidizing agent. For example, although Cr^{2+} is just stable toward oxidation in pure water, in air-saturated water it is rapidly oxidized; Fe^{2+} is oxidized (only slowly in acid, but rapidly in base) to Fe^{3+} in presence of air, although in air-free water Fe^{2+} is quite stable:

$$Fe^{3+} + e = Fe^{2+} \qquad E^0 = +0.77 \text{ V}$$

The slowness of many oxidations by oxygen in acid solution is attributable to the initial reduction to H_2O with HO_2^- as an intermediate if one-electron reducing agents are present:

$$O_2 + 2H^+ + 2e = H_2O_2 \qquad E^0 = +0.682 \text{ V}$$
$$O_2 + H_2O + 2e = OH^- + HO_2^- \qquad E^0 = -0.076 \text{ V}$$

However, the rate of oxidation of various substances, e.g., ascorbic acid, may be vastly increased by catalytic amounts of transition-metal ions, especially Cu^{2+}, where a Cu^I–Cu^{II} redox cycle is involved.

Oxygen is readily soluble in organic solvents, and merely pouring such liquids in air serves to saturate them with oxygen. This should be kept in mind when determining the reactivity of air-sensitive materials in solution in organic solvents.[13]

Measurements of electronic spectra of alcohols, ethers, benzene, and even saturated hydrocarbons show that there is reaction of the charge-transfer type with the oxygen molecule. However, there is no true complex formation since the heats of formation are negligible and the spectral changes are due to contact between the molecules at van der Waals distances. The classic example is that of N,N-dimethylaniline which becomes yellow in air or oxygen but colorless again when the oxygen is removed by nitrogen. Such weak charge-transfer complexes make certain electronic transitions in molecules more intense; they are also a plausible first stage in photo-oxidations.

With certain transition-metal complexes, oxygen adducts may be formed, sometimes reversibly (see Sections 24-A-1, 25-E and 25-F). Although the O_2 entity remains intact, the complexes may be described as having coordinated O_2^- or O_2^{2-} ions; these may be bound to the metal in a three-membered ring or may act as a bridging group. Coordinated molecular oxygen is more reactive than the free molecule, and various substances not directly oxidized under mild conditions can be attacked in presence of metal complexes.

Singlet O_2 and Photochemical Oxidations.[14]   The lowest-energy electron configuration of the O_2 molecule that contains two electrons in π^* orbitals, gives rise to three states, as shown below. Oxygen molecules in excited singlet states, especially the $^1\Delta_g$ state, which has a much longer lifetime than the $^1\Sigma_g^+$ state, react with a variety of unsaturated organic substrates to cause

[13] See D. F. Shriver, *The Manipulation of Air-Sensitive Compounds*, McGraw-Hill, 1969.
[14] C. S. Foote, *Accounts Chem. Res.*, 1968, **1**, 104; D. Valentine, Jr., in *Annual Survey of Photochemistry*, Interscience, 1970, p. 362; *Ann. New York Acad. Sci.*, 1970, **171**, No. 1; D. R. Kearns, *Chem. Rev.*, 1971, **71**, 395.

State	π_a^*	π_b^*	Energy
$^1\Sigma_g^+$	↿	↾	155 kJ ($\sim$13,000 cm^{-1})
$^1\Delta_g$	↿⇂		92 kJ ($\sim$ 8,000 cm^{-1})
$^3\Sigma_g^-$	↿	↾	0 (ground state)

limited, specific oxidations, a very typical reaction being a Diels–Alder-like 1,4-addition to a 1,3-diene:

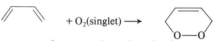

There are three ways of generating the singlet oxygen molecules: (1) photochemically by irradiation in presence of a sensitizer; (2) chemically; (3) in an electrodeless discharge. The last is inefficient and impractical. The photochemical route is believed to proceed as follows, where "sens" represents the photosensitizer (typically a fluorescein derivative, methylene blue, certain porphyrins or certain polycyclic aromatic hydrocarbons):

$$^1Sens \xrightarrow{h\nu} {}^1Sens^*$$
$$^1Sens^* \rightarrow {}^3Sens^*$$
$$^3Sens^* + {}^3O_2 \rightarrow {}^1Sens + {}^1O_2$$
$$^1O_2 + Substrate \rightarrow Products$$

Energy transfer from triplet excited sensitizer, $^3Sens^*$, to $^3O_2$ to give $^1O_2$ is a spin-allowed process. Singlet oxygen (mainly $^1\Delta_g$) is generated chemically in the reaction

$$H_2O_2 + ClO^- \rightarrow Cl^- + H_2O + O_2(^1\Delta_g)$$

and by carrying out this reaction in alcohol in presence of substrates useful amounts of products may be produced. It is also conveniently generated by the thermal decomposition of the solid adducts formed at low temperatures by ozone with triaryl and other phosphites,[15] e.g.,

$$(PhO)_3P + O_3 \xrightarrow{-78°} (PhO)_3PO_3 \xrightarrow{-15°} (PhO)_3PO + O_2(^1\Delta_g)$$

It is possible that singlet oxygen is involved in many biological and other oxidations using O_2, especially in presence of light.

OXYGEN COMPOUNDS

Most oxygen compounds are described in this book during treatment of the chemistry of other elements. Water and the hydroxonium ion have already been discussed (Chapter 5). A few important compounds and classes of compounds will be mentioned here.

14-4. Oxygen Fluorides[16]

Since fluorine is more electronegative than oxygen, it is logical to call its binary compounds with fluorine oxygen fluorides rather than fluorine oxides,

[15] M. E. Brennan, *Chem. Comm.*, **1970**, 956; R. W. Murray, W. C. Jumma and W.-P. Lin, *J. Amer. Chem. Soc.*, 1970, **92**, 3205; P. D. Bartlett and G. D. Mendelhall, *J. Amer. Chem. Soc.*, 1970, **92**, 210.

[16] J. J. Turner, *Endeavour*, 1968, **27**, 42; A. G. Streng, *Chem. Rev.*, 1963, **63**, 607.

although the latter names are sometimes seen. Oxygen fluorides have been intensively studied as potential rocket fuel oxidizers.[17]

Oxygen Difluoride, OF_2. This is prepared by passing fluorine rapidly through 2% sodium hydroxide solution, by electrolysis of aqueous HF–KF solutions, or by action of F_2 on moist KF.[18] It is a pale yellow poisonous gas, b.p. $-145°$. It is relatively unreactive and can be mixed with H_2, CH_4 or CO without reaction, although sparking causes violent explosion. Mixtures of OF_2 with Cl_2, Br_2 or I_2 explode at room temperature. It is fairly readily hydrolyzed by base:

$$OF_2 + 2OH^- \rightarrow O_2 + 2F^- + H_2O$$

It reacts more slowly with water, but explodes with steam:

$$OF_2 + H_2O \rightarrow O_2 + 2HF$$

and it liberates other halogens from their acids or salts:

$$OF_2 + 4HX(aq) \rightarrow 2X_2 + 2HF + H_2O$$

Metals and non-metals are oxidized and/or fluorinated, and in an electric discharge even Xe reacts to give a mixture of fluoride and oxide fluoride.

Dioxygen Difluoride, O_2F_2. The difluoride is a yellow-orange solid (m.p. $109.7°K$), obtained by high-voltage electric discharges on mixtures of O_2 and F_2 at 10–20 mm pressure and temperatures of $77–90°K$. It decomposes into O_2 and F_2 in the gas at $-50°$ with a half-life of about 3 hr. It is an extremely potent fluorinating and oxidizing agent and under controlled conditions OOF groups may be transferred to a substrate.[19] Many substances explode on exposure to oxygen difluoride at low temperatures, and even C_2F_4 is converted into COF_2, CF_4, CF_3OOCF_3, etc. With Cl_2 a purple, fairly stable intermediate $(O_2ClF_3)_n$ can be isolated. O_2F_2 has been used for oxidizing primary aliphatic amines to the corresponding nitroso compounds.

The structures of OF_2 and O_2F_2 are known (14-V, 14-VI). That of O_2F_2 is notable for the shortness of the O—O bond (1.217 Å, cf. 1.48 Å in H_2O_2 and 1.49 in O_2^{2-}) and the relatively long O—F bonds (1.575 Å) compared with those in OF_2 (1.409 Å). A plausible but distinctly *ad hoc* explanation which has been offered for this is that each singly occupied π^* orbital of the O_2 molecule interacts with a singly occupied fluorine σ orbital to form two

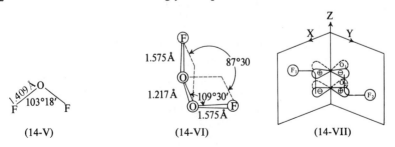

(14-V)   (14-VI)   (14-VII)

[17] *Advanced Propellant Chemistry*, Advances in Chemistry Series No. 54, Amer. Chem. Soc., 1966.
[18] A. H. Borning and K. E. Pullen, *Inorg. Chem.*, 1969, **8**, 1791.
[19] See e.g., I. J. Solomon *et al.*, *J. Amer. Chem. Soc.*, 1968, **90**, 6557.

$3c$—$2e$ OOF bonds in roughly perpendicular planes, as illustrated in (14-VII). Thus a strong, essentially double O—O bond is retained while relatively weak O—F bonds are formed.

Other oxygen fluorides, have been reported, but "O_3F_2" now appears[20] to be a mixture of other substances, e.g., O_2F_2 and $(OOF)_n$, while "O_4F_2," described as a red-brown solid at 77°, has not been satisfactorily confirmed.

14-5. The Dioxygenyl Cation

Compounds containing discrete O_2^+ ions are known. Thus PtF_6 reacts with O_2 to give the orange solid O_2PtF_6, isomorphous with $KPtF_6$. The O_2MF_6 compounds in which M = P, As, or Sb can be prepared in several ways,[21] viz..

$$O_2F_2 + MF_5 \rightarrow O_2MF_6 + \tfrac{1}{2}F_2$$
$$2O_2 + F_2 + 2MF_5 \xrightarrow{h\nu} 2O_2MF_6 \qquad \text{(M = As or Sb only)}$$
$$4OF_2 + 2MF_5 \rightarrow 2O_2MF_6 + 3F_2 \qquad \text{(M = As or Sb only)}$$

The O_2^+ ion is found, as expected, to be paramagnetic and the O—O stretching frequency is $1860 \pm 2 \text{ cm}^{-1}$. Spectroscopic study of gaseous O_2^+ gives an O—O distance of 1.12 Å (cf. 1.09 Å in isoelectronic NO).

14-6. Hydrogen Peroxide, H_2O_2 [22]

Hydrogen peroxide is obtained by electrolytic processes that involve the formation of peroxodisulfate ion and its subsequent hydrolysis. Sulfuric acid or ammonium sulfate–sulfuric acid solutions are electrolyzed at high current density (~ 1 amp/dm^2) at electrodes (usually Pt) that have high overvoltages for O_2 evolution. Although the detailed mechanism of the process is not quite certain, stoichiometrically we have

$$S_2O_8^{2-} + 2H^+ + 2e = 2HSO_4^- \qquad E^0 = 2.18 \text{ V (acid)}$$
$$S_2O_8^{2-} + 2e = 2SO_4^{2-} \qquad E^0 = 2.06 \text{ V (neutral)}$$

An optimum residence time for the solution and low temperature ($-20°$) are used to minimize the hydrolytic reaction (eq. 14-1) in the cell and conse-

$$H_2S_2O_8 + H_2O \rightarrow \quad H_2SO_5 \quad + \quad H_2SO_4 \qquad (14\text{-}1)$$

Peroxodi-   Peroxomono-
sulfuric   sulfuric acid;
acid   "Caro's acid"

quent loss of product by decomposition reactions (eqs. 14-2a, b):

$$2H_2O_2 \rightarrow 2H_2O + O_2 \qquad (14\text{-}2a)$$
$$H_2SO_5 + H_2O_2 \rightarrow H_2SO_4 + H_2O + O_2 \qquad (14\text{-}2b)$$

The peroxodisulfuric acid solution is hydrolyzed separately:

$$H_2S_2O_8 + H_2O \rightarrow H_2SO_5 + H_2SO_4 \qquad \text{(Fast)}$$
$$H_2SO_5 + H_2O \rightarrow H_2O_2 + H_2SO_4 \qquad \text{(Slow)}$$

[20] I. J. Solomon et al., J. Amer. Chem. Soc., 1968, 90, 5408.
[21] R. J. Gillespie and J. Passmore, Accounts Chem. Res., 1971, in the press.
[22] R. Powell, Hydrogen Peroxide Manufacture, Noyes Devel. Corp., Park Ridge, New Jersey, 1968; W. C. Schumb, C. N. Satterfield and R. L. Wentworth, Hydrogen Peroxide, A. C. S. Monograph No. 128, Reinhold, 1955 (a comprehensive reference work).

and the H_2O_2 is rapidly removed by distillation at high temperature and low pressure. Dilute solutions of H_2O_2 so obtained are then concentrated by vacuum distillation to 28–35% by weight. Higher concentrations, 90–99%, are commercially achieved by further, multistage fractionation. Such concentrated materials are very susceptible to decomposition under catalysis by metal ions and it is necessary to add inhibitors such as sodium pyrophosphate or stannate and to store the hydrogen peroxide in pure aluminum ($> 99.6\%$) containers.

Hydrogen peroxide is also produced on a large scale by autoxidation of an anthraquinol, such as 2-ethylanthraquinol (14-VIII), in a cyclic continuous

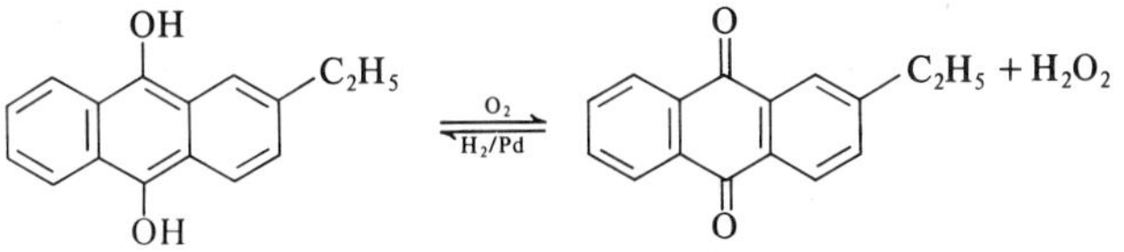

(14-VIII)

process. Hydrogen from the cracking of butane is used to reduce the quinone with Pd on an inert support in free suspension. H_2O_2 is extracted from the oxygenated organic solution by countercurrent columns, whereafter the aqueous product contains about 20% H_2O_2. The process needs only H_2, atmospheric oxygen and water as major raw materials; it is cheaper to operate than the electrolytic method.

Pure H_2O_2 is a pale blue, sirupy liquid, boiling at 152.1° and freezing at $-0.89°$. It resembles water in many of its physical properties. The pure liquid has a dielectric constant at 25° of 93 and a 65% solution in water has a dielectric constant of 120. Thus both the pure liquid and its aqueous solutions are potentially excellent ionizing solvents, but its utility in this respect is limited by its strongly oxidizing nature and its ready decomposition in the presence of even traces of many heavy-metal ions according to the equation:

$$2H_2O_2 = 2H_2O + O_2 \qquad \Delta H = -99 \text{ kJ mol}^{-1} \tag{14-3}$$

In dilute aqueous solution it is more acidic than water:

$$H_2O_2 = H^+ + HO_2^- \qquad K_{20°} = 1.5 \times 10^{-12}$$

The molecule H_2O_2 has a skew, chain structure (Fig. 14-2). There is only a low barrier to internal rotation about the O—O bond. In the liquid state H_2O_2 is even more highly associated via hydrogen bonding than is H_2O.

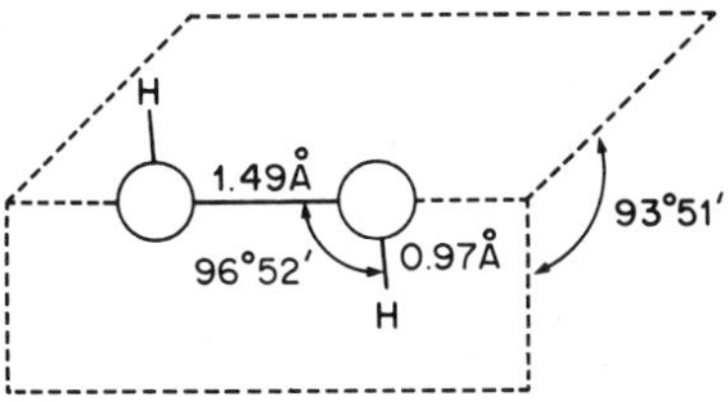

Fig. 14-2. The structure of hydrogen peroxide.

Its oxidation-reduction chemistry in aqueous solution is summarized by the following potentials:

$$H_2O_2 + 2H^+ + 2e = 2H_2O \qquad E^0 = 1.77 \text{ V}$$
$$O_2 + 2H^+ + 2e = H_2O_2 \qquad E^0 = 0.68 \text{ V}$$
$$HO_2^- + H_2O + 2e = 3OH^- \qquad E^0 = 0.87 \text{ V}$$

from which it can be seen that hydrogen peroxide is a strong oxidizing agent in either acid or basic solution; only toward very strong oxidizing agents such as MnO_4^- will it behave as a reducing agent.

Dilute or 30% hydrogen peroxide solutions are widely used as oxidants. In acid solution oxidations with hydrogen peroxide are most often slow, whereas in basic solution they are usually fast. Decomposition of hydrogen peroxide according to reaction 14-3, which may be considered a self-oxidation, occurs most rapidly in basic solution; hence an excess of H_2O_2 may best be destroyed by heating in basic solution.

The oxidation of H_2O_2 in aqueous solution by Cl_2, MnO_4^-, Ce^{4+}, etc., and the catalytic decomposition caused by Fe^{3+}, I_2, MnO_2, etc., have been studied. In both cases, by using labeled H_2O_2, it has been shown that the oxygen produced is derived entirely from the peroxide and not from water. This suggests that oxidizing agents do not break the O—O bond, but simply remove electrons. In the case of oxidation by chlorine, a mechanism of the following kind is consistent with the lack of exchange of ^{18}O between H_2O_2 and H_2O:

$$Cl_2 + H_2{}^{18}O_2 \rightarrow H^+ + Cl^- + H^{18}O^{18}OCl$$
$$H^{18}O^{18}OCl \rightarrow H^+ + Cl^- + {}^{18}O_2$$

It is important to recognize, however, that very many reactions involving H_2O_2 (and also O_2) in solutions are free-radical ones. Metal-ion-catalyzed decomposition of H_2O_2 and other reactions can give rise to radicals of which HO_2 and OH are most important. HO_2 has been detected in ice irradiated at low temperature and also in aqueous solutions where H_2O_2 interacts with Ti^{3+}, Fe^{2+} or Ce^{IV} ions.

Hydrogen peroxide has been estimated to be more than 10^6 times less *basic* than H_2O. However, on addition of concentrated H_2O_2 to tetrafluoroboric acid in tetrahydrothiophene 1,1-dioxide (sulfolane) the conjugate cation $H_3O_2^+$ can be obtained. The solutions are very powerful, but unselective, oxidants for benzene, cyclohexane and other organic materials.

14-7. Peroxides, Superoxides and Ozonides[23]

Ionic Peroxides. Peroxides that contain O_2^{2-} ions are known for the alkali metals, Ca, Sr and Ba. Sodium peroxide is made commercially by air oxidation of Na, first to Na_2O, then to Na_2O_2; it is a yellowish powder, very hygroscopic though thermally stable to 500°, which contains also, according to esr studies, about 10% of the superoxide. Barium peroxide, which was

[23] I. I. Vol'nov, *Peroxides, Superoxides, and Ozonides of Alkali and Alkaline Earth Metals,* Consultants Bureau-Plenum Press, 1966; N-G. Vannerberg, *Progr. Inorg. Chem.,* 1962, **4**, 125; A. W. Petrocelli *et al., J. Chem. Educ.,* 1962, **39**, 557; 1963, **40**, 146.

originally used for making dilute solutions of hydrogen peroxide by treatment with dilute sulfuric acid, is made by action of air or O_2 on BaO; the reaction is slow below $500°$ and BaO_2 decomposes above $600°$.

The ionic peroxides with water or dilute acids give H_2O_2, and all are powerful oxidizing agents. They convert all organic materials into carbonate even at moderate temperatures. Na_2O_2 also vigorously oxidizes some metals; e.g., Fe violently gives FeO_4^{2-}; and Na_2O_2 can be generally employed for oxidizing fusions. The alkali peroxides also react with CO_2:

$$2CO_2(g) + 2M_2O_2 \rightarrow 2M_2CO_3 + O_2$$

Peroxides can also serve as reducing agents for such strongly oxidizing substances as permanganate.

A number of other electropositive metals such as Mg, the lanthanides or uranyl ion also give peroxides; these are intermediate in character between the ionic ones and the essentially covalent peroxides of metals such as Zn, Cd and Hg. The addition of H_2O_2 to solutions of, e.g., Zn^{2+} or UO_2^{2+} gives impure peroxides.

A characteristic feature of the ionic peroxides is the formation of well-crystallized hydrates and H_2O_2 adducts. Thus $Na_2O_2 \cdot 8H_2O$ can be obtained by adding ethanol to 30% H_2O_2 in concentrated NaOH at $15°$ or by rapid crystallization of Na_2O_2 from iced water. The alkaline earths all form the octahydrates, $M^{II}O_2 \cdot 8H_2O$. They are isostructural, containing discrete peroxide ions to which the water molecules are hydrogen-bonded, giving chains of the type $\cdots O_2^{2-} \cdots (H_2O)_8 \cdots O_2^{2-} \cdots (H_2O)_8 \cdots$.

Superoxides. The action of oxygen at pressures near atmospheric on K, Rb or Cs gives yellow to orange crystalline solids of formula MO_2. NaO_2 can be obtained only by reaction of Na_2O_2 with O_2 at 300 atm and $500°$. LiO_2 cannot be isolated and the only evidence for it is the similarity in the absorption spectra of the pale yellow solutions of Li, Na and K on rapid oxidation of the metals in liquid ammonia at $-78°$ by oxygen. Alkaline-earth, Mg, Zn and Cd superoxides occur only in small concentrations as solid solutions in the peroxides. Tetramethylammonium superoxide has been obtained as a yellow solid (m.p. $97°$), which dissolves in water with evolution of O_2. There is clearly a direct correlation between superoxide stability and electropositivity of the metal concerned.

The paramagnetism of the compounds corresponds to one unpaired electron per two oxygen atoms, consistent with the existence of O_2^- ions, as first suggested for these oxides by Pauling. Crystal-structure determinations show the existence of such discrete O_2^- ions. The compounds KO_2, RbO_2 and CsO_2 crystallize in the CaC_2 structure (Fig. 10-3), which is a distorted NaCl structure. NaO_2 is cubic owing to the disorder in the orientation of the O_2^- ions. The superoxides are very powerful oxidizing agents. They react vigorously with water:

$$2O_2^- + H_2O = O_2 + HO_2^- + OH^-$$
$$2HO_2^- = 2OH^- + O_2 \qquad \text{(Slow)}$$

The reaction with CO_2, which involves peroxocarbonate intermediates, is of some technical use for removal of CO_2 and regeneration of O_2 in closed systems. The over-all-reaction is:

$$4MO_2(s) + 2CO_2(g) = 2M_2CO_3(s) + 3O_2(g)$$

Ozonides. The interaction of O_3 with hydroxides of K, Rb and Cs has long been known to give materials that are neither peroxides nor superoxides. These are ozonides:

$$3KOH(s) + 2O_3(g) = 2KO_3(s) + KOH \cdot H_2O(s) + \tfrac{1}{2}O_2(g)$$

KO_3 gives orange-red crystals; it decomposes to KO_2 and O_2 slowly. NH_4O_3 has also been reported.

The ozonide ion is paramagnetic with one unpaired electron and is apparently bent ($\sim 100°$; O—O ~ 1.2 Å) (cf. ClO_2). There is evidence that O_3^- occurs as a reaction intermediate in the decomposition of H_2O_2 in alkaline solution[24] and in radiolytic reactions.[25]

Materials of composition approximating to M_2O_3 (M = alkali metal) are almost certainly mixtures of peroxide and superoxide and there is no evidence for the existence of an O_3^{2-} ion.

The various $O_2^{n\pm}$ species, from O_2^+ to O_2^{2-}, provide an interesting illustration of the effect of varying the number of antibonding electrons on the length and stretching frequency of a bond, as the data in Table 14-2 show.[26]

TABLE 14-2

Various Bond Values for Oxygen Species

Species	O—O Dist. (Å)	Number of $n*$ Electrons	ν_{OO} (cm^{-1})
O_2^+	1.12	1	1860
O_2	1.21	2	1556
O_2^-	1.33	3	1145
O_2^{2-}	1.49	4	~ 770

14-8. Other Peroxo Compounds

A large number of *organic peroxides* and *hydroperoxides* are known. Peroxo carboxylic acids, e.g., peracetic acid, $CH_3CO \cdot OOH$, can be obtained by action of H_2O_2 on acid anhydrides. Peracetic acid is commercially made as 10–55% aqueous solutions containing some acetic acid by interaction of 50% H_2O_2 and acetic acid, with H_2SO_4 as catalyst at 45–60°; the dilute acid is distilled under reduced pressure; it is also made by air oxidation of acetaldehyde. The peroxo acids are useful oxidants and sources of free radicals, e.g., by treatment with Fe^{2+}(aq). Benzoyl peroxide and cumyl hydroperoxide are moderately stable and widely used as polymerization initiators and for other purposes where free-radical initiation is required.

[24] L. J. Heidt, *J. Phys. Chem.*, 1969, **73**, 2361.
[25] B. L. Galf and L. M. Dortman, *J. Amer. Chem. Soc.*, 1969, **91**, 2199.
[26] L. Andrews, *J. Amer. Chem. Soc.*, 1968, **90**, 7368; J. C. Evans, *Chem. Comm.*, **1969**, 682.

Organic peroxo compounds are also obtained by *autoxidation* of ethers, unsaturated hydrocarbons and other organic materials on exposure to air. The autoxidation is a free-radical chain reaction which is initiated almost certainly by radicals generated by interaction of oxygen and traces of metals such as Cu, Co, or Fe.[27] The attack on specific reactive C—H bonds by a radical, X$^\bullet$, gives first R$^\bullet$ and then hydroperoxides which can react further:

$$RH + X^\bullet \rightarrow R^\bullet + HX$$
$$R^\bullet + O_2 \rightarrow RO_2^\bullet$$
$$RO_2^\bullet + RH \rightarrow ROOH + R^\bullet$$

Peroxide formation can lead to explosions if oxidized solvents are distilled. Peroxides are best removed by washing with acidified $FeSO_4$ solution or, for ethers and hydrocarbons, by passage through a column of activated alumina. Peroxides are absent when $Fe^{2+} + SCN^-$ reagent gives no red color.

There are a great variety of inorganic peroxo compounds where —O— is replaced by —O—O— groups. Some of these are discussed elsewhere in this book. Typical are peroxo anions such as peroxo sulfates (14-IX) and (14-X). All peroxo acids yield H_2O_2 on hydrolysis. Peroxodisulfate, as the

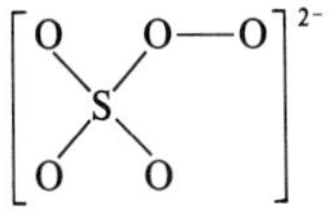

(14-IX)	(14-X)
Peroxomonosulfate	Peroxodisulfate

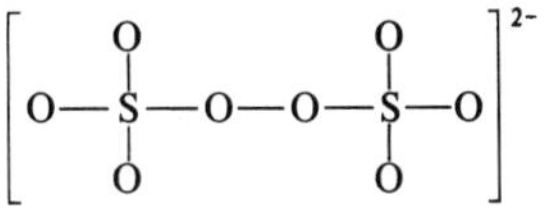

ammonium salt, is commonly used as a strong oxidizing agent in acid solution, for example, to convert C into CO_2, Mn^{2+} into MnO_4^-, or Ce^{3+} into Ce^{4+}. The last two reactions are slow and normally incomplete in the absence of silver ion as a catalyst.

It is important to make the distinction between true peroxo compounds, which contain —O—O— groups, and compounds that contain hydrogen peroxide of crystallization such as $2Na_2CO_3 \cdot 3H_2O_2$ or $Na_4P_2O_7 \cdot nH_2O_2$. Esr studies of peroxoborates and blue peroxocarbonates have shown the presence of free radicals, but it is not yet certain what species are responsible.

[27] J. Betts, *Quart. Rev.*, 1971, **25**, 265.

Further Reading

General:

Ardon, M., *Oxygen: Elementary Forms and Hydrogen Peroxide*, Benjamin, 1965 (a short text).

Hoare, P. J., *The Electrochemistry of Oxygen*, Interscience, 1968.

Perst, H., *Oxonium Ions in Organic Chemistry*, Academic Press, 1971.

Yosk, D. M., and H. Russell, *Systematic Inorganic Chemistry (of the 5th and 6th Group Elements)*, Prentice-Hall, 1946 (excellent on selected aspects).

Oxidation, peroxo species:

Brilkina, T. G., and V. A. Shuschunov, *Reactions of Organometallic Compounds with Oxygen and Peroxides*, Iliffe, 1969.

Connor, J. A., and E. A. V. Ebsworth, *Adv. Inorg. Chem. Radiochem.*, 1964, 6, 280 (peroxo compounds of transition metals).

Edwards, J. O., ed., *Peroxide Reaction Mechanisms*, Interscience, 1962.
Hawkins, E. G. E., *Organic Peroxides; their Formation and Reactions*, Spon, 1961.
Stewart, R., *Oxidation Mechanisms: Applications to Organic Chemistry*, Benjamin, 1964.
Wiberg, K., *Oxidation in Organic Chemistry*, Academic Press, 1966.

Oxides (see also under appropriate elements):

Alper, A. M., *High Temperature Oxides*, 2 vols., Academic Press, 1970.
Brewer, L., *Chem. Rev.*, 1952, **52**, 1 (a comprehensive survey of thermodynamic properties of oxides).
Brewer, L., and G. M. Rosenblatt, *Chem. Rev.*, 1961, **61**, 257 (dissociation energies of gaseous oxides MO_2).
Carrington, A., and M. C. R. Symons, *Chem. Rev.*, 1963, **61**, 443 (structure and reactivity of oxo anions of transition metals).
Gimblett, F. G. R., *Inorganic Polymer Chemistry*, Butterworth, 1963 (aggregation of hydroxo ions and compounds).
Howe, A. T., and P. J. Fenshaw, *Quart. Rev.*, 1967, **21**, 507 (electronic properties of oxides and other binary compounds of first-row transition metals).
Mackenzie, J. D., *Adv. Inorg. Chem. Radiochem.*, 1962, **4**, 293 (nature of oxides in melts).
Wadsley, A. D., in *Non-Stoichiometric Compounds*, L. Mandelcorn, ed., Academic Press, 1964 (binary and ternary oxides).
Ward, R., *Progr. Inorg. Chem.*, 1959, **1**, 465 (mixed metal oxides).

15

The Group VI Elements: S, Se, Te, Po

GENERAL REMARKS

15-1. Electronic Structures, Valences, and Stereochemistries

Some properties of the elements are given in Table 15-1.

TABLE 15-1

Some Properties of the Group VI Elements

Element	Electronic structure	Melting point (°C)	Boiling point (°C)	Radius X^{2-}	Covalent radius —X—	Electro-negativity
S	$[Ne]3s^23p^4$	119^a	444.6	1.90	1.03	2.44
Se	$[Ar]3d^{10}4s^24p^4$	217	684.8	2.02	1.17	2.48
Te	$[Kr]4d^{10}5s^25p^4$	450	990	2.22	1.37	2.01
Po	$[Xe]4f^{14}5d^{10}6s^26p^4$	254	962	2.30		1.76

^{a} For monoclinic S (see text).

The atoms are two electrons short of the configuration of the next noble gas, and the elements show essentially non-metallic covalent chemistry except for polonium and to a very slight extent tellurium. They may complete the noble-gas configuration by forming (a) the *chalconide* ions S^{2-}, Se^{2-}, and Te^{2-}, although these ions exist only in the salts of the most electropositive elements, (b) two electron pair bonds, e.g., $(CH_3)_2S$, H_2S, SCl_2, etc., (c) ionic species with one bond and one negative charge, e.g., RS^-, or (d) three bonds and one positive charge, e.g., R_3S^+.

In addition to such divalent species, the elements form compounds in *formal* oxidation states IV and VI with 4, 5 or 6 bonds; tellurium may give an 8-coordinate ion, TeF_8^{2-}. Some examples of compounds of Group VI elements and their stereochemistries are listed in Table 15-2.

TABLE 15-2

Compounds of Group VI Elements and their Stereochemistries

Valence	Number of bonds	Geometry	Examples
II	2	Angular	Me_2S, H_2Te, S_n
	3	Pyramidal	Me_3S^+
	4	Square-planar	$Te[SC(NH_2)_2]_2Cl_2$
IV	2	Angular	SO_2
	3	Pyramidal	SF_3^+, OSF_2, SO_3^{2-}
		Trigonal-planar	$(SeO_2)_n$
	4	ψ-Trigonal-bipyramidal	SF_4, RSF_3, Me_2TeCl_2
		Tetrahedral	Me_3SO^+
	5	ψ-Octahedral (square-pyramidal)	$SeOCl_2py_2$, SF_5^-, TeF_5^-
	6	Octahedral	$SeBr_6^{2-}$, PoI_6^{2-}, $TeBr_6^{2-}$
VI	3	Trigonal-planar	$SO_3(g)$
	4	Tetrahedral	SeO_4^{2-}, $SO_3(s)$, SeO_2Cl_2
	5	Trigonal-bipyramidal	SOF_4
	6	Octahedral	RSF_5, SeF_6, $Te(OH)_6$
	8(?)	?	$TeF_8^{2-}(?)$

15-2. Group Trends

There are great differences between the chemistry of oxygen and that of sulfur, with more gradual variations through the sequence S, Se, Te, Po. Differences from oxygen are attributable, among other things, to the following:

1. The lower electronegativities of the S–Po elements lessens the ionic character of those of their compounds that are formally analogous to those of oxygen, alters the relative stabilities of various kinds of bonds, and drastically lessens the importance of hydrogen bonding, although weak S···H—S bonds do indeed exist.

2. The maximum coordination number is not limited to four, nor is the valence limited to two, as in the case of oxygen, since d orbitals may be utilized in bonding. Thus sulfur forms several hexacoordinate compounds, for example, SF_6, and for tellurium six is the characteristic coordination number.

3. Sulfur has a strong tendency to catenation, so that it forms compounds having no oxygen, selenium, or tellurium analogs, for example: polysulfide ions, S_n^{2-}; sulfanes, XS_nX (where X may be H, halogen, —CN, or —NR_2); and the polysulfuric acids, $HO_3SS_nSO_3H$ and their salts. Although selenium and tellurium have a smaller tendency to catenation they form rings (Se only) and long chains in their elemental forms. None of these chains is branched, because the valence of the element is only two.

Gradual changes of properties are evident with increasing size, decreasing electronegativity, etc., such as:

1. Decreasing thermal stability of the H_2X compounds. Thus H_2Te is considerably endothermic.

2. Increasing metallic character of the elements.

3. Increasing tendency to form anionic complexes such as $SeBr_6^{2-}$, $TeBr_6^{2-}$, PoI_6^{2-}.

4. Decreasing stability of compounds in high formal positive oxidation states.

5. Emergence of cationic properties for Po and, very marginally, for Te. Thus TeO_2 and PoO_2 appear to have ionic lattices and they react with hydrohalic acids to give Te^{IV} and Po^{IV} halides, and PoO_2 forms a hydroxide $Po(OH)_4$. There are also some ill-defined "salts" of Te and Po, such as $Po(SO_4)_2$, $TeO_2 \cdot SO_3$, etc.

Use of d Orbitals.[1] In addition to the ability of the S–Po elements to utilize d orbitals in hybridization with s and p orbitals so as to form more than four σ bonds to other atoms, sulfur particularly and also selenium appear to make frequent use of $d\pi$ orbitals to form multiple bonds. Thus, for example, in the sulfate ion, where the s and p orbitals are used in σ bonding, the shortness of the S—O bonds suggests that there must be considerable multiple-bond character. The only likely explanation for this is that empty $d\pi$ orbitals of sulfur accept electrons from filled $p\pi$ orbitals of oxygen (see page 143). Similar $d\pi$-$p\pi$ bonding occurs in some phosphorus compounds, but it seems to be more prominent with sulfur, and many instances will be cited later in this Chapter.

THE ELEMENTS

15-3. Occurrence

Sulfur occurs widely in Nature as the element, as H_2S and SO_2, in innumerable sulfide ores of metals, and in the form of various sulfates such as *gypsum* and *anhydrite* ($CaSO_4$), magnesium sulfate, etc. Selenium and tellurium are much less abundant than sulfur and frequently occur as selenide and telluride impurities in metal sulfide ores. They are often recovered from flue dusts of combustion chambers for sulfur ores, particularly those of Ag and Au, and from lead chambers in sulfuric acid manufacture.

Polonium occurs in U and Th minerals as a product of radioactive decay series. It was first isolated from pitchblende which contains less than 0.1 mg of Po per ton. The most accessible isotope is ^{210}Po (α, 138.4d) obtained in gram quantities by irradiation of bismuth in nuclear reactors:

$$^{209}Bi(n, \gamma)^{210}Bi \rightarrow \, ^{210}Po + \beta^-$$

Polonium is separated from Bi by sublimation or in a variety of chemical ways. The study of polonium chemistry is difficult owing to the intense α radiation, which causes damage to solutions and solids, evolves much heat, and makes necessary special handling techniques for protection of the chemist.

---

[1] For discussion see D. W. J. Cruickshank and B. C. Webster in *Inorganic Sulfur Compounds*, G. Nickless, ed., Elsevier, 1968; W. G. Salmond, *Quart. Rev.*, 1968, **22**, 253; L. D. Pettit, *Quart. Rev.*, 1971, **25**, 1.

15-4. The Structures of Elemental Sulfur[2]

The structural relationships of sulfur in all three phases are exceedingly complex and there has been considerable confusion concerning new allotropes, which subsequently turned out to be mixtures or to be impure, and also concerning nomenclature. We shall deal only with the main, well established species.

Solid Sulfur. All modifications of crystalline sulfur contain either (a) sulfur rings, which may have 6, 8, 10 or 12 sulfur atoms and are referred to as cyclohexa-, cycloocta-, etc., -sulfur, or (b) chains of sulfur atoms, referred to as catenasulfur, S_∞.

(a) *Cyclooctasulfur*, S_8. This is the most common form and has three main allotropes (crystal forms):

Orthorhombic sulfur, S_α, is thermodynamically the most stable form and occurs in large yellow crystals in volcanic areas. It can be grown from solutions although the crystals then usually contain solvent. Its structure is shown in Fig. 15-1. At 368.46 °K (95.5 °C) S_α transforms to the high-temperature form, *monoclinic sulfur*, S_β. The enthalpy of the transition is small (0.4 kJ g-atom^{-1} at 95.5 °C) and the process is slow so that it is possible by rapid heating of S_α to attain the melting point of S_α, 112.8°; S_β melts at 119°. Monoclinic S_β crystallizes from sulfur melts and, although slow conversion to S_α occurs, the crystals can be preserved for weeks. Its structure contains S_8 rings as in S_α, but differently packed.

Monoclinic, S_γ, m.p. 106.8 °C, is obtained by slow crystallization of sulfur from ethanolic ammonium polysulfide solutions. It transforms slowly into S_β and/or S_α.

(b) *Cyclohexasulfur*, S_6. This is rhombohedral sulfur (S_ρ) and is obtained by addition of concentrated HCl to a solution of $Na_2S_2O_3$ at $-10°$. The polythionate chains (page 452) initially produced undergo ring closure to give S_6. Extraction of the precipitate with benzene and crystallization give orange crystals, but extremely complicated procedures are required to obtain

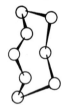 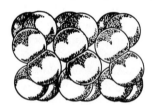

S–S = 2.059Å
∠S–S–S = 107°54′
Dihedral angle = 98.9°

Fig. 15-1. The structure of orthorhombic sulfur in which layers of cyclic S_8 molecules are stacked together.

[2] (a) B. Meyer, *Chem. Rev.*, 1964, **64**, 429; in *Inorganic Sulphur Chemistry*, G. Nickless, ed., Elsevier, 1968; *Elemental Sulphur Chemistry*, Interscience, 1965; (b) F. Tuinstra, *Structural Aspects of the Allotropy of Sulphur and the Other Divalent Elements*, Waltman, 1967 (a critical survey of structures of S, Se, Te and Po); (c) R. Rahman, S. Safe and A. Taylor, *Quart. Rev.*, 1970, **24**, 208 (stereochemistry of sulfur and polysulfides).

high purity.[3] S_6 can also be made by the reaction

$$H_2S_2 + S_4Cl_2 \rightarrow S_6 + 2HCl$$

It decomposes quite rapidly and is chemically very much more reactive than S_8 as the ring is considerably more strained; the reactions may be profoundly affected by impurities and light.

(c) *Other cyclosulfurs.* By controlled reactions of sulfur chlorides with sulfanes or with the compound $(h^5\text{-}C_5H_5)_2TiS_5$, which contains a five-sulfur chain, thermodynamically unstable allotropes containing 7-, 9-, 10- and 12-membered rings can be obtained,[4] e.g.,

$$2H_2S_4 + 2S_2Cl_2 \rightarrow S_{12} + 4HCl$$
$$(h^5\text{-}C_5H_5)_2TiS_5 + S_nCl_2 \xrightarrow{CS_2} (h^5\text{-}C_5H_5)_2TiCl_2 + S_{5+n}$$

These sulfurs have the properties:

S_7, intense yellow, m.p. 39°, polymerizes to a viscous liquid ($\sim 45°$).

S_9, intense yellow, more stable than S_6.

S_{10}, intense yellow, polymerizes above 60° to a viscous liquid.

S_{12}, light yellow, m.p. 145°.

Although these cyclosulfurs are unstable at 25° and are light-sensitive, they can be preserved for weeks at low temperatures, and S_{12} can be stored for months.

(d) *Catenasulfur.*[5] When molten sulfur is poured into ice-water, the so-called plastic sulfur is obtained; although normally this has S_8 inclusions it can be obtained as long fibres by heating S_α in nitrogen at 300° for 5 min. and quenching a thin stream in ice-water. These fibres can be stretched under water and appear to contain helical chains of sulfur atoms with about 3.5 atoms per turn. Unlike the other sulfur allotropes, catenasulfur is insoluble in CS_2; it transforms slowly to S_α.

Liquid Sulfur.[6] The properties of cyclooctasulfur are unusual; the other cyclosulfurs undergo similar changes on melting but have been less well studied. On melting, S_8 first gives a yellow transparent mobile liquid which becomes brown and increasingly viscous above about 160°. The viscosity reaches a maximum at about 200°, and thereafter falls until, at the b.p. of 444.60°, the sulfur is again a rather mobile, dark red liquid. Fig. 15-2(a) shows the viscosity and specific heat as a function of temperature. Although S_8 rings persist in the liquid up to about 193°, the changes in viscosity are due to ring cleavage and the formation of catenasulfur species. The average degree of polymerization is shown in Fig. 15-2(b). Such sulfur chains must have radical ends, and these radicals will in turn attack other rings and chains so that at any temperature an equilibrium between rings and chains of many lengths

[3] R. E. Davis in *Inorganic Sulfur Chemistry*, G. Nickless, ed., Elsevier, 1968, p. 108.
[4] M. Schmidt *et al.*, *Angew. Chem. Internat. Edn.*, 1968, **7**, 632; *Chem. Ber.*, 1968, **101**, 381; U.-I. Zahorsky, *Angew. Chem. Internat. Edn.*, 1968, **7**, 633; I. Kawada and E. Hellner, *Angew. Chem. Internat. Edn.*, 1970, **9**, 379.
[5] A. V. Tobolsky and J. MacKnight, *Polymeric Sulfur and Related Polymers*, Vol. 13 of *Polymer Reviews*, Interscience-Wiley, 1965; J. C. Koh and R. W. Klement, Jr., *J. Phys. Chem.*, 1970, **74**, 4280.
[6] B. Meyer *et al.*, *J. Amer. Chem. Soc.*, 1971, **93**, 1034; *J. Phys. Chem.*, 1971, **75**, 912; R. E. Harris, *J. Phys. Chem.*, 1970, **74**, 3102.

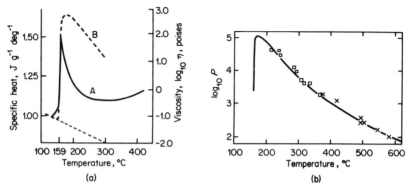

Fig. 15-2. (a) Specific heat (A) and viscosity (B) of liquid sulfur. (b) Chain length (*P*) as a function of temperature, × from magnetic susceptibility measurements and □ from esr measurements. [Reproduced by permission from *Elemental Sulfur—Chemistry and Physics*, B. Meyer, ed., Interscience, 1965.]

will eventually be set up. Esr studies have shown the presence of the radical ends of these chains in molten sulfur above 160°. Their concentration is about 6×10^{-3} mol l^{-1} at 300°. It is presumed that the chains reach their greatest average length, $5–8 \times 10^5$ atoms, at about 200° where the viscosity is highest. The quantitative behavior of the system is sensitive to certain impurities, such as iodine, which can stabilize chain ends by formation of S—I bonds. In the formation of polymers, practically every S—S bond of an S_8 ring broken is replaced by an S—S bond in a linear polymer and the over-all heat of the polymerization is thus expected to be close to zero. An enthalpy of 13.4 kJ mol^{-1} of S_8 converted into polymer has been found at the critical polymerization temperature (159°).

The color changes on melting are due to an increase in the intensity and a shift of an absorption band to the red. This is associated with the formation above ca. 250° of the red species S_3 and S_4, which comprise 1–3% of sulfur at its boiling point.

Sulfur Vapor. In addition to S_8, sulfur vapor contains S_n species in a temperature-dependent equilibrium. S_2 molecules predominate at higher temperatures; above $\sim 2200°$ and at pressures below 10^{-7} cm sulfur atoms predominate. The S_2 species can be rapidly quenched in liquid nitrogen to give a highly colored solid, unstable above $-80°$, which contains S_2 molecules. These have two unpaired electrons (cf. O_2). The electronic absorption bands of S_2 in the visible region account for the deep blue color of hot sulfur vapor.

15-5. The Structures of Elemental Selenium, Tellurium and Polonium[7]

Two thermodynamically unstable allotropes of *selenium* which are soluble in CS_2 and contain cyclooctaselenium, Se_8, can be obtained by evaporation

[7] D. M. Chizhikov and P. Schastlivy, *Selenium and Selenides*, Collet, 1968; *Tellurium and Tellurides*, Collet, 1970.

of solutions below about 72°. However, the stable form consists of grey trigonal, metal-like crystals which may be grown from hot solutions of Se in aniline or from melts. The structure,[8] which has no sulfur analog, contains infinite spiral chains of selenium atoms. Although there are fairly strong single bonds between adjacent atoms in each chain, there is evidently weak interaction of a metallic nature between the neighboring atoms of different chains. Selenium is not comparable with most true metals in its electrical conductivity in the dark, but it is markedly photoconductive and is widely used in photoelectric devices.

The one form of *tellurium* is silvery-white, semimetallic and isomorphous with gray Se. Like the latter it is virtually insoluble in all liquids except those with which it reacts. Gray Se and Te form a continuous range of solid solutions that appear to contain chains in which Se and Te atoms alternate more or less randomly. With sulfur, selenium can give mixed rings S_nSe_{8-n} ($n = 4$–7).[9]

In vapors the concentration of paramagnetic Se_2 and Te_2 molecules and Se and Te atoms is evidently much higher under comparable conditions of temperature and pressure than for sulfur, indicating decreased tendency toward catenation.

The trend towards greater metallic character in the elements is complete at polonium. Whereas sulfur is a true insulator (specific resistivity in $\mu\Omega$-cm $= 2 \times 10^{23}$), selenium (2×10^{11}) and tellurium (2×10^5) are intermediate in their electrical conductivities, and the temperature coefficient of resistivity in all three cases is negative, which is usually considered characteristic of non-metals. Polonium in each of its two allotropes has a resistivity typical of true metals (~ 43 $\mu\Omega$-cm) and a positive temperature coefficient. The low-temperature allotrope, which is stable up to about 100°, has a cubic structure, and the high temperature form is rhombohedral. In both forms the coordination number is six.

15-6. Reactions of the Elements

The allotropes of S and Se containing cyclo-species are soluble in CS_2 and other non-polar solvents such as benzene and cyclohexane. The solutions are light-sensitive, becoming cloudy on exposure, and may also be reactive towards air; unless very special precautions are taken sulfur contains traces of H_2S and other impurities that can have substantial effects on rates of reactions. The nature of the sulfur produced on photolysis is not well established, but such material reverts to S_8 slowly in the dark or rapidly in presence of triethylamine. From the solvent CHI_3, sulfur crystallizes as a charge-transfer compound, $CHI_3 \cdot 3S_8$, with $I \cdots S$ bonds; isomorphous compounds with PI_3, AsI_3, and SbI_3 are also known. It is probable that similar charge-transfer complexes wherein the S_8 ring is retained are first

[8] P. Cherin and P. Ungar, *Inorg. Chem.*, 1967, **6**, 1589.
[9] A. T. Ward, *J. Phys. Chem.*, 1968, **72**, 4133.

formed in reactions of sulfur with, e.g., bromine (cf. page 463). When heated, sulfur, selenium and tellurium burn in air to give the dioxides, MO_2, and the elements react when heated with halogens, most metals, and non-metals. They are not affected by non-oxidizing acids, but the more metallic polonium will dissolve in concentrated HCl as well as in H_2SO_4 and HNO_3.

It has long been known that sulfur, selenium and tellurium will dissolve in oleums to give blue, green and red solutions, respectively, which are unstable and change in color when kept or warmed. The nature of the colored species has been controversial but it is now clear that they are cyclic polycations in which the element is formally in a fractional oxidation state.[10]

It is difficult to isolate solids from the oleum solutions but crystalline salts have been obtained by selective oxidations of the elements with SbF_5 or AsF_5 in liquid HF or with $S_2O_6F_2$ in HSO_3F,[11] or as tetrachloroaluminates, by interaction of the element and its halide in molten $AlCl_3$.[12] Representative reactions are

$$S_8 + 3SbF_5 \rightarrow S_8^{2+} + 2SbF_6^- + SbF_3$$
$$Se_8 + 3AsF_5 \rightarrow Se_8^{2+} + 2AsF_6^- + AsF_3$$
$$2S_8 + S_2O_6F_2 \rightarrow S_{16}^{2+} + 2SO_3F^-$$
$$7Te + TeCl_4 + 4AlCl_3 \rightarrow 2Te_4^{2+} + 4AlCl_4^-$$

X-ray study of salts of the yellow Se_4^{2+} and red Te_4^{2+} ions show that they are square (15-I) and MO considerations suggest that there is a six π-electron quasi-aromatic system. The green Se_8^{2+} ion has a structure (15-II) which is rather similar to that of S_4N_4 (Section 15-9), with a long central Se—Se bond,

(15-I)   (15-II)

except that one Se atom is "up" or *syn* and the other "down" or *anti*. The S_{16}^{2+} and Se_{16}^{2+} ions probably have two M_8 rings joined together. By Raman and other spectroscopic studies the identity of the well-characterized species with those present in solutions has been confirmed. Tellurium forms Te_6^{2+} and Te_n^{n+} in addition to Te_4^{2+} but the structures of these ions are not yet certain.

The blue S_8^{2+} and red S_{16}^{2+} ions are feebly paramagnetic, probably owing to dissociation to form radical cations, e.g.,

$$S_{16}^{2-} \rightleftharpoons 2S_8^+$$

The difference between oxygen which forms only O_2 and O_2^+ and the heavier elements is evidently due to the inability of the latter to form π-bonds.

[10] R. J. Gillespie and J. Passmore, *Accounts Chem. Res.*, 1971, **4**, 413.
[11] R. J. Gillespie *et al.*, *Inorg. Chem.*, 1971, **10**, 1327.
[12] N. J. Bjerrum, *Inorg. Chem.*, 1970, 9, 1965; D. J. Prince, J. D. Corbett and B. Garbisch, *Inorg. Chem.*, 1970, 9, 2730; R. C. Paul, J. K. Puri and K. C. Malhotra, *Chem. Comm.*, **1970**, 776.

Sulfur is also soluble, with reaction, in organic amines such as piperidine, to give colored solutions containing N,N'-polythiobisamines in which there are free radicals (about 1 per 10^4 S atoms),

$$2RR'NH + S_n \rightarrow (RR'N)_2S_{n-1} + H_2S$$

Many sulfur reactions are catalyzed by amines, and such S—S bond-breaking reactions to give free radicals may be involved.

Sulfur and selenium react with many organic molecules. For example, saturated hydrocarbons are dehydrogenated. The reaction of sulfur with alkenes and other unsaturated hydrocarbons is of enormous technical importance: hot sulfurization results in the vulcanization (formation of S bridges between carbon chains) of natural and synthetic rubbers.

It is clear that all reactions of S_8 or other cyclo-species require that the initial attack open the ring to give sulfur chains or chain compounds. Many common reactions can be rationalized by considering a nucleophilic attack on S—S bonds. Some typical reactions are

$$S_8 + 8CN^- \rightarrow 8SCN^-$$
$$S_8 + 8Na_2SO_3 \rightarrow 8Na_2S_2O_3$$
$$S_8 + 8Ph_3P \rightarrow 8Ph_3PS$$

Such reactions cannot possibly proceed by what, according to the stoichiometry, would be ninth-order reactions. It appears that the rate-determining step is the initial attack on the S_8 ring and that subsequent steps proceed very rapidly, so that the reactions can be assumed to proceed as follows:

$$S_8 + CN^- \rightarrow SSSSSSSSCN^-$$
$$S_6\text{—}S\text{—}SCN^- + CN^- \rightarrow S_6SCN^- + SCN^-, \text{ etc.}$$

Sulfur—sulfur bonds occur in a wide variety of compounds, and disulfide bridges are especially important in certain proteins and enzymes. Sulfur bonds can be cleaved homolytically by dissociation or by radical attack on S and heterolytically by nucleophilic or electrophilic attack.

The activation energy for S_N2 attack can be correlated with bond distances—the shorter the bond, the higher the activation energy, which is consistent with poor acceptor property of an S antibonding orbital for the incoming nucleophile in reactions such as

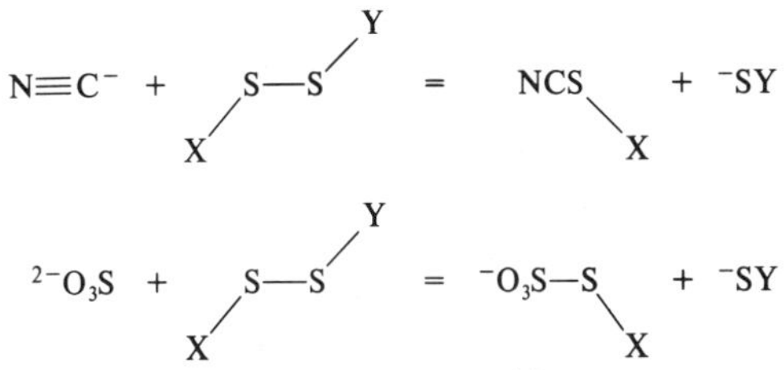

Further discussion of the extensive literature[13] on sulfur reactions is beyond the scope of this text.

[13] See W. A. Pryor, *Mechanism of Sulfur Reactions*, McGraw Hill, 1962; R. E. Davis in *Inorganic Sulfur Chemistry*, R. Nickless, ed., Elsevier, 1968; *Mechanism of Reactions of Sulfur Compounds*, Annual Reports, Intra-Science Research Foundation, Santa Monica, Calif.; J. L. Kice, *Accounts Chem. Res.*, 1968, **1**, 58.

BINARY COMPOUNDS

15-7. Hydrides

The dihydrides, H_2S, H_2Se and H_2Te are extremely poisonous gases with revolting odors; the toxicity of H_2S far exceeds that of HCN. They are readily obtained by the action of acids on metal chalconides. H_2Po has been prepared only in trace quantities, by dissolving magnesium foil plated with Po in $0.2M$ HCl. The thermal stability and bond strengths decrease from H_2S to H_2Po. Although pure H_2Se is thermally stable to 280°, H_2Te and H_2Po appear to be thermodynamically unstable with respect to their constituent elements. All behave as very weak acids in aqueous solution, and the general reactivity and also the dissociation constants increase with increasing atomic number. *Hydrogen sulfide* dissolves in water to give a solution about $0.1\ M$ under 1 atm pressure. The dissociation constants[14] are

$$H_2S + H_2O = H_3O^+ + HS^- \qquad K = 1 \times 10^{-7}$$
$$HS^- + H_2O = H_3O^+ + S^{2-} \qquad K = \sim 10^{-17}$$

In acid solution, H_2S is also a mild reducing agent.

Sulfanes. The compounds H_2S_2 through H_2S_6 have been isolated in pure states whereas higher members are so far known only in mixtures. All are reactive yellow liquids whose viscosities increase with chain length. They may be prepared in large quantities by reactions[15] such as

$$Na_2S_n(aq) + 2HCl(aq) \rightarrow 2NaCl(aq) + H_2S_n(l) \ (n = 4-6)$$
$$S_nCl_2(l) + 2H_2S(l) \rightarrow 2HCl(g) + H_2S_{n+2}(l)$$
$$S_nCl_2(l) + 2H_2S_2(l) \rightarrow 2HCl(g) + H_2S_{n+4}(l)$$

The oils from the first reaction can be cracked and fractionated to give pure H_2S_2 through H_2S_5, whereas the higher sulfanes are obtained from the other reactions. Although the sulfanes are all thermodynamically unstable with respect to the reaction

$$H_2S_n(l) = H_2S(g) + (n-1)S(s)$$

these reactions, which are believed to be free-radical in nature,[15a] are sufficiently slow for the compounds to be stable for considerable periods.

15-8. Metal Chalconides

Most metallic elements react directly with S, Se, Te and, so far as is known, Po. Often they react very readily, mercury and sulfur, for example, at room temperature. Binary compounds of great variety and complexity of structure can be obtained. The nature of the products usually also depends on the ratios of reactants, the temperature of reaction, and other conditions. Many elements form several compounds and sometimes long series of compounds with a given chalconide. We shall give here only the briefest account

[14] W. Giggenbach, *Inorg. Chem.*, 1971, **10**, 1333.
[15] E. Muller and J. B. Hyne, *Canad. J. Chem.*, 1968, **46**, 2341; T. K. Wiewioroski, *Endeavour*, 1970, **29**, 9 (the $S-H_2S-H_2S_n$ system).
[15a] E. Muller and J. B. Hyne, *J. Amer. Chem. Soc.*, 1969, **91**, 1907.

of some of the most important types and shall, for the most part, deal only with the sulfides[16]: the selenides and tellurides[7] are often similar.

Ionic Sulfides; Sulfide Ions.   Only the alkalis and alkaline earths form sulfides that appear to be mainly ionic. They are the only sulfides that dissolve in water and they crystallize in simple ionic lattices, for example, an anti-fluorite lattice for the alkali sulfides and a rock salt lattice for the alkaline-earth sulfides. It is not absolutely certain that they contain S^{2-} and not SH^- ions. Essentially only SH^- ions are present in aqueous solution owing to the low second dissociation constant of H_2S. Although S^{2-} is present in concentrated alkali solutions[14] it cannot be detected below about $8M$ NaOH owing to the reaction

$$S^{2-} + H_2O = SH^- + OH^-   K = \sim 1$$

Aqueous solutions of *polysulfide* ions can be obtained by boiling solutions of the sulfides with sulfur. From such solutions, and also in some other ways, crystalline polysulfides[2c] may be obtained. The structures of three such ions are shown in Fig. 15-3. Only M_2S_4 and M_2S_5 give stable solutions in water; M_2S_2 and M_2S_3 disproportionate to M_2S and M_2S_4 or M_2S_5.

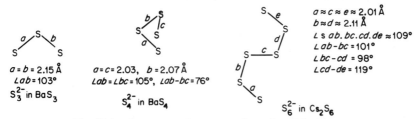

Fig. 15-3.   Structures of representative polysulfide anions.

When alkali polysulfides are dissolved in polar solvents such as acetone, ethanol or Me_2SO, deep-blue solutions may be obtained. Similar blue colors are obtained when S_8 is dissolved in fused alkali halides or sulfates above 400°, but these colors disappear on cooling. The blue species is the paramagnetic disulfide ion[17] S_2^- (cf. O_2^-, page 417) formed by dissociation

$$S_4^{2-} \rightleftarrows 2S_2^-$$

The blue color of the sulfide-containing silicate ultramarine (page 324) is probably due to S_2^- ions trapped in the lattice.

Polysulfide ions can also behave as dinegative chelating ligands, to give what can be considered as heterocyclic sulfur rings. The red salts of the ion $[Pt(S_5)_3]^{2-}$ were obtained in 1903 by boiling H_2PtCl_6 with NH_4S_x, but the constitution was only recently proved.[18] On reduction with CN^- at 60° the ion $[Pt^{II}(S_5)_2]^{2-}$ is obtained. Other complexes with TiS_5 and MoS_4 rings have been obtained, e.g., by interaction of $(h^5\text{-}C_5H_5)_2TiCl_2$ with NaS_x.[19]

[16] A. V. Tobolsky, ed., *The Chemistry of Sulfides*, Interscience-Wiley, 1968.
[17] W. Giggenbach, *Inorg. Chem.*, 1971, **10**, 1306, 1308.
[18] A. E. Wickenden and R. A. Krause, *Inorg. Chem.*, 1969, **8**, 779; P. E. Jones and L. Katz, *Acta Cryst.*, 1969, **B**, **25**, 745.
[19] H. Köpf and B. Block, *Chem. Ber.*, 1969, **102**, 1504; H. Köpf, *Chem. Ber.*, 1969, **102**, 1509; *Angew. Chem. Internat. Edn.*, 1971, **10**, 137.

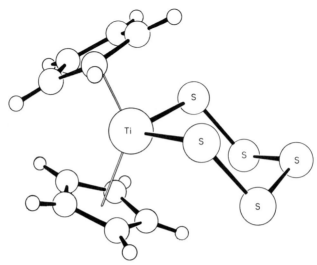

Fig. 15-4. Structure of di-h^5-cyclopentadienyltitanium pentasulfide. [Reproduced by permission from ref. 20.]

The TiS_5 ring has a chair conformation,[20] as shown in Fig. 15-4. A chelate sulfur complex of nickel has also been made[21] by interaction of a macrocyclic sulfide complex with S_2Cl_2.

The S_2 unit can also be bound in transition-metal complexes, e.g., $[Ir(diphos)_2S_2]^+$, in which it is sideways to the metal[22] as in O_2 compounds (Section 21-6). More common are bridge groups between two metal atoms.

Other Metallic Sulfides.[23] Metal sulfides frequently have peculiar stoichiometries, are often non-stoichiometric phases rather than compounds in a classical sense, and are often polymorphic, and many of them are alloy-like or semimetallic in behavior. Sulfides tend to be much more covalent than the corresponding oxides, with the result that quite often there is only limited and occasionally no stoichiometric analogy between the oxides and the sulfides of a given metal. Very often, indeed possibly in most cases, when there is a sulfide and an oxide of identical empirical formula they will have different structures. A few examples will be considered.

Several transition-metal sulfides, for example, FeS, CoS and NiS, adopt a structure called the *nickel arsenide structure*, illustrated in Fig. 15-5.

[20] I. Bernal *et al.*, *Angew. Chem. Internat. Edn.*, 1971, **10**, 921.
[21] N. B. Egan and R. A. Krause, *J. Inorg. Nuclear Chem.*, 1969, **31**, 127.
[22] A. P. Ginsberg and W. E. Lindsell, *Chem. Comm.*, **1971**, 232.
[23] *High Pressure Research*, Vol. 2, *Chalconides*, R. S. Bradley, ed., Academic Press, 1969; A. D. Wadsley in *Non-Stoichiometric Compounds*, L. Mandelcorn, ed., Academic Press, 1964; T. A. Bitter *et al.*, *Inorg. Chem.*, 1968, **7**, 2208.

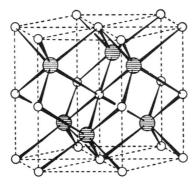

Fig. 15-5. Structure of NiAs (As atoms shaded). The Ni atom in the center of the diagram is surrounded octahedrally by six atoms and has also two near Ni neighbors which are coplanar with four of the As atoms. [Reproduced by permission from A. F. Wells *Structural Inorganic Chemistry*, Clarendon Press, 1945, p. 387.]

In this structure each metal atom is surrounded octahedrally by six sulfur atoms, but also approached fairly closely by two other metal atoms. These metal–metal distances are 2.60–2.68 Å in FeS, CoS and NiS, and at such distances there must be a considerable amount of metal–metal bonding, thus accounting for their alloy-like or semimetallic character. Note that such a structure is not in the least likely for a predominantly ionic salt, requiring as it would the close approach of dispositive ions.

Another class of sulfides of considerable importance is the *disulfides*, represented by FeS_2, CoS_2 and others. All of these contain discrete S_2 units with an S—S distance almost exactly equal to that to be expected for an S—S single bond. These assume one of two closely related structures. First there is the *pyrite structure* named after the polymorph of FeS_2 that exhibits it. This structure may be visualized as a distorted NaCl structure. The Fe atoms occupy Na positions and the S_2 groups are placed with their centers at the Cl positions but turned in such a way that they are not parallel to any of the cube axes. The *marcasite structure* is very similar but somewhat less regular.

FeS is a good example of a well-characterized non-stoichiometric sulfide. It has long been known that a sample with an Fe/S ratio precisely unity is rarely encountered, and in the older literature such formulae as Fe_6S_7 and $Fe_{11}S_{12}$ have been assigned to it. The iron–sulfur system assumes the nickel arsenide structure over the composition range 50–55.5 atom % of sulfur, and, when the S/Fe ratio exceeds unity, some of the iron positions in the lattice are vacant in a random way. Thus the very attempt to assign stoichiometric formulae such as Fe_6S_7 is meaningless. We are dealing not with *one* compound, in the classical sense, but with a *phase* which may be perfect, that is, FeS, or may be deficient in iron. The particular specimen that happens to have the composition Fe_6S_7 is better described as $Fe_{0.858}S$.

An even more extreme example of non-stoichiometry is provided by the Co–Te (and the analogous Ni–Te) system. Here, a phase with the nickel arsenide structure is stable over the entire composition range CoTe to

$CoTe_2$. It is possible to pass continuously from the former to the latter by progressive loss of Co atoms from alternate planes (see Fig. 15-5) until, at $CoTe_2$, every other plane of Co atoms present in CoTe has completely vanished.

Typical of a system in which many different phases occur (each with a small range of existence so that each may be encountered in non-stoichiometric form) is the Cr–S system, where six phases occur in the composition range $CrS_{0.95}$ to $CrS_{1.5}$.

Although there are differences, the chemistry of selenides and tellurides is generally similar to that of sulfides.

15-9. Other Binary Sulfides

Most non-metallic (or metalloid) elements form sulfides that, if not molecular, have polymeric structures involving sulfide bridges. Thus silicon disulfide (15-III) consists of infinite chains of SiS_4 tetrahedra sharing edges, while Sb_2S_3 and Bi_2S_3 are isomorphous (15-IV), forming infinite bands which are then held in parallel strips in the crystal by weak secondary bonds.

(15-III)   (15-IV)

Sulfur–Nitrogen Compounds.[24]   There is an extensive chemistry of S–N compounds comprising species of the types SN^+, S_2N_2, $S_3N_3^+$, S_4N_4, $S_5N_5^+$, $S_6N_6\cdots(SN)_x$ and various derivatives of them. The cationic species are known as *thiazyl ions*, e.g., pentathiazyl $S_5N_5^+$, thiotrithiazyl $S_4N_3^+$, etc. Among the most important and best studied are S_4N_4 and its derivatives.

Tetrasulfur tetranitride. N_4S_4 is made by interaction of S and NH_3 in CCl_4 or by passing S_2Cl_2 over heated pellets of NH_4Cl. The orange crystals are stable in air but may be detonated by shock. They are thermochromic, being nearly colorless at $-190°$, orange at $25°$ and blood-red at $100°$.

The structure of S_4N_4 is a cage with a square set of N atoms and a bisphenoid of S atoms (Fig. 15-6), which is in interesting contrast to the structure of As_4S_4 (realgar) also shown in Fig. 15-6. The $S\cdots S$ distance, 2.59 Å, is longer than the normal S—S single-bond distance, ~ 2.08 Å, but short enough to indicate significant interaction; even the S to S (linked by N) distance, 2.71 Å, is indicative of direct $S\cdots S$ interaction. The N—S distances and the angles in N_4S_4 and also in $N_3S_4^+$ described below suggest the presence of lone pairs on N and S atoms. A recent MO study[25] in which no d orbital was used

[24] H. G. Heal in *Inorganic Sulfur Chemistry*, G. Nickless, ed., Elsevier, 1968; C. W. Allen, *J. Chem. Educ.*, 1967, **44**, 38; M. Becke-Goehring, *Inorg. Macromol. Rev.*, 1969, Vol. 1; R. L. Patton and W. L. Jolly, *Inorg. Chem.*, 1970, **9**, 1079.
[25] R. Gleiter, *J. Chem. Soc.*, A, **1970**, 3174.

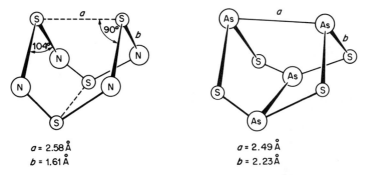

$a = 2.58 \text{ Å}$
$b = 1.61 \text{ Å}$

$a = 2.49 \text{ Å}$
$b = 2.23 \text{ Å}$

Fig. 15-6.   Structures of N_4S_4 and As_4S_4. Both have D_{2d} symmetry.

has given a satisfactory account of both the geometric and the electronic structure. There are long single S—S bonds, lone pairs on the N atoms and some electrons delocalized over the cyclic $(NS)_4$ framework.

Other species.   A variety of other N–S compounds can be prepared, some of them from N_4S_4 as shown in Fig. 15-7.

S_4N_4 undergoes two main types of reaction. (a) Reactions in which the N–S ring is preserved, as in additions of BF_3 or $SbCl_5$,[26] or the reduction to $S_4N_4H_2$. (b) Reactions in which ring cleavage occurs, with reorganization to give other ring systems, examples being the reaction with HCl or $SOCl_2$ to give the thiotrithiazyl ion, $S_4N_3^+$, and with H_2SO_4 [26a] to give $S_2N_2^+$. The ring structures of three derivatives of N_4S_4 are shown in Fig. 15-8. In $S_4N_4H_4$ the S atoms form a square and all the N atoms on the same side

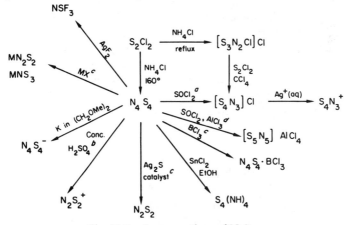

Fig. 15-7.   Some reactions of N_4S_4.
(a) A. J. Bannister and J. S. Padley, *J. Chem. Soc., A,* **1969,** 658.
(b) D. A. C. McNeil, M. Murray and M. C. R. Symons, *J. Chem. Soc., A,* **1967,** 1019.
(c) R. L. Patton and W. L. Jolly, *Inorg. Chem.,* 1969, **8,** 1389, 1392.
(d) A. J. Bannister *et al., Chem. Comm.,* **1969,** 1187.

[26] R. L. Patton and W. L. Jolly, *Inorg. Chem.,* 1969, **8,** 1389, 1392; M. Becke-Goehring and D. Schläfer, *Z. anorg. Chem.,* 1968, **356,** 234.
[26a] S. A. Lipp and W. L. Jolly, *Inorg. Chem.,* 1971, **10,** 33.

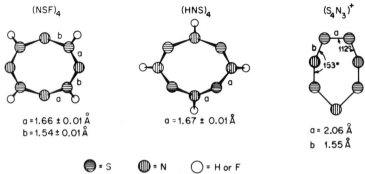

Fig. 15-8.   Structures of some sulfur–nitrogen compounds derived from S_4N_4.

form another square, with the H atoms attached to N (C_{4v}). $S_4N_4F_4$ is also cyclic but differs from $S_4N_4H_4$ in having a bisphenoidal arrangement of N atoms about a square array of S atoms, the F atoms attached to S, and alternating unequal S—N distances (S_4); this implies that there is $d\pi$-$p\pi$ character in four of these S—N bonds but an absence of delocalization. $(NSF)_4$ is isoelectronic with $(NPF_2)_4$ in which delocalization, at least to the extent that there are 3-center N⋯P⋯Nπ systems, is thought to occur.

Finally, the very stable $S_4N_3^+$ ion has a planar structure and absorption and nmr spectra that imply a well-defined, delocalized aromatic π-system.[27] Delocalization effects are also observed in the esr spectrum of radical ions[28] such as $N_4S_4^-$ and $N_2S_2^+$. The latter and also N_2S_2 have planar rings.

Reaction of S_2Cl_2 with NH_3 gives several imides, viz. S_7NH, three isomers of $S_6(NH)_2$, and two isomers of $S_5(NH)_3$, all of which, and $S_4(NH)_4$, have 8-membered rings with NH groups replacing S in S_8.[29]

Finally there is an extensive chemistry of compounds with alkyl or other groups bound to S or to N and in which bonds to other elements may also be present. We quote only sulfaminic chloride, $N_3S_3O_3Cl_3$ (α-form) which has a slightly puckered alternate ring with the N—S bonds approximately equal; each S is bound to O and to an axial chloride.[30]

15-10. Halides

The halides are listed in Table 15-3; only certain of them are discussed.

Sulfur Fluorides.[31]   Some reactions of S—F compounds are shown in Fig. 15-9. Direct fluorination of sulfur yields principally SF_6 and only traces of SF_4 and S_2F_{10}. Reaction of AgF with sulfur in a vacuum produces two

[27] P. Friedman, *Inorg. Chem.*, 1969, **8**, 692.
[28] R. A. Meinzer, D. W. Pratt and R. J. Myers, *J. Amer. Chem. Soc.*, 1969, **91**, 6623; S. A. Lipp, J. J. Chang and W. L. Jolly, *Inorg. Chem.*, 1970, **9**, 1970.
[29] J. C. van der Grampel and A. Vos, *Acta Cryst.*, 1969, *B*, **25**, 611.
[30] A. C. Hazell, G. A. Wiegers and A. Vos, *Acta Cryst.*, 1966, **20**, 186.
[31] S. M. Williamson, *Progr. Inorg. Chem.*, 1966, **7**, 39; H. L. Roberts, *Quart. Rev.*, 1961, **15**, 30.

TABLE 15-3

The Group VI Binary Halides

(m = m.p.; b = b.p.; d = decomposes; sub = sublimes; all in °C)

Fluorides	Chlorides	Bromides	Iodides
	Sulfur		
S_2F_2,[a] m -165, b -10.6	S_2Cl_2,[c] m -80, b 138	S_2Br_2,[c] m -46, d 90	
$[SF_2]$[b]	SCl_2, m -78, b 59		
SF_4, m -121, b -40	SCl_4, d -31		
SF_6, sub -65, m -51			
S_2F_{10}, m -53, b 29			
	Selenium		
	Se_2Cl_2	Se_2Br_2, d in vapor	
	$SeCl_2$, d in vapor	$SeBr_2$, d in vapor	
SeF_4, m -10, b 106	$SeCl_4$, sub 191	$SeBr_4$	
SeF_6, sub -47, m -35			
	Tellurium[d]		
TeF_4, m 130	$TeCl_4$, m 225, b 390	$TeBr_4$, m 380, b 414	TeI_4 m 259[e],
		(d in vapor)	d 100
TeF_6, sub -39, m -38			
Te_2F_{10}, m -34, b 53			

[a] Isomeric mixture of FSSF, m $-133°$, and F_2SS, b $-10.6°$.

[b] Detected by microwave spectroscopy among gaseous products of radio-frequency discharge in SF_6. Molecular parameters (S—F = 1.589 Å, $\angle$ FSF = $98°16'$; $\mu = 1.05$ D), but not bulk properties, are known (D. R. Johnson and F. X. Powell, *Science*, 1969, **164**, 950).

[c] Also the dichlorosulfanes, S_nCl_2, $2 < n < 100(?)$ (see, e.g., F. Fetier and M. Kulus, *Z. anorg. Chem.*, 1969, **364**, 241), and dibromosulfanes, S_nBr_2, $n > 2$.

[d] Lower halides, Te_3Cl_2, Te_2Br and TeI exist in the solid state and $TeCl_2$ and $TeBr_2$ in the gas phase (A. Rabenau, H. Rau and E. Rosenstein, *Angew. Chem. Internat. Edn.*, 1970, **9**, 802).

[e] Melting point obtained in a sealed tube to prevent decomposition: $TeI_4 = Te + 2I_2$.

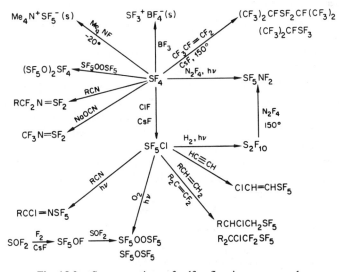

Fig. 15-9. Some reactions of sulfur–fluorine compounds.

isomers, the more stable, volatile and abundant being $F_2S{=}S$, the other F—S—S—F.

Sulfur tetrafluoride.[32] SF_4 is best made by reaction of SCl_2 with NaF in acetonitrile at 70–80°. SF_4 is an extremely reactive substance, instantly hydrolyzed by water to SO_2 and HF, but its fluorinating action is remarkably selective. It will convert C=O and P=O groups smoothly into CF_2 and PF_2, and COOH and P(O)OH groups into CF_3 and PF_3 groups, without attack on most other functional or reactive groups that may be present. Compounds of the type $ROSF_3$, which may be intermediates in the reaction with keto groups, have been prepared.[33] SF_4 is also quite useful for converting metal oxides into fluorides, which are (usually) in the same oxidation state.

Aryl-substituted fluorides can be readily obtained by the reaction

$$(C_6H_5)_2S_2 + 6AgF_2 \rightarrow 2C_6H_5SF_3 + 6AgF$$

which is carried out in trichloro- or trifluoro-methane. The arylsulfur trifluorides are more convenient laboratory fluorinating agents than SF_4 in that they do not require pressure above atmospheric. The structure of SF_4 and of substituted derivatives, RSF_3, is that of a trigonal bipyramid with an equatorial position occupied by the lone-pair. As expected, some donor compounds of SF_4 exist.

Sulfur hexafluoride.   The hexafluoride is normally very resistant to attack and extreme conditions are often required. Thus SF_6 resists molten KOH and steam at 500°. It reacts with O_2 when a platinum wire is exploded electrically,[34] also with some red-hot metals and with alkali metals in liquid ammonia.[35] Because of its inertness, high dielectric strength and molecular weight, it is used as a gaseous insulator in high-voltage generators and other electrical equipment.

SF_6 and its substituted derivatives (see below) have, or may be presumed to have, octahedrally bonded sulfur. In SF_6 the S—F bonds are about 0.2 Å shorter than expected for S—F single bonds. The low reactivity, particularly toward hydrolysis, which contrasts with the very high reactivity of SF_4, is presumably due to a combination of factors including high S—F bond strength and the facts that sulfur is both coordinately saturated and sterically hindered, augmented in the case of SF_6 by the lack of polarity of the molecule. The low reactivity is due to kinetic factors and not to thermodynamic stability, since the reaction of SF_6 with H_2O to give SO_3 and HF would be decidedly favorable ($\Delta F = -460$ kJ mol^{-1}), and the average bond energy in SF_4 (326 kJ mol^{-1}) is slightly higher than that of SF_6. The possibility of electrophilic attack on SF_6 has been confirmed by its reactions with certain Lewis acids. Thus Al_2Cl_6 at 180–200° gives AlF_3, Cl_2 and sulfur chlorides, while the thermodynamically allowed reaction

$$SF_6 + 2SO_3 \rightarrow 3SO_2F_2$$

[32] D. G. Martin, *Ann. New York Acad. Sci.*, 1967, **145**, 161; W. C. Smith, *Angew. Chem. Internat. Edn.*, 1962, **1**, 467.
[33] K. Baum, *J. Amer. Chem. Soc.*, 1969, **91**, 4594.
[34] B. Siegel and P. Breisacher, *J. Inorg. Nuclear Chem.*, 1970, **32**, 1469.
[35] L. Brewer, C.-A. Chang and B. King, *Inorg. Chem.*, 1970, **9**, 814.

proceeds slowly at 250°. SF_6 also reacts rapidly and quantitatively with sodium in ethylene glycol dimethyl ether containing biphenyl at room temperature:

$$8Na + SF_6 \rightarrow Na_2S + 6NaF$$

Electron-transfer from a biphenyl radical ion to an SF_6 molecule to give an unstable SF_6^- ion is probably involved.

Disulfur decafluoride. This compound is best obtained by the photochemical reaction

$$2SF_5Cl + H_2 \xrightleftharpoons{h\nu} S_2F_{10} + 2HCl$$

It is extremely poisonous (the reason for which is not clear), being similar to phosgene in its physiological action. It is not dissolved or hydrolyzed by water or alkalis and is not very reactive. Its structure is such that each sulfur atom is surrounded octahedrally by five fluorine atoms and the other sulfur atom. The S—S bond is unusually long, 2.21 Å, as compared with about 2.08 Å expected for a single bond, whereas the S—F bonds are, as in SF_6, about 0.2 Å shorter than expected for an S—F single bond. At room temperature it shows scarcely any chemical reactivity, though it oxidizes the iodide in an acetone solution of KI. At elevated temperatures, however, it is a powerful oxidizing agent, generally causing destructive oxidation and fluorination, presumably owing to initial breakdown to free radicals:

$$S_2F_{10} \rightarrow 2SF_5\cdot$$
$$SF_5\cdot \rightarrow SF_4 + F\cdot$$

Substituted Sulfur Fluorides.   There is an extensive chemistry of substituted sulfur fluorides of the types RSF_3 and RSF_5, examples of the former having been mentioned above. The SF_5 derivatives bear considerable resemblance to CF_3 derivatives with the principal difference that in reactions with organometallic compounds the SF_5 group is fairly readily reduced, whereas the CF_3 group is not.

The mixed halide SF_5Cl is an important intermediate (Fig. 15-9). Although it can be made by interaction of S_2F_{10} with Cl_2 at 200–250° it is best made by the CsF-catalyzed reaction:[36]

$$SF_4 + ClF \xrightarrow{25°, \ 1 \ hr., \ CsF} SF_5Cl$$

A probable intermediate is the salt $CsSF_5$, which dissociates significantly above 150°:

$$CsF + SF_4 \xrightleftharpoons[150°]{100°} CsSF_5$$

SF_5Cl is a colorless gas, b.p. $-15.1°$, m.p. $-64°$, which is more reactive than SF_6, being rapidly hydrolyzed by alkalis, though it is inert to acids. Its hydrolysis and its powerful oxidizing action towards many organic substances are consistent with the charge distribution $F_5S^{\delta-}$—$Cl^{\delta+}$. Its radical reactions with olefins and fluoroolefins resemble those of CF_3I.

The very reactive yellow *pentafluorosulfur hypofluorite* is one of the few known hypofluorites; it is obtained by the catalytic reaction

$$SOF_2 + 2F_2 \xrightarrow{CsF, \ 25°} SF_5OF$$

[36] C. J. Schack, R. D. Wilson and M. G. Warmer, *Chem. Comm.*, **1969**, 1110.

SOF_4, which is obtained in absence of CsF, is also converted into SF_5OF by CsF.

Sulfur Chlorides.   The chlorination of molten sulfur gives S_2Cl_2, an orange liquid of revolting smell. By using an excess of chlorine and traces of $FeCl_3$, $SnCl_4$, I_2, etc., as catalyst at room temperature, an equilibrium mixture containing ca 85% of SCl_2 is obtained. The dichloride readily dissociates within a few hours:

$$2SCl_2 \rightleftharpoons S_2Cl_2 + Cl_2$$

but it can be obtained pure as a dark-red liquid by fractional distillation in presence of some PCl_5, small amounts of which will stabilize SCl_2 for some weeks.

Sulfur chlorides are used as a solvent for sulfur (giving dichlorosulfanes up to about $S_{100}Cl_2$), in the vulcanization of rubber, as chlorinating agents, and as intermediates.[37] Specific higher chlorosulfanes can be obtained by reactions such as

$$2SCl_2 + H_2S_4 \xrightarrow{\ -80°\ } S_6Cl_2 + 2HCl$$

The sulfur chlorides are readily hydrolyzed by water. In the vapor S_2Cl_2 has a Cl—S—S—Cl structure with S—S = 1.93, S—Cl = 2.06 Å and $\angle$ SSCl = 108° and is twisted out of plane in the same way as H_2O_2 and FSSF.[38]

On treatment of sulfur chlorides with Cl_2 at $-80°$, SCl_4 is obtained as yellow crystals. It dissociates above $-31°$; it may be $SCl_3^+Cl^-$.

Selenium and Tellurium Halides.   The *fluorides*, MF_4, are very reactive. Although in the gaseous state SeF_4 and TeF_4 are ψ-trigonal-bipyramidal with an equatorial lone pair (C_{2v}), yet crystalline TeF_4 has a chain structure with distorted square-pyramidal TeF_5 units linked by *cis*-Te—F—Te single bridges.[39]

The hexafluorides and Te_2F_{10} (Se_2F_{10} is not known) are also considerably more reactive than the sulfur analogs, and TeF_6 is completely hydrolyzed by water in 24 hours; compounds containing the TeF_5 group, e.g., TeF_5Cl are also known.[40]

The *chlorides* and *bromides* are similar to those of sulfur but are thermally more stable, though readily hydrolyzed. In the vapor, only $TeCl_4$ is stable, $SeCl_4$ dissociating to $SeCl_2(g)$ and Cl_2; $SeBr_4$ dissociates even in solution.[41]

There has been much controversy concerning the structures of the tetra-halides.[42] In the vapor state, $TeCl_4$ has the ψ-trigonal-bipyramidal structure

[37] See L. A. Wiles and Z. S. Ariyan, *Chem. and Ind.*, **1962**, 2102, for reactions of S_2Cl_2 with organic compounds.
[38] P. J. Hendra and P. J. D. Parks, *J. Chem. Soc.*, A, **1968**, 908 (for S_2X_2 and Se_2X_2); B. Beagley *et al.*, *Trans Faraday Soc.*, 1969, **65**, 2300 (parameters for other S—S and S—X bonds).
[39] A. J. Edwards and F. I. Hewaidy, *J. Chem. Soc.*, A, **1968**, 2977.
[40] G. W. Fraser, R. D. Peacock and P. M. Watkins, *Chem. Comm.*, **1968**, 1257; A. Engelbrecht, W. Loreck and W. Nehoda, *Z. anorg. Chem.*, 1968, **360**, 88.
[41] G. A. Ozin and A. Vander Voet, *Chem. Comm.*, **1970**, 896; N. Katsaros and J. W. George, *Inorg. Chem.*, 1969, **8**, 759.
[42] B. Buss and B. Krebs, *Inorg. Chem.*, 1971, **10**, 2795; I. R. Beattie, J. R. Horder and P. J. Jones, *J. Chem. Soc.*, A, **1970**, 329; K. J. Wynne and P. S. Pearson, *Inorg. Chem.*, 1970, **9**, 106; R. Ponsionen and D. J. Stufkens, *Rec. Trav. chim.*, 1971, **90**, 521.

(C_{2v}); in the solid state it is Te_4Cl_{16} with a cubane-like structure, the Te atom being bound to three near chlorine atoms and three more distant bridging ones. This structure could be described as an array of $TeCl_3^+$ and Cl^- which would account for the electrical conductivity of the melt. In benzene solution trimers and tetramers of $TeCl_4$ are present. The halides $SeCl_4$ and $TeBr_4$ are similar to $TeCl_4$.

The halides form adducts with both Lewis acids and bases, and halogeno complex ions with halide ions; there is also a distinct tendency to form cationic species.

Complexes. The fluorides are comparatively feeble bases but combine with CsF. The tetrafluorides give MF_5^- only, while TeF_6 gives a product of stoichiometry Cs_2TeF_8 but unknown structure. Salts of the TeF_5^- ion are more readily obtained by adding Cs^+ to solutions of TeO_2 in 40% HF. The ion is ψ-octahedral with a lone pair.[43]

The tetrahalides MCl_4 and MBr_4 also form complex anions, normally MX_6^{2-}, although there is evidence for MX_5^-. Thus by dissolving H_2SeO_3 in strong HCl, adding KCl and saturating the mixture with gaseous HCl the salt K_2SeCl_6 is obtained. Despite the presence of a putative lone pair, a great variety of evidence shows that the ions $SeCl_6^{2-}$, $TeCl_6^{2-}$, and $TeBr_6^{2-}$ have regular octahedral (O_h) symmetry[44] (cf. discussion on page 140).

The tendency of MX_4 to form cationic species is shown in several ways: (a) Molten $TeCl_4$ has salt-like conductivity and in polar solvents such as CH_3NO_2 or DMF the halides behave as 1:1 electrolytes. (b) A number of adducts of the tetrahalides or substituted alkyl halides, notably $(CH_3)_3TeCl$,[45] give conducting solutions and may in addition be ionic in the solid state. Thus $MCl_4 \cdot AlCl_3$ is best formulated as $[MCl_3]^+[AlCl_4]^-$; the SeF_3^+ ion can be distinguished in the complicated structure of $SeF_4 \cdot 2NbF_5$;[46] adducts of TeF_4 with Me_3N, and other ligands, L, are of the type $[L_2TeF_3]^+[TeF_5]^-$.[47] Further, in ionizing solvents, the halides give ionic species such as $[TeX_3S_2]^+$ where S is a solvent molecule such as CH_3CN or C_2H_5OH. The alleged conductivity of $SeCl_4py_2$ in CH_3CN appears to be in error, and both this complex and $TeCl_4py_2$ are probably molecular with *trans*-pyridines.[48]

Other tellurium complex halides. Tellurium forms a number of complex halides in the II as well as the IV state, some of the best-known being those with thiourea (tu) or substituted thioureas as ligands. The red Te^{IV} compounds are made by treating TeO_2 in concentrated hydrochloric acid solution with, e.g., tetramethylthiourea (Me_4tu):

$$TeO_2 + 4HCl + 2Me_4tu \rightarrow trans\text{-}Te^{VI}(Me_4tu)_2Cl_4 + 2H_2O$$

[43] S. H. Mastin, R. R. Ryan and L. B. Asprey, *Inorg. Chem.*, 1970, **9**, 2100.
[44] T. C. Gibb, R. Greatorex, N. N. Greenwood and A. C. Sarma, *J. Chem. Soc.*, A, **1970**, 213 (references to Mössbauer spectra and structure of Te halides, oxides and oxo acids).
[45] M. T. Chen and J. W. George, *J. Amer. Chem. Soc.*, 1968, **90**, 4580.
[46] A. J. Edwards and G. R. Jones, *J. Chem. Soc.*, A, **1970**, 1891.
[47] N. N. Greenwood, A. C. Sarma and B. P. Straughan, *J. Chem. Soc.*, A, **1968**, 1561.
[48] I. R. Beattie *et al.*, *J. Chem. Soc.*, A, **1969**, 482; N. Katsaros and J. W. George, *J. Inorg. Nuclear Chem.*, 1969, **3**, 3503.

The structures of these compounds are octahedral with *trans* sulfur atoms showing no evidence (cf. TeX_6^{2-}) of stereochemical influence of the lone pair.[49]

This ligand can further act as a reducing agent in methanolic $4M$ hydrochloric acid:

$$Te(Me_4tu)_2Cl_4 \xrightarrow{Heat} Te^{II}(Me_4tu)Cl_2 + (Me_4tu)^{2+} + 2Cl^-$$

$$Te^{II}(Me_4tu)Cl_2 + Me_4tu \xrightarrow[Heat]{MeOH} Te^{II}(Me_4tu)_2Cl_2$$
$$\text{red} \qquad\qquad\qquad\qquad\qquad\qquad \text{yellow}$$

The Te^{II} complexes may be either *cis*- or *trans*-TeL_2Cl_2 where L is a thiourea, or, in excess L, of the type TeL_4^{2+}. They appear to be planar or nearly so. The tetramethylthiourea complexes, $SeCl_2L$ and $TeCl_2L$, appear to be 3-coordinate, i.e. *tbp* with axial Cl atoms and two equatorial lone pairs.[50]

Polonium Halides. Polonium halides are similar to those of tellurium, being volatile above 150° and soluble in organic solvents. They are readily hydrolyzed and form complexes, for example, $Na_2[PoX_6]$, isomorphous with those of tellurium. There is tracer evidence for the existence of a volatile polonium fluoride. The metal is also soluble in hydrofluoric acid, and complex fluorides exist.

15-11. Oxides

The principal oxides are given in Table 15-4.

TABLE 15-4

Oxides[a] of S, Se, Te[b] and Po

$\begin{cases} S_2O \\ SO \end{cases}$[c]			
SO_2	SeO_2	TeO_2	PoO_2 [$PoO(OH)_2$]
b.p. $-10.07°$	sub. 315°	m.p. 733°	
m.p. $-75.5°$			
SO_3	SeO_3	TeO_3	
m.p. 16.8° (γ)	m.p. 120°	dec. 400°	
b.p. 44.8°		Te_2O_5	
		dec. $>400°$	

[a] TeO(g), PoO, and S_2O_3 have also been reported.
[b] See W. A. Dutton and W. C. Cooper, *Chem. Rev.*, 1966, **66**, 657 (oxides and oxo acids of Te).
[c] Unstable, see text.

Disulfur Monooxide. This is produced when a glow discharge is passed through SO_2 and in other ways and was long thought to be SO. However, gases of this composition are equimolar mixtures of S_2O and SO_2. The extremely reactive biradical, SO, can be detected as an intermediate in the reactions, however, and $(SO)_2$ has also been detected by mass spectrometry.

[49] S. Husebye and J. W. George, *Inorg. Chem.*, 1969, **8**, 313.
[50] P. J. Hendra and Z. Jovic, *J. Chem. Soc.*, A, 1968, 911; O. Foss in *Selected Topics in Structure Chemistry*, Univ. Press, Oslo, 1967, p. 145; K. J. Wynne and P. S. Pearson, *Chem. Comm.*, **1971**, 293.

S_2O is believed to have the structure SSO. It is unstable at ordinary temperatures, decomposing to SO_2 and polymeric oxides; it is orange-red when condensed from the gas phase at $-196°$.

Dioxides. The dioxides are obtained by burning the elements in air, though small amounts of SO_3 also form in the burning of sulfur. Sulfur dioxide is also produced when many sulfides are heated in air. Selenium and tellurium dioxides are also obtained by treating the metals with hot nitric acid to form H_2SeO_3 and $2TeO_2 \cdot HNO_3$, respectively, and then heating these to drive off water or nitric acid.

The dioxides differ considerably in structure. SO_2 is a gas, SeO_2 is a white volatile solid, and TeO_2 is a non-volatile white solid. Gaseous SO_2 and SeO_2 are bent symmetrical molecules; in each case the short S—O and Se—O bond distances imply that there is considerable multiple bonding. There may be $p\pi–p\pi$ bonding as well as $p\pi–d\pi$ bonding due to the overlap of filled $p\pi$ orbitals of oxygen with vacant $d\pi$ orbitals of sulfur. SO_2 solidifies to form a molecular lattice as far as is known, but SeO_2 forms infinite chains (Fig. 15-10). As the values of the angles (see Figure) imply, these chains are not planar. TeO_2 and PoO_2 each exist in two forms. α-TeO_2 has Te at the apex of a square pyramid with four O atoms at the base.

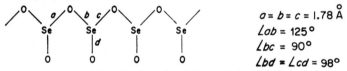

$$a = b = c = 1.78 \text{ Å}$$
$$\angle ab = 125°$$
$$\angle bc = 90°$$
$$\angle bd = \angle cd = 98°$$

Fig. 15-10. Section of an infinite chain of SeO_2.

Sulfur dioxide. This is a weak reducing agent in acid solution, but a stronger one in basic solution, where the sulfite ion is formed.

Liquid SO_2 is a useful non-aqueous solvent.[51] Although it is a relatively poor ionizing medium ($\varepsilon = 15.1$ at $0°$) it dissolves many organic and inorganic substances and is often used as a solvent for nmr studies as well as in preparative reactions. There is little evidence for self-ionization of liquid SO_2, and the conductivity (3×10^{-8} to 2×10^{-7} ohm cm^{-1}) is mainly a reflection of the purity.

Sulfur dioxide has lone pairs and can act as a Lewis base; it can also act as a Lewis acid. With certain amines, crystalline 1:1 charge-transfer complexes are formed in which electrons from nitrogen are presumably transferred to antibonding acceptor orbitals localized on sulfur. One of the most stable[52] is $Me_3N \cdot SO_2$ whose structure is shown in formula (15-V); here the dimensions of the SO_2 molecule appear to be unchanged by complex formation.

[51] N. H. Lichtin, *Progr. Phys. Org. Chem.*, 1963, **1**, 75; T. C. Waddington in *Non-Aqueous Solvent Systems*, T. C. Waddington, ed., Academic Press, 1965; C. C. Addison, W. Karcher and H. Hecht, *Chemistry in Liquid Dinitrogen Tetroxide and Sulphur Dioxide*, Vol. III, *Chemistry in Non-aqueous Ionizing Solvents*, Pergamon Press, 1968; D. F. Burrow in J. J. Lagowski, ed., *The Chemistry of Non-Aqueous Solvents*, Vol. 3, Academic Press, 1970.

[52] J. Grundnes and S. D. Christian, *J. Amer. Chem. Soc.*, 1968, **90**, 2239; D. van der Helm, J. D. Childs and S. D. Christian, *Chem. Comm.*, **1969**, 887.

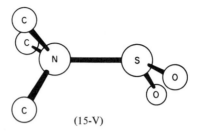

(15-V)

It also forms complexes with halide ions in both aqueous and non-aqueous media; the ion SO_2F^- (page 456) can be regarded as an especially stable solvate.[53]

Although crystals are formed with quinol and some other hydrogen-bonding compounds, these are clathrates or, in the case of $SO_2 \cdot \sim 7H_2O$, a clathrate hydrate (page 160).

Sulfur dioxide also forms complexes with a number of metal species and may be bound in different ways. In the crystalline compound $SbF_5 \cdot SO_2$, which is of interest because of the use of SO_2 as a solvent for super acid systems (page 174), the SO_2 is bound as in (15-VI).[54] Sulfur dioxide also forms numerous complexes with transition metals, as well as undergoing "insertion" reactions with M—M and M—C bonds which are discussed

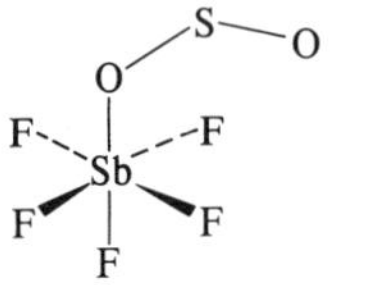

(15-VI) 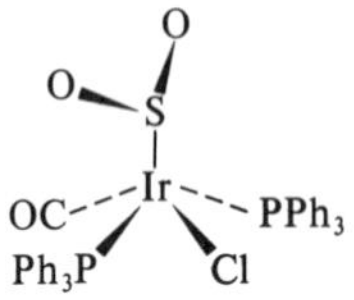   (15-VII)

in Section 24-A-3. The latter reaction is not confined to transition metals, however, and SO_2 can be inserted into Sn—C and Hg—C bonds, for example:[55]

$$RCH_2HgOAc + SO_2 \rightarrow RCH_2SO_2HgOAc$$
$$(CH_3)_4Sn + SO_2 \rightarrow (CH_3)_3SnSO_2CH_3$$

The bonding in $MCl(CO)(SO_2)(PPh_3)_2$, M = Rh or Ir (15-VII), differs from that in the SbF_5 compound in that the S atom is bound to the metal.[56] Metal–sulfur bonding appears to be general in transition-metal species, although there are notable differences in different compounds. In the Ir compound the $IrSO_2$ group is non-planar with a long Ir—S distance (2.49 Å); the geometry resembles that in $Me_3N \cdot SO_2$, and the metal may be considered as an electron-donor. In the octahedral complex $[RuCl(SO_2)(NH_3)_4]Cl$

[53] E. J. Woodhouse and T. H. Norris, *Inorg. Chem.*, 1971, **10**, 614; A. Salama *et al.*, *J. Chem. Soc., A*, **1971**, 1112.

[54] J. W. Moore, H. W. Baird and H. B. Miller, *J. Amer. Chem. Soc.*, 1968, **90**, 1359.

[55] W. Kitching, C. W. Fong and A. J. Smith, *J. Amer. Chem. Soc.*, 1969, **91**, 767; C. W. Fong and W. Kitching, *J. Amer. Chem. Soc.*, 1969, **91**, 767; C. W. Fong and W. Kitching, *J. Amer. Chem. Soc.*, 1971, **93**, 3790.

[56] S. J. La Placa and J. A. Ibers, *Inorg. Chem.*, 1966, **5**, 405; K. W. Muir and J. A. Ibers, *Inorg. Chem.*, 1969, **8**, 1920.

the Ru—S distance is 2.07 Å and the $RuSO_2$ group is planar,[57] suggesting π-bonding by overlap of filled ruthenium d orbitals with antibonding π orbitals on sulfur.

Sulfur trioxide. This is the only important trioxide in this Group. It is obtained by reaction of sulfur dioxide with molecular oxygen, a reaction that is thermodynamically very favorable but extremely slow in the absence of a catalyst. Platinum sponge, V_2O_5, and NO serve as catalysts under various conditions. SO_3 reacts vigorously with water to form sulfuric acid. Commercially, for practical reasons, SO_3 is absorbed in concentrated sulfuric acid, to give oleum (page 181), which is then diluted. SO_3 is also used as such for preparing sulfonated oils and alkyl arenesulfonate detergents. It is also a powerful but generally indiscriminate oxidizing agent; it will, however, selectively oxidize pentachlorotoluene and similar compounds to the alcohol.[58]

The free molecule, in the gas phase, has a planar, triangular structure which may be considered to be a resonance hybrid involving $p\pi$–$p\pi$ S—O bonding, as in (15-VIII), with additional π bonding via overlap of filled oxygen $p\pi$ orbitals with empty sulfur $d\pi$ orbitals, in order to account for the very short S—O distance of 1.41 Å.

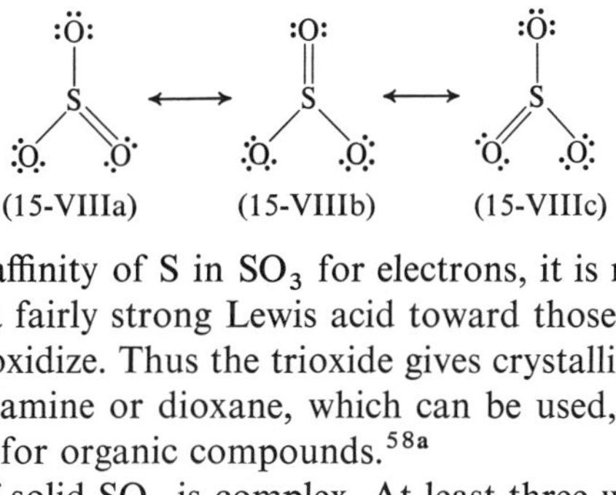

(15-VIIIa)   (15-VIIIb)   (15-VIIIc)

In view of this affinity of S in SO_3 for electrons, it is not surprising that SO_3 functions as a fairly strong Lewis acid toward those bases that it does not preferentially oxidize. Thus the trioxide gives crystalline complexes with pyridine, trimethylamine or dioxane, which can be used, like SO_3 itself, as sulfonating agents for organic compounds.[58a]

The structure of solid SO_3 is complex. At least three well-defined phases are known. There is first γ-SO_3, formed by condensation of vapors at $-80°$ or below. This ice-like solid contains cyclic trimers with structure (15-IX).

A more stable, asbestos-like phase, β-SO_3, has infinite helical chains of linked SO_4 tetrahedra (15-X), and the most stable form, α-SO_3, which also

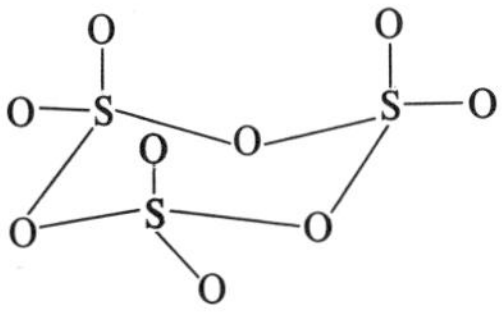

(15-IX) 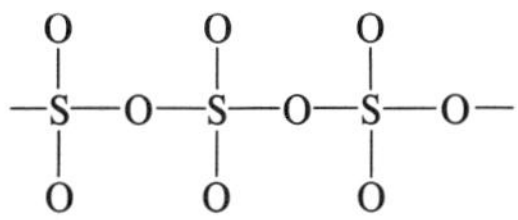   (15-X)

[57] L. H. Voigt, J. L. Katz and S. E. Wiberley, *Inorg. Chem.*, 1965, **4**, 1157.
[58] V. Mark *et al.*, *J. Amer. Chem. Soc.*, 1971, **93**, 3538.
[58a] E. E. Gilbert, *Chem. Rev.*, 1962, **62**, 549 (reactions of SO_3 and its adducts with organic compounds).

has an asbestos-like appearance, presumably has similar chains cross-linked into layers.

Liquid γ-SO_3, which is a monomer–trimer mixture, can be stabilized by the addition of boric acid. In the pure state it is readily polymerized by traces of water.

Selenium trioxide. This trioxide is made by dehydration of H_2SeO_4 by P_2O_5 at 150–160°; it is a strong oxidant and is also rapidly rehydrated by water. SeO_3 dissolves in liquid HF to give fluoroselenic acid, $FSeO_3H$ (cf. FSO_3H, page 181) as a viscous fuming liquid.[59]

Tellurium trioxide. Made by dehydration of $Te(OH)_6$, this orange compound reacts only slowly with water but dissolves rapidly in bases to give tellurates.

OXO ACIDS

15-12. General Remarks

S, Se and Te form oxo acids. Those of sulfur are by far the most important and the most numerous. Some of the acids are not actually known as such, but like phosphorus oxo acids occur only in the form of their anions and salts. In Table 15-5 the various oxo acids of sulfur are grouped according to structural type. This classification is to some extent arbitrary, but it corresponds with the order in which we shall discuss these acids. None of the oxo acids in which there are S—S bonds has any known Se or Te analog.

TABLE 15-5

Principal Oxo Acids of Sulfur

Name	Formula	Structure[b]
Acids Containing One Sulfur Atom		
Sulfurous	$H_2SO_3{}^a$	SO_3^{2-} (in sulfites)
Sulfuric	H_2SO_4	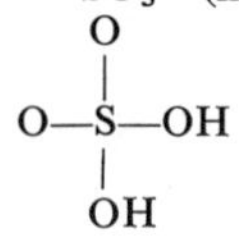 $\begin{array}{c} O \\ \| \\ O-S-OH \\ \| \\ OH \end{array}$
Acids Containing Two Sulfur Atoms		
Thiosulfuric	$H_2S_2O_3$	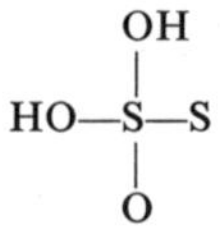 $\begin{array}{c} OH \\ \| \\ HO-S-S \\ \| \\ O \end{array}$
Dithionous	$H_2S_2O_4{}^a$	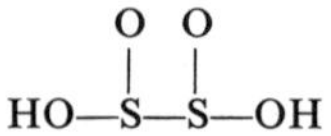 $\begin{array}{c} O \quad O \\ \| \quad \| \\ HO-S-S-OH \end{array}$

[59] A. J. Edwards, M. A. Monty and R. D. Peacock, *J. Chem. Soc., A*, **1967**, 557.

Name	Formula	Structure[b]
Pyrosulfurous (disulfurous)	$H_2S_2O_5$[a]	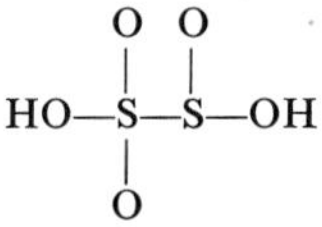 $\begin{array}{cc} O & O \\ \| & \| \\ HO-S-S-OH \\ \| \\ O \end{array}$
Dithionic	$H_2S_2O_6$	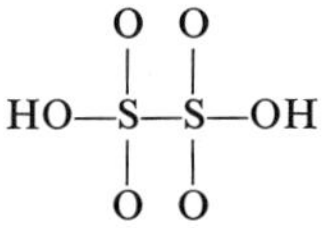 $\begin{array}{cc} O & O \\ \| & \| \\ HO-S-S-OH \\ \| & \| \\ O & O \end{array}$
Pyrosulfuric (disulfuric)	$H_2S_2O_7$	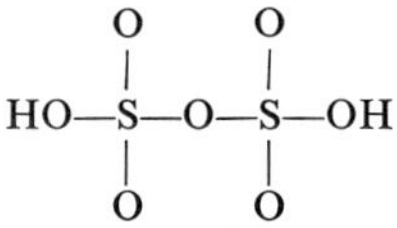 $\begin{array}{cc} O & O \\ \| & \| \\ HO-S-O-S-OH \\ \| & \| \\ O & O \end{array}$

Acids Containing Three or More Sulfur Atoms

Polythionic	H_2	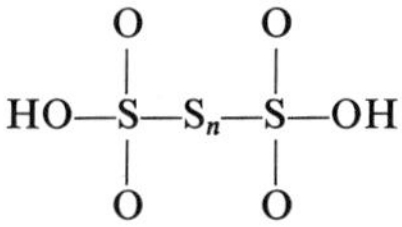 $\begin{array}{cc} O & O \\ \| & \| \\ HO-S-S_n-S-OH \\ \| & \| \\ O & O \end{array}$

Peroxo Acids

Peroxomonosulfuric	H_2SO_5	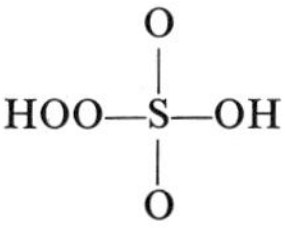 $\begin{array}{c} O \\ \| \\ HOO-S-OH \\ \| \\ O \end{array}$
Peroxodisulfuric	$H_2S_2O_8$	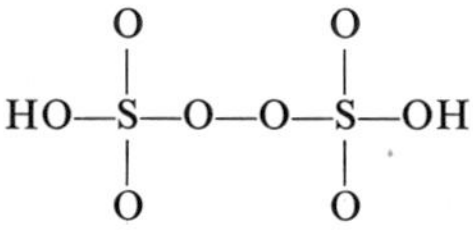 $\begin{array}{cc} O & O \\ \| & \| \\ HO-S-O-O-S-OH \\ \| & \| \\ O & O \end{array}$

[a] Free acid unknown.
[b] In most cases the structure given is inferred from the structure of anions in salts of the acid.

15-13. Sulfurous Acid

SO_2 is quite soluble in water; such solutions, which possess acidic properties, have long been referred to as solutions of sulfurous acid, H_2SO_3. However, H_2SO_3 is either not present or present only in infinitesimal quantities in such solutions. The so-called hydrate, $H_2SO_3 \cdot \sim 6H_2O$, is the gas hydrate $SO_2 \cdot \sim 7H_2O$. The equilibria in aqueous solutions of SO_2 are best represented as

$$SO_2 + xH_2O = SO_2 \cdot xH_2O \text{ (hydrated } SO_2)$$
$$[SO_2 \cdot xH_2O = H_2SO_3 \quad K \lll 1]$$
$$SO_2 \cdot xH_2O = HSO_3^- \text{(aq)} + H_3O^+ + (x-2)H_2O$$

and the first acid dissociation constant for "sulfurous acid" is properly defined as follows:

$$K_1 = \frac{[HSO_3^-][H^+]}{[\text{Total dissolved } SO_2] - [HSO_3^-] - [SO_3^{2-}]} = 1.3 \times 10^{-2}$$

Although sulfurous acid does not exist, two series of salts, the bisulfites, containing HSO_3^-, and the sulfites, containing SO_3^{2-}, are well known. The SO_3^{2-} ion in crystals is pyramidal. In bisulfite solutions there appear to be four species. At low concentrations there are the tautomers (15-XIa, b), which interact by hydrogen bonding at higher concentrations ($10^{-2}M$) to give (15-XIc), which in turn is in equilibrium with the disulfite ion (15-XId).

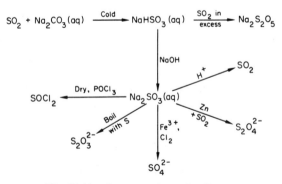

Only the alkali sulfites and bisulfites are commonly encountered; these are water-soluble. Heating solid bisulfites or passing SO_2 into their aqueous solutions affords pyrosulfites:

$$2MHSO_3 \xrightleftharpoons{Heat} M_2S_2O_5 + H_2O$$
$$HSO_3^- (aq) + SO_2 = HS_2O_5^- (aq)$$

Whereas pyro acids, for example, pyrosulfuric, $H_2S_2O_7$, discussed in Chapter 5 (page 181), usually have oxygen bridges, the pyrosulfite ion has an S—S bond and hence an unsymmetrical structure, $O_2S—SO_3(C_s)$.[60] Some important reactions of sulfites are shown in Fig. 15-11.

Solutions of SO_2 and of sulfites possess reducing properties and are often used as reducing agents:

$$SO_4^{2-} + 4H^+ + (x-2)H_2O + 2e = SO_2 \cdot xH_2O \qquad E^0 = 0.17 \text{ V}$$
$$SO_4^{2-} + H_2O + 2e = SO_3^{2-} + 2OH^- \qquad E^0 = -0.93 \text{ V}$$

The strong reducing action of basic solutions of sulfites may be attribut-

Fig. 15-11. Some reactions of sulfites.

[60] A. W. Herlinger and T. V. Long, *Inorg. Chem.*, 1969, **8**, 2661; R. Naylor, J. B. Gill and D. C. Goodall, *Chem. Comm.*, **1971**, 671.

able to the tautomeric species with an S—H bond which is known to exist in aqueous solution (15-XIb). Tautomeric forms of diesters of sulfurous acids, dialkyl sulfites, $OS(OR)_2$ and alkanesulphonic esters, $RSO_2(OR)$, occur.

15-14. Selenous and Tellurous Acids

SeO_2 dissolves readily in water to give solutions that do contain selenous acid with the $OSe(OH)_2$ structure;[61] Raman spectra show that it is negligibly dissociated in aqueous solution, while in half and fully neutralized solutions the ions $HSeO_3^-$ and SeO_3^{2-} are formed, salts of which can be isolated. Above $\sim 4M$, pyroselenate ions are formed:

$$2HSeO_3^- = Se_2O_5^{2-} + H_2O$$

The solid acid, though efflorescent, can be isolated and X-ray studies have shown the presence of layers of pyramidal SeO_3 groups connected by hydrogen bonds. The acid and its salts are moderately strong oxidizing agents:

$$H_2SeO_3 + 4H^+ + 4e = Se + 3H_2O \qquad E^0 = 0.74 \text{ V}$$

and they oxidize SO_2, HI, H_2S, etc.

TeO_2 is virtually insoluble in water. There is no information as to the species actually present in the saturated solution ($\sim 10^{-5}M$), and no hydrated form of TeO_2 has been isolated. The dioxide dissolves in strong bases to give solutions from which crystalline tellurites, bitellurites, and various polytellurites may be isolated.

15-15. Selenic and Telluric Acids (for Sulfuric Acid see page 180)

Vigorous oxidation of selenites or fusion of selenium with potassium nitrate gives *selenic acid* (or its salts). The free acid forms colorless crystals, m.p. 57°. It is very similar to sulfuric acid in its formation of hydrates, in acid strength, and in the properties of its salts, most of which are isomorphous with the corresponding sulfates and bisulfates.

Raman spectra indicate that the aqueous selenic acid system closely resembles aqueous H_2SO_4 in the types of species, i.e., H_2SeO_4, $HSeO_4^-$ and SeO_4^{2-}, rather than the aqueous telluric acid system. It differs mainly in being less stable. It evolves oxygen when heated above about 200° and is a strong, though usually not kinetically fast, oxidizing agent:

$$SeO_4^{2-} + 4H^+ + 2e = H_2SeO_3 + H_2O \qquad E^0 = 1.15 \text{ V}$$

Pyroselenates exist and appear to contain the ion $[O_3SeOSeO_3]^{2-}$ in crystals, but in solution the species appears to be $[SeO_3(OH)]^-$.

Telluric acid is very different from sulfuric and selenic acids and has hydrogen-bonded octahedral molecules, $Te(OH)_6$, in the crystal.

The acid or its salts may be prepared by oxidation of tellurium or TeO_2 by

[61] G. E. Walrafen, *J. Chem. Phys.*, 1967, **46**, 1870.

H_2O_2, Na_2O_2, CrO_3, or other powerful oxidizing agents. It is a moderately strong, but, like selenic acid, kinetically slow oxidizing agent ($E^0 = 1.02$ V). It is a very weak dibasic acid with $K_1 \approx 10^{-7}$. Tellurates of various stoichiometries are known and most, if not all of them contain TeO_6 octahedra.[62] Examples are $K[TeO(OH)_5] \cdot H_2O$, $Ag_2[TeO_2(OH)_4]$ and Hg_3TeO_6.

15-16. Peroxo Acids

No peroxo acid containing Se or Te is known.

Peroxodisulfuric acid can be obtained from its NH_4^+ or Na^+ salts which can be crystallized from solutions after electrolysis of the corresponding sulfates at low temperatures and high current densities. The $S_2O_8^{2-}$ ion has the structure O_3S—O—O—SO_3, with approximately tetrahedral angles about each S atom.

The peroxodisulfate ion is one of the most powerful and useful of oxidizing agents:

$$S_2O_8^{2-} + 2e = 2SO_4^{2-} \qquad E^0 = 2.01 \text{ V}$$

However, the reactions may be complicated mechanistically and in many of them there is good evidence[63] for the formation of the radical ion $\dot{S}O_4^-$ by one-electron reduction:

$$S_2O_8^{2-} + e = SO_4^{4-}$$

The oxidations by $S_2O_8^{2-}$ often proceed slowly but become more rapid in presence of catalysts, the silver ion being commonly used for this purpose. The precise mechanism is not quite clear,[64] but it appears that a weak 1:1 complex is first formed between Ag^+ and $S_2O_8^{2-}$, the rapidly reacting oxidizing species being Ag^{II}.

Peroxomonosulfuric acid (Caro's acid) is obtained by hydrolysis of peroxodisulfuric acid:

$$O{=}\underset{\underset{O}{\|}}{\overset{\overset{OH}{|}}{S}}{-}O{-}O{-}\underset{\underset{O}{\|}}{\overset{\overset{OH}{|}}{S}}{=}O + H_2O \rightarrow O{=}\underset{\underset{O}{\|}}{\overset{\overset{OH}{|}}{S}}{-}OH + O{=}\underset{\underset{O}{\|}}{\overset{\overset{OH}{|}}{S}}{-}OOH$$

and also by the action of concentrated hydrogen peroxide on sulfuric acid or chlorosulfuric acid:

$$H_2O_2 + H_2SO_4 \rightarrow HOOSO_2OH + H_2O$$
$$H_2O_2 + HSO_3Cl \rightarrow HOOSO_2OH + HCl$$

The salts such as $KHSO_5$ can be obtained only impure, admixed with K_2SO_4 and $KHSO_4$; aqueous solutions decompose to give mainly O_2 and SO_4^{2-} with small amounts of H_2O_2 and $S_2O_8^{2-}$.

[62] O. Lindqvist, *Acta Chem. Scand.*, 1970, **24**, 3178; N. E. Erickson and A. G. Maddock, *J. Chem. Soc.*, A, **1970**, 1665.

[63] W. K. Wilmarth and A. Haim in *Peroxide Reaction Mechanisms*, J. O. Edwards, ed., Interscience, 1962; D. A. House, *Chem. Rev.*, 1962, **62**, 185; D. E. Pennington and A. Haim, *J. Amer. Chem. Soc.*, 1968, **90**, 3700; F. Secco and S. Celsi, *J. Chem. Soc.*, A, **1971**, 1092.

[64] Cf. J. D. Miller, *J. Chem. Soc.*, A, **1969**, 2348.

15-17. Thiosulfuric Acid

Thiosulfates are readily obtained by boiling solutions of sulfites with S and in the decomposition of dithionites. The free acid is unstable at ordinary temperatures, but it has been isolated as an etherate at $-78°$ from the reaction

$$SO_3 + H_2S \rightarrow H_2S_2O_3$$

or, free from solvent, by the reaction

$$HSO_3Cl + H_2S \rightarrow H_2S_2O_3 + HCl$$

The alkali thiosulfates are manufactured for use mainly in photography where they are used to dissolve unreacted silver bromide from emulsion by formation of the complexes $[Ag(S_2O_3)]^-$ and $[Ag(S_2O_3)_2]^{3-}$; the thiosulfate ion also forms complexes with other metal ions.

The thiosulfate ion has the structure[65] S—SO_3 with an S—S distance of 2.013 ± 0.003 Å and S—O distances of 1.468 ± 0.004 Å. These distances suggest some S—S π bonding and considerable S—O π bonding similar to that in SO_4^{2-}, for which the S—O distance is 1.44 Å.

The reaction of SO_3 with H_2S above can be extended to give a series of sulfanemonosulfonic acids, HS_xSO_3H:

$$HSSH + SO_3 \rightarrow HSSSO_3H$$

They are stable only in ether solution at low temperatures and their salts are thermally unstable and reactive to water.

15-18. Dithionous (Hypo- or Hydrosulfurous) Acid

The reduction of sulfites in aqueous solutions containing an excess of SO_2, usually by zinc dust, gives the dithionite ion, $S_2O_4^{2-}$. Solutions of this ion are not very stable and decompose in a complex way[66] according to the stoichiometry

$$2S_2O_4^{2-} + H_2O \rightarrow S_2O_3^{2-} + 2HSO_3^-$$

Oxygen-free dithionite solutions show a strong esr signal due to the radical ion $\dot{S}O_2^-$ and there is the reversible dissociation

$$S_2O_4^{2-} \rightleftharpoons 2SO_2^-$$

The solutions are oxidized by air, and in acid solution decomposition, producing some elemental sulfur, is rapid. The Zn^{2+} and Na^+ salts are commonly used as powerful and rapid reducing agents in alkaline solution:

$$2SO_3^{2-} + 2H_2O + 2e = 4OH^- + S_2O_4^{2-} \qquad E^0 = -1.12 \text{ V}$$

and in the presence of 2-anthraquinonesulfonate as catalyst (Fieser's solution) aqueous $Na_2S_2O_4$ efficiently removes oxygen from inert gases.

The structure of the dithionite ion (Fig. 15-12) has several remarkable features. The oxygen atoms, which must bear considerable negative charge, are closely juxtaposed by the eclipsed (C_{2v}) configuration and by the small value of the angle α, which would be 35° for sp^3 tetrahedral hybridization

[65] S. Baggio, Acta Cryst., 1969, A, 25, 5119.
[66] L. Burlamacchi, G. Guarini and E. Tiezzi, Trans. Faraday Soc., 1969, 65, 496; W. J. Lem and M. Wayman, Canad. J. Chem., 1970, 48, 2778.

Fig. 15-12.   Structure of the dithionite ion, $S_2O_4^{2-}$, in $Na_2S_2O_4$.

at the sulfur atom. Secondly, the S—S distance is much longer than S—S bonds in disulfides, polysulfides, etc., which are in the range ca 2.0–2.15 Å. The long bond is believed to be due to weakening by repulsion of lone pairs on sulfur resulting from dp hybridized bonding. The weak bonding is consistent with the strong reducing properties and with the rapid exchange between $S_2O_4^{2-}$ and labeled $^*SO_2$.

15-19. Dithionic Acid

Although $H_2S_2O_6$ might at first sight appear to be the simplest homolog of the polythionates ($S_nO_6^{2-}$) to be discussed in the next Section, dithionic acid and its salts do not behave like the polythionates. Furthermore, from a structural point of view, dithionic acid should not be considered as a member of the polythionate series, since dithionates contain no sulfur atom bound only to other sulfur atoms as do $H_2S_3O_6$ and all higher homologs, $H_2S_nO_6$. The dithionate ion has the structure O_3S—SO_3 with approximately tetrahedral bond angles about each sulfur, and the shortness of the S—O bonds (1.43 Å; cf. 1.44 Å in SO_4^{2-}) again suggests considerable double-bond character.[67]

Dithionate is usually obtained by oxidation of sulfite or SO_2 solutions with manganese(IV) oxide:

$$MnO_2 + 2SO_3^{2-} + 4H^+ = Mn^{2+} + S_2O_6^{2-} + 2H_2O$$

Other oxo acids of sulfur that are formed as by-products are precipitated with barium hydroxide, and $BaS_2O_6 \cdot 2H_2O$ is then crystallized. Treatment of aqueous solutions of this with sulfuric acid gives solutions of the free acid which may be used to prepare other salts by neutralization of the appropriate bases. Dithionic acid is a moderately stable strong acid which decomposes slowly in concentrated solutions and when warmed. The ion itself is completely stable; solutions of its salts may be boiled without decomposition. Although it contains sulfur in an intermediate oxidation state, it resists most oxidizing and reducing agents, presumably for kinetic reasons.

15-20. Polythionates[68]

These anions have the general formula $[O_3SS_nSO_3]^{2-}$. The free acids are not stable, decomposing rapidly into S, SO_2 and sometimes SO_4^{2-}. Also, no acid salt is known. The well-established polythionate anions are those with

[67] I. R. Beattie, M. J. Gall and G. A. Ozin, *J. Chem. Soc., A*, **1969**, 1001.
[68] J. Janickis, *Accounts Chem. Res.*, 1969, **2**, 316.

$n = 1-4$. They are named according to the total number of sulfur atoms and are thus called: trithionate $S_3O_6^{2-}$, tetrathionate $S_4O_6^{2-}$, etc. In all these anions there are sulfur chains, thus disposing of older proposals that there might be $S \rightarrow S$ linkages. The conformations of the chains in these anions are very similar to those of segments of the S_8 ring, whereas the configurations about the end sulfur atoms, $-S-SO_3$, are approximately tetrahedral. They can be regarded as derivatives of sulfanes, hence the name sulfanedisulfonic acids; e.g., tetrathionates can be called disulfanedisulfonates. In addition to these structurally well-characterized polythionates, others containing up to 20 sulfur atoms have been prepared.

Polythionates can be prepared in various ways. Mixtures are obtained by reduction of thiosulfate solutions with SO_2 in presence of As_2O_3 and also by the reaction of H_2S with an aqueous solution of SO_2 which produces a solution called Wackenroder's liquid. A general reaction said to produce polythionates up to very great chain lengths is

$$6S_2O_3^{2-} + (2n-9)H_2S + (n-3)SO_2 \rightarrow 3S_nO_6^{2-} + (2n-12)H_2O + 6OH^-$$

Many polythionates are best made by selective preparations, e.g., the action of H_2O_2 on cold saturated sodium thiosulfate:

$$2S_2O_3^{2-} + 4H_2O_2 \rightarrow S_3O_6^{2-} + SO_4^{2-} + 4H_2O$$

Tetrathionates are obtained by treatment of thiosulfates with iodine in the reaction used in the volumetric determination of iodine:

$$2S_2O_3^{2-} + I_2 \rightarrow 2I^- + S_4O_6^{2-}$$

Various species containing Se and Te are also known such as $Se_nS_2O_6^{2-}$ ($2 \leqslant n \leqslant 6$), $O_3S_2SeS_2SO_3^{2-}$ and $O_3S_2TeS_2O_3^{2-}$.

OXO HALIDES AND HALOOXO ACIDS

15-21. Oxo Halides

Only S and Se are known to form well-defined oxo halides. These are of the three main types: (a) the thionyl and selenyl halides, SOX_2 and $SeOX_2$; (b) the sulfuryl halides, SO_2X_2, and their one selenium analog, SeO_2F_2; and (c) a number of more complex sulfur oxo chlorides and oxo fluorides.

The *thionyl* and *selenyl halides* are:

SOF_2	$SOCl_2$	$SOBr_2$	$SOFCl$
$SeOF_2$	$SeOCl_2$	$SeOBr_2$	

With the exception of thionyl fluoride, all these compounds react rapidly, and sometimes violently, with water, being completely hydrolyzed:

$$SOCl_2 + H_2O \rightarrow SO_2 + 2HCl$$

Thionyl fluoride reacts only slowly with water.

These halides may be prepared in many ways. $SOCl_2$ is usually prepared by the following reaction:

$$PCl_5 + SO_2 \rightarrow SOCl_2 + POCl_3$$

SOF_2 is obtained from $SOCl_2$ by reaction with SbF_3 in presence of $SbCl_5$ (Swarts reagent):

$$3SOCl_2 + 2SbF_3 \rightarrow 3SOF_2 + 2SbCl_3$$

On a large scale, $SOCl_2$ is treated with anhydrous hydrogen fluoride. $SOBr_2$ is also prepared from $SOCl_2$ by treatment of the latter with HBr at $0°$. SOClF is usually a by-product in the preparation of SOF_2. Selenyl chloride may be obtained by reaction of SeO_2 with $SeCl_4$ in carbon tetrachloride, and the fluoride is obtained therefrom by halogen exchange with AgF or HgF_2. $SeOBr_2$ may be obtained by treating a mixture of Se and SeO_2 with bromine or by reaction of SeO_2 with $SeBr_2$.

The thionyl and selenyl halides are stable in a vacuum at ordinary temperatures and below, but when strongly heated they decompose, usually to a mixture of products including the dioxide, free halogen, and lower halides. The structures are pyramidal (Fig. 15-13), the S and Se atoms using sets of, roughly, sp^3 hybrid orbitals, one of which holds the unshared pair. The S—O bonds in the thionyl halides are evidently resonance hybrids of the canonical structures (15-XIIa, b, c). In b and c the multiple bonding results

$$\overset{+}{S}{-}\overset{-}{O} \longleftrightarrow S{=}O \longleftrightarrow \overset{-}{S}{\equiv}\overset{+}{O}$$

$$\text{(15-XIIa)} \qquad \text{(15-XIIb)} \qquad \text{(15-XIIc)}$$

from overlap of filled $p\pi$ orbitals of oxygen with empty $d\pi$ orbitals of sulfur. The net bond order appears to be about 2, as indicated by the bond distances which are ~ 1.45 Å as compared to ~ 1.7 Å expected for an S—O single bond. The bond order also increases in the series $OSBr_2 < OSCl_2 < OSF_2$, the more electronegative halogen causing the greater amount of oxygen-to-sulfur dative π bonding.

Thionyl chloride finds use in preparing anhydrous metal halides from oxides, hydroxides, and hydrated chlorides.

The thionyl and selenyl halides can function as weak Lewis bases, using lone pairs on oxygen and also, more surprisingly, as weak Lewis acids, using vacant d orbitals. The structure of the compound $SeOCl_2 \cdot 2py$ is shown in Fig. 15-13(b).

Oxo halides of the type SO_2X_2 are called *sulfuryl halides*. Those known are SO_2F_2, SO_2Cl_2, SO_2FCl and SO_2FBr, of which sulfuryl chloride and fluoride are most important. The chloride is formed by direct reaction of SO_2 with chlorine in the presence of a catalyst, and the fluoride by fluorination of SO_2Cl_2 or by thermal decomposition of barium fluorosulfate:

$$Ba(SO_3F)_2 \xrightarrow{500°} SO_2F_2 + BaSO_4$$

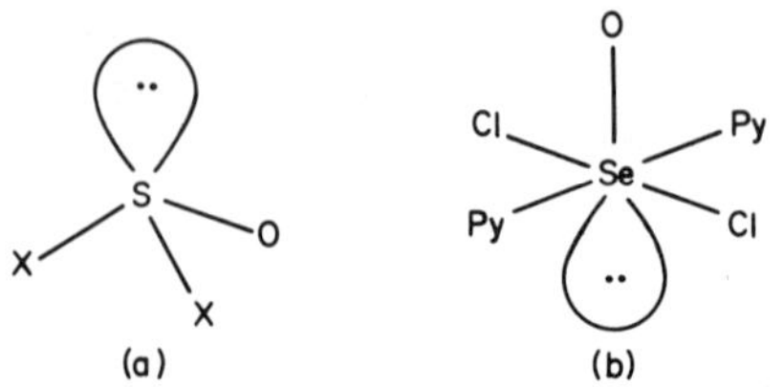

(a)   (b)

Fig. 15-13.   Structures of (a) thionyl halides, SOX_2, and (b) $SeOCl_2 \cdot 2py$.

Sulfuryl fluoride is a chemically inert gas, unaffected by water even at 150°, but slowly hydrolyzed by strong aqueous alkali. Sulfuryl chloride is much less stable than the fluoride, decomposing thermally below 300° and reacting fairly rapidly with water. It fumes strongly in moist air. It can be used as a chlorinating agent.

SeO_2F_2 has been prepared by warming a mixture of barium selenate and fluorosulfonic acid. It is a rather reactive gas, readily hydrolyzed by water.

The sulfuryl halides are known to have distorted tetrahedral structures with S—O bonds similar to those in the thionyl halides.

Some of the more complex oxo chlorides and oxo fluorides of sulfur are those shown in Table 15-6 along with their structures, where known. These structures were determined, in large part, from ^{19}F-nmr studies.

Peroxodisulfuryl difluoride, b.p. 67°, m.p. −51.6°, is obtained by fluorination of SO_3:

$$2SO_3 + F_2 \xrightarrow{170°, \ 3 \ h} S_2O_6F_2$$

At 120°, the O—O bond breaks and $S_2O_6F_2$ dissociates,[69] forming an intense brown paramagnetic species:

$$S_2O_6F_2 = 2FSO_3^{\cdot} \qquad \Delta H_{diss} = 94 \text{ kJ mol}^{-1}$$

The compound is a strong oxidant and is a versatile reagent for the preparation of other sulfur oxo fluorides;[70] thus F_2CO gives CF_3OOSO_2F, while Br^- and I^- give $X(SO_3F)_4^-$.

TABLE 15-6

Some Complex Oxo Halides of Sulfur

Compound	Structure	Compound	Structure
$S_2O_5F_2$	$\begin{array}{ccc} O & & O \\ \| & & \| \\ F-S-O-S-F \\ \| & & \| \\ O & & O \end{array}$	SOF_6	$\begin{array}{c} F \ F \\ F-S-OF \\ F \ F \end{array}$
$S_2O_5Cl_2$	Presumably analogous to that of $S_2O_5F_2$	$S_2O_6F_2$	$\begin{array}{ccc} O & & O \\ \| & & \| \\ F-S-O-O-S-F \\ \| & & \| \\ O & & O \end{array}$
SOF_4[a]	$\begin{array}{c} F \\ \| \ F \\ O-S \\ \| \ F \\ F \end{array}$	$\left. \begin{array}{c} S_3O_8F_2 \\ S_3O_8Cl_2 \end{array} \right\}$	Structures not known, but probably
SO_3F_2	$\begin{array}{c} O \\ \| \\ F-S-OF \\ \| \\ O \end{array}$		$\begin{array}{ccccc} O & & O & & O \\ \| & & \| & & \| \\ X-S-O-S-O-S-X \\ \| & & \| & & \| \\ O & & O & & O \end{array}$

[a] Adducts of SOF_4 and SbF_5 or AsF_5 contain the ion SOF_3^+ (M. Browstein, P. A. W. Dean and R. G. Gillespie, *Chem. Comm.*, **1970**, 9).

[69] P. M. Nutkowitz and G. Vincow, *J. Amer. Chem. Soc.*, 1969, **91**, 5956.
[70] J. K. Ruff and R. F. Merritt, *Inorg. Chem.*, 1968, **7**, 1219.

15-22. Halooxo Acids

Fluorosulfurous acid exists only as salts which are formed by the action of SO_2 on alkali fluorides, e.g.,

$$KF + SO_2 \rightleftharpoons KSO_2F$$

The salts have a measurable dissociation pressure at normal temperatures but are useful and convenient mild fluorinating agents, e.g.,

$$(PNCl_2)_3 + 6KSO_2F \rightarrow (PNF_2)_3 + 6KCl + 6SO_2$$
$$C_6H_5COCl + KSO_2F \rightarrow C_6H_5COF + KCl + SO_2$$

The sulfuryl halides may be considered, formally, as derivatives of sulfuric acid in which both OH groups have been replaced by halogen atoms. If only one OH group be replaced, the acids, FSO_3H (page 181), $ClSO_3H$ and $BrSO_3H$, are obtained.

Chlorosulfuric acid, a colorless fuming liquid, explosively hydrolyzed by water, forms no salts. It is made by treating SO_3 with dry HCl. Its main use is for the sulfonation of organic compounds.

Bromosulfuric acid, prepared from HBr and SO_3 in liquid SO_2 at $-35°$, decomposes at its melting point ($8°$) into Br_2, SO_2 and H_2SO_4.

Further Reading

Abrahams, S. C., *Quart. Rev.*, 1956, **10**, 407 (an excellent review of the stereochemistry of S, Se, Te, Po and O).

Bagnall, K. W., *The Chemistry of* Se, Te *and* Po, Elsevier, 1966 (general inorganic chemistry).

Banks, R. E., and R. B. Haszeldine, *Adv. Inorg. Chem. Radiochem.*, 1961, **3**, 408 (polyfluoroalkyl derivatives of S and Se).

Becke-Goehring, M., *Adv. Inorg. Chem. Radiochem.*, 1960, **2**, 159 (sulfur nitride chemistry).

——, *Quart. Rev.*, 1956, **10**, 437 (sulfur nitride chemistry).

Breslow, D. S., and H. Skolnik, *Multisulfur and Sulfur–Oxygen, Five- and Six-Membered Heterocycles*, Wiley, 1967.

Cady, G. H., *Adv. Inorg. Chem. Radiochem.*, 1960, **2**, 105 (fluorine-containing sulfur compounds).

Cilento, G., *Chem. Rev.*, 1960, **60**, 147 (comprehensive survey of effects of utilization of *d* orbitals in organic sulfur chemistry).

Cooper, W., ed., *The Physics of Selenium and Tellurium*, Pergamon Press, 1969.

—, ed., *Tellurium*, van Nostrand, 1971 (extraction, chemistry and applications).

Cruickshank, D. W. J., *J. Chem. Soc.*, **1961**, 5486 (a detailed discussion of $d\pi–p\pi$ bonding in S—O bonds).

Foss, O., *Adv. Inorg. Chem. Radiochem.*, 1960, **2**, 237 (a comprehensive account of compounds with S—S bonds).

George, J. W., *Progr. Inorg. Chem.*, 1960, **2**, 33 (halides and oxo halides of the elements of Groups Vb and VIb).

Gosselck, J., *Angew. Chem. Intern. Ed.*, 1963, **2**, 660 (review of organoselenium compounds).

Janssen, M., ed., *Organosulfur Chemistry*, Wiley, 1967.

Karchner, J. H., *Analytical Chemistry of Sulfur and its Compounds*, Part 1, Interscience–Wiley, 1970.

Kharasch, N., and C. Y. Meyers, eds., *The Chemistry of Organosulfur Compounds*, Pergamon Press, several volumes (includes some inorganic topics, e.g., acids).

Liler, M. *Reaction Mechanisms in Sulfuric Acid and Other Strong Acid Solutions*, Academic Press, 1971.

Milligan, B., and J. M. Swan, *Rev. Pure Appl. Chem.*, 1962, **12**, 73 (salts of *S*-aryl and alkyl thiosulfuric esters (Bunte salts)).

Nickless, G., ed., *Inorganic Sulfur Chemistry*, Elsevier, 1968 (an extensive treatise).

Parker, A. J., and N. Kharasch, *Chem. Rev.*, 1959, **59**, 583 (scission of S—S bonds).

Price, C. C., and S. Oae, *Sulfur Bonding*, Ronald Press, 1962 (chemical properties and bonding in sulfides, sulfoxides, sulfones, etc.).

Quarterly Reports on Sulfur Chemistry, and *Annual Reports on Mechanism and Reactions of Sulfur Compounds*, Intra Science Research Foundation, Santa Monica, Calif.

Rao, S. R., *Xanthates and Related Compounds*, Dekker, 1971.

Reid, E. E., *Organic Chemistry of Bivalent Sulfur*, Chemical Publishing Co., 1963 (Vol. V covers CS_2, thiourea, etc.).

Roy, A. B. and P. A. Trudinger, *The Biochemistry of Inorganic Compounds of Sulphur*, Cambridge Univ. Press, 1970.

Schenk, P. W., and R. Steudel, *Angew. Chem. Intern. Ed.*, 1965, **4**, 402 (a review of lower oxides and polyoxides of S).

Schmidt, M., *Inorg. Macromol. Rev.*, Vol. 1, 1969 (sulfur-containing polymers).

Senning, A., ed., *Sulfur in Organic and Inorganic Chemistry*, Dekker (volumes in the press).

Specialist Periodical Reports, Chemical Society, London; *Organic Compounds of Sulphur, Selenium and Tellurium*, D. H. Reid, Senior Reporter, Vol. 1, 1970.

Sulphur Manual, Texas Gulf Sulphur Co., New York, 1959.

Symposium on the Inorganic Chemistry of Sulfur, Chem. Soc. (London), Spec. Publ. No. 12, 1958.

Taller, W. N., ed., *Sulphur Data Book*, McGraw-Hill, New York, 1954.

Thorn, G. D., and R. A. Ludwig, *The Dithiocarbamates and Related Compounds*, Elsevier, 1962.

Tobolsky, A. V., ed., *The Chemistry of Sulfides*, Wiley, 1968 (symposium on various topics, mostly polymers and biological systems, but includes N—S chemistry).

van der Heijde, H. B., in *Organic Sulfur Compounds*, N. Kharasch, ed., Pergamon Press, London, 1961 (inorganic sulfur acids).

Yost, D. M., and H. Russell, *Systematic Inorganic Chemistry (of the 5th and 6th Group Elements)*, Prentice-Hall, 1946 (selected aspects of chemistry of S, Se and Te).

16

The Group VII Elements: F, Cl, Br, I, At

GENERAL REMARKS

16-1. Electronic Structures and Valences

Some important properties of the Group VII elements (halogens) are given in Table 16-1. Since the atoms are only one electron short of the noble gas configuration, the elements readily form the anion X^- or a single covalent bond. Their chemistries are essentially completely non-metallic and, in general, the properties of the elements and their compounds change progressively with increasing size. As in other Groups there is a much greater change between the first-row element, fluorine, and the second-row element, chlorine, than between other pairs, but with the exception of the Li–Cs group there are closer similarities within the Group than in any other in the Periodic Table.

The exceptional reactivity of fluorine, discussed below, is a reflection of the low bond energy, while the principal further differences between fluorine and the other halogens arise from its small size and high electronegativity.

TABLE 16-1

Some Properties of the Halogens

Ele-ment	Electronic structure	Ion-ization poten-tial (eV)	Electron affinity[a] (kJ g-atom^{-1}) (298°)	Dissocia-tion energy (kJ mol^{-1})	B.p. (°C)	M.p. (°C)	Crys-tal radius, X^-(Å)[c]	Co-valent radius, X (Å)
F	$1s^2 2s^2 2p^5$	17.42	339	153[b]	−118	−233	1.19	0.71
Cl	[Ne]$3s^2 3p^5$	12.96	355	242	−34.6	−103	1.70	0.99
Br	[Ar]$3d^{10} 4s^2 4p^5$	11.81	331	193	58.76	−7.2	1.87	1.14
I	[Kr]$4d^{10} 5s^2 5p^5$	10.45	302	150	184.35	113.5	2.12	1.33
At	[Xe]$4f^{14} 5d^{10} 6s^2 6p^5$	—	—	—	—	—	—	—

[a] From values, in eV, due to R. S. Berry and C. W. Reimann, *J. Chem. Phys.*, 1963, **38**, 1541, Allred–Rochow give *electronegativity values* 4.10, 2.83, 2.74, 2.21, 1.96.
[b] Generally accepted value. Photoionization with mass spectral analysis gives $D_0 = 129.5$ kJ mol^{-1}, but the reasons for this very low value are not yet apparent (H. Dibler, J. A. Walker and K. E. McCulloh, *J. Chem. Phys.*, 1969, **50**, 4593).
[c] Ladd radii.

The coordination number of the halides in the -1 state is normally only one, but, in certain metal compounds, bridging halides can be obtained (see page 468) where the coordination number is 2 and in some metal atom cluster compounds (see page 547) triply bridging halide ions with coordination number 3 are found.

In the halogen fluorides such as ClF_3, ClF_5, BrF_5 and IF_7 the formal oxidation states and coordination numbers are, of course, higher but only in these compounds and in oxo compounds such as Cl_2O_7 or I_2O_5 is this the case. In the oxo ions the halogen atom can scarcely be said to be in a positive oxidation state except in a formal sense of oxidation number (e.g., $+7$ for Cl in ClO_4^-) for convenience in balancing oxidation–reduction equations. Such formal oxidation numbers bear no relation to actual charge distributions.

For fluorine there is little evidence for positive behavior, even formally, except perhaps in the ion FCl_2^+; in oxygen fluorides the F atom is probably somewhat negative with respect to oxygen, while, in ClF, evidence from chlorine nuclear quadrupole coupling shows that the actual charge distribution involves partial positive charge on Cl.

Bond polarities in other halogen compounds indicate the importance of forms such as I^+Cl^- in ICl or I^+CN^- in ICN. In general, when a halogen atom forms a bond to another atom more electronegative than itself, the bond will be polar with a partial positive charge on the halogen. Nevertheless, there are a few compounds of Cl, Br and I that can be regarded as having positive halogen, such as the ions Cl_3^+, Br_3^+, I_5^+, discussed below.

Even for the heaviest member of the group, astatine, there is little evidence for any extensive "metallic" behavior.

THE ELEMENTS

None of the halogens occur in the elemental state in Nature; this is because of their high reactivity. All exist as diatomic molecules, which, being homonuclear, are without permanent electrical polarity. In the condensed phases, only weak van der Waals forces operate, hence the trend in melting and boiling points of the halogens parallels that in the noble gases. In both cases the dominating factor is the increasing magnitude of the van der Waals forces as size and polarizability of the atoms or molecules increase. The increase in color of the elements and of their covalent compounds with increasing size is due in general to a progressive shift of electronic absorption bands to longer wavelengths in the absorption spectrum.

16-2. Fluorine

Fluorine occurs widely in Nature, notably as *fluorspar*, CaF_2; *cryolite*, Na_3AlF_6; and *fluorapatite*, $3Ca_3(PO_4)_2Ca(F,Cl)_2$. It is more abundant (0.065%) than chlorine (0.055%) in the earth's crust.

The estimated standard potential of fluorine ($E^0 = +2.85$ V) clearly indicates why early attempts to prepare the element by electrolytic methods in aqueous solution suitable for chlorine ($E^0 = +1.36$ V) failed. The element was first isolated in 1886 by Moissan, who pioneered the chemistry of fluorine and its compounds. The greenish gas is obtained by electrolysis. Although anhydrous HF is non-conducting, the addition of anhydrous KF gives conducting solutions. The most commonly used electrolytes are $KF \cdot 2$–$3HF$, which is molten at 70–100°, and KF–HF, which is molten at 150–270°. When the melting point begins to be too high, the electrolyte can be regenerated by resaturation with HF from a storage tank. There have been many designs for fluorine cells; these are constructed of steel, copper, or Monel metal, which become coated with an unreactive layer of fluoride, and steel or copper cathodes with ungraphitized carbon anodes are used. Although fluorine is often handled in metal apparatus, it can be handled in the laboratory in glass apparatus provided traces of HF, which attacks glass rapidly, are removed. This is achieved by passing the gas through sodium or potassium fluoride with which HF forms the bifluorides, MHF_2.

Fluorine is the most chemically reactive of all the elements and combines directly at ordinary or elevated temperatures with all the elements other than oxygen and the lighter noble gases, often with extreme vigor. It also attacks many other compounds, particularly organic compounds, breaking them down to fluorides; organic materials often inflame and burn in the gas.

The great reactivity of the element is in part attributable to the low dissociation energy (Table 16-1) of the F—F bond in the fluorine molecule, and the fact that reactions of atomic fluorine are strongly exothermic. The generally accepted explanation for this anomalous value is that it is due to repulsion between non-bonding electrons. However, it has been recently pointed out[1] that destabilizing effects of comparable magnitude per F atom are found in bond energies of alkali-metal and some covalent fluorides; the electron affinity of fluorine is also ca. 110 kJ mol^{-1} lower than expected from the trend for the other halogens.* Hence the anomaly in F—F bond energy may arise from the addition of an electron to the F atom. Because of the exceptionally small size (atomic radii: F, 0.57; Cl, 0.98; Br, 1.12; I, 1.32) of the F atom, the entering electron causes an anomalously large increase in the repulsive energy among all the valence-shell electrons. The energy of binding the additional electron is thus less by ca. 110 kJ mol^{-1} than that predicted. The low dissociation energy of F_2 can thus be considered to arise from the interaction of each atom and the electron provided to the bond by the other fluorine. A similar effect may account for the low bond energies in H_2O_2 and N_2H_4.

---

* Note that there is *no* anomaly in the ionization potential.
[1] P. Politzer, *J. Amer. Chem. Soc.*, 1969, **91**, 6235.

16-3. Chlorine

Chlorine occurs in Nature mainly as sodium chloride in sea water and in various inland salt lakes, and as solid deposits originating presumably from the prehistoric evaporation of salt lakes. Chlorine is prepared industrially, almost entirely by electrolysis of brine:[2]

$$Na^+ + Cl^- + H_2O \rightarrow Na^+ + OH^- + \tfrac{1}{2}Cl_2 + \tfrac{1}{2}H_2$$

Disadvantages of brine electrolysis are the simultaneous production of sodium hydroxide and the use of mercury as the electrode; the inevitable loss of mercury constitutes a major pollution hazard. A process producing chlorine independently would be advantageous. A recent modification of the old Deacon process:

$$2HCl + \tfrac{1}{2}O_2 \rightleftharpoons Cl_2 + H_2O$$

which has an unfavorable equilibrium, may become economic when nitrogen oxides are used as catalyst and the equilibrium is shifted by removal of water by sulfuric acid.[2a] A related process, where the equilibrium is shifted by removal of chlorine as dichloroethylene, is the oxychlorination reaction:

$$2HCl + \tfrac{1}{2}O_2 + C_2H_4 \rightarrow C_2H_4Cl_2 + H_2O$$

The dichloroethylene is pyrolysed to vinyl chloride, which in turn is used for polymerization.

Chlorine and hydrogen can also be recovered by electrolysis of warm 22% hydrochloric acid which is obtained as a by-product in chlorination processes.

Chlorine is a greenish gas. It is moderately soluble in water with which it reacts (see page 476). When chlorine is passed into dilute solutions of $CaCl_2$ at $0°$, feathery crystals of "chlorine hydrate," $Cl_2 \cdot 7.3H_2O$, are formed. This substance is a clathrate of the gas-hydrate type (see page 160), having all medium holes and $\sim 20\%$ of the small holes in the structure filled with chlorine molecules.

16-4. Bromine

Bromine occurs principally as bromide salts of the alkalis and alkaline earths in much smaller amounts than, but along with, chlorine. Bromine is obtained from brines and sea water by chlorination at a pH of ~ 3.5 and is swept out in a current of air.

Bromine is a dense, mobile, dark-red liquid at room temperature. It is moderately soluble in water (33.6 g l^{-1} at $25°$) and miscible with non-polar solvents such as CS_2 and CCl_4. Like Cl_2 it gives a crystalline hydrate, which is, however, structurally different from that of chlorine.

16-5. Iodine

Iodine occurs as iodide in brines and in the form of sodium and calcium iodates. Also, various forms of marine life concentrate iodine. Production

---

[2] *Chem. Eng. News*, Nov. 9, 1970, p. 32.
[2a] *Chem. Eng. News*, May 5, 1969, 14.

of iodine involves either oxidizing I^- or reducing iodates to I^- followed by oxidation to the elemental state. Exact methods vary considerably depending on the raw materials. A commonly used oxidation reaction, and one suited to laboratory use when necessary, is oxidation of I^- in acid solution with MnO_2 (also used for preparation of Cl_2 and Br_2 from X^-).

Iodine is a black solid with a slight metallic luster. At atmospheric pressure it sublimes (violet vapor) without melting. Its solubility in water is slight (0.33 g l^{-1} at $25°$). It is readily soluble in non-polar solvents such as CS_2 and CCl_4 to give violet solutions; spectroscopic studies indicate that "dimerization" occurs in solutions to some extent:[3]

$$2I_2 \rightleftharpoons I_4$$

Iodine solutions are brown in solvents such as unsaturated hydrocarbons, liquid SO_2, alcohols and ketones, and pinkish brown in benzene (see below).

Iodine forms the well-known blue complex with starch, where the iodine atoms are aligned in channels in the polysaccharide amylose.[3a]

16-6. Astatine, At, Element 85[4]

Isotopes of element 85 have been identified as short-lived products in the natural decay series of uranium and thorium. The element was first obtained in quantities sufficient to afford a knowledge of some of its properties by the cyclotron reaction: $^{209}Bi(\alpha,2n)^{211}At$. The element was named astatine from the Greek for "unstable." About 20 isotopes are known, the longest lived being ^{210}At with a half-life of only 8.3 hours; consequently, macroscopic quantities cannot be accumulated, although some compounds, HAt, CH_3At, AtI, AtBr and AtCl have been detected mass-spectroscopically.[5] Our knowlege of its chemistry is based mainly on tracer studies, which show that it behaves about as one might expect by extrapolation from the other halogens. The element is rather volatile. It is somewhat soluble in water from which it may, like iodine, be extracted into benzene or carbon tetrachloride. It cannot, like iodine, be extracted from basic solutions.

The At^- ion is produced by reduction with SO_2 or zinc but not ferrous ion (which gives some indication of the oxidation potential of At^-). This ion is carried down in AgI or TlI precipitates. Bromine and, to some extent, ferric ion oxidizes it to what appears to be AtO^- or HAtO. HClO or hot $S_2O_8^{2-}$ oxidize it to an anion carried by IO_3^- and therefore probably AtO_3^-. Astatine is also carried when $[I py_2]^+$ salts are isolated, indicating that $[At py_2]^+$ can exist. In $0.1 M$ acid, the potentials appear to be:

$$AtO_3^- \xrightarrow{1.5} HOAt(?) \xrightarrow{1.0} At \xrightarrow{0.3} At^-$$

There is some evidence for the ion At^+.[6]

[3]   D. D. Eley, F. L. Isack and C. H. Rochester, *J. Chem. Soc., A*, **1968**, 1651.
[3a] H. J. Keller and K. Seibold, *J. Amer. Chem. Soc.*, 1971, **93**, 1310.
[4]   V. D. Nefedov, Yu. V. Norseyev, M. A. Toropuva and V. A. Khalkin, *Russ. Chem. Rev.*, 1968, **37**, 87; A. H. W. Aten, Jr., *Adv. Inorg. Chem. Radiochem.*, 1964, **6**, 207.
[5]   E. H. Appelman, E. N. Sloth and M. H. Studier, *Inorg. Chem.*, 1966, **5**, 766.
[6]   Yu. V. Norseyev and V. A. Khalkin, *J. Inorg. Nuclear Chem.*, 1968, **30**, 3239.

16-7. Charge-transfer Compounds of Halogens[7]

As noted above, iodine gives solutions in organic solvents whose color depends on the nature of the solvent. It has been shown that for the brown solutions in donor solvents, solvation and 1:1 complex formation, $I_2 \cdots S$, occur; such interactions are said to be of the "charge-transfer" type and the complexes are called charge-transfer complexes. This name derives from the nature of the interaction, in which the bonding energy is attributable to a partial transfer of charge. The ground state of the system can be described as a resonance hybrid of (16-I) and (16-II), with the former predominating. An electronic transition to an excited state, which is also a resonance hybrid of (16-I) and (16-II) with the latter predominating, is characteristic of these complexes. It usually occurs near or in the visible region and is the cause of their typically intense colors. This transition is called a charge-transfer transition.

$$X_2 \cdots S \quad \leftrightarrow \quad X_2^- S^+$$

$$(16\text{-I}) \qquad\qquad (16\text{-II})$$

Although iodine has been most extensively studied, chlorine and bromine show similar behavior. For a given group of donors, the frequency of the intense charge-transfer absorption band in the ultraviolet is dependent upon the ionization potential of the donor solvent molecule, and electronic charge can be transferred either from a π-electron system as in benzene or from lone-pairs as in ethers or amines. Charge-transfer spectra and complexes are of importance elsewhere in chemistry.

For halogens, and interhalogens such as ICl, charge-transfer compounds can in fact be isolated in the crystalline state, though low temperatures are often required. Thus dioxane with Br_2 gives a compound with a chain structure (16-III) where the Br—Br distance (2.31 Å) is only slightly longer than

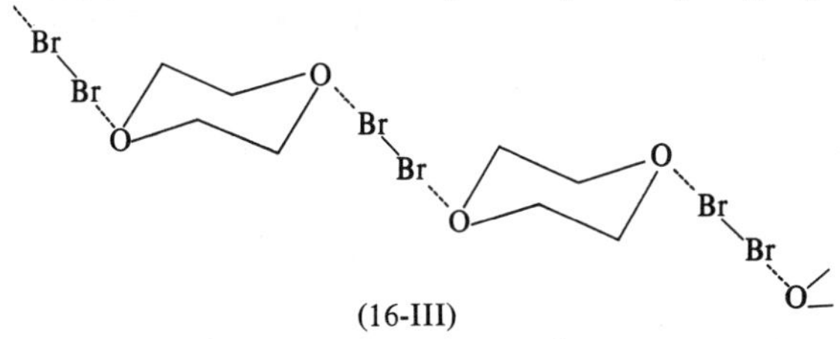

(16-III)

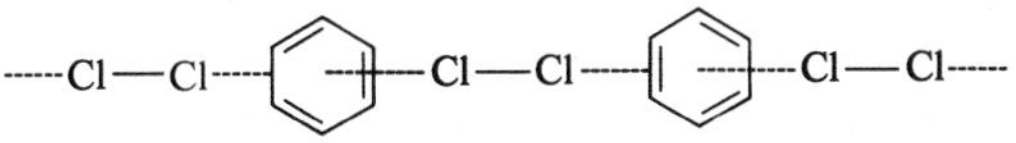

(16-IV)

[7] L. J. Andrews and R. M. Keefer, *Molecular Complexes in Organic Chemistry*, Holden-Day, 1964; C. K. Prout and J. D. Wright, *Angew. Chem. Internat. Edn.*, 1968, **7**, 659 (structures of solid donor–acceptor complexes); H. A. Bent, *Chem. Rev.*, 1968, **68**, 587 (general donor–acceptor interactions); R. S. Mullikan and W. B. Person, *Molecular Complexes*, Interscience-Wiley, 1969; and in *Physical Chemistry*, Vol. 3, D. Henderson, ed., Academic Press, 1969; M. J. Blandamer and M. F. Fox, *Chem. Rev.*, 1970, **70**, 59; R. Foster, *Organic Charge Transfer Complexes*, Academic Press, 1969; for a good example, I_2–C_2H_5OH, see L. M. Julien, W. E. Bennett and W. B. Person, *J. Amer. Chem. Soc.*, 1969, **91**, 6195.

in Br_2 itself (2.28 Å). In the benzene compound (16-IV) the halogen molecules lie along the axis perpendicular to the center of the ring. Other crystals such as $(CH_3)_3NI_2$ and $(CH_3)_2COI_2$ all have one halogen linked to the donor atom and the second pointing away, as in $N\cdots X\!-\!X$. In many of the compounds, especially with O and N donors, there is considerable similarity to hydrogen-bonding interactions.

In certain cases, interaction of the halogen and the donor may become sufficiently strong for the $X\!-\!X$ bond to be broken. Thus, for adducts of the type $R_2Y\cdot X_2$, where Y is O, S, Se or Te, the components R_2Y and X_2 retain their identities when Y is more electronegative than X. Otherwise there is an oxidative addition (Section 24-A-2) of X_2 to Y, and the resulting complex, e.g., $(p\text{-}ClC_6H_4)_2SCl_2$ or Me_2TeCl_2, has a trigonal ψ-bipyramidal structure with axial halide atoms.[8] For pyridine the final crystalline product is $[py\!-\!I\!-\!py]I$; of course, the latter could be regarded as a coordination complex of I^+.

In certain cases, charge-transfer complexes may be intermediates in halogen reactions.[9] Thus, I_2 first reacts with KCNS in water to give a yellow solution, after which there is a slow reaction in neutral or basic solution:

$$I_2 + SCN^- \xrightarrow{\text{Rapid}} I_2NCS^- \xrightarrow{\text{Slow}} ICN + SO_4^{2-}$$

HALIDES

With the exception of He, Ne and Ar, all the elements in the Periodic Table form halides, often in several oxidation states, and halides generally are among the most important and common compounds. The ionic and covalent radii of the halogens are shown in Table 16-1.

There are almost as many ways of classifying halides as there are types of halide—and this is many. There are not only binary halides that can range from simple molecules with molecular lattices to complicated polymers and ionic arrays, but also oxo halides, hydroxo halides, and other complex halides of various structural types.

16-8. Preparation of Anhydrous Halides

Although preparations of individual halogen compounds are mentioned throughout the text, anhydrous halides are of such great importance in chemistry that a few of the more important general methods of preparation can be noted.

1. *Direct interaction.* This method is perhaps the most important preparative method for all halides, and the halogens or their acids where appropriate are employed.

---

[8] N. C. Baenziger, R. E. Buckles, R. J. Maner and T. D. Simpson, *J. Amer. Chem. Soc.*, 1969, **91**, 5749; G. C. Haywood and P. J. Hendra, *J. Chem. Soc., A*, **1969**, 1760.
[9] Cf. E. M. Kosower, *Progr. Phys. Org. Chem.*, 1965, **3**, 81.

Many simple fluorides of metals in lower oxidation states are obtained by dissolving the oxides, carbonates, etc., in HF and drying the product, or by dry reaction. Higher fluorides, such as AgF_2 or CrF_4, usually require the use of elemental fluorine, or a powerful fluorinating compound such as ClF_3, with the metal or a lower fluoride or other salt. For chlorides, bromides, and iodides of transition metals, elevated temperatures are usually necessary in dry reactions. Reaction with Cl_2 or Br_2 is often more rapid when tetrahydrofuran or some other ether is used as the medium, the halide being obtained as a solvate. Where different oxidation states are possible, F_2 and Cl_2 usually give a higher state than bromine or iodine does. Non-metals, such as phosphorus, usually react readily without heating, and their reaction with fluorine may be explosive.

2. *Halogen exchange.* This method is especially important for fluorides, many of which are normally obtained from the chlorides by action of various metal fluorides such as CoF_3, AsF_3, AsF_5, etc.; this type of replacement is much used for organic fluorine compounds (page 489).

Many Cl, Br and I compounds undergo rapid halogen exchange with either the elements or the acids HX. An excess of reagent is usually required since equilibrium mixtures are normally formed.

3. *Halogenation by halogen compounds.* This is an important method, particularly for metal fluorides and chlorides. The reactions involve treatment of anhydrous compounds, often oxides, with halogen compounds such as BrF_3, CCl_4, hexachlorobutadiene, and hexachloropropene at elevated temperatures:

$$NiO + ClF_3 \rightarrow NiF_2$$
$$UO_3 + CCl_2{=}CCl{-}CCl{=}CCl_2 \xrightarrow{\text{Reflux}} UCl_4$$
$$Pr_2O_3 + 6NH_4Cl(s) \xrightarrow{300°} 2PrCl_3 + 3H_2O + 6NH_3$$
$$Sc_2O_3 + CCl_4 \xrightarrow{600°} ScCl_3$$

4. *Dehydration of hydrated halides.* Hydrated halides are usually obtained easily from aqueous solutions. They can sometimes be dehydrated by heating them in a vacuum, but often this leads to oxo halides or impure products. Various reagents can be used to effect dehydration. For example, $SOCl_2$ is often useful for chlorides. Another fairly general reagent is 2,2-dimethoxypropane.

$$[Cr(H_2O)_6]Cl_3 + 6SOCl_2 \xrightarrow{\text{Reflux}} CrCl_3 + 12HCl + 6SO_2$$
$$MX_n \cdot mH_2O + mCH_3C(OCH_3)_2CH_3 \rightarrow MX_n + m(CH_3)_2CO + 2mCH_3OH$$

In many cases the acetone and/or methanol becomes coordinated to the metal, but gentle heating or pumping usually gives the solvate-free halide.

16-9. Binary Ionic Halides

Most metal halides are substances of predominantly ionic character,[10] although partial covalence is important in some. Actually, of course, there is a uniform gradation from halides that are for all practical purposes purely ionic, through those of intermediate character, to those that are essentially

[10] R. G. Pearson and H. B. Gray, *Inorg. Chem.*, 1963, **2**, 358 (discussion of ionic character).

covalent. As a rough guide we can consider those halides in which the lattice consists of discrete ions rather than definite molecular units to be basically ionic, although there may still be considerable covalence in the metal–halogen interaction, and the description "ionic" should never be taken entirely literally. As a borderline case *par excellence*, which clearly indicates the danger of taking such rough classifications as "ionic" and "covalent" or even "ionic" and "molecular" too seriously, we have $AlCl_3$ (page 264); this has an extended structure in which aluminum atoms occupy octahedral interstices in a close-packed array of chlorine atoms; this kind of non-molecular structure could accomodate an appreciably ionic substance. Yet $AlCl_3$ melts at a low temperature (193 °C) to a molecular liquid containing Al_2Cl_6, these molecules being much like the Al_2Br_6 molecules which occur in both solid and liquid states of the bromide. Thus, while $AlCl_3$ cannot be called simply a molecular halide, it is an over-simplification to call it an ionic one.

The relatively small radius of F^-, 1.19 Å, is almost identical with that of the oxide, O^{2-}, ion (1.25 Å); consequently, many fluorides and oxides are ionic with similar formulae and crystal structures, for example, CaO and NaF. The compounds of the other halogens with the same formula usually form quite different lattices and may even give molecular lattices. Thus chlorides and other halides often resemble sulfides, just as the fluorides often resemble oxides. In several cases the fluorides are completely ionic, whereas the other halides are covalent; for example, CdF_2 and SrF_2 have the CaF_2 lattice (nearly all difluorides have the fluorite or rutile structure), but $CdCl_2$ and $MgCl_2$ have layer lattices with the metal atoms octahedrally surrounded by chlorine atoms.

Many metals show their highest oxidation state in the fluorides. Let us consider the Born–Haber cycle in equation (16-1):

$$M(s) \xrightarrow{S} M(g) \xrightarrow{I_4} M^{4+}(g)$$
$$\searrow$$
$$MX_4(s) \qquad (16\text{-}1)$$
$$\nearrow$$
$$2X_2(g) \xrightarrow{2D} 4X(g) \xrightarrow{4A} 4X^-(g)$$

The value of $(A - D/2)$, the energy change in forming 1 g-ion of X^- from $\frac{1}{2}$ mole of X_2, is ~ 250 kJ for all the halogens, and S is small compared to I_4 in all cases. Although the structure of MX_4 and hence the lattice energy may not be known to allow us to say whether $4(A - D/2)$ plus the lattice energy will compensate for $(I_4 + S)$, we can say that the lattice energy and hence the potential for forming an ionic halide in a high oxidation state will be greatest for fluoride, since, generally, for a given cation size the greatest lattice energy will be available for the smallest anion, that is, F^-.

However, for very high oxidation states, which are formed notably with transition metals, for example, WF_6 or OsF_6, the energy available is quite insufficient to allow ionic crystals with, say, W^{6+} or Os^{6+} ions; consequently such fluorides are gases, volatile liquids or solids resembling closely the covalent fluorides of the non-metals. The question as to whether a metal fluoride will be ionic or molecular cannot be reliably predicted, and the distinction between the types is not always sharp.

As noted previously, ionic oxides can also be obtained in high oxidation states, sometimes even higher than the fluorides, for example, RuF_6 and RuO_4. Fluorides in high oxidation states are often hydrolyzed by water, the important factors here being the even greater stability of the ionic and covalent oxides and also the low dissociation of HF in aqueous solution. Thus, for example,

$$4RuF_5 + 10H_2O \rightarrow 3RuO_2 + RuO_4 + 20HF$$

The halides of the alkali metals, except perhaps lithium, of the alkaline earths, with the definite exception of beryllium, and of most of the lanthanides and a few halides of the d-group metals and actinides can be considered as mainly ionic materials. As the charge/radius ratio of the metal ions increases, however, covalence increases. Consider, for instance, the sequence KCl, $CaCl_2$, $ScCl_3$, $TiCl_4$. KCl is completely ionic, but $TiCl_4$ is an essentially covalent molecular compound. Similarly, for a metal with variable oxidation state, the lower halides will tend to be ionic, whereas the higher ones will tend to be covalent. As examples we can cite $PbCl_2$ and $PbCl_4$, and UF_4, which is an ionic solid, while UF_6 is a gas.

The size and polarizability of the halide ion are also important in determining the character of the halide. Thus we have the classic case of the aluminum halides, where AlF_3 is basically ionic, $AlCl_3$ has a layer lattice, while $AlBr_3$ and AlI_3 exist as covalent dimers.

Most ionic halides dissolve in water to give hydrated metal ions and halide ions. However, the lanthanide and actinide elements in the $+3$ and $+4$ oxidation states form fluorides insoluble in water. Fluorides of Li, Ca, Sr and Ba also are sparingly soluble, the lithium compound being precipitated by ammonium fluoride. Lead gives a sparingly soluble salt, PbClF, which can be used for gravimetric determination of F^-. The chlorides, bromides and iodides of Ag^I, Cu^I, Hg^I and Pb^{II} are also quite insoluble. The solubility through a series of mainly ionic halides of a given element, $MF_n \rightarrow MI_n$ may vary in either order. In cases where all four halides are essentially ionic, the solubility order will be iodide > bromide > chloride > fluoride, since the governing factor will be the lattice energies which increase as the ionic radii decrease. This order is found among the alkali, alkaline-earth, and lanthanide halides. On the other hand, if covalence is fairly important, it can invert the trend, making the fluoride most and the iodide least soluble, as in the familiar cases of silver and mercurous halides.

It should be clear that no broad simple generalization is possible, and the properties of metal halides are determined by the interplay of a number of factors.

16-10. Molecular Halides

A solid consisting of separate molecules held together by van der Waals forces and perhaps dipole–dipole and dipole–induced dipole forces will have a low lattice energy; therefore molecular halides are generally considerably more volatile than ionic halides. There is probably a rough correlation between increasing metal-to-halogen covalence and increasing tendency to the formation of molecular compounds. Thus the molecular halides are sometimes also called the covalent halides. The designation molecular is preferable since it states a fact.

Most of the electronegative elements, and the metals in very high oxidation states, form molecular halides. Among the most important compounds are the hydrogen halides (page 177).

The acid strengths increase in the order $HCl < HBr < HI$. The reasons for this, as well as the weakness of HF, have been considered on page 168. The H—X bond energies and the thermal stabilities decrease markedly in the order $HF > HCl > HBr > HI$, that is, with increasing atomic number of the halogen. The same trend is found, in varying degrees, among the halides of all elements giving a set of molecular halides, such as those of C, B, Si, P, etc. Interhalogen compounds are discussed below.

The formation of halide bridges between two or, less often, three other atoms is an important structural feature. Between two metal atoms, the most common situation involves two halogen atoms as in (16-V), but examples with one and three bridge atoms are known. In older literature such bridges were often depicted as involving a covalent bond to one metal atom and donation of an electron pair to the other as in 16-V. In virtually every case,

(16-V)

however, structural data show that the two bonds to each bridging halogen atom are equivalent, and a representation such as (16-V) is thus inappropriate. Molecular-orbital theory provides a simple, flexible formulation.

As indicated in Fig. 16-1a, each metal atom presents an empty σ orbital directed more or less toward the bridging halide, X^-; these orbitals are ϕ_1 and ϕ_2. The X^- ion has four filled valence-shell orbitals. One of these (which may be taken as an s orbital, a p orbital, or a hybrid), ϕ_3, will be directed down, as shown; the other, ϕ_4, is a pure p orbital. The metal orbitals may be combined (Fig. 16-1b) into a symmetric combination, Φ_1, and an antisymmetric combination, Φ_2, which may interact with ϕ_3 and ϕ_4, respectively, to form bonding and antibonding orbitals, as shown in Fig. 16-1c. The four electrons then occupy the two bonding MO's, giving a mean M—X bond order of 1. It is also evident that the M—X—M angle is not sharply limited. In reality, angles ranging from 60° to 180° are found, though the majority for Cl and Br are in the range of 70–100°.

With Cl^- and Br^-, bridges are characteristically bent, whereas fluoride bridges may be either bent or linear. Thus, in BeF_2 there are infinite chains, $\cdots BeF_2BeF_2\cdots$, with bent bridges, similar to the situation in $BeCl_2$. On the other hand, transition-metal pentahalides afford a notable contrast. While

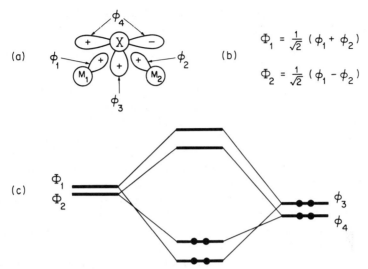

Fig. 16-1.   Molecular orbital formulation of a bridging halide system, MXM.

the pentachlorides dimerize [see (16-VI)] with bent M—Cl—M bridges, the pentafluorides form cyclic tetramers (16-VII), with linear M—F—M bridges.

(16-VI)

(16-VII)

The fluorides probably adopt the tetrameric structures with linear bridges, in part because the smaller size of F than of Cl would introduce excessive M···M repulsion in an M$\diagdown^{F}_{F}\diagup$M system. Another notable example of a linear M—F—M bridge occurs in $K[(C_2H_5)_3Al—F—Al(C_2H_5)_3]$; the linearity of this Al—F—Al chain may be due in part to overlap of filled $2p\pi$ orbitals of fluorine with empty $3d\pi$ orbitals of the aluminum atoms, thus giving some π character to the AlF bonds.

Molecular Fluorides.   Many molecular fluorides exist, but it is clear that because of the high electronegativity of fluorine the bonds in such compounds tend to be very polar. Because of the low dissociation energy of F_2 and the relatively high energy of many bonds to F (e.g., C—F, 486; N—F, 272; P—F, 490 kJ mol^{-1}), molecular fluorides are often formed very exothermically; this is just the opposite of the situation with nitrogen where the great strength of the bond in N_2 makes nitrogen compounds mostly endothermic. Interestingly, in what might be considered a direct confrontation between these two effects, the tendency of fluorine to exothermicity wins. Thus, for NF_3 we have:

$$\begin{aligned}
\tfrac{1}{2}N_2(g) &= N(g) & \Delta H &\approx 475 \text{ kJ mol}^{-1} \\
\tfrac{3}{2}F_2(g) &= 3F(g) & \Delta H &\approx 232 \text{ kJ mol}^{-1} \\
N(g)+3F(g) &= NF_3(g) & -3E_{N-F} &\approx -3(272) = -816 \text{ kJ mol}^{-1} \\
\text{Therefore, } \tfrac{1}{2}N_2(g)+\tfrac{3}{2}F_2(g) &= NF_3(g) & \Delta H &\approx -109 \text{ kJ mol}^{-1}
\end{aligned}$$

The high electronegativity of fluorine often has a profound effect on the properties of molecules in which several F atoms occur. Representative are facts such as (a) CF_3COOH is a strong acid, (b) $(CF_3)_3N$ and NF_3 have no basicity and (c) CF_3 derivatives in general are attacked much less readily by electrophilic reagents in anionic substitutions than are CH_3 compounds. The CF_3 group may be considered as a kind of large pseudohalogen with an electronegativity about comparable to that of Cl.

The molecular fluorides of both metals and non-metals are usually gases or volatile liquids; analogs with the other halogens are often lacking, the reason being partly the size factor, which with fluorine permits higher coordination numbers, and partly the factors discussed above concerning the stability of higher oxidation states and covalent bond formation. Where the central atom has suitable vacant orbitals available, and especially if the polarity of the single bonds M—F would be such as to leave a considerable charge on M, as in, say, SF_6, multiple bonding can occur using filled p orbitals of fluorine for overlap with vacant orbitals of the central atom. This multiple bonding is probably the major factor in the shortness and high strength of many bonds to fluorine. Ionic–covalent resonance may also be a contributing factor—if it is not, in fact, only another way of expressing the same thing. An example is provided by the hexafluorides of the 2nd and 3rd row transition metals where the metal–fluorine stretching frequencies and the stability trends are most readily explained by some $(F)p\pi \rightarrow d\pi(M)$ bonding. Thus, as expected, the stability is least with the electron-rich metals, e.g., Pd and Pt, at the heavy ends of the Groups. With the chloro complexes, MCl_6^{2-}, the opposite holds, as would be expected if $(M)d\pi \rightarrow d\pi(Cl)$ bonding is involved in addition to σ bonding (cf. page 599).

The volatility of molecular fluorides is due to the absence of intermolecular bonding other than van der Waals forces, since the polarizability of fluorine is very low and no suitable outer orbitals exist for other types of attraction.

Reactivity. The detailed properties of a given molecular halide depend on the particular elements involved and these are discussed where appropriate in other Chapters. However, a fairly general property of molecular halides is their easy hydrolysis to produce the hydrohalic acid and an acid of the other element. Typical examples are

$$\begin{aligned}
BCl_3+3H_2O &\rightarrow B(OH)_3+3H^++3Cl^- \\
PBr_3+3H_2O &\rightarrow HPO(OH)_2+3H^++3Br^- \\
SiCl_4+4H_2O &\rightarrow Si(OH)_4+4H^++4Cl^-
\end{aligned}$$

Where the maximum covalency is attained, as in CCl_4 or SF_6, the halides may be quite inert towards water. However, this is a result of kinetic and not thermodynamic factors. Thus for CF_4 the equilibrium constant for the reaction

$$CF_4(g) + 2H_2O(l) = CO_2(g) + 4HF(g)$$

is ca. 10^{23}. The necessity for a means of attack is well illustrated by the fact that SF_6 is not hydrolyzed, whereas SeF_6 and TeF_6 are hydrolyzed at 25° through expansion of the coordination sphere which is possible only for selenium and tellurium.

16-11. Halide Complexes

All the halide ions have the ability to function as ligands and form, with various metal ions or covalent halides, complexes such as SiF_6^{2-}, $FeCl_4^-$, HgI_4^{2-}, etc., as well as mixed complexes along with other ligands, for example, $[Co(NH_3)_4Cl_2]^+$. We merely make some general remarks and cite some typical characteristics of such complexes, reserving detailed discussions for other places in connection with the chemistries of the complexed elements.

One of the important general questions that arise concerns the relative affinities of the several halide ions for a given metal ion. There is no simple answer to this however. For crystalline materials it is obvious that lattice energies play an important role and there are cases, such as BF_4^-, BCl_4^-, BBr_4^-, in which the last two are known only in the form of crystalline salts with large cations, where lattice energies are governing. In considering the stability of the complex ions *in solution* it is important to recognize that (*a*) the stability of a complex involves, not only the absolute stability of the M—X bond, but also its stability relative to the stability of ion—solvent bonds, and (*b*) in general an entire series of complexes will exist, $M^{n+}(aq)$, $MX^{(n-1)+}(aq)$, $MX_2^{(n-2)+}(aq)$, ..., $MX_x^{(n-x)+}(aq)$, where x is the maximum coordination number of the metal ion. Of course, these two points are of importance in all types of complexes in solution.

A survey of all of the available data on the stability of halide complexes shows that generally the stability decreases in the series $F > Cl > Br > I$, but with some metal ions the order is the opposite, namely, $F < Cl < Br < I$. No rigorous theoretical explanation for either sequence or for the existence of the two classes of acceptors relative to the halide ions has been given. It is likely that charge/radius ratio, polarizability, and the ability to use empty outer d orbitals for back-bonding are significant factors. From the available results it appears that for complexes where the replacement stability order is $Cl < Br < I$, the actual order of M—X bond strength is $Cl > Br > I$, so that ionic size and polarizability appear to be the critical factors.

The limiting factor in the formation of fluoro complexes for cations of small size and high charge is competitive hydrolysis; even at high concentrations many fluoro complexes are hydrolyzed and particularly so where the oxidation state is high. There is an empirical relation[11] of wide applicability:

$$\log Q = -1.56 + 0.48 Z_+^2 / r_+$$

[11] R. E. Mesmer and C. F. Baes, Jr., *Inorg. Chem.*, 1969, **8**, 618.

where Q is the formation constant for the reaction

$$M^{n+} + F^- = MF^{(n-1)+}$$

and Z_+ and r_+ are the charge and radius of the cation. It is to be emphasized that all complex fluoro "acids" such as HBF_4 and H_2SiF_6 are necessarily strong since the proton can be bound only to a solvent molecule.

There are many references to the effect of steric factors in accounting for such facts as the existence of $FeCl_4^-(aq)$ as the highest ferric complex with Cl^- whereas FeF_6^{3-} is rather stable, and similar cases such as $CoCl_4^{2-}$, SCl_4, $SiCl_4$ as the highest chloro species compared with the fluoro species CoF_6^{3-}, SF_6, SiF_6^{2-}. In many such cases thorough steric analysis, considering the probable bond lengths and van der Waals radii of the halide ions, shows that this steric factor alone cannot account for the differences in maximum coordination number, and this point requires further study.

Although iodo complexes are generally the least stable, and are dissociated or unstable in aqueous solution, a large number of complex anions can be made even where the metal M^{n+} is oxidizing towards I^-, provided that non-aqueous media, e.g., nitriles, CH_3NO_2, or liquid HI, are used.[12] The latter possibility arises since HCl has a free energy of formation ca. 85 kJ mol^{-1} higher than that of HI so that in the anhydrous reaction

$$MCl_6^{(6-n)-} + 6HI(l) = MI_6^{(6-n)-} + 6HCl$$

the equilibrium lies to the right and the greater volatility of HCl provides additional driving force.

Finally, it may be mentioned that, in effecting the separation of metal ions, one can take advantage of halide complex formation equilibria in conjunction with anion-exchange resins. To take an extreme example, Co^{2+} and Ni^{2+}, which are not easily separated by classical methods, can be efficiently separated by passing a strong hydrochloric acid solution through an anion-exchange column. Co^{2+} forms the anionic complexes $CoCl_3^-$ and $CoCl_4^{2-}$ rather readily, whereas it does not seem that any anionic chloro complex of nickel is formed in aqueous solution even at the highest attainable activities of Cl^-; however, tetrachloronickelates can be obtained in fused salt systems or in non-aqueous media. More commonly, effective separation depends on properly exploiting the *difference* in complex-formation between two cations *both* of which have some tendency to form anionic halide complexes.

OXIDES, OXO ACIDS AND THEIR SALTS

16-12. Oxides

The oxides of chlorine, bromine and iodine are listed in Table 16-2. Oxides of *fluorine* were discussed in Chapter 14 (page 412); while these were called oxygen fluorides because of the greater electronegativity of fluorine,

---

[12] J. L. Ryan, *Inorg. Chem.*, 1969, **8**, 2053, 2058.

TABLE 16-2

Oxides of the Halogens

Fluorine[a]	B.p. (°C)	M.p. (°C)	Chlorine	B.p. (°C)	M.p. (°C)	Bromine[b]	B.p. (°C)	Iodine[b]
F_2O	−145	−224	Cl_2O	∼4[c]	−116	Br_2O	−18	I_2O_4
			Cl_2O_3[d]					
F_2O_2	−57	−163	ClO_2	∼10[c]	−5.9	Br_3O_8 or BrO_3		I_4O_9
			Cl_2O_4	44.5	−117			
			Cl_2O_6		3.5	BrO_2		I_2O_5
			Cl_2O_7	82[c]	−91.5	Br_2O_7(?)		I_2O_7

[a] See page 412.   [b] Decompose on heating.   [c] Explodes.   [d] Explodes below 0°.

those of the remaining halogens are conventionally and properly called *halogen oxides* since oxygen is the more electronegative element, although not by a very great margin relative to chlorine. All the oxides may be formally considered as anhydrides or mixed anhydrides of the appropriate oxo acids, but this aspect of their chemistry is of little practical consequence. For the most part they are neither common nor especially important.

Chlorine Oxides.   All the chlorine oxides are highly reactive and unstable, tending to explode under various conditions. Probably the best characterized is *chlorine monoxide*, Cl_2O. It is a yellowish-red gas at room temperature. It explodes rather easily to Cl_2 and O_2 when heated or sparked. It dissolves in water, forming an orange-yellow solution that contains some HOCl, of which it is formally the anhydride; hypochlorite is formed in alkali. The molecule is angular (111°) and symmetrical with Cl—O = 1.71 Å.[13] It is prepared by treating freshly prepared mercuric oxide with chlorine gas or with a solution of chlorine in carbon tetrachloride:

$$2Cl_2 + 2HgO \rightarrow HgCl_2 \cdot HgO + Cl_2O$$

Chlorine dioxide is also highly reactive and is liable to explode very violently;[14] apparently mixtures with air containing less than ∼ 50 mm partial pressure of ClO_2 are safe. ClO_2 is useful as a very active oxidizing agent in certain processes and is made on a fairly large scale, but it is always produced where and as required. The best preparation is the reduction of $KClO_3$ by moist oxalic acid at 90°, since the CO_2 liberated also serves as a diluent for the ClO_2. Commercially, the gas is made by the exothermic reaction of sodium chlorate in 4–4.5M sulfuric acid containing 0.05–0.25M chloride ion with sulfur dioxide:

$$2NaClO_3 + SO_2 + H_2SO_4 \rightarrow 2ClO_2 + 2NaHSO_4$$

ClO_2 is a yellowish gas at room temperature. The molecule is angular (118°) with Cl—O = 1.47 Å.[15] Although ClO_2 is an odd molecule, it has no marked tendency to dimerize, perhaps because the electron is more effectively

[13] B. Beagley, A. H. Clark and T. G. Hewitt, *J. Chem. Soc.*, A, **1968**, 658.
[14] E. T. McHale and G. von Elbe, *J. Amer. Chem. Soc.*, 1967, **89**, 2795; *J. Chem. Phys.*, 1968, **72**, 1849.
[15] A. H. Clark and B. Beagley, *J. Chem. Soc.*, A, **1970**, 46.

delocalized than in other odd molecules such as NO_2. It is soluble in water and solutions with up to 8 g l^{-1} are stable in the dark, but in light decompose slowly to HCl and $HClO_3$. In alkaline solution a mixture of chlorite and chlorate ions is formed fairly rapidly. Acid solutions are much more stable,[16] but reduction to $HClO_2$ occurs first, followed by decomposition to $HCl+ HClO_3$. The photolysis of ClO_2 at low temperature affords what is believed to be Cl_2O_3 as a dark brown solid that explodes readily.[14]

Dichlorine tetroxide, or chlorine perchlorate, $ClOClO_3$, is obtained by the reaction

$$CsClO_4 + ClSO_3F \rightarrow CsSO_3F + ClOClO_3$$

It is stable for short periods at room temperature and is more stable than its analogs $FOClO_3$ and $BrOClO_3$.[17]

Dichlorine hexoxide, Cl_2O_6, is obtained as a red oily liquid by the action of ozone on ClO_2. It is unstable, decomposing even at its melting point into ClO_2 and O_2, and it reacts explosively with organic matter and other reducing agents. With water or alkalis a mixture of chlorate and perchlorate ions is formed. Its structure is unknown.

Dichlorine heptoxide is the most stable of the chlorine oxides. It is an oily liquid formed by dehydration of perchloric acid with P_2O_5 at $-10°$, followed by vacuum-distillation, with precautions against explosions. It reacts with water and OH^- to regenerate perchlorate ion. Electron-diffraction shows the structure $O_3ClOClO_3$ with a ClOCl angle of $118.6°$.[17a]

Bromine Oxides.   The bromine oxides are all of very low thermal stability. Br_2O, a dark-brown liquid, decomposes at an appreciable rate above $-50°$. Br_3O_8 (also claimed to be BrO_3) is a white solid unstable above $-80°$ except in an atmosphere of ozone. BrO_2 is a yellow solid unstable from about $-40°$; under certain conditions it decomposes in a vacuum, evolving Br_2O, to a white solid that may be Br_2O_7.

Iodine Oxides.   Of these, white crystalline iodine pentoxide is the most important and is made by the reaction

$$2HIO_3 \xrightleftharpoons{240°} I_2O_5 + H_2O$$

It has IO_3 pyramids sharing one oxygen to give O_2IOIO_2 units, but quite strong intermolecular I$\cdots$O interactions lead to a three-dimensional network.[17b] This compound is stable up to about $300°$ where it melts with decomposition to iodine and oxygen. It is the anhydride of iodic acid and reacts immediately with water. It reacts as an oxidizing agent with various substances such as H_2S, HCl and CO. One of its important uses is as a reagent for the determination of CO, the iodine which is produced quantitatively being then determined by standard iodometric procedures:

$$5CO + I_2O_5 \rightarrow I_2 + 5CO_2$$

The other oxides of iodine, I_2O_4, I_4O_9 and I_2O_7 are of less certain nature. They decompose when heated at $\sim 100°$ to I_2O_5 and iodine, or to iodine and

[16] R. K. Murmann and R. C. Thompson, *J. Inorg. Nuclear Chem.*, 1970, **32**, 1405.
[17] C. J. Schack *et al.*, *Inorg. Chem.*, 1970, **9**, 1387; 1971, **10**, 1278, 1589.
[17a] J. D. Witt and R. M. Hammaker, *Chem. Comm.*, **1970**, 667.
[17b] K. Selte and A. Kjekshus, *Acta Chem. Scand.*, 1970, **24**, 1912.

oxygen. The yellow solid I_2O_4, which is obtained by partial hydrolysis of $(IO)_2SO_4$ (discussed below), appears to have a network built up of polymeric I—O chains that are cross-linked by IO_3 groups. I_4O_9, which can be made by treating I_2 with ozonized oxygen, can be regarded as $I(IO_3)_3$, similarly cross-linked. I_2O_7 has been obtained as an orange polymeric solid by the action of 65% oleum on HIO_4, despite the fact that, in 100% H_2SO_4, HIO_4 decomposes to O_2, O_3 and $H_2IO_3^+$.

16-13. Oxo Acids

The known oxo acids of the halogens are listed in Table 16-3. The chemistry of these acids and their salts is very complicated. Solutions of all the acids and of several of the anions can be obtained by reaction of the free halogens with water or aqueous bases. We shall discuss these reactions first. In the following, the term halogen refers to chlorine, bromine, and iodine only.

TABLE 16-3

Oxo Acids of the Halogens

Fluorine	Chlorine	Bromine	Iodine
FOH	ClOH[a]	BrOH[a]	IOH[a]
	$HClO_2$[a]	$HBrO_2$(?)[a]	—
	$HClO_3$[a]	$HBrO_3$[a]	HIO_3
	$HClO_4$	$HBrO_4$[a]	HIO_4, H_5IO_6, $H_4I_2O_9$

[a] Not obtainable in pure state.

Reaction of Halogens with H_2O and OH^-. A considerable degree of order can be found in this area if full and proper use is made of thermodynamic data in the form of oxidation potentials and equilibrium constants and if the relative rates of competing reactions are also considered. The basic thermodynamic data are given in Table 16-4. From these, all necessary potentials and equilibrium constants can be derived.

TABLE 16-4

Standard Potentials (in Volts) for Reactions of the Halogens

Reaction	Cl	Br	I
(1) $H^+ + HOX + e = \frac{1}{2}X_2(g,l,s) + H_2O$	1.63	1.59	1.45
(2) $3H^+ + HXO_2 + 3e = \frac{1}{2}X_2(g,l,s) + 2H_2O$	1.64	—	—
(3) $6H^+ + XO_3^- + 5e = \frac{1}{2}X_2(g,l,s) + 3H_2O$	1.47	1.52	1.20
(4) $8H^+ + XO_4^- + 7e = \frac{1}{2}X_2(g,l,s) + 4H_2O$	1.42	1.59[b]	1.34
(5) $\frac{1}{2}X_2(g,l,s) + e = X^-$	1.36	1.07	0.54[a]
(6) $XO^- + H_2O + 2e = X^- + 2OH^-$	0.89	0.76	0.49
(7) $XO_2^- + 2H_2O + 4e = X^- + 4OH^-$	0.78	—	—
(8) $XO_3^- + 3H_2O + 6e = X^- + 6OH^-$	0.63	0.61	0.26
(9) $XO_4^- + 4H_2O + 8e = X^- + 8OH^-$	0.56	0.69[b]	0.39

[a] Indicates that I^- can be oxidized by oxygen in aqueous solution.
[b] Calc. from data of G. K. Johnson et al., Inorg. Chem., 1970, 9, 119.

The halogens are all to some extent soluble in water. However, in all such solutions there are species other than solvated halogen molecules, since a disproportionation reaction occurs *rapidly*. Two equilibria serve to define the nature of the solution:

$$X_2(g,l,s) = X_2(aq) \qquad\qquad K_1$$
$$X_2(aq) = H^+ + X^- + HOX \qquad K_2$$

The values of K_1 for the various halogens are: Cl_2, 0.062; Br_2, 0.21; I_2, 0.0013. The values of K_2 can be computed from the potentials in Table 16-4 to be 4.2×10^{-4} for Cl_2, 7.2×10^{-9} for Br_2 and 2.0×10^{-13} for I_2. We can also estimate from

$$\tfrac{1}{2}X_2 + e = X^-$$

and

$$O_2 + 4H^+ + 4e = 2H_2O \qquad E^0 = 1.23$$

that the potentials for the reactions

$$2H^+ + 2X^- + \tfrac{1}{2}O_2 = X_2 + H_2O$$

are -1.62 for fluorine, -0.13 for chlorine, 0.16 for bromine and 0.69 for iodine.

Thus for saturated solutions of the halogens in water at 25° we have the results shown in Table 16-5. There is an appreciable concentration of hypochlorous acid in a saturated aqueous solution of chlorine, a smaller concentration of HOBr in a saturated solution of Br_2, but only a negligible concentration of HOI in a saturated solution of iodine.

TABLE 16-5

Equilibrium Concentrations in Aqueous Solutions of the Halogens, 25°, mol l^{-1}

	Cl_2	Br_2	I_2
Total solubility	0.091	0.21	0.0013
Concentration X_2(aq), mol l^{-1}	0.061	0.21	0.0013
$[H^+] = [X^-] = [HOX]$	0.030	1.15×10^{-3}	6.4×10^{-6}

Hypohalous acids. The colorless compound FOH, m.p. $-117°$, is thermally unstable with a half-life less than an hour at 25°. It is made by passing F_2 over ice and collecting the gas in a trap.[18] It reacts rapidly with water. The other XOH compounds are also unstable and cannot be obtained pure. In water their dissociation constants are: HOCl, 3.4×10^{-8}; HOBr, 2×10^{-9}, HOI, 1×10^{-11}. As can be readily seen, reaction of halogens with water does not constitute a suitable method for preparing aqueous solutions of the hypohalous acids owing to the unfavorable equilibria. A useful general method is interaction of the halogen and a well-agitated suspension of mercuric oxide:

$$2X_2 + 2HgO + H_2O \rightarrow HgO \cdot HgX_2 + 2HOX$$

The similar reaction of HgO and I_2 can be used to form organic hypoiodates from alcohols; these can be used as oxidants.[18a] The hypohalous acids are good oxidizing agents, especially in acid solution (see Table 16-4).

[18] H. M. Studier and E. H. Appelman, *J. Amer. Chem. Soc.*, 1971, **93**, 2349.
[18a] A. Goosen and H. A. H. Lane, *J. Chem. Soc.*, B, **1969**, 995.

In the vapor phase, HOCl is formed in the equilibrium

$$H_2O(g) + Cl_2O(g) \rightleftharpoons 2HOCl(g)$$

A microwave study[19] of the vapor yields the parameters H—O = 0.97 Å, O—Cl = 1.69 Å, $\angle$ HOCl = $103 \pm 3°$.

The hypohalite ions can all be produced in principle by dissolving the halogens in base according to the general reaction

$$X_2 + 2OH^- \rightarrow X^- + XO^- + H_2O$$

and for these reactions the equilibrium constants are all favorable—7.5×10^{15} for Cl_2, 2×10^8 for Br_2 and 30 for I_2—and the reactions are rapid.

However, the situation is complicated by the tendency of the hypohalite ions to disproportionate further in basic solution to produce the halate ions:

$$3XO^- = 2X^- + XO_3^-$$

For this reaction, the equilibrium constant is in each case very favorable, that is, 10^{27} for ClO^-, 10^{15} for BrO^- and 10^{20} for IO^-. Thus the actual products obtained on dissolving the halogens in base depend on the rates at which the hypohalite ions initially produced undergo disproportionation, and these rates vary from one to the other and with temperature.

The disproportionation of ClO^- is slow at and below room temperature. Thus, when chlorine reacts with base "in the cold," reasonably pure solutions of Cl^- and ClO^- are obtained. In hot solutions, $\sim 75°$, the rate of disproportionation is fairly rapid and, under proper conditions, good yields of ClO_3^- can be secured.

The disproportionation of BrO^- is moderately fast even at room temperature. Consequently solutions of BrO^- can only be made and/or kept at around 0°. At temperatures of 50–80° quantitative yields of BrO_3^- are obtained:

$$3Br_2 + 6OH^- \rightarrow 5Br^- + BrO_3^- + 3H_2O$$

The rate of disproportionation of IO^- is very fast at all temperatures, so that it is unknown in solution. Reaction of iodine with base gives IO_3^- quantitatively according to an equation analogous to that for Br_2.

It remains now to consider the equilibria of the oxo anions not yet mentioned and their kinetic relations to those we have discussed. *Halite ions* and *halous acids* do not arise in the hydrolysis of the halogens. HIO_2 apparently does not exist, $HBrO_2$ is doubtful, while $HClO_2$ is not formed by disproportionation of ClOH if for no other reason than that the equilibrium constant is quite unfavorable:

$$2ClOH = Cl^- + H^+ + HClO_2 \qquad K \sim 10^{-5}$$

The reaction

$$2ClO^- = Cl^- + ClO_2^- \qquad K \sim 10^7$$

is favorable, but the disproportionation of ClO^- to ClO_3^- and Cl^- (see above) is so much more favorable that the first reaction is not observed.

Finally, we must consider the possibility of production of *perhalate ions*

[19] D. C. Lindsey, D. G. Lister and D. J. Millen, *Chem. Comm.*, **1969**, 950.

by disproportionation of the halate ions. Since the acids HXO_3 and HXO_4 are all strong, these equilibria are independent of pH. The reaction

$$4ClO_3^- = Cl^- + 3ClO_4^-$$

has an equilibrium constant of 10^{29}, but it takes place only very slowly in solution even near $100°$; hence perchlorates are not readily produced. Neither perbromate nor periodate can be obtained in comparable disproportionation reactions because the equilibrium constants are 10^{-33} and 10^{-53}, respectively.

The only definitely known halous acid, *chlorous acid*, is obtained in aqueous solution by treating a suspension of barium chlorite with sulfuric acid and filtering off the precipitate of barium sulfate. It is a relatively weak acid ($K_a \approx 10^{-2}$) and cannot be isolated in the free state. *Chlorites*, $MClO_2$, themselves are obtained by reaction of ClO_2 with solutions of bases:

$$2ClO_2 + 2OH^- \rightarrow ClO_2^- + ClO_3^- + H_2O$$

Chlorites are used as bleaching agents. In alkaline solution the ion is stable to prolonged boiling and up to a year at $25°$ in absence of light. In acid solutions, however, the decomposition is rapid and is catalysed by Cl^-:

$$5HClO_2 \rightarrow 4ClO_2 + Cl^- + H^+ + 2H_2O$$

but the reaction sequence is complicated.[20]

Halic Acids. Only *iodic acid* is known in the free state. This is a stable white solid obtained by oxidizing iodine with concentrated nitric acid, hydrogen peroxide, ozone or various other strong oxidizing agents. It can be dehydrated to its anhydride, I_2O_5, as already noted. Salts such as KHI_2O_6 exist in the solid state, probably owing to favorable lattice energies and low solubility. In aqueous solutions the predominant species is IO_3^-.[21] *Chloric* and *bromic acids* are best obtained in solution by treating the barium halates with sulfuric acid.

All the halic acids are strong acids and are powerful oxidizing agents; the mechanisms of reduction, e.g., by I^- are very complicated.[21a] The halate ions, XO_3^-, are all pyramidal, as is to be expected from the presence of an octet, with one unshared pair, in the halogen valence shell.

Iodates of certain $+4$ metal ions—notably those of Ce, Zr, Hf and Th—can be precipitated from $6M$ nitric acid to provide a useful means of separation.

Perchlorates. Although disproportionation of ClO_3^- to ClO_4^- and Cl^- is thermodynamically very favorable, the reaction occurs only very slowly in solution and does not constitute a useful preparative procedure. Perchlorates are commonly prepared by electrolytic oxidation of chlorates. The properties of perchloric acid have been discussed on page 180.

Perchlorates of practically all electropositive metals are known. Except

[20] R. G. Keefer and G. Gordon, *Inorg. Chem.*, 1968, **7**, 239.
[21] R. A. Direk *et al.*, *J. Amer. Chem. Soc.*, 1971, **93**, 77.
[21a] See, e.g., A. F. M. Barton and G. A. Wright, *J. Chem. Soc., A*, **1968**, 1747, 2096.

for a few with large cations of low charge, such as $CsClO_4$, $RbClO_4$ and $KClO_4$, they are readily soluble in water. Solid perchlorates containing the tetrahedral ClO_4^- ion are often isomorphous with salts of other tetrahedral anions, e.g., MnO_4^-, SO_4^{2-}, BF_4^-. A particularly important property of the perchlorate ion is its slight tendency to serve as a ligand in complexes. Thus perchlorates are widely used in studies of complex ion formation, the *assumption* being made that no appreciable correction for the concentration of perchlorate complexes need be considered. While this may often be true for aqueous solutions, there are indications that, when no other donor is present to compete, perchlorate ion exercises a donor capacity. The structures of $(CH_3)_3SnClO_4$ and $[Co(MeSC_2H_4SMe)_2(ClO_4)_2]$ provide particularly convincing instances of this, but in certain cases, even in aqueous solutions, e.g., of $Copy_4(ClO_4)_2$, some complexing is possible (see Section 21-8).[22] The use of perchlorate as an ion for the isolation of crystalline salts of organometallic ions such as $(h^5\text{-}C_5H_5)_2Fe^+$ is to be avoided since such salts are often dangerously explosive: the use of the trifluoromethanesulfonate ion, $CF_3SO_3^-$, PF_6^{2-}, or of BF_4^- which behave very much like ClO_4^-, is preferable, but even BF_4^- and PF_6^{2-} may act as ligands.[22a] Although ClO_4^- is potentially a good oxidant:

$$ClO_4^- + 2H^+ + 2e = ClO_3^- + H_2O \qquad E^0 = 1.23 \text{ V}$$

in aqueous solution it is reduced only by Ru^{II}, V^{II}, V^{III} and Ti^{III}.[23] Despite the more favorable potential for reduction by Eu^{2+} or Cr^{2+} no reaction occurs, for reasons that are not entirely clear.

Perbromic Acid and Perbromates. Perbromates have only recently been prepared and previously there have been many papers justifying theoretically their non-existence.[24] This provides an excellent example of the folly of concluding the non-existence of certain compounds until all conceivable preparative methods have been exhausted.

The potential

$$BrO_4^- + 2H^+ + 2e = BrO_3^- + H_2O \qquad E^0 = +1.76 \text{ V}$$

shows that only the strongest oxidants can form perbromate.[24a] The failure of ozone ($E = +2.07$ V) and $S_2O_8^{2-}$ ($E = +2.01$ V) to cause oxidation is probably due to kinetic reasons. Perbromate is a stronger oxidant than ClO_4^- (1.23 V) or IO_4^- (1.64 V); there is no convincing explanation of this anomaly, and similar instability of the high oxidation state occurs in other first longperiod elements, e.g., for Se in the S, Se, Te group.

Small amounts of perbromic acid or perbromates can be obtained by oxidation of BrO_3^- electrolytically or by the action of XeF_2. The best

[22] W. C. Jones and W. E. Bull, *J. Chem. Soc., A*, **1968**, 1849.
[22a] H. G. Mayfield and W. E. Bull, *J. Chem. Soc., A*, **1971**, 2279; P. Legzdins *et al.*, *J. Chem. Soc., A*, **1970**, 3322.
[23] K. W. Kallen and J. E. Early, *Inorg. Chem.*, **1971**, **10**, 1152; E. Bishop and N. Adams, *Talanta*, 1970, **17**, 1125.
[24] For references see M. M. Cox and J. W. Moore, *J. Phys. Chem.*, 1970, **74**, 627.
[24a] G. K. Johnson *et al.*, *Inorg. Chem.*, 1970, **9**, 119.

preparation involves oxidation of BrO_3^- by fluorine in $5M$ NaOH solution; by a rather complicated procedure, pure solutions can be obtained:[25]

$$BrO_3^- + F_2 + 2OH^- \rightarrow BrO_4^- + 2F^- + H_2O$$

Solutions of $HBrO_4$ can be concentrated up to $6M$ (55%) without decomposition and are stable indefinitely even at 100°. More concentrated solutions, up to 83%, can be obtained but these are unstable; the hydrate, $HBrO_4 \cdot 2H_2O$, can be crystallized. The ion is tetrahedral[26] with Br—O = 1.61 Å (cf. Cl—O in ClO_4^-, 1.45 Å; and I—O in IO_4^-, 1.79 Å).

In dilute solution, perbromate is a sluggish oxidant at 25° and is slowly reduced by I^- or Br^- but not by Cl^-. However, the $3M$ acid readily oxidizes stainless steel, and the $12M$ acid rapidly oxidizes Cl^- and will explode in contact with tissue paper. Above $6M$ the solutions are erratically, but not explosively, unstable.

Pure potassium perbromate is stable up to 275°, where it decomposes to $KBrO_3$ and even NH_4BrO_4 is stable[26a] to around 170°.

Periodic Acid and Periodates. Periodic acid exists in solution as the tetrahedral ion, IO_4^-, as well as in several hydrated forms. The complexity of the periodates is similar to that found for the oxo acids of Sb and Te, and periodates often resemble tellurates in their stoichiometries.

The main equilibria in acid solutions are:

$$
\begin{aligned}
H_5IO_6 &= H^+ + H_4IO_6^- & K &= 1 \times 10^{-3} \\
H_4IO_6^- &= IO_4^- + 2H_2O & K &= 29 \\
H_4IO_6^- &= H^+ + H_3IO_6^{2-} & K &= 2 \times 10^{-7}
\end{aligned}
$$

In aqueous solutions at 25° the periodate ion, IO_4^-, predominates; the hydrated species are termed orthoperiodates. There is evidence that in very concentrated acids, e.g., $10M$ $HClO_4$, the ion $I(OH)_6^+$ is formed. The various pH-dependent equilibria are established rapidly; kinetic studies of the hydration of IO_4^- suggest either one-step or two-step paths (Fig. 16-2), the latter being more likely.

Salts of periodic acid are of several types but in general there is an IO_6 octahedron present in the structure. The commonest are the acid salts such as $NaH_4IO_6 \cdot H_2O$, $Na_2H_3IO_6$ and $Na_3H_2IO_6$; on addition of CsOH to H_5IO_6 solutions, however, the periodate $CsIO_4$ is precipitated.[27] The free acid H_5IO_6 can be dehydrated to $H_4I_2O_9$ at 80° and to HIO_4 at 100°.

In alkaline solution periodate dimerizes:[27,28]

$$2IO_4^- + 2OH^- \rightarrow H_2I_2O_{10}^{4-}$$

The dimeric ion in solution is presumably the same as that found as a discrete ion $[O_3(OH)IO_2IO_3(OH)]^{4-}$ in the crystalline salt $K_4H_2I_2O_{10} \cdot 8H_2O$; the ion has two IO_6 octahedra sharing one edge. The Cs^+ salt, which can be

[25] E. H. Appelman, *Inorg. Chem.*, 1969, **8**, 223.
[26] S. Siegel, B. Tain and E. H. Appelman, *Inorg. Chem.*, 1969, **8**, 1190.
[26a] J. N. Keith and I. J. Solomon, *Inorg. Chem.* 1970, **9**, 1560.
[27] R. M. Kress, H. W. Dogden and C. J. Nyman, *Inorg. Chem.*, 1968, **7**, 446.
[28] J. Aveston, *J. Chem. Soc., A*, **1969**, 273; G. J. Buist, W. C. Hipperson and J. D. Lewis, *J. Chem. Soc., A*, **1969**, 307.

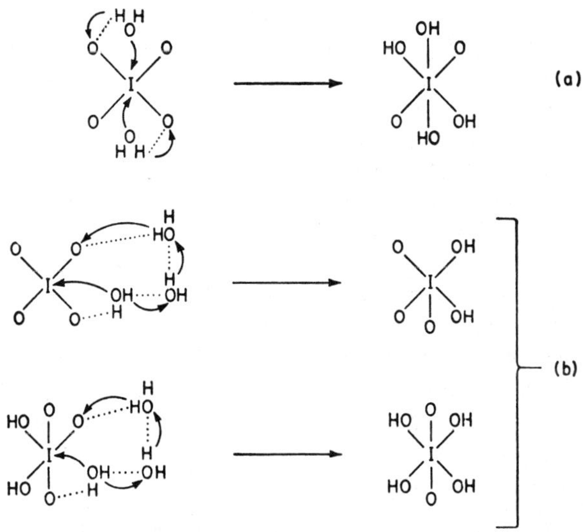

Fig. 16-2. Schematic representation of (a) the one-step and (b) the two-step mechanism for aquation of IO_4^- to $IO_2(OH)_4^-$. Dotted lines represent hydrogen bonds.

made by heating $CsIO_4$ with an excess of $CsOH$, is dehydrated to $Cs_4I_2O_9$ at $60°$.[27]

The chief characteristic of periodic acids is that they are powerful oxidizing agents that usually react smoothly and rapidly. They are thus useful for analytical purposes, for example, to oxidize manganous ion to permanganate. Ozone (derived from O atoms) may be liberated in the reactions, but not hydrogen peroxide.

Periodic acid or its salts are also commonly used in organic chemistry.[29]

INTERHALOGEN COMPOUNDS AND IONS

16-14. General Survey

The halogens form compounds that are binary and ternary combinations of themselves. With the exception of BrCl, ICl, ICl_3 and IBr, the compounds are all halogen fluorides (Table 16-6) (below) such as ClF, BrF_3, IF_5 and IF_7. Ternary compounds occur only as polyhalide ions, the principal types of which are listed in Tables 16-7 and 16-8 (page 485).

All the interhalogen compounds are of the type XX_n' where n is an odd number and X' is always the lighter halogen when n is greater than 1. Because n is always odd it follows that all interhalogen compounds are diamagnetic, having all valence electrons present either as shared (bonding) or unshared pairs. The general scarcity and instability of odd molecules

[29] B. Sklarz, *Quart. Rev.*, 1967, **21**, 3; G. Dryhurst, *Periodate Oxidations of Diols and Other Functional Groups*, Pergamon Press, 1970.

makes this seem reasonable, and it is to be expected that any further inter-halogens, if discovered, will also contain an even number of atoms. No ternary interhalogen compounds are known, although attempts have been made to prepare them. This is probably because any ternary molecules formed can readily redistribute to form a mixture of the (presumably) more stable binary compounds and/or elemental halogens. Another general observation is that stability of the compounds with higher n increases as X becomes larger and X' smaller.

The structures of the interhalogen compounds are all known, to varying degrees of accuracy. These structures and the reasons for them in terms of the electronic configuration of the molecules[30] are discussed in Chapter 4.

Chemically, the interhalogens are all rather reactive. They are corrosive oxidizing substances and attack most other elements, producing mixtures of the halides. They are all more or less readily hydrolyzed (some such as BrF_3 being dangerously explosive in this respect), in some cases according to the equation

$$XX' + H_2O = H^+ + X'^- + HOX$$

The diatomic compounds often add to ethylenic double bonds and may react with the heavier alkali and alkaline-earth metals to give polyhalide salts.

The *diatomic compounds* are ClF, BrF, BrCl, IBr and ICl. In their physical properties, they are usually intermediate between the constituent halogens. They are of course polar, whereas the halogen molecules are not. ClF is colorless; BrF, BrCl, ICl and IBr are red or reddish-brown. IF is unknown except in minute amounts observed spectroscopically: it is apparently too unstable with respect to disproportionation to IF_5 and I_2 to permit its isolation. The other isolable diatomic compounds have varying degrees of stability with respect to disproportionation and fall in the following stability order where the numbers in parentheses represent the disproportionation constants for the gaseous compounds and the elements in their standard states at 25°: ClF $(2.9 \times 10^{-11}) >$ ICl $(1.8 \times 10^{-3}) >$ IBr $(4.9 \times 10^{-2}) >$ BrCl (0.34). BrF is omitted since it is extremely unstable and its characteristic disproportionation is to Br_2 and BrF_3: this is due, not to any particular weakness of the BrF bond (210 kJ mol^{-1}), but to the even greater stabilities of the products of disproportionation.

ClF may be prepared by direct interaction at 220–250° and it is readily freed from ClF_3 by distillation, but it is best prepared by interaction of Cl_2 and ClF_3 at 250–350°. BrF also results on direct reaction of Br_2 with F_2, but it has never been obtained in high purity because of its ready dispro-portionation. Iodine monochloride is obtained as brownish-red tablets (β-form) by treating liquid chlorine with solid iodine in stoichiometric amount, and cooling to solidify the liquid product. It readily transforms to the α-form, ruby-red needles. BrCl is unstable:[30a]

$$2BrCl \rightleftharpoons Br_2 + Cl_2 \qquad K = 0.145 \ (25° \ in \ CCl_4)$$

[30] B. M. Deb and C. A. Coulson, *J. Chem. Soc.*, A, **1971**, 958.
[30a] T. Surles and A. I. Popov, *Inorg. Chem.*, 1969, **8**, 2049.

IBr as a solid results from direct combination; it is endothermic and extensively dissociated in the vapor. Despite the instability of the BrX compounds, the fluorosulfonate $BrOSO_2F$, obtained[30b] by treating Br_2 with $S_2O_6F_2$, is stable to 150°.

Iodine trichloride, ICl_3, is also formed (like ICl) by treatment of liquid chlorine with the stoichiometric quantity of iodine, or with a deficiency of iodine followed by evaporation of the excess of chlorine. It is a fluffy orange powder, unstable much above room temperature.

The most important and most intensively studied compounds are the halogen fluorides.

16-15. The Halogen Fluorides[31]

These compounds and some of their important physical properties are listed in Table 16-6.

TABLE 16-6

Some Physical Properties of Halogen Fluorides

	M.p. (°C)	B.p. (°C)	Specific conductivity[a] at 25 °C (ohm^{-1}cm^{-1})	Structure
ClF	−156.6	−100.1	—	
ClF$_3$	−76.3	11.75	3.9×10^{-9}	Planar; distorted "T"
ClF$_5$	−103	−14	—	Square-pyramidal
BrF	−33	20	—	
BrF$_3$	9	126	$>8.0 \times 10^{-3}$	Planar; distorted "T"
BrF$_5$	−60	41	9.1×10^{-8}	Square-pyramidal
IF$_3$[b]	—	—	—	
IF$_5$	10	101	5.4×10^{-6}	Square-pyramidal
IF$_7$	6.45[c]	—	—	Pentagonal-bipyramidal

[a] Values in the literature may be very inaccurate in view of the possibility of hydrolysis by traces of water.
[b] Alleged to be a yellow powder obtained by fluorination of I_2 in Freon at −78° and to decompose to I_2 and IF_5 above −35°.
[c] Triple point; sublimes 4.77° at 1 atm.

The preparations of ClF and BrF have already been mentioned. ClF_3 may be prepared by direct combination of the elements at 200–300° and is available commercially. It is purified by converting it into $KClF_4$ by the action of KF and thermally decomposing the salt at 130–150°. ClF_5 can be made by interaction of F_2 and ClF_3 above 200° but is best prepared by the reaction:

$$KCl + 3F_2 \xrightarrow[\text{bomb}]{200°} KF + ClF_5$$

ClF_5 is a colorless gas; it is somewhat less stable than ClF_3 and above 165° there is the equilibrium[32]

$$ClF_5 \rightleftharpoons ClF_3 + F_2$$

[30b] A. M. Qureshi and F. Aubke, *Inorg. Chem.*, 1971, **10**, 1116.
[31] L. Stein in *Halogen Chemistry*, V. Gutmann, ed., Vol. 1, Academic Press, 1967.
[32] G. A. Hyde and M. M. Bondakian, *Inorg. Chem.*, 1968, 7, 2648.

The other halides are best prepared by the reactions:

$$Br_2 + 3F_2 \xrightarrow{200^\circ} 2BrF_3$$
$$BrF_3 + F_2 \longrightarrow BrF_5$$
$$I_2 + 5F_2 \xrightarrow{25^\circ} 2IF_5$$
$$PdI_2 + 8F_2 \xrightarrow{200^\circ} PdF_2 + 2IF_7$$

The halogen fluorides are very reactive and with water or organic substances react vigorously or explosively. They are powerful fluorinating agents for inorganic compounds or, when diluted with nitrogen, for organic compounds. The most useful compounds are ClF, ClF_3, and BrF_3.[33] Although only qualitative data are usually available, the order of reactivity is approximately $ClF_3 > BrF_5 > IF_7 > ClF > BrF_3 > IF_5 > BrF$.

Certain compounds, notably ClF, BrF_3 and IF_5 have high entropies of vaporization, suggesting some association in the liquid state; BrF_3 and IF_5 also show appreciable electrical conductance. To account for these observations, association by fluorine bridging in addition to self-dissociation,

$$2BrF_3 \rightleftharpoons BrF_2^+ + BrF_4^-$$
$$2IF_5 \rightleftharpoons IF_4^+ + IF_6^-$$

have been postulated. An analogy with other solvent systems can be made—in liquid BrF_3, for example, the "acid" would be BrF_2^+ and the "base" BrF_4^-. Indeed, suitable compounds such as $BrF_3 \cdot SbF_5$ (or $BrF_2^+ \cdot SbF_6^-$) and $KBrF_4$ dissolve in BrF_3 to give highly conducting solutions. In liquid HF, the equilibrium

$$HF + BrF_3 \rightleftharpoons BrF_2^+ + HF_2^-$$

has been conclusively proved,[34] while in ClF_3–BrF_3 mixtures, there is the equilibrium

$$ClF_3 + BrF_3 \rightleftharpoons ClF_2^+ + BrF_4^- \qquad K = \text{ca. } 10^{-4}$$

A substantial number of salts containing *fluorohalogen cations* or *anions* (Table 16-7) have been prepared[34–37] by reaction of the halogen fluoride with Lewis acids such as BF_3, SbF_5, etc., or with Lewis bases such as F^- and NOF. Although some of the products are evidently ionic, it does not follow that they all are, and some of them may best be regarded as adducts of the halogen fluoride acting as a Lewis base. Examples of such reactions are

$$2ClF + AsF_5 \rightarrow FCl_2^+ AsF_6^-$$
$$ClF + CsF \rightarrow Cs^+ ClF_2^-$$
$$ClF_3 + CsF \rightarrow Cs^+ ClF_4^-$$
$$ClF_5 + PtF_6 \rightarrow ClF_4^+ PtF_6^- + \tfrac{1}{2}F_2$$
$$BrF_3 + ClF_3 \rightarrow ClF_2^+ BrF_4^-$$
$$ClF_3 + AsF_5 \rightarrow ClF_2^+ AsF_6^-$$
$$IF_5 + CsF \rightarrow Cs^+ IF_6^-$$

Salts may also be obtained by other reactions such as

$$CsCl + 2F_2 \rightarrow CsClF_4$$

Although X-ray study should answer the problem of ionicity, the results are not always clear cut. Thus, in the SbF_6^- salts of ClF_2^+ and BrF_2^+ the Cl and Br atoms have two close and two distant fluorine neighbors (belonging

[33] T. Surles *et al.*, *Inorg. Chem.*, 1971, **10**, 611, 913.
[34] R. J. Gillespie and M. J. Morton, *Quart. Rev.*, 1971, **25**, 553.
[35] K. O. Christie and W. Sawodny, *Inorg. Chem.*, 1969, **8**, 212.
[36] S. P. Beaton *et al.*, *Inorg. Chem.*, 1969, **8**, 2175.
[37] F. O. Roberts and G. Mamantov, *Inorg. Chim. Acta*, 1968, **2**, 317.

TABLE 16-7

Halogen Fluoro Cations and Anions

Parent	Cation	Anion	Parent	Cation	Anion
ClF	FCl_2^+	ClF_2^-	BrF_5	BrF_4^+	BrF_6^-
ClF_3	ClF_2^+	ClF_4^-	IF_5	IF_4^+	IF_6^-
ClF_5	ClF_4^+	—	IF_7	IF_6^+	—
BrF_3	BrF_2^+	BrF_4^-			

to SbF_6^- ions) in a much distorted square.[38] Hence, at least approximately, the solid can be said to contain XF_2^+ ions. A similar situation arises for the adduct $ICl_3 \cdot SbCl_3$ which also has a distorted square of Cl about I.

The IF_6^- ion, as its Cs^+ salt, appears to have a symmetry lower than octahedral (cf. XeF_6 and discussion page 140) and may be polymeric in the solid[36]. Note that although BrF_6^- is well-defined, there is no evidence[35] for ClF_6^-, suggesting that ClF_5 is coordinatively saturated with one localized lone pair and five F atoms; in agreement with this, ClF_5 appears not to associate.

There is good evidence for some of the anionic species, e.g., for $KBrF_4$ in the crystal and in liquid BrF_3 there are square BrF_4^- ions.[39]

The reaction of ClF_3 or ClF_5 with SbF_5 gives, respectively, the ClF_2^+ or ClF_4^+ salt of SbF_6^- when the reactants are carefully purified, but otherwise unidentified paramagnetic species are formed.[40]

16.16. Other Interhalogen Ions[34,41]

In addition to the fluoro ions discussed above, several other types are known, examples of which are listed in Table 16-8.

TABLE 16-8

Principal Types of Polyhalide Ions[a] (Other than Fluoro Ions)

Cations	Anions[b]		Cations	Anions[b]	
X_n^+	X_3^-	X_5^-	X_n^+	X_3^-	X_5^-
Br_2^+	Cl_3^-		I_5^+	I_2Cl^-	I_5^-
I_2^+	Br_3^-	ICl_4^-	I_4^{2+}	IBr_2^-	
Cl_3^+	I_3^-	$IBrCl_3^-$	ICl_2^+	ICl_2^-	
Br_3^+ [c]	Br_2Cl^-	$I_2Br_2Cl^-$	$I_2Cl^+(?)$	$IBrCl^-$	
I_3^+	$BrCl_2^-$	I_4Cl^-		$IBrF^-$	

[a] Also I_7^-, I_6Br^-, Br_6Cl^-, I_9^-; for fluoro species, see Table 16-7.
[b] Usually as salts of large univalent cations, e.g., Cs^+, Et_4N^+, etc.
[c] As $Sb_3F_{16}^-$ salt (A. J. Edwards and E. R. Jones, *J. Chem. Soc.*, A, **1971**, 2318).

[38] A. J. Edwards and R. J. C. Sills, *J. Chem. Soc.*, A, **1970**, 2697; K. O. Christie and C. J. Schack, *Inorg. Chem.*, 1970, **9**, 2296.
[39] K. O. Christie and C. J. Schack, *Inorg. Chem.*, 1970, **9**, 1852; T. Surles *et al.*, *Inorg. Chem.*, 1970, **9**, 2726.
[40] K. O. Christie and J. S. Muirhead, *J. Amer. Chem. Soc.*, 1969, **91**, 7777.
[41] A. I. Popov in *Halogen Chemistry*, V. Gutman, ed., Vol. I, p. 225, Academic Press, 1967.

Anionic Species. One of the earliest to be recognized is I_3^- whose formation accounts for the increased solubility of I_2 in water on addition of KI. Few of the other ions are stable in aqueous solution, the most stable being I_3^-. In non-aqueous media such as CH_3OH or CH_3CN, stabilities are substantially higher. Many of the ions, especially the larger ones, exist only in crystalline salts with large cations such as Cs^+ or R_4N^+. The ions or salts are usually prepared by direct interaction between X^- and X_2, e.g.,

$$KI + Cl_2 \rightarrow KICl_2 \leftarrow KCl + ICl$$

Of the simple X_3^- ions, Cl_3^- is the least stable but is formed when concentrated solutions of Cl^- are saturated with chlorine:

$$Cl^-(aq) + Cl_2 \rightleftharpoons Cl_3^-(aq) \qquad K \approx 0.2$$

There is slight evidence in the bromine system for Br_3^-, but the series of polyions I_5^-, I_7^- and I_9^- is well-established for iodine.[42]

The X_3^- species are linear but not necessarily symmetrical. In solution, I_3^- appears to be symmetrical, as it is in $Ph_4As^+I_3^-$, but in CsI_3 there are two I—I distances (2.83 and 3.03 Å), while long polymeric chains are present in benzamide hydrogen triiodide. The anions in $CsBr_3$, CsI_2Br and $CsIBr_2$ are also unsymmetrical.[42a]

The I_5^- ion (Fig. 16-3) can best be described as I^- with two I_2 molecules fairly weakly coordinated to it; and I_7^- appears to be similar but with even weaker bonds between I_3^- and two I_2 molecules—whether such an arrangement really constitutes a discrete "ion" is questionable.

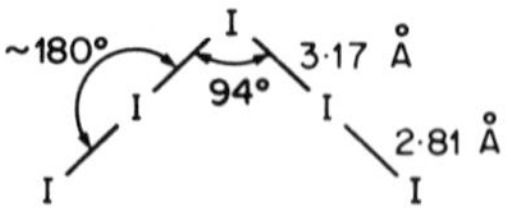

Fig. 16-3.   Structure of the pentaiodide ion, I_5^- in $[Me_4N]I_5$.

Cationic Species. There is no evidence for any ion of the type X^+ but Br_2^+ and I_2^+ are known. The Cl_3^+ and Br_3^+ ions are made by the reactions.[34, 43]

$$ClF + Cl_2 + AsF_5 \rightarrow Cl_3^+ AsF_6^-$$
$$O_2^+ AsF_6^- + \tfrac{3}{2}Br_2 \rightarrow Br_3^+ AsF_6^- + O_2$$

The iodine-containing cations are better established, but they exist only in media with low nucleophilic properties since in more strongly donor solvents the ions are either attacked or are complexed to give species such as $[py_2 I]^+$.

When iodine is dissolved in fluorosulfuric acid and is oxidized by addition of $S_2O_6F_2$, the species $I(SO_3F)_3$, I_2^+, I_3^+ and I_5^+ can be obtained depending on the amount of oxidant added.[44] Thus with equimolar amounts we have:

$$5I_2 + 5S_2O_6F_2 \xrightarrow{\text{FSO}_3\text{H}} 4I_2^+ + 4SO_3F^- + 2I(SO_3F)_3$$

The blue color produced and the spectral properties, now shown to be due to I_2^+, were previously attributed to I^+.

[42] L. E. Topol, *Inorg. Chem.*, 1971, **10**, 736.
[42a] J. E. Davies and E. K. Nunn, *Chem. Comm.*, 1969, 1274.
[43] O. Glemser and A. Sinali, *Angew. Chem. Internat. Edn.*, 1969, **8**, 517.
[44] R. J. Gillespie, J. B. Milne and M. J. Morton, *Inorg. Chem.*, 1968, **7**, 2221.

On cooling, the solution, which is paramagnetic with $\mu_{\text{eff}} = 2.0$ B.M. corresponding to one unpaired electron per I_2^+ ion as expected, becomes diamagnetic and brown. This is attributed to dimerization:

$$2I_2^+ \rightleftharpoons I_4^{2+}$$

The cation as a blue solution and solid has also been obtained by the reaction of I_2 with SbF_5.[45]

Solutions of I_2^+ and I_3^+ can be obtained by oxidation in oleum:

$$2I_2 + 6H_2S_2O_7 \rightarrow 2I_2^+ + 2HS_3O_{10}^- + 5H_2SO_4 + SO_2$$
$$3I_2 + 6H_2S_2O_7 \rightarrow 2I_3^+ + 2HS_3O_{10}^- + 5H_2SO_4 + SO_2$$

while ICl_3 in oleum gives ICl_2^+.[46]

Finally, molten iodine is electrically conducting and this can be ascribed to self-ionization:

$$3I_2 \rightleftharpoons I_3^+ + I_3^-$$

Complexes of Cationic Halogens. Although the I^+ ion appears not to exist, compounds of the types $[I\,py_2]^+X^-$ ($X = NO_3^-$, RCO_2^-, or ClO_4^-), or $I\,py\,X$ where the anion is also coordinated as in $I\,py\,ONO_2$ or $I\,py\,OCOR$, are known.

Such compounds are generally prepared by treatment of a silver salt with the stoichiometric amount of iodine and an excess of pyridine in $CHCl_3$, e.g.,

$$AgNO_3 + I_2 + 2py = [I\,py_2]NO_3 + AgI\downarrow$$

After removal of AgI the complexes can be isolated from the solution. Electrolysis of $[I\,py_2]^+NO_3^-$ in chloroform yields iodine at the cathode; however, compounds of the type $py\,IOCOR$ give only feebly conducting solutions in acetone. Direct interaction of I_2 and pyridine gives a compound that X-ray studies show to contain the planar ion $[py\,I\,py]^+$ along with I_3^- and I_2 molecules. The only other ion of this type is $[(NH_2)_2CS]_2I^+ \cdot I^-$ obtained by grinding iodine and thiourea together.[47]

Similar but less stable complexes of chlorine and bromine are known.

OTHER COMPOUNDS

16-17. Oxo Halogeno Fluorides[48]

The most important compound of this type is *perchloryl fluoride*, ClO_3F. It can be prepared by the action of F_2 or FSO_3H on $KClO_4$, but is best made by the solvolytic reaction of $KClO_4$ with a super acid,[49] e.g., a mixture of HF and SbF_5:

$$KClO_4 + 2HF + SbF_5 \xrightarrow{40-50°} FClO_3 + KSbF_6 + H_2O$$

The toxic gas (m.p. $-147.8°$, b.p. $-46.7°$) is thermally stable to 500° and

[45] R. D. W. Kemmitt et al., J. Chem. Soc., A, **1968**, 682.
[46] R. J. Gillespie and K. C. Malhotra, Inorg. Chem., 1969, **8**, 1751.
[47] H. Hope and G. H. Y. Lin, Chem. Comm., **1970**, 169.
[48] V. M. Khutsoretskii, L. V. Okhlobystina and A. A. Fainzil'berg, Russ. Chem. Rev., 1967, **36**, 145 (a comprehensive review).
[49] C. A. Wamser, W. B. Fox, D. Gould, and B. Sukornick, Inorg. Chem., 1968, **7**, 1933.

resists hydrolysis. At elevated temperatures it is a powerful oxidizing agent and has selective fluorinating properties, especially for replacement of H by F in CH_2 groups.[50] It can also be used to introduce ClO_3 groups into organic compounds, e.g., C_6H_5Li gives $C_6H_5ClO_3$. In these reactions it appears that the nucleophile attacks the chorine atom,[51] e.g.,

$$RO^- + FClO_3 \longrightarrow \begin{matrix} O \\ O \end{matrix} \overset{\overset{F}{|}}{\underset{\underset{O-R}{|}}{Cl}} - O \longrightarrow ROClO_3 + F^-$$

Perbromyl fluoride,[52] m.p. $-110°$, is made similarly, but it is more reactive than ClO_3F and is hydrolyzed by base:

$$BrO_3F + 2OH^- = BrO_4^- + H_2O + F^-$$

presumably by initial associative attack of OH^- on Br.

Other compounds are the explosive ClO_2F that is formed by hydrolysis of chlorine fluorides and reacts with Lewis bases to give salts of the ClO_2^+ ion,[53] $ClOF_3$ and IO_2F_3. The latter is made by the reactions

$$Ba_3H_4(IO_6)_2 + 14HSO_3F \rightarrow 2HOIOF_4 + 8H_2SO_4 + 3Ba(SO_3F)_2$$
$$HOIOF_4 + SO_3 \rightarrow IO_2F_3 + FSO_3H$$

The white solid (m.p. 41°) appears to exist as two trigonal-bipyramidal isomers in solutions according to ^{19}F-nmr spectra.[54]

16-18. Compounds of Iodine(III)

A large number of these compounds are known, very many containing organic groups.[55] The only bromine compounds are $K[X(OSO_2F)_4]$ and $X(OSO_2F)_3$, where $X = Br$ or I.[55a] If $AgClO_4$ is added to an excess of I_2 in ether at $-85°$ a yellow solution containing $I(ClO_4)_3$ is obtained:

$$I_2 + 2AgClO_4 \rightarrow AgI + AgClO_4 \cdot IClO_4$$
$$AgClO_4 \cdot IClO_4 + IClO_4 \rightarrow AgI + I(ClO)_3$$

Fuming nitric acid in the presence of acetic anhydride oxidizes iodine, producing the compound $I(OCOCH_3)_3$. On oxidation of iodine with concentrated nitric acid in the presence of acetic anhydride and phosphoric acid, the compound IPO_4 is obtained. No direct structural information is available for these compounds, but they are probably best regarded as covalent. Nevertheless, electrolytic dissociation can occur and when a saturated solution of $I(OCOCH_3)_3$ in acetic anhydride is electrolyzed the quantity of silver iodide formed at a silvered platinum gauze *cathode* is in

[50] M. Schlosser and G. Heinz, *Chem. Ber.*, 1969, **102**, 1944.
[51] W. A. Sheppard, *Tetrahedron Letters*, **1969**, 83.
[52] H. H. Claasen and E. H. Appelman, *Inorg. Chem.*, 1970, **9**, 622.
[53] K. O. Christie, C. J. Schack, D. Pilipovitch and W. Sawodny, *Inorg. Chem.*, 1969, **8**, 2489.
[54] A. Engelbrecht and P. Peterfy, *Angew. Chem. Internat. Edn.*, 1969, **8**, 768.
[55] D. F. Banks, *Chem. Rev.*, 1966, **66**, 243 (an extensive review).
[55a] H. A. Carter *et al.*, *Inorg. Chem.*, 1970, **9**, 2485.

good agreement with calculations according to Faraday's law if the presence of I^{3+} is assumed:

$$I^{3+} + Ag + 3e \rightarrow AgI$$

The compounds are sensitive to moisture and are not stable much above room temperature. They are hydrolyzed with disproportionation of the I^{III}, as illustrated for IPO_4 thus:

$$5IPO_4 + 9H_2O \rightarrow I_2 + 3HIO_3 + 5H_3PO_4$$

Covalent I^{III} is known also in the compound triphenyliodine, $(C_6H_5)_3I$, and a large number of diaryliodonium salts, such as $(C_6H_5)_2I^+X^-$, where X may be one of a number of common anions. Aryl compounds such as $C_6H_5ICl_2$ are also well known and can be prepared by direct interaction; they can be regarded as trigonal-bipyramidal with axial chlorine atoms and equatorial phenyl group and lone-pairs. The compounds can be used for chlorination of alkenes.

Oxo Compounds. The so-called iodosyl sulfate, $(IO)_2SO_4$, which is a yellow solid obtained by the action of H_2SO_4 on I_2O_5 and I_2, has polymeric $(I-O)_n$ chains cross-linked by the anion. Similarly, HIO_3 in H_2SO_4 gives monomeric IO_2HSO_4 in dilute solution, but at high concentrations polymerization occurs and a white solid, $I_2O_5SO_3$, can be obtained.

ORGANIC COMPOUNDS OF FLUORINE[56]

Although the halogens form innumerable organic compounds, the properties of certain organic fluorine compounds, and especially the unusual ways in which they are prepared, call for their brief inclusion in this text.

16-19. Preparative Methods

The main preparative methods are the following:

1. *Replacement of other halogens by means of metal fluorides.* The driving force for a reaction

$$R-Cl + MF = R-F + MCl$$

depends in part on the free-energy difference of MF and MCl, which is approximately equal to the difference in lattice energies. Since lattice energies are proportional to the reciprocal of the interionic distance, the increase in free energy when MCl is formed from MF is proportional to the difference of the reciprocals of the sums of the ionic radii in MF and MCl. Thus the larger the cation, M, the more favorable tends to be the free energy for the above reaction, and the fluorine-exchanging ability therefore increases with increasing metal ion radius in isomorphous compounds. Thus, among the alkalis, LiF is the poorest and CsF the most effective. For AgF the difference in lattice energies is small owing to contributions of non-ionic bonding in AgCl, so that AgF is a very powerful fluorinating agent.

[56] W. A. Sheppard and C. M. Sharts, *Organic Fluorine Chemistry*, Benjamin, 1970 (a comprehensive monograph).

Other fluorinating agents, each having particular advantages under given conditions, are AgF_2, CoF_3, SbF_3 ($+SbCl_5$ catalyst), HgF_2, KHF_2, ZnF_2, AsF_3, etc. Examples of some fluorinations are:

$$C_6H_5PCl_2 + AsF_3 \xrightarrow{25°} C_6H_5PF_2 + AsCl_3$$
$$C_6H_5CCl_3 + SbF_3 \longrightarrow C_6H_5CF_3 + SbCl_3$$

2. *Replacement of other halogens using hydrogen fluoride.* It is clearly very desirable commercially to replace Cl by F using a method that does not require a metal fluoride. Anhydrous HF is relatively inexpensive and can be employed in a number of cases, e.g.,

$$2CCl_4 + 3HF \rightarrow \underset{\text{b.p.} -29.8°}{CCl_2F_2} + \underset{\text{b.p. } 23.7°}{CCl_3F} + 3HCl$$
$$CHCl_3 + 2HF \rightarrow CHClF_2 + 2HCl$$

where the products from CCl_4 are widely used aerosol propellants. Such reactions require $SbCl_5$ as a catalyst, temperatures of 50–150°, and pressures of 50–500 lb in^{-2}; the steps appear to be

$$SbCl_5 + 3HF \rightarrow SbCl_2F_3 + 3HCl$$
$$SbCl_2F_3 + 2CCl_4 \rightarrow SbCl_5 + CCl_3F + CCl_2F_2$$

There are other catalysts that will operate under milder conditions, and an important example is the use of Cr^{IV} or Cr^{V} fluoride in reactions such as

$$CCl_3COCCl_3 \xrightarrow[\text{Cr}]{\text{HF}} CF_3COCF_3$$
$$\text{catalyst}$$

3. *Electrolytic replacement of hydrogen by fluorine.* One of the most important preparative methods, both in the laboratory and industrially, is the electrolysis[57] of organic compounds in liquid HF at voltages (~ 4.5–6) below that required for the liberation of fluorine. Steel cells with Ni anodes and steel cathodes are used, and fluorination occurs at the anode. Although many organic compounds give conducting solutions in liquid HF, a conductivity additive may be required. Examples of such fluorinations are

$$(C_2H_5)_2O \rightarrow (C_2F_5)_2O$$
$$C_8H_{18} \rightarrow C_8F_{18}$$
$$(CH_3)_2S \rightarrow CF_3SF_5 + (CF_3)_2SF_4$$
$$(C_4H_9)_3N \rightarrow (C_4F_9)_3N$$
$$CH_3COOH \rightarrow CH_3COOF \xrightarrow{H_2O} CF_3COOH$$

4. *Direct replacement of hydrogen by fluorine.* Although most organic compounds inflame or explode when mixed with fluorine in normal circumstances, direct fluorination of many compounds is possible under appropriate conditions. There are two main techniques.

Catalytic fluorination involves mixing the reacting compound and fluorine diluted with nitrogen *in the presence* of the catalyst.[58] The catalyst may be copper gauze, silver-coated copper gauze, or cesium fluoride. An example is

$$C_6H_6 + 9F_2 \xrightarrow{Cu, 265°} C_6F_{12} + 6HF$$

A recently developed procedure[59] giving high yields involves the reaction

[57] S. Nagase, *Fluorine Chem. Rev.*, 1967, **1**, 77.
[58] For references see M. Wechsberg and G. H. Cady, *J. Amer. Chem. Soc.*, 1969, **91**, 443.
[59] J. L. Margrave and R. J. Lagow, *Chem. Eng. News*, 1970, Jan. 12, p. 40.

of the substrate in the *solid* state (liquids or gases are frozen) with fluorine diluted with helium over a rather long period (12–36 hours) at a low temperature in presence of a heat sink in the form of the reactor and containers. The purpose is to allow the heat generated in the exothermic reaction (over-all for replacement of H by F, ca. 420 kJ mol^{-1}), which could lead to C—C bond breaking, to be slowly dissipated. The replacement reaction appears to proceed by several steps, each less exothermic than the C—C average bond strength, so that, provided the reaction time allows completion of individual reactions, fluorination without degradation is possible. Examples of materials that can be fluorinated in this way are polystyrene, anthracene and other polynuclear hydrocarbons, and compounds containing atoms other than carbon such as carboranes, β-trichloroborazines, phthalocyanine, etc.

Inorganic fluorides such as cobaltic fluoride have also commonly been used for the vapor-phase fluorination of organic compounds, e.g.,

$$(CH_3)_3N \xrightarrow{\text{CoF}_3} (CF_3)_3N + (CF_3)_2NF + CF_3NF_2 + NF_3$$

5. *Replacement of oxygen by fluorine.*   A particularly useful and selective fluorinating agent for oxygen compounds is SF_4 (page 438); e.g., ketones $RR'CO$ may be converted to $RR'CF_2$.

Carbonyl compounds can also be catalytically fluorinated to give fluorohypochlorites $RR'CF(OCl)$ by using ClF in presence of CsF.[60] Similarly the CsF-catalysed interaction of CO_2 and F_2 gives difluorobis(fluorooxy)-methane, $CF_2(OF)_2$.[61]

6. *Other methods.*   The use of CsF as a catalyst in various fluorination reactions has been noted; another example is

$$R_FCN + F_2 \xrightarrow{\text{CsF}, -78°} R_FCF_2NF_2 \qquad (R_F = \text{perfluoralkyl})$$

Towards unsaturated fluorocarbons the F^- ion has a high order of nucleophilicity and adds to the positive center of a polarized multiple bond. The carbanion so produced may then undergo double-bond migration or may act as a nucleophile leading to the elimination of F^- or another ion by an S_N2 mechanism. Fluoride-initiated reactions of these types have wide scope in synthesis.[62] The reactions can be carried out in DMF or diglyme by using either the sparingly soluble CsF or the more soluble Et_4NF. KSO_2F (page 456), which is soluble in tetramethylene sulfone (thiolane dioxide), can also be used.

As examples we have

$$CF_2{=}CFCF_3 \xrightarrow{F^-} (CF_3)_2CF^- \xrightarrow{I_2} (CF_3)_2CFI + I^-$$
$$C_6F_8Fe(CO)_3 \xrightarrow{F^-} C_6F_9Fe(CO)_3$$

Finally, note that thermal decomposition of aromatic diazonium fluoroborates may give fluoroaromatic compounds:

$$C_6H_5N_2Cl \xrightarrow{\text{NaBF}_4} C_6H_5N_2BF_4 \xrightarrow{\text{Heat}} C_6H_5F + N_2 + BF_3$$

[60] D. E. Gould *et al.*, *J. Amer. Chem. Soc.*, 1970, **92**, 2313.
[61] F. A. Hohorst and J. M. Shreeve, *Inorg. Chem.*, 1968, **7**, 624.
[62] J. A. Young, *Fluorine Chem. Rev.*, 1967, **1**, 359; J. S. Johar and R. D. Dresdner, *Inorg. Chem.*, 1968, **7**, 683; see also L. L. German and I. L. Knunyantz, *Angew. Chem. Internat. Edn.*, 1969, **8**, 349 (conjugated additions in liquid HF).

16-20. Properties of Organofluorine Compounds

The C—F bond energy is indeed very high (486 kJ mol^{-1}; cf. C—H 415, and C—Cl 332 kJ mol^{-1}), but organic fluorides are not necessarily particularly stable thermodynamically; rather, the low reactivities of fluorine derivatives must be attributed to the impossibility of expansion of the octet of fluorine and the inability of, say, water to coordinate to fluorine or carbon as the first step in hydrolysis, whereas with chlorine this may be possible using outer d orbitals. Because of the small size of the F atom, H atoms can be replaced by F atoms with the least introduction of strain or distortion, as compared with replacement by other halogen atoms. The F atoms also effectively shield the carbon atoms from attack. Finally, since C bonded to F can be considered to be effectively oxidized (whereas in C—H it is reduced), there is no tendency for oxidation by oxygen. Fluorocarbons are attacked only by hot metals, e.g., molten sodium. When pyrolyzed, fluorocarbons tend to split at C—C rather than C—F bonds.

The replacement of H by F leads to increased density, but not to the same extent as by other halogens. Completely fluorinated derivatives, C_nF_{2n+2}, have very low boiling points for their molecular weights and low intermolecular forces; the weakness of these forces is also shown by the very low coefficient of friction for polytetrafluoroethylene, $(CF_2—CF_2)_n$.[63]

Commercially important organic fluorine compounds are chlorofluorocarbons, which are used as non-toxic, inert refrigerants, aerosol bomb propellants, and heat-transfer agents, and fluoroolefins, which are used as monomers for free-radical-initiated polymerizations to oils, greases, etc., and also as chemical intermediates. $CF_3CHBrCl$ and some similar compounds are safe and useful anaesthetics.[63a] $CHClF_2$ is used for making tetrafluoroethylene:

$$2CHClF_2 \xrightarrow{500-1000°} CF_2{=}CF_2 + 2HCl$$

C_2F_4 is also made by quenching CF_2 radicals formed at temperatures above 1500° by the reaction of almost any source of fluorine with carbon. Tetrafluoroethylene, b.p. $-76.6°$, can be polymerized thermally or in aqueous emulsion by use of oxygen, peroxides, etc., as free-radical initiators. A convenient laboratory source of C_2F_4 is thermal cracking of polymer at 500–600°.

The fluorinated carboxylic acids are notable first for their strongly acid nature—for example, for CF_3COOH, $K_a = 5.9 \times 10^{-1}$, whereas CH_3COOH has $K_a = 1.8 \times 10^{-5}$. Secondly, many standard reactions of carboxylic acids can be made leaving the fluoroalkyl group intact:

$$C_3F_7COOH \xrightarrow[C_2H_5OH]{H_2SO_4} C_3F_7COOC_2H_5 \xrightarrow{NH_3} C_3F_7CONH_2 \begin{array}{l} \xrightarrow{P_2O_5} C_3F_7CN \\ \xrightarrow{LiAlH_4} C_3F_7CH_2NH_2 \end{array}$$

[63] For physical properties of fluoro compounds see C. R. Patrick, *Chem. in Britain*, **1971**, 154.

[63a] E. R. Larsen, *Fluorine Chem. Rev.*, 1969, **3**, 1.

Trifluoroacetic anhydride obtained by action of P_2O_5 on the acid is widely used in organic chemistry as an acylating agent in presence of acid.

The $C_nF_{2n+1}COOH$ acids can be converted into *perfluoroalkyl halides* by, for example, the action of I_2 on the silver salt. These halides are relatively reactive, undergoing free-radical reactions when heated or irradiated, although, because of the very strong electron-attracting nature of the perfluoroalkyl groups, they fail to show many of the common nucleophilic reactions of alkyl halides. They do not, for example, readily form Grignard reagents. The bond dissociation energy for the reaction

$$CF_3I = CF_3{}^{\bullet} + I^{\bullet}$$

is only 115 kJ mol^{-1}. CF_3I is an important intermediate for preparation of trifluoromethyl derivatives. Thus the reaction of CF_3I and similar compounds with metals and non-metals has led to an extensive range of CF_3 derivatives, for example,

$$CF_3I + P \xrightarrow{\text{Heat}} (CF_3)_nPI_{3-n}$$

Of trifluoromethyl derivatives, two are worth special comment. *Trifluoronitrosomethane*, CF_3NO, b.p. $-84°$, and related compounds, generally R_FNO, can be made by the reaction

$$CF_3I + NO \xrightarrow[\text{uv}]{\text{Hg}} CF_3NO$$

or by photolytic or pyrolytic reactions

$$R_FCOOAg + NOCl \xrightarrow{-AgCl} R_FCOONO \xleftarrow{NOCl} (R_FCO)_2O$$

$$\text{uv or}\ \Big\downarrow\ \text{heat}$$

$$R_FNO + CO_2$$

The nitroso compounds[64] can be co-polymerized with C_2F_4 at low temperatures in aqueous media to form very stable rubber-like materials of the type

$$\left[\begin{array}{c} -N-O-CF_2-CF_2- \\ | \\ CF_3 \end{array} \right]_n$$

A second important and useful compound is trifluoro(fluorooxo)methane, CF_3OF (b.p. $-95°$), which is obtained by fluorination of CH_3OH or CO by F_2 in presence of AgF_2. It is stable to 450° and is a reactive, strong oxidant which undergoes a wide variety of reactions, e.g.:

$$CF_3OF + SO_2 \rightarrow CF_3OSO_2F$$
$$CF_3OF + SO_3 \rightarrow CF_3OOSO_2F$$
$$CF_3OF + C_2F_4 \rightarrow CF_3OCF_2CF_3$$
$$CF_3OF + \tfrac{1}{2}N_2F_2 \rightarrow CF_3ONF_2$$
$$CF_3OF + SF_4 \rightarrow CF_3OSF_5$$

It can also fluorinate activated olefins, including steroids.[65] The corresponding bis(fluorooxo) compound, $CF_2(OF)_2$, which is also a strong oxidant, is obtained by fluorination of CO_2 by CsF and F_2.[61]

[64] M. C. Henry, C. B. Griffis and E. C. Stump, *Fluorine Chem. Rev.*, 1967, **1**, 3.
[65] D. H. R. Barton *et al.*, *Chem. Comm.*, **1968**, 804.

Further Reading

Advances in Fluorine Chemistry, J. C. Tatlow *et al.*, eds. (review volumes on all aspects of fluorine chemistry).

Brown, D., *Halides of Lanthanides and Actinides*, Interscience-Wiley, 1968.

Canterford, J. H., and R. Colton, *Halides of First Row Transition Metals*, 1969; *Halides of Second and Third Row Transition Metals*, Interscience-Wiley, 1968.

Dehnicke, K., *Angew. Chem. Internat. Edn.*, 1965, **4**, 22 (synthetic methods for oxo halides).

Eagers, R. Y., *Toxic Properties of Inorganic Fluorine Compounds*, Elsevier, 1969.

Fluorine Chem. Rev., P. Tarrant, ed., Dekker, Vol. 1, 1967 (various topics).

Halogen Chemistry, V. Gutmann, ed., Academic Press, 1967 (three volumes on various aspects of chemistry).

Jolles, Z. E., ed., *Bromine and its Compounds*, Benn, London, and Academic Press, New York, 1966 (reference book).

Journal of Fluorine Chemistry, H. J. Eeméleus and J. C. Tatlow, eds., Vol. 1, 1971.

Kubo, M., and D. Nakamura, *Adv. Inorg. Chem. Radiochem.*, 1966, **8**, 257 (nuclear quadrupole resonance studies, especially of halogen compounds).

Lawless, E. W., and I. C. Smith, *Inorganic High Energy Oxidizers*, Dekker, 1969 (properties of F_2, fluorohalogen and fluoronitrogen compounds).

Mellor, J. W., *Comprehensive Treatise on Inorganic Chemistry*, Suppl. II, Pt. I, Longmans, Green, New York–London, 1956 (contains all the Group VII elements).

Neumark, H. R., *et al.*, *The Chemistry and Chemical Technology of Fluorine*, Interscience-Wiley, 1967.

Novikov, S. S., V. V. Sevost'yanova and A. A. Fainzil'berg, *Russ. Chem. Rev.*, **1962**, 671 (positive halogen organic compounds).

Pattison, F. L. M., *Toxic Aliphatic Fluorine Compounds*, Elsevier, 1959.

Pavpath, A. E., and A. L. Leffler, *Aromatic Fluorine Compounds*, Reinhold, New York, 1962 (a comprehensive reference text).

Pennman, R. A., *Inorg. Chem.*, 1969, **8**, 1379 (use of molar refractivity for ascertaining composition of transition-metal fluoro complexes).

Rudge, A. J., *Manufacture and Use of Fluorine and its Compounds*, Oxford Univ. Press, 1962.

Scherer, O., *Technische organische Fluor Chemie, Topics in Current Chemistry*, Vol. 2, Part 13/14, Springer Verlag, 1970.

Schmeisser, M., and K. Brändle, *Adv. Inorg. Chem. Radiochem.*, 1963, **5**, 41 (oxides and oxo fluorides of halogens; a comprehensive review).

Schumaker, J. C., ed., *Perchlorates* (A. C. S. Monograph, No. 146), Reinhold, New York, 1960 (an extensive review of perchloric acid and perchlorates).

Sconce, J. S., ed., *Chlorine: its Manufacture, Properties, and Uses* (A. C. S. Monograph, No. 154), Reinhold, New York, 1962.

Simons, J. H., ed., *Fluorine Chemistry*, Vols. I–V, Academic Press, New York, 1954 (comprehensive reference books on special topics in fluorine chemistry).

Treichel, P. M., and F. G. A. Stone, *Adv. Organometallic Chem.*, 1964, **1**, 143 (fluorocarbon derivatives of metals).

Zinov'ev, A. A., *Russ. Chem. Rev.*, **1963**, 268 (perchloric acid and Cl_2O_7).

The Noble Gases

THE ELEMENTS

17-1. Group Trends

The closed-shell electronic structures of the noble-gas atoms are completely stable, as shown by the high ionization potentials, especially of the lighter members (Table 17-1). The elements are all low-boiling gases whose physical properties vary fairly systematically with atomic number. The boiling point of helium is the lowest of any known substance. The boiling points and heats of vaporization increase monotonically with increasing atomic number.

TABLE 17-1

Some Properties of the Noble Gases

Outer shell configuration		Atomic number	1st IP (eV)	Normal b.p., °K	ΔH_{vap}, (kJ mol^{-1})	% by volume in the atmosphere	Promotion energy (eV), $ns^2np^6 \rightarrow$ $ns^2np^5(n+1)s$
He	$1s^2$	2	24.58	4.18	0.09	5.24×10^{-4}	—
Ne	$2s^22p^6$	10	21.56	27.13	1.8	1.82×10^{-3}	16.6
Ar	$3s^23p^6$	18	15.76	87.29	6.3	0.934	11.5
Kr	$4s^24p^6$	36	14.00	120.26	9.7	1.14×10^{-3}	9.9
Xe	$5s^25p^6$	54	12.13	166.06	13.7	8.7×10^{-6}	8.3
Rn	$6s^26p^6$	86	10.75	208.16	18.0		6.8

The heats of vaporization are measures of the work that must be done to overcome interatomic attractive forces. Since there are no ordinary electron-pair interactions between noble-gas atoms, these weak forces must be of the van der Waals or London type; such forces are proportional to the polarizability and inversely proportional to the ionization potentials of the atoms; they increase therefore as the size and diffuseness of the electron clouds increase.

The ability of the noble gases to enter into chemical combination with other atoms is very limited, only Kr, Xe and Rn having so far been induced to do so, and only bonds to F and O are stable. This ability would be expected to increase with decreasing ionization potential and decreasing energy of promotion to states with unpaired electrons. The data in Table 17-1 for ionization potentials and for the lowest-energy promotion process show that chemical activity should increase down the group. According to present knowledge, the threshold of actual chemical activity is reached only at Kr. The chemical activity of Xe is markedly greater. That of Rn is presumably still greater, but it is difficult to assess because the half-life of the longest-lived isotope ^{222}Rn, is only 3.825 days so that only tracer studies can be made.

17-2. Occurrence, Isolation and Applications

The noble gases occur as minor constituents of the atmosphere (Table 17-1). Helium is also found as a component (up to $\sim 7\%$) in certain natural hydrocarbon gases in the United States. This helium undoubtedly originated from decay of radioactive elements in rocks, and certain radioactive minerals contain occluded helium which can be released on heating. All isotopes of radon are radioactive and are occasionally given specific names (e.g., actinon, thoron) derived from their source in the radioactive decay series; ^{222}Rn is normally obtained by pumping off the gas from radium chloride solutions. Ne, Ar, Kr and Xe are obtainable as products of fractionation of liquid air.

The main uses of the gases are in welding (argon provides an inert atmosphere), in gas-filled electric light bulbs and radio tubes (argon), and in discharge tubes (neon); radon has been used therapeutically as an α-particle source in the treatment of cancer. Helium as liquid is extensively used in cryoscopy and as an inert protective gas in chemical reactions.

The amounts of He and Ar formed by radioactive decay in minerals can be used to determine the age of the specimen. For example, in the course of the decay of ^{238}U, eight α-particles are produced; these acquire electrons to form He atoms by oxidizing other elements present. If the rock is sufficiently impermeable, the total He remains trapped therein. If the amounts of trapped helium and remaining ^{238}U are measured, the age of the specimen can be calculated, for one-eighth of the atoms of He represent the number of ^{238}U atoms that have decayed. A correction must be applied for thorium, which also decays by α emission and generally occurs in small amounts with uranium. Argon arises in potassium-containing minerals by electron-capture of ^{40}K; a complication arises here since ^{40}K also decays by β-emission to ^{40}Ca, and the accuracy of the age determination in this case depends on accurate determination of the branching ratio of ^{40}K.

The trapping of Ar, Kr and Xe in clathrate compounds has been discussed in Chapter 5.

17-3. Special Properties of Helium[1]

Naturally occurring helium is essentially all ^{4}He, although ^{3}He occurs to the extent of $\sim 10^{-7}$ atom %. ^{3}He can be made in greater quantities by nuclear reactions and by β^- decay of tritium.

The pressure–temperature diagram near absolute zero for helium is shown in Fig. 17-1. Its most remarkable feature is that helium has no triple point; that is, there is *no* combination of temperature and pressure at which solid, liquid, and gas coexist in equilibrium. Helium is the only known substance lacking a triple point. It is also the only one that cannot be solidified at atmospheric pressure. These departures of helium from the universal pattern are due to a quantum effect. Its zero-point energy is so high that it outweighs the weak interatomic forces which, without the application of external pressure, are not strong enough to bind the helium atoms into the crystalline state. Still more remarkable, however, is the transition that takes place across the line $\lambda-\lambda' \sim 2.2\,°$K from He$_I$ to He$_{II}$. This is marked by a huge specific heat anomaly, He$_{II}$ having much lower entropy. Nevertheless, He$_{II}$ has the random ordering of a true liquid. It is, however, a unique liquid in exhibiting the phenomenon of superconductivity. It has immeasurably low viscosity and readily forms films only a few hundred atoms thick which flow apparently without friction, even up over the edges of a vessel. No fully satisfactory explanation of these properties has yet been devised.

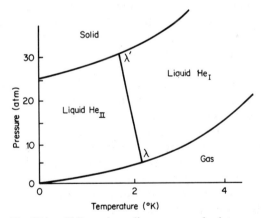

Fig. 17-1. Helium phase diagram near absolute zero.

[1] J. Wilks, *The Properties of Liquid and Solid Helium*, Oxford University Press, 1967; *The Technology of Liquid Helium*, R. H. Kropschot, B. W. Birmingham and D. B. Mann, ed., Nat. Bur. Stand. Monograph No. 111, U. S. Govt. Printing Office, Washington D.C., 1968; S. J. Putterman and I. Rudnick, *Physics Today*, 1971, **24**, 39.

THE CHEMISTRY OF THE NOBLE GASES[2]

After his observation that O_2 reacts with PtF_6 to give the compound $[O_2^+][PtF_6^-]$ (page 414), N. Bartlett in 1962 recognized that, since the ionization potential of xenon is almost identical with that of the oxygen molecule, an analogous reaction should occur with xenon. He then confirmed this prediction by obtaining a red crystalline solid, originally believed to be "$XePtF_6$" (see below), by direct interaction of Xe with PtF_6. This discovery led to rapid and extensive developments in xenon chemistry. The structures and related problems of bonding in xenon compounds have been discussed in Chapter 4.

17-4. The Chemistry of Xenon

Xenon reacts directly only with fluorine, but compounds in oxidation states from II to VIII are known, some of which are exceedingly stable and can be obtained in large quantities. The more important compounds and some of their properties are given in Table 17-2.

Fluorides. The equilibrium constants for the reactions

$$Xe + F_2 \rightarrow XeF_2$$
$$XeF_2 + F_2 \rightarrow XeF_4$$
$$XeF_4 + F_2 \rightarrow XeF_6$$

have been either measured or calculated for the range 25–500°. The studies show unequivocally that only these three fluorides exist. The equilibria are established rapidly only above 250° and this is the lower limit for thermal preparative methods. Although in principle knowledge of the constants should allow the choice of reaction conditions to give a particular compound, in practice the methods were developed empirically. All three fluorides are volatile, readily subliming at room temperature. They can be stored indefinitely in nickel or Monel metal containers, but XeF_4 and XeF_6 are particularly susceptible to hydrolysis and traces of water must be rigorously excluded.

Xenon difluoride. This is best obtained by interaction of F_2 and an excess of Xe at high pressure, but there are other methods such as the interaction of Xe and O_2F_2 at $-118°$ and procedures in which XeF_2 is trapped from mixtures of F_2 and Xe at low pressures. It is soluble in water, giving solutions $0.15M$ at 0°, that evidently contain XeF_2 molecules. The hydrolysis is slow in dilute acid but rapid in basic solution:

$$XeF_2 + 2OH^- \rightarrow Xe + \tfrac{1}{2}O_2 + 2F^- + H_2O$$

The solutions, which have a pungent odor due to XeF_2, are powerful oxi-

[2] J. G. Malm and E. H. Appelman, *Atomic Energy Rev.* 1969, **7** (3), 3, Internat. Atomic Energy Agency, Vienna (an exhaustive and critical review including references up to late 1969); J. H. Holloway, *Noble Gas Chemistry*, Methuen, 1968 (includes history); J. G. Malm, H. Selig, J. Jortner and S. A. Rice, *Chem. Rev.*, 1965, **65**, 119 (bonding in xenon compounds); H. Selig in *Halogen Chemistry*, V. Gutman, ed., Vol. 1, Academic Press, 1967 (xenon fluorides); *Gmelins Handbuch der anorganischen Chemie*, Main Suppl. to 8th ed., Vol. 1, *Noble Gas Compounds*, 1970 (excellent account including clathrates); L. Stein, *Yale Sci. Mag.*, 1970, **44**, 2 (radon chemistry).

TABLE 17-2

Principal Xenon Compounds[a]

Oxidation state	Compound	Form	M.p. (°C)	Structure	Remarks
II	XeF_2	Colorless crystals	129	Linear	Hydrolyzed to $Xe + O_2$; v. soluble in HF(l)
	$XeF_2 \cdot 2SbF_5$	Yellow solid	63		
IV	XeF_4	Colorless crystals	117	Square	Stable, $\Delta H_f^{298°} = -284$ kJ mol^{-1}
	$XeOF_2$	Colorless crystals	31		Barely stable
VI	XeF_6	Colorless crystals	49.6	Distorted octahedral	Stable, $\Delta H_f^{298°} = -402$ kJ mol^{-1}
	$CsXeF_7$	Colorless solid			Decomp. $> 50°$
	Cs_2XeF_8	Yellow solid		Archim. antiprism[c]	Stable to 400°
	$XeOF_4$	Colorless liquid	-46	ψ-Octa-hedral[b]	Stable
	XeO_3	Colorless crystals		ψ-Tetra-hedral[b]	Explosive, $\Delta H_f^{298°} = +402$ kJ mol^{-1} hygroscopic; stable in solution
	$nK^+[XeO_3F^-]_n$	Colorless crystals		Square-pyramidal (F bridges)	Very stable
VIII	XeO_4	Colorless gas		Tetrahedral	Explosive
	XeO_6^{4-}	Colorless salts		Octahedral	Anions $HXeO_6^{3-}$, $H_2XeO_6^{2-}$, $H_3XeO_6^-$ also exist

[a] Other species have been observed in mass spectra, e.g., XeO_3F_2, or in emission spectra. Salts of the square pyramidal XeF_5^+ and T-shaped XeF_3^+ ions are known.

[b] Lone pair present.

[c] In the salt $(NO^+)_2[XeF_8]^{2-}$ from XeF_6 and NOF (F. W. Petersen *et al.*, *Science*, 1971, **173**, 1238).

dizing agents, e.g., HCl gives Cl_2, Ce^{III} gives Ce^{IV}, and the estimated potential is

$$XeF_2(aq) + 2H^+ + e = Xe + 2HF(aq) \qquad E_0 = +2.64 \text{ V}$$

XeF_2 also acts as a mild fluorinating agent for organic compounds; for example, in solution or in the vapor phase benzene is converted into C_6H_5F.[3]

Xenon tetrafluoride. XeF_4 is the easiest fluoride to prepare, and essentially quantitative conversion is obtained when a 1:5 mixture of Xe and F_2 is heated in a nickel vessel at 400° and ca. 6 atm pressure for a few hours. Its properties are similar to those of XeF_2 except regarding its hydrolysis, as discussed below. With hydrogen, XeF_4 rapidly gives Xe and HF, and with fluorine under pressure it forms XeF_6. XeF_4 specifically fluorinates the ring in substituted arenes such as toluene.[3a]

[3] M. J. Shaw *et al.*, *J. Amer. Chem. Soc.*, 1970, **92**, 5096, 6498; D. R. MacKenzie and J. Fajer, *J. Amer. Chem. Soc.*, 1970, **92**, 4994.

[3a] N. C. Chang *et al.*, *J. Org. Chem.*, 1970, **35**, 4020.

Xenon hexafluoride. This preparation requires more severe conditions, but at high pressures (> 50 atm) and temperature ($> 250°$) quantitative conversion may be obtained. The solid is colorless but becomes yellow when heated and gives a yellow liquid and vapor. It reacts rapidly with quartz:

$$2XeF_6 + SiO_2 \rightarrow 2XeOF_4 + SiF_4$$

and it is extremely readily hydrolyzed.

Structural evidence[4] that XeF_6 is not precisely octahedral and may be considered to have a lone-pair is supported by the formation of adducts with Lewis acids (see below). A cubic phase of XeF_6 studied at $-80°$ contains XeF_5^+ and F^- ions that are associated in tetrameric and hexameric rings.[5]

Xenon oxo fluorides. The tetrafluoride has just been noted; it is best made by partial hydrolysis of XeF_6:

$$XeF_6 + H_2O \rightarrow XeOF_4 + 2HF$$

The interaction of $XeOF_4$ and XeO_3 gives XeO_2F_2.

Xenon Fluoride Complexes. The di-, tetra- and hexafluorides of xenon can act as fluoride-donors, while XeF_6 appears to be able to act as an acceptor also. Thus the fluorides react with fluoro-Lewis acids such as AsF_5, SbF_5 and RuF_5 to give adducts.[6] XeF_2 gives complexes of stoichiometry $XeF_2 \cdot 2MF_5$ and $2XeF_2 \cdot MF_5$ while XeF_4 gives $XeF_3^+Sb_2F_{11}^-$. Although $XeF_2 \cdot IF_5$ has a molecular lattice, the other adducts are recognizably ionic, containing XeF^+ and $Xe_2F_3^+$ ions, e.g.,

$$2XeF_2 + AsF_5 \xrightarrow{\text{BrF}_5} Xe_2F_3^+ AsF_6^-$$

In the XeF^+ compounds there is considerable distortion of the anion,[6,7] probably due to a strong covalent contribution of the type $FXe^+ \cdots FMF_5^-$. The X-ray studies show that the $Xe_2F_3^+$ ion is planar with the structure (17-I).

(17-I)

From these studies it is evident that the original reaction of Xe and PtF_6 noted earlier is best written:

$$Xe + PtF_6 \xrightarrow{25°} XePtF_6$$
$$XePtF_6 + PtF_6 \xrightarrow{25°} XeF^+PtF_6^- + PtF_5^- \xrightarrow{60°} XeF^+Pt_2F_{11}^-$$

The product made at $25°$ appears to contain both $XeFPtF_6$ and PtF_5^-.

The hexafluoride also gives complexes such as $2XeF_6 \cdot SbF_5$, and the adduct of PtF_5 has been shown to be $XeF_5^+PtF_6^-$ in which there is interaction between the cation and the anion via fluorine bridges. XeF_6 also appears to

[4] L. S. Bartell and R. M. Gavin, Jr., *J. Chem. Phys.*, 1966, **48**, 2466.
[5] G. R. Jones and R. D. Burbank, Abstr. Amer. Cryst. Assoc. Meeting, March, 1970.
[6] F. O. Sladky, P. A. Bulliner and N. Bartlett, *J. Chem. Soc.*, A, **1969**, 2179, 2188; J. H. Holloway and J. G. Knowles, *J. Chem. Soc.*, A, **1969**, 756; G. R. Jones, R. D. Burbank and N. Bartlett, *Inorg. Chem.*, 1970, **9**, 2264; R. J. Gillespie, B. Landa and G. J. Schrobilgen, *Chem. Comm.*, **1971**, 1543.
[7] V. M. McRae, R. D. Peacock and D. R. Russell, *Chem. Comm.*, **1969**, 62.

act as a fluoride ion acceptor since it reacts with alkali-metal fluorides (other than LiF) to give heptafluoro- or octafluoro-xenates(vi):

$$XeF_6 + RbF \rightarrow RbXeF_7$$

The Rb and Cs salts are well-characterized; above 20° and 50°, respectively, they decompose:

$$2MXeF_7 \rightarrow XeF_6 + M_2XeF_8$$

This formation of XeF_7^- and XeF_8^{2-} salts from XeF_6 resembles the behavior of UF_6. The Rb and Cs octafluoroxenates are the most stable xenon compounds yet made and decompose only above 400°; they hydrolyze in the atmosphere to give xenon-containing oxidizing products. The sodium fluoride adduct of XeF_6 decomposes below 100° and can be used to purify XeF_6.

Oxo fluoro anions can also be obtained, but not directly from $XeOF_4$. The very stable salts $MXeO_3F$ are obtained when a solution of XeO_3 is treated with KF or CsF; X-ray study shows that the anion is polymeric, i.e., $(XeO_3F^-)_n$ with chains of XeO_3 groups linked by an angular Xe—F—Xe bridge,[8] giving distorted square-pyramidal coordination of Xe. An apparent 10-valence-electron shell in XeO_3F^- is really, because of polymerization, a 12-electron shell (cf. BrF_3 and I_2Cl_6).

Xenon–Oxygen Compounds. Both XeF_4 and XeF_6 are violently hydrolyzed by water to give Xe^{VI}, evidently in the form of undissociated XeO_3:

$$3XeF_4 + 6H_2O \rightarrow XeO_3 + 2Xe + \tfrac{3}{2}O_2 + 12HF$$
$$XeF_6 + 3H_2O \rightarrow XeO_3 + 6HF$$

Aqueous solutions, colorless, odorless, and stable, as concentrated as $11M$ in Xe^{VI}, have been obtained. They are non-conducting. On evaporation, XeO_3 is obtained.

Xenon trioxide is a white deliquescent solid and is dangerously explosive; its formation is one of the main reasons why great care must be taken to avoid water in studies of XeF_6. The molecule is pyramidal (C_{3v}). XeO_3 can be quantitatively reduced by iodide:

$$XeO_3 + 6H^+ + 9I^- \rightarrow Xe + 3H_2O + 3I_3^-$$

There is some evidence that xenate esters are formed in the violent reactions with alcohols.

In water, XeO_3 appears to be present as XeO_3 molecules, but in base solutions we have

$$XeO_3 + OH^- \rightleftharpoons HXeO_4^- \qquad K = 1.5 \times 10^{-3}$$

The main species[9] in solutions, $HXeO_4^-$, slowly disproportionates to produce Xe^{VIII} and Xe:

$$2HXeO_4^- + 2OH^- \rightarrow XeO_6^{4-} + Xe + O_2 + 2H_2O$$

Aqueous Xe^{VIII} arises, not only in the above disproportionation, but also when ozone is passed through a dilute solution of Xe^{VI} in base. These yellow *perxenate* solutions are powerful and rapid oxidizing agents.

Stable, insoluble perxenate salts can be precipitated from Xe^{VIII} solutions, e.g., $Na_4XeO_6 \cdot 8H_2O$, $Na_4XeO_6 \cdot 6H_2O$, $Ba_2XeO_6 \cdot 1.5H_2O$; the first two contain XeO_6 octahedra. The solutions of sodium perxenate are alkaline

[8] D. J. Hodgson and J. A. Ibers, *Inorg. Chem.*, 1969, **8**, 326.
[9] J. L. Peterson, H. H. Claasen and E. H. Appelman, *Inorg. Chem.*, 1970, **9**, 619.

owing to hydrolysis, and the following equilibrium constants have been estimated:

$$HXeO_6^{3-} + OH^- = XeO_6^{4-} + H_2O \qquad K < 3$$
$$H_2XeO_6^{2-} + OH^- = HXeO_6^{3-} + H_2O \qquad K \sim 4 \times 10^3$$

so that even at pH 11–13 the main species is $HXeO_6^{3-}$.[10] From the above equilibria it follows that for H_4XeO_6, pK_3 and pK_4 are $\sim 4 \times 10^{-11}$ and $< 10^{-14}$, respectively. Hence, by comparison with H_6TeO_6 and H_5IO_6, H_4XeO_6 appears to be an anomalously weak acid.

Solutions of perxenates are reduced by water, at pH 11.5 at a rate of about 1% per hour, but in acid solutions almost instantaneously:

$$H_2XeO_6^{2-} + H^+ \rightarrow HXeO_4^- + \tfrac{1}{2}O_2 + H_2O$$

This reduction appears to proceed almost entirely through the formation of OH radicals according to the scheme

$$Xe^{VIII} + H_2O \rightleftharpoons Xe^{VII} + OH^{\cdot}$$
$$Xe^{VII} + H_2O \rightleftharpoons Xe^{VI} + OH^{\cdot}$$
$$2OH^{\cdot} \rightarrow H_2O_2$$
$$Xe^{VIII} + H_2O_2 \rightarrow Xe^{VI} + O_2$$

Xenon tetroxide is a highly unstable and explosive gas formed by action of concentrated H_2SO_4 on barium perxenate. It is tetrahedral (Xe—O = 1.736 Å),[11] according to electron-diffraction studies.

The aqueous chemistry of xenon is briefly summarized by the potentials:

Acid solution: $H_4XeO_6 \xrightarrow{2.36\,V} XeO_3 \xrightarrow{2.12\,V} Xe$
$XeF_2 \xrightarrow{2.64\,V} Xe$

Alkaline solution: $HXeO_6^{3-} \xrightarrow{0.94\,V} HXeO_4^- \xrightarrow{1.26\,V} Xe$

17-5. Other Noble Gas Chemistry

The bonds from xenon to elements other than F and O are highly unstable. There is some evidence for the existence of *xenon dichloride*, $XeCl_2$, made by the action of electric discharges on a mixture of Xe, F_2 and CCl_4.

However, ion cyclotron resonance spectroscopy at 70 eV of a mixture of CH_3F, Xe and H_2 shows that the ion CH_3Xe^+ is stable. The estimated Xe—C bond energy is about 180 kJ mol^{-1} [12]

Krypton difluoride is obtained when an electric discharge is passed through Kr and F_2 at $-183°$, or when the gases are irradiated with high-energy electrons or protons. It is a volatile white solid that decomposes slowly at room temperature. It is a highly reactive fluorinating agent. With SbF_5 a complex $KrF_2 \cdot 2SbF_5$, which perhaps is $KrF^+Sb_2F_{11}^-$, is formed; this is stable below its melting point (50°).

Electron-diffraction confirms the linearity of KrF_2.

[10] G. D. Downey, H. H. Claassen and E. H. Appelman, *Inorg. Chem.*, 1971, **10**, 1817.
[11] G. Gundersen, K. Hedberg and J. L. Huston, *Acta Cryst.*, 1969, *A*, **25**, 5124.
[12] D. Holtz and J. C. Beauchamp, *Science*, 1971, **173**, 1237.

Further Reading

Cook, G. A., ed., *Argon, Helium, and the Rare Gases*, 2 Vols., Interscience-Wiley, 1961 (comprehensive, exclusive of chemical behavior).
Dalrymple, G. B. and M. A. Lanphere, *Potassium–Argon Dating*, Freeman, 1969.

18

Zinc, Cadmium and Mercury

GENERAL REMARKS

18-1. Position in the Periodic Table: Group Trends

These three elements follow copper, silver and gold and have two s electrons outside filled d shells. Some of their properties are given in Table 18-1. Whereas in Cu, Ag and Au the filled d shells may lose one or

TABLE 18-1

Some Properties of the Group IIb Elements

	Zinc	Cadmium	Mercury
Outer configuration	$3d^{10}4s^2$	$4d^{10}5s^2$	$5d^{10}6s^2$
Ionization potentials, eV			
1st	9.39	8.99	10.43
2nd	17.89	16.84	18.65
3rd	40.0	38.0	34.3
Melting point, °C	419	321	−38.87
Boiling point, °C	907	767	357
Heat of vaporization, kJ mol⁻¹	131	112	61.5
E^0 for $M^{2+} + 2e = M$, V	−0.762	−0.402	0.854
Radii of divalent ions, Å	0.69	0.92	0.93

two d electrons to give ions or complexes in the II and III oxidation states, this is no longer possible for the Group II elements and there is no evidence for oxidation states higher than II. This follows from the fact that the third ionization potentials are extremely high for Zn, Cd and Hg, and energies of solvation or lattice formation cannot suffice to render the III oxidation states chemically stable.

Divergence from the Group valence does occur, however, in the ions M_2^{2+}, of which only Hg_2^{2+} is ordinarily stable, and Hg_n^{2+} ($n = 3$–6), as discussed below. There are no simple M^+ ions with a single s electron in this Group or for any of the elements of the first, second, or third long Periods. Where such ions might have been expected, there is either disproportionation as in

$Ga^{II} \rightarrow Ga^{I} + Ga^{III}$ (page 279) or formation of a metal—metal bond as in Ga_2S_2 and Hg_2^{2+}.

Since these elements form no compound in which the d shell is other than full, they are regarded as non-transition elements, whereas by the same criteria Cu, Ag and Au are considered as transition elements. Also the metals are softer and lower-melting, and Zn and Cd are considerably more electropositive, than their near neighbors in the transition groups. However, there is some resemblance to the d-group elements in their ability to form complexes, particularly with ammonia, amines, halide ions and cyanide. For complexes, even with CN^-, it must be borne in mind that the possibility of $d\pi$ bonding between the metal and the ligand is very much lowered compared to the d-transition elements, owing to the electronic structure, and no carbonyl, nitrosyl, olefin complex, etc., of the type given by transition metals is known.

The chemistries of Zn and Cd are very similar, but that of Hg differs considerably and cannot be regarded as homologous. As examples we quote the following. The hydroxide $Cd(OH)_2$ is more basic than $Zn(OH)_2$, which is amphoteric, but $Hg(OH)_2$ is an extremely weak base. The chlorides of Zn and Cd are essentially ionic, whereas $HgCl_2$ gives a molecular lattice. Zinc and cadmium are electropositive metals, but Hg has a high positive standard potential; further, the Zn^{2+} and Cd^{2+} ions somewhat resemble Mg^{2+} (see Section 18-9), but Hg^{2+} does not. Though all the M^{2+} ions readily form complexes, those of Hg^{2+} have formation constants greater by orders of magnitude than those for Zn^{2+} or Cd^{2+}.

All three elements form a variety of covalently bound compounds, and the polarizing ability of the M^{2+} ions is larger than would be predicted by comparing the radii with those of the Mg–Ra group, a fact that can be associated with the greater ease of distortion of the filled d shell. Compared to the noble gas-like ions of the latter elements the promotional energies, $ns^2 \rightarrow nsnp$ (433, 408 and 524 kJ mol^{-1} for Zn, Cd and Hg, respectively), involved in the formation of two covalent bonds are also high and this has the consequence, particularly for Hg, that further ligands can be added only with difficulty. This is probably the main reason for the fact that two-coordination is the commonest for Hg.

18-2. Stereochemistry

The stereochemistry of the elements in the II state is summarized in Table 18-2; the nature of Hg_n^{2+} compounds is discussed below. Since there is no ligand field stabilization effect in Zn^{2+} and Cd^{2+} ions because of their completed d shells, their stereochemistry is determined solely by considerations of size, electrostatic forces, and covalent bonding forces. The effect of size is to make Cd^{2+} more likely than Zn^{2+} to assume a coordination number of six. For example, ZnO crystallizes in lattices where the Zn^{2+} ion is in tetrahedral holes surrounded by four oxide ions, whereas CdO has the rock

TABLE 18-2

Stereochemistry of Divalent Zinc, Cadmium and Mercury

Coordination Number	Geometry	Examples
2	Linear	$Zn(CH_3)_2$, HgO, $Hg(CN)_2$
3	Planar	$[Me_4N]^+HgX_3^-$ [a]
4	Tetrahedral	$[Zn(CN)_4]^{2-}$, $ZnCl_2(s)$, ZnO
		$[Cd(NH_3)_4]^{2+}$, $HgCl_2(OAsPh_3)_2$
	Planar	Bis(glycinyl)Zn
5	Trigonal-bipyramidal	Terpy $ZnCl_2$; $[Zn(SCN)$ tren$]^+$;
		$[Co(NH_3)_6][CdCl_5]^b$
	Square-pyramidal	$Zn(acac)_2 \cdot H_2O$
6	Octahedral	$[Zn(NH_3)_6]^{2+}$ (solid only), CdO
		$CdCl_2$, $[Hg\,en_3]^{2+}$
8	Dist. square-antiprism	$[Hg(NO_2)_4]^{2-}$
	Dodecahedral	$(Ph_4As)_2Zn(NO_3)_4^c$

[a] Na^+, NH_4^+ and Cs^+ salts do not contain HgX_3^-, while adducts such as $HgX_2\cdot$ dioxan contain only HgX_2 and dioxan molecules in chains (T. Brill, *J. Inorg. Nuclear Chem.*, 1970, **32**, 1868; T. Brill and Z. Z. Hugus, Jr., *Inorg. Chem.*, 1970, **9**, 984).

[b] E. F. Epstein and I. Bernal, *J. Chem. Soc, A.*, **1971**, 3628.

[c] J. Drummond and J. S. Wood, *J. Chem. Soc., A*, **1970**, 226.

salt structure. Similarly, $ZnCl_2$ crystallizes in at least three polymorphs, two or more of which have tetrahedrally coordinated zinc atoms; $CdCl_2$, on the other hand, has only one form, involving octahedral coordination.

In their complexes, Zn, Cd and Hg commonly have coordination numbers four, five and six, with five especially common for zinc.[1] While the coordination number for all three elements in organo compounds is usually two, only Hg commonly forms linear bonds in other cases, for example in HgO. Indeed, linear 2-coordination is more characteristic for Hg(II) than for any other metal species.

THE ELEMENTS

18-3. Occurrence, Isolation and Properties

The elements have relatively low abundance in Nature (of the order 10^{-6} of the earth's crust for Zn and Cd), but have long been known because they are easily obtained from their ores.

Zinc occurs widely in a number of minerals, but the main source is *sphalerite*, (ZnFe)S, which commonly occurs with galena, PbS; cadmium minerals are scarce but, as a result of its chemical similarity to Zn, Cd occurs by isomorphous replacement in almost all zinc ores. There are numerous

[1] For references see D. P. Madden, M. M. da Mota and S. M. Nelson, *J. Chem. Soc., A*, **1970**, 790.

methods of isolation, initially involving flotation and roasting; Zn and Pb are commonly recovered simultaneously by a blast furnace method.[2] Cadmium is invariably a by-product and is usually separated from zinc by distillation or by precipitation from sulfate solutions by zinc dust:

$$Zn + Cd^{2+} = Zn^{2+} + Cd \qquad E = +0.36 \text{ V}$$

The only important ore of mercury is *cinnabar*, HgS; this is roasted to the oxide which decomposes at ca. 500°, the mercury vaporizing.

Properties. Some properties of the elements are listed in Table 18-1. Zinc and cadmium are white, lustrous, but tarnishable metals. Like Be and Mg, with which they are isostructural, their structures deviate from perfect hexagonal close packing by elongation along the six-fold axis. Mercury is a shiny liquid at ordinary temperatures. All are remarkably volatile for heavy metals, mercury, of course, uniquely so.[3] Mercury gives a monoatomic vapor and has an appreciable vapor pressure (1.3×10^{-3} mm) at 20°. It is also surprisingly soluble in both polar and non-polar liquids;[4] for example, a saturated solution in air-free water at 25° has 6×10^{-8} g/g. Because of its high volatility and toxicity, mercury should always be kept in stoppered containers and handled in well-ventilated areas. Mercury is readily lost from dilute aqueous solutions and even from solutions of mercuric salts[4a] owing to reduction of these by traces of reducing materials and by disproportionation of Hg_2^{2+}.

Both Zn and Cd react readily with non-oxidizing acids, releasing hydrogen and giving the divalent ions, whereas Hg is inert to non-oxidizing acids. Zinc also dissolves in strong bases because of its ability to form zincate ions (see below), commonly written ZnO_2^{2-}:

$$Zn + 2OH^- \rightarrow ZnO_2^{2-} + H_2$$

Cadmium does not react with base since cadmiate ions are of negligible stability.

Zinc and cadmium react readily when heated with oxygen, to give the oxides. Although mercury and oxygen are unstable with respect to HgO at room temperature, their rate of combination is exceedingly slow; the reaction proceeds at a useful rate at 300–350°, but, around 400° and above, the stability relation reverses and HgO decomposes rapidly into the elements:

$$HgO(s) = Hg(s) + \tfrac{1}{2}O_2 \qquad \Delta H_{diss} = 160 \text{ kJ mol}^{-1}$$

This ability of mercury to absorb oxygen from air and regenerate it again in pure form was of considerable importance in the earliest studies of oxygen by Lavoisier and Priestley.

All three elements react directly with halogens and with non-metals such as S, Se, P, etc.

[2] A. W. Richards, *Chem. in Britain*, **1969**, 203; *Non-Ferrous Extractive Metallurgy in the United Kingdom*, W. Ryan, ed., Institution of Mining and Metallurgy, London, 1968.

[3] For physical properties see A. V. Grosse, *J. Inorg. Nuclear Chem.*, 1965, **27**, 773.

[4] See, e.g., J. N. Spencer and A. F. Voigt, *J. Phys. Chem.*, 1968, **72**, 464, 471.

[4a] T. Y. Toribara, C. P. Shields and L. Koval, *Talanta*, 1970, **17**, 1025.

Zinc and cadmium form many alloys, some, such as brass, being of technical importance. Mercury combines with many other metals, sometimes with difficulty but sometimes, as with sodium or potassium, very vigorously, giving *amalgams*. Some amalgams have definite compositions; that is, they are compounds, such as Hg_2Na. Some of the transition metals do not form amalgams, and iron is commonly used for containers of mercury. Sodium amalgams and amalgamated zinc are frequently used as reducing agents for aqueous solutions.

THE UNIVALENT STATE

This state is of importance only for mercury, but unstable species of Zn^I and Cd^I exist. Although the last two have the formula M_2^{2+} there is evidence that highly unstable, strongly reducing Zn^+ and Cd^+ ions can be obtained when aqueous solutions of Zn^{2+} and Cd^{2+} are irradiated.[5]

When zinc is added to molten $ZnCl_2$ at 500–700° a yellow, diamagnetic glass is obtained on cooling. According to Raman and other spectra,[6] this glass contains Zn_2^{2+}. It is soluble in warm saturated $ZnCl_2$ solution, and the resulting greenish-yellow solution is stable for some days but, on dissolution in CH_3OH or acetone, decomposition and precipitation of Zn occur in less than a minute.

Similarly, when cadmium is dissolved in molten cadmium halides, very dark red melts are obtained. The high color may be due to the existence of both Cd^I and Cd^{II} joined by halide bridges, since mixed valence states in complexes are known to give intense colors in many other cases. If, for example, aluminum chloride is added to the Cd–$CdCl_2$ melt, only a greenish-yellow melt is obtained and phase studies here and for the bromide have shown the presence of Cd^I. Yellow solids can be isolated; they are diamagnetic and can be formulated $(Cd_2)^{2+}(AlCl_4^-)_2$.[7] When such solids are added to donor solvents or to water, cadmium metal is at once formed, together with Cd^{2+}, so that it is not surprising that there is no evidence for Cd^I in aqueous solution. The stabilization by the tetrahaloaluminate ions is presumably due to lowering of the difference between the lattice energies of the two oxidation states and lessening of the tendency to disproportionate. A similar case of this type of stabilization has been noted for Ga^I, as in $Ga^+AlCl_4^-$.

The force constants for the M—M bonds, obtained from Raman spectra, clearly show the stability order: Zn_2^{2+}, 0.6; Cd_2^{2+}, 1.1; Hg_2^{2+}, 2.5 mdyne $Å^{-1}$, which may be compared with values of 0.98 for K_2 and 1.7 for Na_2 and I_2.

---

[5] D. Meyerstein and W. A. Mulac, *Inorg. Chem.*, 1970, **9**, 1762; R. S. Eachus and M. C. R. Symons, *J. Chem. Soc.*, A, **1970**, 3080.
[6] D. H. Kerridge and S. A. Tariq, *J. Chem. Soc.*, A, **1967**, 1122.
[7] R. A. Potts, R. D. Barnes and J. D. Corbett, *Inorg. Chem.*, 1968, **7**, 2558.

18-4. The Mercurous Ion, Mercurous–Mercuric Equilibria and Mercurous Compounds

The mercurous ion, Hg_2^{2+}, is readily obtained from mercuric salts by reduction in aqueous solution and is readily re-oxidized to Hg^{2+}.

There have been many lines of evidence showing the binuclear nature of Hg_2^{2+}. A few of these may be noted:

1. Mercurous compounds are diamagnetic both as solids and in solution, whereas Hg^+ would have an unpaired electron.

2. X-ray determination of the structures of several mercurous salts shows the existence of individual Hg_2^{2+} ions. The Hg—Hg distances are far from constant (Table 18-3) and show no correlation with the electronegativity of the anion.

TABLE 18-3

Mercury–Mercury Bond Lengths in Mercurous Compounds[a]

Salt	Hg—Hg (Å)	Salt	Hg—Hg (Å)
Hg_2F_2	2.51	$Hg_2(NO_3)_2 \cdot 2H_2O$	2.54
Hg_2Cl_2	2.53	$Hg_2(BrO_3)_2$	2.51
Hg_2Br_2	2.49	$Hg_2SO_4^{b}$	2.50
Hg_2I_2	2.69		

[a] E. Dorm, *Chem. Comm.*, **1971**, 466.

[b] Contains chains, $-O-Hg-Hg-O-\overset{\overset{O}{\|}}{\underset{\underset{O}{\|}}{S}}-O-$ (E. Dorm, *Acta Chem. Scand.*, **1969**, 23, 1607).

3. The Raman spectrum of an aqueous solution of mercurous nitrate contains a strong line which can only be attributed to an Hg—Hg stretching vibration.

4. There are various kinds of equilibria for which constant equilibrium quotients can be obtained only by considering the mercurous ion to be Hg_2^{2+}. For example, suppose we add an excess of mercury to a solution initially X molar in mercuric nitrate. An equilibrium between Hg, Hg^{2+} and mercurous ion will be reached (see below); depending on the assumed nature of mercurous ion, the following equilibrium quotients can be written:

$$Hg(l) + Hg^{2+} = Hg_2^{2+} \qquad K = [Hg_2^{2+}]/[Hg^{2+}] = f/(1-f)$$
$$Hg(l) + Hg^{2+} = 2Hg^+ \qquad K' = [Hg^+]^2/[Hg^{2+}] = (2fX)^2/(1-f)X = 4f^2X/(1-f)$$

where f represents the fraction of the initial Hg^{2+} determined by analysis or otherwise to have disappeared when equilibrium is reached. It is found that, when values of K and K' are calculated from experimental data at different values of X, the former are substantially constant while the latter are not.

5. The electrical conductances of solutions of mercurous salts resemble closely, in magnitude and variation with concentration, the conductances of uni-divalent rather than uni-univalent electrolytes.

The greater strength of the Hg—Hg bond in Hg_2^{2+} compared with Cd—Cd in Cd_2^{2+} is reflected also on comparison of the bond energies $HgH^+ > CdH^+$ in the spectroscopic ions, and qualitatively the stability of Hg_2^{2+} is probably related to the large electron affinity of Hg^+. The electron affinity of M^+ (equal to the first ionization potential of the metal) is 1.4 eV greater for Hg^+ than for Cd^+. This results from the fact that the $4f$ shell in Hg shields the $6s$ electrons relatively poorly. The high ionization potential of Hg also accounts for the so-called "inert pair" phenomenon, namely, the exceptionally noble character of mercury and its low energy of vaporization.

Hg^I–Hg^{II} Equilibria. An understanding of the thermodynamics of these equilibria is essential to an understanding of the chemistry of the mercurous state. The important thermodynamic values are the potentials

$$Hg_2^{2+} + 2e = 2Hg(l) \qquad E^0 = 0.789 \text{ V} \qquad (18\text{-}1)$$
$$2Hg^{2+} + 2e = Hg_2^{2+} \qquad E^0 = 0.920 \text{ V} \qquad (18\text{-}2)$$
$$Hg^{2+} + 2e = Hg(l) \qquad E^0 = 0.854 \text{ V} \qquad (18\text{-}3)$$

For the disproportionation equilibrium

$$Hg_2^{2+} = Hg(l) + Hg^{2+} \qquad E^0 = -0.131 \text{ V} \qquad (18\text{-}4)$$

we then have
$$K = [Hg^{2+}]/[Hg_2^{2+}] = 6.0 \times 10^{-3}$$

The implication of the standard potentials is clearly that only oxidizing agents with potentials in the range -0.79 to -0.85 V can oxidize mercury to Hg^I but not to Hg^{II}. Since no common oxidizing agent meets this requirement, it is found that when mercury is treated with an excess of oxidizing agent it is entirely converted into Hg^{II}. However, when mercury is in at least 50% excess only Hg^I is obtained since, according to equation 18-4 Hg(l) readily reduces Hg^{2+} to Hg_2^{2+}.

The equilibrium constant for reaction 18-4 shows that, although Hg_2^{2+} is stable with respect to disproportionation, it is only so by a small margin. Thus any reagents that reduce the activity (by precipitation or complexation) of Hg^{2+} to a significantly greater extent than they lower the activity of Hg_2^{2+} will cause disproportionation of Hg_2^{2+}. There are many such reagents, so that the number of stable Hg^I compounds is rather restricted.

Thus, when OH^- is added to a solution of Hg_2^{2+}, a dark precipitate consisting of Hg and HgO is formed; evidently mercurous hydroxide, if it could be isolated, would be a stronger base than HgO. Similarly, addition of sulfide ions to a solution of Hg_2^{2+} gives a mixture of Hg and the extremely insoluble HgS. Mercurous cyanide does not exist because $Hg(CN)_2$ is so slightly dissociated though soluble. The reactions in these cited cases are

$$Hg_2^{2+} + 2OH^- \rightarrow Hg(l) + HgO(s) + H_2O$$
$$Hg_2^{2+} + S^{2-} \rightarrow Hg(l) + HgS$$
$$Hg_2^{2+} + 2CN^- \rightarrow Hg(l) + Hg(CN)_2(aq)$$

Mercurous Compounds. As indicated above, no hydroxide, oxide, or sulfide can be obtained by addition of the appropriate anion to aqueous Hg_2^{2+}, nor have these compounds been otherwise made.

Among the best known of the few mercurous compounds are the *halides.* The fluoride is unstable toward water, being hydrolyzed to hydrofluoric acid and unisolable mercurous hydroxide (which disproportionates as above). The other halides are highly insoluble, which thus precludes the possibilities of hydrolysis or disproportionation to give Hg^{II} halide complexes. *Mercurous nitrate* is known only as the dihydrate $Hg_2(NO_3)_2 \cdot 2H_2O$, which X-ray studies have shown to contain the ion $[H_2O—Hg—Hg—OH_2]^{2+}$; a *perchlorate*, $Hg_2(ClO_4)_2 \cdot 4H_2O$, is also known; both are very soluble in water, and the halides and other relatively insoluble salts of Hg_2^{2+} may conveniently be prepared by adding the appropriate anions to their solutions. Other known mercurous salts are the sparingly soluble sulfate, chlorate, bromate, iodate and acetate.

Mercurous ion forms few *complexes*; this may in part be due to a low tendency for Hg_2^{2+} to form coordinate bonds, but is probably due mainly to the fact that mercuric ion will form even more stable complexes with most ligands, for example, CN^-, I^-, amines and alkyl sulfides, so that the Hg_2^{2+} disproportionates. Nitrogen ligands of low basicity tend to favor Hg_2^{2+} and there are relatively stable complexes with aniline, $[Hg_2(PhNH_2)]^{2+}$, and with 1,10-phenanthroline. In $[Hg_2 \text{ phen}] (NO_3)_2$ the ligand is bound to only one metal atom.[8]

Complexes can readily be obtained in solution with ligands that form essentially ionic metal—ligand bonds and hence no strong complexes with mercury(II). Such ligands are oxalate, succinate, pyrophosphate and tripolyphosphate. Pyrophosphate gives the species $[Hg_2(P_2O_7)_2]^{6-}$ (pH range 6.5–9) and $[Hg_2(P_2O_7)OH]^{3-}$ for which stability constants have been measured.

18-5. Polymercury Cations

It has long been known that mercury dissolves slowly in fluorosulfuric acid to give a yellow solution. The color has been shown to be due to the trimercury ion, Hg_3^{2+}. Golden salts of the ion are obtained by oxidation reactions[8a] similar to those used to make salts of the S_n^{2+} ions (page 428):

$$3Hg + 3AsF_5 \xrightarrow{\text{SO}_2(l)} Hg_3(AsF_6)_2 + AsF_3$$
$$Hg + Hg_2Cl_2 + 2AlCl_3 \xrightarrow{\text{NaCl melt}} Hg_3(AlCl_4)_2$$

X-ray study of $Hg_3(AsF_6)_2$ shows that the ion is linear with $Hg—Hg = 2.55$ Å.

By using different stoichiometries of the reactants, the ions Hg_4^{2+}, Hg_5^{2+} and Hg_6^{2+} may be obtained, e.g.,

$$6Hg + 3AsF_5 \xrightarrow{\text{SO}_2(l)} Hg_6(AsF_5)_2$$

DIVALENT ZINC AND CADMIUM COMPOUNDS

18-6. Oxides and Hydroxides

The oxides, ZnO and CdO, are formed on burning the metals in air or by pyrolysis of the carbonates or nitrates; oxide smokes can be obtained by

---

[8] R. C. Elder, J. Halpern and J. S. Bond, *J. Amer. Chem. Soc.*, 1967, **89**, 6877.

[8a] G. Torsi and G. Mamantov, *Inorg. Nuclear Chem. Lett.*, 1970, **6**, 843; R. J. Gillespie *et al.*, *Chem. Comm.*, **1971**, 782, 1168.

combustion of the alkyls, cadmium oxide smokes being exceedingly toxic. Zinc oxide is normally white but turns yellow on heating; cadmium oxide varies in color from greenish-yellow through brown to nearly black, depending on its thermal history. These colors are the result of various kinds of lattice defects. Both oxides sublime without decomposition at very high temperatures.

The hydroxides are precipitated from solutions of salts by addition of bases. The solubility products of $Zn(OH)_2$ and $Cd(OH)_2$ are about 10^{-11} and 10^{-14}, respectively, but $Zn(OH)_2$ is more soluble than would be expected from this constant owing to the equilibrium

$$Zn(OH)_2(s) = Zn(OH)_2(aq) \qquad K = 10^{-6}$$

Further, $Zn(OH)_2$ readily dissolves in an excess of alkali bases to give zincate ions, probably of the type $[Zn(OH)_3(H_2O)]^-$ or $[Zn(OH)_4]^{2-}$. At high hydroxide concentrations the only species observed in solution is the latter;[9] solid zincates such as $NaZn(OH)_3$ and $Na_2[Zn(OH)_4]$ can be crystallized from concentrated solutions.

$Cd(OH)_2$ is insoluble in bases. Both Zn and Cd hydroxide readily dissolve in an excess of strong ammonia to form the ammine complexes, for example, $[Zn(NH_3)_4]^{2+}$.

18-7. Sulfides, Selenides and Tellurides

These are all crystalline substances, insoluble in water. Three structures are represented among the eight compounds as shown in Table 18-4. The

TABLE 18-4

Structures[a] of Zn and Cd Oxides and Chalconides[b]

Metal	O	S	Se	Te
Zn	W, Z	Z, W	Z	Z, W
Cd	NaCl	W, Z	W, Z	Z

[a] W = wurtzite structure; Z = zinc blende structure; NaCl = rock salt structure.
[b] Where two polymorphs occur, the one stable at lower temperatures is listed first.

rock salt and zinc blende structures have been described on pages 50 to 53. In the former, the cation is octahedrally surrounded by six anions, whereas in the latter the cation is tetrahedrally surrounded by anions. The *wurtzite structure* (from the mineral wurtzite, which is the stable high-temperature modification of ZnS) also gives the cations tetrahedral coordination and is shown in Fig. 18-1. It will be seen from Table 18-4 that zinc and cadmium prefer tetrahedral coordination in their chalconides.

[9] J. S. Fordyce and R. L. Baum, *J. Chem. Phys.*, 1965, **43**, 843.

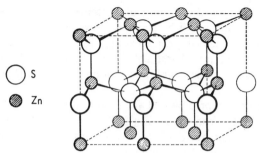

Fig. 18-1.   The wurtzite structure.

18-8. Halides

All four halides of both zinc and cadmium are known. Some of their relevant properties are given in Table 18-5.

TABLE 18-5

Some Properties of the Zinc and Cadmium Halides

	Solubility in water (mol l^{-1})	M.p. (°C)	B.p. (°C)	Structure
ZnF_2	1.57 (20°)	872	1502	Rutile
$ZnCl_2$	31.8 (25°)	275	756	Three forms; see text
$ZnBr_2$	20.9 (25°)	394	697	*ccp* anions with Zn in tetrahedral
ZnI_2	13 (25°)	446	(Sublimes)	interstices
CdF_2	0.29 (25°)	1110	1747	Fluorite
$CdCl_2$	7.7 (20°)	868	980	Close-packed anions with Cd in
$CdBr_2$	4.2 (20°)	568	1136	octahedral interstices
CdI_2	2.3 (20°)	387	(Sublimes)	

Both ZnF_2 and CdF_2 show distinct evidence of being considerably more ionic than the other halides of the same element. Thus they have higher melting and boiling points, and they are considerably less soluble in water. The latter fact is attributable, not only to the high lattice energies of the fluorides, but also to the fact that the formation of halo complexes in solution, which enhances the solubility of the other halides, does not occur for the fluorides (see below).

The structures of the chlorides, bromides and iodides may be viewed as close-packed arrays of halide ions, but there is a characteristic difference in that zinc ions occupy tetrahedral interstices whereas cadmium ions occupy octahedral ones. Actually, there are at least three polymorphs of $ZnCl_2$ known, two of which are similar and in which the presence of zinc ions in tetrahedral holes has been demonstrated. Raman spectra show that, depending on the concentration, the species present in aqueous solutions of $ZnCl_2$ are $[Zn(H_2O)_6]^{2+}$, $ZnCl^+(aq)$, $ZnCl_2(aq)$ and $[ZnCl_4(H_2O)_2]^{2-}$, but no indication was found of $[ZnCl_3]^-$ or $[ZnCl_4]^{2-}$ (see below).

Zinc chloride is so soluble in water that mole ratios $H_2O–ZnCl_2$ can easily be less than 2:1. Both zinc and cadmium halides are quite soluble in alcohol, acetone and similar donor solvents, and in some cases adducts can be obtained.

Aqueous solutions of cadmium halides appear, superficially, to be incompletely dissociated, that is, to be weak electrolytes. Although there are significant amounts of the undissociated halides, CdX_2, and polymeric species[10] present in moderately concentrated solutions, there are other species also present as shown in Table 18-6. Thus the solutions are best regarded as systems containing all possible species in equilibrium rather than simply as solutions of a weak electrolyte.

TABLE 18-6

Approximate Concentrations of Dissociated and Undissociated Species in $0.5M$ $CdBr_2$ Solution at 25°

	Concentration (M)		Concentration (M)
Cd^{2+}	0.013	Br^-	0.200
$CdBr^+$	0.259	$CdBr_2$	0.164
$CdBr_3^-$	0.043	$CdBr_4^{2-}$	0.021

18-9. Oxo Salts and Aquo Ions

Salts of oxo acids such as the nitrate, sulfate, sulfite, perchlorate and acetate are soluble in water. The Zn^{2+} and Cd^{2+} ions are rather similar to Mg^{2+}, and many of their salts are isomorphous with magnesium salts, for example, $Zn(Mg)SO_4·7H_2O$ and $M_2^ISO_4·Hg(Mg)SO_4·6H_2O$. The aquo ions are quite strong acids, and aqueous solutions of salts are hydrolyzed. In perchlorate solution the only species for Zn, Cd (and Hg) below 0.1 M are the MOH^+ ions, e.g.,

$$Zn^{2+}(aq) + H_2O \rightleftharpoons ZnOH^+(aq) + H^+$$

For more concentrated cadmium solutions, the principal species is Cd_2OH^{3+}:

$$2Cd^{2+}(aq) + H_2O \rightleftharpoons Cd_2OH^{3+}(aq) + H^+$$

In presence of complexing anions, e.g., halide, species such as $Cd(OH)Cl$ or $CdNO_3^+$ may be obtained.[10, 11]

Zinc forms a basic acetate, $Zn_4O(OCOCH_3)_6$, isomorphous with the oxoacetate of beryllium, on distillation of the normal acetate in a vacuum. This is a crystalline solid rapidly hydrolyzed by water, unlike the beryllium compound, the difference being due to the possibility of coordination numbers exceeding four for zinc.

Carbonates and hydroxo carbonates are also known.

[10] J. W. Macklin and R. A. Plane, *Inorg. Chem.*, 1970, **9**, 821.
[11] A. R. Davis and R. A. Plane, *Inorg. Chem.*, 1968, **7**, 2565.

18-10. Complexes of Zinc and Cadmium

Complex anions with halides are formed by both metals but the formation constants differ widely; they are also many orders of magnitude smaller than those of Hg^{2+}, as is clear from Tables 18-7 and 18-8. The exact values are

TABLE 18-7

Equilibrium Constants for Some Typical Complexes of
Zn, Cd and Hg
$(M^{2+} + 4X = [MX_4]; \; K = [MX_4]/[M^{2+}][X]^4)$

X	K		
	Zn^{2+}	Cd^{2+}	Hg^{2+}
Cl^-	1	10^3	10^{16}
Br^-	10^{-1}	10^4	10^{22}
I^-	10^{-2}	10^6	10^{30}
NH_3	10^9	10^7	10^{19}
CN^-	10^{21}	10^{19}	10^{41}

TABLE 18-8

Some Formation Constants of Zinc and Cadmium Halide Complexes (at 25°)

Halogen		Log K_1	Log K_2	Log K_3	Log K_4	Medium
Zn	F	0.75	Not obs.	Not obs.	Not obs.	0.5–1.0M NaClO$_4$
	Cl	−1.0 to +1.0	−1.0 to +1.0	−1.0 to +1.0	−1.0 to +1.0	Variable
	Br	−0.60	−0.37	−0.73	0.44	Ionic str. = 4.5
	I	−2.93	1.25	−0.07	−0.59	Ionic str. = 4.5
Cd	F	0.46	0.07	Not obs.	Not obs.	1.0M NaClO$_4$
	Cl	1.77	1.45	−0.25	−0.05	2.1M KNO$_3$
	Br	1.97	1.25	0.24	0.15	1M KNO$_3$
	I	2.96	1.33	1.07	1.00	1.6M KNO$_3$

not important since ionic strength effects are rather large, but certain qualitative features can be discerned. The formation of fluoro complexes is restricted, and none has been isolated as a solid. There is evidence for the attainment of all four stages of complexation by both zinc and cadmium with Cl^-, Br^- and I^-, the cadmium complexes being moderately stable while those of zinc are of rather low stability. The ZnX_4^{2-} complexes can be isolated as salts of large cations but cadmium[11a] can also form $CdCl_5^{3-}$. Zn^{2+} tends to form stronger bonds to F and O, whereas Cd^{2+} is more strongly bound to Cl, S and P ligands.

Complex cations with ammonia and amine ligands are well-defined and can be obtained as crystalline salts.

Zinc complexes of dithiocarbamates and of other sulfur compounds are important accelerators in the vulcanization of rubber by sulfur. The isostructural Zn and Cd compounds, $[M(S_2CNEt_2)_2]_2$, achieve 5-coordination

[11a] T. V. Long *et al.*, *Inorg. Chem.*, 1970, **9**, 459.

by dimerizing as shown for cadmium in (18-I); one metal—sulfur bond is considerably longer than the others.[12] These dimers may be split by amines to form 5-coordinate 1:1 adducts.[13] Zinc β-diketonates also form 5-coordinate adducts with water,[14] alcohols, and nitrogen bases. Five-coordination is, in fact, now being often found for zinc, a recent example being the hydrazinecarboxylato complex, $Zn(NH_2NHCOO)_2$.[15]

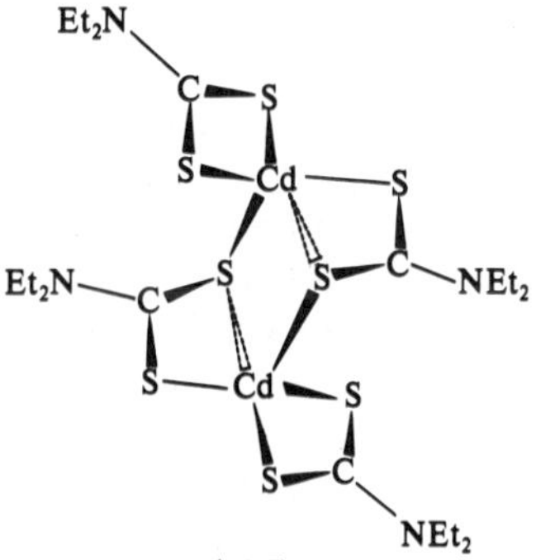

(18-I)

18-11. The Biological Role of Zinc

Zinc appears to be one of the most important metals biologically. It is probably second only to iron among the heavy metals (that is excluding Na^+, Ca^{2+} and Mg^{2+}). In the past 15–20 years more than 25 zinc-containing proteins have been identified, most of them enzymes. The Zn^{2+} ion is contained in several dehydrogenases, aldolases, peptidases, phosphatases, an isomerase of yeast, a transphosphorylase, and a phospholipase, attesting to its importance in carbohydrate, lipid and protein metabolism in virtually all organisms.[16]

Table 18-9 lists a few of the enzymes[17] known to contain zinc as an

TABLE 18-9

Some Metalloenzymes of Zinc

Name	Mol. wt.	Zn atoms per mol	Sources
Carbonic anhydrases	~30,000	1	Erythrocytes
Carboxypeptidases	~35,000	1	Pancreas
Dehydrogenases	~85,000*	$\geqslant 2$	Yeast, liver
Alkaline phosphatases	89,000	4	E. coli

* Some are about twice this. Sub-units of ~42,000 are apparently involved in all cases.

[12] J. F. Villa et al., Chem. Comm., 1971, 307.
[13] S. K. Gupta and T. S. Srivastava, J. Inorg. Nuclear Chem., 1970, 32, 1611.
[14] R. L. Belford et al., Inorg. Chem., 1969, 8, 1312.
[15] F. Bigoli et al., Chem. Comm., 1970, 120.
[16] B. L. Vallee and W. E. C. Wacker in The Proteins, H. Neurath, ed., Vol. V, 2nd edn., Academic Press, 1970.
[17] See A. F. Paresi and B. L. Vallee, Amer. J. Clinical Nutrition, 1969, 22, 1222.

essential constituent. Among the enzymes most thoroughly studied are the following:

Aldol dehydrogenases, from diverse species including yeast, horses and humans, catalyze the oxidation of ethanol or the reduction of acetaldehyde, using diphosphopyridine nucleotide (DPN) as a co-factor. Crystalline yeast alcohol dehydrogenase has a molecular weight of 150,000 and contains four Zn^{2+} ions and binds four DPN molecules per mole. Its structure and chemistry are not yet known in detail.

Alkaline phosphatases of bacteria catalyze hydrolyses of a variety of phosphomonoesters and also act as phosphotransferases. The one from *E. coli* contains four Zn^{2+} ions per unit of molecular weight of 89,000. Much still remains to be learned of the chemistry, and the structure is unknown.

Carboxypeptidase is specific for hydrolyses of *C*-terminal amino acids of proteins, peptides and corresponding esters. Carboxypeptidase from beef pancreas has a molecular weight of 34,300 and contains one zinc atom per mole. Chemical and X-ray crystallographic studies show that the Zn atom is at the active site and give a moderately detailed idea of how the enzyme functions and what role is played by the metal:[18] the Zn^{2+} ion is bound to the enzyme by two N and one O atom, a distorted tetrahedron being completed by a water molecule. A peptide carbonyl group displaces the water molecule when the substrate is bound, and hydrolysis of the peptide bond containing the bound carbonyl group then occurs by a series of proton transfers.

Metallothioneins are curious proteins, whose function is still unknown. They have been found in kidneys, liver and gut of horses, rabbits and humans. They have molecular weights of ca. 10,000, they contain ca. 9% by weight of metal, and nearly 1/3 of their amino acid residues are cysteines

$$—NH—CH—CO—$$
$$\qquad\quad |$$
$$\qquad CH_2SH$$

(18-II)

(18-II). The metal content is mainly zinc, but an appreciable fraction is cadmium. Metallothionein is the only biological material known to contain cadmium.

DIVALENT MERCURY COMPOUNDS[19]

18-12. Mercuric Oxide and Sulfide

Red mercuric oxide is formed on gentle pyrolysis of mercurous or mercuric nitrate, by direct interaction of mercury and oxygen at 300–350°, or as red

[18] R. E. Dickerson and I. Geis, The *Structure and Action* of *Proteins*, Harper and Row, New York, 1969, pp. 87–94.
[19] K. Aurivillius, *Arkiv Kemi*, 1965, **24**, 151; D. Grdenič, *Quart. Rev.*, 1965, **19**, 303.

crystals by heating of an alkaline solution of K_2HgI_4. Addition of OH^- to aqueous Hg^{2+} gives a yellow precipitate of HgO; the yellow form differs from the red only in particle size. The usual form of the oxide has a structure with zigzag chains —Hg—O—Hg— with Hg—O = 2.03 Å, $\angle$ HgOHg = 109°, and $\angle$ OHgO = 179°; there is only weak bonding between the chains, the shortest Hg—O distance here being 2.82 Å.

No hydroxide has been obtained, but the oxide is soluble (10^{-3} to 10^{-4} mol l^{-1}) in water, the exact solubility depending on particle size, to give a solution of what is commonly assumed to be the hydroxide, although there is no proof for such a species. This "hydroxide" is an extremely weak base:

$$K = [Hg^{2+}][OH^-]^2/[Hg(OH)_2] = 1.8 \times 10^{-22}$$

and is somewhat amphoteric, though more basic than acidic. The equilibria involved in a solution of red HgO in aqueous in $HClO_4$ have been interpreted in terms of the species Hg^{2+}, $HgOH^+$ and $Hg(OH)_2$. There is, however, no evidence for any hydroxo complex even in $2M$ NaOH. On the other hand, in the solid state mixed oxides such as $Hg_2Nb_2O_7$ are known; all of these contain linear O–Hg–O groups.[20]

Mercuric sulfide, HgS, is precipitated from aqueous solutions as a black, highly insoluble compound. The solubility product is 10^{-54}, but the sulfide is somewhat more soluble than this figure would imply because of some hydrolysis of Hg^{2+} and S^{2-} ions. The black sulfide is unstable with respect to a red form identical with the mineral cinnabar and changes into it when heated or digested with alkali polysulfides or mercurous chloride. The red form has a distorted sodium chloride lattice with Hg—S chains similar to those in HgO. Another form, occurring as the mineral metacinnabarite, has a zinc blende structure, as have the selenide and telluride.

18-13. Mercuric Halides

The *fluoride* is essentially ionic and crystallizes in the fluorite structure; it is almost completely decomposed, even by cold water, as would be expected for an ionic compound that is the salt of a weak acid and an extremely weak base. Not only does mercury(II) show no tendency to form covalent Hg—F bonds, but no fluoro complex is known.

In sharp contrast to the fluoride, the other halides show marked covalent character. *Mercuric chloride* crystallizes in an essentially molecular lattice, the two short Hg—Cl distances being about the same length as the Hg—Cl bonds in gaseous $HgCl_2$, while the next shortest distances are much longer (see Table 18-10).

In the larger lattice of $HgBr_2$ each Hg atom is surrounded by six Br atoms, but two are so much closer than the other four that it can be considered that perturbed $HgBr_2$ molecules are present. The normal red form of HgI_2

[20] A. W. Sleight, *Inorg. Chem.*, 1968, **7**, 1704.

TABLE 18-10

Hg—X Distances in Mercuric Halides (in Å)

Compound	Solid			Vapor
	Two at	Two at	Two at	
HgF_2		Eight at 2.40		—
$HgCl_2$	2.25	3.34	3.63	2.28 ± 0.04
$HgBr_2$	2.48	3.23	3.23	2.40 ± 0.04
HgI_2 (red)		Four at 2.78		2.57 ± 0.04

has a layer structure with HgI_4 tetrahedra linked at some of the vertices. However, at 126° it is converted into a yellow molecular form.[21]

In the vapor all three halides are distinctly molecular, as they are also in solutions. Relative to ionic HgF_2, the other halides have very low melting and boiling points (Table 18-11). They also show marked solubility in many organic solvents. In aqueous solution they exist almost exclusively (~99%) as HgX_2 molecules, but some hydrolysis occurs, the principal equilibrium being, e.g.,[22]

$$HgCl_2 + H_2O \rightleftharpoons Hg(OH)Cl + H^+ + Cl^-$$

TABLE 18-11

Some Properties of Mercuric Halides

Halide	M.p. (°C)	B.p. (°C)	Solubility, moles/100 moles at 25°			
			in H_2O	in C_2H_5OH	in $C_2H_5OCOCH_3$	in C_6H_6
HgF_2	645 d.	—	Hydrolyzes	Insol.	Insol.	Insol.
$HgCl_2$	280	303	0.48	8.14	9.42	0.152
$HgBr_2$	238	318	0.031	3.83	—	—
HgI_2	257	351	0.00023	0.396	0.566	0.067

18-14. Mercuric Oxo Salts

Among the mercuric salts that are totally ionic and hence highly dissociated in aqueous solution are the nitrate, sulfate, and perchlorate. Because of the great weakness of mercuric hydroxide, aqueous solutions of these salts tend to hydrolyze extensively and must be acidified to be stable.

In aqueous solutions of $Hg(NO_3)_2$ the main species are $Hg(NO_3)_2$, $HgNO_3^+$ and Hg^{2+}, but at high concentrations of NO_3^- the complex anion $[Hg(NO_3)_4]^{2-}$ is formed.[23]

Mercuric carboxylates, especially the acetate and the trifluoroacetate, are of considerable importance because of their utility in attacking unsaturated hydrocarbons, as discussed below. They are made by dissolving HgO in the

[21] A. J. Melveger et al., Inorg. Chem., 1968, 7, 1630.
[22] L. Ciavatta and M. Grimaldi, J. Inorg. Nuclear Chem., 1968, 30, 563.
[23] A. R. Davis and D. E. Irish, Inorg. Chem., 1969, 8, 1699.

hot acid and crystallizing. The trifluoroacetate[24] is also soluble in benzene, acetone and tetrahydrofuran, which increases its utility, while the acetate is soluble in water and alcohols.

Other salts such as the oxalate and phosphates are sparingly soluble in water.

Mercuric ions catalyze a number of reactions of complex compounds[25] such as the aquation of $[Cr(NH_3)_5X]^{2+}$. Bridged transition states, e.g.,

$$[(H_2O)_5CrCl]^{2+} + Hg^{2+} = [(H_2O)_5Cr\!-\!Cl\!-\!Hg]^{4+}$$

are believed to be involved.

18-15. Mercuric Complexes

A number of these have been mentioned above. The Hg^{2+} ion has indeed a strong tendency to complex formation, and the characteristic coordination numbers and stereochemical arrangements are two-coordinate, linear, and four-coordinate, tetrahedral. Octahedral coordination is less common; a few three- and five-coordinate complexes are also known. There appears to be considerable covalent character in the mercury—ligand bonds, especially in the two-coordinate compounds. The most stable complexes are those with C, N, P and S as ligand atoms.

Halide and pseudohalide complexes.[26] In the halide systems, depending on the concentration of halide ions, there are equilibria such as[27]

$$HgX^+ \rightleftharpoons HgX_2 \rightleftharpoons HgX_3^- \rightleftharpoons HgX_4^{2-}$$

At $1M$ Cl^- the main species is $[HgCl_4]^{2-}$ but in tributyl phosphate as solvent the most stable ion is $[HgCl_3]^-$; at $10^{-1}M$ Cl^- the concentrations of $HgCl_2$, $HgCl_3^-$ and $HgCl_4^{2-}$ are about equal.

Mercuric cyanide, $Hg(CN)_2$, which contains discrete molecules with linear C—Hg—C bonds,[28] is soluble in CN^-, to give $Hg(CN)_3^-$ and $Hg(CN)_4^{2-}$ only.[29] SCN^- is similar to CN^- in its complexing behavior.

Oxo ion complexes. Several of these exist, e.g., $[Hg(SO_3)_2]^{2-}$ and $[Hg\ ox_2]^{2-}$. The yellow crystals formed by adding KNO_2 to $Hg(NO_3)_2$ solution contain $K_2[Hg(NO_2)_4]\cdot KNO_3$, where the nitrite ion is bidentate, giving an eight-coordinate very distorted square antiprism.[30]

Other ligands. The acetylacetonate contains mercury bound to the central (or γ) carbon atom.[31] It also appears that mercurous perchlorate and nitrate react with aqueous acetone to give complex species which have the acetone bound to mercury as the enolate anion.

[24] H. C. Brown and M.-H. Rei, *J. Amer. Chem. Soc.*, 1968, **90**, 5647.
[25] See, e.g., D. A. Buckingham *et al.*, *Inorg. Chem.*, 1970, **9**, 11; J. N. Armor and A. Haim, *J. Amer. Chem. Soc.*, 1971, **93**, 861; S. W. Foong, B. Kipling and A. G. Sykes, *J. Chem. Soc., A.*, **1971**, 118.
[26] G. B. Deacon, *Rev. Pure Appl. Chem.* (*Australia*), 1963, **13**, 189.
[27] J. E. D. Davis and D. A. Long, *J. Chem. Soc., A*, **1968**, 2564.
[28] R. C. Seccombe and C. H. L. Kennard, *J. Organometallic Chem.*, 1969, **13**, 243.
[29] K. G. Ashurst, N. P. Finkelstein and L. A. Goold, *J. Chem. Soc., A*, **1971**, 1899.
[30] D. Hall and R. V. Holland, *Inorg. Chim. Acta*, 1969, **3**, 235.
[31] F. Bonati and G. Minghetti, *J. Organometallic Chem.*, 1970, **22**, 5.

Phosphine, arsine, sulfide and bi- and ter-pyridine[32] generally give mercuric halide complexes that are either monomeric, L_2HgX_2, or dimeric as in (18-III). More highly bridged structures such as (18-IV) may be formed with PR_3 and AsR_3.

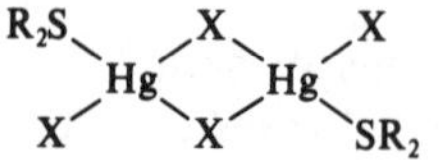

(18-III) 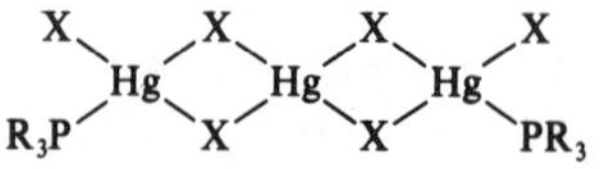   (18-IV)

Although there is a tendency to form ammonobasic compounds, a variety of amines form complexes with Hg^{II} and the affinity of Hg^{II} for nitrogen ligands in aqueous solution exceeds that of the transition metals. In addition to the ammonia and amine complexes of the type $Hg(NH_3)_2X_2$, tetra-ammines such as $[Hg(NH_3)_4](NO_3)_2$ can be prepared in saturated aqueous ammonium nitrate. The ion $[Hg\ en_3]^{2+}$ has octahedral Hg^{II}, as have complexes of the type $[HgL_6](ClO_4)_2$ obtained when suitable oxygen donors are added to Hg^{II} perchlorate in ethanol.

18-16. Novel Compounds of Mercury(II) with Nitrogen[33]

It has been known since the days of alchemy that when Hg_2Cl_2 is treated with aqueous ammonia a black residue is formed, and this reaction is still used in qualitative analysis to identify Hg_2Cl_2. Only relatively recently has the nature of the reaction been clarified. These residues contain nitrogen compounds of Hg^{II} plus metallic mercury, and the Hg^{II} compounds can be obtained directly from Hg^{II} salts.

There are three known products of the reaction of $HgCl_2$ with ammonia, the proportion of any one of them depending on the conditions. The possible products are $Hg(NH_3)_2Cl_2$, $HgNH_2Cl$ and $Hg_2NCl \cdot H_2O$ and they are formed according to the following equations:

$$HgCl_2 + 2NH_3 \rightleftharpoons Hg(NH_3)_2Cl_2(s)$$
$$HgCl_2 + 2NH_3 \rightleftharpoons HgNH_2Cl(s) + NH_4^+ + Cl^-$$
$$2HgCl_2 + 4NH_3 + H_2O \rightleftharpoons Hg_2NCl \cdot H_2O + 3NH_4^+ + 3Cl^-$$

The equilibria represented here seem to be labile so that the product obtained can be controlled by varying the concentrations of NH_3 and NH_4^+. In concentrated NH_4Cl solution, the diammine $Hg(NH_3)_2Cl_2$ is precipitated, whereas, with dilute ammonia and no excess of NH_4^+, the amide $HgNH_2Cl$ is formed. The compound $Hg_2NCl \cdot H_2O$ is probably not produced in a pure state by the above reaction, but it can be obtained by treating the compound $Hg_2NOH \cdot 2H_2O$ (Millon's base) with hydrochloric acid. Millon's base itself is made by the action of aqueous ammonia on yellow mercuric oxide.

The diammine has been shown to consist of discrete tetrahedral molecules.

[32] L. W. Houk and G. R. Dobson, *J. Chem. Soc.*, A, **1968**, 1846; A. R. Davis, C. J. Murphy and R. A. Plane, *Inorg. Chem.*, 1970, 9, 423.
[33] D. Breitinger and K. Broderson, *Angew. Chem. Internat. Edn.*, 1970, 9, 357 (an extensive review).

The amide has infinite chains —Hg—NH$_2$—Hg—NH$_2$—, where the N—Hg—N segments are linear while the bonds about nitrogen are tetrahedral; the chloride ions lie between the chains. The analogous bromide has the same structure.

Millon's base has a three-dimensional framework of composition Hg$_2$N with the OH$^-$ ions and water molecules occupying rather spacious cavities and channels. Many salts of Millon's base are known, for example, Hg$_2$NX·nH$_2$O (X = NO$_3^-$, ClO$_4^-$, Cl$^-$, Br$^-$ or I$^-$; $n = 0-2$). In these the framework appears to remain essentially unaltered, and thus it is an ion-exchanger similar to a zeolite.

Returning to the dark residues given by mercurous chloride, one or both of HgNH$_2$Cl and Hg$_2$NCl·H$_2$O are present together with free metal. The insolubility of these compounds causes the disproportionation of the Hg$_2^{2+}$, for example

$$Hg_2Cl_2(s) + 2NH_3 = Hg(l) + HgNH_2Cl(s) + NH_4^+ + Cl^-$$

The action of aqueous ammonia on organomercury compounds such as C$_6$H$_5$HgCl also gives amido derivatives, (C$_6$H$_5$Hg)$_2$NH$_2^+$, that are ionic in nature.[34]

There is no evidence for any intermediate ammonobasic or ammine compound of mercury(I).

18-17. Compounds with Metal-to-Mercury Bonds

A wide variety of compounds are known in which the Hg atom is bound to other metals, very commonly transition-metal atoms, in complexes. These are all best formulated as Hg^{II} compounds, and where the structures are known the compounds have linear bonds M—Hg—M or M—HgX.[35] Not only Hg, but also Zn and Cd, form similar compounds.[36] Some typical examples are (CO)$_4$CoMCo(CO)$_4$ (M = Zn, Cd or Hg), Fe(CO)$_4$(HgCl)$_2$, h^5-C$_5$H$_5$Fe(CO)$_2$–Hg–Co(CO)$_4$. The general nature of compounds with metal–metal bonds has been outlined elsewhere.

Some of these compounds are made by action of HgCl$_2$ on hydrido complexes or on carbonylate anions (Section 22-7), e.g.,

$$2C_5H_5Mo(CO)_3^- Na^+ + HgCl_2 \rightarrow [C_5H_5Mo(CO)_3]_2Hg + 2NaCl$$
$$(Ph_2MeAs)_3RhHCl_2 + HgCl_2 \rightarrow (Ph_2MeAs)_3Rh(HgCl)Cl_2 + HCl$$

For other transition-metal complexes, HgCl$_2$ may undergo the oxidative-addition reaction (Section 24-A-2), adding as HgCl and Cl, or it may add as the HgCl$_2$ *molecule*.[37] Examples of the latter type of addition are not confined to transition metals, but sulfides R$_2$S may also form "oxidative adducts"

[34] J. B. Deacon and J. M. S. Green, *J. Chem. Soc.*, A, **1968**, 1182.
[35] W. P. Griffith, *J. Chem. Soc.*, A, **1969**, 834; B. Lee, J. M. Burlitch and J. L. Hoard, *J. Amer. Chem. Soc.*, 1967, **85**, 6363; A. R. Manning, *J. Chem. Soc.*, A, **1968**, 1018; G. M. Sheldrick and R. N. F. Simpson, *J. Chem. Soc.*, A, **1968**, 1005.
[36] See, e.g., J. M. Burlitch and A. Ferrari, *Inorg. Chem.*, 1970, **9**, 563.
[37] D. M. Adams, D. J. Cook and R. D. W. Kemmitt, *J. Chem. Soc.*, A, **1968**, 1067.

such as $(R_2S)HgCl_2$. The adduct $(h^5\text{-}C_5H_5)(CO)_2Co \cdot HgCl_2$ has the structure (18-V) in which the axial positions of the trigonal-bipyramid are occupied by weakly coordinated chlorine atoms of adjacent molecules.

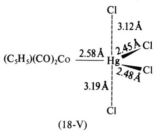

(18-V)

ORGANOMETALLIC COMPOUNDS

18-18. Organozinc and Organocadmium Compounds[38]

Organozinc compounds are historically important[39] since they were the first organometallic compounds to be prepared; their discovery by Frankland in 1849 played a decisive part in the development of modern ideas of chemical bonding. The zinc and cadmium compounds are also of interest since their mild reactivities toward certain organic functional groups give them unique synthetic potentialities.

Organozinc compounds of the types "RZnX" and R_2Zn are known, whereas except for Bu^nCdCl^{40} only R_2Cd compounds have been isolated. The constitution of RMX in solution has presented a problem similar to that for Grignard reagents. Spectroscopic studies[40a] show that RMX species predominate in ethers, e.g.,

$$Me_2Cd + CdI_2 \xrightleftharpoons{\text{THF}} 2MeCdI \qquad K \geqslant 100$$

There is self exchange in Me_2Cd and with Zn, Ga and In alkyls *via* alkyl bridged species.[41]

The zinc alkyls can be obtained by thermal decomposition of RZnI, which is prepared by the reaction of alkyl iodides with a zinc–copper couple:

$$C_2H_5I + Zn(Cu) \rightarrow C_2H_5ZnI \xrightarrow{\text{Heat}} \tfrac{1}{2}(C_2H_5)_2Zn + \tfrac{1}{2}ZnI_2$$

The alkyls may also be prepared, and the diaryls most conveniently obtained, by the reaction of zinc metal with an organomercury compound:

$$R_2Hg + Zn \rightarrow R_2Zn + Hg$$

or by reaction of zinc chloride with organolithium or Grignard reagents.

[38] I. Sheverdina and K. A. Kocheskov, *The Organic Compounds of Zinc and Cadmium*, North Holland, 1967; G. E. Coates and K. Wade, *Organometallic Compounds*, Vol. 1, Methuen, 1967.
[39] See J. S. Thayer, *J. Chem. Educ.*, 1969, **46**, 764.
[40] J. R. Sanders and E. C. Ashby, *J. Organometallic Chem.*, 1970, **25**, 277.
[40a] D. F. Evans and G. Fazakerley, *J. Chem. Soc.*, A, **1971**, 182.
[41] W. Bremser, M. Winokur and J. D. Roberts, *J. Amer. Chem. Soc.*, 1970, **92**, 1080.

The best preparation of R_2Cd compounds is by treatment of the anhydrous cadmium halide with RLi or RMgX. The reaction of cadmium metal with alkyl iodides in dimethylformamide or $(CH_3)_2SO$ gives RCdI in solution.

The R_2Zn and R_2Cd compounds are non-polar liquids or low-melting solids, soluble in most organic liquids. The lower alkyl zinc compounds are spontaneously flammable, and all react vigorously with oxygen and with water. The cadmium compounds are less sensitive to oxygen but are less stable thermally.

Both zinc and cadmium compounds react readily with compounds containing active hydrogen, such as alcohols:

$$R_2M + R'OH \rightarrow RMOR' + RH$$

and are generally similar to RLi or RMgX, although their lower reactivity allows selective alkylations not possible with the more standard reagents. An important example is the use of the cadmium compounds in the synthesis of ketones from acyl chlorides:

$$2RCOCl + R_2Cd \rightarrow 2RCOR' + CdCl_2$$

With lithium alkyls and aryls, complexes such as $Li[ZnPh_3]$ and $Li_2[CdMe_4]$ may be formed,[42] while hydrido complexes such as $Li[(Et_2Zn)_2H]$ appear to have Zn—H—Zn bridges.[43]

18-19. Organomercury Compounds[44]

A vast number of organomercury compounds are known, some of which have useful physiological properties. They are of the types RHgX and R_2Hg. They are commonly made by the interaction of $HgCl_2$ and RMgX, but Hg—C bonds can also be made in other ways discussed below.

The *RHgX compounds* are crystalline solids whose properties depend on the nature of X. When X is an atom or group that can form covalent bonds to mercury, for example, Cl, Br, I, CN, SCN or OH, the compound is a covalent non-polar substance more soluble in organic liquids than in water. When X is SO_4^{2-} or NO_3^-, the substance is salt-like and presumably quite ionic, for instance, $[RHg]^+NO_3^-$. Acetates behave as weak electrolytes. For iodides or thiocyanates, complex anions, e.g., $RHgI_2^-$ and $RHgI_3^{2-}$, may be formed.

The *dialkyls and diaryls* are non-polar, volatile, toxic, colorless liquids or low-melting solids. Unlike the Zn and Cd alkyls, they are much less affected by air or water, presumably because of the low polarity of the Hg—C bond and the low affinity of mercury for oxygen. However, they are photochemically and thermally unstable, as would be expected from the low bond strengths, which are of the order 50–200 kJ mol^{-1}. In the dark, mercury

[42] L. M. Sutz and B. F. Little, *J. Organometallic Chem.*, 1969, **16**, 227; J. Yamamoto and C. A. Wilkie, *Inorg. Chem.*, 1971, **10**, 1129.
[43] G. J. Kubas and D. F. Shriver, *J. Amer. Chem. Soc.*, 1970, **92**, 1949.
[44] L. G. Makarova and A. N. Nesmeyanov, *The Organic Compounds of Mercury*, North Holland, 1967 (comprehensive monograph).

compounds can be easily kept for months. The decomposition generally proceeds by homolysis of the Hg—C bond and free-radical[45] reactions.

All RHgX and R_2Hg compounds have linear bonds, but deviation from linearity has been claimed in a few cases, and, in solution in particular, solvation effects may contribute to non-linearity. The cyclopentadienyl, $(C_5H_5)_2Hg$,[46] C_5H_5HgX[46] and indenyl, $(C_9H_7)_2Hg$,[47] compounds are fluxional (see Sects. 1-9 and 23-7).

The principal utility of dialkyl- and diaryl-mercury compounds, and a very valuable one, is in the preparation of other organo compounds by inter-change, e.g.,

$$\frac{n}{2}R_2Hg + M \rightarrow R_nM + \frac{n}{2}Hg$$

This reaction proceeds essentially to completion with the Li and Ca groups, and with Zn, Al, Ga, Sn, Pb, Sb, Bi, Se and Te, but with In, Tl and Cd reversible equilibria are established. Partial alkylation of reactive halides can be achieved, e.g., $AsCl_3 + Et_2Hg \rightarrow EtHgCl + EtAsCl_2$

The special class of perhalogenomercury compounds, which may be used as sources of dihalogenocarbenes for transfer reactions to organic sub-strates, has been mentioned on page 285. A convenient preparation of such compounds[48] is

$$PhHgCl + CHX_3 + t\text{-}C_4H_9OK \xrightarrow{\text{Benzene}} PhHgCX_3 + KCl + t\text{-}C_4H_9OH$$

There has been very extensive study of the mechanisms of reaction of organomercury compounds, but only brief mention can be made here.[49] Exchange reactions of the type:

$$MeHgBr + {}^*HgBr_2 \rightleftharpoons Me^*HgBr + HgBr_2$$

have been extensively studied using tracer mercury. It was shown that the electrophilic substitution, S_E2, proceeds with full retention of configuration when an optically active group, sec-butyl, is present. The reaction, which is also catalyzed by anions, is believed to proceed through a cyclic transition state such as (18-VI). Reactions such as

$$R_2Hg + HgX_2 = 2RHgX$$

have equilibrium constants of 10^5–10^{11} and proceed at rates that are slow

[45] K. Bass, Organometallic Chem. Rev., 1966, 1, 391.

[46] E. Maslowsky and K. Nakamoto, Inorg. Chem., 1969, 8, 1108; W. Kitching and B. F. Hegarty, J. Organometallic Chem., 1969, 16, P39; F. A. Cotton and T. J. Marks, J. Amer. Chem. Soc., 1969, 91, 7281; P. West et al., J. Amer. Chem. Soc., 1969, 91, 5649.

[47] F. A. Cotton and T. J. Marks, J. Amer. Chem. Soc., 1969, 91, 3178.

[48] D. Seyferth and R. L. Lambert, J. Organometallic Chem., 1969, 16, 21; D. C. Muller and D. Seyferth, J. Amer. Chem. Soc., 1969, 91, 1754, 5027.

[49] F. R. Jensen and B. Rickborn, Electrophilic Substitution of Organomercurials, McGraw-Hill, 1968; O. A. Reutov and I. P. Beletskaya, Reactions of Organometallic Compounds, North Holland, 1968 (despite the title, this is mainly concerned with mercury compounds); D. E. Matteson, Organometallic Chem. Rev., A, 1969, 4, 263 (a good critical review of electrophilic displacements by mercury compounds); R. E. Dessy and W. Kitching Adv. Organometallic Chem., 1966, 4, 268 (reactions mechanism involving Hg–C bonds, including mercuration and oxomercuration); W. Kitching, Organometallic Chem. Rev., 1968, 3, 35, 61 (mercuration and oxomercuration).

and solvent-dependent. Nmr studies have also shown that, in solutions of RHgI compounds, there is relatively fast exchange of R groups and, again, a mechanism involving a cyclic intermediate or transition state (18-VII) has been postulated.

$$
\begin{array}{cc}
\text{(18-VI)} & \text{(18-VII)}
\end{array}
$$

Finally, we note that mercury released to the environment, as metal, e.g., by losses from electrolytic cells used for NaOH and Cl_2 production, or as compounds such as alkylmercury seed dressings or fungicides, constitutes a serious hazard.[50] This is a result of biological methylation to give $(CH_3)_2Hg$ or CH_3Hg^+. It is known that models for vitamin B_{12} such as methylcobaloximes or methylpentacyanocobaltate (Section 25-F) which have $Co—CH_3$ bonds will transfer the CH_3 to Hg^{2+}. There are a number of microorganisms that can perform the same function, possibly by similar routes.

Mercuration and Oxomercuration. An important reaction for the formation of Hg—C bonds, and one that can be adapted to the synthesis of a wide variety of organic compounds, is the addition of mercuric salts, notably the acetate, trifluoroacetate or nitrate to unsaturated compounds.[51]

The simplest reaction is the mercuration reaction of aromatic compounds, which is commonly achieved by the action of mercuric acetate in methanol, e.g.,

$$
\text{C}_6\text{H}_6 + \text{Hg(OCOCH}_3)_2 \longrightarrow \text{C}_6\text{H}_5\text{HgOCOCH}_3 + \text{CH}_3\text{COOCH}_3
$$

Even aromatic organometallic compounds such as tricarbonylcyclobutadienyliron, $C_4H_4Fe(CO)_3$ (Section 23-3), can be mercurated.[51a]

By use of aryl compounds prepared in this way, the arylation of alkenes and a variety of other organic unsaturated compounds can be achieved.[52] A palladium compound, usually Li_2PdCl_4, is used as a transfer agent and organopalladium species are believed to be unstable intermediates; by use of air and a Cu^{II} salt the reaction can be made catalytic, since the Pd metal formed in the reaction is dissolved by Cu^{II} and the Cu^I so produced is oxidized by air (cf. the Wacker process, Section 24-B-5). A typical reaction is

$$
\text{C}_6\text{H}_5\text{HgCl} + \!\!\!\!\!\!\!\!\!\!\!\!\!\begin{array}{c}\backslash\ \ /\\ \text{C}=\text{C}\\ /\ \ \backslash\end{array}\!\!\!\!\!\!\!\!\!\!\!\! \xrightarrow{\text{Li}_2\text{PdCl}_4} \begin{array}{c}\text{C}_6\text{H}_5\backslash\ \ /\\ \text{C}=\text{C}\\ /\ \ \backslash\end{array}
$$

[50] L. Dunlap, *Chem. Eng. News*, July 5th, 1971, p. 22.
[51] See L. F. Fieser and M. Fieser, *Reagents for Organic Synthesis*, Wiley, 1967, pp. 644–660.
[51a] G. Amiet, K. Nicholas and R. Pettit, *Chem. Comm.*, **1970**, 161.
[52] R. F. Heck, *J. Amer. Chem. Soc.*, 1968, **90**, 5518, 5542; 1969, **91**, 6707.

The interaction of mercuric salts is not confined to arenes, but with alkenes there appears to be a general reversible reaction

$$\underset{}{\overset{}{>}}C=C\overset{}{\underset{}{<}} \;+\; HgX_2 \;\;\rightleftharpoons\;\; \overset{X}{\underset{}{\overset{}{|}}}C-C\overset{}{\underset{HgX}{<}} \qquad (18\text{--}5)$$

In most cases, the reactions have to be carried out in an alcohol or other protic medium, so that further reaction with the solvent is normally complete and the reaction is called *oxomercuration*, e.g.,

$$\overset{}{\underset{}{>}}C=C\overset{}{\underset{}{<}} \;+\; Hg(OCOCH_3)_2 \;+\; C_2H_5OH \;\longrightarrow\; \overset{C_2H_5O}{\underset{}{|}}C-C\overset{}{\underset{HgOCOCH_3}{<}} \;+\; CH_3COOH$$

The evidence that HgX_2 adds across the bond is usually indirect, often, by observing the products on hydrolysis, e.g.,

$$CH_2{=}CH_2 + Hg(NO_3)_2 \xrightarrow{\;OH^-\;} HOCH_2CH_2Hg^+ + 2NO_3^-$$

or by removal of Hg as $HgCl_2$ by action of HCl, which reverses the addition:

$$\overset{}{\underset{}{>}}C=C\overset{}{\underset{}{<}} + Hg(OCOCH_3)_2 \longrightarrow \overset{CH_3COO}{\underset{}{|}}C-C\overset{}{\underset{HgOCOCH_3}{<}} \xrightarrow{\;HCl\;} \overset{}{\underset{}{>}}C=C\overset{}{\underset{}{<}} + HgCl_2 + 2CH_3COOH$$

The reversibility of the initial addition is readily established by using $Hg(OCOCF_3)_2$ and, since the latter is soluble in non-polar solvents, the equilibrium constants for the addition reaction (18-5) where $X = OCOCF_3$ can readily be measured and the dependence on the nature of the alkene and the solvent studied.[53]

In reactions such as the above, mercurinium ions of the type (18-VIII) and (18-IX) are believed to be intermediates.[54] Indeed in $FSO_3H\text{--}SbF_5\text{--}SO_2$ at $-70°$ long-lived ions have been obtained[55] by reactions such as

$$CH_3OCH_2CH_2HgCl \xrightarrow{\;H^+\;} CH_3OH_2^+ + CH_2CH_2Hg^{2+} + HCl$$

$$\overset{Hg^{2+}}{\underset{}{\triangle}}\quad \overset{X}{\underset{Hg}{\triangle}}$$

$$\qquad (18\text{-VIII}) \qquad (18\text{-IX})$$

The above type of addition has been used for the simple preparation, from alkenes and other unsaturated substances, of alcohols, ethers and amines—the additions of HgX_2 are carried out in water, alcohols or acetonitrile, respectively. The mercury is removed from the intermediate by reduction with

[53] H. C. Brown and K.-T. Liu, *J. Amer. Chem. Soc.*, 1970, **92**, 1760; R. D. Bach and H. F. Henneike, *J. Amer. Chem. Soc.*, 1970, **92**, 5589; J. E. Byrd and J. Halpern, *J. Amer. Chem. Soc.*, 1970, **92**, 6967.
[54] See, e.g., D. J. Pasto and J. A. Gontarz, *J. Amer. Chem. Soc.*, 1970, **92**, 7480.
[55] G. A. Olah and P. R. Clifford, *J. Amer. Chem. Soc.*, 1971, **93**, 1262, 2320.

sodium borohydride.[56] Such additions are also useful in that the products are those in the Markownikoff direction. Two examples are the following:

$$RCH{=}CH_2 + CH_3CN + Hg(NO_3)_2 = \underset{\underset{\underset{ONO_2}{|}}{\underset{N\,=\,CCH_3}{|}}}{RCHCH_2HgNO_3} \xrightarrow[\text{NaOH}]{\text{NaBH}_4} \underset{\underset{\underset{O}{\|}}{\underset{NHCCH_3}{|}}}{RCHCH_3}$$

The catalytic activity of mercuric salts in sulfuric acid solutions for hydration of acetylenes doubtless proceeds by routes similar to the above; the overall reaction is

$$RC{\equiv}CR' + H_2O \xrightarrow{H^+} \underset{\underset{OH\ H}{|\quad |}}{R{-}C{=}C{-}R'} \rightarrow \underset{\underset{O}{\|}}{R{-}C{-}CH_2R'}$$

Acetylene itself gives acetaldehyde.

Finally it is of interest that methanolic solutions of Hg^{II} acetate readily absorb carbon monoxide at atmospheric pressure[57] and the resulting compound can be converted by halide salts into compounds of the type $XHgOCOCH_3$. It has been shown that carbon monoxide is, in effect, inserted between the Hg and O of a solvolyzed mercuric ion, though the mechanism is not established in detail:

$$Hg(OCOCH_3)_2 + CH_3OH \rightleftharpoons CH_3COOHgOCH_3 + CH_3COOH$$
$$CH_3COOHg{-}OCH_3 + CO \rightleftharpoons CH_3COOHgC(O)OCH_3$$

The CO can be regenerated from the compounds by heating or by action of concentrated hydrochloric acid. Under pressures of 25 atm, reactions such as

$$RHgNO_3 + CO + CH_3OH \rightarrow RCOOCH_3 + Hg + HNO_3$$

can be carried out.

[56] H. C. Brown *et al.*, *J. Amer. Chem. Soc.*, 1969, **91**, 5646, 5647; D. J. Pasto and J. A. Gontarz, *J. Amer. Chem. Soc.*, 1969, **91**, 719; G. M. Whitesides and J. S. Fillipo, *J. Amer. Chem. Soc.*, 1970, **92**, 6611; H. Arzoumanian and J. Metzger, *Synthesis*, 1971, **10**, 527.

[57] See, e.g., L. R. Barlow and J. M. Davidson, *J. Chem. Soc.*, A, **1968**, 1609: J. M. Davidson, *J. Chem. Soc.*, A, **1969**, 193.

Further Reading

Bidstrup, P. L., *Toxicity of Mercury and its Compounds*, Elsevier, 1964.
Chizhikov, D. M., *Cadmium*, Pergamon Press, 1966 (mainly the technology of production).
Fleischer, A., and J. J. Lander, *Zinc–Silver Oxide Batteries*, Wiley, 1971.
Roberts, H. L., *Adv. Inorg. Chem. Radiochem.*, 1968, **11**, 309 (an excellent account of mercury chemistry containing useful thermodynamic and other data).

PART THREE

Chemistry of the Transition Elements

19

Introduction to the Transition Elements

The purpose of this Chapter is to provide brief general discussions of a few topics that pertain to the transition elements as a class, rather than to any particular one or group within the class. Of course, some of these topics, such as magnetism or optical rotation, have indeed still greater generality, but they are discussed here because of their special pertinence among the transition elements.

ELECTRONIC STRUCTURES

19-1. Definition and General Characteristics of Transition Elements

The transition elements may be strictly defined as those that, *as elements*, have partly filled d or f shells. Here we shall adopt a slightly broader definition and include also elements that have partly filled d or f shells in any of their commonly occurring oxidation states. This means that we treat the coinage metals, Cu, Ag and Au, as transition metals, since Cu^{II} has a $3d^9$ configuration, Ag^{II} a $4d^9$ configuration and Au^{III} a $5d^8$ configuration. From a purely chemical point of view it is also appropriate to consider these elements as transition elements since their chemical behavior is, on the whole, quite similar to that of other transition elements.

With the above broad definition in mind, one finds that there are at present some 56 transition elements, counting the heaviest elements through the one of atomic number 104. Clearly, the majority of all known elements are transition elements. All these transition elements have certain general properties in common:

1. They are all metals.

2. They are practically all hard, strong, high-melting, high-boiling metals that conduct heat and electricity well. In short, they are "typical" metals of the sort we meet in ordinary circumstances.

3. They form alloys with one another and with other metallic elements.

4. Many of them are sufficiently electropositive to dissolve in mineral acids, although a few are "noble"—that is, they have such low electrode potentials that they are unaffected by simple acids.

5. With very few exceptions, they exhibit variable valence, and their ions and compounds are colored in one if not all oxidation states.

6. Because of partially filled shells they form at least some paramagnetic compounds.

This large number of transition elements is subdivided into three main groups: (a) the main transition elements or d-block elements, (b) the lanthanide elements and (c) the actinide elements.

The main transition group or d block includes those elements that have partially filled d shells only. Thus, the element scandium, with the outer electron configuration $4s^2 3d$, is the lightest member. The eight succeeding elements, Ti, V, Cr, Mn, Fe, Co, Ni and Cu, all have partly filled 3d shells either in the ground state of the free atom (all except Cu) or in one or more of their chemically important ions (all except Sc). This group of elements is called the *first transition series*. At zinc the configuration is $3d^{10}4s^2$, and this element forms no compound in which the 3d shell is ionized, nor does this ionization occur in any of the next nine elements. It is not until we come to yttrium, with ground-state outer electron configuration $5s^2 4d$, that we meet the next transition element. The following eight elements, Zr, Nb, Mo, Tc, Ru, Rh, Pd and Ag, all have partially filled 4d shells either in the free element (all but Ag) or in one or more of the chemically important ions (all but Y). This group of nine elements constitutes the *second transition series*.

Again there follows a sequence of elements in which there are never d-shell vacancies under chemically significant conditions until we reach the element lanthanum, with an outer electron configuration in the ground state of $6s^2 5d$. Now, if the pattern we have observed twice before were to be repeated, there would follow eight elements with enlarged but not complete sets of 5d electrons. This does not happen, however. The 4f shell now becomes slightly more stable than the 5d shell, and, through the next fourteen elements, electrons enter the 4f shell until at lutetium it becomes filled. Lutetium thus has the outer electron configuration $4f^{14}5d6s^2$. Since both La and Lu have partially filled d shells and no other partially filled shells, it might be argued that both of these should be considered as d-block elements. However, for chemical reasons, it would be unwise to classify them in this way, since all of the fifteen elements La $(Z = 57)$ through Lu $(Z = 71)$ have very similar chemical and physical properties, those of lanthanum being in a sense prototypal; hence, these elements are called the *lanthanides*, and their chemistry will be considered separately in Chapter 27. Since the properties of Y are extremely similar to, and those of Sc mainly like, those of the lanthanide elements proper, and quite different from those of the regular d-block elements, we shall treat them also in Chapter 27.

For practical purposes, then, the *third transition series* begins with hafnium, having the ground-state outer electron configuration $6s^2 5d^2$, and embraces the elements Ta, W, Re, Os, Ir, Pt and Au, all of which have partially filled $5d$ shells in one or more chemically important oxidation states as well as (excepting Au) in the neutral atom.

Continuing on from mercury, which follows gold, we come via the noble-gas radon and the radioelements Fr and Ra to actinium, with the outer electron configuration $7s^2 6d$. Here we might expect, by analogy to what happened at lanthanum, that in the following elements electrons would enter the $5f$ orbitals, producing a lanthanide-like series of fifteen elements. What actually occurs is, unfortunately, not so simple. Although, immediately following lanthanum, the $4f$ orbitals become decisively more favorable than the $5d$ orbitals for the electrons entering in the succeeding elements, there is apparently not so great a difference between the $5f$ and $6d$ orbitals until later. Thus, for the elements immediately following Ac, and their ions, there may be electrons in the $5f$ or $6d$ orbitals or both. Since it appears that later on, after four or five more electrons have been added to the Ac configuration, the $5f$ orbitals do become definitely the more stable, and since the elements from about americium on do show moderately homologous chemical behavior, it has become accepted practice to call the fifteen elements beginning with Ac the *actinide elements*.

There is an important distinction, based upon electronic structures, between the three classes of transition elements. For the d-block elements the partially filled shells are d shells, $3d$, $4d$ or $5d$. These d orbitals project well out to the periphery of the atoms and ions so that the electrons occupying them are strongly influenced by the surroundings of the ion and, in turn, are able to influence the environments very significantly. Thus, many of the properties of an ion with a partly filled d shell are quite sensitive to the number and arrangement of the d electrons present. In marked contrast to this, the $4f$ orbitals in the lanthanide elements are rather deeply buried in the atoms and ions. The electrons that occupy them are largely screened from the surroundings by the overlying shells ($5s$, $5p$) of electrons, and therefore reciprocal interactions of the $4f$ electrons and the surroundings of the atom or the ion are of relatively little chemical significance. This is why the chemistry of all the lanthanides is so homologous, whereas there are seemingly erratic and irregular variations in chemical properties as one passes through a series of d-block elements. The behavior of the actinide elements lies between those of the two types described above because the $5f$ orbitals are not so well shielded as are the $4f$ orbitals, although not so exposed as are the d orbitals in the d-block elements.

19-2. Position in the Periodic Table

Fig. 19-1 shows in a qualitative way the relative variations in the energies of the atomic orbitals as a function of atomic number in neutral atoms.

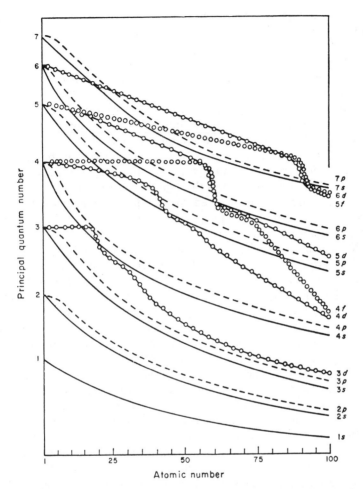

Fig. 19-1. The variation of the energies of atomic orbitals with increasing atomic number in neutral atoms (energies not strictly to scale).

It is well to realize that in a multielectron atom, one with say twenty or more electrons, the energies of all of the levels are more or less dependent on the populations of all the other levels. Hence the diagram is rather complicated.

We see that in hydrogen all the subshells of each principal shell are equienergic. As we proceed to more complex atoms, these various subshells, *s, p, d, f, g,* etc., split apart and at the same time drop to lower energies. This descent in energy occurs because the degree to which an electron in a particular orbital is shielded from the nuclear charge by all the other electrons in the atom is insufficient to prevent a steady increase in the *effective nuclear charge* felt by that electron with increasing atomic number. In other words, each electron is imperfectly shielded from the nuclear charge by the other electrons. The energy of an electron in an atom is given by

$$E = -\frac{2\pi^2 \mu e^4 (Z^*)^2}{n^2 h^2}$$
(19-1)

where Z^* is the effective nuclear charge, and the energy of the electron falls as Z^* increases.

The reason why the diagram is so complicated, however, is that all sub-shells do not drop in parallel fashion, but rather in varying and somewhat irregular ways. This is because the several subshells of the same principal shell are shielded to different degrees by the core of electrons beneath.

From Fig. 19-1 we see that the $1s$, $2s$, $2p$, $3s$ and $3p$ levels occur in that sequence in all known atoms. Thus through those atoms (H to Ar) in which this sequence of orbitals is being filled, they are filled in that order. While the filling of this set of orbitals is taking place, the energies of the higher and as yet unfilled orbitals are being variously affected by the screening power of these first eighteen electrons. In particular, the $3d$ levels, which penetrate the argon core rather little, have scarcely dropped in energy when we reach argon ($Z = 18$), whereas the $4s$ and $4p$ levels, especially the former, which penetrate the argon core quite a bit, have dropped rather steeply. Thus, when two more electrons are added to the argon configuration to give the potassium and calcium atoms, they enter the $4s$ orbital, which has fallen below the $3d$ orbitals. As these two electrons are added, the nuclear charge is also increased by two units. Since the $3d$ orbitals penetrate the electron density in the $4s$ orbitals very considerably, the net result is that the effective nuclear charge for the $3d$ orbitals increases rather abruptly, and they now drop well below the $4p$ orbitals to about the level of the $4s$ orbital. The next electron therefore enters the $3d$ shell, and scandium has the configuration $[Ar]4s^2 3d$. This $3d$ electron screens the $4p$ levels more effectively than it screens the remaining $3d$ orbitals, so the latter remain the lowest available orbitals and the next electron is also added to the $3d$ shell to give Ti with the configuration $[Ar]4s^2 3d^2$. This process continues in a similar way until the entire $3d$ shell is filled. Thus at Zn we have the configuration $[Ar]4s^2 3d^{10}$, and the $4p$ orbitals, now the lowest available ones, become filled in the six succeeding elements.

This same sequence of events is repeated again in the elements following krypton, which has the electron configuration $[Ar]3d^{10}4s^2 4p^6$. Because of the way in which the shielding varies, the $4d$ levels, which in a one-electron atom would be next in order of stability, are higher in energy than the $5s$ and $5p$ orbitals, so that the next two electrons added go into the $5s$ orbitals, giving the alkali and alkaline-earth elements Rb and Sr. But the shielding of the $4d$ orbitals by these $5s$ electrons is very poor, so that the $4d$ orbitals feel strongly the increase of two units of nuclear charge and take a sharp drop, becoming appreciably more stable than the $5p$ orbitals, and the next electron added becomes a $4d$ electron. Thus the next element, Y, is the first member of the second transition series. This series is completed at Ag, configuration $[Kr]4d^{10}5s^2$, and then six $5p$ electrons are added to make Xe, the next noble gas.

At Xe ($Z = 54$) the next available orbitals are the $6s$ and $6p$ orbitals. The $4f$ orbitals are so slightly penetrating with respect to the Xe core that they have scarcely gained any stability, while the more penetrating $6s$ and $6p$ levels have gained a good deal. Hence the next two electrons added are $6s$ electrons, giving again an alkali and an alkaline-earth element, Cs and Ba, respectively. However, the $6s$ shell scarcely shields the $4f$ orbitals, so the latter abruptly feel an increase in effective nuclear charge and thus suffer a steep drop in energy. At the same time, however, the energy of the $5d$ levels also drops abruptly, just as did that of $(n-1)d$ levels previously as electrons are added to the ns level, and the final situation is one in which, at Ba, the $6s$, $5d$ and $4f$ levels are all of about the same energy. The next entering electron, in the element lanthanum, enters a $5d$ orbital, but the following element, cerium, has the configuration $6s^2 4f^2$. Through the next twelve elements electrons continue to enter the $4f$ orbitals, and it is likely that even at cerium they are intrinsically more stable than the $5d$'s. Certainly they are so by the time we reach ytterbium, with the configuration $6s^2 4f^{14}$. Now, with the $6s$ and $4f$ shells full, the next lowest levels are unequivocally the $5d$'s, and from lutetium, with the configuration $6s^2 4f^{14} 5d$, through mercury, with the configuration $[Xe]6s^2 4f^{14} 5d^{10}$, the ten $5d$ electrons are added. Chemically, lanthanum and lutetium, each of which has a single $5d$ electron, are very similar to one another, and all the elements in between, with configurations $[Xe]4f^n 6s^2$, have chemical properties intermediate between those of lanthanum and lutetium. Consequently these fifteen elements are all considered as members of one class, the lanthanides. Hafnium, $[Xe]4f^{14} 5d^2 6s^2$, through gold are the eight elements that we regard as the members of the third transition series.

Following mercury there are six elements in which electrons enter the $6p$ orbitals until the next noble gas, radon, is reached. Its configuration is $[Xe]4f^{14} 5d^{10} 6s^2 6p^6$. The $5f$ orbitals have dropped so much more slowly, because of their relatively non-penetrating character, than have the $7s$ and $7p$ orbitals that the next two electrons beyond the radon core are added to the $7s$ level, and again an alkali and an alkaline-earth element are formed, namely, Fr, $[Rn]7s$, and Ra, $[Rn]7s^2$. But, again in analogy to the situation one row up in the Periodic Table, both the $5f$ and the $6d$ orbitals penetrate the $7s$ orbitals very considerably; they are thus abruptly stabilized relative to the $7p$ orbitals, and the next electrons added enter them. It appears that as we proceed through actinium and the following elements, the energies of the $6d$ and $5f$ orbitals remain for a while so similar that the exact configuration is determined by interelectronic forces of the sort discussed in Section 19-3. In the case of protactinium it is not certain whether the ground state is $[Rn]7s^2 6d^3$, $[Rn]7s^2 6d^2 5f$, $[Rn]7s^2 6d 5f^2$ or $[Rn]7s^2 5f^3$. These four configurations doubtless differ very little in energy, and for chemical purposes the question which is actually the lowest is not of great importance. The next element, uranium, appears definitely to have the configuration $[Rn]7s^2 5f^3 6d$, and the elements thereafter are all believed to have the configurations $[Rn]7s^2 5f^n 6d$. The important point is that, around actinium, the $6d$ and $5f$ levels are of practically the same energy, with the $5f$'s probably becoming slowly more stable later on.

19-3. Electron Configurations of the Atoms and Ions

In this Section we shall look more closely at the factors determining the electron configurations of transition-metal atoms and ions. The discussion in the preceding Section is not entirely adequate or accurate because it takes account only of the shielding of a given electron from the nuclear charge by other electrons in the atom. One electron may help to determine the orbital occupied by another electron not only in this indirect way but also because of direct interactions between the electrons. It is these direct interactions that cause the differences in energy between different states derived from the same configuration, as explained in more detail in Section 3-3. In cases where the energies of two orbitals differ by an amount comparable to or less than the energies arising from electron–electron interactions, it is not possible to predict electron configurations solely by consideration of the order of orbital energies, and the problem requires a more searching analysis.

An especially conspicuous and important example of the dominance of interelectronic interactions over orbital-energy differences is the "special stability" of half-filled shells. Examples are found in the first transition series and in the lanthanides, notably the boxed items in the annexed series.

	Sc	Ti	V	Cr	Mn	Fe	Co	Ni	Cu	Zn
$4s$	2	2	2	1	2	2	2	2	1	2
$3d$	1	2	3	5	5	6	7	8	10	10

	Sm	Eu	Gd	Tb
$6s$	2	2	2	2
$5d$	0	0	1	0
$4f$	6	7	7	9

Half-filled shells have an amount of exchange energy considerably greater than would be interpolated from the energies of the configurations to either side of them. Hence there is a driving force either to take an electron "out of turn," as with Cr and Cu, or to shunt an excess electron to another shell of similar energy in order to achieve or maintain the half-filled arrangement. All spins are parallel, giving maximum spin multiplicity in these half-filled shells.

In the second transition series the irregularities become more complex, as shown in the annexed series Y to Cd. No simple analysis is possible here;

	Y	Zr	Nb	Mo	Tc	Ru	Rh	Pd	Ag	Cd
$5s$	2	2	1	1	1	1	1	0	1	2
$4d$	1	2	4	5	6	7	8	10	10	10

both nuclear–electron and electron–electron forces play their roles in determining these configurations. Although a preference for the filled $4d$ shell is evident at the end of the series and the elements Nb and Mo show a preference for the half-filled shell, the configuration of Tc shows that this preference is not controlling throughout this series.

It is also well to point out that the interelectronic forces and variations in total nuclear charge play a large part in determining the configurations of ions. We cannot say that because $4s$ orbitals became occupied before $3d$ orbitals they are always more stable. If this were so, then we should expect the elements of the first transition series to ionize by loss of $3d$ electrons, whereas, in fact, they ionize by loss of $4s$ electrons first. Thus it is the net effect of all the forces—nuclear-electronic attraction, shielding of one electron by others, interelectronic repulsions and the exchange forces—that determines the stability of an electron configuration; and, unfortunately, there are many cases in which the interplay of these forces and their sensitivity to changes in nuclear charge and the number of electrons present cannot be simply described.

MAGNETIC PROPERTIES OF CHEMICAL SUBSTANCES

19-4. The Importance of Magnetism in Transition Element Chemistry

Many—indeed, most—compounds of the transition elements are paramagnetic, and much of our understanding of transition metal chemistry has been derived from magnetic data. Consequently, it is necessary to explain the salient facts and principles of magnetism, from a chemical viewpoint, before proceeding to detailed discussion of chemistry.

All the magnetic properties of substances in bulk are ultimately determined by the electrical properties of the subatomic particles, electrons, and nucleons. Because the magnetic effects due to nucleons and nuclei are some 10^{-3} times those due to electrons, they ordinarily have no detectable effect on magnetic phenomena of direct chemical significance. This is not to say that chemical phenomena do not have significant effects upon nuclear magnetism; it is just such effects that make nuclear magnetic resonance spectroscopy an extremely useful tool for the chemist. Thus we shall concentrate our attention entirely on the properties of the electron and on the magnetic properties of matter that result therefrom. We shall see that there are direct and often sensitive relationships between the magnetic properties of matter in bulk and the number and distribution of unpaired electrons in its various constituent atoms or ions.

There are several kinds of magnetism, qualitatively speaking; the salient features of each are summarized in Table 19-1. In the following Sections we shall consider first *paramagnetism*. A paramagnetic substance is attracted into a magnetic field with a force proportional to the field strength times the field gradient. Paramagnetism is generally caused by the presence in the substance of ions, atoms or molecules having unpaired electrons. Each of these has a definite paramagnetic moment that exists in the absence of any external magnetic field. A *diamagnetic* substance is repelled by a magnetic

TABLE 19-1

Main Types of Magnetic Behavior

Type	Sign of χ_M	Magnitude[a] of χ_M in cgs units	Dependence of χ_M on H	Origin
Diamagnetism	−	1–500×10^{-6}	Independent	Electron charge
Paramagnetism	+	0–10^{-2}	Independent	Spin and orbital motion of electrons on individual atoms
Ferromagnetism	+	10^{-2}–10^6	Dependent	Cooperative interaction between magnetic moments of individual atoms
Antiferromagnetism	+	0–10^{-2}	May be dependent	

[a] Assuming molecular or ionic weights in the range of about 50–1000. χ_M is the susceptibility per mole of substance, as explained on page 541.

field. All matter has this property to some extent. Diamagnetic behavior is due to small magnetic moments that are induced by the magnetic field but do not exist in the absence of the field. Moments so induced are in opposition to the inducing field, thus causing repulsion. Finally, there are the more complex forms of magnetic behavior known as ferromagnetism and antiferromagnetism, and still others that will not be discussed here.

19-5. Origin of Paramagnetic Moments

Electrons determine the magnetic properties of matter in two ways. First, each electron is, in effect, a magnet in itself. From a pre-wave-mechanical viewpoint, the electron may be regarded as a small sphere of negative charge spinning on its axis. Then, from completely classical considerations, the spinning of charge produces a magnetic moment. Secondly, an electron traveling in a closed path around a nucleus, again according to the pre-wave-mechanical picture of an atom, will also produce a magnetic moment, just as does an electric current traveling in a loop of wire. The magnetic properties of any individual atom or ion will result from some combination of these two properties, that is, the inherent *spin moment* of the electron and the *orbital moment* resulting from the motion of the electron around the nucleus. These physical images should not, of course, be taken too literally, for they have no place in wave mechanics, nor do they provide a basis for quantitatively correct predictions. They are qualitatively useful conceptual aids, however.

The magnetic moments of atoms, ions, and molecules are usually expressed in units called *Bohr magnetons*, abbreviated B.M. The Bohr magneton is defined in terms of fundamental constants as

$$1 \text{ B.M.} = \frac{eh}{4\pi mc} \qquad (19\text{-}2)$$

where e is the electronic charge, h is Planck's constant, m is the electron mass and c is the speed of light. This is *not*, however, the moment of a single electron. Because of certain features of quantum theory, the relationship is a little more complicated.

The magnetic moment, μ_s, of a single electron is given, according to wave mechanics, by the equation

$$\mu_s \text{ (in B.M.)} = g\sqrt{s(s+1)} \qquad (19\text{-}3)$$

in which s is simply the absolute value of the spin quantum number and g is the gyromagnetic ratio, more familiarly known as the "g factor." The quantity $\sqrt{s(s+1)}$ is the value of the angular momentum of the electron, and thus g is the ratio of the magnetic moment to the angular momentum, as its name is intended to suggest. For the free electron, g has the value 2.00023, which may be taken as 2.00 for most purposes. From equation 19-3 we can calculate the spin magnetic moment of one electron as

$$\mu_s = 2\sqrt{\tfrac{1}{2}(\tfrac{1}{2}+1)} = \sqrt{3} = 1.73 \text{ B.M.}$$

Thus any atom, ion or molecule having one unpaired electron (e.g., H, Cu^{2+}, ClO_2) should have a magnetic moment of 1.73 B.M. from the electron spin alone. This may be augmented or diminished by an orbital contribution as will be seen below.

There are transition-metal ions having one, two, three... up to seven unpaired electrons. As indicated in Section 3-3, the spin quantum number for the ion as a whole, S, is the sum of the spin quantum numbers, $s = \tfrac{1}{2}$, for the individual electrons. For example, in the manganese(II) ion with five unpaired electrons, $S = 5(\tfrac{1}{2}) = \tfrac{5}{2}$; and in the gadolinium(III) ion with seven unpaired electrons, $S = 7(\tfrac{1}{2}) = \tfrac{7}{2}$. Thus we can use equation 19-3, substituting S for s, to calculate the magnetic moment due to the electron spins alone, the so-called "spin-only" moment, for any atom or ion provided we know the total spin quantum number, S. The results are summarized in Table 19-2 for all possible real cases.

In the two examples chosen above, namely, Mn^{II} and Gd^{III}, the observed values of their magnetic moments agree very well with the spin-only values in Table 19-2. Generally, however, experimental values differ from the spin-only ones, usually being somewhat greater. This is because the orbital motion of the electrons also makes a contribution to the moment. The theory by which the exact magnitudes of these orbital contributions may be calculated is by no means simple and we shall give here only a superficial and pragmatic account of the subject. More detailed discussion of a few specific cases will be found in several places later in the text.

For Mn^{II}, Fe^{III}, Gd^{III} and other ions whose ground states are S states there is no orbital angular momentum even in the free ion. Hence there

TABLE 19-2

"Spin-only" Magnetic Moments for Various Numbers
of Unpaired Electrons

No. of unpaired electrons	S	μ_S (B.M.)
1	$\frac{1}{2}$	1.73
2	1	2.83
3	$\frac{3}{2}$	3.87
4	2	4.90
5	$\frac{5}{2}$	5.92
6	3	6.93
7	$\frac{7}{2}$	7.94

cannot be any orbital contribution to the magnetic moment, and the spin-only formula applies exactly.* In general, however, the transition-metal ions in their ground states, D or F being most common, do possess orbital angular momentum. Wave mechanics shows that for such ions, if the orbital motion makes its full contribution to the magnetic moments, they will be given by

$$\mu_{S+L} = \sqrt{4S(S+1) + L(L+1)} \qquad (19\text{-}4)$$

in which L represents the orbital angular momentum quantum number for the ion.

In Table 19-3 are listed magnetic moments actually observed for the common ions of the first transition series together with the calculated values of μ_S and μ_{S+L}. It will be seen that observed values of μ frequently exceed μ_S, but seldom are as high as μ_{S+L}. This is because the electric fields of other atoms, ions and molecules surrounding the metal ion in its compounds restrict the orbital motion of the electrons so that the orbital angular momen-

TABLE 19-3

Theoretical and Experimental Magnetic Moments for Various Transition Metal Ions
(in Bohr magnetons)

Ion	Ground state quantum numbers		Spectroscopic symbol	μ_S	μ_{S+L}	Observed moments
	S	L				
V^{4+}	$\frac{1}{2}$	2	^2D	1.73	3.00	1.7–1.8
Cu^{2+}	$\frac{1}{2}$	2	^2D	1.73	3.00	1.7–2.2
V^{3+}	1	3	^3F	2.83	4.47	2.6–2.8
Ni^{2+}	1	3	^3F	2.83	4.47	2.8–4.0
Cr^{3+}	$\frac{3}{2}$	3	^4F	3.87	5.20	~3.8
Co^{2+}	$\frac{3}{2}$	3	^4F	3.87	5.20	4.1–5.2
Fe^{2+}	2	2	^5D	4.90	5.48	5.1–5.5
Co^{3+}	2	2	^5D	4.90	5.48	~5.4
Mn^{2+}	$\frac{5}{2}$	0	^6S	5.92	5.92	~5.9
Fe^{3+}	$\frac{5}{2}$	0	^6S	5.92	5.92	~5.9

* Because of certain high-order effects and also, in part, because of covalence in metal–ligand bonds, slight departures (i.e. a few tenths of a B.M.) from the spin-only moments are sometimes observed.

tum and hence the orbital moments are wholly or partially "*quenched.*" In some cases, e.g., d^3 and d^8 ions in octahedral environments and d^7 ions in tetrahedral ones, the quenching of orbital angular momentum in the ground state is expected to be complete according to the simplest arguments, and yet such systems deviate from spin-only behavior. However, when the effect of spin–orbit coupling is considered, it is found that orbital angular momentum is mixed into the ground state from the first excited state of the system. This phenomenon is discussed quantitatively for the d^7 ion Co^{II} in Sect. 25-F-3. In the case of a d^3 ion in an octahedral environment, the orbital contribution is introduced in *opposition* to the spin contribution and moments slightly below the spin-only value are therefore observed, as for Cr^{III}.

Finally, it is noteworthy that in many systems that contain unpaired electrons, as well as in a few, e.g., CrO_4^{2-}, that do not, weak paramagnetism that is *independent* of temperature can arise by a coupling of the ground state of the system with excited states of high energy under the influence of the magnetic field. This *temperature-independent paramagnetism*, TIP, thus resembles diamagnetism in that it is not due to any magnetic dipole existing in the molecule but is induced when the substance is placed in the magnetic field. It also resembles diamagnetism in its temperature-independence and in its order of magnitude, viz., 0–500×10^{-6} cgs units per mole (cf. Table 19-1). It is often ignored in interpreting the paramagnetic behavior of ions with unpaired electrons, but in work that pretends to accuracy it should not be. Certainly, when measured susceptibilities are corrected for diamagnetism (see next Section) it is illogical not to correct them also for TIP if this is known to occur in the system concerned.

19-6. Diamagnetism

Diamagnetism is a property of all forms of matter. All substances contain at least some, if not all, electrons in closed shells. In closed shells the electron-spin moments and orbital moments of individual electrons balance one another out so that there is no net magnetic moment. However, when an atom or molecule is placed in a magnetic field, a small magnetic moment directly proportional to the strength of the field is induced. The electron spins have nothing to do with this induced moment; they remain tightly coupled together in antiparallel pairs. However, the planes of the orbitals are tipped slightly so that a small net orbital moment is set up in opposition to the applied field. It is because of this opposition that diamagnetic substances are repelled from magnetic fields.

Even an atom with a permanent magnetic moment will have diamagnetic behavior working in opposition to the paramagnetism when placed in a magnetic field, provided only that the atom has one or more closed shells of electrons. Thus, the net paramagnetism measured is slightly less than the true paramagnetism because some of the latter is "canceled out" by the diamagnetism.

Since diamagnetism is usually several orders of magnitude weaker than paramagnetism, substances with unpaired electrons almost always have a net paramagnetism. Of course, a very dilute solution of a paramagnetic ion in a diamagnetic solvent such as water may be diamagnetic because of the large ratio of diamagnetic to paramagnetic species in it. Another important feature of diamagnetism is that its magnitude does not vary with temperature. This is because the moment induced depends only on the sizes and shapes of the orbitals in the closed shells and these are not temperature-dependent.

19-7. Magnetic Susceptibility

Chemically useful information is obtained by proper interpretation of measured values of magnetic moments. However, magnetic moments are not measured directly. Instead, one measures the magnetic susceptibility of a material from which it is possible to calculate the magnetic moment of the paramagnetic ion or atom therein.

Magnetic susceptibility is defined in the following way. If a substance is placed in a magnetic field of magnitude H, the flux B, within the substance, is given by

$$B = H + 4\pi I \tag{19-5}$$

I is called the intensity of magnetization. The ratio B/H, called the magnetic permeability of the material, is given by

$$B/H = 1 + 4\pi(I/H) = 1 + 4\pi\kappa \tag{19-6}$$

κ is called the magnetic susceptibility per unit volume, or simply the volume susceptibility. The physical significance of equation 19-6 is easily seen. The permeability, B/H, is just the ratio of the density of lines of force within the substance to the density of such lines in the same region in the absence of the specimen. Thus, the volume susceptibility of a vacuum is by definition zero, since in a vacuum it must be that $B/H = 1$. The susceptibility of a diamagnetic substance is negative because lines of force from induced dipoles cancel out some lines of force due to the applied field. For paramagnetic substances the flux is greater within the substance than it would be in a vacuum, and thus paramagnetic substances have positive susceptibilities.

There are numerous methods for measuring magnetic susceptibilities, all of which depend on measuring the force exerted upon a body when it is placed in an inhomogeneous magnetic field. The more paramagnetic the body is, the more strongly will it be drawn toward the more intense part of the field.

19-8. Magnetic Moments from Magnetic Susceptibilities

It is generally more convenient to discuss magnetic susceptibility on a weight basis than on a volume basis and thus the following relations are used:

$$\kappa/d = \chi \tag{19-7a}$$
$$M\chi = \chi_M \tag{19-7b}$$

In these equations d is the density in g cm^{-3} and M is the molecular weight. χ is called the gram susceptibility and χ_M is called the molar susceptibility. When a value of χ_M is obtained from the measured volume susceptibility, κ, it can be corrected for the diamagnetic contribution and for the TIP to give a "corrected" molar susceptibility, χ_M^{corr}, which is the most useful quantity in drawing conclusions about electronic structure.

In his classic studies, Pierre Curie showed that paramagnetic susceptibilities depend inversely on temperature and often follow or closely approximate the behavior required by the simple equation

$$\chi_M^{corr} = C/T \tag{19-8}$$

Here T represents the absolute temperature, and C is a constant that is characteristic of the substance and known as its Curie constant. Equation 19-8 expresses what is known as *Curie's law.**

Now, on theoretical grounds, just such an equation is to be expected. The magnetic field in which the sample is placed tends to align the moments of the paramagnetic atoms or ions; at the same time, thermal agitation tends to randomize the orientations of these individual moments. The situation is entirely analogous to that encountered in the electric polarization of matter containing electric dipoles, with which the student is probably already familiar from a standard physical chemistry course. Applying a straight-forward statistical treatment, one obtains the following equation showing how the molar susceptibility of a substance containing independent atoms, ions or molecules, each of magnetic moment μ (in B.M.), will vary with temperature:

$$\chi_M^{corr} = \frac{N\mu^2/3k}{T} \tag{19-9}$$

where N is Avogadro's number, and k is the Boltzmann constant. Obviously, by comparison of equations 19-8 and 19-9,

$$C = N\mu^2/3k \tag{19-10}$$

and at any given temperature

$$\mu = \sqrt{3k/N} \cdot \sqrt{\chi_M^{corr} T} \tag{19-11}$$

which, on evaluating $\sqrt{3k/N}$ numerically, becomes

$$\mu = 2.84\sqrt{\chi_M^{corr} T}. \tag{19-12}$$

Thus, to recapitulate, one first makes a direct measurement of the volume susceptibility of a substance from which χ_M is calculated, and in accurate work is corrected for diamagnetism and TIP. From this corrected molar susceptibility and the temperature of the measurement, equation 19-12 enables one to calculate the magnetic moment of the ion, atom or molecule responsible for the paramagnetism.

From equation 19-8 we should expect that, if we measure χ_M for a substance at several temperatures and plot the reciprocals of the χ_M^{corr} values against T, we shall obtain a straight line of slope C which intersects the origin.

* Actually Curie's law was originally based on χ; that is, the effects of diamagnetism and TIP were neglected. However, its significance and utility are enhanced when these are taken into account.

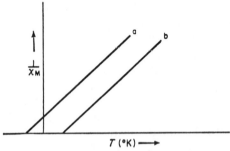

Fig. 19-2. Some deviations from the Curie law that may be fitted to the Curie–Weiss law.

Although there are many substances that, within the limits of experimental error, do show this behavior, there are also many others for which the line does not go through the origin, but instead looks somewhat like one of those shown in Fig. 19-2, cutting the T axis at a temperature below $0°K$ as in (a) or above $0°K$ as in (b). Obviously, such a line can be represented by a slight modification of the Curie equation,

$$\chi_M^{corr} = \frac{C}{T - \theta} \tag{19-13}$$

Here θ is the temperature at which the line cuts the T axis. This equation expresses what is known as the *Curie–Weiss law*, and θ is known as the *Weiss constant*. Actually, just such an equation can be derived if one assumes, not that the dipoles in the various ions, atoms or molecules of a solid are completely independent—as was assumed in deriving equation 19-9—but that instead the orientation of each one is influenced by the orientations of its neighbors as well as by the field to which it is subjected. Thus the Weiss constant can be thought of as taking account of the interionic or intermolecular interactions, thereby enabling us to eliminate this extraneous effect by computing the magnetic moment from the equation

$$\mu = 2.84\sqrt{\chi_M^{corr}(T - \theta)} \tag{19-14}$$

instead of from equation 19-12. Unfortunately, there are also cases in which magnetic behavior appears to follow the Curie–Weiss equation without the Weiss constant having this simple interpretation. In such cases, it is often quite wrong to use equation 19-14. In cases where the Curie law does not accurately fit the data, and where the applicability of the Curie–Weiss law is in doubt (even though it *may* fit the data), the best practice is to compute a magnetic moment at a given temperature using the Curie law, e.g., 19-12, and call this an *effective magnetic moment*, μ_{eff}, at the specified temperature. In this way no possibly unjustified implications are attached to empirically sound facts.

19-9. Ferromagnetism and Antiferromagnetism

In addition to the simple paramagnetism we have discussed, where the Curie or Curie–Weiss law is followed and the susceptibility shows no depen-

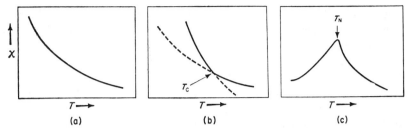

Fig. 19-3. Diagrams indicating the qualitative temperature-dependence of magnetic susceptibility for (a) simple paramagnetism, (b) ferromagnetism, and (c) antiferromagnetism.

dence on field strength, there are other forms of paramagnetism in which the dependence on both temperature and field strength is complicated. Two of the most important of these are ferromagnetism and antiferromagnetism. No attempt will be made to explain either of these in detail, either phenomenologically or theoretically, but it is important for the student to recognize their salient features. Fig. 19-3 compares the qualitative temperature-dependence of the susceptibility for (a) simple paramagnetism, (b) ferromagnetism, and (c) antiferromagnetism. Of course, (a) is just a rough graph of Curie's law. In (b) it should be noted that there is a discontinuity at some temperature, T_C, called the Curie temperature. Above the Curie temperature the substance follows the Curie or the Curie–Weiss law; that is, it is a simple paramagnetic. Below the Curie temperature, however, it varies in a different way with temperature and is also field-strength dependent. For antiferromagnetism there is again a characteristic temperature, T_N, called the Néel temperature. Above T_N the substance has the behavior of a simple paramagnetic, but below the Néel temperature the susceptibility *drops* with decreasing temperature.

These peculiarities in the behavior of ferromagnetic and antiferromagnetic substances below their Curie or Néel points are due to interionic interactions, which have magnitudes comparable to the thermal energies at the Curie or Néel temperature and thus become progressively greater than thermal energies as the temperature is further lowered. In the case of antiferromagnetism, the moments of the ions in the lattice tend to align themselves so as to cancel one another out. Above the Néel temperature thermal agitation prevents very effective alignment, and the interactions are manifested only in the form of a Weiss constant. However, below the Néel temperature this antiparallel aligning becomes effective and the susceptibility is diminished. In ferromagnetic substances the moments of the separate ions tend to align themselves parallel and thus to reinforce one another. Above the Curie temperature, thermal energies are more or less able to randomize the orientations; below T_C, however, the tendency to alignment becomes controlling, and the susceptibility increases much more rapidly with decreasing temperature than it would if the ion moments behaved independently of one another.

Presumably, even in those substances we ordinarily regard as simple paramagnetics there are some interionic interactions, however weak, and therefore there must be some temperature, however low, below which they will show ferromagnetic or antiferromagnetic behavior, depending on the sign of the interaction. The question why such interactions are so large in some substances that they have Curie or Néel temperature near and even above room temperature is still something of an unsolved problem. Suffice it to say here that in many cases it is certain that the magnetic interactions cannot be direct dipole–dipole interactions but instead the dipoles are coupled through the electrons of intervening atoms in oxides, sulfides, halides and similar compounds.

In general, ferro- and antiferromagnetic interactions are decreased when the magnetic species are separated from one another physically. Thus, when the magnetic behavior of a solid shows the effects of interionic coupling, solutions of the same substance will be free from such interactions. This includes solid solutions; for example, when K_2OsCl_6, which has μ_{eff} per Os atom at $300\,°K$ of 1.44 B.M., is contained at a level of $\leqslant 10$ mole percent in diamagnetic and isomorphous K_2PtCl_6, its μ_{eff} value at the same temperature rises to 1.94 B.M. owing to the elimination of antiferromagnetic coupling between the Os^{IV} ions through intervening chlorine atoms.

There are special cases, of considerable interest, in which antiferromagnetic coupling occurs between a few, say 2 or 3, paramagnetic ions which are held together in a polynuclear complex. Of course such interactions correspond to incipient bond formation and when they become strong enough they lead to a situation in which the state with paired electron spins, the bonded state, is so stable that the substance is entirely diamagnetic at normal temperatures. However, in some cases, as illustrated by the dimeric carboxylates of Cu^{II} (see Sect. 25-H-3), the energy of interaction between the unpaired electrons is small relative to thermal energy at room temperature and the compound remains paramagnetic. However, μ_{eff} is less than that for isolated ions, and it decreases markedly with decreasing temperature. For example, μ_{eff} for Cu^{II} is normally about 1.8–1.9 B.M. at $300\,°K$, but in $Cu_2(CH_3COO)_4 \cdot 2H_2O$, it is ~ 1.4 B.M. and drops to lower values at lower temperatures. Presumably at $0\,°K$, or close to it, where thermal energy becomes unavailable, the moment would become zero.

19-10. Electron Spin Resonance, esr

This phenomenon opens an extra dimension in the inference of chemically important features of molecular electronic structure by magnetic measurements. Because its proper understanding requires an extensive knowledge of wave mechanics, beyond the scope of this text, we shall present here only a brief and heuristic account, intended to draw attention to the kinds of useful result that may be obtained.

Electron spin resonance may be observed when molecules or ions con-

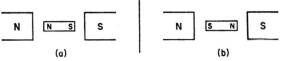

Fig. 19-4.   Diagram indicating (a) unfavorable and (b) favorable alignments of a small magnet (e.g., an electron) in the field of a larger one.

taining one or more unpaired electrons are placed in a magnetic field. In a molecule containing a single unpaired electron in an S state ($L = 0$), the effect of the magnetic field is to lift the spin degeneracy, that is, to make the energy of the electron different for its two M_S values, $+\frac{1}{2}$ and $-\frac{1}{2}$. This effect is easily understood by thinking of the electron classically as a small magnet in the field of a larger one. When its field is lined up with that of the larger one (Fig. 19-4a), there is an increase in potential energy, whereas the opposite alignment (Fig. 19-4b) decreases the potential energy. A quantitative treatment shows that the energy difference between these two electron spin alignments is equal to $g\beta H$, where g is the same gyromagnetic ratio as was discussed earlier, β is the Bohr magneton and H is the strength of the magnetic field. The lower state is slightly more populated than the upper one at thermal equilibrium. Thus, when radiation of frequency ν such that $h\nu = g\beta H$ is applied to the system, there is a net absorption because absorptive transitions upward are more numerous than radiative transitions downward. By sweeping the frequency of an oscillator (in the microwave region) through the appropriate frequency range, ν is observed as the frequency of maximum absorption. From this the g value may be calculated. In this simple case, it would be 2.00, but in other cases, more complex behavior is observed. It is the added complexities from which detailed information on electronic structure can often be obtained. The three main types of more complex behavior and their significance will now be mentioned briefly.

1. It will often happen that the observed g value will differ from 2.00. This deviation can be attributed to orbital contributions to the magnetism, and, from the very precise data afforded by esr measurements, these orbital contributions can be evaluated quite precisely. With such information, fairly detailed knowledge of orbital populations, degrees of hybridization, etc., may often be obtained.

In some substances there are two (or more) identical ions in different chemical environments or differently oriented with respect to the crystallographic axes. In a bulk-susceptibility measurement, only the average magnetic properties of both ions could be determined, but even slight differences between them will result in their resonance frequencies being detectably different in an esr measurement.

2. Magnetic anisotropy can often be observed by making measurements on small, oriented, single crystals. The anisotropy means that the g value and hence the resonance frequency vary according to the orientation of the crystal with respect to the direction of the external magnetic field, H. The g value in any particular arbitrary direction can be expressed as the tensor

resultant of three tensor components, g_x, g_y, g_z, in mutually perpendicular directions. There are many cases in which two of these tensor components are equal, and the two separate g values are called $g_{\parallel}$ and $g_{\perp}$; $g_{\parallel}$ is the value in the unique direction, while $g_{\perp}$ is the value in any direction in the plane perpendicular to this direction. From the properties of tensors, it follows that the g value averaged over all directions, g_{av}, is given by

$$g_{av}^2 = \tfrac{1}{3}(g_x^2 + g_y^2 + g_z^2) = \tfrac{1}{3}(g_{\parallel}^2 + 2g_{\perp}^2) \qquad (19\text{-}15)$$

Thus, although a bulk-susceptibility measurement made in the usual way on a powdered sample could provide only a value of g_{av}, and give no hint about the individual g_x, g_y, g_z, or $g_{\parallel}$ and $g_{\perp}$ values, a relatively simple esr measurement using a small single crystal can provide detailed knowledge of anisotropy. Anisotropy measurements can be and have been made by making bulk susceptibility measurements on quite large single crystals, but the experimental problems are formidable and the results of relatively low accuracy. Anisotropy data can usually be interpreted to give quite detailed information on metal–ligand bonding in complexes.

3. It often happens that the small magnetic fields of atomic nuclei cause splittings (fine structure) of esr lines. From the magnitudes of such splittings, semiquantitative information as to "electron densities" of the unpaired electrons in particular orbitals of particular atoms can be obtained.

In conclusion, one other great advantage and one disadvantage of esr measurements relative to bulk-susceptibility measurements should be recorded. Esr is extremely sensitive, responding, under optimum conditions (of line width, instrumental sensitivity, and signal-to-noise ratio), to $\sim 10^{-12}$ paramagnetic species per liter, whereas bulk-susceptibility measurements can be accurate, as a rule, only when the net paramagnetism is greatly in excess of the diamagnetism in the sample and when the available quantity is relatively large ($\geqslant 100$ mg). Thus, esr can be used to study minute samples or larger samples containing the paramagnetic species, either ions or free radicals, at very low concentrations. A frequent disadvantage of the esr method is that very low temperatures, for example, liquid nitrogen or even liquid helium temperatures, are often required in order to reduce spin–lattice relaxation effects sufficiently to make lines observable. In many instances, especially with paramagnetic ions having even numbers of electrons, there has been only slight success at observing resonance under any experimental conditions. Also, the apparatus for esr measurements is expensive and complex, and the interpretation of the spectra often requires considerable skill. Bulk susceptibility measurements do have the advantage that they can always be made and used to obtain some idea of the magnetic moment of the paramagnetic ion if it is present in sufficient concentration. Esr measurements have by no means supplanted bulk susceptibility measurements except for special cases and special purposes. The bulk susceptibility measurement still remains one of the chemist's most valuable and frequently used tools in his efforts to understand the electronic structures of transition metal compounds.

19-11. Metal-to-Metal Bonds and Metal Atom Clusters

Within approximately the past decade it has been realized, and abundantly demonstrated, that the transition metals have a marked propensity to form homo- and hetero-nuclear bonds among themselves. There is now a large number of molecules and ionic species containing such bonds, and detailed discussions will be found in succeeding Chapters. At this point, however, we present an introductory overview with cross references to the later, detailed presentations.

Structurally, compounds containing metal-to-metal (M—M) bonds may be divided into two broad classes: (1) those with only two-center bonds, which may, however, be of multiple character,[1] and (2) those containing three or more metal atoms, like or unlike, arranged in a polygonal or polyhedral array with, in some cases, considerable delocalization of the bonding electrons.[2,3] Some of the principal polyhedral arrangements in such *metal atom cluster* compounds* have already been mentioned and illustrated in Chapter 1, pages 30–36. To summarize briefly, the most common ones are equilateral triangles, octahedra and tetrahedra, but the trigonal bipyramid, square pyramid and cube also occur. An unusual situation, so far found only for some gold species,[4] involves a polyhedron of metal atoms surrounding a central metal atom which is bonded to all the peripheral ones. A typical one of these, $Au_{11}I_3(PR_3)_7$, contains an Au_{11} cluster in which ten Au atoms surround a central one and to each peripheral Au atom is bound either iodine or PR_3 (Fig. 26-I-2).

It is important to stress that the term metal atom cluster was coined[5] for those systems in which M—M bonding is strong enough to make a significant contribution to the heat of formation. Such substances are non-classical or non-Werner complexes. Of course, many polynuclear complexes were known when Werner formulated his coordination theory, and many more have since been discovered. In such a complex each of the metal atoms interacts with its own set of ligands, some of which are shared, but they do not interact, or interact only slightly, with any other metal atom. In many of these cases there are interactions that have a marked effect upon magnetic properties but where no thermodynamically significant M—M bond occurs. These classical polynuclear complexes require neither a new nomenclature nor any new concept for their description, and they should not be (but, unfortunately, sometimes have been) called cluster compounds. Not sur-

[1] F. A. Cotton, *Rev. Pure Appl. Chem.*, 1967, **17**, 25; *Accounts Chem. Res.*, 1969, **2**, 240.
[2] B. R. Penfold, *Perspectives in Structural Chemistry*, J. D. Dunitz and J. A. Ibers, eds., Vol. 2, Wiley, 1968, p. 71.
[3] M. C. Baird, *Progr. Inorg. Chem.*, 1968, **9**, 1; D. L. Kepert and K. Vrieze in *Halogen Chemistry*, Vol. 3, V. Gutmann, ed., Academic Press, 1967, p. 1; R. D. Johnston, *Adv. Inorg. Chem. Radiochem.*, 1970, **13**, 471; P. Chini, *Inorg. Chim. Acta Rev.*, 1968, **2**, 31.
[4] (a) M. McPartlin, R. Mason and L. Malatesta, *Chem. Comm.*, **1969**, 334; (b) V. G. Albano *et al.*, *Chem. Comm.*, **1970**, 1210.
[5] J. A. Bertrand, F. A. Cotton and W. A. Dollase, *Inorg. Chem.*, 1963, **2**, 1166.
* The term "staphylonuclear" has also been used to describe such entities.

prisingly, there are borderline cases where the M—M interaction is weak but cannot be said with certainty to be unworthy of the designation "bond."

The surest indication of the existence of a metal—metal bond is provided by the molecular structure, when this is fully known. In cases such as $Mn_2(CO)_{10}$ or $Re_2Cl_8^{2-}$, where the metal atoms are adjacent to each other and there are no bridging groups, the existence of metal—metal bonds is self-evident. More generally, short distances between metal atoms are indicative of the existence of a bond, even if bridging groups are present, but the distance criterion must be used cautiously, since great variations are possible. For example, in $[h^5\text{-}C_5H_5Mo(CO)_3]_2$ there is necessarily an Mo—Mo bond because the dimer has adjacent Mo atoms and no bridging group; the Mo—Mo distance is 3.22 Å. In MoO_2, on the other hand, where Mo—Mo bonding is also postulated, the Mo atoms lie in chains, bridged by pairs of oxygen atoms, with the alternating Mo—Mo distances of 2.50 and 3.10 Å; the shorter ones are considered to correspond to bonds and the latter not, and yet the latter is shorter than the bond distance in $[h^5\text{-}C_5H_5Mo(CO)_3]_2$.

Undoubtedly, the length of a bond of given multiplicity between a given pair of metal atoms is a sensitive function of oxidation states, nature of additional ligands and other aspects of the molecular structure, and the use of interatomic distances to infer the existence and strength of metal—metal bonds requires caution.

Another kind of evidence frequently adduced to show the existence of metal—metal bonds is the lowering (even to zero) of magnetic moments compared with the values expected for isolated metal ions of the kinds present. The lowering is assumed to be due to pairing of spins in metal—metal interactions. Because magnetic susceptibilities are relatively easy to measure, this indirect criterion has often been used, but in fact the greatest care is required if erroneous conclusions are to be avoided, as the following observations will show.

(1) Where ions with an even number of electrons are concerned, pairing of spins may often be due to distortions or irregularities in the environment about the isolated ion, which split apart orbitals that would otherwise be degenerate. This possibility is particularly relevant for the heavier transition elements, where intraionic spin pairing is easiest.

(2) Interionic spin pairing may occur by means of electron interactions through intervening anions. Good examples of this in binuclear complexes are provided by $[(EtOCS_2)_2MoO]_2O$ and $[(RuCl_5)_2O]^{4-}$, both of which contain linear M—O—M groups, as described in detail at appropriate places in Chapter 26. A more elaborate case is that of RuO_2. This compound has the rutile structure (Fig. 2–3). Each Ru^{IV} ion is in the center of a practically regular octahedron of oxide ions, while each of the oxide ions is shared between three Ru^{IV} ions. Thus there is no localized two-center interaction between the ruthenium ions, nor do the rather long Ru—Ru distances seem consistent with any strong direct metal—metal interaction. Nonetheless,

this compound, which contains octahedrally coordinated d^4 ions, is essentially diamagnetic. The most likely explanation here is the formation of extended molecular orbitals—in effect, energy bands—by extended overlap of metal and oxygen orbitals. On the other hand, MoO_2 has a distorted rutile structure in which the $Mo^{IV}(d^2)$ ions are drawn together in pairs, as noted already. This compound is also essentially diamagnetic, but here at least part, if not all, of the pairing of electron spins is attributable to metal—metal bonding. The point to be emphasized here is that, from the magnetic data alone, the difference between the two cases could not have been appreciated and the magnetic properties might have been used as a basis for postulating direct metal—metal interactions in both compounds whereas in only one of them does it necessarily occur.

(3) The high values of the spin–orbit coupling constants in the heavier transition elements can often lead to very low magnetic susceptibilities in the absence of metal—metal bonds. This matter is discussed in Chapter 26.

Two types of binuclear system in which M—M interactions can run the gamit from very strong and multiple to weak or even entirely repulsive are shown in Fig. 19-5. The first of these is the *confacial bioctahedron*, in which the nine ligand atoms define two octahedra with a common face. The $M_2Cl_9^{3-}$ ions formed by the Group VI metals Cr, Mo and W illustrate the manner in which the strength of M—M interaction can be estimated by considering the structure parameters;[1] we know from the magnetic and spectral properties of $Cr_2Cl_9^{3-}$ that no M—M bond exists in this system. The Cr^{3+} ions have high-spin d^3 configurations with magnetic properties and spectra essentially typical for such ions in isolated octahedral complexes; if there is no attractive interaction between the metal atoms, there must be a net repulsive one at the distances involved and this shows up very clearly in the structure; the

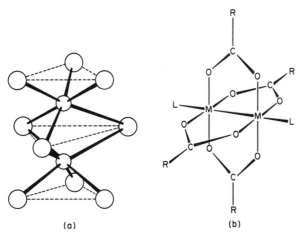

Fig. 19-5. Two types of structure in which a full range of M—M interaction from strongly bonding to net repulsive can be observed. (a) The confacial bioctahedron. (b) The carboxylato-bridged dinuclear molecules.

chromium atoms are displaced from the centers of their octahedra so as to be further away from each other, and the various angles and Cr—Cl distances also reflect the repulsive force. Thus the Cr—Cl—Cr angles exceed the value for an ideal confacial bioctahedron (that is, one in which two perfect octahedra are fused on a common face), the Cl_{bridge}—Cr—Cl_{bridge} angles are < 90° and the $Cl_{term.}$—Cr—$Cl_{term.}$ angles are > 90° and the Cl_{bridge}—Cr distances are greater than the $Cl_{term.}$—Cr distances.

In $W_2Cl_9^{3-}$, which has no unpaired electron, the structure by itself provides evidence that a strong W—W bond exists. The W atoms are markedly displaced from the centers of their octahedra *toward* each other and the various angles and W—Cl distances all reflect a compression caused by this strong W—W bond. The $Mo_2Cl_9^{3-}$ ion has magnetic, spectral and structural properties indicating an M—M bond of only medium strength.

In the case of the tetracarboxylato-bridged species, Fig. 19-5b, there is again considerable freedom for the M—M distance to adjust itself according to the strength of the M—M interaction. In the case of $Cu_2(O_2CCH_3)_4(H_2O)_2$ there is only a weak antiferromagnetic coupling of spins; the Cu—Cu distance is very large, ~ 2.65 Å. At the other extreme we have $Mo_2(O_2CCF_3)_4$ in which there is a very strong (quadruple) Mo—Mo bond and the distance is only 2.109 Å.

Factors Favoring M—M Bonding.   The most important single factor appears to be low formal oxidation state. Thus the overwhelming majority of species containing M—M bonds have metal atoms in a formal oxidation state of II or less. The importance of this factor is probably connected with the dependence of M—M orbital overlaps on the size of the orbitals. In order to attain a sufficient overlap of the metal valence-shell orbitals at distances that do not introduce too much repulsive interaction of the cores, the formal charge has to be low, thus permitting expansion of the valence shell orbitals.

A second factor is the suitability of the valence-shell configuration and the metal—ligand bonding system. Thus a strong, quadruple Re—Re bond exists in $Re_2Cl_8^{2-}$ and a number of related dinuclear rhenium species, despite the relatively high formal oxidation number (+3), because the electron configuration is optimal for strong (quadruple) bonding.

Obviously, too many electrons in the valence shell will militate against the formation of both M—M multiple bonds and metal atom clusters since the electrons would have to occupy antibonding as well as bonding orbitals, even if good orbital overlaps can occur. Thus, among lower halides and chalcogenides it is only in the early groups, i.e. Nb, Ta, Cr, Mo, W, Tc and Re that M—M bonded compounds are common. However, for metals in the group Mn, Fe, Co and Ni metal atom clusters are formed extensively when CO groups or other ligands capable of drawing electrons out of the antibonding orbitals are present. Thus the polynuclear metal carbonyls constitute one of the largest classes of M—M bonded compounds.

Molecular-orbital Descriptions of M—M Bonding.   In some cases M—M bonding can be adequately described in terms of *2c-2e* bonds. Obviously all

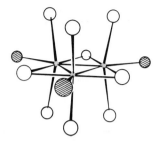

Fig. 19-6. The structure of $Re_3X_9L_3$ molecules. The small circles represent rhenium atoms; the hatched circles represent L groups; the large open circles represent the X groups.

isolated M—M single bonds, e.g., in $(OC)_5Mn—Mn(CO)_5$, can be so described, and even a few clusters, such as the triangular one in $Os_3(CO)_{12}$ or the octahedral one in $Mo_6X_8^{4+}$, can be satisfactorily treated as a collection of $2c$-$2e$ M—M bonds. There are, however, many cases where it would be extremely cumbersome to construct such a scheme, since resonance hybridization of many canonical forms would have to be invoked. For this and other reasons, the most general and uniform basis on which to discuss bonding in these compounds is a set of molecular orbitals that are symmetry-adapted to the case in question. For illustration we shall mention here just two examples.

The trinuclear $Re_3Cl_9L_3$ group, which has the type of structure shown in Fig. 19-6, will illustrate how delocalized MO's are used to treat the bonding in clusters. The simplest approach is to note that there is a set of five ligand atoms around each metal atom, approximately at five of the vertices of an octahedron. If, from the valence shell of each metal atom (which consists of five d, one s, and three p orbitals), an appropriate set of orbitals is reserved for these bonds, there will remain four unused atomic orbitals or combinations of atomic orbitals. If one further assumes that these are pure or nearly pure d orbitals, it is not difficult, using the requisite theory, to determine qualitatively how they will interact to produce a set of MO's delocalized over the set of rhenium atoms.

The approximate ordering of the MO's is shown in Fig. 19-7. It is seen that there are four bonding MO's, two of which (e', e'') are doubly degenerate. Thus altogether the bonding MO's can hold twelve electrons. Each rhenium atom has four electrons in its valence orbitals after allowance for

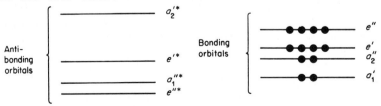

Fig. 19-7. The qualitative ordering of molecular orbital energies for M—M bonding in the Re_3 cluster of $Re_3X_9L_3$ species.

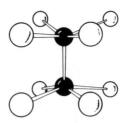

Fig. 19-8.   The structure of $M_2X_8^{n-}$ species such as $Re_2Cl_8^{2-}$ and $Mo_2Cl_8^{4-}$.

metal—ligand bonds, so that the bonding orbitals become fully occupied. The cluster is therefore diamagnetic with maximal M—M bonding.

Since there are six electron pairs distributed over the three Re—Re edges of the Re_3 triangle, or two pairs per Re—Re interaction, the MO picture is formally equivalent to describing each Re—Re bond as a double bond, consisting of a σ bond and a π bond perpendicular to the Re_3 plane.

Similar MO treatments of the bonding in other clusters have been described. In many cases the procedure and results are more complex than those used here for illustration, and there has been some controversy over differences in results,[6] but the basic approach is generally accepted.

Another reason why a description of the bonding in terms of MO's delocalized over the entire cluster is preferable to a system of localized bonds is that certain clusters can be oxidized and reduced so as to retain their integrity and symmetry. Thus, for example, we have the redox series

$$Ta_6Cl_{12}^{2+} \rightleftarrows Ta_6Cl_{12}^{3+} \rightleftarrows Ta_6Cl_{12}^{4+}$$

Such processes are most conveniently understood in terms of MO's. In the particular case cited, the MO pattern suggested[6b] has a singly degenerate orbital as the highest filled one. Thus, the three species mentioned should have 0, 1, and 0 unpaired electrons. This is in agreement with experiment.

To illustrate the treatment[7] of very strong multiple M—M bonds, we consider species such as $Re_2X_8^{2-}$ and $Mo_2Cl_8^{4-}$. The structure of these is shown in Fig. 19-8. The M—M bonds are very short (Re—Re, ~ 2.24 Å; Mo—Mo, ~ 2.13 Å) and the two MX_4 units have an *eclipsed* rotational orientation. After allowance is made for net charge and M—X bonds each metal atom remains with four d (or predominantly d) orbitals and four electrons. One of these orbitals has σ character, two of them form a degenerate pair with π character, and the fourth has δ character with respect to the common four-fold axis of the $ReCl_4$ groups. It is therefore possible for two such groups to interact forming one σ, two π, and one δ bonding orbitals (and, of course, antibonding orbitals corresponding to each). These four bonding orbitals are then occupied by the four pairs of electrons, and the quadruple

[6]   (a) L. D. Crossway, D. P. Olsen and G. H. Duffey, *J. Chem. Phys.*, 1963, **38**, 73; (b) F. A. Cotton and T. E. Haas, *Inorg. Chem.*, 1964, **3**, 10; (c) M. L. Robin and N. A. Kuebler, *Inorg. Chem.*, 1965, **4**, 978; S. F. A. Kettle, *Theor. Chim. Acta*, 1965, **3**, 211; 1966, **4**, 150; *J. Chem. Soc., A*, **1967**, 314.

[7]   F. A. Cotton, *Inorg. Chem.*, 1965, **4**, 334; F. A. Cotton and C. B. Harris, *Inorg. Chem.*, 1967, **6**, 924.

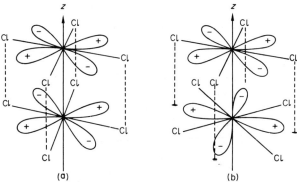

(a)   (b)

Fig. 19-9. A sketch showing how δ overlap depends on rotational angle. (a) Eclipsed configuration having maximum overlap. (b) Staggered configuration having zero overlap.

bond is formed. The δ component restricts rotation in just such a way as to favor the eclipsed configuration: the δ overlap is maximal for this configuration and goes to *zero* for the staggered configuration, as shown in Fig. 19-9.

In addition to the bonding (σ, π, δ) and antibonding (σ^*, π^*, δ^*) orbitals just mentioned, there are two approximately non-bonding orbitals of σ type, namely, $\sigma_n(1)$, $\sigma_n(2)$, with maximum amplitude along the four-fold axis but projecting away from the two rhenium atoms. A simple energy-level diagram for $Re_2Cl_8^{2-}$ based on semiquantitative MO theory[7] (omitting, for clarity, all MO's concerned predominantly with Re—Cl bonding) is given in Fig. 19-10.

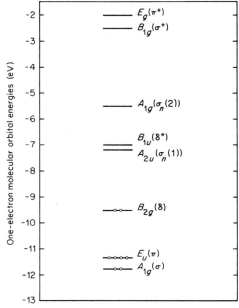

Fig. 19-10. An energy-level diagram for the MO's involved in Re—Re bonding in $Re_2Cl_8^{2-}$.

Heteronuclear M—M Bonds. The transition metals, especially in their carbonyl-type compounds, form many bonds to the non-transitional metal or metalloidal atoms. This is particularly true for the elements Zn, Cd, Hg, Cu, Ag, Au, Tl, Ge, Sn and Pb. These are nearly always $2c$-$2e$ bonds which require no special comment. However, there are cases where mixed metal clusters of a more complex nature are formed. Examples of these are trigonal-bipyramidal Sn_2Pt_3 species and tetrahedral Ge_2Co_2 species.

Guide to Detailed Discussion of M—M Bonded Species. Explicit discussions of particular M—M bonded compounds will be found mainly in the following places:

(1) Polynuclear metal carbonyls: Section 22-3.

(2) Cluster compounds of Nb, Ta, Mo, W and Re under the chemistries of those elements: Chapter 26.

(3) Species with quadruple and other multiple M—M bonds under the chemistry of Tc, Re, Mo, Cr, Ru and Rh: Chapters 25 and 26.

Further Reading

Specialist Periodical Reports, The Chemical Society, London: *Spectroscopic Properties of Inorganic and Organometallic Compounds*, Vol. 1, 1968 (includes esr); *Electronic Structure and Magnetism of Inorganic Compounds*, Vol. 1, 1971; *Inorganic Chemistry of Transition Elements*, Vol. 1, 1972.

Atkins, P. W., and M. C. R. Symons, *The Structure of Inorganic Radicals*, Elsevier, 1967.

Ayscough, P. B., *Electron Spin Resonance in Chemistry*, Methuen, 1967.

Bersohn, M., and J. C. Baird, *Introduction to Electron Paramagnetic Resonance*, Benjamin, 1966.

Carrington, A., and A. D. McLachlan, *Introduction to Electron Paramagnetic Resonance*, Benjamin, 1967.

Earnshaw, A., *Introduction to Magnetochemistry*, Academic Press, 1968.

Figgis, B. N., and J. Lewis, *Progr. Inorg. Chem.*, 1964, **6**, 37 (an extended discussion of both theory and experimental results concerning magnetochemistry of the transition elements).

Figgis, B. N., and J. Lewis, *Technique of Inorganic Chemistry*, H. B. Jonassen and A. Weissberger, eds., Volume IV, Interscience-Wiley, 1965 (detailed introduction to experimental techniques of magnetochemistry).

Goodenough, J. B., *Magnetism and the Chemical Bond.*, Interscience-Wiley, **1963** (deals principally with magnetic properties of extended arrays, e.g., metal oxides).

Hill, H. A. O., and P. Day, eds., *Physical Methods in Advanced Inorganic Chemistry*, Interscience-Wiley, 1968 [contains good articles on electron paramagnetic resonance (by E. König) and other forms of spectroscopy as applied mainly to transition elements].

Hochstrasser, R. M., *J. Chem. Educ.*, 1965, **42**, 154 (detailed but non-mathematical discussion of electron configurations when interelectronic interaction energies and orbital energy differences are comparable).

König, E., Landoldt-Bornstein, Vol. 2, *Magnetic Properties of Coordination and Organometallic Transition Metal Compounds*, Springer, 1966 (comprehensive compilation of data).

McGarvey, B. R., *Transition Metal Chemistry*, 1966, **3**, 89 (esr spectra).

McMillan, J. A., *Electron Paramagnetism*, Reinhold, 1968.

Mulay, L. N., *Magnetic Susceptibility*, Interscience-Wiley, 1963 (good coverage of experimental methods).

Nyholm, R. S., and M. L. Tobe, *Adv. Inorg. Chem. Radiochem.*, 1963, **5**, 1 (general discussion on stability of oxidation states in transition metal compounds).

Pake, G. E., *Paramagnetic Resonance*, W. A. Benjamin, Inc., 1962 (good introduction with due attention to transition-metal ions).

20

The Electronic Structures of Transition Metal Complexes

INTRODUCTION

20-1. Genealogy of the Several Theories

The studies of Werner and his contemporaries followed by the ideas of Lewis and Sidgwick on electron-pair bonding led to the idea that ligands are groups that can in some way donate electron pairs to metal ions or other acceptors, thus forming the so-called coordinate link. This approach to bonding in complexes was extended by Pauling and developed into the *valence bond theory* of metal–ligand bonding. This theory enjoyed great and virtually exclusive popularity among chemists through the 1930's and 1940's, but during the 1950's it was supplemented by the *ligand field theory*. This was developed between 1930 and 1940 by physicists, mainly J. H. Van Vleck and his students, and rediscovered in the early 1950's by several theoretical chemists. The ligand field theory as we have it today evolved out of a purely electrostatic theory called the *crystal field theory* which was first expounded in 1929 by H. Bethe.

The crystal field theory, CFT, as we shall see, treats the interaction between the metal ion and the ligands as a purely electrostatic problem in which the ligand atoms are represented as *point* charges (or as point dipoles). At the opposite extreme, so to speak, the metal–ligand interaction can be described in terms of *molecular orbitals* formed by overlap of ligand and metal orbitals. Although these two methods certainly use different physical representations of the problem and are, at least superficially, very different in their algebraic form, there is a close fundamental relationship between them, as Van Vleck pointed out long ago, because both make explicit and rigorous use of the symmetry properties of the complex. More recently, this relationship has been explored further and CFT has been described as an "operator-equivalent" formalism.[1]

[1] J. S. Griffith, *J. Chem. Phys.*, 1964, **41**, 516.

The basic difficulty with the CFT treatment is that it takes no account of the partly covalent nature of the metal–ligand bonds, and therefore whatever effects and phenomena stem directly from covalence are entirely inexplicable in simple CFT. On the other hand, CFT provides a very simple and easy way of treating numerically many aspects of the electronic structures of complexes. MO theory, in contrast, does not provide numerical results in such an easy way. Therefore, a kind of modified CFT has been devised in which certain parameters are empirically adjusted to allow for the effects of covalence without explicitly introducing covalence into the CFT formalism. This modified CFT is often called ligand field theory, LFT. However, LFT is sometimes also used as a general name for the whole gradation of theories from the electrostatic CFT to the MO formulation. We shall use LFT in the latter sense in this Chapter, and we introduce the name *adjusted crystal field theory*, ACFT, to specify the form of CFT in which some parameters are empirically altered to allow for covalence without explicitly introducing it.

Ligand field theory, in the sense of the term indicated above, can be defined as the theory of (1) the origin and (2) the consequences of the splitting of inner orbitals of ions by their chemical environments. The inner orbitals of usual interest in this connection are partly filled ones, i.e. d or f orbitals. However, we shall restrict this discussion to d orbitals. To a considerable degree, it is possible to consider the two parts of LFT separately, which has the important consequence that many phenomena that are due to the existence of d orbital splitting can be understood pragmatically even if an exact explanation of why the splitting exists is not available. Of course, a truly rigorous discussion of all the consequences of d orbital splittings cannot be divorced from a discussion of the forces responsible for the splittings.

We shall begin by outlining the CFT formalism. It is extremely important for the reader to bear in mind while reading Section 20-2, however, that this is a sheer formalism, devoid of physical meaning because *ligand atoms are not points*. On the contrary, they are bodies with about the same size and structure as the metal atom itself. However, the CFT formalism is actually the historical origin of ligand field theory, it does provide useful results and it is absolutely necessary to be conversant with it in order to read the literature.

THE ELECTROSTATIC CRYSTAL FIELD THEORY, CFT

20-2. The Splitting of d Orbitals by Electrostatic Fields

Let us consider a metal ion, M^{m+}, lying at the center of an octahedral set of point charges, as shown in Fig. 20-1. Let us suppose that this metal ion has a single d electron outside of closed shells; such an ion might be Ti^{III}, V^{IV}, etc. In the free ion, this d electron would have had equal probability of being in any one of the five d orbitals, since all are equivalent. Now, however,

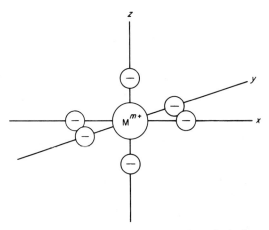

Fig. 20-1. Sketch showing six negative charges arranged octahedrally around a central M^{m+} ion with a set of Cartesian axes for reference.

the d orbitals are not all equivalent. Some are concentrated in regions of space closer to the negative ions than are others, and the electron will obviously prefer to occupy the orbital(s) in which it can get as far as possible from the negative charges. By examining the shapes of the d orbitals (see Fig. 20-2) and comparing them with Fig. 20-1, we see that both the d_{z^2} and $d_{x^2-y^2}$ orbitals have lobes that are heavily concentrated in the vicinity of the charges, whereas the d_{xy}, d_{yz} and d_{zx} orbitals have lobes that project between the charges. It can also be seen that each of the three orbitals in the latter group, namely, d_{xy}, d_{yz}, d_{zx}, is equally favorable for the electron; these three orbitals have entirely equivalent environments in the octahedral com-

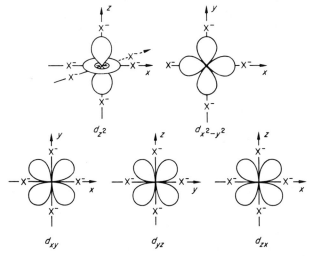

Fig. 20-2. Sketches showing the distribution of electron density in the five d orbitals with respect to a set of six octahedrally arranged negative charges (cf. Fig. 20-1).

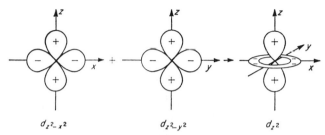

Fig. 20-3.   Sketches of $d_{z^2-x^2}$ and $d_{z^2-y^2}$ orbitals which are usually combined to make the d_{z^2} orbital.

plex. The two relatively unfavorable orbitals, d_{z^2} and $d_{x^2-y^2}$, are also equivalent; this is not obvious from inspection of Fig. 20-2, but Fig. 20-3 shows why it is so. As indicated, the d_{z^2} orbital can be resolved into a linear combination of two orbitals, $d_{z^2-x^2}$ and $d_{z^2-y^2}$, each of which is obviously equivalent to the $d_{x^2-y^2}$ orbital. It is to be stressed, however, that these two orbitals do not have separate existences, and the resolution of the d_{z^2} orbital in this way is only a device to persuade the reader *pictorially* that d_{z^2} is equivalent to $d_{x^2-y^2}$ in relation to the octahedral distribution of charges. Straightforward, though not simple, mathematical arguments can be used to prove this rigorously.

Thus, in the octahedral environment of six negative charges, the metal ion now has two kinds of d orbitals: three of one kind, equivalent to one another and conventionally labeled t_{2g} (sometimes $d\varepsilon$ or γ_5), and two of another kind, equivalent to each other, conventionally labeled e_g (sometimes $d\gamma$ or γ_3); furthermore, the e_g orbitals are of higher energy than the t_{2g} orbitals. These results may be expressed in an energy-level diagram as shown in Fig. 20-4a.

In Fig. 20-4a it will be seen that we have designated the energy difference between the e_g and the t_{2g} orbitals as Δ_o, where the subscript o stands for octahedral. The additional feature of Fig. 20-4a—the indication that the e_g levels lie $\frac{3}{5}\Delta_o$ above and the t_{2g} levels lie $\frac{2}{5}\Delta_o$ below the energy of the unsplit d orbitals—will now be explained. Let us suppose that a cation containing ten d electrons, two in each of the d orbitals, is first placed at the center of a hollow sphere whose radius is equal to the M—X internuclear distance, and

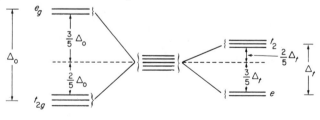

(a) Octahedral complex   (b) Tetrahedral complex

Fig. 20-4.   Energy-level diagrams showing the splitting of a set of d orbitals by octahedral and tetrahedral electrostatic crystal fields.

that charge of total quantity $6e$ is spread uniformly over the sphere. In this spherically symmetric environment the d orbitals are still five-fold degenerate.* The entire energy of the system, that is, the metal ion and the charged sphere, has a definite value. Now suppose the total charge on the sphere is caused to collect into six discrete point charges, each of magnitude e, and each lying at a vertex of an octahedron but still on the surface of the sphere. Merely redistributing the negative charge over the surface of the sphere in this manner cannot alter the total energy of the system when the metal ion consists entirely of spherically symmetrical electron shells, and yet we have already seen that as a result of this redistribution electrons in e_g orbitals now have higher energies than those in t_{2g} orbitals. It must, therefore, be that the total increase in energy of the four e_g electrons equals the total decrease in energy of the six t_{2g} electrons. This then implies that the rise in the energy of the e_g orbitals is $\frac{6}{4}$ times the drop in energy of the t_{2g} orbitals, which is equivalent to the $\frac{3}{5}:\frac{2}{5}$ ratio shown.

This pattern of splitting, in which the algebraic sum of all energy shifts of all orbitals is zero, is said to "preserve the center of gravity" of the set of levels. This center of gravity rule is quite general for any splitting pattern when the forces are purely electrostatic and where the set of levels being split is well removed in energy from all other sets with which they might be able to interact.

By an analogous line of reasoning it can be shown that the electrostatic field of four charges surrounding an ion at the vertices of a tetrahedron causes the d shell to split up as shown in Fig. 20-4b. In this case the d_{xy}, d_{yz} and d_{zx} orbitals are less stable than the d_{z^2} and $d_{x^2-y^2}$ orbitals. This may be appreciated qualitatively if the spatial properties of the d orbitals are considered with regard to the tetrahedral array of four negative charges as depicted in Fig. 20-5. If the cation, the anions and the cation–anion distance are the same in both the octahedral and tetrahedral cases, it can be shown that

$$\Delta t = \tfrac{4}{9}\Delta_o$$

In other words, other things being about equal, the crystal field splitting in a

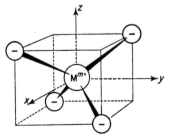

Fig. 20-5.   Sketch showing the tetrahedral arrangement of four negative charges around a cation, M^{m+}, with respect to coordinate axes that may be used in identifying the d orbitals.

* The energies of all orbitals are, of course, greatly but *not equally* raised when the charged sphere encloses the ion.

tetrahedral complex will be about half the magnitude of that in an octahedral complex.

The above results have been derived on the assumption that ionic ligands, such as F^-, Cl^- or CN^-, may be represented by point negative charges. Ligands that are neutral, however, are dipolar (e.g., 20-I and 20-II), and they

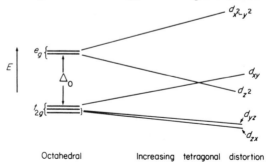

(20-I)   (20-II)

approach the metal ion with their negative poles. Actually, in the field of the positive metal ion such ligands are further polarized. Thus, in a complex such as a hexammine, the metal ion is surrounded by six dipoles with their negative ends closest; this array has the same general effects upon the d orbitals as an array of six anions, so that all the above results are valid for complexes containing neutral, dipolar ligands.

We next consider the pattern of splitting of the d orbitals in tetragonally distorted octahedral complexes and in planar complexes. We begin with an octahedral complex, MX_6, from which we slowly withdraw two *trans*-ligands. Let these be the two on the z-axis. As soon as the distance from M^{m+} to these two ligands becomes greater than the distance to the other four, new energy differences among the d orbitals arise. First of all, the degeneracy of the e_g orbitals is lifted, the z^2 orbital becoming more stable than the $(x^2 - y^2)$ orbital. This happens because the ligands on the z-axis exert a much more direct repulsive effect on a d_{z^2} electron than upon a $d_{x^2-y^2}$ electron. At the same time the three-fold degeneracy of the t_{2g} orbitals is also lifted. As the ligands on the z-axis move away, the yz and zx orbitals remain equivalent to one another, but they become more stable than the xy orbital because their spatial distribution makes them more sensitive to the charges along the z-axis than is the xy orbital. Thus for a small tetragonal distortion of the type considered, we may draw the energy level diagram shown in Fig. 20-6. It

Fig. 20-6. Energy-level diagram showing the further splitting of the d orbitals as an octahedral array of ligands becomes progressively distorted by the withdrawal of two *trans*-ligands, specifically those lying on the z-axis.

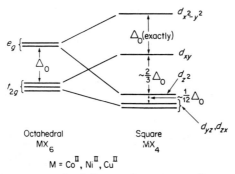

Fig. 20-7.   Approximate energy-level diagram for corresponding octahedral and square complexes of some metal ions in the first transition series.

should be obvious that for the opposite type of tetragonal distortion, that is, one in which two *trans*-ligands lie closer to the metal ion than do the other four, the relative energies of the split components will be inverted.

As Fig. 20-6 shows, it is in general *possible* for the tetragonal distortion to become so large that the z^2 orbital eventually drops below the xy orbital. Whether this will *actually happen* for any particular case, even when the two *trans*-ligands are completely removed so that we have the limiting case of a square, four-coordinated complex, depends upon quantitative properties of the metal ion and the ligand concerned. Semiquantitative calculations with parameters appropriate for square complexes of Co^{II}, Ni^{II} and Cu^{II} lead to the energy-level diagram shown in Fig. 20-7, in which the z^2 orbital has dropped so far below the xy orbital that it is nearly as stable as the (yz, zx) pair. As Fig. 20-6 indicates, the d_{z^2} level might even drop below the (d_{xz}, d_{yz}) levels and in fact experimental results suggest that in some cases (e.g., $PtCl_4^{2-}$) it does.

For a square pyramidal (*spy*) set of ligands, the splitting diagram has to be qualitatively similar to that for the square set, because removal of only one z-axis ligand from an octahedron introduces a perturbation of the same qualitative nature as does the removal, or partial removal, of both. Hence, Fig. 20-6 (and also Fig. 20-7, though not quantitatively) are applicable to this case.

Trigonal-bipyramidal Complexes.   A d orbital splitting diagram for an ion in a trigonal-bipyramidal (*tbp*) ligand field can also be deduced by an electrostatic argument. The *tbp* has D_{3h} symmetry. Taking the three-fold axis as the z-axis, we first note that the d orbitals are divided into three groups, those in each group having a different spatial relationship to the ligand set: (1) d_{z^2}; (2) d_{xy}, $d_{x^2-y^2}$; (3) d_{xz}, d_{yz}. The pairs of orbitals in each of the last two sets are indeed equivalent through this may not be immediately obvious by inspection.

All lobes of the d_{z^2} orbital lie in the region of ligands; hence, an electron in this orbital should be very unstable. All lobes of the d_{xz} and d_{yz} orbitals lie between ligands; hence, this should be a relatively stable orbital. An

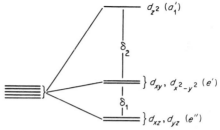

Fig. 20-8.   The splitting pattern of the d orbitals in a trigonal-bipyramidal crystal field.

electron in a d_{xy} or $d_{x^2-y^2}$ orbital is remote from the axial ligands but is in proximity to the equatorial ones. An electron in one of these orbitals should therefore be more stable than one in d_{z^2} but less stable than one in d_{xz} or d_{yz}. Thus, as a qualitative expectation, when the axial and equatorial ligands are the same or similar and lie about equal distances from the metal ion, the d orbital pattern should be as shown in Fig. 20-8.

From the foregoing qualitative arguments it is not possible to estimate the relative magnitudes of the two splittings; however, quantitative treatments [2,3] and experimental data both indicate that in general δ_2 is 2–3 times greater than δ_1. For $TiX_3(NR_3)_2$ (X = Cl or Br) molecules, for example, $\delta_1 \approx 5500 \text{ cm}^{-1}$ and $\delta_2 \approx 9000 \text{ cm}^{-1}$.

SOME CONSEQUENCES AND APPLICATIONS OF ORBITAL SPLITTING

We have now shown how the d orbitals are split by octahedral, tetrahedral, and other arrays of ligands according to the CFT formalism. Regardless of what may be said concerning the artificiality of the model, it is a fact that, qualitatively, these splittings are correctly predicted. We shall now examine some important consequences of the splittings, intending to return later to a discussion of more realistic explanations of the origin of the splittings.

20-3. Magnetic Properties from Crystal Field Theory

In a study of the magnetic properties of a transition-metal complex, our first concern will be to know how many unpaired electrons are present. We shall now see how this property may be understood in terms of the orbital splittings described in the preceding Section.

It is a general rule that if a group of n or less electrons occupies a set of n degenerate orbitals, they will spread themselves among the orbitals and give n unpaired spins. This is Hund's first rule, or the rule of maximum multiplicity. It means that pairing of electrons is an unfavorable process; energy must be expended in order to make it occur. If two electrons are not only to

[2]   C. A. L. Becker, D. W. Meek and T. M. Dunn, *J. Phys. Chem.*, 1968, **72**, 3588.
[3]   J. S. Wood, *J. Chem. Soc., A*, **1969**, 1582 and references cited therein.

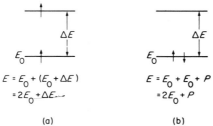

$$E = E_0 + (E_0 + \Delta E)$$
$$= 2E_0 + \Delta E$$

$$E = E_0 + E_0 + P$$
$$= 2E_0 + P$$

(a)   (b)

Fig. 20-9. A hypothetical two-orbital system in which two possible distributions of two electrons and the resulting total energies are as shown.

have their spins paired but also to be placed in the same orbital, there is a further unfavorable energy contribution because of the increased electrostatic repulsion between electrons that are compelled to occupy the same regions of space. Let us suppose now that in some hypothetical molecule we have two orbitals separated by an energy ΔE and that two electrons are to occupy these orbitals. Referring to Fig. 20-9, we see that when we place one electron in each orbital, their spins will remain uncoupled and their combined energy will be $(2E_0 + \Delta E)$. If we place both of them in the lower orbital, their spins will have to be coupled to satisfy the exclusion principle, and the total energy will be $(2E_0 + P)$, where P stands for the energy required to cause pairing of two electrons in the same orbital. Thus, whether this system will have distribution (a) or (b) for its ground state depends on whether ΔE is greater or less than P. If $\Delta E < P$, the triplet state (a) will be the more stable; if $\Delta E > P$, the singlet state (b) will be the more stable.

Octahedral Complexes. We shall first apply an argument of the type outlined above to octahedral complexes, using the d-orbital-splitting diagram previously deduced from CFT. As indicated in Fig. 20-10, we may place one, two and three electrons in the d orbitals without any possible uncertainty about how they will occupy the orbitals. They will naturally enter the more stable t_{2g} orbitals with their spins all parallel, and this will be true

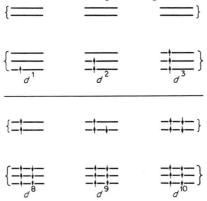

Fig. 20-10. Sketches showing the unique ground-state occupancy schemes for d orbitals in octahedral complexes with d configurations $d^1, d^2, d^3, d^8, d^9, d^{10}$.

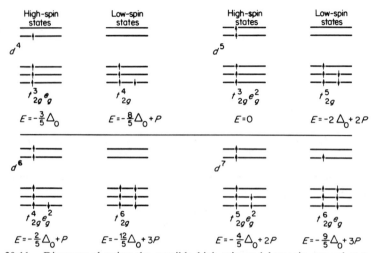

Fig. 20-11. Diagrams showing the possible high-spin and low-spin ground states for d^4, d^5, d^6 and d^7 ions in octahedral crystal fields, including the notation for writing out the configurations and expressions for their energies, derived as explained in the text.

irrespective of the strength of the crystal field as measured by the magnitude of Δ. Further, for ions with eight, nine, and ten d electrons, there is only one possible way in which the orbitals may be occupied to give the lowest energy (see Fig. 20-10). For each of the remaining configurations, d^4, d^5, d^6 and d^7, two possibilities exist, and the question of which represents the ground state can only be answered by comparing the values of Δ_0 and P, an average pairing energy. The two configurations for each case, together with simple expressions for their energies, are set out in Fig. 20-11. The configurations with the maximum possible number of unpaired electrons are called the *high-spin* configurations, and those with the minimum number of unpaired spins are called the *low-spin* or *spin-paired* configurations. These configurations can be written out in a notation similar to that used for electron configurations of free atoms, whereby we list each occupied orbital or set of orbitals, using a right superscript to show the number of electrons present. For example, the ground state for a d^3 ion in an octahedral field is t_{2g}^3; the two possible states for a d^5 ion in an octahedral field are t_{2g}^5 and $t_{2g}^3 e_g^2$. This notation is further illustrated in Fig. 20-11. The energies are referred to the energy of the unsplit configuration (the energy of the ion in a spherical shell of the same total charge) and are simply the sums of $-\frac{2}{5}\Delta_0$ for each t_{2g} electron, $+\frac{3}{5}\Delta_0$ for each e_g electron and P for every pair of electrons occupying the same orbital.

For each of the four cases where high- and low-spin states are possible, we may obtain from the equations for the energies which are given in Fig. 20-11 the following expression for the relation between Δ_0 and P at which the high- and low-spin states have equal energies:

$$\Delta_0 = P$$

The relationship is the same in all cases and means that the spin state of any ion in an octahedral electrostatic field depends simply upon whether the magnitude of the field as measured by the splitting energy, Δ_0, is greater or less than the mean pairing energy, P, for the particular ion. For a particular ion, of the d^4, d^5, d^6 or d^7 type, the stronger the crystal field, the more likely it is that the electrons will crowd as much as possible into the more stable t_{2g} orbitals, whereas in the weaker crystal fields, where $P > \Delta_0$, the electrons will remain spread out over the entire set of d orbitals as they do in the free ion. For ions of the other types, d^1, d^2, d^3, d^8, d^9 and d^{10}, the number of unpaired electrons is fixed at the same number as in the free ion irrespective of how strong the crystal field may become.

Approximate theoretical estimates of the mean pairing energies for the relevant ions of the first transition series have been made from spectroscopic data. In Table 20-1 these energies, along with Δ_0 values for some complexes

TABLE 20-1

Crystal Field Splittings, Δ_0, and Mean Electron Pairing Energies, P, for Several Transition-metal Ions
(energies in cm^{-1})

Config- uration	Ion	P^a	Ligands	Δ	Spin state	
					Predicted	Observed
d^4	Cr^{2+}	23,500	$6H_2O$	13,900	High	High
	Mn^{3+}	28,000	$6H_2O$	21,000	High	High
d^5	Mn^{2+}	25,500	$6H_2O$	7,800	High	High
	Fe^{3+}	30,000	$6H_2O$	13,700	High	High
d^6	Fe^{2+}	17,600	$6H_2O$	10,400	High	High
			$6CN^-$	33,000	Low	Low
	Co^{3+}	21,000	$6F^-$	13,000	High	High
			$6NH_3$	23,000	Low	Low
d^7	Co^{2+}	22,500	$6H_2O$	9,300	High	High

[a] It will be shown later that owing to the so-called nephelauxetic effect these energies in the complexes should probably be $\sim20\%$ lower than the free-ion values given. It can be seen, however, that even if they are decreased by this amount the correct spin states are still predicted.

(derived by methods to be described in the next Section), are listed. It will be seen that the theory developed above affords correct predictions in all cases. It will further be noted that the mean pairing energies vary irregularly from one metal ion to another as do the values of Δ_0 for a given set of ligands. Thus, as Table 20-1 shows, the d^5 systems should be exceptionally stable in their high-spin states, whereas the d^6 systems should be exceptionally stable in their low-spin states. These expectations are in excellent agreement with the experimental facts.

Finer details of magnetic properties such as orbital contributions, unusual temperature-dependencies of magnetic moments and magnetic anisotropies can also be calculated by using CFT or, better, ACFT, and the results are generally good approximations. They are not perfect because the basic premise in the theory—namely, that the interaction between the metal ion and its surroundings is a purely electrostatic perturbation by nearest neighbors—is not perfect.

Tetrahedral Complexes. It is found that for the d^1, d^2, d^7, d^8 and d^9 cases only high-spin states are possible, whereas for d^3, d^4, d^5 and d^6 configurations both high-spin and low-spin states are in principle possible. The existence of low-spin states would require that $\Delta_t > P$. Since Δ_t values are only about half as great as Δ_0 values, it is to be expected that low-spin tetrahedral complexes of ions of the first transition series with d^3, d^4, d^5 and d^6 configurations would be scarce or unknown. None has so far been shown to exist, and there seems little chance that any will be found.

Square and Tetragonally-distorted Octahedral Complexes. These two cases may be considered together because, as noted earlier, they merge into one another.

Let us consider as an example the d^8 system in an octahedral environment which is then subjected to a tetragonal distortion. We have already seen (Fig. 20-6) how a decrease in the electrostatic field along the z-axis splits apart the $(x^2 - y^2)$ and z^2 orbitals. We have also seen that, if the tetragonal distortion, that is, the disparity between the contributions to the electrostatic potential of the two z-axis ligands and the other four becomes sufficiently great, the z^2 orbital may fall below the xy orbital. In either case, the two least stable d orbitals are now no longer degenerate but are separated by some energy, Q. Now the question whether the tetragonally distorted d^8 complex will have high or low spin depends on whether the pairing energy, P, is greater or less than the energy Q. Fig. 20-12a shows the situation for the case of a "weak" tetragonal distortion, that is, for one in which the second highest d orbital is still d_{z^2}.

Fig. 20-12b shows a possible arrangement of levels for a strongly tetragonally distorted octahedron, or for the extreme case of a square, four-coordinate complex (compare with Fig. 20-7), and the low-spin form of occupancy of these levels for a d^8 ion. In this case, owing to the large sepa-

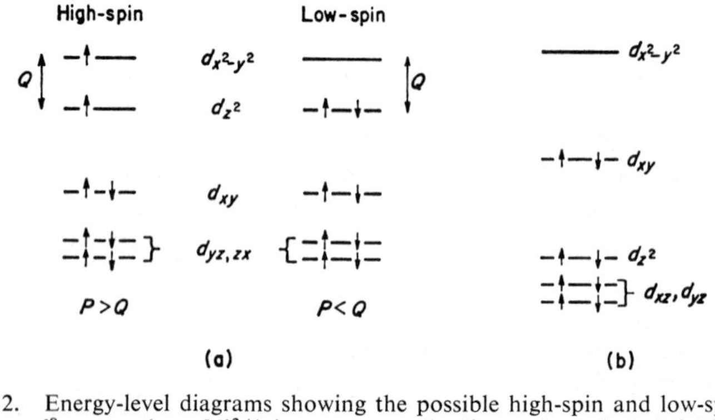

Fig. 20-12. Energy-level diagrams showing the possible high-spin and low-spin ground states for a d^8 system (e.g., Ni^{2+}) in a tetragonally distorted octahedral field. (a) Weak tetragonal distortion; (b) strong distortion or square field.

ration between the highest and the second highest orbitals the high-spin configuration is impossible of attainment with the pairing energies of the real d^8 ions, e.g., Ni^{II}, Pd^{II}, Pt^{II}, Rh^I, Ir^I and Au^{III}, which normally occur, and all square complexes of these species are diamagnetic (unless the ligands have unpaired electrons). Similarly, for a d^7 ion in a square complex, as exemplified by certain Co^{II} complexes, only the low-spin state with one unpaired electron should occur, and this is in accord with observation.

Five-coordinate Complexes. We shall consider only the two regular geometries : *spy* and *tbp*. It should be stressed, however, that in practice distortions of these limiting geometries are more the rule than the exception and may have a significant effect on the magnetic properties. By far the majority of cases studied are for $Co^{II}(d^7)$ and $Ni^{II}(d^8)$. Reviews of the data are available.[4]

Spy. Employing a diagram similar to that in Fig. 20-7, we first conclude that only the two larger splittings can ever be large enough to lead to spin-pairing. Thus d^2 and d^3 configurations will always be high-spin; d^4–d^8 configurations can either be high- or low-spin depending on the magnitude of these larger splittings. Experimental data are relatively scarce since most work on 5-coordination has employed polydentate ligands predisposed to favor a *tbp* configuration. For d^6, d^7 and d^8 configurations in complexes of the type $[M(OER_3)_4(ClO_4)]^+$ (E = P or As) the spin states are high, while the spin states are low for essentially all other ligand sets having, at least roughly, *spy* geometry.

Tbp. From the orbital splitting diagram, Fig. 20-8, where δ_1 is small, comparable to Δ_t values, while δ_2 is larger, comparable to Δ_0 values, we can make the following inferences:

d^2 complexes must be high-spin;

d^3 and d^4 complexes are likely to be high-spin since δ_1 is unlikely to exceed spin-pairing energies;

d^5–d^8 complexes may have either high- or low-spin configurations depending on whether δ_2 is smaller or greater than the mean spin-pairing energy.

The possibilities and some examples are summarized in Table 20-2.

High-spin, Low-spin Crossovers. One may envisage the occurrence of critical orbital splittings of a magnitude comparable to P so that high- and low-spin states will have about the same energy.

Though such situations occur infrequently, there are several well documented cases.[5] Certain complexes of the iron(III) ion, d^5, containing dithiocarbamate ligands (20-III) were the first to be discovered (by Cambi and co-workers in 1931–1933) and have been very extensively studied in

[4] L. Sacconi, *Pure Appl. Chem.*, 1968, **17**, 97; *J. Chem. Soc. (A)*, **1970**, 248; M. Ciampolini, *Structure and Bonding*, 1969, **6**, 52.

[5] E. K. Barefield, D. H. Busch and S. M. Nelson, *Quart. Rev.*, 1968, **22**, 457; R. L. Martin and A. H. White, *Transition Metal Chemistry*, Vol. 4, R. L. Carlin, ed., Marcel Dekker, Inc., **1968**, p. 113.

TABLE 20-2

Some High- and Low-spin *tbp* Complexes

	High-spin				Low-spin			
d^n	Configuration and ground state	Number of unpaired electrons	Examples[a,b] Formula	Examples[a,b] Observed mag. moment (B.M., 25°)	Configuration	Number of unpaired electrons	Examples[a,b] Formula	Examples[a,b] Observed mag. moment (B.M., 25°)
d^2	$e''^2\ (^3A_2')$	2	$VCl_3(NMe_3)_2$	2.76				
d^3	$e''^2e'\ (^4E')$	3	$CrCl_3(NMe_3)_2$	3.88	$e''^3\ (^2E'')$	1	None	
d^4	$e''^2e'^2\ (^5A_1')$	4	$Cr(Me_6tren)Cl^+$	4.85	$e''^4\ (^1A_1')$	0	None	
d^5	$e''^2e'^2a_1'\ (^5A_1')$	5	$[(MeSalam)_2Mn]_2$	~5.9	$e''^4e'\ (^2E')$ or $e''^3e'^2\ (^4E'')$	1 or 3	None	
d^6	$e''^3e'^2a_1'\ (^5E'')$	4	$Fe(Me_6tren)Cl^+$	5.34	$e''^4e'^2\ (^3A_2')$	2	None	
d^7	$e''^4e'^2a_1'\ (^4A_2')$	3	$Co(Me_6tren)Cl^+$ $[(MeSalam)_2Co]_2$[c]	4.43	$e''^4e'^3\ (^1E')$	1	None	
d^8	$e''^4e'^3a_1'\ (^3E')$	2	$Ni(Me_6tren)Cl^+$	3.42	$e''^4e'^4\ (^1A_1')$	0	$[Co(CNCH_3)_5]^+$ $[Ni(TAP)CN]^+$	diamag. diamag.

[a] Only complexes of first transition series.
[b] $Me_6tren = (Me_2NCH_2CH_2)_3N$; TAP $= (Me_2AsCH_2CH_2CH_2)_3As$; MeSalam = *N*-methylsalicylaldiminato ion.
[c] Compound reported only as "high-spin".

$$R_2N—C\Big\langle{}^{S}_{S}$$

(20-III)

recent years.[5,6] In the immediately following discussion of these and a few other systems, a notation for the *states* arising from electron configurations (e.g., the $^6A_{1g}$ state of the $t_{2g}^3 e_g^2$ configuration) will be used, in advance of an explanation of the notation. All that needs to be understood about it for the moment is that the left superscripts denote the spin multiplicities of the states.

The tris(dithiocarbamato)iron(III) molecules, Fe(dtc)$_3$, have trigonally distorted octahedral configurations of the six sulfur donor atoms; as a first approximation, however, splittings due to the trigonal distortion of octahedral symmetry may be neglected and the orbitals and states denoted according to ideal O_h symmetry. On this basis then, the ground states for high- and low-spin configurations are, respectively, $^6A_{1g}(t_{2g}^3 e_g^2)$ and $^2T_{2g}(t_{2g}^5)$. With most R groups the $^2T_{2g}$ state lies several hundred cm^{-1} below the $^6A_{1g}$ state. Thus, at low temperatures, the effective magnetic moment, μ_{eff}, tends toward a value of ~ 2.1 B.M., which is characteristic of a t_{2g}^5 configuration. As the temperature increases, molecules begin to populate the high-spin state and the average effective magnetic moment rises, following a sigmoidal curve that appears to be approaching an asymptotic limit. This limiting value must, of course, be less than the μ_{eff} for a pure high-spin complex, since it will never be possible to excite thermally all the molecules into the high-spin state. This behavior is shown in Fig. 20-13, for a typical case where the energy difference between the low- and high-spin states, ΔE, is within the range 50–250 cm^{-1}; this is the crucial range since thermal energies vary from ~ 70 cm^{-1} at 100°K to ~ 200 cm^{-1} at room temperature. If ΔE becomes greater than ~ 29 kJ mol^{-1} (> 10 times RT at the highest temperature of measurement) thermal population of the high-spin excited state is negligible—or nearly so—and simple, temperature-independent low-spin behavior is ob-

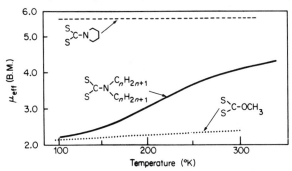

Fig. 20-13.   Variation of μ_{eff} with temperature for some FeL$_3$ complexes, where L is a dithiocarbamate or xanthate.

[6] A. H. Ewald, R. L. Martin, E. Sinn and A. H. White, *Inorg. Chem.*, 1969, **8**, 1837.

served. This is the case when the R, R' groups are cyclohexyl, or when NRR' is replaced by OR (to give xanthates instead of dithiocarbamates). Conversely, the replacement of NRR' by pyrrolidino causes the high-spin state to become more stable by an energy much in excess of RT and we have simple, temperature-independent high-spin behavior.

It should be noted that, while the spin crossover behavior just discussed is relatively simple for solutions (it is formally no different from any ordinary chemical equilibrium, A $\rightleftarrows$ B, between isomers), it becomes much more complex in the solid state. The experimental results can be fitted only approximately by considering a simple Boltzmann distribution of molecules between two states. However, when the theory is refined by introducing interactions between the two crossing states and higher ones, and particularly by allowing for the fact that the vibrational contributions to the total free energies of the two states will differ because the transfer of electrons from e orbitals (which are antibonding, see Sections 20-14 and 20-17) to non-bonding t_2 orbitals markedly affects bond strengths, the experimental data can be fitted quantitatively. A variety of other methods (e.g., Mössbauer, nmr, infrared and electronic spectra) have been used to study these systems and a very detailed and generally satisfactory understanding of their behavior has been achieved.[5] Among other d^5 cases that have been studied are various metmyoglobin complexes.

For d^6 systems, crossover phenomena involving the $^5T_{2g}(t_2^4e^2)$ and $^1A_{1g}(t_2^6)$ states have been observed mainly with Fe^{II} complexes such as $Fe(phen)_2(NCS)_2$, the aquo Co^{III} ion and Co^{III} in certain mixed oxides. Only for a few iron(II) compounds, especially $Fe(phen)_2(NCS)_2$, have fairly detailed studies been made.

Several Co^{II} complexes afford examples of crossover equilibria in a d^7 system where $^4T_1(t_2^5e^2)$ and $^2E(t_2^6e)$ states are concerned. For about a dozen complexes of the type $[CoL_2]X_2$, where L is a tridentate ligand of the 2,6-pyridinedialdehyde dihydrazone type (20-IV) or a closely related one, there is a ^2E ground state with the $^4T_1$ state lying only a few hundred cm^{-1} higher. More recently the complex Co[tri-(2-pyridyl)amine]$_2$ perchlorate has been shown to behave similarly.[7]

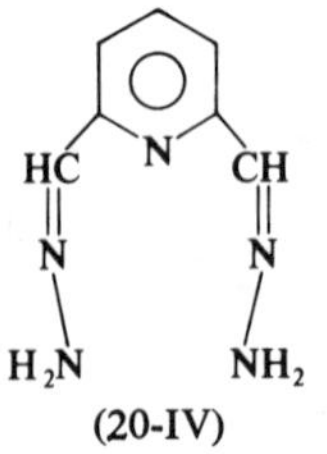

(20-IV) 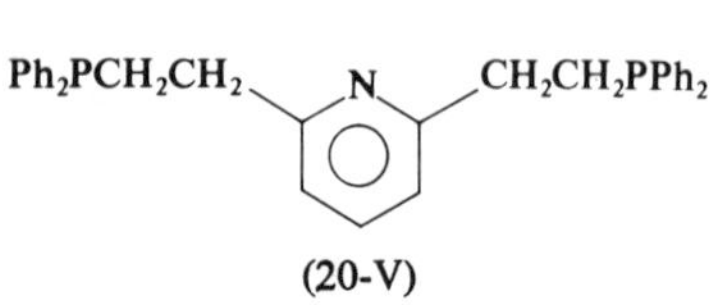   (20-V)

In addition to the spin-crossover equilibria occurring in octahedral complexes, several cases are known[8] for the 5-coordinate Co^{II} and Ni^{II} complexes

[7] P. E. B. Barnard et al., Chem. Comm., 1970, 520.
[8] S. M. Nelson and W. J. S. Kelly, Chem. Comm., 1968, 436; 1969, 94.

$M(PpyP)X_2$, where PpyP is (20-V), with $M = Co$, $X = Br$ or I and with $M = Ni$, $X = Cl$.

ELECTRONIC ABSORPTION SPECTRA*

20-4. Octahedral and Tetrahedral Complexes

d^1 and d^9 Systems. Let us first consider the simplest possible case, an ion with a d^1 configuration, lying at the center of an octahedral field, for example, the Ti^{III} ion in $[Ti(H_2O)_6]^{3+}$. The d electron will occupy a t_{2g} orbital. Upon irradiation with light of frequency v, equal to Δ_0/h, where h is Planck's constant and Δ_0 is the energy difference between the t_{2g} and the e_g orbitals, it should be possible for such an ion to capture a quantum of radiation and convert that energy into energy of excitation of the electron from the t_{2g} to the e_g orbital. The absorption band which results from this process is found in the visible spectrum of the hexaquotitanium(III) ion, shown in Fig. 20-14, and is responsible for its violet color. Three features of this absorption band are of importance here : its position, its intensity and its breadth.

In discussing the positions of absorption bands in relation to the splittings of the d orbitals, it is convenient and common practice to use the same unit, the reciprocal centimeter or wave number, abbreviated cm^{-1}, for both the unit of frequency in the spectra and the unit of energy for the orbitals. With this convention, we see that the spectrum of Fig. 20-14 tells us that Δ_0 in $[Ti(H_2O)_6]^{3+}$ is 20,000 cm^{-1}. Since there are 83.7 cm^{-1} per kJ, this means that the splitting energy is ~ 240 kJ mol^{-1}, which is comparable with the usual

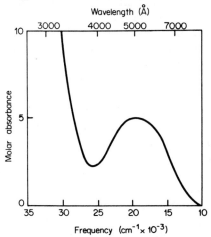

Fig. 20-14.   The visible absorption spectrum of $[Ti(H_2O)_6]^{3+}$.

* Only the d–d or crystal-field transitions are discussed here. Discussion of charge-transfer transitions, which require an MO description, is postponed to Section 20-18.

values of chemical bond energies. In general, we shall see that the crystal field theory enables us to calculate the energy separations between various states of the d electrons from the frequencies of the absorption bands in the visible spectra. In the present case the relationship is the simplest possible one, namely, the observed frequency is identical with the d orbital splitting.

Turning now to the intensity of this absorption band in the $[Ti(H_2O)_6]^{3+}$ ion, we note that it is extremely weak by comparison with absorption bands found in many other systems. The reason for this is that the electron is jumping from one orbital that is centrosymmetric to another that is also centrosymmetric, and that all transitions of this type are nominally "forbidden" by the rules of quantum mechanics. One-electron transitions which are "allowed" have intensities that give molar absorbance values at the absorption peaks of $\sim 10^4$. If the postulate of the crystal field theory, that in both the ground and the excited states the electrons of the metal ion occupy completely pure d orbitals that have no other interaction than a purely coulombic one with the environment of the ion, were precisely correct, the intensity of this band would be precisely zero. It gains a little intensity because the postulate is not perfectly valid in ways that will be discussed on page 578. It will also be noted that the band is several thousand cm^{-1} broad, rather than a sharp line at a frequency precisely equivalent to Δ_0. This too is a general phenomenon that will be discussed in detail below.

It is also possible to interpret the d–d spectrum of a d^9 ion as simply as we have done with the spectrum of a d^1 ion. This is another application of the *hole formalism* (see page 84) according to which a d^{10-n} configuration will have the same behavior in a crystal field, except for certain changes in the signs of energy terms, as a d^n configuration. The former has as many holes in its d shell as the latter has electrons. According to the hole formalism, which is perfectly rigorous within the limits of the electrostatic crystal field theory, n holes in the d shell may be treated like n positrons. Now all of the d level splittings that we have deduced for the static fields will be quantitatively the same for one positron, *except* that the patterns will be inverted because a positron will be most electrostatically stable in just those regions where an electron is least electrostatically stable, and *vice versa*. We can thus look upon a Cu^{II} ion in an octahedral environment as a one-positron ion in an octahedral field and deduce that in the ground state the positron will occupy an e_g orbital from which it may be excited by radiation providing energy Δ_0 to a t_{2g} orbital.

Experimentally it is found, however, that the absorption band of the Cu^{II} ion in aqueous solution is not a simple, symmetrically shaped band but instead appears to consist of several nearly superposed bands. The observant reader may have noticed that the absorption band of the $[Ti(H_2O)_6]^{3+}$ ion is not quite a simple, symmetrical band either. In each case these complications are traceable to distortions of the octahedral environment that are required by the Jahn–Teller theorem. We shall discuss this theorem on page 590.

d^2–d^8 **Ions; Energy-level Diagrams.** In order to interpret the spectra of complexes in which the metal ions have more than one but less than nine d electrons, we must employ an energy-level diagram based upon the Russell–Saunders states of the relevant d^n configuration in the free (uncomplexed) ion. It can be shown that, just as the set of five d orbitals is split apart by the electrostatic field of surrounding ligands to give two or more sets of lower degeneracy, so also are the various Russell–Saunders states of a d^n configuration. The number and types of the components into which an octahedral or tetrahedral field will split a state of given L is the same regardless of the d^n configuration from which it arises, and these facts are summarized in Table 20-3. The designations of the states of the ion in the crystal field are

TABLE 20-3

Splitting of Russell–Saunders States in Octahedral and Tetrahedral Electrostatic Fields

State of free ion	States in the crystal field
S	A_1
P	T_1
D	$E + T_2$
F	$A_2 + T_1 + T_2$
G	$A_1 + E + T_1 + T_2$
H	$E + 2T_1 + T_2$

the *Mulliken* symbols; their origin is in group theory, but they may be regarded simply as labels. The main letters in these symbols have the following meanings: A and B designate singly degenerate states, E designates doubly degenerate states, and T indicates triple degeneracy. Left superscripts can be applied just as they are to the letter symbols for free ion states (i.e., S, P, D, etc.) to indicate spin multiplicity.

Although the states into which a given free ion state is split are the same in number and type in both octahedral and tetrahedral fields, the pattern of energies is inverted in one case relative to the other. This is quite analogous to the results in the d^1 case, as we have already seen (Fig. 20-4).

A discussion of the way in which the energies of the crystal field states are calculated would be beyond the scope of this book. For the inorganic chemist it is not essential to know how the energy level diagrams are obtained in order to use them; it is, of course, necessary to know how to interpret them properly. At this point we shall examine several of them in some detail in order to explain their interpretation. Others will be introduced subsequently in discussing the chemistry of particular transition elements. For the convenience of the reader, a complete set of diagrams is presented in the Appendix. The diagrams used in the body of the text will generally represent the relative energies only qualitatively, whereas those in the Appendix (the Tanabe and Sugano diagrams) are semiquantitative.

We shall look first at the energy level diagram for a d^2 system in an octa-

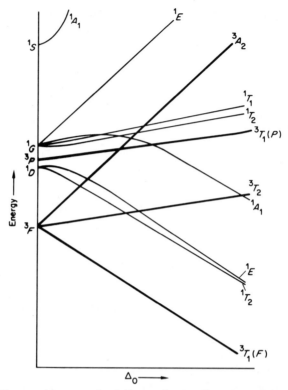

Fig. 20-15. The complete energy-level diagram for the d^2 configuration in an octahedral crystal field. The heavier lines are those for the triplet states.

hedral field, as shown in Fig. 20-15. The ordinate is in energy units, usually cm^{-1}, and the abscissa is in units of crystal field splitting energy as measured by Δ_0, the splitting of the one-electron orbitals. At the extreme left are the Russell–Saunders states of the free ion. It may be seen that each of these splits in the crystal field into the components specified in Table 20-3. Three features of this energy level diagram particularly to be noted because they will be found in all such diagrams are:

1. States with identical designations never cross.

2. The crystal field states have the same spin multiplicity as the free ion states from which they originate.

3. States that are the only ones of their type have energies that depend linearly on the crystal field strength, whereas, when there are two or more states of identical designation, their lines will in general show curvature. This is because such states interact with one another as well as with the crystal field.

It is interesting to note in this diagram that a triplet state lies lowest at all field strengths shown, and, since its slope is as steep as the slope of any other state, it will continue to be the lowest state no matter how intense the crystal field may become. This is in complete agreement with our previous conclusion, based on the simple splitting diagram for the d orbitals, which showed that the two d electrons would have their spins parallel in an octahedral field irrespective of how strong the field might be.

In order to use this energy level diagram to predict or interpret the spectra of octahedral complexes of d^2 ions, for example, the spectrum of the $[V(H_2O)_6]^{3+}$ ion, we first note that there is a quantum-mechanical selection rule that forbids transitions between states of different spin multiplicity. This means that in the present case only three transitions, those from the $^3T_1$ ground state to the three triplet excited states, $^3T_2$, $^3A_2$ and $^3T_1(P)$, will occur. Actually, spin-forbidden transitions, that is, those between levels of different spin multiplicity, do occur very weakly because of weak spin–orbit interactions, but they are several orders of magnitude weaker than the spin-allowed ones and are ordinarily not observed.

Experimental study of the $[V(H_2O)_6]^{3+}$ ion reveals just three absorption bands with energies of about 17,000, 25,000 and 38,000 cm^{-1}. Using an energy level diagram like that in Fig. 20-15, in which the separations of the free ion states are adjusted to match exactly those appropriate for V^{III}, one finds that, at a Δ_0 of 21,500 cm^{-1}, the three transitions are expected at 17,300, 25,500 and 38,600 cm^{-1}, in excellent agreement with observation. For high-spin complexes of the first transition series metals in their normal oxidation states, quantitative agreement of this sort, or nearly as good, is not always attained unless the adjusted CFT (ACFT) treatment is used in which the actual energies of the free ion states are somewhat altered as will be explained in Section 20-16.

We shall look next at the energy level diagram for a d^8 ion in an octahedral field (Fig. 20-16), restricting our attention to the triplet states except for the lowest energy singlet state which comes from the ^1D state of the free ion. This has been included to show that for this system the ground state will always be a spin triplet no matter how strong the crystal field, a result in agreement with the conclusion previously drawn from consideration of the distribution of eight electrons among a set of five d orbitals. On comparing the arrangements of the three components derived from the ^3F ground state for the d^2 and the d^8 cases, we note that one pattern is the inverse of the other. This is a manifestation of the hole formalism for the d^2-d^{10-2} configurations which is fundamentally analogous to the d^1-d^{10-1} example we previously examined.

Similar energy level diagrams may be drawn for d^n systems in tetrahedral crystal fields. There is an interesting relationship between these and the ones for certain systems in octahedral fields. We have already seen that the splitting pattern for the d orbitals in a tetrahedral field is just the inverse of that for the d orbitals in an octahedral field. A similar inverse relationship exists between the energy level diagrams of d^n systems in tetrahedral and octahedral fields. The components into which each Russell–Saunders state is split are reversed in their energy order in the tetrahedral compared to the octahedral

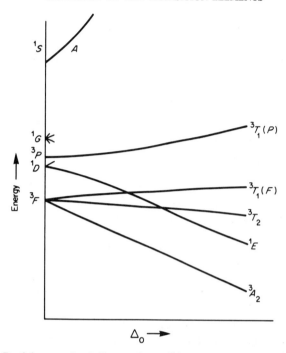

Fig. 20-16.   Partial energy-level diagram for a d^8 ion in an octahedral field, showing the triplet states and only the lowest singlet state.

cases. Furthermore, as in the one-electron case, a purely electrostatic inter-action of metal ion with the ligands will produce only $\frac{4}{9}$ the splittings in the tetrahedral case as in the octahedral case, all other factors such as metal—ligand distance being kept constant.

Finally, there are rather extensive qualitative similarities between the energy level diagrams of groups of the d^n systems because of the combined effects of reversals in the splitting patterns by changing from an octahedral to a tetrahedral field and by changing from a d^n to a d^{10-n} configuration. When we go from d^n in an octahedral field to d^n in a tetrahedral field, all splittings of the Russell–Saunders states are inverted. But the same inversions occur on changing from the d^n configuration in an octahedral (tetrahedral) field to the d^{10-n} configuration in an octahedral (tetrahedral) field. These relations, combined with the fact that, for the free ions, the Russell–Saunders states of the pairs of d^n and d^{10-n} systems are identical in number, type and relative (though certainly not absolute) energies, mean that various pairs of configuration–environment combinations have qualitatively identical energy level diagrams in crystal fields, and that these differ from others only in the reversal of the splittings of the individual free ion states. These relations are set out in Table 20-4.

It will be evident from the foregoing description and illustrations of energy level diagrams that they may be used to determine from observed spectral

TABLE 20-4

Relations between Energy Level Diagrams for Various d^n Configurations
in Octahedral and Tetrahedral Crystal Fields

Octahedral d^1 and tetrahedral d^9	Reverse[a] of	Octahedral d^9 and tetrahedral d^1
Octahedral d^2 and tetrahedral d^8	Reverse of	Octahedral d^8 and tetrahedral d^2
Octahedral d^3 and tetrahedral d^7	Reverse of	Octahedral d^7 and tetrahedral d^3
Octahedral d^4 and tetrahedral d^6	Reverse of	Octahedral d^6 and tetrahedral d^4
Octahedral d^5	Identical with	Tetrahedral d^5

[a] "Reverse" means that the order of levels coming from each free ion state is reversed; it does *not* mean that the diagram as a whole is reversed.

bands the magnitudes of Δ_o and Δ_t in complexes. It may be noted in the diagrams for the d^2 and d^8 systems that there are three spin-allowed absorption bands whose positions are all determined by the one* parameter, Δ_0 or Δ_t. Thus in these cases the internal consistency of the theory may be checked.

Certain generalizations may be made about the dependence of the magnitudes of Δ values on the valence and atomic number of the metal ion, the symmetry of the coordination shell and the nature of the ligands. For octahedral complexes containing high-spin metal ions, it may be inferred from the accumulated data for a large number of systems that:

1. Δ_o values for complexes of the first transition series are 7500–12,500 cm^{-1} for divalent ions and 14,000–25,000 cm^{-1} for trivalent ions.

2. Δ_o values for corresponding complexes of metal ions in the same Group and with the same valence increase by 25–50% on going from the first transition series to the second and by about this amount again from the second to the third. This is well illustrated by the Δ_o values for the complexes $[Co(NH_3)_6]^{3+}$, $[Rh(NH_3)_6]^{3+}$ and $[Ir(NH_3)_6]^{3+}$, which are, respectively, 23,000, 34,000 and 41,000 cm^{-1}.

3. Δ_t values are about 40–50% of Δ_o values for complexes differing as little as possible except in the geometry of the coordination shell, in agreement with theoretical expectation.

4. The dependence of Δ values on the identity of the ligands follows a regular order known as the spectrochemical series which will now be explained.

20-5. The Spectrochemical Series

It has been found by experimental study of the spectra of a large number of complexes containing various metal ions and various ligands, that ligands may be arranged in a series according to their capacity to cause d orbital splittings. This series, for the more common ligands, is: $I^- < Br^- < Cl^- < F^- < OH^- < C_2O_4^{2-} \sim H_2O < -NCS^- < py \sim NH_3 < en < bipy < o$-phen

---

* This is only approximately true because the separation between the ^3F and ^3P states is not the same in the complexed ion as it is in the free ion, and this separation therefore becomes a second parameter, in addition to Δ, to be determined from experiment. We shall return to this point on page 601.

$< NO_2^- < CN^-$. The idea of this series is that the d orbital splittings and hence the relative frequencies of visible absorption bands for two complexes containing the same metal ion but different ligands can be predicted from the above series whatever the particular metal ion may be. Naturally, one cannot expect such a simple and useful rule to be universally applicable. The following qualifications must be remembered in applying it:

1. The series is based upon data for metal ions in common oxidation states. Because the nature of the metal–ligand interaction in an unusually high or unusually low oxidation state of the metal may be in certain respects qualitatively different from that for the metal in a normal oxidation state, striking violations of the order shown may occur for complexes in unusual oxidation states.

2. Even for metal ions in their normal oxidation states inversions of the order of adjacent or nearly adjacent members of the series are sometimes found.

20-6. Further Remarks on Intensities and Line Widths, illustrated by Manganese(II)

Fig. 20-17 shows the spectra of octahedrally coordinated Mn^{2+}, in $[Mn(H_2O)_6]^{2+}$, and tetrahedrally coordinated Mn^{2+}, in the $[MnBr_4]^{2-}$ ion. Let us look first at the $[Mn(H_2O)_6]^{2+}$ spectrum. The spectra of other octahedrally coordinated Mn^{2+} complexes are quite similar.[9] The most striking features are (a) the weakness of the bands, (b) the large number of bands and (c) the great variation in the widths of the bands, with one extremely narrow

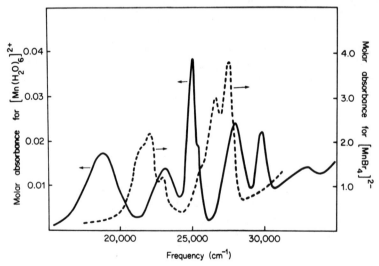

Fig. 20-17. The d–d absorption spectra of the $[Mn(H_2O)_6]^{2+}$ ion (——, molar absorbance scale to the left) and the $[MnBr_4]^{2-}$ ion (- - -, molar absorbance to the right).

[9] J. J. Foster and N. S. Gill, *J. Chem. Soc.*, *A*, **1968**, 2625.

indeed. All these main features of the spectra are easily understood in terms of ligand field theory.

The bands in $[Mn(H_2O)_6]^{2+}$ are more than 10^2 times weaker than those normally found for $d–d$ crystal field transitions (cf. Fig. 20-14). It is because of their extreme weakness that the ion has such a pale color, and the many other salts and complexes of Mn^{II} in which the ion finds itself in octahedral surroundings are also very pale pink in color; finely ground solids often appear to be white. The reason for the weakness of the bands is very simple. The ground state of the d^5 system in a weak octahedral field has one electron in each d orbital, and their spins are parallel, making it a spin sextuplet. This corresponds to the ^6S ground state of the free ion, which is not split by the ligand field. This, however, is the only sextuplet state possible, for every conceivable alteration of the electron distribution $t_{2g}^3 e_g^2$ results in the pairing of two or four spins, thus making quartet or doublet states. Hence, all excited states of the d^5 system have different spin multiplicity from the ground state, and transitions to them are spin forbidden. Because of weak spin–orbit interactions, such transitions are not totally absent, but they give rise only to very weak absorption bands. As a rough rule such spin-forbidden transitions give absorption bands ~ 100 times weaker than those for similar but spin-allowed transitions.

In order to understand the number and widths of the spin-forbidden bands, we must refer to an energy level diagram. Fig. 20-18 shows a simplified one for the d^5 system in which all spin-doublet states are omitted. Most of these are of very high energy, and transitions to them from the sextuplet ground state are doubly spin-forbidden and hence never observed. It is seen that there are four Russell–Saunders states of the free ion that are quartets, and their splittings as a function of ligand field strength are shown. The observed bands of $[Mn(H_2O)_6]^{2+}$ can be fitted by taking Δ equal to about 8600 cm^{-1}, as indicated by the vertical dashed line in the diagram. The diagram shows that, to the approximation used to calculate it, the ^4E and $^4A_1$ states arising from the ^4G term are degenerate. This is very nearly but not exactly so, as the slight shoulder on the sharp band at $\sim 25,000$ cm^{-1} shows.

It will also be noted that there are three states, the $^4A_2$ state from ^4F, the ^4E state from ^4D, and the $(^4E_1, ^4A_1)$ state from ^4G, whose energies are independent of the strength of the ligand field. Such a situation, which never occurs for upper states of the same spin multiplicity as the ground state, makes it unusually easy to measure accurately the decrease in the interelectronic repulsion parameters.

Theoretical considerations show that the widths of spectral bands due to $d–d$ transitions should be proportional to the slope of the upper state relative to that of the ground state. In the present case, where the ground-state energy is independent of the ligand field strength, this means that the band widths should be proportional to the slopes of the lines for the respective upper states as they are seen in Fig. 20-18. Comparison of the spectrum of $[Mn(H_2O)_6]^{2+}$ with the energy-level diagram shows that this expectation is

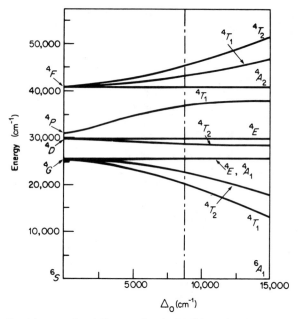

Fig. 20-18. Partial energy-level diagram for the Mn^{II} ion, showing only the ^6S state and the quartet states. The separations of the Russell–Saunders states at $\Delta = 0$ are those appropriate for the $[Mn(H_2O)_6]^{2+}$ ion (*not* the actual free Mn^{2+} ion) and the vertical line (–.–.–) is at the Δ value (8600 cm^{-1}) for this species.

very well fulfilled indeed. Thus the narrowest bands are those at $\sim 25,000$ and $\sim 29,500$ cm^{-1}, which correspond to the transitions to upper states with zero slope. The widths of the other lines are also seen to be greater in proportion to the slopes of the upper state energy lines.

The reason why the band widths are proportional to the slopes is easy to grasp in a qualitative way. As the ligand atoms vibrate back and forth, the strength of the ligand field, Δ, also oscillates back and forth about a mean value corresponding to the mean position of the ligands. Now, if the separation between the ground and excited states is a sensitive function of Δ, the energy difference will vary considerably over the range in Δ that corresponds to the range of metal—ligand distances covered in the course of the vibrational motion. If, on the other hand, the energy separation of the two states is rather insensitive to Δ, only a narrow range of energy will be encompassed over the range of the vibration. This argument is illustrated in Fig. 20-19.

Tetrahedral complexes of Mn^{II} are yellow-green, and the color is more intense than that of the octahedral complexes. A typical spectrum is shown in Fig. 20-17. First it will be noted that the molar absorbance values are in the range 1.0–4.0, whereas for octahedral Mn^{II} complexes (see Fig. 20-17), they are in the range 0.01–0.04. This increase by a factor of about 100 in the intensities of tetrahedral complexes over octahedral ones is entirely typical. The reasons for it are not known with complete certainty, but it is

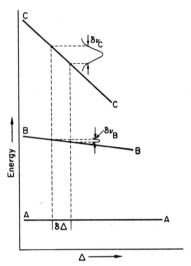

Fig. 20-19. Diagram showing how band width is related to the slope of the upper state relative to that of the ground state. A–A gives the energy of the ground state, B–B, and C–C the energies of two upper states as functions of the ligand field strength, Δ. $\delta\Delta$ represents the range of variation of Δ due to the ligand vibrations, and $\delta\nu_B$ and $\delta\nu_C$ are the widths of the bands due to the transitions $A \rightarrow B$ and $A \rightarrow C$.

thought to be due in part to mixing of metal p and d orbitals in the tetrahedral environment which is facilitated by overlap of metal d orbitals with ligand orbitals in the tetrahedral complexes.

It may also be seen in Fig. 20-17 that there are six absorption bands in two groups of three, just as there are in $[Mn(H_2O)_6]^{2+}$, but they are here much closer together. This is to be expected since the Δ value for the tetrahedral complex, $[MnBr_4]^{2-}$, should be less than that for $[Mn(H_2O)_6]^{2+}$. From the energy level diagram (Fig. 20-18) it can be seen that the uppermost band in the group at lower energy should be due to the transition to the field-strength-independent $(^4E_1, {}^4A_1)$ level, and this band does seem to be quite narrow, although it is partly overlapped by the other two in the group.

OPTICAL ACTIVITY

The most important optically active inorganic compounds, or at least those most intensively studied in recent years, are complexes of transition metals containing two or three chelate rings. The special importance of transition-metal complexes in the study of optical activity is due to two things. First, there are several transition-metal ions, especially Co^{III}, Cr^{III}, Rh^{III}, Ir^{III} and Pt^{IV}, that give such kinetically inert complexes that resolution of optical isomers is possible and the rates of racemization are slow enough for spectroscopic studies to be carried out conveniently. Secondly, detailed studies of optical activity require as a starting point a reasonably clear understanding of the energy levels and electronic spectra of the compounds to be studied, and the studies are greatly facilitated if the compounds have easily observable absorption bands in the visible region, both of which requirements are uniquely met by transition-metal compounds.

20-7. Basic Principles and Definitions

Optical activity in a molecule can be expected when and only when the molecule is so structured that it cannot be superposed on its mirror image. Such a molecule is said to be *dissymmetric*, or *chiral*. Chiral, from the Greek word for hand, is useful in emphasizing the left-hand to right-hand type of relationship between non-superposable mirror images or enantiomorphs.

The general conditions for the existence of chirality have been specified earlier (page 21). In short, a dissymmetric or chiral molecule must have either no element of symmetry or, at most, only proper axes of symmetry.

The six-coordinate chelate complexes of the types M(bidentate ligand)$_3$ and *cis*-M(bidentate ligand)$_2$X$_2$, which have symmetries D_3 and C_2, respectively, fulfill this condition and are the commonest cases in which the "center of dissymmetry" is the metal ion itself.

The simplest way in which optical activity may be observed is doubtless already familiar. It consists in the observation that the plane of polarization of plane-polarized monochromatic light is rotated upon passing through a solution containing one or the other—or an excess of one or the other—of two enantiomorphic molecules.

In order to appreciate more fully this phenomenon and some others closely related to it, the nature of plane-polarized light must be considered in more detail. When observed along the direction of propagation, a beam of plane-polarized light appears to have its electric vector, which oscillates as a sine wave with the frequency of the light, confined to one plane. There is also an oscillating magnetic vector confined to a perpendicular plane, but we shall not be specifically interested in this.

It is useful to think of a beam of plane-polarized light as the resultant of two coterminous beams of right- and left-circularly polarized light that have equal amplitudes and are in phase. A circularly polarized beam is one in which the electric vector rotates uniformly about the direction of propagation by 2π during each cycle. Fig. 20-20 shows how two such beams, circularly polarized in opposite senses, give a plane-polarized resultant. The most important property of the two circularly polarized components to be noted here is that they are enantiomorphous to one another, that is, one is the non-superposable mirror image of the other.

Now, just as there are molecules AB consisting of two separately dissymmetric halves, say A(+),A(−), and B(+),B(−), which are different substances, for example, the diastereoisomers A(+)B(+) and A(−)B(+), having numerically different physical properties, so the physical interactions of the two circularly polarized beams with a given enantiomorph of a dissym-

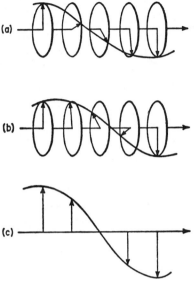

Fig. 20-20. (a) Right-circularly polarized light. (b) Left-circularly polarized light. (c) The plane-polarized resultant of (a) and (b). The horizontal arrow gives the direction of propagation, and the arrows perpendicular to this direction denote the instantaneous spatial direction of the electric vector. [Reproduced by permission from S. F. Mason, *Chem. in Britain*, **1965**, 245.]

metric molecule will be quantitatively different. The two important differences are (1) the refractive indices for left- and right-circularly polarized light, n_l and n_r, respectively, will be different, and (2) the molar absorbances, ε_l and ε_r, will be different.

If only the refractive-index difference existed, the rotation of the plane of polarization would be explained as shown in Fig. 20-21 since the retarding of one circularly polarized component relative to the other can be seen to have this net effect.

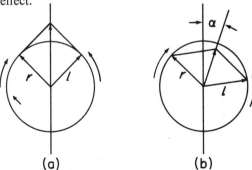

Fig. 20-21. (a) An instantaneous view along the direction of propagation of the two vectors r and l of the circularly polarized beams and their resultant which lies in a vertical plane. (b) If $n_l > n_r$, the beam l is retarded relative to beam r, thus causing the plane of the resultant to be tilted by the angle α.

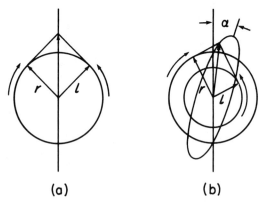

(a)   **(b)**

Fig. 20-22.   (a) Same as Fig. 20-21(a). (b) If $n_l > n_r$, and also $\varepsilon_l > \varepsilon_r$, the vectors r and l will be affected, qualitatively, as shown, and they will then give rise to a resultant which traces out the indicated ellipse. Note that the quantity $\varepsilon_l - \varepsilon_r$ is here vastly exaggerated compared to real cases in order to make the diagram clear.

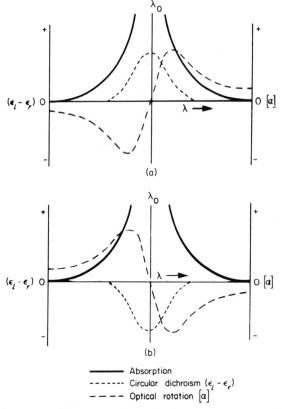

Absorption
Circular dichroism $(\epsilon_l - \epsilon_r)$
Optical rotation $[\alpha]$

Fig. 20-23.   The Cotton effect, as manifested in circular dichroism, $\varepsilon_l - \varepsilon_r$, and optical rotatory dispersion, $[\alpha]$, as would be given by a dissymmetric compound with an absorption band centered at λ_0, on the assumption that there are no other absorption bands close by. (a) A positive Cotton effect. (b) A negative Cotton effect.

Actually, the simultaneous existence of a difference between ε_l and ε_r means that the rotated "plane" is no longer strictly a plane. This can be seen in Fig. 20-22; since one rotating electric vector is not exactly equal in length to the other after the two components have traversed the optically active medium, their resultant describes an ellipse, whose principal axis defines the "plane" of the rotated beam and whose minor axis is equal to the absolute difference $|\varepsilon_l - \varepsilon_r|$. This difference is usually very small so that it is a very good approximation to speak of rotating "the plane"; however, it can be measured and constitutes the *circular dichroism*. It is most important that both the optical rotation and the circular dichroism are dependent on wavelength, especially in the region of an electronic absorption band of the atom or ion lying at the "center of dissymmetry." Moreover, at a given wavelength the values of $n_l - n_r$ and $\varepsilon_l - \varepsilon_r$ for one enantiomorph are equal and opposite to those for the other enantiomorph. The variations of $n_l - n_r$ and $\varepsilon_l - \varepsilon_r$ with wavelength for a pair of enantiomorphs in the region of an absorption band with a maximum at λ_0 are illustrated schematically in Fig. 20-23. The variation of the angle of rotation with wavelength is called *optical rotatory dispersion*, ORD. This, together with the circular dichroism, CD, and the attendant introduction of ellipticity into the rotated beam, are, all together, called the *Cotton effect*, in honor of the French physicist, Aimé Cotton, who made pioneering studies of the wavelength-dependent aspects of these phenomena in 1895.

20-8. Applications

Cotton effects are studied today by inorganic chemists for two principal purposes. First, they can be used, in a fairly empirical way, to correlate the configurations of related dissymmetric molecules and thus to follow the steric course of certain reactions. Secondly, both theoretical and experimental work is in progress to establish generally reliable criteria for determining spectroscopically the absolute configurations of molecules.

Before summarizing the results in each of these areas, it will be well to mention the types of optically active complexes that have been most studied. These are shown in Fig. 20-24. Unfortunately the notation for the chirality

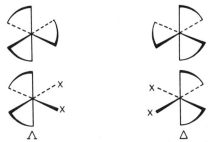

Fig. 20-24.   The principal types of optically active chelate molecules and the IUPAC nomenclature (IUPAC Bulletin No. 33, page 68, 1968).

of such molecules is at present in great confusion, owing to a proliferation of symbols and different rationales for their use. The symbols for absolute chirality recommended (but not yet adopted) by a Commission of IUPAC are also defined in Fig. 20-24 and will be used here.

The use of ORD and CD data in making empirical correlations of configurations is increasingly important. The basic idea is simply that electronic transitions of similar nature in similar molecules should have the same signs for CD and ORD effects when the molecules have the same absolute chirality. The chief uncertainty in the conclusions arises from failure to satisfy adequately the criteria of "similarity" in the nature of the electronic transitions and in the structures of the molecules themselves. As an illustration, we use a case where the required similarities are obviously present. Fig. 20-25 shows the CD curves for $(+)$ [Co en$_3$]$^{3+}$ and $(+)$ [Co(l-pn)$_3$]$^{3+}$, where the $(+)$ signs indicate that these are the enantiomers having positive values of [α] at the sodium D line. Clearly these two complex ions must have the same absolute configuration. That of the $(+)$ [Co en$_3$]$^{3+}$ ion has been determined by means of anomalous X-ray scattering as Λ.

It must not be thought that the identical rotational directions at the sodium D line (or at any other single wavelength) could itself have been taken as a criterion of identical absolute configuration. Such relationships fail so frequently as to make the "criterion" useless: for example, $(+)$[Co en$_2$(NH$_3$)Cl]$^{2+}$ and $(-)$[Co en$_2$(NCS)Cl]$^{+}$ have the same configuration.

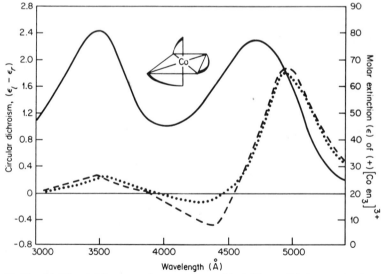

Fig. 20-25.   (a) The visible absorption spectrum of $(+)$ [Co en$_3$]$^{3+}$ (———). (b) The circular dichroism of $(+)$ [Co en$_3$]$^{3+}$ (. . . .). (c) The circular dichroism of $(+)$ [Co(l-pn)$_3$]$^{3+}$ (– – –). The molar absorbance scale is on the right, the CD scale on the left. The absorption spectrum of $(+)$ [Co(l-pn)$_3$]$^{3+}$ is practically identical with that of $(+)$ [Co en$_3$]$^{3+}$. The small sketch shows the absolute configuration (Λ) of $(+)$ [Co en$_3$]$^{3+}$ as determined by anomalous scattering of X-rays.

Attempts to develop general, non-empirical relations between the signs of CD or ORD effects and absolute configurations have thus far been unsuccessful when dealing with *d–d* transitions.

In cases where the ligands themselves have electronic transitions in the ultraviolet region that persist in perturbed but identifiable form in the complexes, there is a reliable method of assigning absolute configurations directly from observed ORD or, better, CD data without employing any reference compound of known configuration.[10-12] This is based on the fact that when two or three such ligands are in close proximity in the complex their individual electric dipole transition moments can couple to produce exciton splittings. Detailed analysis shows that each component of the split band will have a different sign for its circular dichroism and that this sign may be predicted from the absolute configuration by an argument not dependent on numerical accuracy.

The types of ligand that lend themselves to this treatment are bipyridine and phenanthroline, which have strong near-uv transitions. In the case of $[Fe(phen)_3]^{2+}$ the correctness of the method has been confirmed by an X-ray crystallographic determination of absolute configuration.[13]

The Faraday Effect.[14] All substances rotate the plane of polarization of light and exhibit ORD and CD effects when placed in a magnetic field that has a component in the direction of propagation of the polarized radiation. Phenomenologically these effects, collectively called the *Faraday effect*, are analogous to ordinary optical activity and the *Cotton effect*, but their interpretation and application to chemical problems are more complicated and less advanced and have only rather recently received detailed study. Work to date has been concerned with non-transition-metal complexes and organic compounds to about the same extent as with transition-metal complexes, so that mention of this topic here rather than elsewhere is purely a matter of juxtaposing it conveniently with the preceding discussion of natural optical activity.

Because all substances, including solvents, cell windows, etc., exhibit the Faraday effect, it is, in general, only possible to get interpretable results by measuring the magnetic circular dichroism (MCD) through the electronic absorption bands of the complex of interest. This can be done with apparatus normally used for conventional CD measurements by surrounding the sample cell with a small superconducting solenoid.

MCD spectra are chiefly useful in testing or confirming the assignments of orbitally allowed electronic transitions such as charge-transfer bands (see Section 20-18). Space does not permit an account of the theoretical basis here, but one illustration may suffice to indicate the utility of the method in a

[10] A. J. McCaffery, S. F. Mason and B. J. Norman, *J. Chem. Soc., A,* **1969,** 1428.
[11] I. Hanazaki and S. Nagakura, *Inorg. Chem.,* 1969, **8,** 654.
[12] B. Bosnich, *Inorg. Chem.,* 1968, 7, 2379.
[13] D. H. Templeton, A. Zalkin and T. Ueki, *Acta Cryst.,* 1966, **21,** *A,* 154.
[14] P. N. Schatz and A. J. McCaffery, *Quart. Rev.,* 1969, **23,** 552.

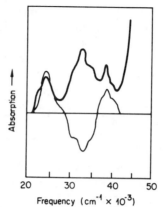

Fig. 20-26.   The absorption spectrum (bold curve) and MCD spectrum (thinner curve) of
$K_3Fe(CN)_6$ in the region of charge-transfer bands.

favorable case. For the $Fe(CN)_6^{3-}$ ion there are three charge-transfer bands
in visible and near-uv spectra as shown in Fig. 20-26. Theory suggests
that all of them must involve transitions from orbitals of t_{1u} and t_{2u} symmetry on the ligands to the vacancy in the metal d orbitals of t_{2g} symmetry.
However, theory could not directly show with certainty which transition was
which. However, it was possible to show theoretically that a $t_{1u} \rightarrow t_{2g}$ transition should have a positive MCD effect while a $t_{2u} \rightarrow t_{2g}$ transition should
have a negative one. From the observed MCD spectrum (Fig. 20-26), it is
then clear that the $t_{2u} \rightarrow t_{2g}$ transition lies between the two $t_{1u} \rightarrow t_{2g}$ transitions.
In favorable circumstances the shape and/or magnitude of the MCD effect
can show that a transition involves an excited state that is degenerate.

STRUCTURAL AND THERMODYNAMIC EFFECTS
OF INNER ORBITAL SPLITTINGS

We have so far considered how the electrostatic effects of ligands cause the
d electrons to prefer certain regions of space (i.e. certain orbitals) to others.
We shall now look briefly at some ways in which the non-spherical distribution of the d electrons, caused by the environment, reacts back upon the
environment.

20-9. Ionic Radii

We consider first the effect of d-orbital splittings on the variation of ionic
radii with atomic number in a series of ions of the same charge. We shall
use as an example the octahedral radii of the divalent ions of the first transition series. Fig. 20-27 shows a plot of the experimental values. The points
for Cr^{2+} and Cu^{2+} are indicated with open circles because the Jahn–Teller
effect, to be discussed below, makes it impossible to obtain these ions in truly

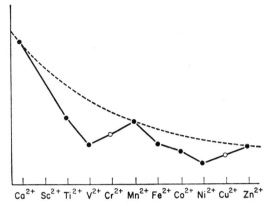

Fig. 20-27. The relative ionic radii of divalent ions of the first transition series. The broken line is a theoretical curve explained in the text.

octahedral environments, thus rendering the assessment of their "octahedral" radii somewhat uncertain. A smooth curve has also been drawn through the points for Ca^{2+}, Mn^{2+} and Zn^{2+} ions which have the electron configurations $t_{2g}^0 e_g^0$, $t_{2g}^3 e_g^2$, and $t_{2g}^6 e_g^4$, respectively. In these three cases the distribution of d electron density around the metal ion is spherical because all d orbitals are either unoccupied or equally occupied. Because the shielding of one d electron by another from the nuclear charge is imperfect, there is a steady contraction in the ionic radii. It is seen that the radii of the other ions are all below the values expected from the curve passing through Ca^{2+}, Mn^{2+} and Zn^{2+}. This is because the d electrons in these ions are not distributed uniformly (i.e. spherically) about the nuclei, as we shall now explain.

The Ti^{2+} ion has the configuration t_{2g}^2. This means that the negative charge of two d electrons is concentrated in those regions of space away from the metal–ligand bond axes. Thus, compared to the effect that they would have if distributed spherically around the metal nucleus, these two electrons provide abnormally little shielding between the positive metal ion and the negative ligands; therefore the ligand atoms are drawn in closer than they would be if the d electrons were spherically distributed. Thus, in effect, the radius of the metal ion is smaller than that for the hypothetical, isoelectronic spherical ion. In V^{2+} the same effect is found in even greater degree because there are now three t_{2g} electrons providing much less shielding between metal ion and ligands than would three spherically distributed d electrons. For Cr^{2+} and Mn^{2+}, however, we have the configurations $t_{2g}^3 e_g$ and $t_{2g}^3 e_g^2$, in which the electrons added to the t_{2g}^3 configuration of V^{2+} go into orbitals that concentrate them mainly between the metal ion and the ligands. These e_g electrons thus provide a great deal more screening than would be provided by spherically distributed electrons, and indeed the effect is so great that the radii actually increase. The same sequence of events is repeated in the second half of the series. The first three electrons added to the spherical $t_{2g}^3 e_g^2$ configuration of Mn^{2+} go into the t_{2g} orbitals where the screening power is abnormally low and the radii therefore decrease abnormally rapidly. On going from Ni^{2+}, with the configuration $t_{2g}^6 e_g^2$, to Cu^{2+} and Zn^{2+}, electrons are added to the e_g orbitals where their screening power is abnormally high, and the radii again cease to decrease and actually show small increases. Similar effects may be expected with trivalent ions, with ions of other transition series and in tetrahedral environments as well, although in these other circumstances fewer experimental data are available to verify the predictions that may be made by reasoning similar to that used above.

20-10. Jahn–Teller Effects

In 1937 Jahn and Teller proved a rather remarkable theorem which states that any non-linear molecular system in a degenerate electronic state will be unstable and will undergo some kind of distortion that will lower its symmetry and split the degenerate state. Although it may sound somewhat abstract, this simple theorem has great practical importance in understanding the structural chemistry of certain transition-metal ions. In order to illustrate this, we shall begin with the Cu^{2+} ion. Suppose this ion finds itself in the center of an octahedron of ligands. As shown on page 572, this ion may be thought of as possessing one hole in the e_g orbitals, and the electronic state of the ion is hence a degenerate, E_g, state. According to the Jahn–Teller theorem then, the octahedron cannot remain perfect at equilibrium but must become distorted in some way.

The dynamic reason for the distortion is actually rather easy to appreciate in terms of simple physical reasoning. Let us suppose that, of the two e_g orbitals, it is the $(x^2 - y^2)$ orbital that is doubly occupied while the z^2 orbital is only singly occupied. This must mean that the four negative charges or the negative ends of dipoles in the xy-plane will be more screened from the electrostatic attraction of the Cu^{2+} ion than will the two charges on the z-axis. Naturally, then, the latter two ligands will be drawn in somewhat more closely than the other four. If, conversely, the z^2 orbital is doubly occupied and the $(x^2 - y^2)$ orbital only singly occupied, the four ligands in the xy-plane will be drawn more closely to the cation than will the other two on the z-axis. It is also possible that the unpaired electron could be in an orbital that is some linear combination of $(x^2 - y^2)$ and z^2, in which case the resulting distortion would be some related combination of the simple ones considered above. These simple considerations call attention to several important facts relating to the operation of the Jahn–Teller theorem:

1. The theorem only predicts that for degenerate states a distortion must occur. It does not give any indication of what the geometrical nature of the distortion will be or how great it will be.

2. In order to make a prediction of the nature and magnitude of the distortion, detailed calculations must be made of the energy of the entire complex as a function of all possible types and degrees of distortion. The configuration having the lowest over-all energy may then be predicted to be the

equilibrium one. Such *a priori* calculations, however, are extremely laborious and few have been attempted.

3. It may be noted that there is one general restriction on the nature of the distortions, namely, that if the undistorted configuration has a center of symmetry, so also must the distorted equilibrium configuration.

In order to give a little more insight into the energy problem noted under paragraph 2 above, let us consider what happens to the d orbital energies when there occurs a small distortion of the type in which the octahedron becomes stretched along its z-axis. The effects are shown in Fig. 20-28. In this diagram the various splittings are not drawn to scale, in the interest of clarity. Both the splittings due to the distortion are much smaller than Δ_0 and, as noted below, δ_2 is much smaller than δ_1. It should also be noted that each of the splittings obeys a center-of-gravity rule. The two e_g orbitals separate so that one goes up as much as the other goes down; the t_{2g} orbitals separate so that the doubly degenerate pair goes down only half as far as the single orbital goes up. It can be seen that, for the d^9 case, there is no net energy change for the t_{2g} electrons, since four are stabilized by $\delta_2/3$ while two are destabilized by $2\delta_2/3$. For the e_g electrons, however, a net stabilization occurs, since the energy of one electron is raised by $\delta_1/2$, but two electrons have their energies lowered by this same amount; the net lowering of the electronic energy is thus $\delta_1/2$. It is this stabilization that provides the driving force for the distortion.

It is easy to see from Fig. 20-28 that, for both the configurations $t_{2g}^6 e_g$

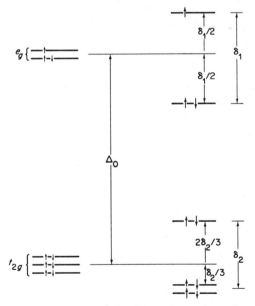

Fig. 20-28.   Schematic diagram of the splittings caused by an elongation of an octahedron along one axis. The various splittings are not to the same scale, δ_1 and δ_2 being much smaller relative to Δ_0 than indicated.

and $t_{2g}^6 e_g^3$, distortion of the octahedron will cause stabilization; thus we predict, as could also be done directly from the Jahn–Teller theorem, that distortions are to be expected in the octahedral complexes of ions with these configurations, but not for ions having t_{2g}^6, $t_{2g}^6 e_g^2$ or $t_{2g}^6 e_g^4$ configurations. In addition, it should also be obvious from the foregoing considerations that a high-spin d^4 ion, having the configuration $t_{2g}^3 e_g$, will also be subject to distortion. Some real ions having those configurations which are subject to distortion are:

$t_{2g}^3 e_g$: high-spin Cr^{II} and Mn^{III}
$t_{2g}^6 e_g$: low-spin Co^{II} and Ni^{III}
$t_{2g}^6 e_g^3$: Cu^{II}

For low-spin Co^{II} no satisfactory structural detail is available, but in the other four cases there are ample data to show that distortions do occur and that they take the form of elongation of the octahedron along one axis. Indeed, in a number of Cu^{II} compounds the distortions of the octahedra around the cupric ion are so extreme that the coordination is best regarded as virtually square and, of course, Cu^{II} forms many square complexes. Specific illustrations of distortions in the compounds of these several ions will be mentioned when the chemistry of the elements is described in Chapter 25.

It may also be noted that the Jahn–Teller theorem applies to excited states as well as to ground states, although in such cases the effect is a complicated dynamic one because the short life of an electronically excited state does not permit the attainment of a stable equilibrium configuration of the complex. To illustrate the effect on excited states, we may consider the $[Ti(H_2O)_6]^{3+}$, $[Fe(H_2O)_6]^{2+}$, and $[CoF_6]^{3-}$ ions. The first of these has an excited state configuration e_g. The presence of the single e_g electron causes the excited state to be split, and it is this that accounts for the broad, flat contour of the absorption band of $[Ti(H_2O)_6]^{3+}$ as seen in Fig. 20-14. In both $[Fe(H_2O)_6]^{2+}$ and $[CoF_6]^{3-}$, the ground state has the configuration $t_{2g}^4 e_g^2$, and the excited state with the same number of unpaired electrons has the configuration $t_{2g}^3 e_g^3$. Thus, the excited states of these ions are subject to Jahn–Teller splitting into two components, and this shows up very markedly in their absorption spectra, as Fig. 20-29 shows for $[CoF_6]^{3-}$.

Jahn–Teller distortions can also be caused by the presence of 1, 2, 4 or 5

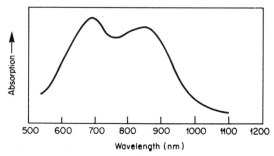

Fig. 20-29. The absorption spectrum of the $[CoF_6]^{3-}$ ion in $K_2Na[CoF_6]$, showing the splitting due to a Jahn–Teller distortion of the excited state with the configuration $t_{2g}^3 e^3$.

electrons in the t_{2g} orbitals of an octahedrally coordinated ion. This can easily be seen by referring to the lower part of Fig. 20-28. If one t_{2g} electron is present, distortion by elongation on one axis will cause stabilization by $\delta_2/3$. Distortion by flattening along one axis would produce a splitting of the t_{2g} orbitals which is just the reverse of that shown in Fig. 20-28 for elongation and thus would cause stabilization by twice as much, namely, $2\delta_2/3$. The same predictions can obviously be made for the t_{2g}^4 case. For a t_{2g}^2 configuration—if we assume, with good reason since δ_2 will be much less than the electron-pairing energy, that pairing of electrons will not occur— the elongation distortion would be favored since it will provide a total stabilization of

$$2 \times \delta_2/3 = 2\delta_2/3$$

whereas the flattening would give a net stabilization energy of only

$$2\delta_2/3 - \delta_2/3 = \delta_2/3$$

For the t_{2g}^5 case, flattening is again predicted to cause the greater stabilization.

There is, however, little experimental confirmation of these predictions of Jahn–Teller effects for partially filled t_{2g} shells. This is mainly due to the fact that the effects are expected, theoretically, to be much smaller than those for partially filled e_g orbitals. In terms of Fig. 20-28, theory shows that for a given amount of distortion δ_2 is much smaller than δ_1. Thus the stabilization energies, which are the driving forces for the distortions, are evidently not great enough to cause well-defined, clearly observable distortions in cases of partially occupied t_{2g} orbitals. From the CFT point of view the relation $\delta_2 \ll \delta_1$ is easily understood. Since the e_g orbitals are directed right at the ligands, the presence of an electron in one e_g orbital but not in the other will cause a much larger disparity in the metal—ligand distances than will non-uniform occupancy of t_{2g} orbitals, which concentrate their electrons between the metal—ligand bonds where their effect on metal—ligand distances is much less (cf. the discussion of ionic radii in Section 20-9).

In terms of MO theory, as given below, the reason for $\delta_2 \ll \delta_1$ is also very simple, The e_g orbitals are antibonding with respect to the metal—ligand σ bonds, and thus a change in the population of these orbitals should strongly affect the metal—ligand bond strength. On the other hand, t_{2g} orbitals are non-bonding in respect to metal—ligand σ interaction, though they may have antibonding or bonding character in respect to metal—ligand π bonding. However, since σ bonding is usually far more important than π bonding, changes in the t_{2g} population have much less influence on the metal—ligand bond strengths.

20-11. Thermodynamic Effects of Crystal Field Splittings

We have seen in Section 20-2 that the d orbitals of an ion in an octahedral field are split so that three of them become more stable (by $2\Delta_0/5$) and two of them less stable (by $3\Delta_0/5$) than they would be in the absence of the splitting. Thus, for example, a d^2 ion will have each of its two d electrons stabilized by $2\Delta_0/5$, giving a total stabilization of $4\Delta_0/5$. Recalling from Section 20-4 that Δ_0 values run about 10,000 and 20,000 cm^{-1} for di- and tri-valent ions of the first transition series, we can see that these "extra" stabilization energies—extra in the sense that they would not exist if the d shells of the metal ions were symmetrical as are the other electron shells of the ions—will amount to ~ 100 and ~ 200 kJ mol^{-1}, respectively, for di- and tri-valent d^2 ions. These *ligand field stabilization energies*, LFSE's, are of course of the same order of magnitude as the energies of most chemical changes, and they will therefore play an important role in the thermodynamic properties of transition-metal compounds.

Let us first of all consider high-spin octahedral complexes. Every t_{2g} electron represents a stability increase (i.e. energy lowering) of $2\Delta_0/5$, whereas every e_g electron represents a stability decrease of $3\Delta_0/5$. Thus, for any configuration $t_{2g}^p e_g^q$, the net stabilization will be given by $(2p/5 - 3q/5)\Delta_0$. The results* obtained for all the ions, that is, d^0 to d^{10}, using this formula are collected in Table 20-5. Since the magnitude of Δ_0 for any particular complex can be obtained from the spectrum, it is possible to determine the magnitudes

TABLE 20-5

Ligand Field Stabilization Energies, LFSE's, for Octahedrally and Tetrahedrally Coordinated High-spin Ions

Number of d electrons	Stabilization energies		Difference, Octa. – Tetra.[b]
	Octa.	Tetra.	
1, 6	$2\Delta_0/5$	$3\Delta_t/5$	$\Delta_0/10$
2, 7[a]	$4\Delta_0/5$	$6\Delta_t/5$	$2\Delta_0/10$
3, 8	$6\Delta_0/5$	$4\Delta_t/5$	$8\Delta_0/10$
4, 9	$3\Delta_0/5$	$2\Delta_t/5$	$4\Delta_0/10$
0, 5, 10	0	0	0

[a] For the d^2 and d^7 ions, the figure obtained in this way and given above is not exactly correct because of the effect of configuration interaction.

[b] On the assumption that $\Delta_0 = 2\Delta_t$.

of these crystal field stabilization energies independently of thermodynamic measurements and thus to see what part they play in the thermodynamics of the transition-metal compounds.

Hydration, Ligation and Lattice Energies. As a first example, let us consider the heats of hydration of the divalent ions of the first transition series. These are the energies of the processes

$$M^{2+}(gas) + \infty H_2O = [M(H_2O)_6]^{2+}(aq)$$

and they can be estimated by using thermodynamic cycles. The energies calculated are shown by the filled circles in Fig. 20-30. It will be seen that a smooth curve, which is nearly a straight line, passes through the points for

* The same results are obtained, a little less easily, from a molecular-orbital treatment, and thus the concept of LFSE's is valid regardless of the model, CFT or otherwise, used to derive the d orbital splittings (cf. F. A. Cotton, *J. Chem. Educ.*, 1964, **41**, 466 for details).

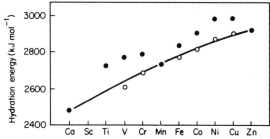

Fig. 20-30. Hydration energies of some divalent ions. Solid circles are experimentally derived hydration energies. Open circles are energies corrected for LFSE.

the three ions, $Ca^{2+}(d^0)$, $Mn^{2+}(d^5)$ and Zn^{2+} (d^{10}), which have no LFSE while the points for all other ions lie above this line. On subtracting the LFSE from each of the actual hydration energies, the values shown by open circles are obtained, and these fall on the smooth curve. It may be noted that, alternatively, LFSE's could have been estimated from Fig. 20-30 and used to calculate Δ_0 values. Either way, the agreement between the spectrally and thermodynamically assessed Δ_0 values provides evidence for the fundamental correctness of the idea of d orbital splitting.

Two more examples of these thermodynamic consequences of crystal field splittings are shown in Figs. 20-31 and 20-32. In Fig. 20-31 the lattice energies of the dichlorides of the metals from calcium to zinc are plotted versus atomic number. Once again they define a curve with two maxima and a minimum at Mn^{2+}. As before, the energies for all the ions having LFSE's lie above the curve passing through the energies of the three ions that have no ligand field stabilization energy. Similar plots are obtained for the lattice energies of other halides and of the chalconides of di- and tri-valent metals. In Fig. 20-32 are plotted the gas-phase dissociation energies, estimated by means of thermodynamic cycles, for the hexammine ions of some divalent metals. These energies are for the process

$$[M(NH_3)_6]^{2+}(g) = M^{2+}(g) + 6NH_3(g)$$

and thus are equal to six times the mean $M-NH_3$ bond energy. The hydration energies discussed above do not have quite this simple interpretation

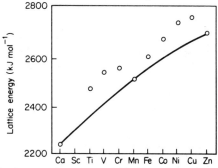

Fig. 20-31. The lattice energies of the dichlorides of the elements from Ca to Zn.

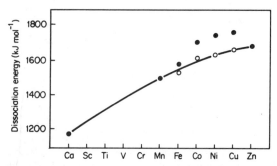

Fig. 20-32. Dissociation energies of the hexammines of some divalent metal ions. Solid circles are the total energies. Open circles are the total energies minus the ligand field stabilization energies estimated from spectroscopically derived Δ_0 values.

because they include the energy of further hydration of the hexaquo ions in addition to the energy of the process

$$M^{2+}(g) + 6H_2O(g) = [M(H_2O)_6]^{2+}(g)$$

decreased by the heat of vaporization of six moles of water. Although the data for the dissociation of the hexammines are limited, they show the same trend as do the hydration and lattice energies and, as shown by the open circles of Fig. 20-32, the deviations from the smooth curve are equal within experimental error to the spectroscopically estimated LFSE's.

It will be noted in all three of the Figs. 20-30 to 20-32, that the smooth curves from Ca^{2+} through Mn^{2+} to Zn^{2+}, on which the corrected energies do, or presumably would, fall, rise with increasing atomic number. This is to be expected, since the smooth curve on which the radii of (real or hypothetical) spherical ions lie, as shown in Fig. 20-27, falls from Ca^{2+} to Zn^{2+}. A steady decrease in ionic radius naturally leads to a steady increase in the electrostatic interaction energy between the cation and the ligand anions or dipoles. A feature of particular importance, which is most directly evident in Fig. 20-32, is that the LFSE's, critical as they may be in explaining the *differences* in energies between various ions in the series, make up only a small fraction, 5–10%, of the *total* energies of combination of the metal ions with the ligands. In other words, the LFSE's, though crucially important in many ways, are not by any means major sources of the binding energies in complexes.

Formation Constants of Complexes.   It is a fairly general observation that the equilibrium constants for the formation of analogous complexes of the divalent-metal ions of Mn through Zn with ligands that contain nitrogen as the donor atom fall in the following order of the metal ions: $Mn^{2+} < Fe^{2+} < Co^{2+} < Ni^{2+} < Cu^{2+} > Zn^{2+}$. There are occasional exceptions to this order, sometimes called the Irving–Williams order, which may be attributed to the occurrence of spin-pairing in strong crystal fields. Spin-pairing, naturally, affects the relative energies in a different way. The generality of the above order of stability constants receives a natural explanation in terms of LFSE's. Since the magnitudes of stability constants are proportional to the antilogarithms of standard free energy changes, the above order is also that of the $-\Delta G^0$ values for the formation reactions. The standard free energies of formation are related to the enthalpies by the relation

$$-\Delta G^0 = -\Delta H^0 + T\Delta S^0$$

and there are good reasons for believing that entropies of complex formation are substantially constant in the above series of ions. We thus come finally to the conclusion that the above order of formation constants is also the order of $-\Delta H^0$ values for the formation reactions. Indeed, in a few cases direct measurements of the ΔH^0 values have shown this to be true.

The formation of a complex in aqueous solution involves the displacement of water molecules by ligands. If the metal ion concerned is subject to crystal field stabilization, as is, for example, Fe^{2+}, this stabilization will be greater in the complex than in the aquo ion, since the nitrogen-containing ligand will be further along in the spectrochemical series than is H_2O (see page 577). For Mn^{2+}, however, there is no LFSE in the hexaquo ion or in the complex, so that complexation cannot cause any increased stabilization. Thus the Fe^{2+} ion has more to gain by combining with the ligands than does Mn^{2+}, and it accordingly shows a greater affinity for them. Similarly, of two ions, both of which experience crystal field stabilization, the one that experiences the greater amount from both the ligand and from H_2O will also experience the larger increase on replacement of H_2O by the ligand. Thus, in general, the order of the ions in the stability series follows their order in regard to crystal field stabilization energies.

Octahedral vs. Tetrahedral Coordination.   Finally, in this Section we shall consider a phenomenon that is structural in nature but depends directly on LFSE's, namely, variation in the relative stability of octahedral and tetrahedral complexes for various metal ions. It should be clearly understood that the ΔH^0 of transformation of a tetrahedral complex of a given metal ion to an octahedral complex of the same ion, as represented, for instance, by the equation

$$[MCl_4]^{2-} + 6H_2O = [M(H_2O)_6]^{2+} + 4Cl^- \qquad (20\text{-}1)$$

is a quantity to which the difference in LFSE's of the octahedral and tetrahedral species makes only a small contribution. The metal—ligand bond energies, polarization energies of the ligands, hydration energies, and other contributions all play larger roles, and the calculation of ΔH^0 for a particular metal, M, is a difficult and, as yet, insuperable problem. However, if a reaction of this type occurs (actually or hypothetically as a step in a thermodynamic cycle) for a series of metal ions increasing regularly in atomic number, say the ions Mn^{2+}, Fe^{2+}, ..., Cu^{2+}, Zn^{2+}, it is reasonable to suppose that the various factors contributing to ΔH^0 will all change uniformly *except* the differences in the LFSE's. The latter therefore might be expected to play a decisive role, despite being inherently small parts of the entire ΔH^0 in each individual case, in determining an irregular variation in the equilibrium constants for such reactions from one metal ion to another. There are two cases in which experimental data corroborate this expectation.

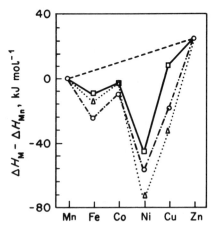

Fig. 20-33.   Enthalpies of reaction 20-1 expressed as differences between ΔH for a particular metal and ΔH for the Mn^{2+} compounds. Squares and circles are "experimental"
values derived in different ways from thermodynamic data, while triangles are values calculated from LFSE differences. In each case the ΔH values are plotted relative to the
interpolated value between Mn^{2+} and Zn^{2+} as indicated by the straight line between the
points for these two ions. [Reproduced with permission from A. B. Blake and F. A. Cotton,
Inorg. Chem., 1964, **3**, 9.]

For reaction 20-1 mentioned above, carried out hypothetically in the vapor
phase, the enthalpies have been estimated from thermodynamic data for the
metals, M, in the series Mn^{2+}, Fe^{2+}, ..., Cu^{2+}, Zn^{2+}. At the same time,
from the spectra of the $[M(H_2O)_6]^{2+}$ and $[MCl_4]^{2-}$ ions the values of Δ_0 and
Δ_t have been evaluated and the differences between the two LFSE's calculated. Fig. 20-33 shows a comparison between these two sets of quantities.
It is evident that the qualitative relationship is very close even though some
quantitative discrepancies exist. The latter may well be due to inaccuracies
in the ΔH values since these are obtained as net algebraic sums of the independently measured enthalpies of several processes. The qualitatively close
agreement between the variation in the enthalpies and the LFSE difference
justifies the conclusion that it is the variations in LFSE's that account for
gross qualitative stability relations such as the fact that tetrahedral complexes
of Co^{II} are relatively stable while those of Ni^{II} are not.

A second illustration of the importance of LFSE's in determining stereochemistry is provided by the *site preference* problem in the mixed metal
oxides which have the spinel or inverse spinel structures. These structures
have been described on page 54, but the reason for the occurrence of inverse
spinels was not given. In every case where inversion occurs, an explanation
in terms of LFSE's can be given. For example, $NiAl_2O_4$ is inverted, that is
Ni^{2+} ions occupy octahedral interstices and half the aluminum ions occupy
tetrahedral ones. It could *not* be predicted that this would necessarily occur
simply because the LFSE of Ni^{2+} is much greater in octahedral than in
tetrahedral environments, because there are other energy differences that
oppose exchanging the sites of Ni^{2+} ions and Al^{3+} ions. However, it can be

said that Ni^{2+} is the ion most likely to participate in such an inversion and that, if inversion is to occur at all, it must occur with $NiAl_2O_4$. For $FeAl_2O_4$, in contrast, LFSE differences would again dictate a qualitative preference by Fe^{2+} for the octahedral site, but inversion does not occur. However, as Table 20-5 shows, the magnitude of the preference energy may be around an order of magnitude smaller than for Ni^{2+}.

Another example of the role of LFSE's in determining site preferences is provided by the series of oxides Fe_3O_4, Mn_3O_4 and Co_3O_4, of which only the first is inverted. All energy changes connected with inversion should be similar in the three compounds except the differences in LFSE's, and these are just such as to favor inversion in Fe_3O_4 but not in the others. Thus, transfer of the d^5 Fe^{3+} ion involves no change in LFSE, but transfer of the high-spin d^6 Fe^{2+} ion from a tetrahedral to an octahedral hole produces a net gain in LFSE. For Mn_3O_4, transfer of the d^5 Mn^{2+} ion makes no change in LFSE, but transfer of the d^4 Mn^{3+} ion from an octahedral to a tetrahedral hole would decrease the LFSE, so that the process of inverting Mn_3O_4 is disfavored. For Co_3O_4 the transfer of the Co^{2+} ions to octahedral holes would be only slightly favored by the LFSE's, whereas the transfer of a low-spin d^6 Co^{3+} ion from an octahedral hole to a tetrahedral one where it would presumably become high-spin would cause an enormous net decrease in LFSE, so that here, even more than in the case of Mn_3O_4, we do not expect inversion.

ADJUSTMENTS TO THE CFT TO ALLOW FOR COVALENCE

20-12. Experimental Evidence for Metal–Ligand Orbital Overlap

Electron Spin Resonance Spectra.   One source of such evidence, probably the most direct we possess, is electron spin resonance (esr) data. The nature of electron spin resonance has been briefly described on page 544. In many cases it has been found that, instead of the single absorption band expected for a group of d electrons localized on a particular metal atom, there is observed a complex pattern of sub-bands, as shown in Fig. 20-34 for the now classic case of the $[IrCl_6]^{2-}$ ion. The pattern of sub-bands, called the hyperfine structure, has been satisfactorily explained by assuming that certain of the iridium orbitals and certain orbitals of the surrounding chloride ions overlap to such an extent that the single unpaired electron is not localized entirely on the metal ion but instead is about 5% localized on each Cl^- ion. The hyperfine structure is caused by the nuclear magnetic moments of the chloride ions, and the hyperfine splittings are proportional to the fractional extent to which the unpaired electron occupies the orbitals of these chloride ions. The electron is thus only 70% an "iridium(IV) $4d$ electron," instead of the 100% that is assumed in the purely electrostatic crystal field theory. Another similar example is that of the $[Mo(CN)_8]^{3-}$ ion. The esr spectrum

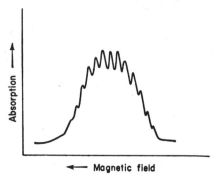

Fig. 20-34. The esr spectrum of the $[IrCl_6]^{2-}$ ion, obtained with the applied magnetic field aligned along one of the Cl—Ir—Cl axes of the complex ion in a single crystal of $Na_2PtCl_6 \cdot 6H_2O$ containing 0.5% of Ir^{IV} substitutionally replacing Pt^{IV}.

of this ion when it is enriched in ^{13}C, which has a nuclear spin (^{12}C does not), exhibits marked hyperfine structure showing that the unpaired electron is significantly delocalized on to the carbon atoms of the CN^- ions. There are numerous other examples.

Nuclear Magnetic Resonance Spectra. Closely related to esr experiments of the sort just mentioned are nuclear magnetic resonance (nmr) experiments in which the nuclear resonances of atoms in ligands are found to be affected by unpaired electrons in a manner that can only be explained by assuming that electron spin density is transferred from metal orbitals into orbitals of ligand atoms. Thus, for example, the resonance frequency of the ring proton, H_α, in tris(acetylacetonato)vanadium(III) (Fig. 20-35) is considerably shifted from its position in a comparable diamagnetic compound, say its Al^{III} analog. In order to account for the magnitude of the shift, it is necessary to assume that the spin density of unpaired electrons, *formally* restricted to t_{2g} metal orbitals in the crystal field treatment, actually moves out into the π electron system of the ligand to a significant extent and eventually into the $1s$ orbitals of the hydrogen atoms. Perhaps the most extensive and detailed studies of this phenomenon are those made on the Ni^{II} complexes of amino-tropone imines (see Section 25-G-4). Even in MF_6^{2-} octahedra, where we should certainly expect the metal—ligand bonding to be as electrostatic as any-where, fluorine nuclear resonance spectra have shown that delocalization of the spin density of metal-ion d electrons takes place to the extent of 2–5%.

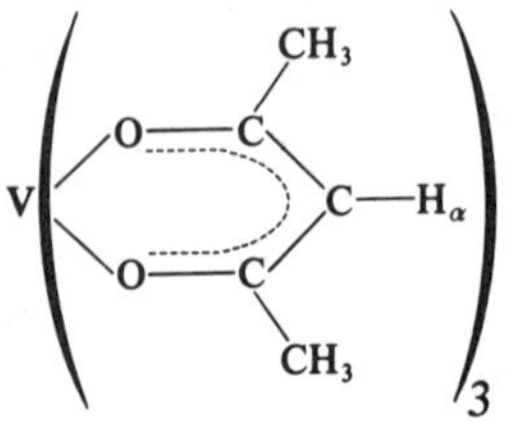

Fig. 20-35. Tris(acetylacetonato)vanadium(III), indicating the ring hydrogen, H_α, whose nuclear resonance frequency is strongly affected by the unpaired electrons on the V^{III} ion.

Intensities of "d–d" Transitions. Another indication that metal-ion and ligand orbitals overlap with the result that the "d orbitals" of the metal ions are not pure metal-ion d orbitals is given by the intensities of the optical absorption bands due to "d–d" transitions. If the crystal field approximation were perfect, the only mechanisms by which these absorptions could gain intensity would be by interactions of the d orbital wave functions with vibrational wave functions of the complex ion and by mixing of d orbitals with other *metal-ion* orbitals in those complexes (e.g., tetrahedral ones) where there is no center of symmetry. There are, however, cases in which it is fairly certain that these two processes are insufficient to account for the intensities observed, and it must be assumed then that the additional process of overlap and mixing of the metal d orbitals with various ligand atom orbitals, which is a powerful mechanism for enhancing the intensities, occurs to a significant degree.

The Nephelauxetic Effect.[15] As noted on page 575, if the energy-level diagrams for transition-metal ions with two to eight d electrons are calculated, with the assumption that the separations between the various Russell–Saunders states are exactly the same in the complexed ion as in the free, gaseous ion (which leaves Δ as the only variable parameter), the fitting of experimental data is not exact. In some cases the discrepancies are quite marked. It is invariably found that the fit can be improved by assuming that the separations between the Russell–Saunders states are smaller in the complexed than in the free ion. Now the separations between these states are attributable to the repulsions between the d electrons in the d^n configuration, so that the decrease in the energy separations between the states suggests that the d electron cloud has expanded in the complex, thus increasing the mean distance between d electrons and decreasing the interelectronic repulsions. It is now generally believed that this expansion of the d electron cloud occurs at least partly because the metal-ion d orbitals overlap with ligand atom orbitals, thus providing paths by which d electrons can, and do, escape to some extent from the metal ion. This effect of ligands in expanding the d electron clouds has been named the *nephelauxetic* (from the Greek, meaning "cloud expanding") effect, and it has been found that the common ligands can be arranged in order of their ability to cause cloud expansion. This order, which is more or less independent of the metal ion, similarly to the spectrochemical series, is in part: $F^- < H_2O < NH_3 < $ oxalate $\sim$ ethylenediamine $< $ —$NCS^- < Cl^- \sim CN^- < Br^- < I^-$.

The magnitude of the nephelauxetic effect may be conveniently expressed as the nephelauxetic ratio, β. This is the ratio of a given interelectronic repulsion parameter for the metal ion in a complex to its value in the gaseous ion. It is sometimes possible to evaluate different β's for different sets of orbitals, e.g., the e_g and the t_{2g} orbitals of an octahedral complex, reflecting the different amounts of σ and π covalence. Thus, from spin-forbidden

[15] C. K. Jørgensen, *Progr. Inorg. Chem.*, 1962, **4**, 23; *Helv. Chim. Acta*, 1967, **21**, 131.

$t_{2g} \rightarrow t_{2g}$ or $e_g \rightarrow e_g$ transitions the $\beta(t_{2g}, t_{2g})$ or β_π and $\beta(e_g, e_g)$ or β_σ values can be obtained, while the usual $t_{2g} \rightarrow e_g$ type of transition gives a $\beta(t_{2g}, e_g)$ or $\beta_{\sigma, \pi}$ value. Since most ordinary ligands engage in more σ bonding than π bonding, the order of β values should, in general, be: $\beta_\pi > \beta_{\sigma, \pi} > \beta_\sigma$. Some representative examples to bear this out are: $Cr(H_2O)_6^{3+}$, $\beta_\pi = 0.91$, $\beta_{\sigma, \pi} = 0.79$; $Ni(NH_3)_6^{2+}$, $\beta_{\sigma, \pi} = 0.85$, $\beta_\sigma = 0.77$.

Antiferromagnetic Coupling. Still another evidence of some overlap between metal-ion d orbitals and ligand orbitals in compounds that are usually described as "ionic" comes from detailed consideration of anti- ferromagnetism as observed in, for example, the oxides MnO, FeO, CoO and NiO. As we have already noted (page 542), an antiferromagnetic substance is one that follows a Curie or Curie–Weiss law at high temperatures but below a certain temperature (the Néel temperature) shows decreasing rather than increasing magnetic susceptibility as the temperature is lowered further. It has been conclusively shown by neutron-diffraction studies that this effect is not due to pairing of electron spins within individual ions but is due rather to a tendency of half of the ions to have their magnetic moments lined up in the opposite direction to those of the other half of the ions. Such antiparallel aligning, in which nearest-neighbor metal ions separated by an oxide ion collinear with them have opposed moments, cannot be explained merely by the direct effect, over the intervening distance, of one magnetic dipole on another; their separation is too great to permit an effect of the observed magnitude. Instead, the oxide ions are assumed to participate in the following way. Let us consider an M^{2+}—O^{2-}—M^{2+} set in which each metal ion pos- sesses an unpaired electron. The oxide ion also has pairs of electrons occupy- ing π orbitals. If there is overlap between that d orbital of one metal ion which contains its unpaired electron and a π orbital of the oxide ion, an electron from the oxide ion will move so as to occupy partially the d orbital. In so doing, however, it must have its spin opposed to that of the d electron because of the exclusion principle. The other π electron then has its spin aligned parallel to that of the d electron on the first metal ion. If it moves to some extent into the d orbital of the second metal ion which already contains that metal ion's unpaired d electron, the spin of that d electron will have to be aligned opposite to that of the entering π electron and, hence, opposite to that of the d electron on the first metal ion. The net result is that by this intervention of the oxide ion, which can only occur because there is some finite though not necessarily large degree of overlap between metal d and oxygen π orbitals, we obtain from a system in which the two metal-ion d electrons were free to orient their spins independently one in which they are coupled together with their spins antiparallel. If this latter state has slightly lower energy at low temperatures than does the former, then as the tempera- ture is lowered the entire metal oxide lattice will tend to drop into it, and antiferromagnetism will be exhibited. This, in somewhat simplified form, is the currently accepted explanation of antiferromagnetic behavior in most "ionic" salts, oxides and chalconides, and its key assumption is that these substances are not in fact completely ionic but instead involve some significant degree of overlap between metal-ion d orbitals and anion orbitals.

20-13. The Theoretical Failure of an Ionic Model

It is obvious, as noted already, that the CFT model cannot have any *physical* value, because it does not attempt to represent the ligands as they really are, even in purely electrostatic particulars. However, the question naturally arises as to what result would be obtained if we were to represent the ligands in a realistic manner, that is, as finite spheres of negative charge with a positive charge located at the center of the sphere, but still use only coulombic forces between these realistic ligand atoms and the d electrons of the metal ion. This question has been investigated quantitatively,[16] with results that will now be summarized in a simplified, qualitative form. Fig. 20-36 shows schematically the spacial relationship of metal-ion d orbitals and the charge clouds and nuclei of ligand atoms in a complex.

An electron occupying an orbital that has lobes such as A and A' in Fig. 20-36 directed at the ligand atoms will feel the positive charge of the nucleus (or the net positive charge of the compact core, in a heavier atom) rather strongly, thus off-setting appreciably the repulsive effect it feels from the diffuse electron cloud into which it penetrates. On the other hand, an electron occupying an orbital that has lobes, such as B and B' in Fig. 20-36, directed between ligand atoms feels the repulsive effect of the electron cloud nearly as much as the electron in A but feels the countervailing effect of the nucleus less. Thus, instead of concluding that the electron must be much more stable in orbital B than in A, as we do in the point-charge model, we

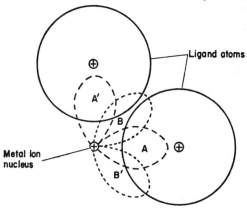

Fig. 20-36.   A drawing of part of a complex, showing two lobes, A, A' of an e_g type metal orbital, two lobes, B, B' of a t_{2g} type metal orbital and two ligand atoms. The spheres enclosing most of the electron clouds of the ligand atoms are also shown.

[16] W. H. Kleiner, *J. Chem. Phys.*, 1952, **20**, 1784; A. J. Freeman and R. E. Watson, *Phys. Rev.*, 1960, **120**, 1254. The latter paper should be read by all serious students of ligand field theory.

conclude that the stability difference will be either only slightly in favor of B, or possibly, even *slightly in favor of orbital A*. Actual calculations, using realistic (i.e. Hartree–Fock) orbitals for metal and ligand atoms, have shown this to be true quantitatively.

20-14. The Adjusted Crystal Field Theory, ACFT (also called Ligand Field Theory)

We have now recognized that the central assumption of crystal field theory, namely, that the metal ion and the surrounding ligand atoms interact with one another in a purely electrostatic way and do not mix their orbitals or share electrons, is never strictly true. The question then is whether we may still use the crystal field theory, perhaps with certain modifications and adjustments, as a *formalism* to make predictions and calculations even though we do not take its assumptions literally. The answer to this question is in the affirmative provided the degree of orbital overlap is not too great, and experience has now shown that, for most complexes of metals in their normal oxidation states, the amount of overlap is small enough to be manageable in such a way. The crystal field theory so modified to take account of the existence of moderate amounts of orbital overlap we shall call the adjusted crystal field theory, although the term ligand field theory has often been used to designate this particular form of CFT. When the amount of orbital overlap is excessive—and this is likely to happen for complexes that contain the metals in unusual oxidation states—we must have recourse to the molecular orbital theory which is outlined in the next Section.

The most straightforward modifications of simple crystal field theory that make allowance for orbital overlap involve using all parameters of interelectronic interactions as variables rather than taking them equal to the values found for the free ions. Of these parameters, three are of decisive importance, namely, the spin–orbit coupling constant, λ, and the interelectronic repulsion parameters, which may be the Slater integrals, F_n, or certain usually more convenient linear combinations of these called the Racah parameters, B and C.

The spin–orbit coupling constant plays a considerable role in determining the detailed magnetic properties of many ions in their complexes, for example, the deviations of some actual magnetic moments from spin-only values and inherent temperature-dependence of some moments. All studies to date show that in ordinary complexes the values of λ are 70–85% of those for the free ions. It is possible to get excellent agreement between crystal field theory predictions and experimental observations simply by using these smaller λ values.

The Racah parameters are measures of the energy separations of the various Russell–Saunders states of an atom. The energy differences between states of the same spin multiplicity are, in general, multiples of B only, whereas the differences between states of different multiplicity are expressed as sums of multiples of both B and C. To illustrate their use, let us take the d^8 (2 positron) system as it occurs in some tetrahedral nickel(II) complexes. As a 2-positron system in a tetrahedral field, it has, qualitatively, the same energy level diagram as a 2-electron system in an octahedral field which we have already shown in Fig. 20-30. Now an approximate calculation of the energy of the transition, v_3, from the $^3T_1(F)$ ground state to the $^3T_1(P)$ state gives the result

$$v_3 = (E_P - E_F) + \tfrac{3}{5}\Delta_t$$

In the $[NiX_4]^{2-}$ complexes ($X = Cl^-$, Br^- or I^-), this transition is observed at $\sim 14,000\ cm^{-1}$. It is completely impossible to account for this result if we assume that the energy difference $(E_P - E_F)$ has the same value in these complexes as it has in the free ion, for in the free ion it is about $16,000\ cm^{-1}$, which is greater than v_3. The only way out of this paradox is to assume that $(E_P - E_F)$ shrinks to $\sim 70\%$ of the free-ion value. Now theory expresses $(E_P - E_F)$ as $15B$, so that this is equivalent to saying that B', the value of this Racah parameter for the Ni^{2+} ion in the complexes, is $\sim 70\%$ of B, the value in the free ion. Similarly, it is found that the observed energies of several transitions from the $^3T_1(F)$ ground state to excited singlet states require that their separations from the ground state be reduced to about 70% of the free-ion values. This implies that the Racah parameter C is also diminished and by about the same amount as B. Indeed it is a general rule that

$$B'/B \approx C'/C \approx 0.7$$

Moreover, for a series of analogous complexes with different ligands, the B'/B ratios will be in the order required by the nephelauxetic series.

Thus in order to calculate an energy-level diagram and/or details of magnetic behavior in ligand field theory, one proceeds in the same manner as in crystal field theory except that, instead of assuming the free-ion values for λ, B and C, one either assumes somewhat smaller ones or leaves them as parameters to be evaluated from the experimental observations. In this way all the computational and conceptual advantages of the simple electrostatic theory are preserved while allowance is made—in an indirect and admittedly artificial way—for the consequences of finite orbital overlap. One also bears in mind that there are other consequences of the overlap, for example, electron delocalization.

THE MOLECULAR ORBITAL THEORY

20-15. Qualitative Principles

The molecular orbital theory starts with the premise that overlap of atomic orbitals occurs, where permitted by symmetry, to an extent determined by the spacial nature of the orbitals. All degrees of overlap, including the electrostatic situation, come within its scope.

The first task in working out the MO treatment for a particular type of complex is to find out which orbital overlaps are or are not possible because of the inherent symmetry requirements of the problem. This can be done quite elegantly and systematically by using some principles of group theory, but such an approach is outside the scope of this discussion. Instead we shall simply present the results that are obtained for octahedral complexes, illustrating them pictorially. It may be noted that, ultimately, for the experimental inorganic chemist, this pictorial representation is much more important than mathematical details, for it provides a basis for visualizing the bonding and thinking concretely about it.

The molecular orbitals we shall use here will be of the LCAO type. Our method for constructing them, which we shall apply specifically to octahedral complexes, will take the following steps:

1. We note that there are nine valence shell orbitals of the metal ion to be considered. Six of these—d_{z^2}, $d_{x^2-y^2}$, s, p_x, p_y and p_z—have lobes lying along the metal—ligand bond directions (i.e. are suitable for σ bonding), whereas three, namely, d_{xy}, d_{yz}, d_{zx}, are so oriented as to be suitable only for π bonding.

2. We shall assume initially that each of the six ligands possesses one σ orbital. These individual σ orbitals must then be combined into six "symmetry" orbitals, each constructed so as to overlap effectively with a particular one of the six metal-ion orbitals which are suitable for σ bonding. Each of the metal orbitals must then be combined with its matching symmetry orbital of the ligand system to give a bonding and an antibonding molecular orbital.

3. If the ligands also possess π orbitals, these too must be combined into "symmetry" orbitals constructed so as to overlap effectively with the metal-ion π orbitals, and the bonding and antibonding MO's then formed by overlap.

Complexes with No π Bonding. The six σ symmetry orbitals are indicated in Fig. 20-37, in which they are illustrated pictorially, expressed algebraically as normalized linear combinations of the individual ligand σ orbitals, and juxtaposed with the metal-ion orbitals with which they are matched by symmetry. On the left side of Fig. 20-37 are the symmetry symbols, A_{1g}, E_g and T_{1u}, for these orbitals. These symbols are of group theoretical origin, and they stand for the symmetry class to which belong the metal orbital, the matching symmetry orbital of the ligand system, and the molecular orbitals that will result from the overlap of these two. They are very commonly used simply as convenient labels, but they also carry information. The symbol A_{1g} always represents a single orbital that has the full symmetry of the molecular system. The symbol E_g represents a pair of orbitals that are equivalent except for their orientations in space, whereas T_{1u} represents a set of three orbitals that are equivalent except for their orientations in space. The subscripts g and u are used to indicate whether the orbital(s) is centrosymmetric (g from the German *gerade* meaning even) or anticentrosymmetric (u from the German *ungerade* meaning uneven).

The final step now needed to obtain the molecular orbitals themselves is to allow each metal orbital to overlap with its matching symmetry orbital

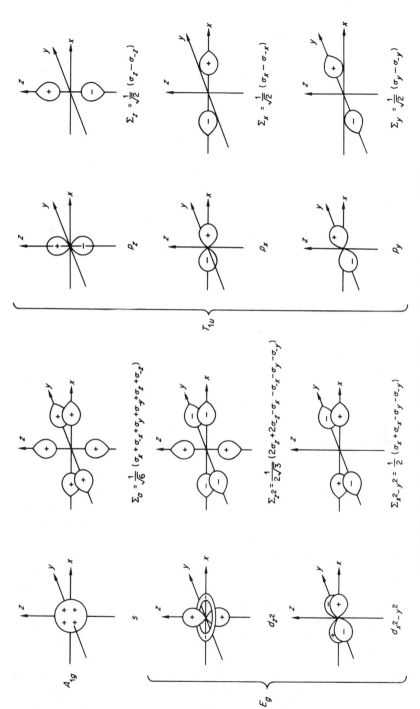

Fig. 20-37.   The six metal ion σ orbitals and their matching ligand symmetry orbitals.

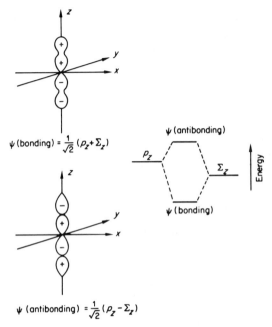

Fig. 20-38. To the left are orbital pictures showing the bonding and antibonding molecular orbitals that are the z components of the T_{1u} sets. To the right is an energy-level diagram showing how the energies of the various orbitals are related.

of the ligand system. As usual two combinations are to be considered: one in which the matched orbitals unite with maximum positive overlap, thus giving a bonding MO, and the other in which they unite with maximum negative overlap to give the corresponding antibonding MO. This process is illustrated for the pair p_z and Σ_z in Fig. 20-38. From the energy point of view, these results may be expressed in the usual type of MO energy-level diagram (see page 100), as shown at the right side of Fig. 20-38. It will be noted there that the p_z and Σ_z orbitals are not assumed to have the same energies, for in general they do not. To a first approximation, the energies of the bonding and antibonding MO's lie equal distances below and above, respectively, the mean of the energies of the combining orbitals.

In just the same way, the other metal-ion orbitals combine with the matching symmetry orbitals of the ligand system to form bonding and antibonding MO's. The MO's of the same symmetry class—which are equivalent except for their spacial orientations—have the same energies, but orbitals of different symmetry classes do not in general have the same energies, since they are not equivalent. The energy-level diagram that results when all the σ interactions are considered is shown in Fig. 20-39. Here we name the orbitals only by their symmetry designations, using the asterisk to signify that a molecular orbital is antibonding. It should be noted in Fig. 20-39 that the three metal-ion d orbitals that are suitable for forming π bonds but not σ

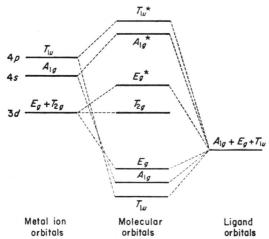

Fig. 20-39. The molecular orbital energy-level diagram, qualitative, for an octahedral complex between a metal ion of the first transition series and six ligands that do not possess π orbitals.

bonds are given their appropriate symmetry label, T_{2g}, and shown as remaining unchanged in energy, since we are considering ligands having no π orbitals with which they might interact.

There are certain implications of this energy-level diagram deserving special attention. In general, in molecular orbital diagrams of this type, it may be assumed that, if a molecular orbital is much nearer in energy to one of the atomic orbitals used to construct it than to the other one, it has much more the character of the first one than of the second. On this basis then, Fig. 20-39 implies that the six bonding σ MO's, three T_{1u}'s, the A_{1g} and the two E_g's, have more the character of ligand orbitals than they do of metal orbitals. It can then be said that electrons occupying these orbitals will be mainly "ligand electrons" rather than "metal electrons," though they will partake of metal-ion character to some significant extent. Conversely, electrons occupying any of the antibonding MO's are to be considered as predominantly metal electrons. Any electrons in the T_{2g} orbitals will be *purely* metal electrons when there are no ligand π orbitals as in the case being considered.

Let us now look at the center of the MO diagram, where we see the T_{2g} orbitals and, somewhat higher in energy, the E_g^* orbitals. The latter, as noted, are predominantly of metal-ion d orbital character though with some ligand orbital character mixed in. Is this not the same situation, qualitatively speaking, as we obtained from the electrostatic arguments of crystal field theory? Indeed it is, and, moreover, it is the same result we get from the adjusted crystal field theory, where we allow for the occurrence of orbital overlap that destroys to some extent the "purity" of the metal-ion d orbitals.

Complexes with π Bonding. If the ligands have π orbitals, filled or unfilled, it is necessary to consider their interactions with the T_{2g} d orbitals, that is, the d_{xy}, d_{yz} and d_{zx} orbitals. The simplest case is the one where each ligand has a pair of π orbitals mutually perpendicular, making $6 \times 2 = 12$ altogether. From group theory it is found that these may be combined into four triply degenerate sets belonging to the symmetry classes T_{1g}, T_{2g}, T_{1u} and T_{2u}. Those in the classes T_{1g} and T_{2u} will remain rigorously non-bonding. (We use the terms bonding, non-bonding, and antibonding with reference to the metal—ligand interactions regardless of the character of the orbitals in respect to bonding between atoms within polyatomic ligands.) This is because the metal ion does not possess any orbitals of these symmetries with which they might interact. The T_{1u} set can interact with the metal-ion p orbitals, which are themselves a set with T_{1u} symmetry, and in a quantitative discussion it would be necessary to make allowance for this. However, in a qualitative treatment we may assume that, since the p orbitals are already required for the σ bonding, we need not consider π bonding by means of T_{1u} orbitals, which are thus non-bonding in character. This then leaves us with only the T_{2g} set of symmetry orbitals to overlap with the metal ion T_{2g} d orbitals.

The ligand π orbitals may be simple $p\pi$ orbitals, as in the Cl^- ion, simple $d\pi$ orbitals as in phosphines or arsines, or molecular orbitals of a polyatomic ligand as in CO, CN^- or pyridine. When they are simple $p\pi$ or $d\pi$ orbitals, it is quite easy to visualize how they combine to form the proper symmetry orbitals for overlapping with the metal-ion orbitals. This is illustrated for $p\pi$ orbitals in Fig. 20-40.

The effects of π bonding via molecular orbitals of the T_{2g} type upon the energy levels must now be considered. These effects will vary depending on the energy of the ligand π orbitals relative to the energy of the metal T_{2g} orbitals and on whether the ligand π orbitals are filled or empty. Let us consider first the case where there are empty π orbitals of higher energy than the metal T_{2g} orbitals. This situation is found in complexes where the ligands are phosphines or arsines, for example. As shown in Fig. 20-41a, the net result of the π interaction is to stabilize the metal T_{2g} orbitals (which, of course, also acquire some ligand orbital character in the process) relative

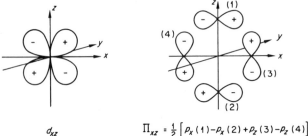

$$d_{xz} \qquad \Pi_{xz} = \tfrac{1}{2}\left[p_x(1) - p_x(2) + p_z(3) - p_z(4) \right]$$

Fig. 20-40.   At the right is the symmetry orbital, made up of ligand p orbitals, that has the proper symmetry to give optimum interaction with the metal ion d_{xz} orbital shown at the left. There are analogous symmetry orbitals, π_{xy} and π_{yz}, that are similarly related to the metal ion d_{xy} and d_{yz} orbitals.

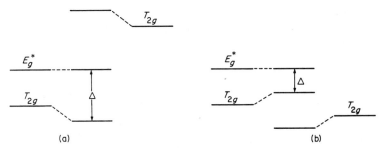

Fig. 20-41. Energy-level diagrams showing how π interactions can affect the value of Δ. (a) Ligands have π orbitals of higher energy than the metal T_{2g} orbitals; (b) ligands have π orbitals of lower energy than the metal T_{2g} orbitals.

to the metal E_g^* orbitals. In effect, the π interaction causes the Δ value for the complex to be greater than it would be if there were only σ interactions.

A second important case is the one in which the ligands possess only filled π orbitals of lower energy than the metal T_{2g} orbitals. As shown in Fig. 20-41b, the interaction here destabilizes the T_{2g} orbitals relative to the E_g^* orbitals and thus diminishes the value of Δ. This is probably the situation in

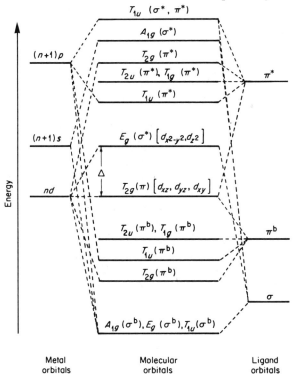

Fig. 20-42. A qualitative molecular orbital diagram for an $M(CO)_6$ or $[M(CN)_6]^{n-}$ compound. [Adapted from H. B. Gray and N. A. Beach, *J. Amer. Chem. Soc.*, 1963, **85**, 2922.]

complexes of metal ions in their normal oxidation states, especially the lower ones, with the ligand atoms oxygen and fluorine.

There are also important cases in which the ligands have both empty and filled π orbitals. In some, such as the Cl^-, Br^- and I^- ions, these two types are not directly interrelated, the former being outer d orbitals and the latter valence-shell p orbitals. In others, such as CO, CN^- and pyridine, the empty and filled π orbitals are the antibonding and bonding $p\pi$ orbitals. In such cases the net effect is the result of competition between the interaction of the two types of ligand π orbitals with the metal T_{2g} orbitals, and simple predictions are not easily made. Fig. 20-42 shows a diagram for octahedral metal cyano complexes and carbonyls that is based on some rough calculations of overlaps and comparisons with absorption spectra. While it has only a qualitative value, and even in this sense may not be entirely correct, most workers would probably consider it to be essentially correct for the Group VI hexacarbonyls and cyano complexes of Fe^{II}, Ru^{II} and Os^{II}. However, the order of some of the MO's may well be changed by large changes in the metal orbital energies, as, for example, on going from a normal oxidation state of the metal to an unusually low or high one.

20-16. Quantitative Calculations

In recent years considerable effort has been devoted to devising and testing schemes by which quantitative molecular-orbital energy-level diagrams might be obtained without excessive computation. Virtually all of this work has used the LCAO approximation together with the assumption that the energies of the MO's can be estimated from the magnitudes of the overlaps. This general approach is called an extended Hückel or, alternatively, a Wolfsberg–Helmholz calculation. There are a number of variations in detail, depending on how the values of certain initial parameters (orbital energies, radial wave functions, the proportionality factor relating overlaps to energies) are chosen, on the number of orbitals and overlaps included in the calculation, and on the manner in which initial parameters are permitted to vary during the calculation so as to lead to a final result that is "self-consistent," i.e. so that the parameter values used in the final cycle of calculation are consistent with the electron distribution that is calculated in that cycle. The main steps in the method, with some comments on the alternative procedures for handling the details, will now be sketched.

1. The ligand orbitals are combined into symmetry-correct combinations. The general nature of these combinations has been indicated already in Section 20-15. Although the form of the combinations of a given set of ligand orbitals is rigorously fixed by symmetry requirements, there have been some differences in the choice of ligand orbitals to be used.

2. The overlaps are calculated. First, the overlaps between metal orbitals and individual ligand orbitals are calculated, using, preferably, the best available wave functions, namely, the Hartree or Hartree–Fock types,

although for the heavier metal atoms these are not yet available. Then, by using the expressions for the symmetry-correct combinations of ligand orbitals, the so-called group overlaps are calculated. These group overlaps are those between a ligand orbital combination as a whole and the metal orbital (or orbitals) with which it gives non-zero overlap. In some calculations the overlap of the ligand orbitals with one another is also considered.

3. The energies of the metal orbitals and the ligand orbital combinations are estimated. This is a complex and non-rigorous procedure, subject to numerous variations in the hands of different workers. For both metal and ligand orbitals, the energies vary with the state of ionization of the atoms and it is necessary to determine the required relationship, so that as the charge distribution varies through successive cycles of calculation (see step 6), the energies may be appropriately reassigned. The rate of variation of energy of an orbital with state of ionization is, however, difficult to estimate accurately, because *in a molecule* it varies less than in the free atoms. In most calculations, however, the variation used has been that for free atoms. This can be done in such a way as to obtain a reasonable fit to observed spectral bands, but then the ground-state electron distribution is rather unrealistic in that the degree of covalence is markedly overestimated.

4. The values of the integrals of the type

$$\int \psi_i \mathscr{H} \Sigma_i \, d\tau$$

where ψ_i and Σ_i represent, respectively, a metal orbital and a combination of ligand orbitals with the same symmetry, are evaluated by assuming that they are proportional to the corresponding overlap integrals times the mean energy of the overlapping orbitals and times a proportionality constant. The constant has usually been taken as 2.00, but the possibilities that it should be varied with orbital type (i.e. σ or π) and perhaps from one metal to another have been suggested in certain cases.

5. An initial charge distribution is chosen, the orbital energies and the interaction energies are calculated and the secular equation is solved to give the energies and the compositions (i.e. metal and ligand orbital coefficients) of the molecular orbitals. The choice of initial charge distribution is such that the metal ion carries only a small charge, say, ca. $+0.5$, since experience shows that the final cycle of calculation (see next step) will lead to only a small net charge on the metal, whatever its *formal* oxidation state.

6. With the results of the preceding step, electrons are assigned to the MO's and, by using a method of "population analysis" due to Mulliken, the net charge on each atom is assessed. If these differ from the charge distribution initially assumed, steps 5 and 6 are repeated with the new charge distribution (or a compromise between this and the original one). Once again, if a discrepancy between the assumed and the calculated charge distribution is found, another cycle of steps 5 and 6 is carried out. This practice is repeated until only a negligible difference is obtained.

The procedure just outlined is relatively quick and easy and gives results that are helpful in the interpretation of spectra. The energy-level diagrams so obtained have value for discussing the bonding in broad qualitative terms. However, the details and the exact ordering of energy levels obtained by this procedure must generally be taken *con granulo salis*.

There have been a number of attempts to carry out more thorough calculations, in which the approximation of step 4 is replaced by more rigorous evaluation of the various interaction integrals. Discussion of these calculations would involve a level of quantum-mechanical discussion inappropriate in this book. The reader is therefore referred to the literature[17] for further information.

20-17. MO Diagrams for Some Other Types of Complex

The other two main types of complex, tetrahedral and square, have also been frequently discussed within the framework of MO theory. Tetrahedral species can be divided roughly into two broad classes: (1) Oxo species in which the formal oxidation number of the metal is high (≥ 6) and where there must be very extensive π bonding. Examples, including some which have been subject to prolonged controversy, are MnO_4^-, MnO_4^{2-}, CrO_4^{2-} and MoO_4^{2-}. (2) Complexes in which the metals are in lower oxidation states,

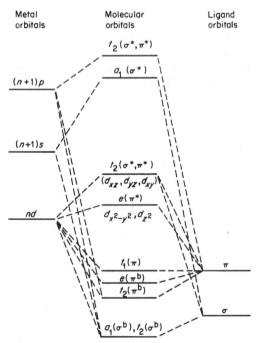

Fig. 20-43. A molecular orbital diagram that would be applicable to most tetrahedral complexes of transition-metal ions in lower oxidation states.

[17] J. P. Dahl and C. J. Ballhausen, *Adv. Quantum Chem.*, 1968, **4**, 170.

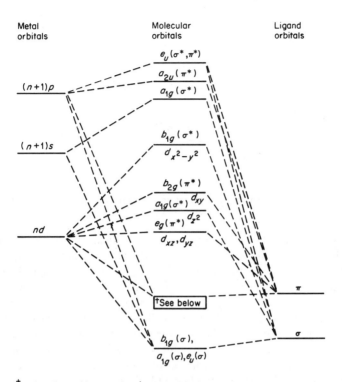

Metal orbitals

Molecular orbitals

Ligand orbitals

$e_u(\sigma^*, \pi^*)$

$a_{2u}(\pi^*)$

$(n+1)p$

$a_{1g}(\sigma^*)$

$(n+1)s$

$b_{1g}(\sigma^*)$

$d_{x^2-y^2}$

$b_{2g}(\pi^*)$

$a_{1g}(\sigma^*)\ d_{xy}$

$e_g(\pi^*)\ d_{z^2}$

nd

d_{xz}, d_{yz}

†See below

π

$b_{1g}(\sigma),$

$a_{1g}(\sigma), e_u(\sigma)$

σ

$^{\dagger}\pi$ bonding orbitals: $a_{2g}, b_{2u}, a_{2u}, b_{2g}, e_g, e_u$

Fig. 20-44. A qualitative molecular orbital diagram for a square ML_4 complex. The arrangement here should be essentially correct for $PtCl_4^{2-}$.

such as $+2$ or $+3$, and the ligands are halide ions, amine-N atoms or RO^- ions. A diagram that would be qualitatively applicable to most complexes of the second type is given in Fig. 20-43.

For most square complexes, the general form of the MO diagram is as shown in Fig. 20-44.

For substances such as $(h^5\text{-}C_5H_5)_2M$ and $M(CO)_6$ compounds, the molecular-orbital approach has been extensively used. It is more appropriate, however, to postpone consideration of these to Chapters 22 and 23 where such compounds are discussed comprehensively.

The Two-dimensional Spectrochemical Series. As shown in Fig. 20-42, the extent of the energy separation between the e_g and t_{2g} orbitals depends, not only on how much the former are raised in energy by the σ bonding, but also on how much the latter are influenced by π interactions. It is thus possible to envision a two-dimensional plot in which ligands are ordered along one axis according to their tendency to render the e_g orbitals anti-bonding (only positive values of a σ parameter possible) and along the other axis according to their ability to shift the t_{2g} orbitals either up or down (both positive and negative π parameters possible). The first attempt to implement

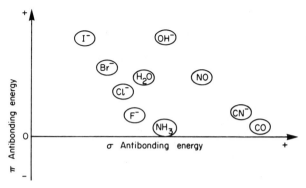

Fig. 20-45. A qualitative two-dimensional spectrochemical series of some ligands, as suggested by McClure.

this idea was made by McClure,[18] who developed a formalism whereby splittings of the bands in substituted octahedral complexes, e.g., *trans*-CoA_4B_2, are expressed as functions of σ_A, σ_B, π_A and π_B. There have been further efforts to develop the scheme,[19] but there is a need for more extensive low-temperature polarized crystal spectra in order to have enough accurate data on splittings to cover all important ligands. The qualitative ordering originally suggested by McClure is shown in Fig. 20-45. It must be noted here that the position of CO seems highly doubtful since a vast amount of other evidence implies that a strong π interaction (negative in the sense of the diagram) is of the essence in M—CO bonding.

Even in qualitative form the idea of the two-dimensional series is valuable in emphasizing the dual origin of *d* orbital splittings.

20-18. Charge-transfer Spectra

We have so far (Sections 20-4 to 20-6) considered only those electronic transitions in which electrons move between orbitals that have predominantly metal *d* orbital character. Thus, the charge distribution in the complex is about the same in the ground and the excited state. There is another important class of transitions in which the electron moves from a molecular orbital centered mainly on the ligands to one centered mainly on the metal atom, or *vice versa*. In these, the charge distribution is considerably different in ground and the excited states (at least we naïvely assume it is), and so they are called *charge-transfer transitions*.

As just implied, there are two broad classes: ligand-to-metal (L→M) and metal-to-ligand (M→L). In general, the former are better understood. In most cases charge-transfer (CT) processes are of higher energy than *d–d* transitions; thus they usually lie at the extreme blue end of the visible spec-

[18] D. S. McClure in *Advances in the Chemistry of the Coordination Compounds*, S. Kirschner, ed., McMillan Company, 1961.
[19] L. Dubicki and R. L. Martin, *Austral. J. Chem.*, 1969, **22**, 839.

trum, or in the ultraviolet region. Also, nearly all observed CT transitions are fully allowed (that is, they are $g \leftrightarrow u$ transitions, with $\Delta S = 0$) and hence the CT bands are strong. Extinction coefficients are typically 10^3–10^4, or more. There are, of course, many forbidden CT transitions which give rise to weak bands; these are seldom observed because they are covered up by the strong CT bands.

L→M Transitions in Octahedral Complexes. The first CT spectra to be systematically studied, and still among the best understood, are the L→M transitions in hexahalo complexes. Fig. 20-46 shows a partial MO diagram for such complexes, and the four types of transition that would be expected are indicated. Actually, each of the transitions shown is a group of transitions, since the excited orbital configuration (and in many cases also the ground configuration) gives rise to several different states of similar but not identical energies.

Transitions of the v_1 type will obviously be of lowest energy. Secondly, since the π and π^* orbitals involved are both approximately non-bonding, they will not vary steeply with M—L distance as the ligands vibrate. Thus, according to the same type of argument already applied to the d–d transitions (page 578), the bands for these transitions should be relatively narrow. A third factor that should assist in identifying the v_1 set of bands is that they will be missing in those cases where the $\pi^*(t_{2g})$ orbitals are filled, i.e. in d^6 complexes. Table 20-6 indicates how observed transitions in a variety of complexes have been assigned to the v_1 class. Two trends in these bands support the assigned L→M nature of these transitions. For a given metal ion (e.g., Os^{4+}), their energies decrease in the sequence MCl_6, MBr_6, MI_6, which is the order of decreasing ionization potentials (i.e. easier oxidizability) of the halogen atoms. As the oxidation state of the metal increases (e.g.,

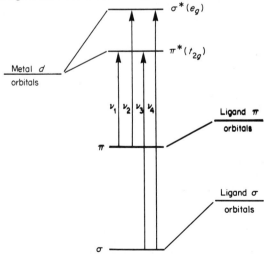

Fig. 20-46.   Partial MO diagram for an octahedral MX_6 complex, showing the four main classes of L→M charge-transfer transitions.

TABLE 20-6

$L \to M$ CT Transitions (cm^{-1} $\times 10^{-3}$) in some Hexahalo Complexes[a,b]

d^n	Complex	v_1 Set[c]	v_2 Set	v_4 Set
$4d^4$	$RuCl_6^{2-}$	17.0–24.5 (0.65–3.0)	36.0–41.0 (12–18)	—
$4d^5$	$RuCl_6^{3-}$	25.5–32.5 (0.60–2.1)	43.6(16)	—
$5d^4$	$OsCl_6^{2-}$	24.0–30.0 (1.0–8.0)	47.0(20)	—
	$OsBr_6^{2-}$	17.0–25.0 (1.6–7.5)	35.0–41.0 (10–15)	—
	OsI_6^{2-}	11.5–18.5 (2.5–6.0)	27.0–35.5 (8.0–9.2)	44.6(41)
$5d^6$	$PtBr_6^{2-}$	—	27.0–33.0 (7.0–18)	44.2(70)
	PtI_6^{2-}	—	20.0–30.0 (8.0–13)	40.0–43.5 (40–60)

[a] Data and assignments from C. K. Jørgensen, *Mol. Phys.*, 1959, **2**, 309; *Adv. Chem. Phys.*, 1963, **5**, 33.
[b] Numbers in parentheses are molar extinction coefficients ($\times 10^{-3}$).
[c] Band half-widths, 400–1000 cm^{-1}.

$RuCl_6^{3-}$, $RuCl_6^{2-}$) its orbitals should be deeper and thus the transition should go to lower energy.

Transitions of the v_2 type should give the lowest-energy CT bands in t_{2g}^6 complexes (e.g., those of the PtX_6^{2-} type). Since the transition is from a mainly non-bonding level to a distinctly antibonding one, the bands should be fairly broad. The transitions assigned to the v_2 sets all have half widths of 2000–4000 cm^{-1}. The shifts in energy of these bands with change of halogen and change of metal oxidation state are again as expected for $L \to M$ transitions.

Transitions of the v_3 set are all expected to be broad and weak and are not observed. The v_4 transitions have been observed in a few cases, but in many cases they must lie beyond the range of observation.

$L \to M$ Transitions in Tetrahedral Complexes. For the tetrahalo complexes, e.g., the NiX_4^{2-}, CoX_4^{2-} and MnX_4^{2-} species, strong $L \to M$ CT spectra can be observed and assigned in much the same way as for the octahedral MX_6^{n-} complexes.[20]

Of course, CT spectra are not restricted, like d–d spectra, to transition-metal complexes. For example, the series of complexes $HgCl_4^{2-}$, $HgBr_4^{2-}$ and HgI_4^{2-} have absorptions at 43,700, 40,000 and 31,000 cm^{-1} which may be assigned as $L \to M$ CT transitions.

$M \to L$ Transitions. Transitions of this type can only be expected when the ligands possess low-lying empty orbitals and the metal ion has filled orbitals lying higher than the highest filled ligand orbitals. The best examples are

[20] P. Day and C. K. Jørgensen, *J. Chem. Soc.*, **1964**, 6226.

provided by complexes containing CO, CN or aromatic amines (e.g. pyridine, bipyridine or phenanthroline) as ligands.

In the case of the octahedral metal carbonyls $Cr(CO)_6$ and $Mo(CO)_6$, pairs of intense bands at 35,800 and 44,500 cm^{-1} for the former and 35,000 and 43,000 cm^{-1} for the latter have been plausibly assigned to transitions from the bonding (mainly metal) to the antibonding (mainly ligand) components of the metal—ligand π bonding interactions.[21]

For $Ni(CN)_4^{2-}$ there are three medium-to-strong bands at 32,000, 35,200 and 37,600 cm^{-1} which have been assigned as transitions from the three types of filled metal d orbitals, d_{xy}, d_{z^2} and (d_{xz}, d_{yz}) to the lowest-energy orbital formed from the π^* orbitals of the set of CN groups.[22]

[21] H. B. Gray and N. A. Beach, *J. Amer. Chem. Soc.*, 1963, **85**, 2923.
[22] H. B. Gray and C. J. Ballhausen, *J. Amer. Chem. Soc.*, 1963, **85**, 260.

Further Reading

Ballhausen, C. J., *Introduction to Ligand Field Theory*, McGraw-Hill, 1962 (an excellent book for the inorganic chemist to learn the quantitative aspects of LFT; also reviews experimental data).

Cotton, F. A., *Chemical Applications of Group Theory*, 2nd ed., Wiley, 1971 (Chapter 9 discusses the symmetry basis of LFT).

Dieke, G. H., *Spectra and Energy Levels of Rare Earth Ions in Crystals*, Interscience-Wiley, 1968 (application of CFT to the complex spectra of the lanthanide ions).

Dunitz, J. D., and L. E. Orgel, *Adv. Inorg. Chem. Radiochem.*, 1960, **2**, 1 (effects of electronic structures on molecular structures).

Dunn, T. M., in *Modern Coordination Chemistry*, J. Lewis and R. G. Wilkins, eds., Interscience, 1960, p. 229 (survey of LFT).

Ferguson, J., *Progr. Inorg. Chem.*, 1970, **12**, 159 (a review of experimental data and assignments of *d–d* spectra for complexes of the 1st transition series).

Figgis, B. N., *Introduction to Ligand Fields*, Wiley, 1966 (good introduction to mathematical LFT).

George, P., and D. S. McClure, *Progr. Inorg. Chem.*, 1959, **1**, 38 (effects of crystal field splittings on thermodynamic properties of compounds).

Griffith, J. S., *The Theory of Transition Metal Ions*, Cambridge University Press, 1961 (a comprehensive mathematical treatise).

Jørgensen, C. K., *Modern Aspects of Ligand Field Theory*, North Holland, 1971.

Jørgensen, C. K., *Progr. Inorg. Chem.*, 1970, **12**, 101 (a good survey of the author's very elaborate methods of analyzing and correlating CT spectra).

Lever, A. B. P., *Inorganic Electronic Spectroscopy*, Elsevier Publishing Co., 1968 (semiquantitative introduction).

McClure, D. S., *Solid State Phys.*, 1959, **9**, 399.

Orgel, L. E., *An Introduction to Transition Metal Chemistry, Ligand Field Theory*, 2nd ed., Methuen and John Wiley, 1966 (a comprehensive, non-mathematical introduction to LFT).

Owen, J., and J. H. M. Thornley, *Reports of Progress in Physics*, 1966, **29**, 675 (a literature review on crystal field and MO theory of transition-metal complexes at a rigorous level).

Schläfer, H. L., and G. Gliemann, *Basic Principles of Ligand Field Theory*, Wiley-Interscience, 1969 (quantitative introduction; references only to about 1965, however).

Theissing, H. H., and P. J. Caplan, *Spectroscopic Calculations for a Multielectron Ion*, Interscience-Wiley, 1966 (an exhaustive discussion of the energy levels and spectra of the Cr^{3+} ion).

21

Classical Complexes

COMPONENTS AND STRUCTURES

21-1. Introduction

The term "classical complexes" is used here for want of a better one. We are attempting to designate those complexes that can, for the most part, be described in terms of (1) a set of ligands with a discrete electron population, and (2) a metal atom with a well-defined oxidation number. In a general way classical complexes are of the type that Werner dealt with in laying the foundations of coordination chemistry, as contrasted with other types that have more complicated behavior.

We specifically wish to exclude here (but only until Chapter 22) complexes in which metal—ligand bonding is highly covalent and/or multiple, thus tending to blur, if not entirely to abolish, the significance of formal oxidation numbers. Ligands that themselves can both donate and accept electrons, so that their own oxidation state is indistinct, make it difficult if not impossible to attribute an unambiguous oxidation number to the metal atom. The extreme examples of such ligands are the 1,2-dithiolene ligands (page 724). Many others, such as NO, CO, CN^- and various phosphines are also capable of varying their net balance of donor and acceptor behavior toward a given metal atom. Because of this the *physical significance* of the oxidation number assigned to the metal atom surrounded by such ligands becomes vague. This will be true even when the conventional choice of an oxidation number is still unambiguous.

A very cogent example is provided by the pair of ions $Fe(CN)_6^{3-}$ and $Fe(CN)_6^{4-}$. The formal oxidation states of the iron atom are obviously $+3$ and $+2$, respectively, but detailed studies of Mössbauer and esr spectra, coupled with an MO treatment of the bonding,[1] indicate that the effective charge residing on the metal atom is practically the same in the two cases. In other words, the conventional practice of attributing the difference of one in the electron populations of the two species to the metal atoms is not

---

[1] R. G. Shulman and S. Sugano, *J. Chem. Phys.*, 1965, **42**, 39.

merely inaccurate, it is totally false. It is the electron density on the ligands, not the metal, that differs.

In addition to excluding here complexes in which the redox properties of the ligands are as important as those of the metal (such ligands have been aptly called *non-innocent* ones by Jørgensen), we exclude complexes in which direct metal—metal bonding is significant (see pages 547–554 for discussion of these) and organometallic compounds which are fully discussed in Chapters 23 and 24.

Naturally, the boundaries between "classical" and "non-classical" complexes and between innocent and non-innocent ligands are imperfectly defined, both logically and historically. After all, Werner and his contemporaries dealt with cyano and phosphine complexes, and pyridine is not as innocent as ammonia but much more so than, say, cyanide.

Another way, though again not a hard-and-fast one, in which the "classical" complexes we wish to discuss here may usually be distinguished from those we shall deal with in Chapter 22 has to do with the formal oxidation numbers of the metal atoms. By and large, the classical complexes are those in which the oxidation numbers are $+2$, $+3$ or $+4$, whereas the non-classical complexes tend to have metal atoms with low ($+1$, 0 and even negative) formal oxidation numbers.

As a concluding observation, the types of ligand found in "classical" complexes of the transition metals are generally found also in complexes of the non-transition metals. On the other hand, ligands such as NO, CO, dithiolenes and CN^- form few complexes with non-transition metal ions.

21-2. Types of Ligand

The points to be covered here do not all pertain exclusively to transition-metal complexes. They are discussed at this place in the text because even those that are relevant to complexes of non-transition metals have their greatest significance with respect to transition-metal complexes.

Among the most important "classical" ligands are the halide ions, F^-, Cl^-, Br^-, I^-, the anions of various oxo acids, such as NO_3^-, NO_2^-, RCO_2^-, SO_4^{2-} and neutral molecules in which the donor atoms are usually N or O, examples being NH_3, RNH_2, H_2O, MeOH, R_3PO, R_2SO and CH_3CN.

The simplest role that each of these ligands can play is that of an electron-pair donor to a single cation. This is illustrated by (21-I) to (21-IV).

$$M:\ddot{\underset{..}{X}}: \quad (X = F^-, Cl^-, Br^-, \text{ or } I^-) \qquad\qquad M:\ddot{O}SO_3$$

$$(21\text{-I}) \qquad\qquad\qquad\qquad (21\text{-II})$$

$$\qquad\qquad\qquad\qquad\qquad\qquad M:\ddot{O}NO_2$$

$$(21\text{-III}) \qquad\qquad\qquad\qquad\qquad (21\text{-IV})$$

Some of these donors occur in sets of two, or more, in a complex ligand that is structurally capable of permitting all the donors simultaneously to form bonds to the same metal atom. This is illustrated for several important types of donor groups in (21-V) to (21-VII). Such ligands are called *polydentate* (or multidentate) ligands and also *chelating* ligands. They will be discussed in detail in Section 21-3.

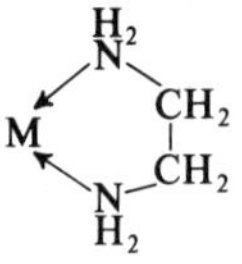

(21-V) 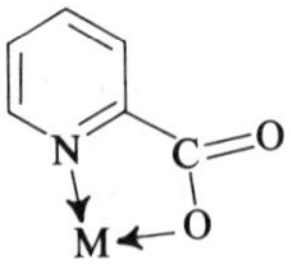   (21-VI) 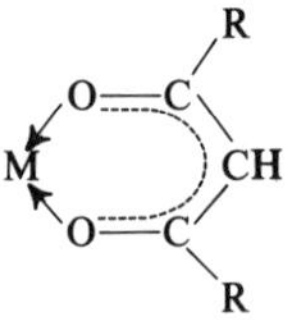   (21-VII)

Another important role of ligands is as bridging groups. In many cases they serve as *monodentate* bridging ligands. This means that there is only *one* ligand atom which forms two (or even three) bonds to different metal atoms. For monoatomic ligands, such as the halide ions, and those containing only one possible donor atom, this monodentate form of bridging is, of course, the only possible one. A few examples are shown in (21-VIII) to (21-X). Ligands having more than one atom that can be an electron donor often function as *bidentate* bridging ligands. Examples are shown in (20-XI) and (21-XII).

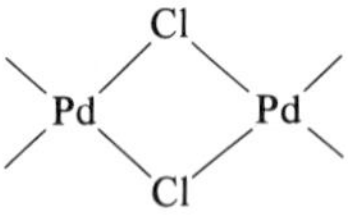

(21-VIII) 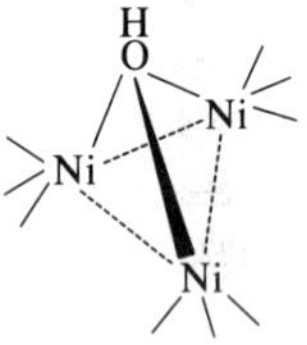   (21-IX) 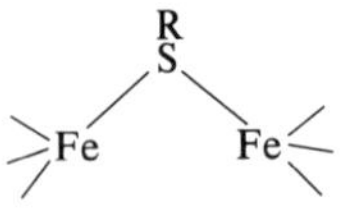   (21-X)

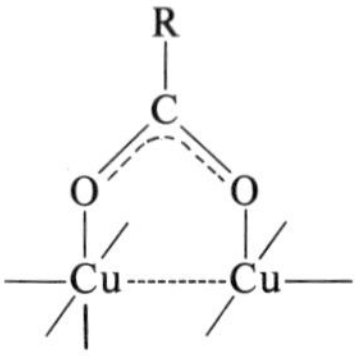

(21-XI) 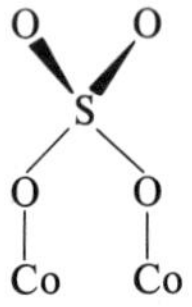   (21-XII)

In the next few Sections, various structural and chemical classes of ligands will be considered in more detail.

21-3. A Survey of Polydentate Ligands

A few important examples of the basic types of polydentate ligand will be cited for each order of "denticity."

Bidentate Ligands.   There are an enormous number of these.[2] They may be classified according to the size of the chelate ring formed, as in the following list of some of the more important ones:

[2] Cf. C. M. Harris and S. E. Livingstone, in *Chelating Agents and Metal Chelates*, F. P. Dwyer and D. P. Mellor, eds., Academic Press, 1964, p. 95.

4-rings:

dithiocarbamate, R_2dtc: R_2NC

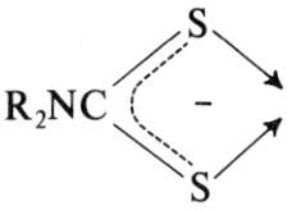

xanthates: ROC

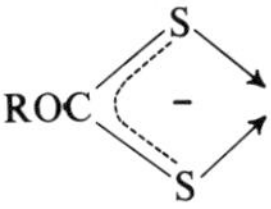

carboxylates: RC

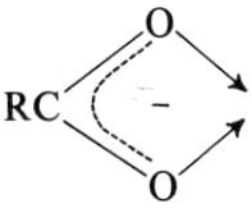

5-rings:

ethylenediamine, en: $H_2NCH_2CH_2NH_2$

2,2′-bipyridine, bipy:

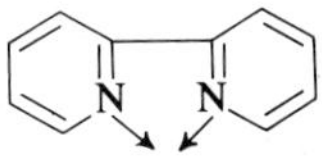

1,10-phenanthroline, phen:

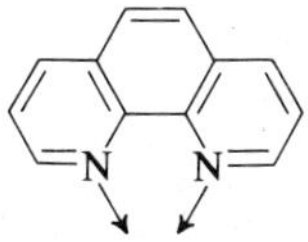

o-phenylenebis(dimethylarsine), diars:

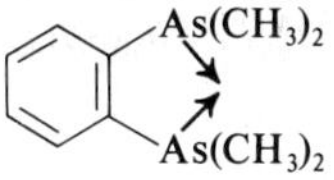

1,2-bis(diphenylphosphino)ethane, diphos:

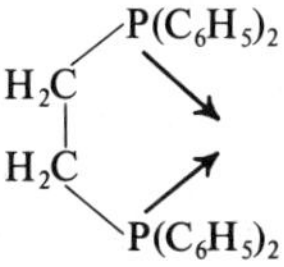

6-rings:

β-diketonates, e.g., acetylacetonate, acac⁻: HC

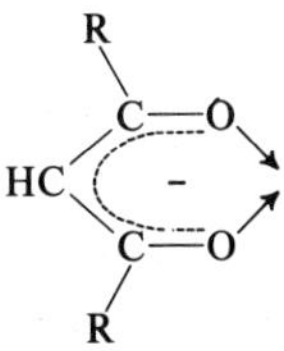

salicylaldiminato, sal⁻:

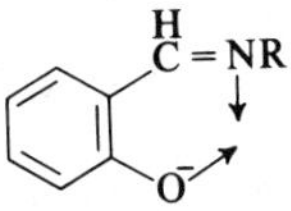

Tridentate Ligands. Some are *obligate planar* such as

terpyridine, terpy: 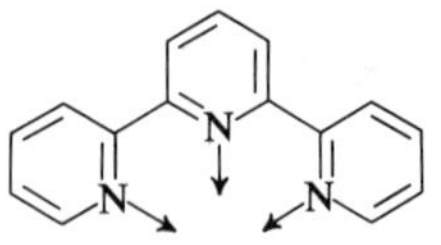

acylhydrazones of salicylaldehyde: 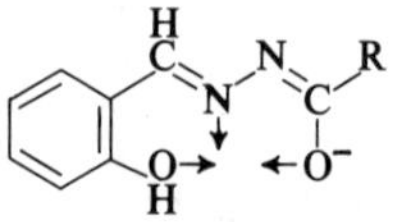

and many similar ones where maintenance of the π-conjugation markedly favors planarity. Such ligands must form complexes of the types (21-XIII) or (21-XIV).

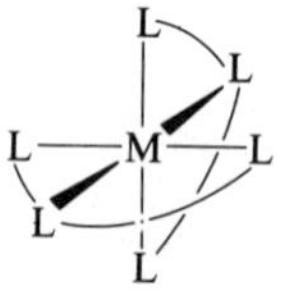

(21-XIII)

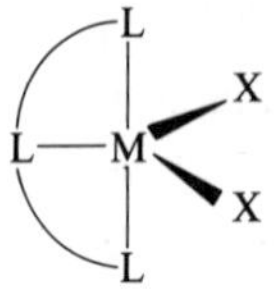

(21-XIV)

There are also many flexible tridentate ligands such as

diethylenetriamine, dien:

$$HN\begin{matrix} \diagup CH_2CH_2NH_2 \\ \diagdown CH_2CH_2NH_2 \end{matrix}$$

bis(3-dimethylarsinylpropyl)methylarsine, triars:

$$CH_3As\begin{matrix} \diagup (CH_2)_3\text{-}As(CH_3)_2 \\ \diagdown (CH_2)_3\text{-}As(CH_3)_2 \end{matrix}$$

which are about equally capable of meridional (21-XIII, XIV) and facial (21-XV) coordination.

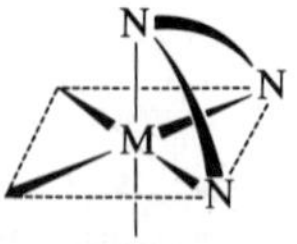

(21-XV)

Quadridentate Ligands. There are three main types:

Open-chain, unbranched:

Triethylenetetramine, trien:

$$H_2N(CH_2)_2NH(CH_2)_2NH(CH_2)_2NH_2$$

Schiff bases derived from acac; e.g., (21-XVI)

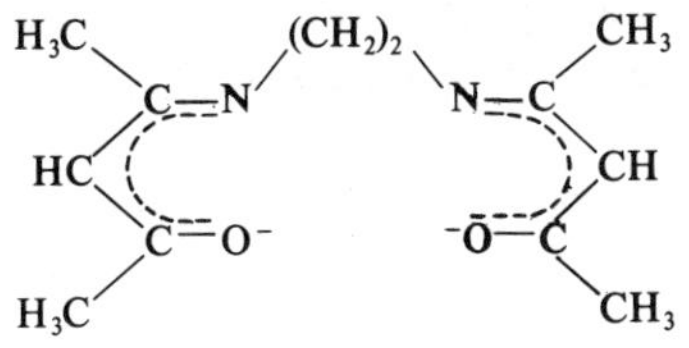

(21-XVI)

Macrocyclic. These are almost exclusively obligate planar, as in all the following examples, though some puckering does occur. Further discussion of some of these will be found in Section 21-4.

porphyrins, substitution products of (21-XVII)
phthalocyanine (21-XVIII)
Schiff bases; e.g., (21-XIX)

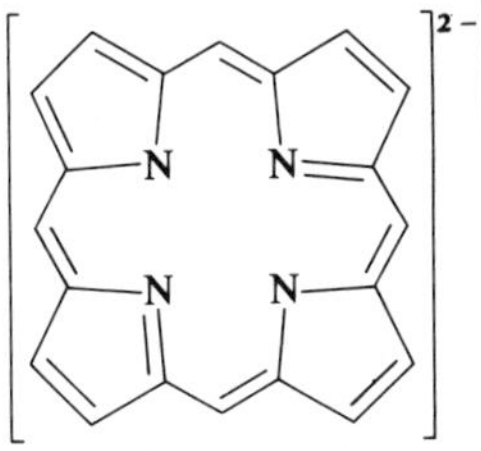

(21-XVII) 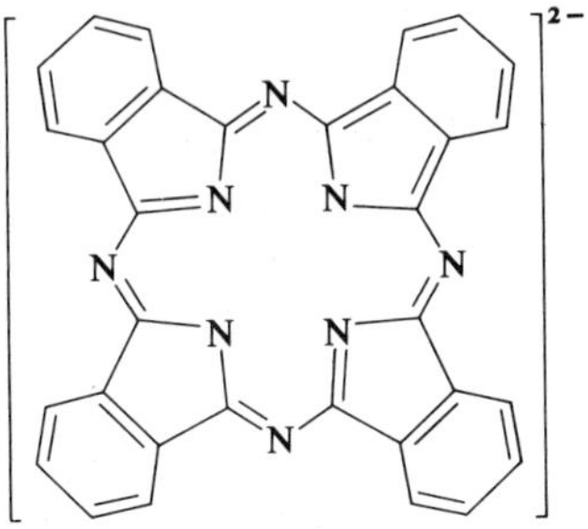   (21-XVIII) 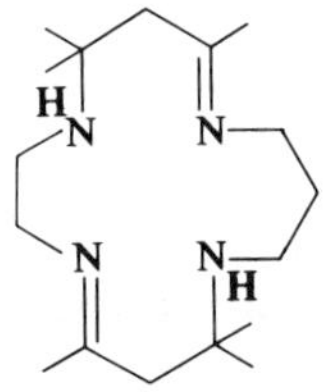   (21-XIX)

Tripod ligands. These are of the type X(——Y)$_3$, where X is N, P or As, the Y groups are R$_2$N, R$_2$P, R$_2$As, RS or RSe, and the connecting chains, ——, are (CH$_2$)$_2$, (CH$_2$)$_3$ or *o*-phenylene. Some common ones are

N(CH$_2$CH$_2$NH$_2$)$_3$	tren
N[CH$_2$CH$_2$N(CH$_3$)$_2$]$_3$	Me$_6$tren
N[CH$_2$CH$_2$P(C$_6$H$_5$)$_2$]$_3$	TPN
P[*o*-C$_6$H$_4$P(C$_6$H$_5$)$_2$]$_3$	QP
N(CH$_2$CH$_2$SCH$_3$)$_3$	TSN
As[*o*-C$_6$H$_4$As(C$_6$H$_5$)$_2$]$_3$	QAS

The tripod ligands are used particularly to favor formation of trigonal-bipyramidal complexes of divalent metal ions, as shown schematically in (21-XX), but they do not invariably give this result. For instance, while Ni(TPN)I$_2$ is trigonal-bipyramidal, Co(TPN)I$_2$ is square pyramidal (21-XXI).

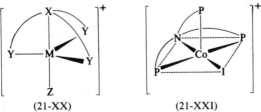

(21-XX)   (21-XXI)

Penta- and Hexa-dentate Ligands.   Perhaps the best known hexadentate
ligand is ethylenediaminetetraacetate, $EDTA^{4-}$, which can also be penta-
dentate as $EDTAH^{3-}$.

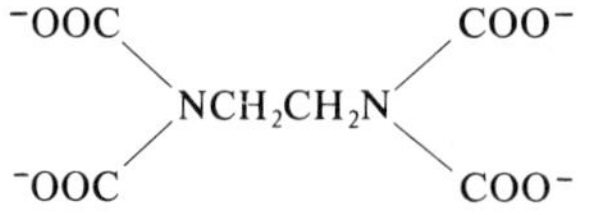

(EDTR) 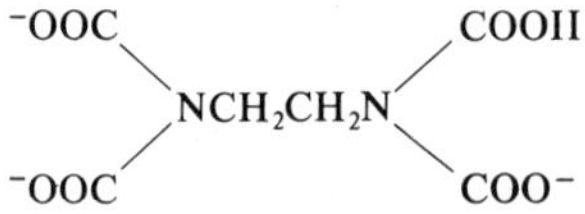   (EDTAH)

Among others is the annexed, (21-XXII), which is specifically designed to
result in trigonal-prismatic coordination.[2a]

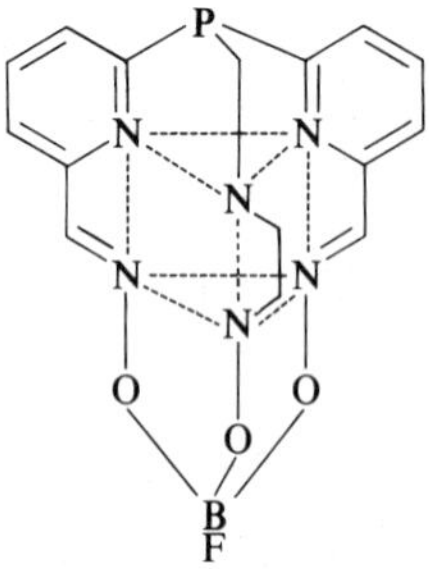

(21-XXII)

21-4. Macrocyclic Ligands

These are large rings so constituted as to bring three or more donor atoms
into contact with one metal ion. The donor atoms are most commonly nitrogen atoms, though oxygen or sulfur atoms, or a mixed set, also occur. Macrocycles giving a set of four essentially coplanar nitrogen atoms[3] have been
most extensively studied, in part because of their relevance, or assumed
relevance, to naturally occurring systems[3] such as the porphyrins and the
corrins (Section 25-F-1).

The phthalocyanines were one of the earliest classes of synthetic macrocyclic tetraamines to be discovered. They are prepared by the annexed type

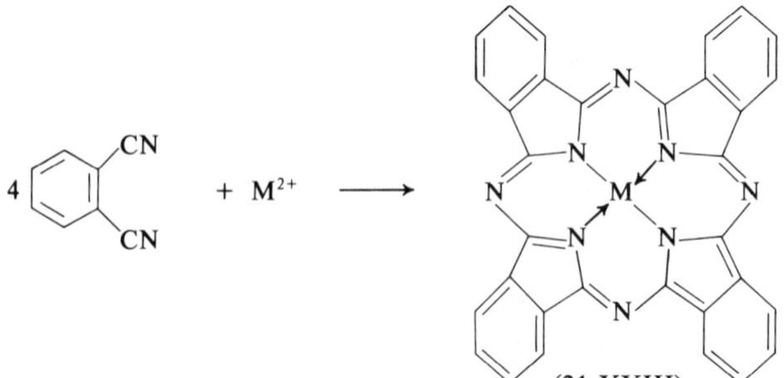

(21-1)

(21-XXIII)

[2a] J. E. Parks, B. E. Wagner and R. H. Holm, *Inorg. Chem.*, 1971, **10**, 2472.
[3] D. St. C. Black and E. Markham, *Rev. Pure Appl. Chem.*, 1965, **15**, 109.

of reaction (21-1), in which the metal ion plays an essential role as a template. These complexes such as (21-XXIII) characteristically have exceptional thermal stability, subliming in a vacuum at temperatures around 500 °C; they are an important commercial class of pigments. They are usually rather insoluble, but sulfonated derivatives are soluble in polar solvents. The conjugated π systems give a pronounced ring current, which can be exploited in studying nmr spectra.[4]

A great variety of nitrogen-containing macrocycles[5] can be made by employing the Schiff base condensation reaction (21-2), often (but not

$$\begin{array}{c} R \\ \diagdown \\ \diagup \\ R' \end{array} C{=}O \ + \ H_2NR'' \ \longrightarrow \ \begin{array}{c} R \\ \diagdown \\ \diagup \\ R' \end{array} C{=}N \diagdown_{R''} \ + H_2O \qquad (21\text{-}2)$$

necessarily) with a metal ion as a template, and with subsequent hydrogenation to obtain a saturated system not subject to hydrolytic degradation by reversal of reaction (21-2). Some representative preparative reactions are (21-3) to (21-6).

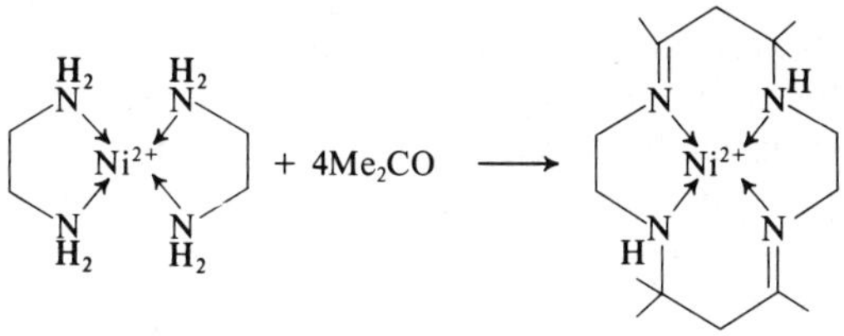

$$(21\text{-}3)$$

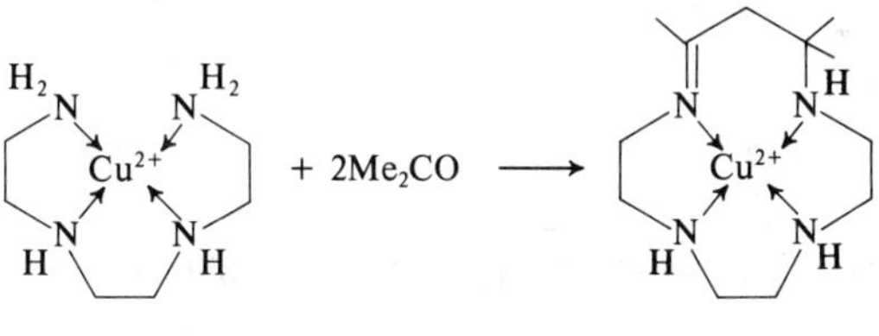

$$(21\text{-}4)$$

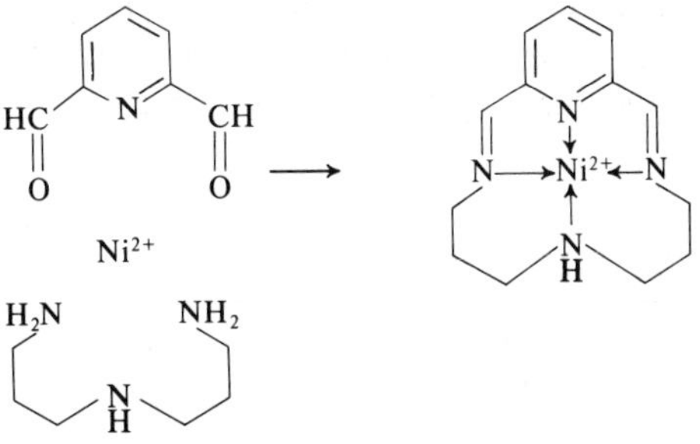

$$(21\text{-}5)$$

[4] J. N. Esposito, L. E. Sutton and M. E. Kenney, *Inorg. Chem.*, 1967, **6**, 1116.
[5] N. F. Curtis, *Coordination Chem. Rev.*, 1968, **3**, 3; D. H. Busch, *Helv. Chim. Acta, Fasc. Extraordinarius*, Alfred Werner Commemoration Volume, 1967.

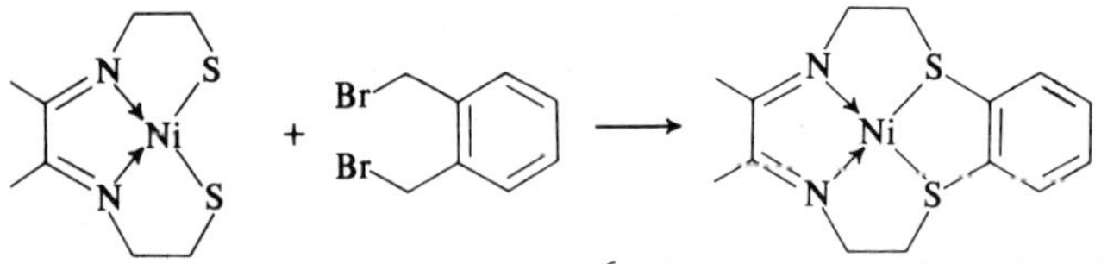

By conventional organic reactions[6] the macrocycles (21-XXIV) and (21-XXV) can be prepared and then used to form complexes.

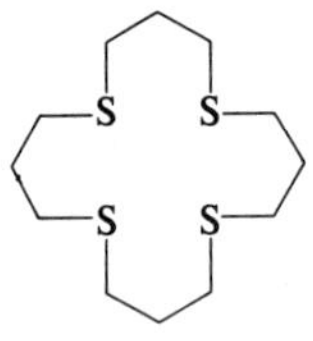

(21-XXIV) 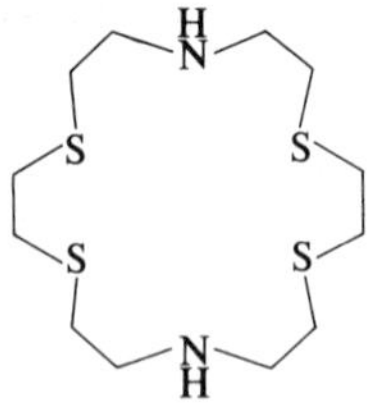   (21-XXV)

Although they have been developed primarily for use with alkali and alkaline-earth ions (see page 200), it is appropriate to mention here the polyether macrocycles. These are of two main types: (1) monocyclic ones,[7] such as (21-XXVI), and (2) the bicyclic molecules,[8] such as (21-XXVII). These compounds are remarkably powerful complexing agents for alkali and alkaline-earth ions, giving unusual coordination geometries.[9] There is as yet no detailed report of their complexes with other metal ions.

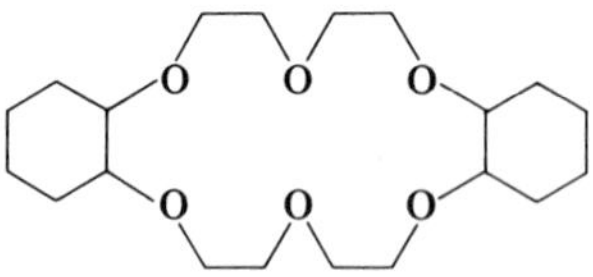

(21-XXVI) 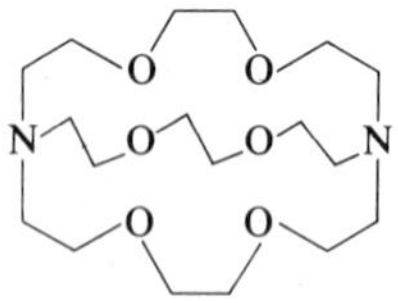   (21-XXVII)

21-5. Some Other Observations on Polydentate Ligands

β-Ketoenolate Complexes.[10] β-Diketones have the property of forming stable anions as a result of enolization followed by ionization, as illustrated in (21-7). These β-ketoenolate ions form very stable chelate complexes with

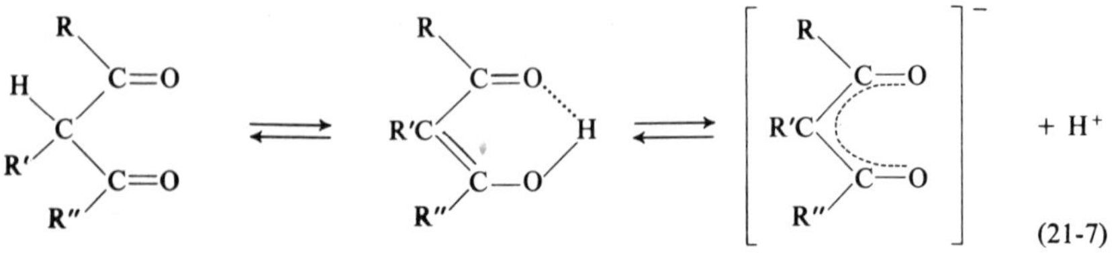

[6] W. Rosen and D. A. Busch, *Chem. Comm.*, **1969**, 148; D. St. C. Black and I. A. McLean, *Chem. Comm.*, **1968**, 1004.
[7] C. J. Pederson, *J. Amer. Chem. Soc.*, 1967, **89**, 7017; 1970, **92**, 386, 391.
[8] J. M. Lehn, J. P. Sauvage and B. Dietrich, *J. Amer. Chem. Soc.*, 1970, **92**, 2916.
[9] D. Bright and M. R. Truter, *Nature*, 1970, **225**, 176.
[10] J. P. Fackler, Jr., *Progr. Inorg. Chem.*, 1966, **7**, 361.

a great range of metal ions. The commonest such ligand is the acetylacetonate ion, acac⁻, in which $R = R'' = CH_3$ and $R' = H$. A general abbreviation for β-ketoenolate ions in general is diketo.

Among the commonest types of diketo complex are those with the stoichiometries $M(diketo)_3$ and $M(diketo)_2$. The former all have structures based on an octahedral disposition of the six oxygen atoms. The tris-chelate molecules then actually have D_3 symmetry and exist as enantiomers. In cases where there are unsymmetrical diketo ligands (i.e., those with $R \neq R''$) geometrical isomers also exist, as indicated in (21-XXVIII). Such compounds have been of value in investigations of the mechanism of racemization of tris-chelate complexes, which are discussed in detail in Section 21-17.

21-XXVIIIa	21-XXVIIIb
cis	*trans*

Substances of composition $M(diketo)_2$ are almost invariably oligomeric, unless the R groups are very bulky ones, such as $(CH_3)_3C—$. Thus, for example, the acetylacetonates of nickel(II), cobalt(II) and zinc(II) are trimeric, tetramic and trimeric, respectively (see appropriate Sections of Chapter 25 for structures and further discussion), while the complexes containing the hindered β-diketonate with $R = R'' = (CH_3)_3C—$ are monomeric. These facts show that in $M(diketo)_2$ molecules the M atoms are coordinately unsaturated;[11] they prefer a coordination number of 6 (or at least 5) and attain such coordination numbers (usually 6) by sharing of oxygen atoms. The presence of bulky R groups sterically impedes such oligomerization. In the presence of good donors such as H_2O, ROH or py, the metal atoms expand their coordination numbers from 4 to 5 or 6 by binding such donors. Thus complexes of the type *trans*-$M(diketo)_2L_{1,2}$ are formed, instead of oligomers.

Tropolonato Complexes.[12] A system similar to the β-diketonates is provided by tropolone and its anion (21-XXIX). The tropolonato ion gives many complexes that are broadly similar to analogous β-diketonate complexes, although there are often very significant differences.

$$\text{(21-XXIX)}$$

It should be noted that the tropolonato ion forms a five-membered chelate ring and that the "bite," that is, the O-to-O distance, is smaller than for β-diketonato ions. This leads in tristropolonato complexes to considerable

[11] D. P. Graddon, *Coordination Chem. Rev.*, 1969, **4**, 1.
[12] E. L. Muetterties and C. M. Wright, *Quart. Rev.*, 1967, **21**, 109.

distortion from an octahedral set of oxygen atoms. Thus, in $Fe(O_2C_7H_5)_3$[13] the O—Fe—O ring angles are only 78° and, the entire configuration is twisted about the three-fold axis towards a more prismatic structure. The upper set of Fe—O bonds is twisted only 40° instead of 60° relative to the lower set.

Polypyrazolylborate Ligands.[14] Anionic ligands of the types (21-XXX) and (21-XXXI) are relatively easy to prepare and have some interesting and

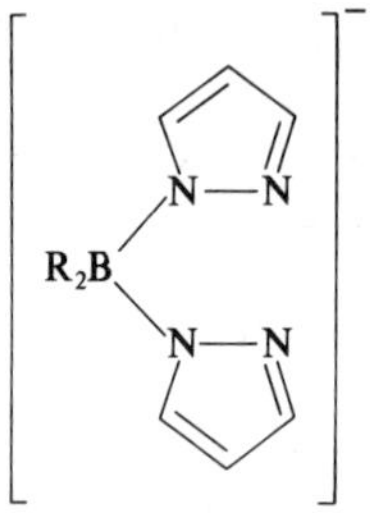

(21-XXX) 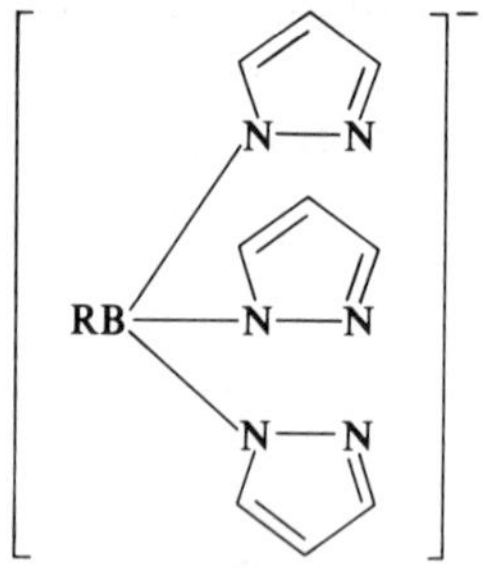   (21-XXXI)

unusual properties. The groups R may be H or alkyl groups and the pyrazole rings may also bear substituents. The dipyrazolyl anions, $R_2Bpz_2^-$, have a formal analogy to the β-diketonate ions and, like them, form complexes of the type $(R_2Bpyz)_2M$. However, because of the much greater steric requirements of the $R_2Bpz_2^-$ ligand such compounds are always strictly monomeric. Again, for steric reasons no tris complex, i.e. $(R_2Bpz_2)_3M$, is known.

The $RBpz_3^-$ ligands give a number of unusual complexes. These ligands themselves are unique in being the only known trigonally tridentate, uninegative ligands. They form trigonally distorted octahedral complexes, $(RBpz_3)_2M^{0,+}$, with di- and tri-valent metal ions, most of which are exceptionally stable. At least to a degree, an analogy can be made between $RBpz_3^-$ and the cyclopentadienyl anion, $C_5H_5^-$; both are 6-electron, uninegative ligands. There are some mono-$RBpz_3^-$ complexes that bear considerable resemblance to half-sandwich complexes, $C_5H_5ML_x$ (see Section 23-2).

$Mo(CO)_6$ reacts with $R_2Bpz_2^-$ and $RBpz_3^-$ to give the anions $[(R_2Bpz_2)-Mo(CO)_4]^-$ and $[(RBpz_3)Mo(CO)_3]^-$ which then form a number of interesting but not, as yet, fully characterized derivatives. One of the more uncommon reactions is (21-8). When R = H in these compounds they are stable towards

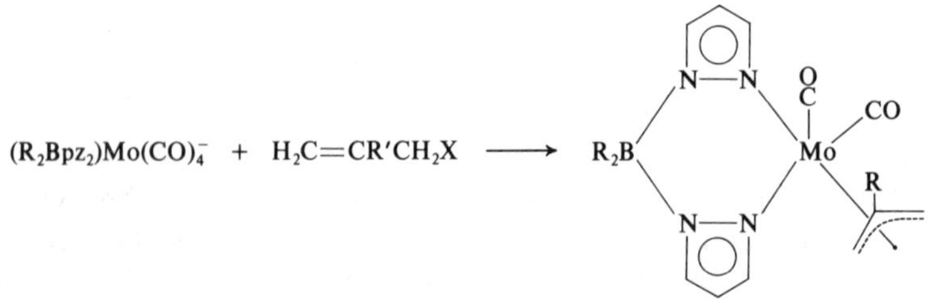

$$(R_2Bpz_2)Mo(CO)_4^- + H_2C{=}CR'CH_2X \longrightarrow \qquad\qquad (21\text{-}8)$$

[13] T. A. Hamor and D. J. Watkin, *Chem. Comm.*, **1969**, 440.
[14] S. Trofimenko, *Accounts Chem. Res.*, 1971, **4**, 17.

air and against attack by a variety of nucleophiles, which is remarkable since the Mo atom has a valence-shell electron configuration of only 16 electrons. The compounds $(HBpz_3)Mo(CO)_2$(allyl) are also very stable. It is not known whether the latter have tridentate $HBpz_3^-$ ligands and hence 18-electron configurations or bidentate $HBpz_3^-$ ligands and 16-electron configurations.

The pyrazolylborate ligands appear to have many still unexplored applications in the design of unusual and informative coordination compounds.

Conformations of Chelate Rings. For many purposes, simple diagrams of chelate rings in which the ring is schematically represented as though all atoms are coplanar are adequate. Of course, in some cases, such as β-diketonate complexes, this is actually true. However, it is well known that the rings are often puckered and, as a result, an understanding of the relative stabilities and certain spectroscopic properties of many chelate complexes can only be achieved by considering carefully the effects of the ring conformations.[15] This can be illustrated by considering the important case of 5-membered rings such as those formed by ethylenediamine.

Fig. 21-1 shows three ways of viewing the puckered rings, and identifies the absolute configurations in the λ, δ notation. As indicated clearly in the

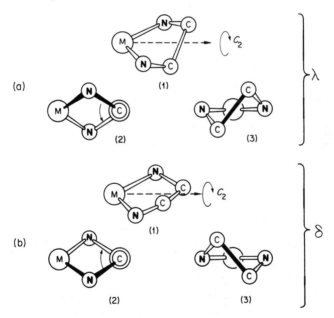

Fig. 21-1. Diagrams showing different ways of viewing the puckering of ethylenediamine chelate rings. The absolute configurations, λ and δ, are defined. [Reproduced by permission from C. J. Hawkins, ref. 15.]

[15] For a detailed discussion and literature references see C. J. Hawkins, *Absolute Configuration of Metal Complexes*, Wiley-Interscience, 1971, Chapter 3. See also J. K. Beattie, *Accounts Chem. Res.*, 1971, **4**, 253. Note that there are some differences in notation in these two articles.

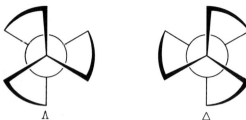

Fig. 21-2. Diagrams of tris-chelate octahedral complexes (actual symmetry: D_3), showing how the absolute configurations Λ and Δ are defined according to the translation (twist) of the helices.

Figure, the-chelate ring has as its only symmetry element a C_2 axis. It must therefore (see Section 1-6) be chiral, and the two forms of a given ring are enantiomorphs. When this source of enantiomorphism is combined with the two enantiomorphous ways, Λ and Δ, of orienting the chelate rings about the metal atom, Fig. 21-2, a number of diastereomeric molecules become possible, specifically, the following eight:

$$\Lambda(\delta\delta\delta) \qquad \Delta(\lambda\lambda\lambda)$$
$$\Lambda(\delta\delta\lambda) \qquad \Delta(\lambda\lambda\delta)$$
$$\Lambda(\delta\lambda\lambda) \qquad \Delta(\lambda\delta\delta)$$
$$\Lambda(\lambda\lambda\lambda) \qquad \Delta(\delta\delta\delta)$$

The two columns are here arranged so as to place an enantiomorphous pair on each line. In the following discussion we shall mention only members of the Λ series; analogous energy relationships must of course exist among corresponding members of the Δ series.

The relative stabilities of the four diastereomers has been extensively investigated. First, it can easily be shown that the diastereomers must, in principle, differ in stability because there are different non-bonded (repulsive) interactions between the rings in each case. Fig. 21-3 shows these differences for any two rings in the complex. When any reasonable potential function is used to estimate the magnitudes of the repulsive energies it is concluded that the order of decreasing stability is

$$\Lambda(\delta\delta\delta) > \Lambda(\delta\delta\lambda) > \Lambda(\delta\lambda\lambda) > \Lambda(\lambda\lambda\lambda)$$

However, this is not the actual order because enthalpy differences between diastereomers are rather small (2–3 kJ mol^{-1}), and an entropy factor must also be considered. Entropy favors the $\delta\delta\lambda$ and $\delta\lambda\lambda$ species because they are three times as probable as the $\delta\delta\delta$ and $\lambda\lambda\lambda$ ones. Hence, the best estimate of relative stabilities, which in fact agrees with all experimental data, becomes

$$\Lambda(\delta\delta\lambda) > \Lambda(\delta\delta\delta) \approx \Lambda(\delta\lambda\lambda) \gg \Lambda(\lambda\lambda\lambda).$$

In crystalline compounds, the $\Delta(\delta\delta\delta)$ isomer (or its enantiomorph) has been found most often, but the other three have also been found. These crystallographic results probably prove nothing about the intrinsic relative stabilities since hydrogen bonding and other intermolecular interactions can easily outweigh the small intrinsic energy differences.

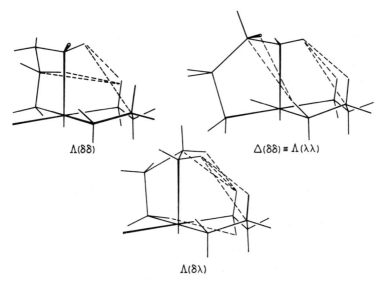

$$\Lambda(\delta\delta) \qquad \Delta(\delta\delta) \equiv \Lambda(\lambda\lambda)$$

$$\Lambda(\delta\lambda)$$

Fig. 21-3. Diagrams showing the different sets of repulsive interactions that exist between the three different pairs of ring conformations in octahedral ethylenediamine complexes. Broken lines represent the significant repulsive interactions. [Reproduced by permission from C. J. Hawkins, ref. 15.]

Nmr studies of solutions[16] of Ru^{II}, Pt^{IV}, Ni^{II}, Rh^{III}, Ir^{III} and Co^{III} [M en$_3$]$^{n+}$ complexes have yielded the most useful data and the general conclusions seem to be that the order of stability suggested above is correct and that ring inversions are very rapid. Both experiment and theory suggest that the barrier to ring inversion is only about 25 kJ mol^{-1}. Thus the four diastereomers of each overall form (Λ or Δ) are in labile equilibrium.

One of the interesting and important applications of the foregoing type of analysis is to the determination of absolute Λ or Δ configurations by using substituted ethylenediamine ligands of known absolute configuration. This is nicely illustrated by the [Co(l-pn)$_3$]$^{3+}$ isomers. The absolute configuration of l-pn [pn = 1,2-diaminopropane, $NH_2CH(CH_3)$—CH_2NH_2] is known. It would also be expected from consideration of repulsions between rings in the tris complex (as indicated in Fig. 21-3) that pn chelate rings would always take a conformation that puts the CH$_3$ group in an equatorial position. Hence, an l-pn ring can be confidently expected to have the δ conformation shown in Fig. 21-4. Note that, because of the extreme unfavor-

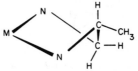

Fig. 21-4. The absolute configuration and expected conformation (i.e. with an equatorial CH$_3$ group) for an M(l-pn) chelate ring.

[16] See e.g., J. L. Sudmeier and G. L. Blackmer, *Inorg. Chem.* 1971, **10**, 2010.

ability of having axial CH_3 groups, only two tris complexes are expected to occur, viz., $\Lambda(\delta\delta\delta)$ and $\Delta(\delta\delta\delta)$. But, by the arguments already advanced for en rings, the Λ isomer should be the more stable of these two, by 5–10 kJ mol^{-1}. Thus, we predict that the most stable $[Co(l\text{-}pn)_3]^{3+}$ isomer must have the absolute configuration Λ about the metal.

In fact, the most stable $[Co(l\text{-}pn)_3]^{3+}$ isomer is the one with $+$ rotation at the sodium-D line and, as shown in Fig. 20-25, it has the same circular dichroism spectrum and hence the same absolute configuration as $(+)$-$[Co\ en_3]^{3+}$. The absolute configuration of the latter has been determined and it is indeed Λ. Thus the argument based on conformational analysis is validated.

Ambidentate Ligands.   These are ligands with two or more different donor sites only one of which is attached to a single metal atom at a given time.[17] Some important or common ligands of this type are:

$$\left[\begin{array}{c} N \diagdown^{\textstyle O}_{\textstyle O} \end{array}\right]^{-}$$

  M—NO_2, nitro
  M—ON=O, nitrito

CN^-:   Cyano (MCN) and isocyano (MNC)

NCS^-:   Thiocyanato (MSCN) and isothiocyanato (MNCS); also, analogous MOCN, MNCO and MSeCN, MNCSe possibilities exist

R_2SO: S-bonded or O-bonded

In a few cases ambidentate ligands give rise to *linkage isomers*, that is, complexes that differ only in the manner in which one or more ambidentate ligands are attached to the metal atom. The first example, discovered by S. M. Jørgensen in 1894, involved the nitro and nitrito isomers of $[Co(NH_3)_5NO_2]Cl_2$. The nitro isomer here, and apparently in all other cases, is the stable one. With NO_2^-, however, many further complexities arise because of the several other (bidentate and bridging) modes of attachment that can occur (see Section 21-8).

The NCS^- ion forms linkage isomers fairly often. The stable form of linkage of SCN^- to a given metal depends markedly on at least three factors: (1) the position of the metal in the Periodic Table and its oxidation state; (2) the nature of the other ligands (simple donors vs. π-acidic ones) also bound to the metal; and (3) the character of the solvent in which the complex may be dissolved. It happens frequently that the balance of these factors makes two isomeric configurations close enough in energy for both to be isolable.

Recent examples[18] of the ambidentate character of the thiocyanate

[17] J. L. Burmeister, *Coordination Chem. Rev.*, 1966, **1**, 205; A. H. Norburg and A. I. P. Sinha, *Quart. Rev.*, 1970, **24**, 69.

[18] G. R. Clark, G. J. Palenik and D. W. Meek, *J. Amer. Chem. Soc.*, 1970, **92**, 1077; U. A. Gregory *et al.*, *J. Chem. Soc.*, A, **1970**, 2770; R. L. Burmeister, R. L. Hassell and R. J. Phelan, *Inorg. Chem.*, 1971, **10**, 2030.

ion are depicted in (21-XXXII) and (21-XXXIIIa,b). The latter is a case of linkage isomerism that involves bridging SCN groups.

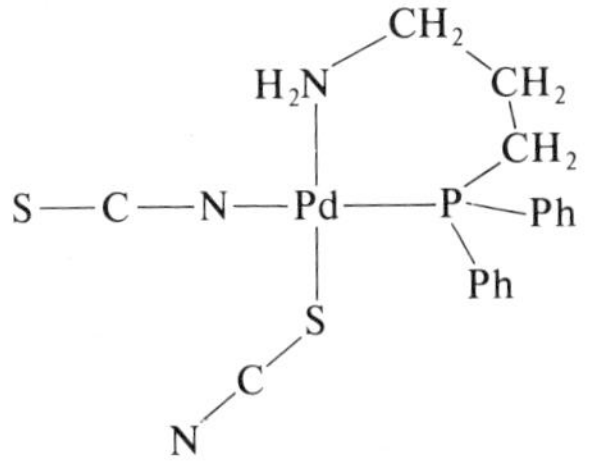

(21-XXXII)

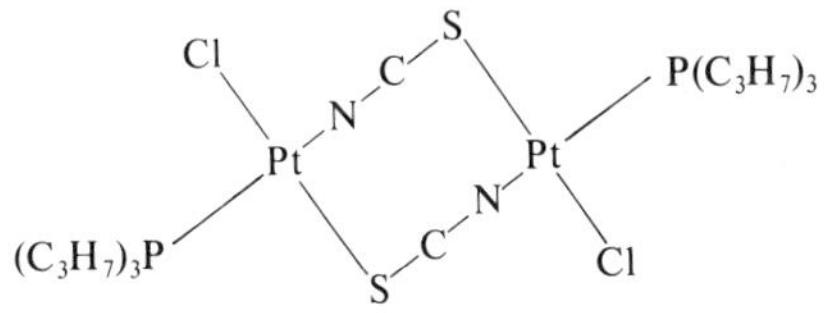

(21-XXXIIIa) 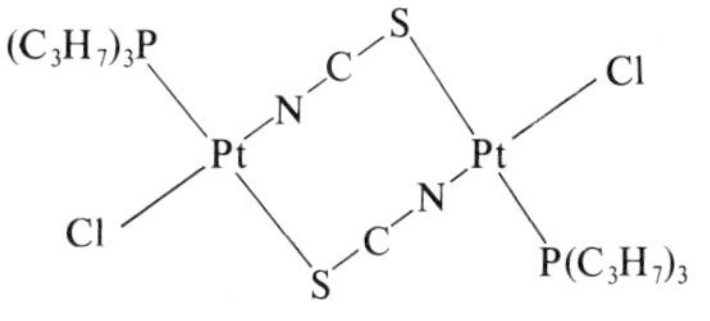   (21-XXXIIIb)

21-6. Dioxygen as a Ligand

Although the most common mode of reaction of molecular oxygen with transition-metal complexes is oxidation, that is, extraction of electrons from the metal (or, on occasion, from the ligand system), it has been recognized in recent years that in appropriate circumstances the oxygen molecule, which we shall call *dioxygen*, can become a ligand. The reaction of dioxygen with a complex so as to incorporate the dioxygen ligand intact is called *oxygenation*, as contrasted to oxidation, in which O_2 loses its identity.

Oxygenation reactions are commonly though not invariably reversible. That is, on increasing temperature and/or reducing the partial pressure of O_2, the dioxygen ligand is lost by dissociation or transferred to another acceptor (which may become oxidized). The process of reversible oxygenation plays an essential role in life processes. The best known examples involve the hemoglobin and myoglobin molecules of higher animals; they are discussed in Section 25-E-7. There are synthetic oxygen-carrying cobalt complexes that have recently been characterized, though it is not yet firmly established how the oxygen is bound to the cobalt. These are discussed in Section 25-F-7.

Our principal concern here is with a class of compounds that first came to light in 1963, with Vaska's discovery[19] of the reaction (21-9). As indicated,

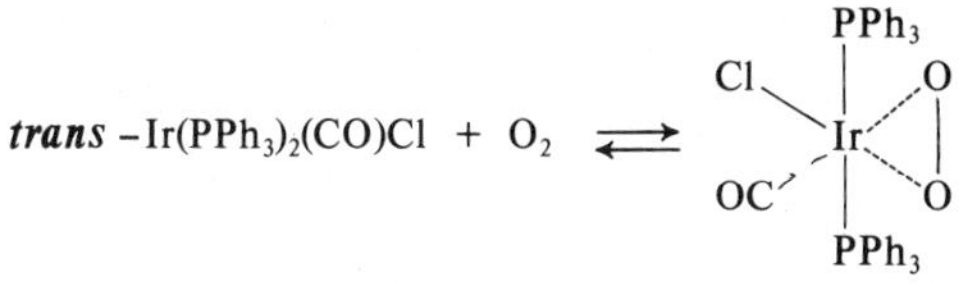

$$\textit{trans} - Ir(PPh_3)_2(CO)Cl + O_2 \rightleftharpoons \quad \text{(21-9)}$$

[19] L. Vaska, *Science*, 1963, **140**, 809.

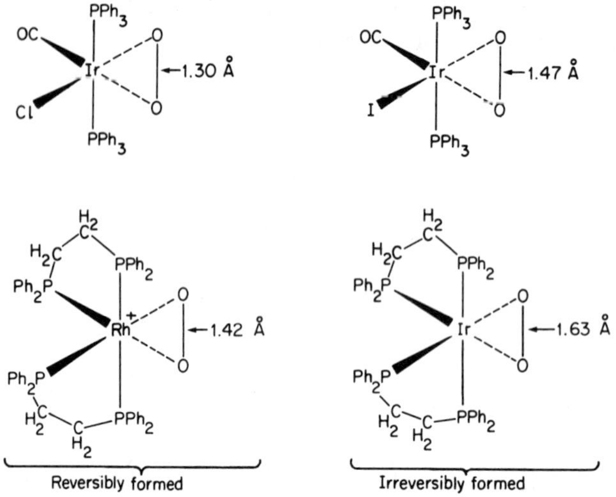

Fig. 21-5. Four representative dioxygen complexes, illustrating the correlation between O—O distances and the reversibility of their formation reactions.

this reaction is reversible. Since then, diamagnetic dioxygen complexes of Fe, Ru, Rh, Ir, Ni, Pd and Pt have been prepared.[20] In all those complexes that have been studied by X-ray diffraction the metal atom and the dioxygen ligand form an isosceles triangle (21-XXXIV). However, the O—O distances vary greatly, from 1.31 Å to 1.63 Å, as illustrated by structures shown

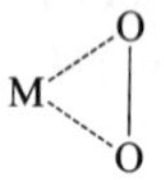

(21-XXXIV)

in Fig. 21-5. This variation seems to depend on the electron density at the metal atom, which in turn depends markedly on the other ligands present. In addition, there is a close correlation between the O—O bond length and the degree of reversibility of the reaction: the compounds with the longest O—O bonds are formed irreversibly.

The nature of the metal-to-dioxygen bonding is not well understood.[21] Undoubtedly both σ and π orbitals of the oxygen atoms play some role. In the most irreversibly formed complexes, i.e. those with the longest O—O bonds, the electronic structure can perhaps be described fairly accurately by postulating a set of three single bonds, two M—O and one O—O. However, this may be too simple a picture, since it has been claimed that electron-spectroscopic results imply the transfer of ca. 1.4 electrons to O_2 in

[20] See R. W. Horn, E. Weissberger and J. P. Collman, Inorg. Chem., 1970, 9, 2367.
[21] C. D. Cook et al., J. Amer. Chem. Soc., 1971, 93, 1904; A. Nakamura et al., J. Amer. Chem. Soc., 1971, 93, 6052; L. Vaska et al., J. Amer. Chem. Soc., 1971, 93, 6672.

$(Ph_3P)_2PtO_2$, which is an irreversibly formed compound. At least in a formal sense, the formation of a dioxygen complex can be viewed as an oxidative addition reaction (see Section 24-A-1).

Dioxygen complex formation has been shown to provide a path for catalytic oxidation of substrates such as SO_2 and organic unsaturated compounds. These reactions are discussed in Section 24-A-4.

12-7. Nitrogen Ligands

The nitrogen atom serves as the donor atom in a great variety of ligands some of which are

:N≡N: :N≐O: :NO₂⁻   :NH₃   :NR₃, etc.

:N≡C—S⁻   :N≡C—R,   R—N̈=N̈—R,   ⟨N⟩:

N₃⁻, etc.

It forms multiple bonds to metal atoms in groupings such as R—N=M N≡M and M⋯N⋯M.

In the conventional cases (e.g., with RCN, O_2N^-, SCN^-, R_3N) the N atom is mainly a donor of σ electrons and no further discussion is required. The cases of N_2 and NO are discussed in the next Chapter along with CO and related ligands. The N atom (or nitride ion) and a few other special cases will be discussed below.

The Nitride Ion. The N^{3-} ion resembles the oxide ion in being a strong π-donor as well as a σ-donor and, as such, it is found with metals in high oxidation states, where extensive charge transfer from ligand to metal is appropriate. It may occur as a terminal ligand (M≡N), as a linearly bridging one (M⋯N⋯M) and as a triple bridge (21-XXXV). Terminal nitride ligands[22] occur in compounds of Re^V, Os^{VI} and Os^{VIII}, e.g.,

$Cl_2(PhEt_2P)_3ReN$ $[X_5OsN]^{2-}$, (X = Cl, Br or CN)
$Cl_2(Ph_3P)_2ReN$ $[O_3OsN]^-$

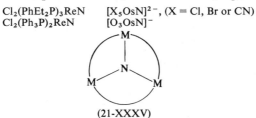

(21-XXXV)

Further discussion will be found under these metals in Chapter 26. Formally, the M—N bonds should be triple. They are very short (Re—N = 1.60 Å in $Cl_2(Ph_3P)_2ReN$; Os—N = 1.61 Å in $[Cl_5OsN]^{2-}$) and M—N stretching frequencies are rather high, viz. 950–1180 cm⁻¹.

[22] D. Bright and J. A. Ibers, *Inorg. Chem.*, 1969, **8**, 709.

Bridging nitride ligands have been found in several dinuclear ruthenium(IV) and osmium(IV) complexes,[23] e.g.,

$$[M_2NCl_8(H_2O)_2]^{3-}, \qquad [M_2N(NH_3)_8Cl_2]^{3+}$$
$$(M = Ru\ or\ Os) \qquad\qquad (M = Ru\ or\ Os)$$

The bridges appear to be linear and symmetrical in all cases; for $[Ru_2NCl_8$-$(H_2O)_2]^{3-}$ an X-ray study proves this and gives an Ru—N distance of 1.72 Å. The triply-bridging nitrogen ligand, as in (21-XXXV), is rather rare but does occur in species such as $[Ir_3N(SO_4)_6(H_2O)_3]^{4-}$.

Imino Complexes. Primary arylamines can condense with some M=O groups (much as they do with keto groups in the formation of so-called Schiff bases) to form M=NR groups, which may also be obtained in other ways. These formal double bonds are quite short,[22] viz. 1.69–1.71 Å in a series of $ReCl_3(PR_3')_2NR$ compounds with R = aryl or CH_3. No simple correlation exists between bond distances and bond orders for the formal double bonds in the imino complexes and the triple bonds in the nitrido ones, apparently because there are several overriding steric factors. For example, the formal triple Re≡N bond in the five-coordinate complex $ReNCl_2(PPh_3)_2$ is shorter (1.60 Å) than the formal double bonds in the several imino complexes; however, in the six-coordinate nitrido complex, $ReNCl_2(PEt_2Ph)_3$, the formal triple bond, Re≡N, is 1.79 Å. This is thought[22] to be due to extreme crowding about the Re atom in the last case.

The RN ligand is often referred to as a nitrene, just as RR'C ligands can be considered as carbenes (Section 23-8). In addition to the reaction of M=O with H_2NR, there are other routes to the formation of nitrene complexes,[24] e.g.,

$$Ir(Ph_2MeP)_2(CO)Cl + \tfrac{1}{2}CF_3N=NCF_3 \rightarrow Ir(Ph_2MeP)_2(CO)Cl(NCF_3)$$

The splitting of the N=N bond may be favored by the electronegative CF_3 group, since, as will be mentioned next, azo compounds often maintain their integrity as ligands.

Miscellaneous Nitrogen Donors. Azo compounds, R—N=N—R, which have both σ and π electrons, characteristically use their σ lone pairs,[25] as in the examples (21-XXXVI) and (21-XXXVII). Coordinated azo groups play an important role in the complexes formed by many azo dyes and in the formation of metal—carbon bonds to phenyl groups, as in the compound (21-XXXVIII). See also Sect. 24-A-4.

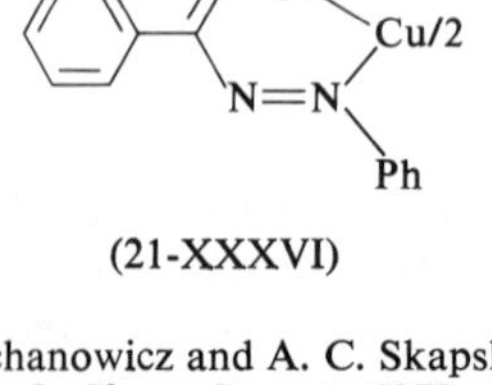

(21-XXXVI) 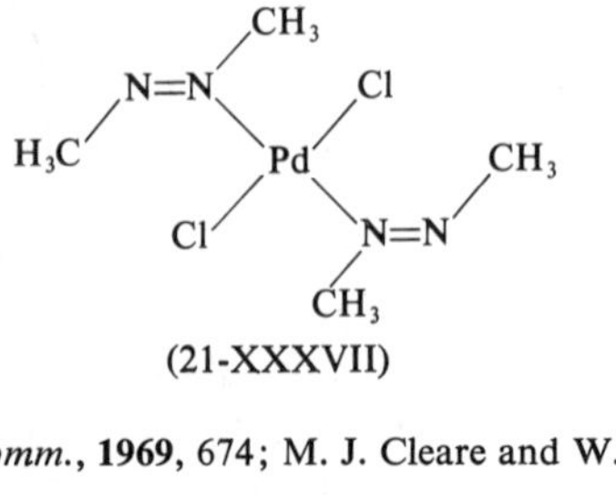   (21-XXXVII)

[23] M. Ciechanowicz and A. C. Skapski, *Chem. Comm.*, **1969**, 674; M. J. Cleare and W. P. Griffith, *J. Chem. Soc., A*, **1970**, 1117.

[24] J. Ashley Smith *et al.*, *Chem. Comm.*, **1969**, 409.

[25] A. L. Balch and D. Petridis, *Inorg. Chem.*, 1969, **8**, 2247.

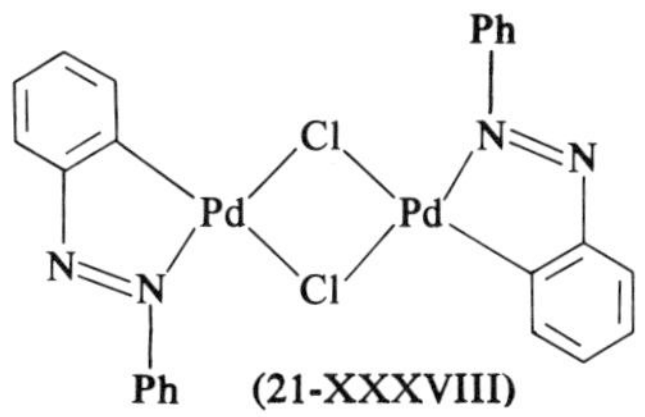

(21-XXXVIII) 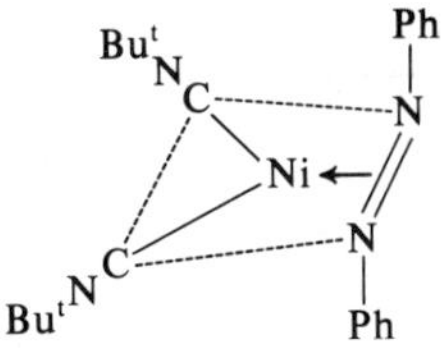   (21-XXXIX)

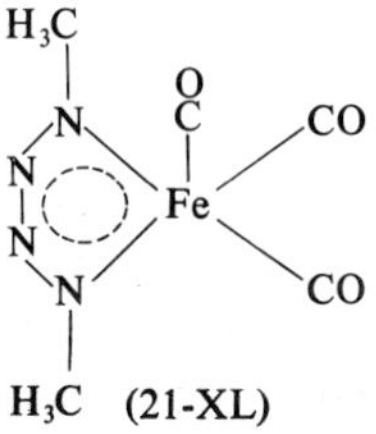

(21-XL)

Only recently has the first proven example (21-XXXIX), of an azo ligand coordinated through the π electrons been reported.[26]

An unusual example of π-donation as well as catenation of nitrogen is afforded by the compound (21-XL).[27] Since the $Fe(CO)_3$ group is normally an acceptor of four electrons, it may be assumed that two π electrons as well as two σ electrons are donated by the N_4 chain. The short (1.83 Å) Fe—N distances are indicative of considerable N—Fe multiple bonding.

21-8. Oxo Anions as Ligands; Infrared Criteria of Structure

Many of the most important ligands having oxygen donor atoms are oxo anions, such as NO_2^-, NO_3^-, SO_4^{2-} and ClO_4^-. The complexes are often structurally diverse because these ligands can all be bound in more than one way. We have already mentioned (Section 21-5) that NO_2^- is ambidentate; in addition, it can be a bridging ligand. While none of the remaining oxo anions listed is ambidentate, all of them can be either uni- or bi-dentate and all of them can serve as bridges. The structural role played by one of these ligands in any given complex can usually be clarified, at least in part, by the use of infrared spectra and thus, in discussion each of these ligands, the infrared structural criteria will be summarized. The usual infrared criteria pertain to the number and positions of the X—O stretching bands, although more elaborate vibrational analysis has in some cases been carried out.[28]

The Nitrite Ion. Aside from its function as a nitro ligand (i.e. as an N-donor) this ion is known to play at least four other structural roles, as shown in (21-XLI) to (21-XLIV). In one case three of these (21-XLI), (21-XLIII) and (21-XLIV) have been found in the same compound.[29] The nitrite ion itself

[26] R. S. Dickson et al., J. Amer. Chem. Soc., 1971, 93, 4636.
[27] R. J. Doedens, Chem. Comm., 1968, 1271.
[28] K. Nakamoto, Infrared Spectra of Inorganic and Coordination Compounds, 2nd edn. Wiley-Interscience, 1970.
[29] D. M. L. Goodgame, M. A. Hitchman and D. F. Marsham, J. Chem. Soc., A, 1971, 259.

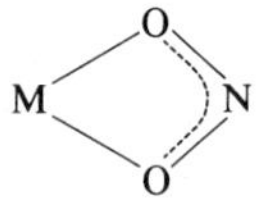

(21-XLI)

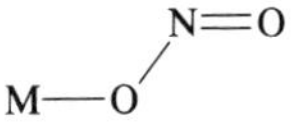

(21-XLII)

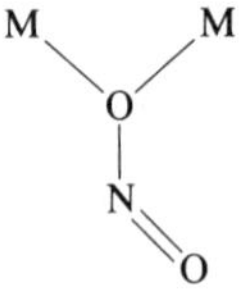

(21-XLIII)

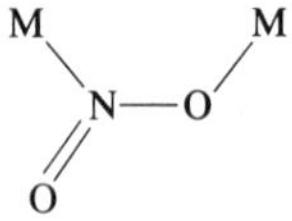

(21-XLIV)

has low symmetry (C_{2v}) and its three vibrational modes, symmetric N—O stretching, ν_s, antisymmetric N—O stretching, ν_{as}, and bending, δ, are all infrared-active to begin with. Thus, the number of bands cannot change upon coordination and the use of ir spectra to infer structure must depend upon interpretation of shifts in the frequencies. The δ vibration is rather insensitive to coordination geometry, but there are characteristic shifts of ν_s and ν_{as} which can often distinguish reliably between the nitro and the nitrito structure. Thus, for nitro complexes the two frequencies are similar, typical values being 1300–1340 cm^{-1} for ν_s and 1360–1430 cm^{-1} for ν_{as}. This is in keeping with the equivalence of the N—O bond orders in the nitro case. For nitrito bonding, the two N—O bonds have very different strengths (as indicated by (21-XLII) and the two N—O stretching frequencies are typically in the ranges 1400–1500 cm^{-1} for N=O and 1000–1100 cm^{-1} for N—O. Criteria for distinguishing among the bridge structures and between such structures as (21-XLII) on the one hand and (21-XLIII) and (21-XLIV) on the other are, as might be expected, more equivocal.[30] Both ir and X-ray results indicate that the bridging geometry (21-XLIV) is more common than (21-XLIII).

The Nitrate Ion. This ion is known to have four structural roles (21-XLV) to (21-XLVIII). A great many nitrato complexes have been studied[31] and the symmetrical bidentate structure (21-XLVI) seems definitely to be the preferred one, with the unidentate form (21-XLV) a less common alternative.

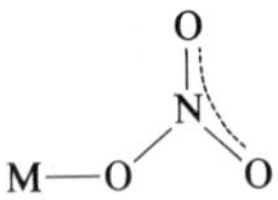

(21-XLV)

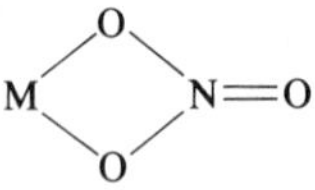

(21-XLVI)

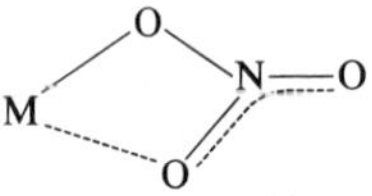

(21-XLVII)

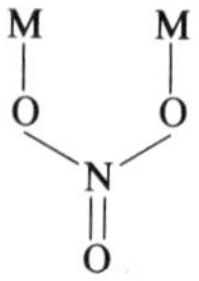

(21-XLVIII)

[30] Ref. 28, page 163.
[31] C. C. Addison et al., Quart. Rev., 1971, 25, 289.

The free nitrate ion has relatively high symmetry (D_{3h}) and thus its infrared spectrum is fairly simple. The totally symmetric N—O stretching mode is not ir-active, but the doubly degenerate N—O stretching mode gives rise to a strong band at ~ 1390 cm^{-1}. There are also two ir-active deformation modes, one of which is doubly degenerate, at 830 cm^{-1} and 720 cm^{-1}. When a nitrate ion becomes coordinated in any of the ways shown in (21-XLV), (21-XLVI) or (21-XLVIII), its effective symmetry is reduced to only C_{2v}. This causes the degeneracies to split and all modes (six) to be ir-active. Hence, it is easy to distinguish between ionic and coordinated nitrate groups.

Because the two commonest forms of coordinated nitrate ion have the same effective symmetry and hence the same number of ir-active vibrational modes, criteria for distinguishing between them must be based on the positions of the bands rather than their number. In practice, the situation is quite complex and there are no entirely straightforward criteria. This is because the array of frequencies depends not only on the geometry of coordination, but also on the strength of coordination. While it might be supposed that for (21-XLVI), the uncoordinated N=O group would give rise to a band at higher frequency than any band for (21-XLV), this is not necessarily the case if the bidentate group is less strongly polarized by the binding to metal than is a unidentate group.

All the modes of coordinated nitrate groups are also active in the Raman spectrum, and it appears that the relative *intensities* of the Raman bands due to N—O stretches, together with their depolarization ratios, may prove to be a generally reliable criterion.[31]

Carboxylate Ions. Discussion will be limited to the most important of these, the acetate ion; others behave similarly. In "ionic" acetates or in aqueous solution, the "free" $CH_3CO_2^-$ ion has symmetric and antisymmetric C—O stretching modes at ~ 1415 and ~ 1570 cm^{-1}. These frequencies can vary by ± 20 cm^{-1}. Since the symmetry of even the free ion is low and it gives two ir-active bands, evidence for the mode of coordination must be derived from the positions rather than the number of bands. When the carboxyl group is unidentate, one of the C—O bonds should have enhanced double bond character and give rise to a high-frequency band. Such bands are observed in the region 1590-1650 cm^{-1}, and are considered diagnostic of unidentate coordination.

Symmetrical bidentate coordination (as in $Zn(CH_3CO_2)_2 \cdot 2H_2O$ and $Na[UO_2(CH_3CO_2)_3]$), and symmetrical bridging (as in the $M_2(O_2CCH_3)_4L_2$ and $M_3O(O_2CCH_3)_6L_3$ types of molecules) leaves the C—O bonds still equivalent, and its effect on the frequencies is not easily predictable. In fact, no criteria for distinguishing these cases has been found. In general, multiple bands appear between 1400 and 1550 cm^{-1}, the multiplicity being attributable to coupling between CH_3CO_2 groups bonded to the same metal atom(s).

The Sulfate Ion. This ligand can be unidentate, bidentate-chelate or bidentate-bridging, and infrared spectra can usually distinguish quite well between these roles.

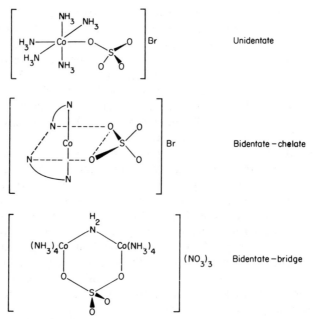

Fig. 21-6. Examples of the three structural roles of the sulfate ion as ligand.

The free sulfate ion has a regular tetrahedral structure, with T_d symmetry. When it functions as a unidentate ligand, the coordinated oxygen atom is no longer equivalent to the other three and the effective symmetry is lowered to C_{3v}. Since the M—O—S chain is normally bent, the actual symmetry is even lower, but this perturbation of C_{3v} symmetry does not measureably affect the infrared spectra. When two oxygen atoms become coordinated, either to the same metal ion or to different ones, the symmetry is lowered still further, to C_{2v}. Examples of each of these structural situations are provided by the compounds in Fig. 21-6.

The distinction between uncoordinated, unidentate and bidentate SO_4^{2-} by infrared spectra is very straightforward. Table 21-1 summarizes the selec-

TABLE 21-1

Correlation of the Types and Activities of S—O Stretching Modes of SO_4^{2-}

State of SO_4^{2-}	Effective symmetry	Types and activities of modes* (R = Raman; I = ir)			
Uncoord.	T_d	$\nu_1(A_1, R)$		$\nu_3(T_2, I, R)$	
Unidentate	C_{3v}	$\nu_1(A_1, I, R)$		$\nu_{3a}(A_1, I, R)$	$\nu_{3b}(E, I, R)$
Bidentate	C_{2v}	$\nu_1(A_1, I, R)$	$\nu_{3a}(A_1, I, R)$	$\nu_{3b}(B_1, I, R)$	$\nu_{3c}(B_2, I, R)$

* ν_2 and ν_4 are not listed because they are O—S—O bending modes. Note also that the arrows drawn have only rough qualitative significance since all modes of the same symmetry will be of mixed parentage in the higher symmetry.

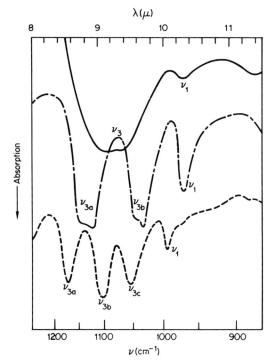

Fig. 21-7. Infrared spectra of $[Co(NH_3)_6]_2(SO_4)_3 \cdot 5H_2O$ (———); $[Co(NH_3)_5SO_4]$ Br (–·–·–); $[(NH_3)_4 Co(\mu\text{-}NH_2)(\mu\text{-}SO_4) Co(NH_3)_4](NO_3)_3$ (– – – –). [Reproduced by permission from K. Nakamoto, ref. 28.]

tion rules for the S—O stretching modes in the three cases. It can be seen that uncoordinated SO_4^{2-} should have one, unidentate SO_4^{2-} three and bidentate SO_4^{2-} four S—O stretching bands in the infrared. Note that the appearance of four bands for the bidentate ion is expected regardless of whether it is chelating or bridging.

These predictions are compared with experimental data in Fig. 21-7. It can be seen that the agreement is excellent, except that v_1 does appear weakly in the spectrum of the uncoordinated SO_4^{2-} ion. This is due to non-bonded interactions of SO_4^{2-} with its neighbors in the crystal, which perturb the T_d symmetry; the same environmental effects also cause the v_3 band to be very broad.

Even though bridging and chelating sulfate cannot be distinguished on the basis of the number of bands they give, it appears the former have bands at lower frequencies than the latter, typical values being

Bridging	Chelating
1160–1200	1210–1240
~1110	1090–1176
~1130	995–1075
960–1000	930–1000

Perchlorate Ion.   As noted above, the perchlorate ion has relatively little tendency to serve as a ligand and is often used where an anion unlikely to coordinate is required. However, there are cases, verified by X-ray crystallography, in which ClO_4^- is coordinated. Examples are: $(CH_3)_3SnClO_4$,[32] $Co(CH_3SCH_2CH_2SCH_3)_2(ClO_4)_2$[33] and $Co(Ph_2MeAsO)_4(ClO_4)_2$.[34] The perchlorate ion has T_d symmetry just as the sulfate ion does and therefore changes in its spectrum upon coordination are qualitatively the same as those just discussed for the sulfate ion.

The characteristic band of the ClO_4^- ion, namely, the triply degenerate Cl—O stretching mode, occurs at about 1110 cm^{-1} and is usually observed as a very broad band. A weak band is often found at about 980 cm^{-1} due to the infrared-forbidden totally symmetric stretching frequency, which, as in the case of the sulfate ion, can acquire a little intensity due to perturbation by an unsymmetrical environment. In compounds such as the last two mentioned above, which contain unidentate ClO_4^-, there are three bands, at about 1120 cm^{-1}, 1040 cm^{-1} and 920 cm^{-1}, in accord with expectation for C_{3v} symmetry.

In compounds containing bidentate bridging perchlorate groups, such as $(CH_3)_3SnClO_4$, which has infinite —Sn—ClO_4—Sn—ClO_4— chains, there are four Cl—O stretching bands, at ~ 1200 cm^{-1}, 1100 cm^{-1}, 1000 cm^{-1} and 900 cm^{-1}. No bidentate chelate perchlorate complex has yet been reported.

STABILITY OF COMPLEX IONS IN AQUEOUS SOLUTION

21-9. Aquo Ions

In a fundamental sense metal ions simply dissolved in water are already complexed—they have formed aquo ions. The process of forming what we more conventionally call complexes is really one of displacing one set of ligands, which happen to be water molecules, by another set. Thus the logical place to begin a discussion of the formation and stability of complex ions in aqueous solution is with the aquo ions themselves.

From thermodynamic cycles the enthalpies of plunging gaseous metal ions into water can be estimated and the results, $2 \cdot 10^2$–$4 \cdot 10^3$ kJ mol^{-1} (see Table 21-2), show that these interactions are very strong indeed. It is of importance in understanding the behavior of metal ions in aqueous solution to know how many water molecules each of these ions binds by direct metal—oxygen bonds. To put it another way, if we regard the ion as being an aquo complex, $[M(H_2O)_x]^{n+}$, which is then further and more loosely solvated, we wish to know the coordination number x and also the manner in which

[32] H. C. Clark and R. J. O'Brien, *Inorg. Chem.*, 1963, **2**, 740.
[33] F. A. Cotton and D. L. Weaver, *J. Amer. Chem. Soc.*, 1965, **87**, 4189.
[34] P. Pauling, G. B. Robertson and G. A. Rodley, *Nature*, 1965, **205**, 73.

TABLE 21-2

Enthalpies of Hydration* of Some Ions, kJ mol^{-1}

H^+	-1091	Ca^{2+}	-1577	Cd^{2+}	-1807
Li^+	-519	Sr^{2+}	-1443	Hg^{2+}	-1824
Na^+	-406	Ba^{2+}	-1305	Sn^{2+}	-1552
K^+	-322	Cr^{2+}	-1904	Pb^{2+}	-1481
Rb^+	-293	Mn^{2+}	-1841	Al^{3+}	-4665
Cs^+	-264	Fe^{2+}	-1946	Fe^{3+}	-4430
Ag^+	-473	Co^{2+}	-1996	F^-	-515
Tl^+	-326	Ni^{2+}	-2105	Cl^-	-381
Be^{2+}	-2494	Cu^{2+}	-2100	Br^-	-347
Mg^{2+}	-1921	Zn^{2+}	-2046	I^-	-305

* Absolute values are based on the assignment of -1091 ± 10 kJ mol^{-1} to H^+ (Cf. H. F. Halliwell and S. C. Nyburg, *Trans. Faraday Soc.*, 1963, **59**, 1126). Each value probably has an uncertainty of at least $10n$ kJ mol^{-1}, where n is the charge of the ion.

the x water molecules are arranged around the metal ion. Classical measurements of various types—for example, ion mobilities, apparent hydrated radii, entropies of hydration, etc.—fail to give such detailed information because they cannot make any explicit distinction between those water molecules directly bonded to the metal—the x water molecules in the inner coordination sphere—and additional molecules that are held less strongly by hydrogen bonds to the water molecules of the inner coordination sphere. There are, however, ways of answering the question in many instances, ways depending, for the most part, on modern physical and theoretical developments. A few illustrative examples will be considered here.

For the transition-metal ions, the spectral and, to a lesser degree, magnetic properties depend upon the constitution and symmetry of their surroundings. As a favorable but not essentially atypical example, the Co^{II} ion is known to form both octahedral and tetrahedral complexes. Thus, we might suppose that the aquo ion could be either $[Co(H_2O)_6]^{2+}$ with octahedral symmetry, or $[Co(H_2O)_4]^{2+}$ with tetrahedral symmetry. It is found that the spectrum and magnetism of Co^{II} in pink aqueous solutions of its salts with non-coordinating anions such as ClO_4^- or NO_3^- are very similar to the corresponding properties of octahedrally coordinated Co^{II} in general, and virtually identical with those of Co^{II} in such hydrated salts as $Co(ClO_4)_2 \cdot 6H_2O$ or $CoSO_4 \cdot 7H_2O$ where octahedral $[Co(H_2O)_6]^{2+}$ ions are known from X-ray studies definitely to exist. Complementing this, we have the fact that the spectral and magnetic properties of the many known tetrahedral Co^{II} complexes, such as $[CoCl_4]^{2-}$, $[CoBr_4]^{2-}$, $[Co(NCS)_4]^{2-}$ and $[py_2CoCl_2]$, which have intense green, blue or purple colors, are completely different from those of Co^{II} in aqueous solution. Thus, there can scarely be any doubt that aqueous solutions of otherwise uncomplexed Co^{II} contain predominantly[35] well-defined, octahedral $[Co(H_2O)_6]^{2+}$ ions, further hydrated, of course.

[35] However, there are also *small* quantities of tetrahedral $[Co(H_2O)_4]^{2+}$; see T. J. Swift, *Inorg. Chem.*, 1964, **3**, 526.

Evidence of similar character can be adduced for many of the other transition-metal ions. For all the di- and tri-positive ions of the first transition series, it is certain that the aquo ions are octahedral $[M(H_2O)_6]^{2(\text{or } 3)+}$ species, although in those of Cr^{II}, Mn^{III} and Cu^{II} there are definite distortions of the octahedra because of the Jahn–Teller effect (see Section 20-11). Information on aquo ions of the second and third transition series, of which there are only a few, however, is not so certain. It is probable that the coordination is octahedral in many, but higher coordination numbers may occur.

For ions that do not have partly filled d shells, evidence of the kind mentioned is lacking, since such ions do not have spectral or magnetic properties related in a straightforward way to the nature of their coordination spheres. We are therefore not sure about the state of aquation of many such ions, although nmr and other relaxation techniques have now supplied some such information. It should be noted that, even when the existence of a well-defined aquo ion is certain, there are vast differences in the average length of time which a water molecule spends in the coordination sphere, the so-called mean residence time. For Cr^{III} and Rh^{III} this time is so long that, when a solution of $[Cr(H_2O)_6]^{3+}$ in ordinary water is mixed with water enriched in ^{18}O, many hours are required for complete equilibration of the enriched solvent water with the coordinated water. From a measurement of how many molecules of H_2O in the Cr^{III} and Rh^{III} solutions fail immediately to exchange with the enriched water added, the coordination numbers of these ions by water were shown to be 6. These cases are exceptional, however. Most other aquo ions are far more labile, and a similar equilibration would occur too rapidly to permit the same type of measurement. This particular rate problem is only one of several which will be discussed more fully in Section 21-14.

Aquo ions are all more or less acidic; that is, they dissociate in a manner represented by the equation

$$[M(H_2O)_x]^{n+} = [M(H_2O)_{x-1}(OH)]^{(n-1)+} + H^+ \qquad K_A = \frac{[H^+][M(H_2O)_{x-1}(OH)]}{[M(H_2O)_x]}$$

The acidities vary widely, as the following K_A values show:

M in $[M(H_2O)_6]^{n+}$	K_A
Al^{III}	1.12×10^{-5}
Cr^{III}	1.26×10^{-4}
Fe^{III}	6.3×10^{-3}

Coordinated water molecules in other complexes also dissociate in the same way, for example,

$$[Co(NH_3)_5(H_2O)]^{3+} = [Co(NH_3)_5(OH)]^{2+} + H^+ \qquad K \approx 10^{-5.7}$$
$$[Pt(NH_3)_4(H_2O)_2]^{4+} = [Pt(NH_3)_4(H_2O)(OH)]^{3+} + H^+ \qquad K \approx 10^{-2}$$

21.10. The "Stepwise" Formation of Complexes

The thermodynamic stability of a species is a measure of the extent to which this species will form from, or be transformed into, other species under certain conditions *when the system has reached equilibrium.* The kinetic stability of a species refers to the speed with which transformations leading to the attainment of equilibrium will occur. In this and the next several Sections we consider problems of thermodynamic stability, that is, the nature of equilibria once they are established. Kinetic problems are discussed on page 652.

If, in a solution containing metal ions, M, and monodentate ligands, L, only soluble mononuclear complexes are formed, the system at equilibrium may be described by the following equations and equilibrium constants:

$$M + L = ML \qquad K_1 = \frac{[ML]}{[M][L]}$$

$$ML + L = ML_2 \qquad K_2 = \frac{[ML_2]}{[ML][L]}$$

$$ML_2 + L = ML_3 \quad . \quad K_3 = \frac{[ML_3]}{[ML_2][L]}$$

$$\begin{matrix} \cdot & \cdot & \cdot & \cdot \\ \cdot & \cdot & \cdot & \cdot \\ \cdot & \cdot & \cdot & \cdot \end{matrix}$$

$$ML_{N-1} + L = ML_N \qquad K_N = \frac{[ML_N]}{[ML_{N-1}][L]}$$

There will be N such equilibria, where N represents the maximum coordination number of the metal ion M for the ligand L. N may vary from one ligand to another. For instance, Al^{3+} forms $AlCl_4^-$ and AlF_6^{3-} and Co^{2+} forms $CoCl_4^{2-}$ and $Co(NH_3)_6^{2+}$, as the highest complexes with the ligands indicated.

Another way of expressing the equilibrium relations is the following:

$$M + L = ML \qquad \beta_1 = \frac{[ML]}{[M][L]}$$

$$M + 2L = ML_2 \qquad \beta_2 = \frac{[ML_2]}{[M][L]^2}$$

$$M + 3L = ML_3 \qquad \beta_3 = \frac{[ML_3]}{[M][L]^3}$$

$$\begin{matrix} \cdot & \cdot & \cdot & \cdot \\ \cdot & \cdot & \cdot & \cdot \\ \cdot & \cdot & \cdot & \cdot \end{matrix}$$

$$M + NL = ML_N \qquad \beta_N = \frac{[ML_N]}{[M][L]^N}$$

Since there can be only N independent equilibria in such a system, it is clear that the K_i's and the β_i's must be related. The relationship is indeed rather obvious. Consider, for example, the expression for β_3. Let us multiply both numerator and denominator by $[ML][ML_2]$ and then rearrange slightly:

$$\beta_3 = \frac{[ML_3]}{[M][L]^3} \cdot \frac{[ML][ML_2]}{[ML][ML_2]}$$

$$= \frac{[ML]}{[M][L]} \cdot \frac{[ML_2]}{[ML][L]} \cdot \frac{[ML_3]}{[ML_2][L]}$$

$$= K_1 K_2 K_3$$

It is not difficult to see that this kind of relationship is perfectly general, namely,

$$\beta_k = K_1 K_2 K_3 \dots K_k = \prod_{i=1}^{i=k} K_i$$

The K_i's are called the *stepwise formation constants* (or stepwise stability constants), and the β_i's are called the *over-all formation constants* (or over-all stability constants); each type has its special convenience in certain cases.

In all the above equilibria we have written the metal ion without specifying charge or degree of solvation. The former omission is obviously of no importance, for the equilibria may be expressed as above whatever the charges. Omission of the water molecules is a convention which is usually convenient and harmless. It must be remembered when necessary. See, for example, the discussion of the chelate effect, in the next Section.

With only a few exceptions, there is generally a slowly descending progression in the values of the K_i's in any particular system. This is illustrated by the data* for the Cd^{II}–NH_3 system where the ligands are uncharged and by the Cd^{II}–CN^- system where the ligands are charged.

$$Cd^{2+}+NH_3 = [Cd(NH_3)]^{2+} \qquad K = 10^{2.65}$$
$$[Cd(NH_3)]^{2+}+NH_3 = [Cd(NH_3)_2]^{2+} \qquad K = 10^{2.10}$$
$$[Cd(NH_3)_2]^{2+}+NH_3 = [Cd(NH_3)_3]^{2+} \qquad K = 10^{1.44}$$
$$[Cd(NH_3)_3]^{2+}+NH_3 = [Cd(NH_3)_4]^{2+} \qquad K = 10^{0.93} \; (\beta_4 = 10^{7.12})$$

$$Cd^{2+}+CN^- = [Cd(CN)]^+ \qquad K = 10^{5.48}$$
$$[Cd(CN)]^++CN^- = [Cd(CN)_2] \qquad K = 10^{5.12}$$
$$[Cd(CN)_2]+CN^- = [Cd(CN)_3]^- \qquad K = 10^{4.63}$$
$$[Cd(CN)_3]^-+CN^- = [Cd(CN)_4]^{2-} \qquad K = 10^{3.65} \; (\beta_4 = 10^{18.8})$$

Thus, typically, as ligand is added to the solution of metal ion, ML is first formed more rapidly than any other complex in the series. As addition of ligand is continued, the ML_2 concentration rises rapidly, while the ML concentration drops, then ML_3 becomes dominant, ML and ML_2 becoming unimportant, and so forth until the highest complex, ML_N, is formed to the nearly complete exclusion of all others at very high ligand concentrations. These relationships are conveniently displayed in diagrams such as those shown in Fig. 21-8.

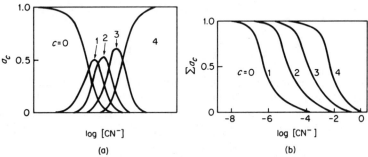

Fig. 21-8.   Plots of the proportions of the various complexes $[Cd(CN)_c]^{(2-c)+}$ as a function of the ligand concentration:

$$\alpha_c = [Cd(CN)_c]/\text{total Cd} \qquad \Sigma\alpha_c = \sum_{c=0}^{4} [Cd(CN_c)]$$

[Reproduced by permission from F. J. C. Rossetti in *Modern Coordination Chemistry*, J. Lewis and R. G. Wilkins, eds., Interscience, 1960, p. 10.]

* Cd–NH_3 constants determined in $2M$ NH_4NO_3; Cd–CN^-constants determined in $3M$ $NaClO_4$.

A steady decrease in K_i values with increasing i is to be expected provided there are only slight changes in the metal—ligand bond energies as a function of i, which is usually the case. For example, in the Ni^{2+}–NH_3 system to be discussed below, the enthalpies of the successive reactions $Ni(NH_3)_{i-1} + NH_3 = Ni(NH_3)_i$ are all within the range 16.7–18.0 kJ mol^{-1}.

There are several reasons for a steady decrease in K_i values as the number of ligands increases: (1) statistical factors; (2) increased steric hindrance as the number of ligands increases if they are bulkier than the H_2O molecules they replace; (3) coulombic factors, mainly in complexes with charged ligands. The statistical factors may be treated in the following way. Suppose, as is almost certainly the case for Ni^{2+}, that the coordination number remains the same throughout the series $[M(H_2O)_N] \cdots [M(H_2O)_{N-n}L_n] \cdots$ $[ML_N]$. The $[M(H_2O)_{N-n}L_n]$ species has n sites from which to lose a ligand, whereas the species $[M(H_2O)_{N-n+1}L_{n-1}]$ has $(N-n+1)$ sites at which to gain a ligand. Thus the relative probability of passing from $[M(H_2O)_{N-n+1}L_{n-1}]$ to $[M(H_2O)_{N-n}L_n]$ is proportional to $(N-n+1)/n$. Similarly, the relative probability of passing from $[M(H_2O)_{N-n}L_n]$ to $[M(H_2O)_{N-n-1}L_{n+1}]$ is proportional to $(N-n)/(n+1)$. Hence, on the basis of these statistical considerations alone, we expect

$$K_{n+1}/K_n = \frac{(N-n)}{n+1} \div \frac{N-n+1}{n} = \frac{n(N-n)}{(n+1)\ (N-n+1)}$$

In the Ni^{2+}–NH_3 system ($N = 6$), we find the comparison between experimental ratios of successive constants and those calculated from the above formula to be as shown in Table 21-3. The experimental ratios are consistently smaller than the statistically expected ones, which is typical and shows that other factors are also of importance.

TABLE 21-3

Comparison of Experimental and Statistical
Formation Constants of Ni^{2+}–NH_3 Complexes

	Experimental	Statistical
K_2/K_1	0.28	0.417
K_3/K_2	0.31	0.533
K_4/K_3	0.29	0.562
K_5/K_4	0.36	0.533
K_6/K_5	0.2	0.417

There are cases where the experimental ratios of the constants do not remain constant or change monotonically; instead, one of them is singularly large or small. There are several reasons for this: (1) an abrupt change in coordination number and hybridization at some stage of the sequence of complexes; (2) special steric effects which become operative only at a certain stage of coordination; and (3) an abrupt change in electronic structure of the metal ion at a certain stage of complexation. Each of these will now be illustrated.

Values of K_3/K_2 are anomalously low for the halogeno complexes of mercury(II); HgX_2 species are linear, whereas $[HgX_4]^{2-}$ species are tetrahedral. Presumably the change from sp to sp^3 hybridization occurs on going from HgX_2 to $[HgX_3]^-$. K_3/K_2 is anomalously small for the ethylenediamine complexes of Zn^{II}, and this is believed to be due to the change from sp^3 to sp^3d^2 hybridization if it is assumed that $[Znen_2]^{2+}$ is tetrahedral. For the Ag^+–NH_3 system $K_2 > K_1$, indicating that the linear, sp-hybridized structure is probably attained with $[Ag(NH_3)_2]^+$ but not with $[Ag(NH_3)(H_2O)_{3(or\ 5)}]^+$.

With 6,6′-dimethyl-2,2′-bipyridine (21-XLIX), many metal ions that form tris-2,2′-bipyridine complexes form only bis or mono complexes, or, in some cases, no isolable complexes at all, because of the steric hindrance between the methyl groups and other ligands attached to the ion.

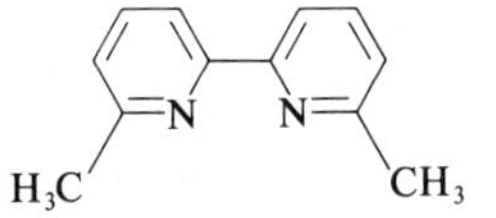

(21-XLIX)

In the series of complexes of Fe^{II} with 1,10-phenanthroline (and also with 2,2′-bipyridine), K_3 is greater than K_2. This is because the tris complex is diamagnetic (i.e. the ferrous ion has the low-spin state t_{2g}^6), whereas in the mono and bis complexes, as in the aquo ion, there are four unpaired electrons. This change from the $t_{2g}^4 e_g^2$ to the t_{2g}^6 causes the enthalpy change for addition of the third ligand to be anomalously large because the e_g electrons are antibonding.

21-11. The Chelate Effect

This term refers to the enhanced stability of a complex system containing chelate rings as compared to the stability of a system which is as similar as possible but contains none or fewer rings. As an example, consider the following equilibrium constants:

$$Ni^{2+}(aq) + 6NH_3(aq) = [Ni(NH_3)_6]^{2+}(aq) \qquad \log\beta = 8.61$$
$$Ni^{2+}(aq) + 3en(aq) = [Ni\ en_3]^{2+}(aq) \qquad \log\beta = 18.28$$

The system $[Ni\ en_3]^{2+}$ in which three chelate rings are formed is nearly 10^{10} times as stable as that in which no such ring is formed. While the effect is not always as pronounced as this, such a chelate effect is a very general one.

In order to understand this effect we must invoke the following well-known thermodynamic relationships:

$$\Delta G° = -RT \ln \beta$$
$$\Delta G° = \Delta H° - T\Delta S°$$

Thus β increases as $\Delta G°$ becomes more negative. A more negative $\Delta G°$ can result from making $\Delta H°$ more negative or from making $\Delta S°$ more positive.

As a very simple case, consider the reactions, and the pertinent thermodynamic data[36] for them, which are shown in Table 21-4. In this case the enthalpy difference is well within experimental error; the chelate effect can thus be traced entirely to the entropy difference.

TABLE 21-4

Two Reactions Illustrating a Purely Entropy-based Chelate Effect

$Cd^{2+}(aq) + 4CH_3NH_2(aq) = [Cd(NH_2CH_3)_4]^{2+}(aq)$ $\log \beta = 6.52$				
$Cd^{2+}(aq) + 2H_2NCH_2CH_2NH_2(aq) = [Cd\ en_2]^{2+}(aq)$ $\log \beta = 10.6$				
Ligands	ΔH°(kJ mol^{-1})	ΔS°(kJ mol^{-1} deg^{-1})	$-T\Delta S$(kJ mol^{-1})	ΔG°(kJ mol^{-1})
$4CH_3NH_2$	-57.3	-67.3	20.1	-37.2
$2\ en$	-56.5	$+14.1$	-4.2	-60.7

In the example first cited, the enthalpies make a slight favorable contribution, but the main source of the chelate effect is still to be found in the entropies. We may look at this case in terms of the following metathesis:

$$[Ni(NH_3)_6]^{2+}(aq) + 3\ en(aq) = [Ni\ en_3]^{2+}(aq) + 6NH_3(aq) \qquad \log \beta = 9.67$$

for which the enthalpy change is -12.1 kJ mol^{-1}, whereas $-T\Delta S^\circ = -55.1$ kJ mol^{-1}. The enthalpy change corresponds very closely to that expected from the increased LFSE of $[Ni\ en_3]^{2+}$ which is estimated from spectral data to be -11.5 kJ mol^{-1} and can presumably be so explained.

As a final example, which illustrates the existence of a chelate effect despite an unfavorable enthalpy term, we may use the reaction

$$[Ni\ en_2(H_2O)_2]^{2+}(aq) + tren(aq) = [Ni\ tren(H_2O)_2]^{2+}(aq) + 2\ en(aq) \qquad \log \beta = 1.88$$
$$[tren = N(CH_2CH_2NH_2)_3]$$

For this reaction we have $\Delta H^\circ = +13.0$, $-T\Delta S^\circ = -23.7$ and $-\Delta G^\circ = -10.7$ (all in kJ mol^{-1}). The positive enthalpy change can be attributed both to greater steric strain resulting from the presence of three fused chelate rings in Ni tren, and to the inherently weaker M—N bond when N is a tertiary rather than a primary nitrogen atom. Nevertheless, the greater number of chelate rings (3 vs. 2) leads to greater stability, owing to an entropy effect that is only partially cancelled by the unfavorable enthalpy change.

Probably the main cause of the large entropy increase in each of the three cases we have been considering is the net increase in the number of unbound molecules—ligands *per se* or water molecules. Thus, while six NH_3 displace six H_2O, making no net change in the number of independent molecules, it takes only three en molecules to displace six H_2O. Another more pictorial way to look at the problem is to visualize a chelate ligand with one end attached to the metal ion. The other end cannot then get very far away, and the probability of it, too, becoming attached to the metal atom is greater

[36] S. J. Ashcroft and C. T. Mortimer, *Thermochemistry of Transition Metal Complexes*, Academic Press, 1970 (a comprehensive review, from which all data used in this Section are taken).

than if this other end were instead another independent molecule which would then have access to a much larger volume of the solution.

The latter view provides an explanation for the decreasing magnitude of the chelate effect with increasing ring size, as illustrated by data such as those shown below for copper complexes of $H_2N(CH_2)_2NH_2$ and $H_2N(CH_2)_3NH_2(tn)$:

$$[Cu\ en_2]^{2+}(aq) + 2tn(aq) = [Cu\ tn_2]^{2+}(aq) + 2\ en(aq) \qquad \log \beta = -2.86$$

Of course, when the ring that must be formed becomes sufficiently large (seven-membered or more), it becomes more probable that the other end of the chelate molecule will contact another metal ion than that it will come around to the first one and complete the ring.

KINETICS AND MECHANISMS IN REACTIONS OF COMPLEX IONS

21-12. Introduction

There are many reactions of complexes in which the composition of the coordination sphere changes. Included in this category are those in which complexes are formed from the metal ions and the ligands, since the "uncomplexed" metal ions are actually aquo complexes. The ability of a particular complex ion to engage in reactions that result in replacing one or more ligands in its coordination sphere by others is called its *lability*. Those complexes for which reactions of this type are very rapid are called *labile*, whereas those for which such reactions proceed only slowly or not at all are called *inert*. It is important to emphasize that these two terms refer to rates of reactions and should not be confused with the terms stable and unstable which refer to the thermodynamic tendency of species to exist under equilibrium conditions. A simple example of this distinction is provided by the $[Co(NH_3)_6]^{3+}$ ion which will persist for days in an acid medium because of its kinetic inertness or lack of lability despite the fact that it is thermodynamically unstable, as the following equilibrium constant shows:

$$[Co(NH_3)_6]^{3+} + 6H_3O^+ = [Co(H_2O)_6]^{3+} + 6NH_4^+ \qquad K \sim 10^{25}$$

In contrast, the stability of $Ni(CN)_4^{2-}$ is extremely high,

$$[Ni(CN)_4]^{2-} = Ni^{2+} + 4CN^- \qquad K \sim 10^{-22}$$

but the rate of exchange of CN^- ions with isotopically labeled CN^- added to the solution is immeasurably fast by ordinary techniques. Of course this lack of any necessary relation between thermodynamic stability and kinetic lability is to be found generally in chemistry, but its appreciation here is especially important.

In the first transition series, virtually all octahedral complexes save those of Cr^{III} and Co^{III} are normally labile; that is, ordinary complexes come to equilibrium with additional ligands, including H_2O, so rapidly that the reactions appear instantaneous by ordinary techniques of kinetic measurement.

Complexes of Cr^{III} and Co^{III} ordinarily undergo ligand replacement reactions with half-times of the order of hours, days or even weeks at $25°$, thus making them convenient systems for detailed kinetic and mechanistic study. Some of the factors responsible for the great range of reaction rates will be discussed in the following pages.

In recent years increasing attention has been devoted to detailed kinetic studies of ligand replacement reactions with the objective of learning details of the mechanisms by which such reactions take place. Although a great deal remains to be done, some important advances have been made. The range of systems accessible to study depends on the experimental techniques available. At present the techniques can be classified into three broad categories; reaction rates for which they are generally used are indicated by the half times:

1. Static methods ($t_{\frac{1}{2}} \geqslant 1$ min)
2. Flow or rapid-mixing techniques (1 min $\geqslant t_{\frac{1}{2}} \geqslant 10^{-3}$ sec)
3. Relaxation methods ($t_{\frac{1}{2}} \leqslant 10^{-1}$ sec).

As a more explicit definition of the terms inert and labile, we may adopt an operational definition (due in a slightly different form to Taube) which says that complexes whose reactions may be studied by static methods are inert and the faster ones labile. The static methods are the classical ones in which reactants are mixed simply by pouring them both into one vessel and the progress of the reaction is then followed by observation of the time variation of some physical or chemical observable (e.g., light absorption, gas evolution, pH, isotopic exchange). Flow and rapid-mixing techniques differ mainly in achieving rapid mixing (in $\sim 10^{-3}$ sec) of the reactants, but use many of the same observational techniques as in static measurements. Relaxation methods are relatively new and have enormously increased the field accessible to study. They depend either (*a*) on creating a single disturbance in a state of equilibrium in a very short period of time (usually by a temperature or pressure jump) and following the process of relaxation to an equilibrium state by a combination of spectrophotometric and fast electronic recording devices, or (*b*) upon continuous disturbances by ultrasonic waves or radiofrequency signals in presence of a magnetic field (i.e., nmr). The latter methods are capable of following the very fastest reactions, and in many cases rate constants up to the limit ($\sim 10^{10}$ sec^{-1}) set by diffusion processes have been measured by ultrasonic methods.

The *direct* result of a kinetic study can at best be a *rate law*, that is, an equation showing how the velocity, v, of a reaction at a given temperature and in a given medium, varies as a function of the concentration of the reactants. Certain constants, k_i, called *rate constants*, will appear in the rate law. For example, a rate law for the reaction

$$A + B = C + D$$

might be

$$v = k_a[A] + k_{ab}[A][B] + k_{ab}[A][B][H^+]^{-1}$$

This would mean that the reaction occurs by three detectable paths, one influenced only by [A], a second influenced by [A] and [B] and a third that depends also on pH. The third term shows that not only A and B, but also [OH$^-$] (since this is related inversely to [H$^+$]), participate in the activated complex when this path is followed.

The ultimate purpose of a rate and mechanism study is usually to *interpret* the rate law correctly so as to determine the correct *mechanism* for the reaction. By mechanism, we mean a specification of what species actually combine to produce activated complexes, and what steps occur before and/or after the formation of the activated complex.*

21-13. Possible Mechanisms for Ligand-replacement Reactions

Two extreme mechanistic possibilities may be considered for such reactions. First, there is the S_N1 mechanism, in which the complex dissociates, losing the ligand to be replaced, the vacancy in the coordination shell then being taken by the new ligand. This path may be represented as follows:

$$[L_5MX]^{n+} \xrightarrow{\text{Slow}} X^- + \underset{\substack{\text{Five-coordinated}\\\text{intermediate}}}{[L_5M]^{(n+1)+}} \xrightarrow[\text{fast}]{+Y^-} [L_5MY]^{n+}$$

The important feature here is that the first step, in which X$^-$ is lost, proceeds *relatively* slowly and thus determines the rate at which the complete process can proceed. In other words, once it is formed, the intermediate complex, which is only five coordinated, will react with the new ligand, Y$^-$, almost instantly. The rate law for such a process is

$$v = k[L_5MX] \tag{21-10}$$

When this mechanism is operative, the rate of the reaction necessarily is directly proportional to the concentration of $[L_5MX]^{n+}$ but independent of the concentration of the new ligand, Y$^-$. The symbol S_N1 stands for *substitution, nucleophilic, unimolecular*. The other extreme pathway for a ligand exchange is the S_N2 mechanism. In this case the new ligand attacks the original complex directly to form a seven-coordinated activated complex which then ejects the displaced ligand, as indicated in the following scheme:

$$[L_5MX]^{n+} + Y^- \xrightarrow{\text{Slow}} \left\{ \begin{bmatrix} L_5M \begin{smallmatrix} X \\ \\ Y \end{smallmatrix} \end{bmatrix}^{(n-1)+} \right\} \xrightarrow{\text{Fast}} [L_5MY]^{n+} + X^- \tag{21-11}$$

When this mechanism is operative, the rate of the reaction will be proportional to the concentration of $[L_5MX^n]^+$ times that of Y$^-$, the rate law being

$$v = k[L_5MX][Y^-] \tag{21-12}$$

The symbol S_N2 stands for *substitution, nucleophilic, bimolecular*.

---

* We do not attempt here to explain or justify most of the reaction rate theory which must underlie any discussion of kinetics and mechanisms. The few brief definitions have been given only as reminders, or where the concepts are particularly important in following discussion. Readers lacking the necessary elementary understanding of chemical kinetics are referred to standard physical chemistry tests.

Unfortunately, these two extreme mechanisms are just that—extremes—and real mechanisms are seldom so simple. It is more realistic to recognize that it is likely that some degree of bond formation will occur before bond breaking is complete, that is, that the transition state may not be either the truly 5-coordinate species or the one in which the *leaving* and *entering* groups are both strongly bound at once. Subsequently, we shall use the terms S_N1 and S_N2 not to imply necessarily the extremes, but to describe mechanisms which may only approximate to these extremes.

To complicate matters further, a rate law of type (21-10) or (21-12) does not *prove* that the reaction proceeds by an S_N1 or S_N2 mechanism, even approximately. The three most important cases in illustration of this are (1) solvent intervention, (2) ion-pair formation and (3) conjugate-base formation.

(1) *Solvent intervention.* Most reactions of complexes have been studied in water, which is itself a ligand and which is present in high and effectively constant concentration ($\sim 55.5M$). Thus, the rate law (21-10) might be observed even if the actual course of the reaction were

$$[L_5MX] + H_2O \rightarrow [L_5MH_2O] + X \quad \text{slow} \tag{21-13a}$$
$$[L_5MH_2O] + Y \rightarrow [L_5MY] + H_2O \quad \text{fast} \tag{21-13b}$$

Moreover, we should not be able to tell from the rate law alone whether either (21-13a) or (21-13b) proceeded by S_N1 or S_N2 type processes.

(2) *Ion-pair formation.* When the reacting complex is a cation and the entering group is an anion, especially when one or both have high charges, ion pairs (or *outer sphere complexes*, as they are also called) will form to some extent,

$$[L_5MX]^{n+} + Y^{m-} = \{[L_5MX]Y\}^{n-m} \tag{21-14}$$

with an equilibrium constant K. These equilibrium constants can be estimated from theory or by comparison with measurements on systems where no subsequent reaction occurs, and they are generally in the range of 0.1–20. Now if the only path by which $[L_5MX]^{n+}$ and Y^{m-} can react with significant velocity involves preliminary formation of the ion pair, then the rate law might* be

$$v = k'K[L_5MX][Y] = k''[L_5MX][Y] \tag{12-15}$$

But the only *kinetic* observation will be a rate law of type (21-12) and only by additional experiments can we learn whether the reaction is truly an S_N2 type in the sense of equation (21-11) or whether the ion pair is involved. Even if it can be shown that the ion pair is involved (and this is frequently determinable) the question of how the ion pair transforms itself into the

---

* Rigorously, the velocity for such a process is given by the equation

$$v = \frac{k'K[L_5MX][Y]}{1 + K[Y]}$$

which reduces to equation (21-15) when $K[Y] \ll 1$. Thus, if [Y] can be made sufficiently large (say $\sim 1M$) and if K exceeds ~ 0.1, careful measurements can provide kinetic evidence against a simple bimolecular mechanism. Such experiments are not often possible in aqueous solution, but in some non-aqueous solvents, where K values can become large, they are generally feasible.

products remains unanswered, because S_N1, S_N2 or solvent participation processes might all occur.

(3) *Conjugate-base formation.* Whenever a rate law involving [OH⁻] is found, there is the question whether OH⁻ actually attacks the metal giving an S_N2 reaction in the sense of (21-11), or whether it appears in the rate law because it first reacts rapidly to remove a proton from a ligand, forming the conjugate base (CB), which then reacts, as in the following sequence:

$$[Co(NH_3)_5Cl]^{2+} + OH^- = [Co(NH_3)_4(NH_2)Cl]^+ + H_2O \qquad \text{fast} \qquad (21\text{-}16a)$$

$$[Co(NH_3)_4(NH_2)Cl]^+ \xrightarrow[\text{then}]{+Y} \xrightarrow{} [Co(NH_3)_5Y]^{2+} + Cl^- \qquad \text{slow} \qquad (21\text{-}16b)$$

In cases where (21-16b) proceeds by an S_N1 mechanism the overall mechanism represented by (21-16a) and (21-16b) is called an S_N1CB mechanism. Of course, in cases where there is no protonic hydrogen atom available (or if it is known that a process like (21-16a) is too slow) the appearance of [OH]⁻ in the rate law probably does indicate an authentic S_N2 process.

It may be noted here that rate laws are often written with terms $[H^+]^{-n}$, like the third term in the law for the hypothetical reaction $A + B = C + D$ above. It should be recognized that this is equivalent to using $[OH^-]^n$ with a different numerical value for the rate constant, since

$$k[H^+]^{-n} = k' K_w [OH^-]^n$$

21-14. Water-exchange and Formation of Complexes from Aquo Ions

Our knowledge of this subject depends largely upon the results of relaxation measurements, since nearly all the reactions are very fast. As an important special case, we shall first consider the rates at which aquo ions exchange water molecules with solvent water. Except for $Cr(H_2O)_6^{3+}$, with a half time $\sim 3.5 \times 10^5$ sec and an activation energy of 112 kJ mol⁻¹ and $Rh(H_2O)_6^{3+}$, which is still slower ($E_{act} \approx 137$ kJ mol⁻¹), these reactions are all fast as shown in Fig. 21-9. It will be seen that though they are all "fast" a range of some 10 orders of magnitude is spanned. These data therefore provide a good base for tackling the question of what factors influence the rates of reaction of similar complexes of different metal ions.

First, by considering the alkali and alkaline-earth ions, the influence of size and charge may be seen. Within each group the rate of exchange increases with size, and for M^+ and M^{2+} ions of similar size, the one of lower charge exchanges most rapidly. Since $M-OH_2$ bond strength should increase with charge and decrease with size of the metal ion, these correlations suggest that the transition state for the exchange reaction is attained by breaking an existing $M-OH_2$ bond to a much greater extent than a new one is formed; that is, that the mechanism is essentially dissociative.

Referring again to Figure 21-9 it will be seen that other series, i.e. (Al^{3+}, Ga^{3+}, In^{3+}), (Sc^{3+}, Y^{3+}), and (Zn^{2+}, Cd^{2+}, Hg^{2+}) also obey the radius rule. There are, however, several cases in which two ions of about the same size disobey the charge rule, i.e. the more highly charged ion exchanges faster.

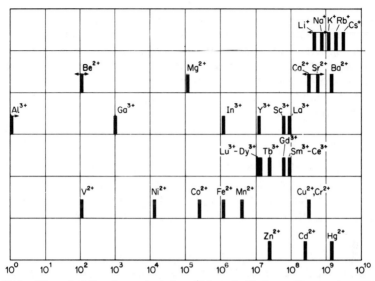

Fig. 21-9. Characteristic rate constants (sec^{-1}) for substitutions of inner sphere H_2O of various aquo ions. [Adapted from M. Eigen, *Pure Appl. Chem.*, 1963, **6**, 105, with revised data kindly provided by M. Eigen. See also H. D. Bennett and B. F. Caldin, *J. Chem. Soc. A*, **1971**, 2198.]

Such exceptions are thought to be due to differences in coordination numbers, but this is not quite certain.

It will also be noted that the divalent transition-metal ions do not follow the charge and radius rules very well. There are at least two additional factors involved here. With Cu^{2+}, the coordination polyhedron is not a regular octahedron, but, rather, two bonds are very much longer and weaker than the other four (see Section 25-H-4). By way of these, the rate of exchange is thus increased. Secondly, for most transition-metal ions the rates of ligand-exchange reactions are influenced by the changes in d electron energies as the coordination changes from that in the reactant to that in the transition state. This will always increase the activation energy and hence decrease the rate, but the magnitude of the effect is not monotonically related to the atomic number; rather, it varies irregularly from ion to ion as explained in Chapter 20.

Extensive studies of the rates at which an aquo ion combines with a ligand to form a complex have revealed the following remarkable general rules:

(1) The rates for a given ion show little or no dependence (less than a factor of 10) on the identity of the ligand.

(2) The rates for each ion are practically the same as the rate of water exchange for that ion, usually ~10 times slower.

It is believed that the only reasonable explanation for these observations is that the formation reactions proceed in two steps, the first being formation of the aquo ion–ligand outer sphere complex, followed by elimination of H_2O from the aquo ion in the same manner as in the water-exchange process.

When the observed rate constants are divided by estimated ion-pair formation constants, numbers very close to the water-exchange rate constants are obtained, generally to within the uncertainties in the estimated ion-pair formation constants. Activation energies and entropies, where available, are essentially the same for the two processes.

One slow complex formation reaction which has been studied in detail is that of equation 21-17:

$$Cr(H_2O)_6^{3+} + NCS^- = Cr(H_2O)_5NCS^{2+} + H_2O \qquad (21\text{-}17)$$

The rate law was found to be

$$v = [Cr(H_2O)_6^{3+}][NCS^-] (k_1 + k_2[H^+]^{-1} + k_3[H^+]^{-2}) \qquad (21\text{-}18)$$

From this it is concluded that there are three important paths, involving $Cr(H_2O)_6^{3+}$, $Cr(H_2O)_5(OH)^{2+}$ and $Cr(H_2O)_4(OH)_2^+$ as reactants. The dependence on $[NCS^-]$ does not by itself prove that the mechanism is S_N2, since the path

$$Cr(H_2O)_6^{3+} \underset{k_{-a}}{\overset{k_a}{\rightleftharpoons}} Cr(H_2O)_5^{3+} + H_2O$$

$$Cr(H_2O)_5^{3+} + NCS^- \overset{k_b}{\rightleftharpoons} Cr(H_2O)_5NCS^{2+}$$

will lead to the same rate law (using the appropriate hydroxo complexes for the second and third terms) provided that* $k_{-a} \gg k_b[NCS^-]$. In fact, this *is* true since the rate of water exchange is about 25 times faster than the rate of reaction (21-17). Thus the rate law does not, from a purely algebraic point of view, distinguish between dissociative and associative mechanisms. However, it is also found that the relative values of the rate constants in (21-18) is $k_3 > k_2 > k_1$. This order is consistent with an associative process, but inconsistent with a dissociative process.

The metal ion–water exchange (or elimination) process must also be the principal rate-determining feature in the early stages of hydrolytic polymerization of many metal ions. For instance, the reaction

$$2Fe(OH)^{2+} = Fe_2(OH)_2^{4+}$$

has a rate constant of $4.5 \times 10^2 M^{-1} \sec^{-1}$ at 25°, and considering the adverse effect of the like charges of the combining species, this is in reasonable accord with the water exchange rate for $Fe(OH)^{2+}$, namely $\sim 3 \times 10^4 \sec^{-1}$ at 25°.

21-15. Ligand-displacement Reactions in Octahedral Complexes

A general equation for a ligand-displacement reaction is

$$[L_nMX] + Y = [L_nMY] + X$$

In aqueous solution the special case in which Y is H_2O (or OH^-) is of overwhelming importance. It appears that there are few, if any, reactions in which X is not first replaced by H_2O, and only then does the other ligand, Y, enter the complex by displacing H_2O. Thus our discussion will be restricted almost entirely to the subject of the *aquation* or *hydrolysis* reaction.

---

* This well-known result for this type of reaction sequence is proved in textbooks on chemical kinetics.

The rates of hydrolyses of cobalt(III) ammine complexes are pH-dependent and generally follow the rate law

$$v = k_A[L_5CoX] + k_B[L_5CoX][OH^-] \qquad (21\text{-}19)$$

In general, k_B (for *base hydrolysis*) is some 10^4 times k_A (for *acid hydrolysis*). The interpretation of this rate law has occasioned an enormous amount of experimental study and discussion, but as yet there is nothing approaching a complete and generally accepted interpretation. Here, we can but touch on a few main aspects of the problem.

Acid Hydrolysis. We turn first to the term $k_A[L_5CoX]$. Since the entering ligand is H_2O, which is present in high ($\sim 55.5M$) and effectively constant concentration, the rate law tells us *nothing* as to the order in H_2O; the means for deciding whether this is an associative (S_N1) or a dissociative (S_N2) process must be sought elsewhere.

Among the most thoroughly studied systems are those involving $[Co(NH_3)_5X]$. There are various kinds of data, some of which favor an essentially dissociative mechanism, but the question can perhaps most safely be described as unresolved. To illustrate the work that has been done, the following points may be mentioned:

(1) The variation of rates with the identity of X correlates well with the variation in thermodynamic stability of the complexes. This indicates that breaking the Co—X bond is at least important in reaching the transition state.

(2) In a series of complexes where X is a carboxylate ion, there is not only the correlation of higher rates with lower basicity of the $RCOO^-$ group, but an *absence* of any slowing down due to increased size of R, after due allowance for the basicity effect. For an S_N2 mechanism, increased size of R should decrease the rate, at least if the attack were on the same side as X, although an attack on the *opposite* side is not excluded by these data.

(3) In the case where X is H_2O, that is, for the water-exchange reaction, the pressure-dependence of the rate has been measured and the volume of activation found to be $+1.2$ ml per mole. This result definitely excludes a predominantly S_N2 mechanism, but it does not agree satisfactorily with an extreme S_N1 mechanism either. It is most consistent with a transition state in which the initial Co—OH_2 bond is stretched quite far while formation of a new Co—$*OH_2$ bond is only beginning to occur, that is, a predominantly dissociative mechanism.

(4) For an extreme S_N1 mechanism, the five-coordinate intermediate $Co(NH_3)_5^{3+}$ would be generated and the behavior of this would be independent of its source. In studies where the ions Hg^{2+}, Ag^+ and Tl^{3+} were used to assist in removal of Cl^-, Br^- and I^- because of their high affinity for these halide ions, the ratio $H_2^{18}O/H_2^{16}O$ in the product was studied. For a genuine $Co(NH_3)_5^{3+}$ intermediate this ratio should be > 1 and constant regardless of the identity of X. When the assisting cation was Hg^{2+} the ratio 1.012 was observed for all three $[Co(NH_3)_5X]^{3+}$ ions, indicating the existence of $Co(NH_3)_5^{3+}$ as an intermediate. However, with Ag^+ the ratio varied (1.009,

1.007, 1.010) indicating that $Co(NH_3)_5^{3+}$ does not have a completely independent existence in this case. For Tl^{3+} the ratios were 0.996, 0.993 and 1.003, showing considerable deviation from a pure dissociative mechanism. If one assumes that with no assisting cation present bond breaking would proceed less far in the transition state, it could then be argued that these experiments favor an S_N2 mechanism, but there is also the unresolved question as to whether the entering water molecule in the assisted aquations comes from the bulk of the solvent or from the coordination sphere of the assisting metal ion. Thus, like many another mechanistic study, this one is tricky to interpret.

(5) A means of generating $Co(NH_3)_5^{3+}$ has been found by using the reaction (21-20) where azide is the sixth ligand:

$$[Co(NH_3)_5N_3]^{2+} + HNO_2 = [Co(NH_3)_5N_3NO]^{3+} = Co(NH_3)_5^{3+} + N_2 + N_2O \quad (21\text{-}20)$$

The relative rates of reaction of this with various anions, e.g., Cl^-, Br^-, SCN^-, F^-, HSO_4^-, $H_2PO_4^-$ and with H_2O were studied. The agreement between these results and those in the reaction (21-21)

$$Co(NH_3)_5(H_2O)^{3+} + X^- = Co(NH_3)_5X^{2+} + H_2O \quad (21\text{-}21)$$

was close, thus indicating that (21-21) also involves the intermediate $Co(NH_3)_5^{3+}$ or something of similar reactivity. By the principal of microscopic reversibility, this intermediate must also participate in the reverse of (21-21), that is, in the hydrolysis reaction itself. However, there are other experiments which are considered to show that the usual aquation reactions (e.g., that of $[Co(NH_3)_5NO_3]^{2+}$) *cannot* proceed through the same intermediate as that generated by oxidation of $[Co(NH_3)_5N_3]^{2+}$.

For the reaction (21-22), where L–L represents a bidentate amine, it has

$$[Co(L\text{--}L)_2Cl_2]^+ + H_2O = [Co(L\text{--}L)_2Cl(H_2O)]^{2+} + Cl^- \quad (21\text{-}22)$$

been found that the rate is increased by increasing bulk of the ligands, a result not in agreement with an S_N2 mechanism but consistent with an S_N1 mechanism.

Another class of hydrolyses which have been extensively studied are those of *trans*-Co^{III} en_2AX species, in which the leaving group is X^-. The variation in rates and stereochemistry (*cis* or *trans*) of products as functions of the nature of A have been examined, and certain informative correlations established. When A is NH_3 or NO_2^- the data indicate that the mechanism is S_N2, whereas for $A = OH^-$, Cl^-, N_3^-, and NCS^- an S_N1 mechanism is postulated. The assignments of mechanism in these cases depend heavily upon detailed consideration of the stereochemical possibilities for the intermediates or activated complexes and are thus of an indirect though apparently reliable nature.

Although, as indicated in the preceding discussion, the majority of substitution reactions in octahedral complexes appear to proceed by an essentially dissociative pathway, there may be exceptions.

As already noted the volume of activation, ΔV^*, for the reaction

$$[Co(NH_3)_5(H_2O^*)]^{3+}(aq) + H_2O(l) \rightarrow [Co(NH_3)_5(H_2O)]^{3+}(aq) + H_2O^*(l)$$

was found to be $+1.2 \text{ cm}^3 \text{ mol}^{-1}$. There is independent evidence that this particular process occurs dissociatively, and a positive ΔV^* is consistent with this. However, for the water exchange

$$[Cr(H_2O)_6]^{3+}(aq) + H_2O^*(l) \rightarrow [Cr(H_2O)_5(H_2O^*)]^{3+}(aq) + H_2O(l)$$

ΔV^* is $-9.3 \text{ cm}^3 \text{ mol}^{-1}$. This seems quite incompatible with a dissociative mode of reaction, but appears to be consistent with a rate-limiting step that can be represented[37] as

$$[\{Cr(H_2O)_6\}(H_2O)_x]^{3+}(aq) \rightarrow [\{Cr(H_2O)_7\}(H_2O)_{x-1}]^{3+}(aq)$$

The idea here is that one of the x water molecules in the secondary coordination sheath, immediately surrounding the six directly coordinated water molecules, slips into the primary shell, momentarily increasing the primary coordination number to 7. The x secondary water molecules are, of course, in rapid exchange with bulk solvent.

Base Hydrolysis.[38] The interpretation of a term of the type $k_B[ML_5X][OH^-]$ in a rate law for base hydrolysis has long been disputed. It could, of course, be interpreted as representing a genuine S_N2 process, OH^- making a nucleophilic attack on Co^{III}. However, the possibility of an S_N1CB mechanism, discussed above, must also be considered. There are arguments on both sides and it is of course possible that the mechanism may vary for different complexes. Studies of base hydrolysis in octahedral complexes have so far dealt mainly with those of Co^{III} and it is now reasonably sure that for these the predominant mechanism is, indeed, S_N1CB, with an intermediate which is probably of *tbp* geometry. The following discussion will largely center around Co^{III} complexes, but it should be pointed out that when the metal is changed from Co^{III} the reactivity patterns and, presumably, the mechanistic aspects are considerably changed. Detailed investigations of these other systems are still to be made.

It may be noted first, that base hydrolysis of Co^{III} complexes is generally very much faster than acid hydrolysis, i.e. $k_B \gg k_A$ in equation (21-19). This, in itself, provides evidence against a simple S_N2 mechanism, and therefore in favor of the S_N1CB mechanism because there is no reason to expect OH^- to be uniquely capable of electrophilic attack on the metal. In the reactions of square complexes (see below) it turns out to be a distinctly inferior nucleophile toward Pt^{II}.

The validity of an S_N1CB mechanism can be examined in terms of three aspects of the overall process: (1) the acid–base behavior of the reacting complex. (2) The structure of the 5-coordinate intermediate. (3) The ability of the amido or hydroxo group, which results from deprotonation of an amino or H_2O ligand, to stabilize such an intermediate.

The S_N1CB mechanism, of course, requires that the reacting complex have at least one protonic hydrogen atom on a non-leaving ligand, and that the rate of reaction of this hydrogen be fast compared to the rate of ligand

[37] D. R. Stranks and T. W. Swaddle, *J. Amer. Chem. Soc.*, 1971, **93**, 2783.
[38] M. L. Tobe, *Accounts Chem. Res.*, 1970, **3**, 377.

displacement. It has been found that the rates of proton exchange in many complexes subject to rapid base hydrolysis are in fact some 10^5 times faster than the hydrolysis itself [e.g., in $Co(NH_3)_5Cl^{2+}$ and $Co\ en_2NH_3Cl^{2+}$]. Such observations are in keeping with the S_N1CB mechanism but afford no positive proof of it. In the case of the complex (21-L) the rates of proton

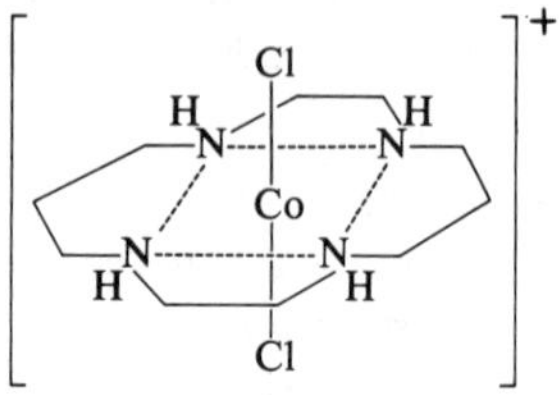

(21-L)

exchange and base hydrolysis were found to be similar. It was further found that there was more exchange in the product than in the reactant and that this additional exchange did not occur subsequent to the act of base hydrolysis. Therefore, it was concluded that the proton exchange formed an integral part of the base hydrolysis reaction.

There have been a number of experiments supporting the idea that the conjugate base does react dissociatively to produce a reactive, 5-coordinate intermediate with a lifetime sufficient for its reactivity to be independent of its origin, and for its characteristic pattern of discrimination among available entering ligands to be manifested. It is a common feature of base hydrolysis of the simpler Co^{III} amine systems, e.g., $Co\ en_2AX$, that there is considerable stereochemical change. For example,

$$\Lambda\text{-}cis\text{-}[Co\ en_2Cl_2]^+ + OH^- \rightarrow [Co\ en_2Cl(OH)]^+ + Cl^-$$

63% *trans*
21% Λ-*cis*
16% Δ-*cis*

This sort of result is best accommodated by postulating a *tbp* intermediate, upon which the attack of H_2O can occur in several ways, each leading to one of these isomers. One possible intermediate, and the lines of attack on it, are shown in Fig. 21-10. If loss of X were to leave a square pyramidal intermediate, it seems more likely that there would be complete retention of stereochemistry. In the case of (21-L) where the rigidity of the macrocyclic ligand would favor the formation of a square-pyramidal intermediate, the *trans*-configuration is, in fact, fully retained in the product.

Finally, there is the question of why the conjugate base so readily dissociates to release the ligand X. In view of the very low acidity of coordinated amines, the concentration of the conjugate base is a very small fraction of the total concentration of the complex. Thus, its reactivity is enormously greater, by a factor far in excess of the mere ratio of k_B/k_A. It can be estimated that the ratio of the rates of aquation of $[Co(NH_3)_4NH_2Cl]^+$ and $[Co(NH_3)_5Cl]^{2+}$ must be greater than 10^6. Two features of the conjugate base have been considered in efforts to account for this reactivity. First, there is the obvious

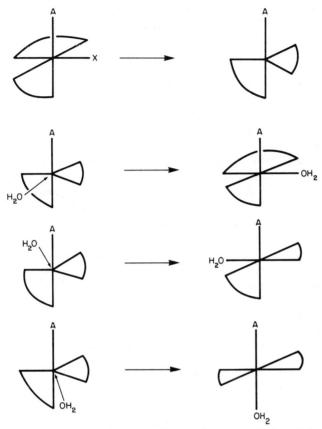

Fig. 21-10. Diagrams showing how a *tbp* intermediate in base hydrolysis by the S_N1CB mechanism could yield three isomeric products.

charge effect. The conjugate base has a charge which is one unit less positive than the complex from which it is derived. Though it is difficult to construct a rigorous argument, it seems entirely unlikely that the charge effect, in itself, can account for the enormous rate difference involved. It has been proposed that the amide ligand could labilize the leaving group X by a combination of electron repulsion in the ground state and a π-bonding contribution to the stability of the five-coordinate intermediate, as suggested in Fig. 21-11. However, there are observations that some workers consider to contradict this explanation also, and thus the question of why the conjugate base is hyperreactive remains unsettled.

Fig. 21-11. A sketch showing how an amide group could promote the dissociation of another ligand, X.

It should also be noted that some results obtained in non-aqueous solvents also support the S_N1CB mechanism. Thus, in dimethylsulfoxide, reactions of the type

$$Co\ en_2NO_2Cl^+ + Y^- = Co\ en_2NO_2Y^+ + Cl^-$$
$$(Y = N_3^-, NO_2^-\ or\ SCN^-)$$

are slow, with half times in hours, but when traces of OH^- or piperidine are added the half times are reduced to minutes. Since it was also shown that reaction of $Co\ en_2NO_2OH^+$ with Y^- is slow, an S_N2 mechanism with this as an intermediate is ruled out, and a genuine conjugate-base mechanism must prevail here.

Further interesting evidence for the conjugate base (though it does not bear on whether the actual aquation is S_N1 or S_N2) comes from a study[39] of the activity of OOH^- in base hydrolysis. Since OOH^- is a weaker base but a better nucleophile toward metal ions than OH^-, base hydrolysis by OOH^- compared to OH^- should proceed more slowly if its function is to form the conjugate base by removing a proton, but faster if it attacks the metal in a genuine S_N2 process. Experimental data are in agreement with the former.

Finally, in complexes having no protonic hydrogen, acceleration by base should not be observed according to the S_N1CB mechanism. This is in general true (as for 2,2'-bipyridine complexes, for instance), but there are a few cases in which a reaction of the first-order in OH^- is observed nonetheless. One of these is the hydrolysis of $Co(EDTA)^-$ by OH^-. The formation of the 7-coordinate intermediate (21-LI) has been proposed, since it is also found

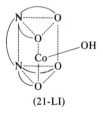

(21-LI)

that the complex racemizes with a first-order dependence on OH^-, but with a rate faster than that of hydrolysis. The 7-coordinate species could revert to $Co(EDTA)^-$ again without hydrolysis occurring, but with concomitant racemization.

Anation Reactions. These are reactions in which an anion displaces H_2O from the coordination sphere. In general, attempts to distinguish between S_N1 and S_N2 mechanisms have been unsuccessful because of complications such as ion pairing or the slow rate of anation compared to water exchange. In order to avoid the ion-pairing problem, an anionic complex may be used, and $Co(CN)_5H_2O^{2-}$ has proved very suitable. In a classical study[40] it was shown that the reaction (21-23) proceeds by an essentially limiting S_N1 mechanism with the intermediate, $Co(CN)_5^{2-}$, having a long

[39] R. G. Pearson and D. N. Edgington, *J. Amer. Chem. Soc.*, 1962, **84**, 4607.
[40] A. Haim and W. K. Wilmarth, *Inorg. Chem.*, 1962, **1**, 573.

enough lifetime to discriminate between various ligands present in the solution.

$$Co(CN)_5H_2O^{2-} = Co(CN)_5^{2-} + H_2O \qquad (21\text{-}23a)$$
$$Co(CN)_5^{2-} + X^- = Co(CN)_5X^{3-} \qquad (21\text{-}23b)$$

Electrophilic Attack on Ligands. There are some reactions known where ligand exchange does not involve the breaking of metal—ligand bonds, but instead bonds within the ligands themselves are broken and re-formed. One well-known case is the aquation of carbonato complexes. When isotopically labeled water, H_2*O, is used, it is found that no $*O$ gets into the coordination sphere of the ion during aquation,

$$[Co(NH_3)_5OCO_2]^+ + 2H_3*O^+ \rightarrow [Co(NH_3)_5(H_2O)]^{3+} + 2H_2*O + CO_2$$

The most likely path for this reaction involves proton attack on the oxygen atom bonded to Co followed by expulsion of CO_2 and then protonation of the hydroxo complex (eqn. 21-24). Similarly, in the reaction of NO_2^- with

$$\left\{ \begin{array}{c} Co(NH_3)_5\text{---}O\cdots C \diagdown^{O}_{\diagup O} \\ \overset{..}{H^+} \\ \overset{..}{O} \\ \diagup \diagdown \\ H \quad H \end{array} \right\} \longrightarrow [Co(NH_3)_5\text{---}O]^{2+}_{|\;H} \xrightarrow{+H^+} [Co(NH_3)_5(H_2O)]^{3+}$$

Transition state

$$(21\text{-}24)$$

pentaamnmineaquocobalt(III) ion, isotopic labeling studies show that the oxygen originally in the bound H_2O turns up in the bound NO_2^-. This remarkable result is explained by the reaction sequence (21-25):

$$2NO_2^- + 2H^+ = N_2O_3 + H_2O$$

$$[Co(NH_3)_5*OH]^{2+} + N_2O_3 \longrightarrow \left[\begin{array}{c} (NH_3)_5CO\text{---}*O\cdots H \\ \overset{..}{ON}\cdots\overset{..}{O}NO \end{array} \right] \xrightarrow{Fast} \qquad (21\text{-}25a)$$

Transition state

$$HNO_2 + [Co(NH_3)_5*ONO]^{2+} \xrightarrow{Slow} [Co(NH_3)_5(NO*O)]^{2+} \quad (21\text{-}25b)$$

21-16. Ligand-displacement Reactions in Square Complexes

Mechanism of Ligand-displacement Reactions.[41] For square complexes, the mechanistic problem is more straightforward and hence better understood. One might expect that four-coordinate complexes would be more likely than octahedral ones to react by an S_N2 mechanism, and extensive studies of Pt^{II} complexes have shown that this is so.

For reactions in aqueous solution of the type (21-26) the rate law takes the general form (21-27). It is believed that the second term corresponds to

$$PtL_nCl_{4-n} + Y = PtL_nCl_{3-n}Y + Cl^- \qquad (21\text{-}26)$$
$$v = k[PtL_nCl_{4-n}] + k'[PtL_nCl_{4-n}][Y] \qquad (21\text{-}27)$$

[41] L. Cattalini, in *Inorganic Reaction Mechanisms*, J. O. Edwards, ed., Wiley-Interscience, 1970 (earlier reviews and subsequent journal references can be found here).

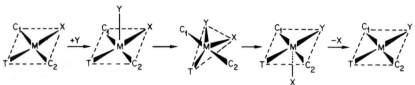

Fig. 21-12.   The course of ligand displacement at a planar complex and the trigonal-bipyramidal five-coordinate structure.

a genuine S_N2 reaction of Y with the complex, while the first term represents a two-step path in which one Cl^- is first replaced by H_2O (probably also by an S_N2 mechanism) as the rate-determining step followed by relatively fast replacement of H_2O by Y.

It has been found that the rates (21-26) of reaction for the series of four complexes in which $L = NH_3$ and $Y = H_2O$ vary by only a factor of 2. This is a remarkably small variation since the charge on the complex changes from -2 to $+1$ as n goes from 0 to 3. Since Pt—Cl bond breaking should become more difficult in this series, while the attraction of Pt for a nucleophile should increase in the same order, the virtual constancy in the rate argues for an S_N2 process in which both Pt—Cl bond breaking and Pt$\cdots$OH$_2$ bond formation are of comparable importance.

A general representation of the stereochemical course of displacement reactions of square complexes is given in Fig. 21-12. It should be carefully noted that this process is entirely stereospecific: *cis* and *trans* starting materials lead, respectively, to *cis* and *trans* products. Whether any of the three intermediate configurations possess enough stability to be regarded as actual intermediates rather than merely phases of the activated complex remains uncertain. Since the starting complex possesses an empty valence shell orbital with which a fifth Pt—ligand bond could be formed (see Section 26-H-1 for a discussion of isolable 5- and 6-coordinated Pt^{II} complexes) the first alternative requires consideration.

It is interesting that the rates of reaction of the series of complexes [MCl(o-tolyl)(PtEt$_3$)$_2$] with pyridine vary enormously with change in the metal, M. The relative rates for Ni, Pd and Pt are $5 \times 10^6 : 10^5 : 1$, which seems to be in accord with the relative ease with which these metal ions increase their coordination numbers from 4 to 6, as this is inferred from their general chemical behavior.

Although the evidence is less than complete, it appears likely that the S_N2 mechanism is valid for the reactions of square complexes other than those of Pt^{II}, such as those of Ni^{II}, Pd^{II}, Rh^{I}, Ir^{I} and Au^{III}.

The order of nucleophilic strength of entering ligands [i.e. the order of the rate constants k' in equation (21-27)] for substitution reactions on Pt^{II} is

$$F^- \sim H_2O \sim OH^- < Cl^- < Br^- \sim NH_3 \sim \text{olefins} < C_6H_5NH_2 < C_5H_5N$$
$$< NO_2^- < N_3^- < I^- \sim SCN^- \sim R_3P$$

This order of nucleophilicity toward Pt^{II} has been the subject of much discussion.

There is no correlation with the order of the ligands in terms of basicity, redox potentials or other forms of reactivity. The order found is remarkably consistent for a variety of substrates and can be expressed in the form of a linear-free energy relationship, similar to those employed for many types of organic reactions. One first defines the quantity n^0:

$$n^0 = \log(k'/k)$$

where k' and k are as defined in eqn. (21-27) when the complex in eqn. (21-26) is *trans*-Pt py$_2$Cl$_2$ in methanol at 30°. We then write

$$\log K' = sn^0 + \log k$$

Log k, the "intrinsic reactivity," varies from one reacting complex to another, as does s, the "discrimination factor", but n^0 values are practically invarient with substrate.

It was observed long ago that *cis–trans* isomerization in planar complexes is catalyzed by traces of free ligands. Since a single displacement reaction is, as mentioned above, stereospecific and conserves stereochemistry, this is best explained in terms of the two stage mechanism shown in Fig. 21-13.

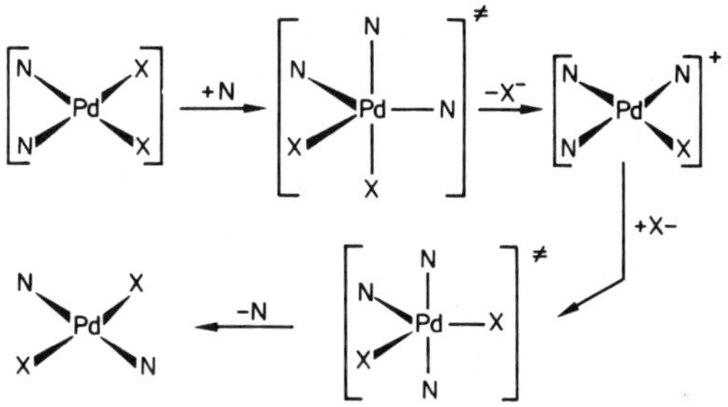

Fig. 21-13. Two-stage mechanism for the catalytic isomerization of *cis*-[Pd(amine)$_2$X$_2$] complexes to the corresponding *trans*-isomers.

The *trans* Effect.   This is a particular feature of ligand-replacement reactions in square complexes which is of less importance in reactions of octahedral complexes except in some special cases where CO (or NO) is present as a ligand, or where M=O or M≡N bonds are present (see, e.g., Section 26-D-9). Most work has been done with Pt^{II} complexes, which are numerous and varied and have fairly convenient rates of reaction. Consider the general reaction (21-28):

$$[PtLX_3]^- + Y^- \rightarrow [PtLX_2Y]^- + X^- \qquad (21\text{-}28)$$

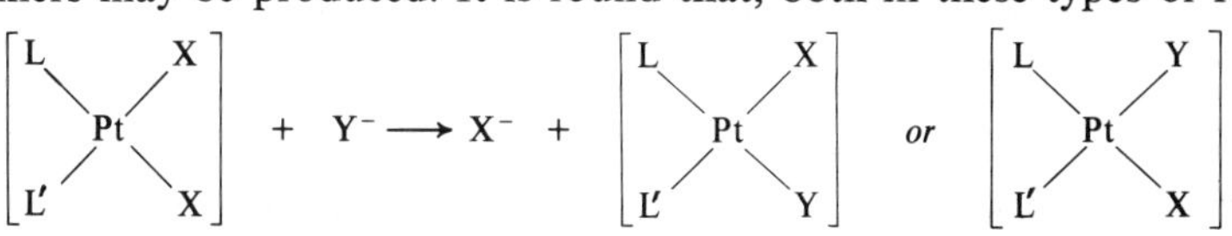

Sterically, there are two possible reaction products, with *cis*- and *trans*-orientation of Y with respect to L. It has been observed that the relative proportions of the *cis*- and *trans*-products vary appreciably with the ligand L. Moreover, in reactions of the type (21-29) either or both of the indicated isomers may be produced. It is found that, both in these types of reaction and in others, a fairly extensive series of ligands may be arranged in the same order with respect to their ability to facilitate substitution in the position *trans* to themselves. This phenomenon is known as the *trans effect*. The approximate order of increasing *trans* influence is:

$$H_2O, OH^-, NH_3, py < Cl^-, Br^- < -SCN^-, I^-, NO_2^-,$$
$$C_6H_5^- < SC(NH_2)_2, CH_3^- < H^-, PR_3 < C_2H_4, CN^-, CO$$

It is to be emphasized that the *trans* effect is here defined solely as a kinetic phenomenon. It is the effect of a coordinated group upon the rate of substitution at the position *trans* to itself in a square or octahedral complex.

The *trans* effect series has proved very useful in rationalizing known synthetic procedures and in devising new ones. As an example we may consider the synthesis of the *cis-* and *trans*-isomers of $[Pt(NH_3)_2Cl_2]$. The synthesis of the *cis*-isomer is accomplished by treatment of the $[PtCl_4]^{2-}$ ion with ammonia (reaction 21-30). Since Cl^- has a greater *trans*-directing

$$\text{(reaction 21-30 structures)} \tag{21-30}$$

influence than does NH_3, substitution of NH_3 into $[Pt(NH_3)Cl_3]^-$ is least likely to occur in the position *trans* to the NH_3 already present and thus the *cis*-isomer is favored. The *trans*-isomer is made by treating $[Pt(NH_3)_4]^{2+}$ with Cl^- (reaction 21-31). Here the superior *trans*-directing influence of Cl^- causes the second Cl^- to enter *trans* to the first one, producing *trans*-$[Pt(NH_3)_2Cl_2]$.

$$\text{(reaction 21-31 structures)} \tag{21-31}$$

All theorizing about the *trans* effect must recognize the fact that since it is a kinetic phenomenon, depending upon activation energies, the stabilities of both the ground state and the activated complex are relevant. It is in principle possible that the activation energy can be affected by changes in one or the other of these energies or by changes in both.

The earliest attempt to explain the *trans* effect was the so-called polarization theory of Grinberg which is primarily concerned with effects in the ground state. This theory deals with a postulated charge distribution as shown in Fig. 21-14. The primary charge on the metal ion induces a dipole in the ligand L which in turn induces a dipole in the metal. The orientation

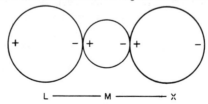

Fig. 21-14.   Distribution of dipoles according to the polarization theory of the *trans* effect.

of this dipole on the metal is such as to repel negative charge in the *trans*-ligand X. Hence X is less attracted by the metal atom because of the presence of L. This theory would lead to the expectation that the magnitude of the *trans* effect of L and its polarizability should be monotonically related, and for some ligands in the *trans* effect series, e.g., H^-, $I^- > Cl^-$, such a correlation is observed. In effect this theory says that the *trans* effect is attributable to a ground-state weakening of the bond to the ligand which is to be displaced.

An alternative theory of the *trans* effect was developed with special reference to the activity of ligands such as phosphines, CO and olefins which are known to be strong π acids (see Chapters 22 and 23 for further details). This model attributes their effectiveness primarily to their ability to stabilize a 5-coordinate transition state or intermediate. This model is, of course, only relevant if the reactions are bimolecular; there is good evidence that this is so in the vast majority, if not all cases. Fig. 21-15 shows how the ability of a ligand to withdraw metal $d\pi$ electron density into its own empty π or π^* orbitals could enhance the stability of a species in which both the incoming ligand, Y, and the outgoing ligand, X, are simultaneously bound to the metal atom.

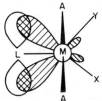

Fig. 21-15. Postulated *tbp* 5-coordinate activated complex for reaction of Y with *trans*-MA$_2$LX, to displace X.

Very recently, evidence has been presented to show that even in cases where stabilization of a 5-coordinate activated complex may be important there is still a ground-state effect—a weakening and polarization of the *trans* bond.[42] In the anion $C_2H_4PtCl_3^-$ the Pt—Cl bond *trans* to ethylene is slightly longer than the *cis* ones, the Pt—*trans*-Cl stretching frequency is lower than the average of the two Pt—*cis*-Cl frequencies, and the nqr spectrum indicates that the *trans*-Cl atom is more ionically bonded.

The present consensus of opinion among workers in the field appears to be that, in each case over the entire series of ligands whose *trans* effect has been studied, both the ground-state bond weakening and the activated-state stabilizing roles may be involved to some extent. For a hydride ion or a methyl group it is probable that we have the extreme of pure, ground-state bond weakening. With the olefins the ground-state effect may play a secondary role compared to activated-state stabilization, although the relative importance of the two effects in such cases remains a subject for speculation and further studies are needed.

[42] J. P. Yesinowski and T. L. Brown, *Inorg. Chem.*, 1971, **10**, 1097.

21-17. Isomerization and Racemization of Tris-chelate Complexes

As already noted, tris-chelate complexes exist in enantiomeric configurations, Λ and Δ, about the metal atom (Fig. 21-2) and, when the chelating ligand is unsymmetrical, that is, when it has two different ends, there are also geometrical isomers, *cis* and *trans*, as shown in (XXVIIIa, b). Each geometrical isomer exists in enantiomeric forms, and thus there are four different molecules.

In the case of tris complexes with symmetrical ligands, the process of inversion (interconversion of enantiomers) is of considerable interest. When the metal ions are of the inert type, it is often possible to resolve the complex and then the process of racemization can be followed by measurement of optical rotation as a function of time. Possible pathways for racemization fall into two broad classes: (1) Those without bond rupture. (2) Those with bond rupture.

There are two pathways without bond rupture that have been widely discussed. One is the trigonal, or Bailar, twist and the other is the rhombic, or Ray–Dutt, twist. They are shown as (a) and (b) in Fig. 21-16.

The simplest dissociative pathways, in which one end of one ligand becomes detached from the metal atom are shown as (c)–(f) in Fig. 21-16. The intermediate may be five-coordinate, with either *tbp* or *spy* geometry, and the dangling ligand may occupy either an axial or an equatorial (or basal) position. In the case of the *spy* intermediates, it is possible that a solvent molecule might temporarily occupy a position in the coordination shell.

It has proved extremely difficult to determine unequivocally which of the various pathways shown is the principal one in a particular case. One of the earliest studies dealt with $[Cr(C_2O_4)_3]^{3-}$. For racemization of this complex it is likely that the mechanism is of a ring-opening type since it has been shown that *all* oxalate oxygen atoms exchange with solvent water at a rate faster than that for oxalate exchange but almost equal to that of racemization.

A considerable amount of effort has been devoted to M(diketo)$_3$ complexes, the reason being that by using unsymmetrical diketonate ligands the processes of isomerization and racemization can be studied simultaneously. Since isomerization can occur *only* by a dissociative pathway, it is often possible to exploit well designed experiments to yield information on the pathways for both isomerization and racemization.[43,44]

To illustrate the approach, let us consider some of the data and deductions for the system $Co(CH_3COCHCOCH(CH_3)_2)_3$, measured in C_6H_5Cl.[43] It was found that the isomerization and racemization are both intramolecular processes, and that they occur at approximately the same rate and with activation energies that are identical within experimental error. It thus appears likely that the two processes have the same transition state. This

[43] J. G. Gordon, II, and R. H. Holm, *J. Amer. Chem. Soc.*, 1970, **92**, 5319.
[44] J. R. Hutchison, J. G. Gordon, II, and R. H. Holm, *Inorg. Chem.*, 1971, **10**, 1004.

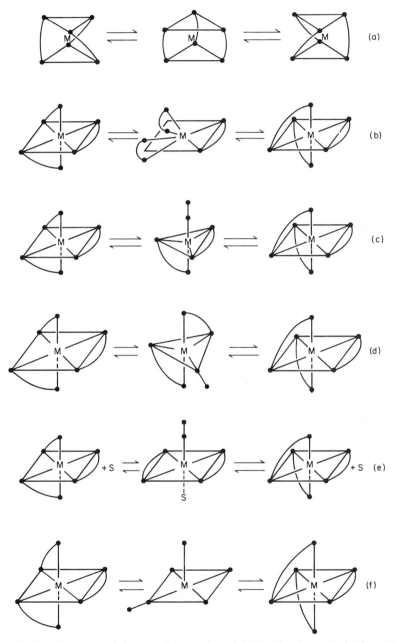

Fig. 21-16. Diagrams of five possible modes of intramolecular racemization of a tris-chelate complex. (a) The trigonal shift, (b) the rhombic shift, (c, d) pathways with trigonal-bipyramidal intermediates, (e, f) ring-opening pathways with square-pyramidal intermediates.

excludes a twist mechanism as the principal pathway for racemization. Moreover, it was found that isomerization occurs mainly with inversion of configuration. This imposes a considerable restriction on the acceptable pathways. Detailed consideration of the stereochemical consequences of the various dissociative pathways, and combinations thereof, lead to the conclusion that for this system the major pathway is through a *tbp* intermediate with the dangling ligand in an axial position.

In the case of the much more labile tris-diketonate complexes of aluminum and gallium, the techniques of study are more complex since it is impossible to isolate even partially resolved samples. The most probable mechanisms for these labile complexes appear to be certain twist processes along with bond rupture to give *spy* transition states.

The best available evidence[45] for a trigonal twist mechanism has been obtained for the mixed ligand complex (21-LII). It is proposed that the

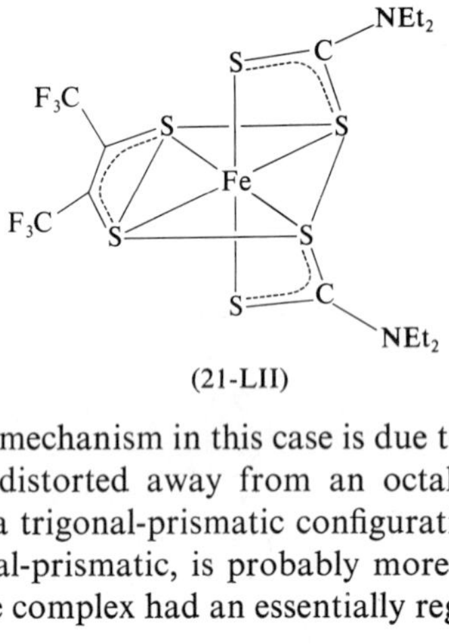

(21-LII)

preference for a twist mechanism in this case is due to the fact that the structure is considerably distorted away from an octahedral configuration of sulfur atoms toward a trigonal-prismatic configuration. Thus the transition state, which is trigonal-prismatic, is probably more energetically accessible than it would be if the complex had an essentially regular octahedral ground configuration.

21-18. Electron-transfer Reactions

These can be divided into two main classes: (1) those in which the electron transfer effects no net chemical change and (2) those in which there is a chemical change. The former, called *electron-exchange* processes, can be followed only indirectly, as by isotopic labelling or by nmr. The latter are the usual oxidation–reduction reactions and can be followed by many standard chemical and physical methods. The electron-exchange processes are of interest because of their particular suitability for theoretical study.

There are two well established general mechanisms for electron-transfer processes. In the first, called the *outer-sphere mechanism*, each complex retains its own full coordination shell, and the electron must pass through

[45] L. H. Pignolet, R. A. Lewis and R. H. Holm, *J. Amer. Chem. Soc.*, 1971, **93**, 360.

both. This, of course, is a purely formal statement in that we do not imply that the "same" electron leaves one metal atom and arrives at the other. In the second case, the *inner-sphere mechanism*, the two complexes form an intermediate in which at least one ligand is shared, i.e. belongs simultaneously to both coordination shells.

The Outer-sphere Mechanism. This mechanism is certain to be the correct one when *both* species participating in the reaction undergo ligand exchange reactions more slowly than they participate in the electron-transfer process. An example is the reaction

$$[Fe^{II}(CN)_6]^{4-} + [Ir^{IV}Cl_6]^{2-} \rightarrow [Fe^{III}(CN)_6]^{3-} + [Ir^{III}Cl_6]^{3-}$$

where both reactants are classified as inert ($t_{1/2}$ for aquation of $0.1M$ solution > 1 ms) but the redox reaction has a rate constant of $\sim 10^5$ l mol^{-1} sec^{-1} at 25°.

For reactions of the electron-exchange type a sketch of the energy vs. reaction coordinate takes the symmetrical form shown in Fig. 21-17. The energy of activation, E_{act}, is made up of three parts: (1) the electrostatic energy (repulsive for species of like charge), (2) the energy required to distort the coordination shells of both species and (3) the energy required to modify the solvent structure about each species. There have been various attempts to compute each of these terms accurately and thus to provide a quantitative theory of electron-exchange reactions, and reasonable success has been achieved in some instances. In this discussion, however, we shall take only a qualitative approach.

Table 21-5 lists some electron-exchange reactions believed to proceed by the outer-sphere mechanism, though for the Co^{II}—Co^{III} reactions this might not be correct since one of the reactants (i.e. the Co^{II} partner) undergoes ligand substitution rapidly. The range covered by the rate constants is very large, extending from $\sim 10^{-4}$ up to perhaps nearly the limit of diffusion control ($\sim 10^9$). It is possible to account qualitatively for the observed variation in rates in terms of the second contribution mentioned to the activation energy. The transition state for electron exchange will be one in which each species has the same dimensions. This is so because a transition state for a process in which there is no adjustment of bond lengths prior to the electron jump would necessarily have a much higher energy. Suppose for an M^{II}–M^{III} exchange the electron jumped while both ions were in their

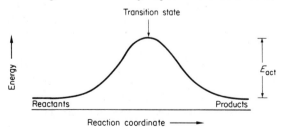

Fig. 21-17. Graph of the energy vs. reaction coordinate for an electron-exchange reaction in which reactants and products are identical.

TABLE 21-5

Rates of Some Electron-exchange Reactions
with Outer-sphere Mechanisms

Reactants	Rate constants (l mole^{-1} sec^{-1})
$[Fe(bipy)_3]^{2+}$, $[Fe(bipy)_3]^{3+}$	
$[Mn(CN)_6]^{3-}$, $[Mn(CN)_6]^{4-}$	
$[Mo(CN)_8]^{3-}$, $[Mo(CN)_8]^{4-}$	
$[W(CN)_8]^{3-}$, $[W(CN)_8]^{4-}$	$>10^6$ at $25°$
$[IrCl_6]^{2-}$, $[IrCl_6]^{3-}$	
$[Os(bipy)_3]^{2+}$, $[Os(bipy)_3]^{3+}$	
$[Fe(CN)_6]^{3-}$, $[Fe(CN)_6]^{4-}$	Second order, $\sim 10^5$ at $25°$
$[MnO_4]^-$, $[MnO_4]^{2-}$	Second order, $\sim 10^3$ at $0°$
$[Co\ en_3]^{2+}$, $[Co\ en_3]^{3+}$	
$[Co(NH_3)_6]^{2+}$, $[Co(NH_3)_6]^{3+}$	Second order, $\sim 10^{-4}$ at $25°$
$[Co(C_2O_4)_3]^{3-}$, $[Co(C_2O_4)_3]^{4-}$	

normal configurations. This would produce an M^{II} complex with bonds compressed *all the way* to the length appropriate to an M^{III} complex and an M^{III} complex with the bonds lengthened *all the way* to the length of those in the M^{II} complex. This would be the zenith of energy, and, as the bonds readjusted, the energy of the exchanging pair would drop to the initial energy of the systems. However, this zenith of energy is obviously higher than it would be if the reacting ions first adjusted their configurations so that each met the other one *only halfway* and then exchanged the electron. The more the two reacting species differ initially in their sizes the higher will be the activation energy.

In the seven cases at the top of Table 21-5, the two species differ by only one electron in an orbital which is approximately non-bonding with respect to the metal–ligand interaction (see Chapter 20). Therefore the lengths of the metal—ligand bonds should be practically the same in the two participating species and the contribution to the activation energy of stretching and contracting bonds should be small. For the $MnO_4^- - MnO_4^{2-}$ case, the electron concerned is not in a strictly non-bonding orbital. In the three cases of slow electron exchange, there is a considerable difference in the metal—ligand bond lengths. However, there is also a change in the extent of electron spin pairing among the non-exchanging electrons on each metal ion. Since it is possible that this could affect the rate of the process either through the activation energy or by influencing the frequency factor (the transmission coefficient, in terms of the absolute theory of rate processes), the significance of the Co^{II}–Co^{III} results is not entirely clear.

The importance of the energy required to change metal—ligand bond distances is suggested by the fact that V^{2+} and Cr^{2+} both appear to react with substitution-inert $[Co(NH_3)_6]^{3+}$ by an outer-sphere mechanism. However, V^{2+} reacts the faster, despite the fact that the redox potential is more favorable with Cr^{2+}. This can be understood since the $Cr^{II}(t_{2g}^3 e_g)$ to $Cr^{III}(t_{2g}^3)$ oxidation presumably requires more reorganization of bond lengths than does the $V^{II}(t_{2g}^3)$ to $V^{III}(t_{2g}^2)$ oxidation. It is also possible, however, that the rate difference is due to the different degrees of orbital overlap by e_g and t_{2g} orbitals with the oxidant.

In electron-transfer reactions between two dissimilar ions, in which there is a net decrease in free energy, the rates are generally higher than in comparable electron-exchange processes. In other words, one factor favoring rapid electron transfer is the thermodynamic favorability of the overall reaction. This generalization seems to apply not only to the outer-sphere processes now under discussion but also to the inner-sphere mechanism to be discussed shortly.

In several cases the rate constants for reactions in Table 21-5 have been found to depend on the identity and concentration of cations present in the solution. The general effect is an increase in rate with an increase in concentration of the cations, but certain cations are particularly effective. The general effect can be attributed to the formation of ion pairs which then decrease the electrostatic contribution to the activation energy. Certain specific effects found for example in the $MnO_4^- - MnO_4^{2-}$ and $[Fe(CN)_6]^{4-} - [Fe(CN)_6]^{3-}$ systems are less easily interpreted with certainty. The effect of $[Co(NH_3)_6]^{3+}$ on the former is thought to be due to ion pairing, greatly enhanced by the high charge. There is no evidence that the cations participate in the actual electron transfer, though this may be so in some cases.

Ligand-bridged or Inner-sphere Processes. Ligand-bridged transition states have been shown to occur in a number of reactions, mainly through the elegant experiments devised and executed by H. Taube and his school. He has demonstrated that the following general reaction occurs:

$$[Co(NH_3)_5X]^{2+} + Cr^{2+}(aq) + 5H^+ = [Cr(H_2O)_5X]^{2+} + Co^{2+}(aq) + 5NH_4^+ \quad (21\text{-}32)$$

$$(X = F^-, Cl^-, Br^-, I^-, SO_4^{2-}, NCS^-, N_3^-, PO_4^{3-}, P_2O_7^{4-}, CH_3COO^-,$$
$$C_3H_7COO^-, \text{crotonate, succinate, oxalate, maleate})$$

The significance and success of these experiments rest on the following facts. The Co^{III} complex is not labile while the Cr^{II} aquo ion is, whereas, in the products, the $[Cr(H_2O)_5X]^{2+}$ ion is not labile while the Co^{II} aquo ion is. It is found that the transfer of X from $[Co(NH_3)_5X]^{2+}$ to $[Cr(H_2O)_5X]^{2+}$ is quantitative. The most reasonable explanation for these facts is a mechanism such as that illustrated in (21-33).

$$Cr^{II}(H_2O)_6^{2+} + Co^{III}(NH_3)_5Cl^{2+} = [(H_2O)_5Cr^{II}ClCo^{III}(NH_3)_5]^{4+}$$

$$\updownarrow \begin{array}{l} \text{electron} \\ \text{transfer} \end{array} \quad (21\text{-}33)$$

$$Cr(H_2O)_5Cl^{2+} + Co(NH_3)_5(H_2O)^{2+} = [(H_2O)_5Cr^{III}ClCo^{II}(NH_3)_5]^{4+}$$

$$\downarrow H^+, H_2O$$

$$Co(H_2O)_6^{2+} + 5NH_4^+$$

Since all Cr^{III} species, including $Cr(H_2O)_6^{3+}$ and $Cr(H_2O)_5Cl^{2+}$, are substitution-inert, the quantitative production of $Cr(H_2O)_5Cl^{2+}$ must imply that electron transfer, $Cr^{II} \rightarrow Co^{III}$, and Cl^- transfer from Co to Cr are mutually interdependent acts, neither possible without the other. Postulation of the binuclear, chloro-bridged intermediate appears to be the only chemically credible way to explain this. As the general equation (21-32) above implies, many (though not all) ligands can serve as bridges.

In reactions between Cr^{2+} and CrX^{2+} and between Cr^{2+} and $Co(NH_3)_5X^{2+}$, which are inner sphere, the rates decrease as X is varied in the order $I^- > Br^- > Cl^- > F^-$. This seems a reasonable one if ability to "conduct" the transferred electron is associated with polarizability of the bridging group, and it appeared that this order might even be considered diagnostic of the mechanism. However, the opposite order[46] is found for $Fe^{2+}/Co(NH_3)_5X^{2+}$ and $Eu^{2+}/Co(NH_3)_5X^{2+}$ reactions; the $Eu^{2+}/Cr(H_2O)_5X^{2+}$ reactions give the order first mentioned, thus showing that the order is not simply a function of the reducing ion used.[47] The order must, of course, be determined by the relative stabilities of transition states with different X and the variation in reactivity order has been rationalized on this basis.[48]

There are now a number of cases, for example those of $Co(NH_3)_5X^{2+}$ with $Co(CN)_5^{3-}$, where $X = F^-$, CN^-, NO_3^- and NO_2^-, and that [49] of Cr^{2+} with $IrCl_6^{2-}$ in which the electron transfer takes place by both inner- and outer-sphere pathways.

As indicated, evidence for the inner-sphere mechanism is usually indirect, though very persuasive. In rare cases, the lifetime of the bridged intermediate allows direct observation of it. Thus for the reaction

$$cis\text{-}[Ru(NH_3)_4Cl_2]^+ + Cr^{2+}(aq) = cis\text{-}[Ru(NH_3)_4(H_2O)Cl]^+ + CrCl^{2+}(aq)$$

the participation of the bridged intermediate

$$[Ru(NH_3)_4Cl-Cl-Cr(H_2O)_5]^{3+}$$

is indicated by direct spectroscopic evidence, which also suggests that the formation of this intermediate and the electron transfer are relatively rapid so that this intermediate spends most of its lifetime in the $Ru^{II}-Cr^{III}$ form, whilst the dissociation of this form of the intermediate is rate-determining.[50]

There are some cases in which an inner-sphere mechanism involves multiple bridges. Examples are

$$cis\text{-}[Cr(H_2O)_4(N_3)_2]^+ + {}^*Cr^{2+} = cis\text{-}[{}^*Cr(H_2O)_4(N_3)_2]^+ + Cr^{2+}$$
$$cis\text{-}[Co(NH_3)_4(OOCR)_2]^+ + Cr^{2+} = cis\text{-}[Cr(H_2O)_4(OOCR)_2]^+ + Co^{2+}$$
$$[Co(EDTA)]^- + Cr^{2+} = Co^{2+} + [Cr(H_2O)_3(EDTA)]^-$$

In the last case, three oxygen atoms of EDTA evidently serve as bridges.

In considering the inner-sphere mechanism, the question naturally arises: are we really dealing with bridge-facilitated *electron* transfer, or can the process be regarded as an *atom* transfer? For example, in the $Cr^{2+}/Co(NH_3)_5Cl^{2+}$ reaction, the net effect could be described simply as transfer of a Cl *atom* from Co to Cr. While there is no experimental way to settle the

[46] H. Diebler and H. Taube, *Inorg. Chem.*, 1965, **4**, 1029; J. P. Candlin, J. Halpern and D. L. Trimm, *J. Amer. Chem. Soc.*, 1964, **86**, 1089.
[47] A. Adin and A. G. Sykes, *J. Chem. Soc.*, A, **1968**, 354.
[48] A. Haim, *Inorg. Chem.*, 1968, **7**, 1475.
[49] R. N. F. Thorneley and A. G. Sykes, *J. Chem. Soc.*, A, **1970**, 232.
[50] W. G. Movius and R. G. Linck, *J. Amer. Chem. Soc.*, 1969, **91**, 5395.

question in such simple cases, there are observations in more elaborate cases that militate against the atom transfer idea at least as a general one. Thus, in reactions with two bridging groups, two ligands but only one electron are transferred. Conversely, in various Pt^{II}/Pt^{IV} exchanges two electrons but only one ligand is transferred. There is even a case in which electron transfer by an inner-sphere mechanism is not accompanied by ligand transfer,[51] viz.

$$Co(EDTA)^{2-} + Fe(CN)_6^{3-} = Co(EDTA)^- + Fe(CN)_6^{4-}$$

There are also indirect but very convincing theoretical arguments for rejecting the atom-transfer concept as an adequate "intimate" mechanism. These have been summarized in detail for the $Cr(H_2O)_5F^{2+}/Cr^{2+}$ case.[52]

There is also the subtle question of the "intimate" mechanism[52,53] of electron transfer by the inner-sphere path, that is a detailed idea of how electron density is shifted from the reductant to the oxidant, once the bridged binuclear intermediate is formed, but recently some progress has been made, especially by the adroit use of various organic ligands as bridging groups. Basically, two types of "intimate" mechanism have been considered: (1) A "chemical" mechanism, in which an electron is transferred to the bridging group, thus reducing it to a radical anion, whereupon, an electron hopping process eventually carries the electron to the oxidant metal ion. (2) A tunnelling mechanism, where by the electron simply passes from reductant to oxidant by quantum-mechanical tunnelling through the barrier constituted by the bridging ligand.

In using organic bridging groups to investigate this question, the problem early arises of distinguishing between adjacent and remote attack by the reductant on the potential bridging group. In the case of benzoate ion as bridging group, attack must be on the coordinated carboxyl group, and there is evidence to show that it actually occurs on the carbonyl oxygen atom, as shown below.

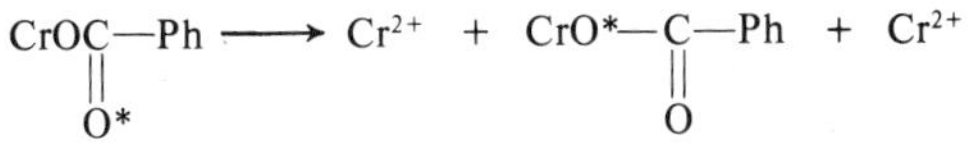

$$CrO\underset{\underset{O^*}{\|}}{C}-Ph \longrightarrow Cr^{2+} + CrO^*\underset{\underset{O}{\|}}{-C}-Ph + Cr^{2+}$$

A definitive example of remote attack is provided by the reaction (21-34). The evidence required to prove remote attack here is more elaborate than

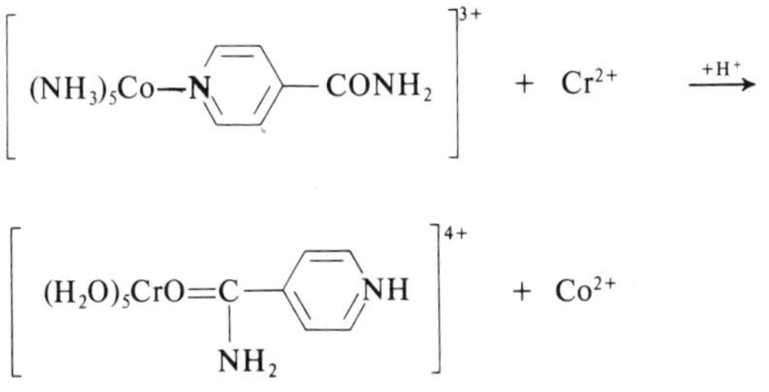

$$\left[(NH_3)_5Co-N\!\!\!\bigcirc\!\!\!-CONH_2\right]^{3+} + Cr^{2+} \xrightarrow{+H^+}$$

$$\left[(H_2O)_5CrO=C\underset{NH_2}{\overset{|}{-}}\!\!\!\bigcirc\!\!\!NH\right]^{4+} + Co^{2+}$$

(21-34)

[51] D. H. Huchital and R. G. Wilkins, *Inorg. Chem.*, 1976, **6**, 1022.
[52] H. Taube and E. S. Gould, *Accounts Chem. Res.*, 1969, **2**, 321.
[53] J. Halpern and L. E. Orgel, *Discuss. Faraday Soc.*, 1960, **29**, 32; P. V. Manning, R. C. Jarnagin and M. Silver, *J. Phys. Chem.*, 1964, **68**, 265.

might at first sight be suspected. The mere fact that the Cr(III) product contains the amide-bound ligand does not assure that remote attack occurred as the rate-determining step. It is necessary to exclude the possibility that (21-LIII) might initially be formed and then isomerized by unreacted Cr^{2+},

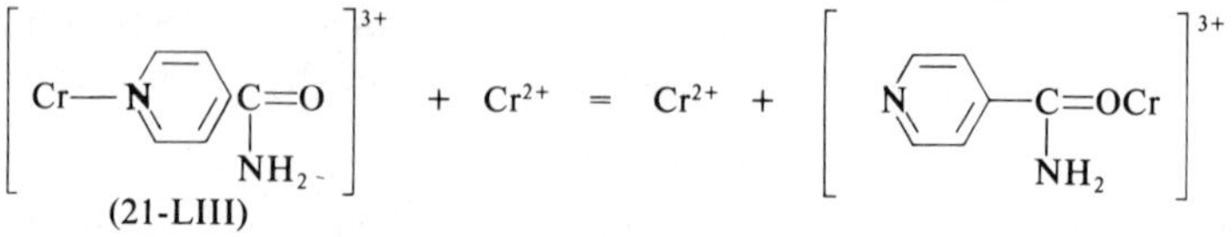

$$\left[Cr-N \overset{}{\underset{}{\bigcirc}} C=O \atop NH_2^- \right]^{3+} + Cr^{2+} = Cr^{2+} + \left[N \overset{}{\underset{}{\bigcirc}} -C=OCr \atop NH_2 \right]^{3+}$$
(21-LIII)

as illustrated. In fact, the equilibrium represented is established only very slowly, and it lies well to the left (pyridine being a much better ligand than an amide). In addition, the reaction (21-35) proceeds much more slowly

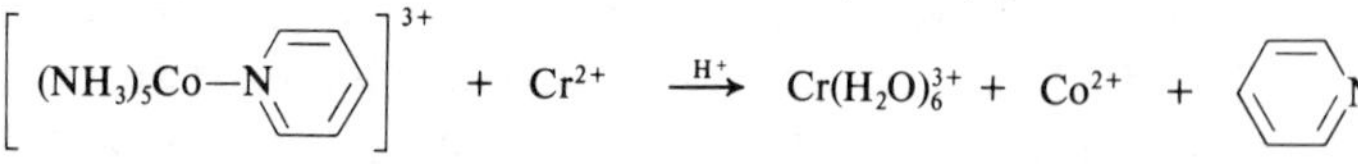

$$\left[(NH_3)_5Co-N \overset{}{\underset{}{\bigcirc}} \right]^{3+} + Cr^{2+} \xrightarrow{H^+} Cr(H_2O)_6^{3+} + Co^{2+} + \overset{}{\underset{}{\bigcirc}} NH^+ \quad (21\text{-}35)$$

than reaction (21-34), and exclusively by an outer-sphere mechanism; all chromium appears as $Cr(H_2O)_6^{3+}$. Thus, direct remote attack seems certain in the case of the p-amido ligand.

An indication that the chemical intimate mechanism can be operative in remote attack is afforded by the annexed rate data. The first two pairs of

Reactants	Rate ratio, Co/Cr
$Co(NH_3)_5F^{2+}/Cr^{2+}$ $Cr(NH_3)_5F^{2+}/Cr^{2+}$ }	$\sim 10^6$
$Co(NH_3)_5OH^{2+}/Cr^{2+}$ $Cr(NH_3)_5OH^{2+}/Cr^{2+}$ }	$\sim 10^6$
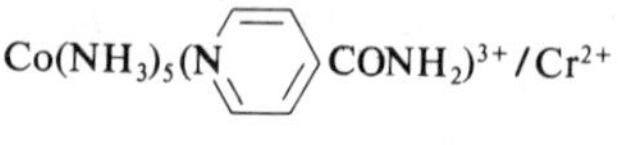 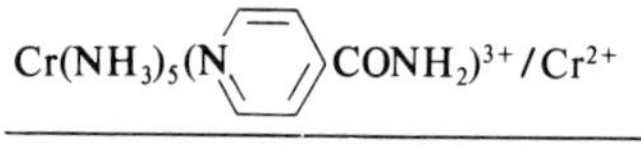 $Co(NH_3)_5(N \bigcirc CONH_2)^{3+}/Cr^{2+}$ $Cr(NH_3)_5(N \bigcirc CONH_2)^{3+}/Cr^{2+}$ }	~ 10

reactions are inner sphere but presumably involve tunnelling as the intimate mechanism. They indicate that tunnelling to Co^{3+} is characteristically a million times faster than to Cr^{3+}. The small rate ratio in the last pair of reactions then strongly implies that the rates are primarily set by the rate of reduction of the bridging ligand, this rate being only a second-order function of what metal ion is attached to the far end.

In conclusion, it should be said that space has permitted only a fragmentary and somewhat superficial presentation of the extensive and subtle studies that have been carried out, mostly by Taube and his coworkers, on the nature of inner-sphere reactions. Various references already cited as well as Taube's recent book (see reading list) must be consulted for further information.

Two-electron Transfers and Non-complementary Reactions. There are some elements that have stable oxidation states differing by two electrons, without a stable state in between. It has been shown that in the majority of these cases, if not in all, two-electron transfers occur. The Pt^{II}–Pt^{IV} system (to be discussed briefly below) and the Tl^{I}–Tl^{III} system have been studied in some detail. For the latter in aqueous perchlorate solution the rate law is

$$v = k_1[Tl^+][Tl^{3+}] + k_2[Tl^+][TlOH^{2+}]$$

In presence of other anions, more complicated rate laws are found indicating that two-electron transfers occur through various Tl^{3+} complexes.

There are a number of other redox reactions which also appear to proceed by 2-electron transfers, examples being

$$Sn^{II} + Tl^{III} \rightarrow Sn^{IV} + Tl^{I}$$
$$Sn^{II} + Hg^{II} \rightarrow Sn^{IV} + Hg^{0}$$
$$V^{II} + Tl^{III} \rightarrow V^{IV} + Tl^{I}$$

All these reactions are *complementary*, meaning that in the overall stoichiometry the oxidant gains and the reductant loses two electrons.

Reactions of non-complementary type must in general have more complex, multistep mechanisms, since ternary activated complexes are improbable. Thus, for example, the reaction of Sn^{II} with Fe^{III} proceeds by the two-step mechanism:

$$Sn^{II} + Fe^{III} \rightarrow Sn^{III} + Fe^{II}$$
$$Sn^{III} + Fe^{III} \rightarrow Sn^{IV} + Fe^{II}$$

Similarly, the reaction of Tl^{III} with Fe^{II} involves a reactive intermediate Tl^{II} species.

Ligand Exchange via Electron Exchange. When a metal atom forms cations in two oxidation states, one giving labile complexes and the other inert complexes, substitution reactions of the latter can be accelerated by the presence of trace quantities of the former. For example, the reactions of type (21-36) that are catalyzed by a trace of Cr^{2+} must occur as shown in

$$[Cr(NH_3)_5X]^{2+} + 5H^+ = [Cr(H_2O)_5X]^{2+} + 5NH_4^+ \qquad (21\text{-}36)$$
$$(X = F^-, Cl^-, Br^-, I^-)$$

equation (21-37), in view of the complete retention of X by Cr^{III} while the NH_3's are completely lost.

$$[Cr(NH_3)_5X]^{2+} + {}^*Cr^{2+}(aq) \rightarrow$$
$$\{[(H_3N)_5Cr\!-\!X\!-\!{}^*Cr(H_2O)_5]^{4+}\} \rightarrow [{}^*Cr(H_2O)_5X]^{2+} + \{Cr(NH_3)_5^{2+}\}$$
$$\text{Transition state} \qquad\qquad\qquad\qquad \underset{+\,5H^+}{\overset{\text{rapidly}}{\big\downarrow}} \qquad (21\text{-}37)$$
$$5NH_4^+ + Cr^{2+}(aq)$$

There are various other cases, especially the $[Cr(H_2O)_5X]^{2+}$–$[Cr(H_2O)_6]^{2+}$ exchanges, in which retention of the X groups shows that they must be bridges in the activated complex. Also, when Fe^{3+} is reduced by Cr^{2+} in the presence of halide ions, the chromium(III) is produced as $[Cr(H_2O)_5X]^{2+}$. Similar phenomena are found also in the Co^{II}–Co^{III} aquo system and fairly generally in Pt^{II}–Pt^{IV} systems.

Pt^{II} catalyzes the exchange of chloride ion with $[Pt(NH_3)_4Cl_2]^{2+}$, the rate law being

$$v = k[Pt^{II}][Pt^{IV}][Cl^-]$$

The mechanism proposed to explain this is the following:

$Pt(NH_3)_4^{2+}$ + $*Cl^-$ = $Pt(NH_3)_4*Cl^+$ (fast pre-equilibrium)

$$= Pt(NH_3)_4Cl^+ + Pt(NH_3)_4*ClCl^{2+}$$

The structure of the proposed activated complex, or intermediate, is very plausible, being quite comparable to the structures found in crystals of several compounds containing equal molar quantities of Pt^{II} and Pt^{IV} (see Section 27-H-1). There is also considerable kinetic evidence corroborating this mechanism.[54] It is likely that traces of Pt^{II} generated by adventitious reducing agents (traces of other metal ions, organic matter, etc.) or photochemically play a role in many reactions of Pt^{IV} complexes. It is also known that traces of other metals, notably Ir, which have several oxidation states can catalyze Pt^{IV} reactions.

[54] R. R. Rettew and R. C. Johnson, *Inorg. Chem.*, 1965, **4**, 1968.

Further Reading

For references to complexes of specific ligands see Chapter 25, Introduction.

Bailar, J. C., Jr., ed., *The Chemistry of Coordination Compounds*, Reinhold, 1956.
Basolo, F., and R. G. Pearson, *Mechanisms of Inorganic Reactions*, 2nd edn., John Wiley, 1967.
Benson, D., *Mechanisms of Inorganic Reactions in Solution*, McGraw-Hill, 1968.
Chaberek, S., and A. E. Martell, *Sequestering Agents*, Wiley-Interscience, 1959.
Clifford, A. F., *Inorganic Chemistry of Qualitative Analysis*, Prentice-Hall, 1961 (contains much general information on complex formation and equilibria).
Collman, J. P., *Angew. Chem. Internat. Edn.*, 1965, **4**, 132 (substitution reactions of metal acetylacetonates).
Dwyer, F. P., and D. P. Mellor, eds., *Chelating Agents and Metal Chelates*, Academic Press, 1964 (contains articles on chelates, nature of metal—ligand bond, redox potentials, chelates in biochemistry, etc.).
Edwards, J. O., *Inorganic Reaction Mechanisms*, Benjamin, 1964 (an excellent introduction to basic kinetic theory and applications to substitution, electron transfer, free radical and other inorganic reactions).
Edwards, J. O., ed., *Inorganic Reaction Mechanisms*, Vol. 13 of *Progr. Inorg. Chem.*, Interscience, 1970 (authoritative reviews on cobalt binuclear complexes, fast reactions, peroxide reactions, redox processes, electron transfer and substitution in square d^8 complexes).
Eigen, M., and R. G. Wilkins, in *Mechanisms of Inorganic Reactions*, Adv. in Chem. Series, No. 49, p. 55, American Chemical Society (a recent survey of complex formation studies, with extensive references to literature on these and other fast reactions of complexes).
Fronaeus, S., in *Technique of Inorganic Chemistry*, Vol. I, Interscience-Wiley, 1963 (methods for determination of stability constants).

Grinberg, A. A., *The Chemistry of Complex Compounds*, Pergamon Press, 1962 (a translation of a Russian text which is good on classical complex chemistry and the historical background).

Halpern, J., *Quart. Rev.*, 1961, **15**, 207 (electron-transfer reactions).

Halpern, J., *J. Chem. Educ.*, 1968, **45**, 372 (electron-transfer reactions).

Hill, H. A. O. and P. Day, *Physical Methods in Advanced Inorganic Chemistry*, Interscience, 1968 (includes much material pertaining to complexes).

Hinton, J. F., and E. S. Amis, *Chem. Rev.*, 1971, **71**, 627 (an extensive review on solvation numbers of ions).

Hunt, J. P., *Coord. Chem. Rev.*, 1971, **7**, 1 (H_2O exchanges using ^{17}O).

Hunt, J. P., *Metal Ions in Aqueous Solution*, Benjamin, 1963 (structures of water and ionic solutions, equilibria involving complex ions, rates and mechanisms and redox reactions).

Inorganic Reaction Mechanisms, Vol. 1, Specialist Periodical Report, The Chemical Society, London, 1971 (this is the first of a continuing series of literature reviews on the title topic).

Lewis, J., and R. G. Wilkins, eds., *Modern Coordination Chemistry*, Interscience, 1960.

Lincoln, S. F., *Coord. Chem. Rev.*, 1971, **6**, 309 (solvent coordination numbers of metal ions in solution).

McAuley, A., and J. Hill, *Quart. Rev.*, 1969, **23**, 18 (electron-transfer reactions).

Perrin, D. D., *Organic Complexing Reagents*, Interscience-Wiley, 1964 (a useful survey of organic ligands and their applications in analysis).

Ringbom, A., *Complexation in Analytical Chemistry*, Vol. 16, *Chemical Analysis*, Interscience-Wiley, 1963 (critical guide for selection of analytical methods based on complexes).

Rossotti, F. J. C., and H. Rossotti, *The Determination of Stability Constants*, McGraw-Hill, 1961 (very thorough treatment).

Sillén L. G., and A. E. Martell, *Stability Constants of Metal–Ion Complexes*, Special Publication No. 17, Chemical Society, London, 1964; also Suppl. No. 1, Special Publication No. 25, 1971 (authoritative and critical compilations of stability constants of metal complexes for both inorganic and organic ligands; includes useful definitions and examples).

Sutin, N., *Accounts Chem. Res.*, 1968, **1**, 225 (electron-transfer reactions).

Sykes, A. G., *Chem. in Britain*, 1970, **6**, 159 (redox reactions).

Sykes, A. G., *Kinetics of Inorganic Reactions*, Pergamon Press, 1966 (a good short introduction).

Taube, H., *Electron Transfer Reactions of Complex Ions in Solution*, Academic Press, 1970.

Wells, A. F., *Structural Inorganic Chemistry*, 3rd edn., Oxford University Press, 1962 (authoritative book with references on structural aspects including complexes).

22

Complexes of π-Acceptor (π-Acid) Ligands

A characteristic feature of the d group transition metals is their ability to form complexes with a variety of neutral molecules such as carbon monoxide, isocyanides, substituted phosphines, arsines, stibines or sulfides, nitric oxide, various molecules with delocalized π orbitals, such as pyridine, 2,2'-bipyridine, 1,10-phenanthroline, and with certain ligands containing 1,2-dithioketone or 1,2-dithiolene groups, such as the dithiomaleonitrile anion. Very diverse types of complex exist, ranging from binary molecular compounds such as $Cr(CO)_6$ or $Ni(PF_3)_4$ through mixed species such as $Co(CO)_3NO$ and $(C_6H_5)_3PFe(CO)_4$, to complex ions such as $[Fe(CN)_5CO]^{3-}$, $[Mo(CO)_5I]^-$, $[Mn(CNR)_6]^+$, $[Vphen_3]^+$, and $\{Ni[S_2C_2(CN)_2]_2\}^{2-}$.

In many of these complexes, the metal atoms are in low-positive, zero or negative formal oxidation states. It is a characteristic of the ligands now under discussion that they can stabilize low oxidation states; this property is associated with the fact that these ligands possess vacant π orbitals in addition to lone-pairs. These vacant orbitals accept electron density from filled metal orbitals to form a type of π bonding that supplements the σ bonding arising from lone-pair donation; high electron density on the metal atom—of necessity in low oxidation states— can thus be delocalized onto the ligands. The ability of ligands to accept electron density into low-lying empty π orbitals can be called π-*acidity*, the word acidity being used in the Lewis sense.

There are many unsaturated organic molecules and ions capable of forming more or less stable complexes with transition metals in low oxidation states besides those to be discussed in this Chapter. There are those that form the so-called π *complexes*: they will be discussed in the following Chapter. The separation is justified, because there is a qualitative difference in the bonding, from a structural standpoint. The ligands discussed here form bonds to the metal by using σ orbitals and exercise their π acidity by using π orbitals whose nodal planes include the axis of the σ bond. In π complexes, on the other hand, both donation and back-acceptance by the ligand are accomplished by use of ligand π orbitals. For the π complexes, therefore, the metal atom lies out of the molecular plane of the ligand, whereas for the complexes discussed here the metal lies along the axes of linear ligands or in the plane of the planar ones.

We can note at this point that the stoichiometries of many, though not all, of the complexes can be predicted by use of the *noble-gas formalism*. This requires that the number of valence electrons possessed by the metal atom plus the number of pairs of σ electrons contributed by the ligands be equal to the number of electrons in the succeeding noble-gas atom.

As explained later, this is simply a phenomenological way of formulating the tendency of the metal atom to use its valence orbitals, nd, $(n+1)s$ and $(n+1)p$, as fully as possible, in forming bonds to ligands. While it is of considerable utility in the design of new compounds, particularly of metal carbonyls, nitrosyls and isocyanides, and their substitution products, it is by no means infallible. It fails altogether for the bipyridine and dithioolefin type of ligand and there are numerous exceptions even among carbonyls, such as $V(CO)_6$ and the stable $[M(CO)_2(diphos)_2]^+$ (M = Mo or W) ions.

In general these compounds have to be prepared by indirect methods from other compounds, although it is sometimes possible to combine metal and ligand directly. Ni is most reactive, combining directly with CO, CH_3PCl_2, and 1,2-bis(diethylphosphino)benzene. Co and Pd also combine with the last of these, and the metals Fe, Co, Mo, W, Rh and Ru also combine with CO, but with the exceptions of Ni and Fe the reactions are too sluggish to be of practical value.

CARBON MONOXIDE COMPLEXES

The most important π-acceptor ligand is carbon monoxide. Many of its complexes are of considerable structural interest as well as of importance industrially and in catalytic and other reactions. Carbonyl derivatives of at least one type are known for all of the transition metals. A few examples of complexes with the unknown CS as a ligand are known, one example being *trans*-RhCl(CS) (PPh$_3$)$_2$ (page 708).

22-1. Mononuclear Metal Carbonyls

The simplest metal carbonyls are of the type $M(CO)_x$. The known ones and some of their properties are listed in Table 22-1. The compounds are all hydrophobic but soluble to varying degrees in non-polar liquids.

In all cases the M—C—O groups are linear. Thus the octahedral, *tbp* and tetrahedral structures have, respectively, strict O_h, D_{3h} and T_d symmetries.

TABLE 22-1

Mononuclear, Binary Metal Carbonyls

Compound	Color and form	Structure[a]	Comments
V(CO)$_6$	Black crystals; dec. 70°; sublimes *in vacuo*	Octahedral[b]	Yellow-orange in solution; paramagnetic (1e$^-$)
Cr(CO)$_6$ Mo(CO)$_6$ W(CO)$_6$	Colorless crystals; all sublime *in vacuo*	Octahedral; M—C = 1.92(4), Cr 2.06(2), Mo 2.06(4), W	Stable to air; dec. 180–200°
Fe(CO)$_5$	Yellow liquid; m.p. −20° b.p. 103°	*tbp*; Fe—C^d = 1.810(3) axial, 1.833(2) eq.	Action of uv gives Fe$_2$(CO)$_9$
Ru(CO)$_5$^{c}	Colorless liquid; m.p. −22°	*tbp* (by ir)	Very volatile and difficult to obtain pure
Os(CO)$_5$^{c}	Colorless liquid; m.p. ∼ −15°	*tbp* (by ir)	Very volatile and difficult to obtain pure
Ni(CO)$_4$	Colorless liquid; m.p. −25° b.p. 43°	Tetrahedral Ni—C = 1.84(4)	Very toxic; musty smell; flammable; decomposes readily to metal

[a] M—C bond lengths are in Ångstroms. Figures in parentheses are error estimates, occurring in least significant digit.

[b] This structure is not known with certainty; there may be some distortion.

[c] F. Calderazzo and F. L'Epplattenier, *Inorg. Chem.*, 1967, **6**, 1220. This is only the second time Ru(CO)$_5$ or Os(CO)$_5$ has been reported.

[d] B. Beagley *et al.*, *Acta Cryst.*, 1969, *B*, **25**, 737.

22-2. Bonding in Linear M—C—O Groups

The fact that refractory metals, with high heats of atomization (~ 400 kJ mol^{-1}), and a generally inert molecule such as CO are capable of uniting to form stable, molecular compounds must certainly be considered, at face value, surprising, especially when it is noted that the CO molecules remain as individuals in the resulting molecules. Moreover, it is known that the simple Lewis basicity (donor ability) of CO is negligible. However, the explanation lies in the multiple nature of the M—CO bond, for which there is much evidence, some of it semiquantitative.

While it is possible to formulate the bonding in terms of a resonance hybrid of (22-Ia) and (22-Ib), a molecular-orbital formulation is more

$$\bar{M}\!-\!\overset{+}{C}\!\equiv\!O\!: \qquad \longleftrightarrow \qquad M\!=\!C\!=\!\overset{..}{O}\!:$$

$$(22\text{-Ia}) \qquad\qquad\qquad (22\text{-Ib})$$

detailed, more graphic and probably more accurate. The MO picture is as follows: There is first a dative overlap of the filled carbon σ orbital (Fig. 22-1a) and second a dative overlap of a filled $d\pi$ or hybrid $dp\pi$ metal orbital with an empty antibonding $p\pi$ orbital of the carbon monoxide (Fig. 22-1b). This bonding mechanism is *synergic*, since the drift of metal electrons into CO orbitals will tend to make the CO as a whole negative and hence to

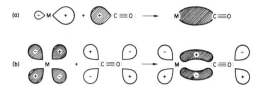

Fig. 22-1. (a) The formation of the metal←carbon σ bond using an unshared pair on the C atom. (b) The formation of the metal→carbon π bond. The other orbitals on the CO are omitted for clarity.

increase its basicity via the σ orbital of carbon; at the same time the drift of electrons to the metal in the σ bond tends to make the CO positive, thus enhancing the acceptor strength of the π orbitals. Thus, up to a point, the effects of σ bond formation strengthen the π bonding and vice versa. It may be noted here that dipole moment studies indicate that the moment of an M—C bond is only very low, about 0.5 D, suggesting a close approach to electroneutrality.

The main lines of physical evidence showing the multiple nature of the M—CO bonds are bond lengths and vibrational spectra. According to the preceding description of the bonding, as the extent of back-donation from M to CO increases, the M—C bond becomes stronger and the C≡O bond becomes weaker. Thus, the multiple bonding should be evidenced by shorter M—C and longer C—O bonds as compared to M—C single bonds and C≡O triple bonds, respectively. Actually very little information can be obtained from the CO bond lengths, because in the range of bond orders (2–3) concerned, CO bond length is relatively insensitive to bond order, as shown in Fig. 3-19 (page 123). The bond length in CO itself is 1.128 Å, while the bond lengths in metal carbonyl molecules are ∼1.15 Å, a shift in the proper direction but of little quantitative significance owing to its small magnitude and the uncertainties (∼0.02 Å) in the individual distances. For M—C distances, the sensitivity to bond order in the range concerned (1–2) is relatively high, probably about 0.3–0.4 Å per unit of bond order, and good evidence for multiple bonding can therefore be expected from such data. However, there is a difficulty in applying this criterion, in that the estimation of the length of an M—C single bond is difficult because zero-valent metals do not form such bonds.

In order to estimate[1] the extent to which the metal—carbon bonds are "shortened" we measure the lengths of M—CO bonds in the same molecule in which some other bond, M—X, exists, such that this bond must be single. Then, using the known covalent radius for X, estimating the single bond covalent radius of C to be 0.70 Å when an sp hybrid orbital is used (the greater s character makes this ∼0.07 Å shorter than that for sp³ carbon), the length for a single M—CO bond in this molecule can be estimated and compared with the observed value. Relatively few data suitable for this purpose are

[1] F. A. Cotton and R. M. Wing, Inorg. Chem., 1965, 4, 314.

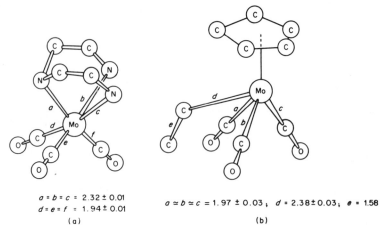

$a = b = c = 2.32 \pm 0.01$
$d = e = f = 1.94 \pm 0.01$

(a)

$a \simeq b \simeq c = 1.97 \pm 0.03$; $d = 2.38 \pm 0.03$; $e = 1.58$

(b)

Fig. 22-2. The molecular structures of (a) *cis*-Mo(dien)(CO)$_3$ and (b) h^5-C$_5$H$_5$Mo(CO)$_3$-C$_2$H$_5$. [Reproduced by permission from ref. 1 and R. Mason *et al.*, *Proc. Chem. Soc.*, **1963**, 273, respectively.]

available, but the data shown in Fig. 22-2 can be used to illustrate the argument. From Mo(dien)(CO)$_3$ [where dien = NH(CH$_2$CH$_2$NH$_2$)$_2$] we take the mean Mo—N bond length to represent an Mo—N single bond, since the amine N atom has no orbitals available for π bonding. When the covalent radius of sp^3 N ($\sim$0.70 Å) is subtracted and the covalent radius of sp C (also, by chance, 0.70 Å) is added, the length of a single Mo—CO bond should be 2.32 ± 0.02 Å. The observed length, 1.94 ± 0.01 Å, shows that there is extensive Mo—C π bonding. Similarly, in (h^5-C$_5$H$_5$)Mo(CO)$_3$C$_2$H$_5$ the Mo—C$_2$H$_5$ distance, less 0.77 (the radius for sp^3 carbon), plus 0.70 gives 2.32 ± 0.04 Å for the single-bond length. That observed is 1.97 ± 0.03 Å; again, there is a large difference indicating appreciable π bonding. The good agreement in the numbers for these two cases is very satisfying and lends strength to the argument.

From the vibrational spectra of metal carbonyls, to be discussed more fully in Section 22-5, it is possible to infer very directly, and even semiquantitatively, the existence and extent of M—C multiple bonding. This is most easily done by studying the CO stretching frequencies rather than the MC stretching frequencies, since the former give rise to strong sharp bands well separated from all other vibrational modes of the molecules. MC stretching frequencies, on the other hand, are in the same range with other types of vibration (e.g., MCO bends) and therefore assignments are not easy to make, nor are the so-called "MC stretching modes" actually pure MC stretching motions. The inferring of M—C bond orders from the behavior of C—O vibrations depends on the assumption that the valence of C is constant, so that a given increase in the M—C bond order must cause an equal decrease in the C—O bond order; this, in turn, will cause a drop in the CO vibrational frequency.

From the direct comparison of CO stretching frequencies in carbonyl molecules with the stretching frequency of CO itself, certain useful qualitative conclusions can be drawn. The CO molecule has a stretching frequency of 2143 cm^{-1}. Terminal CO groups in neutral metal carbonyl molecules are found in the range 2125–1850 cm^{-1}, showing the reduction in CO bond orders. Moreover, when changes are made that should increase the extent of M—C back-bonding, the CO frequencies are shifted to even lower values. Thus, if some CO groups are replaced by ligands with low or negligible back-accepting ability, those CO groups that remain must accept $d\pi$ electrons from the metal to a greater extent in order to prevent the accumulation of negative charge on the metal atom. Thus, the frequencies for $Cr(CO)_6$ are ~ 2100, ~ 2000 and ~ 1985 (exact values vary with phase and solvent) whereas, when three CO's are replaced by amine groups which have essentially no ability to back-accept, as in the Cr analog of $Mo(dien)(CO)_3$ (Fig. 22-2a), there are two CO stretching modes with frequencies of ~ 1900 and ~ 1760 cm^{-1}. Similarly, when we go from $Cr(CO)_6$ to the isoelectronic $V(CO)_6^-$, when more negative charge must be taken from the metal atom, a band is found at ~ 1860 cm^{-1} corresponding to the one found at ~ 2000 cm^{-1} in $Cr(CO)_6$. A series of these isoelectronic species illustrating this trend, with their ir-active CO stretching frequencies (cm^{-1}) is: $Ni(CO)_4$ (~ 2060); $Co(CO)_4^-$ (~ 1890); $Fe(CO)_4^{2-}$ (~ 1790). (The anionic species will be described in detail in Section 22-7.) Conversely, a change that would tend to inhibit the shift of electrons from metal to CO π orbitals, such as placing a positive charge on the metal, should cause the CO frequencies to *rise*, and this effect has been observed in several cases, the following being representative:

$Mn(CO)_6^+$, ~ 2090	$Mndien(CO)_3^+$, ~ 2020, ~ 1900
$Cr(CO)_6$, ~ 2000	$Crdien(CO)_3$, ~ 1900, ~ 1760
$V(CO)_6^-$, ~ 1860	

In order to obtain semiquantitative estimates of M—C π bonding from vibrational frequencies, it is first necessary to carry out an approximate dynamical analysis of the CO stretching modes and thus derive force constants for the CO groups. This procedure will be discussed in Section 22-5. A procedure such as the following may then be used.[2]

For molecules of the type cis-$ML_3(CO)_3$, of which cis-$Mo(dien)(CO)_3$ is an example, each CO group shares the two metal $d\pi$ orbitals from which it receives electrons with the ligand, L, which is *trans* to it. In the hexacarbonyl itself (L = CO) each $d\pi$ orbital is thus contributing equally to all six ligands. Since there are three $d\pi$ orbitals (see page 609), each containing one electron pair, there can be, at most, $\frac{1}{2}$ of a π bond for each M—C pair, if the $d\pi$ electrons are fully used. Now in the case where L represents a ligand incapable of accepting $d\pi$ electrons, each of the three CO groups has access to $\frac{1}{3}$ of the three pairs of $d\pi$ electrons. Thus, if the $d\pi$ electrons enter fully into the

[2] F. A. Cotton, *Helv. Chim. Acta*, Fasc. Extraordinarius, Alfred Werner Commemoration Volume, Verlag Helvetica Chimica Acta, Basle, 1967, p. 117.

back-bonding, each M—C pair will have a full π bond. Thus, if we assume full participation by the metal $d\pi$ electrons in each case, the M—C bond orders go from 1.5 in $M(CO)_6$ to 2.0 in cis-$ML_3(CO)_3$ where L is a ligand atom without π-acceptor orbitals, such as an aliphatic amino nitrogen atom. The CO bond orders in these two cases should then be, respectively, 2.5 and 2.0. From data for CO groups in organic molecules, it can be estimated that for a change of 0.5 in CO bond order, the bond force constant should change by ~ 3.4 md/Å. It has been found that the CO force constants in $M(dien)(CO)_3$ molecules are 3.4 ± 0.4 md/Å less than those in the corresponding $M(CO)_6$ molecules. Thus, the extensive, essentially complete participation of the metal $d\pi$ electrons in M—C π bonding is nicely demonstrated by these data.

Other types of data that have been used, though less extensively, to verify the multiple nature of M—CO bonds are the intensities of carbonyl stretching modes,[3] the frequencies of metal—carbon stretching modes,[4] equilibrium constants for ligand exchange reactions,[5] molecular-orbital calculations[6] and electronic spectra.[7]

It is important when seeking quantitative correlations between CO stretching frequencies and the strength of M—CO bonding to observe that the σ electrons on carbon occupy an MO that is slightly antibonding. Thus, the donation of these electrons will by itself cause some increase in the $C{\equiv}O$ bond strength.

A final word concerning the actual linearity of "linear" M—C—O groups is appropriate. Except when the M—C—O group lies along a molecular symmetry axis of order three or higher, it will generally be at least slightly bent. Angles of 165° to 179° are common. This results simply because the π bonding is of somewhat different strength in the two mutually perpendicular directions.

The CO groups in linear M—C—O chains are frequently designated *terminal* CO groups, to distinguish them from the bridging ones, which will be discussed in the next Section.

22-3. Polynuclear Metal Carbonyls

As mentioned in Section 19-5, one of the largest classes of M—M-bonded compounds, especially of the metal atom cluster type, is that in which the

[3] D. J. Darensbourg and T. L. Brown, *Inorg. Chem.*, 1968, **7**, 959.
[4] G. R. Dobson and L. W. Houk, *Inorg. Chim. Acta*, 1967, **1**, 287; R. J. H. Clark and B. C. Crosse, *J. Chem. Soc., A*, **1969**, 224.
[5] R. J. Angelici and C. M. Ingemanson, *Inorg. Chem.*, 1969, **8**, 83.
[6] N. A. Beach and H. B. Gray, *J. Amer. Chem. Soc.*, 1968, **90**, 5713; K. G. Caulton and R. F. Fenske, *Inorg. Chem.*, 1968, **7**, 1273; A. F. Schreiner and T. L. Brown, *J. Amer. Chem. Soc.*, 1968, **90**, 3366, 5947; M. Dartiguenave, Y. Dartiguenave and H. B. Gray, *Bull. Soc. Chim. France*, **1969**, 4223; D. A. Brown and R. M. Rawlinson, *J. Chem. Soc., A*, **1969**, 1530.
[7] E. W. Abel *et al.*, *J. Mol. Spectroscopy*, 1969, **30**, 20; P. S. Braterman and A. P. Walker, *Discuss. Faraday Soc.*, No. 47, **1969**, 121.

ligands are strong π-acids, such as CO, NO, PR_3 and a few organo groups. By far the most important ligand in this respect is CO. The number of compounds encompassed by the title of this Section is so vast that only a few typical and specially important ones can be discussed here.

Table 22-2 lists some well-characterized di- and poly-nuclear, neutral, unsubstituted metal carbonyls. In all these compounds there are M—M bonds but in many cases they are supplemented by bridging CO groups of several types. The principal structural types of cluster are shown in Fig. 22-3.

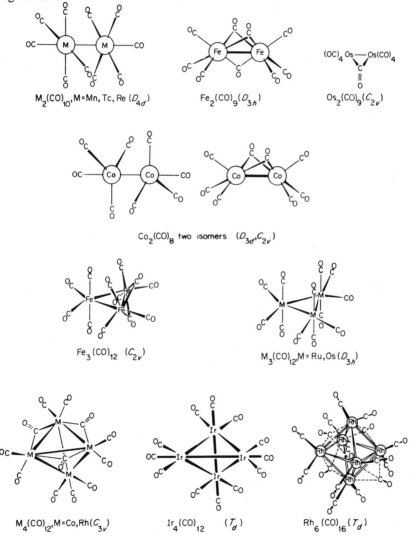

Fig. 22-3.   The structures of some polynuclear metal carbonyls. The detailed structure of $Os_2(CO)_9$ is not known.

TABLE 22-2

Unsubstituted Neutral Polynuclear Metal Carbonyls

Dinuclear		
$Mn_2(CO)_{10}$	$Fe_2(CO)_9$	$Co_2(CO)_8$
$Tc_2(CO)_{10}$		$[Rh_2(CO)_8]^a$
$Re_2(CO)_{10}$	$Os_2(CO)_9$	
$ReMn(CO)_{10}$		
Polynuclear		
	$Fe_3(CO)_{12}$	$Co_4(CO)_{12}, Co_6(CO)_{16}$
	$Ru_3(CO)_{12}$	$Rh_4(CO)_{12}, Rh_6(CO)_{16}$
$[Re(CO)_4]_3(?)$	$Os_3(CO)_{12}$	$Ir_4(CO)_{12}$
	$M_2M'(CO)_{12}$	
	(M, M' = Fe, Ru, Os)	

^{a} Contrary to earlier reports of its isolation, it now appears that $Rh_2(CO)_8$ exists only in equilibrium with $Rh_4(CO)_{12}$ at low temperatures and under high pressure of CO [R. Whyman, *Chem. Comm.*, **1970**, 1194].

Considerable difficulty was encountered in determining the structures of $Fe_3(CO)_{12}$, $Co_4(CO)_{12}$ and $Rh_4(CO)_{12}$ because of disorder or twinning of the crystals, but unequivocal results have recently been obtained for these important molecules. As shown in Fig. 22-3, $Fe_3(CO)_{12}$ has a somewhat different structure[8] from its Ru and Os analogs; however, infrared spectra show that this structure rearranges to some extent in an unknown way in solution.[9] In an argon matrix at 20°K, $Fe_3(CO)_{12}$ exhibits an ir spectrum in accord with the C_{2v} structure which the molecule has in the crystal.[10] The $Co_4(CO)_{12}$ and $Rh_4(CO)_{12}$ structures[11] are the same, but differ from that of $Ir_4(CO)_{12}$.

It can be seen in the various structures depicted in Fig. 22-3 that CO groups fulfil three different structural functions. The terminal groups have already been discussed. The two principal types of bridging group are depicted in Fig. 22-4. The doubly bridging types are formally similar to keto groups. They occur fairly frequently and, apparently, always in conjunction with an M—M bond. This is presumably because the keto bridge could not by itself keep the metal atoms sufficiently close together. For example, addition of another bridging CO group to the bridged form of $Co_2(CO)_8$ to give $(OC)_3CO(\mu\text{-}CO)_3Co(CO)_3$ does not occur (except, perhaps, under

(a)   (b)

Fig. 22-4.   The two main types of bridging CO groups:
(a) Doubly bridging ("ketonic"). (b) Triply bridging.

[8]   C. H. Wei and L. F. Dahl, *J. Amer. Chem. Soc.*, 1969, **91**, 1351.
[9]   J. Knight and M. J. Mays, *Chem. Comm.*, **1970**, 1006.
[10] M. Poliakoff and J. J. Turner, *Chem. Comm.*, **1970**, 1008.
[11] C. H. Wei, *Inorg. Chem.*, 1969, **8**, 2384.

very high pressure of CO); this additional bridge could be added only at the expense of the Co—Co bond but evidently the latter is essential to stability.

Bridging CO groups very often occur in pairs, as in (22-IIa). Any pair of bridging CO groups can always be regarded as an alternative to a non-bridged arrangement with two terminal groups, as in (22-IIb):

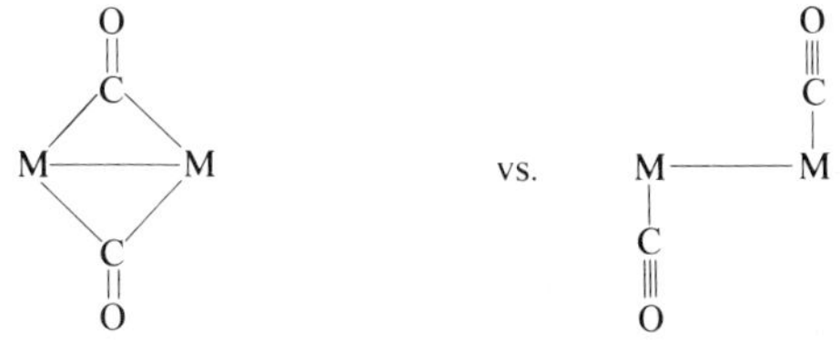

$$\text{(22-IIa)} \qquad\qquad \text{vs.} \qquad\qquad \text{(22-IIb)}$$

The relative stabilities of the alternatives appear to depend primarily upon the size of the metal atoms. The larger the metal atoms the greater is the preference for a non-bridged structure. Thus in any group the relative stability of non-bridged structures increases as the group is descended. Good examples of this are found in Table 22-2 and Fig. 22-3. Thus, $Fe_3(CO)_{12}$ has two bridging CO's while $Ru_3(CO)_{12}$ and $Os_3(CO)_{12}$ have none. Again, $Co_4(CO)_{12}$ and $Rh_4(CO)_{12}$ have three bridging CO's, while $Ir_4(CO)_{12}$ has none. Recently it has been shown[12] that, although $Fe_2(CO)_9$ has three bridging CO's, $Os_2(CO)_9$ has only one. The generalization concerning metal atom size also covers the trend horizontally in the Periodic Table. Thus the large Mn atoms form only the non-bridged $(OC)_5Mn—Mn(CO)_5$ molecule whereas the dinuclear cobalt carbonyl, $Co_2(CO)_8$, exists as an equilibrium mixture of the bridged and non-bridged structures shown in Fig. 22-3, with a very small enthalpy difference between them.

It has recently been shown that the rate at which interconversion of a pair of structures of the types (22-IIa ↔ 22-IIb) occurs may be quite high,[13] thus providing the basis for stereochemically non-rigid or fluxional behavior.

In general, the doubly bridging, or "ketonic," CO groups are symmetrical (equal M—C distances), but unsymmetrical ones occur in a few cases, such as $[Fe_4(CO)_{13}]^{2-}$ [14] and $Fe_3(CO)_{11}PPh_3$.[15]

Carbonyl groups can also bridge triangular arrays of three metal atoms (Fig. 22-4b). Although this occurs *relatively* infrequently, quite a few cases are now known. One example is provided by $Rh_6(CO)_{16}$, Fig. 22-3, and two more are shown in Fig. 22-5.

The presence of bridging CO groups can often be recognized from the infrared spectra of metal carbonyl compounds. As already noted, and further discussed in Section 22-5, terminal M—CO groups usually have CO stretch-

[12] J. R. Moss and W. A. G. Graham, *Chem. Comm.*, **1970**, 835.
[13] J. G. Bullitt, F. A. Cotton and T. J. Marks, *J. Amer. Chem. Soc.*, 1970, **92**, 2155.
[14] R. J. Doedens and L. F. Dahl, *J. Amer. Chem. Soc.*, 1966, **88**, 4847.
[15] D. J. Dahm and R. A. Jacobson, *J. Amer. Chem. Soc.*, 1968, **90**, 5106.

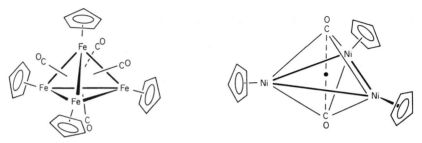

Fig. 22-5. Two examples of triply bridging CO groups.

ing frequencies in the region 1900–2150 cm^{-1} in neutral molecules. In accord with the formal double-bond character of "ketonic" bridging CO groups, their CO stretching frequencies generally fall in the lower range 1750–1850 cm^{-1}. Triply bridging CO groups have still lower CO stretching frequencies; these vary from 1620 to 1730 cm^{-1} in neutral molecules.

There are several examples of polynuclear carbonyls containing single carbon atoms intimately buried in the metal atom cluster. There is as yet no definite idea as to the origin of these carbon atoms. The compounds are usually obtained by refluxing simpler polynuclear carbonyls in hydrocarbons. The two best characterized examples are $Fe_5(CO)_{15}C$ and $Ru_6(CO)_{17}C$, whose structures are shown in Fig. 22-6.

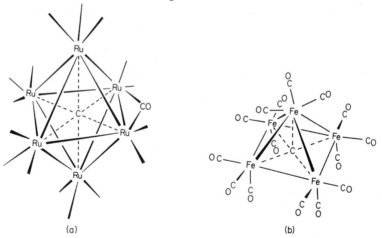

(a)   (b)

Fig. 22-6. The structures of two metal cluster-type carbonyl molecules that contain incorporated carbon atoms. (a) $Ru_6(CO)_{17}C$ (CO groups omitted for clarity). (b) $Fe_5(CO)_{15}C$.

22-4. Preparation of Metal Carbonyls

Only for nickel and iron is direct reaction of metal with carbon monoxide feasible. Finely divided nickel will react at room temperature; an appreciable rate of reaction with iron requires elevated temperatures and pressures. In all other cases the carbonyls are prepared from metal compounds under reductive conditions.

Dicobalt octacarbonyl is prepared by the reaction at 250–300 atm pressure and 120–200°:

$$2CoCO_3 + 2H_2 + 8CO \rightarrow Co_2(CO)_8 + 2CO_2 + 2H_2O$$

Other binary carbonyls are made from the metal halides. The general method is to treat them, usually in suspension in an organic solvent such as tetrahydrofuran, with carbon monoxide at 200–300 atm pressure and temperatures up to 300° in presence of a reducing agent. A variety of reducing agents have been employed—electropositive metals like Na, Al or Mg, trialkylaluminums, copper or the sodium ketyl of benzophenone (Ph_2CONa). The detailed course of the reactions is not well known, but when organometallic reducing agents are employed it is likely that unstable organo derivatives of the transition metal are formed as intermediates. Vanadium carbonyl is most easily obtained by the reaction

$$VCl_3 + CO + Na \text{ (excess)} \xrightarrow[\text{120° 5000 psi}]{\text{diglyme}} [Na \text{ diglyme}_2][V(CO)_6] \xrightarrow[\text{then sublime 50°}]{\text{H}_3\text{PO}_4} V)CO)_6$$

Some carbonyls, e.g., $Os(CO)_5$, $Tc_2(CO)_{10}$ and $Re_2(CO)_{10}$, may be prepared from the oxides (OsO_4, Tc_2O_7, Re_2O_7) by action of CO at high temperatures ($\sim 300°C$) and pressures (~ 300 atm).

Metal acetylacetonates in organic solvents often form suitable starting materials. For example,

$$Ru(C_5H_7O_2)_3 \xrightarrow[\text{150°}]{\text{CO + H}_2 \text{ at 200 atm}} Ru_3(CO)_{12}$$

The binuclear carbonyl $Fe_2(CO)_9$ is obtained as orange mica-like plates by photolysis of $Fe(CO)_5$ in hydrocarbon solvents. The green $Fe_3(CO)_{12}$ is best made by acidification of a polynuclear carbonylate anion (see below) which in turn is obtained from $Fe(CO)_5$ by the action of organic amines such as triethylamine.

The bimetallic carbonyls are generally obtained by the method used for the first known example, h^5-$C_5H_5(CO)_3MoW(CO)_3h^5$-C_5H_5, namely, the interaction of a carbonyl halide with the sodium salt of a carbonylate anion.

22-5. Vibrational Spectra

The vibrational spectra, especially the infrared spectra, of metal carbonyls, have proved to be a rich and convenient source of information concerning both structure and bonding. Some aspects of both these applications have already been mentioned. Here we shall extend the discussion and present an overall summary.

Structural Diagnosis. Perhaps the most important day-to-day use of infrared spectra in the inorganic research laboratory is in deducing the structures of molecules containing carbonyl groups. There are a number of different ways in which this may be done.

1. *Detecting bridging CO groups.* We have already noted that for neutral molecules, bridging CO groups absorb in the range 1700–1850 cm^{-1} while

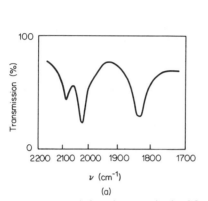
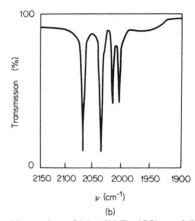

Fig. 22-7.   The infrared spectra in the CO stretching region of (a) solid $Fe_2(CO)_9$ and (b) $Os_3(CO)_{12}$ in solution. Note the greater sharpness of the solution spectra. The most desirable spectra are those obtained in non-polar solvents or in the gas phase. [Data from F. A. Cotton in *Modern Coordination Chemistry*, J. Lewis and R. G. Wilkins, eds., Wiley-Interscience, New York, 1960, and D. K. Huggins, N. Flitcroft and H. D. Kaesz, *Inorg. Chem.*, 1965, **4**, 166.]

terminal ones generally absorb at higher frequencies, 1850–2125 cm^{-1}. Fig. 22-7 illustrates how these facts may be used to infer structures. It is evident that $Fe_2(CO)_9$ has strong absorption bands in both the terminal and the bridging regions. From this alone it could be inferred that the structure must contain both types of CO groups; X-ray study shows that this is so. For $Os_3(CO)_{12}$ several structures consistent with the general rules of valence can be envisioned; some of these would have bridging CO groups, while one (that shown in Fig. 22-3, which is the actual one) does not. The infrared spectrum alone, Fig. 22-7(b), shows that no structure with bridging CO groups is acceptable, since there is no absorption band below 2000 cm^{-1}.

In the case of $Co_2(CO)_8$ (and several other molecules) the infrared spectrum has been used to demonstrate and study the temperature-dependent equilibrium between bridged and non-bridged structures (see Fig. 22-3) in solution. By measuring the spectrum at various temperatures, it is possible to divide the entire spectrum into two sets of bands, one of which increases in intensity while the other decreases, with increasing temperature. It is further found that absorption in the bridging region occurs in only one of these sets, thus showing that one structure has bridges, while the other does not.

In using the positions of CO stretching bands to infer the presence of bridging CO groups it is necessary to keep certain conditions in mind. The frequencies of terminal CO stretches can be quite low if (*a*) there are a number of ligands present that are good donors but poor π-acceptors, or (*b*) there is a net negative charge on the molecule. In either case, back-donation to the CO groups becomes very extensive, thus increasing the M—C bond orders, decreasing the C—O bond orders and driving the CO stretching frequencies down. Thus in $Mo(dien)(CO)_3$ one of the CO stretching bands is as low as $1760 \, \text{cm}^{-1}$ and in the $Fe(CO)_4^{2-}$ ion there is a band at $1790 \, \text{cm}^{-1}$.

2. *Molecular symmetry from the number of bands.* It is often possible to infer the symmetry of the arrangement of the CO groups from the number of CO stretching bands that are found in the infrared spectrum, though a certain amount of judgment and experience is necessary in order consistently to avoid errors. The procedure consists in first determining from the mathematical and physical requirements of symmetry how many CO stretching bands ought to appear in the infrared spectrum for each of several possible structures.[16] The experimental observations are then compared with the predictions and those structures for which the predictions disagree with observation are considered to be eliminated. In favorable cases there will be only one possible structure remaining. In carrying out this procedure, due regard must be given to the possibilities of bands being weak or superposed and, of course, the correct model must be among those considered. The reliability of the procedure can usually be increased if the behavior of approximate force constants and the relative intensities of the bands are also considered.

To illustrate the procedure, consider the *cis-* and *trans-*isomers of an $ML_2(CO)_4$ molecule. Fig. 22-8 shows the approximate forms of the CO stretching vibrations and also indicates those that are expected to absorb infrared radiation, when only the symmetry of the $M(CO)_4$ portion of the molecule is considered. When $L = (C_2H_5)_3P$, the two isomeric compounds can be isolated. One has four infrared bands (2016, 1915, 1900, 1890 cm^{-1})

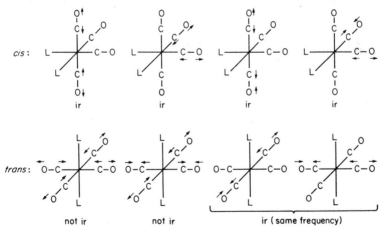

Fig. 22-8. Schematic indication of the forms of the CO stretching vibrations of *cis-* and *trans-*$ML_2(CO)_4$ molecules. For the *cis-*isomer, all four are distinct and can absorb infrared radiation. For the *trans-*isomer, two are equivalent and have the same frequency, forming a degenerate vibration; only this one can absorb infrared radiation.

[16] See, for example, K. Nakamoto, *Infrared Spectra of Inorganic and Coordination Compounds*, 2nd edn., Wiley, 1970, or F. A. Cotton, *Chemical Applications of Group Theory*, 2nd edn., Wiley, 1971, for explanations of how this is done.

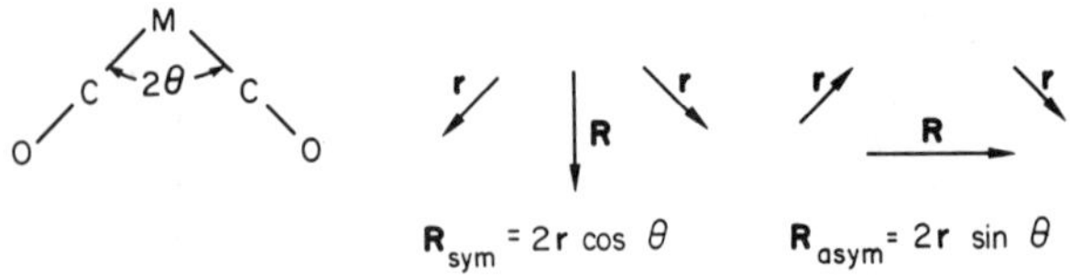

$$R_{sym} = 2r \cos \theta \qquad R_{asym} = 2r \sin \theta$$

Fig. 22-9. Diagrams showing how the dipole vectors, r, of individual CO groups combine to give the dipole vectors **R** for the symmetric and the antisymmetric mode of vibration of an $M(CO)_2$ moiety.

and is thus the *cis*-isomer, while the other shows only one strong band (1890 cm^{-1}) and is thus the *trans*-isomer.

It may also be noted that since no major interaction is to be expected between the CO stretching motions in two $M(CO)_4$ groups if they are connected only through the two heavy metal atoms, $Os_3(CO)_{12}$ should have the four-band spectrum of a *cis*-$ML_2(CO_4$ molecule, and as seen in Fig. 22-7 it does.

3. *Bond angles from relative intensities.* To a very good approximation the relative intensities of different CO stretching modes can be calculated by using a very simple model. In this model, each CO oscillator is treated as a dipole vector and the total dipole vector for the entire vibrational mode is taken to be the vector sum of these individual vectors. Since the intensities

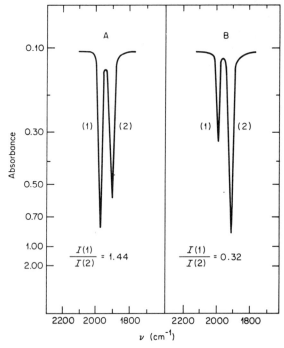

Fig. 22-10. The CO stretching bands and their relative intensities for the two cations of Figure 22-11. Data from F. A. Cotton and C. M. Lukehart, *J. Amer. Chem. Soc.*, 1971, **93**, 2672, and unpublished work.

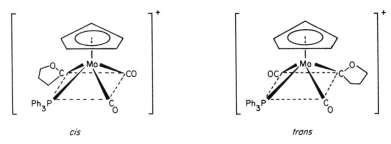

cis   trans

Fig. 22-11.   The two isomers of $[h^5\text{-}C_5H_5Mo(CO)_2PPh_3C_4H_6O]^+$. The C_4H_6O ligand is a "coordinated carbene," discussed in Section 23-8.

are proportional to the squares of the dipole vectors there is a relationship of the intensities to the angles between C—O bond directions.

For the case of two CO groups, as shown in Fig. 22-9, the ratio of the intensities of the symmetric and antisymmetric bands is given by

$$R_{sym}/R_{asym} = \left\{\frac{2r\cos\theta}{2r\sin\theta}\right\}^2 = \cotan^2\theta$$

Fig. 22-10 shows the ir spectra of the two isomeric ions of Fig. 22-11. Both, of course, have two CO stretching bands, so that no decision between them can be made on this simple basis. However, the intensity ratios are quite different in the two cases. As will be discussed below, the band of higher frequency can be assigned in both cases to the symmetric mode. From careful measurement of the areas, the intensity ratios given on the spectra are obtained. From each of these the angle 2θ can be obtained, viz.,

A. $2\theta = 2\sqrt{\text{arccotan } 1.44} = 79°$

B. $2\theta = 2\sqrt{\text{arccotan } 0.32} = 121°$

It is obvious that the angles 79° and 121° can be associated only with the cis- and trans-isomer, respectively.

Another illustration of this type of argument is provided by the $[RhCl(CO)_2]_2$ molecule (Section 26-G-2), which has the bent structure shown schematically in (22-III) in the crystal and illustrated in Fig. 26-G-2. Since the

(22-III)

reason for bending is not obvious, there might be a temptation to attribute it to intermolecular forces in the crystal. The infrared spectrum of the molecule in solution (Fig. 22-12), however, shows qualitatively and quantitatively, that the structure is essentially identical with that observed in the crystal.[17]

[17] J. G. Bullitt and F. A. Cotton, Inorg. Chim. Acta, 1971, 5, 637.

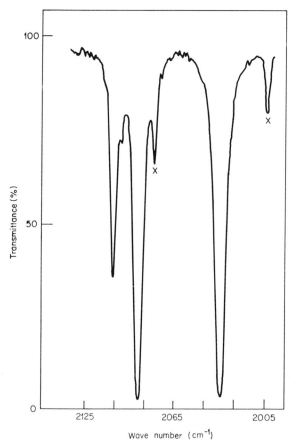

Fig. 22-12.   The infrared spectrum of $Rh_2Cl_2(CO)_4$. The small peaks marked X are ^{13}CO satellites. The large peaks are the three infrared-active fundamentals.

The appearance of three bands shows that the molecule is bent. Moreover, from the ratios of the intensities of these bands, the dihedral angle is calculated to be $58 \pm 2°$, which is not significantly different from the value of $56°$ found for ϕ in the crystal.

    4. *Detecting conformers.*   The number of bands in the carbonyl stretching region for certain molecules can often be used to show that two conformational isomers are present. Thus, in the molecule $(h^5\text{-}C_5H_5)Fe(CO)_2(h^1\text{-}C_5H_5)$ (see Chapter 23 for further details) one would tend to expect only two bands, due to the symmetric and the antisymmetric mode. In fact, there are four bands, as shown in Fig. 22-13a. This shows that there must be nearly equal populations of two conformations, presumably those shown below the spectrum. Similarly, for $(h^5\text{-}C_5H_5)Mo(CO)_2C_3H_5$, the appearance of two sets of bands indicates the presence of the two conformers shown below the spectrum (Fig. 22-13b).

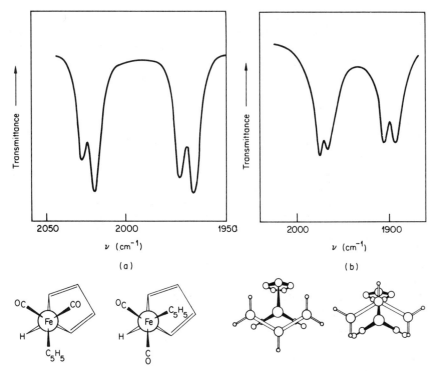

Fig. 22-13. Infrared spectra that demonstrate the presence of conformational isomers. (a) For the molecule $(h^5\text{-}C_5H_5)Fe(CO)_2(h^1\text{-}C_5H_5)$; data from F. A. Cotton and T. J. Marks, *J. Amer. Chem. Soc.*, 1969, **91**, 7523. (b) For the molecule $(h^5\text{-}C_5H_5)Mo(CO)_2C_3H_5$; data from R. B. King, *Inorg. Chem.*, 1966, **5**, 2242, and A. Davison and W. C. Rode, *Inorg. Chem.*, 1967, **6**, 2124.

Force Constants and Bonding. In order to make the fullest use of vibrational spectra, Raman as well as infrared data should be obtained and mathematical analysis should be conducted to obtain values of the force constants. The main use for force constants is in seeking a quantitative understanding of bonding relationships.

A rigorous vibrational analysis of even a simple mononuclear metal carbonyl molecule is complex. It is for the Group VI carbonyls, $M(CO)_6$, where M = Cr, Mo or W, and $Ni(CO)_4$ that the most sophisticated treatments have been reported.[18] The C—O force constants, in md/Å, found in these four cases (for the molecules in the gas phase) are: Cr, 17.2; Mo, 17.3; W, 17.2; and Ni, 17.9. The force constant for the CO triple bond in CO^+ is 19.8 and the value for a double bond must be 12–13. Thus the CO bond orders in the $M(CO)_6$ molecules and $Ni(CO)_4$ must be about 2.65 and 2.75, respectively. In the same four molecules the M—C bond stretching force constants are: Cr, 2.08; Mo, 1.96; W, 2.36; Ni, 2.02. Since the CO force

[18] L. H. Jones, R. S. McDowell and M. Goldblatt, *J. Chem. Phys.*, 1968, **48**, 2663; *Inorg. Chem.*, 1969, **8**, 2349.

constants suggested that there may be less π bonding for $Ni(CO)_4$ than for $Cr(CO)_6$, the similarity of the M—C force constants might appear to indicate more σ bonding to Ni than to Cr. However, unless the differences involved are very large, arguments of this kind must be regarded circumspectly, because CO bond strength is influenced not only by the π back-bonding but also by the σ bonding, since, as already noted, the σ donor orbital is somewhat CO antibonding.

The effort required to obtain results of the kind just discussed is so great that there have been many attempts to devise some simple method of calculating useful force constants from limited data and by simple means. The most widely used of these simple methods is that commonly called the Cotton–Kraihanzel (CK) method,[19] although very similar approximations have been suggested by others.[20] The most important approximation is that CO force constants can be calculated from the CO stretching frequencies alone because these are at very much higher frequencies (> 1850 cm^{-1}) than all other vibrations (< 700 cm^{-1}) in simple metal carbonyls as well as most substituted carbonyls. The main effect of this approximation is that the force constants so obtained are not "absolute;" but in a series of related molecules the shift from absolute values will probably be essentially constant. Hence, relative values, and, thus the differences between force constants for different but similar molecules should be rather reliably given.

Beyond this, there are other assumptions in the CK method that are of doubtful validity in the light of the results of the rigorous vibrational analyses. In order to simplify the equations and to make the application simple, two further approximations are used. (1) The actual fundamental frequencies are used, neglecting the fact that these vibrations are not truly harmonic. (2) The interaction constants between *cis*- and *trans*-CO groups are assumed to be related in a simple way. This assumed relationship is predicated on the idea that the coupling between CO groups is due entirely to the electronic effects of interaction with the metal d orbitals. The first assumption is certainly not true; the effects of anharmonicity of the vibrations amount to 20–30 cm^{-1}, which is comparable to the separations between bands due to different modes of vibration, typically $\leqslant 100$ cm^{-1}. As for the causes of the couplings, they are probably due only in part to electronic effects and undoubtedly arise in part from direct electrostatic interaction between the oscillating dipoles of the different CO groups.

Nevertheless, simplified treatments, such as the CK approximation (but not necessarily this particular one), have proved useful in correlating data and provide a much better approach than the direct use of observed frequencies without even empirical allowance for the effects of interaction between the stretching of the two or more CO groups in the molecule.

[19] F. A. Cotton and C. S. Kraihanzel, *J. Amer. Chem. Soc.*, 1962, **84**, 4432; *Inorg. Chem.*, 1963, **2**, 533; F. A. Cotton, *Inorg. Chem.*, 1964, **3**, 702.
[20] L. H. Jones, *Spectrochim. Acta*, 1963, **19**, 329; H. D. Kaesz *et al.*, *J. Amer. Chem. Soc.*, 1967, **89**, 2844.

One of the ideas inherent in the CK method, empirically valid in a qualitative sense and extremely useful in interpreting the spectra of $M(CO)_n$ moieties, is that the coupling constant for stretching of two CO groups on the same metal atom is positive. Physically, this means that modes in which CO groups stretch in phase have higher frequencies than those in which they stretch out of phase. Another way to state this is that the stretching of one CO group makes it harder to stretch another at the same time. This is to be expected from either the CK approximation, where only the influence of back-bonding is considered, or from a consideration of dipolar interactions; therefore, even if both factors contribute, this is an expected and understandable rule. It was on this basis that in assigning the spectra in Fig. 22-10 we could safely assume that the upper bands were due to the symmetric modes (those in which the two CO groups stretch or contract simultaneously).

22-6. Reactions

The number of carbonyls and the variety of their reactions is so enormous that only a few types of reaction can be mentioned. For $Mo(CO)_6$ and $Fe(CO)_5$, Fig. 22-14 gives a suggestion of the extensive chemistry that any individual carbonyl typically has. Other examples will be encountered in succeeding Chapters.

The most important general reactions of carbonyls are those in which CO groups are displaced by other ligands. These may be individual donor molecules, with varying degrees of back-acceptor ability themselves, e.g., PX_3, PR_3, $P(OR)_3$, SR_2, NR_3, OR_2, RNC, etc., or unsaturated organic molecules such as C_6H_6 or cycloheptatriene. Derivatives of organic molecules are discussed separately in the next Chapter.

Another important general reaction is that with bases (OH^-, H^-, NH_2^-) leading to the carbonylate anions that are discussed in the next Section.

Although many of the substitution reactions with other π-acid ligands proceed thermally (temperatures up to 200° in some cases being required for the less reactive carbonyls) it is sometimes more convenient to obtain a particular product by photochemical methods; in some cases, substitution proceeds readily *only* under irradiation. For example, the thermal reactions of $Fe(CO)_5$ and triphenylphosphine or triphenylarsine (L) give mixtures, whereas, photochemically, $Fe(CO)_4L$ and $Fe(CO)_3L_2$ can be obtained quite simply. Manganese carbonyl and $h^5\text{-}C_5H_5Mn(CO)_3$ are usually quite resistant to substitution reactions, but the former under irradiation gives $[Mn(CO)_4PR_3]_2$. In the very rapid photochemical production of acetylene and olefin complexes from $Mo(CO)_6$ and $W(CO)_6$ it is believed that $M(CO)_5$ radicals are the initiating species; even in absence of other ligands, bright yellow solutions are produced when Cr, Mo and W hexacarbonyls are irradiated in various solvents. Metal carbonyls in presence of organic halogen compounds such as CCl_4 have been found to act as initiators for the free-radical polymerization of methyl methacrylate and other monomers.

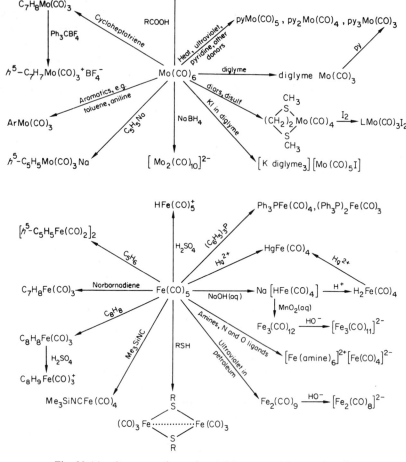

Fig. 22-14.   Some reactions of molybdenum and iron carbonyls.

Kinetic and mechanistic studies of metal carbonyl reactions have been limited mainly to CO exchange processes.[21] In general, these reactions seem to take place by an S_N1 mechanism, although some specific exceptions are known.

22-7. Carbonylate Anions and Carbonyl Hydrides

It was first shown by W. Hieber, who has made many notable contributions to carbonyl chemistry, that if iron pentacarbonyl is treated with

[21] R. J. Angelici, *Organometallic Chem. Rev.*, 1968, **3**, 173; D. A. Brown, *Inorg. Chim. Acta Rev.*, 1967, **1**, 35; H. Werner, *Angew. Chem. Internat. Edn.*, 1968, **7**, 930; W. Strohmeier, *Angew. Chem. Internat. Edn.*, 1964, **3**, 730; F. Basolo and R. G. Pearson, *Mechanisms of Inorganic Reactions*, 2nd ed., Wiley, 1967.

aqueous alkali it dissolves to give an initially yellow solution containing the ion $HFe(CO)_4^-$, which on acidification gives a thermally unstable gas $H_2Fe(CO)_4$. Carbonylate anions have been obtained from most of the carbonyls. Some do not give hydrides on acidification; however, the carbonylate ions of Mn, Re, Fe and Co certainly do.

The carbonylate anions can be obtained in a number of ways—by treating carbonyls with aqueous or alcoholic alkali hydroxide or with amines, sulfoxides or other Lewis bases, by cleaving metal—metal bonds with sodium, or in special cases by refluxing carbonyls with salts in an ether medium. Illustrative examples are

$$Fe(CO)_5 + 3NaOH(aq) \rightarrow Na[HFe(CO)_4](aq) + Na_2CO_3(aq) + H_2O$$
$$Co_2(CO)_8 + 2Na/Hg \xrightarrow{THF} 2Na[Co(CO)_4]$$
$$Mn_2(CO)_{10} + 2Li \xrightarrow{THF} 2Li[Mn(CO)_5]$$
$$Mo(CO)_6 + KI \xrightarrow{diglyme} [K(diglyme)_3]^+[Mo(CO)_5I]^- + CO$$
$$2Co^{2+}(aq) + 11CO + 12OH^- \xrightarrow{KCN\ (aq)} 2[Co(CO)_4]^- + 3CO_3^{2-} + 6H_2O$$
$$TaCl_5 + CO + Na(excess) \xrightarrow[3000-5000\ psi]{diglyme\ \sim 100^\circ} [Na(diglyme)_2]^+[Ta(CO)_6]^-$$

The simpler carbonylate ions, like the binary carbonyls, obey the noble-gas formalism, and their stoichiometries can thus be predicted. The ions are usually fairly readily oxidized by air; the alkali-metal salts are soluble in water, from which they can be precipitated by large cations such as $[Co(NH_3)_6]^{3+}$ or $[Ph_4As]^+$.

In addition to mononuclear carbonylate anions, a variety of polynuclear species have been obtained. The iron carbonylate ions, which have been much studied, are obtained by the action of aqueous alkali or Lewis bases on binary carbonyls or in other ways, for example,

$$Fe_2(CO)_9 + 4OH^- \rightarrow [Fe_2(CO)_8]^{2-} + CO_3^- + 2H_2O$$
$$2Fe_3(CO)_{12} + 7OH^- \rightarrow [Fe_3(CO)_{11}]^{2-} + [HFe_3(CO)_{11}]^- + 2CO_3^{2-} + 3H_2O$$
$$Fe(CO)_5 + Et_3N \xrightarrow{H_2O,\ 80^\circ} [Et_3NH][HFe_3(CO)_{11}]$$
$$[HFe_3(CO)_{11}]^- + Fe(CO)_5 \rightarrow [HFe_4(CO)_{13}]^- + 3CO$$

The $Fe_2(CO)_8^{2-}$ ion has a direct Fe—Fe bond with no bridges and is thus analogous to one isomer of the isoelectronic $Co_2(CO)_8$. The $HFe_3(CO)_{11}^-$ ion has the structure shown in Fig. 22-15, which can be considered analogous to that of $Fe_3(CO)_{12}$ with one bridging CO group replaced by a bridging H^-. The structure of the $Fe_4(CO)_{13}^{2-}$ ion, also shown in Fig. 22-15, provided one of the earliest examples of unsymmetrical, doubly bridging CO groups.

Some other mono- and poly-nuclear carbonylate anions are obtained by reactions such as

$$Cr(CO)_6 + 3KOH \rightarrow K[HCr(CO)_5] + K_2CO_3 + H_2O$$
$$Cr(CO)_6 + BH_4^- \rightarrow HCr_2(CO)_{10}^-$$
$$Ni(CO)_4 + Na \text{ in } NH_3(l) \rightarrow Ni_4(CO)_9^{2-}$$

The structure of the $[HCr_2(CO)_{10}]^-$ ion (22-IV) is of interest because it appears to contain a linear Cr—H—Cr bridge system.[22]

Cobalt and rhodium form some very elaborate metal cluster-type car-

[22] L. B. Handy et al., J. Amer. Chem. Soc., 1966, 88, 366.

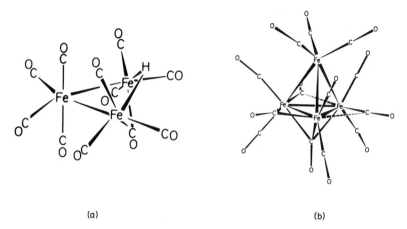

(a)   (b)

Fig. 22-15.   The structures of (a) the HFe$_3$(CO)$_{11}^-$ and (b) the Fe$_4$(CO)$_{13}^{2-}$ ion.

(22-IV)

bonylate anions. In each of the two cobalt anions,[23] Co$_6$(CO)$_{14}^{4-}$ and Co$_6$(CO)$_{15}^{2-}$ there is an octahedral set of metal atoms. In the former there is a terminal CO group on each metal atom and a triply bridging CO group on each octahedral face; the O_h symmetry of the octahedron is thereby maintained. In Co$_6$(CO)$_{15}^{2-}$ the symmetry is reduced to C_{3v} by the location of three triply bridging CO's, six doubly bridging CO's and a terminal CO on each metal atom.

Rhodium, in addition to forming analogs of the cobalt species just mentioned, also forms the more complex species Rh$_7$(CO)$_{16}^{3-}$ and Rh$_{12}$(CO)$_{30}^{2-}$. The former[24] has a distorted octahedron of Rh atoms with the seventh Rh lying over one splayed face; there are doubly and triply bridging as well as terminal CO groups. The Rh$_{12}$(CO)$_{30}^{2-}$ ion consists of two centrosymmetrically joined Rh$_6$(CO)$_{15}^{2-}$ units with two electrons deleted so as to permit formation of a CO-bridged Rh—Rh bond.[25]

There is also a polynuclear carbonylate anion, [Fe$_6$(CO)$_{16}$C]$^{2-}$, which incorporates a carbon atom.[26] The C atom resides in the center of an octahedron of iron atoms.

An important general reaction of carbonylate anions   or of closely

[23] V. G. Albano et al., J. Organometallic Chem., 1969, 16, 461.
[24] V. G. Albano, P. L. Bellon and G. F. Ciani, Chem. Comm., 1969, 1024.
[25] V. G. Albano and P. L. Bellon, J. Organometallic Chem., 1969, 19, 405.
[26] M. R. Churchill et al., J. Amer. Chem. Soc., 1971, 93, 3073.

related ions such as h^5-$C_5H_5Fe(CO)_2^-$ is that with a halide, generally RX. Some examples where R is an alkyl or aryl group will be discussed in the next Chapter, but R can also be groups such as $R_3'Si$, $R_2'P$, $R'S$, etc. The reactions are usually carried out in tetrahydrofuran solution in which the sodium carbonylate salts are soluble. Not all halides react with every carbonylate ion, of course, but a great variety of derivatives have been made. Two typical examples are

$$(CO)_5Mn^- + ClCH_2 - CH = CH_2 \rightarrow (CO)_5MnCH_2 - CH = CH_2 + Cl^-$$
$$h^5\text{-}C_5H_5(CO)_3W^- + ClSiMe_3 \rightarrow h^5\text{-}C_5H_5(CO)_3WSiMe_3 + Cl^-$$

A special case of this general reaction occurs when R is a complex of another transition (or non-transition) metal, in which case compounds with metal—metal bonds are obtained. Some examples are[27]

$$h^5\text{-}C_5H_5(CO)_3Mo^- + ClW(CO)_3 h^5\text{-}C_5H_5 \rightarrow h^5\text{-}C_5H_5(CO)_3MoW(CO)_3 h^5\text{-}C_5H_5 + Cl^-$$
$$Ta(CO)_6^- + C_2H_5HgCl \rightarrow C_2H_5HgTa(CO)_6 + Cl^-$$
$$Fe(CO)_4^{2-} + 2Ph_3PAuCl \rightarrow (Ph_3Au)_2Fe(CO)_4 + 2Cl^-$$
$$Mn(CO)_5Br + Co(CO)_4^- \rightarrow (OC)_4CoMn(CO)_5 + Br^-$$

Carbonyl hydrides. In some cases hydrides corresponding to the carbonylate anions can be isolated. A few of these are listed in Table 22-3 along with their main properties.

TABLE 22-3

Some Carbonyl Hydrides and their Properties

Compound	Form	M.p. (°C)	M—H stretch (cm^{-1})	τ value[a]	Comment
$HMn(CO)_5$	Colorless liquid	−25	1783	17.5	Stable liquid at 25°; weakly acidic
$H_2Fe(CO)_4$	Yellow liquid, colorless gas	−70	?	21.1	Decomposes at −10° giving H_2 + red $H_2Fe_2(CO)_8$
$H_2Fe_3(CO)_{11}$	Dark red liquid		?	25	
$HCo(CO)_4$	Yellow liquid, colorless gas	−26	~1934	20	Decomp. above m.p. giving H_2 + $Co_2(CO)_8$
$HW(CO)_3(h^5\text{-}C_5H_5)$	Yellow crystals	69	1854	17.5	Stable short time in air
$HFe(CO)_3(PPh_3)_2^+$	Yellow	—	?	17.6	Formed by protonation of $Fe(CO)_3(PPh_3)_2$ in H_2SO_4

[a] τ value is position of high-resolution proton magnetic resonance line in parts per million referred to tetramethylsilane reference as 10.00.

In general the carbonyl hydrides are rather unstable substances. They can be obtained by acidification of the appropriate alkali carbonylates or in other ways. Examples of the preparations are

$$NaCo(CO)_4 + H^+(aq) \rightarrow HCo(CO)_4 + Na^+(aq)$$
$$Fe(CO)_4I_2 \xrightarrow{\text{NaBH}_4 \text{ in THF}} H_2Fe(CO)_4$$
$$Mn_2(CO)_{10} + H_2 \xrightarrow[200°]{200 \text{ atm}} 2HMn(CO)_5$$
$$Co + 4CO + \tfrac{1}{2}H_2 \xrightarrow[150°]{50 \text{ atm}} HCo(CO)_4$$

[27] C. E. Coffey, J. Lewis and R. S. Nyholm, *J. Chem. Soc.*, **1964**, 1741; A. S. Kasenally, R. S. Nyholm and M. B. H. Stiddard, *J. Chem. Soc.*, **1965**, 5343; K. A. Keblys and M. Dubeck, *Inorg. Chem.*, 1964, 3, 1646; J. Lewis and S. B. Wild, *J. Chem. Soc.*, A, **1966**, 69.

The iron and cobalt carbonyl hydrides form pale yellow solids or liquids at low temperatures and in the liquid state begin to decompose above about $-10°$ and $-20°$, respectively; they are relatively more stable in the gas phase, however, particularly when diluted with carbon monoxide. They both have revolting odors and are readily oxidized by air. $HMn(CO)_5$ is appreciably more stable.

The hydrides are not very soluble in water but in water they behave as acids, ionizing to give the carbonylate ions:

$$HMn(CO)_5 = H^+ + [Mn(CO)_5]^- \qquad pK \sim 7$$
$$H_2Fe(CO)_4 = H^+ + [HFe(CO)_4]^- \qquad pK_1 \sim 4$$
$$[HFe(CO)_4]^- = H^+ + [Fe(CO)_4]^{2-} \qquad pK_2 \sim 13$$
$$HCo(CO)_4 = H^+ + [Co(CO)_4]^- \qquad \text{strong acid}$$

Like hydride complexes in general, the carbonyl hydrides generally exhibit sharp M—H stretching bands in the infrared and proton nuclear resonance absorptions at very high τ values as shown in Table 22-3.

Although there was considerable controversy for many years about the structural role of the hydrogen atom in metal carbonyl hydrides and related compounds,[28] careful structural studies, especially those employing neutron diffraction,[29] have clarified the situation. It appears that, in general, the hydrogen atom occupies a regular place in the coordination polyhedron and that the M—H distances are approximately equal to the values expected from the sum of single-bond covalent radii. A good example is afforded by the structure[29] of $HMn(CO)_5$, shown in Fig. 22-16.

In the case of polynuclear carbonyl hydride species the location of the hydrogen atoms has generally proved elusive. In no case has this problem yet been definitively solved by neutron diffraction. In some cases, their probable positions have been inferred indirectly. For example, in

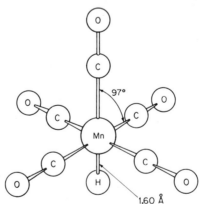

Fig. 22-16. The structure of the $HMn(CO)_5$ molecule, showing both the stereochemical activity of the hydrogen atom and the metal-to-hydrogen distance which approximates to the sum of normal covalent radii.

[28] J. A. Ibers, *Ann. Rev. Phys. Chem.*, 1965, **16**, 375.
[29] S. J. LaPlaca et al., *Inorg. Chem.*, 1969, **8**, 1928.

$H_2Ru_6(CO)_{18}$, X-ray crystallography shows a distorted octahedral array of Ru atoms, with three terminal CO groups on each;[30] the distortion of the octahedron is such as to suggest that the hydrogen atoms function as triple bridges on two opposite triangular faces.

22-8. Carbonyl Halides and Related Compounds

Carbonyl halides, $M_x(CO)_yX_z$, are known for most of the elements forming binary carbonyls but also for Pd, Pt and Au which do not form binary carbonyls; Cu^I and Ag^I carbonyl complexes also exist.

The carbonyl halides are obtained either by the direct interaction of metal halides and carbon monoxide, usually at high pressure, or in a few cases by the cleavage by halogens of polynuclear carbonyls, for example,

$$Mn_2(CO)_{10} + Br_2(l) \xrightarrow{40°} 2Mn(CO)_5Br \underset{CO, 150 \text{ atm.}}{\overset{\text{in petrol at } 120°}{\rightleftharpoons}} [Mn(CO)_4Br]_2 + 2CO$$

$$RuI_3 + 2CO \xrightarrow{220°} [Ru(CO)_2I_2]_n + \tfrac{1}{2}I_2$$

$$2PtCl_2 + 2CO \longrightarrow [Pt(CO)Cl_2]_2$$

A few examples of the halides and some of their properties are listed in Table 22-4. Carbonyl halide anions are also known; they are often derived

TABLE 22-4

Some Examples of Carbonyl Halide Complexes

Compound	Form	M.p. (°C)	Comment
$Mn(CO)_5Cl$	Pale yellow crystals	Sublimes	Loses CO at 120° in organic solvents; can be substituted by pyridine, etc.
$[Re(CO)_4Cl]_2$	White crystals	Dec. >250	Halogen bridges cleavable by donor ligands or by CO (pressure)
$[Ru(CO)_2I_2]_n$	Orange powder	Stable >200	Halide bridges cleavable by ligands
$[Pt(CO)Cl_2]_2$	Yellow crystals	195; sublimes	Hydrolyzed H_2O; PCl_3 replaces CO

by reaction of ionic halides with metal carbonyls or substituted carbonyls:

$$M(CO)_6 + R_4N^+X^- \xrightarrow{\text{diglyme}} R_4N^+[M(CO)_5X]^- + CO \quad M = Cr, Mo, W$$

$$Mn_2(CO)_{10} + 2R_4N^+X^- \longrightarrow (R_4N^+)_2[Mn_2(CO)_8X_2]^{2-} + 2CO$$

$$(R_4N^+)_2[Mn_2(CO)_8X_2]^{2-} + 2R_4N^+Y^- \longrightarrow 2(R_4N^+)_2[Mn(CO)_4XY]^{2-}$$

$$M(CO)_4(bipy) + 2KCN \longrightarrow K_2[M(CO)_4(CN)_2] + bipy \quad M = Cr, Mo, W$$

The structures of the carbonyl halides present little problem; where they are dimeric or polymeric they are invariably bridged through the halogen atoms and *not* by carbonyl bridges, for example, in (22-V) and (22-VI). The

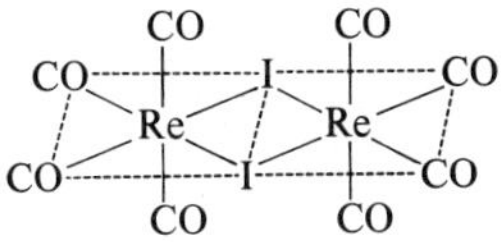

(22-V) 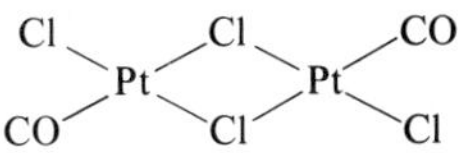   (22-VI)

[30] M. R. Churchill *et al.*, *Chem. Comm.*, **1970**, 458.

halogen bridges can be broken by numerous donor ligands such as pyridine, substituted phosphines, isocyanides, etc. The breaking of halogen bridges by other donor ligands is not of course confined to the carbonyl halides, and other bridged halides such as those given by olefins (Sect. 23-1) can be cleaved. As an example we may cite the reaction

$$[Mn(CO)_4I]_2 + 4py \rightarrow 2Mn(CO)_3Ipy_2 + 2CO \qquad (22\text{-}1)$$

The initial product of the cleavage in reaction (22-1) is (22-VII) but the reaction can proceed further and the product (22-VIII) is isolated. This occurs

(22-VII)   (22-VIII)

because in (22-VII) two of the CO groups are *trans* to each other and thus will be competing across the metal atom for the same metal π bonding orbitals. Hence in the presence of any ligand like a nitrogen, phosphorus or arsenic donor, of *lower* π bonding requirement or capacity compared to CO, one of the *trans* CO groups will be displaced. It follows that the two pyridine (or other) ligands inserted must appear in the *cis*-position to each other. This type of labilization of groups in certain stereochemical situations has been discussed elsewhere under the *trans effect* (see page 667). We can also note that in (22-VIII), which is resistant to further displacement of CO, three of the octahedral positions are occupied by essentially non-π bonding ligands, so that the remaining three CO groups must be responsible for the delocalization of the negative charge on the metal atom; they will, however, now have the exclusive use of the electrons in the d_{xy}, d_{yz}, and d_{xz} metal orbitals for π bonding, and hence the metal—carbon bonding is about at a maximum in (22-VIII) and similar derivatives.

We may finally note that, although $Ni(CO)_4$ exists, there is no corresponding palladium or platinum carbonyl, while, on the other hand, carbonyl halides of Pd^{II} and Pt^{II} exist but there is no carbonyl halide of Ni^{II} (substituted phosphine halides of Ni^{II} do exist, however). The reasons for these differences probably lie in the electronic structures (Ni, d^8s^2; Pd, d^{10}; Pt, d^9s) and promotional energies involved in forming the complexes in the zero and II oxidation states; it would appear that Ni^0 can form π bonds more readily than Pd^0 or Pt^0, but Pd^{II} and Pt^{II} can form π bonds more readily than Ni^{II}. All three derivatives, $M(PF_3)_4$ (M = Ni, Pd or Pt) and complexes such as $PtCO(PPh_3)_3$ and $Pt_3(CO)_4(PMe_2Ph)_3$ are known, however.

22.9 Thiocarbonyl and related complexes

Although, unlike carbon monoxide, CS does not exist under ordinary conditions, a number of complexes containing the thiocarbonyl ligand have been made, all of them, so far, having additional ligands present. The complexes have been prepared mainly from carbon disulfide.[31] Compounds in which CS_2 itself acts as a ligand are intermediates and in these the CS_2 can be bound either end on through the sulfur atom or in a three-membered ring (see Sect. 24-A-2). Thus $RhCl(PPh_3)_3$ reacts with CS_2 to give $RhCl(h^1\text{-}CS_2)$ $(h^2\text{-}CS_2)$ $(PPh_3)_2$. In presence of polar solvents, e.g. methanol, a cationic species containing the $h^2\text{-}CS_2$ group is formed and this is then attacked by triphenylphosphine to form the thiocarbonyl complex $trans\text{-}RhCl(CS)$ $(PPh_3)_2$ and triphenylphosphine sulfide:

$$(Ph_3P)_2Rh\underset{S^+}{\overset{S}{\diagdown}}C \quad +: PPh_3 \rightarrow SPPh_3 + (Ph_3P)_2Rh(CS)^+ \xrightarrow{\ Cl^-\ } trans\text{-}RhCl(CS)(PPh_3)_2$$

In other reactions the sulfur atom is removed from CS_2 by olefins probably as an alkene episulfide but this is not certain,[31b] e.g.,

$$h^5\text{-}C_5H_5Mn(CO)_2 \text{ (cyclooctene)} + CS_2 = h^5\text{-}C_5H_5Mn(CO)_2CS + ?$$

The isoelectronic iron cation has been made by a more rational synthesis via the ethylchlorothioformate,[32a] viz.,

$$h^5\text{-}C_5H_5Fe(CO)_2Na \xrightarrow{\text{EtOC(S)Cl}} h^5\text{-}C_5H_5Fe(CO)_2C(S)OEt \xrightarrow[-\text{EtOH}]{\text{HCl}} [h^5\text{-}C_5H_5Fe(CO)_2(CS)]^+$$

Thiocarbonyl complexes have CS stretches in the region 1270-1360 cm^{-1}, depending on the oxidation state of the metal, charge on the complex, etc., whereas the stretch for CS trapped in a matrix at $-190°$ is at 1274 cm^{-1}. The $d\pi\text{-}p\pi$ bonding is similar to that for the carbonyls. The CS group in $h^5\text{-}C_5H_5Fe(CO)_2(CS)^+$ can undergo nucleophilic attacks, e.g., with azide ion similar to those undergone by coordinated CO (see Sect. 24-A-2).[32b]

It may be noted that there is an extensive chemistry of CS_2 compounds and that CS_2 can undergo insertion reactions (Sect. 24-A-3) with metal-hydrogen, metal-carbon and other bonds.[33] The structure of two complexes made by interaction of CS_2 with $HRe(CO)_2$ $(PPh_3)_3$[34a] and $HPtCl(PPh_3)_2$[34b] respectively confirms that in the first case there is a chelate dithioformate group, ReS_2CH, whereas in the second a $Pt\,SC(S)H$ group is formed. Carbon dioxide also undergoes similar insertion reactions and a few CO_2 complexes of transition metals have also been claimed.[35]

[31] (a) M. C. Baird, G. Hartwell and G. Wilkinson, *J. Chem. Soc. (A)*, **1967**, 865, 2037; J. D. Gilbert, M. C. Baird and G. Wilkinson, *J. Chem. Soc. (A)*, **1968**, 2198; M. Yagupsky and G. Wilkinson, *J. Chem. Soc. (A)*, **1968**, 2813; M. J. Mays and F. P. Stefanini, *J. Chem. Soc. (A)*, **1971**, 2747.
(b) I. S. Butler and A. E. Fenster, *Chem. Comm.*, 1970, 933.
[32] (a) L. Busetto and R. J. Angelici, *J. Amer. Chem. Soc.*, 1968, **90**, 3283; (b) L. Busetto, M. Graziani and U. Belluco, *Inorg. Chem.*, 1971, **10**, 78.
[33] D. Commereuc, l. Douek and G. Wilkinson, *J. Chem. Soc. (A)*, **1970**, 1771.
[34] (a) V. G. Albano, P. L. Bellon and G. Ciani, *J. Organometal. Chem.*, 1971, **31**, 75; (b) A. Palazzi, L. Busetto and M. Graziani, *J. Organometal. Chem.*, 1971, **30**, 273.
[35] I. S. Kolomnikov *et al.*, *Chem. Comm.*, **1971**, 972.

22-10. Dinitrogen (N_2) Complexes

The fact that CO and N_2 (hereafter called *dinitrogen*) are isoelectronic has for years led to speculation as to the possible existence of M—NN bonds analogous to M—CO bonds, but it was only in 1965 that the first example, $[Ru(NH_3)_5N_2]Cl_2$, was reported.[36a] Subsequent work[36b] has shown that the $[Ru(NH_3)_5N_2]^{2+}$ cation can be obtained in a number of ways, e.g.,

by reaction of N_2H_4 with aqueous $RuCl_3$;
by reaction of NaN_3 with $[Ru(NH_3)_5(H_2O)]^{3+}$;
by reaction of N_2 (100 atm) with $[Ru(NH_3)_5H_2O]^{3+}$;
by reaction of $RuCl_3(aq)$ with Zn in $NH_3(aq)$.

Of these the direct reaction with N_2 to displace H_2O is perhaps most notable. Despite early reports to the contrary, no way of reductively transforming the coordinated N_2 in $[Ru(NH_3)_5N_2]^{2+}$ into NH_3 has yet been found. In fact, reduction of coordinated N_2 to NH_3 in a stable complex has not yet been reported. However, there are several systems (page 345) in which reduction of N_2 to NH_3 is catalyzed[37] by low-valent metal atoms, presumably *via* transient M–N_2 complexes.

A bridging N_2 ligand was first observed in the product of the following reaction:

$$[Ru(NH_3)_5Cl]^{2+} \xrightarrow[N_2]{Zn/Hg} \{[Ru(NH_3)_5]_2N_2\}^{4+}$$

The Raman spectrum of this dinuclear ion has a strong line at 2100 cm^{-1}, indicative of a linear $Ru^{II}N{\equiv}NRu^{II}$ rather than a bent Ru^{III}—N$=$N—Ru^{III} structure.

The terminal-type N_2 ligands have strong infrared bands in the range 1930–2230 cm^{-1} (100–400 cm^{-1} below that of free N_2, 2331 cm^{-1}), and these bands may be used diagnostically.

The formation of N_2 complexes by direct uptake of N_2 gas at atmospheric pressure has been observed in reactions such as

$$CoH_3(Ph_3P)_3 + N_2 \rightarrow Co(H)(N_2)(Ph_3P)_3$$

$$Co(acac)_3 + 3Ph_3P + N_2 \xrightarrow{Al(iso-C_4H_9)_3} Co(H)(N_2)(Ph_3P)_3$$

$$FeCl_2 + 3PEtPh_2 + N_2 \xrightarrow{NaBH_4,\ EtOH} FeH_2(N_2)(PEtPh_2)_3$$

$$2h^5\text{-}C_5H_5Fe(diphos)I + N_2 + 2TlBF_4 \xrightarrow{acetone} [h^5\text{-}C_5H_5(diphos)Fe—N{=}N—Fe(diphos)\text{-} h^5\text{-}C_5H_5]^{2+}(BF_4^-)_2 + 2TlI$$

There are numerous other metal complexes with tertiary phosphines or arsines, examples being *cis* and *trans* $Mo(N_2)_2(PR_3)_4$, $ReX(N_2)(PR_3)_4$ and $OsX_2(N_2)(AsR_3)_3$. These are readily obtained by reduction of the

[36] (a) A. D. Allen and F. Bottomly, *Accounts Chem. Res.*, 1968, **1**, 360 (review of this and early work); R. Murray and D. C. Smith, *Coordination Chem. Rev.*, 1968, **3**, 429; J. E. Fergusson and J. I. Love, *Rev. Pure Appl. Chem.*, 1970, **20**, 33; (b) J. Chatt *et al.*, *J. Chem. Soc., A*, **1970**, 1479.
[37] E. E. van Tamelen, H. Rudler and C. Bjorkland, *J. Amer. Chem. Soc.*, 1971, **93**, 3526.
[38] W. E. Silverthorne, *Chem. Comm.*, **1971**, 1310; M. Aresta *et al.*, *Inorg. Chim. Acta.*, 1971, **5**, 203; *Chem. Comm.*, **1971**, 781.

halides with Na or Zn amalgams under nitrogen, often even at 1 atm,[39] e.g.,

$$MoCl_4(PPhMe_2)_2 + N_2 + 2PPhMe_2 \xrightarrow{\text{Na/Hg in THF}} cis\text{-}Mo(N_2)_2(PPhMe_2)_4$$

$$mer\text{-}OsCl_3(AsR_3)_3 + N_2 \xrightarrow{\text{Zn/Hg, in THF}} OsCl_2(N_2)(AsR_3)_3$$

Several dinitrogen complexes have been structurally characterized.[40] The main structural features are presented in Table 22-5. The most important

TABLE 22-5

Structural Parameters of Some M–N≡N Groups[40]

Compound	M–N (Å)	N≡N (Å)	MNN
[Ru(NH$_3$)$_5$N$_2$]Cl$_2$^{a}	2.10	1.12	180
CoH(PPh$_3$)$_3$N$_2$·Et$_2$O	1.80	1.16	175
CoH(PPh$_3$)$_3$N$_2$^{b}	1.78	1.10	178
	1.89	1.12	178
[Ru en$_2$(N$_3$)(N$_2$)]PF$_6$	1.89(1)	1.11(1)	179(1)

^{a} Errors uncertain owing to disorder in the crystal.
^{b} There are two independent molecules in the unit cell.

points are the essential linearity of the M—N—N groups, the very small increase in the N—N distance (1.098 Å in free N_2) and the fact that the M—N distances are somewhat shorter than would be expected for single bonds.

The bonding in M—N_2 groups is, as might be anticipated, qualitatively similar to that in terminal M—CO groups; the same two basic components, M←N_2 σ-donation and M→N_2 π-acceptance, are involved. The major quantitative differences, which account for the lower stability of N_2 complexes, appear to arise from small differences in the energies of the MO's of CO and N_2.[41] For CO the σ donor orbital is weakly antibonding, whereas the corresponding orbital for N_2 is of bonding character. Thus, N_2 is a significantly poorer σ donor than is CO. Now, it is observed that in pairs of N_2 and CO complexes where the metal and other ligands are identical, the fractional lowerings of N_2 and CO frequencies are nearly identical. For the CO complexes, weakening of the CO bond, insofar as electronic factors are concerned, is due entirely to back-donation from metal $d\pi$ orbitals to CO π^* orbitals, with the σ-donation slightly cancelling some of this effect. For N_2 complexes, on the other hand, N≡N bond weakening results from both σ-donation and π-back-acceptance. The very similar changes in stretching frequencies for these two ligands suggests then that N_2 is weaker than CO in both its σ-donor and π-acceptor functions. This in turn would account for the poor stability of N_2 complexes in general.

[39] T. A. George and C. D. Seibold, *J. Organometal. Chem.*, 1971, **30**, C13; J. Chatt *et al.*, *J. Chem. Soc. (A)*, **1970**, 842; **1971**, 702, 895.
[40] F. Bottomly and S. J. Nyburg, *Acta Cryst.*, **1968**, B, **24**, 1289; J. H. Enemark *et al.*, *Chem. Comm.*, **1968**, 96; B. R. Davis, N. C. Payne and J. A. Ibers, *J. Amer. Chem. Soc.*, 1969, **91**, 1240; B. R. Davis and J. A. Ibers, *Inorg. Chem.*, 1970, **9**, 2768.
[41] J. Chatt, D. P. Melville and R. L. Richards, *J. Chem. Soc. (A)*, **1969**, 2841.

Electron emission spectra suggest that the N≡N bonds in both [ReCl(diphos)$_2$(N$_2$)] and [ReCl(diphos)$_2$(N$_2$)]$^+$ are appreciably polar.[42]

22-11. Isocyanide Complexes

Isocyanide complexes can be obtained by direct substitution reactions of the metal carbonyls and in other ways. They include such crystalline, air-stable compounds as red Cr(CNPh)$_6$, white [Mn(CNCH$_3$)$_6$]I and orange Co(CO)(NO)(CNC$_7$H$_7$)$_2$, all of which are soluble in benzene.

Isocyanides generally appear to be stronger σ donors than CO, and various complexes such as [Ag(CNR)$_4$]$^+$, [Fe(CNR)$_6$]$^{2+}$ and [Mn(CNR)$_6$]$^{2+}$ are known where π bonding is of relatively little importance; derivatives of this type are not known for CO. However, the isocyanides are capable of extensive back-acceptance of π electrons from metal atoms in low oxidation states. This is indicated qualitatively by their ability to form compounds such as Cr(CNR)$_6$ and Ni(CNR)$_4$, analogous to the carbonyls and more quantitatively by comparison of CO and CN stretching frequencies. As shown in Table 22-6 the extent to which CN stretching frequencies in

TABLE 22-6

Lowering of CO and CN Frequencies in Analogous Compounds, Relative to Values for Free CO and CNAr[a]

Molecule[b]	Δv (cm^{-1}) for each fundamental mode		
Cr(CO)$_6$	43	123	160
Cr(CNAr)$_6$	68	140	185
Ni(CO)$_4$	15	106	
Ni(CNAr)$_4$	70	125	

[a] Ar represents C$_6$H$_5$ and p-CH$_3$OC$_6$H$_4$.
[b] Data for isonitriles from F. A. Cotton and F. Zingales, J. Amer. Chem. Soc., 1961, 83, 351.

Cr(CNAr)$_6$ and Ni(CNAr)$_4$ molecules are lowered relative to the frequencies of the free CNAr molecules exceeds that by which the CO modes of the corresponding carbonyls lie below the frequency of CO. While these results are not to be taken in a literally quantitative sense, they do show that the back-acceptor capacity of isocyanides rivals that of CO. Various other infrared studies[43] have led to this or a similar conclusion.

Another indication of extensive back-donation in isocyanide complexes of low-valent metals comes from the Co—C distances found[44] in the Co^I complex, [Co(CNCH$_3$)$_5$]$^+$; the mean value, 1.87 Å, is at least 0.17 and

[42] G. J. Leigh et al., Chem. Comm., 1970, 1661.
[43] M. Bigorgne, J. Organometallic Chem., 1963, 1, 101; K. K. Joshi, P. L. Pauson and W. H. Stubbs, J. Organometallic Chem., 1963, 1, 51; R. C. Taylor and W. D. Horrocks, Inorg. Chem., 1964, 3, 584.
[44] F. A. Cotton, T. G. Dunne and J. S. Wood, Inorg. Chem., 1965, 4, 318.

probably ~ 0.25 Å shorter than that for a Co—C single bond and indicates a bond order of 1.5–2.0.

While relatively few compounds containing bridging isocyanide groups, as in (22-IX), have actually been reported

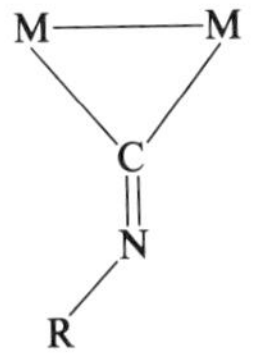

(22-IX)

they appear to constitute a perfectably stable and potentially widespread structural unit. Examples are

$(h^5\text{-}C_5H_5)Fe(CO)(\mu\text{-}CO)(\mu\text{-}CNPh)Fe(h^5\text{-}C_5H_5)$

and

$(h^5\text{-}C_5H_5)Ni(\mu\text{-}CNR)_2Ni(h^5\text{-}C_5H_5).$

22-12. Nitric Oxide Complexes[45]

The NO molecule is closely akin to CO except that it contains one more electron, which occupies a π^* orbital (cf. page 107). Consistently with the general similarity of CO and NO, they form many comparable complexes, although, as a result of the presence of the additional electron, NO also forms a class (bent MNO) with no carbonyl analogs.

Linear, Terminal MNO Groups. Just as the CO group reacts with a metal atom that presents an empty σ orbital and a pair of filled $d\pi$ orbitals, as illustrated in Fig. 22-1, to give a linear MCO grouping with a C → M σ bond and a significant degree of M→C π bonding, so the NO group engages in a structurally and electronically analogous reaction with a metal atom that may be considered, at least formally, to present an empty σ orbital and a pair of $d\pi$ orbitals containing only three electrons. The full set of four electrons for the $Md\pi \rightarrow \pi^*(NO)$ interactions is thus made up of three electrons from M and one from NO. In effect, NO contributes three electrons to the total bonding configuration under circumstances where CO contributes only two. Thus, for purposes of formal electron "book-keeping", the ligand NO can be regarded as a three-electron donor in the same sense as the ligand CO is considered a two-electron donor. This leads to the following very useful general rules concerning stoichiometry, which may be applied without specifically allocating the difference in the number of electrons to any particular (i.e., σ or π) orbitals:

(1) Compounds isoelectronic with one containing an $M(CO)_n$ grouping are those containing $M'(CO)_{n-1}(NO)$, $M''(CO)_{n-2}(NO)_2$, etc., where M',

[45] B. F. G. Johnson and J. A. McCleverty, *Progr. Inorg. Chem.*, 1966, 7, 277; W. P. Griffith, *Adv. Organometallic Chem.*, 1968, 7, 211.

M″, etc. have atomic numbers that are 1, 2, ..., etc. less than M. Some examples are: $(h^5\text{-}C_5H_5)CuCO$, $(h^5\text{-}C_5H_5)NiNO$; $Ni(CO)_4$, $Co(CO)_3NO$, $Fe(CO)_2(NO)_2$, $Mn(CO)(NO)_3$; $Fe(CO)_5$, $Mn(CO)_4NO$.

(2) Three CO groups can be replaced by two NO groups. Examples of pairs of compounds so related are

$$Fe(CO)_5, \qquad Fe(CO)_2(NO)_2$$
$$Mn(CO)_4NO, \qquad Mn(CO)(NO)_3$$
$$Co(CO)_3NO, \qquad Co(NO)_3$$

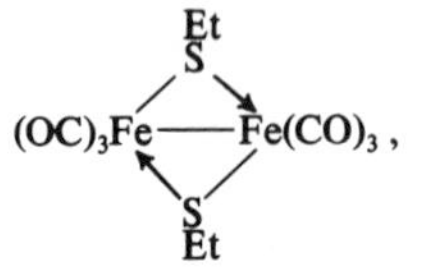

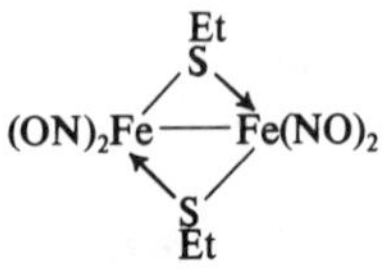

(Roussin's red ester)

The compound $Co(NO)_3$ is the only pure nitrosyl [i.e., $M(NO)_n$ type] compound that is well established.[46] $Fe(NO)_4$ and $Ru(NO)_4$ have been reported but their true structural natures have never been unambiguously established. For the iron compound an ionic structure, $NO^+[Fe(NO)_3]^-$, has been suggested; the anion is plausible, being isoelectronic with $Co(NO)_3$. The rules of stoichiometry given above would lead one to propose that $Cr(NO)_4$ might exist, isostructural and isoelectronic with $Ni(CO)_4$, $Co(CO)_3NO$ and $Fe(CO)_2(NO)_2$, but there is no report of it as yet.

It should be noted that the designation "linear MNO group" does not disallow a small amount of bending in cases where the group is not in an axially symmetric environment, just as with terminal MCO groups. Thus MNO angles as low as 161° in $[(ON)_2FeI]_2$[47] and 167° in the similar Roussin red ester, $[(ON)_2FeS(C_2H_5)]_2$, and quite commonly 170–175° may be found in "linear" MNO groups. As noted later, "bent MNO groups" have angles of 120–140°.

In compounds containing both MCO and linear MNO groups, the M—C and M—N bond lengths differ by a fairly constant amount, ~0.07 Å, which is approximately equal to the expected difference in the C and N radii. Thus, structural data suggest that under comparable circumstances M—CO and M—NO bonds are about equally strong.[48] In a chemical sense the M—N bonds appear to be stronger, since substitution reactions on mixed carbonyl nitrosyl compounds typically result in displacement of CO in preference to NO. For example, $Co(CO)_3NO$ reacts with a variety of R_3P, X_3P, amine and RNC compounds, invariably to yield the $Co(CO)_2(NO)L$ product.

The NO vibration frequencies for linear MNO groups substantiate the idea of extensive M to N π-bonding, leading to appreciable population of

[46] I. H. Sabherwal and A. B. Burg, Chem. Comm., 1970, 1001.
[47] L. F. Dahl, E. Rodulfo de Gil and R. D. Feltham, J. Amer. Chem. Soc., 1969, 91, 1653.
[48] B. A. Frenz, J. H. Enemark and J. A. Ibers, Inorg. Chem., 1969, 8, 1288.

NO π^* orbitals. Both the NO and O_2^+ species contain one π^* electron and their stretching frequencies are 1860 and 1876 cm^{-1}, respectively. Thus the observed frequencies in the range 1800–1900 cm^{-1}, which are typical of linear MNO groups in molecules with small or zero charge, indicate the presence of approximately one electron pair shared between metal $d\pi$ and NO π^* orbitals.

The compound $[(C_2H_5)_2NCS_2]_2FeNO$ contains a linear FeNO group,[49] but requires a word of comment because of its electronic structure.[50] The $[(C_2H_5)_2NCS_2]_2Fe$ group does not present a half-filled $d\pi$ orbital along with an empty σ orbital according to the idealized prescription presented above for forming linear MNO bonds. In this case, the interaction of the two filled $d\pi$ orbitals with the π^*(NO) orbitals is very strong ($\nu_{NO} = 1735$ cm^{-1}). The antibonding component is thus raised to such a high energy that the odd electron is transferred to a metal orbital, probably the $\sigma^*(d_{x^2-y^2})$ orbital.

Bridging NO Groups. These appear to occur somewhat less commonly than bridging CO groups, but there are the same two types, doubly and triply bridging. A triply bridging NO group occurs in the compound $(h^5\text{-}C_5H_5)_3Mn_3(NO)_4$, which also contains three doubly bridging NO's.[51] Unfortunately disorder in the crystal prevented the determination of accurate molecular dimensions. A crystallographically characterized symmetrical doubly bridging NO group[52a] occurs in

$$(h^5\text{-}C_5H_5)(NO)Cr(\mu\text{-}NO)(\mu\text{-}NH_2)Cr(NO)(h^5\text{-}C_5H_5).$$

and quite unsymmetrical doubly bridging NO groups occur[52b] in $h^5\text{-}C_5H_5(NO_2)Mn(\mu\text{-}NO)_2Mn(h^5\text{-}C_5H_5)NO$.

From infrared spectroscopy, however, it is certain that bridging NO groups occur in various other compounds, e.g., in $[h^5\text{-}C_5H_5CoNO]_2$.

Just as with the corresponding types of bridging CO groups, the NO stretching frequencies decrease with the extent of the bridging. Thus, in $(h^5\text{-}C_5H_5)_3Mn_3(NO)_4$ there are two bands due to the doubly bridging NO groups at 1543 and 1481 cm^{-1} and one from the triply bridging group at 1320 cm^{-1}. In $(h^5\text{-}C_5H_5)(NO)Cr(\mu\text{-}NO)(\mu\text{-}NH_2)Cr(NO)(h^5\text{-}C_5H_5)$ the terminal NO groups absorb at 1644 cm^{-1} while the bridging group has a frequency of 1505 cm^{-1}.

Bridging NO groups are also to be regarded as three-electron donors. The doubly bridging ones may be represented as

$$\ddot{N}\!::\!\ddot{O}$$

where the additional electron required to form two metal-to-nitrogen single bonds is supplied by one of the metal atoms. The situation is formally quite analogous to that for bridging halogen atoms.

[49] G. R. Davies *et al.*, *J. Chem. Soc.*, B, **1970**, 1275.
[50] H. B. Gray, I. Bernal and E. Billig, *J. Amer. Chem. Soc.*, 1962, **84**, 3404.
[51] R. C. Elder, F. A. Cotton and R. A. Schunn, *J. Amer. Chem. Soc.*, 1967, **89**, 3645.
[52] (a) L. Y. Y. Chan and F. W. B. Einstein, *Acta Cryst.*, 1970, B, **26**, 1899; (b) J. L. Calderon *et al.*, *Chem. Comm.*, **1971**, 1476.

Bent, Terminal MNO Groups.   It has long been known that NO can form single bonds to univalent groups such as halogens and alkyl radicals, affording the bent species

$$\begin{array}{cc} \overset{..}{N}=\overset{..}{O}: & \overset{.}{N}=\overset{..}{O}: \\ \diagup & \text{and} \quad \diagup \\ X & R \end{array}$$

Relatively recent studies[53] have shown that metal atoms with suitable electron configurations and partial coordination shells may bind NO in a similar way. It appears that this type of NO complex is formed when the incompletely coordinated metal ion, L_nM, would have a $t_{2g}^6 e_g$ configuration, thus being prepared to form one more single σ bond. Table 22-7 lists some compounds in which this type of M—NO structure has been demonstrated by X-ray crystallography.

TABLE 22-7
Compounds for which Bent MNO Groups have been Proved

	< MNO (deg.)	ν_{NO} (cm^{-1})
$[Co\ en_2Cl(NO)]^+$	121	1611
$[IrCl(CO)\ (PPh_3)_2NO]^+$	124	1680
$IrCl_2(NO)\ (PPh_3)_2$	123	1560
$IrI(CH_3)\ (NO)\ (PPh_3)_2$	120	1525
$[RuCl(NO)\ (PPh_3)_2NO]^+$	136	1687[a]
$[Co(NH_3)_5NO]^{2+\ b}$	119	1610
$Co[S_2CN(CH_3)_2]_2(NO)$	139[c]	1626

[a] The other NO group is of the linear MNO type and its stretching frequency is 1845 cm^{-1}.

[b] This mononuclear cation occurs in the so-called "black" $[Co(NH_3)_5NO]Br_2$. The "red isomer" is evidently not an isomer but a binuclear species, $[(NH_3)_5Co(N_2O_4)Co(NH_3)_5]Br_4$, and thus not a nitrosyl complex at all.

[c] Structure imprecise owing to crystal twinning.

The NO stretching frequencies for the authenticated cases fall in the range 1525–1690 cm^{-1}, that is, generally lower than those for linear MNO systems, except perhaps when the latter occur in anionic complexes, such as $[Cr(CN)_5(NO)]^{4-}$ ($\nu_{NO} = 1515$ cm^{-1}). Tentatively, at least, this may be used as a criterion of structure type. If this is, indeed, a valid criterion, then several NO complexes currently in the literature may be strongly suspected of having bent MNO groups, e.g., $Cr(S_2CNR_2)_2(NO)_2$ and others, where very low ν_{NO} are found in neutral or cationic complexes. In organic nitroso compounds, RNO, ν_{NO} is generally found in the 1500–1600 cm^{-1} range.

Nitrosyl Cyano Complexes.   Although they are not basically different from other nitrosyl complexes, $[M(CN)_5NO]^{n-}$ species have generally received special attention. Those studied by X-ray crystallography have octahedrally coordinated metal atoms and linear MNO groups. These

[53] D. J. Hodgson et al., J. Amer. Chem. Soc., 1968, **90**, 4486; D. A. Snyder and D. L. Weaver, Inorg. Chem., 1970, **9**, 2760; C. G. Pierpont et al., J. Amer. Chem. Soc., 1970, **92**, 4760; J. A. Ibers et al., Inorg. Chem., 1971, **10**, 1035; 1043; J. Chem. Soc., A, 1971, 2146.

are $[Fe(CN)_5NO]^{2-}$ (the so-called nitroprusside ion), $[Mn(CN)_5NO]^{3-}$, $[Cr(CN)_5NO]^{4-}$, $[V(CN)_5NO]^{3-}$ and $[Mo(CN)_5NO]^{4-}$. In each of these cases, except $[V(CN)_5NO]^{3-}$, the $M(CN)_5^{n-}$ moiety would have a t_{2g}^5 electron configuration and an empty σ orbital, so that the formation of linear MNO groups is to be expected. The linearity of the CrNO group in $[Cr(CN)_5NO]^{4-}$ has not yet been demonstrated, but this may reasonably be assumed by analogy with the isoelectronic $[Mo(CN)_5NO]^{4-}$ ion where an angle of $175 \pm 3°$ has been found.[54] Recent work shows that the $[V(CN)_5NO]^{5-}$ ion, which would be isoelectronic with the others just mentioned, probably does not exist. The alleged compound $K_5[V(CN)_5NO]\cdot 2H_2O$ is actually $K_3[V(CN)_5NO]\cdot 2H_2O$.[55] The $[V(CN)_5NO]^{3-}$ ion has two less electrons than the others. It is reasonable to suppose that the d_{xy} orbital in this case is vacant so that the two t_{2g} orbitals which are π orbitals with respect to the V—N—O axis (d_{xz}, d_{yz}) are populated by three electrons in the $V(CN)_5$ fragment, just as in the other $M(CN)_5$ fragments, before reaction with NO. Thus, qualitatively, the same sort of metal—NO π system can be established in all cases.

TABLE 22-8

Some Properties of $[M(CN)_5NO]^{n-}$ Ions

Ion	ν_{NO} (cm^{-1})	d_{N-O} (Å)
$[Fe(CN)_5NO]^{2-}$	1939	1.13
$[Mn(CN)_5NO]^{3-}$	1725	1.21
$[Cr(CN)_5NO]^{4-}$	1515	—
$[V(CN)_5NO]^{3-}$	1530	1.29

From the trends in ν_{NO} and d_{N-O} shown in Table 22-8 it can be argued[56a] that the extent and strength of π bonding increases substantially in the order Fe < Mn < Cr ~ V. As the effective nuclear charge for the $d\pi$ orbitals increases (from V to Fe) the π orbitals drop in energy and contract, and therefore overlap less well with the NO π^* orbitals. The composition of the occupied, bonding pair of π MO's that extends over the M—N—O group has been estimated to vary from > 50% NO π^* in the vanadium compound to only about 25% in the iron compound.

The ion $[Co(CN)_5NO]^{3-}$ has often been discussed along with those just mentioned, but it very probably differs structurally since the $Co(CN)_5^{3-}$ group, as an incomplete octahedron, would have filled t_{2g} orbitals and one electron in a σ orbital; it should thus combine with NO to form a bent Co—N—O group. An X-ray study of this would be of much interest.

Ruthenium Nitrosyls. These nitrosyls are more numerous than those of any other element and are, in fact, a dominant feature of ruthenium chemistry (see Section 26-F-5).

[54] D. H. Svedung and N.-G. Vannerberg, *Acta Chem. Scand.*, 1968, **22**, 1551.
[55] S. Jagner and N.-G. Vannerberg, *Acta Chem. Scand.*, 1970, **24**, 1988.
[56] (a) P. T. Manoharan and H. B. Gray, *Inorg. Chem.*, 1966, **5**, 823; (b) D. B. Brown, *Inorg. Chim. Acta*, 1971, **5**, 314.

Ambiguities in Assigning Oxidation States. In both linear and bent terminal NO complexes there is ambiguity about how to assign an oxidation state to the metal.

For the linear type, a formalism found commonly in the recent literature regards the NO as an NO^+ ion. For example, in the well-known nitroprusside ion $[Fe(CN)_5NO]^{2-}$, the metal is assigned an oxidation number of $+2$, which implies that it has a d^6 electron configuration. While this is a plausible value, Mössbauer and other physical evidence[56b] indicates that the effective oxidation number is closer to $+3$. This is the number obtained if NO is considered to interact as an essentially neutral ligand. The idea that NO first transfers its π^* electron to iron in order to become NO^+, that this *cation* then serves as an electron-*donor*, and that finally $d\pi$ electron density is then transferred back to the very π^* orbitals from which an electron was initially removed seems to be an unnecessarily complicated way of describing the net electron distribution. Its only virtue seems to be that in this and certain other cases it leads to an intuitively satisfactory oxidation number for the metal. However, there are many cases where it leads to intuitively *un*satisfactory oxidation numbers, viz., in $Fe(NO)_2(CO)_2$ or $Co(NO)_3$ where oxidation numbers of -2 and -3 are obtained.

Since the only thing that is physically and chemically important is the actual electron distribution, it seems that in essentially covalent bonding situations ionic formulations with their attendant necessity for, or consequence of, assigning oxidation numbers are best avoided. By treating the M—NO bonds as we have here, namely, in terms of a pairing of the π^* electron of NO with a $d\pi$ electron of the metal we describe an essentially electroneutral electron distribution and at the same time achieve a correct specification of the electron configuration on the metal atom.

For the bent M—N—O groups the literature already contains two completely opposed formulations. It has been suggested by some[57] that such compounds be treated as NO^- complexes, whereas other workers have proposed that they be considered as NO^+ complexes wherein the metal atom is donating a pair of σ electrons to an sp^2 σ orbital on the nitrogen atom. It seems unlikely that either of these extreme ionic descriptions is physically realistic. Probably, in each case there is an essentially covalent M—N σ bond which may be more or less polarized, one way or the other depending on the exact nature of the metal and its attached ligands. Probably the NO^- description comes close to the truth in some cases where it also gives intuitively reasonable formal oxidation numbers. In $[Coen_2Cl(NO)]^+$ and $[Co(NH_3)_5NO]^{2+}$, for example, if NO is treated as a coordinated NO^- ion, the presence of $Co(III)$ is then implied and this seems quite consistent with the large body of cobalt(III) ammine chemistry. The idea that these species contain $Co(I)$ donating to NO^+ appears a little bizarre.

To summarize the position with respect to both the linear and the bent

[57] R. D. Feltham, W. Silverthorn and G. McPherson, *Inorg. Chem.*, 1969, **8**, 344, and earlier papers in that series.

terminal MNO groups, the electron distribution associated with each can perhaps best be stated as follows. In linear MNO groups the basic interaction is by donation of a σ pair from N along with a covalent $N\pi^*-d\pi$ bond, often with some additional bonding due to $d\pi \rightarrow N\pi^*$ interaction. In bent MNO groups the interaction is basically a single covalent bond formed by a σ electron on the metal and a σ electron in an sp^2 orbital on nitrogen, with a range of bond polarity possible.

22-13. Donor Complexes of Group V and Group VI Ligands

A variety of trivalent phosphorus, arsenic, antimony and bismuth compounds, as well as divalent oxygen, sulfur and selenium compounds, can also give complexes with transition metals. These donor molecules are, of course, quite strong Lewis bases and give complexes with acceptors such as BR_3 compounds where d orbitals are not involved. However, the donor atoms do have empty $d\pi$ orbitals and back-acceptance into these orbitals is possible, as shown in Fig. 22-17.

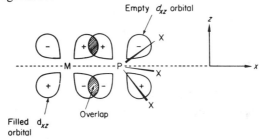

Fig. 22-17.   Diagram showing the back-bonding from a filled metal d orbital to an empty phosphorus $3d$ orbital in the PX_3 ligand, taking the internuclear axis as the z-axis. An exactly similar overlap occurs in the yz plane using the d_{yz} orbitals.

The extent to which back-donation occurs depends on the identity of the donor atom and on the electronegativity of the groups attached to it. Analogous PX_3, AsX_3 and SbX_3 compounds differ very little, while the ligands having a nitrogen atom, which lacks π orbitals, cause significantly lower frequencies for the CO vibrations, as indicated by the CO stretching frequencies (cm^{-1}) in the following series of compounds:

$(PCl_3)_3Mo(CO)_3$	2040, 1991
$(AsCl_3)_3Mo(CO)_3$	2031, 1992
$(SbCl_3)_3Mo(CO)_3$	2045, 1991
dien $Mo(CO)_3$	1898, 1758

A similar situation prevails with the Group VI ligands. The electronegativity of the groups X has a pronounced effect, as shown by the following CO stretching frequencies:

$[(C_2H_5)_3P]_3Mo(CO)_3$	1937	1841
$[(C_6H_5O)_3P]_3Mo(CO)_3$	1994	1922
$[Cl_2(C_2H_5O)P]_3Mo(CO)_3$	2027	1969
$(Cl_3P)_3Mo(CO)_3$	2040	1991
$(F_3P)_3Mo(CO)_3$	2090	2055

Based on data such as these an extensive series of ligands involving Group V and Group VI donor atoms can be arranged in the following order of decreasing π acidity:[58]

$$CO \sim PF_3 > PCl_3 \sim AsCl_3 \sim SbCl_3 > PCl_2(OR) > PCl_2R > PCl(OR)_2 > PClR_2 \sim P(OR)_3$$
$$> PR_3 \sim AsR_3 \sim SbR_3 \sim SR_2 > RCN > NR_3 \sim OR_2 \sim ROH > H_2NCOR$$

Within an extensive series of phosphorus ligands, PX_3, involving 47 different X groups and 70 different ligands (including mixed ones, i.e. PXX_2' and $PXX'X''$ types), a ranking was achieved based on the observed CO stretching frequencies of $PX_3Ni(CO)_3$ molecules.[59] It was found that the ordering of the X groups according to their effect in increasing the electron-acceptor power of the PX_3 or PXX_2' molecules correlated extremely well with an ordering of X groups according to their influence on the ionization constants of acids of the type $XX'P(O)OH$.

Correlations such as those just mentioned, extensive and consistent as they are, do not in themselves unequivocally prove that PX_3 compounds have significant π acidity. There have, in fact, been several attempts[60] to attribute all variations in CO stretching frequencies as a function of X in PX_3 substituents to σ bonding, with the conclusion that there is little if any π component in any of the M—PX_3 bonds. According to this view, an increase in the electronegativity of X causes PX_3 to donate less well to M, thus causing M to engage in less $d\pi \rightarrow \pi^*$ dative bonding to the CO groups.

A way to distinguish objectively between these views is to examine the effect of changing X on the length of the M—PX_3 bond. If the change to a more electronegative X leads to greater $Md\pi \rightarrow Pd\pi$ bonding, while not greatly affecting the σ bond, a strengthening and, hence, shortening of the M—PX_3 bond should result. If only σ bonding is of importance and the effect of increasing the electronegativity of X on ν_{CO} is due solely to a weakening of the M—PX_3 σ bond then, indubitably, the M—PX_3 bond must get weaker and longer. Precisely the required experiment has been done,[61] in which the structures of $Ph_3PCr(CO)_5$ and $(PhO)_3PCr(CO)_5$ were determined. The P—Cr bond in the latter compound is 0.11 Å *shorter* than that in the former. The role of $Md\pi \rightarrow Pd\pi$ bonding in such compounds would seem then to be indisputable.

It is noteworthy that ir spectral evidence as well as photoelectron spectroscopy[62] show that PF_3 is as good or better than CO as a π acid. It is not then surprising that PF_3 can form an extensive group of $M_x(PF_3)_y$ com-

[58] W. D. Horrocks, Jr., and R. C. Taylor, *Inorg. Chem.*, 1963, **2**, 723; F. A. Cotton, *Inorg. Chem.*, 1964, 3, 702; W. Strohmeier and F. J. Müller, *Chem. Ber.*, 1967, **100**, 2812.

[59] C. A. Tolman, *J. Amer. Chem. Soc.*, 1970, **92**, 2953.

[60] M. Bigorgne, *J. Inorg. Nuclear Chem.*, 1964, **26**, 107; R. J. Angelici and M. D. Malone, *Inorg. Chem.*, 1967, 6, 1731; L. M. Venanzi, *Chem. in Britain*, 1968, 4, 162; see also R. J. P. Williams, *Chem. in Britain*, 1968, **4**, 277, and S. O. Grim and D. A. Wheatland, *Inorg. Chem.*, 1969, **8**, 1716, for rebuttals.

[61] H. J. Plastas, J. M. Stewart and S. O. Grim, *J. Amer. Chem. Soc.*, 1969, **91**, 4326.

[62] J. C. Green, D. I. King and J. H. D. Eland, *Chem. Comm.*, 1970, 1121.

pounds, many of which are analogs of corresponding $M_x(CO)_y$ compounds and some of which, e.g., $Pd(PF_3)_4$ and $Pt(PF_3)_4$, actually have no carbonyl analogs. Even anions, such as $Co(PF_3)_4^-$, and hydrides, such as $HCo(PF_3)_4$, are known.[63] Photoelectron spectroscopy has also indicated that in $Ni(PF_3)_4$ and $Pt(PF_3)_4$ the charge shifts due to σ and π bonding are practically compensatory,[64] which is similar to the situation in M—CO bonding. Thus the ability of PX_3 ligands, especially PF_3, to stabilize low oxidation states of metals[65] is understandable.

The PF_2 group can serve as a bridging group, but differs from doubly bridging CO in being a 3-electron ligand. The compounds $(PF_3)_3Co(PF_2)_2$-$Co(PF_3)_3$[66], and $(PF_3)_3Fe(PF_2)_2Fe(PF_3)_3$[67] have been reported; the latter presumably contains a metal—metal bond while the former does not.

When it comes to the actual stability of PX_3-substituted carbonyl molecules, especially those with two or more substituents, steric effects become very important.[68] For highly substituted molecules, such as $Ni(PX_3)_4$, steric effects become much more important than electronic effects in determining the position of exchange equilibria such as

$$NiL_4 + 4L' \rightleftharpoons NiL_3L' + 3L' + L \rightleftharpoons NiL_2L'_2 + 2L' + 2L$$
$$\rightleftharpoons NiLL'_3 + L' + 3L \rightleftharpoons NiL'_4 + 4L$$

and partial dissociation equilibria, e.g.,

$$Ni(PPh_3)_4 \rightleftharpoons Ni(PPh_3)_3 + PPh_3$$

It has been shown that these steric effects can be correlated with an easily measured parameter, the angle of the ligand cone. The ligand cone is defined by a conic surface with apex at the metal atom which can just enclose the van der Waals surface of all ligand atoms over all rotational orientations about the M—P bond. The smaller the ligand cone angle the greater the competitive binding ability of the ligand in situations where steric interference between ligands becomes significant. In terms of the ligand cone angle, PPh_3 with a cone angle of $184 \pm 2°$ is the poorest of common ligands while $P(OCH_3)_3$ $(107 \pm 2°)$ is one of the very best.

22-14. Cyanide Complexes

The formation of cyanide complexes is restricted almost entirely to the transition metals of the d block and their near neighbours Zn, Cd and Hg. This appears to indicate that metal—CN π bonding is of importance in the stability of cyanide complexes and, as shown below, there is evidence of various types to support this. However, the π-accepting tendency of CN^-

[63] T. Kruck, Angew. Chem. Internat. Edn., 1967, 6, 53.
[64] I. H. Hillier et al., Chem. Comm., 1970, 1316.
[65] L.-Malatesta, R. Ugo and S. Cenini, Advances in Chemistry Series, No. 62, American Chemical Society, Washington, D.C., page 318.
[66] T. Kruck and W. Lang, Angew. Chem. Internat. Edn., 1967, 6, 454.
[67] P. L. Timms, Chem. Comm., 1969, 1033.
[68] C. A. Tolman, J. Amer. Chem. Soc., 1970, 92, 2956.

does not seem to be nearly as high as for CO, NO^+ or RNC, which is, of course, reasonable in view of its negative charge. CN^- is a strong nucleophile so that back-bonding need not be invoked to explain the stability of its complexes with metals in normal (i.e. II, III) oxidation states. Nonetheless, because of the formal similarity of CN^- to CO, NO and RNC, it is convenient to discuss its complexes in this Chapter.

Types of Cyano Complex. The majority of cyano complexes have the general formula $[M^{n+}(CN)_x]^{(x-n)-}$ and are anionic, such as $[Fe(CN)_6]^{4-}$, $[Ni(CN)_4]^{2-}$, $[Mo(CN)_8]^{3-}$. Mixed complexes, particularly of the type $[M(CN)_5X]^{n-}$, where X may be H_2O, NH_3, CO, NO, H or a halogen, are also well known.

Although bridging cyanide groups might be expected in analogy with those formed by CO, none has been definitely proved. However, linear bridges, M—CN—M, are well known and play an important part in the structures of many crystalline cyanides and cyano complexes. Thus AuCN, $Zn(CN)_2$ and $Cd(CN)_2$ are all polymeric with infinite chains.

The free anhydrous acids corresponding to many cyano anions can be isolated, examples being $H_3[Rh(CN)_6]$ and $H_4[Fe(CN)_6]$. These acids are thus different from those corresponding to many other complex ions, such as $[PtCl_6]^{2-}$ or $[BF_4]^-$, which cannot be isolated except as hydroxonium (H_3O^+) salts, and they are also different from metal carbonyl hydrides in that they contain no metal—hydrogen bonds. Instead, the hydrogen atoms are situated in hydrogen bonds between anions, i.e. $MCN\cdots H\cdots NCM$. Different sorts of structures arise depending on the stoichiometry. For example, in $H[Au(CN)_4]$ there are chains, while in $H_2[Pd(CN)_4]$ there are sheets. For octahedral anions there is a difference in structure depending on whether or not the number of protons equals half the number of cyanide ions. For $H_3[M(CN)_6]$ compounds an infinite, regular three-dimensional array is formed in which the hydrogen bonds are perhaps symmetrical, whereas in other cases the structures appear to be more complicated.

Metal—Cyanide Bonding. The cyanide ion occupies a very high position in the spectrochemical series, gives rise to large nephelauxetic effects and produces a strong *trans*-effect. All these properties are accounted for most easily by postulating M—CN π bonding, and semiempirical MO calculations support this. From close analysis of the vibrational spectra of cyanide complexes, the existence of π bonding has been confirmed more directly, but it does not appear to be nearly as extensive as in carbonyls.

However, the cyanide ion does have the ability to stabilize metal ions in low formal oxidation states, and it presumably does this by accepting electron density into its π^* orbitals. The fact that cyano complexes of zerovalent metals are generally much less stable (in a practical as opposed to a well-defined thermodynamic or chemical sense) than similar metal carbonyls has often been taken to show the poor π-acidity of CN^-, but it should be noted that the cyano compounds, e.g., $[Ni(CN)_4]^{4-}$, are anionic and might thus tend to be more reactive for this reason alone. In some instances cyano complexes are known in two or even three successive oxidation states, $[M(CN)_n]^{x-}$, $[M(CN)_n]^{(x+1)-}$, $[M(CN)_n]^{(x+2)-}$, and in this respect they resemble the complexes to be discussed next.

LIGANDS WITH EXTENDED π SYSTEMS

22-15. Bipyridine and Similar Amines[69]

The three ligands bipyridine (bipy) (22-X), 1,10-phenthroline (phen) (22-XI), and terpyridine (terpy) (22-XII) form complexes with a variety of

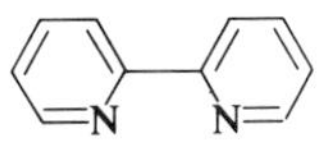

(22-X) 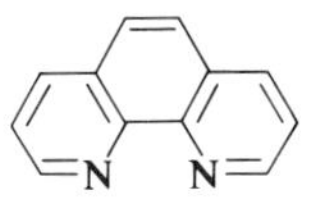   (22-XI) 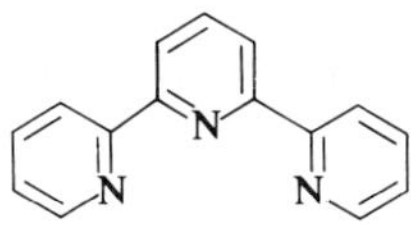   (22-XII)

metal atoms in a great range of oxidation states. For metal ions in "normal" oxidation states, the interaction of metal $d\pi$ orbitals with the ligand π^* orbitals is significant, but not exceptional. However, these ligands can stabilize metal atoms in very low formal oxidation states and in such complexes it is believed that there is extensive occupation of the ligand π^* orbitals, so that the compounds can often be best formulated as having radical anion ligands, $L^{\cdot}$. Most work has been carried out on bipy complexes, but it is apparent that phen and terpy afford very similar ones.

The methods of preparation are varied. Complexes involving transition-metal ions in "normal" oxidation states can usually be obtained by conventional reactions and then reduced with a variety of reagents such as Na/Hg, Mg or BH_4^-. The most general method employs Li_2bipy:

$$MX_y + yLi_2bipy + n(bipy) \xrightarrow{\text{THF}} M(bipy)_n + yLiX + yLi(bipy)$$

It is also noteworthy that many highly reactive organometallic compounds can be stabilized against hydrolysis, for example, by addition of these ligands. This is particularly true of R_2Zn, R_2Cd and R_2Hg species.

The low-valent metal complexes are invariably colored, and usually intensely so. For those containing transition metals, the bands responsible are believed to be mainly $d \to \pi^*$ charge-transfer bands. In other cases $\pi \to \pi^*$ ligand bands may also be active. For the ML_2 complexes of Be, Mg, Ca and Sr, esr spectra show the presence of a ground, or low-lying excited, state that is a spin triplet. This can be best explained by postulating an M^{2+} cation and two radical anion ligands, $L^{\cdot}$. Also for the tris-bipy complexes $[Cr(bipy)_3]^+$, $[V(bipy)_3]$ and $[Ti(bipy)_3]^-$, esr data indicate that there is strong σ interaction with metal 4s orbitals, while the unpaired electrons are extensively delocalized on the ligands.

[69] W. R. McWhinnie and J. D. Miller, *Adv. Inorg. Chem. Radiochem.*, 1969, **12**, 135.

22-16. 1,2-Dithiolene Ligands[70]

Although intensive study of complexes of these ligands began only in 1957, an enormous body of very detailed literature has been built up. The discussion here will be limited mainly to the basic types encompassed by formula (22-XIII), although several other classes of complexes such as (22-XIV), (22-XV) and even (22-XVI) are doubtless related. Moreover, there are many

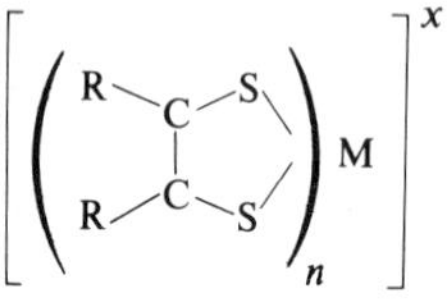

R = H, alkyl, C_6H_5, CF_3, CN
$n = 2$; $x = 0, -1, -2$
$n = 3$; $x = 0, -1, -2, -3$

(22-XIII)

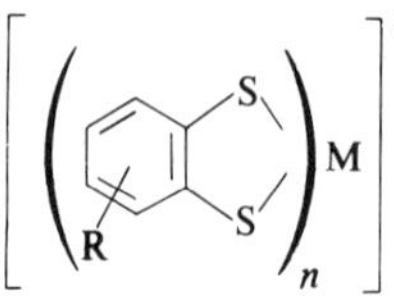

R = alkyl
$n = 2$; $x = 0, -1, -2$

(22-XIV)

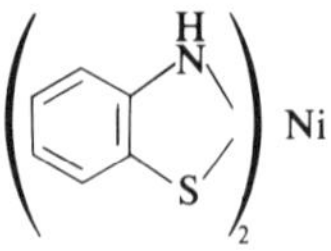

(22-XV)

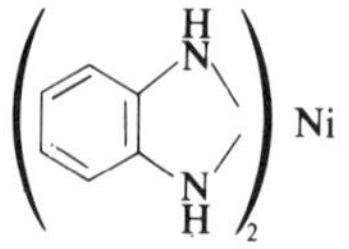

(22-XVI)

mixed ligand complexes in which 1,2-dithiolene ligands are present along with others such as h^5-C_5H_5, NO, olefins, CN^-, O, etc. In all formulae and structures presented here, lines drawn between atoms are for guidance but do not necessarily indicate valences or bond orders, since the chief property of these systems is the lack of simple descriptions of the bonding.

These compounds, particularly those of types (22-XIII) and (22-XIV), can be prepared in various ways. A few representative preparative reactions are shown in (22-2), (22-3) and (22-4).

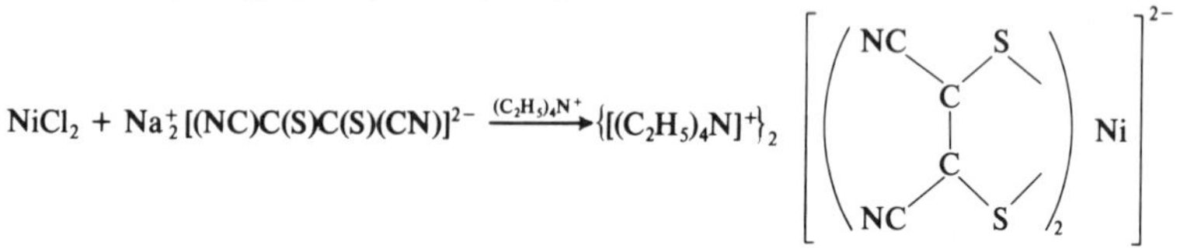

$$NiCl_2 + Na_2^+[(NC)C(S)C(S)(CN)]^{2-} \xrightarrow{(C_2H_5)_4N^+} \{[(C_2H_5)_4N]^+\}_2$$

(22-2)

$$Ni(CO)_4 + 2(C_6H_5)_2C_2 + 4S \longrightarrow$$

$\xrightarrow[(C_2H_5)_4N^+]{p\text{-}H_2NC_6H_4NH_2}$

$[(C_2H_5)_4N]^+$

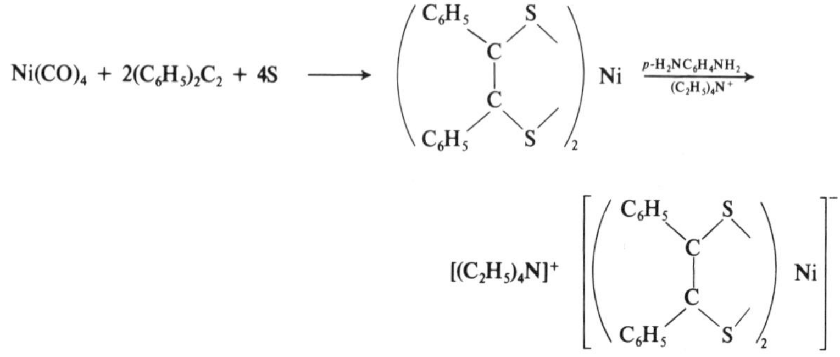

(22-3)

[70] J. A. McCleverty, *Progr. Inorg. Chem.*, 1968, **10**, 49; R. Eisenberg, *Progr. Inorg. Chem.*, 1970, **12**, 295.

$$Ni(CO)_4 + 2 \begin{array}{c} F_3C \\ \diagdown \\ \diagup \\ F_3C \end{array} \begin{array}{c} C-S \\ | \quad | \\ C-S \end{array} \longrightarrow \left(\begin{array}{c} F_3C \quad S \\ \diagdown C \diagup \diagdown \\ | \quad \\ C \\ \diagup \quad \diagdown \\ F_3C \quad S \end{array} \right)_2 Ni \xrightarrow[\text{acetone} + (C_2H_5)_4N^+]{\text{spontaneously in}}$$

$$[(C_2H_5)_4N]^+ \left[\left(\begin{array}{c} F_3C \quad S \\ \diagdown C \diagup \diagdown \\ | \quad \\ C \\ \diagup \quad \diagdown \\ F_3C \quad S \end{array} \right)_2 Ni \right]^- \xrightarrow[+ (C_2H_5)_4N^+]{p\text{-}H_2NC_6H_4NH_2}$$

$$[(C_2H_5)_4N]_2^+ \left[\left(\begin{array}{c} F_3C \quad S \\ \diagdown C \diagup \diagdown \\ | \quad \\ C \\ \diagup \quad \diagdown \\ F_3C \quad S \end{array} \right)_2 Ni \right]^{2-} \qquad (22\text{-}4)$$

For complexes containing only dithiolene ligands, four types of structure have been observed. These are shown in Fig. 22-18. The planar, D_{2h}, structure is found for a majority of the structurally characterized bis complexes. The second structure type, observed in the remaining bis complexes, is dimeric, each metal atom being 5-coordinate. The metal atoms are significantly displaced from the planes of the dithiolene ligands (by 0.2–0.4 Å) but the bridging linkages are relatively weak. The third type of structure is one having trigonal-prismatic, D_{3h}, coordination geometry.

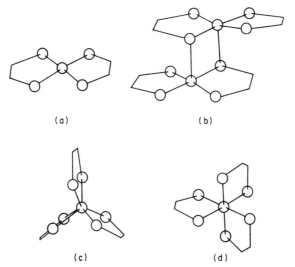

Fig. 22-18.   The four basic structure types for "pure" dithiolene complexes: (a) square coordination; (D_{2h} molecular symmetry); (b) five-coordinate dimer; (c) trigonal-prismatic coordination (D_{3h} symmetry); (d) octahedral-coordination (D_3 symmetry).

The interligand $S \cdots S$ distances in this structure are rather short (3.0–3.1 Å), which suggests that there may be weak interactions directly between the sulfur atoms; this structure is found only in a few of the more highly oxidized or neutral tris complexes, one example being $Mo[Se_2C_2(CF_3)_2]_3$[71]; the remaining ones have the more conventional, octahedral form of 6-coordination.

The most interesting characteristic of the 1,2-dithiolene complexes is their ability, unsurpassed and seldom matched among other complexes, to form series of structurally similar molecules differing in their electron populations. Examples are:

$$\{Ni[S_2C_2(CN)_2]_2\} \underset{-e}{\overset{+e}{\rightleftharpoons}} \{Ni[S_2C_2(CN)_2]_2\}^- \underset{-e}{\overset{+e}{\rightleftharpoons}} \{Ni[S_2C_2(CN)_2]_2\}^{2-}$$

$$[CoL_2]_2 \underset{-e}{\overset{+e}{\rightleftharpoons}} [CoL_2]_2^- \underset{-e}{\overset{+e}{\rightleftharpoons}} [CoL_2]_2^{2-} \underset{-2e}{\overset{+2e}{\rightleftharpoons}} 2[CoL_2]^{2-} \qquad [L = S_2C_2(CF_3)_2]$$

$$[CrL_3]^0 \underset{-e}{\overset{+e}{\rightleftharpoons}} [CrL_3]^{1-} \underset{-e}{\overset{+e}{\rightleftharpoons}} [CrL_3]^{2-} \underset{-e}{\overset{+e}{\rightleftharpoons}} [CrL_3]^{3-} \qquad [L = S_2C_2(CN)_2]$$

The electronic structures of the 1,2-dithiolene complexes have provoked a great deal of controversy. The ring system involved can be written in two extreme forms, (22-XVII) and (22-XVIII); the formal oxidation number of

(22-XVII)   (22-XVIII)

the metal differs by two in these two cases. In molecular-orbital terms the problem is one of the extent to which electrons are in metal d orbitals or delocalized over the ligand. Undoubtedly, in general, considerable delocalization occurs, which accounts for the ability of these complexes to exist with such a range of electron populations. The exact specification of orbital populations in any given case is a difficult and subtle question that we shall not discuss in detail here.

[71] C. G. Pierpoint and R. Eisenberg, *J. Chem. Soc., A,* **1971**, 2285.

Further Reading

π-Bonding

Pettit, L. D., *Quart. Rev.*, 1971, **25**, 1 (a review of multiple bonding and back-donation in complexes of transition and non-transition elements).

Carbonyls

Abel, E. W., and F. G. A. Stone, *Quart. Rev.*, 1969, **23**, 325 (structures of metal carbonyls).

Abel, E. W., and F. G. A. Stone, *Quart. Rev.*, 1970, **24**, 498 (preparations and chemistry of carbonyls; this review gives a valuable list of reviews of individual subtopics).

Abel, E. W., and S. P. Tyfield, *Adv. Organometallic Chem.*, 1970, **8**, 117 (metal carbonyl cations).

Biryukov, B. P., and I. T. Struckhov, *Russ. Chem. Rev.*, 1970, **39**, 9 (metal-metal distances and radii in polynuclear carbonyls and π-complexes).

Calderazzo, F., R. Ercoli and G. Natta in *Organic Syntheses via Metal Carbonyls*, I. Wender and P. Pino, eds., Interscience-Wiley, 1968 (a comprehensive review of many phases of metal carbonyl chemistry with extensive tables).

Haines, L. M., and M. H. B. Stiddard, *Adv. Inorg. Chem. Radiochem.*, 1969, **12**, 53 (vibrational spectra).

Hieber, W., *Adv. Organometallic Chem.*, 1970, **8**, 1 (a review of forty years research by a pioneer).

Johnson, R. D., *Adv. Inorg. Chem. Radiochem.*, 1970, **13**, 471 (transition metal clusters with π-acid ligands including CO, NO, h^5-C_5H_5, etc.).

King, R. B., *Accounts Chem. Res.*, 1970, **3**, 417 (use of carbonylate anions in synthesis).

Ryang, M., and S. Tsutsumi, *Synthesis*, 1971, **2**, 55 (organic syntheses using metal carbonyls).

Cyanides

Britten, D., in *Perspectives in Structural Chemistry*, Vol. 1, Wiley, 1967 (structural chemistry of the cyano group).

Sharpe, A. G., and B. M. Chadwick, *Adv. Inorg. Chem. Radiochem.*, 1966, **8**, 84 (an authoritative review).

Isocyanides

Malatesta, L., and F. Bonati, *Isocyanide Complexes of Metals*, Interscience-Wiley, 1969.

Volger, A., in *Isonitrile Chemistry*, I. Ugi, ed., Academic Press, 1970 [coordinated isonitriles (isocyanides)].

Group V ligands

Booth, G., *Adv. Inorg. Chem. Radiochem.*, 1964, **6**, 1 (comprehensive review of complexes of phosphines, arsines and stibines).

Kruck, Th., *Angew. Chem. Internat. Edn.*, 1967, **6**, 53 (complexes of PF_3).

Nixon, J. F., *Adv. Inorg. Chem. Radiochem.*, 1970, **13**, 364 (fluorophosphines and their metal complexes).

Nyholm, R. S., *Quart. Rev.*, 1970, **24**, 1 (complexes of perfluoro ligands includi ng PF_3)

See also Further Reading for Chapters 23, 24 and 25.

23

Organometallic Compounds of Transition Metals

The ability of transition elements to form organo derivatives only began to be appreciated properly during the 1950's, whereas the organo derivatives of non-transition metals and metalloids had been actively studied for more than a hundred years before that. Nonetheless, the organometallic compounds of the transition metals now constitute an enormous, diversified field of chemistry which is still expanding rapidly. It gains breadth by merging into the field of metal carbonyls and related compounds discussed in Chapter 22. It also affords structural and bonding situations quite different from those generally occurring among the organo derivatives of non-transition elements. For these reasons an entire Chapter is devoted to the topic at this point.

The transition metals form a variety of compounds in which there is a normal σ bond to carbon, although the binary alkyls and aryls are usually less stable thermally and chemically than those in which other ligands, notably π-bonding ligands, are also bound to the metal atom. More important, the unique characteristics of d orbitals also allow certain types of unsaturated hydrocarbons and some of their derivatives to be bound to metals in a non-classical manner to give molecules or ions with structures that have no counterpart elsewhere in chemistry. Not only is a wide range of organo compounds of different types isolable, but labile species play an important role in many reactions of olefins, acetylenes and their derivatives catalyzed by metal complexes. Many of these reactions also involve the incorporation of carbon monoxide and/or hydrogen into unsaturated molecules. These catalytic processes are accorded separate treatment in Chapter 24.

23-1. Olefin Complexes[1a]

About 1830, Zeise, a Danish pharmacist, characterized a compound of stoichiometry $PtCl_2 \cdot C_2H_4$ which is now known to be a dimer with chlorine bridges; he also isolated, from the reaction products of chloroplatinate with ethanol, salts of the ion $[C_2H_4PtCl_3]^-$. The structures of these ethylene com-

[1a] H. W. Quinn and J. H. Tsai, *Adv. Inorg. Chem. Radiochem.*, 1969, **12**, 217.

plexes were fully established only recently although they were the first orga-nometallic derivatives of transition metals to be prepared.

It was later found that certain metal halides or ions other than Pt^{II}, notably Cu^I, Ag^I, Hg^{II} and Pd^{II}, formed complexes when treated with a variety of olefins. Thus cuprous chloride in aqueous suspension absorbs ethylene, both components dissolving well beyond their normal solubilities and in a 1:1 mole ratio. Solid cuprous halides also absorb some gaseous olefins, but the dissociation pressures of the complexes are quite high. The reaction of silver ions especially, with a variety of unsaturated substances, has been studied by physical measurements such as distribution equilibria between an aqueous and an organic solvent phase. The results can be accounted for in terms of equilibria of the type

$$Ag^+(aq) + olefin = [Ag\ olefin]^+(aq)$$

For series of olefins, certain trends can be correlated with steric and induc-tive factors. In some cases, the interaction of hydrocarbons with Ag^+ ions gives crystalline precipitates that are often useful for purification of the olefins. Thus cyclooctatetraene or bicyclo-2,5-heptadiene, when shaken with aqueous silver perchlorate (or nitrate), give white crystals of stoichio-metry olefin·$AgClO_4$ or 2olefin·$AgClO_4$, depending on the conditions.[1b] Benzene with $AgNO_3$, $AgClO_4$ or $AgBF_4$ also gives crystalline complexes. In the structure of $[C_6H_6·Ag]^+ClO_4^-$ the metal ion is asymmetrically located with respect to the ring.

Before proceeding to a more general discussion, let us consider the struc-ture and bonding in the two of the simplest olefin–metal complexes, the monoolefin–metal and the 1,3-diene–metal species. Several fundamental considerations applicable in more complex cases will thereby be brought out.

Monoolefin–Metal Bonding. The basis for discussion of bonding is a clear specification of structure. The structures of two monoolefin complexes, including the anion of Zeise's salt, are shown in Fig. 23-1. The fact that the plane of the olefin, and indeed the C=C axis itself, are perpendicular to

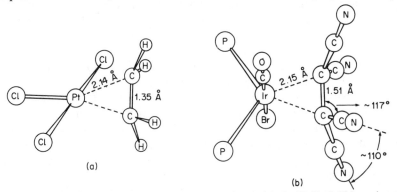

Fig. 23-1. The structures of two monoolefin complexes: (a) the $[PtCl_3C_2H_4]^-$ anion of Zeise's salt; (b) the $(Ph_3P)_2(CO)BrIr[C_2(CN)_4]$ molecule, with Ph groups omitted for clarity.

[1b] E. A. H. Griffith and E. L. Amma, *J. Amer. Chem. Soc.*, 1971, **93**, 3167.

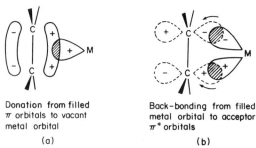

Donation from filled   Back-bonding from filled
π orbitals to vacant   metal orbital to acceptor
metal orbital   π* orbitals

(a)   (b)

Fig. 23-2. Diagrams showing the molecular orbital view of olefin–metal bonding according to Dewar. The donor part of the bond is shown in (a), and the back-bonding part in (b).

one of the expected bond directions from the central metal atom is of key significance. In addition, the expected line of a bond orbital from the metal strikes the C=C bond at its midpoint. The entire $Cl_3PtC_2H_4$ group in Zeise's salt has C_{2v} symmetry.[2]

The most generally useful, and generally accepted, description of the bonding that is fully consistent with the observed structural parameters is the so-called Dewar–Chatt MO description. It is illustrated in Fig. 23-2. The bonding is assumed to consist of two interdependent components: (a) overlap of the π-electron density of the olefin with a σ-type acceptor orbital on the metal atom; and (b) a "back-bond" resulting from flow of electron density from filled metal d_{xz} or other $d\pi$-$p\pi$ hybrid orbitals into *antibonding* orbitals on the carbon atoms. This view is thus similar to that discussed for the bonding of carbon monoxide and similar weakly basic ligands and implies the retention of appreciable "double-bond" character in the olefin. Of course, the donation of π-bonding electrons to the metal σ orbital and the introduction of electrons into the π-antibonding orbital both weaken the π bonding in the olefin, and in every case except the anion of Zeise's salt there is significant lengthening of the olefin C—C bond. There appears to be some correlation between lengthening of the bond and the electron-withdrawing power of the substituents of the olefin. This is exemplified by the two structures shown in Fig. 23-1, where the $C_2(CN)_4$ complex has a C—C bond about as long as a normal single bond.

In the extreme of a very long C—C distance, it is not unreasonable to formulate the bonding as involving two *2e-2c* M—C bonds and a C—C single bond—a kind of metallo-cyclopropane ring. It will be seen in Fig. 23-1b that the bond angles at the two olefin carbon atoms are not inconsistent with such a view. Actually, this representation of the bonding and that embodied in the Dewar–Chatt MO description are neither incompatible nor mutually exclusive; they are, rather, complementary, and there is a smooth gradation of one description into the other. The Dewar–Chatt scheme is, of course, more flexible and is thus more widely useful.

[2] L. Manojlovic-Muir, K. W. Muir and J. A. Ibers, *Discuss. Faraday Soc.*, 1969, **47**, 84 (a comprehensive review of structures of monoolefin–metal complexes).

The important qualitative idea about metal–olefin bonding, which is exemplified by these simple systems, is that the bonding has dual character. There is donation of π-bonding electrons from the highest filled olefin orbitals into metal orbitals of suitable symmetry and there is donation of electrons from filled metal orbitals of suitable symmetry back into the lowest-lying π-antibonding orbitals of the olefin. The two components are synergically related. As one component increases, it tends to promote an increase in the other. There is an obvious similarity to metal—CO bonding.

On both theoretical[3] and experimental[4] grounds, it appears that the metal—olefin bond is essentially electroneutral, with donation and back-acceptance approximately balanced.

Scarcely more complicated, in principle, than simple monoolefin complexes are those in which two unconjugated double bonds form independent linkages to a metal atom. Two of the more important complex-forming non-conjugated diolefins are 1,5-cyclooctadiene and norbornadiene, representative complexes of which are shown as (23-I) and (23-II), respectively. An interesting case of three unconjugated double bonds coordinated to one metal atom is shown in (23-III).

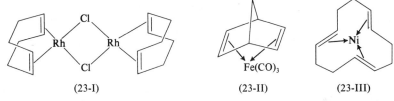

| (23-I) | (23-II) | (23-III) |

When two or more conjugated double bonds are engaged in bonding to a metal atom the interactions become more complex, though qualitatively the two types of basic, synergic components are involved. The case of 1,3-butadiene unit is an important one and shows why it would be a drastic oversimplification to treat such cases as simply collections of separate monoolefin–metal interactions.

1,3-Butadiene–Metal Bonding. Two extreme formal representations of the bonding of a 1,3-butadiene group to a metal atom are possible, as shown in Fig. 23-3. The degree to which individual structures approach either of

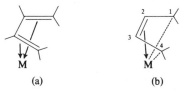

| (a) | (b) |

Fig. 23-3. Two extreme formal representations of the bonding of a 1,3-butadiene group to a metal atom. (a) implies that there are two more or less independent monoolefin–metal interactions, while (b) depicts σ bonds to C-1 and C-4 coupled with a monoolefin–metal interaction to C-2 and C-3.

[3] J. W. Moore, *Acta Chem. Scand.*, 1966, **20**, 1154.
[4] J. P. Yesinowski and T. L. Brown, *Inorg. Chem.*, 1971, **10**, 1097.

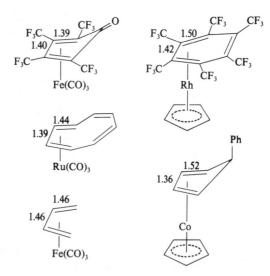

Fig. 23-4. Some structures in which butadiene units are bonded to metal atoms. The numbers give the average lengths, in Å, of the C-1/C-2 and C-3/C-4 bonds and the C-2/C-3 bond length, in each case.

these extremes can be judged by the lengths of the C—C bonds; a short-long-short pattern is indicative of (a) while a long-short-long pattern is indicative of (b). Fig. 23-4 shows a selection of structures covering a range of such distance ratios. Actually, in no case has a pronounced short-long-short pattern been established with precision. The actual variation seems to lie between approximate equality of all three bond lengths and the long-short-long pattern. This can be understood in the following way.

In the ground state of butadiene the bonding MO's are occupied in such a way that the bond orders and bond lengths are as shown in Fig. 23-5. To promote butadiene from its ground state to the first excited state, an electron is removed from the highest occupied bonding orbital and placed in the lowest-lying antibonding orbital. It is seen in Fig. 23-5 that this has the effect of weakening (hence lengthening) the C-1/C-2 and C-3/C-4 bonds while strengthening (hence shortening) the C-2/C-3 bond.* When a butadiene

	Bond orders			Bond lengths (Å)		
(a) Ground state	1.89	1.45	1.89	1.36	1.45	1.36
(b) First excited state	1.45	1.67	1.45	1.45	1.39	1.45

Fig. 23-5. Bond orders and bond lengths for 1,3-butadiene in its ground state and the first excited state, based on Hückel calculations.

* It is entirely understandable that just these bond-order changes occur, if the nature of the orbitals is considered in detail. The highest filled orbital in the ground state has the form $\phi_1 + \phi_2 - (\phi_3 + \phi_4)$. It is markedly bonding for C-1/C-2 and C-3/C-4 but antibonding for C-2/C-3. The lowest unoccupied MO in the ground state of butadiene has the form $\phi_1 - \phi_2 - \phi_3 + \phi_4$. It is antibonding for C-1/C-2 and C-3/C-4 but bonding for C-2/C-3.

unit becomes bonded to a metal atom we expect that there will be two parts to the interaction (just as in the case of monoolefin–metal interactions): donation of electron density from the highest filled orbital of the butadiene to the metal and donation of metal electron density to the lowest antibonding orbital of the butadiene. Clearly these two shifts in electron density have much the same effect on the π-bonding and hence on the C—C bond lengths in the butadiene component as does promotion of the isolated butadiene molecule from its ground configuration to the first excited configuration. Thus, the more strongly the butadiene is bonded to the metal the more pronounced should be the long-short-long pattern of the C—C bond lengths. Apparently, in order to create a stable molecule the bonding must be at least strong enough to cause approximate equalization of the C—C bond lengths.

It may also be noted that when the butadiene group competes with CO groups for metal-atom electron density, back-donation into its antibonding orbitals will be restricted. However, when it competes with the h^5-C_5H_5 group (to be discussed on page 736), which is not itself a strong back-acceptor, a considerable flow of metal electron density into the butadiene antibonding orbital can occur. These considerations lead to structural predictions, namely, that for (butadiene)$M(CO)_3$ molecules the long-short-long pattern will be less pronounced than for (butadiene)$M(h^5$-$C_5H_5)$ molecules. Fig. 23-4 suggests that these predictions are qualitatively correct.

In the case of $C_4H_6Fe(CO)_3$, where the C—C distances are about equal, an nmr study[5] in which the ^{13}C–H coupling constants were measured indicates that C—C and C—H bonds continue to be formed with carbon sp^2 hybrid orbitals. This supports the foregoing description in which it is essentially pure carbon $p\pi$ orbitals that are employed in binding the olefin to the metal. Thus, it is probably unwise to refer to Fig. 23-3b as a π–σ bonding scheme although in the extreme such a designation might become partly true.

It is noteworthy that the $Fe(CO)_3$ moiety has a pronounced affinity for 1,3-diene groups and irradiation of 1,4-dienes in presence of $Fe(CO)_5$ and heating with $Fe_3(CO)_{12}$ are efficient methods of effecting 1,4- to 1,3-diene isomerization.[6] Examples are illustrated in the annexed formulae.

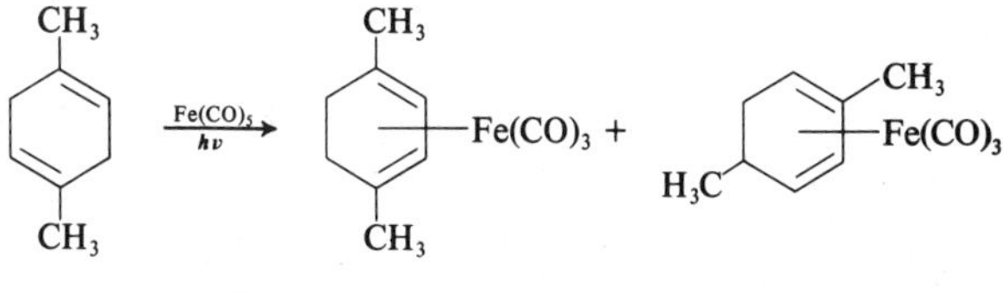

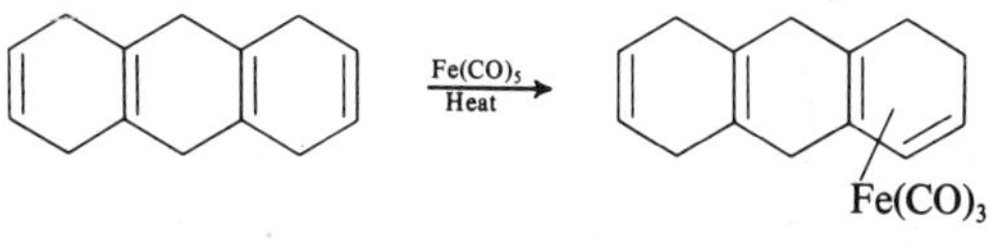

[5] H. L. Retcofsky, E. N. Frankel and H. S. Gutowsky, *J. Amer. Chem .Soc.*, 1966, **88**, 2710.
[6] A. J. Birch *et al.*, *J. Chem. Soc., A*, **1968**, 332.

A System of Notation.   Before the discussion proceeds further, and more complex structures are considered, a system of notation[7] will be outlined. The number of carbon atoms that are attached to a metal atom is specified by a prefix such as *trihapto-, tetrahapto-, pentahapto-,* etc. These prefixes can be abbreviated as h^3-, h^4-, h^5-, etc. If necessary, the bound carbon atoms may be specified by numbers, using a conventional naming and numbering scheme for the organic group. The annexed illustrations should make clear the use of the notation.

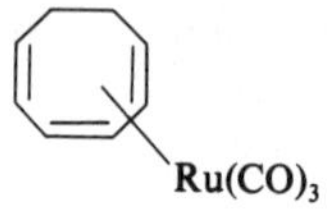

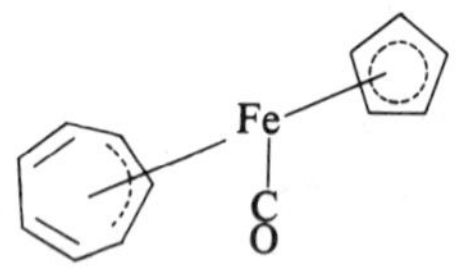

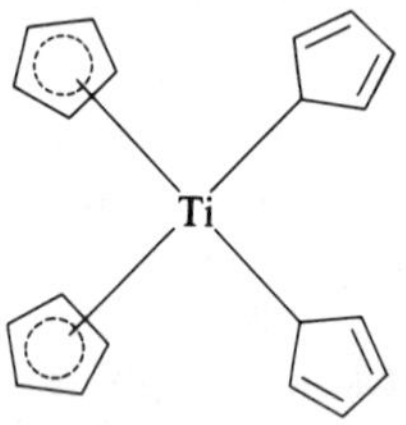

(1,2,3,4-*tetrahapto*-1,3,5-cyclooctatriene) (tricarbonyl)ruthenium
(1,2,3-*trihapto*cycloheptatrienyl) (*pentahapto*cyclopentadienyl)carbonyliron
$(h^5\text{-}C_5H_5)_2(h^1\text{-}C_5H_5)_2Ti$

In general, an incompletely coordinated (coordinately unsaturated) metal atom in an $M(CO)_n$ group will combine with a polyolefinic compound so as to complete its normal coordination requirement. It does so on the basis that each double bond is, formally, the equivalent of one 2-electron donor. Several categories may be distinguished among the many compounds of this type:

(1) $M(CO)_x$ *lacks n electron pairs.* In this case $M(CO)_x$ (or some similar group with an incompletely coordinated metal atom) will combine with an *n*-ene olefin, or part of an olefin with more than *n* π bonds. For example, the $Cr(CO)_3$, $Mo(CO)_3$ and $W(CO)_3$ groups require three double bonds to complete their coordination. Thus they combine with a 1,3,5-triene, such as cycloheptatriene as in (23-IV). $Fe(CO)_3$, on the other hand, has unused binding capacity sufficient to interact with only two double bonds; it therefore combines with 1,3-dienes, as already noted, or with 1,3-diene portions of more extended olefin systems (see 23-V).

[7] F. A. Cotton, *J. Amer. Chem. Soc.*, 1968, **90**, 6230.

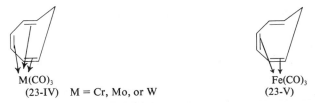

M(CO)₃
(23-IV)   M = Cr, Mo, or W

Fe(CO)₃
(23-V)

The cyclooctatetraene molecule, with four essentially unconjugated double bonds interacts with $M(CO)_x$ or other incompletely coordinated components in various ways, depending on the specific requirements of that component. For instance, with $(h^5\text{-}C_5H_5)Co$ or $PtCl_2$, it uses its 1- and 5-olefinic linkages, as in (23-VI) and (23-VII), while $Fe(CO)_3$, with its predilection for binding 1,3-diolefins, attaches to two adjacent double bonds, as in (23-VIII).

(23-VI)   (23-VII)   (23-VIII)

(2) $M(CO)_x$ *lacks an odd number of electrons.* The observant reader will note that all cases so far considered involve metal-containing components in which the metal requires only an even number of electrons to complete its coordination. There are, of course, about an equal number of instances in which an odd number of electrons are required. In those cases the metal will combine with some sort of *polyenyl* (as opposed to polyene) group. The important cases are those in which 3 or 5 electrons are required. In these the organic moieties are

$$R_2\underset{\text{Allyl}}{C\text{---}\overset{R}{C}\text{---}CR_2}$$
$$R_2\underset{\text{Dienyl}}{C\text{---}\overset{R}{C}\text{---}\overset{R}{C}\text{---}\overset{R}{C}\text{---}CR_2}$$

which can be considered as neutral radicals capable of donating 3 and 5 electrons, respectively, to a metal atom. The *allyl* compounds are so numerous and important that they are accorded special treatment in Section 23-5, where additional comments will be made on the chemistry of dienyl species as well. It will suffice here to give a few representative examples of polyenylmetal complexes, namely, (23-IX) to (23-XI). It will be noted

(23-IX)   (23-X)   (23-XI)

in these examples that here again a bound enyl (allyl) or dienyl fragment may be part of a more extended olefinic chain.

As a general rule, polyolefins do not tend to form very stable compounds in which only one double bond is coordinated to a given metal atom. There is a strong tendency for at least a second double bond to become coordinated *via* displacement of some other ligand. Thus we have the annexed reactions, where species in square brackets are not isolable:

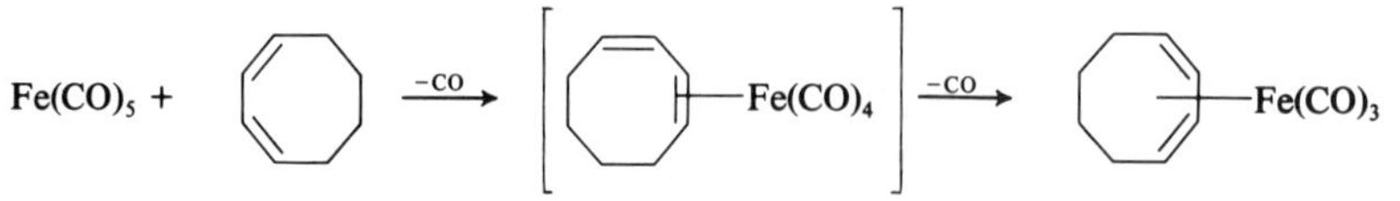

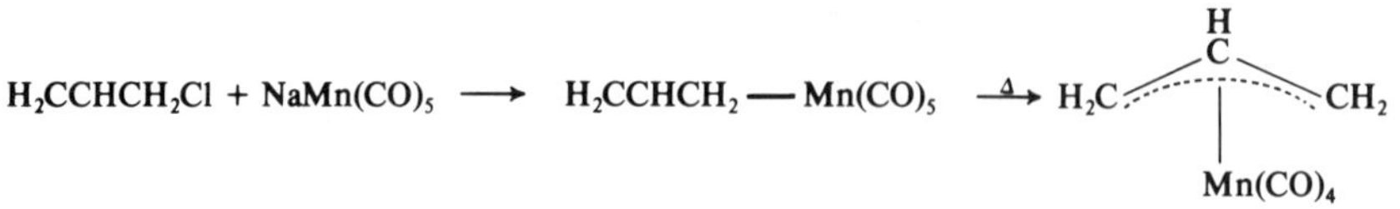

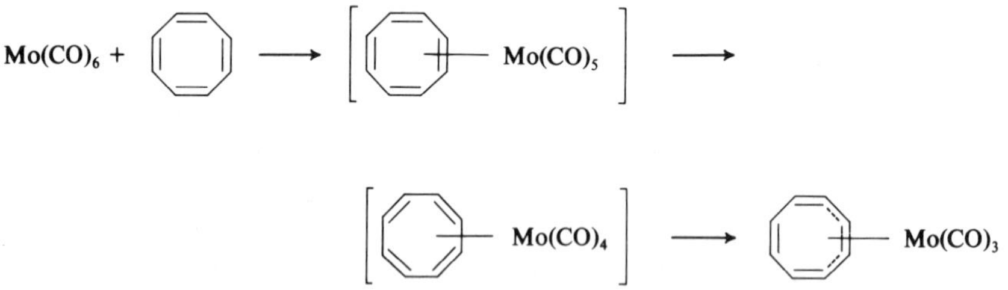

$Fe(CO)_5$ + [cyclooctatetraene] $\xrightarrow{-CO}$ [cyclooctatetraene—$Fe(CO)_4$] $\xrightarrow{-CO}$ cyclooctatetraene—$Fe(CO)_3$

H_2CCHCH_2Cl + $NaMn(CO)_5$ $\longrightarrow$ H_2CCHCH_2—$Mn(CO)_5$ $\xrightarrow{\Delta}$ $H_2C\cdots C(H)\cdots CH_2$ / $Mn(CO)_4$

$Mo(CO)_6$ + [cyclooctatetraene] $\longrightarrow$ [cyclooctatetraene—$Mo(CO)_5$] $\longrightarrow$

[cyclooctatetraene—$Mo(CO)_4$] $\longrightarrow$ cyclooctatetraene—$Mo(CO)_3$

In cases where the polyene or polyenyl group forms a closed cycle, e.g., the dienyl group (A) or the triene (B), we have a special type of compound,

(A)   (B)

often with exceptional stability. The next few Sections will be devoted to such compounds.

23-2. Cyclopentadienyl Compounds

Although the cyclopentadienyl group, C_5H_5, has long been known to form compounds with metal or metalloidal atoms, no such compound was considered noteworthy until the first cyclopentadienyl compound of a transition metal was reported in 1951. This compound was bis(*pentahapto*cyclopentadienyl)iron, $(h^5\text{-}C_5H_5)_2Fe$, commonly called *ferrocene*. X-ray studies, as well as a variety of other physical and chemical evidence, showed that ferrocene has a "sandwich" structure (23-XII). The remarkable structure of this molecule as well as its extraordinary thermal stability (to above 500°C) provided the initial impetus for the immense and continuing research effort on ferrocene[8] and a host of related compounds.

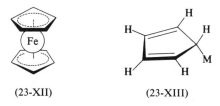

       (23-XII)   (23-XIII)

In terms of structure and bonding, nearly all non-ionic cyclopentadienyl-metal compounds can be placed in one of two main classes. (1) Those containing one or more $(h^5\text{-}C_5H_5)M$ groups, as in ferrocene; and (2) those containing $(h^1\text{-}C_5H_5)M$ groups (23-XIII). In this Section we shall be mainly concerned with the former, since the $(h^1\text{-}C_5H_5)M$ compounds are merely one of many M—C single-bond systems (see Section 23-6). There are also some essentially ionic cyclopentadienyl compounds which will be mentioned later in this Section.

Cyclopentadiene is a weak acid ($pK_a \sim 20$) and with a variety of bases can give salts of the symmetrical cyclopentadienide ion, $C_5H_5^-$; like other ring systems that give "sandwich" compounds this ion has the "aromatic" sextet of π electrons. Because of the intrinsic stability of the $C_5H_5^-$ anion, it has long been customary to treat cyclopentadienylmetal compounds as though they contain $C_5H_5^-$ anions and metal cations, much as metal halides are considered to contain X^- ions and cations. Thus, ferrocene is usually regarded as an iron(II) compound, made up of Fe^{2+} and $2C_5H_5^-$. As always, such an assignment of formal oxidation numbers need not be taken too literally.

A general method for the preparation of all *pentahapto*cyclopentadienyl compounds is the reaction of sodium cyclopentadienide with a metal halide or complex halide in tetrahydrofuran, ethylene glycol dimethyl ether, dimethylformamide or similar solvent. A solution of the sodium salt is obtained by treating dispersed sodium with the hydrocarbon in tetrahydrofuran:

$$2C_5H_6 + Na \rightarrow C_5H_5^- + Na^+ + H_2 \text{ (main reaction)}$$
$$3C_5H_6 + 2Na \rightarrow 2C_5H_5^- + 2Na^+ + C_5H_8$$
$$C_5H_5^- + MR^+ \rightarrow C_5H_5MR$$

An alternative method useful in some cases employs a strong base, preferably diethylamine, for example,

$$2C_5H_5 + 2(C_2H_5)_2NH + FeCl_2 \xrightarrow[\text{of amine}]{\text{in excess}} (h^5\text{-}C_5H_5)_2Fe + 2(C_2H_5)_2NH_2Cl$$

Since the $C_5H_5^-$ anion functions as a uninegative ligand, the di-h^5-cyclopentadienylmetal complexes are of the type $[(h^5\text{-}C_5H_5)_2M]X_{n-2}$,

[8] M. Rosenblum, *The Iron Group Metallocenes; Ferrocene, Ruthenocene, Osmocene*, John Wiley and Sons, 1965.

where the oxidation state of the metal M is n and X is a uninegative ion. Hence in the II oxidation state we obtain neutral, sublimable, and organic solvent-soluble molecules like $(h^5\text{-}C_5H_5)_2Fe$ and $(h^5\text{-}C_5H_5)_2Cr$, and in III, IV and V oxidation states species such as $(h^5\text{-}C_5H_5)_2Co^+$, $(h^5\text{-}C_5H_5)_2TiCl_2$ and $(h^5\text{-}C_5H_5)_2NbBr_3$, respectively.

All the $3d$ elements have been obtained in neutral molecules, and, with the exceptions of the manganese and titanium compounds discussed below, they appear to have the same structure and essentially the same bonding as in ferrocene; however, only ferrocene is air-stable, the others being sensitive to destruction or oxidation by air; the stability order is $Ni > Co > V \gg Cr$. It is still uncertain whether $(h^5\text{-}C_5H_5)_2Ti$ actually exists. Apparently, there are at least two substances, one metastable relative to the other, of stoichiometry $C_{10}H_{10}Ti$, as well as various derivatives such as H_2 and N_2 complexes. The chemistry is complex and knowledge of structures virtually non-existent as yet[9] (see also Section 25-A-4).

The cationic species, several of which can exist in aqueous solutions provided these are acidic, behave like other large unipositive ions, e.g., Cs^+, and can be precipitated by silicotungstate, $PtCl_6^{2-}$, BPh_4^- and other large anions. The $(h^5\text{-}C_5H_5)_2Co^+$ ion is remarkably stable and is unaffected by concentrated sulfuric and nitric acids, even on heating.

Some typical h^5-cyclopentadienyl compounds are given in Table 23-1.

TABLE 23-1

Some Di-h^5-cyclopentadienylmetal Compounds

Compound	Appearance; m.p. (°C)	Unpaired electrons	Other properties[a]
$(h^5\text{-}C_5H_5)_2Fe$	Orange crystals; 174	0	Oxidized by Ag^+(aq), dil. HNO_3; $h^5\text{-}Cp_2Fe^+ = h^5\text{-}Cp_2Fe$, $E^0 = -0.3$ V (vs. SCE); stable thermally to $> 500°$
$(h^5\text{-}C_5H_5)_2Cr$	Scarlet crystals; 173	2	Veryair- sensitive; soluble HCl giving C_5H_6 and blue cation, probably $[h^5\text{-}C_5H_5CrCl(H_2O)_n]^+$
$(h^5\text{-}C_5H_5)_2Ni$	Bright green; 173 (d.)	2	Fairly air-stable as solid; oxidized to Cp_2Ni^+; NO gives $h^5\text{-}CpNiNO$; Na/Hg in C_2H_5OH gives $h^5\text{-}CpNiC_5H_7$
$(h^5\text{-}C_5H_5)_2Co^+$	Yellow ion in aqueous solution	0	Forms numerous salts and a stable strong base (absorbs CO_2 from air); thermally stable to $\sim400°$
$(h^5\text{-}C_5H_5)_2TiCl_2$	Bright red crystals; 230	0	Sl. sol. H_2O giving $h^5\text{-}Cp_2TiOH^+$; C_6H_5Li gives $h^5\text{-}Cp_2Ti(C_6H_5)_2$; reducible to $h^5\text{-}Cp_2TiCl$; Al alkyls give polymerization catalyst
$(h^5\text{-}C_5H_5)_2WH_2$	Yellow crystals; 163	0	Moderately stable in air, soluble benzene, etc.; soluble in acids giving $h^5\text{-}Cp_2WH_3^+$ ion

[a] $Cp = C_5H_5$.

9   H. H. Brintzinger and J. E. Bercaw, *J. Amer. Chem. Soc.*, 1970, **92**, 6182; 1971, **93**, 2045; R. H. Marvich and H. H. Brintzinger, *J. Amer. Chem. Soc.*, 1971, **93**, 2046.

In addition to the *di(pentahapto)*cyclopentadienyls, many other sandwich-bonded compounds exist in which only one h^5-C_5H_5 ring is present, together with other ligands. A few of these compounds, their methods of preparation and reactions are shown in Fig. 23-6.

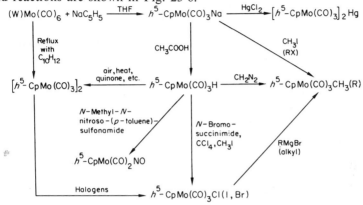

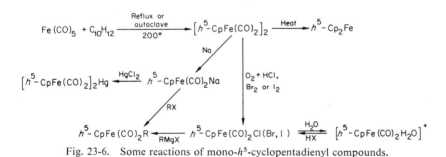

Fig. 23-6.   Some reactions of mono-h^5-cyclopentadienyl compounds.

The h^5-C_5H_5 ring has a resemblance to benzene in its C—C bond order and it was expected that the ring would have aromatic character. For compounds whose properties allow them to survive the reaction conditions, this possibility has been amply realized. The most extensive study has been made with ferrocene, which has been shown to undergo a large number of reactions such as Friedel–Crafts acylation, metalation by butyllithium, sulfonation, etc.; there is now an exceedingly extensive organic chemistry of ferrocene.[10] A monocyclopentadienyl compound, h^5-$C_5H_5Mn(CO)_3$, behaves similarly. It is now known that in certain organic reactions of the molecules the metal atom is directly involved. One example is the intramolecular hydrogen bonding of ferrocene alcohols; another is the protonation of ferrocene (see below) in Friedel–Crafts reactions.

Bonding in (h^5-C_5H_5)Metal Compounds.   The basic qualitative features of this bonding, especially in ferrocene itself and other very symmetrical compounds, can be considered well understood, although much effort is still

[10] D. E. Bublitz and K. L. Rhinehart, Jr., *Organic Reactions*, 1969, **17**, 1.

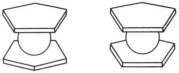

Fig. 23-7. Staggered and eclipsed configurations of a $(h^5\text{-}C_5H_5)_2M$ compound.

devoted to refining the picture quantitatively.[11] Before proceeding to details, it should be noted that the discussion of bonding does not depend critically on whether the preferred rotational orientation of the rings (see Fig. 23-7) in a $(h^5\text{-}C_5H_5)_2M$ compound is staggered (D_{5d}) or eclipsed (D_{5h}); nor is that question unequivocally settled. It is experimentally certain

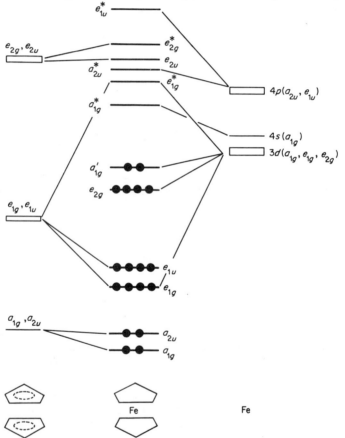

Fig. 23-8. An approximate MO diagram for ferrocene. Different workers often disagree about the exact order of the MO's; the order shown here, especially for the antibonding MO's, may be incorrect in detail, but the general pattern is widely accepted.

[11] R. Prins, *Mol. Phys.*, 1970, **19**, 603; *J. Chem. Phys.*, 1969, **50**, 4804; *Chem. Comm.*, **1970**, 280; C. B. Harris, *Inorg. Chem.*, 1968, **7**, 1517; M. D. Fayer and C. B. Harris, *Inorg. Chem.*, 1969, **8**, 2792; Y. S. Sohn, D. N. Hendrickson and H. B. Gray, *J. Amer. Chem. Soc.*, 1971, **93**, 3603.

that the barrier to rotation is very low ($\leqslant 5$ kJ mol^{-1}). The most recent evidence seems to suggest that the eclipsed configuration may be the more stable,[12] but in condensed phases, especially crystals, where there are intermolecular energies of the same or greater magnitude than the barrier, either configuration may be found.

The bonding is best treated in the LCAO-MO approximation. A semiquantitative energy-level diagram is given in Fig. 23-8. Each C_5H_5 ring, taken as a regular pentagon, has five π MO's, one strongly bonding (a), a degenerate pair which are weakly bonding (e_1) and a degenerate pair which are markedly antibonding (e_2), as shown in Fig. 23-9. The pair of rings taken together then has ten π orbitals and, if D_{5d} symmetry is assumed, so that there is a center of symmetry in the (h^5-C_5H_5)$_2$M molecule, there will be centrosymmetric (g) and antisymmetric (u) combinations. This is the origin of

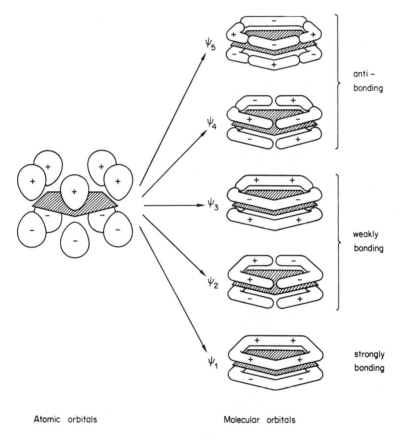

Atomic orbitals   Molecular orbitals

Fig. 23-9.   The π molecular orbitals formed from the set of $p\pi$ orbitals of the C_5H_5 ring.

[12] R. K. Bohn and A. Haaland, *J. Organometallic Chem.*, 1966, **5**, 470; G. J. Palenik, *Inorg. Chem.*, 1970, **9**, 2424.

the set of orbitals shown on the left of Fig. 23-8. On the right are the valence shell ($3d$, $4s$, $4p$) orbitals of the iron atom. In the center are the MO's formed when the ring π orbitals and the valence orbitals of the iron atom interact.

For $(h^5\text{-}C_5H_5)_2Fe$, there are 18 valence electrons to be accommodated: five π electrons from each C_5H_5 ring and eight valence shell electrons from the iron atom. It will be seen that the pattern of MO's is such that there are exactly nine bonding or non-bonding MO's and ten antibonding ones. Hence, the 18 electrons can just fill the bonding and non-bonding MO's, giving a closed configuration. Since the occupied orbitals are either of a type (which are each symmetric around the 5-fold molecular axis) or they are *pairs* of e_1 or e_2 type which are also, *in pairs*, symmetrical about the axis, no intrinsic barrier to internal rotation is predicted. The very low barriers observed may be attributed to van der Waals forces directly between the rings.

It will be seen in Fig. 23-8 that among the principal bonding interactions is that giving rise to the strongly bonding e_{1g} and strongly antibonding e_{1g}^* orbitals. In order to give one concrete example of how ring and metal orbitals overlap, the nature of this particular important interaction is illustrated in Fig. 23-10. This particular interaction is in general the most

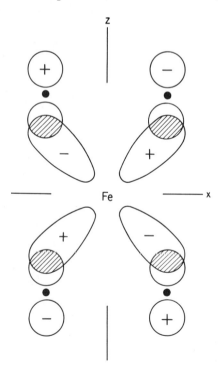

Fig. 23-10.   A sketch showing how one of the e_1-type d orbitals d_{xz} overlaps with an e_1-type ring π orbital to give a delocalized metal–ring bond. The view is a cross-section taken in the xz plane.

important single one because the directional properties of the e_1-type d orbitals (d_{xz} and d_{yz}) give excellent overlap with the e_1-type ring π orbitals, as Fig. 23-10 clearly shows.

Ionic Compounds; Cyclopentadienides.   As noted above, cyclopentadiene is weakly acidic, dissociating to form H^+ and the pentagonal ion $C_5H_5^-$. With the cations of very electropositive metals this ion forms essentially ionic compounds which may appropriately be called *cyclopentadienides*. The principal ones are formed by the alkali metals, MC_5H_5, the alkaline-earth metals, $M(C_5H_5)_2$, the lanthanides and the actinides,[13] $M(C_5H_5)_3$. Europium gives $Eu(C_5H_5)_2$. In addition, $Mn(C_5H_5)_2$ also seems to be ionic, since the $Mn(II)$ therein retains its spin sextuplet d^5 configuration, and the compound is strongly antiferromagnetic, much like other ionic MnX_2 compounds. The ionic cyclopentadienides are typically very reactive toward air and water and react readily with ferrous chloride in tetrahydrofuran to give ferrocene.

Some of the ionic $M(C_5H_5)_2$ compounds have sandwich structures, like that of ferrocene. This is not surprising since such an arrangement is electrostatically the most favorable; structure is not a criterion of bond type in cyclopentadienyl–metal systems. The beryllium compound, $Be(C_5H_5)_2$ has a most interesting unsymmetrical sandwich structure (23-XIV), in which

(23-XIV)

the Be atom presumably oscillates between two positions. The Be radius is so small that even at the closest distance of approach of the two C_5H_5 rings the Be atom cannot make good bonds to both simultaneously.

Although it was originally thought that there was some special feature of the bonding that gave the parallel ring, or ferrocene-like, structure, this view had to be modified to account for the monocyclopentadienyl compounds and the subsequent observation that in several di-h^5-cyclopentadienyl compounds the rings are *not* parallel.

Systems containing only one h^5-C_5H_5 ring include $(h^5$-$C_5H_5)Mn(CO)_3$, $(h^5$-$C_5H_5)Co(CO)_2$, $(h^5$-$C_5H_5)NiNO$ and $(h^5$-$C_5H_5)CuPR_3$.[14] The ring-to-metal bonding in these cases can be accounted for by a conceptually simple modification of the picture given above for $(h^5$-$C_5H_5)_2M$ systems. In each case a principal axis of symmetry can be chosen so as to pass through the metal atom and intersect the ring plane perpendicularly at the ring center;

[13] L. J. Nugent *et al.*, *J. Organometallic Chem.*, 1971, **27**, 365.
[14] F. A. Cotton and J. Takats, *J. Amer. Chem. Soc.*, 1970, **92**, 2353; L. T. J. Delbaere, D. W. McBride and R. B. Ferguson, *Acta Cryst.*, 1970, *B*, **26**, 518.

in other words, the C_5H_5M group is a pentagonal pyramid, symmetry C_{5v}. The single ring may then be considered to interact with the various metal orbitals in about the same way as do each of the rings in the sandwich system. The only difference is that opposite to this single ring is a different set of ligands which interact with the opposite lobes of, for example, the de_1 orbitals, to form their own appropriate bonds to the metal atom.

Compounds with two h^5-C_5H_5 rings that are not parallel are also numerous. They include a number of $(C_5H_5)_2MX_2$ compounds in which M = Ti, Zr or Mo and X represents a univalent group such as a halogen, H or an R group, as well as others such as $(C_5H_5)_2TaH_3$ and $(C_5H_5)_2WH_3^+$. The angle subtended at the metal atom by the centroids of the two rings is generally 130–135°. A detailed description of the metal orbitals employed in these cases is difficult to give since the low symmetry permits a great variety of overlaps. Empirically, however, there is nothing to suggest that the rings in these compounds are bonded significantly differently from those in ferrocene.

It is also possible to have covalent $(h^5$-$C_5H_5)M$ groups even when the metal atom has no valence-shell d orbitals provided it has p orbitals of suitable energy and size. As shown in Fig. 23-11, a pair of p_x and p_y orbitals can overlap with the e_1 π orbitals of C_5H_5 in much the same way as do d_{xz} and d_{yz} orbitals. The C_5H_5In and C_5H_5Tl molecules are the best documented cases of this type of bonding.

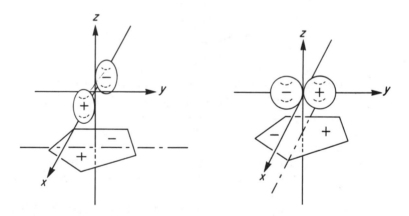

Fig. 23-11.   Diagrams showing how p_x and p_y orbitals are symmetry-adapted to overlap the e_1 π orbitals of a C_5H_5 ring.

23-3. Other Arene–Metal Compounds

The h^5-cyclopentadienyl group is only one of several delocalized carbocyclic groups that can be symmetrically bound to metal atoms, although it has special importance because it gives a greater variety of stable species than any other group. Other symmetric ring systems that form complexes are C_3Ph_3, C_4H_4, C_6H_6, C_7H_7 and C_8H_8. There is a formalism of describing these as if they assume the charge required to achieve an aromatic electron configuration. Since the "magic numbers" for aromaticity are 2, 6 and 10, the carbocycles would be written as

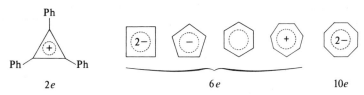

| 2e | 6e | 10e |

The charges thus assumed are often used in assigning formal oxidation numbers to metal atoms, e.g., $(h^5\text{-}C_5H_5)_2Fe$ is said to contain Fe^{II}. Like all assignments of oxidation numbers, except in genuinely ionic compounds, these should not be treated too literally.

Benzenoid–Metal Complexes. Of these other carbocycles, those containing benzene and substituted benzenes are perhaps the most important.[15] Curiously enough the first $(h^6\text{-}C_6H_6)M$-containing compounds were prepared as long ago as 1919, but their true identities were only recognized 35 years later. A series of chromium arene compounds was first obtained by Hein by the reaction of $CrCl_3$ with C_6H_5MgBr; they were formulated by Hein as "polyphenylchromium" compounds, viz., $(Ph)_nCr^{0,1+}$ where $n = 2$, 3 or 4. They actually contain "sandwich"-bonded C_6H_6 and C_6H_5—C_6H_5 groups, as, for example, in (23-XV).

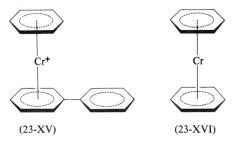

| (23-XV) | (23-XVI) |

The prototype neutral compound, dibenzenechromium, $(C_6H_6)_2Cr$ (23-XVI), has also been obtained from the Grignard reaction of $CrCl_3$ but a more effective method of wider applicability to other metals is the direct interaction of an aromatic hydrocarbon and a transition-metal halide in presence of Al powder as a reducing agent and halogen acceptor and $AlCl_3$ as a Friedel–Crafts-type activator. Although the neutral species are formed directly in the case of chromium, the usual procedure is to hydrolyze the reaction mixture with dilute acid which gives the cations $(C_6H_6)_2Cr^+$,

[15] H. Zeiss, P. J. Wheatley and H. J. S. Winkler, *Benzenoid–Metal Complexes*, Ronald Press, 1966.

(mesitylene)$_2$Ru^{2+}, etc. In several cases these cations can be reduced to the neutral molecules by reducing agents such as hypophosphorous acid.

Dibenzenechromium, which forms dark brown crystals, is much more sensitive to air than is ferrocene, with which it is isoelectronic; it does not survive the reaction conditions of aromatic substitution. X-ray and electron-diffraction measurements on dibenzenechromium and other arene complexes show that the carbon—carbon bond lengths are equivalent.[16] As with the h^5-C$_5$H$_5$ compounds, a variety of complexes with only one arene ring have been prepared, for example,

$$C_6H_5CH_3 + Mo(CO)_6 \xrightarrow{\text{Reflux}} C_6H_5CH_3Mo(CO)_3 + 3CO$$
$$C_6H_6 + Mn(CO)_5Cl + AlCl_3 \rightarrow C_6H_6Mn(CO)_3^+ AlCl_4^-$$

A third method by which some benzenoid complexes may be obtained involves cyclization of doubly substituted acetylenes, viz.,

$$(C_6H_5)_2Mn + 6CH_3C{\equiv}CCH_3 \rightarrow [C_6(CH_3)_6]_2Mn$$

Bisbenzenoid complexes are now known for V, Cr, Mo, W, Mn, Tc, Re, Fe, Ru, Os, Co, Rh and Ir. None, however, is sufficiently stable to have any significant chemistry in which they remain intact. The monobenzenoid metal tricarbonyl species, such as (C$_6$H$_6$)Mo(CO)$_3$ (23-XVII) are generally more stable.

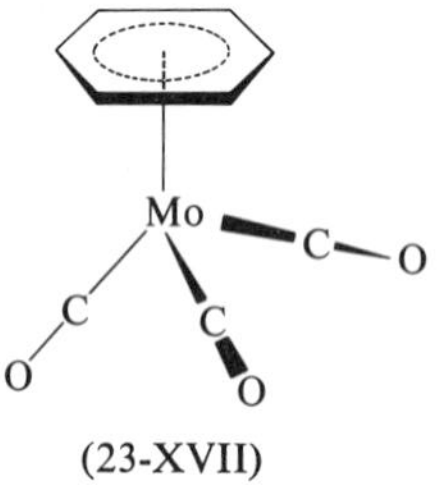

(23-XVII)

Other Carbocyclic Derivatives. The only (C$_3$R$_3$)M compounds yet prepared are those containing (C$_3$Ph$_3$) combined with nickel.[17] They are obtained by the reactions

$$Ni(CO)_4 + C_3Ph_3X \rightarrow (h^3\text{-}C_3Ph_3)Ni(CO)_2X \xrightarrow{\text{py}} (h^3\text{-}C_3Ph_3)NiXpy_2 \cdot py$$
$$(h^3\text{-}C_3Ph_3)NiXpy_2 \cdot py \xrightarrow{C_5H_5Tl} (h^3\text{-}C_3Ph_3)(h^5\text{-}C_5H_5)Ni$$

X-ray crystallographic studies have confirmed the symmetrical *trihapto* character of the (C$_3$Ph$_3$)Ni bonding.

It is well known that cyclobutadiene, C$_4$H$_4$, is antiaromatic and unstable in the free state, but it can be stabilized by bonding to a metal atom with

[16] A. Haaland, *Acta Chem. Scand.*, 1965, **19**, 41; M. F. Bailey and L. F. Dahl, *Inorg. Chem.*, 1965, **4**, 1299, 1314; E. Keulen and F. Jellinek, *J. Organometallic Chem.*, 1966, **5**, 490.

[17] R. N. Tuggle and D. L. Weaver, *J. Amer. Chem. Soc.*, 1969, **91**, 6506; *Inorg. Chem.*, 1971, **10**, 1504; M. D. Rausch, R. M. Tuggle and D. L. Weaver, *J. Amer. Chem. Soc.*, 1970, **92**, 4981.

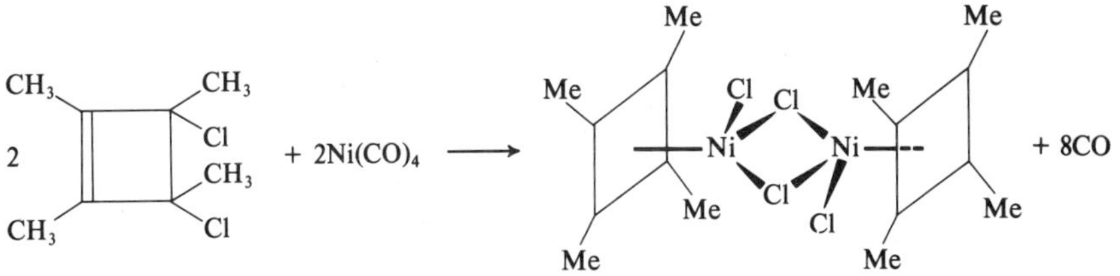

Fig. 23-12.   The preparation and structure of $[(CH_3C)_4NiCl_2]_2$.

suitable electron configuration. The first such compound prepared was the methyl-substituted nickel compound obtained as in Fig. 23-12.

The interaction of diphenylacetylene and $Fe_3(CO)_{12}$ under specified conditions gives a complex $Ph_4C_4Fe(CO)_3$ in which there is an essentially square carbon ring bound to iron.[18] Still another well-documented example is the dinuclear molybdenum species (23-XVIII).[19]

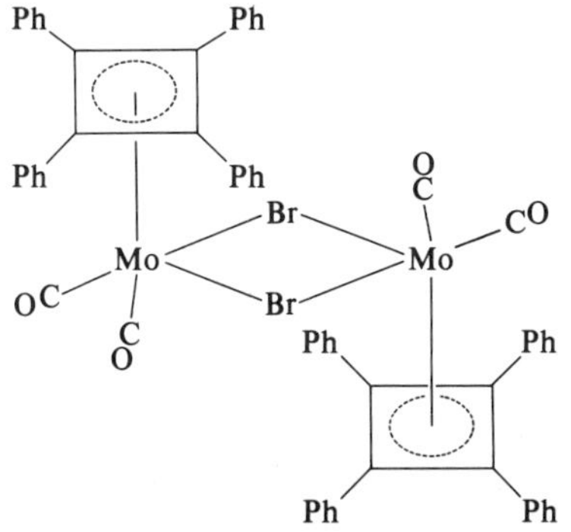

(23-XVIII)

Seven-membered ring complexes are few and not too stable. They are prepared indirectly rather than from $C_7H_7^+$. For example,

$$C_7H_8Mo(CO)_3 + (C_6H_5)_3C^+BF_4^- \rightarrow h^7\text{-}C_7H_7Mo(CO)_3^+ BF_4^- + (C_6H_5)_3CH$$
$$h^5\text{-}C_5H_5V(CO)_4 + C_7H_8 \xrightarrow{\text{Reflux}} (h^5\text{-}C_5H_5)(h^7\text{-}C_7H_7)V$$

Nmr studies show the equivalence of the seven hydrogen atoms in $[C_7H_7Mo(CO)_3]^+$. In the case of $h^7\text{-}C_7H_7Mo(CO)_2(C_6F_5)$ the symmetric bonding of the ring has been demonstrated crystallographically.[20]

As already noted, the cyclooctatetraene molecule is not itself aromatic and it forms a number of complexes in which it retains its polyolefin nature. However, there are a few molecules in which it acts as a planar, aromatic ligand, namely, when it can be considered as a $C_8H_8^{2-}$ ion (aromatic, 10-electron system) bound to a metal atom large enough to overlap effectively with such a large ring. The known compounds are $(C_8H_8)_2U$ and $(C_8H_8)_2Th$, and the arene–metal character has been demonstrated by X-ray diffraction.[21]

[18] R. P. Dodge and V. Schomaker, *Acta Cryst.*, 1965, **18**, 614.
[19] M. Mathew and G. J. Palenik, *Canad. J. Chem.*, 1969, **47**, 705.
[20] M. R. Churchill and T. A. O'Brien, *J. Chem. Soc.*, **1969**, 1110.
[21] A. Streitwieser, Jr., and N. Yoshida, *J. Amer. Chem. Soc.*, 1969, **91**, 7528.

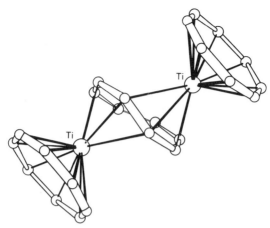

Fig. 23-13. The structure of $(C_8H_8)_3Ti_2$. [Reproduced by permission from Reference 22a.]

For these actinide complexes the participation of $5f$ orbitals has been invoked to explain the metal-to-ring bonding. The only example of symmetrical C_8H_8—M bonding for a d-block metal is found in $(C_8H_8)_3Ti_2$, the structure[22a] of which is shown in Fig. 23-13. The bonding to the central C_8H_8 ring here is quite peculiar and has not been analyzed in detail.

Heterocyclic Sandwich Compounds. There is no reason, in principle, why certain heterocyclic rings that have a significant degree of aromatic character should not be able to form "sandwich"-type bonds to metals. Few such compounds are known and in some cases the failures are not explained, though in others, e.g., of pyridine, they are due to the fact that the heterocyclic atom has lone-pairs that are more basic than the π electrons. Compounds containing thiophene, e.g., $C_4H_4SCr(CO)_3$ and the anion of pyrrole, viz., $C_4H_4NMn(CO)_3$, are analogous to $C_6H_6Cr(CO)_3$ and h^5-$C_5H_5Mn(CO)_3$, respectively. Azaferrocene (h^5-cyclopentadienyl-h^5-pyrrolyliron) has also been obtained as red crystals, m.p. 114°, of lower thermal stability than ferrocene:

$$h^5\text{-}C_5H_5Fe(CO)_2I + C_4H_4NK \xrightarrow{\text{benzene}}$$
$$h^5\text{-}C_5H_5Fe\text{-}h^5\text{-}C_4H_4N$$
$$FeCl_2 + C_5H_5Na + C_4H_4NNa \xrightarrow{\text{THF}}$$

Azaferrocene appears to be isomorphous with ferrocene so that in the crystal the N atoms must be randomly placed in any of the ten positions occupied by carbon in the ferrocene molecule.

Neutral pyrrole compounds, e.g., $(C_4H_4NH)Cr(CO)_3$, can be made by treating $Cr(CO)_3(MeCN)_3$ with pyrrole[22b].

While they can scarcely be called *organo*metallic compounds, the numer-

[22a] H. Dierks and H. Dietrich, *Acta Cryst.*, 1968, *B*, **24**, 58.
[22b] K. Öfele and E. Dotzaner, *J. Organometal. Chem.*, 1971, **30**, 211.

ous carborane-metal complexes (page 253) and $[B_3N_3(CH_3)_6]Cr(CO)_3$ (page 257) may be noted here as closely related, structurally and electronically, to the true arene–metal compounds.

23-4. Alkyne Complexes[23]

By analogy with the bonding of olefins, it might have been expected that for acetylenes, where there are two π bonds at right angles to each other, a metal atom could be bound to each. This possibility has been realized, for example, in the reaction of cobalt carbonyl with acetylenes of various types:

$$Co_2(CO)_8 + RC{\equiv}CR \rightarrow Co_2(CO)_6(RC{\equiv}CR) \qquad (R = CF_3, C_6H_5, \text{ etc.})$$

The diphenylacetylene derivative has been studied by X-ray diffraction and it is indeed found (23-XIX) that two cobalt atoms (which also are bound to each other by a metal—metal bond) are linked to the acetylene; the angle between the cobalt atoms and the C—C axis of the acetylene is about 90° as expected. A similar case is the nickel complex (23-XX). A complex (23-XXI)

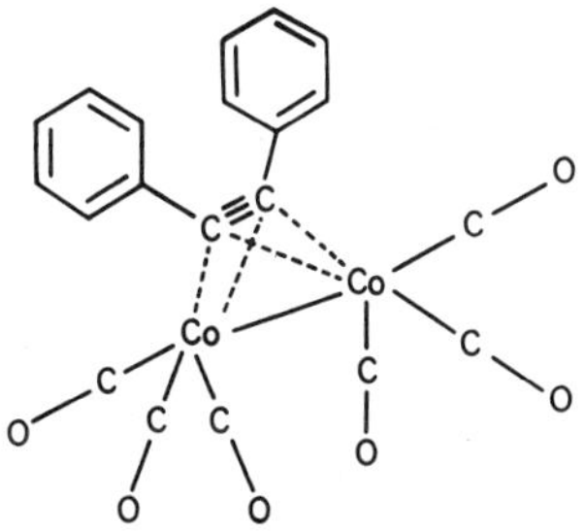

(23-XIX) 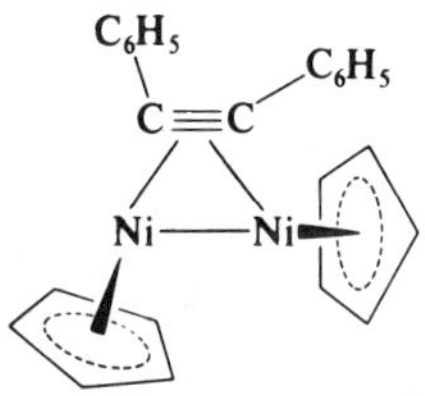   (23-XX)

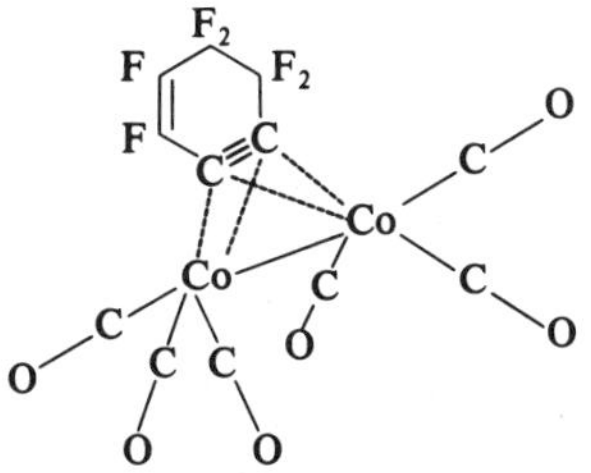

(23-XXI)

which has a bridging acetylene-containing carbocyclic ring isomeric with perfluorobenzene (a kind of "benzyne") is obtained by interaction of $Co_2(CO)_8$ with perfluoro-1,3-cyclohexadiene, which is partially defluorinated in the reaction.

In complexes of this type, as indicated in the drawings, the R—C≡C—R groups are appreciably bent at the acetylenic carbon atoms. Experimental

[23] W. Hübel in *Organic Syntheses via Metal Carbonyls*, I. Wender and P. Pino, eds., Interscience-Wiley, 1968, p. 273; C. Hoogzand and W. Hübel, *ibid.*, p. 343.

studies by nmr[24] and electron spectroscopy,[25] as well as MO calculations,[26] all indicate that there is a qualitative, if not a quantitative, relation between the degree of bending and the extent of both acetylene-to-metal σ orbital donation and back-donation from metal π-orbitals to the acetylene π^* orbitals. The bending appears to increase in proportion to the excess of π back-donation (metal to acetylene) over the σ forward-donation (acetylene to metal). There is, however, a need for more extensive physical studies to test this correlation more thoroughly.

There are a number of alkyne complexes in which an alkyne is coordinated to only one metal atom and serves simply as the equivalent of an olefin or carbon monoxide ligand. Thus we have the annexed reactions.

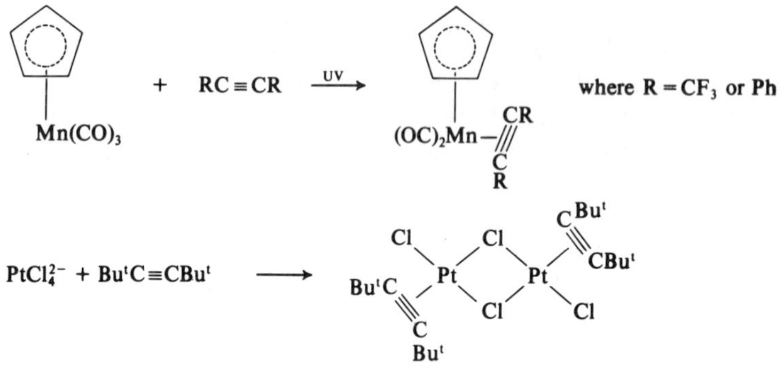

$$PtCl_4^{2-} + Bu^tC\equiv CBu^t \longrightarrow$$

In these cases, the alkyne occupies only one of the conventional number of coordination positions for the metal atom in question. The structural nature of this type of interaction is well represented[27] by

$$PtCl_2(p\text{-}CH_3C_6H_4NH_2)(Bu^tC\equiv CBu^t),$$

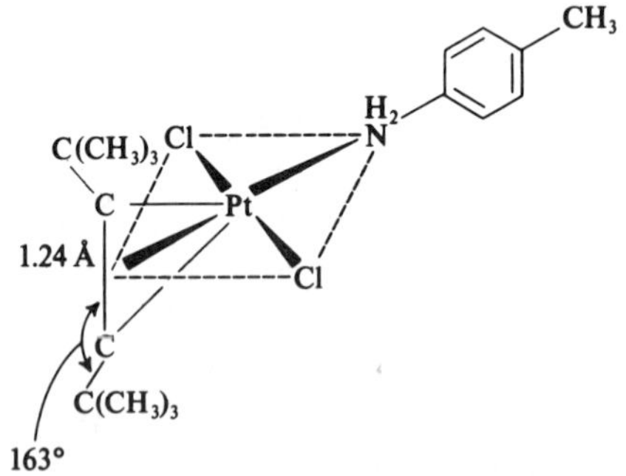

Fig. 23-14.   The structure of a complex in which the alkyne fills essentially the same role as an olefin (cf. Fig. 23-1).

[24] C. D. Cook and K. Y. Wan, *J. Amer. Chem. Soc.*, 1970, **92**, 2595.
[25] C. D. Cook *et al.*, *J. Amer. Chem. Soc.*, 1971, **93**, 1904.
[26] A. C. Blizzard and D. P. Santry, *J. Amer. Chem. Soc.*, 1968, **90**, 5749; J. H. Nelson *et al.*, *J. Amer. Chem. Soc.*, 1969, **91**, 7005.
[27] G. R. Davies *et al.*, *J. Chem. Soc., A*, **1970**, 1873.

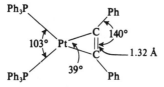

Fig. 23-15. The structure of $(Ph_3P)_2Pt(PhC_2Ph)$, in which diphenylacetylene is most simply formulated as a divalent, bidentate ligand. The C—C bond length is 1.32 Å.

which is shown in Fig. 23-14. The distortion of the alkyne from linearity is real but not great and the C—C bond remains quite short.

Arylalkynes can also form, with Pt, Pd and Ir, complexes such as that[28] shown in Fig. 23-15, in which they can be considered to serve as bidentate, divalent ligands. In these the C—C stretching frequency is lowered considerably, to the range 1750–1770 cm^{-1}, indicative of a C—C double bond. The C—C bond length of 1.32 Å is consistent with this view, as is the relatively large distortion from linearity.

There are a number of more complex structures in which, however, the integrity of the individual alkyne is retained. In (23-XXII), the structure has been confirmed by X-ray crystallography;[29] a satisfactory description of the bonding, which can adequately employ all metal valence orbitals, can be given, employing molecular orbital theory. Another compound for which no simple representation of the bonding appears possible is (23-XXIII). The

(23-XXII)

(23-XXIII)

bonding in the complex molecule depicted in (23-XXIV) can be described by supposing that each carbon atom forms a σ C—Co bond and that the remaining pair of π electrons is employed in a bridge bond between the other two Co atoms.

There are a large number of compounds formed from alkynes and metal carbonyl molecules in which the several alkynes link up among themselves

[28] J. O. Glanville, J. M. Stewart and S. O. Grim, *J. Organometallic Chem.*, 1967, 7, 9.
[29] L. F. Dahl and R. Bau, personal communications.

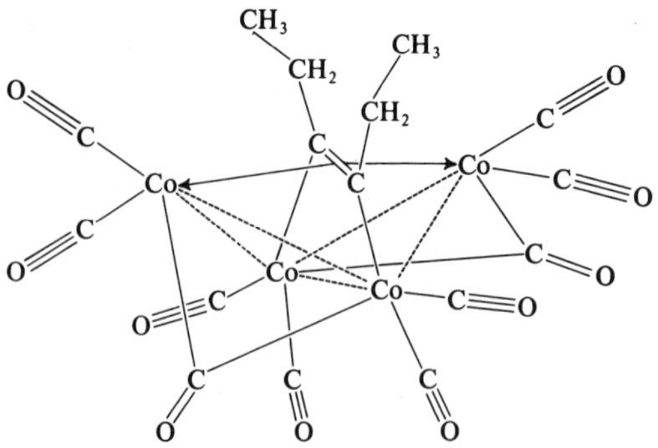

(23-XXIV)

and with CO groups to form new organic ligands. We have already noted in Section 23-3 that RC≡CR molecules form substituted benzenes. Those in which CO is incorporated lead to cyclopentadienone and quinone complexes, especially the former. A few typical reactions are annexed.

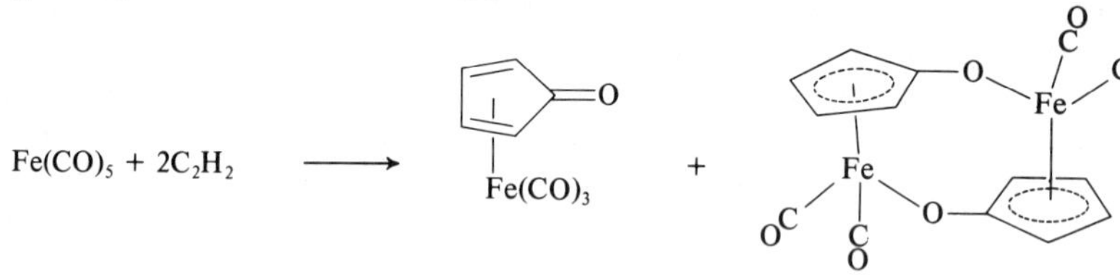

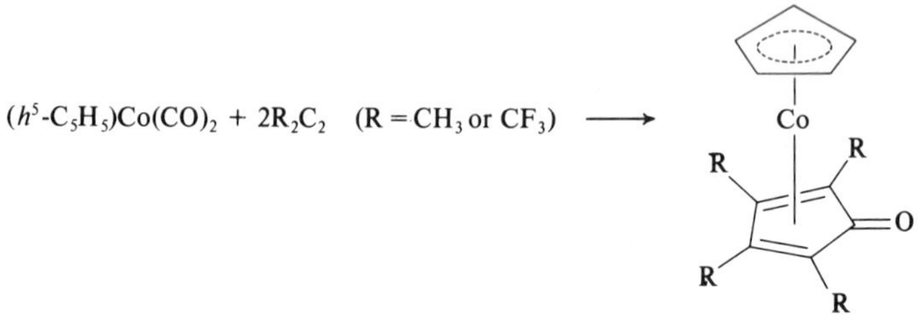

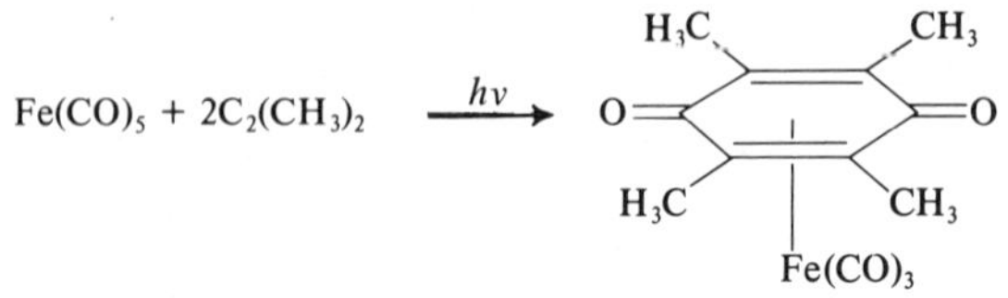

$$Fe(CO)_5 + 2C_2H_2 \longrightarrow$$

$$(h^5\text{-}C_5H_5)Co(CO)_2 + 2R_2C_2 \quad (R = CH_3 \text{ or } CF_3) \longrightarrow$$

$$Fe(CO)_5 + 2C_2(CH_3)_2 \xrightarrow{\ hv\ }$$

Many of these cyclopentadienone and quinone complexes can, of course, be made directly from a metal carbonyl and the cyclic ketone itself.

In still other cases rings that include metal atoms are formed, as illustrated by reaction (23-1).

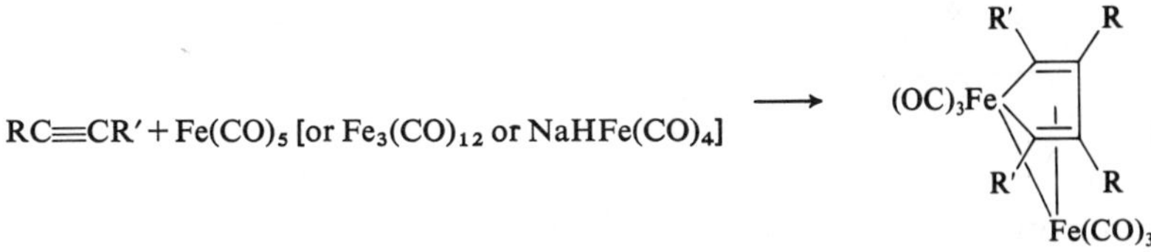

$$RC \equiv CR' + Fe(CO)_5 \text{ [or } Fe_3(CO)_{12} \text{ or } NaHFe(CO)_4]$$

(23-1)

R, R' = H, CH$_3$, Ph or OH

There are many cases in which still more complex organic ligands are elaborated from alkyne starting materials. A particularly striking example is shown in reaction (23-2).

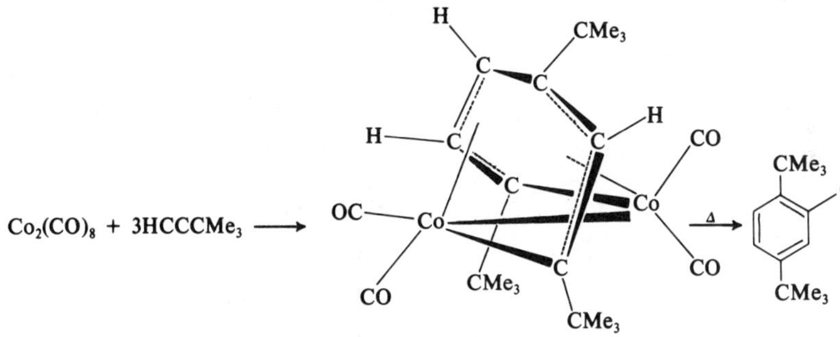

$$Co_2(CO)_8 + 3HCCCMe_3 \longrightarrow$$

(23-2)

23-5. Allyl and Other Enyl Complexes

The allyl group, bound in a *trihapto* manner (23-XXV), is a very important

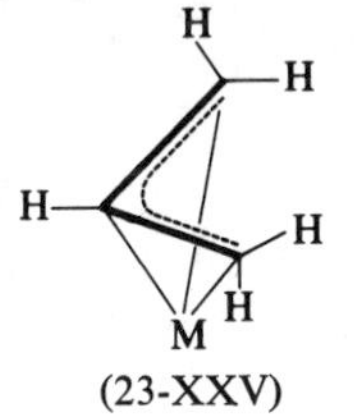

(23-XXV)

3-electron ligand. Some typical reactions for preparation of simple h^3-allyl complexes are represented by the reactions (23-3) to (23-5).

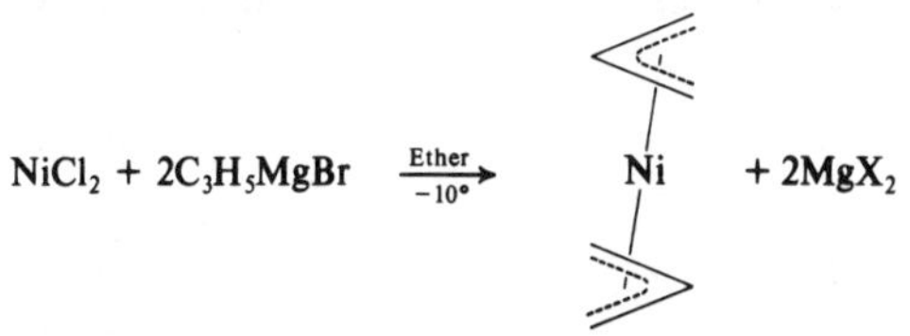

$$NiCl_2 + 2C_3H_5MgBr \xrightarrow[-10°]{Ether} Ni + 2MgX_2 \qquad (23-3)$$

$$Na^+ \left[\begin{array}{c} \\ Mo(CO)_3 \end{array} \right]^- + C_3H_5Cl \longrightarrow \begin{array}{c} \\ Mo(CO)_3 \\ H_2C \\ CH = CH_2 \end{array} \xrightarrow[uv]{\Delta \text{ or }} \begin{array}{c} \\ Mo \\ O^C {}_O{}^C \end{array}$$

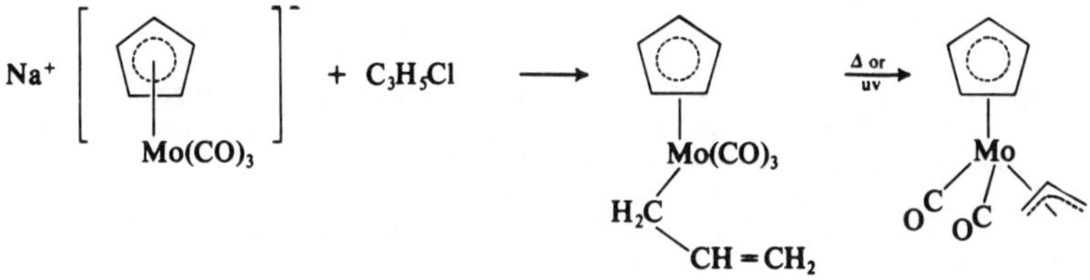

(23-4)

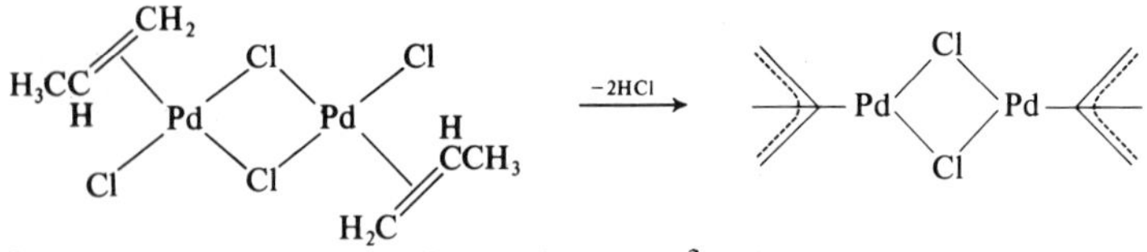

$$(23\text{-}5)$$

An important pathway to formation of h^3-allyl complexes is by protonation of (*tetrahapto*-1,3-butadiene)metal species. This may occur in a variety of ways: (1) By addition of an acid such as HCl to a butadiene complex[30] (reaction 23-6); (2) by reversible addition of a proton only[31] (reaction 23-7); or (3) by reaction of a diene with a metal complex containing a proton ligand[32] (reaction 23-8). The weight of evidence suggests that the proton

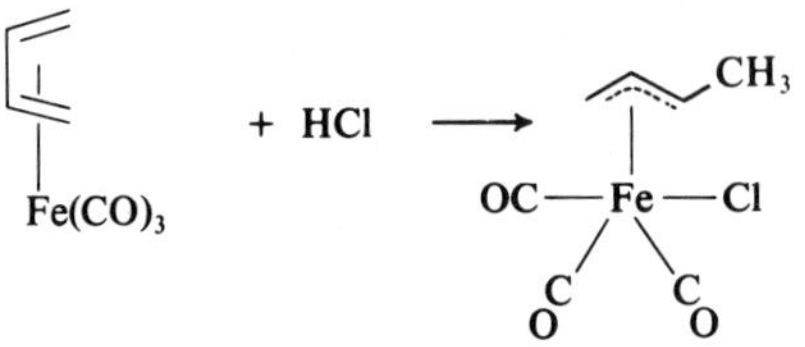

$$(23\text{-}6)$$

$$(23\text{-}7)$$

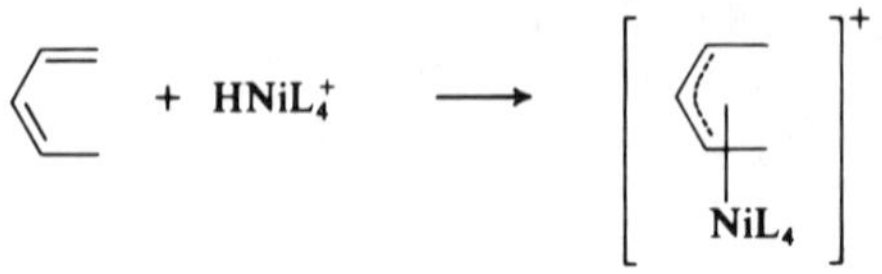

$$(23\text{-}8)$$

attacks from the *endo* (metal) side of the olefin,[31,33] but this has not been definitely proved (as it has in the case of proton attack on various non-coordinated double bonds).[34]

trihapto-Allylmetal groups are also found in a wide variety of compounds containing more complex ligands, as building blocks in more elaborate structures. Several examples illustrating this are (23-XXVI) to (23-XXIX).

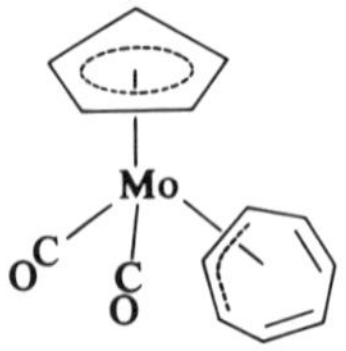

(23-XXVI)

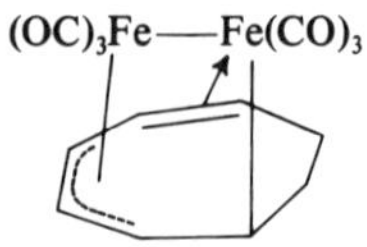

(23-XXVII)

[30] G. F. Emerson, J. E. Mahler and R. Pettit, *Chem. and Ind. (London)*, **1964**, 836.
[31] D. A. T. Young, J. R. Holmes and H. D. Kaesz, *J. Amer. Chem. Soc.*, 1969, **91**, 6968.
[32] C. A. Tolman, *J. Amer. Chem. Soc.*, 1970, **92**, 6785.
[33] M. A. Haas, *Organometallic Chem. Rev., A*, 1969, **4**, 307.
[34] A. Davison *et al.*, *J. Chem. Soc.*, **1962**, 4821; S. Winstein *et al.*, *J. Amer. Chem. Soc.*, 1965, **87**, 3267.

(23-XXVIII)   (23-XXIX)

The structures of a great many complexes containing h^3-allylmetal components have been determined. A representative one is shown in Fig. 23-16. The M—C distances usually vary about as they do in this structure, though often the difference between those to the central and terminal atoms is less than 0.1 Å.

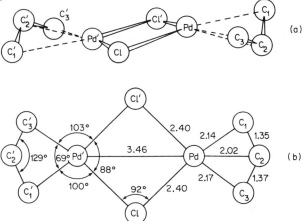

Fig. 23-16.   Structure of allylpalladium(II) chloride dimer. Bond lengths are in Å. [Reprinted by permission from *J. Organometallic Chem.*, 1965, **3**, 43.]

Allylmetal complexes play a role in many syntheses of oligomers of small olefins; nickel has been used particularly often in such syntheses.[35] For example, $Ni(CO)_4$ effects the reaction (23-9). The participation of an

$$BrCH_2CH{=}CH(CH_2)_2CH{=}CHCH_2Br \xrightarrow[-NiBr_2]{+Ni(CO)_4}$$

(23-9)

42%   5%

h^3-allylnickel intermediate is strongly supported by the isolation of (23-XXX) and its reaction with CO as in (23-10).

$$+ 3CO \longrightarrow \qquad + R_3PNi(CO)_3 \qquad (23\text{-}10)$$

(23-XXX)

[35] P. Heimbach, P. W. Jolly and G. Wilke, *Adv. Organometallic Chem.*, 1970, **8**, 29.

pentahapto-Dienyl Complexes. *pentahapto*-Cyclohexadienyl systems can be generated by addition or deletion of H^- from suitable systems as indicated by reactions (23-11) and (23-12). A considerable number of other complexes

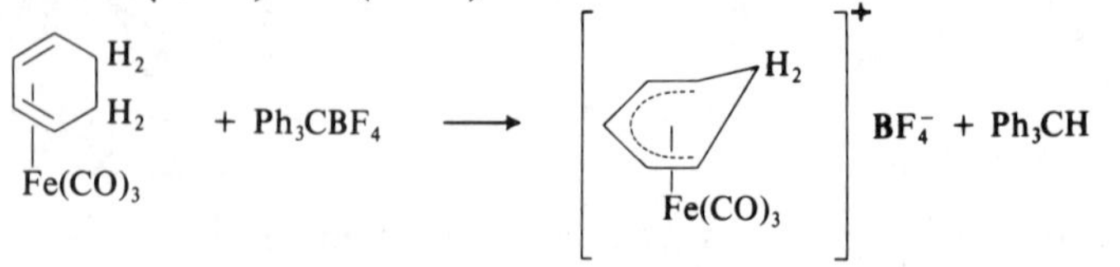

$$+ \; Ph_3CBF_4 \longrightarrow \qquad\qquad BF_4^- + Ph_3CH \qquad (23\text{-}11)$$

$$+ \; H^- \longrightarrow \qquad\qquad\qquad (23\text{-}12)$$

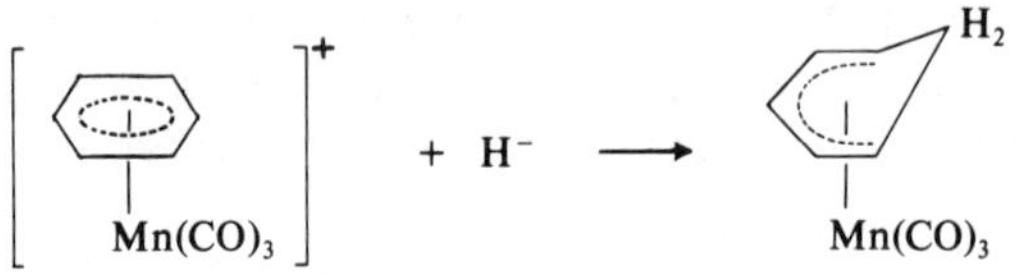

of the C_6H_7 group and substituted derivatives thereof are known.[36] The structure of the manganese complex formed in reaction (23-12) has been confirmed crystallographically.[36]

23-6. Transition Metal to Carbon σ-Bonds[37]

Early efforts to prepare simple alkyls or aryls of transition metals,[38] such as diethyl-iron or -nickel, showed that such compounds were generally unstable under ordinary conditions although they might be present in solutions at low temperatures. In view of this it has often been assumed that bonds from transition metals to carbon are weak. However, it was found that, when π-acid ligands such as CO, h^5-C_5H_5 or PR_3 are present, thermally quite stable σ-alkyls and σ-aryls can be obtained. The first extensive series to be prepared—by standard methods such as the action of metal complex halides on RMgX or RLi or of sodium salts of metal anions on organic halides—consisted of compounds such as h^5-$C_5H_5W(CO)_3CH_3$. The idea arose that the presence of such ligands stabilized the σ-bond, possibly increasing its strength. However, in view of the lack of bond-energy data, and for other reasons, this view cannot be sustained. There are indeed only three estimates of transition-metal-to-carbon σ-bond energies: for Pt—C_6H_5 in $Pt(C_6H_5)_2(PEt_3)_2$, ca. 250 kJ mol^{-1},[39] for Pt—CH_3 in h^5-$C_5H_5Pt(CH_3)_3$, ca. 164 kJ mol^{-1},[40] and Ti—CH_3 and Ti—C_6H_5 in $(h^5$-$C_5H_5)_2TiR_2$ ca. 250 and 350 kJ mol^{-1}, respectively,[41] all of which it will be noted have π-acid ligands attached.

It is now evident that the main reason for the instability of transition-metal alkyls is a kinetic one—there are easy pathways for decomposition. There

[36] M. R. Churchill and F. R. Scholer, *Inorg. Chem.*, 1969, **8**, 1950.
[37] G. W. Parshall and J. J. Mrowca, *Adv. Organometallic Chem.*, 1968, **7**, 157; M. R. Churchill in *Perspectives in Structural Chemistry*, J. D. Dunitz and J. A. Ibers, eds., John Wiley and Sons, 1970, Vol. III, p. 91.
[38] See F. A. Cotton, *Chem. Rev.*, 1955, **55**, 551.
[39] S. J. Ashcroft and C. T. Mortimer, *J. Chem. Soc.*, A, **1967**, 930.
[40] K. W. Eggar, *J. Organometallic Chem.*, 1970, **24**, 501.
[41] V. I. Tel'noi *et al.*, *Dokl. Akad. Nauk S.S.S.R.*, 1967, **174**, 1374.

are several possible modes of decomposition, such as homolysis of the bond and transfers of hydrogen from the alkyl ligand to the metal. A particularly common and well-studied reaction[42,43] is the transfer from the β-carbon of the alkyl chain:

$$M-CH_2-CH_2-R \rightleftharpoons MH + CH_2=CHR$$

resulting in the elimination of olefin and formation of an M—H bond. The reverse of this reaction, i.e. the formation of alkyls by addition of olefins to M—H bonds, is of very great importance in a number of catalytic reactions discussed in the next Chapter. Once the hydrogen has been transferred to metal, further reaction to give the metal, hydrogen or hydrogenation of olefin to saturated hydrocarbon can occur.

If this hydrogen transfer could be hindered, the possibility of obtaining stable alkyls would exist. Thus, if we have a group of the type $-CH_2-XR_n$ where X is an atom that can form a single but *not* a multiple bond to carbon, olefin elimination cannot occur. The CH_3 group, of course, also cannot be eliminated easily by hydrogen transfer, and a number of methyls such as $Ti(CH_3)_4$ are more stable than the corresponding ethyls; stable benzyls[44] of Ti and Zr, $M(CH_2C_6H_5)_4$, also meet the condition, but the best examples of elimination-stabilized alkyls are the trimethylsilylmethyls[45] $M^n(CH_2SiMe_3)_n$ and neopentyls $M^n(CH_2CMe_3)_n$ (see Table 23-2). Isolable

TABLE 23-2
Some Representative Transition-metal Alkyls

Compound	Properties	Comment
$Ti(CH_2C_6H_5)_4$	Red crystals, m.p. 70°	Catalyzes polymerization of ethylene (see Chapter 24).[a]
Me_3NbCl_2[b]	Yellow crystals	Sublimes in vac. at 25°; gives CH_4 on standing.
$VO(CH_2SiMe_3)_3$[c]	Yellow needles, m.p. 75°	Air-stable.
$Cr(CH_2SiMe_3)_4$[c]	Red needles, m.p. 40°	Paramagnetic, tetrahedral.
$Mo_2(CH_2SiMe_3)_6$[c]	Yellow plates, m.p. 99°	Mo—Mo triple bond 2.167 Å; staggered configuration.
$Mo_2(CH_2CMe_3)_6$[c]	Yellow plates, d. 135°	Contains Mo—Mo triple bond.
$Co(salen)C_2H_5$	Red brown crystals	Co—C = 1.99Å.[e]
$[Rh(NH_3)_5C_2H_5]\,(ClO_4)_2$	White crystals	Loses 1 NH_3 in water.
$C_2H_5Mn(CO)_5$	Yellow oil, m.p. −30°	Dec. in vac. at −10°; inserts CO to give acyl.
$PtCH_3Br(PEt_3)_2$	White crystals	Square, *trans*-PEt$_3$ groups.
$(h^5-C_5H_5)Fe(CO)_2(t-C_4H_9)$[d]	Orange, m.p. 72°	Only known t-butyl.

[a] U. Giannini, U. Zucchini and E. Albizzati, *J. Polymer. Sci. Part B.*, 1970, **8**, 405.
[b] G. J. Juvinall, *J. Amer. Chem. Soc.*, 1964, **86**, 4202.
[c] F. Huq *et al.*, *Chem. Comm.*, **1971**, 1079, 1477; and ref. 45.
[d] W. P. Giering and M. Rosenblum, *J. Organometallic Chem.*, 1970, **25**, C71.
[e] M. Calligari *et al.*, *J. Chem. Soc., A,* **1971**, 2720.

[42] See, e.g., R. P. A. Sneeden and H. H. Zeiss, *J. Organometallic Chem.*, 1970, **22**, 713; G. M. Whitesides *et al.*, *J. Amer. Chem. Soc.*, 1970, **92**, 1426; 1971, **93**, 1379.
[43] M. Tamura and J. Kochi, *J. Organometallic Chem.*, 1971, **29**, 111.
[44] U. Zucchini, E. Albizzati and U. Giannini, *J. Organometallic Chem.*, 1971, **26**, 357; G. R. Davies *et al.*, *Chem. Comm.*, **1971**, 677; I. W. Bassi *et al.*, *J. Amer. Chem. Soc.*, 1971, **93**, 3787.
[45] G. Yagupsky *et al.*, *Chem. Comm.*, **1970**, 1369.

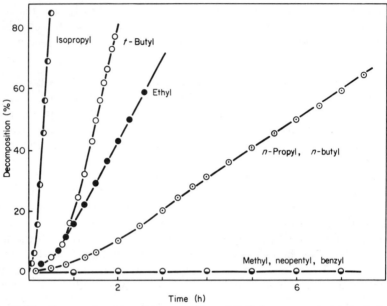

Fig. 23-17. Comparative rates of decomposition of dialkylmanganese in tetrahydrofuran at 2° (0.048M MnCl₂, 0.27M Grignard reagent), the alkyl group being as stated in the diagram. [Reprinted by permission from M. Tamura and J. Kochi, *J. Organometallic Chem.*, 1971, **21**, 111.]

and thermally stable compounds are known for V, Nb, Ta, Cr, Mo, W, etc. A good illustration also of the increased stability of alkyls that cannot readily eliminate is provided by decomposition studies[43] of manganese dialkyls prepared *in situ* (Fig. 23-17) which shows that the stability order is

$$Me_3CCH_2 \approx C_6H_5CH_2 \approx CH_3 \gg n\text{-}C_3H_7, \quad n\text{-}C_4H_9 > C_2H_5 > t\text{-}C_4H_9 > iso\text{-}C_3H_7$$

Alkyls of completely different type, which again meets the criterion, are the anionic chelate carborane derivatives of Co, Ni and Cr,[46] e.g., $\{Cu^{II}[B_{10}C_2H_{10})_2]\}^{2-}$ shown in Fig. 23-18 (see also page 253).

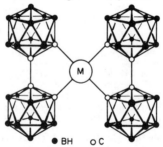

● BH   ○ C

Fig. 23-18. The proposed structure for $M^{(4-n)+}[B_{10}C_2H_{10})^{n-}$ (M = Ni or Cu, n = 1 or 2; M = Co, n = 1 or 3) derivatives. [Reproduced by permission from D. A. Owen and M. F. Hawthorne, *J. Amer. Chem. Soc.*, 1971, **93**, 873.]

[46] M. F. Hawthorne *et al.*, *J. Amer. Chem. Soc.*, 1971, **93**, 873, 1362.

While the presence of π-acid ligands does, as noted above, often allow stable alkyls to be isolated, their presence does not necessarily guarantee stability if a ligand can dissociate, thereby providing a vacant site for the H-transfer–olefin-elimination reaction to proceed. Indeed such instability of the alkyl is essential to the functioning of hydrogenation and isomerization catalysts discussed in Chapter 24. One representative example is a ruthenium complex where the ethyl compound can be detected by nmr spectroscopy only under ca. 30 atm ethylene pressure since in the reaction

$$RuHCl(PPh_3)_3 + C_2H_4 \rightleftharpoons RuC_2H_5Cl(PPh_3)_3$$

the equilibrium at 25° and 1 atm lies well to the left.

Thus the prime function of ligands, whether π-acid or otherwise, in stabilizing the metal alkyls is merely to occupy coordination positions firmly rather than to produce any significant electronic change in the nature of the metal-to-carbon bond. This is clearly illustrated in two ways:

(a) The alkyl $Ti(CH_3)_4$ decomposes well below 0° but on addition of, e.g., bipyridine, the adduct is stable to $+30°$.[47]

(b) It is possible to have quite stable alkyls with even water or ammonia as ligand provided that the species are substitution-inert. Thus there are a number of complex ions of Cr^{III}, Co^{III} and Rh^{III} such as $[Rh(NH_3)_5C_2H_5]^+$.[48] Particularly important examples of such M—C bonds are the cobalt complexes of the vitamin B_{12} type and their synthetic analogues discussed in Section 25-F-7. One example is the dimethylglyoxime complex (23-XXXI)

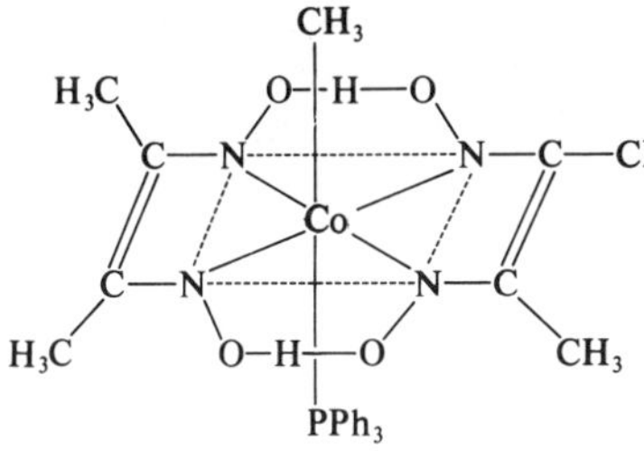

(23-XXXI)

where the methyl may be replaced by a variety of other alkyl groups. Some stable transition-metal alkyls and their properties are listed in Table 23-2.

It may be noted also that there are a number of reasonably stable *lithium alkyl anions*, which are made by interaction of lithium alkyls with metal halides. Some of these, notably the copper(I) alkyls, $Li[CuR_2]$, are important reagents in organic synthesis (see Section 25-H-2). Some chromium alkyl salts have anions with multiple metal—metal bonds,[49] e.g., $Li_4[Cr_2Me_8] \cdot 4C_4H_8O$.

Perfluoroalkyls of various types, commonly of carbonyls or h^5-cyclopentadienyls, are known and are usually more stable thermally than their hydro-

[47] R. Tabacchi and A. Jacot-Guillard, *Chimia*, 1970, **24**, 271; cf. G. W. A. Fowles, D. A. Rice and J. D. Williams, *J. Chem. Soc., A*, **1971**, 1920.

[48] M. D. Johnson and N. Winterton, *J. Chem. Soc., A*, **1970**, 507; K. Thomas *et al.*, *J. Chem. Soc., A*, **1968**, 1801.

[49] J. Krausse and G. Schödel, *J. Organometallic Chem.*, 1971, **27**, 59.

carbon analogs. One example, obtained by interaction of $Fe(CO)_5$ with C_2F_4, is (23-XXXII). A few binary alkyls, e.g., $Ti(C_6F_5)_4$ and $[Cu(C_6F_5)]_8$ are known.

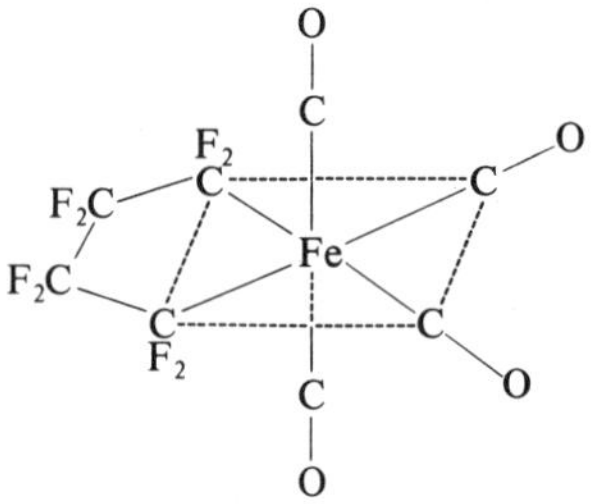

(23-XXXII)

X-ray studies have suggested that the metal-to-carbon bond distance is shorter than expected in the fluoroalkyls relative to hydrocarbon alkyls.[50] This shortening has been attributed to some π contribution to the $M—R_F$ bond by flow of electrons from filled metal orbitals to orbitals on carbon, partially made available by withdrawal of electron density by the highly electronegative fluorine atoms. Although bond lengths are related to bond strengths, it is, of course, impossible to correlate M—C bond lengths directly with thermal stabilities or chemical reactivities.

Considerable stability of M—C bonds results when, in addition to excluding the possibility of β-elimination of hydrogen, resistance to homolysis of the M—C bond is enhanced by chelation.[51] In some cases such compounds are formed by attack of the metal atom on an *ortho*-C—H bond (see Section 24-A-4), as in reaction (23-13), but in others prior lithiation is employed, as in reactions (23-14) and (23-15).

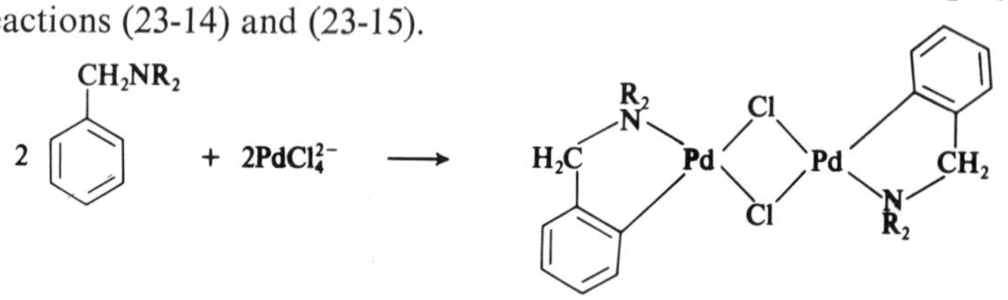

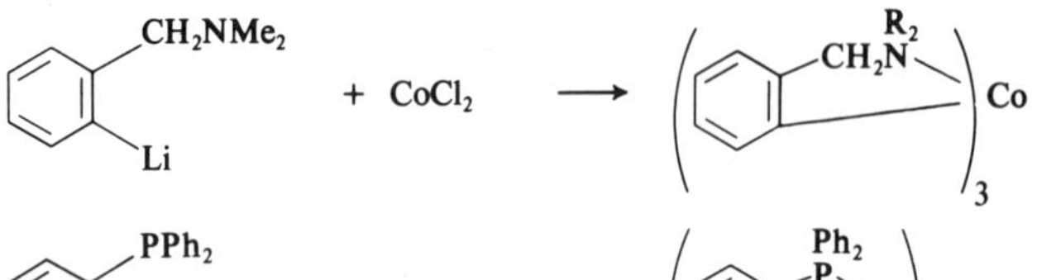

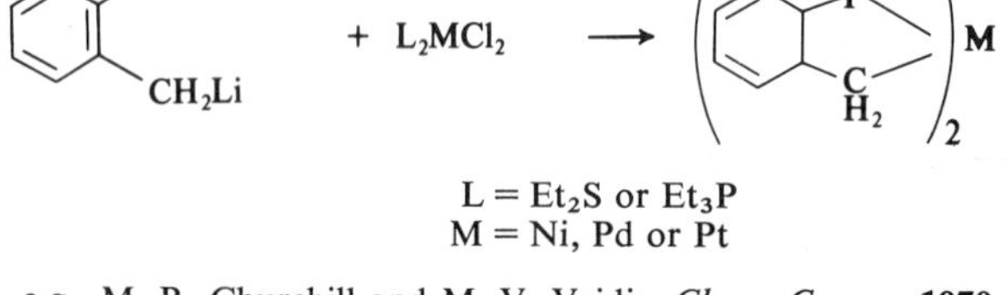

L = Et$_2$S or Et$_3$P
M = Ni, Pd or Pt

[50] See, e.g., M. R. Churchill and M. V. Veidis, *Chem. Comm.*, **1970**, 1099.
[51] G. W. Parshall, *Accounts Chem. Res.*, 1970, **3**, 139; G. Longoni *et al.*, *Chem. Comm.*, **1971**, 471; A Tzachach and H. Nindel, *J. Organometallic Chem.*, 1970, **24**, 155.

Finally, it may be mentioned that a number of stable h^1-C_5H_5—M or σ-cyclopentadienyl—metal bonds are known and were, in fact, among the earliest stable σ bonds from transition metals to carbon to be discovered. Their stability may be attributed both to their inability to decompose by β-hydrogen elimination and to the coordinately saturated nature of the metal atoms in most of the compounds that contain them. *monohapto*-cyclopentadienylmetal compounds are noteworthy and are discussed further in the next Section, because they are usually fluxional.

23-7. Fluxional Organometallic Molecules

Fluxional behavior is characteristic of certain classes of organometallic compound.[52] A few examples have already been considered in Section 1-9 as illustrations; a more organized account will be given here.

h^1-**Cyclopentadienyl Compounds.** It is very common for these compounds to be fluxional.[53] At higher temperatures, usually even at room temperature, the five protons of the ring give rise to a single sharp nmr signal. This indicates that there is some rapid rearrangement that causes each proton to move rapidly among the three types of environment (sites) available; these are designated α, β, and γ, in (23-XXXIII). At lower temper-

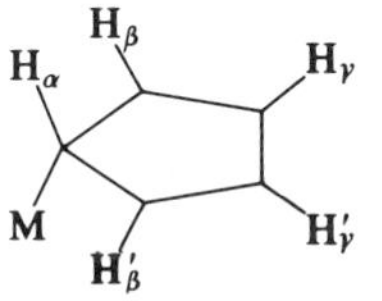

(23-XXXIII)

atures, this single line broadens, then collapses, and the complex spectrum expected for the $(h^1$-$C_5H_5)$M structure finally appears and becomes sharp at still lower temperatures. Obviously, at very low temperatures the $(h^1$-$C_5H_5)$M group is stereochemically fairly rigid; the process by which hydrogen atoms exchange sites being too slow to affect the nmr spectrum. The rate of rearrangement is so rapid at room temperature that all protons pass among the three environments at a rate of $> 10^3 \text{ sec}^{-1}$.

Many studies have been carried out to determine the detailed pathway of rearrangement. In many cases experimental evidence[54,55] eliminates the possibility that intermolecular exchange, either by unimolecular dissociation or by some other more complex path, is responsible for the site exchange. Unimolecular, non-dissociative pathways must therefore be considered, and the four most obvious ones are shown in Fig. 23-19. Pathway (1),

[52] F. A. Cotton, *Accounts Chem. Res.*, 1968, **1**, 257.
[53] G. Wilkinson and T. S. Piper, *J. Inorg. Nuclear Chem.*, 1955, **2**, 32; **3**, 104.
[54] F. A. Cotton and T. J. Marks, *J. Amer. Chem. Soc.*, 1969, **91**, 7523 and prior references cited therein.
[55] C. H. Campbell and M. L. H. Green, *J. Chem. Soc.*, A, **1970**, 1318.

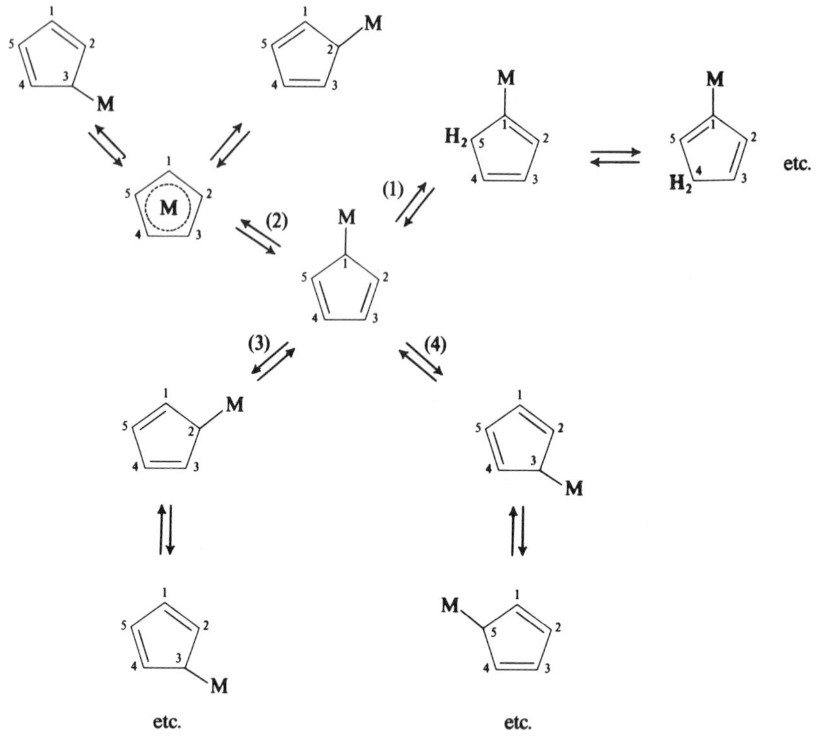

Fig. 23-19. Four theoretically possible pathways for intramolecular rearrangement of an
$(h^1\text{-}C_5H_5)M$ component.

in which the M—C bond remains intact and hydrogen atoms migrate, is
actually quite unlikely at the low temperatures involved although it could
conceivably become important in some special cases at higher temperatures.
The best direct evidence against it is the fact that pentamethylcyclopenta-
dienyl compounds can be fluxional—methyl migrations are highly unlikely.
Of the remaining three pathways, (2) can be conclusively eliminated from
direct evidence since it would cause a uniform rate of exchange for all sites;
this would lead to symmetrical collapse of the entire low-temperature spectrum, whereas, in fact, different portions of this spectrum collapse at
different rates.[56] This leads to elimination of all the possibilities except (3)
and (4). Detailed studies[54, 55] have established that pathway (3), a sequence
of 1,2-shifts is the correct one. The activation energies[55] for such shifts
are in the range of 30–40 kJ mol^{-1} for the $(h^5\text{-}C_5H_5)M(CO)_2(h^1\text{-}C_5H_5)$
(M = Fe or Ru) compounds.

Cycloheptatrienyl Complexes.   There are a number of cases in which
cycloheptatrienyl rings are bound in *trihapto* fashion, i.e. use an allyllic portion
of the ring. In the three cases (23-XXXIV), (23-XXXV) and (23-XXXVI)

[56] M. J. Bennett *et al., J. Amer. Chem. Soc.*, 1966, **88**, 4371.

it has been shown[57-59] that this type of compound is fluxional and that the rearrangement pathway is a sequence of 1,2-shifts. Even the molecule (23-XXXVII) is fluxional[60] and again the pathway is a sequence of 1,2-shifts.

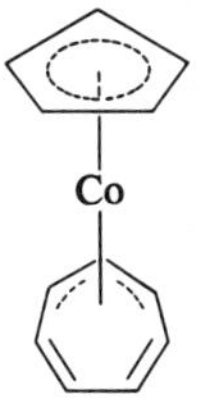

(23-XXXIV) 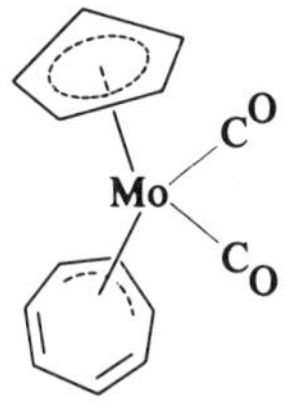   (23-XXXV) 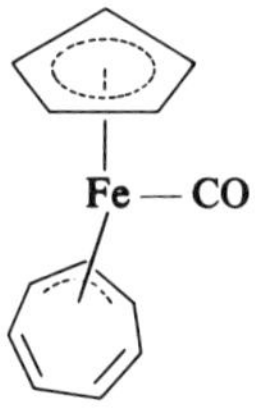   (23-XXXVI)

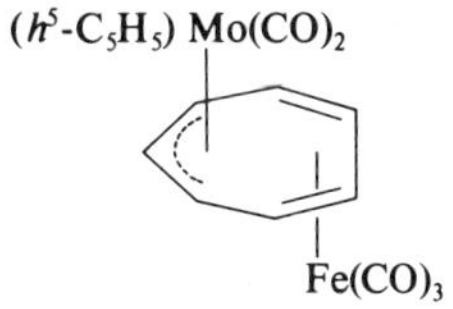

$(h^5\text{-}C_5H_5) Mo(CO)_2$

Ph_3Sn

Fe(CO)_3

(23-XXXVII) 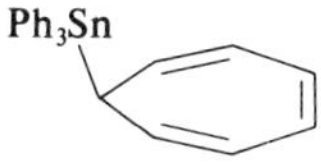   (23-XXXVIII)

There is only one certain example of a *monohapto*-cycloheptatrienylmetal complex. This does not involve a transition metal but deserves mention here because it provides a clear example of electronic control of a rearrangement pathway. It has been shown[61] that cycloheptatrienyl triphenyltin, (23-XXXVIII), rearranges by a sequence of 1,5-shifts; this is in agreement with theoretical predictions for so-called sigmatropic shifts, where orbital symmetry allows a 1,5-shift but forbids the 1,2- and 1,3-shifts which are also conceivable in this system (a 1,4-shift is equivalent to a 1,5-shift in this case). In the case of $(h^1\text{-}C_5H_5)M$ rearrangements, the observed 1,2-shifts are equivalent to 1,5-shifts which are favored by orbital symmetry rules. It is not clear, therefore, whether the 1,3-shifts are excluded by the orbital symmetry rules or by a preference for 1,2-shifts just because the latter are the shortest. Thus, the significance of the results for $Ph_3SnC_7H_7$ is that the distinction is made and orbital symmetry rules are shown to be relevant to these processes.

h^3-**Allyl Complexes.** These are characteristically fluxional[62] unless the allyl group is incorporated in a ring, which restricts its mobility. The simplest

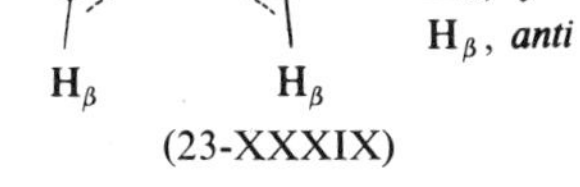

H_α, *syn*
H_β, *anti*

(23-XXXIX)

[57] M. A. Bennett, R. Bramley and R. Watt, *J. Amer. Chem. Soc.*, 1969, **91**, 3089.
[58] J. W. Faller, *Inorg. Chem.*, 1969, **8**, 767.
[59] D. Ciappenelli and M. Rosenblum, *J. Amer. Chem. Soc.*, 1969, **91**, 6876.
[60] F. A. Cotton and C. R. Reich, *J. Amer. Chem. Soc.*, 1969, **91**, 847.
[61] R. B. Larrabee, *J. Amer. Chem. Soc.*, 1971, **93**, 1510.
[62] K. Vrieze, H. C. Volger and P. W. N. M. van Leeuwen, *Inorg. Chem. Acta Rev.*, 1969, **3**, 109.

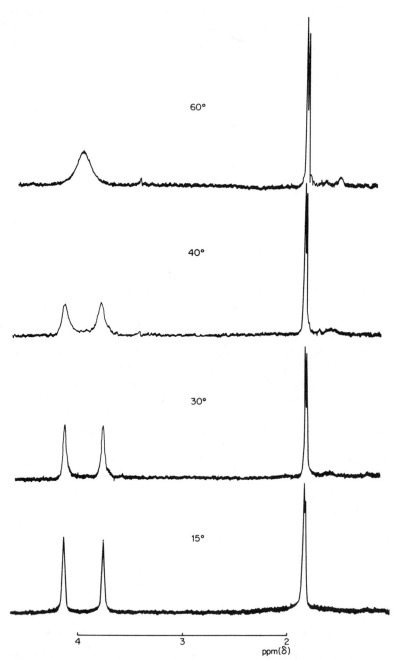

Fig. 23-20.   Temperature-dependence of the nmr spectrum of $(Ph_3As)_2Cl_2Rh(C_4H_7)$ at 100 MHz. The signal at 2 ppm is due to the 2-methyl group. The others are due to the *syn-* and *anti-*protons on C-1 and C-3. [Reproduced by permission from K. Vrieze and H. C. Volger, *J. Organometallic Chem.*, 1967, **9**, 537.]

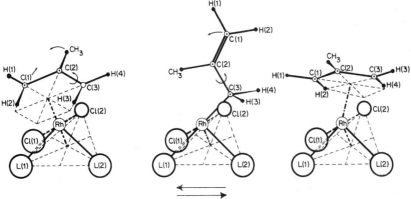

Fig. 23-21. Interconversion of the *syn-* and *anti-*proton at C-3 via a σ-allyl form (either transition state or intermediate). Obviously the same process occurs also at C-1. [Reproduced by permission from K. Vrieze and H. C. Volger, *J. Organometallic Chem.*, 1967, **9**, 537.]

and perhaps commonest sort of fluxional behavior is exchange of *syn-* and *anti-*protons (see 23-XXXIX). Fig. 23-20 presents a set of nmr spectra at several temperatures, showing how this exchange is observed;[63] many similar cases could be cited. Separate resonances for the *syn-* and *anti-* protons are seen at low temperature for $(Ph_3As)_2Cl_2RhC_4H_7$ (where C_4H_7 is a 2-methylallyl group); as the temperature is raised these signals broaden and eventually collapse to give a single signal for all four protons. The most probable general explanation for this sort of behavior (although there may be exceptions) is that the h^3-allyl-metal group rearranges to a short-lived h^1-allyl-metal intermediate, in which rotation takes place about the C-2/C-3 bond, after which an h^3-allyl-metal structure is recovered.[63,64] This is shown in detail for the $(Ph_3As)_2Cl_2RhC_4H_7$ case in Fig. 23-21.

There are also cases in which the orientation of an h^3-allyl group changes without *syn–anti* exchange. Two of these are represented in reactions (23-16) and (23-17). For the former case it has been proposed[65] that the

$$\text{(23-16)}$$

$$\text{(23-17)}$$

[63] K. Vrieze and H. C. Volger, *J. Organometallic Chem.*, 1967, **9**, 537.
[64] J. W. Faller and M. E. Thomsen, *J. Amer. Chem. Soc.*, 1969, **91**, 6871.
[65] J. W. Faller, M. J. Incorvia and M. E. Thomson, *J. Amer. Chem. Soc.*, 1969, **91**, 518.

mechanism is a dissociation of the amine ligand, intermediate formation of the $C_3H_5PdCl_2PdC_3H_5$ dimer and attack on the dimer by amine to regenerate $C_3H_5PdClAm$. In (23-17), on the other hand, the allyl group apparently undergoes hindered internal rotation which directly inter-converts the two isomers.[66]

An elegant study[67] of the system (23-XL) has shown that in the absence of added base it is non-fluxional. Addition of pyridine causes *syn–anti* proton exchange to occur in such a way that C-1 and C-3 of the allyl group do not interchange positions. The only plausible explanation of this result appears to be that pyridine attacks Pd, displacing one terminal carbon atom of the allyl group, whereupon rotation about the C—C single bond occurs. The allyl group then recovers the *trihapto*-form of bonding, with expulsion of the pyridine.

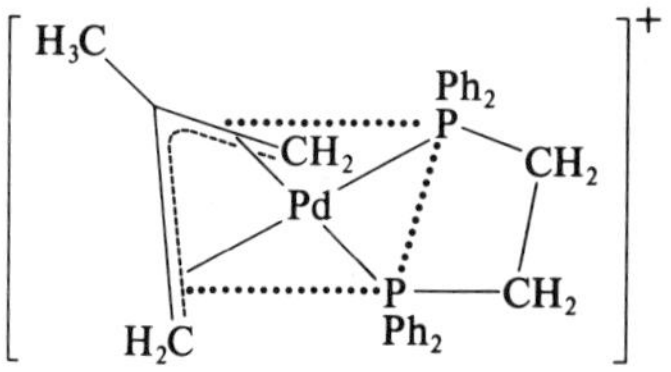

(23-XL) 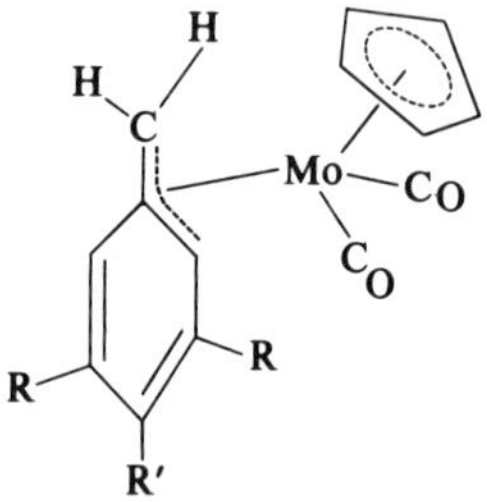   (23-XLI)

The fluxional behavior of systems of the type (23-XLI) has been studied in intimate detail and here again the results strongly imply that the *tri-*, *mono-*, *trihapto*-pathway is operative.[68]

Cyclic Polyolefin Complexes. The principal examples of these are the cyclooctatetraene complexes (23-XLII) and (23-XLIII). In both classes it has

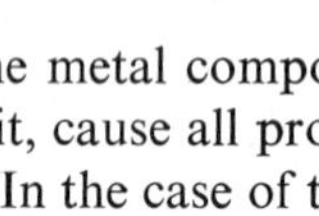

M(CO)$_3$ M(CO)$_3$
M = Fe, Ru or Os   M = Cr, Mo or W
(23-XLII)   (23-XLIII)

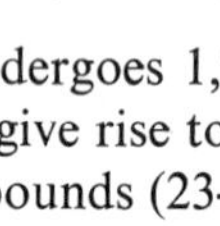

been shown[69] that the metal component undergoes 1,2-shifts which, in the high-temperature limit, cause all protons to give rise to a single sharp signal in the nmr spectrum. In the case of the compounds (23-XLII), the activation energy is so small that temperatures as low as $-150°$ or lower are required to yield the spectrum characteristic of the instantaneous structure. A num-

[66] J. W. Faller and M. J. Incorvia, *Inorg. Chem.*, 1968, **7**, 840.
[67] D. L. Tibbetts and T. L. Brown, *J. Amer. Chem. Soc.*, 1970, **92**, 3031.
[68] F. A. Cotton and T. J. Marks, *J. Amer. Chem. Soc.*, 1969, **91**, 1339.
[69] F. A. Cotton *et al.*, *J. Amer. Chem. Soc.*, 1969, **91**, 6598; F. A. Cotton, J. W. Faller and A. Musco, *J. Amer. Chem. Soc.*, 1968, **90**, 1438.

ber of other fluxional cyclooctatetraene complexes such as $(C_8H_8)_2Fe$,[70] $C_8H_8Fe_2(CO)_5$,[71] $(C_8H_8)_2Ru_3(CO)_4$,[72] and $(C_8H_9)(C_8H_8)Co$[73] have been examined though the pathways have not been definitely established for these compounds.

23-8. Carbene Complexes

A carbene (carbenoid) is a species of general formula (23-XLIV), in which X and X′ represent a great variety of nominally univalent groups. As compounds of divalent carbon, carbenes are short-lived in the free state, but they have been studied by special experimental techniques, as noted in Chapter 10. The more stable carbenes are those in which the electronic structure can be represented by the resonance (23-XLV). When this is com-

(23-XLIV)

(23-XLV)

$$:C{=}O \longleftrightarrow :\bar{C}{\equiv}\overset{+}{O}$$

(23-XLVI)

(23-XLVIIa) (23-XLVIIb) (23-XLVIIc) (23-XLVIId)

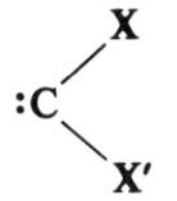

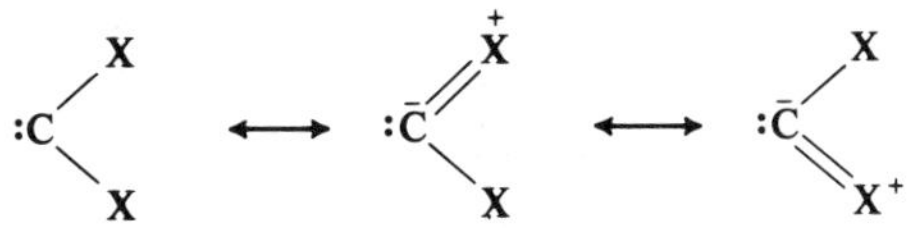

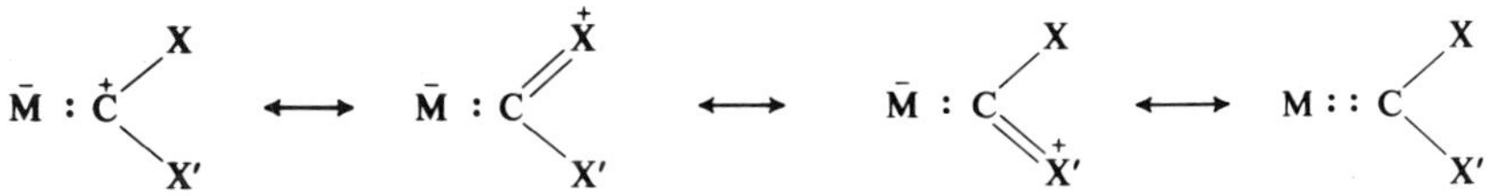

pared with the resonance representation of carbon monoxide (23-XLVI) the idea is suggested that complexes of carbenes might exist, stabilized by resonance of the type (23-XLVII). A number of such complexes can in fact be prepared in which X, X′ or both are RO, RHN or RR′N.

One general method of preparation employs the addition of organolithium reagents to coordinated carbon monoxide[74] (see also Chapter 24, Section A-4), followed by alkylation, viz.,

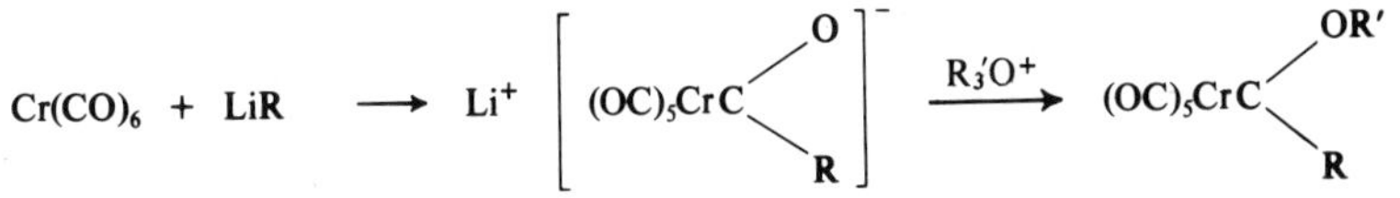

R, R′ = Me, Et

[70] A. Carbonaro *et al.*, *J. Amer. Chem. Soc.*, 1969, **91**, 4453.
[71] C. E. Keller, G. F. Emerson and R. Pettit, *J. Amer. Chem. Soc.*, 1965, **87**, 1388; E. B. Fleischer *et al.*, *J. Amer. Chem. Soc.*, 1966, **88**, 3158.
[72] M. J. Bennett, F. A. Cotton and P. Legzdins, *J. Amer. Chem. Soc.*, 1968, **90**, 6335.
[73] A. Greco, M. Green and F. G. A. Stone, *J. Chem. Soc., A*, **1971**, 285.
[74] E. O. Fischer and A. Massböl, *Chem. Ber.*, 1967, **100**, 2445; E. O. Fischer and H. J. Kollmeier, *Angew. Chem. Internat. Edn.*, 1970, **9**, 309.

Another method of some generality which begins with coordinated iso-cyanides[75] is illustrated in (23-18).

$$cis\text{-}PtCl_2(Et_3P)(CNPh) + EtOH \rightarrow cis\text{-}Cl_2(Et_3P)PtC \overset{OEt}{\underset{NHPh}{<}} \tag{23-18}$$

It is also possible to carry out exchange reactions[76] whereby OR is replaced with NRR′; a representative example is (23-19).

$$(OC)_5CrC \overset{OMe}{\underset{Me}{<}} + NH_2Et \longrightarrow (OC)_5CrC \overset{NHEt}{\underset{Me}{<}} + MeOH \tag{23-19}$$

A ready intramolecular cyclization (23-20) leading to a carbene complex has been recently reported.[77]

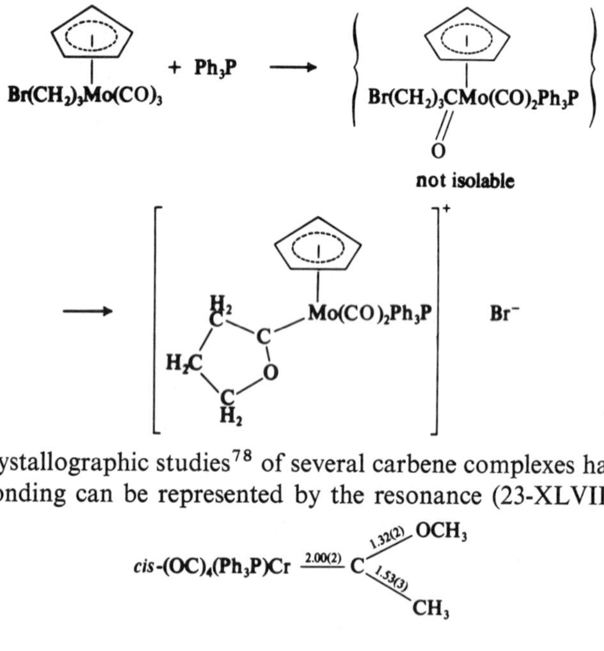

$$\tag{23-20}$$

X-ray crystallographic studies[78] of several carbene complexes have shown that the bonding can be represented by the resonance (23-XLVII) with all

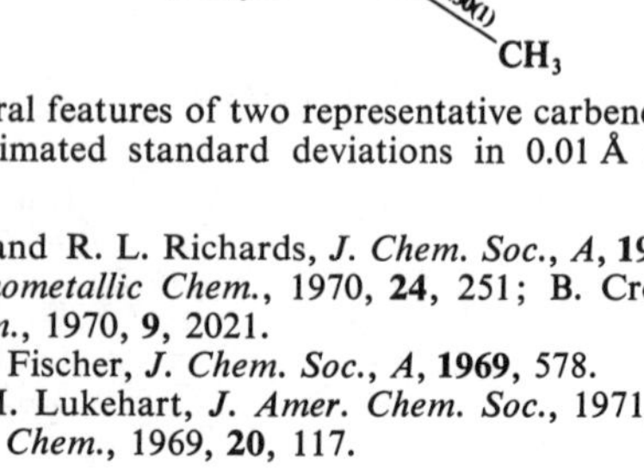

Fig. 23-22. Some structural features of two representative carbene complexes. Distances are in Å, with estimated standard deviations in 0.01 Å in parentheses.

[75] E. M. Badley, J. Chatt and R. L. Richards, *J. Chem. Soc., A,* **1971**, 21; F. Bonati and G. Minghetti, *J. Organometallic Chem.,* 1970, **24**, 251; B. Crociani, T. Boschi and U. Belluco, *Inorg. Chem.,* 1970, **9**, 2021.
[76] J. A. Connor and E. O. Fischer, *J. Chem. Soc., A,* **1969**, 578.
[77] F. A. Cotton and C. M. Lukehart, *J. Amer. Chem. Soc.,* 1971, **93**, 2672.
[78] O. S. Mills, *Pure Appl. Chem.,* 1969, **20**, 117.

canonical forms participating significantly. The M—CXX' skeleton is always found to be planar. Fig. 23-22 shows two examples[79] indicating how the shortened M—C, C—O, and C—N distances reflect the presence of the multiple-bond character implied by the resonance (23-XLVII).

[79] O. S. Mills and A. D. Redhouse, *J. Chem. Soc., A,* **1969**, 1274; J. A. Connor and O. S. Mills, *J. Chem. Soc., A,* **1969**, 334.

Further Reading

Annual Reports and *Specialist Periodical Reports*, The Chemical Society, London. Sections of the former cover organo derivatives of the transition elements. Among the latter there are currently annual volumes on *Spectroscopic Properties of Inorganic and Organometallic Compounds* and *Inorganic Reaction Mechanisms*, the latter covering organometallic compounds.

Organometallic Chemistry Reviews constitute one of the most important items of secondary literature. There are two series: The *A* series comprises topical articles that appear on the initiative of their authors. The *B* series presents regular, annual reviews of the organo chemistry of each element in the Periodic Table.

Advances in Organometallic Chemistry, F. G. A. Stone and R. West, eds., Vols. 1 (1964) to 8 (1970), Academic Press (contain many articles on transition-metal compounds).

Birmingham, J. M., *Adv. Organometallic Chem.,* 1965, **2**, 365 (synthesis of h^5-C_5H_5 compounds).

Cais, M., in *The Chemistry of Alkenes,* S. Patai, ed., Interscience-Wiley, 1964 (alkene–metal complexes).

Cotton, F. A., and C. M. Lukehart, *Prog. Inorg. Chem.,* 1972, Vol. 16 (A comprehensive review of all phases of the chemistry of carbenoid complexes).

Cox, J. D., and G. Pilcher, *Thermochemistry of Organic and Organometallic Compounds,* Academic Press, 1970.

Dub, M., *Organometallic Compounds,* 2nd edn., Vol. 1, Springer Verlag (a compendium).

Fischer, E. O., and H. Werner, *Metal π-Komplex mit di- und oligo-olefinischen Liganden,* Verlag Chemie, 1963 (review of olefin complexes).

Fischer, E. O., and H. Werner, *Metal π-Complexes,* Vol. 1, Elsevier, 1966.

Green, M. L. H., *Organometallic Compounds,* Vol. 2, *The Transition Elements,* Methuen and Co. Ltd., 1968.

Jones, R., *Chem. Rev.,* 1968, **68**, 785 (π-complexes of substituted olefins).

King, R. B., *Organometallic Syntheses,* Vol. 1, *Transition Metal Compounds,* Academic Press, 1965 (detailed preparative procedures).

King, R. B., *Transition Metal Organometallic Chemistry: An Introduction.* Academic Press, 1969.

Little, W. F., *Survey of Progress in Chemistry,* 1963, **1**, 133 (review on sandwich compounds, especially organic chemistry of ferrocene).

Rosenblum, M., *The Iron Group Metallocenes: Ferrocene, Ruthenocene, Osmocene,* Vol. 1 of *The Chemistry of Organometallic Compounds,* Wiley, 1965.

Wilke, G., *et al., Angew. Chem. (Internat. Edn),* 1966, **5**, 151 (allyl complexes and their uses in catalysis).

Wilkinson, G. and F. A. Cotton, *Progr. Inorg. Chem.,* 1959, **1**, 1 (a comprehensive review of early work on sandwich compounds).

24

Organometallic Compounds in Homogeneous Catalytic Reactions

Virtually every element in one form or another acts as catalyst for *some* chemical reaction. Since strenuous precautions are seldom taken to remove traces of ions, such as Cu^{2+} and Cl^-, or other contaminants, notably oxygen and water, from reagents commonly used, many quite ordinary reactions may well be catalyzed to some extent by extraneous materials. In this Chapter we discuss certain catalytic reactions involving transition-metal complexes, mainly those in homogeneous solution that involve the formation of metal-to-carbon bonds.

Most organic chemicals produced in bulk quantities, the majority of which are oxygenated compounds, are derived from natural gas or petroleum, usually by conversion of these hydrocarbons into olefins such as propene or butadiene. The use of transition-metal complexes for the conversion of unsaturated substances into polymers, alcohols, ketones, carboxylic acids etc., has been intensively studied and has generated a vast patent as well as scientific-journal literature. In particular, the discovery of the low-pressure polymerization of ethylene and propene by Ziegler and Natta led, not only to technical syntheses of polyalkenes and rubbers, but also to the wide use of aluminum alkyls as alkylating agents and reductants for metal complexes. Similarly, the discovery by Smidt of palladium-catalyzed oxidation of alkenes stimulated an enormous growth in the use of palladium complexes for a variety of catalytic and stoichiometric reactions of organic compounds.

Before discussing specific catalytic reactions it is necessary to discuss a number of stoichiometric ones that are important in themselves as well as in their relevance, actual or potential, to catalyses. Although the principles discussed below doubtless have applicability to heterogeneous catalysis, we shall not discuss such processes specifically. For separation of the products from reactants and catalyst, heterogeneous systems have great practical advantage over homogeneous ones, but it is often difficult to obtain from studies of heterogeneous systems the same insight into reaction mechanisms as can sometimes be obtained from homogeneous systems.

24-A. STOICHIOMETRIC REACTIONS

The transfer of atoms or groups of atoms from a metal atom to a ligand and *vice versa* is one of the fundamental process involved in catalysis by metals. The metal atom or its ligand may be attacked by electrophilic or nucleophilic reagents, but it may be difficult to prove that an attacking reagent is bound to a metal before transfer to a ligand occurs. In certain cases, however, the distinction between direct attack on a ligand and coordination prior to transfer can be made.

24-A-1. Coordinative Unsaturation

If two substances A and B are to react at a metal atom contained in a complex in solution, then clearly there must be vacant sites for their coordination. In heterogeneous reactions, the surface atoms of metals, metal oxides, halides, etc., are necessarily coordinatively unsaturated; but, when even intrinsically coordinatively unsaturated complexes such as square d^8 species are in solution, solvent molecules will occupy the remaining sites and these will have to be displaced by reacting molecules, thus:

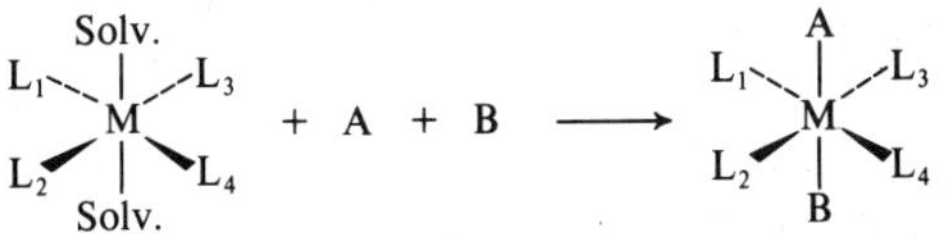

In five- or six-coordinated metal complexes, coordination sites may be made available by dissociation of one or more ligands either thermally or photochemically. Examples of thermal dissociations are

$$RhH(CO)(PPh_3)_3 \underset{}{\overset{-PPh_3}{\rightleftharpoons}} RhH(CO)(PPh_3)_2 \underset{}{\overset{-PPh_3}{\rightleftharpoons}} RhH(CO)(PPh_3)$$
$$Ni[P(o\text{-tolyl})_3]_4 \underset{}{\overset{-PR_3}{\rightleftharpoons}} Ni[P(o\text{-tolyl})_3]_3$$

The iridium analog of the first complex, namely, $IrH(CO)(PPh_3)_3$, does not catalyze the reactions that the Rh species does at $25°$, but does this when dissociation is induced either by heat or by ultraviolet irradiation.

Dissociation may also be promoted by a change in oxidation state as in oxidative addition reactions discussed below. Thus, although the complex $RhCl(PPh_3)_3$ (Section 26-G-2) is not appreciably dissociated in solution,[1] under hydrogen one phosphine is lost:

$$Rh^ICl(PPh_3)_3 + H_2 \rightleftharpoons Rh^{III}Cl(H)(H)(PPh_3)_2 + PPh_3$$

Oxygen also promotes the dissociation. Finally, sites may be made more readily available by the use of ligands with high *trans*-effects. For example, the reaction

$$PtCl_4^{2-} + C_2H_4 \rightarrow [PtCl_3C_2H_4]^- + Cl^-$$

is quite slow, but it is accelerated by addition of tin(II) chloride which forms $PtCl_3(SnCl_3)^{2-}$ where the Cl *trans* to $SnCl_3$ is labilized.

[1]   H. Arai and J. Halpern, *Chem. Comm.*, **1971**, 1571.

24-A-2. The Acid–Base Behavior of Metal Atoms in Complexes[2]

Protonation and Lewis Base Behavior. In electron-rich complexes the metal atom may have substantial non-bonding electron density located on it and consequently may be attacked by the proton or by other electrophilic reagents. An example is $(h^5\text{-}C_5H_5)_2ReH$ which is a base comparable in strength to ammonia, cf.

$$H_3N + H^+ \rightleftharpoons H_4N^+$$
$$(h^5\text{-}C_5H_5)_2HRe + H^+ \rightleftharpoons (h^5\text{-}C_5H_5)_2H_2Re^+$$

While other $h^5\text{-}C_5H_5$ compounds (page 736) with lone pairs can be protonated,[3] ferrocene has its electrons in more or less bonding orbitals and is therefore an exceedingly weak base, being protonated only by superacids. Many metal carbonyls and phosphine or phosphite complexes,[4] including clusters,[5] can be protonated, and in some cases the salts can be isolated:

$$Fe(CO)_5 + H^+ \rightleftharpoons FeH(CO)_5^+$$
$$Ni[P(OEt)_3]_4 + H^+ \rightleftharpoons NiH[P(OEt)_3]_4^+$$
$$Ru(CO)_3(PPh_3)_2 + H^+ \rightleftharpoons [RuH(CO)_3(PR_3)_2]^+$$
$$Os_3(CO)_{12} + H^+ \rightleftharpoons [HOs_3(CO)_{12}]^+$$

The acidification of carbonylate anions (page 702) can be regarded similarly, e.g.,

$$Mn(CO)_5^- + H^+ \rightleftharpoons HMn(CO)_5$$

The more strongly basic compounds also form adducts with Lewis acids,[6] e.g., $IrCl(CO)(PPh_3)_2 \cdot BF_3$.

Acceptor Properties or Lewis Acidity of Complexes. Coordinatively unsaturated compounds, whether transition metal or not, can generally add neutral or anionic nucleophiles, e.g.,

$$PF_5 + F^- \rightleftharpoons PF_6^-$$
$$TiCl_4 + 2OPCl_3 \rightarrow TiCl_4(OPCl_3)_2$$

but it is important to note that, even when they are electron-rich, coordinatively unsaturated species may show Lewis acid *as well as* Lewis base behavior, e.g.,

$$\textit{trans-}IrCl(CO)(PPh_3)_2 + CO \rightleftharpoons IrCl(CO)_2(PPh_3)_2$$
$$PdCl_4^{2-} + Cl^- \rightleftharpoons PdCl_5^{3-}$$

The Oxidative Addition Reaction.[7] When a complex behaves simultaneously as Lewis acid and Lewis base we have the so-called oxidative addition reaction, which can be written generally as

$$L_yM^n + XY \rightarrow L_yM^{n+2}(X)(Y)$$

The reverse reaction can be termed *reductive elimination*.

[2] J. C. Kotz and D. G. Pedrotty, *Organometallic Chem. Rev., A,* 1969, **4**, 479; D. F. Shriver, *Accounts Chem. Res.,* 1970, **3**, 231.

[3] See, e.g., J. C. Kotz and D. G. Pedrotty, *J. Organometallic Chem.,* 1970, **22**, 425.

[4] W. C. Drinkard et al., *Inorg. Chem.,* 1970, **9**, 392; R. A. Schunn, *Inorg. Chem.,* 1970, **9**, 394; C. A. Tolman, *J. Amer. Chem. Soc.,* 1970, **92**, 4217.

[5] See, e.g., J. Knight and M. J. Mays, *J. Chem. Soc., A,* **1970**, 711; A. J. Deeming, B. F. G. Johnson and J. Lewis, *J. Chem. Soc., A,* **1970**, 2967.

[6] R. N. Scott, D. F. Shriver and D. D. Lehman, *Inorg. Chim. Acta,* 1970, **4**, 73.

[7] (a) L. Vaska et al., *Accounts Chem. Res.,* 1968, **1**, 335; *Chem. Comm.,* **1971**, 1080; *Trans. N.Y. Acad. Sci.,* 1971, **33**, 70; *J. Amer. Chem. Soc.,* 1971, **93**, 6671; (b) J. Halpern, *Accounts Chem. Res.,* 1970, **3**, 386; (c) J. P. Collman, *Accounts Chem. Res.,* 1968, **1**, 136; S. Carra and R. Ugo, *Inorg. Chim. Acta, Rev.,* 1967, **1**, 49; J. P. Collman and W. R. Roper, *Adv. Organometallic Chem.,* 1968, **7**, 54.

For the addition reactions to proceed, we must have (*a*) non-bonding electron density on the metal M, (*b*) two vacant coordination sites on the complex L_yM to allow formation of two new bonds to X and Y, and (*c*) a metal M with its oxidation states separated by two units.

Many reactions of compounds even of non-metals, not usually thought of as oxidative additions, may be so designated, e.g.,

$$(CH_3)_2S + I_2 \rightleftharpoons (CH_3)_2SI_2$$
$$PF_3 + F_2 \rightleftharpoons PF_5$$
$$SnCl_2 + Cl_2 \rightleftharpoons SnCl_4$$

For transition metals, the most intensively studied reactions are those of complexes of metals with the d^8 and d^{10} electron configuration, notably, Fe^0, Ru^0, Os^0; Rh^I, Ir^I; Ni^0, Pd^0, Pt^0 and Pd^{II} and Pt^{II}. An especially well-studied complex[7a] is the square *trans*-IrCl(CO) $(PPh_3)_2$ (Section 26-G-2) which undergoes reactions such as

$$\text{trans-}Ir^ICl(CO)\,(PPh_3)_2 + HCl \rightleftharpoons Ir^{III}HCl_2(CO)\,(PPh_3)_2$$

It will be noted that in additions of molecules such as H_2, HCl, or Cl_2, two new bonds to the metal are made and the H—H, H—Cl or Cl—Cl bond is broken. However, molecules that contain multiple bonds may be added oxidatively *without* cleavage to form new complexes which have 3-membered rings (cf. page 729), for example,

The latter reaction also provides an example of the situation where the most stable coordination number in the oxidized state would be exceeded, so that expulsion of a ligand may occur; other examples[8] are

$$Ru^0(CO)_3(PPh_3)_2 + I_2 \xrightarrow{-CO} Ru^{II}I_2(CO)_2(PPh_3)_2$$
$$Mo^0(CO)_4bipy + HgCl_2 \xrightarrow{-CO} Mo^{II}(CO)_3bipy\,Cl(HgCl)$$

In Table 24-A-1 we list types of molecules that have been added oxidatively to at least one complex. So far, the C—H bond in alkanes or alkenes cannot normally be broken under mild conditions in oxidative additions, although the saturated hydrocarbon cubane is isomerized by certain Rh^I complexes and initial breaking of a C—H bond by oxidative addition is involved.[7b, 9]

When no ligand loss is involved, there will be an equilibrium reaction

$$L_yM^n + XY \rightleftharpoons L_yM^{n+2}XY$$

Whether the equilibrium lies on the reduced or the oxidized side depends very

[8] See, e.g., M. Cooke, M. Green and T. A. Kuc, *J. Chem. Soc., A*, **1971**, 1200; J. W. McDonald and F. Basolo, *Inorg. Chem.*, 1971, **10**, 492.

[9] See also J. Wristers, L. Brewer and R. Pettit, *J. Amer. Chem. Soc.*, 1970, **92**, 7499; P. G. Gassman and F. J. Williams, *J. Amer. Chem. Soc.*, 1970, **92**, 7631.

TABLE 24-A-1

Substances that can be Added to Complexes
in Oxidative Addition Reactions[a]

Atoms separate	Atoms remain attached
H_2	O_2
HX^b (X = Cl, Br, I, CN, COOR, ClO_4)	SO_2
H_2S, C_6H_5SH	$CF_2{=}CF_2$, $(CN)_2C{=}C(CN)_2$
RX	$RC{\equiv}CR'$
RCOX ⎫ R = Me, Ph, CF_3, etc.,	RNCS
RSO_2X ⎬	RNCO
R_3SnX ⎪ X = Cl, Br, I	$RN{=}C{=}NR'$
R_3SiX ⎭	$RCON_3$
Cl_3SiH	$R_2C{=}C{=}O$
Ph_3PAuCl	CS_2
HgX_2, CH_3HgX (X = Cl, Br, I)	$(CF_3)_2CO$, $(CF_3)_2CS$, CF_3CN
$C_6H_5N_2^+$ BF_4^-,^{c} Ph_3C^+ BF_4^-	
$C_6H_6^d$	
$CH_3C(CN)_3^e$	

(tetrachloro-o-benzoquinone structure)^f

[a] For specific examples see reactions of complexes of Rh^I, Ir^I, Pd^0 and Pt^0 (Chapter 26). Cationic species such as $[Ir(CO)(diphos)_2]^+$, where diphos = $Me_2P(CH_2)_2PMe_2$, may also undergo oxidative additions.

[b] Organic compounds with acidic hydrogen atoms also may add, e.g., $RC{\equiv}CH$, C_5H_6, succinimide, etc.

[c] Addition of diazonium salts to $IrCl(CO)(PPh_3)_2$ gives an iridium tetrazene heterocycle —N(–N=N–)N— (F. W. B. Einstein et al., J. Amer. Chem. Soc., 1971, 93, 1826).

[d] On addition to $(h^5\text{-}C_5H_5)_2MH$, where M = Nb or Ta (F. N. Tebbe and G. W. Parshall, J. Amer. Chem. Soc., 1971, 93, 3793), and to an unisolated species, probably $(h^5\text{-}C_5H_5)_2W$ (M. L. H. Green and P. J. Knowles, J. Chem. Soc., A, 1971, 1508) to form phenyl hydrido species.

[e] Undergoes C—C bond cleavage with $Pt(PPh_3)_3$ (J. L. Burmeister and L. M. Edwards, J. Chem. Soc., A, 1971, 1663). C—C bond cleavage also occurs on addition of C_6H_5CN to $Pt(PEt_3)_3$ (D. H. Gerlach et al., J. Amer. Chem. Soc., 1971, 93, 3553) and in reactions of strained hydrocarbons such as cyclopropane with some d^8 and d^{10} complexes.

[f] See, e.g., A. L. Balch and Y. S. Sohn, J. Organometallic Chem., 1971, 30, C31.

critically on (a) the nature of the metal and its ligands, (b) the nature of the added molecule XY and of the M—X and M—Y bonds so formed, and (c) on the medium in which the reaction is conducted. At the present time there are insufficient quantitative data to allow predictions to be made, but the various factors have been considered in some detail.[7a] Qualitatively, however, we can say that for the addition to proceed, $E_{MX} + E_{MY}$ must exceed $E_{XY} + P$, where E_{MX} and E_{MY} are the free energies of the new bonds to the metal, E_{XY} is the free energy for bond dissociation of XY, and P is the promotional energy for oxidation of the metal. Usually only E_{XY} is known.

The higher oxidation states are usually more stable for the heavier than for the lighter metals, so that, for example, Ir^{III} species are generally more stable than Rh^{III} species. For the ligands, factors that tend to increase the electron density on the metal make for an increase in oxidizability. For example, in the reactions of square iridium(I) complexes with carboxylic acids,[10] e.g.,

$$\textit{trans-}Ir^IX(CO)L_2 + RCOOH \rightleftharpoons Ir^{III}HX(O_2CR)(CO)L_2$$

the equilibria lie further to the Ir^{III} side in the order $X = Cl < Br < I$ and in the orders $L = PPh_3 < PMePh_2 < PMe_2Ph < PMe_3$ and $L = P(p\text{-}FC_6H_4)_3 < P(p\text{-}MeC_6H_4)_3$. There is no direct correlation to be expected between the pK_a of various acids, HX, *in water* and the propensity to add to metals. For example, HCOOH (pK_a 3.75) and CH_3COSH (pK_a 3.33) have similar acidities, but on addition to $Pt(PPh_3)_4$ the formic acid adduct is very unstable whereas the thioacetic acid adduct is quite stable. This difference is evidently due to the greater affinity towards platinum of sulfur than of oxygen.[11] Hence both the acidity and the nephelauxetic effect (page 601) of the conjugate base are important.

Stereochemistry and Mechanism of Addition.   When the XY molecule adds without severance of X from Y the two new bonds to the metal are necessarily in *cis*-positions, but when X and Y are separated the product may be one or more of several isomers with either *cis*- or *trans*-MX and -MY groups, e.g.,

The final product will be the isomer or isomer mixture that is the most stable thermodynamically under the pertaining conditions. The ligands, solvent, temperature, pressure, etc., will have a decisive influence on this. The nature of the final product does not necessarily give a guide to the initial product of the reaction, since isomerization of the initial product may also occur. The following observations on the direction of addition and on the addition mechanism have been made:

(a) When solid *trans-*$IrCl(CO)(PPh_3)_2$ reacts with HCl gas the product has H and Cl in *cis*-positions.

(b) In the addition of HCl or HBr to *trans-*$IrCl(CO)(PPh_3)_2$ in non-polar solvents such as benzene, *cis*-addition is also found. If wet solvents or the polar solvent dimethylformamide or benzene–methanol are used, *cis–trans*-mixtures are obtained.[12]

In polar media, initial protonation of a square complex will produce first a cationic 5-coordinate complex which may then isomerize by an intramolecular mechanism (page 140). Coordination of halide ion would finally give the oxidized product:

$$MXL_3 + H^+(solv) \rightarrow MHXL_3^+$$
$$MHXL_3^+ + Cl^-(solv) \rightarrow MHClXL_3$$

Isomers could also result from rapid exchange of halide ions through 5-coordinate intermediates.

[10] B. L. Shaw and R. E. Stainbank, *J. Chem. Soc., A,* **1971,** 3716.
[11] D. M. Roundhill, P. G. Tripathy and B. W. Renoe, *Inorg. Chem.,* 1971, **10,** 727.
[12] D. M. Blake and M. Kubota, *Inorg. Chem.,* 1970, **9,** 989; *J. Amer. Chem. Soc.,* 1970, **92,** 2578.

(c) The addition of alkyl halides to compounds such as trans-IrCl(CO)-$(PR_3)_2$ in polar media may well be of the S_N2 type,[13] as is usually the case for the much studied attacks on alkyl halides in organic chemistry, viz.,

$$L_yM\!:\ \overset{\frown}{C}R^1R^2R^3X \rightarrow L_yM^{\delta+}\text{---}\underset{\underset{R^3}{|}}{\overset{\overset{R^1\ \ R^2}{\diagdown\!\diagup}}{C}}\text{---}X^{\delta-} \rightarrow [L_yM-CR^1R^2R^3]^{+} + X^-$$

$$\downarrow$$

$$L_yMX(CR^1R^2R^3)$$

Such reactions should proceed with inversion of configuration if an optically active halide is used. However, a claim to have established inversion in one case has been shown to be erroneous[14] and definitive proof is as yet lacking.

(d) In the absence of solvent, the addition of the optically active halide $H^*C(CH_3)BrCOOC_2H_5$ (and also of an optically active silane $R^1R^2R^3SiH$) to trans-IrCl(CO)$(PPh_3)_2$ has been claimed to proceed with retention of the configuration at carbon.[15] This conclusion, which was based on recovery of the optically active halide after cleavage from the complex with bromine, may not be correct since brominative cleavage of M—C bonds may occur with either inversion or retention of configuration.[16]

At the present time, therefore, it appears that the oxidative addition reaction may proceed by the following routes: (1) a purely ionic mechanism involving a 5-coordinate cationic intermediate, especially in polar solvents and with ionizable molecules such as hydrogen halides; (2) by the classical type of S_N2 attack of nucleophiles on alkyl halides; or (3) under non-polar conditions, particularly with molecules that have little or no polarity—such as hydrogen—by one-step concerted processes, which may give products with the new bonds formed in cis-positions,[15] viz.,

$$\oplus X\text{---}Y\ominus$$

$$L\overset{+}{\underset{-}{\rightleftharpoons}}Ir\overset{-}{\underset{+}{\rightleftharpoons}}L \longrightarrow \underset{L}{\overset{X}{\diagdown}}Ir\underset{L}{\overset{Y}{\diagup}}$$

The fact that many of the d^8 complexes react with molecular hydrogen might seem surprising in view of the high energy (ca. 450 kJ mol^{-1}) of the H—H bond. The attack on the H_2 molecule probably results from electron density of the metal entering the hydrogen $1s^*$ antibonding orbitals, thus leading to bond weakening. Where two coordination sites are available on the metal atom two cis-M—H bonds result (cf. page 788). An alternative reaction is heterolysis of H_2 by the removal of H^+ in the presence of strong bases, e.g.,

$$RuCl_2(PPh_3)_3 + H_2 + Et_3N \rightarrow RuHCl(PPh_3)_3 + Et_3NH^+Cl^-$$

[13] A. J. Hart-Davis and W. A. G. Graham, Inorg. Chem., 1970, 9, 2658.
[14] F. R. Jensen and B. Knickel, J. Amer. Chem. Soc., 1971, 93, 6339.
[15] R. G. Pearson and W. R. Muir, J. Amer. Chem. Soc., 1970, 92, 5519.
[16] G. M. Whitesides and D. J. Boschetto, J. Amer. Chem. Soc., 1971, 93, 1529; D. Dodd and M. D. Johnson, Chem. Comm., 1971, 571.

24-A-3. Migration of Atoms or Groups from Metal to Ligand; the "Insertion Reaction"[17]

The concept of "insertion" is of wide applicability in chemistry when defined as a reaction wherein any atom or group of atoms is inserted between two atoms initially bound together:

$$L_nM—X + YZ \rightarrow L_nM—(YZ)—X$$

Some representative examples are

$$R_3SnNR_2 + CO_2 \rightarrow R_3SnOC(O)NR_2$$
$$Ti(NR_2)_4 + 4CS_2 \rightarrow Ti(S_2CNR_2)_4$$
$$R_3PbR' + SO_2 \rightarrow R_3PbOS(O)R'$$
$$[(NH_3)_5RhH]^{2+} + O_2 \rightarrow [(NH_3)_5RhOOH]^{2+}$$
$$(CO)_5MnCH_3 + CO \rightarrow (CO)_5MnCOCH_3$$

An early example (Berthelot, 1869) is

$$SbCl_5 + 2HC≡CH \rightarrow Cl_3Sb(CH=CHCl)_2$$

For transition metals, which are our main concern, the most detailed studies have been made on the "insertion" of CO into metal-to-carbon bonds, but other representative insertions are listed in Table 24-A-2.

Mechanistic studies have been made using mainly $CH_3Mn(CO)_5$ or related compounds.[18] With ^{14}CO as tracer, it has been shown that (*a*) the CO molecule that becomes the acyl-carbonyl is not derived from external CO but is one already coordinated to the metal atom, (*b*) the incoming CO is added *cis* to the acyl group, i.e.

(*c*) the conversion of alkyl into acyl can be effected by addition of ligands other than CO, e.g.,

Kinetic studies of the reactions show that the first step involves an equilibrium between the octahedral alkyl and a five-coordinate acyl species, e.g.,

$$CH_3Mn(CO)_5 \rightleftharpoons CH_3COMn(CO)_4$$

[17] M. F. Lappert and B. Prokai, *Adv. Organometallic Chem.*, 1967, **5**, 225; A. Wojcicki, *Accounts Chem. Res.*, 1971, **4**, 344 (SO₂); R. F. Heck in *Mechanisms of Inorganic Reactions*, Advances in Chem. Series, No. 49, Amer. Chem. Soc., Washington D.C., 1965; W. Kitching and C. W. Fong, *Organometallic Chem. Rev. A.*, 1970, **5**, 281 (SO₂ and SO₃ insertions).

[18] See, e.g., P. K. Maples and C. S. Kraihanzel, *J. Amer. Chem. Soc.*, 1968, **90**, 6645; W. D. Bannister *et al.*, *J. Chem. Soc.*, *A*, **1969**, 698; B. L. Booth *et al.*, *J. Chem. Soc.*, *A*, **1970**, 308, 1979; K. Noack and F. Calderazzo, *J. Organometallic Chem.*, 1967, **10**, 101; *Inorg. Chem.*, 1968, **7**, 345; M. Green and D. J. Westlake, *J. Chem. Soc.*, *A*, **1971**, 367; R. W. Glyde and R. J. Mawby, *Inorg. Chem.*, 1971, **10**, 854; M. Kubota, D. M. Blake and S. A. Smith, *Inorg. Chem.*, 1971, **10**, 1430.

TABLE 24-A-2

Some Representative Insertion or Group-transfer Reactions

"Inserted" molecule	Bond	Product
CO	M—CR$_3$	MCOCR$_3$
	M—H	MCHO
	M—OH	MCOOH
	M—NR$_2$	MCONR$_2$
SO$_2$	M—C	MS(R)O$_2$ or MOS(O)R^a
	M—M	MOS(O)M
	M—(h^3-C$_3$H$_5$)	MSO$_2$CH$_2$CH=CH$_2$
CO$_2$	M—H	MCOOH
	M—C^b	MC(O)OR
	M—NR$_2$	MOC(O)NR$_2$
CS$_2$	M—M	MSC(S)M
	M—H	MS$_2$CH and MSC(S)H
C$_2$H$_4$	M—H	MC$_2$H$_5$
C$_2$F$_4$	M—H	MCF$_2$CF$_2$H
CH$_2$=C=CH$_2$	M—R	M(h^3-allyl)
SnCl$_2$	M—M	MSn(Cl$_2$)M
RNC	M—R'	MC(R')=NR
	M—h^3-C$_3$H$_5$	MC(=NR) (CH$_2$CH=CH$_2$)

^{a} (h^5-C$_5$H$_5$)Fe(CO)$_2$(CH$_2$C≡CMe) gives a hetero ring Fe—C=CMe–S(O)–OCH$_2$ but usually the S-sulfinate, MS(R)O$_2$ is formed (M. R. Churchill and J. Wormald, *J. Amer. Chem. Soc.*, 1971, **93**, 354; *Inorg. Chem.*, 1971, **10**, 572; A. Wojcicki *et al.*, *J. Amer. Chem. Soc.*, 1971, **93**, 2535, *Inorg. Chem.*, 1971, **10**, 2130; C. W. Fong and W. Kitching, *J. Amer. Chem. Soc.*, 1971, **93**, 3791).

^{b} (h^5-C$_5$H$_5$)$_2$TiPh$_2$ gives

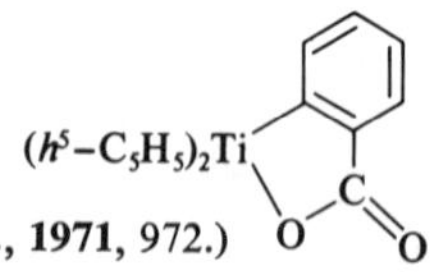

$(h^5$–C$_5$H$_5)_2$Ti

(M. E. Vol'pin *et al.*, *Chem. Comm.*, **1971**, 972.)

The incoming ligand (L=CO, Ph$_3$P, etc.) then adds to the 5-coordinate species:

$$CH_3COMn(CO)_4 + L \rightleftharpoons CH_3COMn(CO)_4L$$

Since 5-coordinate species can undergo intramolecular rearrangements (page 140), more than one isomer of the final product may be formed.

The insertion reaction is thus best considered as an alkyl migration to a coordinated carbon monoxide ligand in a *cis*-position, and the migration probably proceeds through a three-center transition state:

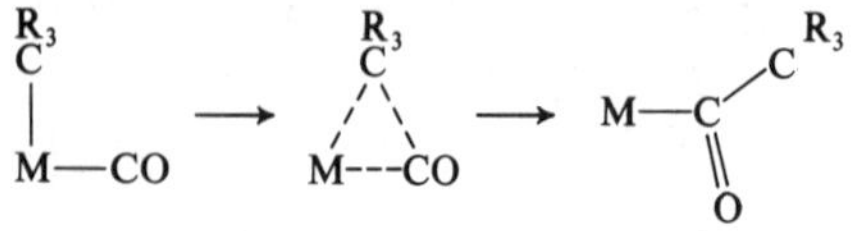

For a PPh$_3$-promoted insertion of CO into the Fe—C bond in h^5-C$_5$H$_5$Fe(CO)$_2$R, where R=CHDCHDCMe$_3$, to give h^5-C$_5$H$_5$Fe(CO)-(PPh$_3$)COR it was shown that the reaction proceeds—as expected from this mechanism—with complete *retention* of configuration.[19]

In more complicated molecules, four-center transition states may be in-

[19] G. M. Whitesides and D. J. Boschetto, *J. Amer. Chem. Soc.*, 1969, **91**, 4313.

volved. For example, there is the important hydride transfer[20] to alkenes to form an alkyl, e.g.,

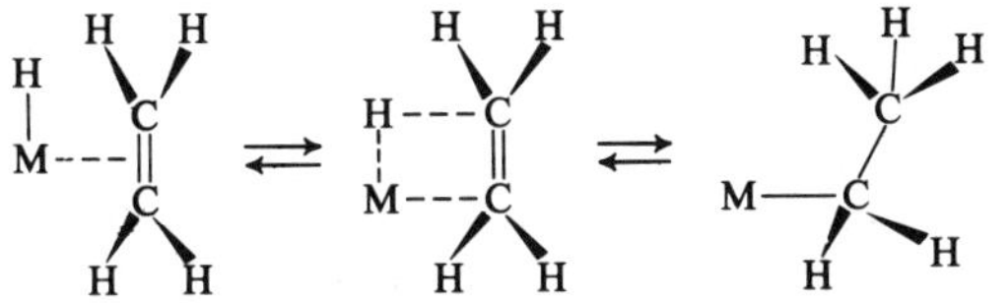

It may be noted that this reaction involves the transfer of hydrogen to the second (β) carbon atom of the resulting alkyl. For hydrocarbons, this has so far not been proved, mainly because such reactions are very readily reversible, as discussed below. For fluoroolefins such as tetrafluoroethylene, stable alkyls result and here transfer to the β-carbon atom has been proved, e.g.,

$$RhH(CO)(PPh_3)_3 + C_2F_4 \rightarrow Rh(CF_2CF_2H)(CO)(PPh_3)_2 + PPh_3$$

24-A-4. Reactions of Coordinated Ligands[21]

In Section 24-A-3, we discussed *intra*molecular attack on a coordinated ligand promoted by an incoming new ligand. Here we discuss attack by an external reagent directly upon a coordinated ligand, specifically those involving transition metal-to-carbon bonds, and also intramolecular hydrogen transfer. It is not always easy to *prove* that the ligand reacts while coordinated; or that prior coordination of the reagent is not involved in which case the reaction would be in effect intramolecular.

Reactions involving H^+ and H^-.[21b,c] These are the simplest types of electrophilic and nucleophilic reactions of organotransition-metal complexes, and among the most important are those involving alkyl and alkene compounds (cf. Chapter 23, page 754).

Hydride ion addition to certain h^5-C_5H_5 compounds produces cyclopentadiene olefin complexes (equations 24-A-1, 24-A-2); addition to arene complexes gives h^5-cyclohexadienyls (24-A-3):

$$(h^5\text{-}C_5H_5)_2Co^+ + H^- \xrightarrow[\text{in THF}]{\text{NaBH}_4} h^5\text{-}C_5H_5CoC_5H_6 \qquad (24\text{-}A\text{-}1)$$

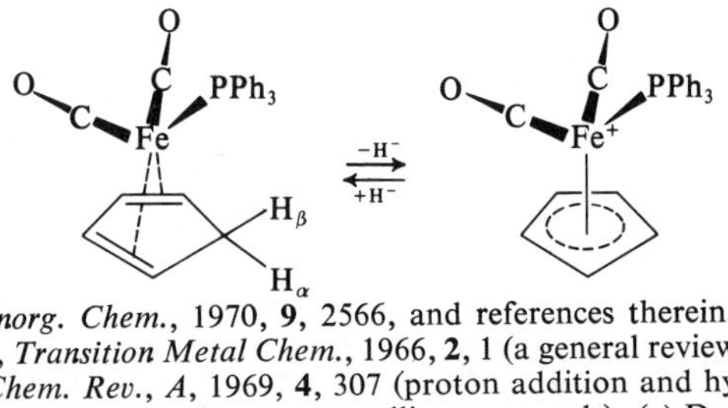

(24-A-2)

[20] R. A. Schunn, *Inorg. Chem.*, 1970, **9**, 2566, and references therein.
[21] (a) J. P. Collman, *Transition Metal Chem.*, 1966, **2**, 1 (a general review); (b) M. A. Haas, *Organometallic Chem. Rev.*, A, 1969, **4**, 307 (proton addition and hydride abstractions from ligands in transition-metal organometallic compounds); (c) D. A. White, *Organometallic Chem. Rev.*, A, 1968, **3**, 497 (general review of electrophilic and nucleophilic attack on ligands in transition-metal organometallic compounds); (d) F. R. Hartley, *Chem. Rev.*, 1969, **69**, 799 (reactions of Pd and Pt olefin and acetylene complexes); (e) J. Mašek, *Inorg. Chim. Acta*, 1969, **3**, 99 (electrophilic behavior of coordinated NO).

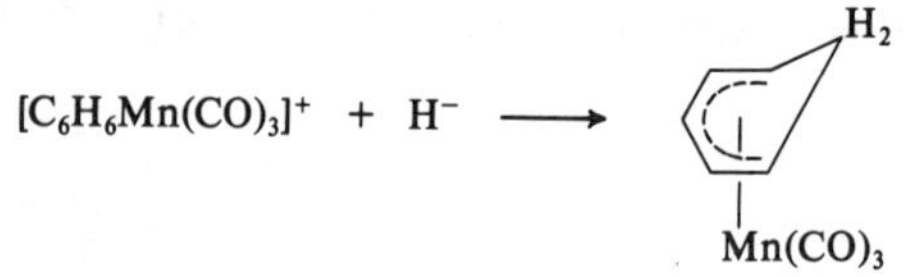

$$[C_6H_6Mn(CO)_3]^+ \; + \; H^- \; \longrightarrow \qquad\qquad\qquad (24\text{-}A\text{-}3)$$

The cyclopentadiene derivatives obtained in this way are noteworthy in that it has been shown that one of the hydrogen atoms (H_α on the *exo*-side of the ring) gives an intense and unusually low ($\sim 2750 \text{ cm}^{-1}$) C—H stretching frequency. In addition, this particular hydrogen atom is the labile one. It is removed by chemical agents that can remove hydride ion, for example, acids, $(C_6H_5)_3C^+BF_4^-$, N-bromosuccinimide, etc.

Similar hydride transfer reactions are possible for certain complex alkyls, where conversion into the olefin complex can be achieved by abstraction of H^- by triphenylmethyl tetrafluoroborate:

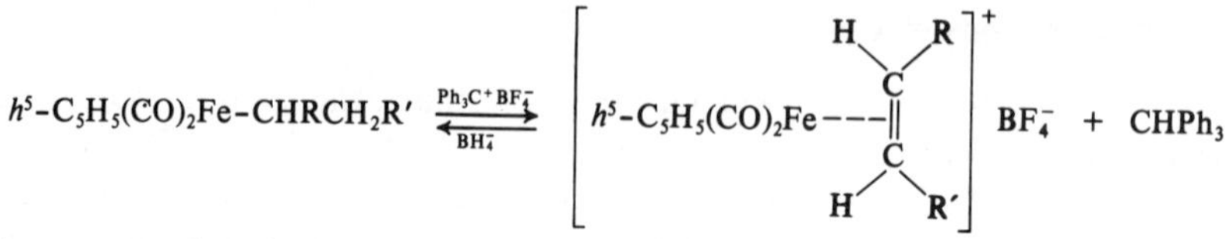

$$h^5\text{-}C_5H_5(CO)_2Fe\text{-}CHRCH_2R' \; \underset{BH_4^-}{\overset{Ph_3C^+BF_4^-}{\rightleftharpoons}} \; \left[h^5\text{-}C_5H_5(CO)_2Fe\text{---}\|\begin{array}{c}\end{array} \right]^+ BF_4^- \; + \; CHPh_3$$

There are related reactions involving the addition to, or abstraction of *protons* from the organic group. Thus acetonyl and other oxoalkyl compounds give olefin-coordinated ions in which the enol forms, e.g., of acetone, are stabilized by bonding to the metal atom:

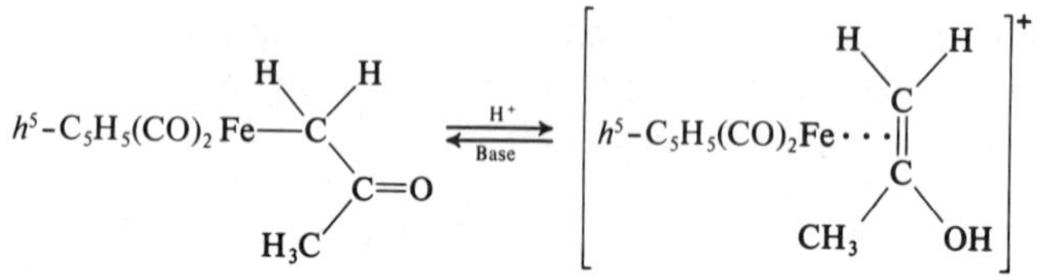

$$h^5\text{-}C_5H_5(CO)_2Fe\text{---}C \underset{Base}{\overset{H^+}{\rightleftharpoons}} \left[h^5\text{-}C_5H_5(CO)_2Fe\cdots\| \right]^+$$

A similar case occurs with cyanoalkyls, which may be reversibly protonated to ketene imine complexes:

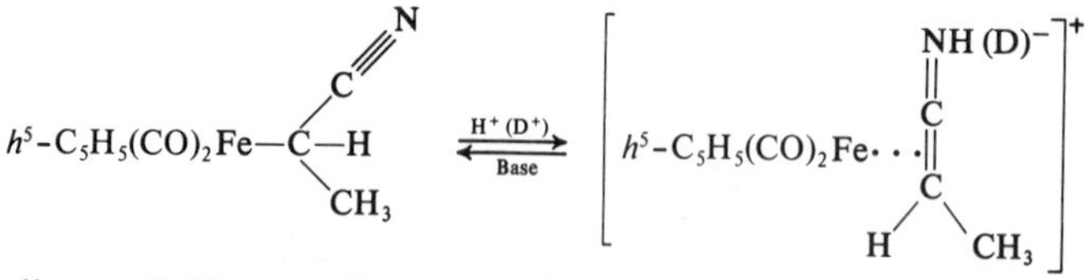

$$h^5\text{-}C_5H_5(CO)_2Fe\text{---}\overset{}{\underset{CH_3}{C}}\text{---}H \underset{Base}{\overset{H^+(D^+)}{\rightleftharpoons}} \left[h^5\text{-}C_5H_5(CO)_2Fe\cdots\| \right]^+$$

and, finally, σ-allylic complexes may be protonated:

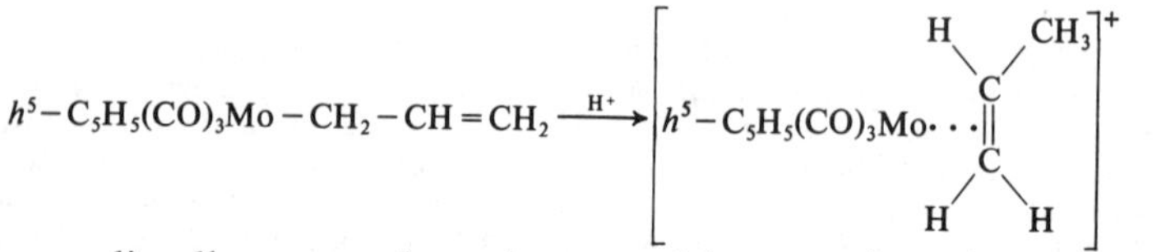

$$h^5\text{-}C_5H_5(CO)_3Mo\text{-}CH_2\text{-}CH{=}CH_2 \overset{H^+}{\longrightarrow} \left[h^5\text{-}C_5H_5(CO)_3Mo\cdots\| \right]^+$$

Where cyclic alkenes are bound to transition metals and not all of the double bonds are involved in bonding to the metal, we may have also reversible hydrogen transfer reactions, e.g.,

As noted earlier, protonations of this type may give metal-stabilized carbonium ions,[22] such as $C_6H_7^+$.

Nucleophilic Attack. There are innumerable reactions involving anions or bases such as OH^-, OR^-, $OCOR^-$, N_3^-, R^-, NR_3, N_2H_4, etc. and the ligands attacked may be CO, NO, RCN, RNC, alkenes, etc. It is not always certain that attack is direct, and prior coordination may well occur in some cases, so that then the reactions could be considered as intramolecular transfers.

Some reactions have long been known, such as attack on coordinated NO^{23} and CO by OH^- ion, e.g.,

$$[Fe(CN)_5NO]^{2-} + HO^- \xrightarrow{\text{Slow}} \left[Fe(CN)_5N\begin{matrix} O \\ OH \end{matrix} \right]^{3-} \xrightarrow[\text{Fast}]{HO^-} [Fe(CN)_5NO_2]^{4-} + H_2O$$

$$Fe(CO)_5 + HO^- \longrightarrow \left[(CO)_4Fe-C\begin{matrix} O \\ OH \end{matrix} \right]^- \xrightarrow{HO^-} (CO)_4FeH^- + HCO_3^-$$

The attack of alkoxide ions on CO to give M—COOR groups is especially common and has been observed for complexes of Mn, Re, Fe, Ru, Os, Co, Rh, Ir, Pd, Pt and Hg,[24] e.g.,

$$[Ir(CO)_3(PPh_3)_2]^+ \xrightarrow[H^+]{CH_3O^-} Ir(CO)_2(COOCH_3)(PPh_3)_2$$

The reaction is important in the synthesis of carboxylic acids and esters from olefins, CO and water or alcohols, while similar attacks on CO by OH^- or H_2O are involved in reductions of Co^{2+} or Rh^{3+} by CO to form CO_2.[25]

Coordinated carbon monoxide can also be attacked by lithium alkyls or dialkylamides,[26] e.g.,

$$LiCH_3 + W(CO)_6 \rightarrow Li^+[(CO)_5W-C(O)CH_3]^-$$

These anions may in turn be converted into coordinated carbenes (see page 767). Carbon monoxide may also be attacked by NH_3, primary amines, or

[22] See, e.g., R. Aumann and S. Winstein, *Tetrahedron Letters*, 1970, **12**, 903; A. Eisenstadt and S. Winstein, *Tetrahedron Letters*, 1971, **13**, 613.

[23] J. Mašek and H. Wendt, *Inorg. Chim. Acta*, 1969, **3**, 455; E. J. Baran and A. Müller, *Chem. Ber.*, 1969, **102**, 3915.

[24] See, e.g., S. D. Ibekwe and K. A. Taylor, *J. Chem. Soc., A*, **1970**, 1; J. E. Byrd and J. Halpern, *J. Amer. Chem. Soc.*, 1971, **93**, 1634.

[25] H. C. Clark and W. J. Jacobs, *Inorg. Chem.*, 1970, **9**, 1229.

[26] D. J. Darensbourg and M. Y. Darensbourg, *Inorg. Chem.*, 1970, **9**, 1691; E. O. Fischer et al., *Chem. Ber.*, 1970, **103**, 1262; *Angew. Chem. Internat. Edn.*, 1970, **9**, 309; *J. Organometallic Chem.*, 1970, **23**, 215.

hydrazine;[27] attack by azide ion[27a] affords the group $MC(O)N_3$, which readily loses nitrogen and rearranges to give MNCO.

Alkene and dienyl complexes are also susceptible to attack by O, N and C nucleophiles, and many examples are known,[28a,b] e.g.,

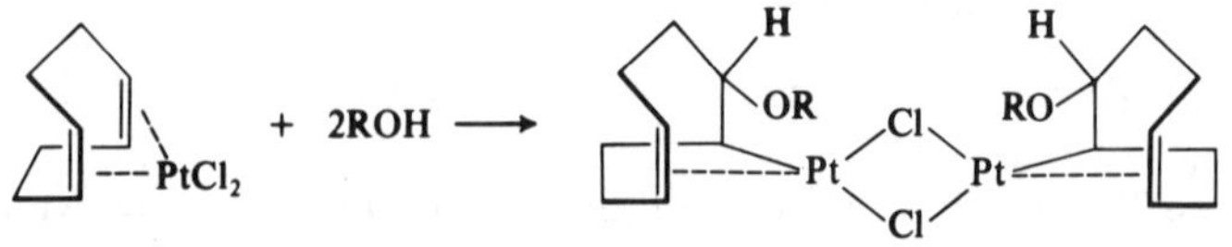

Studies[28b] of the attack of diethylamine on a coordinated alkene have shown unequivocally that there is a direct attack on the alkene and that there is no intramolecular rearrangement of an intermediate involving coordinated amine.

Finally nitrile or isocyanide complexes may be attacked. Thus nitrile complexes with aromatic amines and alcohols give complexes of amidines and imidate esters respectively,[29a] e.g.,

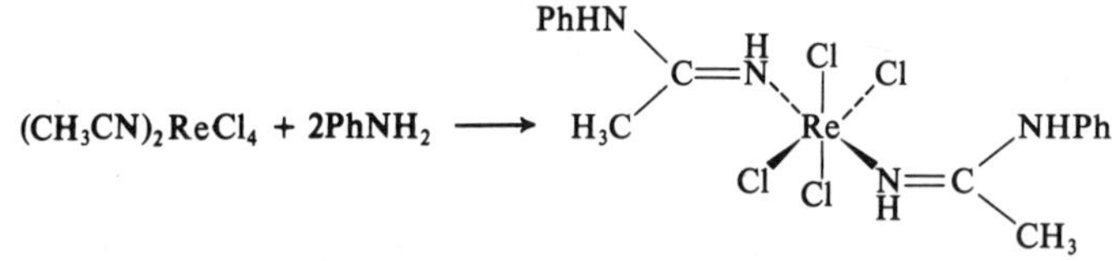

Isocyanide complexes[29b] on the other hand are attacked to form carbene complexes (page 767), e.g.,

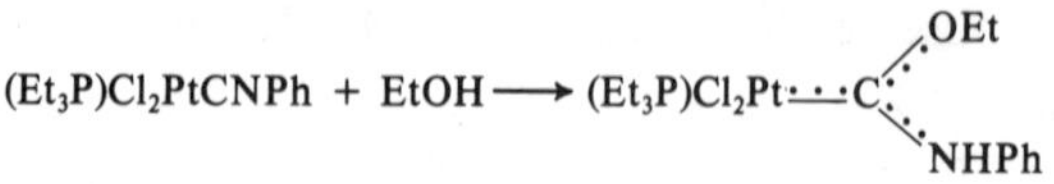

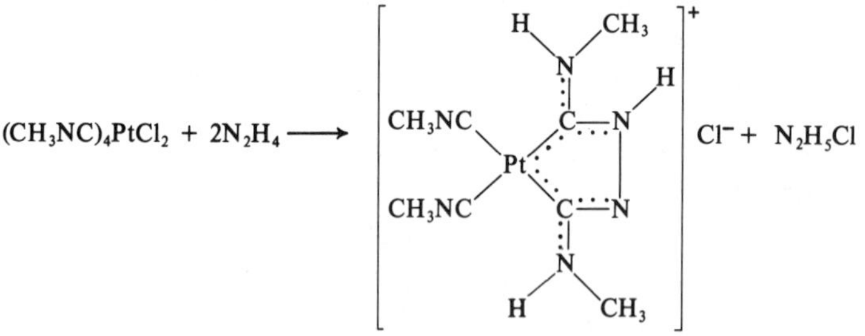

[27] J. T. Moelwyn-Hughes, A. W. B. Garner and A. S. Howard, *J. Chem. Soc., A*, **1971**, 2361, 2370.

[27a] L. M. Charley and R. J. Angelici, *Inorg. Chem.*, 1971, **10**, 868.

[28a] L. A. P. Kane-McGuire, *J. Chem. Soc., A*, **1971**, 1602.

[28b] A. De Renzie, R. Palumbo and G. Paiaro, *J. Amer. Chem. Soc.*, 1971, **93**, 880.

[29a] G. Rouschias and G. Wilkinson, *J. Chem. Soc.*, **1968**, 489; H. C. Clark, L. E. Manzer, *Inorg. Chem.*, 1971, **10**, 2699.

[29b] G. Rouschias and B. L. Shaw, *J. Chem. Soc., A*, **1971**, 2097, W. M. Butler and J. H. Enemark, *Inorg. Chem.*, 1971, **10**, 2416.

Intramolecular Hydrogen Transfer.   It is possible to transfer groups from the metal to ligand by insertion reactions, but a rather special case of transfer reactions is one between certain ligands and the metal atom in which a hydrogen atom is initially transferred and is then subsequently lost. Such reactions are especially important for triarylphosphines and triaryl phosphites.[30] Some examples are the following:

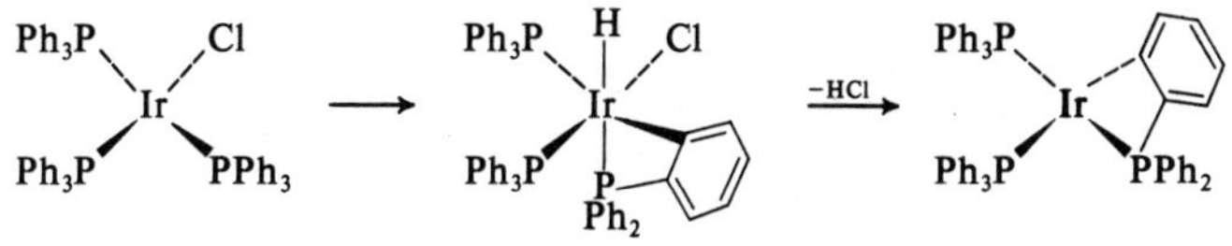

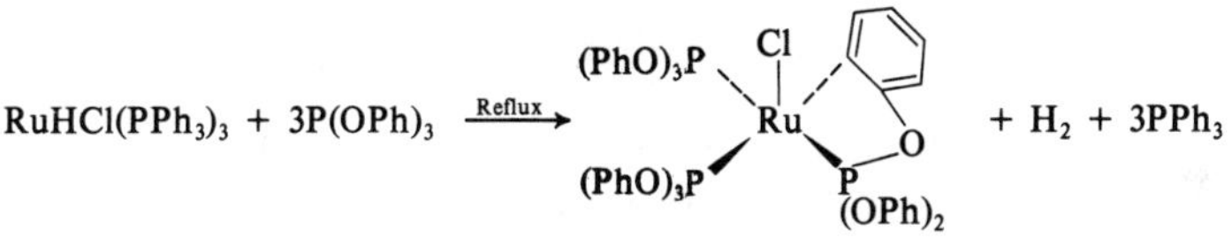

$$RuHCl(PPh_3)_3 + 3P(OPh)_3 \xrightarrow{Reflux}$$

$$+ H_2 + 3PPh_3$$

$$CH_3Rh(PPh_3)_3 \xrightarrow{Heat} CH_4 + Rh(C_6H_4PPh_2)(PPh_3)_2$$

In a few cases, the reactions may be reversible.[31] The hydrido complex formed may be characterized by nmr spectra; an example of such a case[31] is

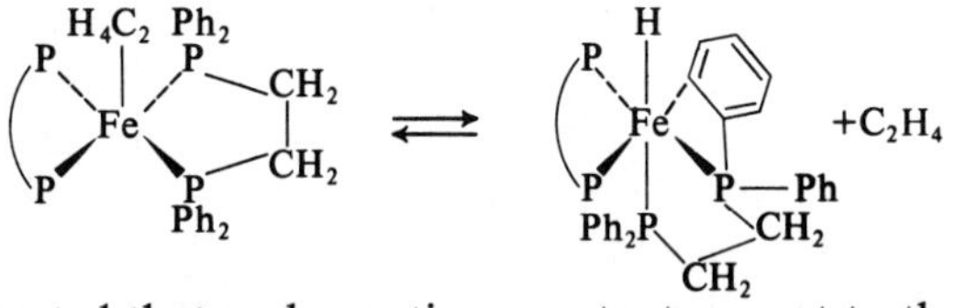

$$+C_2H_4$$

It will be noted that such reactions are tantamount to the oxidative addition of C—H of a phenyl ring to a metal atom; a few cases of the direct addition of benzene are known (Table 24-A-1).

The formation of M—C bonds in this way is also observed in reactions with azobenzene, N,N-dimethylbenzylamine and similar compounds,[32] e.g.,

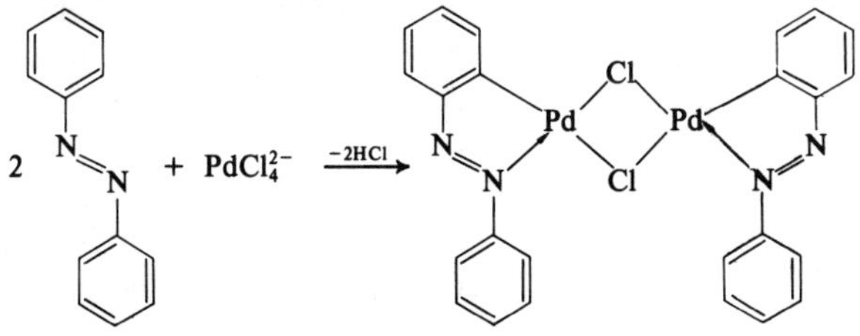

Reactions of Coordinated Molecular Oxygen.   Several complexes, notably tertiary phosphine complexes of Ru^0, Rh^I, Ir^I and Pt^0 react with oxygen to

[30] G. W. Parshall, *Accounts Chem. Res.*, 1970, **3**, 139; S. D. Robinson *et al.*, *J. Chem. Soc.*, A, **1970**, 639; **1971**, 3413; A. J. Cheney *et al.*, *Chem. Comm.*, **1970**, 1176.

[31] U. A. Gregory *et al.*, *J. Chem. Soc.*, A., **1971**, 1118.

[32] R. F. Heck, *J. Amer. Chem. Soc.*, 1968, **90**, 313; J. M. Kliegman and A. C. Cope, *J. Organometallic Chem.*, 1969, **16**, 309; B. Crociani *et al.*, *J. Chem. Soc.*, A **1970**, 531; D. L. Weaver, *Inorg. Chem.*, 1970, **9**, 2250; M. I. Bruce, M. Z. Iqbal and F. G. A. Stone, *J. Chem. Soc.*, A, **1970**, 3204; G. Pini *et al.*, *Chem. Comm.*, **1971**, 470.

give adducts, sometimes reversibly. The crystal structures of several iridium complexes e.g., $IrCl(CO)(O_2)(PPh_3)_2$ (24-A-I), have been studied.[33] The O—O bond distances in irreversible systems are longer than those in the reversible ones. Thus for the above chloride, where the oxygen addition is reversible, the O—O distance is 1.30 Å, whereas for the iodide, where oxygen is added irreversibly at 25°/1 atm, the O—O distance is 1.51 Å (see also page 635).

(24-A-I)   (24-A-II)   (24-A-III)

The bonding of oxygen in such complexes, and indeed also the bonding of acetylenes, tetracyanoethylene and some other molecules, has given rise to some discussion,[33,34] since between known extremes of the type (24-A-II) and (24-A-III), where there is a real difference in oxidation state of the metal, there are molecules in which the bond distances suggest an intermediate bonding situation (see discussion Chapter 23, page 730).

Although an MO approach deals with the intermediate situations, a useful idea is that of a partial oxidation state for the metal.[7a] Thus, comparing the CO stretches in trans-$Ir^ICl(CO)(PPh_3)_2$, defined as containing Ir^I, and in $IrCl_3(CO)(PPh_3)_2$, defined as Ir^{III}, with those in other complexes, e.g., the O_2 or SO_2 adducts, we can say that in $IrCl(CO)O_2(PPh_3)_2$ the Ir has "oxidation state" 1.89 whereas in the SO_2 adduct it is 2.0. There is a rough correlation of this definition of oxidation state with the reversibility of the reaction; thus reactions of C_2F_4 (2.57) and I_2 (2.85) are also reversible, whereas that of Br_2 (2.95) is not.

When coordinated, oxygen is kinetically more reactive than in the free molecule, and numerous substrates may be oxidized. In stoichiometric reactions the oxidized entity commonly remains bound to the metal, but some catalytic reactions are known. Thus triphenylphosphine is oxidized to Ph_3PO and CO to CO_2.[35]

However, catalytic oxidations of cyclohexene or ethylbenzene in the presence of $RhCl(PPh_3)_3$ give the products characteristic of free-radical reactions:[36] thus cyclohexene gives cyclohexenone, cyclohexenol, cyclohexyl hydroperoxide and unidentified products.

For some complexes, the reaction proceeds through peroxo intermediates

[33] J. A. McGinnety, N. C. Paine and J. A. Ibers, J. Amer. Chem. Soc., 1969, 91, 6301.

[34] C. Bombieri et al., J. Chem. Soc., A, 1970, 1313; D. M. P. Mingos, Nature, 1971, 229, 193; 230, 154.

[35] B. W. Graham et al., Chem. Comm., 1970, 1272; J. Halpern and A. L. Pickard, Inorg. Chem., 1970, 9, 2798; J. Kiji and J. Furakawa, Chem. Comm., 1970, 977.

[36] V. P. Kurkov, J. Z. Pasky and J. B. Lavigne, J. Amer. Chem. Soc., 1968, 90, 4743; L. W. Fine, M. Grayson and V. H. Suggs, J. Organometallic Chem., 1970, 22, 219; see also Section 26-G-2.

which may be isolated and characterized,[37] e.g., the platinum peroxocarbonate:

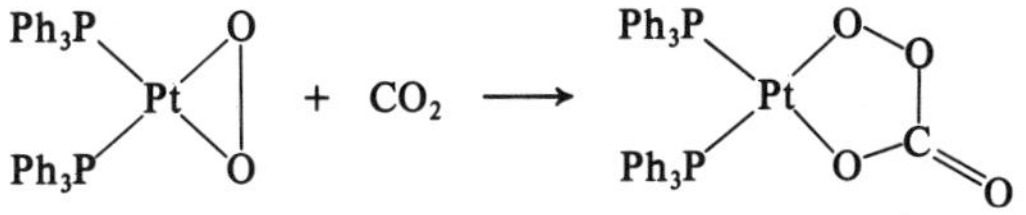

The mechanism of oxidation of SO_2 by $IrCl(CO)O_2(PPh_3)_2$ to give a sulfato complex has been shown[38] by use of ^{18}O tracer to be similar, namely,

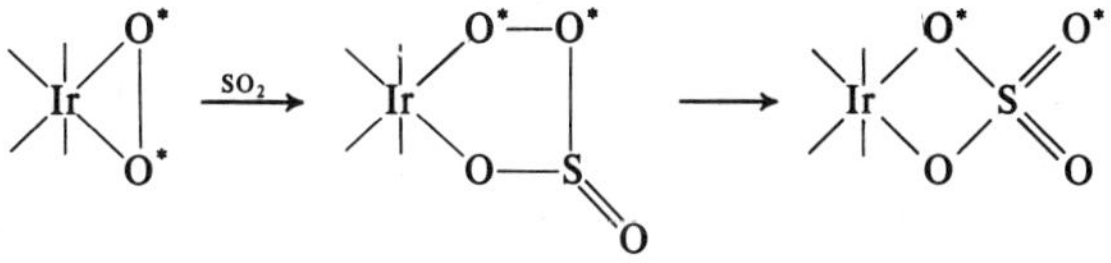

24-B. CATALYTIC REACTIONS OF ALKENES

The term catalyst is an ambiguous one and requires careful use. In heterogeneous reactions, where for example a gas mixture is passed over a solid that evidently undergoes no change, the term, meaning a substance added to accelerate a reaction, may have some point. However, homogeneous catalytic reactions in solution are commonly very complex and proceed by way of linked chemical reactions involving different metal species. The concept of one particular species being "the catalyst," even if it is the one added to initiate or accelerate the reaction, has no validity. It is necessary to think in terms of catalytic intermediates involved in the various chemical reactions of a catalytic cycle. Examples of such cycles will be given later.

24-B-1. Isomerization[39]

Many transition-metal ions and complexes, especially those of Group VIII metals, promote double-bond migration—that is, isomerization—in alkenes, to give invariably the thermodynamically most stable isomeric mixture. Thus 1-alkenes give (cis + trans)-2-alkenes. The mechanism of isomerization involves the reaction discussed above (page 779), namely, transfer of a hydrogen atom from the metal to the coordinated alkene, thus forming an alkyl. The reaction is characteristic for many transition-metal hydrido species; and, in addition, many complexes that do not have M—H bonds, e.g., $(Et_3P)_2NiCl_2$, will isomerize alkenes provided that a source of hydride ion such as molecular hydrogen is present.

The first step in the reaction[20,40] must be the coordination of the olefin, e.g.,

$$L_nMH + RCH=CH_2 \rightleftharpoons L_nMH(RCH=CH_2)$$

[37] C. J. Nyman et al., J. Amer. Chem. Soc., 1971, 93, 617; Inorg. Chem., 1971, 10, 1311.
[38] J. P. Collman et al., Inorg. Chem., 1970, 9, 2367; 1971, 10, 219; J. J. Levison and S. D. Robinson, J. Chem. Soc., A, 1971, 762.
[39] M. Orchin, Adv. Catalysis, 1966, 16, 1; R. F. Hartley, Chem. Rev., 1969, 69, 799; A. J. Hubert and U. Reimlinger, Synthesis, 1970, 405.
[40] H. C. Clark and H. Kurosawa, Chem. Comm., 1971, 957.

which is followed by the transfer reaction

$$L_nMH(RCH{=}CH_2) \rightleftharpoons L_nM{-}CH_2CH_2R$$

So far, it has not been possible to measure equilibrium constants for the first of these reactions, although they will probably be rather similar to those for non-hydridic complexes (page 729) which can often be measured. The equilibrium constants will depend on the steric and electronic nature of the olefin as well as on the nature of the metal complex. 1-Alkenes appear to have constants ca. 50 times those for 2-alkenes. Also, owing to the rapidity of the hydrogen transfer reaction, it is not possible in systems that isomerize olefins even to show the presence of the hydridoalkene species; a few model compounds are known with non-conjugated dienes and some substituted unsaturated compounds such as fumaronitrile,[41] but the best example[42] is the hydridoethylene complex trans-$[PtH(C_2H_4)(PEt_3)_2]BPh_4$.*

With alkenes other than ethylene, there is, of course, the possibility of addition of M—H to the double bond in either the Markownikoff or the anti-Markownikoff direction, just as with the addition, say, of hydrogen chloride. Thus we may have reactions that give either product (A) or product (B). Since in the anti-Markownikoff addition the H atom is transferred from

$$L_nMH + RCH_2CH{=}CH_2$$

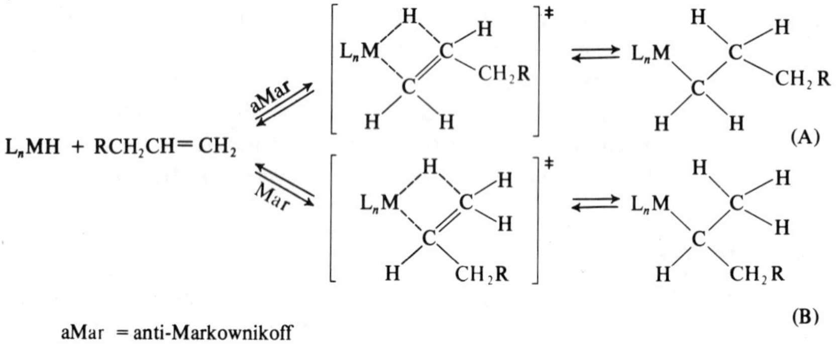

aMar = anti-Markownikoff
Mar = Markownikoff

(A)

(B)

the metal to the β-carbon of the chain, to give the primary alkyl derivative (A), the reverse reaction must re-form the original alkene. There is thus *no* isomerization in this case; note, however, that because of rotation about the C—C bond the *same* H atom need not necessarily be removed, so that *hydrogen atom exchange* can occur. On the other hand, there are *two* possibilities for the secondary alkyl derivative (B); if the H atom is transferred from the CH_3 group, again the original 1-alkene is formed, but if it is transferred from the methylene of the CH_2R group, then a 2-alkene is formed. Thus isomerizations can occur only on Markownikoff addition, and it should be noted that either *cis*- or *trans*-isomers or both may be formed.

---

*The *cis*-isomer would probably rapidly undergo hydrogen transfer reaction to form a Pt—C_2H_5 group.
[41] See, e.g., W. H. Baddley and M. S. Frazer, *J. Amer. Chem. Soc.*, 1969, **91**, 3661.
[42] A. J. Deeming, B. F. G. Johnson and J. Lewis, *Chem. Comm.*, 1970, 598.

It is often difficult to prove the presence of metal–alkyl species in the solutions although in certain cases this can be achieved by nmr methods. Thus the reaction

$$RuHCl(PPh_3)_3 + C_2H_4 \rightleftharpoons Ru(C_2H_5)Cl(PPh_3)_3$$

can be observed, although only under pressure.[43] In other cases, reversibility may be possible only under drastic conditions,[44] e.g.,

$$\textit{trans-}PtHCl(PEt_3)_2 + C_2H_4 \underset{180°}{\overset{80\ atm,\ 95°}{\rightleftharpoons}} \textit{trans-}Pt(C_2H_5)Cl(PEt_3)_2$$

But the forward reaction is catalyzed in polar media by anions such as $SnCl_3^-$, NO_3^- and PF_6^-, which are good leaving groups and readily provide a site for alkene coordination in solvated species of the type $[PtH(PR_3)_2S]^+$ where S is a solvent molecule.[40, 45]

It is also possible to show the reality of hydrogen transfer by exchange studies, e.g.,

$$RhD(CO)(PPh_3)_2 + RCH{=}CH_2 \rightleftharpoons RhH(CO)(PPh_3)_2 + RCD{=}CH_2$$

where the appearance of the Rh—H group can be followed by the growth of the characteristic proton resonance line.[46] Here hydrogen atom exchange is much faster than isomerization, and exchange reactions with 1-alkenes are much faster than those with *cis-* or *trans-*2-alkenes, probably for the same steric reasons that result in high selectivity for hydrogenation of 1-alkenes as discussed below.

There are many problems still to be solved for this quite simple reaction. For the reaction to be reversible and fast the alkyls must be unstable, but the criteria for stability of different types of alkyls are not fully established. It must be noted that the alkyl group occupies only one coordination site, whereas in the transition state and in the hydridoalkene complex, *two* sites must be available, so that clearly a prime factor is coordinative unsaturation which allows decomposition of the alkyl by elimination of alkene. The factors affecting the direction of addition, or the selectivity in elimination from secondary alkyls to give *cis-* or *trans-*isomers, are similarly not well understood. For the first, it appears that the greater the polarity of the M—H bond in the direction $M^{\delta-}{-}H^{\delta+}$, i.e., the greater the acidity as shown by dissociation in a solvent such as water, the greater is the tendency to Markownikoff addition.[47]

24-B-2. Hydrogenation of Alkenes[48]

It has long been known that molecular hydrogen can be activated by transition-metal ions such as Ag^+ or MnO_4^-, or by complexes such as those

[43] P. S. Hallman, B. R. McGarvey and G. Wilkinson, *J. Chem. Soc., A,* **1968**, 3143.
[44] J. Chatt *et al., J. Chem. Soc., A,* **1968**, 190.
[45] R. Cramer and R. V. Lindsay, Jr., *J. Amer. Chem. Soc.,* 1966, **88**, 3534.
[46] M. Yagupsky and G. Wilkinson, *J. Chem. Soc., A,* **1970**, 941.
[47] D. Evans, J. A. Osborn, and G. Wilkinson, *J. Chem. Soc., A,* **1968**, 3133.
[48] R. S. Coffey, A. Andreeta, F. Conti and G. F. Ferrari, in *Aspects of Homogeneous Catalysis,* Vol. 1, R. Ugo, *ed.,* Carlo Manfredi, Milan, 1970.

given by Cu^{2+} in quinoline or Co^{2+} in aqueous cyanide,[49] and certain of these species can be used as catalysts for the slow reduction of unsaturated substances. Although the intermediary of species with M—H bonds was often postulated, proof was not obtained and indeed in the Co^{2+}—CN^- system free radicals appear to be involved.[50]

The first rapid and practical system for the homogeneous reduction of alkenes, alkynes and other unsaturated substances at 25° and 1 atm pressure used the complex $RhCl(PPh_3)_3$ (Sect. 26-G-2) in benzene solution.[51] Because of differences in rates of hydrogenation depending on the nature of groups at the double bond, selective reductions are possible, for example,[52]

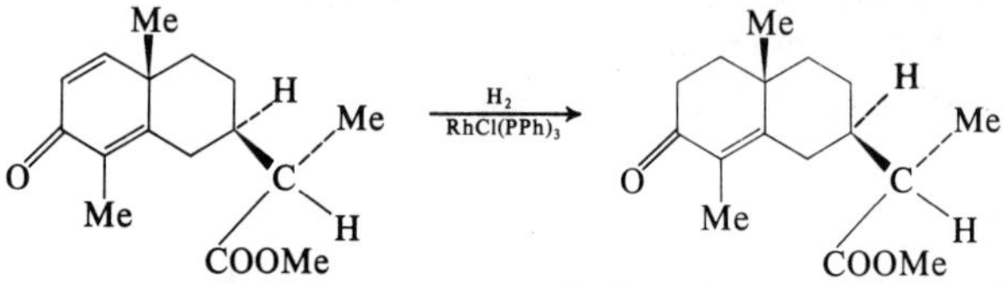

Further, in contrast to heterogeneous catalysis, where scattering of deuterium throughout the molecule usually results, selective addition of D_2 to a double bond occurs. Finally, asymmetric hydrogenation has been achieved by use of complexes with phosphines that are optically active either at the phosphorus atom or at a carbon atom on the group attached to P.[53] The mechanism of reduction probably involves the following steps:

(a) As noted earlier, $RhCl(PPh_3)_3$ is essentially undissociated in benzene. However, in oxidative addition reactions, not only with hydrogen but also with other molecules such as CH_3I, it appears that the resulting 6-coordinate species, unlike their iridium analogs (Sect. 26-G-2), readily lose a phosphine to give 5-coordinate species; thus we have

$$RhCl(PPh_3)_3 + H_2 \rightleftharpoons RhH_2Cl(PPh_3)_2 + PPh_3$$

The intermediacy of a bisphosphine complex is confirmed by the fact that generation of such a species *in situ* by addition of PPh_3 to the cyclooctene complex, viz., $[(C_8H_{14})_2RhCl]_2 + 4PPh_3 \rightarrow 4C_8H_{14} + 2RhCl(PPh_3)_2$

gives the maximum rate of hydrogenation when the ratio $Rh:PPh_3$ is 1:2. Nmr studies indicate that the stereochemistry of the dihydrido complex is that shown in (24-B-I). In benzene, the species is probably 5-coordinate; but from

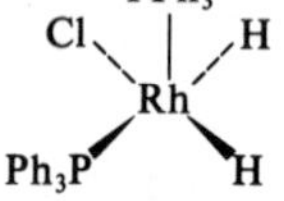

(24-B-I)

[49] J. Kwiatek, *Catalyst Revs.*, 1968, **1**, 37; J. Halpern and L.-Y. Wong, *J. Amer. Chem. Soc.*, 1968, **90**, 6065.
[50] L. M. Jackman, J. A. Hamilton and J. M. Lawlor, *J. Amer. Chem. Soc.*, 1968, **90**, 1915.
[51] G. Wilkinson *et al.*, *J. Chem. Soc.*, A, **1966**, 1711, 1736; **1967**, 1574; **1968**, 1054; S. Siegel and D. W. Ohrt, *Chem. Comm.*, **1971**, 1529.
[52] See M. Fieser and L. F. Fieser, *Reagents for Organic Synthesis*, Vol. 2, Wiley 1969; P. N. Rylander and L. Hasbrouck, *Engelhard Tech. Bull.*, **1970**, 85.
[53] W. S. Knowles, M. J. Sabecky and B. D. Vineyard, *Chem. Comm.*, **1972**, 10.

donor solvents such as acetic acid crystalline solvates may be isolated, so that in such solvents the sixth position is probably occupied by the solvent molecule.

(b) In benzene or other weakly bound solvents, coordination of the alkene to the coordinatively unsaturated complex then occurs:

$$RhH_2Cl(PPh_3)_2 + RCH{=}CH_2 \rightleftharpoons RhH_2Cl(PPh_3)_2(RCH{=}CH_2)$$

However, when a stronger donor is present, the alkene cannot displace the donor molecule and, for example, the pyridine complex $RhH_2Cl(PPh_3)_2py$ will not act as a catalyst for hydrogenation.

(c) The final step then involves hydrogen transfer to the coordinated olefin, as discussed above. This probably occurs by a two-step process, to give first an alkyl (the addition could be either Markownikoff or anti-Markownikoff), followed by a second rapid H transfer to give alkane:

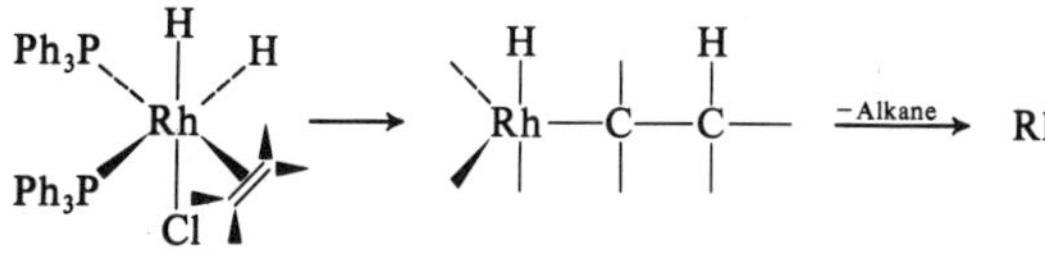

In benzene solution the amount of isomerization is small, but addition of alcohols or traces of oxygen appears to increase the rate of isomerization.[54] Further work on these effects is desirable.

A number of other reducing systems are now known, most of them differing from $RhCl(PPh_3)_3$ in that only one M—H bond is involved. Two complexes, $RuHCl(PPh_3)_3$[43] and $RhH(CO)(PPh_3)_3$,[55] are unusual in that they are highly selective for the reduction of 1-alkenes for reasons which are probably steric, as discussed below.

Consider the rhodium compound, as this has been mentioned in connexion with isomerization (page 785) and is important in hydroformylation discussed below. The hydrogenation is believed to follow the path shown in Fig. 24-B-1. The complex $RhH(CO)(PPh_3)_3$ dissociates in benzene, and both the isomerization discussed above and hydrogenation are suppressed by the addition of an excess of PPh_3. In contrast, $CoH(CO)(PPh_3)_3$ and $IrH(CO)(PPh_3)_3$ do not dissociate except when heated or irradiated; thus the former causes hydrogenation only at ca. 150° and 150 atm of hydrogen.[56] The high selectivity for hydrogenation of 1-alkenes is probably due to the fact that in the square species [(A), (B) in Fig. 24-B-I] steric effects of the bulky trans-PPh_3 groups allow the alkyl group to have some degree of stability only when it is primary, i.e. as in Rh—CH_2—R. In a secondary alkyl complex, e.g., Rh—$CH(CH_3)CH_2R$, with a substituent on the carbon atom adjacent to the metal (formed by Markownikoff addition or from a 2-alkene) there would be much greater steric repulsion. The loss of stability

[54] G. C. Bond and R. A. Hillyard, Discuss. Faraday Soc., No. 46, 1968; A. S. Hussey and Y. Takeuchi, J. Amer. Chem. Soc., 1969, 91, 672; R. L. Augustine and J. F. van Peppen, Chem. Comm., 1970, 495.

[55] C. O'Connor and G. Wilkinson, J. Chem. Soc., 1968, 2665; see also M. G. Burnett and R. J. Morrison, J. Chem. Soc., A, 1971, 2325.

[56] M. Hidai et al., Tetrahedron Lett., 1970, 20, 1715.

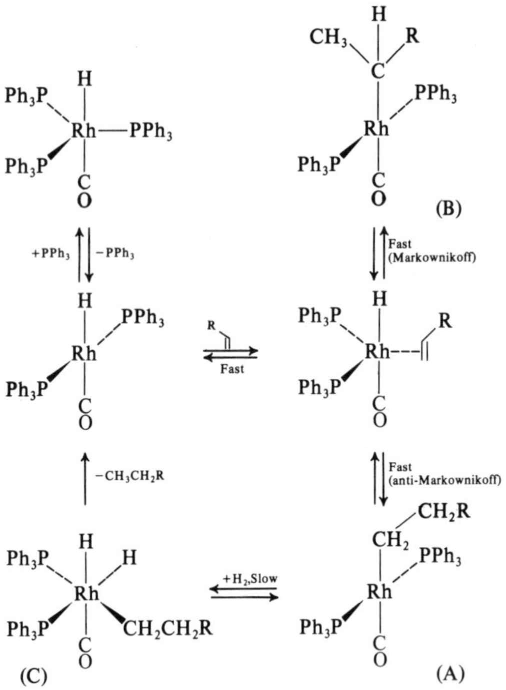

Fig. 24-B-1. Catalytic cycle of hydrogenation and isomerization of 1-alkenes by RhH(CO)(PPh₃)₃ at 25° and 1 atm pressure.

of the alkyl complex would mean that it would be too short-lived to undergo the slow oxidative addition of H_2 to form the dihydrido species [(C) in Fig. 24-B-1]. That the secondary alkyl complex is formed simultaneously from 1-alkenes by Markownikoff addition is shown by the fact that the rates of hydrogenation and of isomerization of 1-alkenes are comparable.[46]

Systems that reduce unsaturated groupings in addition to C=C or C≡C, e.g., >C=O, —N=N—, —CH=N—, —NO₂, are $[RhH_2(PMe_3)_2L_2]^+$, where L=solvent, and the complex $RhCl_3py_3$ in dimethylformamide solution in the presence of sodium borohydride.[57] A phosphorus-containing polystyrene polymer forms a complex with rhodium on treatment with $RhCl(PPh_3)_3$ and can act as an efficient heterogeneous catalyst.[58]

24-B-3. Hydroformylation of Alkenes[59]

The hydroformylation reaction, which was discovered by O. Roelen in Germany, is the addition of H_2 and CO (or formally of H and the formyl

[57] R. R. Schrock and J. A. Osborn, *Chem. Comm.*, **1970**, 567; J. C. Orr *et al.*, *Chem. Comm.*, **1970**, 162; P. Abley, J. Jardine and F. J. McQuillan, *J. Chem. Soc., C*, **1971**, 480.
[58] R. H. Grubbs and L. C. Kroll, *J. Amer. Chem. Soc.*, 1971, **93**, 3062.
[59] J. Falbe, *Synthesen mit Kohlenmonoxyd*, Springer Verlag, 1967, translated by C. R. Adams as *Carbon Monoxide in Organic Synthesis*, 1970; A. J. Chalk and R. F. Harrod, *Adv. Organometallic Chem.*, 1968, **6**, 119; I. Wender and P. Pino, eds., *Organic Synthesis via Metal Carbonyls*, Vol. 1, Interscience-Wiley, 1968.

group, HCO) to an alkene, usually a 1-alkene, to form an aldehyde, which may be further reduced to the alcohol:

$$RCH{=}CH_2 + H_2 + CO \rightarrow RCH_2CH_2CHO \xrightarrow{H_2} RCH_2CH_2CH_2OH$$

The reaction as originally used employed cobalt compounds as catalysts at temperatures of ca. 150° and >300 psi pressure, and some three million tons a year of alcohols, usually C_7–C_9, are produced in this way. The process ordinarily gives both straight- and branched-chain products in the ratio ca. 3 : 1, but considerable efforts have been made to improve the yield of the linear product; patents describe the use of tributylphosphine-substituted cobalt carbonyls for this purpose.[60]

The main features of the reaction mechanism, which include H transfer and CO insertion reactions, were elucidated mainly by Orchin and by Heck and Breslow through studies on the reaction of the hydridocarbonyl, $HCo(CO)_4$ (page 705), with olefins.[61] However, the cobalt system is very difficult to study and further information on the catalyst cycle is provided by work with $RhH(CO)(PPh_3)_3$. This complex is catalytically active even at 25° and 1 atm pressure[62] and, in contrast to the cobalt system, produces only aldehyde. On use of high concentrations of PPh_3 or $P(OPh)_3$, and especially of molten PPh_3 (m.p. 80°) as solvent, high yields of linear aldehyde can be obtained with little or no loss of alkene as alkane (which is a disadvantage of the cobalt systems). The proposed reaction cycle is shown in Fig. 24-B-2. The initial step is associative attack of the alkene on the species $RhH(CO)_2(PPh_3)_2$ [(A) in Fig. 24-B-2], which leads to the alkyl complex (B). The latter then undergoes CO insertion to form the acyl derivative (C) which subsequently undergoes oxidative addition of molecular hydrogen to give the dihydridoacyl complex (D).

The last of these steps, which is the only one in the cycle that involves a change in oxidation state of the metal, is probably rate-determining. The final steps are another H transfer to the carbon atom of the acyl group in (D), followed by loss of aldehyde and regeneration of the 4-coordinate species (E). An excess of CO over H_2 inhibits the hydroformylation reaction, probably through the formation of 5-coordinate dicarbonyl acyls (F), which cannot react with hydrogen.

The high PPh_3 concentrations that are essential to provide high yields (>95%) of linear aldehyde are probably required in order to suppress dissociation and the formation of monophosphine species, and also to force the associative attack of olefin on the bisphosphine species (A), for which the specificity of anti-Markownikoff addition is high.

In absence of an excess of phosphine, attack probably occurs on a species

[60] W. Rupilius, J. J. McCoy and M. Orchin, *Ind. Eng. Chem. (Prod. Res. Devel.)*, 1971, **10**, 142; G. F. Pregaglia, A. Andreetta and G. F. Ferrari, *J. Organometal. Chem.*, 1971, **30**, 387.

[61] R. F. Heck, *Adv. Organometallic Chem.*, 1966, **4**, 243 (reactions of alkyl and acyl cobalt carbonyls).

[62] C. K. Brown and G. Wilkinson, *J. Chem. Soc., A*, **1970**, 2753.

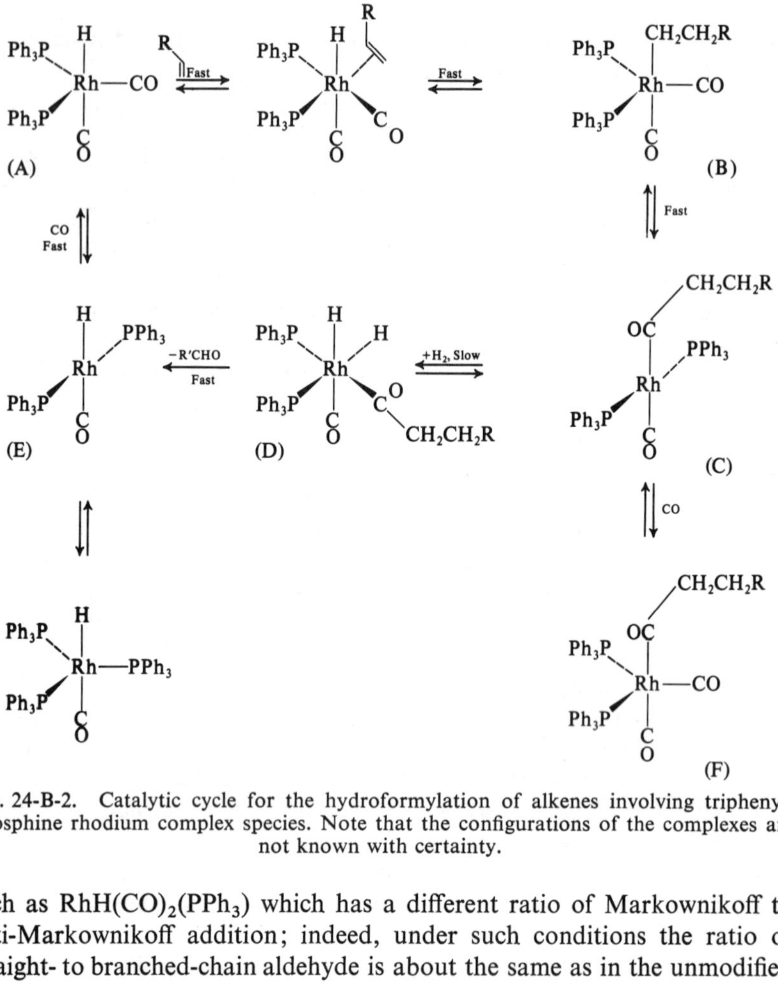

Fig. 24-B-2. Catalytic cycle for the hydroformylation of alkenes involving triphenyl-phosphine rhodium complex species. Note that the configurations of the complexes are not known with certainty.

such as $RhH(CO)_2(PPh_3)$ which has a different ratio of Markownikoff to anti-Markownikoff addition; indeed, under such conditions the ratio of straight- to branched-chain aldehyde is about the same as in the unmodified cobalt system, namely, 3:1.

Although the intermediates shown in Fig. 24-B-2 are too unstable for isolation it has been possible,[63] by using alkenes such as C_2F_4, which gives stable species such as the square alkyl $Rh(C_2F_4H)(CO)(PPh_3)_2$, and by using the similar but much more stable iridium complexes derived from $IrH(CO)_2(PPh_3)_2$, to characterize analogs for all the species with the ex- ception of the final dihydridoacyl complex; a comparable complex, $IrHCl(COCH_2R)(CO)(PPh_3)_2$, may however be isolated.

A wide variety of unsaturated substances can be hydroformylated by cobalt or rhodium catalysts, but conjugated alkenes, e.g., butadiene, may give a number of products including hydrogenated monoaldehydes. The mechanism[64] is then quite different from that for monoalkenes, since addi-

[63] G. Yagupsky, C. K. Brown and G. Wilkinson, *J. Chem. Soc., A*, **1970**, 1392.
[64] C. K. Brown *et al.*, *J. Chem. Soc., A*, **1971**, 850.

tion of M—H to conjugated dienes[65] leads to allylic species, which may be present as σ-bonded intermediates or, more commonly as h^3-allyls (cf. page 753), e.g.,

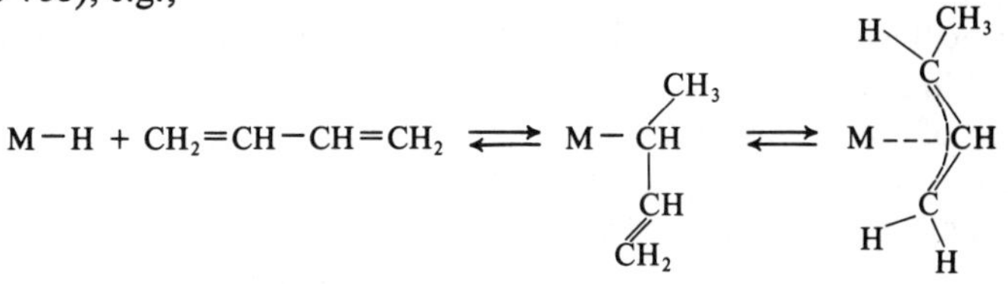

$$M-H + CH_2\!=\!CH-CH\!=\!CH_2 \rightleftharpoons M-\underset{\underset{CH_2}{\overset{\|}{CH}}}{\overset{CH_3}{\underset{|}{CH}}} \rightleftharpoons M\text{---}CH$$

In addition to hydroformylation, there are a large number of other carbonylation reactions.[59] One reaction of industrial importance is the synthesis of acetic acid by the reaction[66]

$$CH_3OH + CO \rightarrow CH_3COOH$$

which is catalyzed by rhodium complexes such as $[RhCl(CO)_2]_2$ or $RhCl(CO)(PPh_3)_2$ in presence of HI, I_2 or CH_3I as an activator. Other important reactions of alcohols or water[67] are

$$CH_2\!=\!CH_2 + CO + H_2O \xrightarrow{\text{PdCl}_2,\ \text{Cu}^{2+}} CH_3CH_2COOH$$

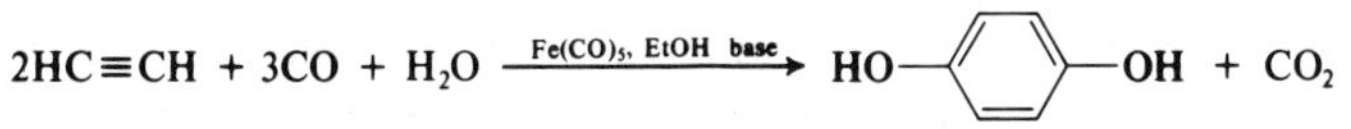

$$2HC\!\equiv\!CH + 3CO + H_2O \xrightarrow{\text{Fe(CO)}_5,\ \text{EtOH base}} HO-\!\!\!\!\bigcirc\!\!\!\!-OH + CO_2$$

while the reduction of nitrobenzene to aniline or to phenyl isocyanate is also of commercial importance,[68] e.g.,

$$C_6H_5NO_2 + 2CO + H_2 \xrightarrow[\substack{200\ \text{atm}\\135-160°}]{\text{Ru}_3(\text{CO})_{12}} C_6H_5NH_2 + 2CO_2$$

24-B-4. Hydrosilylation of Alkenes[69]

The hydrosilylation reaction (often termed hydrosilation) of alkenes is

$$RCH\!=\!CH_2 + HSiR_3 \rightarrow RCH_2CH_2SiR_3$$

Although the commercial Speier reaction uses hexachloroplatinic acid as the catalyst, phosphine complexes of cobalt, rhodium, palladium or nickel can also be used.[70a] The addition of silanes to trans-$IrCl(CO)(PPh_3)_2$ or $IrH(CO)(PPh_3)_3$ provides a model[70b] for the first step, namely, the oxidative addition $IrCl(CO)(PPh_3)_2 + R_3SiH \rightarrow IrHCl(SiR_3)(CO)(PPh_3)_2$

[65] C. A. Tolman, J. Amer. Chem. Soc., 1970, 92, 6785.
[66] J. F. Roth et al., Chem. Tech., 1971, 600.
[67] W. Reppe, N. V. Kutepow and A. Magin, Angew. Chem. Internat. Edn., 1969, 8, 727.
[68] F. L'Epplattenier, P. Matthys and F. Calderazzo, Inorg. Chem., 1970, 9, 342; A. F. M. Iqbal, Tetrahedron Letters, 1971, 3385.
[69] C. Eaborn and R. W. Bott in Organometallic Compounds of Group IV Elements, Vol. I, Part 1, 1968, A. G. McDiarmid, ed., Dekker; P. N. Rylander, Engelhard Tech. Bull., 1970, 10, 130; A. J. Chalk, J. Organometallic Chem., 1970, 21, 207.
[70a] M. Hara, K. Ohno and J. Tsuji, Chem. Comm., 1971, 247; A. J. Archer, R. N. Haszeldine and R. V. Parish, Chem. Comm., 1971, 524.
[70b] J. F. Harrod and C. A. Smith, Can. J. Chem., 1970, 48, 870.

but for further additions to alkenes to be possible the catalytic species must have additional vacant coordination sites for alkene coordination and activation. The SiR_3 grouping is known to have a very strong *trans*-effect, and this will also act to labilize coordination sites.[71] It may be noted further that addition of optically active silanes to Pt, Co and Ir species proceeds with retention of configuration (cf. discussion on page 776).[72]

Before leaving the subject of additions, we may note that HCN can be added to alkenes to give nitriles:[73]

$$RCH{=}CH_2 + HCN \rightarrow RCH_2CH_2CN$$

The mechanism, which involves nickel– or palladium–phosphite complexes, is probably similar to those discussed above.

24-B-5. Alkene Polymerization and Oligomerization[74]

Most unsaturated substances such as alkenes, alkynes, aldehydes, acrylonitrile, epoxides, isocyanates, etc., can be converted into polymeric materials of some sort—either very high polymers, or low-molecular-weight polymers, or oligomers such as linear or cyclic dimers, trimers, etc. In addition, copolymerization of several components, e.g., styrene–butadiene–dicyclopentadiene, is very important in the synthesis of rubbers. Not all such polymerizations, of course, require transition-metal catalysts and we consider here only a few examples that do. The most important is Ziegler–Natta polymerization of ethylene and propene.

Ziegler–Natta Polymerization.[74, 75]   The discovery by Ziegler that hydrocarbon solutions of $TiCl_4$ in presence of triethylaluminum polymerize ethylene at 1 atm pressure has led to an extremely diverse chemistry in which aluminum alkyls are used to generate transition metal–alkyl species. At about the same time, it was found by the Phillips Petroleum Co. that, when suspended in inert hydrocarbon solvents, specially activated chromium oxides on an alumina support also polymerize ethylene. However, much polyethylene is still produced by the original Imperial Chemical Industries method, which is a thermally induced free-radical polymerization operating at very high pressures. Neither of the last two processes will produce stereoregular polypropene, which was Natta's development of the Ziegler method.

The Ziegler–Natta system is heterogeneous, and the active metal species is

[71] F. F. Saalfield, M. V. McDowell and A. G. MacDiarmid, *J. Amer. Chem. Soc.*, 1970, **92**, 2324; J. Chatt *et al.*, *J. Chem. Soc.*, A, **1970**, 1343.

[72] L. M. Sommer *et al.*, *J. Amer. Chem. Soc.*, 1969, **91**, 7051, 7061.

[73] E. S. Brown and E. A. Rick, *Chem. Comm.*, **1969**, 112.

[74] A. D. Ketley, ed., *The Stereochemistry of Macromolecules*, Vols. I–III, Arnold (London), Dekker (New York), 1967; L. S. Reich and A. Schindler, *Polymerisation by Organometallic Compounds*, Interscience-Wiley, 1966; G. Natta, ed., *Stereoregular Polymers*, Volts. I, II, Pergamon Press.

[75] J. Boor, Jr., *Ind. Eng. Chem. (Prod. Res. Devel.)*, 1970, **9**, 437; W. Cooper, *ibid*, 1970, **9**, 457 (synthetic rubbers).

a fibrous form of $TiCl_3$ formed *in situ* from $TiCl_4$ and $AlEt_3$, but preformed $TiCl_3$ can be used. The second function of the aluminum alkyl appears to be replacement of one of the chloride ions in a Ti^{3+} ion at the surface by an alkyl radical derived from it; the surface Ti atom has one of its six coordination sites vacant. An ethylene molecule is then bound to the vacant site. After promotion of an electron from the Ti—alkyl bond to a molecular orbital of the complex, a four-center transition state is produced which enables an alkyl group to transfer to coordinated ethylene. A further molecule of ethylene is then bound to the vacant site and the process repeated. The mechanism is then as follows:

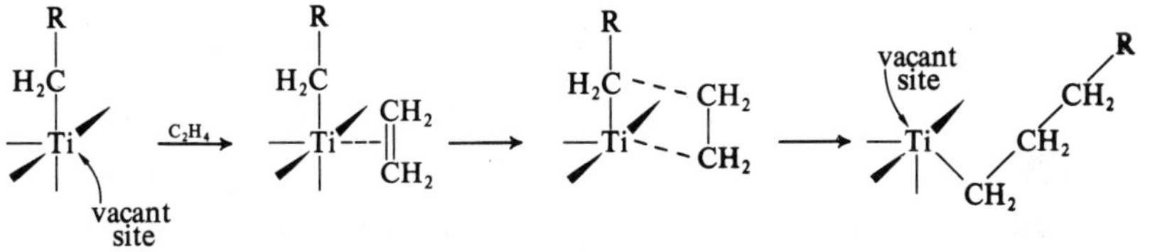

The stereoregular polymerization of propene may arise because of the nature of the sterically hindered surface sites on the $TiCl_3$ lattice.

An important extension of Ziegler–Natta polymerization is the copolymerization of styrene, butadiene and a third component such as dicyclopentadiene or 1,4-hexadiene (see below) to give synthetic rubbers. Vanadyl halides rather than titanium halides are then used as the metal catalyst.

Finally, we note that Wilke and his coworkers[76] have shown that zerovalent complexes, especially of nickel, obtained by reduction with aluminum alkyls can be used in a wide variety of polymerizations such as trimerization of butadiene to *trans,trans,trans*-cyclododecatriene.

Alkene Metathesis.[77]   Although these reactions are not polymerizations, it is convenient to deal with them here. The olefin metathesis or dismutation is of the following type:

$$CH_2=CHR + CH_2=CHR \rightleftharpoons \begin{matrix} CH_2 \\ \| \\ CH_2 \end{matrix} + \begin{matrix} CHR \\ \| \\ CHR \end{matrix}$$

Under certain conditions, with the same catalysts, isomerizations and polymerizations also occur. The reaction can be carried out heterogeneously at 150–500° over catalysts such as WO_3, transition metals on silica supports, etc., or homogeneously in solution. Homogeneous catalysts that have been used are WCl_6 plus $EtAlCl_2$ in ethanol, and Mo $py_2(NO)_2Cl_2$ plus aluminum alkyl in chlorobenzene.[78] However, lithium alkyls can also be employed and even combinations of $AlCl_3$ with WCl_6. Hence it appears that the formation of transition-metal alkyl species is not an essential feature, but that the Li or

[76] G. Wilke *et al.*, *Angew. Chem. Internat. Edn.*, 1970, **9**, 367; 1965, **4**, 327; 1963, **2**, 105; *Adv. Organometallic Chem.*, 1970, **8**, 29.
[77] G. C. Bailey, *Catalyst Rev.*, 1970, **3**, 37.
[78] See, e.g., W. B. Hughes *et al.*, *J. Amer. Chem. Soc.*, 1970, **92**, 528, 532.

Al alkyls act partly to reduce the halide, and partly to remove Cl from the halide, thus giving a coordinatively unsaturated species. Aluminum chloride alone can do this, e.g.,

$$WCl_6 + AlCl_3 \rightleftharpoons WCl_5^+ + AlCl_4^-$$

but the nature of the species involved is not yet established.

The coordinatively unsaturated metal halide species is required to provide two sites for coordination of two alkene molecules. It has been proposed that the reaction proceeds through a cyclobutane-like intermediate:

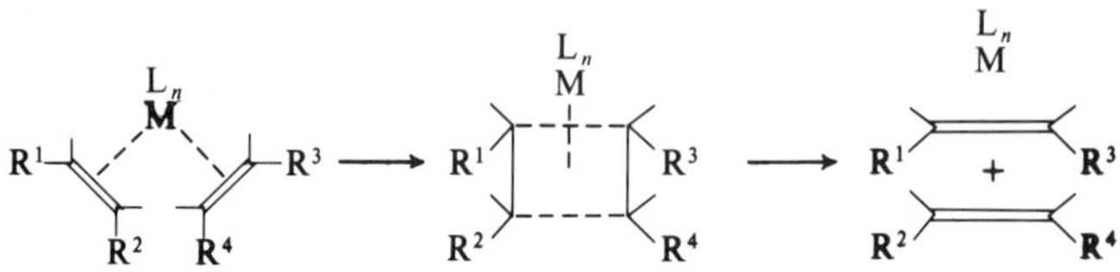

but no experimental support has been provided and other explanations have been devised.[79] A metallocycle of the type (24-B-II) may be involved.

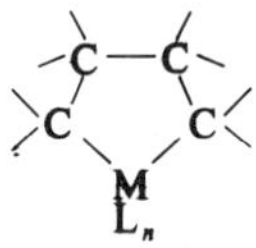

(24-B-II)

The du Pont 1,4-Hexadiene Synthesis. An important industrial process for the synthesis of 1,4-hexadiene, a component of ternary rubbers, illustrates a different type of mechanism, which is more closely related to the processes discussed in the previous Sections. The synthesis involves the reaction of ethylene and butadiene and may be carried out by using rhodium chloride in ethanolic hydrogen chloride solution or by nickel(0) phosphite complexes in acid solution.[80]

The proposed reaction cycle for NiL_4, where $L = P(OEt)_3$, is shown in Fig. 24-B-3. The partial sequence of steps definitely established is

$$NiL_4 \xrightleftharpoons{H^+} HNiL_4^+ \xrightleftharpoons[+L]{-L} HNiL_3^+ \xrightarrow{C_4H_6} h^3\text{-}C_4H_7NiL_3^+$$

$$h^3\text{-}C_4H_7NiL_3^+ \xrightleftharpoons[+L]{-L} h^3\text{-}C_4H_7NiL_2^+$$

$$h^3\text{-}C_4H_7NiL_2(C_2H_4) \xrightleftharpoons{C_2H_4} h^3\text{-}C_4H_7NiL_2^+$$

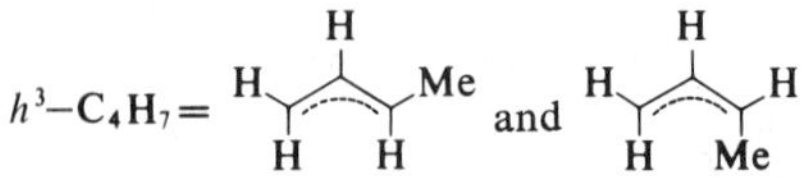

Coupling of the h^3-crotonyl [1-methylallyl, $CH_2 = CH—CH(CH_3)—$] group and ethylene, followed by a β-hydrogen abstraction from the resulting alkyl, gives 1,4-hexadiene and regenerates the hydridonickel complex.

[79] G. S. Lewandos and R. Pettit, *J. Amer. Chem. Soc.*, 1971, **93**, 7087.
[80] C. A. Tolman, *J. Amer. Chem. Soc.*, 1970, **92**, 6777.

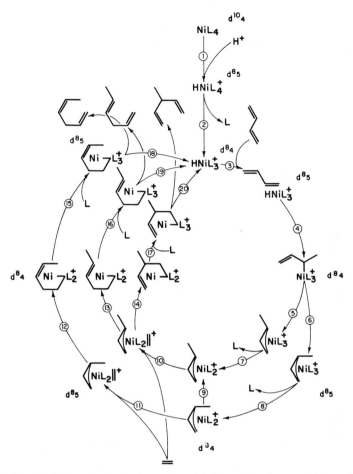

Fig. 24-B-3. Catalytic cycle for the formation of 1,4-hexadiene from ethylene and buta-diene by use of nickel complexes. The numbers 4 or 5 after the electron configurations denote the coordination number of the species. [Reproduced by permission from a diagram provided by Dr. C. A. Tolman, Central Research Laboratory, E. I. du Pont de Nemours and Co., and published in *J. Amer. Chem. Soc.*, 1970, **92**, 6777].

24-B-6. Palladium-catalyzed Reactions[81]

It has long been known that ethylene compounds of palladium, e.g. $[C_2H_4PdCl_2]_2$, are rapidly decomposed in aqueous solution to form

[81] P. M. Maitlis, *Organic Chemistry of Palladium*, Academic Press, 1971 (an authoritative monograph on organo compounds and catalytic uses); A. Aguilo, *Adv. Organometallic Chem.*, 1967, **5**, 321 (extensive review of oxypalladation reactions); J. Tsuji, *Accounts Chem. Res.*, 1969, **2**, 144 (carbonylation and other reactions); E. W. Stern, *Catalyst Rev.*, 1968, **1**, 74 (reactions of alkenes, CO, etc., with palladium complexes); F. R. Hartley, *Chem. Rev.*, 1969, **9**, 799 (exhaustive review of Pd and Pt alkene and alkyne complexes, including catalytic uses); *Amer. Chem. Soc.*, Div. Petrol Chem., Preprints, Vol. 14, No. 2, Minneapolis Meeting, April 1969 (detailed symposium papers); R. Hüttel, *Synthesis*, 1970, 225; *Brennstoff-Chem.*, 1969, **50**, 331 (preparative organic reactions).

acetaldehyde and Pd metal. The conversion of this stoichiometric reaction into a cyclic one was achieved by Smidt and his co-workers (at Wacker Chemie), their main contribution being the linking together of the known individual reactions:

$$C_2H_4 + PdCl_2 + H_2O \rightarrow CH_3CHO + Pd + 2HCl$$
$$Pd + 2CuCl_2 \rightarrow PdCl_2 + 2CuCl$$
$$2CuCl + 2HCl + \tfrac{1}{2}O_2 \rightarrow 2CuCl_2 + H_2O$$
$$\overline{C_2H_4 + \tfrac{1}{2}O_2 \rightarrow CH_3CHO}$$

The oxidation of ethylene by palladium(II)–copper(II) chloride solution is essentially quantitative and only low Pd concentrations are required; the process can proceed either in one stage or in two stages; in the latter the reoxidation by O_2 is done separately.[82]

Oxidation of alkenes of the type $RCH=CHR'$ or $RCH=CH_2$ gives ketones; an important example is the formation of acetone from propene.

When media other than water are used, different but related processes operate. Thus in acetic acid ethylene gives vinyl acetate,[83] while vinyl ethers may be formed in alcohols; there are usually competing reactions giving unwanted side products. There has been much industrial study of such processes, but although the oxidations to aldehyde are successful a plant to produce vinyl acetate has been abandoned as uneconomic owing to corrosion problems and difficulty in catalyst recovery.

The mechanism of the reaction has led to much speculation, but the following appears to satisfy the observed facts. Since the reaction proceeds in Pd^{II} solutions of chloride ion concentration $>0.2M$, the metal is most likely present as $[PdCl_4]^{2-}$. The following reactions then proceed:

$$[PdCl_4]^{2-} + C_2H_4 \rightleftharpoons [PdCl_3(C_2H_4)]^- + Cl^- \quad \text{(Fast)}$$
$$[PdCl_3(C_2H_4)]^- + H_2O \rightleftharpoons [PdCl_2(H_2O)(C_2H_4)] + Cl^-$$
$$[PdCl_2(H_2O)C_2H_4] + H_2O \rightleftharpoons [PdCl_2(OH)(C_2H_4)]^- + H_3O^+$$

The *trans*-isomer of this hydroxo species is doubtless more stable than the *cis*-isomer (see *trans*-effect, page 667), but kinetically significant amounts of the latter will be present so that further reaction occurs by *cis*-transfer. This transfer may well be Cl^- ion- or solvent-assisted in a 5-coordinate solvated species, e.g.,

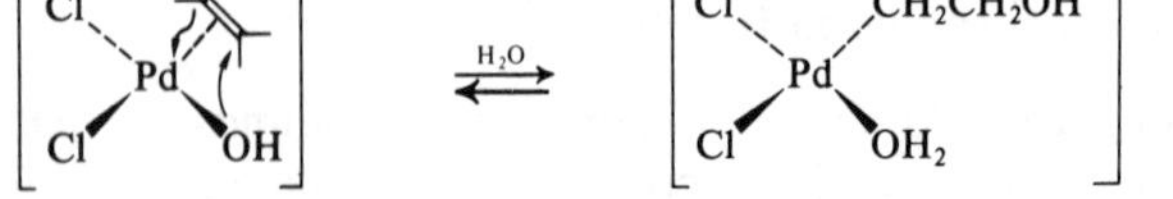

This reaction followed by three further steps, namely (a) a fast hydrogen transfer from the β-carbon of the chain to the metal, (b) hydride transfer

[82] For details see G. Szonyi, *Homogeneous Catalysis*, Adv. in Chem. Series, No. 70, Amer. Chem. Soc., 1968, p. 53.
[83] R. van Helden *et al.*, *Rec. Trav. Chim.*, 1968, **87**, 961.

from metal to the α-CH_2 group (as in hydrogenation), and finally (*c*) reductive elimination of Pd metal:

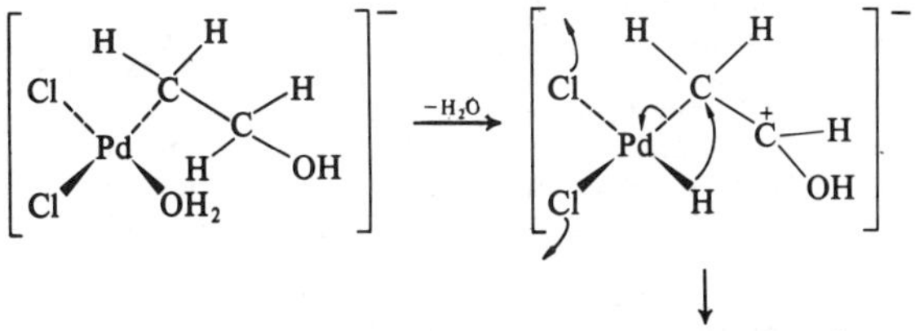

$$CH_3CHO + H^+ \longleftarrow CH_3CHOH^+ + Pd^0 + 2Cl^-$$

The sequence accounts not only for the rate laws and dependence on (inhibition by) Cl^- and H^+, but also for results of deuteration studies which show that in D_2O *no* deuterium is incorporated into the acetaldehyde.

In acetic acid media, the acetate ion can be regarded as the nucleophile comparable to OH^-, and in alcohols the alkoxide ions.

The mechanism for the oxidation of Pd metal by Cu^{II} chloro complexes is not well understood, but electron transfer *via* halide bridges (cf. page 675) is probably involved. The extremely rapid air-oxidation of Cu^I chloro complexes is better known and probably proceeds through an initial oxygen complex:

$$CuCl_2^- + O_2 \rightleftharpoons ClCuO_2 + Cl^-$$

followed by formation of radicals such as O_2^-, OH, or HO_2:

$$ClCuO_2 + H_3O^+ \rightarrow CuCl^+ + HO_2 + H_2O$$

As with Ziegler and Natta's discovery, recognition of the reactivity of palladium complexes has led to a deluge of patents and papers involving all manner of organic substances. All reactions appear to involve the formation of Pd—C and possibly Pd—H bonds, and to proceed by insertions or transfers; where conjugated alkenes or allyl compounds are involved, palladium allylic species are doubtless intermediates.

A few examples[84] are the following:

$$RC{\equiv}CH \xrightarrow{(CH_3CN)_2PdCl_2} \text{arenes or Dewar arenes}$$

$$RNH_2 + CO + PdCl_2 \rightarrow RNCO + Pd + 2HCl$$

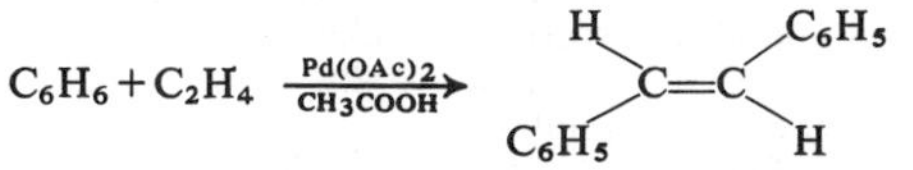

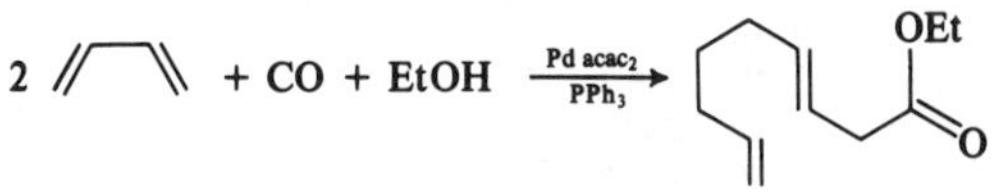

[84] M. Avram *et al.*, *Chem. Ber.*, 1969, **103**, 3996; P. M. Maitlis *et al.*, *J. Amer. Chem. Soc.*, 1970, **92**, 2276, 2285; J. K. Stille and L. F. Hines, *J. Amer. Chem. Soc.*, 1970, **92**, 1798. W. E. Billups, W. E. Walker and T. C. Shields, *Chem. Comm.*, **1971**, 1067; R. S. Shue, *J. Amer. Chem. Soc.*, 1971, **93**, 7116.

Further Reading

Elastomer Stereospecific Polymerisation, Adv. in Chem. Series, No. 52, Amer. Chem. Soc., 1966 (includes transition-metal systems).

Homogeneous Catalysis: Industrial Applications and Implications, Adv. in Chem. Series, No. 70. Amer. Chem. Soc., 1968 (various aspects including general reviews and useful summary of commercial homogeneous processes).

Organometallic Compounds in Industry, Topics in Current Chemistry, Vol. 16, Issue 3-4, Springer-Verlag [reactions of palladium compounds with olefins (R. F. Heck); organometallic catalysts for olefin polymerization and rubber production (A. Gumbolt and H. Weber)].

Discussions of the Faraday Society, No. 46, Butterworths 1968 (homogeneous catalysis, particularly of hydrogenation and oxidation by Pd^{II}, Mn^{III}, Co^{III}, etc.).

Augustine, E. L., ed., *Oxidation—Techniques and Application in Organic Synthesis*, Dekker, 1969 (use of transition metal and other compounds for oxidation of hydrocarbons, O- and N-containing functional groups).

Bird, C. W., *Transition Metal Intermediates in Organic Synthesis*, Logos Press (London), Academic Press (New York), 1967 (a comprehensive discussion of a variety of reactions including oligomerization, hydroformylation, carboxylation, etc. with many collected data and references).

Candlin, J. P., K. A. Taylor and D. T. Thompson, *Reactions of Transition Metal Complexes*, Elsevier, 1968 (general review on all types of reaction with copious references).

Chalk, A. J., and J. F. Harrod, *Adv. Organometallic Chem.*, 1968, **6**, 119 (catalyses by cobalt carbonyls: hydrogenation, hydroformylation, carboxylation, hydrosilylation, etc.).

Collman, J. P., *Trans. Metal Chem.*, 1966, **2**, 1 [reactions of coordinated ligands, of σ-bonded atoms, π-bonded ligands and quasiaromatic systems (h^5-C_5H_5, arene and acac complexes); an extensive review].

Collman, J. P., *Accounts Chem. Res.*, 1968, **1**, 136 (general discussion of basic reactions involved in catalytic reactions of additive nature).

Falbe, J., *Synthesen mit Kohlenmonoxyd*, Springer-Verlag, 1967 (translated as *Carbon Monoxide in Organic Synthesis*, 1970) (comprehensive account of all types of carbonylation reactions).

Halpern, J., and R. S. Nyholm, *Proc. 3rd Int. Conf. Catal.*, Vol. 1, W. H. Sachter *et al.*, eds., Butterworths, 1965 (general reviews).

Halpern, J., *Ann. Rev. Phys. Chem.*, 1965, **16**,103 (homogeneous hydrogenation, acetylene hydrations, etc.).

Heck, R. F., *Accounts Chem. Res.*, 1969, **2**, 10 (general review on addition reactions: hydrogenation, dimerization, carbonylation, etc.).

Heck, R. F., *Adv. Organometallic Chem.*, 1966, **4**, 243 (synthesis of alkyl and aryl cobalt carbonyls and their reactions; insertion and other reactions).

Henrici-Olivé, G., and S. Olivé, *Angew. Chem. Internat. Edn.*, 1971, **10**, 105 (general review on principles of homogeneous catalysis).

James, B. R., *Inorg. Chim. Acta Rev.*, 1970, **4**, 73 (comprehensive review of catalyses by Ru complexes).

Kwiatek, J., *Catalyst Rev.*, 1968, **1**, 37 (reactions of pentacyanocobaltate).

Muetterties, E. L., *Transition Metal Hydrides*, Dekker, 1971.

Ochiai, E. I., *Co-ordination Chem. Rev.*, 1968, **3**, 49 (a general review of all aspects of catalysis by metal ions and their complexes).

Paquette, L. A., *Accounts Chem. Res.*, 1971, **4**, 280 (catalysis of strained σ-bond rearrangements by Ag^+ ion).

Parshall, G. W., and J. J. Mrowca, *Adv. Organometallic Chem.*, 1968, **7**, 157 (reactions of σ-alkyls and aryls of transition metals).

Ryang, M., *Organometallic Chem. Rev.*, A, 1970, **5**, 67 (extensive review of stoichiometric reactions of metal carbonyls in organic syntheses).

Schrautzer, G. N., ed., *Transition Metals in Homogeneous Catalysis*, Dekker, 1971.

Ugo, R., ed., *Aspects of Homogeneous Catalysis*, 1970, Vol. 1, Carlo Manfredi, Milan (a new review series).

Wender, I., and P. Pino, eds., *Organic Syntheses via Metal Carbonyls*, Wiley, Vol. 1, 1968 (cyclic polymerization of acetylenes; organic synthesis *via* alkyl and acyl Co carbonyls; reactions of nitrogen compounds).

25

The Elements of the First Transition Series

GENERAL REMARKS

The general features of the transition elements and of their complex chemistry have been presented in the preceding Chapters. We discuss in this Chapter the elements of the first transition series, titanium through copper. There are two main reasons for considering these elements apart from their heavier congeners of the second and third transition series: (1) in each group, for example, V, Nb, Ta, the first-series element always differs appreciably from the heavier elements, and comparisons are of limited use; and (2) the aqueous chemistry of the first-series elements is much simpler, and the use of ligand-field theory in explaining both the spectra and magnetic properties of compounds has been far more extensive. The ionization potentials for the first-series atoms are listed in Table 25-1. In the separate Sections A–H for each element the oxidation states and stereochemistries are summarized; we do not specify, except in cases of special interest, distortions from perfect geometries which can be expected in octahedral d^1 and d^2 (slight), high-spin octahedral d^4 (two long coaxial bonds), low-spin octahedral d^4 (slight),

TABLE 25-1
Ionization Potentials of the Elements of the First Transition Series

Element	Configuration	Ionization potential (eV)							
		1st	2nd	3rd	4th	5th	6th	7th	8th
Sc	$3d^14s^2$	6.54	12.80	24.75	73.9	92	111	138	159
Ti	$3d^24s^2$	6.83	13.57	27.47	43.24	99.8	120	141	171
V	$3d^34s^2$	6.74	14.65	29.31	48	65.2	128.9	151	174
Cr	$3d^54s^1$	6.76	16.49	30.95	49.6	73.2	90.6	161	185
Mn	$3d^54s^2$	7.43	15.64	33.69	52	76	98	119	196
Fe	$3d^64s^2$	7.90	16.18	30.64	57.1	78	102	128	151
Co	$3d^74s^2$	7.86	17.05	33.49	53	83.5	106	132	161
Ni	$3d^84s^2$	7.63	18.15	35.16	56	78	110	136	166
Cu	$3d^{10}4s^1$	7.72	20.29	36.83	58.9	82	106	140	169
Zn	$3d^{10}4s^2$	9.39	17.96	39.7	62	86	112	142	177

high-spin octahedral d^6, d^7 (slight) or low-spin octahedral d^7, d^8 molecules (two long coaxial bonds). A few other general features of the elements can be mentioned here.

The energies of the $3d$ and $4s$ orbitals in the neutral atoms are quite similar, and their configurations are $3d^n4s^2$ except for Cr, $3d^54s^1$, and Cu, $3d^{10}4s^1$, which are attributable to the stabilities of the half-filled and the filled d shells, respectively. Since the d orbitals become stabilized relative to the s orbital when the atoms are charged, the predominant oxidation states in ionic compounds and complexes of non-π-bonding ligands are II or greater. Owing to its electronic structure, copper has a higher second ionization potential than the other elements and the Cu^I state is important. The high values of third ionization potentials also indicate why it is difficult to obtain oxidation states for nickel and copper greater than II. Although ionization potentials give some guidance concerning the relative stabilities of oxidation states, this problem is a very complex one and not amenable to ready generalization. Indeed it is often futile to discuss relative stabilities of oxidation states, as some oxidation states may be perfectly stable under certain conditions, for example, in solid compounds, in fused melts, in the vapor at high temperatures, in absence of air, etc., but non-existent in aqueous solutions or in air. Thus there is no aqueous chemistry of Ti^{2+}, yet crystalline $TiCl_2$ is stable up to about $400°$ in absence of air; also, in fused potassium chloride, titanium and titanium trichloride give Ti^{II} as the main species and Ti^{IV} is in vanishingly small concentrations; on the other hand, in aqueous solutions in air only Ti^{IV} species are stable.

However, it is sometimes profitable to compare the relative stabilities of ions differing by unit charge when surrounded by similar ligands with similar stereochemistry, as in the case of the Fe^{3+}–Fe^{2+} potentials (page 861), or with different anions. In these cases, as elsewhere, many factors are usually involved; some of these have already been discussed, but they include (a) ionization potentials of the metal atoms, (b) ionic radii of the metal ions, (c) electronic structure of the metal ions, (d) the nature of the anions or ligands involved with respect to their polarizability, donor $p\pi$- or acceptor $d\pi$-bonding capacities, (e) the stereochemistry either in a complex ion or a crystalline lattice, and (f) nature of solvents or other media. In spite of the complexities there are a few trends to be found, namely:

1. From Ti to Mn the highest valence, which is usually found only in oxo compounds or fluorides or chlorides, corresponds to the total number of d and s electrons in the atom. The stability of the highest state decreases from Ti^{IV} to Mn^{VII}. After Mn, that is, for Fe, Co and Ni, the higher oxidation states are difficult to obtain.

2. In the characteristic oxo anions of the valence states IV to VII, the metal atom is tetrahedrally surrounded by oxygen atoms, whereas in the oxides of valences up to IV the atoms are usually octahedrally coordinated.

3. The oxides of a given element become more acidic with increasing oxidation state and the halides more covalent and susceptible to hydrolysis by water.

4. In the II and III states, complexes in aqueous solution or in crystals are usually either in four or six coordination and, across the first series, generally of a similar nature in respect to stoichiometry and chemical properties.

5. The oxidation states less than II, with the exception of Cu^I, are found only with π-acid type ligands or in organometallic compounds.

Finally we re-emphasize that the occurrence of a given oxidation state as well as its stereochemistry depend very much on the experimental conditions and that species that cannot have independent existence under ordinary conditions of temperature and pressure in air may be the dominant species under others. In this connection we may note that transition-metal ions may be obtained in a particular configuration difficult to produce by other means through incorporation by isomorphous substitution in a crystalline host lattice, for example, tetrahedral Co^{3+} in other oxides, tetrahedral V^{3+} in the $NaAlCl_4$ lattice, as well as by using ligands of fixed geometry such as phthalocyanins.

Although some discussion of the relationships between the first, second and third transition series is useful, we defer this until the next Chapter.

In the discussion of individual elements we have kept to the traditional order, i.e. elemental chemistries are considered separately, with reference to their oxidation state. However, it is possible to organize the subject matter from the standpoint of the d^n electronic configuration of the metal. While this can bring out useful similarities in spectra and magnetic properties in certain cases, and has a basis in theory (Chapter 20) nevertheless the differences in chemical properties of d^n species due to differences in the nature of the metal, its energy levels and especially the charge on the ion, often exceed the similarities. Nonetheless, such cross-considerations, as for example in the d^6 series V^{-1}, Cr^0, Mn^I, Fe^{II}, Co^{III}, Ni^{IV}, can provide a useful exercise for students.

Before we consider the chemistries, a few general remarks on the various oxidation states of the first-row elements are pertinent.

The II state. All the elements Ti–Cu inclusive form well-defined binary compounds in the divalent state, such as oxides and halides, which are essentially ionic in nature. Except for Ti, they form well-defined aquo ions, $[M(H_2O)_6]^{2+}$; the potentials are summarized in Table 25-2.

In addition, all the elements form a wide range of complex compounds, which may be cationic, neutral or anionic depending on the nature of the ligands.

TABLE 25-2

Standard Potentials (Acid Solution) for +2 and +3 States[a] (in volts)

	Ti	V	Cr	Mn	Fe	Co	Ni	Cu[b]
$M^{2+} + 2e = M$	−1.6	−1.18	−0.91	−1.18	−0.44	−0.28	−0.24	+0.34
$M^{3+} + e = M^{2+}$	−0.37	−0.25	−0.41	+1.54	+0.77	+1.84	—	—

[a] Some potentials depend on acidity and complexing anions, e.g., for Fe^{3+}–Fe^{2+} in $1M$ acids: HCl, +0.70; $HClO_4$, +0.75; H_3PO_4, +0.44; $0.5M$ H_2SO_4, +0.68 V.

[b] $Cu^{2+} + e = Cu^+$, $E_0 = +0.15$ V; $Cu^+ + e = Cu$, $E_0 = +0.52$ V.

The III state. All the elements form at least some compounds in this state, which is the highest known for copper and then only in certain complex compounds. The fluorides and oxides are again generally ionic in nature, although the chlorides may have considerable covalent character, as in $FeCl_3$.

The elements Ti–Co form aquo ions although the Co^{III} and Mn^{III} ones are readily reduced. In aqueous solution certain anions readily form complex species and for Fe^{3+}, for example, one can be sure of obtaining the $Fe(H_2O)_6^{3+}$ ion only at high acidity (to prevent hydrolysis) and when non-complexing anions such as ClO_4^- or $CF_3SO_3^-$ are present. There is an especially extensive aqueous complex chemistry of the substitution-inert octahedral complexes of Cr^{III} and Co^{III}.

The trivalent halides, and indeed also the halides of other oxidation states, generally act readily as Lewis acids and form neutral compounds with donor ligands, e.g., $TiCl_3(NMe_3)_2$, and anionic species with corresponding halide ions, e.g., VCl_4^-, FeF_6^{3-}.

The IV state. This is the most important oxidation state of Ti where the main chemistry is that of TiO_2 and $TiCl_4$ and its derivatives. This is also an important state for vanadium which forms the vanadyl ion VO^{2+} and many derivatives, cationic, anionic, and neutral containing the VO group. For the remaining elements, Cr–Ni, the IV state is found mainly in fluorides, fluoro complex anions, and cation complexes; however, an important class of compounds are the salts of the oxo ions and other oxo species.

The V, VI and VII states. These occur only as $Cr^{V,VI}$, $Mn^{V,VI,VII}$ and Fe^V, and apart from the fluorides CrF_5, CrF_6 and oxofluorides, MnO_3F, the main chemistry is that of the oxo anions $M^nO_4^{(8-n)-}$. All the compounds in these oxidation states are powerful oxidizing agents.

The lower oxidation states I, 0, −I. All the elements form some compounds, at least, in these states but only with ligands of π-acid type. There are few such compounds for Ti and they are confined essentially to bipyridine complexes, e.g., $Ti(bipy)_3$. An exception, of course, is the Cu^I state for copper, where some insoluble binary Cu^I compounds such as CuCl are known, as well as complex compounds. In the absence of complexing ligands the Cu^+ ion has only a transitory existence in water, although it is quite stable in CH_3CN. Finally, it should be noted that it is also possible to classify the chemistry according to the nature of the ligands present, and again this can provide a useful exercise for students. The following reference list provides a guide to particular ligands in their transition-metal compounds; for references to carbonyls and π-acid ligands and organo compounds see Chapters 22, 23 and 24.

A. *Binary Compounds.*

General: Samsonov, G. V., ed., *Refractory Transition Metal Compounds*, Academic Press, 1964.

Carbides: Frad, W. A., *Adv. Inorg. Chem. Radiochem.*, 1968, **11**, 153.

Storms, E. K., *The Refractory Carbides*, Academic Press, 1967 (Ti, Zr, Hf, V, Nb, Ta, Cr, Mo, W, Th, U).

Toth, L. E., *Transition Metal Carbides and Nitrides*, Academic Press, 1971.

Nitrides: Juza, R., *Adv. Inorg. Chem. Radiochem.*, 1966, **9**, 81 (nitrides of first transition series).

Oxides: Ward, R., *Progr. Inorg. Chem.*, 1959, **1**, 465 (mixed oxides).

Sulfides: Jellinek, F., in *Inorganic Sulphur Chemistry*, D. Nickless, ed., Elsevier, 1968.

Electronic Properties: Howe, T. A., and P. J. Frensham, *Quart. Rev.*, 1967, **21**, 507 (electronic properties of oxides, nitrides, etc.).

B. Group IV — Ligands.

Baird, M. C., *Progr. Inorg. Chem.*, 1968, **9**, 1.

Brooks, E. H., and R. J. Cross, *Organometallic Chem. Rev.*, A, 1970, **6**, 227 (transition-metal compounds with bonds to Group IV elements Si, Ge, Sn, Pb).

Young, J. F., *Adv. Inorg. Chem. Radiochem.*, 1968, **11**, 91.

Group V — Nitrogen Ligands.

Bailey, R. A. *et al.*, *Coordination Chem. Rev.*, 1971, **6**, 407 (infrared spectra of SCN⁻ and related complexes).

Bottomley, F., *Quart. Rev.*, 1970, **24**, 617 (hydrazine as ligand and reactant).

Curtis, N. F., *Coordination Chem. Rev.*, 1968, **3**, 3 (macrocyclic complexes from condensation of metal amine complexes with aliphatic carbonyl compounds).

Griffith, W. P., in C. B. Colburn, ed., *Developments in Inorganic Nitrogen Chemistry*, Vol. 1, Elsevier, 1966, pp. 241–306 (transition-metal complexes of N-donor ligands).

Hambright, P., *Coordination Chem. Rev.*, 1971, **6**, 247 (metalloporphyrins).

Holm, R. H., G. W. Everett and A. Chakravorty, *Progr. Inorg. Chem.*, 1966, **7**, 83 (metal complexes of Schiff bases and β-oxo amines).

Lever, A. P. B., *Adv. Inorg. Chem. Radiochem.*, 1965, **7**, 28 (phthalocyanin complexes).

Lindsay, L. F., and S. E. Livingstone, *Coordination Chem. Rev.*, 1967, **2**, 173 (complexes of diimines, etc., with Fe, Co and Ni).

McKenzie, E. D., *Coordination Chem. Rev.*, 1971, **6**, 187 (bipy and phen complexes).

McWhinnie, W. R., and J. D. Miller, *Adv. Inorg. Chem. Radiochem.*, 1969, **12**, 135 (bipy, phen and terpy complexes).

Reedijk, J., *Methyl Cyanide as a Ligand*, Bronder-Offset, Rotterdam, 1968.

Sinn, E., and C. M. Harris, *Coordination, Chem. Rev.*, 1969, **4**, 391 (Schiff base-metal complexes as ligands).

Walton, R. A., *Quart. Rev.*, 1965, **19**, 126 (nitrile complexes).

Wendlandt, W. W., and T. P. Smith, *Thermal Properties of Transition Metal Amine Complexes*, Elsevier, 1967.

Yamada, S., *Coordination Chem. Rev.*, 1966, **1**, 415 (stereochemistry of Schiff base complexes).

Group V — P, As and Sb Ligands.

Booth, G., *Adv. Inorg. Chem. Radiochem.*, 1964, **6**, 1.

Group VI — Oxygen Ligands.

Oxo Compounds: Baran, V., *Coordination Chem. Rev.*, 1971, **6**, 65 (hydroxyl ion as ligand; hydroxo species; hydrolysis of metal aquo ions).

Griffith, W. P., *Coordination Chem. Rev.*, 1970, **5**, 459 (compounds with M—O and M=O bonds, especially complex ions).

Peroxo Compounds: Connor, J. A., and E. A. V. Ebsworth, *Adv. Inorg. Chem. Radiochem.*, 1964, **6**, 280 (peroxo species of transition metals including lanthanides and actinides).

Alkoxides: Bradley, D. C., *Progr. Inorg. Chem.*, 1960, **2**, 203; *Coordination Chem. Rev.*, 1967, **2**, 299.

Nitrate Compounds and Complexes: Addison, C. C., and N. Logan, *Adv. Inorg. Chem. Radiochem.*, 1964, **6**, 72 (anhydrous metal nitrates).

Addison, C. C., and D. Sutton, *Progr. Inorg. Chem.*, 1967, **8**, 195 (nitrato complexes); C. C. Addison *et al.*, *Quart. Rev.*, 1971, **25**, 289.

Addison, C. C. *et al.*, *Quart. Rev.*, 1971, **25**, 289 (structural aspects).

RNO Complexes: Garvey, R. G., J. H. Nelson and R. O. Ragsdale, *Coordination Chem. Rev.*, 1968, **3**, 375 (aromatic N oxide complexes).

β-Oxoenolate Complexes: Fackler, J. P., *Progr. Inorg. Chem.*, 1966, **7**, 361 (general review).

Fortman, J. J., and R. E. Sievers, *Coordination Chem. Rev.*, 1971, **6**, 331 (optical and geometric isomerism).

Gibson, D., *Coordination Chem. Rev.*, 1969, **4**, 225 (carbon-bonded β-diketonate complexes).

Graddon, D. P., *Coordination Chem. Rev.*, 1969, **4**, 1 (Lewis acid behavior of M^{II} β-diketonate complexes).

Musso, H. *et al.*, *Angew. Chem. Internat. Edn.*, 1971, **10**, 225 (spectroscopic properties).

Carboxylate Complexes: Herzog, S., and W. Kalies, *Z. Chem.*, 1968, **8**, 81 (review of structures of carboxylate complexes). Oldham, C., *Progr. Inorg. Chem.*, 1968, **10**, 223 (complexes of simple carboxylic acids). Tapscott, R. E., R. L. Belford and I. C. Paul, *Coord. Chem. Rev.*, 1969, **4**, 323 (tartrate complexes).

Polyethers: Petersen, C. J., and K. K. Frensdorff, *Angew. Chem. Internat. Edn.*, 1972, **11**, 16.

Group VI — S, Se, Te Ligands.

Coucouvanis, D., *Progr. Inorg. Chem.*, 1970, **11**, 233 [dithiocarbamate, dithiolene (see page 724) and related complexes].

Eisenberg, R., *Progr. Inorg. Chem.*, 1970, **12**, 295 (dithiolato complexes).

Lindoy, L. F., *Coordination Chem. Rev.*, 1969, **4**, 41 (reactions of sulfur ligand complexes).

Livingstone, S. E., *Quart. Rev.*, 1965, **19**, 386 (extensive review of S, Se and Te ligands).

Livingstone, S. E., *Coordination Chem. Rev.*, 1971, **9**, 59; Cox, M., and J. Darken, *Coordination Chem. Rev.*, 1971, **7**, 29 (thio-β-diketone complexes).

Reynolds, W. L., *Progr. Inorg. Chem.*, 1970, **12**, 1 (Me_2SO complexes).

Vitzthum, G., and E. Lindner, *Angew. Chem. Internat. Edn.*, 1971, **10**, 315 (sulfinate complexes).

Group VII — Halogen Ligands.

Colton, R., and J. H. Canterford, *Halides of the First Row Transition Metals*, Interscience-Wiley, 1969 (an authoritative reference text on halides, oxo halides, halide adducts, complex anions, etc.; only more recent references are quoted in this book).

Fowles, G. W. A., *Progr. Inorg. Chem.*, 1964, **6**, 1 (reactions of metal halides with NH_3 and RNH_2).

Marcus, Y., *Coordination Chem. Rev.*, 1967, **2**, 195, 257 (metal chloride complexes studied by ion-exchange and solvent-extraction).

O'Donnell, T. A., *Rev. Pure Appl. Chem.*, 1970, **20**, 159 (reactivity of higher fluorides).

Hydrido Complexes.

Green, M. L. H., and D. J. Jones, *Adv. Inorg. Chem. Radiochem.*, 1965, **7**, 115.

Ginsberg, A. P., *Transition Metal Chem.*, 1965, **1**, 111.

Muetterties, E. L., ed., *Transition Metal Hydrides*, Vol. 1, Dekker, 1971.

Kaesz, H. D. and R. B. Saillant, *Chem. Rev.*, 1972, **72** (in press).

Mixed Valence States.

Robin, M. B., and P. Day, *Adv. Inorg. Chem. Radiochem.*, 1967, **10**, 247 (extensive survey of compounds containing metals in two or more oxidation states; includes binary compounds as well as complexes).

Allen, G. C., and N. S. Hush, *Progr. Inorg. Chem.*, 1967, **8**, 357 (similar survey).

Spectra, Photochemistry and Thermochemistry.

Specialist Periodical Reports, The Chemical Society, London:

Spectroscopic Properties of Inorganic and Organometallic Compounds, Vol. 1, 1968 (authoritative and comprehensive reviews of nmr, nqr, esr, vibrational, electronic and Mössbauer spectra).

Adamson, A. W. *et al.*, *Chem. Rev.*, 1968, **68**, 541 (photochemical behavior of transition-metal complexes).

Ashcroft, S. J., and C. T. M. Mortimer, *Thermochemistry of Transition Metal Complexes*, Academic Press, 1970.

Balzani, V., and V. Carassiti, *Photochemistry of Coordination Compounds*, Academic Press, 1970.

Brunner, H., *Angew. Chem. Internat. Edn.*, 1971, **10**, 249 (optical activity due to asymmetric metal atoms).

Hester, R. E., *Coordination Chem. Rev.*, 1967, **2**, 319 (Raman spectra).

McAuley, A., *Coordination Chem. Rev.*, 1970, **5**, 245 (metal ion oxidation in solution).

Nakagawa, I., *Coordination Chem. Rev.*, 1969, **4**, 423 (far ir spectra and lattice vibrations of complex salts).

Wehry, E. L., *Quart. Rev.*, 1967, **21**, 213 (photochemical behavior of transition-metal complexes).

25-A. TITANIUM[1]

Titanium is the first member of the d-block transition elements and has four valence electrons, $3d^2 4s^2$. Titanium(IV) is the most stable and common oxidation state; compounds in lower oxidation states, $-I$, 0, II and III, are quite readily oxidized to Ti^{IV} by air, water or other reagents. The energy for removal of four electrons is high, so that the Ti^{4+} ion does not have a real

---

[1] R. J. H. Clark, *The Chemistry of Titanium and Vanadium*, Elsevier, 1968 (mainly chemistry of compounds); J. Barksdale, *Titanium: Its Occurrence, Chemistry and Technology*, 2nd edn., Ronald Press; R. I. Jaffee and N. E. Promisel, eds., *The Science, Technology and Applications of Titanium*, Pergamon Press, 1970.

existence and Ti^{IV} compounds are generally covalent in nature. In this IV state, there are some resemblances to the elements Si, Ge, Sn and Pb, especially Sn. The estimated ionic radii ($Sn^{4+} = 0.71$, $Ti^{4+} = 0.68$ Å) and the octahedral covalent radii ($Sn^{IV} = 1.45$, $Ti^{IV} = 1.36$ Å) are similar; thus TiO_2 (rutile) is isomorphous with SnO_2 (cassiterite) and is similarly yellow when hot. Titanium tetrachloride, like $SnCl_4$, is a distillable liquid readily hydrolyzed by water, behaving as a Lewis acid and giving adducts with donor molecules; $SiCl_4$ and $GeCl_4$ do not give stable, solid, molecular addition compounds with ethers although $TiCl_4$ and $SnCl_4$ do so, a difference that may be attributed to the ability of the halogen atoms to fill the coordination sphere of the smaller Si and Ge atoms. There are also similar halogeno anions such as TiF_6^{2-}, GeF_6^{2-}, $TiCl_6^{2-}$, $SnCl_6^{2-}$ and $PbCl_6^{2-}$, some of whose salts are isomorphous, while the Sn and Ti nitrates $M(NO_3)_4$ are also isomorphous. There are other similarities such as the behavior of the tetrachlorides on ammonolysis to give amido species. It is a characteristic of Ti^{IV} compounds that they undergo hydrolysis to species with Ti—O bonds, in many of which there is octahedral coordination by oxygen; Ti—O—C bonds are well known, and compounds with Ti—O—Si and Ti—O—Sn bonds are known.

The stereochemistry of titanium compounds is summarized in Table 25-A-1.

TABLE 25-A-1

Oxidation States and Stereochemistry of Titanium

Oxidation state	Coordination number	Geometry	Examples
Ti^{-1}	6	Octahedral	Ti bipy$_3^-$
Ti^0	6	Octahedral	Ti bipy$_3$
Ti^{II}, d^2	4	Distorted tetrahedral	$(h^5\text{-}C_5H_5)_2Ti(CO)_2$
	6	Octahedral	$TiCl_2$
Ti^{III}, d^1	3	Planar	$Ti[N(SiMe_3)_2]_3$
	5	tbp	$TiBr_3(NMe_3)_2$
	6^a	Octahedral	TiF_6^{3-}, $Ti(H_2O)_6^{3+}$, $TiCl_3 \cdot 3THF$
Ti^{IV}, d^0	4^a	Tetrahedral	$TiCl_4$
		Distorted tetrahedral	$(h^5\text{-}C_5H_5)_2TiCl_2$
	5	Distorted tbp	$K_2Ti_2O_5$
		?	$TiX_4(NMe_3)$
	6^a	Octahedral	TiF_6^{2-}, $Ti(acac)_2Cl_2$
			TiO_2,^{b} $[Cl_3POTiCl_4]_2$
	7	ZrF_7^{3-} type	$[Ti(O_2)F_5]^{3-}$
		Pentagonal bipyramid	$Ti_2(ox)_3 \cdot 10H_2O$
	8	Dist. dodecahedral	$TiCl_4$ diars$_2$^{c}
			$Ti(S_2CNEt_2)_4$

^{a} Most common state.

^{b} Distortions occur in some forms of TiO_2 and in $BaTiO_3$.

^{c} diars = o-phenylenebis(dimethylarsine). The As atoms form the elongated tetrahedron and Cl atoms the flattened one.

25-A-1. The Element

Titanium is relatively abundant in the earth's crust (0.6%). The main ores are *ilmenite*, $FeTiO_3$, and *rutile*, one of the several crystalline varieties of

TiO_2. It is not possible to obtain the metal by the common method of reduction with carbon because a very stable carbide is produced and, moreover, the metal is rather reactive toward oxygen and nitrogen at elevated temperatures. However, because the metal appears to have certain uniquely useful metallurgical properties, the following rather expensive process (Kroll process) is used, although an electrolytic procedure of unspecified nature has been developed.[2]

Ilmenite or rutile is treated at red heat with carbon and chlorine to give $TiCl_4$, which is fractionated to free it from impurities such as $FeCl_3$. The $TiCl_4$ is then reduced with molten magnesium at $\sim 800°$ in an atmosphere of argon. This gives metallic titanium as a spongy mass from which the excess of Mg and $MgCl_2$ is removed by volatilization at $\sim 1000°$. The sponge may then be fused in an atmosphere of argon or helium in an electric arc and cast into ingots.

Extremely pure titanium can be made on the laboratory scale by the van Arkel–de Boer method (also used for other metals) in which TiI_4 that has been carefully purified is vaporized and decomposed on a hot wire in a vacuum.

The metal has a hexagonal close-packed lattice and resembles other transition metals such as iron and nickel in being hard, refractory (m.p. $1680° \pm 10°$, b.p. $3260°$), and a good conductor of heat and electricity. It is, however, quite light in comparison to other metals of similar mechanical and thermal properties and unusually resistant to certain kinds of corrosion and has therefore come into demand for special applications in turbine engines and industrial chemical, aircraft, and marine equipment.

Although rather unreactive at ordinary temperatures, titanium combines directly with most non-metals, for example, hydrogen, the halogens, oxygen, nitrogen, carbon, boron, silicon, and sulfur, at elevated temperatures. The resulting nitride, TiN, carbide, TiC and borides, TiB and TiB_2, are interstitial compounds which are very stable, hard and refractory.

The metal is not attacked by mineral acids at room temperature or even by hot aqueous alkali. It dissolves in hot HCl, giving Ti^{III} species, whereas hot nitric acid converts it into a hydrous oxide which is rather insoluble in acid or base. The best solvents are HF or acids to which fluoride ions have been added. Such media dissolve titanium and hold it in solution as fluoro complexes.

TITANIUM COMPOUNDS

25-A-2. The Chemistry of Titanium(IV), d^0

This is the most important oxidation state for titanium and we consider its chemistry first.

---

[2] *Chem. Eng. News.*, **1968**, Dec. 23, p. 32.

Binary Compounds. *Halides.* The tetrachloride, $TiCl_4$, is one of the most important titanium compounds since it is the usual starting point for the preparation of most other Ti compounds. It is a colorless liquid, m.p. $-23°$, b.p. $136°$, with a pungent odor. It fumes strongly in moist air and is vigorously, though not violently, hydrolyzed by water:

$$TiCl_4 + 2H_2O \rightarrow TiO_2 + 4HCl$$

With some HCl present or a deficit of H_2O, partial hydrolysis occurs, giving oxo chlorides; Raman spectra of the yellow solution of $TiCl_4$ in aqueous HCl indicate that the species present is $[TiO_2Cl_4]^{4-}$ or $[TiOCl_5]^{3-}$ but not $[TiCl_6]^{2-}$, since the oxo chloride, $TiOCl_2$, in HCl gives the same spectrum.[3]

$TiBr_4$ and TiI_4 are similar to $TiCl_4$, but they are crystalline at room temperature and are isomorphous with SiI_4, GeI_4 and SnI_4, having molecular lattices. The fluoride is obtained as a white powder by action of F_2 on Ti at $200°$; it sublimes readily and is hygroscopic; its structure is not known. All the halides behave as Lewis acids; with neutral donors such as ethers they give adducts, and with halide ions give the respective halogeno complex anions (see below).

Titanium Oxide; Complex Oxides; Sulfide. The dioxide TiO_2, has three crystal modifications, rutile, anatase, and brookite, all of which occur in Nature. In rutile, the commonest, the titanium is octahedrally coordinated, and this structure has been discussed on page 51, as it is a common one for MX_2 compounds. In anatase and brookite there are very distorted octahedra of oxygen atoms about each titanium, two being relatively close. Although rutile has been assumed to be the most stable form because of its common occurrence, thermochemical data indicate that anatase is $8-12$ kJ mol^{-1} more stable than rutile.

The dioxide is used as a white pigment. Naturally occurring forms are usually colored, sometimes even black, owing to the presence of impurities such as iron. Pigment-grade material is generally made by hydrolysis of $TiOSO_4$ or vapor-phase oxidation of $TiCl_4$ with oxygen. The solubility of TiO_2 depends considerably on its chemical and thermal history. Strongly roasted specimens are chemically inert.

The precipitates obtained on adding base to Ti^{IV} solutions are best regarded as hydrous TiO_2. This substance dissolves in concentrated alkali hydroxide, to give solutions from which hydrated "titanates" having formulas such as $M_2^I TiO_3 \cdot nH_2O$ and $M_2^I Ti_2O_5 \cdot nH_2O$ but of unknown structure may be obtained.

A considerable number of materials called "titanates" are known, some of which are of technical importance. Nearly all of them have one of the three major mixed metal oxide structures (page 54), and indeed the names of two of the structures are those of the titanium compounds that were the first found to possess them, namely, $FeTiO_3$, *ilmenite*, and $CaTiO_3$, *perovskite*. Other titanites with the ilmenite structure are $MgTiO_3$, $MnTiO_3$, $CoTiO_3$

[3] J. E. D. Davies and D. A. Long, *J. Chem. Soc.*, A, **1968**, 2560.

and $NiTiO_3$, while others with the perovskite structure are $SrTiO_3$, and $BaTiO_3$. There are also titanates with the spinel structure such as Mg_2TiO_4, Zn_2TiO_4 and Co_2TiO_4.

Barium titanate is of particular interest, since it shows remarkable ferroelectric behavior. The reason for this is understood in terms of the structure. Here the ion, Ba^{2+}, is so large relative to the small ion, Ti^{4+}, that the latter can literally "rattle around" in its octahedral hole. When an electric field is applied to a crystal of this material, it can be highly polarized because each of the Ti^{4+} ions is drawn over to one side of its octahedron thus causing an enormous electrical polarization of the crystal as a whole.

The compound Ba_2TiO_4 has discrete, somewhat distorted TiO_4 tetrahedra and the structure is related to that of β-K_2SO_4 or β-Ca_2SiO_4.

The *sulfide*,[4] TiS_2, in common with the disulfides of Zr, Hf and Sn, has a metallic luster and is a semi-conductor; it has a CdI_2-type structure with planar sheets of TiS_6 octahedra joined by edges.

Titanium(IV) Complexes. *Aqueous Chemistry; Oxo Salts.* There is no simple aquated Ti^{4+} ion because of the high charge-to-radius ratio, and in aqueous solutions hydrolyzed species occur and basic oxo salts or hydrated oxides may be precipitated. Although there have been claims for a titanyl ion, TiO^{2+}, this ion appears *not* to exist either in solutions or in crystalline salts such as $TiOSO_4 \cdot H_2O$. The latter has been shown to have $(TiO)_n^{2n+}$ chains:

These are joined together in the crystal by sulfate groups, each of which is in contact with three metal ions; the water molecule is associated with the titanium atoms, so that the latter are approximately octahedrally coordinated by oxygen.

In dilute perchloric acid solutions, there appears to be an equilibrium between the main species:

$$Ti(OH)_3 + H^+ \rightarrow Ti(OH)_2 + H_2O$$

each of which is almost certainly octahedrally coordinated, as, for example, $[Ti(OH)_2(H_2O)_4]^{2+}$. In sulfuric acid, these and other species such as $Ti(OH)_3HSO_4$ and $Ti(OH)_2HSO_4^+$ have been invoked. On increase in the pH, polymerization and further hydrolysis eventually give colloidal or precipitated hydrous TiO_2.

Anionic Complexes. The solutions obtained by dissolving Ti, TiF_4 or hydrous oxides in aqueous HF contain various fluoro complex ions but predominantly the very stable TiF_6^{2-} ion, which can be isolated as crystalline salts.

In aqueous HCl, $TiCl_4$ gives oxo chloro complexes, but from solutions saturated with gaseous HCl salts of the $[TiCl_6]^{2-}$ ion may be obtained.

4 L. E. Conroy and K. C. Park, *Inorg. Chem.*, 1968, 7, 459.

These are better made by interaction of $TiCl_4$ with KCl or of $TiCl_4$ and $Me_4N^+Cl^-$ in $SOCl_2$ solution. The green or yellow salts are hydrolyzed in water to oxo species.

Adducts of TiX_4. All the halides form adducts, which may be of the type TiX_4L or TiX_4L_2. They are normally crystalline solids often soluble in organic solvents, so that their spectroscopic and other properties have been well studied.[5] With certain donors such as R_3P or R_3As, very sparingly soluble adducts may be formed; tetrahydrofuran is especially strongly coordinated.

In general, the adducts appear to involve octahedral coordination of Ti^{IV}. Thus $[TiCl_4(OPCl_3)]_2$ and $[TiCl_4(MeCOOEt)]_2$ are dimeric, with two chlorine bridges, while $TiCl_4(OPCl_3)_2$ has octahedral coordination with *cis*-$OPCl_3$ groups. There are only a few eight-coordinate species and these either have chelate P or As donors[6] or are dithiocarbamates.

Peroxo Complexes. One of the most characteristic reactions of aqueous Ti^{IV} solutions is the development of an intense orange color on addition of hydrogen peroxide, and this reaction can be used for the colorimetric determination of either Ti or of H_2O_2. Detailed studies of the system[7] show that below pH 1, the main peroxo species is probably $[Ti(O_2)(OH)aq]^+$; in less acid solutions, rather complex polymerization processes lead eventually to a precipitate of a peroxohydrate. Various crystalline salts can be isolated,[7,8] e.g., of the ions $[Ti(O_2)F_5]^{3-}$, $[Ti(O_2)(SO_4)_2]^{2-}$ and $[Ti_2O(O_2)_2(dipic)_2]^{2-}$ where dipic = 2,6-pyridinedicarboxylate. The latter has a μ-oxo bridge, and all species appear to have bidentate peroxo groups.

Solvolytic Reactions of $TiCl_4$; *alkoxides and related compounds.* Titanium tetrachloride reacts with a wide variety of compounds containing active hydrogen atoms, such as those in OH groups, with removal of HCl. The replacement of chloride is usually incomplete in absence of an HCl acceptor such as an amine or the ethoxide ion.

Alkoxides.[9] These have been much studied and are generally typical of other transition-metal alkoxides, such as those of Hf, Ce, V, Nb, Fe and U, which will not be discussed in detail. The compounds can be obtained by reactions such as

$$TiCl_4 + 4ROH + 4NH_3 \rightarrow Ti(OR)_4 + 4NH_4Cl$$
$$TiCl_4 + 3EtOH \rightarrow 2HCl + TiCl_2(OEt)_2 \cdot EtOH$$

The titanium alkoxides are liquids or solids that can be distilled or sublimed and are soluble in organic solvents such as benzene. They are exceedingly readily hydrolyzed by even traces of water, the ease decreasing with increasing chain length of the alkyl group; such reactions give polymeric species

5   See, e.g., C. E. Michelson, D. S. Dyer and R. O. Ragsdale, *J. Chem. Soc., A,* **1970,** 2296, for TiF_4 adducts; B. Hessett and P. G. Perkins, *J. Chem. Soc., A,* **1970,** 3229, 3331; R. S. Borden and R. N. Hammer, *Inorg. Chem.,* 1970, **9,** 2005.
6   W. P. Crisp, R. L. Deutscher and D. L. Kepert, *J. Chem. Soc., A,* **1970,** 2199.
7   J. Muhlebach, K. Müller and G. Swartzenbach, *Inorg. Chem.* 1970, **9,** 2381; D. Swartzenbach, *Inorg. Chem.,* 1970, **9,** 2391.
8   W. P. Griffith and T. D. Wickens, *J. Chem. Soc., A,* **1967,** 590.
9   R. Feld and P. L. Cowe, *The Organic Chemistry of Titanium,* Butterworth, 1965 (includes also Ti—C compounds).

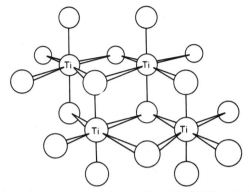

Fig. 25-A-1.   The tetrameric structure of crystalline $Ti(OC_2H_5)_4$. Only Ti and O atoms are shown.

with —OH— or —O— bridges. The initial hydrolytic step probably involves coordination of water to the metal; a proton on H_2O could then interact with the oxygen of an OR group through hydrogen bonding, leading to hydrolysis:

$$\underset{H}{\overset{H}{\diagdown}}\overset{+}{O}\text{—}\bar{M}(OR)_x \longrightarrow \underset{H\cdots\cdots :O}{\overset{H}{\diagdown}}\overset{+}{O}\text{—}\bar{M}(OR)_{x-1} \longrightarrow M(OH)(OR)_{x-1} + ROH$$
$$\underset{R}{\diagdown}$$

Probably the most important structural feature of the titanium and other alkoxides is that, although monomeric species can in certain cases exist, especially in very dilute solution, these compounds are in general polymers. Solid $Ti(OC_2H_5)_4$ is a tetramer, with the structure shown in Fig. 25-A-1. This compact structure neatly allows each Ti atom to attain octahedral coordination. However, in benzene solution, $Ti(OR)_4$ compounds are trimeric for primary alkoxides[10] but are unassociated when made from secondary and tertiary alcohols.[11] The alkoxides are often referred to as "alkyl titanates" and under this name are used in heat-resisting paints, where eventual hydrolysis to TiO_2 occurs.

β-Diketonates.   These are made by treating $TiCl_4$ with the β-diketone in an inert solvent.[12] The acetylacetonate, $TiCl_2acac_2$, which has *cis*–Cl groups and is a fluxional molecule, is readily hydrolyzed to an oxo-bridged species where the Ti—O—Ti group is almost linear, suggesting some $p\pi$–$d\pi$ bonding (see page 407).

Nitrogen compounds.   Nitrogen compounds with N—H bonds appear to react with titanium halides to give initially an adduct, from which hydrogen halide is eliminated by base catalysis. Thus the action of diluted gaseous

[10] D. C. Bradley and C. E. Holloway, *J. Chem. Soc.*, A, **1968**, 3116; W. R. Russo and W. H. Nelson, *J. Amer. Chem. Soc.*, 1970, **92**, 1521.

[11] L. H. Thomas and G. H. Davies, *J. Chem. Soc.*, A, **1969**, 1271.

[12] D. W. Thompson, W. A. Somers and M. O. Workman, *Inorg. Chem.*, 1970, 9, 1252; R. C. Fay and R. N. Lowry, *Inorg. Chem.*, 1970, **9**, 2048.

ammonia on $TiCl_4$ gives the addition product, but with an excess of ammonia ammonolysis occurs and up to three Ti—Cl bonds are converted into Ti—NH_2 bonds. With increasing replacement, the remaining Ti—Cl bonds become more ionic and even liquid ammonia ammonolyzes only three bonds. Primary and secondary amines react in a similar way to give orange or red solids such as $TiCl_2(NHR)_2$ and $TiCl_3NR_2$ which can be further solvated by the amine.

The action of lithium alkylamides, $LiNR_2$, on $TiCl_4$ leads to liquid or solid compounds of the type $Ti[N(C_2H_5)_2]_4$, which, like the alkoxides, are readily hydrolyzed by water with liberation of amine.[13] Similar dialkylamides are known also for both Ti^{III} and Ti^{II}.[14]

Such amides undergo a wide range of "insertion" reactions (Chapter 24); thus with CS_2 they form dithiocarbamates, $Ti(S_2CNR_2)_4$.[14]

Other Titanium(IV) Compounds.   The anhydrous nitrate is a very interesting volatile compound (m.p. 58°), made by the action of N_2O_5 on hydrated Ti^{IV} nitrate.[15] It is very reactive toward organic substances, often causing inflammation or explosion; it probably reacts by releasing the very reactive NO_3 radical. The structure shown in Fig. 25-A-2, is a special case of the dodecahedral structure as explained on page 29.

The sulfate, of unknown structure is made by the reaction

$$TiCl_4 + 6SO_3 \rightarrow Ti(SO_4)_2 + 2S_2O_5Cl_2$$

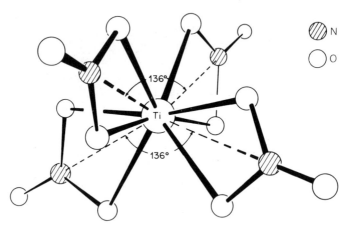

Fig. 25-A-2.   The structure of titanium(IV) nitrate. Each NO_3 group is bidentate and they are disposed so that the N atoms form a slightly distorted tetrahedron with D_{2d} symmetry. [Reproduced by permission from C. D. Garner and F. C. Wallwork, *J. Chem. Soc., A,* **1966,** 1496.]

[13] D. C. Bradley and K. J. Chivers, *J. Chem. Soc., A,* **1968,** 1965; G. Chandra and M. F. Lappert, *J. Chem. Soc., A,* **1968,** 1940; E. C. Alyea *et al., Chem. Comm.,* **1969,** 1064.

[14] M. F. Lappert *et al., J. Chem. Soc., A,* **1970,** 2550; **1971,** 874, 1314; M. Colapietro *et al., Chem. Comm.,* **1970,** 743.

[15] C. C. Addison *et al., J. Chem. Soc., A,* **1967,** 808.

25-A-3. The Chemistry of Titanium(III), d^1

There is an extensive chemistry of solid compounds and Ti^{III} species in solution.

Binary Compounds. The most important compound is the chloride, $TiCl_3$, which has several crystalline forms.[16] It can be made by H_2 reduction of $TiCl_4$ vapor at 500–1200°; this and other high-temperature methods give the violet α-form. The reduction of $TiCl_4$ by aluminum alkyls in inert solvents gives a brown β-form which is converted into the α-form at 250–300°. Two other forms are known, but these and the α-form have layer lattices containing $TiCl_6$ groups whereas β-$TiCl_3$ is fibrous with single chains of $TiCl_6$ octahedra sharing edges. The latter is of particular importance because the stereospecific polymerization of propene (page 794) depends critically on the structure of the β-form.

The trichloride is oxidized by air and reacts with donor molecules to give adducts of general formula $TiCl_3 \cdot nL$ ($n = 1$–6). When heated above 500°, $TiCl_3$ disproportionates (see below).

The *oxide*, Ti_2O_3 (corundum structure), is obtained by reducing TiO_2 at 1000° in a stream of H_2. It is rather inert and is attacked only by oxidizing acids. The addition of OH^- ions to aqueous Ti^{III} solutions gives a purple precipitate of the hydrous oxide.

Aqueous Chemistry and Complexes of Titanium(III). Aqueous solutions of the $[Ti(H_2O)_6]^{3+}$ ion can be readily obtained by reducing aqueous Ti(IV) either electrolytically or with zinc. The violet solutions reduce oxygen and hence must be handled in a nitrogen or hydrogen atmosphere:

$$\text{``}TiO^{2+}\text{''(aq)} + 2H^+ + e = Ti^{3+} + H_2O \quad E^0 = \text{ca. } 0.1 \text{ V}$$

The Ti^{3+} solutions are commonly used as fairly rapid mild reducing agents, e.g., in volumetric analysis.

In dilute $HClO_4$, H_2SO_4 or HCl solutions, $[Ti(H_2O)_6]^{3+}$ is the main species. It hydrolyzes as follows:

$$[Ti(H_2O)_6]^{3+} = [Ti(OH)(H_2O)_5]^{2+} + H^+ \quad K = 1.3 \times 10^{-4}$$

In more concentrated HCl solutions the predominant complex is $[TiCl(H_2O)_5]^{3+}$ although on crystallization *trans*-$[TiCl_2(H_2O)_4]Cl \cdot 2H_2O$ is obtained. The hexaquo ion also occurs in other salts such as alums, e.g., $CsTi(SO_4)_2 \cdot 12H_2O$.

Electronic Structure. The Ti^{III} ion is a d^1 system, and in an octahedral ligand field the configuration must be t_{2g}. One absorption band is expected ($t_{2g} \rightarrow e_g$ transition), and has been observed in several compounds. The spectrum of the $[Ti(H_2O)_6]^{3+}$ ion is discussed on page 571. The violet color of the hexaquo ion is attributable to the band that is so placed as to permit some blue and most red light to be transmitted.

Although a d^1 ion in an electrostatic field of perfect O_h symmetry should

[16] See D. F. Hoeg in A. D. Ketley, ed., *The Stereochemistry of Macromolecules*, Vol. 1, Dekker, 1967.

show a highly temperature-dependent magnetic moment as a result of spin–orbit coupling, with μ_{eff} becoming 0 at $0°K$, the combined effects of distortion and covalence (which causes delocalization of the electron) cause a leveling out of μ_{eff}, which has in general been found to vary from not less than about 1.5 B.M. at $80°K$ to ~ 1.8 B.M. at about $300°K$. Room temperature values of μ_{eff} are generally close to 1.7 B.M.

Other Titanium(III) Compounds.   As noted above, titanium halides form a wide variety of adducts with donor molecules. The acceptor behavior is also shown by the formation of anionic species[17] such as $[TiCl_6]^{3-}$, $[TiCl_5(H_2O)]^{2-}$, $[TiF_6]^{3-}$ and $[Ti_2Cl_9]^{3-}$. The last of these, like its V and Cr analogs, can be made by interaction of fused $TiCl_3$ and Et_2NH_2Cl; it has the confacial bioctahedron structure (page 31). The adducts may be neutral or ionic depending on the circumstances. Examples[18] are $TiBr_3$ $bipy_2$, $[TiBr_2$ $bipy_2]^+[TiBr_4$ $bipy]^-$ and $[Ti\{OC(NH_2)_2\}_6]^{3+}$.

25-A-4. The Chemistry of Titanium(II), d^2

Compounds of divalent titanium are few, and Ti^{II} has no aqueous chemistry because of its oxidation by water, although it has been reported that ice-cold solutions of TiO in dilute HCl contain Ti^{II} ions which persist for some time. The well-defined compounds are $TiCl_2$, $TiBr_2$, TiI_2 and TiO. The halides are best obtained by reduction of the tetrahalides with titanium:

$$TiX_4 + Ti \rightarrow 2TiX_2$$

or by disproportionation of the trihalides:

$$2TiX_3 \rightarrow TiX_2 + TiX_4$$

(the volatile tetrahalides are readily removed). The equilibria and thermodynamics of these interactions have been studied in detail.

The oxide, which is made by heating Ti and TiO_2, has the NaCl structure but is normally non-stoichiometric.

There are a few complexes such as the halides $[TiCl_5]^{3-}$, $[TiCl_4]^{2-}$ and adducts with CH_3CN, pyridine or other ligands.[19]

25-A-5. Organometallic Compounds[9,20]

The formation of Ti—C bonds on the surface of crystalline $TiCl_3$ by reaction of $TiCl_4$ with aluminum alkyls, and the discovery by Ziegler and Natta that the system polymerizes ethylene and propene (Chapter 24), have led to intense study of organotitanium chemistry and the use of organo species in

[17] See, e.g., P. J. Nassiff *et al.*, *Inorg. Chem.*, 1971, **10**, 368; T. J. Kistenmacher and G. D. Stucky, *Inorg. Chem.*, 1971, **10**, 122.
[18] G. W. A. Fowles, B. J. Russ and J. E. Lester, *J. Chem. Soc., A*, **1968**, 805, 1180; P. H. Davis and J. S. Wood, *Inorg. Chem.*, 1970, 9, 1111.
[19] G. W. A. Fowles, T. E. Lester and R. A. Walton, *J. Chem. Soc., A*, **1968**, 1081.
[20] G. A. Razuvaev and V. N. Latyaeva, *Organometallic Chem. Rev.*, 1967, **2**, 349; R. S. P. Coutts and P. C. Wailes, *Adv. Organometallic Chem.*, 1970, 9, 136 (compounds of Ti^{II} and Ti^{III}, including alkyls, halide adducts, etc.); see also *Organometallic Chem. Rev.*, Section B, Annual Surveys.

a wide variety of catalytic reactions of unsaturated substances. Virtually all compounds with one or more alkyl groups on titanium polymerize alkenes.

There is good evidence for simple molecular alkyls of titanium(III) and titanium(IV) but the only ones stable under ordinary conditions are CH_3TiCl_3 and the benzyl, $Ti(CH_2Ph)_4$;[21] the yellow $Ti(CH_3)_4$ is unstable above $-20°$. However, both $Ti(CH_3)_4$ and CH_3TiCl_3 form thermally stable adducts[22] with donor ligands, although even these are sensitive to air and water.

One of the most important simple compounds is $(h^5\text{-}C_5H_5)_2TiCl_2$, which forms red crystals (m.p. 230°) and is readily made from $TiCl_4$ by action of C_5H_5Na (Chapter 23); it has a distorted tetrahedral structure (25-A-I). The compound behaves as a homogeneous catalyst for alkene polymerization in

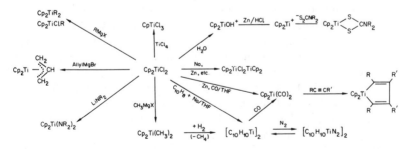

(25-A-I)   (25-A-II)

presence of aluminum alkyls. It has an extremely varied chemistry involving reduction to Ti^{III} and Ti^{II} species, loss of one ring to give $C_5H_5TiX_3$ compounds and replacement of halogen by other unidentate ligands. Some reactions are shown in Fig. 25-A-3. The alkyl and aryl derivatives, e.g., $(C_5H_5)_2TiClR$,[23] may be quite stable. The tetracyclopentadienyl $(C_5H_5)_4Ti$, which has the alkyl-like structure (25-A-II), is fluxional in two senses:[23a] (a) as in other cases (page 761) the σ-bonded rings undergo rapid shifts, and (b) the two types of ring interchange their roles rapidly, so that at 25° all twenty protons give only a single broad line.

The reduction of $(C_5H_5)_2TiCl_2$ to Ti^{III} species can be carried out in solu-

Fig. 25-A-3. Some reactions of dicyclopentadienyl compounds of titanium(II), -(III), and -(IV) (Cp = $h^5\text{-}C_5H_5$).

[21] I. W. Bassi *et al.*, *J. Amer. Chem. Soc.*, 1971, **93**, 3787.
[22] G. W. A. Fowles, D. A. Rice and J. D. Wilkins, *J. Chem. Soc. A*, **1971**, 1920.
[23] J. A. Waters and G. A. Mortimer, *J. Organometallic Chem.*, 1970, **22**, 417.
[23a] J. L. Calderon *et al.*, *J. Amer. Chem. Soc.*, 1971, **93**, 3587, 3592.

tions or in solid-state reactions, and the cation $(C_5H_5)_2Ti^+$ can be isolated as salts.

Of more interest is the reduction to Ti^{II} species (Fig. 25-A-3). The nature of the compound of stoichiometry $C_{10}H_{10}Ti$ has been a matter of prolonged dispute, but it now appears[24] that there are two isomers, a dark green one being dimeric with hydride bridges and a carbene-like C_5H_4 group (cf. carbene ligands, page 767) as in (25-A-III); a dark metastable form may be

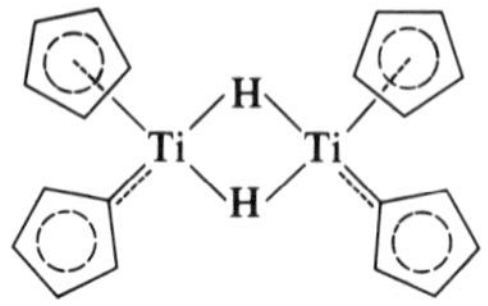

(25-A-III)

$[C_{10}H_{10}Ti]_2$. The compound catalyzes a number of reactions, including hydrogenation, but its most unusual property is to absorb molecular nitrogen reversibly to give a "nitride" of unknown structure. The latter on hydrolysis by a proton source yields about 65% of its N as NH_3.[25] Other reduced titanium systems have also been shown to absorb nitrogen, and it was previously known that traces of nitrogen compounds were found in Ziegler–Natta polymers.

Compounds containing CO or NO are poorly established, presumably because of the lack of π-bonding electron density on titanium. The dicarbonyl $(h^5\text{-}C_5H_5)_2Ti(CO)_2$[26] is reasonably stable; like $[C_{10}H_{10}Ti]_2$ it undergoes, with loss of CO, a number of oxidative addition (Chapter 24) reactions. The lower valence states for titanium of 0 and -1 are known only in the bipyridine complexes such as Ti bipy$_3$.[27]

25-B. VANADIUM

The maximum oxidation state of vanadium is V. But for this there is little similarity, other than in some of the stoichiometry, to the chemistry of elements of the P group. The chemistry of V^{IV} is dominated by the formation of oxo species, and a wide range of compounds with VO^{2+} groups is known. There are four well-defined cationic species, $[V^{II}(H_2O)_6]^{2+}$, $[V^{III}(H_2O)_6]^{3+}$, $V^{IV}O^{2+}$aq and $V^{V}O_2^+$aq, and none of these disproportionates because the ions become better oxidants as the oxidation state increases; both V^{II} and V^{III} ions are oxidized by air. As with Ti, and in common with other transition elements, the vanadium halides and oxo halides behave as Lewis acids, forming adducts with neutral ligands and halogeno complex ions with halide ions.

[24] H. H. Brintzinger et al., J. Amer. Chem. Soc., 1970, **92**, 6182; 1971, **93**, 2045, 2046.
[25] E. E. van Tamelen et al., Accounts Chem. Res., 1970, **3**, 361; J. Amer. Chem. Soc., 1970, **92**, 5251, 5253.
[26] F. Calderazzo, J. J. Salzman, and P. Mosimann, Inorg. Chim. Acta, 1967, **1**, 65.
[27] R. Pappalardo, Inorg. Chim. Acta, 1968, **2**, 209.

The oxidation states and stereochemistries for vanadium are summarized in Table 25-B-1.

TABLE 25-B-1

Oxidation States and Stereochemistry of Vanadium

Oxidation state	Coordination number	Geometry	Examples
V^{-I}	6	Octahedral	$V(CO)_6^-$, $Li[V(bipy)_3] \cdot 4C_4H_8O$
V^0	6	Octahedral	$V(CO)_6$, $V(bipy)_3$, $V[C_2H_4(PMe_2)_2]_3$
	7	?	$V(CO)_6AuPPh_3$
V^I, d^4	6	Octahedral	$[V(bipy)_3]^+$
		Tetragonal pyramidal	$h^5\text{-}C_5H_5V(CO)_4$
V^{II}, d^3	6	Octahedral	$[V(H_2O)_6]^{2+}$, $[V(CN)_6]^{4-}$
V^{III}, d^2	3	Planar	$V[N(SiMe_3)_2]_3$
	4	Tetrahedral	$[VCl_4]^-$
	5	*tbp*	*trans*-$VCl_3(SMe_2)_2$, $VCl_3(NMe_3)_2$
	6^a	Octahedral	$[V(NH_3)_6]^{3+}$, $[V(C_2O_4)_3]^{3-}$, VF_3
V^{IV}, d^1	4	Tetrahedral	VCl_4, $V(NEt_2)_4$, $V(CH_2SiMe_3)_4$
	5	Tetragonal pyramidal	$VO(acac)_2$
		?	$[VO(SCN)_4]^{2-}$, VCl_5^-
		tbp	$VOCl_2$ *trans*-$(NMe_3)_2$
	6^a	Octahedral	VO_2 (rutile), K_2VCl_6, $VO(acac)_2py$
	8	Dodecahedral	$VCl_4(diars)_2$
V^V, d^0	4	Tetrahedral(C_{3v})	$VOCl_3$
	5	*tbp*	$VF_5(g)$
		spy	$CsVOF_4$
	6^a	Octahedral	$VF_5(s)$, VF_6^-, V_2O_5 (very distorted, almost *tbp* with one distant O); $[VO_2ox_2]^{3-}$
	7	Pentagonal bipyramidal	$VO(NO_3)_3 \cdot CH_3CN^b$

[a] Most important states.
[b] Contains both mono- and bi-dentate NO_3 groups (F. B. W. Einstein *et al.*, *Inorg. Chem.*, 1971, **10**, 678).

25-B-1. The Element

Vanadium has an abundance in Nature of about 0.02%. It is widely spread but there are few concentrated deposits. Important minerals are *patronite* (a complex sulfide), vanadinite [$Pb_5(VO_4)_3Cl$], and *carnotite* [$K(UO_2)VO_4 \cdot \frac{3}{2}H_2O$]. The last of these is more important as a uranium ore, but the vanadium is usually recovered as well. Vanadium also occurs widely in certain petroleums, notably those from Venezuela, and it can be isolated from them as oxovanadium(IV) porphyrins.[1] V_2O_5 is recovered from flue dusts after combustion.

Very pure vanadium is rare because, like titanium, it is quite reactive toward oxygen, nitrogen and carbon at the elevated temperatures used in conventional thermometallurgical processes. Since its chief commercial use is in alloy steels and cast iron, to which it lends ductility and shock resistance,

[1] See M. Zerner and M. Gouterman, *Inorg. Chem.*, 1966, **5**, 1699; R. Bonnett and P. Brewer, *Tetrahedron Letters*, 1970, **30**, 2579.

commercial production is mainly as an iron alloy, *ferrovanadium*. The very pure metal can be prepared by the de Boer–van Arkel process (page 809). It is reported to melt at ~1700°, but addition of carbon (interstitially) raises the melting point markedly: vanadium containing 10% of carbon melts at ~2700°. The pure, or nearly pure, metal resembles titanium in being corrosion-resistant, hard and steel-grey. In the massive state it is not attacked by air, water, alkalies or non-oxidizing acids other than HF at room temperature. It dissolves in nitric acid, concentrated sulfuric acid and aqua regia.

At elevated temperatures it combines with most non-metals. With oxygen it gives V_2O_5 contaminated with lower oxides, and with nitrogen the interstitial nitride, VN. Arsenides, silicides, carbides, and other such compounds, many of which are definitely interstitial and non-stoichiometric, are also obtained by direct reaction of the elements.

VANADIUM COMPOUNDS[2,3]

25-B-2. Vanadium Halides

These are listed in Table 25-B-2 together with some of their reactions.

TABLE 25-B-2

The Halides of Vanadium

$$VF_5{}^a \xrightarrow{PCl_3} VF_4 \xrightarrow{\sim 150° \, b} VF_3 \xrightarrow[115°]{H_2 + HF} VF_2$$

$VF_5{}^a$ colorless m.p. 19.5° b.p. 48°		VF_4 lime-green sub. >150°		VF_3 yellow-green		VF_2 blue

$$VCl_4{}^a \underset{Cl_2}{\overset{Reflux}{\rightleftharpoons}} VCl_3 \xrightarrow{>450° \, b} VCl_2$$

$-{}^d$ (below VF_5)

$VCl_4{}^a$ red-brown b.p. 154° ; VCl_3 violet ; VCl_2 pale green

25° ↑ HF in $CClF_3$; 600° ↑ HF(g) ; 600° ↑ HF(g)

$[VBr_4{}^c]$ magenta $\underset{Br_2}{\overset{>-23°}{\rightleftharpoons}}$ $VBr_3{}^a$ black $\xrightarrow{>280°}$ VBr_2 red-brown

$[VI_4(g)]$ ← $VI_3{}^e$ brown $\xrightarrow{>280°}$ VI_2 dark violet

---

[a] Made by direct interaction at elevated temperatures, F_2, 300°; Cl_2, 500°; Br_2, 150°.
[b] Disproportionation reaction, e.g., $2VCl_3 = VCl_2 + VCl_4$.
[c] Isolated from vapor at ~550° by rapid cooling; decomposes above −23°.
[d] The alleged VCl_5 is $PCl_4 \cdot VCl_5$ (I. M. Griffiths, D. Nicholls and K. R. Seddon, *J. Chem. Soc., A*, **1971**, 1513).
[e] Made in a temperature gradient with V at >400°, I_2 at 250–300°.

2 R. J. H. Clark, *The Chemistry of Titanium and Vanadium*, Elsevier, 1968.
3 D. Nicholls, *Coordination Chem. Rev.*, 1966, **1**, 379 (extensive review on the coordination chemistry of vanadium compounds); J. O. Hill, I. G. Worsley and L. G. Hepler, *Chem. Rev.*, 1971, **71**, 127 (thermodynamic properties and oxidation potentials).

The *tetrachloride* is obtained not only from $V + Cl_2$ but also by the action of CCl_4 on red-hot V_2O_5 and by chlorination of ferrovanadium (followed by distillation to separate VCl_4 from Fe_2Cl_6). It is an oil which is violently hydrolyzed by water to give solutions of oxovanadium(IV) chloride; its magnetic and spectral properties confirm its non-associated tetrahedral nature. It has a high dissociation pressure and loses chlorine slowly when kept, but rapidly on boiling, leaving VCl_3. The latter may be decomposed to VCl_2 which is then stable (m.p. 1350°):[4]

$$2VCl_3(s) \rightarrow VCl_2(s) + VCl_4(g)$$
$$VCl_3(s) \rightarrow VCl_2(s) + \tfrac{1}{2}Cl_2(g)$$

The bromide system is similar but there is only indirect evidence for VI_4 in the vapor phase.[5] The trihalides have the BI_3 structure in which each metal atom is at the center of a nearly perfect octahedron of halogen atoms.

Liquid VF_5 has an uncommonly high viscosity (cf. SbF_5). The polymeric structure of the solid, which has endless chains of VF_6 octahedra linked by V—F—V *cis* bridges,[6] presumably persists before breaking down to give monomeric VF_5 in the vapor.

The halides form adducts with donor molecules, and anionic species with the appropriate halide ions. The oxo halides are discussed below.

25-B-3. The Chemistry of Vanadium(V)

Vanadium(V) Oxide. Vanadium(V) oxide is obtained on burning the finely divided metal in an excess of oxygen, although some quantities of lower oxides are also formed. The usual method of preparation is by heating so-called ammonium metavanadate:

$$2NH_4VO_3 \rightarrow V_2O_5 + 2NH_3 + H_2O$$

It is thus obtained as an orange powder which melts at about 650° and solidifies on cooling to orange, rhombic needle crystals. Addition of dilute H_2SO_4 to solutions of NH_4VO_3 gives a brick-red precipitate of V_2O_5. This has slight solubility in water (~ 0.007 g/l) to give pale yellow acidic solutions. Although mainly acidic in character, and hence readily soluble in bases, V_2O_5 also dissolves in acids. That the V^V species so formed are moderately strong oxidizing agents is indicated by the fact that chlorine is evolved when V_2O_5 is dissolved in hydrochloric acid, V^{IV} being produced. It is also reduced by warm sulphuric acid. The following standard potential has been estimated:

$$VO_2^+ + 2H^+ + e = VO^{2+} + H_2O \qquad E^0 = 1.0 \text{ V}$$

Vanadates.[7] Vanadium pentoxide dissolves in sodium hydroxide to give colorless solutions. On addition of acid to about pH 6.5 the solutions become bright orange, and continue so until, at around pH 2, a brown precipitate of

[4] S. S. Kim, S. A. Reed and J. W. Stout, *Inorg. Chem.*, 1970, **9**, 1584.
[5] K. O. Berry, R. R. Smardzewski and R. E. McCarley, *Inorg. Chem.*, 1969, **8**, 1994.
[6] A. J. Edwards and G. R. Jones, *J. Chem. Soc.*, A, **1969**, 1651.
[7] M. T. Pope and B. W. Dale, *Quart. Rev.*, 1968, **22**, 527 (isopolyvanadates).

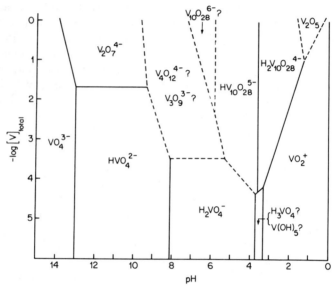

Fig. 25-B-1. Approximate conditions of pH and vanadium concentration under which a given species is the major component at 25°. [Reproduced by permission from M. T. Pope and B. W. Dale.[7]]

V_2O_5 is formed. This redissolves in more acid to form the dioxovanadium(v) ion, VO_2^+.

These reactions have been studied intensively by a variety of physical methods including Raman spectra,[8] but the nature of the species formed has been, and to some extent still is, controversial. The available data seem most consistent with the following main equilibria in alkaline solutions:

$$VO_4^{3-} + H^+ \rightleftharpoons [VO_3(OH)]^{2-}$$
$$2[VO_3(OH)]^{2-} + H^+ \rightleftharpoons [V_2O_6(OH)]^{3-} + H_2O$$
$$[VO_3(OH)]^{2-} + H^+ \rightleftharpoons [VO_2(OH)_2]^-$$
$$3[VO_3(OH)]^{2-} + 3H^+ \rightleftharpoons V_3O_9^{3-} + 3H_2O$$

and in acid solutions:

$$[V_{10}O_{28}]^{6-} + H^+ \rightleftharpoons [HV_{10}O_{28}]^{5-}$$
$$[HV_{10}O_{28}]^{5-} + H^+ \rightleftharpoons [H_2V_{10}O_{28}]^{4-}$$
$$[H_2V_{10}O_{28}]^{4-} + 14H^+ \rightleftharpoons 10VO_2^+ + 8H_2O$$

The ranges of existence of the various species at different V concentrations and pH are summarized in Fig. 25-B-1. It must be noted that equilibrium measurements in dilute aqueous solutions cannot establish the extent of hydration of the various species, so that the formulae given are arbitrary in this regard. In general, though not always, we write the simplest possible formula, that is, the one containing no water. Thus $[VO_3(OH)]^{2-}$ probably is $[VO_2(OH)_3]^{2-}$ or $[VO_2(OH)_3(H_2O)]^{2-}$, etc.

The above equilibria show that in the most basic solutions, mononuclear, tetrahedral vanadate ions, VO_4^{3-}, are formed and that, as the basicity is

[8] W. P. Griffith and P. J. B. Lesniak, J. Chem. Soc., A, 1969, 1066.

reduced, these first protonate and then aggregate into dinuclear and tri-nuclear species, written above as $[V_2O_6(OH)]^{3-}$ and $[V_3O_9]^{3-}$. These anions may be built up from dioxovanadium ions VO_2^+ (which are discussed further below) and OH^- ions as indicated in (25-B-I, 25-B-II and 25-B-III), but this

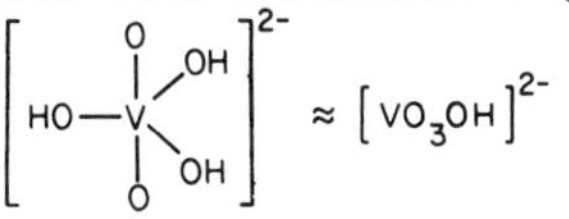

(25-B-I) 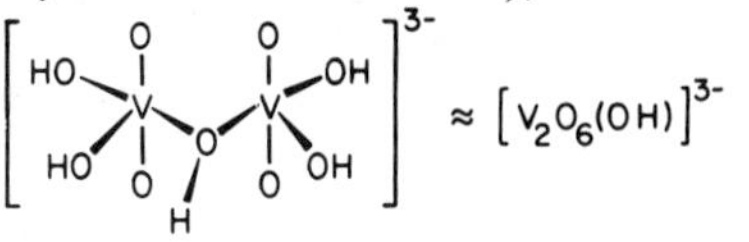   (25-B-II)

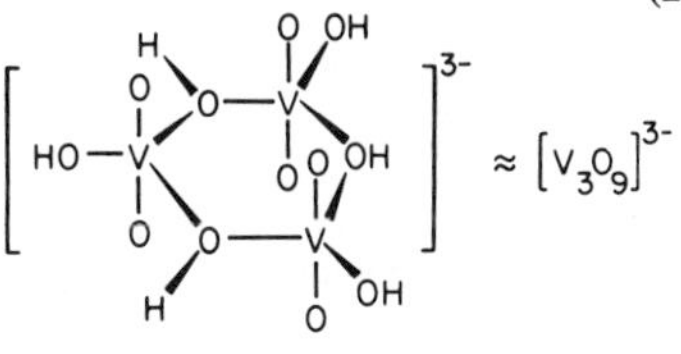

(25-B-III)

is only speculative. Although crystalline vanadates may be obtained from the solutions, their stoichiometries do not necessarily provide information con-cerning the nature of species in solution. Failure to recognize that an equilib-rium can exist between such solids and the solutions without the same spe-cies being present in both phases is responsible for a great deal of confusion in the older literature in this field. And, in addition, there is the complication that equilibria may be reached only very slowly.

However, the orange solutions readily deposit beautiful large orange crys-tals that are evidently hydrates of the orange species in solution. The X-ray structures of the salts $K_2Zn_2V_{10}O_{28} \cdot 16H_2O$, $Na_6V_{10}O_{28} \cdot 18H_2O$ and $Ca_3V_{10}O_{28} \cdot 16H_2O$ have been determined. When solutions of the orange crystals are warmed, more sparingly soluble vanadates such as $K_3V_5O_{14}$ and KV_3O_8 are precipitated and these also have been studied. Other crystal-line salts such as KVO_3 and $Na_4V_2O_7 \cdot 18H_2O$ can be prepared.[9]

The structures of three anions are shown in Fig. 25-B-2. KVO_3 has infinite chains of VO_4 tetrahedra sharing corners; $KVO_3 \cdot H_2O$ has chains of VO_5 polyhedra; and the decavanadates $V_{10}O_{28}^{6-}$ have ten VO_6 octahedra linked together. Although there may be several species in solution containing the V_{10} unit, it now seems certain that, at least in this case, the basic struc-ture in solution and in the crystals is the same.

Finally, it is to be noted that the decavanadate ion is only one example of the type of polyanion generally called isopolyanions, which are a charac-teristic feature of other of the lighter d group transition elements, notably Nb, Ta, Mo and W (Chapter 26). It is also possible that the linking of MO_6 octahedra can occur around another, different atom, in which case the polyanions are called heteropolyanions. For vanadium, only few examples are known,[10] e.g., $[Mn^{IV}V_{13}O_{39}]^{9-}$.

[9] C. M. Flynn, J. V. Pope and M. T. Pope, *J. Amer. Chem. Soc.*, 1970, **92**, 85.
[10] C. M. Flynn and M. T. Pope, *J. Amer. Chem. Soc.*, 1970, **92**, 85, *Inorg. Chem.*, 1971, **10**, 2745; G. A. Tsigidmos and C. J. Hallada, *Inorg. Chem.*, 1968, **7**, 437.

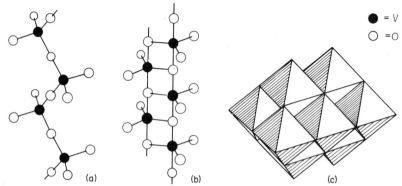

● = V

○ = O

Fig. 25-B-2.   Structures in the crystalline state of anions in (a) KVO_3, (b) $KVO_3 \cdot H_2O$, and (c) decavanadates, $[V_{10}O_{28}]^{16-}$.

Vanadium(V) Oxo Halides.   These are $VOX_3 (X = F, Cl$ or $Br)$, VO_2F and VO_2Cl. VOF_3, made by the action of F_2 on V_2O_5 at 450°, has a sheet structure with both VFV and VF_2V bridges.[11] The oxotrichloride, is made by the action of Cl_2 on $V_2O_5 + C$ at ca. 300° and is a yellow, readily hydrolyzed liquid.

The Dioxovanadium(V) Ion; Vanadium(V) Complexes.   As noted above, in strong acid solutions the ion VO_2^+ is formed, which is doubtless complexed where anions other than ClO_4^- are involved. Unlike the oxotitanium compounds where there is little evidence for multiple bonding, M=O, the oxovanadium (V or IV) compounds show the infrared and Raman bands that are characteristic for M=O groups. The VO_2^+ group in solution is angular since the spectra[12] are similar to those found for octahedral complexes[12a] such as cis-$[VO_2Cl_4]^{3-}$, cis-$[VO_2EDTA]^{3-}$ and cis-$[VO_2ox_2]^{3-}$. The cis arrangement for dioxo compounds of metals with no d electrons is preferred over the $trans$-arrangement found in some other metal dioxo systems, e.g., RuO_2^{2+} because the strongly π-donating O ligands then have exclusive share of one $d\pi$ orbital each (d_{xz}, d_{yz}) and share a third one (d_{xy}), whereas in the $trans$-configuration they would have to share two $d\pi$ orbitals and leave one unused.

The addition of hydrogen peroxide to acidic V^V solutions gives a red color due to the formation of peroxo complexes, where oxygen atoms in VO_4^{3-} are replaced by one or more O_2^{2-} groups; several peroxovanadates have been isolated,[12b] e.g., $K[V(O_2)_3$ bipy]$\cdot 4H_2O$.

The complex compounds formed under non-aqueous conditions are largely those derived from Lewis acid behaviour of VF_5 and the oxo halides. Thus addition of KF to VF_5 in ligand HF gives KVF_6 which is hydrolyzed

[11] A. J. Edwards and P. Taylor, *Chem. Comm.*, **1970**, 1474.
[12] W. P. Griffith and T. D. Wickens, *J. Chem. Soc.*, A, **1968**, 400.
[12a] W. R. Scheidt *et al.*, *J. Amer. Chem. Soc.*, **1971**, **93**, 3867; R. R. Ryan, S. H. Mastin and M. J. Reisfeld, *Acta Cryst.*, 1971, B, **25**, 1270.
[12b] J. Sala-Pala and J. E. Guèrchais, *J. Chem. Soc.*, A, **1971**, 1132; I.-B. Svensson and R. Stomberg, *Acta Chem. Scand.*, 1971, **25**, 989.

by moisture. Other examples are the donor adducts $VOCl_3(NEt_3)_2$, $VOCl_3(MeCN)_2$, the ions VOF_4^- and $VOCl_4^-$,[13] and compounds such as alkoxides derived from them by solvolytic reactions, e.g., $VO(OR)_2Cl$.

25-B-4. The Chemistry of Vanadium(IV), d^1

This important oxidation state is the most stable one under ordinary conditions. Thus aqueous solutions of V^{3+} are oxidized by air to V^{IV}, and V^V is readily reduced to V^{IV} by mild reducing agents:

$$VO^{2+} + 2H^+ + e = V^{3+} + H_2O \qquad E_0 = +0.34 \text{ V}$$
$$VO_2^+ + 2H^+ + e = VO^{2+} + H_2O \qquad E_0 = +1.0 \text{ V}$$

Oxovanadium(IV) Compounds.[14] The most important V^{IV} compounds are those containing the VO unit, which can persist through a variety of chemical reactions.

The oxovanadium(IV) ion, $[VO(H_2O)_5]^{2+}$ is blue and occurs in a number of salts. An extremely wide range of compounds, most of which are blue, can be obtained from the aqueous solutions. An alternative source is the solution of the oxochloride, $VOCl_2$, readily formed by heating V_2O_5 with ethanolic hydrochloric acid; the green deliquescent solid $VOCl_2$ can be made by reducing $VOCl_3$ with hydrogen. Addition of base to $[VO(H_2O)_5]^{2+}$ gives the yellow hydrous oxide $VO(OH)_2$ which redissolves in acids giving the cation.

Oxovanadium(IV) compounds may, depending on the nature of the ligands, be cationic, neutral or anionic and be either 5-coordinate, where the stereochemistry is that of the square pyramid, or 6-coordinate, containing a distorted octahedron; examples are $[VO\, bipy_2Cl]^+$, $VO\, acac_2$ and $[VO(NCS)_4]^{2-}$. Most oxovanadium compounds are blue-green, but a salicylaldimine complex is yellow and appears to be polymeric with donation from the oxygen of one VO group to the vacant site of the other.[15]

The VO bond is essentially a double bond. (a) Electronic, esr and vibrational spectra of the aquo ion are all consistent with the formulation $[VO(H_2O)_5]^{2+}$. The water in the position *trans* to O appears to be labile according to ^{17}O exchange studies.[16] (b) All oxovanadium(IV) compounds have infrared and Raman bands characteristic of a $V{=}O$ group; this is so even for $VO(OH)_2$. (c) X-ray crystallographic studies on several compounds confirm the VO formulation. Thus the hydrate $VOSO_4 \cdot 5H_2O$ has an octahedral structure with one H_2O *trans* to O at 2.22 Å, three other H_2O molecules *cis* at 2.04 Å and the oxygen of the monodentate OSO_3^{2-} at 1.98 Å in the fourth *cis*-position.[17] The acetylacetonate has the structure shown in

[13] D. Nicholls and D. N. Wilkinson, *J. Chem. Soc., A*, **1970**, 1103; J. A. S. Howell and K. C. Moss, *J. Chem. Soc., A*, **1971**, 270.
[14] J. Selbin, *Coordination Chem. Rev.*, 1966, **1**, 293; *Chem. Rev.*, 1965, **65**, 153; *Angew. Chem. Internat. Edn.*, 1966, **5**, 712.
[15] M. Mathew, A. J. Carty and G. J. Palenik, *J. Amer. Chem. Soc.*, 1970, **92**, 3197.
[16] K. Wuthrich and R. E. Connick, *Inorg. Chem.*, 1968, **7**, 1377.
[17] C. J. Ballhausen, B. F. Djurinskij and K. J. Watson, *J. Amer. Chem. Soc.*, 1968, **90**, 3305.

(25-B-IV). In all cases the VO bond is short, e.g., in the sulfate the V—O bond length is 1.591 Å and in (25-B-IV) it is 1.56 Å, so that VO bonds can

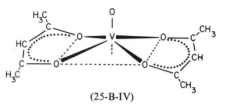

(25-B-IV)

properly be regarded as multiple ones, the π-component arising from electron flow $O(p\pi) \rightarrow V(d\pi)$. Even in VO_2, which has a distorted rutile structure, one bond (1.76 Å) is conspicuously shorter than the others in the VO_6 unit (note that in TiO_2 all Ti—O bonds are substantially equal).

All of the five-coordinate complexes such as (25-B-IV) take up a sixth ligand quite readily,[18] becoming octahedral, with donors such as pyridine or triphenylphosphine. The V=O stretching frequency is quite sensitive to the nature of the *trans*-ligand, and donors that increase the electron density on the metal thereby reduce its acceptor properties toward O, and hence lower the V—O multiple-bond character and the stretching frequency.[19] The esr and electronic spectra of VO^{2+} species are sensitive also to solvents for the same reason, although H-bonding solvents appear to form hydrogen bonds to the V=O group rather than occupy the sixth site.[20]

Because of the strong VO π bonding in oxovanadium(IV) compounds, interpretation of the electronic spectra is not as simple as it would be for an ordinary octahedral complex and at present there are differences of opinion about the exact ordering of the orbitals.

Other Vanadium(IV) Complexes. These can, in the main, be considered to be derived from VX_4 or VOX_2 by addition of other neutral ligands or halide ions. The compounds are normally made by direct interaction in a non-aqueous solvent.

Thus VCl_4 and Me_4NCl react in $SOCl_2$ to give[21] the salt $[Me_4N]_2[VCl_6]$, while adducts $VCl_4 \cdot L_2$ are formed[22] with PR_3 and other ligands. Salts of oxovanadium(IV) are usually made from V_2O_5 in HCl solutions.[13,23]

Finally, vanadium alkoxides and dialkylamides, such as $[V(OEt)_4]_2$ or $V(NR_2)_4$, can be obtained by methods similar to those used for titanium; the dialkylamides react with CS_2, undergoing a V—N insertion reaction (see page 777) to give eight-coordinate V^{IV} compounds, $V(S_2CNR_2)_4$.[24]

[18] See, e.g., K. Dichmann *et al.*, *Chem. Comm.*, **1970**, 1295.
[19] For examples and references see: C. J. Popp, J. H. Nelson and R. O. Ragsdale, *J. Amer. Chem. Soc.*, 1969, **91**, 610; G. Vigee and J. Selbin, *J. Inorg. Nuclear Chem.*, 1968, **30**, 2273; L. J. Boucher and T. F. Yen, *Inorg. Chem.*, 1969, **8**, 639.
[20] C. M. Guzy, J. B. Raynor and M. C. R. Symons, *J. Chem. Soc., A*, **1969**, 2791.
[21] R. D. Bereman and C. H. Brubaker, *Inorg. Chem.*, 1969, **8**, 2480.
[22] R. E. Bridgland and W. R. McGregor, *J. Inorg. Nuclear Chem.*, 1969, **31**, 43.
[23] J. E. Drake, J. E. Vekris and J. S. Wood, *J. Chem. Soc., A*, **1969**, 345.
[24] O. Piovesana and C. Furlani, *J. Chem. Soc., A*, **1970**, 2929; *Chem. Comm.*, **1971**, 256; M. Bonamico *et al.*, *Chem. Comm.*, **1971**, 365.

Vanadium(IV) Oxide and Oxoanions. The dark blue oxide, VO_2, is obtained by mild reduction of V_2O_5, a classic method being by fusion with oxalic acid; it is amphoteric, being about equally readily soluble in both acids and bases. In strongly basic solutions vanadate(IV) ions, VO_4^{4-}, probably not of this simple nature, exist and from these solutions, and less basic ones, various vanadate(IV) compounds, also called *hypovanadates*, are obtainable. They are of the type $M_2^IV_4O_9 \cdot 7H_2O$. By fusion of VO_2 with alkaline-earth oxides, other vanadate(IV) compounds can be formed, e.g., $M^{II}VO_3$, $M^{II}_2VO_4$. Little is known of their structures. A deep violet polyanion, $[V_{10}O_{28}H_4]^{4-}$, appears, however,[25] to contain both V^{IV} and V^V.

25-B-5. The Chemistry of Vanadium(III), d^2

Vanadium(III) Oxide. This is a black, refractory substance made by reduction of V_2O_5 with hydrogen or carbon monoxide. It has the corundum structure but is difficult to obtain pure since it has a marked tendency to become oxygen-deficient without change in structure. Compositions as low in oxygen as $VO_{1.35}$ are reported to retain the corundum structure.

V_2O_3 is entirely basic in nature and dissolves in acids to give solutions of the V^{III} aquo ion or its complexes. From these solutions addition of OH^- gives the hydrous oxide, which is very easily oxidized in air.

The Aquo Ion and Complexes. The blue aquo ion $[V(H_2O)_6]^{3+}$ can be obtained as above or by electrolytic or chemical reduction of V^{IV} or V^V solutions. Such solutions, and also others, of V^{III} are subject to aerial oxidation in view of the potential

$$VO^{2+} + 2H^+ + e = V^{3+} + H_2O \qquad E^0 = 0.36 \text{ V}$$

When solutions of V^{2+} and VO^{2+} are mixed, V^{3+} is formed but a brown intermediate species which has an oxo bridge, VOV^{4+}, occurs; this is similar to a chromium(III) species $CrOCr^{4+}$ obtained when Cr^{2+} is oxidized under conditions where a Cr^{IV} complex might be expected—by the two-electron oxidant Tl^{3+}.

Vanadium(III) forms a number of complex ions, mostly anionic, e.g., $[V(C_2O_4)_3]^{3-}$, $[V(CN)_6]^{3-}$ and $[V(NCS)_6]^{3-}$, but some are neutral,[26] e.g., $V[S_2P(OEt)_2]_3$. In addition to its occurrence in aqueous solutions, where partial hydrolysis to $V(OH)^{2+}$ and VO^+ occurs, the $[V(H_2O)_6]^{3+}$ ion occurs in the vanadium alums, $M^IV(SO_4)_2 \cdot 12H_2O$. The ammonium alum is obtained as air-stable blue-violet crystals by electrolytic reduction of NH_4VO_3 in H_2SO_4. However, for the hydrated halides $VX_3 \cdot 6H_2O$, nqr spectroscopy confirms the structure $[VCl_2(H_2O)_4]Cl \cdot 2H_2O$ as found in similar hydrates of Fe^{III} and Cr^{III}.[27] The bromide and some bromo complexes can be made

[25] C. Heitner-Wirguin and J. Selbin, *J. Inorg. Nuclear Chem.*, 1969, **31**, 3181.

[26] C. Furlani *et al.*, *J. Chem. Soc., A*, **1970**, 2929.

[27] L. Podimore, P. W. Smith and B. Stoessiger, *Chem. Comm.*, **1970**, 221; cf. $[VCl_2(MeOH)_4]^+$, A. T. Casey and R. J. M. Clark, *Inorg. Chem.*, 1969, **8**, 1216.

by heating V_2O_5 with ethanolic HBr^{28} (as noted above, HCl gives VO^{2+} species).

The electronic structure of an octahedrally coordinated d^2 ion, of which V^{III} is the example *par excellence*, has been discussed in Chapter 21. It need only be added here that data for a number of V^{III} octahedral complexes, such as $V(H_2O)_6^{3+}$, VF_6^{3-}, $V(C_3H_2O_4)_3^{3-}$ and V^{3+} substituted into α-Al_2O_3 have been interpreted satisfactorily in terms of the ligand-field model, although in general it has been found necessary to take account of the effects of a trigonal distortion (to D_{3d}) of the basically octahedral field.

Other Vanadium(III) Complexes. As with other halides, there are adducts[29] such as $VX_3(NMe_3)_2$ and VX_3py_3, and anionic species,[30] e.g., VCl_4^- and $V_2Cl_9^{3-}$.

25-B-6. The Chemistry of Vanadium(II)

This is the least known of the oxidation states. The black *oxide*, VO, has a rock-salt type lattice but it shows a marked tendency to non-stoichiometry, being obtainable with anywhere from ~ 45 to ~ 55 atom % oxygen. It has a metallic luster and rather good electrical conductivity of a metallic nature. There is probably considerable V—V bonding. The oxide is basic and dissolves in mineral acids, giving V^{1i} solutions.

Aqueous Solutions, Salts and Complexes. Electrolytic or zinc reduction of acidic solutions of V^V, V^{IV} or V^{III} produces violet air-sensitive solutions containing the $[V(H_2O)_6]^{2+}$ ion. These are strongly reducing (Table 25-2, page 803) and are oxidized by water with evolution of hydrogen, despite the fact that the standard potential V^{3+}/V^{2+} would indicate otherwise. The oxidation of V^{2+} by air is complicated and appears to proceed in part by direct oxidation to VO^{2+} and in part by way of an intermediate species of type VOV^{4+}. There is also some evidence to suggest that at 5mM concentration, the V^{II} ion is a hydrated dimer VOV^{2+}.

Several crystalline salts containing $[V(H_2O)_6]^{2+}$ ions are known,[31] although the hydrate $VCl_2 \cdot 4H_2O$ is actually *trans*-$VCl_2(H_2O)_4$. The most important are the sulfate $VSO_4 \cdot 6H_2O$, which is formed as violet crystals on addition of ethanol to reduced sulfate solutions, and the double sulfates (Tutton salts), $M_2[V(H_2O)_6](SO_4)_2$ where $M = NH_4^+$, K^+, Rb^+ or Cs^+. From the V^{2+} solutions amine complexes, e.g., $[V\,en_3]Cl_2 \cdot H_2O$, can be isolated.

The electronic absorption spectra are consistent with octahedral aquo ions both in crystals and in solution, and the energy-level diagram is analogous to that for Cr^{III} (page 839). The magnetic moments of the sulfates lie close to the spin only value (3.87 B.M.).

[28] D. Nicholls and D. N. Wilkinson, *J. Chem. Soc.*, A, **1969**, 1233.
[29] G. W. A. Fowles *et al.*, *J. Chem. Soc.*, A, **1967**, 1592, 1869.
[30] A. T. Casey and R. J. H. Clark, *Inorg. Chem.*, 1968, **7**, 1598.
[31] L. F. Larkworthy *et al.*, *J. Chem. Soc.*, A, **1968**, 2936; **1970**, 1095.

In aqueous solution, $[V(H_2O)_6]^{2+}$ is kinetically inert because of its d^3 configuration (cf. Cr^{3+}) and substitution reactions are relatively slow. Although F^- and SCN^- form weak complexes, there is little evidence for complexing with Cl^-, Br^-, I^- or SO_4^{2-}. Reorganization of the precursor complex is believed to be the rate-determining step in many reductions by V^{2+} since the rates of redox reactions are similar to those of substitution, e.g.,

$$V(H_2O)_6^{2+} + SCN^- \rightarrow (H_2O)_5VNCS^+ + H_2O$$

and reduction reactions appear to proceed by a substitution-controlled inner-sphere mechanism (page 675).[32] However, in the reduction of $IrCl_6^{2-}$ by V^{2+}, the rate constant is higher by ca. 10^5, so that here an outer-sphere mechanism is involved.[33]

25-B-7. Carbonyl and Organovanadium Compounds[34]

Compounds are known in the -1 to $+5$ oxidation states, but the most extensive chemistry is that of h^5-cyclopentadienyl, arene and carbonyl complexes.

Unlike titanium, vanadium forms a simple, octahedral carbonyl:

$$VCl_3 + 4Na + 6CO \xrightarrow[\substack{160° \\ 200\ atm.}]{diglyme} [Na\ diglyme_2] [V(CO)_6] + 3NaCl$$
$$\downarrow HCl-Et_2O$$
$$V(CO)_6$$

This green-black compound is unusual in that it is the only paramagnetic carbonyl; it undergoes substitution reactions typical of other metal carbonyls (Chapter 23). Substituted anions such as $[V(CO)_4diphos]^-$ and $[V(CO)_5PPh_3]^-$ have been prepared.[35]

The chemistry of σ-alkyls and aryls is less well known than that of Ti, and such bonds are usually unstable, although $V(CH_2SiMe_3)_4$ and $VO(CH_2SiMe_3)_3$ are isolable. Unstable alkyls are present in the solutions of V halides and Al alkyls, which are used in Ziegler–Natta type reactions for the copolymerisation of styrene, butadiene, and dicyclopentadiene to give synthetic rubbers. The dicyclopentadienyl derivatives of V^{II}, V^{III} and V^{IV} are well-established, and, unlike the unusual Ti^{II} compound (page 818), the V^{II} complex is the simple paramagnetic molecule $(h^5\text{-}C_5H_5)_2V$. This undergoes oxidative additions with many compounds (cf. Chapter 24) to give V^{III} or V^{IV} compounds, e.g., Cp_2VCS_2, Cp_2VCl_2, etc.

Mixed complexes such as $h^5\text{-}C_5H_5V(CO)_4$, $h^5\text{-}C_5H_5V(h^7\text{-}C_7H_7)$ and the diarenes, e.g., $V(C_6H_6)_2$, are known, as well as some alkene complexes.

[32] N. Sutin, *Accounts Chem. Res.*, 1968, **1**, 225; J. M. Malin and J. H. Swinehart, *Inorg. Chem.*, 1969, **8**, 1407; M. Green, R. S. Taylor and A. G. Sykes, *J. Chem. Soc., A*, **1971**, 509.

[33] R. N. F. Thornley and A. G. Sykes, *J. Chem. Soc., A*, **1970**, 1036.

[34] *Organometallic Chem. Rev.*, Section B, Annual Surveys; *Gmelin's Handbuch der anorganischen Chemie*, 8th edn., Main Suppl. Vol. 2, *Organometallic Compounds of Vanadium*, Verlag Chemie, 1971.

[35] A. Davison and J. E. Ellis, *J. Organometallic Chem.*, 1971, **31**, 239.

25-C. CHROMIUM

For chromium, as for Ti and V, the highest oxidation state is that corresponding to the total number of $3d$ and $4s$ electrons. Although Ti^{IV} is the most stable state for titanium and V^V is only mildly oxidizing, chromium(VI), which exists only in oxo species such as CrO_3, CrO_4^{2-}, and CrO_2F_2, is strongly oxidizing. Apart from stoichiometric similarities, chromium resembles the Group VI elements of the sulfur group only in the acidity of the trioxide and the covalent nature and ready hydrolysis of CrO_2Cl_2.

Although Cr^V and Cr^{IV} are formed as transient intermediates in the reduction of Cr^{VI} solutions, these oxidation states have no stable aqueous chemistry except as peroxo complexes because of their ready disproportionation to Cr^{III} and Cr^{VI}. Some solid and gaseous compounds do, however, exist.

The most stable and important state is Cr^{III}, d^3, which in an octahedral complex has each t_{2g} level singly occupied, giving a sort of half-filled shell stability. The lower oxidation states are strongly reducing; in aqueous solution only the divalent state, Cr^{2+}, is known. Since for Cr and also for the following elements of the first transition series the more important oxidation states are the lower ones, we discuss them first.

The oxidation states and stereochemistry are summarized in Table 25-C-1.

TABLE 25-C-1

Oxidation States and Stereochemistry of Chromium

Oxidation state	Coordination number	Geometry	Examples
Cr^{-II}		?	$Na_2[Cr(CO)_5]$
Cr^{-I}		Octahedral	$Na_2[Cr_2(CO)_{10}]$
Cr^0	6	Octahedral	$Cr(CO)_6$, $[Cr(CO)_5I]^-$, $Cr(bipy)_3$
Cr^I, d^5	6	Octahedral	$[Cr(bipy)_3]^+$
Cr^{II}, d^4	4	Distorted tetrahedral	$CrCl_2(MeCN)_2$, $CrI_2(OPPh_3)_2$
	5	Trigonal bipyramidal	$[Cr(Me_6 tren)Br]^+$
	6	Distorted[b] octahedral	CrF_2, $CrCl_2$, CrS
	7	?	$[Cr(CO)_2(diars)_2X]X$
Cr^{III}, d^3	3	Planar	$Cr(NPr_2)_3$
	4	Distorted tetrahedral	$[PCl_4]^+$ $[CrCl_4]^-$, $[Cr(CH_2SiMe_3)_4]^-$
	5	Trigonal bipyramidal	$CrCl_3(NMe_2)_2$
	6^a	Octahedral	$[Cr(NH_3)_6]^{3+}$, $Cr(acac)_3$, $K_3[Cr(CN)_6]$
Cr^{IV}, d^2	4	Tetrahedral	$Cr(OC_4H_9)_4$, Ba_2CrO_4, $Cr(CH_2SiMe_3)_4$
	6	Octahedral	K_2CrF_6, $[Cr(O_2)_2en] \cdot H_2O$
Cr^V, d^1	4	Tetrahedral	CrO_4^{3-}
	5	?	CrF_5, $CrOCl_4^-$
	6	Octahedral	$K_2[CrOCl_5]$
	8	Quasi-dodecahedral	K_3CrO_8 (see text)
Cr^{VI}, d^0	4	Tetrahedral	CrO_4^{2-}, CrO_2Cl_2, CrO_3

[a] Most stable state.
[b] Four short and two long bonds.

25-C-1. The Element

The chief ore is *chromite*, $FeCr_2O_4$, which is a spinel with Cr^{III} on octahedral sites and Fe^{II} on the tetrahedral ones. If pure chromium is not required —as for use in ferrous alloys—the chromite is reduced with carbon in a furnace affording the carbon-containing alloy ferrochromium:

$$FeCr_2O_4 + 4C \rightarrow Fe + 2Cr + 4CO$$

When pure chromium is required, the chromite is first treated with molten alkali and oxygen to convert the Cr^{III} to chromate(VI) which is dissolved in water and eventually precipitated as sodium dichromate. This is then reduced with carbon to Cr^{III} oxide:

$$Na_2Cr_2O_7 + 2C \rightarrow Cr_2O_3 + Na_2CO_3 + CO$$

This oxide is then reduced with aluminum:

$$Cr_2O_3 + 2Al \rightarrow Al_2O_3 + 2Cr$$

Chromium is a white, hard, lustrous and brittle metal, m.p. $1903° \pm 10°$. It is extremely resistant to ordinary corrosive agents, which accounts for its extensive use as an electroplated protective coating. The metal dissolves fairly readily in non-oxidizing mineral acids, for example, hydrochloric and sulfuric acids, but not in cold aqua regia or nitric acid, either concentrated or dilute. The last two reagents passivate the metal in a manner which is not well understood. The electrode potentials of the metal are

$$Cr^{2+} + 2e = Cr \qquad E^0 = -0.91 \text{ V}$$
$$Cr^{3+} + 3e = Cr \qquad E^0 = -0.74 \text{ V}$$

so that it is rather active when not passivated. Thus it readily displaces copper, tin and nickel from aqueous solutions of their salts.

At elevated temperatures, chromium unites directly with the halogens, sulfur, silicon, boron, nitrogen, carbon and oxygen.

CHROMIUM COMPOUNDS

25-C-2. Binary Compounds

Halides. These are listed in Table 25-C-2. The anhydrous Cr^{II} halides are obtained by action of HF, HCl, HBr or I_2 on the metal at 600–700° or by reduction of the trihalides with H_2 at 500–600°. $CrCl_2$ is the most common and most important of these halides, dissolving in water to give a blue solution of Cr^{2+} ion.

Of the Cr^{III} halides the red-violet chloride, which can be prepared in a variety of ways, e.g., by the action of $SOCl_2$ on the hydrated chloride, is singularly important. It can be sublimed in a stream of chlorine at about 600°, but if heated to such a temperature in the absence of chlorine it decomposes to Cr^{II} chloride and chlorine. The flaky or leaflet form of $CrCl_3$ is a consequence of its crystal structure, which is of an unusual type. It consists of a cubic close-packed array of chlorine atoms in which two-thirds of the octahedral holes between *every other* pair of Cl planes are occupied by metal

TABLE 25-C-2

Halides of Chromium

	Cr^{II}	Cr^{III}	Higher and mixed oxidation states		
F	CrF_2	$CrF_3{}^a$ green, m.p. 1404°	$CrF_4{}^b$ green, subl. 100° $Cr_2F_5{}^d$	CrF_5 red, m.p. 30°	$CrF_6{}^c$ yellow
Cl	$CrCl_2$	$CrCl_3$ violet, m.p. 1150°	$CrCl_4{}^e$		
Br	$CrBr_2$	$CrBr_3$ black, subl.	$CrBr_4{}^e$		
I	CrI_2	CrI_3 black, decomp.			

^a Melts only in a closed system; in an open system disproportionates above 600° to give CrF_5.

^b Becomes brown on slightest contact with moisture.

^c Unstable above −100°.

^d Often non-stoichiometric; contains regular $Cr^{III}F_6$ and highly distorted $Cr^{II}F_6$ octahedra sharing corners and edges.

^e Not known as solids; appear to exist in vapors formed when the trihalides are heated in an excess of the halogen.

atoms. Those alternate layers of chlorine atoms with no metal atoms between them are held together only by van der Waals' forces and thus the crystal has pronounced cleavage parallel to the layers. $CrCl_3$ is the only substance known to have this exact structure, but $CrBr_3$, as well as $FeCl_3$ and triiodides of As, Sb and Bi, have a structure that differs only in that the halogen atoms are in hexagonal rather than cubic close packing.

Chromic chloride does not dissolve at a significant rate in pure water, but it dissolves readily in presence of Cr^{II} ion or reducing agents such as $SnCl_2$ that can generate some Cr^{II} from the $CrCl_3$. This is because the process of solution can then take place by electron-transfer from Cr^{II} in solution *via* a Cl bridge to the Cr^{III} in the crystal. This Cr^{II} can then leave the crystal and act upon a Cr^{III} ion elsewhere on the crystal surface, or perhaps it can act without moving. At any rate, the "solubilizing" effect of reducing agents must be related in this or some similar way to the mechanism by which chromous ions cause decomposition of otherwise inert Cr^{III} complexes in solution (page 679).

Chromic chloride forms adducts with a variety of donor ligands. The tetrahydrofuranate, $CrCl_3 \cdot 3THF$, which is obtained as violet crystals by action of a little zinc on $CrCl_3$ in THF, is a particularly useful material for the preparation of other chromium compounds such as carbonyls or organo compounds, as it is soluble in organic solvents.

Chromium(IV) fluoride is made by fluorination of the metal at ca. 350°, while at 350–500° CrF_5 is obtained.

The hexafluoride is formed in low yield with fluorine at 200 atm pressure and 400° in a bomb.

Oxides. Only Cr_2O_3, CrO_2 and CrO_3 are of importance. The green oxide, $\alpha\text{-}Cr_2O_3$, which has the corundum structure (page 54), is formed on burning the metal in oxygen, on thermal decomposition of Cr^{VI} oxide or ammonium dichromate or on roasting the hydrous oxide, $Cr_2O_3 \cdot nH_2O$. The latter, commonly called chromic hydroxide, although its water content is variable, is precipitated on addition of hydroxide to solutions of Cr^{III} salts. The oxide, if ignited too strongly, becomes inert toward both acid and base, but otherwise it and its hydrous form are amphoteric, dissolving readily in acid to give aquo ions, $[Cr(H_2O)_6]^{3+}$, and in concentrated alkali to form "chromites".

Fusing Cr_2O_3 with the oxides of a number of bivalent metals gives well-crystallized compounds $M^{II}O \cdot Cr_2O_3$ having the spinel structure with the Cr^{III} ions occupying the octahedral interstices.

Chromium oxide and chromium supported on other oxides such as Al_2O_3 are important catalysts for a wide variety of reactions.[1]

Chromium(IV) oxide, CrO_2, is normally synthesized by hydrothermal reduction of CrO_3. It has an undistorted rutile structure (i.e., no M—M bonds as in MoO_2). It is ferromagnetic and has metallic conductance presumably due to delocalization of electrons into energy bands formed by overlap of metal d and oxygen $p\pi$ orbitals.

Chromium(VI) oxide, CrO_3, can be obtained as an orange-red precipitate on adding sulfuric acid to solutions of Na or K dichromate. The red solid is thermally unstable above its melting point, 197°, losing oxygen to give Cr_2O_3 after various intermediate stages. It is readily soluble in water and highly poisonous.

Interaction of CrO_3 and organic substances is vigorous and may be explosive. However, CrO_3 is widely used in organic chemistry as an oxidant, commonly in acetic acid as solvent.[2] The mechanism has been much studied and is believed to proceed initially by the formation of chromate esters (when pure, they are highly explosive) which undergo C—H bond cleavage as the rate-determining step to give Cr^{IV} as the first product; the general scheme appears to be

$$H_2A + Cr^{VI} \rightleftharpoons Cr^{IV} + A \text{ (slow)}$$
$$Cr^{IV} + Cr^{VI} \rightleftharpoons 2Cr^V$$
$$Cr^V + H_2A \rightleftharpoons Cr^{III} + A$$

The crystal structure of CrO_3 consists of infinite chains of CrO_4 tetrahedra sharing corners.[3]

Other Binary Compounds. The chromium sulfide system is very complex, with two forms of Cr_2S_3 and several intermediate phases between these and CrS.[4] Rhombohedral Cr_2S_3 has complex electrical and magnetic properties.[5]

[1] C. P. Poole, Jr., and D. S. MacIver, *Adv. Catalysis*, 1967, **17**, 223; R. L. Burwell *et al.*, *Adv. Catalysis*, 1969, **20**, 2; P. W. Selwood, *J. Amer. Chem. Soc.*, 1970, **92**, 39.
[2] See, e.g., K. B. Wiberg and S. K. Mukherjee, *J. Amer. Chem. Soc.*, 1971, **93**, 2543; A. K. Aswathy and J. Rocek, *J. Amer. Chem. Soc.*, 1969, **91**, 991; L. F. Fieser and M. Fieser, *Reagents for Organic Synthesis*, Wiley, Vols. 1 and 2, 1967, 1969.
[3] J. S. Stephens and D. W. J. Cruickshank, *Acta Cryst.*, 1970, *B*, **26**, 222.
[4] A. W. Sleight and T. A. Bither, *Inorg. Chem.*, 1969, **8**, 566; T. J. A. Popma and C. F. van Bruggen, *J. Inorg. Nuclear Chem.*, 1969, **31**, 73; F. Hulliger, *Structure and Bonding*, 1968, **4**, 83.
[5] C. F. van Bruggen, M. B. Vellinga, and C. Haas, *J. Solid State Chem.*, 1970, **2**, 303.

25-C-3. The Chemistry of Chromium(II), d^4

The Chromous Ion.   Aqueous solutions of the Cr^{2+} ion, which is sky-blue in color, are best prepared by dissolving electrolytic Cr metal in dilute mineral acids, but they can be prepared by reducing Cr^{III} solutions with zinc amalgam or electrolytically. Various hydrated salts can be crystallized from these solutions, examples being $Cr(ClO_4)_2 \cdot 6H_2O$, $CrCl_2 \cdot 4H_2O$ and $CrSO_4 \cdot 5H_2O$.

The Cr^{2+} ion is readily oxidized:

$$Cr^{3+} + e = Cr^{2+} \qquad E^0 = -0.41 \text{ V}$$

and the solutions must be protected from air—even then, they decompose at rates varying with the acidity and the anions present, by reducing water with liberation of hydrogen.

The mechanisms of reductions of other ions by Cr^{2+} have been extensively studied especially by H. Taube and his co-workers, since the resulting Cr^{3+} complex ions are substitution-inert. Much information regarding ligand-bridged transition states (page 675) has been obtained in this way. The products of oxidation of Cr^{2+} depend on the nature of the oxidant. With one-electron oxidants, the ions $[Cr(H_2O)_6]^{3+}$ or $[Cr(H_2O)_5X]^{2+}$, where X is derived from the oxidant, are the usual products. With two-electron oxidants there is the possibility of oxidation by the route:

$$Cr^{II} + 2e \rightarrow Cr^{IV}$$
$$Cr^{II} + Cr^{IV} \rightarrow Cr^{III}$$

and in some cases it is known that intermediate species (possibly binuclear) are formed.[6] Thus in the action of O_2 on perchlorate solutions of Cr^{2+} an initial CrO_2Cr group may be formed which is protonated to give $[(H_2O)_4Cr(OH)_2Cr(H_2O)_4]^{4+}$; this in turn is converted into the final product $[Cr(H_2O)_6]^{3+}$; all the O_2 atoms consumed appear in the latter.

The use of Cr^{2+} in ClO_4^- or BF_4^- solutions for the reduction of alkyl halides and other organic substances has been extensively studied, in part because Cr^{III} species with a Cr—C bond may be thereby obtained.[7] Thus reduction of $CHCl_3$ gives $[Cr(CHCl_2)(H_2O)_5]^{2+}$; it appears that in such reductions a free-radical rather than an electron-transfer mechanism may be operating. Alkyl halides can additionally be reduced to alkanes, particularly if aqueous DMF is used as solvent and ethylenediamine is present;[8] the intermediate alkyl, e.g., $[Cr^{III}R\ en_2(H_2O)]^{2+}$, is hydrolytically and proto-lytically unstable, giving rise to RH.

Another example of trapping is the reduction of polysulfide ion by Cr^{2+} whereby the green ion $[CrSH(H_2O)_5]^{2+}$ is formed.[9]

[6]   See, e.g., R. C. Thompson and G. Gordon, *Inorg. Chem.*, 1966, **5**, 557, 562.
[7]   J. R. Hanson and E. Premuzic, *Angew. Chem. Internat. Edn.*, 1968, **7**, 247 (review); M. D. Johnson *et al.*, *J. Chem. Soc.*, A, **1970**, 507, 511, 517, 523; D. D. Davis and W. B. Bigelow, *J. Amer. Chem. Soc.*, 1970, **92**, 5127; W. Schmidt, J. H. Swinehart and H. Taube, *J. Amer. Chem. Soc.*, 1971, **93**, 1117.
[8]   J. K. Kochi and J. W. Powers, *J. Amer. Chem. Soc.*, 1970, **92**, 137.
[9]   M. Ardon and H. Taube, *J. Amer. Chem. Soc.*, 1967, **89**, 3661.

Chromous Complexes. Comparatively few Cr^{II} halide complexes are known. The halides form adducts with ammonia, e.g., $CrCl_2 \cdot nNH_3$ ($n = 1$–6) and with nitriles, e.g., $CrCl_2 \cdot 2CH_3CN$, and the halogeno complex ions, $KCrF_3$ (distorted perovskite structure) and K_2CrCl_4 have also been obtained.

Reaction of ethanolic solutions of $CrCl_2$ with ethylenediamine[10] leads to salts of $[Cr\ en_3]^{2+}$ and $[Cr\ en_2Cl_2]^+$, but the diamine is readily lost in aqueous solutions. Other complexes are $K_4[Cr(NCS)_6]$, $K_4[Cr(CN)_6]$ and $[Cr(CO)_2diars_2X]X$.

Five-coordinate complexes with distorted trigonal-bipyramidal geometry are formed with several tripod ligands (page 625), examples[11] being $[Cr(Me_6tren)Br]^+$, $[Cr(PN_3)I]^+$ and

$$[Cr(PN_3)Br]^+ \quad \{PN_3 = (Et_2NCH_2CH_2)_2N(CH_2CH_2PPh_2)\}.$$

These are high-spin complexes with the two expected d—d transitions $e'' \to a_1$ and $e' \to a_1$.

Chromium(II) forms several dinuclear species with quadruple interactions between the metal atoms. Among these is the acetate, $Cr_2(O_2CCH_3)_4(H_2O)_2$, one of the most familiar and stable of chromous compounds, which is precipitated as a red solid when a Cr^{2+} solution is added to a solution of sodium acetate. Its structure is typical of carboxylato-bridged dinuclear complexes (page 550) and has the H_2O molecules as end groups.[12] The Cr—Cr distance, 2.36 Å, suggests that there is a strong, multiple interaction, involving one σ, two π, and one δ component. Such a quadruple interaction between the two d^4 metal ions accounts for the diamagnetism of the compound. A number of other essentially diamagnetic Cr^{II} carboxylates are known and presumably have the same dinuclear structure.

Perhaps the most remarkable dinuclear chromous compounds are the allyl, $Cr_2(C_3H_5)_4$, with a reported Cr—Cr distance[13] of 1.97 Å, and the anion in $Li_4Cr_2(CH_3)_8 \cdot 4C_4H_8O$ which is reported to have the $Re_2X_8^{2-}$ type of structure, with a Cr—Cr distance of 1.98 Å.[14] Evidently the d^4 Cr^{II} atoms have again combined *via* a quadruple bond, but the shortness of the distances is certainly arresting.

Electronic Structure of Cr^{II} Compounds. In an octahedral environment two electron distributions, $t_{2g}^3 e_g$ and t_{2g}^4, are possible. The magnetic data available show that in general chromium(II) compounds are of the high-spin type. The Curie–Weiss law is usually obeyed and the moments are ~ 4.95 B.M., i.e., close to the spin-only value. Aside from the alkanoates and the benzoate, there is a red form of the formate which is nearly diamagnetic, but a blue paramagnetic formate exists also; the structures are probably different from that of the acetate (see above). The $[Cr(CN)_6]^{4-}$ ion has a moment of only ~ 3.2 B.M. and is thus a low-spin complex.

[10] A. Earnshaw, L. F. Larkworthy and K. C. Patel, *J. Chem. Soc., A*, **1969**, 1339.
[11] F. Mani and L. Sacconi, *Inorg. Acta Chim.*, **4**, 365.
[12] F. A. Cotton *et al.*, *J. Amer. Chem. Soc.*, 1970, **92**, 3801.
[13] T. Aoki *et al.*, *Bull. Chem. Soc. Japan*, 1969, **42**, 545.
[14] J. Krausse and G. Schödl, *J. Organometallic Chem.*, 1971, **27**, 59.

For the mononuclear high-spin complexes only one spin-allowed absorption band, an $^5E_g \rightarrow {}^5T_{2g}(t_{2g}{}^3 e_g \rightarrow t_{2g}{}^2 e_g^2)$ transition, is to be expected in O_h symmetry. The blue color of the aquo Cr^{II} ion is attributable to the existence of such a band, which is rather broad, at about 700 mμ. However, because of the distortion of the octahedron, to be discussed below, the band is actually attributable to several nearly superposed transitions, and there is also another band in the near infrared.

As noted above (page 592), an ion with a d^4 high-spin configuration should cause Jahn–Teller distortion of an octahedral environment. In several instances precise X-ray studies have shown marked distortions, of the type found so commonly in Cu^{II} compounds (page 912) where two ligands are much farther from the metal ion than are the other four. For example, in $CrCl_2$ there are four Cl^- at 2.39 Å and two at 2.90 Å, and quite similar distortions have been observed in CrF_2, $CrBr_2$, and CrS. The compound Cr_2F_5 contains both Cr^{2+} and Cr^{3+} ions in octahedral environments, but the octahedra about the Cr^{2+} ion are highly distorted with four short (1.96–2.01 Å) and two long (2.57 Å) bonds.

25-C-4. The Chemistry of Chromium(III), d^3

Chromium(III) Complexes.[15]   There are literally thousands of chromium(III) complexes which, with a few exceptions, are all hexacoordinate. The principal characteristic of these complexes in aqueous solutions is their relative kinetic inertness.

Ligand-displacement reactions of Cr^{III} complexes are only about 10 times faster than those of Co^{III}, with half-times in the range of several hours. It is largely because of this kinetic inertness that so many complex species can be isolated as solids and that they persist for relatively long periods of time in solution, even under conditions where they are thermodynamically quite unstable.

The hexaquo ion, $[Cr(H_2O)_6]^{3+}$, which is regular octahedral, occurs in aqueous solution and in numerous salts such as the violet hydrate, $[Cr(H_2O)_6]Cl_3$, and other similar salts and in an extensive series of alums, $M^ICr(SO_4)_2 \cdot 12H_2O$. The chloride has three isomers, the others being the dark green *trans*-$[CrCl_2(H_2O)_4]Cl \cdot 2H_2O$, which is the normal commercially available salt, and pale green $[CrCl(H_2O)_5]Cl_2 \cdot H_2O$. The aquo ion is acidic (pK=4), and the hydroxo ion condenses to give a dimeric hydroxo bridged species:

$$[Cr(H_2O)_6]^{3+} \underset{H^+}{\overset{-H^+}{\rightleftharpoons}} [Cr(H_2O)_5OH]^{2+} \rightleftharpoons [(H_2O)_5Cr\underset{OH}{\overset{OH}{<>}}Cr(H_2O)_5]^{4+}$$

On addition of further base, soluble polymeric species of high molecular weight and eventually dark green gels are formed.

[15] J. E. Earley and R. D. Cannon, *Transition Metal Chem.*, 1965, **1**, 34 (aqueous chemistry of Cr^{III}).

Ammonia and Ammine Complexes[16] are the most numerous chromium derivatives and the most extensively studied. They include the pure ammines, $[CrAm_6]^{3+}$, the mixed ammine–aquo types, that is, $[CrAm_{6-n}(H_2O)_n]^{3+}$ ($n = 0$–4, 6), the mixed ammine–acido types, that is, $[CrAm_{6-n}R_n]^{(3-n)+}$ ($n = 1$–4, 6), and mixed ammine–aquo–acido types, for example, $[CrAm_{6-n-m}(H_2O)_nR_m]^{(3-m)+}$. In these general formulae, Am represents the monodentate ligand NH_3 or half of a polydentate amine such as ethylenediamine, and R represents an acido ligand such as a halide, nitro or sulfate ion. These ammine complexes provide examples of virtually all the kinds of isomerism possible in octahedral complexes.

The preparation of polyamine complexes sometimes presents difficulties in part due to the fact that in neutral or basic solution hydroxo or oxo-bridged polynuclear complexes are often formed. Such polyamines are often conveniently prepared from the Cr^{IV} peroxo species[17] noted below; thus the action of HCl on $[Cr^{IV}en(H_2O)(O_2)_2]\cdot H_2O$ forms the blue salt $[Cr^{III}en(H_2O)_2Cl_2]Cl$.

Some representative polynuclear complexes whose structures are fairly certain[17a] are shown below.

$$[(NH_3)_5Cr(OH)Cr(NH_3)_5]^{5+} \underset{H^+}{\overset{OH^-}{\rightleftharpoons}} [(NH_3)_5CrOCr(NH_3)_5]^{4+}$$

$$\downarrow H_2O \text{ (one day, } 100°) \qquad\qquad \downarrow OH^- \text{ (several days, } 25°)$$

$$[(NH_3)_5Cr(OH)Cr(NH_3)_4(H_2O)]^{5+} \xleftarrow{H^+} [(NH_3)_5Cr(OH)Cr(NH_3)_4(OH)]^{4+}$$

The oxo-bridged complex has a linear Cr—O—Cr group, indicating $d\pi$–$p\pi$ bonding as in other cases of M—O—M groups.

Anionic complexes are also common and are of the type $[CrX_6]^{3-}$ where X may be F^-, Cl^-, NCS^-, CN^-, but they may also have lower charges if neutral ligands are present as in the ion $[Cr(NCS)_4(NH_3)_2]^-$.[18] Complexes of bi- or poly-dentate anions are also known, one example being $[Cr\ ox_3]^{3-}$.

A different type of anionic complex is the dark blue $Cr_2Cl_9^{3-}$ which is made by the procedure used for similar compounds of V, Ti, Tl and W, namely, interaction of $CrCl_3$ and CsCl at 650–850°. The ion has three bridging Cl atoms and is similar to $W_2Cl_9^{3-}$ (page 961) except that the Cr^{3+} ions repel each other from the centers of their octahedra and the magnetic moments are normal, indicating that there is no Cr—Cr bond.[19]

As expected, Cr^{III} can also form *complexes of other types*, including neutral complexes with β-diketonates and similar ligands, e.g., Cr acac$_3$ and $Cr(OCOCF_3)_3$.[20] It also forms *basic acetate* compounds that have an unusual structure. The basic unit is $[Cr_3O(CH_3COO)_6L_3]^+$ in which there is

[16] C. S. Garner and D. A. House, *Transition Metal Chem.*, 1970, **6**, 59; W. W. Fee, J. N. MacB. Harrowfield and W. G. Jackson, *J. Chem. Soc.*, A, **1970**, 2612.

[17] D. A. House, R. G. Hughes and C. S. Garner, *Inorg. Chem.*, 1967, **6**, 1077, 1519.

[17a] M. Yevitz and J. A. Stanko, *J. Amer. Chem. Soc.*, 1971, **93**, 1512.

[18] For other examples see M. A. Bennett, R. J. H. Clark and A. D. J. Goodwin, *Inorg. Chem.*, 1967, **6**, 1621.

[19] R. Saillant and R. A. D. Wentworth, *Inorg. Chem.*, 1969, **8**, 1226.

[20] D. W. A. Sharp *et al.*, *J. Chem. Soc.*, A, **1968**, 3110.

an equilateral triangle of Cr atoms with an O atom at the center. There are two bridging CH_3COO groups across each edge of the triangle. Finally, a molecule L, e.g., H_2O, py, etc., is coordinated to each Cr so that it has distorted octahedral coordination.[21] This oxygen-centered triangular structure appears to be a characteristic feature of M^{III} carboxylates and possibly also of sulfates and certain chloro complexes. It is, moreover, found for M^{III} oxoacetates of Fe, Mn, V, Ru and possibly Co^{III}.

The coordination number 3 occurs in dialkylamides,[22] e.g., Cr[N(iso-$C_3H_7)_2]_3$; a combination of steric factors and multiple bonding has been proposed to explain the stability of such monomers.

Finally, as noted above, $CrCl_3$ forms numerous adducts with ethers, nitriles, amines and phosphines, which have formulas $CrCl_3 \cdot 2L$ or $CrCl_3 \cdot 3L$. The adduct $CrCl_3 \cdot 2NMe_3$ provides one of the very few examples of authenticated non-octahedral Cr^{III} complexes, as here X-ray studies confirm the trigonal-bipyramidal structure with axial amine groups.[23] Halide-bridged complexes such as $[CrCl_3(PR_3)_2]_2$ have been prepared by direct interaction.[24]

Electronic Structures of Chromium(III) Complexes.[25]   The magnetic properties of the octahedral Cr^{III} complexes are uncomplicated. From the simple orbital-splitting diagram (page 563) it follows that all such complexes must have three unpaired electrons irrespective of the strength of the ligand field, and this has been confirmed for all known mononuclear complexes. More sophisticated theory further predicts that the magnetic moments should be very close to, but slightly below, the spin-only value of 3.88 B.M.; this, too, is observed experimentally.

The spectra of Cr^{III} complexes are also well understood in their main features. A partial energy-level diagram is shown in Fig. 25-C-1. It is seen that three spin-allowed transitions are expected, and these have been observed in a considerable number of complexes. Indeed, the spectrochemical series was originally established by Tsuchida using data for Cr^{III} and Co^{III} complexes. In the aquo ion, the bands are found at 17,400, 24,700 and 37,000 cm^{-1}.

Ruby, natural or synthetic, is α-Al_2O_3 containing occasional Cr^{III} ions in place of Al^{III} ions. The environment of the Cr^{III} in ruby is thus a slightly distorted (D_{3d}) octahedron of oxide ions. The frequencies of the spin-allowed bands of Cr^{III} in ruby indicate that the Cr^{III} ions are under considerable compression, since the value of Δ_0 calculated is significantly higher than in the $[Cr(H_2O)_6]^{3+}$ ion or in other oxide lattices and glasses. Also, in ruby, spin-forbidden transitions from the $^4A_2$ ground state to the doublet states arising from the ^2G state of the free ion are observed. The transitions to the ^2E and $^2T_1$ states give rise to extremely sharp lines because the slopes

[21] L. Dubicki and R. L. Martin, *Austral. J. Chem.*, 1969, **22**, 701.

[22] D. C. Bradley *et al.*, *Chem. Comm.*, **1968**, 495; **1971**, 411.

[23] J. S. Wood, *Inorg. Chem.*, 1968, **7**, 852.

[24] M. A. Bennett, R. J. H. Clark and A. D. J. Goodwin, *J. Chem. Soc.*, A, **1970**, 541.

[25] L. S. Forster, *Transition Metal Chem.*, 1969, **5**, 1; J. R. Perumareddi, *Coordination Chem. Rev.*, 1969, **4**, 73; C. Furlani, *Coordination Chem. Rev.*, 1966, **1**, 51.

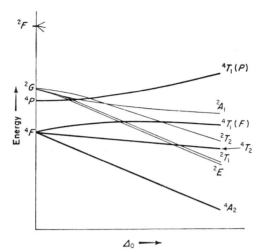

Fig. 25-C-1. Partial energy-level diagram for a d^3 ion in an octahedral field (also for a d^7 ion in a tetrahedral field). The quartet states are drawn with heavier lines.

of the energy lines for these states are the same as that for the ground state (except in extremely weak fields). This relationship is explained more fully on page 581 in connection with Mn^{II} complexes.

The same doublet states play a key role in the operation of the ruby laser. In this device a large single crystal of ruby is irradiated with light of the proper frequency to cause excitation to the $^4T_2(F)$ state. The exact magnitudes of certain energy differences and relaxation times are such, in the ruby, that the system rapidly makes a radiationless transition (i.e., by loss of energy to the crystal lattice in the form of vibrations) to the ^2E and $^2T_1$ states, instead of decaying directly back to the ground state. The systems then return from these doublet states to the ground state by stimulated emission of very sharp lines which are in phase with the stimulating radiation. Thus, bursts of extremely intense, monochromatic and coherent (all emitters in phase) radiation are obtained which are of use in communication and as sources of energy.

Organochromium(III) Complexes.[26] Mention has been made above of complexes with stable Cr—C bonds, but in addition there are a number of other alkyls and aryls of formula CrR_3L_n where L is usually an ether molecule such as tetrahydrofuran. These are obtained by action of lithium alkyls or Grignard reagents on $CrCl_3 \cdot 3THF$. The alkyls are rather unstable and their modes of decomposition have been much studied,[26a] but the aryls such as $Cr(C_6H_5)_3 \cdot 3THF$ are considerably more stable. Some anionic methyl complex anions, e.g., $Li_3[Cr(CH_3)_6]$, are also reasonably stable.[14]

[26] *Gmelin's Handbuch der anorganischen Chemie*, 8th edn., Main Suppl. Vol. 3, *Organometallic Compounds of Chromium*, Verlag Chemie, 1971.
[26a] R. P. A. Sneedon and H. H. Zeiss, *J. Organometallic Chem.*, 1970, **22**, 713; 1971, **26**, 101.

25-C-5. The Chemistry of Chromium(IV), d^2

Chromium(IV) compounds are perhaps slightly more numerous and stable than those of chromium(V), but still relatively rare. Some mixed oxides, $M_2^{II}CrO_4$, $M_3^{II}CrO_5$ and $M_4^{II}CrO_6$ are known but only the blue-black $M_2^{II}CrO_4$ compounds containing Ba and Sr are well characterized. These air-stable compounds contain discrete CrO_4^{4-} ions with magnetic moments of ~ 2.8 B.M.; they are made, e.g., by the reaction

$$SrCrO_4 + Cr_2O_3 + 5Sr(OH)_2 \xrightarrow{1000°} 3Sr_2CrO_4 + 5H_2O$$

Complexes. The most common compounds are peroxo compounds discussed below. Fluorochromates(IV),[27] e.g., K_2CrF_6 and $KCrF_5$ may be obtained by fluorination of stoichiometric amounts of $CrCl_3$ and KCl; both probably contain octahedral Cr^{IV}.

The most unusual Cr^{IV} compounds are the species with Cr—C, Cr—N and Cr—O bonds, exemplified by $Cr(CH_2SiMe_3)_4$,[28] $Cr(NEt_2)_4$ and $Cr(OBu^t)_4$;[29] these are surprisingly stable, blue, volatile, monomeric, paramagnetic substances. The silyl compound is obtained by oxidation of the Cr^{III} species obtained by action of Me_3SiCH_2MgCl on $CrCl_3 \cdot 3THF$, and the dialkyl-amides are similarly obtained by using $LiNR_2$; the alkoxides can be made from the dialkylamides by action of alcohols. The magnetic and electronic absorption properties are consistent with a distorted tetrahedral structure.

25-C-6. The Chemistry of Chromium(V), d^1

This oxidation state is not well known in simple compounds, but electron-spin resonance spectra suggest that many oxide lattices containing chromium may, when suitably oxidized or reduced, contain Cr^{5+}; such ions are believed to be the active sites in polymerization of ethylene to give polyethylene over chromium-containing alumina catalysts (cf. Chapter 24).

The oxofluoride, $CrOF_3$, has been prepared in impure form by the action of ClF_3, BrF_3 or BrF_5 on CrO_3 or $K_2Cr_2O_7$. $CrOCl_3$ is made by reaction of CrO_3 with $SOCl_2$. Some halo and oxohalo complexes are also known. The moisture-sensitive fluorooxochromates(V), $KCrOF_4$ and $AgCrOF_4$, can be obtained by treating CrO_3 mixed with KCl or $AgCl$ with BrF_3. Chlorooxo compounds with the general formula $M_2^I[CrOCl_5]$ are obtained by reduction of CrO_3 with concentrated hydrochloric acid in the presence of alkali metal ions at $0°$.

Alkali and alkaline earth chromates(V) have been prepared. They are dark green hygroscopic solids that hydrolyze with disproportionation to Cr^{III} and Cr^{VI}. Na_3CrO_4 has a magnetic susceptibility corresponding to one unpaired electron, and both Li_3CrO_4 and Na_3CrO_4 as well as the $M_3^{II}(CrO_4)_2$ compounds appear to contain discrete, tetrahedral CrO_4^{3-} ions.

[27] G. Siebert and W. Hoppe, *Naturwiss.*, 1970, **58**, 95.
[28] G. Yagupsky *et al.*, *Chem. Comm.*, **1970**, 1369; J. C. S. Dalton, 1972 (in press).
[29] For references see D. C. Bradley *et al.*, *J. Chem. Soc.*, A, **1971**, 772, 1433.

The only evidence for a persisting Cr^V species in solution was obtained by dissolution of chromates(VI) in 65% oleum. The quantity of O_2 evolved and the magnetic properties of the blue solution were consistent with the formation of Cr^V, but the nature of the species is uncertain.

25-C-7. The Chemistry of Chromium(VI), d^0

Chromate and Dichromate Ions. In basic solutions above pH 6, CrO_3 forms the tetrahedral yellow *chromate* ion, CrO_4^{2-}; between pH 2 and pH 6, $HCrO_4^-$ and the orange-red *dichromate* ion $Cr_2O_7^{2-}$ are in equilibrium; and at pH's below 1 the main species is H_2CrO_4. The equilibria are the following:

$$HCrO_4^- \rightleftharpoons CrO_4^{2-} + H^+ \qquad K = 10^{-5.9}$$
$$H_2CrO_4 \rightleftharpoons HCrO_4^- + H^+ \qquad K = 4.1$$
$$Cr_2O_7^{2-} + H_2O \rightleftharpoons 2HCrO_4^- \qquad K = 10^{-2.2}$$

In addition there are the base-hydrolysis equilibria:

$$Cr_2O_7^{2-} + OH^- \rightleftharpoons HCrO_4^- + CrO_4^{2-}$$
$$HCrO_4^- + OH^- \rightleftharpoons CrO_4^{2-} + H_2O$$

which have been studied kinetically for a variety of bases.[30]

The pH-dependent equilibria are quite labile and on addition of cations that form insoluble chromates, e.g., Ba^{2+}, Pb^{2+}, Ag^+, the chromates and not the dichromates are precipitated. Further, the species present depend on the nature of the acid used, and only for HNO_3 and $HClO_4$ are the equilibria as given above. When hydrochloric acid is used, there is essentially quantitative conversion into the chlorochromate ion, while with sulfuric acid a sulfato complex results:

$$CrO_3(OH)^- + H^+ + Cl^- \rightarrow CrO_3Cl^- + H_2O$$
$$CrO_3(OH)^- + HSO_4^- \rightarrow CrO_3(OSO_3)^{2-} + H_2O$$

Orange potassium chlorochromate can be prepared simply by dissolving $K_2Cr_2O_7$ in hot $6M$ HCl and crystallizing. It can be recrystallized from HCl but is hydrolyzed by water:[31]

$$CrO_3Cl^- + H_2O \rightarrow CrO_3(OH)^- + H^+ + Cl^-$$

The potassium salts of CrO_3F^-, CrO_3Br^- and CrO_3I^- are obtained similarly. They owe their existence to the fact that dichromate, though a powerful oxidizing agent, is kinetically slow in its oxidizing action toward halide ions.

Acid solutions of dichromate are strong oxidants:

$$Cr_2O_7^{2-} + 14H^+ + 6e = 2Cr^{3+} + 7H_2O \qquad E^0 = 1.33 \text{ V}$$

The mechanism of oxidation of Fe^{2+} and other common ions by Cr^{VI} has been studied in detail; with one- and two-electron reductants, respectively, Cr^V and Cr^{IV} are initially formed.[32]

The chromate ion in basic solution, however, is much less oxidizing:

$$CrO_4^{2-} + 4H_2O + 3e = Cr(OH)_3(s) + 5OH^- \qquad E^0 = -0.13 \text{ V}$$

[30] See J. R. Pladziewicz and J. H. Espenson, *Inorg. Chem.*, 1971, **10**, 634.
[31] J. Y. Tong and R. L. Johnson, *Inorg. Chem.*, 1966, **5**, 1902.
[32] J. H. Espenson *et al.*, *J. Amer. Chem. Soc.*, 1970, **92**, 1880, 1884, 1889; G. P. Haight, T. J. Huang and B. Z. Shakhashiri, *J. Inorg. Nuclear Chem.*, 1971, **33**, 2168.

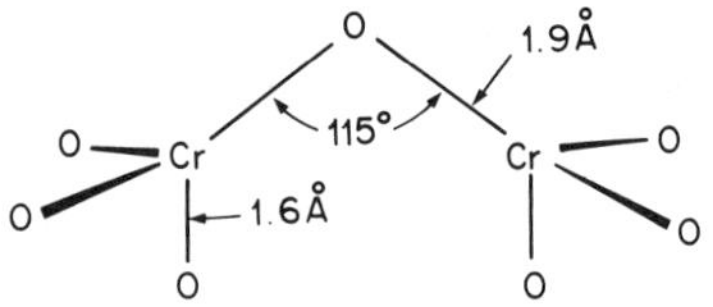

Fig. 25-C-2. The structure of the dichromate ion as found in $(NH_4)_2Cr_2O_7$.

Chromium(VI) does not give rise to the extensive and complex series of polyacids and polyanions characteristic of the somewhat less acidic oxides of V^V, Mo^{VI} and W^{VI}. The reason for this is perhaps the greater extent of multiple bonding, $Cr=O$, for the smaller chromium ion. Other than the chromate and dichromate ions there is no oxo acid or anion of major importance, although trichromates, $M_2^ICr_3O_{10}$, and tetrachromates, $M_2^ICr_4O_{13}$, have been reported. The dichromate ion in the ammonium salt has the structure shown in Fig. 25-C-2.

Oxo Halides. The most important oxo halide is chromyl chloride, CrO_2Cl_2, a deep red liquid (b.p. 117°). It is formed by the action of hydrogen chloride on chromium(VI) oxide:

$$CrO_3 + 2HCl \rightarrow CrO_2Cl_2 + H_2O$$

by warming dichromate with an alkali-metal chloride in concentrated sulfuric acid: $K_2Cr_2O_7 + 4KCl + 3H_2SO_4 \rightarrow 2CrO_2Cl_2 + 3K_2SO_4 + 3H_2O$

and in other ways. It is photosensitive but otherwise rather stable, although it vigorously oxidizes organic matter, sometimes selectively.[33] It is hydrolyzed by water to chromate ion and hydrochloric acid.

Chromyl fluoride, CrO_2F_2, made by fluorination of CrO_2Cl_2 or by the action of anhydrous HF on CrO_3, is a very stable red-brown gas.[34]

Other oxo halides, e.g., $CrOF_4$, are known; and fluorocarboxylates such as $CrO_2(OCOCF_3)_2$ can be made by reaction of CrO_3 with the carboxylic anhydrides.[35]

25-C-8. Peroxo Complexes of Chromium(IV), (V) and (VI)

Like other transition metals, notably Ti, V, Nb, Ta, Mo and W, chromium forms peroxo compounds in the higher oxidation states. They are all more or less unstable, both in and out of solution, decomposing slowly with the evolution of oxygen, and some of them are explosive or flammable in air. The main ones are the adducts of the deep blue chromium peroxide, CrO_5, the violet peroxochromates, the red peroxochromates, and the addition compounds of CrO_4.

When acid dichromate solutions are treated with hydrogen peroxide, a

[33] F. Freeman, P. D. McCart and N. J. Yamachika, *J. Amer. Chem. Soc.*, 1970, **92**, 4621; B. K. Sharpies and T. C. Flood, *J. Amer. Chem. Soc.*, 1971, **93**, 2316.
[34] W. V. Rochat, J. N. Gerlach and G. L. Gard, *Inorg. Chem.*, 1970, **9**, 999.
[35] J. N. Gerlach and G. L. Gard, *Inorg. Chem.*, 1970, **9**, 1565.

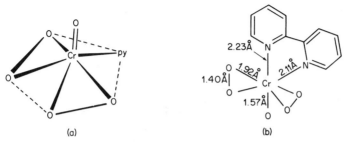

Fig. 25-C-3.   The structures of (a) CrO(O$_2$)$_2$·py and (b) CrO(O$_2$)$_2$·bipy. The coordination polyhedron of (a) is approximately a pentagonal pyramid with the oxide oxygen at the apex.

deep blue color rapidly appears but does not persist long. The overall reaction is

$$2HCrO_4^- + 3H_2O_2 + 8H^+ \rightarrow 2Cr^{3+} + 3O_2 + 8H_2O$$

but, depending on the conditions, the intermediate species may be characterized. At temperatures below 0°, green cationic species are formed:[36]

$$2HCrO_4^- + 4H_2O_2 + 6H^+ \rightarrow Cr_2(O_2)^{4+} + 3O_2 + 8H_2O$$
$$6HCrO_4^- + 13H_2O_2 + 16H^+ \rightarrow Cr_3(O_2)_2^{5+} + 9O_2 + 24H_2O$$

The blue species, which is one of the products at room temperature:

$$HCrO_4^- + 2H_2O_2 + H^+ \rightarrow CrO(O_2)_2 + 3H_2O$$

decomposes fairly readily, giving Cr^{3+}, but it may be extracted into ether where it is more stable and, on addition of pyridine to the ether solution, the compound pyCrO$_5$, a monomer in benzene and essentially diamagnetic, is obtained. These facts lead to the formulation of the blue species as aquo, ether or pyridine adducts of the molecule CrO(O$_2$)$_2$ containing Cr^{VI}. On decomposition or on reaction with organic substrates, the etherate CrO$_5$·Et$_2$O provides a source of oxygen, possibly singlet oxygen.[37] X-ray study[38] of the pyridine, 2,2'-bipyridine and 1,10-phenanthroline adducts confirms the bisperoxo formulation shown in Fig. 25-C-3. The second donor bond in the last two compounds is weak and not essential; when the chelate effect operates it is formed, otherwise not.

The action of H$_2$O$_2$ on neutral or slightly acidic solutions of K, NH$_4$ or Tl dichromate leads to diamagnetic, blue-violet, violently explosive salts. They are believed to contain the ion [Cr^{VI}O(O$_2$)$_2$OH]$^-$ and are thus related to CrO$_5$ since they contain the same number (2) of peroxo groups; CrO$_5$ can be converted into the violet salts merely by addition of OH$^-$.

On treatment of alkaline chromate solutions with 30% hydrogen peroxide —and after further manipulations—the red-brown peroxochromates, M$_3^I$CrO$_8$, can be isolated. They are paramagnetic with one unpaired electron per formula unit, and K$_3$CrO$_8$ forms mixed crystals with K$_3$NbO$_8$ and K$_3$TaO$_8$ in both of which the heavy metals are pentavalent. Thus it may be

[36] A. C. Adams et al., J. Amer. Chem. Soc., 1969, 90, 5761.
[37] H. W. S. Chan, Chem. Comm., 1970, 1550.
[38] R. Stromberg and I. Ainalem, Acta Chem. Scand., 1968, 22, 1439.

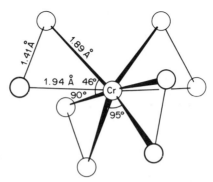

Fig. 25-C-4. The dodecahedral (D_{2d}) structure of the CrO_8^{3-} ion.

formulated as a quasi-dodecahedral (D_{2d}) tetraperoxo complex of Cr^V (Fig. 25-C-4).

When the reaction mixture used in preparing $(NH_4)_3CrO_8$ is heated to 50° and then cooled to 0°, brown crystals of $(NH_3)_3CrO_4$ are obtained. From this, on gentle warming with KCN solutions, $K_3[CrO_4(CN)_3]$ is obtained. An X-ray study of the ammonia compound has revealed the structure shown in Fig. 25-C-5; it has been suggested that it contains Cr^{II} coordinated by two superoxide ions; in view of the magnetic data which show the presence of two unpaired electrons, this would require the rather unlikely assumption that the Cr^{II} is here diamagnetic and it seems more natural to consider that the compound contains Cr^{IV}, with two unpaired electrons, coordinated by peroxide ions, which may, however, be abnormal.

Finally, the action of H_2O_2 on aqueous solutions of CrO_3 containing ethylenediamine or other amines gives chromium(IV) complexes such as the olive-green $[Cr\ en(H_2O)(O_2)_2]\cdot H_2O$. As noted above, these are useful sources for the preparation of Cr^{III} amine complexes.[39]

The bonding of the peroxo group in peroxo compounds of Ti, Cr, Nb and Ta has been discussed in terms of "bent" metal—oxygen bonds. The compounds for which there is information have dodecahedral (K_3CrO_8) or pentagonal-bipyramidal $[K_2W_{12}O_{11}, CrO_4(NH_3)_3, CrO_5py]$ structures with

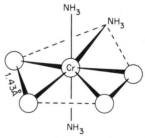

Fig. 25-C-5. The pentagonal bipyramidal structure of $(NH_3)_3CrO_4$.

[39] D. A. House, R. G. Hughes, and C. S. Garner, *Inorg. Chem.*, 1967, **6**, 1077, 1519; S. H. Caldwell and D. A. House, *Inorg. Chem.*, 1969, **8**, 151.

the peroxo group(s) forming three-membered rings with the metal atom. In both structures, the normal angles between metal–ligand orbitals would be 71–72°, while the peroxo group, with the O—M—O angle of 45°, lies between the metal–ligand directions. Thus the metal σ-orbital axes lie ca. 20° outside the M–O lines and the bonds are considered to be "bent." Similar "bent" bonds occur in other compounds containing three-membered rings, notably cyclopropane and ethylene oxide. This bonding implies a somewhat shortened O—O distance—cf. O—O in BaO_2, 1.49 Å, O—O in CrO_5py, 1.40 Å. These figures can be compared with the O—O distance of 1.30 Å found in $[(NH_3)_5CoO_2Co(NH_3)_5]^{5+}$ and $O_2IrClCO(PPh_3)_2$ (see pages 635 and 783) which is, however, characteristic of O_2.

It has been predicted that stable peroxo compounds will be formed where the normal ligand–bond angles are ca. 70°.

25-D. MANGANESE

As with Ti, V and Cr, the highest oxidation state of manganese corresponds to the total number of $3d$ and $4s$ electrons. This VII state occurs only in the oxo compounds MnO_4^-, Mn_2O_7 and MnO_3F, and these compounds show some similarity to corresponding compounds of the halogens, for example,

TABLE 25-D-1

Oxidation States and Stereochemistry of Manganese

Oxidn. state	Coordn. no.	Geometry	Examples
Mn^{-III}	4	Tetrahedral	$Mn(NO)_3CO$
Mn^{-II}	4 or 6	Square	$[Mn(phthalocyanine)]^{2-b}$
Mn^{-I}	5	Trigonal bipyramid	$Mn(CO)_5^-$, $[Mn(CO)_4PR_3]^-$
	4 or 6	Square	$[Mn(phthalocyanine)]^{-b}$
Mn^0	6	Octahedral	$Mn_2(CO)_{10}$
Mn^I, d^6	6	Octahedral	$Mn(CO)_5Cl$, $K_5[Mn(CN)_6]$, $[Mn(CNR)_6]^+$
Mn^{II}, d^5	4	Tetrahedral	$MnCl_4^{2-}$, $MnBr_2(OPR_3)_2$
	4	Square	$[Mn(H_2O)_4]SO_4 \cdot H_2O$, $Mn(S_2CNEt_2)_2$
	6^a	Octahedral	$[Mn(H_2O)_6]^{2+}$, $[Mn(SCN)_6]^{4-}$
	5	?	$[Mn(dienMe)X_2]$
		Trigonal bipyramid	$[Mn(trenMe_6)Br]Br^b$
	7	NbF_7^{2-} structure	$[Mn(EDTA)H_2O]^{2-}$
	8	Dodecahedral	$(Ph_4As)_2Mn(NO_3)_4^c$
Mn^{III}, d^4	5	Square pyramidal	$[bipyH_2][MnCl_5]$
	6^a	Octahedral	$Mn(acac)_3$, $[Mn\ ox_3]^{3-}$, MnF_3(distorted)
	7	?	$[Mn(EDTA)H_2O]^-$
Mn^{IV}, d^3	6	Octahedral	MnO_2, $Mn(SO_4)_2$, $MnCl_6^{2-}$
Mn^V, d^2	4	Tetrahedral	MnO_4^{3-}
Mn^{VI}, d^1	4	Tetrahedral	MnO_4^{2-}
Mn^{VII}, d^0	3	Planar	MnO_3^+
	4^a	Tetrahedral	MnO_4^-, MnO_3F

[a] Most common states.
[b] M. Ciampolini and N. Nardi, *Inorg. Chem.*, 1966, **5**, 1150.
[c] J. Drummond and J. S. Wood, *J. Chem. Soc. A*, **1970**, 226.

in the instability of the oxide. Manganese(VII) is powerfully oxidizing, usually being reduced to Mn^{II}. The intermediate oxidation states are known, but only a few compounds of Mn^V have been characterized; nevertheless, Mn^V species are frequently postulated as intermediates in the reduction of permanganates. Although Mn^{II} is the most stable state, it is quite readily oxidized in alkaline solution. The oxidation states and stereochemistry of manganese are summarized in Table 25-D-1.

25-D-1. The Element

Manganese is relatively abundant, constituting about 0.085% of the earth's crust. Among the heavy metals, only iron is more abundant. Although widely distributed, it occurs in a number of substantial deposits, mainly oxides, hydrous oxides or carbonate, and from all these, or the Mn_3O_4 obtained by roasting them, the metal can be obtained by reduction with aluminum.

Manganese is roughly similar to iron in its physical and chemical properties, the chief difference being that manganese is harder and more brittle but less refractory (m.p. 1247°). It is quite electropositive, and readily dissolves in dilute, non-oxidizing acids. It is not particularly reactive toward non-metals at room temperatures, but at elevated temperatures it reacts vigorously with many. Thus it burns in chlorine to give $MnCl_2$, reacts with fluorine to give MnF_2 and MnF_3, burns in nitrogen above 1200° to give Mn_3N_2, and combines with oxygen, giving Mn_3O_4, at high temperatures. It also combines directly with boron, carbon, sulfur, silicon and phosphorus but not with hydrogen.

MANGANESE COMPOUNDS[1]

In this Section we shall not discuss the extensive chemistry of carbonyl compounds such as $Mn_2(CO)_{10}$ or $Mn(CO)_5Cl$ or organometallic complexes such as $(h^5\text{-}C_5H_5)_2Mn$, or $h^5\text{-}C_5H_5Mn(CO)_3$. These have been dealt with in Chapters 22, 23 and 24.

25-D-2. The Chemistry of Divalent Manganese, d^5

This is the most important and generally is the most stable oxidation state for the element. In neutral or acid aqueous solution it exists as the very pale pink hexaquo ion, $[Mn(H_2O)_6]^{2+}$, which is quite resistant to oxidation as shown by the potentials

$$MnO_4^- \underline{\hspace{4cm}} Mn^{3+} \underline{\quad 1.5\ V \quad} Mn^{2+} \underline{\quad -1.18\ V \quad} Mn$$
$$\underline{\hspace{5cm}}$$
$$1.5\ V$$

---

[1] T. A. Zordan and L. G. Hepler, *Chem. Rev.*, 1968, **68**, 737 (thermodynamic properties and reduction potentials).

In basic media, however, the hydroxide $Mn(OH)_2$ is formed and this is very easily oxidized even by air, as shown by the potentials:

$$MnO_2 \cdot yH_2O \xrightarrow{-0.1 \text{ V}} Mn_2O_3 \cdot xH_2O \xrightarrow{-0.2 \text{ V}} Mn(OH)_2$$

Binary Compounds. Manganese(II) oxide is a grey-green to dark green powder made by roasting the carbonate in hydrogen or nitrogen. It has the rock-salt structure and is insoluble in water. Manganese(II) hydroxide is precipitated from Mn^{2+} solutions by alkali-metal hydroxides as a gelatinous white solid which rapidly darkens because of oxidation by atmospheric oxygen. $Mn(OH)_2$ is a well-defined compound—not an indefinite hydrous oxide—having the same crystal structure as magnesium hydroxide. It is only very slightly amphoteric:

$$Mn(OH)_2 + OH^- = Mn(OH)_3^- \qquad K \approx 10^{-5}$$

Manganous sulfide is a salmon-colored substance precipitated by alkaline sulfide solutions. It has a relatively high K_{sp} (10^{-14}) and redissolves easily in dilute acids. It is a hydrous form of MnS and becomes brown when left in air owing to oxidation. If air is excluded, the salmon-colored material changes on long storage, or more rapidly on boiling, into green, crystalline, anhydrous MnS.

MnS, MnSe and MnTe have the rock-salt structure. They are all strongly antiferromagnetic, as are also the anhydrous halides. The superexchange mechanism (page 602) is believed responsible for their antiferromagnetism.

Manganous Salts. Manganese(II) forms an extensive series of salts with all common anions. Most are soluble in water, although the phosphate and carbonate are only slightly so. Most of the salts crystallize from water as hydrates. With non-coordinating anions the salts contain $[Mn(H_2O)_6]^{2+}$, but $MnCl_2 \cdot 4H_2O$ contains cis-$MnCl_2(H_2O)_4$ units, while $MnCl_2 \cdot 2H_2O$ has polymeric chains.

The anhydrous salts must in general be obtained by dry reactions or by using non-aqueous solvents. Thus $MnCl_2$ is made by reaction of chlorine with the metal or of HCl with the metal, the oxide or the carbonate. The sulfate, $MnSO_4$, is obtained on fuming down sulfuric acid solutions. It is quite stable and may be used for manganese analysis provided no other cations giving non-volatile sulfates are present.

Manganous Complexes. Manganese(II) forms many complexes, but the equilibrium constants for their formation in aqueous solution are not high compared to those for the divalent cations of succeeding elements (Fe^{II}–Cu^{II}), as noted on page 596, because the Mn^{II} ion is the largest of these and it has no ligand-field stabilization energy in its complexes (except in the few of low spin). Many hydrated salts, $Mn(ClO_4)_2 \cdot 6H_2O$, $MnSO_4 \cdot 7H_2O$, etc., contain the $[Mn(H_2O)_6]^{2+}$ ion, and the direct action of ammonia on anhydrous salts leads to the formation of ammoniates, some of which have been shown to contain the $[Mn(NH_3)_6]^{2+}$ ion. Chelating ligands such as ethylenediamine, EDTA, oxalate ions, etc., form complexes isolable from aqueous solution.

Some EDTA complexes, e.g., $[Mn(OH_2)EDTA]^{2-}$, are 7-coordinate, while certain tridentate amines give 5-coordinate species.

In aqueous solution the formation constants for halogeno complexes are very low, e.g.,

$$Mn_{aq}^+ + Cl^- \rightleftharpoons MnCl_{aq}^+ \qquad K \approx 3.85$$

but when ethanol or acetic acid is used as solvent salts of complex anions of varying types may be isolated, such as[2]

MnX_3^- : octahedral with perovskite structure.

MnX_4^{2-} : tetrahedral (green-yellow) or polymeric octahedral with halide bridges (pink).

$MnCl_6^{4-}$: only Na and K salts known; octahedral.

The precise nature of the product obtained depends on the nature of the cation used and also on the halide and the solvent, but MnI_2 gives only MnI_4^{2-}. By contrast, the thiocyanates, $M_4^I[Mn(NCS)_6]$, can be crystallized as hydrates from aqueous solution. Salts of ions such as *trans*-$[MnCl_4(H_2O)_2]^{2-}$ and *trans*-$[Mn_2Cl_6(H_2O)_4]^{2-}$ are also known.

Manganese(II) *acetylacetonate* is trimeric, and octahedral coordination is probably achieved by sharing of oxygen atoms as in the corresponding Co^{II} and Ni^{II} complexes (pages 877, 899). It reacts readily with water and other donors, forming octahedral species, e.g. $Mn(acac)_2(H_2O)_2$.[3]

Although the normal coordination number of Mn^{II} is six, the MnX_4^{2-} ions are tetrahedral. Also Mn^{2+} ions are known to occupy tetrahedral holes in certain glasses and to substitute for Zn^{II} in ZnO. Tetrahedral Mn^{II} has a green-yellow color, far more intense than the pink of the octahedrally coordinated ion, and it very often exhibits intense yellow-green fluorescence. Most commercial phosphors are manganese-activated zinc compounds, wherein Mn^{II} ions are substituted for some of the Zn^{II} ions in tetrahedral surroundings, as for example in Zn_2SiO_4.

A wide variety of adducts of anhydrous manganese(II) halides with ligands such as py, MeCN, R_3P, R_3AsO, etc., mostly of stoichiometry MnX_2I_2, are known. Some are octahedral and some tetrahedral.

Electronic Spectra of Manganese(II) Compounds. As noted in Chapter 20 the high-spin d^5 configuration has certain unique properties; manganese(II), as the most prominent example of this configuration, has thus been discussed at some length above (page 578). Only a few brief remarks need be made here.

The majority of Mn^{II} complexes are high-spin. In octahedral fields, this configuration gives spin-forbidden as well as parity-forbidden transitions, thus accounting for the extremely pale color of such compounds. In tetrahedral environments, the transitions are still spin-forbidden but no longer parity-forbidden; these transitions are therefore about 10 times stronger and the compounds have a noticeable pale yellow-green color. The high-spin d^5

[2] J. J. Foster and N. S. Gill, *J. Chem. Soc.*, A, **1968**, 2625; see also R. Colton and J. H. Canterford, *Halides of the First Row Transition Metals*, Wiley, 1969.

[3] S. Onuma and S. Shibata, *Bull. Chem. Soc. Japan*, 1970, **43**, 2395.

configuration gives an essentially spin-only magnetic moment of ~ 5.9 B.M., which is temperature-independent.

At sufficiently high values of Δ_0 a t_{2g}^5 configuration gives rise to a doublet ground state; for Mn^{II} the pairing energy is high and only a few of the strongest ligand sets, e.g., those in $[Mn(CN)_6]^{4-}$, $[Mn(CN)_5NO]^{3-}$ and $[Mn(CNR)_6]^{2+}$, can accomplish this.

In the square environment provided by phthalocyanine, Mn^{II} has a $^4A_{1g}$ ground state.[4] A quartet ground state also occurs in the planar $Mn(S_2CNEt_2)_2$.[4a]

25-D-3. The Chemistry of Manganese(III), d^4

Binary Compounds. The oxides[5] are the most important compounds and, when any manganese oxide or hydroxide is heated at $1000°$, black crystals of Mn_3O_4, haussmannite, are formed. This is a spinel, $Mn^{II}Mn^{III}_2O_4$. When $Mn(OH)_2$ is allowed to oxidize in air, a hydrous oxide is formed which gives $MnO(OH)$ on drying. Although the mineral manganite, γ-$MnO(OH)$, formally contains Mn^{III}, magnetic measurements indicate that it contains Mn^{II} and Mn^{IV}. The mineral bixbyite, Mn_2O_3, does appear to contain Mn^{III}.

Manganese(III) occurs in a number of other mixed oxide systems. *Manganese(III) fluoride* is obtained on fluorination of $MnCl_2$ or other compounds and is a red-purple solid instantaneously hydrolyzed by water. It has been used as a fluorinating agent. The black trichloride, which decomposes above $-40°$ can be made by the action of HCl on Mn^{III} acetate at $-100°$.

The Manganese(III) Aquo Ion; Complexes.[6] The manganic ion[7] can be obtained by electrolytic or persulfate oxidation of Mn^{2+} solutions or by reduction of MnO_4^-. However, it cannot be obtained in high concentrations as it is reduced by water:

$$Mn^{3+} + e = Mn^{2+} \qquad E_0 = 1.54 \text{ V } (3M \text{ HClO}_4)$$

It also has a strong tendency to disproportionate:

$$2Mn^{3+} + 2H_2O = Mn^{2+} + MnO_2(s) + 4H^+ \qquad K \approx 10^9$$

The latter reaction can be slowed by the presence of an excess of Mn^{2+} and H^+ and is inappreciable for $[H^+] > 3M$. Hydrolysis also occurs at low acidity:

$$Mn^{3+} + H_2O = MnOH^{2+} + H^+ \qquad K \approx 1$$

The Mn^{III} state can also be stabilized in solution by sulfate, oxalate, pyrophosphate and some other complexing anions, but even the most stable of the resulting species, which is the possibly 7-coordinate complex

[4] C. G. Barraclough *et al.*, *J. Chem. Phys.*, 1970, **53**, 1638.
[4a] S. Lahity and V. K. Anand, *Chem. Comm.*, **1971**, 1111.
[5] See M. B. Robin and P. Day, *Adv. Inorg. Chem. Radiochem.*, 1967, **10**, 288.
[6] G. Davies, *Coordination Chem. Rev.*, 1969, **4**, 199.
[7] L. Ciavatta and M. Grimaldi, *J. Inorg. Nuclear Chem.*, 1969, **31**, 3071; G. Davies and K. Kustin, *Inorg. Chem.*, 1969, **8**, 484; C. F. Wells and C. Barnes, *J. Chem. Soc., A*, **1971**, 430.

anion $[MnEDTA(H_2O)]^-$ undergoes slow decomposition due to oxidation of the ligand.[8]

One of the best known compounds is the so-called "manganic acetate." This dark red substance is obtained as a hydrate by action of $KMnO_4$ on a hot solution of Mn^{II} acetate in glacial acetic acid; the "anhydrous" compound can be obtained by crystallization from acetic acid containing acetic anhydride. The structure has only recently been determined and it now appears that the acetate is a basic acetate similar to those of Fe^{III} and Cr^{III} (page 837). Thus it has the stoichiometry

$$[Mn_3O(OCOCH_3)_6]^+[OCOCH_3]^- \cdot CH_3COOH$$

and it has a unit with an oxygen atom in the center of a triangle of Mn atoms linked by acetate bridges.[9] The compound oxidizes olefins (to lactones) and also aromatic hydrocarbons such as toluene;[10] in addition, it catalyzes the decarboxylation of carboxylic acids.[11] The reactions all involve free radicals.

The dark brown crystalline *acetylacetonate*, $Mn(acac)_3$, is readily obtained by oxidation of basic solutions of Mn^{2+} by air or chlorine in the presence of acetylacetone. Like the acetate, it acts as an oxidant, coupling phenols[12] and, in presence of donors such as Me_2SO, initiating the free-radical polymerization of acrylonitrile and styrene.[13] Salts of the *halogeno complexes*, $MnCl_5^{2-}$ and MnF_5^{2-}, are also known; thus dissolution of $MnO(OH)$ in HF followed by addition of NH_4HF gives purple crystals of $(NH_4)_2MnF_5$; this ion is polymeric with distorted octahedra linked by sharing corners to form an infinite linear chain —Mn—F—Mn—, which results in antiferromagnetic behavior.[14]

Finally, it is of note that manganese(III) and manganese(IV) complexes are probably of great importance in photosynthesis, where oxygen evolution is dependent upon manganese. Some porphyrin complexes have been studied as models for the system.[15] There have been interesting findings with some oxygen-containing compounds: like the rather similar cobalt oxygen carriers (page 887), the manganese(II) complexes of the bis(salicylaldehyde) ethylenediiminato anion (salen) and the 3-methoxy derivative of this ligand react with oxygen to produce at least the following compounds:[16]

$[Mn^{III}salen]_2O_2$; peroxo-bridged, $\mu_{eff} = 2.79$ B.M.

$[Mn^{IV}salen\ O]_n$; polymeric with Mn—O—Mn bridges, $\mu_{eff} = 1.97$ B.M.

$[Mn^{IV}O(3\text{-MeO-salen})]$; Mn=O bond; $\mu_{eff} = 3.68$ B.M.

[8] R. E. Hainin and M. A. Suwyn, *Inorg. Chem.*, 1967, **6**, 139, 142.

[9] L. W. Hessel and C. Romers, *Rec. Trav. Chim.*, 1969, **88**, 545.

[10] E. I. Heiba, R. M. Dessau and W. J. Koehl, Jr., *J. Amer. Chem. Soc.*, 1969, **91**, 138, 5905; J. R. Gilmore and J. M. Mellor, *J. Chem. Soc., C*, **1971**, 2355.

[11] J. M. Anderson and J. K. Kochi, *J. Amer. Chem. Soc.*, 1970, **92**, 2450.

[12] M. J. S. Dewar and T. Nakaya, *J. Amer. Chem. Soc.*, 1968, **90**, 7134.

[13] C. H. Bamford and A. N. Ferrar, *Chem. Comm.*, **1970**, 315.

[14] S. Emori *et al.*, *Inorg. Chem.*, 1969, **8**, 1385; A. J. Edwards, *J. Chem. Soc., A.*, **1971**, 2653.

[15] L. J. Boucher and H. K. Garber, *Inorg. Chem.*, 1970, **9**, 2644; L. J. Boucher, *J. Amer. Chem. Soc.*, 1970, **92**, 2725; *Coord. Chem. Rev.*, 1972, **7**, 289.

[16] T. Yarino *et al.*, *Chem. Comm.*, **1970**, 1317.

Electronic Structure of Mn^{III} Compounds.[17]   The ground state, $^5E_g(t_{2g}{}^3e_g)$, for octahedral Mn^{III} is subject to a Jahn–Teller distortion. Because of the odd number of e_g electrons this distortion should be appreciable (page 592), and it might be expected to resemble the distortions in Cr^{II} and Cu^{II} compounds, namely, a considerable elongation of two *trans*-bonds with little difference in the lengths of the other four. This has been observed, for example, in MnF_5^{2-} and in MnF_3 which have the same basic structure as VF_3 where each V^{3+} ion is surrounded by a regular octahedron of F^- ions, except that two Mn—F distances are 1.79 Å, two more are 1.91 Å, and the remaining two are 2.09 Å. There is also distortion of the spinel structure of Mn_3O_4; here Mn^{2+} ions are in tetrahedral interstices and Mn^{3+} ions in octahedral interstices; each of the latter tends to distort its own octahedron and the cumulative effect is that the entire lattice is distorted from cubic to elongated tetragonal. However, in the case of $Mn(acac)_3$ the arrangement of the six oxygen atoms does not show this kind of distortion, or indeed any large deviation from an octahedral disposition; the reason for this is not clear, but in this case the chelate rings with their π systems introduce a significant low symmetry (D_3) component into the ligand field and this may influence the operation of the Jahn–Teller effect here in a manner that remains to be investigated.

A simplified energy level diagram for d^4 systems is shown in Fig. 25-D-1. It is consistent with the existence of both high-spin and low-spin octahedral complexes. Because the next quintet state (^5F, derived from the d^3s configuration) lying $\sim 110,000$ cm^{-1} above the ^5D ground state of the free ion is of such high energy, only one spin-allowed absorption band ($^5E_g \rightarrow {}^5T_{2g}$) is to be expected in the visible region. For $[Mn(H_2O)_6]^{3+}$ and tris(oxalato)- and tris(acetylacetonato)-manganese(III) a rather broad band appears around

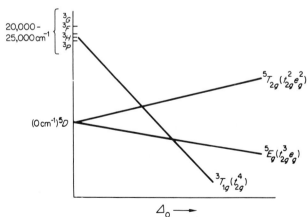

Fig. 25-D-1.   Simplified energy level diagram for the d^4 system, Mn^{III}, in octahedral surroundings.

[17] T. S. Davis, J. P. Fackler and M. J. Weeks, *Inorg. Chem.*, 1968, **7**, 1994.

20,000 cm^{-1} and the red or red-brown colors of high-spin Mn^{III} compounds may be attributed to such absorption bands. However, the spectra of some six-coordinate Mn^{III} complexes are not so simple and they are difficult to interpret in all their details, presumably because both static and dynamic Jahn–Teller effects perturb the simple picture based on fixed and perfect O_h symmetry.

The only low-spin manganese(III) compounds are salts of the $[Mn(CN)_6]^{3-}$ ion. Manganese(II) in the presence of an excess of CN^- is readily oxidized, even by a current of air, with the production of this ion which is first isolated from the solution as the Mn^{II} salt, $Mn_3^{II}[Mn(CN)_6]_2$, from which other salts are obtained. For $[Mn(CN)_6]^{3-}$, there appears to be no transition likely below a frequency where it would be obscured by strong ultraviolet bands, and none has been observed.

25-D-4. The Chemistry of Manganese(IV), d^3

Binary Compounds.   The most important compound is *manganese dioxide*, which is a grey to black solid occurring in ores such as pyrolusite, where it is usually non-stoichiometric. When made by the action of oxygen on manganese at a high temperature it has the rutile structure found for many other oxides, MO_2, e.g., those of Ru, Mo, W, Re, Os, Ir and Rh.[18] However, as normally made by heating $Mn(NO_3)_2 \cdot 6H_2O$ in air ($\sim 530°$) it is non-stoichiometric. A hydrated form is obtained by reduction of aqueous $KMnO_4$ in basic solution. Manganese(IV) occurs in a number of mixed oxides.

Manganese dioxide is inert to most acids except when heated, but it does not dissolve to give Mn^{IV} in solution; instead it functions as an oxidizing agent, the exact manner of this depending on the acid. With HCl, chlorine is evolved:

$$MnO_2 + 4HCl \rightarrow MnCl_2 + Cl_2 + 2H_2O$$

and this reaction is often used for small-scale generation of the gas in the laboratory. With sulfuric acid at 110°, oxygen is evolved and an Mn^{III} acid sulfate is formed. Hydrated manganese dioxide is used in organic chemistry for the oxidation of alcohols and other compounds.[19]

The *tetrafluoride*, MnF_4, obtained by direct interaction, is an unstable blue solid decomposing slowly to MnF_3 and F_2.

Salts and Complexes.   If $MnSO_4$ in sulfuric acid is oxidized with permanganate, black crystals of $Mn(SO_4)_2$ may be obtained on cooling. This substance is quickly hydrolyzed by treatment with water or dilute sulfuric acid, depositing hydrous MnO_2.

Complex salts include the chloro and fluoro salts, $M_2^I[MnX_6]$, which can be obtained by reduction of $KMnO_4$ in fuming HCl or liquid HF, respectively, with ether; there are also iodates, $M_2^I[Mn(IO_3)_6]$, and some curious glycerol complexes such as $Na_2[Mn(C_3H_5O_3)_2]$.

[18] D. B. Rogers *et al.*, *Inorg. Chem.*, 1969, **8**, 841.
[19] O. Meth-Cohn and H. Suschitzky, *Chem. and Ind.*, **1969**, 443 (a review).

Manganese(IV) has also been obtained in heteropolyniobate and hetero-polyvanadate ions, where it is known or assumed to have an octahedral environment.[20]

25-D-5. The Chemistry of Manganese(V), d^2

This is a little-known oxidation state and there are few authenticated examples. The so-called hypomanganates, MnO_4^{3-}, can be obtained as bright blue salts by reduction of MnO_4^- with an excess of SO_3^{2-}:

$$MnO_4^- + e = MnO_4^{2-} \qquad E = +0.56$$
$$MnO_4^{2-} + e = MnO_4^{3-} \qquad E = ca. +0.3$$

Alkali-metal salts are deliquescent and easily hydrolyzed. The MnO_4^{3-} ion is also formed by disproportionation when MnO_2 dissolves in concentrated KOH, and its spectra have been obtained when the ion is trapped in a lattice of Ca_2ClPO_4.[21]

25-D-6. The Chemistry of Manganese(VI), d^1, and -(VII), d^0

Manganese(VI) is known in only one environment, namely, as the deep green *manganate* ion, MnO_4^{2-}. This ion is formed on oxidizing MnO_2 in fused KOH with potassium nitrate, air or other oxidizing agent. Only two salts, K_2MnO_4 and several hydrated forms of Na_2MnO_4, have been isolated in completely pure condition. Both are very dark green, in fact nearly black, in color.

The manganate ion is stable only in very basic solutions. In acid, neutral or only slightly basic solutions it readily disproportionates according to the equation:

$$3MnO_4^{2-} + 4H^+ = 2MnO_4^- + MnO_2(s) + 2H_2O \qquad K \sim 10^{58}$$

Manganese(VII) is best known in the form of salts of the *permanganate* ion, MnO_4^-, of which the potassium salt is by far the commonest, being widely used as a laboratory oxidizing agent. $KMnO_4$ is manufactured on a large scale by electrolytic oxidation of a basic solution of potassium manganate. Aqueous solutions of MnO_4^- may be prepared by oxidation of solutions of Mn^{II} ion with very powerful oxidizing agents such as PbO_2 or $NaBiO_3$. The ion has an intense purple color, and crystalline salts appear almost black. Both crystalline $HMnO_4$ and a dihydrate, probably $H_5O_2^+MnO_4^-$ have been obtained at $-75°C$. The former is a violent oxidant for organic materials and itself decomposes violently above $3°$; it is slightly soluble in perfluorodecalin.[22]

Solutions of permanganate are intrinsically unstable, decomposing slowly but observably in acid solution:

$$4MnO_4^- + 4H^+ \rightarrow 3O_2(g) + 2H_2O + 4MnO_2(s)$$

[20] C. M. Flynn, Jr., and G. D. Stucky, *Inorg. Chem.*, 1969, **8**, 332, 335; C. M. Flynn, Jr., and M. T. Pope, *J. Amer. Chem. Soc.*, 1970, **92**, 85; *Inorg. Chem.*, 1970, **9**, 2009.

[21] J. Milsten and S. L. Holt, *Inorg. Chem.*, 1969, **8**, 1021.

[22] N. A. Frigerio, *J. Amer. Chem. Soc.*, 1969, **91**, 6200.

In neutral or slightly alkaline solutions in the dark, decomposition is immeasurably slow. It is, however, catalyzed by light so that standard permanganate solutions should be stored in dark bottles.

In basic solution permanganate functions as a powerful oxidizing agent:

$$MnO_4^- + 2H_2O + 3e = MnO_2(s) + 4OH^- \qquad E^0 = +1.23 \text{ V}$$

In very strong base and with an excess of MnO_4^-, however, manganate ion is produced:

$$MnO_4^- + e = MnO_4^{2-} \qquad E^0 = +0.56 \text{ V}$$

In acid solution permanganate is reduced to Mn^{2+} by an excess of reducing agent:

$$MnO_4^- + 8H^+ + 5e = Mn^{2+} + 4H_2O \qquad E^0 = +1.51 \text{ V}$$

but because MnO_4^- oxidizes Mn^{2+}:

$$2MnO_4^- + 3Mn^{2+} + 2H_2O = 5MnO_2(s) + 4H^+ \qquad E^0 = +0.46 \text{ V}$$

the product in presence of an excess of permanganate is MnO_2.

The mechanisms of oxidations by permanganate are very complicated and involve several bimolecular steps. One example is the oxidation of chloride ion[23] where MnO_4^- is first reduced to Mn^{3+}:

$$MnO_4^- + 4Cl^- + 8H^+ \rightarrow Mn^{3+} + 2Cl_2 + 4H_2O$$

The Mn^{3+} is only very slowly reduced further to Mn^{2+} unless a catalyst such as Cu^{2+} or Ag^+ is present. The rate law suggests that the rate-determining step involves the interaction of a chloro oxo species (cf. below) and Cl^-:

$$HMnO_3Cl^+ + Cl^- \rightarrow Cl_2 + HMn^VO_3$$

In the manganate ion there is one unpaired electron, but its behavior cannot be predicted by a ligand-field treatment because the overlap of metal and oxygen orbitals is too great. Instead, a molecular-orbital approach must be used. This is, of course, in no way surprising, for we surely could not expect to have an Mn^{6+} ion surrounded by O^{2-} ions without a great deal of electron density being drawn from the oxide ions into the Mn orbitals.

The permanganate ion has no unpaired electron, but it does have a small paramagnetism that is temperature-independent (see page 539).

Attempts to provide a complete description of the electronic structures of MnO_4^{2-} and MnO_4^-, capable of accounting for both their spectra and their magnetic properties have been made by using semi-empirical MO theory (see page 614). While some success has been achieved, there remain certain difficulties, some of which may be due to inherent inconsistencies in the methods of treatment. However, the fact that in MnO_4^{2-} and MnO_4^{3-} the extra electrons occupy antibonding orbitals implies that the Mn—O bond length should increase from Mn^{VI} to Mn^V; this has been shown to be so for $KMnO_4$ and K_2MnO_4 where the bond lengths are 1.63 and 1.66 Å, respectively,[24] the angles in both cases being close to the tetrahedral angle.

[23] K. J. Liv, H. Lester and N. C. Peterson, *Inorg. Chem.*, 1966, **5**, 2128.
[24] G. J. Palenik, *Inorg. Chem.*, 1967, **6**, 503, 507.

Manganese(VII) Oxide and Oxo Halides. The addition of small amounts of $KMnO_4$ to concentrated H_2SO_4 gives a clear green solution where the ionization

$$KMnO_4 + 3H_2SO_4 \rightleftharpoons K^+ + MnO_3^+ + H_3O^+ + 3HSO_4^-$$

appears to occur. The electronic absorption spectra are consistent with a planar trigonal ion MnO_3^+.

With larger amounts of $KMnO_4$, the explosive oil, Mn_2O_7 separates. This can be extracted into CCl_4 or chlorofluorocarbons in which it is reasonably stable and safe. By interaction of Mn_2O_7 and $ClSO_3H$, the green, volatile, explosive liquids MnO_3Cl and $MnOCl_3$ and the highly unstable brown MnO_2Cl_2 can be obtained.[25] Interaction of $KMnO_4$ and FSO_3H gives the green liquid MnO_3F which also explodes at room temperature.

25-E. IRON

With this element, the trends already noted in the relative stabilities of oxidation states continue, except that there is now no compound or chemically important circumstance in which the oxidation state is equal to the total number of valence-shell electrons, which in this case is eight. The highest oxidation state known is VI, and it is rare and of little importance. Even the trivalent state, which rose to a peak of importance at chromium, now loses ground to the divalent state. We shall see below that this trend continues, with the sole exception of Co^{III} which is stable in a host of complexes.

Iron compounds are amenable to study by a type of nuclear resonance spectroscopy that depends on a phenomenon known as the *Mössbauer effect*.[1] Although the Mössbauer effect has been observed for about one-third of the elements, only for iron and to a lesser extent tin has it been a major research tool for the chemist. With reference to iron, the effect depends on the fact that the nuclide ^{57}Fe, which is formed in the decay of ^{57}Co, has an excited state ($t_{1/2} \sim 10^{-7}$ sec) at 14.4 keV above the ground state; this can lead to a very sharp resonance absorption peak. Thus, if γ-radiation from the ^{57}Co source falls on an absorber where the iron nuclei are in an environment identical with that of the source atoms, then resonant absorption of γ-rays occurs. However, if the Fe nuclei are in a different environment, no absorption occurs and the radiation is transmitted and can be measured. In order to obtain resonant absorption, it is then necessary to impart a velocity to the absorber, relative to the source. This motion changes the energy of the incident quanta (Döppler effect), so that at a certain velocity there is correspondence with the excitation energy of the nuclei in the absorber. The shifts in the absorption position relative to stainless steel as arbitrary zero are custom-

[25] T. S. Briggs, *J. Inorg. Nuclear Chem.*, 1968, **30**, 2866.
[1] R. H. Herber, *Prog. Inorg. Chem.*, 1967, **8**, 1; V. I. Goldanskii and R. Herber, *Chemical Applications of Mössbauer Spectroscopy*, Academic Press, 1968; N. N. Greenwood in *Physical Chemistry, An Advanced Treatise*, Vol. 4, Chapter 12, Academic Press, 1970; N. N. Greenwood and T. C. Gibb, *Mössbauer Spectrometry*, Chapman Hall, 1971.

arily expressed in velocities (mm/sec) rather than in energies. The shift in the resonance absorption depends both on the chemical environment and on temperature.

The chemical or isomer shift (δ) is a linear function of electron density (due to electrons occupying s orbitals) at the nucleus. This, in turn, is influenced by many factors (oxidation state, spin state, s-character in the σ bonds, $d\pi$ back-bonding, ionicity), and correlations are not simple. For low-spin complexes δ is rather independent of oxidation state from -2 to $+2$. Even $Fe(CN)_6^{4-}$ and $Fe(CN)_6^{3-}$ have almost identical δ values. For high-spin compounds, however, δ varies markedly with formal oxidation state and indeed provides an excellent means of establishing it. A few examples of how Mössbauer spectra can be employed in studying the chemistry of iron will now be given; others will be mentioned as appropriate below.

In general, the Mössbauer spectra of compounds containing two or more iron atoms can give evidence for the occurrence of structurally non-equivalent iron atoms and in a number of instances there have been valuable applications of this kind. It is important to note that the converse procedure is highly dangerous, namely, concluding that all iron atoms must be equivalent when no resolution into separate peaks is observed. In certain cases, when the environment of the iron atom is unsymmetrical the consequent electric field gradient interacts strongly enough with the nuclear quadrupole to produce substantial splitting of the resonance and this observation can often be informative as to structure.

As implied, though not explicitly stated, in the opening description, Mössbauer spectra can only be recorded on nuclei bound in a rigid solid environment. It is therefore primarily used to study crystalline substances, though solutions that have been frozen to glasses are also suitable.

Prussian blue, formed from Fe^{3+} and $Fe(CN)_6^{4-}$, and Turnbull's blue, formed from Fe^{2+} and $Fe(CN)_6^{3-}$, have been shown to be identical, both having the composition $Fe_4[Fe(CN)_6]_3$, i.e., ferric ferrocyanide. High-spin Fe^{III} and low-spin Fe^{II} were each identified, indicating that individual iron atoms have distinct, well-defined electron configurations with lifetimes of at least 10^{-7} sec.

For a series of ions $[Fe(CN)_5L]^{n-}$ information about the π-acid character of the ligands L has been obtained, since π donation from Fe to L deshields the Fe nucleus. The strongest π acid has the highest s electron density at the nucleus and hence the smallest δ. The order of δ values was found to be $NO < CO < CN^- < SO_3^{2-} < Ph_3P < NO_2^- < NH_3$.

The oxidation states and stereochemistry of iron are summarized in Table 25-E-1.

25-E-1. The Element

Iron is the second most abundant metal, after aluminum, and the fourth most abundant element in the earth's crust. The core of the earth is believed

TABLE 25-E-1

Oxidation States and Stereochemistry of Iron

Oxidn. state	Coordn. no.	Geometry	Examples
Fe^{-II}	4	Tetrahedral	$Fe(CO)_4^{2-}$, $Fe(CO)_2(NO)_2$
Fe^0	5	Trigonal bipyramidal	$Fe(CO)_5$, $(Ph_3P)_2Fe(CO)_3$, $Fe(PF_3)_5$
	6	Octahedral (?)	$Fe(CO)_5H^+$, $Fe(CO)_4PPh_3H^+$
Fe^I, d^7	6	Octahedral	$[Fe(H_2O)_5NO]^{2+}$
Fe^{II}, d^6	4	Tetrahedral	$FeCl_4^{2-}$, $FeCl_2(PPh_3)_2$
	5	Trigonal bipyramidal	$[FeBr(Me_6tren)]Br$
	5	Square pyramidal	$[Fe(ClO_4)(OAsMe_3)_4]ClO_4$
	6^a	Octahedral	$[Fe(H_2O)_6]^{2+}$, $[Fe(CN)_6]^{4-}$
	8	Dodecahedral (D_{2h})	$[Fe(1,8\text{-naphthyridine})_4](ClO_4)_2$
Fe^{III}, d^5	3	Trigonal	$Fe[N(SiMe_3)_2]_3$
	4	Tetrahedral	$FeCl_4^-$, Fe^{III} in Fe_3O_4
	5	Square pyramidal	$FeCl(dtc)_2$,^{b} $Fe(acac)_2Cl$
	5	Trigonal bipyramidal	$Fe(N_3)_5^{2-}$
	6^a	Octahedral	Fe_2O_3, $[Fe(C_2O_4)_3]^{3-}$, $Fe(acac)_3$, $FeCl_6^{3-}$
	7	Approx. pentagonal bipyramidal	$[FeEDTA(H_2O)]^-$
	8	Dodecahedral	$[Fe(NO_3)_4]^-$
Fe^{IV}, d^4	6	Octahedral	$[Fe(diars)_2Cl_2]^{2+}$
	7	?	$FeH_4(PR_3)_3$ (?)
Fe^{VI}, d^2	4	Tetrahedral	FeO_4^{2-}

^{a} Most common states.
^{b} dtc = dithiocarbamate.

to consist mainly of iron and nickel, and the occurrence of many iron meteorites suggests that it is abundant throughout the solar system. The major iron ores are *hematite*, Fe_2O_3, *magnetite*, Fe_3O_4, *limonite*, $FeO(OH)$, and *siderite*, $FeCO_3$.

The technical production and metallurgy of iron will not be discussed here.[2] Chemically pure iron can be prepared by reduction of pure iron oxide (which is obtained by thermal decomposition of ferrous oxalate, carbonate or nitrate) with hydrogen, by electrodeposition from aqueous solutions of iron salts, or by thermal decomposition of iron carbonyl.

Pure iron is a white, lustrous metal, m.p. 1528°. It is not particularly hard, and it is quite reactive. In moist air it is rather rapidly oxidized to give a hydrous oxide which affords no protection since it flakes off, exposing fresh metal surfaces. In a very finely divided state, metallic iron is pyrophoric. It combines vigorously with chlorine on mild heating and also with a variety of other non-metals including the other halogens, sulfur, phosphorus, boron, carbon and silicon. The carbide and silicide phases play a major role in the technical metallurgy of iron.

The metal dissolves readily in dilute mineral acids. With non-oxidizing acids and in absence of air, Fe^{II} is obtained. With air present or when warm dilute nitric acid is used, some of the iron goes to Fe^{III}. Very strongly oxidizing media such as concentrated nitric acid or acids containing dichrom-

[2] See Gmelin-Durrer, *The Metallurgy of Iron*, Vols. 3a,b, Verlag Chemie, 1971.

ate passivate iron. Air-free water and dilute air-free hydroxides have little effect on the metal, but hot concentrated sodium hydroxide attacks it. In presence of air and water iron rusts to give a hydrated ferric oxide.[2a]

At temperatures up to 906° the metal has a body-centered lattice. From 906° to 1401°, it is cubic close-packed, but at the latter temperature it again becomes body-centered. It is ferromagnetic up to its Curie temperature of 768° where it becomes simply paramagnetic.

IRON COMPOUNDS

25-E-2. The Oxides of Iron

Because of the fundamental structural relationships between them, we discuss these compounds together, rather than separately under the different oxidation states. Three iron oxides are known. They all tend to be non-stoichiometric, but the ideal compositions of the phases are FeO, Fe_2O_3 and Fe_3O_4.

Iron(II) oxide is obtained by thermal decomposition of iron(II) oxalate in a vacuum as a pyrophoric black powder that becomes less reactive if heated to higher temperatures. The crystalline substance can only be obtained by establishing equilibrium conditions at high temperature and then rapidly quenching the system, since at lower temperatures FeO is unstable with respect to Fe and Fe_3O_4; slow cooling allows disproportionation. FeO has the rock-salt structure. The FeO referred to thus far is iron-defective (see below), having a typical composition of $Fe_{0.95}O$. Essentially stoichiometric FeO has been prepared from $Fe_{0.95}O$ and Fe at 1050°K and 50 kiloatm; it is about 0.4% less dense.[3]

The brown hydrous ferric oxide, FeO(OH), exists in several forms depending on the method of preparation, e.g., by hydrolysis of iron(III) chloride solutions at elevated temperatures or by oxidation of iron(II) hydroxide. When heated at 200° the final product is the red-brown $\alpha\text{-}Fe_2O_3$. This oxide occurs in Nature as the mineral hematite. It has the corundum structure where the oxide ions form a *hexagonally* close-packed array with Fe^{III} ions occupying octahedral interstices. However, by careful oxidation of Fe_3O_4 or by heating one of the modifications of FeO(OH) (lepidocrocite) one obtains another type of Fe_2O_3 called $\gamma\text{-}Fe_2O_3$. The structure of this phase may be regarded as a *cubic* close-packed array of oxide ions with the Fe^{III} ions distributed randomly over both the octahedral and the tetrahedral interstices.

Finally, there is Fe_3O_4, a mixed $Fe^{II}\text{-}Fe^{III}$ oxide which occurs in Nature in the form of black, octahedral crystals of the mineral magnetite. It can be made by ignition of Fe_2O_3 above 1400°. It has the inverse spinel structure

[2a] U. R. Evans, *Quart. Rev.*, 1967, **21**, 29.
[3] T. Katsura *et al.*, *J. Chem. Phys.*, 1967, **47**, 4559.

(page 55). Thus the Fe^{II} ions are all in octahedral interstices, whereas the Fe^{III} ions are half in tetrahedral and half in octahedral interstices of a cubic close-packed array of oxide ions. The electrical conductivity, which is 10^6 times that of Fe_2O_3 is probably due to rapid valence oscillation between the Fe sites.

25-E-3. Other Binary Compounds

Halides. Only iron(II) (ferrous) and iron(III) (ferric) halides are known. The anhydrous ones are:

FeF_3	$FeCl_3$	$FeBr_3$	—
FeF_2	$FeCl_2$	$FeBr_2$	FeI_2

The three ferric halides can be obtained by direct halogenation of the metal. The iodide does not exist in the pure state though some may be formed in equilibrium with FeI_2 and iodine. In effect, iron(III) is too strong an oxidizing agent to co-exist with such a good reducing agent as I^-. In aqueous solution Fe^{3+} and I^- react quantitatively:

$$Fe^{3+} + I^- \rightarrow Fe^{2+} + \tfrac{1}{2}I_2$$

The fluoride is white, having only spin-forbidden electronic transitions in the visible spectrum (cf. Mn^{II}) and no low-energy charge-transfer band. $FeCl_3$ and $FeBr_3$ are red-brown because of charge-transfer transitions. All three halides have non-molecular crystal structures with Fe^{III} ions occupying two-thirds of the octahedral holes in alternate layers. In the gas phase there are dimeric molecules, presumably consisting of tetrahedra sharing an edge, and monomers. All three will decompose to the Fe^{II} halides on strong heating in a vacuum.

Only iron(III) chloride is commonly encountered and then as the hexahydrate, yellow lumps obtained by evaporation of aqueous solutions on a steam-bath. Hydrates of the fluoride and bromide are also known.

The iron(II) halides are all known in both anhydrous and hydrated forms. The iodide and bromide can be prepared by reaction of the elements though iron must be present in excess in the case of $FeBr_2$. For FeF_2 and $FeCl_2$ it is necessary to use HF or HCl in order to avoid forming the trihalides. $FeCl_3$ may be reduced to $FeCl_2$ by heating it in hydrogen, by treatment of a tetrahydrofuran solution with an excess of iron filings or by refluxing it in chlorobenzene.

Iron dissolves in the aqueous hydrohalic acids and from these solutions the hydrated halides, $FeF_2 \cdot 8H_2O$ (colorless), $FeCl_2 \cdot 6H_2O$ (pale green), $FeBr_2 \cdot 4H_2O$ (pale green), may be crystallized. $FeCl_2 \cdot 6H_2O$ contains *trans*-$[FeCl_2(H_2O)_4]$ units.

Iron forms many binary compounds with the Group V and Group VI elements. Many are non-stoichiometric and/or interstitial. The sulfides are the most common.[4] Iron(II) sulfide (FeS) and FeS_2 have been discussed above (page 433). Iron(III) sulfide is an unstable amorphous black precipitate

[4] J. C. Ward, *Rev. Pure Appl. Chem.*, 1970, **20**, 175.

obtained by addition of sulfide ions to an aqueous solution of iron(III). The crystalline substance[4a] can be made by treating a suspension of iron(III) hydroxide with H_2S under pressure.

25-E-4. Aqueous and Coordination Chemistry of Iron(II), d^6

Iron(II) forms salts with virtually every stable anion, generally as green, hydrated, crystalline substances isolated by evaporation of aqueous solutions. The sulfate and perchlorate contain octahedral $[Fe(H_2O)_6]^{2+}$ ions. An important double salt is Mohr's salt, $(NH_4)_2SO_4 \cdot FeSO_4 \cdot 6H_2O$, which is fairly stable toward both air-oxidation and loss of water. It is commonly used in volumetric analysis to prepare standard solutions of iron(II) and as a calibration substance in magnetic measurements. Many other ferrous compounds are more or less susceptible to superficial oxidation by air and/or loss of water of crystallization, thus making them unsuitable as primary standards. This behavior is particularly marked for $FeSO_4 \cdot 7H_2O$, which slowly effloresces and becomes yellow-brown when kept.

Iron(II) carbonate, hydroxide and sulfide may be precipitated from aqueous solutions of ferrous salts. Both the carbonate and the hydroxide are white, but in presence of air they quickly darken owing to oxidation. The sulfide also undergoes slow oxidation.

$Fe(OH)_2$ is somewhat amphoteric. It readily redissolves in acids, but also in concentrated sodium hydroxide. If 50% NaOH is boiled with finely divided iron and then cooled, fine blue-green crystals of $Na_4[Fe(OH)_6]$ are obtained. The strontium and barium salts may be precipitated similarly.

Iron(II) hydroxide can be obtained as a definite crystalline compound having the brucite, $Mg(OH)_2$, structure.

Aqueous Chemistry. Aqueous solutions of iron(II), not containing other complexing agents, contain the pale blue-green hexaaquoiron(II) ion, $[Fe(H_2O)_6]^{2+}$. The potential of the Fe^{3+}–Fe^{2+} couple, 0.771 V, is such that molecular oxygen can convert ferrous into ferric ion in acid solution:

$$2Fe^{2+} + \tfrac{1}{2}O_2 + 2H^+ = 2Fe^{3+} + H_2O \qquad E^0 = 0.46 \text{ V}$$

In basic solution, the oxidation process is still more favorable:

$$\tfrac{1}{2}Fe_2O_3 \cdot 3H_2O + e = Fe(OH)_2(s) + OH^- \qquad E^0 = -0.56 \text{ V}$$

Thus, ferrous hydroxide almost immediately becomes dark when precipitated in presence of air and is eventually converted into $Fe_2O_3 \cdot nH_2O$.

Neutral and acid solutions of ferrous ion oxidize *less* rapidly with increasing acidity (despite the fact that the potential of the oxidation reaction becomes more positive). This is because Fe^{III} is actually present in the form of hydroxo complexes, except in extremely acid solutions, and there may also be kinetic reasons.

The oxidation of Fe^{II} to Fe^{III} in neutral solutions by molecular oxygen has been the subject of much speculation and may involve[5] a reaction

[4a] R. Schrader and C. Pietzsch, *Z. Chem.*, **1968**, 154.
[5] K. Goto, H. Tamura and M. Nagayama, *Inorg. Chem.*, 1970, **9**, 963

between $FeOH^+$ and O_2OH^-. The related problem of oxidation of Fe^{2+} by H_2O_2 (Fenton's reagent) is complicated and involves radicals generated by the reaction[6]

$$Fe^{II} + H_2O_2 \rightarrow Fe^{III}(OH) + OH$$

Complexes. Iron(II) forms a number of complexes, most of them octahedral. Ferrous complexes can normally be oxidized to ferric complexes and the Fe^{II}–Fe^{III} aqueous system provides a good example of the effect of complexing ligands on the relative stabilities of oxidation states:

$$[Fe(CN)_6]^{3-} + e = [Fe(CN)_6]^{4-} \qquad E^0 = 0.36 \text{ V}$$
$$[Fe(H_2O)_6]^{3+} + e = [Fe(H_2O)_6]^{2+} \qquad E^0 = 0.77 \text{ V}$$
$$[Fe(phen)_3]^{3+} + e = [Fe(phen)_3]^{2+} \qquad E^0 = 1.12 \text{ V}$$

The ferrous halides combine with gaseous ammonia, forming several ammoniates of which the highest are the hexammoniates that contain the ion, $[Fe(NH_3)_6]^{2+}$. Other anhydrous ferrous compounds also absorb ammonia. The ammine complexes are not stable in water, however, except in saturated aqueous ammonia. With chelating amine ligands, many complexes stable in aqueous solution are known. For example, ethylenediamine forms the entire series:

$$[Fe(H_2O)_6]^{2+} + en = [Fe(en)(H_2O)_4]^{2+} + 2H_2O \qquad K = 10^{4.3}$$
$$[Fe(en)(H_2O)_4]^{2+} + en = [Fe(en)_2(H_2O)_2]^{2+} + 2H_2O \qquad K = 10^{3.3}$$
$$[Fe(en)_2(H_2O)_2]^{2+} + en = [Fe(en)_3]^{2+} + 2H_2O \qquad K = 10^2$$

The complexes of 2,2'-bipyridine and 1,10-phenanthroline have been especially well studied.[7] β-Diketones (dike) form stable, inner-salt complexes, $Fe(dike)_2$; these appear to be polymeric with the exception of $Fe(Bu^tCOCHCOBu^t)_2$.

The brown-ring test for nitrates and nitrites depends on the fact that, under the conditions of the test, nitric oxide is generated. This combines with ferrous ion to produce a brown complex[8] $[Fe(H_2O)_5NO]^{2+}$.

The hexacyanoferrate(II) ion, commonly called ferrocyanide, is a very stable and well-known complex of iron(II). The free acid, $H_4[Fe(CN)_6]$, can be precipitated as an ether addition compound (probably containing oxonium ions, R_2OH^+) by adding ether to a solution of the ion in strongly acidic solution; the ether can then be removed to leave the acid as a white powder. It is a strong tetrabasic acid when dissolved in water; in the solid, the protons are bound to the nitrogen atoms of the CN groups with intermolecular hydrogen bonding (see page 722).

Several diphosphine complexes, trans-$[FeCl_2(diphos)_2]$, can be prepared and by treatment of these with $LiAlH_4$ in THF the hydrido complexes, trans-$[FeHCl(diphos)_2]$ and trans-$[FeH_2(diphos)_2]$, can be obtained. These hydrido complexes are easily oxidized by air but have good thermal stability.

Fe^{II} has a lower tendency to form *tetrahedral complexes* than has Co^{II} or Ni^{II}, but a number of these are known, and others doubtless can be made.

[6] For references see C. Walling *et al.*, *Inorg. Chem.* 1970, **9**, 931; *J. Amer. Chem. Soc.*, 1971, **93**, 4275.
[7] E. König, *Coordination Chem. Rev.*, 1968, **3**, 471.
[8] L. Burlamacchi, G. Martini and E. Tiezzi, *Inorg. Chem.*, 1969, **8**, 2021.

The $[FeX_4]^{2-}$ ions exist in salts with several large cations while neutral, FeL_2X_2, and cationic, FeL_4^{2+}, complexes with, e.g., $L = (Me_2N)_3PO$ or Ph_3PO, have been characterized. Iron(II) tetrahedrally coordinated by sulfur atoms occurs in the protein rubredoxin (page 873) and in $Fe(SPMe_2NPMe_2S)_2$.[9]

A moderate number of *five-coordinate complexes* are known. Those with tripod ligands (page 625), e.g., $[FeBr(Me_6tren)]^+$ or $[(Fe(QP)X]^+$ (X = Cl, Br, I, or NO_3), are either known to be or believed to be distorted trigonal-bipyramidal. $[Fe(OAsMe_3)_4ClO_4]ClO_4$ has square-pyramidal coordination. The *spy* (square-pyramidal) complexes are low-spin ($\mu \approx 3.1$ B.M.) while the others are high-spin. Only for the Me_6tren complex have both the structure and the electronic spectrum been well studied.[10]

Electronic Structures of Iron(II) Complexes. The ground state, ^5D, of a d^6 configuration is split by octahedral and tetrahedral ligand fields into $^5T_2$ and ^5E states; there are no other quintet states and hence only one spin-allowed *d–d* transition occurs if one of these is the ground state. All tetrahedral complexes are high-spin and the $^5E \rightarrow {}^5T_2$ band typically occurs at ~ 4000 cm^{-1}. The magnetic moments are normally 5.0–5.2 B.M., owing to the spins of the four unpaired electrons and a small, second-order orbital contribution. For high-spin octahedral complexes, e.g., $Fe(H_2O)_6^{2+}$, the $^5T_{2g} \rightarrow {}^5E_g$ transition occurs in the visible or near-infrared region ($\sim 10,000$ cm^{-1} for the aquo ion) and is broad or even resolvably split owing to a Jahn–Teller effect in the excited state, which derives from a $t_{2g}^3 e_g^3$ configuration. Magnetic moments are around 5.2 B.M. in magnetically dilute compounds.

For Fe^{II} quite strong ligand fields are required to cause spin-pairing but a number of low-spin complexes, such as $Fe(CN)_6^{4-}$, $Fe(CNR)_6^{2+}$ and $[Fe(phen)_3]^{2+}$ are known.[11] Fe phen$_2(CN)_2$ is also diamagnetic though most Fe phen$_2X_2$ complexes are high-spin. When X = SCN or SeCN a spin state cross-over situation occurs (cf. page 567), and the magnetic moment is temperature-dependent, ranging from ~ 5.1 B.M. at 300°K to ~ 1.5 B.M. at $\leqslant 150$°K.[12] Some other cross-over cases that have been well studied are $Fe(HBR_3)_2$, where R = 1-pyrazolyl,[13] and several complexes with 2-pyridylmethylamine and some similar ligands.[14]

While it can be shown that for strict octahedral symmetry no d^6 ion can have a ground state with two unpaired electrons (only 4 or 0), this might be possible in 6-coordinate complexes in which there are significant departures from O_h symmetry in the ligand field. Perhaps the best documented examples are complexes of the type $[Fe(LL)_2ox]$ and $[Fe(LL)_2mal]$ where LL represents a bidentate diamine ligand such as *o*-phen or bipy, and ox, mal represent oxal-

[9] M. R. Churchill and J. Wormald, *Inorg. Chem.*, 1971, **10**, 1778.
[10] M. Ciampolini, *Structure and Bonding*, 1969, **6**, 52.
[11] See G. M. Bancroft, M. J. Mays and B. E. Prater, *J. Chem. Soc., A*, **1970**, 956.
[12] E. König and K. Madeja, *Inorg. Chem.*, 1967, **6**, 48.
[13] J. P. Jesson *et al.*, *J. Amer. Chem. Soc.*, 1967, **89**, 3158.
[14] G. A. Renovitch and W. A. Baker, Jr., *J. Amer. Chem. Soc.*, 1967, **89**, 6377; R. N. Sylva and H. A. Goodwin, *Austral. J. Chem.*, 1967, **20**, 479.

ato and maleato ions. These complexes have magnetic susceptibilities that follow the Curie–Weiss law over a broad temperature range, with $\mu \approx 3.90$ B.M. (part of which is due to a temperature-independent paramagnetism).[15]

In the case of phthalocyanin iron(II) the extreme tetragonality of the ligand field apparently places one d orbital ($d_{x^2-y^2}$) at extremely high energy, and the six electrons adopt a high-spin distribution among the remaining four, thus giving a triplet ground state, independent of temperature.[16]

Iron(II) occurs in square coordination, but with a high-spin configuration, in the mineral gillespite, $BaFeSi_4O_{10}$. Spectroscopic studies[17] show that the energy order of d orbitals is $z^2 < xz, yz < xy \ll x^2 - y^2$; this is of particular interest since such an ordering cannot be accounted for by the point-charge crystal-field model.

25-E-5. Aqueous and Coordination Chemistry of Iron(III)

Iron(III) occurs in salts with most anions, except those that are incompatible with it because of their character as reducing agents. Examples obtained as pale pink to nearly white hydrates from aqueous solutions are $Fe(ClO_4)_3 \cdot 10H_2O$, $Fe(NO_3)_3 \cdot 9(\text{or } 6)H_2O$ and $Fe_2(SO_4)_3 \cdot 10H_2O$.

Aqueous Chemistry. One of the most conspicuous features of ferric iron in aqueous solution is its tendency to hydrolyze and/or to form complexes. It has been established that the hydrolysis (equivalent in the first stage to acid dissociation of the aquo ion) is governed in its initial stages by the following equilibrium constants:

$$[Fe(H_2O)_6]^{3+} = [Fe(H_2O)_5(OH)]^{2+} + H^+ \qquad K = 10^{-3.05}$$
$$[Fe(H_2O)_5(OH)]^{2+} = [Fe(H_2O)_4(OH)_2]^+ + H^+ \qquad K = 10^{-3.26}$$
$$2[Fe(H_2O)_6]^{3+} = [Fe(H_2O)_4(OH)_2Fe(H_2O)_4]^{4+} + 2H^+ \qquad K = 10^{-2.91}$$

In the last of these equations the binuclear species is believed to have the structure (25-E-I). From the constants for these equilibria it can be seen that

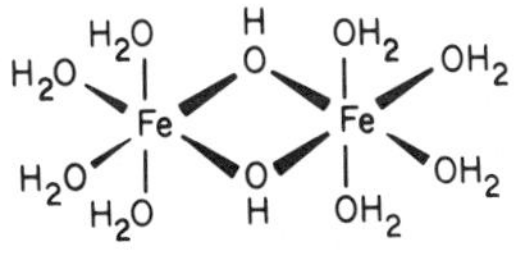

(25-E-I)

even at the rather acid pH's of 2–3, the extent of hydrolysis is very great, and in order to have solutions containing Fe^{III} mainly (say $\sim 99\%$) in the form of the pale purple hexaquo ion the pH must be around zero. As the pH is raised above 2–3, more highly condensed species than the dinuclear one noted above are formed, attainment of equilibrium becomes sluggish, and soon colloidal gels are formed; ultimately, hydrous ferric oxide is precipitated as a red-brown gelatinous mass.

[15] E. König and K. Madeja, *Inorg. Chem.*, 1968, **7**, 1848.
[16] I. Dezsi *et al.*, *J. Inorg. Nuclear Chem.*, 1969, **31**, 1661.
[17] R. G. Burns, M. G. Clark and A. J. Stone, *Inorg. Chem.*, 1966, **5**, 1268.

The kinetics of dissociation of the dinuclear complex has been studied[18] and from the rate law, which includes both acid-dependent and acid-independent terms, and other considerations, possible pathways have been considered.

There is no evidence that any definite hydroxide, $Fe(OH)_3$, exists, and the red-brown precipitate commonly called ferric hydroxide is best described as hydrous ferric oxide, $Fe_2O_3 \cdot nH_2O$. At least a part of such precipitates seems to be the $FeO(OH)$ noted above.

The various hydroxo species, such as $[Fe(OH)(H_2O)_5]^{2+}$, are yellow because of charge-transfer bands in the ultraviolet region which have tails coming into the visible region. Thus aqueous solutions of ferric salts even with non-complexing anions are yellow unless strongly acid.

Hydrous iron(III) oxide is readily soluble in acids but also to a slight extent also in strong bases. When concentrated solutions of strontium or barium hydroxide are boiled with ferric perchlorate, the hexahydroxoferrates(III), $M_3^{II}[Fe(OH)_6]_2$, are obtained as white crystalline powders. With alkali-metal hydroxides, substances of composition M^IFeO_2 can be obtained; these can also be made by fusion of Fe_2O_3 with the alkali-metal hydroxide or carbonate in the proper stoichiometric proportion. Moderate concentrations of what is presumably the $[Fe(OH)_6]^{3-}$ ion can be maintained in strongly basic solutions.

Ferric iron in aqueous solution is rather readily reduced by many reducing agents, such as I^-, as noted above. It also oxidizes sulfide ion, so that ferric sulfide precipitated on addition of H_2S or a sulfide to an Fe^{III} solution rapidly changes to a mixture of iron(II) sulfide and colloidal sulfur. Adding carbonate or hydrogen carbonate to an iron(III) solution precipitates the hydrous oxide.

Iron(III) Complexes. Iron(III) forms a large number of complexes, mostly octahedral ones, and the octahedron may be considered its characteristic coordination polyhedron. Iron(III) does also form a few tetrahedral complexes, of which $FeCl_4^-$ is the most important.

The hexaquo ion exists in very strongly acid solutions of ferric salts, in the several ferric alums, $M^IFe(SO_4)_2 \cdot 12H_2O$, and presumably also in the highly hydrated crystalline salts.

The affinity of iron(III) for amine ligands is very low. No simple ammine complex exists in aqueous solution; addition of aqueous ammonia only precipitates the hydrous oxide. Chelating amines, for example, EDTA, do form some definite complexes among which is the 7-coordinate $[Fe(EDTA)H_2O]^-$ ion. Also, those amines such as 2,2′-bipyridine and 1,10-phenanthroline that produce ligand fields strong enough to cause spin-pairing form fairly stable complexes, isolable in crystalline form with large anions such as perchlorate.

Iron(III) has its greatest affinity for ligands that coordinate by oxygen, viz., phosphate ions, polyphosphates, and polyols such as glycerol, sugars,

[18] B. A. Sommer and D. W. Margerum, *Inorg. Chem.*, 1970, **9**, 2517; H. N. Po and N. Sutin, *Inorg. Chem.*, 1971, **10**, 428.

etc. With oxalate the trisoxalato complex, $[Fe(C_2O_4)_3]^{3-}$, and with β-diket-ones the neutral $[Fe(dike)_3]$ complexes are formed. Formation of complexes with β-diketones is the cause of the intense colors that develop when they are added to solutions of ferric ion, and this serves as a useful diagnostic test for them.

Ferric ion forms complexes with halide ions and SCN^-. Its affinity for F^- is quite high, as shown by the equilibrium constants

$$Fe^{3+} + F^- = FeF^{2+} \qquad K_1 \approx 10^5$$
$$FeF^{2+} + F^- = FeF_2^+ \qquad K_2 \approx 10^5$$
$$FeF_2^+ + F^- = FeF_3 \qquad K_3 \approx 10^3$$

The corresponding constants for chloro complexes are only ~ 30, ~ 5, ~ 0.1, respectively. In very concentrated HCl the tetrahedral $FeCl_4^-$ ion is formed and its salts with large cations may be isolated. The complexes with SCN^- are an intense red and this serves as a sensitive qualitative and quantitative test for ferric ion; $Fe(SCN)_3$ and/or $Fe(SCN)_4^-$ may be extracted into ether.[19] Fluoride ion, however, will discharge this color. In the solid state, FeF_6^{3-} ions are known but in solutions only species with fewer F atoms occur. Other halo complexes that occur in crystalline compounds are $FeBr_4^-$, $FeCl_6^{3-}$ and $Fe_2Cl_9^{3-}$, as well as $Fe(SCN)_6^{3-}$.

With the cyanide ion, salts of the hexacyanoferrate ion and the free acid, $H_3[Fe(CN)_6]$, are well known. In contrast to $[Fe(CN)_6]^{4-}$ the $[Fe(CN)_6]^{3-}$ ion is quite poisonous; for kinetic reasons the latter dissociates and reacts rapidly, whereas the former is not labile. There are a variety of substituted ions $[Fe(CN)_5X]$, where $X = H_2O$, NO_2, etc., of which the best known is the nitroprusside ion[20] $[Fe(CN)_5NO]^{2-}$; this is attacked by OH^- to give $[Fe(CN)_5NO_2]^{2-}$.

It has long been known that treating a solution of Fe^{III} with hexacyano-ferrate(II) yields a blue precipitate called *Prussian blue* and that treating a solution of Fe^{II} with hexacyanoferrate(III) yields a blue precipitate called *Turnbull's blue*. These substances are actually identical, having the formulas $M^IFeFe(CN)_6 \cdot xH_2O$, where M^I is Na, K or Rb, but not Li or Cs. Their structure is closely related to those of brown ferric ferricyanide. $FeFe(CN)_6$, the white, insoluble potassium ferrous ferrocyanide, $K_2FeFe(CN)_6$, and a number of similar compounds such as $KCu^{II}Fe(CN)_6$ and $Cu^{II}_2Fe(CN)_6$. In all cases the basic structural feature seems to be a cubic array of iron ions with CN^- ions along cube edges between them and water molecules in some of the cubes. In $Fe^{III}Fe^{III}(CN)_6$ this is the complete structure. In $M^IFe^{II}Fe^{III}(CN)_6$ every other cube contains an M^I ion at its center, and in $M^I_2Fe^{II}Fe^{II}(CN)_6$ every cube contains an M^I ion at its center. Other com-pounds such as the Cu^{II} salts appear to have the same sort of structure. In Prussian (Turnbull's) blue the Fe^{II} ions are surrounded by the C atoms while

[19] A. G. Maddock and L. O. Medeiros, *J. Chem. Soc., A*, **1969**, 1946.
[20] J. H. Swinehart, *Coordination Chem. Rev.*, 1967, **2**, 385; *Inorg. Chem.*, 1968, **7**, 1855; B. A. Goodman and J. B. Rayner, *J. Chem. Soc., A*, **1970**, 2038.

the Fe^{III} ions are surrounded by N atoms, so that it could be called ferric ferrocyanide.[21]

Iron(III), like Cr^{III}, Mn^{III} and some other trivalent ions, forms basic carboxylates. The $[Fe_3O(CH_3COO)_6(H_2O)_3]^+$ cation[22] is isostructural with that of Cr^{III} (see page 837).

Binuclear, oxygen-bridged Fe^{III} complexes have been actively studied in recent years.[23] Two types, dihydroxo- and monooxo-bridged ones have been recognized. The former are represented by $[Fe(H_2O)_4OH]_2$, mentioned above and by $[Fe(pic)_2OH]_2$ [pic is the anion (25-E-II)]. Neither has been definitively

(25-E-II)

characterized as to structure, but there is reasonable evidence that both are indeed dihydroxo-bridged. The aquo ion, observed only in solution, is reported to be diamagnetic, whereas $[Fe(pic)_2OH]_2$ as a solid has a temperature-dependent magnetic moment that behaves as expected for weak antiferromagnetic coupling ($J = -8$ cm^{-1}) between two high-spin ions. The magnetic moment asymptotically approaches the full high-spin value (~ 5.9 B.M.) at higher temperatures. The monooxo-bridged complexes have very different magnetic behavior: their magnetic moments also approach zero at low temperatures, but they tend towards a high-temperature limit corresponding to only *one* unpaired electron per Fe^{III} atom; an unambiguous explanation for this has not yet been given. These monooxo-bridged complexes are typified by $[Fe(salen)]_2O\cdot CH_2Cl_2$ and $[Fe(HEDTA)]_2O^{2-}$, for both of which the crystal structures are known.[24] In each case the Fe—O distances (~ 1.79 Å) indicate multiple bonding and the Fe—O—Fe group is non-linear, the angles being 142° in the salen complex and 165° in the HEDTA complex. The magnetic behavior has been discussed in terms of two strongly coupled high-spin d^5 ions ($J \approx -90$ cm^{-1}), in terms of two coupled intermediate-spin ($S = 3/2$) ions and from a molecular-orbital point of view in which the bending of the Fe—O—Fe group is considered to cause a very large splitting in the energy of the π orbitals. A definitive description of the electronic structures of these molecules has yet to be given.

The complex $[Fe(EDTA)H_2O]^-$ mentioned above also dimerizes and studies[25] of the spectra, kinetics and redox behavior of the system have been interpreted on the assumption that the dinuclear species is $[Fe(EDTA)]_2O^{4-}$, structurally analogous to the HEDTA complex.

Fe(salen)Cl has the interesting property of forming two differently struc-

[21] For references see R. E. Wilde *et al.*, *Inorg. Chem.*, 1970, **9**, 2512; A. Ludi, H.-U. Güdel and M. Rüegg, *Inorg. Chem.*, 1970, **9**, 2224.
[22] K. Anzenhofer and J. J. de Boer, *Rev. Trav. Chim.*, 1969, **88**, 286.
[23] See R. W. Catterall, K. S. Murray and K. I. Peverill, *Inorg. Chem.*, 1971, **10**, 1301.
[24] See P. Coggon *et al.*, *J. Amer. Chem. Soc.*, 1971, **93**, 1014.
[25] H. J. Schugar *et al.*, *J. Amer. Chem. Soc.*, 1969, **91**, 71; R. G. Wilkins and R. E. Yelin, *Inorg. Chem.*, 1969, **8**, 1470.

tured crystals depending on the solvent from which it is crystallized. One contains 5-coordinate monomers (25-E-III) with a magnetic moment close to 5.9 B.M., while the other contains the bridged binuclear species (25-E-IV) and exhibits marked antiferromagnetic coupling between the iron atoms.[26]

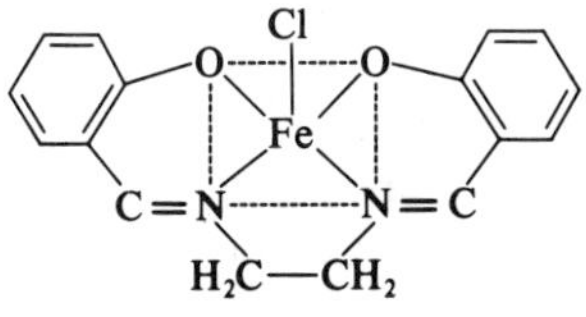

(25-E-III)

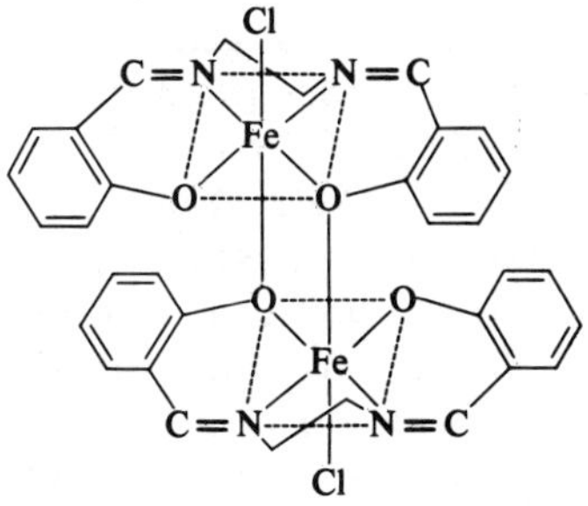

(25-E-IV)

Electronic Structures of Iron(III) Compounds. Iron(III) is isoelectronic with manganese(II), but much less is known of the details of Fe^{III} spectra because of the very much greater tendency of the trivalent ion to have charge-transfer bands in the near-ultraviolet region which have sufficiently strong low-energy wings in the visible to obscure almost completely—or completely in many cases—the very weak, spin-forbidden $d–d$ bands. Insofar as they are known, however, the spectral features of iron(III) ions in octahedral surroundings are in accord with theoretical expectations.

Magnetically, iron(III), like manganese(II), is high-spin in nearly all its complexes, except those with the strongest ligands of which $[Fe(CN)_6]^{3-}$, $[Fe(bipy)_3]^{3+}$ and $[Fe(phen)_3]^{3+}$ are well-known examples. In the high-spin complexes, the magnetic moments are always very close to the spin-only value of 5.9 B.M. because the ground state (derived from the ^5S state of the free ion) has no orbital angular momentum and there is no effective mechanism for introducing any by coupling with excited states. The low-spin complexes, with t_{2g}^5 configurations, usually have considerable orbital contributions to their moments at about room temperature, values of ~2.3 B.M. being obtained. The moments are, however, intrinsically temperature-dependent, and at liquid-nitrogen temperature (77°K) they decrease to ~1.9 B.M.

Certain tris(dithiocarbamate) complexes constitute well-studied examples of spin-crossover equilibria, as discussed elsewhere (page 567), but the ethyl xanthate is essentially low-spin.[27]

25-E-6. Higher Oxidation States

Only for the states IV and VI are there substantiated reports of compounds.

[26] M. Gerloch and F. E. Mabbs, *J. Chem. Soc.*, A, **1967**, 1598, 1900.
[27] B. F. Hoskins and B. P. Kelly, *Chem. Comm.*, **1970**, 45.

Iron(IV). The best known compounds are Sr_2FeO_4 and Ba_2FeO_4, both made by the reaction:

$$M_3^{II}[Fe(OH)_6]_2 + M^{II}(OH)_2 + \tfrac{1}{2}O_2 \xrightarrow{800-900°} 2M_2^{II}FeO_4 + 7H_2O$$

These contain no discrete FeO_4^{4-} ion but are mixed metal oxides, the barium one having the spinel structure.

The cationic species $[Fe(diars)_2X_2]^{2+}$, $X = Cl$ or Br, are obtained by oxidation of the $[Fe(diars)_2X_2]^+$ ions with $15M$ nitric acid. Their magnetic properties ($\mu_{eff} = 2.98$ B.M. at $293°K$) are consistent with there being two unpaired electrons in a tetragonally distorted set of t_{2g} orbitals.[28]

Finally, complexes such as $FeH_4(PEtPh_2)_3$ have been obtained[28a] by action of $NaBH_4$ on ethanol solutions of ferrous chloride containing the phosphine. These lose hydrogen to give dihydrides, e.g., $FeH_2(PEtPh_2)_3$. They are probably borohydrides and not Fe^{IV} species since they have an ir band ca 2400 cm^{-1}.

Iron(VI). Compounds containing the discrete, tetrahedral, red-purple ferrate(VI) ion, FeO_4^{2-}, can be obtained by oxidizing suspensions of $Fe_2O_3 \cdot xH_2O$ in concentrated alkali with chlorine, by anodic oxidation of iron in concentrated alkali, or by oxidation of iron filings with fused KNO_3. The sodium and potassium salts are very soluble, but the barium salt can be precipitated. The ion is paramagnetic with two unpaired electrons. It is relatively stable in basic solution[28b] but decomposes in neutral or acid solution according to the equation:

$$2FeO_4^{2-} + 10H^+ = 2Fe^{3+} + \tfrac{3}{2}O_2 + 5H_2O$$

It is an even stronger oxidizing agent than permanganate and can oxidize NH_3 to N_2, Cr^{II} to CrO_4^{2-}, and arsenite to arsenate. It also oxidizes primary amines and alcohols to aldehydes and epoxidizes squalene.

25-E-7. Biochemistry of Iron

Iron is by far the most widespread and important transition metal with a functional role in living systems. Iron-containing proteins participate in two main processes: oxygen-transport and electron-transfer. There are then other molecules whose function is to store and transport iron itself.

Iron also occurs in conjunction with molybdenum in enzymes that catalyze nitrogen fixation.

Heme Proteins.[29] The chief heme proteins are the hemoglobins, myoglobins, cytochromes and some enzymes such as catalase and peroxidase. These

[28] E. A. Paez, W. T. Oosterhuis and D. L. Weaver, *Chem. Comm.*, **1970**, 506.
[28a] M. Aresta *et al.*, *Inorg. Chim. Acta*, 1971, **5**, 115.
[28b] H. Goff and R. Kent Murinann, *J. Amer. Chem. Soc.*, **1971**, **93**, 6058; R. J. Audette, J. W. Quail and P. J. Smith, *Chem. Comm.*, **1972**, 38; K. B. Sharples and T. C. Flood, *J. Amer. Chem. Soc.*, **1971**, **93**, 2316.
[29] See R. E. Dickerson and I. Geis, *The Structure and Action of Proteins*, Harper and Row, 1969, for further structural details and references to chemical data; for synthetic biochemical models see J. H. Wang, *Accounts Chem. Res.*, **1970**, **3**, 90; T. H. J. Huisman and W. A. Schroeder, eds., *New Aspects of the Structure, Function and Synthesis of Haemoglobins*, Butterworths, 1971.

$$CO_2^- \qquad CO_2^-$$
$$| \qquad\qquad |$$
$$(CH_2)_2 \qquad (CH_2)_2$$

H_3C- $\qquad$ $-CH_3$

Fe

$-CH_3$

H_2C $\qquad$ H_3C $\qquad$ $-CH_3$

$CH_3 \qquad HC$

(25-E-V) $\qquad CH_2$

are characterized by the presence of the heme group (25-E-V) as the iron-containing unit.

Hemoglobin and myoglobin. These are closely related. The former has a molecular weight of 64,500 and consists of four subunits, each containing

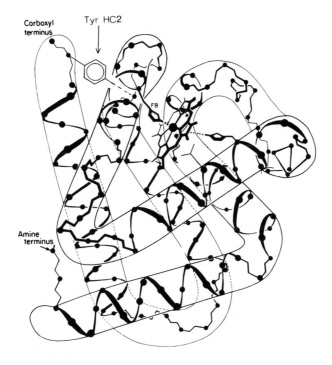

Fig. 25-E-1. A schematic representation of one of the four subunits of hemoglobin. The continuous black band represents the peptide chain and the various sections of helix are evident. Dots on the chain represent the α-carbon atoms. The heme group can be seen at upper right center with the iron as a large dot. The coordinated histidine side chain is labelled F8 (meaning the 8th residue of the F helix). [This Figure was adapted from one kindly supplied by M. Perutz.]

one heme group; myoglobin is very similar to one of the subunits of hemoglobin. One of the hemoglobin subunits is shown in Fig. 25-E-1. Hemoglobin has two functions: (1) It binds oxygen molecules to its iron atoms and transports them from the lungs to muscles where they are delivered to myoglobin molecules which store the oxygen until it is required for metabolic action. (2) The hemoglobin then uses certain amino groups to bind carbon dioxide and carry it back to the lungs.

The heme group is attached to the protein in both hemoglobin and myoglobin through a coordinated histidine-nitrogen atom, F8, shown in Fig. 25-E-1. In the position *trans* to the histidine-nitrogen atom there is a water molecule in the deoxy species or the oxygen molecule in the oxygenated species. The structure of the $Fe—O_2$ grouping is still unknown, but changes in the oxidation state of iron and the introduction of O_2 (and other ligands) cause interesting and important changes in the structure of the iron—heme complex, as will be described below.

Hemoglobin is not simply a passive container for oxygen but an intricate molecular machine. This may be appreciated by comparing its affinity for O_2 to that of myoglobin. For myoglobin (Mb) we have the following simple equilibrium:

$$Mb + O_2 = MbO_2 \qquad K = \frac{[MbO_2]}{[Mb][O_2]}$$

If f represents the fraction of myoglobin molecules bearing oxygen and P represents the equilibrium partial pressure of oxygen, it follows that

$$K = \frac{f}{(1-f)P} \qquad \text{and} \qquad f = \frac{KP}{1+KP}$$

This is the equation for the hyperbolic curve labeled Mb in Fig. 25-E-2. Hemoglobin with its four subunits has more complex behavior; it approximately follows the equation

$$f = \frac{KP^n}{1+KP^n} ; \qquad n \approx 2.8$$

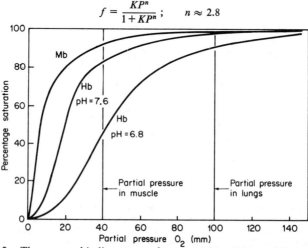

Fig. 25-E-2. The oxygen-binding curves for myoglobin (Mb) and hemoglobin (Hb), showing also the pH-dependence (Bohr effect) for the latter.

where the exact value of n depends on pH. Thus, for hemoglobin (Hb) the oxygen binding curves are sigmoidal as shown in Fig. 25-E-2. The fact that n exceeds unity can be ascribed physically to the fact that attachment of O_2 to one heme group increases the binding constant for the next O_2, which in turn increases the constant for the next one and so on.

It will be seen that, while Hb is about as good an O_2 binder as Mb at high O_2 pressure, it is much poorer at the lower pressures prevailing in muscle and hence passes its oxygen on to the Mb as required. Moreover, the need for O_2 will be greatest in tissues that have already consumed oxygen and simultaneously produced CO_2. The CO_2 lowers the pH, thus causing the Hb to release even more oxygen to the Mb. The pH-sensitivity (called the Bohr effect) as well as the progressive increase of the O_2 binding constants in Hb are due to interactions between the subunits; Mb behaves more simply because it consists of only one unit. It is clear that each of the two is essential in the complete oxygen-transport process. Carbon monoxide, PF_3 and a few other substances are toxic because they become bound to the iron atoms of Hb more strongly than O_2; their effect is one of competitive inhibition.

Detailed elucidation of the way in which subunit interactions give rise to both the cooperativity in oxygen binding and the Bohr effect has recently been accomplished by Perutz.[30] The complete analysis is far too complex to review here, but changes in the coordination of the iron play a crucial role and merit comment. Deoxyhemoglobin contains iron(II) in a high-spin state. With two electrons occupying e_g orbitals, the bonding radius of the iron is so large that it cannot fit into the plane of the four nitrogen atoms of the heme porphyrin. It therefore lies some 0.7–0.8 Å out of that plane, the Fe—N distances being about 2.2 Å. The iron is thus pentacoordinate with square-pyramidal coordination provided by four porphyrin nitrogen atoms in the basal positions and an imidazole-nitrogen atom from histidine F8 (see Fig. 25-E-1) in the apical position. When an oxygen molecule is bound, in the position opposite to this histidine the iron atom goes into a low-spin state; the e_g orbitals are then empty and the radius of the iron decreases so much that it now fits into the plane of the porphyrin system. Thus the iron atom moves some 0.75 Å when deoxyhemoglobin becomes oxygenated. Since it remains attached to the side chain of histidine F8 this shift is transmitted to various parts of the subunit, causing, particularly, important movements of the entire helical section in which F8 occurs, a substantial change in the position of tyrosine HC2 and the other amino acid residues attached to it. These changes in one subunit then cause changes in other subunits since the interface between the subunits is altered. Intricate and subtle though all these changes are, the "trigger" so to speak is a relatively simple and well-defined piece of coordination chemistry.

Cytochrome c. This is a heme protein which serves as an electron-carrier. The heme group here is covalently attached through several side chains to

[30] M. F. Perutz, *Nature*, 1970, **228**, 726.

amino acid residues of the protein and, in addition, an octahedron about the iron atom is completed by coordination of a histidine nitrogen atom and one other ligand not yet identified. The cytochrome c molecules in a great range of organisms are nearly identical, indicating that this type of electron-carrier developed very early in the evolutionary process and has served its purpose well, not requiring much change.

Iron–Sulfur Electron Transfer Proteins.[31] *Ferredoxins.* These relatively small proteins (6,000–12,000) which contain non-heme iron, cysteine-sulfur, and so-called "inorganic" sulfur have redox potentials close to that of the standard hydrogen electrode. They appear to occur in all green plants, including algae, in all photosynthetic bacteria and protozoa and in some fermentative anaerobic bacteria. These molecules play an essential role as electron-transfer agents at the low-potential end of the photosynthetic process, but an exact chemical specification of their activity is still lacking.

While having similar redox properties, the several characterized ferredoxins are diverse chemically in that iron content varies from two to eight atoms per molecule. Recently, the structure of the 8-iron ferredoxin from *M. aerogenes* has been determined by X-ray crystallography.[32] This molecule, with a molecular weight just over 6,000 consists of a protein chain in which there are eight cysteine residues (25-E-VI), eight iron atoms and eight

$$-NH-CH-CO-$$
$$|$$
$$CH_2$$
$$|$$
$$SH$$

(25-E-VI)

inorganic (or labile) sulfur atoms. The iron atoms and labile sulfur atoms form two separate clusters of the same sort (Fig. 25-E-3) already found in a high-potential iron protein (see below), with a cysteine-sulfur atom coordi-

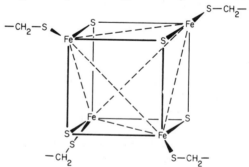

Fig. 25-E-3. The four iron atoms and four "inorganic" sulfur atoms in HiPIP form an eight-atom polyhedra of the type 1-XLVII (page 33). A tetrahedral set of sulfur atoms around each iron atom is completed by cysteine-sulfur atoms.

[31] J. C. M. Tsibris and R. D. Woody, *Coordination Chem. Rev.*, 1970, **5**, 417; R. Malkin and J. C. Rabinowitz, *Ann. Rev. Biochem.*, 1967, *B* **6**, 113; T. Kimura, *Structure and Bonding*, 1968, **5**, 1; D. O. Hall and M. C. W. Evans, *Nature*, 1969, **223**, 1342.
[32] L. H. Jensen, personal communication.

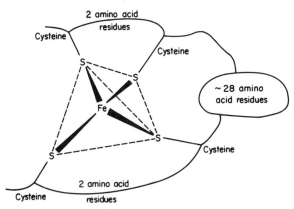

Fig. 25-E-4.   The tetrahedral iron complex found in rubredoxin. The tetrahedron is very
distorted (see text).

nated to each iron atom along a 3-fold axis of the tetrahedron of iron atoms. The centers of the two clusters are about 12 Å apart. Quite different structures had previously been proposed on the basis of indirect evidence (Mössbauer spectra, chemical and magnetic properties) and such data must now be reinterpreted.

Rubredoxin.   This sort of substance was first isolated from *C. pasteurianum* where it appears to participate in a number of biological reactions in which ferredoxin is also active. It contains only one iron atom, four cysteine units, no inorganic sulfur, and has a molecular weight of about 6,000. Its structure in the oxidized form has been investigated in some detail by X-ray crystallography.[33] The iron atom is tetrahedrally coordinated by the four cysteine-sulfur atoms, as shown in Fig. 25-E-4. The tetrahedron is markedly distorted. The Fe—S distances, which should be reliable to within less than 0.05 Å, are 2.39, 2.33, 2.31 and 1.97 Å; the angles range between 101° and 118°.

High-potential Iron Proteins.   These proteins (HiPIP's) have redox potentials some 0.7 V more positive than those of the ferredoxins. *Chromatium* HiPIP, the best characterized one, has a molecular weight of about 10,000, contains four iron atoms, four atoms of inorganic sulfur and four cysteine residues. X-ray investigation[34] has shown that the iron and sulfur atoms form the tetrahedral array (Fig. 25-E-3).

Miscellaneous.   In several classes of invertebrates the oxygen-carrying molecule is a non-heme iron protein called *hemerythrin*. It is presumed that the iron atoms bind the oxygen molecules, but the nature of the coordination sphere which confers this oxygen-binding ability on the iron atoms is unknown, except that a sulfur atom is not involved. The molecule consists of eight apparently identical subunits, each of molecular weight about 12,000 and containing two iron atoms. One molecule of O_2 is bound per subunit.

[33] J. R. Herriott *et al.*, *J. Mol. Biol.*, 1970, **50**, 391; T. V. Long *et al.*, *J. Amer. Chem. Soc.*, 1971, **93**, 1810.
[34] J. Kraut *et al.*, Cold Spring Harbor Symposium on Quantitative Biology, No. 36, 1971.

Iron metabolism[35] requires provision for storing and transporting iron. In man and many other higher animals the storage materials are *ferritin* and *haemosiderin*, which are present in liver, spleen, and bone marrow. Ferritin is a water-soluble, crystalline substance consisting of a roughly spherical protein sheath, of ~ 75 Å inside diameter and ~ 120 Å outside diameter, which in turn is built up of ~ 20 subunits. Within this sheath is a micelle of colloidal Fe_2O_3–H_2O–phosphate. Up to 23% of the dry weight may be iron; the protein portion alone, called apoferritin, is stable and crystallizes and has a molecular weight of about 450,000. Haemosiderin contains even larger proportions of "iron hydroxide" but its constitution is variable and ill-defined in comparison with that of ferritin.

Transferrin is a protein that binds ferric iron very strongly and transports it from ferritin to red cells and *vice versa*. Iron passes between ferritin and transferrin as Fe^{2+}, but the details of the redox process are obscure.

25-F. COBALT

The trend toward decreased stability of the very high oxidation states and the increased stability of the II state relative to the III state, which have been noted through the series Ti, V, Cr, Mn and Fe, persist with cobalt. Indeed, the former trend culminates in the complete absence of oxidation states higher than IV under chemically significant conditions. Even the oxidation state IV is highly uncertain, being represented, if at all, by only a few incompletely characterized compounds, such as Cs_2CoF_6, CoO_2, Ba_2CoO_4 and a heteropolymolybdate. The III state is relatively unstable in simple compounds, but the low-spin complexes are exceedingly numerous and stable, especially where the donor atoms (usually N) make strong contributions to the ligand field. There are also some important complexes of Co^I; this oxidation state is better known for cobalt than for any other element of the first transition series except copper.

The oxidation states and stereochemistry are summarized in Table 25-F-1.

25-F-1. The Element

Cobalt always occurs in Nature in association with nickel and usually also with arsenic. The most important cobalt minerals are *smaltite*, $CoAs_2$, and *cobaltite*, CoAsS, but the chief technical sources of cobalt are residues called "speisses," which are obtained in the smelting of arsenical ores of nickel, copper and lead. The separation of the pure metal is somewhat complicated and of no special relevance here.

Cobalt is a hard, bluish-white metal, m.p. 1493°, b.p. 3100°. It is ferromagnetic with a Curie temperature of 1121°. It dissolves slowly in dilute mineral acids, the Co^{2+}/Co potential being -0.277 V, but it is relatively

[35] *Iron Metabolism*, F. Gross, ed., Springer-Verlag, 1963.

TABLE 25-F-1

Oxidation States and Stereochemistry of Cobalt

Oxidn. state	Coordn. no.	Geometry	Examples
Co^{-I}	4	Tetrahedral	$Co(CO)_4^-$, $Co(CO)_3NO$
Co^0	4	Tetrahedral	$K_4[Co(CN)_4]$, $Co(PMe_3)_4$
Co^I, d^8	4	Tetrahedral	$CoBr(PR_3)_3$
	5^a	Trigonal bipyramidal	$[Co(NCR)_5]^+$, $[Co(CO)_3(PR_3)_2]^+$, $HCo(PF_3)_4^b$
	6	Octahedral	$[Co(bipy)_3]^+$
Co^{II}, d^7	2	Linear	$Co[N(SiMe_3)_2]_2{}^c$
	4^a	Tetrahedral	$[CoCl_4]^{2-}$, $CoBr_2(PR_3)_2$, Co^{II} in Co_3O_4
	5	Trigonal bipyramidal	$[Co(Me_6tren)Br]^+$
	5	Square pyramidal	$[Co(ClO_4)(MePh_2AsO)_4]^+$
	6^a	Octahedral	$CoCl_2$, $[Co(NH_3)_6]^{2+}$
	8	Dodecahedral	$(Ph_4As)_2[Co(NO_3)_4]$
Co^{III}, d^6	4	Tetrahedral	In a 12-heteropolytungstate; in garnets
	6^a	Octahedral	$[Co\,en_2Cl_2]^+$, $[Cr(CN)_6]^{3-}$, $ZnCo_2O_4$, CoF_3, $[CoF_6]^{3-}$
Co^{IV}, d^5	6	Octahedral	$[CoF_6]^{2-}$
Co^V, d^4	4	Tetrahedral?	K_3CoO_4

[a] Most common states.
[b] CoP_4 forms an almost regular tetrahedron.
[c] D. C. Bradley and K. J. Fisher, *J. Amer. Chem. Soc.*, 1971, **93**, 2058.

unreactive. It does not combine directly with hydrogen or nitrogen, and, in fact, no hydride or nitride appears to exist. The metal will combine with carbon, phosphorus and sulfur on heating. It also is attacked by atmospheric oxygen and by water vapor at elevated temperatures, giving CoO.

COBALT COMPOUNDS

25-F-2. Simple Salts and Compounds of Cobalt(II), d^7, and Cobalt(III), d^6

In aqueous solutions containing no complexing agents, the oxidation to Co^{III} is very unfavorable:

$$[Co(H_2O)_6]^{3+} + e = [Co(H_2O)_6]^{2+} \qquad E^0 = 1.84 \text{ V}$$

However, electrolytic or O_3 oxidation of cold acidic perchlorate solutions of Co^{2+} gives $[Co(H_2O)_6]^{3+}$, which is in equilibrium with $[Co(OH)(H_2O)_5]^{2+}$. At $0°$, the half life of these diamagnetic aquo ions is about a month.[1] In the presence of complexing agents, such as NH_3, which form stable complexes with Co^{III} the stability of trivalent cobalt is greatly improved:

$$[Co(NH_3)_6]^{3+} + e = [Co(NH_3)_6]^{2+} \qquad E^0 = 0.1 \text{ V}$$

and in basic media:

$$CoO(OH)(s) + H_2O + e = Co(OH)_2(s) + OH^- \qquad E^0 = 0.17 \text{ V}$$

Water rapidly reduces Co^{3+} at room temperature and this relative insta-

[1] G. Davies and B. Warnquist, *Coordination Chem. Rev.*, 1970, **5**, 349; B. Warnquist *Inorg. Chem.*, 1970, **9**, 682.

bility of uncomplexed Co^{III} is evidenced by the rarity of simple salts and binary compounds, whereas Co^{II} forms such compounds in abundance.

Cobalt(II) oxide, an olive-green substance, is easily prepared by reaction of the metal with oxygen at high temperature, by pyrolysis of the carbonate or nitrate, and in other ways. It has the rock-salt structure and is antiferromagnetic at ordinary temperatures. When cobalt(II) oxide is heated at 400–500° in an atmosphere of oxygen, the oxide Co_3O_4 is obtained. This is a normal spinel containing Co^{II} ions in tetrahedral interstices and diamagnetic Co^{III} ions in octahedral interstices. There is no evidence for the existence of the pure cobaltic oxide Co_2O_3.

Few simple Co^{III} salts are known. The anhydrous fluoride is prepared by fluorination of the metal or of $CoCl_2$ at 300–400°. It is a brown hygroscopic powder, instantly reduced by water and widely used to fluorinate organic compounds. The hydrate, $CoF_3 \cdot 3.5H_2O$, is of uncertain nature. It separates as a green powder on electrolysis of Co^{II} in 40% HF solution. The blue sulfate, $Co_2(SO_4)_3 \cdot 18H_2O$, which is stable when dry but decomposed by water, is precipitated when Co^{II} in $8N$ H_2SO_4 is oxidized either electrolytically or by ozone or fluorine. Co^{III} also occurs in the alums $MCo(SO_4)_2 \cdot 12H_2O$ (M = K, Rb, Cs or NH_4), which are also dark blue and reduced by water.

The action of N_2O_5 on CoF_3 at $-70°$ forms volatile green crystals of $Co(NO_3)_3$, which has a quasi-octahedral (D_3) structure containing equivalent, bidentate nitrate ions. The substance is an active nitrating agent and had a spin-paired electron configuration.[2]

Cobalt(II) forms an extensive group of simple and hydrated salts. The parent base, cobaltous hydroxide, may be precipitated by strong bases as either a blue or a pink solid, depending on conditions, but only the pink form is permanently stable. It is rather insoluble ($K_{sp} = 2.5 \times 10^{-16}$) but somewhat amphoteric, dissolving in very concentrated alkali solution to give a deep blue solution of $[Co(OH)_4]^{2-}$ ions, from which $Na_2Co(OH)_4$ and $Ba_2Co(OH)_6$ may be precipitated. Addition of sulfide ions or H_2S to solutions of Co^{2+} ion causes precipitation of a black solid, usually taken to be CoS and assigned a K_{sp} of about 10^{-22}. However, after storage a short while, this substance becomes far less soluble in acid than the above K_{sp} would indicate since oxidation occurs in air to give the less soluble Co(OH)S.

The anhydrous halides of Co^{II} are all known. CoF_2 is obtained by the action of HF on $CoCl_2$, while the others may be obtained by direct union of the elements at elevated temperatures. All have structures in which the Co^{II} ion is octahedrally coordinated.

Hydrated cobaltous salts with all common anions are known. They are easily obtained by reaction of $Co(OH)_2$ with the appropriate acid or by metathetical reactions. So far as is known, all such hydrated salts are red or pink and contain octahedrally coordinated Co^{II}. In many there are $[Co(H_2O)_6]^{2+}$ ions.

[2] R. J. Fereday, N. Logan and D. Sutton, *J. Chem. Soc.*, A, **1970**, 2699.

25-F-3. Complexes of Cobalt(II), d^7

Divalent cobalt forms numerous complexes of various stereochemical types. Octahedral and tetrahedral ones are most common, but there are a fair number of square ones as well as some which are five-coordinate.[3] Co^{II} forms tetrahedral complexes more readily than any other transition-metal ion. This is in accord with the fact that for a d^7 ion, ligand-field stabilization energies disfavor the tetrahedral configuration relative to the octahedral one to a smaller extent than for any other d^n $(1 \leqslant n \leqslant 9)$ configuration, although it should be carefully noted that this argument is valid only in comparing the behavior of one metal ion to another and not for assessing the absolute stabilities of the configurations for any particular ion (see page 597). Co^{2+} is the only d^7 ion of common occurrence.

Because of the small stability difference between octahedral and tetrahedral Co^{II} complexes, there are several cases in which the two types with the same ligand are both known and may be in equilibrium. The existence of some $[Co(H_2O)_4]^{2+}$ in equilibrium with $[Co(H_2O)_6]^{2+}$ has already been noted (page 645).

Tetrahedral complexes, $[CoX_4]^{2-}$ are generally formed with monodentate anionic ligands, such as Cl^-, Br^-, I^-, SCN^-, N_3^- and OH^-; with a combination of two such ligands and two neutral ones (L), tetrahedral complexes of the type CoL_2X_2 are formed. With ligands that are bidentate mono anions, tetrahedral complexes are formed in some cases, e.g., with N-alkylsalicylaldiminato and bulky β-diketonate anions. With the less hindered ligands of this type association to give a higher coordination number often occurs. Thus in bis-(N-methylsalicylaldiminato)cobalt(II) a dimer with five-coordinate Co atoms is formed (Fig. 25-F-1a), while $Co(acac)_2$

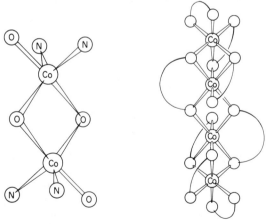

(a)   (b)

Fig. 25-F-1. Schematic representations of the structures of (a) the dimer of bis-(N-methyl-salicylaldiminato)cobalt(II) and (b) the tetramer of bis(acetylacetonato)cobalt(II).

---

[3] L. Sacconi, *J. Chem. Soc., A,* **1970,** 248; R. L. Carlin, *Transition Metal Chem.,* 1965, **1,** 1.

is a tetramer in which each Co atom is six-coordinate, as shown in Fig. 25-F-1b.

Planar complexes are formed with several bidentate monoanions such as dimethylglyoximate, aminooxalate, o-aminophenoxide, dithioacetylacetonate, and the disulfur ligands discussed in Section 22-16. Several neutral bidentate ligands also give planar complexes, although it is either known or reasonable to presume that the accompanying anions are coordinated *to some degree* so that these complexes could also be considered as very distorted octahedral ones. Examples are $[Co en_2] (AgI_2)_2$ and $[Co(CH_3SC_2H_4SCH_3)_2] (ClO_4)_2$. With the tetradentate ligands bis-(salicylaldehydeethylenediiminato) ion and porphyrins, planar complexes are also obtained. The dimethylglyoximate complex is discussed further in Section 25-F-7.

Many Co^{II} complexes, e.g., $Co(NH_3)_6^{2+}$, are readily oxidized by O_2 to give conventional Co^{III} complexes as the ultimate products, e.g., $Co(NH_3)_6^{3+}$, and the overall process is catalyzed by active charcoal. There are interesting binuclear peroxo species that can be observed and isolated in the absence of catalyst. The first in the sequence of steps may involve oxidative addition of O_2 to give a transient Co^{IV} species which then reacts with another Co^{II} species to give a binuclear peroxo-bridged species such as $[(NH_3)_5CoOOCo(NH_3)_5]^{4+}$ or $[(NC)_5CoOOCo(CN)_5]^{6-}$. These species are isolable as moderately stable solid salts but decompose fairly easily in water or acids. The open-chain species $[(NH_3)_5CoO_2Co(NH_3)_5]^{4+}$ can be cyclized in presence of base to

$$[(NH_3)_4Co \overset{O_2}{\underset{NH_2}{<>}} Co(NH_3)_4]^{3+}$$

It seems safe to assume that all such species, open-chain or cyclic, contain low-spin Co^{III} and bridging peroxide, O_2^{2-}, ions; in $[(NH_3)_5CoO_2Co(NH_3)_5]^{4+}$ the O—O distance (1.47 Å) is the same as in H_2O_2.[4]

The above O_2-bridged binuclear complexes can often be oxidized in a one-electron step to species such as $[(NH_3)_5CoO_2Co(NH_3)_5]^{5+}$ and

$$[(NH_3)_4Co \overset{O_2}{\underset{NH_2}{<>}} Co(NH_3)_4]^{4+}$$

These ions were first prepared by Werner who formulated them as peroxo-bridged complexes of Co^{III} and Co^{IV}. Esr data have shown, however, that the single unpaired electron is distributed equally over both cobalt ions, thus ruling out that description. The problem of how best to formulate these complexes has been fairly conclusively settled by X-ray structural data[5] which are summarized in Fig. 25-F-2. The O—O distances are not significantly different from that (1.28 Å) characteristic of superoxide ion, O_2^-.

[4] W. P. Schaefer, *Inorg. Chem.*, 1968, **7**, 125.
[5] G. G. Christoph, R. E. Marsh and W. P. Schaefer, *Inorg. Chem.*, 1969, **8**, 291.

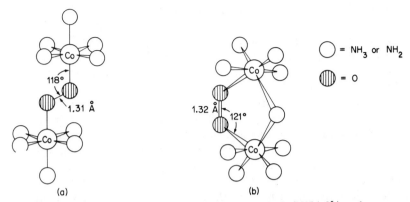

Fig. 25-F-2. The structures of (a) $[(NH_3)_5CoO_2Co(NH_3)_5]^{5+}$ and

(b) $[(NH_3)_4Co\underset{NH_2}{\overset{O_2}{\diagdown\diagup}}Co(NH_3)_4]^{4+}$, showing the octahedral coordination about each

cobalt ion and the angles and distances at the bridging superoxo groups. The five-membered ring in (b) is essentially planar.

The unpaired electron formally belonging to O_2^- resides in a molecular orbital of π symmetry relative to the planar Co—O—O—Co groupings and is delocalized over these four atoms. The cobalt atoms are formally described as Co^{III} ions. The reversible oxygen carriers are discussed below (page 887).

Addition of KCN to an aqueous solution of Co^{II} produces a dark green color, and a purple solid, $K_6Co_2(CN)_{10}$, can be precipitated. The latter contains the $[(NC)_5Co—Co(CN)_5]^{6-}$ ion, similar to the well-characterized, M—M bonded, $[(CH_3NC)_5Co—Co(CNCH_3)_5]^{4+}$ ion (see below). The green solution appears to contain principally $Co(CN)_5^{3-}$, the exact symmetry and solvation of which remain uncertain.[6] The green solution reacts slowly with solvent water, thus:

$$2Co(CN)_5^{3-} + H_2O \rightarrow Co(CN)_5H^{3-} + Co(CN)_5OH^{3-}$$

and also directly with hydrogen, thus:

$$2Co(CN)_5^{3-} + H_2 \rightarrow 2Co(CN)_5H^{3-}$$

possibly through the intermediacy of the $[Co_2(CN)_{10}]^{6-}$ ion.[7] The green solutions provide an effective catalyst for homogeneous hydrogenation of conjugated olefins.[8] The $Co(CN)_5^{3-}$ ion also undergoes bimolecular oxidative insertion reactions with C_2F_4, C_2H_2, SO_2 or $SnCl_2$ to give, for example, $[(NC)_5CoSnCl_2Co(CN)_5]^{6-}$ and reacts[9] with CO to give $[Co(CN)_3(CO)_2]^{2-}$.

With isocyanides, Co(II) forms five-coordinate and six-coordinate complexes as well as Co—Co bonded dinuclear ones.[7,10] The geometry of the

[6] J. P. Maher, J. Chem. Soc., A, 1968, 2918.
[7] J. Halpern and M. Pribanič, Inorg. Chem., 1970, 9, 2616.
[8] J. Kwiatek, Catalyst Rev., 1968, 1, 37.
[9] J. Halpern and M. Pribanič, J. Amer. Chem. Soc., 1971, 93, 96; H. S. Lim and F. C. Anson, Inorg. Chem., 1971, 10, 103.
[10] N. Kataoka and H. Kon, J. Amer. Chem. Soc., 1968, 90, 2978.

$Co(CNR)_5^{2+}$ species in solution has not been established. When R is alkyl these species dimerize in the solid state to give $[(RNC)_5Co\text{—}Co(CNR)_5]^{4+}$ ions with a single bond between the metal atoms. In solution in presence of an excess of RNC, six-coordinate cations $Co(CNR)_6^{2+}$ are formed; their esr spectra indicate that they are low spin $(t_{2g}^6 e_g)$ and axially distorted in accord with the Jahn–Teller theorem.

Cobalt(II) forms a number of well-defined five-coordinate complexes[11] in addition to those somewhat uncertain ones just mentioned, mainly with polydentate ligands such as the quadridentate tripod ligands (page 625) and certain tridentate ligands. The geometry varies, some approaching the trigonal-bipyramidal and others the square-pyramidal limiting cases, while many have an intermediate (C_{2v}) arrangement. Interest in these complexes has centered mainly on correlating their electronic structures with molecular symmetry and the atoms constituting the ligand set; these points will be mentioned below in connection with electronic structures.

Electronic Structures of Co(II) Compounds.[12]   As already noted, cobalt(II) occurs in a great variety of structural environments; because of this the electronic structures and hence the spectral and magnetic properties of the ion are extremely varied. We shall try to mention here each of the principal situations, giving its chief spectral and magnetic characteristics and citing representative cases.

High-spin Octahedral and Tetrahedral Complexes.   For qualitative purposes the partial energy level diagram in Fig. 25-F-3 is useful. In each case

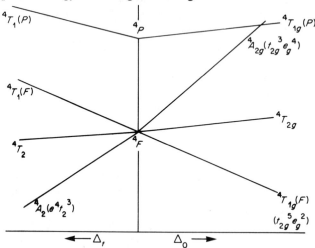

Fig. 25-F-3.   Schematic energy level diagram for quartet states of a d^7 ion in tetrahedral and octahedral ligand fields.

[11] L. Sacconi, *Pure Appl. Chem.*, 1968, **17**, 97; *J. Chem. Soc., A.*, **1970**, 248; P. L. Orioli, *Coordination Chem. Rev.*, 1971, **6**, 285.

[12] (a) R. L. Carlin, *Transition Metal Chem.*, 1965, **1**, 1; J. Ferguson, *Prog. Inorg. Chem.*, 1970, **12**, 249; (b) L. Sacconi, *J. Chem. Soc., A*, **1970**, 248; M. Ciampolini, *Structure and Bonding*, 1969, **6**, 52.

there are a quartet ground state and three spin-allowed electronic transitions to the excited quartet states. Quantitatively the two cases differ considerably, as might be inferred from the simple observation that octahedral complexes are typically pale red or purple, while many common tetrahedral ones are an intense blue. In each case the visible spectrum is dominated by the highest energy transition, $^4A_2 \rightarrow \, ^4T_1(P)$ for tetrahedral and $^4T_{1g}(F) \rightarrow \, ^4T_{1g}(P)$ for octahedral complexes, but in the octahedral systems the $^4A_{2g}$ level is usually close to the $^4T_{1g}(P)$ level and the transitions to these two levels are close together. Since the $^4A_{2g}$ state is derived from a $t_{2g}^3 e^4$ electron configuration, while the $^4T_{1g}(F)$ ground state is derived mainly from a $t_{2g}^5 e_g^2$ configuration, the $^4T_{1g}(F) \rightarrow \, ^4A_{2g}$ transition is essentially a two-electron process and is thus weaker by about a factor of 10^{-2} than the other transitions. In the tetrahedral systems, as illustrated in Fig. 25-F-4 the visible transition is generally about an order of magnitude more intense and displaced to lower energies, in accord with the observed colors mentioned above. For octahedral complexes, there is one more spin-allowed transition $(^4T_{1g}(F) \rightarrow \, ^4T_{2g})$ which generally occurs in the near-infrared region. For tetrahedral complexes there is also a transition in the near-infrared region $[^4A_2 \rightarrow \, ^4T_1(F)]$, as well as one of quite low energy $(^4A_2 \rightarrow \, ^4T_2)$ which is seldom observed because it is in an inconvenient region of the spectrum (1–2 microns) and because it is orbitally forbidden. The visible transitions in both cases, but particularly in the tetrahedral case, generally have complex envelopes because a number of transitions to doublet excited states occur in the same region and these acquire some intensity by means of spin–orbit coupling.

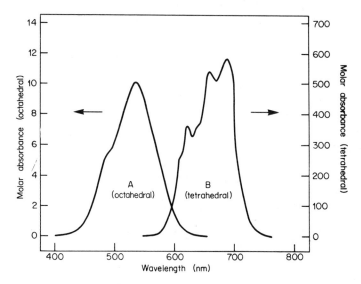

Fig. 25-F-4. The visible spectra of $[Co(H_2O)_6]^{2+}$ (curve A) and $[CoCl_4]^{2-}$ (curve B). The molar absorbance scale at the left applies to curve A, and that at the right applies to curve B.

The octahedral and tetrahedral complexes also differ in their magnetic properties. Because of the intrinsic orbital angular momentum in the octahedral ground state, there is consistently a considerable orbital contribution, and effective magnetic moments for such compounds around room temperature are between 4.7 and 5.2 B.M. For tetrahedral complexes, the ground state acquires orbital angular momentum only indirectly through the mixing in of the $^4T_2$ state by a spin–orbit coupling perturbation. First-order perturbation theory leads to the expression

$$\mu = 3.89 - \frac{15.59\,\lambda'}{\Delta_t}$$

where 3.89 is the spin-only moment for three unpaired electrons and λ' is the effective value of the spin–orbit coupling constant (which is inherently negative). Since λ' varies little from one complex to another, orbital contributions vary inversely with the strength of the ligand field. For example among the tetrahalo complexes we have: $CoCl_4^{2-}$, 4.59 B.M.; $CoBr_4^{2-}$, 4.69 B.M.; CoI_4^{2-}, 4.77 B.M.

Low-spin Octahedral Complexes. As shown in the Tanabe–Sugano diagram, a sufficiently strong ligand field ($\Delta_0 \geqslant 15{,}000\ cm^{-1}$) can cause a ^2E state originating in the ^2G state of the free ion to become the ground state. The electron configuration here is mainly $t_{2g}^6 e_g$ and thus a Jahn–Teller distortion would be expected. Consequently, perfectly octahedral low-spin Co(II) complexes must be rare or non-existent. Very few six-coordinate low-spin Co(II) complexes have actually been reported and none has been structurally characterized. As noted above, the $Co(CNC_6H_5)_6^{2+}$ ion in solution is axially distorted. It appears that ligands tending to give a strong enough field to cause spin pairing give five- rather than six-coordinate complexes, or dinuclear ones such as $Co_2(CNCH_3)_{10}^{4+}$.

Square Complexes. All of these are low-spin with magnetic moments of the 2.2 to 2.7 B.M. at 300°C. Their spectra are complex and neither magnetic nor spectral properties of such compounds have been treated in detail. There are some data to suggest[13] that the unpaired electron occupies the d_{z^2} orbital, as might be expected.

Five-coordinate Complexes.[12b] Both high-spin (three unpaired electrons) and low-spin (one unpaired electron) configurations are found for both trigonal-bipyramidal and square-pyramidal as well as for intermediate configurations. The patterns of orbital energies in the two limiting geometries have been discussed on page 567 and from these we may write the following configurations for the four spin–structure combinations:

Spin	D_{3h}	C_{4v}
High	$(e'')^4(e')^2(a_1)$	$e^4 b_2 a_1 b_1$
Low	$(e'')^4(e')^3$	$e^4 b_2^2 a_1$

[13] R. J. Fitzgerald and G. R. Brubaker, *Inorg. Chem.*, 1969, **8**, 2265.

It appears that the relationships between spin state, geometry, and nature of the ligand atoms are closely interlocked, so that no simple relationship between any two of these factors has been found. However, it does appear that spin state and the nature of the donor-atom set are roughly correlated independently of geometry in such a way that the more heavier-atom (e.g., P, As, Br or S) donors (as compared with O and N) are present the greater is the tendency to spin pairing; this is hardly surprising. For high-spin complexes with fairly regular geometry {e.g., $[Co(Me_6tren)Br]^+$, C_{3v}; $[Co(Ph_2MeAsO)_4(ClO_4)]ClO_4$, C_{4v}} detailed and reasonably convincing spectral assignments have been made. For irregular geometries and for low-spin complexes there are more uncertainties.

25-F-4. Complexes of Cobalt(III), d^6

The complexes of cobalt(III) are exceedingly numerous. Because they generally undergo ligand-exchange reactions relatively slowly, they have, from the days of Werner and Jørgensen, been extensively studied and a large fraction of our knowledge of the isomerism, modes of reaction and general properties of octahedral complexes as a class is based upon studies of Co^{III} complexes. All known discrete Co^{III} complexes are octahedral, though tetrahedral and square-antiprismatic Co^{III} are known in a few solid-state situations.[14]

Co^{III} shows a particular affinity for nitrogen donors, and the majority of its complexes contain ammonia, amines such as ethylenediamine, nitro groups, or nitrogen-bonded SCN groups, as well as halide ions and water molecules. In general, these complexes are synthesized in several steps beginning with one in which the aquo Co^{II} ion is oxidized in solution, typically by molecular oxygen or hydrogen peroxide and often a surface-active catalyst such as activated charcoal, in the presence of the ligands. For example, when a vigorous stream of air is drawn for several hours through a solution of a cobalt(II) salt, CoX_2 ($X = Cl$, Br or NO_3), containing ammonia, the corresponding ammonium salt and some activated charcoal, good yields of the hexammine salts are obtained:

$$4CoX_2 + 4NH_4X + 20NH_3 + O_2 \rightarrow 4[Co(NH_3)_6]X_3 + 2H_2O$$

In the absence of charcoal, replacement usually occurs to give, for example, $[Co(NH_3)_5Cl]^{2+}$ and $[Co(NH_3)_4(CO_3)]^+$. Similarly, on air oxidation of a solution of $CoCl_2$, ethylenediamine and an equivalent quantity of its hydrochloride salt, tris(ethylenediamine)cobalt(III) chloride is obtained.

$$4CoCl_2 + 8en + 4en \cdot HCl + O_2 = 4[Co (en)_3]Cl_3 + 2H_2O$$

However, a similar reaction in acid solution with the hydrochloride gives the green *trans*-dichlorobis(ethylenediamine)cobalt(III) ion as the salt

---

[14] D. L. Wood and J. P. Remeika, *J. Chem. Phys.*, 1967, **46**, 3595; L. C. W. Baker and V. E. Simmons, *J. Amer. Chem. Soc.*, 1959, **81**, 4744; J. M. Pratt and R. G. Thorpe, *Adv. Inorg. Chem. Radiochem.*, 1969, **12**, 375 (*cis–trans* effects in Co^{III} complexes).

trans-[Co en$_2$Cl$_2$] [H$_5$O$_2$]Cl$_2$ which loses HCl on heating. This *trans*-isomer may be isomerized to the red racemic *cis*-isomer on evaporation of a neutral aqueous solution at 90–100°. Both the *cis*- and the *trans*-isomer are aquated when heated in water:

$$[Co\ en_2Cl_2]^+ + H_2O \rightarrow [Co\ en_2Cl(H_2O)]^{2+} + Cl^-$$
$$[Co\ en_2Cl(H_2O)]^{2+} + H_2O \rightarrow [Co\ (en)_2(H_2O)_2]^{3+} + Cl^-$$

and on treatment with solutions of other anions are converted into other [Co en$_2$X$_2$]$^+$ species, for example,

$$[Co\ en_2Cl_2]^+ + 2NCS^- \rightarrow [Co\ en_2(NCS)_2]^+ + 2Cl^-$$

These few reactions are illustrative of the very extensive chemistry of Co^{III} complexes with nitrogen-coordinating ligands.

In addition to the numerous mononuclear ammine complexes of Co^{III}, there are a number of polynuclear ammine complexes in which hydroxo (OH$^-$), peroxo (O$_2^{2-}$), amido (NH$_2^-$), and imido (NH^{2-}) groups function as bridges. Some typical complexes of this class are

$$[(NH_3)_5Co-O-O-Co(NH_3)_5]^{4+} \quad (page\ 878),\ [(NH_3)_3Co(OH)_3Co(OH)_3Co(NH_3)_3]^{3+}$$
$$and\ [(NH_3)_4Co(OH)(NH_2)Co(NH_3)_4]^{4+}$$

Some other Co^{III} complexes of significance are the hexacyano complex [Co(CN)$_6$]$^{3-}$, the oxygen-coordinated complexes such as carbonates,[15] cobalt(III) acetylacetonate and salts of the trisoxalatocobalt(III) anion. The structure of cobaltic acetate is uncertain but this compound possibly contains a Co$_3$O(OCOCH$_3$)$_6$ unit in view of the fact that other trivalent metals, e.g., Cr^{III}, Fe^{III} and Mn^{III}, have such a structure (page 837); it is best made by oxidation of Co^{II} acetate in acetic acid with ozone. It oxidizes alkyl side chains in aromatic hydrocarbons[16] and a cobalt-catalyzed process is used commercially for the oxidation of toluene to phenol. Cobalt(III) in aqueous acid solution commonly oxidizes organic compounds.[16a]

Electronic Structures of Cobalt(III) *Complexes.* The free Co^{III} ion, d^6, has qualitatively the same energy level diagram as does Fe^{II} (see Appendix). However, with Co^{III} the $^1A_{1g}$ state originating in one of the high-energy singlet states of the free ion drops very rapidly and crosses the $^5T_{2g}$ state at a very low value of Δ. Thus all known octahedral Co^{III} complexes, including even [Co(H$_2$O)$_6$]$^{3+}$ and [Co(NH$_3$)$_6$]$^{3+}$, have diamagnetic ground states, with the exceptions of [Co(H$_2$O)$_3$F$_3$] and [CoF$_6$]$^{3-}$ which are paramagnetic with four unpaired electrons.

The visible absorption spectra of Co^{III} complexes may thus be expected to consist of transitions from the $^1A_{1g}$ ground state to other singlet states. Although the entire energy-level pattern for Co^{III} is not known in detail, the two absorption bands found in the visible spectra of regular octahedral Co^{III} complexes represent transitions to the upper states $^1T_{1g}$ and $^1T_{2g}$. In complexes of the type CoA$_4$B$_2$, which can exist in both *cis*- and *trans*-con-

[15] R. P. Perez-MacColl, *Coordination Chem. Rev.*, 1969, **4**, 147.
[16] S. S. Lande, C. D. Falk and J. K. Kochi, *J. Inorg. Nuclear Chem.*, 1971, **33**, 4101.
[16a] A. Meenakshi and M. Santappa, *J. Catalysis*, 1970, **19**, 300.

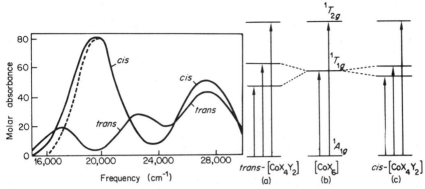

Fig. 25-F-5.   Left: The visible spectra of *cis*- and *trans*-[Co en$_2$F$_2$]$^+$. The broken line shows where the low-frequency side of the $^1A_{1g} \rightarrow {}^1T_{1g}$ band of the *cis*-isomer would be if the band were completely symmetrical. The asymmetry is caused by slight splitting of the $^1T_{1g}$ state. Right: Diagrammatic representation (not to scale) of the energy levels involved in the transitions responsible for the observed bands of octahedral Co^{III} complexes. In the center, (b), are the levels for a regular octahedral complex, [CoX$_6$]. In (a) and (c) the splittings caused by the replacement of two ligands X by two ligands Y are indicated.

figurations, there are certain spectral features which are diagnostic of the *cis*- or *trans*-configuration, as shown in Fig. 25-F-5.

The origin of these features lies in the splitting of the $^1T_{1g}$ state by the environments of lower than O_h symmetry, as also shown diagrammatically in Fig. 25-F-5. Theory shows that splitting of the $^1T_{2g}$ state will always be slight, whereas the $^1T_{1g}$ state will be split markedly in the *trans*-isomer whenever there is a substantial difference in the positions of the ligands, A and B, in the spectrochemical series. Moreover, because the *cis* isomer lacks a center of symmetry it may be expected to have a somewhat more intense spectrum than the *trans*-isomer. These predictions are nicely borne out by the spectra of *cis*- and *trans*-[Co en$_2$F$_2$]$^+$.

25-F-5. Tetravalent Cobalt, d^5; Pentavalent Cobalt, d^4

Compounds in these classes are few and not well characterized. Fluorination of Cs$_2$CoCl$_4$ gives Cs$_2$CoF$_6$, with a crystal structure isomorphous with that of Cs$_2$SiF$_6$ and a magnetic moment rising from 2.46 B.M. at 90° K to 2.97 B.M. at 294° K. The reflectance spectrum has been assigned to an octahedrally coordinated t_{2g}^5 ion.[17] The high magnetic moments could be due to a large orbital contribution in the $^2T_{2g}$ ground state or to partial population of a $^6A_{1g}$ ($t_{2g}^3 e_g^2$) state. The action of oxidizing agents (e.g., Cl$_2$, O$_2$ or O$_3$) on strongly alkaline Co^{II} solutions produces a black material believed to be hydrous CoO$_2$, at least in part, but it is ill-characterized. Ba$_2$CoO$_4$, a red-brown substance obtained by oxidation of 2Ba(OH)$_2$ and 2Co(OH)$_2$ at 1050° and a heteropolymolybdate of Co^{IV}, namely, 3K$_2$O·CoO$_2$·9MoO$_3$·6H$_2$O,

[17] G. C. Allen and K. D. Warren, *Inorg. Chem.*, 1969, **8**, 1902.

have been reported but not further investigated. It would appear at present that Co^{IV} is too unstable to have any very extensive chemistry.

Heating alkali-metal and cobalt oxides in O_2 at 300° gives black substances, e.g., K_3CoO_4.[18] These appear to contain Co^V.

25-F-6. Complexes of Cobalt(I), d^8

With the exception of reduced vitamin B_{12} and cobalt oximes (discussed below), which appear to be Co^I species, all Co^I compounds involve ligands of π-acid type. The commonest coordination number is five, and the trigonal bipyramid appears to be the preferred coordination polyhedron. A few cases of octahedral, square-pyramidal and tetrahedral coordination are also known.

Trigonal-bipyramidal Complexes. Cobalt carbonyl reacts with iso-cyanides, disproportionating to Co^I and Co^{-I}:

$$Co_2(CO)_8 + 5RNC \rightarrow [Co(CNR)_5]^+ [Co(CO)_4]^- + 4CO$$

The ionic nature of the product, as indicated in this equation, is confirmed by the preparation of the same substance by the following reaction:

$$Na[Co(CO)_4] + [Co(CNR)_5]ClO_4 \rightarrow [Co(CNR)_5][Co(CO)_4] + NaClO_4$$

Various salts of the $[Co(CNR)_5]^+$ cations, such as the perchlorate used above, can be prepared by the action of an excess of isocyanide on a Co^{II} salt or by first preparing the $Co(CNR)_4X_2$ compound and then reducing it with RNC or another reducing agent such as N_2H_4, $S_2O_4^{2-}$ or an active metal. In the case of $[Co(CNCH_3)_5]ClO_4$ the cation has been shown crystallographically to be trigonal bipyramidal.

The reduction of Co $acac_3$ by aluminum alkyls under nitrogen gives $CoH(N_2)(PPh_3)_3$.[19] Several similar hydrido species are known, e.g., $CoH(PF_3)_4$ and the unstable $CoH(CO)_4$. Cobalt halide complexes $CoX_2(PR_3)_2$ are reduced by CO to the corresponding Co^I complexes[19a] $CoX(CO)_2(PR_3)_2$. Note, however, that when cobalt halides in tetrahydro-furan containing trimethylphosphine are reduced by sodium amalgam, the paramagnetic cobalt(0) complex $Co(PMe_3)_4$ is obtained.[19b]

In polar solvents and at elevated temperatures triphenylphosphine reacts with cobalt carbonyl to give the cation $[Co(CO)_3(Ph_3P)_2]^+$ in the following disproportionation reaction:

$$Co_2(CO)_8 + 2Ph_3P \rightarrow [Co(CO)_3(Ph_3P)_2]^+ [Co(CO)_4]^- + CO$$

However, at a low temperature ($\sim 0°$) in non-polar solvents a genuine sub-stituted cobalt carbonyl, $[Co_2(CO)_6(Ph_3P)_2]$, is produced. $[Co(CO)_3(Ph_3P)_2]^+$ has been shown by infrared study to be trigonal bipyramidal with the phos-phines occupying the apical positions.

[18] C. Brendel and W. Klemm, *Angew. Chem. Internat. Edn.*, 1970, **9**, 519.
[19] A. Yamamoto *et al.*, *J. Amer. Chem. Soc.*, 1971, **93**, 371.
[19a] M. Bressan, *Inorg. Chem.*, 1970, **9**, 1733.
[19b] A.-F. Klein, *Angew. Chem. Internat. Edn.*, 1971, **10**, 343.

The melon-shaped phosphite, $P(OCH_2)_3CCH_3$, which has good π-acidity, affords several Co^I compounds. When this phosphite reacts with Co^{II} perchlorate, a disproportionation occurs giving $[Co^{III}L_6](ClO_4)_3$ and $[Co^IL_5]ClO_4$, where L represents the phosphite ligand. $[Co^IL_5]NO_3$ has also been prepared. Both are yellow diamagnetic solids and 1:1 electrolytes, but the structures have not been established.

Tetrahedral Complexes. The following general reaction[19c] produces tetrahedral complexes:

$$CoX_2 + 3L \xrightarrow[\text{BH}_4^-\text{ in EtOH}]{\text{Zn dust}} CoXL_3$$

$$[X = Cl, Br, or I; \ L = (C_6H_5)_3P, (C_4H_9)(C_6H_5)_2P, (C_6H_5CH_2)(C_6H_5)_2 P]$$

These compounds are stable, green, crystalline solids, less stable in solution and easily oxidized. Tetrahedral structures are assigned on the basis of magnetism ($\mu_{eff} = 3.0$–3.3 B.M. at $298°$K) and electronic spectra, in which properties they closely resemble tetrahedral nickel(II) species such as $NiX_2(PR_3)_2$ and $[NiX(PR_3)_3]^+$. The Co^I product of the following reaction in strongly alkaline solution may be tetrahedral:

$$2Co^{2+} + 3CO + 6CN^- + 4OH^- \rightarrow 2[Co(CN)_3(CO)]^{2-} + CO_3^{2-} + 2H_2O$$

The reduction of $[Co\ phen_3]\ (ClO_4)_2$ with borohydride in ethanol at $-5°$ produces $[Co\ phen_3]ClO_4$.

25-F-7. Complexes of Biological Interest

Oxygen Carriers. Although no cobalt-containing complex is actually proved to be involved in oxygen metabolism, there are several that are of interest as possible models for the chemistry, especially the metal-to-oxygen binding, that is involved in real biological systems. Of greatest interest would be those that undergo reversible oxygenation and deoxygenation in solution. Though there are fragmentary reports of others, by far the best characterized systems are Schiff base complexes such as Co(acacen) (25-F-I) in dimethylformamide, pyridine, and substituted pyridines;[20] the general reaction, which is reversible at temperatures below $0°$C, is typified by the following:

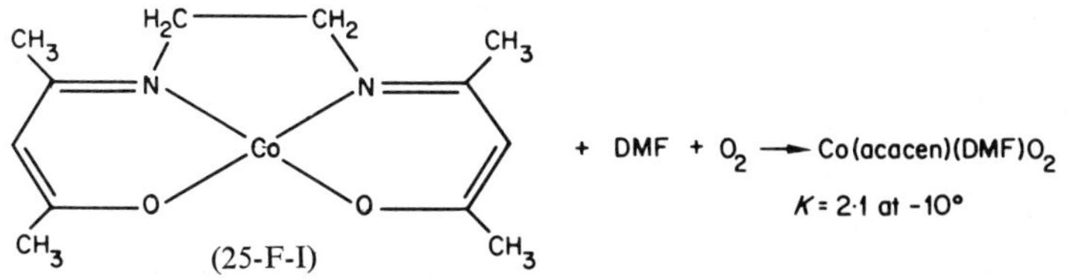

$$+ \ DMF + O_2 \longrightarrow Co(acacen)(DMF)O_2$$

$$K = 2 \cdot 1 \ at \ -10°$$

(25-F-I)

The initial complex has one unpaired electron, and so also do the oxygen adducts, but esr data indicate that in the latter the electron is heavily local-

[19c] M. Aresta, M. Rossi and A. Sacco, *Inorg. Chim. Acta*, 1969, **3**, 227.
[20] F. Basolo *et al.*, *J. Amer. Chem. Soc.*, 1970, **92**, 55, 61; S. Koda, A. Misono and Y. Uchida, *Bull. Chem. Soc. Japan*, 1970, **43**, 3143; C. Busetto *et al.*, *Inorg. Chim. Acta*, 1971, **5**, 129.

ized on the oxygen atoms. There is also an intense absorption band in the infrared region attributable to an O—O stretching vibration. On the basis of these and other observations the adducts have been formulated as octahedral, low-spin Co(III) complexes containing a coordinated superoxide (O_2^-) ion and having a bent Co—O—O chain. This formulation is similar to that proposed by Pauling[21] for oxyhemoglobin. A second type of complex[22] involves the reversible formation of oxygen bridges, Co—O—O—Co, which are similar to those discussed on page 878.

More recently, it has been found[23] that certain porphyrin complexes of Co^{II} in presence of base also have reversible oxygen-binding capacity and, moreover, that these complexes can be introduced into globin to form a cobalt-substituted hemoglobin that has the same type of cooperative oxygen uptake as hemoglobin itself.

Vitamin B_{12}.[24] The best known biological function of cobalt is its intimate involvement in the coenzymes related to vitamin B_{12}, the basic structure of which is shown in Fig. 25-F-6. The macrocyclic ring is the *corrin* system; it is reminiscent of the porphyrin system, the most notable difference being the absence of a methine (CH) bridge between one pair of pyrrole rings.

The term vitamin B_{12} generally means cyanocobalamin which has a Co^{III}—CN group. The best known of the coenzymes also contains Co(III) and a 5'-deoxyadenosyl group (25-F-II) that replaces the CN; this coenzyme was the first organometallic compound discovered in living systems.

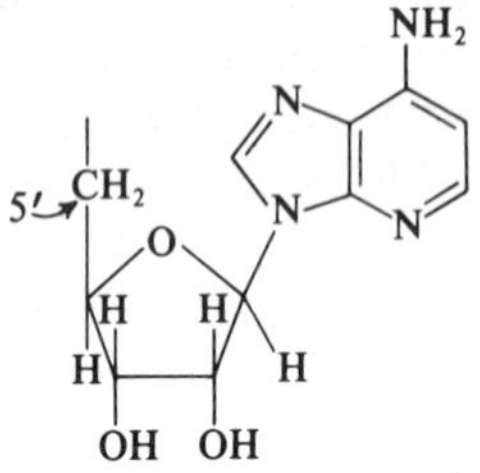

(25-F-II)

The B_{12} coenzymes act in concert with a number of enzymes, but it has so far proved difficult to elucidate their role in detail. Perhaps the best studied systems[25] involve the dioldehydrases, where the following reactions are catalyzed:

$$RCHOHCH_2OH \rightarrow RCH_2CHO + H_2O \quad (R = CH_3 \text{ or } H)$$

From studies of the non-enzymic chemistry of B_{12} coenzymes and of model systems noted below, a body of knowledge about basic B_{12} chemistry

[21] L. Pauling, *Nature*, 1964, **203**, 182.
[22] C. Floriani and F. Calderazzo, *J. Chem. Soc., A*, **1969**, 946; A. Calligaris *et al., J. Chem. Soc., A*, **1970**, 1069; R. D. Gillard and D. A. Phipps, *J. Chem. Soc., A*, **1971**, 1074.
[23] B. M. Hoffman and D. H. Petering, *Proc. Nat. Acad. Sci. USA*, 1970, **67**, 637.
[24] F. Wagner, *Ann. Rev. Biochem.*, 1966, **35**, 405; A. Eschenmoser, *Quart. Rev.*, 1970, **24**, 366 (synthesis of corrin systems and B_{12}).
[25] P. A. Frey *et al., J. Amer. Chem. Soc.*, 1970, **92**, 4488; 1971, **93**, 1242.

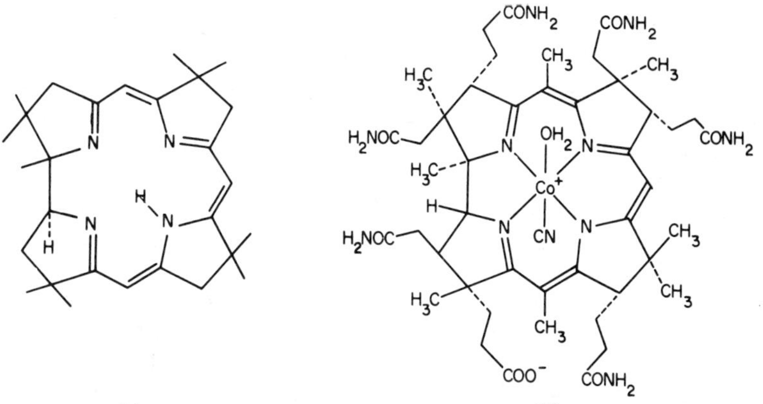

Fig. 25-F-6. (a) The corrin ligand system. (b) Structure of cobyric acid, the simplest known corrinoid natural product. [Reproduced by permission from A. Eschenmoser[24].

has been built up. Some of this chemistry undoubtedly plays a role in its activities as a coenzyme. The cobalamins can be reduced in neutral or alkaline solution to give Co(II) and Co(I) species, often called B_{12r} and B_{12s}, respectively. The latter is a powerful reducing agent, decomposing water to give hydrogen and B_{12r}. These reductions can apparently be carried out *in vivo* by reduced ferredoxin (see page 872). When cyano or hydroxo cobalamin is reduced, the ligand, CN^- or OH^-, is lost and the Co(I) complex is 5-coordinate. There is considerable evidence that these 5-coordinate Co(I) species react with adenosine triphosphate in presence of a suitable enzyme to generate the B_{12} coenzyme. In non-enzymic systems rapid reaction of B_{12s} occurs with alkyl halides, acetylenes, etc., as shown below, where [Co] represents the cobalamin group.

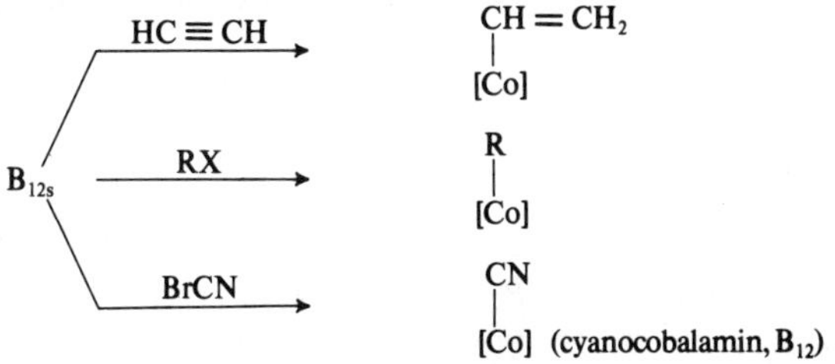

$$B_{12s} \nearrow \xrightarrow{\quad HC \equiv CH \quad} \begin{array}{c} CH = CH_2 \\ | \\ [Co] \end{array}$$

$$\xrightarrow{\quad RX \quad} \begin{array}{c} R \\ | \\ [Co] \end{array}$$

$$\searrow \xrightarrow{\quad BrCN \quad} \begin{array}{c} CN \\ | \\ [Co] \ (\text{cyanocobalamin}, B_{12}) \end{array}$$

Methylcobalamin has an extensive chemistry, some of which is doubtless involved in the metabolism of methane-producing bacteria, and it has been shown that it transfers CH_3 groups to Hg^{II}, Tl^{III}, Pt^{II} and Au^{I}.[26]

Finally there are a considerable number of model systems that consist of a rigid planar ligand system with the axial sites occupied by a base and by an anion that may be a carbanion. Most of these are complexes of Schiff base type, but the best known is cobalt(II) dimethylglyoximate which gives the

---

[26] G. Agnes *et al.*, *Chem. Comm.*, **1971**, 850.

"cobaloximes."[27] These species give reduced Co^I complexes, mostly blue or green, but, in presence of tertiary phosphines, hydrido complexes such as $HCo(DMGH)_2PBu_3^n$ are formed. The cobalt(I) and hydrido complexes are readily alkylated by alkyl halides, epoxides or tosylates, and the reaction is evidently S_N2 in character since inversion of configuration at carbon occurs[28] (cf. page 776). Existence of various compounds with alkyl groups has been established.[29] Alkyls can also be made by interaction of olefins and hydrido-cobalamin species formed on reduction of acetic acid solutions of hydro-oxocobalamin with zinc.[30]

Cobalt has a remarkable facility for replacing zinc in a number of zinc metalloenzymes, presumably because of the similar size of Co^{2+} and Zn^{2+} and the fact that Co^{2+} forms tetrahedral and octahedral complexes with about equal ease. The Co^{2+}/Zn^{2+} replacement often conserves and some-times enhances enzymic function. While the occurrence of Co^{2+} in the en-zymes *in vivo* is not of importance, it has been very useful as a spectrochem-ical probe in studying several enzymes.[31]

25-G. NICKEL

The trend toward decreased stability of higher oxidation states continues with nickel, so that only Ni^{II} occurs in the ordinary chemistry of the element. Even in the few compounds *formally* containing Ni^{III} and Ni^{IV}, especially the latter, there is doubt about the physical significance of these oxidation num-bers. Lower-valent nickel is also uncommon, except for compounds con-taining strongly π-bonding ligands. However, the relative simplicity of nickel chemistry in terms of oxidation number is balanced by considerable com-plexity in coordination numbers and geometries.

The oxidation states and stereochemistry of nickel are summarized in Table 25-G-1.

25-G-1. The Element

Nickel occurs in Nature mainly in combination with arsenic, antimony and sulfur, for example, as *millerite*, NiS, as a red nickel ore that is mainly NiAs, and in deposits consisting chiefly of $NiSb$, $NiAs_2$, $NiAsS$ or $NiSbS$. The most important deposits commercially are *garnierite*, a magnesium—

[27] G. N. Schrauzer, *Accounts Chem. Res.*, 1968, **1**, 97; G. N. Schrauzer *et al.*, *J. Amer. Chem. Soc.*, 1970, **92**, 1551, 2997, 7078; 1971, **93**, 1505; G. Costa *et al.*, *J. Chem. Soc.*, A, **1970**, 2870; B. T. Golding *et al.*, *Angew. Chem. Internat. Edn.*, 1970, **9**, 959; U. Belluco *et al.*, *Inorg. Chim. Acta Rev.*, 1970, **4**, 41.

[28] F. R. Jensen, V. Madan and D. H. Buchanan, *J. Amer. Chem. Soc.*, 1971, **93**, 5283; L. Marzilli, P. A. Marzilli and J. Halpern, *J. Amer. Chem. Soc.*, 1971, **93**, 1374.

[29] M. Calligaris *et al.*, *J. Chem. Soc.*, A, **1971**, 2720; J. Booth *et al.*, *J. Chem. Soc.*, A, **1971**, 1964; G. Costa *et al.*, *Chem. Comm.*, **1971**, 706.

[30] G. N. Schrauzer and R. J. Holland, *J. Amer. Chem. Soc.*, 1971, **93**, 4060.

[31] B. L. Vallee and W. E. C. Wacker, *The Proteins*, 2nd edn., Vol. 5, Academic Press, 1970.

TABLE 25-G-1

Oxidation States and Stereochemistry of Nickel

Oxidn. state	Coordn. no.	Geometry	Examples
Ni^{-I}	4?	?	[Ni$_2$(CO)$_6$]$^{2-}$
Ni^0	3	?	Ni[P(OC$_6$H$_4$-o-Me)$_3$]$_3$
	4	Tetrahedral	Ni(PF$_3$)$_4$, [Ni(CN)$_4$]$^{4-}$, Ni(CO)$_4$
	5	?	NiH[P(OEt)$_3$]$_4^+$
Ni^I, d^9	4	Tetrahedral	Ni(PPh$_3$)$_3$Br
Ni^{II}, d^8 ^{c}	4^a	Square	NiBr$_2$(PEt$_3$)$_2$, [Ni(CN)$_4$]$^{2-}$
	4^a	Tetrahedral	[NiCl$_4$]$^{2-}$, NiCl$_2$(PPh$_3$)$_2$
	5	Square pyramidal	[Ni(CN)$_5$]$^{3-}$, BaNiS$_2$, [Ni$_2$Cl$_8$]$^{4-}$
	5	tbp	[NiX(QAS)]$^+$, [Ni(CN)$_5$]$^{3-}$,
			[NiP{CH$_2$CH$_2$CH$_2$AsMe$_2$}$_3$CN]$^+$
	6^a	Octahedral	NiO, [Ni(NCS)$_6$]$^{4-}$, KNiF$_3$,
			Ni(DMGH)$_2$,^{b} [Ni(bipy)$_3$]$^{2+}$
	6	Trigonal prism	NiAs
Ni^{III}, d^7	5	tbp	NiBr$_3$(PR$_3$)$_2$
	6	Octahedral (distorted)	[Ni(diars)$_2$Cl$_2$]$^+$, [NiF$_6$]$^{3-}$
Ni^{IV}, d^6	6	Octahedral (distorted)	K$_2$NiF$_6$, [Ni(Bu$_2$dtc)$_3$]$^{+d}$

^{a} Most common state.
^{b} Square set of nitrogen atoms about Ni with long Ni—Ni bonds.
^{c} Three-coordinate NiCl$_3^-$ may be present [1] in molten CsAlCl$_4$ at 500°.
^{d} Bu$_2$dtc = N, N'-dibutyldithiocarbamate.

nickel silicate of variable composition, and certain varieties of the iron mineral *pyrrhotite* (Fe$_n$S$_{n+1}$) which contain 3–5% Ni. Elemental nickel is also found alloyed with iron in many meteors, and the central regions of the earth are believed to contain considerable quantities. The metallurgy of nickel is complicated in its details, many of which vary a good deal with the particular ore being processed. In general, the ore is transformed to Ni$_2$S$_3$ which is roasted in air to give NiO, and this is then reduced with carbon to give the metal.[2] Some high-purity nickel is made by the *carbonyl process*: carbon monoxide reacts with impure nickel at 50° and ordinary pressure or with nickel–copper matte under more strenuous conditions, giving volatile Ni(CO)$_4$, from which metal of 99.90–99.99% purity is obtained on thermal decomposition at 200°.

Nickel is silver-white, with high electrical and thermal conductivities (both ~15% of those of silver) and m.p. 1452°, and it can be drawn, rolled, forged and polished. It is quite resistant to attack by air or water at ordinary temperatures when compact and is therefore often electroplated as a protective coating. Because nickel reacts but slowly with fluorine, the metal and certain alloys (Monel) are used to handle F$_2$ and other corrosive fluorides. It is also ferromagnetic, but not so much as iron. The finely divided metal is reactive to air, and it may be pyrophoric under some conditions.

The metal is moderately electropositive:

$$Ni^{2+} + 2e = Ni \qquad E^0 = -0.24 \text{ V}$$

[1] J. Brynestead and G. P. Smith, *J. Amer. Chem. Soc.*, 1970, **92**, 3198.
[2] J. R. Boldt, Jr., *The Winning of Nickel*, Methuen, 1967.

and dissolves readily in dilute mineral acids. Like iron, it does not dissolve in concentrated nitric acid because it is rendered passive by this reagent.

NICKEL COMPOUNDS

25-G-2. The Chemistry of Divalent Nickel, d^8

In the divalent state nickel forms a very extensive series of compounds. This is the only oxidation state of importance in the aqueous chemistry of nickel, and, with the exception of a few special complexes of nickel in other oxidation states, Ni^{II} is also the only important oxidation level in its non-aqueous chemistry.

Binary Compounds. *Nickel*(II) *oxide*, a green solid, with the rock-salt structure is formed when the hydroxide, carbonate, oxalate or nitrate of nickel(II) is heated. It is insoluble in water but dissolves readily in acids.

The *hydroxide*, $Ni(OH)_2$, may be precipitated from aqueous solutions of Ni^{II} salts on addition of alkali-metal hydroxides forming a voluminous green gel which crystallizes [$Mg(OH)_2$ structure] on prolonged storage. It is readily soluble in acid ($K_{sp} = 2 \times 10^{-16}$), and also in aqueous ammonia owing to the formation of ammine complexes. When a concentrated solution of NaOH is added to a considerable molar excess of dilute $Ni(ClO_4)_2$ solution a soluble hydroxo species is formed[3] that is believed, on the basis of equilibrium and kinetic studies, to be $[Ni(OH)]_4$ with a cubic structure consisting of interpenetrating Ni_4 and $(OH)_4$ tetrahedra. $Ni(OH)_2$ is, however, not amphoteric.

Addition of sulfide ions to aqueous solutions of nickel(II) ions precipitates black NiS. This is initially freely soluble in acid, but, like CoS, on exposure to air it soon becomes insoluble owing to oxidation to Ni(OH)S. Fusion of Ni, S and BaS gives $BaNiS_2$, which forms black plates; this product is metallic in nature and has Ni in square-pyramidal coordination.[4]

All four nickel *halides* are known in the anhydrous state. Except for the fluoride, which is best made indirectly, they can be prepared by direct reaction of the elements. All the halides are soluble in water (the fluoride only moderately so), and from aqueous solutions they can be crystallized as the hexahydrates, except for the fluoride which gives $NiF_2 \cdot 3H_2O$. Lower hydrates are obtained from these on storage or heating.

On addition of CN^- ions to aqueous Ni^{II} the *cyanide* is precipitated in a green hydrated form. When heated at 180–200° the hydrate is converted into the yellow-brown, anhydrous $Ni(CN)_2$. The green precipitate readily redissolves in an excess of cyanide to form the yellow $[Ni(CN)_4]^{2-}$ ion, and many hydrated salts of this ion, for example, $Na_2[Ni(CN)_4] \cdot 3H_2O$, may be crystallized from such solutions. In strong cyanide solutions a further CN^- is

[3] G. B. Kolski, N. K. Kildahl and D. W. Margerum, *Inorg. Chem.*, 1969, **8**, 1211.
[4] I. E. Grey and M. Steinfink, *J. Amer. Chem. Soc.*, 1970, **92**, 5093.

taken up to give the red $[Ni(CN)_5]^{3-}$ ion, the structure of which is discussed below. Nickel(II) thiocyanate is also known, as a yellow-brown hydrated solid which reacts with an excess of SCN^- to form complex ions, $[Ni(NCS)_4]^{2-}$ and $[Ni(NCS)_6]^{4-}$.

Other Binary Nickel(II) Compounds. A number of binary nickel compounds, probably all containing Ni^{II} but not all stoichiometric, may be obtained by direct reaction of nickel with various non-metals such as P, As, Sb, S, Se, Te, C and B. Nickel appears to form a nitride Ni_3N. The existence of a hydride is doubtful although the finely divided metal absorbs hydrogen in considerable amounts.

Salts of Oxo Acids. A large number of these are known. They occur most commonly as hydrates, for example, $Ni(NO_3)_2 \cdot 6H_2O$, $NiSO_4 \cdot 7H_2O$, and most of them are soluble in water. Exceptions are the carbonate, $NiCO_3 \cdot 6H_2O$, which is precipitated on addition of alkali hydrogen carbonates to solutions of Ni^{II}, and the phosphate, $Ni_3(PO_4)_2 \cdot 7(?)H_2O$.

Aqueous solutions of Ni^{II} not containing strong complexing agents contain the green hexaquonickel(II) ion, $[Ni(H_2O)_6]^{2+}$, which also occurs in a number of hydrated nickel(II) salts, e.g., $Ni(NO_3)_2 \cdot 6H_2O$, $NiSO_4 \cdot 6H_2O$, $NiSO_4 \cdot 7H_2O$, $Ni(ClO_4)_2 \cdot 6H_2O$.

25-G-3. Stereochemistry and Electronic Structures of Nickel(II) Complexes[5]

Nickel(II) forms a large number of complexes encompassing coordination numbers 4, 5 and 6, and all the main structural types, viz., octahedral, trigonal bipyramidal, square-pyramidal, tetrahedral, and square. Moreover, it is characteristic of Ni^{II} complexes that complicated equilibria, which are generally temperature-dependent and sometimes concentration-dependent, often exist between these structural types. In this Section we shall describe the characteristics of the individual structural types separately, using as examples mainly those compounds that exist completely or almost completely in one form or another. In the next Section we shall discuss the configurational equilibria.

Octahedral Complexes. The maximum coordination number of nickel(II) is 6. A considerable number of neutral ligands, especially amines, displace some or all of the water molecules in the octahedral $[Ni(H_2O)_6]^{2+}$ ion to form complexes such as trans-$[Ni(H_2O)_2(NH_3)_4](NO_3)_2$, $[Ni(NH_3)_6]$-$(ClO_4)_2$, [Ni en$_3$]SO$_4$, etc. Such ammine complexes characteristically have blue or purple colors in contrast to the bright green color of the hexaaquonickel ion. This is because of shifts in the absorption bands when H_2O ligands are replaced by others lying toward the stronger end of the spectrochemical series. This can be seen in Fig. 25-G-1, where the spectra of $[Ni(H_2O)_6]^{2+}$ and [Ni en$_3$]$^{2+}$ are shown. These spectra can readily be interpreted by referring to the energy-level diagram for d^8 ions (page 576). Three spin-allowed transitions are expected, and the three observed bands in

[5] L. Sacconi, *Transition Metal Chem.*, 1968, **4**, 199 (a review).

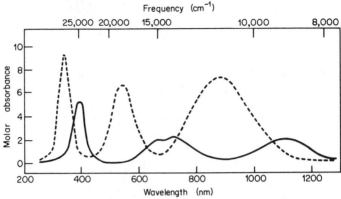

Fig. 25-G-1. Absorption spectra of $[Ni(H_2O)_6]^{2+}$ (——) and $[Ni\ en_3]^{2+}$ (– – –).

each spectrum may thus be assigned as shown in Table 25-G-2. It is a characteristic feature of the spectra of octahedral nickel(II) complexes, exemplified by those of $[Ni(H_2O)_6]^{2+}$ and $[Ni\ en_3]^{2+}$, that molar absorbances of the bands are at the low end of the range (1–100) for octahedral complexes of the first transition series in general, namely, between 1 and 10. The splitting of the middle band in the $[(Ni(H_2O)_6]^{2+}$ spectrum is due to spin-orbit coupling which mixes the $^3T_{1g}(F)$ and ^1E_g states which are very close in energy at the Δ_0 value given by $6H_2O$, whereas in the stronger field of the 3en they are so far apart that no significant mixing occurs.

TABLE 25-G-2

Spectra of Octahedral Nickel(II) Complexes
(approximate band positions in cm^{-1})

Transition	$[Ni(H_2O)_6]^{2+}$	$[Ni\ en_3]^{2+}$
$^3A_{2g}\rightarrow{}^3T_{2g}$	9,000	11,000
$^3A_{2g}\rightarrow{}^3T_{1g}(F)$	14,000	18,500
$^3A_{2g}\rightarrow{}^3T_{1g}(P)$	25,000	30,000

Magnetically, octahedral nickel(II) complexes have relatively simple behavior. From both a simple d-orbital splitting diagram (page 563) and the energy level diagram (page 576), it follows that all of them should have two unpaired electrons, and this is found always to be the case, the magnetic moments ranging from 2.9 to 3.4 B.M. depending on the magnitude of the orbital contribution.

5-Coordinate Nickel(II) Complexes. A considerable number of both trigonal-bipyramidal and square-pyramidal complexes occur and high-$(S=1)$ and low-spin $(S=0)$ examples of each geometry are known.[6] Many of the

[6] (a) M. Ciampolini, *Structure and Bonding*, 1969, **6**, 52; (b) L. Sacconi, *J. Chem. Soc., A*, **1970**, 248; (c) P. L. Orioli and C. A. Ghilardi, *J. Chem. Soc., A*, **1970**, 1511; J. K. Stalick and J. A. Ibers, *Inorg. Chem.*, 1970, **9**, 453; M. Gerloch *et al.*, *J. Chem. Soc., A*, **1970**, 3269; P. Orioli, *Coordination Chem. Rev.*, 1971, **6**, 285.

trigonal bipyramidal complexes are of the type (25-G-I) where the tetra-dentate "tripod" ligand may have X = N, P or As, and Y = N, P, As, S, or Se. A few examples are of such ligands: $Me_6tren = N[CH_2CH_2N(CH_3)_2]_3$, $TPN = N[CH_2CH_2P(C_6H_5)_2]_3$; and $TSP = P(o\text{-}C_6H_4SCH_3)_3$. The ligand Z is generally Cl^-, Br^- or I^- and the complexes (25-G-I) are thus cationic. The

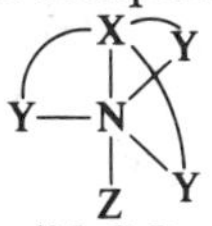

(25-G-I)

actual symmetry is at best C_{3v} rather than the D_{3h} of a true *tbp* and in some cases this is reflected in the complexity of the electronic spectra. Other 5-coordinate complexes which do not contain tripod ligands are several of the $Ni(CN)_2(PR_3)_3$ type,[7] as well as

$$[Ni(Ph_2MeAsO)_4(ClO_4)]ClO_4, \quad Ni(CN)_5^{3-} \quad and \quad NiBr_2NH(CH_2CH_2PPh_2)_2$$

By far the commonest stereochemistry observed is trigonal-bipyramidal (or something approximating thereto), but this may be mainly due to the fact that in most cases a tripod ligand has been used. In the above list only the last three complexes are known in square-pyramidal form, $Ni(CN)_5^{3-}$ being known to occur in also the *tbp* configuration.[8]

Only a few high-spin complexes are known and only $[Ni(Me_6tren)Br]^+$ (*tbp*) and $[Ni(Ph_2MeAsO)_4(ClO_4)]^+$ (*spy*) have been studied in detail spectroscopically. In both cases the observed spectra can be semiquantitatively accounted for by a ligand-field treatment.[6a] Low-spin (i.e. diamagnetic) complexes are numerous and many studies of their spectra have been published. Low-spin complexes occur[6b] in nearly all cases where the ligand set consists mainly of heavier atoms such as P, As, S, Se, Br and I; their spectra have been analyzed by both ligand-field and MO methods with reasonable success.[6a]

Tetrahedral Complexes.   These are mainly of the following stoichiometric types: NiX_4^{2-}, NiX_3L^-, NiL_2X_2, and $Ni(L\text{---}L)_2$, where X represents a halogen, L a neutral ligand such as a phosphine, phosphine oxide or arsine, and L—L is one of several types of bidentate ligand, (25-G-II) to (25-G-IV) being examples. These three bidentate ligands all contain sufficiently bulky substituents on, or adjacent to, the nitrogen atoms to render planarity of the

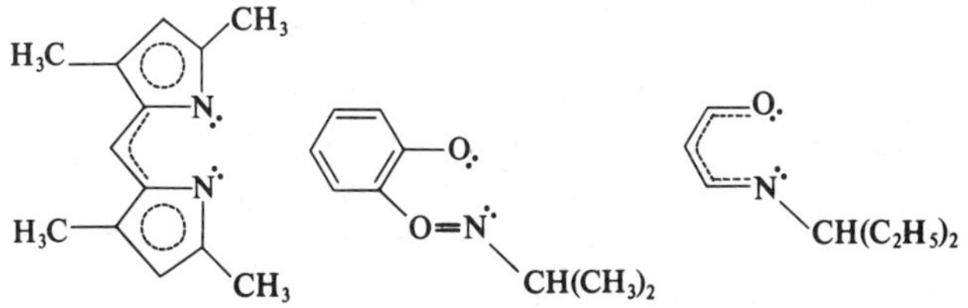

(25-G-II)   (25-G-III)   (25-G-IV)

[7] See, e.g., E. C. Alyea and D. W. Meek, *J. Amer. Chem. Soc.*, 1969, **91**, 5761.
[8] A. Terzis, K. N. Raymond and T. G. Spiro, *Inorg. Chem.*, 1970, 9, 2415.

Ni(L—L)$_2$ molecule sterically impossible. When small substituents are present, planar or nearly planar complexes are formed. It must be stressed that except for the NiX$_4^{2-}$ species a rigorously tetrahedral configuration cannot be expected. However, in some cases there are marked distortions even from the highest symmetry possible, given the inherent shapes of the ligands. Thus, in Ni(L—L)$_2$ molecules the most symmetrical configuration possible would have the planes of the two L—L ligands perpendicular. Most often, however, this dihedral angle differs considerably from 90°; for example when L—L is (25-G-III) the angle is 82° and when L—L is (25-G-II) it is only 76°. Thus the term "tetrahedral" is sometimes used very loosely (i.e. does not imply a regular tetrahedron); since all the so-called tetrahedral species are paramagnetic with two unpaired electrons, it would perhaps be better to simply call them paramagnetic rather than tetrahedral. Indeed, the most meaningful way to distinguish between "tetrahedral" and "planar" 4-coordinate nickel(II) complexes is to consider that for a given ligand set, ABCD, there is a critical value of the dihedral angle between two planes, such as A—Ni—B and C—Ni—D. When the angle exceeds this value the molecule will be paramagnetic; it may be called "tetrahedral" even though the dihedral angle is appreciably less than 90°. Conversely, when the angle is below the critical value the complex will be diamagnetic; it may be called "planar" even if the limit of strict planarity is not actually attained.

For regular or nearly regular tetrahedral complexes there are characteristic spectral and magnetic properties. Naturally the more irregular the geometry of a paramagnetic nickel(II) complex the less likely it is to conform to these specifications. In T_d symmetry the d^8 configuration gives rise to a $^3T_1(F)$ ground state. The transition from this to the $^3T_1(P)$ state occurs in the visible region ($\sim 15,000$ cm^{-1}) and is relatively strong ($\varepsilon \approx 10^2$) compared to the corresponding $^3A_{2g} \rightarrow {}^3T_{1g}$ transition in octahedral complexes. Thus tetrahedral complexes are generally strongly colored and tend to be blue or green unless the ligands also have absorption bands in the visible region. Because the ground state, $^3T_1(F)$, has much inherent orbital angular momentum, the magnetic moment of truly tetrahedral Ni^{II} should be about 4.2 B.M. at room temperature. However, even slight distortions reduce this markedly (by splitting the orbital degeneracy). Thus, fairly regular tetrahedral complexes have moments of 3.5–4.0 B.M., while for the more distorted ones the moments are 3.0–3.5 B.M., i.e. in the same range as for six-coordinate complexes.

Planar Complexes. For the vast majority of 4-coordinate nickel(II) complexes, planar geometry is preferred. This is a natural consequence of the d^8 configuration, since the planar ligand set causes one of the d orbitals ($d_{x^2-y^2}$) to be uniquely high in energy and the eight electrons can occupy the other four d orbitals but leave this strongly antibonding one vacant. In tetrahedral coordination, on the other hand, occupation of antibonding orbitals is unavoidable. With the congeneric d^8 systems Pd^{II} and Pt^{II} this factor becomes so important that no tetrahedral complex is formed.

Planar complexes of Ni^{II} are thus invariably diamagnetic. They are frequently red, yellow or brown owing to the presence of an absorption band of medium intensity ($\varepsilon \approx 60$) in the range 450–600 nm, but other colors do occur when additional absorption bands are present.

As important examples of square complexes, we may mention yellow $Ni(CN)_4^{2-}$, red bis(dimethylglyoximato)nickel(II) (25-G-V), the red β-keto-

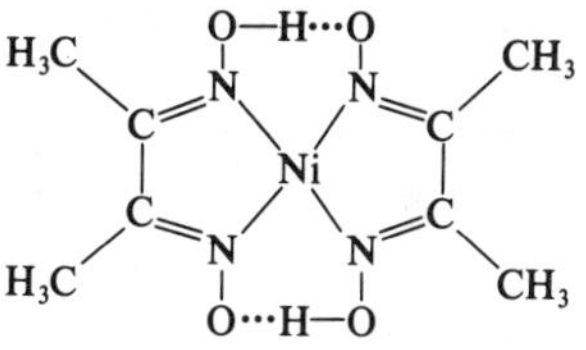

(25-G-V) 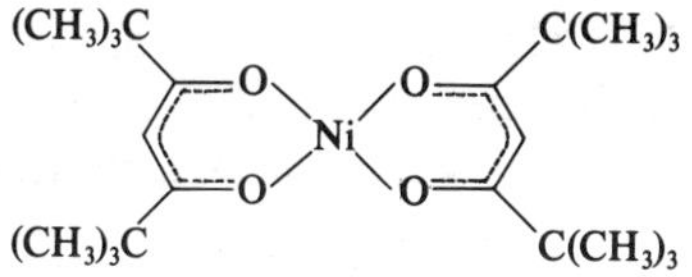   (25-G-VI)

enolate complex (25-G-VI), the yellow to brown $Ni(PR_3)_2X_2$ compounds in which R is alkyl, and complexes containing homologs of the ligands (25-G-III) and (25-G-IV) in which the substituents on nitrogen are small.

25-G-4. "Anomalous" Properties of Nickel(II) Complexes; Conformational Changes

A considerable number of nickel(II) complexes do not behave consistently in accord with expectation for any one of the discrete structural types just described, and they have in the past been termed "anomalous." It now appears that all the "anomalies" can be satisfactorily explained in terms of several types of conformational or other structural change and, ironically, there are so many examples now known that the term "anomalous" can no longer be considered appropriate. The three main structural and conformational changes which nickel(II) complexes undergo are described and illustrated below.

1. *Formation of 5- and 6-Coordinate Complexes by Addition of Ligands to Square Ones.* For any square complex, NiL_4, the following equilibria with additional ligands, L' must in principle exist:

$$ML_4 + L' = ML_4L'$$
$$ML_4 + 2L' = ML_4L'_2$$

In the case where $L = L' = CN$, only the 5-coordinate species is formed, but in most systems studied, in which L' is a good donor such as pyridine, H_2O, C_2H_5OH, etc., the equilibria lie far in favor of the 6-coordinate species, which have a *trans*-structure and a high-spin electron configuration; many may be isolated as pure compounds. Thus, the complex (25-G-VI) is normally prepared in the presence of water and/or alcohol and is first isolated as the green, paramagnetic dihydrate or dialcoholate, from which the red, square complex is then obtained by heating to drive off the H_2O or C_2H_5OH. Similarly, various diamagnetic, square complexes of the salicylaldiminato type become paramagnetic to the degree expected for the presence of two unpaired electrons when dissolved in pyridine, and the dipyridine complexes can be isolated as stable, paramagnetic, octahedral materials. There are, however, cases in which solutions of square complexes attain only a fraction of the paramagnetism expected for complete conversion into octahedral species, thus indicating that the above equilibrium proceeds only part way to the right.

Well-known examples of the square–octahedral ambivalence are provided by the Lifschitz salts, which are complexes of nickel(II) with substituted ethylenediamines, especially the stilbenediamines, one of which is illustrated in (25-G-VII). Many years ago, Lifschitz and others observed that such

$$\left[\begin{array}{c} \text{(C}_6\text{H}_5\text{)HC}\!-\!\overset{H_2}{N}\diagdown \diagup \overset{H_2}{N}\!-\!\text{CH(C}_6\text{H}_5) \\ \text{Ni} \\ \text{(C}_6\text{H}_5\text{)HC}\!-\!\underset{H_2}{N}\diagup \diagdown \underset{H_2}{N}\!-\!\text{CH(C}_6\text{H}_5) \end{array} \right]^{2+}$$

(25-G-VII)

complexes were sometimes blue and paramagnetic and at other times yellow and diamagnetic, depending upon many factors such as temperature, identity of the anions present, the solvent in which they are dissolved or from which they were crystallized, exposure to atmospheric water vapor and the particular diamine involved. The bare experimental facts bewildered chemists for several decades and many hypotheses were promulgated in an effort to explain some or all of the facts. It is now recognized that the yellow species are square complexes, as typified by (25-G-VII), while the blue ones are octahedral complexes, derived from the square ones by coordination of two additional ligands—solvent molecules, water molecules or anions—above and below the plane of the square complex.

2. *Monomer–Polymer Equilibria.* In many cases, four-coordinate complexes associate or polymerize, to give species in which the nickel ions become five- or six-coordinate. In some cases, the association is very strong and the four-coordinate monomers are observed only at high temperatures, while in others the position of the equilibrium is such that both red, diamagnetic monomers and green or blue, paramagnetic polymers are present in a temperature- and concentration-dependent equilibrium around room temperature. A clear example of this situation is provided by various β-ketoenolate complexes. When the β-ketoenolate is the acetylacetonate ion, the trimeric structure shown in Fig. 25-G-2 is adopted. As a result of the sharing of some oxygen atoms, each nickel atom achieves octahedral coordination; the situation is comparable to, but different in detail from, that found for $[\text{Co(acac)}_2]_4$. This trimer is very stable and only at temperatures around 200° (in a non-coordinating solvent) do detectable quantities of monomer appear. It is, however, readily cleaved by donors such as H_2O or pyridine, to give six-coordinate monomers. When the methyl groups of the acetylacetonate ligand are replaced by the very bulky $C(CH_3)_3$ group trimerization is completely prevented and the planar monomer (25-G-VI) results. When groups

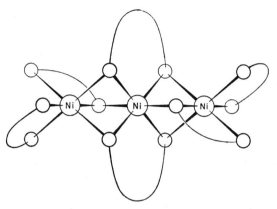

Fig. 25-G-2. Sketch indicating the trimeric structure of nickel acetylacetonate. The unlabelled circles represent oxygen atoms, and the curved lines connecting them in pairs represent the remaining portions of the acetylacetonate rings. [Reproduced by permission from J. C. Bullen, R. Mason and P. Pauling, *Inorg. Chem.*, 1965, **4**, 456.]

sterically intermediate between CH_3 and $C(CH_3)_3$ are used, temperature- and concentration-dependent monomer–trimer equilibria are observed in non-coordinating solvents.

Partial dimerization, presumably to give 5-coordinate, high-spin nickel(II), is known to be the cause of anomalous behavior in some instances. Thus, while *N*-(*n*-alkyl)salicylaldiminato complexes of Ni^{II} are, in general, planar, diamagnetic monomers in chloroform or benzene, when the alkyl group is CH_3 there is an equilibrium between the diamagnetic monomer and a paramagnetic dimer, presumably as shown in (25-G-VIII) and (25-G-IX).

(25-G-VIII)   (25-G-IX)

3. *Square–Tetrahedral Equilibria and Isomerism.* We have already indicated that nickel(II) complexes of certain stoichiometric types, namely, the bishalobisphosphino and bissalicylaldiminato types may have either square or tetrahedral structures, depending on the identity of the ligands. For example, in the NiL_2X_2 cases, when L is triphenylphosphine, tetrahedral structures are found, whereas the complexes with trialkylphosphines generally give square complexes. Perhaps it is then not very surprising that a number

of NiL_2X_2 complexes in which L represents a mixed alkylarylphosphine exist in solution in an equilibrium distribution between the tetrahedral and square forms. Moreover, in some cases it is possible to isolate two crystalline forms of the compound, one yellow to red and diamagnetic, the other green or blue with two unpaired electrons. There is even a case, $Ni[(C_6H_5CH_2)-(C_6H_5)_2P]_2Br_2$, in which both tetrahedral and square complexes are found together in the same crystalline substance.[9]

For a series of NiL_2X_2 complexes in which X = Cl, Br or I, and L is a $(CH_3)(p\text{-}ZC_6H_4)(p\text{-}Z'C_6H_4)P$-type phosphine, nmr studies over a temperature range have demonstrated[10] that at 25° the rate constants for conversion of tetrahedral into planar isomers are in the range 10^5–10^6 sec^{-1} with enthalpies of activation of around 45 kJ mol^{-1}.

The most important and thoroughly examined examples of square–tetrahedral equilibria are provided by the N-(sec-alkyl)salicylaldiminato complexes and the aminotropone iminato complexes (25-G-X).[11] In the tetra-

(25-G-X)

hedral forms unpaired electron spin density from the nickel atoms is introduced into the ligand π system, which results in large shifts in the positions of the various proton nuclear magnetic resonances. The various shifts are in proportion to the spin density at the carbon atom to which the proton is attached, so that from the nmr spectra much can be learned of the nature of the π orbitals into which the spin density is introduced. Such studies have therefore proved to be very important in respect to the electronic structures of a wide variety of aromatic systems which can be attached to the basic ring systems of these complexes, but the detailed results are outside the scope of this book. However, in the course of these studies, the square–tetrahedral equilibria themselves have been studied carefully.

To a certain extent the position of the equilibrium is a function of steric factors, that is of the repulsion between the R groups on the nitrogen atoms of one ligand and various parts of the other ligand, the greater degree of repulsion encountered in the square configuration tending to shift the equilibrium to the tetrahedral side. However, some ring substituents affect the equilibrium by means of electronic effects as well. The ΔH values for the square–tetrahedral conversion are generally of the order of a few kJ per mole and are positive, meaning that the proportion of tetrahedral form in-

[9] B. T. Kilbourn and H. M. Powell, J. Chem. Soc., A, 1970, 1688.
[10] L. H. Pignolet, W. DeW. Horrocks, Jr. and R. H. Holm, J. Amer. Chem. Soc., 1970, 92, 1855.
[11] R. H. Holm and M. J. O'Conner, Progr. Inorg. Chem., 1971, 14, 408.

creases with increasing temperature. In a few cases, however, when extremely bulky substituents are used, the equilibrium lies predominantly on the tetrahedral side. For example in the salicylaldiminato complex with $R = C(CH_3)_3$, ΔH is negative.

25-G-5. Higher Oxidation States of Nickel

Hydroxides and Oxides. There is no evidence for anhydrous oxides of Ni^{III} and Ni^{IV} but there are a number of hydrous oxides and mixed metal oxides, some of considerable complexity, that contain Ni^{III} and Ni^{IV}.

The best defined hydroxide is β-NiO(OH),[12] a black powder obtained by the oxidation of nickel(II) nitrate solutions with bromine in aqueous potassium hydroxide below 25°. It is readily soluble in acids; on ageing, or by oxidation in hot solutions, a Ni^{II}—Ni^{III} hydroxide of stoichiometry $Ni_3O_2(OH)_4$ is obtained. The oxidation of alkaline nickel sulfate solutions by NaOCl gives a black "peroxide," "$NiO_2 \cdot nH_2O$." This is unstable, being readily reduced by water, but is a useful oxidizing agent for organic compounds.[12a]

Electrochemical oxidation of $Ni(OH)_2$ in alkaline solution gives a black oxide that does not have a unique stoichiometry and retains alkali-metal ions; its X-ray pattern is related to that of $MNiO_2$. Further oxidation gives a grey metallic material containing both Ni^{III} and Ni^{IV}.

The Edison or nickel–iron battery, which uses KOH as the electrolyte, is based on the reaction:

$$Fe + 2NiO(OH) + 2H_2O \underset{\text{charge}}{\overset{\text{Discharge}}{\rightleftharpoons}} Fe(OH)_2 + 2Ni(OH)_2 \quad (\sim 1.3 \text{ V})$$

but the mechanism and the true nature of the oxidized nickel species are not fully understood.

There are also various mixed oxides obtained in dry ways. Thus $MNiO_2$ is made by bubbling oxygen through molten alkali-metal hydroxides contained in nickel vessels at about 800°. Other oxides and oxide phases can be made by heating NiO with alkali or alkaline-earth oxides in oxygen. These mixed oxides evolve oxygen on treatment with water or acid.

Complexes of Tetravalent Nickel, d^6. There are several reports of compounds believed to contain Ni(IV) stabilized by coordination with ions of highly electronegative elements, viz., a heteropolymolybdate anion of composition $[NiMo_9O_{32}]^{6-}$, the heteropolyniobate complex[13] $[NiNb_{12}O_{38}]^{12-}$, the periodates $Na(K)NiIO_6 \cdot nH_2O$ and the red K_2NiF_6.[14] Carbollide complexes are also known.[15] All are diamagnetic or only weakly paramagnetic, as would be expected for a low spin t_{2g}^6 configuration.

[12] R. S. McEwen, *J. Phys. Chem.*, 1971, **75**, 1782.
[12a] R. Konaka, S. Terabe and K. Kurama, *J. Org. Chem.*, 1969, **34**, 1334; *J. Amer. Chem. Soc.*, 1969, **91**, 5655; B. T. Golding and D. R. Hall, *Chem. Comm.*, **1970**, 1574.
[13] C. M. Flynn, Jr. and G. D. Stucky, *Inorg. Chem.*, 1969, **8**, 332.
[14] N. A. Matwiyoff *et al.*, *Inorg. Chem.*, 1969, **8**, 750; G. C. Allen and K. D. Warren, *Inorg. Chem.*, 1969, **8**, 753.
[15] M. F. Hawthorne *et al.*, *J. Amer. Chem. Soc.*, 1969, **91**, 758.

Original formulations of sulfur-containing Ni^{IV} complexes with Ni—S—Ni bridges are incorrect and divalent Ni^{II} complexes of the type (25-G-XI)

(25-G-XI)

are present.[16] It also appears that the 1,2-dithiete or 1,2-dithiol complexes of the type $[Ni(S_2C_2R_2)_2]^{2-}$ are most realistically regarded as $Ni(II)$ species,[17] although from a purely formal point of view they might be considered to contain $Ni(IV)$ and $S_2C_2R_2^{2-}$ ligands.

Bromine oxidizes $Ni(Bu_2dtc)_2$ ($Bu_2dtc = N,N'$-dibutyldithiocarbamate) to $Ni(Bu_2dtc)_3Br$, and X-ray study shows this to contain a $[Ni(Bu_2dtc)_3]^+$ cation with a trigonally distorted (D_3) octahedral structure, that might seem rather obviously to contain $Ni(IV)$ coordinated by Bu_2dtc^- ligands. However, in view of the expected oxidizing nature of $Ni(IV)$ and the known ease of oxidation of dtc^- ions this is probably only a naive formalism.

Complexes of Trivalent Nickel, d^7. There are several documented examples of which the $NiX_3(PR_3)_2$ type is longest known. On indirect evidence (e.g., zero dipole moment) these have generally been formulated as trigonal bipyramidal. Recently this was verified crystallographically[18] for $NiBr_3(PMe_2Ph)_2$. The $[Ni(diars)_2Cl_2]^+$ cation is also well-defined but is the subject of a puzzling report in which esr and structural data are said to show[19] that the single unpaired electron occupies a Ni—As σ-*bonding* orbital!

Violet K_3NiF_6, prepared by fluorination of an intimate mixture of $3KCl + NiCl_2$, is reported[20] to contain low-spin nickel(III); the crystals are said to be cubic, though the low-spin d^7 configuration should engender a Jahn–Teller distortion.

A low-spin ($\mu_{eff}^{25°} = 1.7$ B.M.), presumably square, anionic complex containing dicarborane anions as ligands[21] has been reported recently. Earlier, blue-black $KNibi_2$ ($biH_2 = H_2NCONHCONH_2$) with $\mu_{eff}^{25°} = 2.5$ B.M. was mentioned and square coordination was also suggested.[22] X-ray study of these compounds would be desirable.

It has also been reported that $Ni(II)$ complexed with any of several macrocyclic tetraamines (page 627) can be oxidized electrolytically or with $NOBF_4$ to the $Ni(III)$ complex and also reduced to the $Ni(I)$ complex.[23]

[16] A. Avdeef, J. P. Fackler and R. G. Fischer, *J. Amer. Chem. Soc.*, 1970, **92**, 6972.
[17] E. J. Stiefel *et al.*, *J. Amer. Chem. Soc.*, 1967, **89**, 3016; A. L. Balch *et al.*, *J. Amer. Chem. Soc.*, 1967, **89**, 2301.
[18] J. K. Stalick and J. A. Ibers, *Inorg. Chem.*, 1970, **9**, 453.
[19] P. Kreisman *et al.*, *J. Amer. Chem. Soc.*, 1968, **90**, 1067.
[20] L. Stein *et al.*, *Inorg. Chem.*, 1969, **8**, 2472; G. C. Allen and K. D. Warren, *Inorg. Chem.*, 1969, **8**, 1895.
[21] D. A. Owen and M. F. Hawthorne, *J. Amer. Chem. Soc.*, 1970, **92**, 3194.
[22] J. J. Bour and J. J. Steggerda, *Chem. Comm.*, **1967**, 85.
[23] E. K. Barefield *et al.*, *Chem. Comm.*, **1970**, 552, 1718.

25-G-6. Lower Oxidation States of Nickel

The oxidation states $-I$, 0, and I are each represented; the compounds of Ni^0 are decidedly the most numerous. In all cases, however, the ligands involved are of the types with strong π-acid properties and except perhaps for Ni^I the formal oxidation numbers have little physical significance.

The oxidation state $-I$ is represented by the carbonylate anion, $[Ni_2(CO)_6]^{2-}$, of unknown structure. There is also a carbonylate anion $[Ni_4(CO)_9]^{2-}$ in which the formal oxidation state of Ni is $-\frac{1}{2}$, which serves to emphasize that extensive delocalization of electrons and probably Ni—Ni bonds are important in such compounds, thus rendering the classification in terms of oxidation state a pure formality.

The Ni^0 complexes, apart from organo derivatives such as $(PPh_3)_2Ni(H_2C{=}CH_2)$ and $Ni(1,5\text{-cyclooctadiene})_2$, are tetrahedral molecules in which the nickel atom is surrounded by four π-acid type ligands. The pre-eminent example is, of course, nickel carbonyl, $Ni(CO)_4$, which has been described in Chapter 22, along with various derivatives, $NiL_x(CO)_{4-x}$. There are also relatively stable NiL_4 species $[L = PF_3, PCl_3, P(OCN)_3, P(OR)_3]$ and $Ni(diphos)_2$,[24] as well as extremely reactive $K_4[Ni(CN)_4]$ and $K_4[Ni(C{\equiv}CH)_4]$. The phosphite complexes[24a] are of especial interest as they act as bases being readily protonated to give, e.g., $HNi[P(OEt)_3]_4^+$ and these species will catalyze the formation of hexadiene from ethylene and butadiene (page 796).

Nickel(I) complexes are rare. The most stable appear to be the three $Ni(PPh_3)_3X$ (X = Cl, Br and I) compounds[25] which only slowly decompose in air and are stable for long periods in nitrogen. The molecules are tetrahedral and paramagnetic ($\mu_{eff}^{25°} \sim 1.9$ B.M.). They are prepared by a curious process in which $[(h^3\text{-allyl})NiX]_2$ is treated with an excess of PPh_3 in presence of norbornene. By reduction of $(NC)_2(LL)Ni(LL)Ni(LL)(CN)_2$, where $LL = Ph_2P(CH_2)_nPPh_2$ ($n = 3$ or 4), the compounds $(NC)(LL)Ni(LL)Ni(LL)(CN)$ are obtained.[26] The solids are stable in an inert atmosphere; the Ni atoms in these paramagnetic molecules ($\mu_{eff}^{25°} = 2.0$–2.3 B.M.) have been assigned square coordination on indirect evidence.

The dark red cyanide, $K_4[Ni_2(CN)_6]$, made by Na/Hg reduction of $K_2Ni(CN)_4$ has a short Ni—Ni bond.[27]

25-H. COPPER

Copper has a single s electron outside the filled $4d$ shell but cannot be classed in Group I, since it has little in common with the alkalis except formal

[24] R. A. Schunn, *Inorg. Chem.*, 1970, **9**, 394.
[24a] C. A. Tolman, *J. Amer. Chem. Soc.*, 1970, **92**, 4217, 6777, 6785.
[25] L. Porri *et al.*, *Chem. Comm.*, **1967**, 228.
[26] B. Corain, *Chem. Comm.*, **1968**, 509.
[27] D. Bingham and M. G. Burnett, *J. Chem. Soc., A*, **1970**, 2165; O. Jarchow, H. Schultz, and R. Nast, *Angew. Chem., Internat. Edn.*, 1970, **9**, 71.

stoichiometries in the $+I$ oxidation state. The filled d shell is much less effective than is a noble-gas shell in shielding the s electron from the nuclear charge, so that the first ionization potential of Cu is higher than those of the alkalis. Since the electrons of the d shell are also involved in metallic bonding, the heat of sublimation and the melting point of copper are also much higher than those of the alkalis. These factors are responsible for the more noble character of copper, and the effect is to make the compounds more covalent and to give them higher lattice energies, which are not offset by the somewhat smaller radius of the unipositive ion compared to the alkali ions in the same period—Cu^+, 0.93; Na^+, 0.95; and K^+, 1.33 Å.

The second and third ionization potentials of Cu are very much lower than those of the alkalis and account in part for the transition-metal character shown by the existence of colored paramagnetic ions and complexes in the II and III oxidation states. Even in the I oxidation state numerous transition-metal-like complexes, for example, those with olefins, are formed.

There is only moderate similarity between copper and the heavier elements Ag and Au, but some points are noted in the later discussions of these elements (Chapter 26).

The oxidation states and stereochemistry of copper are summarized in Table 25-H-1. While stable copper(0) compounds are not confirmed, reactive intermediates appear to occur in some reactions.

TABLE 25-H-1

Oxidation States and Stereochemistry of Copper

Oxidn. state	Coordn. no.	Geometry	Examples
Cu^I, d^{10}	2	Linear	Cu_2O, $KCuO$, $CuCl_2^-$
	3	Planar	$K[Cu(CN)_2]$, $[Cu(SPMe_3)_3]ClO_4^c$
	4^a	Tetrahedral	CuI, $[Cu(CN)_4]^{3-}$
Cu^{II}, d^9	5	Trigonal bipyramidal	$[Cu(bipy)_2I]^+$, $[CuCl_5]^{3-}$, $[Cu_2Cl_8]^{4-}$
	5	Square pyramidal	$[Cu(DMGH)_2]_2(s)$
	$4^{a,b}$	Tetrahedral (distorted)	(N-isopropylsalicylaldiminato)$_2$Cu $Cs_2[CuCl_4]$
	$4^{a,b}$	Square	CuO, $[Cu py_4]^{2+}$, $(NH_4)_2[CuCl_4]$
	$6^{a,b}$	Distorted octahedral	K_2CuF_4, $K_2[CuEDTA]$, $CuCl_2$
	6	Octahedral	$K_2Pb[Cu(NO_2)_6]$
Cu^{III}, d^8	4	Square	$KCuO_2$
	6	Octahedral	K_3CuF_6

^{a} Most common states.
^{b} These three cases are often not sharply distinguished; see text.
^{c} P. E. Eller and P. W. R. Corfield, *Chem. Comm.*, **1971**, 105.

25-H-1. The Element

Copper is widely distributed in Nature as metal, in sulfides, arsenides, chlorides, and carbonates. It is extracted by oxidative roasting and smelting, or by microbial-assisted leaching, followed by electrodeposition from sulfate solutions.[1]

Copper is a tough, soft, and ductile reddish metal, second only to silver in its high thermal and electrical conductivities. It is used in many alloys such as brasses and is completely miscible with gold. It is only superficially oxidized in air, sometimes giving a green coating of hydroxo carbonate and hydroxo sulfate.

Copper reacts at red heat with oxygen to give CuO and, at higher temperatures, Cu_2O; with sulfur it gives Cu_2S or a non-stoichiometric form of this compound. It is attacked by halogens but is unaffected by non-oxidizing or non-complexing dilute acids in absence of air. Copper readily dissolves in nitric acid and sulfuric acid in presence of oxygen. It is also soluble in ammonia or potassium cyanide solutions in the presence of oxygen, as indicated by the potentials

$$Cu + 2NH_3 \xrightarrow{-0.12 \text{ V}} [Cu(NH_3)_2]^+ \xrightarrow{-0.01 \text{ V}} [Cu(NH_3)_4]^{2+}$$

COPPER COMPOUNDS

25-H-2. The Copper(I) State, d^{10}

Cuprous compounds are diamagnetic and, except where color results from the anion or charge-transfer bands, colorless.

The relative stabilities of the cuprous and cupric states are indicated by the following potential data:

$$Cu^+ + e = Cu \qquad E^0 = 0.52 \text{ V}$$
$$Cu^{2+} + e = Cu^+ \qquad E^0 = 0.153 \text{ V}$$

whence

$$Cu + Cu^{2+} = 2Cu^+ \qquad E^0 = -0.37 \text{ V}; \qquad K = [Cu^{2+}]/[Cu^+]^2 = \sim 10^6$$

The relative stabilities of Cu^I and Cu^{II} in aqueous solution depend very strongly on the nature of anions or other ligands present, and vary considerably with solvent or the nature of neighboring atoms in a crystal.

In aqueous solution only low equilibrium concentrations of Cu^+ ($<10^{-2} M$) can exist (see below) and the only cuprous compounds that are stable to water are the highly insoluble ones such as CuCl or CuCN. This instability towards water is due partly to the greater lattice and solvation energies and higher formation constants for complexes of the cupric ion, so that ionic Cu^I derivatives are unstable.

The equilibrium $2Cu^I \rightleftharpoons Cu + Cu^{II}$ can readily be displaced in either direction. Thus with CN^-, I^- and Me_2S, Cu^{II} reacts to give the Cu^I compound; with anions that cannot give covalent bonds or bridging groups, for example, ClO_4^- and SO_4^{2-}, or with complexing agents that have their greater affinity for Cu^{II}, the Cu^{II} state is favored—thus ethylenediamine reacts with cuprous chloride in aqueous potassium chloride solution:

$$2CuCl + 2en = [Cu \; en_2]^{2+} + 2Cl^- + Cu^0$$

[1] E. G. West, *Chem. in Britain*, **1969**, 199.

That the latter reaction also depends on the geometry of the ligand, that is on its chelate nature, is shown by differences in the $[Cu^{2+}]/[Cu^{+}]^2$ equilibrium with chelating and non-chelating amines. Thus for ethylenediamine, K is $\sim 10^5$, for pentamethylenediamine (which does not chelate) 3×10^{-2} and for ammonia 2×10^{-2}. Hence in the last case the reaction is

$$[Cu(NH_3)_4]^{2+} + Cu^0 = 2[Cu(NH_3)_2]^+$$

The life time of the Cu^+ ion in water depends strongly on conditions. Usually disproportionation is very fast (<1 sec), but ca. $0.01M$ solutions prepared in $0.1M$ $HClO_4$ at $0°$ by the reaction

$$\text{—CuClO}_4(\text{in CH}_3\text{OH}) \xrightarrow[\text{the olefin}]{\text{HClO}_4 \text{ and extract}} \text{Cu}^+(\text{aq})$$

or by reduction of Cu^{2+} with V^{2+} or Cr^{2+}, may last for several hours provided air is excluded.[2]

An excellent illustration of how the stability of the cuprous ion relative to that of the cupric ion may be affected by solvent is the case of acetonitrile. The cuprous ion is very effectively solvated by CH_3CN, and the cuprous halides have relatively high solubilities (e.g., CuI, 35 g/1 kg CH_3CN) vs. negligible solubilities in H_2O. Cu^I is more stable than Cu^{II} in CH_3CN and the latter is, in fact, a comparatively powerful oxidizing agent.

Cuprous Binary Compounds. The *oxide* and *sulfide* are more stable than the corresponding Cu^{II} compounds at high temperatures. Cu_2O is made as a yellow powder by controlled reduction of an alkaline solution of a cupric salt with hydrazine or, as red crystals, by thermal decomposition of CuO. A yellow "hydroxide" is precipitated from the metastable Cu^+ solution mentioned above. Cu_2S is a black crystalline solid prepared by heating copper and sulfur in absence of air.

Nearly colorless KCuO, prepared by heating together K_2O and Cu_2O, contains square $Cu_4O_4^{4-}$ rings with O atoms at the corners;[2a] there are similar Ag and Au compounds.

Cuprous chloride and bromide are made by boiling an acidic solution of the cupric salt with an excess of copper; upon dilution, white CuCl or pale yellow CuBr is precipitated. Addition of I^- to a solution of Cu^{2+} forms a precipitate that rapidly and quantitatively decomposes to CuI and iodine. CuF is unknown. The halides have the zinc blende structure (tetrahedrally coordinated Cu^+). CuCl and CuBr are polymeric in the vapor state, and for CuCl the principal species appears to be a six-ring of alternating Cu and Cl atoms with Cu—Cl~2.16 Å. White CuCl becomes deep blue at $178°$ and melts to a deep green liquid.

The halides are highly insoluble in water, log K_{sp} values at $25°$ being -4.49, -8.23 and -11.96 for CuCl, CuBr and CuI, respectively. Solubility is enhanced by an excess of halide ions (owing to formation of, e.g., $CuCl_2^-$,

[2] J. A. Altermatt and S. E. Manahan, *Inorg. Nuclear Chem. Letters*, 1968, **4**, 1; O. J. Parker and J. H. Esperson, *Inorg. Chem.*, 1969, **8**, 185, 1523; *J. Amer. Chem. Soc.*, 1969, **91**, 1968.

[2a] K. Hesterman and R. Hoppe, *Z. anorg. Chem.*, 1968, **360**, 113.

$CuCl_3^{2-}$ and $CuCl_4^{3-}$) and by other complexing species such as CN^-, NH_3, $S_2O_3^{2-}$.

Other relatively common Cu^I compounds are the cyanide, conveniently prepared by the reaction:

$$2Cu^{2+}(aq) + 4CN^-(aq) \rightarrow 2CuCN(s) + C_2N_2$$

and soluble in an excess of cyanide to give, mainly, $Cu(CN)_4^{3-}$; also common is the sulfate, a greyish solid, stable in absence of moisture, prepared by the reaction

$$Cu_2O + (CH_3)_2SO_4 \xrightarrow{100°} Cu_2SO_4 + (CH_3)_2O$$

Cuprous Complexes.[3] With simple ligands (e.g., halide ions, amines), the coordination is almost invariably tetrahedral. No discrete 2- or 3-coordinate species have been found in solids (contrast with Ag^I and Au^I), although linear 2-coordination is found in Cu_2O and $KCuO$. $KCu(CN)_2$ contains no linear $Cu(CN)_2^-$ ion (as do its Ag^I and Au^I analogs), but instead it has a spiral polymeric structure in which each Cu^I atom is bound to two CN-carbon atoms and one CN-nitrogen atom in a nearly coplanar array (see Fig. 25-H-1). In the (1,3-dimethyltriazeno)copper(I) complex each copper atom is linearly coordinated by two nitrogen atoms (25-H-I) and there is

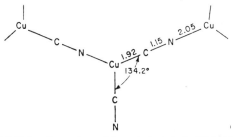

$a \approx b \approx c \approx d \approx 2.66\text{Å}$

$e = 4.42\text{Å}; f = 2.97\text{Å}$

Point symmetry: D_2

(25-H-I)

also one close crosswise Cu—Cu contact (2.97 Å) in the rhombus defined by the four copper atoms.[4]

Among the structurally characterized compounds containing tetrahedrally coordinated Cu^I are those with formulas $M_2^I CuX_3$, $M^I Cu_2X_3$ and

Fig. 25-H-1. A portion of the spiral chain in $K[Cu(CN)_2]$.

[3] W. E. Hatfield and R. Whyman, *Transition Metal Chem.*, 1969, **5**, 47 (a review of Cu^I, Cu^{II} and Cu^{III} complexes).

[4] J. E. O'Connor, G. A. Janusonis and E. R. Corey, *Chem. Comm.*, **1968**, 445.

$[Co(NH_3)_6]_4[Cu_5Cl_{17}]$. Compounds of the first two kinds are built of single and double chains, respectively, of CuX_4 tetrahedra, in which sharing of X ions enables each Cu to complete its tetrahedron. In the last compound[5] there are $[Co(NH_3)_6]^{3+}$, Cl^- and $Cu_5Cl_{16}^{11-}$ ions in a 4:1:1 ratio; the $Cu_5Cl_{16}^{11-}$ ion consists of four $CuCl_4$ tetrahedra each sharing one Cl atom with a central Cu^+ ion so as to give it tetrahedral coordination.

Copper(I) forms several kinds of polynuclear complex in which four Cu atoms lie at the vertices of a tetrahedron. The earliest to be recognized were the $Cu_4I_4L_4$ ($L = R_3P$, R_3As) species, in which there is a triply bridging I atom on each face of the Cu_4 tetrahedron and one ligand, L, is coordinated to a Cu atom at each vertex (25-H-II). The phosphine complexes can be made

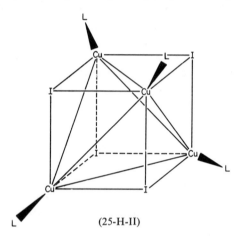

(25-H-II)

from Cu^I or Cu^{II} halides,[6] but with other anions[7] different types may be obtained, e.g., $(R_3P)_3Cu^+PF_6^-$.

The compounds $Cu_4OCl_6(Ph_3PO)_4$, $Cu_4OCl_6py_4$ and $[(CH_3)_4N]_4$-$[Cu_4OCl_{10}]$ all have a tetrahedron of Cu atoms[8] surrounding a central oxygen atom, with a bridging chlorine atom on each edge of the Cu_4 tetrahedron and each of the remaining four ligands coordinated to a Cu atom, as shown in (25-H-III). Each Cu atom thus has distorted trigonal-bipyramidal coordination. There are also distorted Cu_4 tetrahedra in $[CuS_2CNEt_2]_4$; the short (2.71 Å) Cu—Cu distances imply significant metal–metal interactions.

A distorted cubic set of Cu^I atoms occurs in the $[Cu_8(S_2CC(CN)_2)_{12}]^{4-}$ ion,[9] with Cu—Cu distances in the range 2.81–2.87 Å.

[5]   P. Murray-Rust, P. Day and C. K. Prout, *Chem. Comm.*, **1966**, 277.
[6]   F. H. Jardine *et al.*, *J. Chem. Soc., A,* **1970**, 238; *J. Inorg. Nuclear Chem.*, 1971, **33**, 2941.
[7]   E. L. Mutterties and C. W. Alegranti, *J. Amer. Chem. Soc.*, 1970, **92**, 411; S. J. Lippard and G. J. Palenik, *Inorg. Chem.*, 1971, **10**, 1323.
[8]   J. A. Bertrand and J. A. Kelly, *Inorg. Chem.*, 1969, **8**, 1982.
[9]   L. E. McCandlish *et al.*, *J. Amer. Chem. Soc.*, 1968, **90**, 7357.

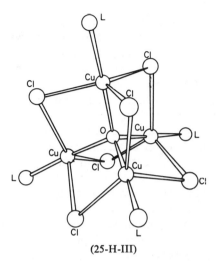

(25-H-III)

The reaction of [PPh$_3$CuCl]$_4$ with Na[HB(OMe)$_3$] yields crystalline [CuH(PPh$_3$)]$_6$, which has the irregularly octahedral heavy atom structure shown in Fig. 25-H-2.[10] The positions of the hydrogen atoms are uncertain; they seem most likely to lie as bridges along the six longer Cu—Cu edges. The hydridic nature of some other complexes is not fully proved (page 187).

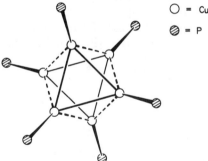

O = Cu

⊘ = P

Fig. 25-H-2. The Cu$_6$P$_6$ group in [CuH(PPh$_3$)]$_6$. The Cu—Cu distances shown as full lines have an average value of 2.65 Å while those shown by broken lines average 2.54 Å.

Organo Derivatives. Only in the Cu^I state does copper form Cu–C bonds; the most stable are (h^5-C$_5$H$_5$)CuPR$_3$, obtained by interaction of Cu$_4$I$_4$(PR$_3$)$_4$ and TlC$_5$H$_5$.[11]

The alkyls and aryls, RCu, obtained by action of RMgX or RLi on cuprous halides, are evidently polymeric. The alkyls are generally unstable, decomposing autocatalytically to alkene and alkane, probably *via* Cu^0 or CuH

[10] S. A. Bezman *et al.*, *J. Amer. Chem. Soc.*, 1971, **93**, 2063.
[11] L. T. J. Delbaere, D. W. McBride and R. B. Fergusen, *Acta. Cryst.*, 1970, *B*, **26**, 515; F. A. Cotton and J. Takats, *J. Amer. Chem. Soc.*, 1970, **92**, 2353.

species.[12] The aryls are more stable, e.g., C_6H_5Cu[13] decomposes above about 80°, possibly because of participation of π as well as σ electrons in Cu—C bonding. The fluoro analogs, both alkyls[14] and aryls,[15] are more stable than the hydrocarbon derivatives. Pentafluorophenylcopper is a tetramer, while $Cu(m\text{-}CF_3C_6H_4)$ is an octamer.

The decarboxylation of $CuOCOC_6F_5$ in quinoline forms CuC_6F_5, and indeed organo coppers appear generally to be intermediates in the decarboxylation of carboxylic acids, and their catalytic action explains the long-known effectiveness of copper metal or copper salts in such reactions.

The reaction of RCu with an excess of lithium alkyls in ether gives anionic complexes, $Li[CuR_2]$, which, however, are not isolated, so that the true constitution is uncertain. These organo cuprates are very useful synthetic agents[16] for alkylations, reacting, for example, with alkyl, vinyl or aryl halides to form C—C bonds, and for syntheses of alkenes and allenes.

Acetylene and Olefin Complexes. Cuprous chloride in concentrated hydrochloric acid absorbs acetylene to give colorless species such as $CuCl(C_2H_2)$ and $[CuCl_2(C_2H_2)]^-$. These halide solutions can also catalyze the conversion of acetylene into vinylacetylene (in concentrated alkali chloride solution) or to vinyl chloride (at high HCl concentration), and the reaction of acetylene with hydrogen cyanide to give acrylonitrile is also catalyzed.

Cuprous ammine solutions react with acetylenes containing the HC≡C-group to give yellow or red precipitates, which are believed to have the structure (25-H-IV). Propynylcopper dissolves in triethylphosphine in toluene to give a cyclic polymer of similar type $[Et_3PCuC≡CMe]_3$.

(25-H-IV)

Cuprous acetylides provide a useful route to the synthesis of a variety of organic acetylenic compounds and heterocycles, by reaction with aryl and other halides.[17] A particularly important indirect use, where acetylides are

[12] K. Wada, M. Tamura and J. Kochi, *J. Amer. Chem. Soc.*, 1970, **92**, 6656; G. M. Whitesides *et al.*, *J. Amer. Chem. Soc.*, 1970, **92**, 1426.

[13] M. Nilsson and O. Wennerstrom, *Acta Chem. Scand.*, 1970, **24**, 482.

[14] V. C. R. McLoughlin and J. Thrower, *J. Organometallic Chem.*, 1969, **25**, 5921.

[15] A. Cairncross *et al.*, *J. Amer. Chem. Soc.*, 1970, **92**, 3187; 1971, **93**, 247, 249; T. Cohen and R. A. Schambach, *J. Amer. Chem. Soc.*, 1970, **92**, 3189.

[16] G. M. Whitesides *et al.*, *J. Amer. Chem. Soc.*, 1969, **91**, 4871; E. J. Corey and I. Kuwajima, *J. Amer. Chem. Soc.*, 1970, **92**, 395; P. Rona and P. Crabbé, *J. Amer. Chem. Soc.*, 1969, **92**, 3289.

[17] C. E. Castro *et al.*, *J. Amer. Chem. Soc.*, 1969, **82**, 6464.

probable intermediates, is the oxidative dimerization of acetylenes.[18] A common procedure is to use the N,N,N',N'-tetramethylethylenediamine complex of CuCl in a solvent, or CuCl in pyridine–methanol, and oxygen as oxidant.

The oxidation can also be done stoichiometrically by cupric acetate in pyridine:

$$2RC\equiv CH + 2Cu^{2+} + 2py \rightarrow RC\equiv C-C\equiv CR + 2Cu^{I} + pyH^{+}$$

There are many complexes of Cu^{I} with olefins or other unsaturated compounds, and these are in general rather similar[19] to those of Ag^{+} (page 1051). Although they are usually made by direct reaction with Cu^{I} halides, they may also be prepared from cupric salts by reduction with trialkyl phosphites in ethanolic solution in presence of the alkene.[19a] The crystalline compounds obtained when using chelating alkenes such as norbornadiene or cyclic polyalkenes usually have polymeric structures, such as that for the cyclo-octatetraene complex (25-H-V), and only one of the double bonds is normally coordinated.

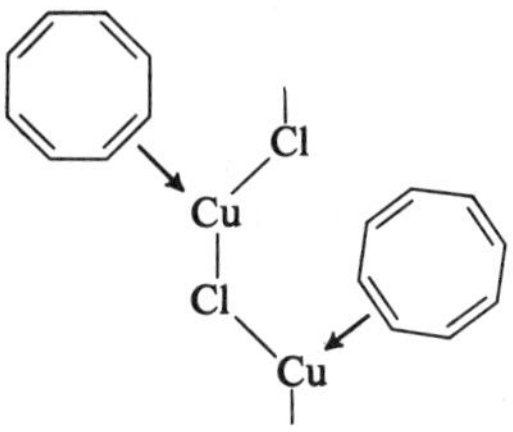

(25-H-V)

Carbon monoxide is absorbed by solutions of chlorocuprates(I), and the halogen bridged dimer $(CuClCO)_2$ can be obtained as crystals. The gas is absorbed quantitatively by aqueous ammonia solutions of Cu^{I} and can be regenerated by acidification. Other unstable copper carbonyls such as $h^5\text{-}C_5H_5Cu(CO)$, $[Cuen(CO)]Cl$ and $CuCl(CO)(Me_2N=CH_2)^+$ can be isolated.[20] A binary carbonyl formed when copper atoms produced thermally are trapped in a CO matrix at $20°K$ can be detected by its infrared spectrum.[20a]

25-H-3. The Copper(II) State, d^9

The dipositive state is the most important one for copper. Most cuprous compounds are fairly readily oxidized to cupric compounds, but further oxidation to Cu^{III} is difficult. There is a well-defined aqueous chemistry of Cu^{2+}, and a large number of salts of various anions, many of which are water-soluble, exist in addition to a wealth of complexes.

[18] L. M. Fieser and M. Fieser, *Reagents for Organic Synthesis*, Wiley, 1967, 1969.
[19] J. M. Harvilchuck, D. A. Aikens and R. C. Murray, *Inorg. Chem.*, 1969, **8**, 539.
[19a] B. W. Cook, R. G. J. Miller and P. F. Todd, *J. Organometallic Chem.*, 1969, **19**, 421.
[20] F. A. Cotton and T. J. Marks, *J. Amer. Chem. Soc.*, 1970, **92**, 5114; G. Rucci *et al.*, *Chem. Comm.*, **1971**, 652, 1132.
[20a] J. S. Ogden, *Chem. Comm.*, **1971**, 978.

Stereochemistry.[3,21]   The d^9 configuration makes Cu^{II} subject to Jahn–Teller distortion (page 592) if placed in an environment of cubic (i.e. regular octahedral or tetrahedral) symmetry, and this has a profound effect on all its stereochemistry. With only one possible exception, to be mentioned below, it is never observed in these regular environments. When 6-coordinate, the "octahedron" is severely distorted, as indicated by the data in Table 25-H-2.

TABLE 25-H-2

Interatomic Distances in Some Cupric Coordination Polyhedra

Compound	Distances (Å)
$CuCl_2$	4Cl at 2.30, 2Cl at 2.95
$CsCuCl_3$	4Cl at 2.30, 2Cl at 2.65
$CuCl_2 \cdot 2H_2O$	2O at 2.01, 2Cl at 2.31, 2Cl at 2.98
$CuBr_2$	4Br at 2.40, 2Br at 3.18
CuF_2	4F at 1.93, 2F at 2.27
$[Cu(H_2O)_2(NH_3)_4]$ in $CuSO_4 \cdot 4NH_3 \cdot H_2O$	4N at 2.05, 1O at 2.59, 1O at 3.37
K_2CuF_4	2F at 1.95, 4F at 2.08

The typical distortion is an elongation along one four-fold axis, so that there is a planar array of four short Cu—L bonds and two *trans* long ones. In the limit, of course, the elongation leads to a situation indistinguishable from square coordination as found in CuO and many discrete complexes of Cu^{II}. Thus the cases of tetragonally distorted "octahedral" coordination and square coordination cannot be sharply differentiated.

In $K_2Pb[Cu(NO_2)_6]$ there is a *regular* octahedron of nitrogen atoms about the Cu^{2+} ion.[22] The structure, based on neutron diffraction data, does not appear to involve either a dynamic Jahn–Teller distortion or a disordering of statically distorted species. It is not known why the Jahn–Teller distortion is too small for detection in this instance.

In addition to the normal square merging into tetragonally distorted octahedral complexes there are other stereochemistries of which the most important is distorted tetrahedral. The $M_2^ICuX_4$ compounds, M^I representing a univalent cation and X representing Cl^- or Br^-, contain non-planar CuX_4^{2-} ions provided the cations are large. Thus $(NH_4)_2CuCl_4$ contains planar $CuCl_4^{2-}$ ions, but Cs_2CuCl_4 and Cs_2CuBr_4 as well as several salts with still larger cations have been shown to contain CuX_4^{2-} ions that are squashed tetrahedra as indicated in Fig. 25-H-3. It appears that an orange color is characteristic of the distorted tetrahedral $CuCl_4^{2-}$ ions, while compounds containing square $CuCl_4^{2-}$ ions are pale yellow.

[21] B. J. Hathaway and D. E. Billing, *Coordination Chem. Rev.*, 1970, **5**, 143 (electronic properties and stereochemistry).
[22] N. W. Isaacs, C. H. L. Kennard and D. A. Wheeler, *J. Chem. Soc.*, A, **1969**, 386.

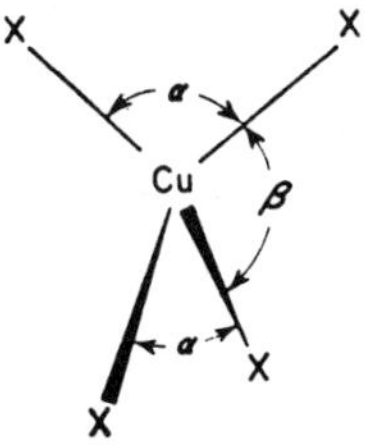

Fig. 25-H-3. Squashed tetrahedral structures of $[CuX_4]^{2-}$ ions in Cs_2CuX_4 salts; $\alpha > \beta$.

Other cases of distorted tetrahedral coordination occur in some bis(salicylaldiminato) complexes[22a] (25-H-VI), where R is bulky.

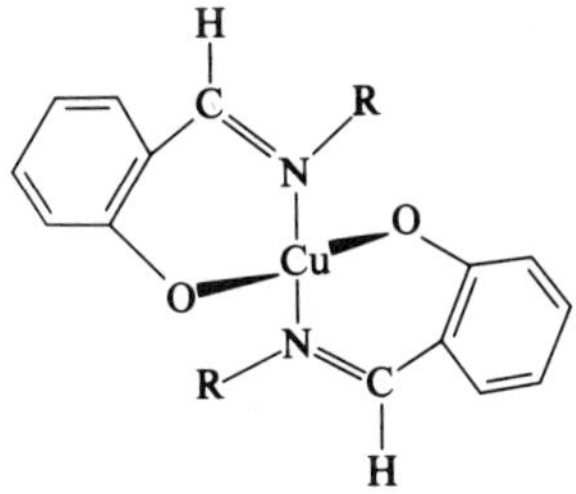

(25-H-VI)

When $R = (CH_3)_2CH$ and $(CH_3)_3C$ the angles between the two chelate ring planes in the crystalline materials are 60° and 54°, respectively, and there are reasons to believe these distortions persist in solution. However, in most cases, with small R groups, the coordination is planar. Again, in the dipyrromethene complex, (25-H-VII), steric interference of methyl groups renders

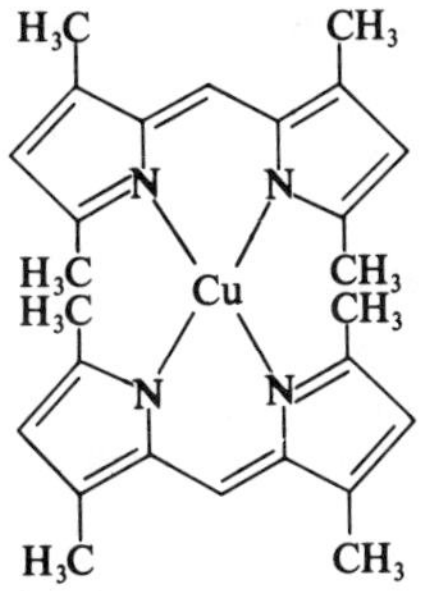

(25-H-VII)

a planar configuration impossible. The angle between the mean planes of the two ligands is 66°.[23]

Numerous planar complexes are known. With a few exceptions such as the salicylaldiminato and dipyrromethene complexes mentioned above, neutral four-coordinate complexes containing chelating ligands have planar

[22a] R. H. Holm and M. S. O'Connor, *Prog. Inorg. Chem.*, 1971, **14**, 325.
[23] M. Elder and B. R. Penfold, *J. Chem. Soc., A*, **1969**, 2556.

coordination. Variants on this include some cases where additional ligands complete a very elongated octahedron and many where there is dimerization of the type shown schematically in (25-H-VIII) for the β-form of bis-

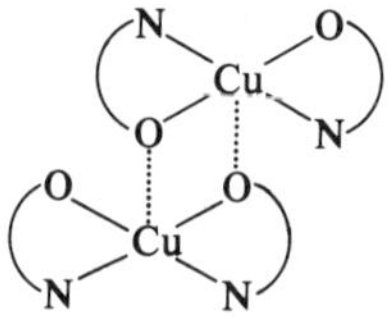

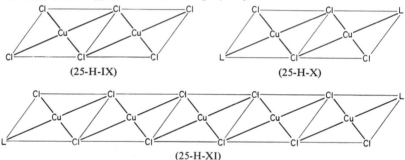

(25-H-VIII)

(8-quinolinolato)copper(II), in which each metal atom becomes 5-coordinate. There are also cases of extended arrays of planar complexes, as in $Cu_2Cl_6^{2-}$ (25-H-IX), $Cu_2Cl_4(CH_3CN)_2$ (25-H-X; $L = CH_3CN$) and $Cu_5Cl_{10}(C_3H_7OH)_2$ (25-H-XI; $L = C_3H_7OH$).

(25-H-IX)   (25-H-X)

(25-H-XI)

Trigonal-bipyramidal coordination is found in several cases. In $Cu(terpy)Cl_2$ and $[Cu(dipy)_2I]I$ the *tbp* symmetry is necessarily imperfect, but the $CuCl_5^{3-}$ ion has full (crystallographic) D_{3h} symmetry in $[Cr(NH_3)_6]$-$[CuCl_5]$, while $Cu_2Cl_8^{4-}$ is similar but with two bridged chlorides.[24] Other compounds containing *tbp* coordination of Cu^{II} are $[Cu(Me_4dien)(N_3)]Br$[25] and $[Cu(NH_3)_2][Ag(SCN)_3]$,[26] while copper(II) dimethylglyoximate is dimeric with a distorted square-pyramidal configuration[27] in which the fifth position is bound to the oxygen of one of the NO groups.

Spectral and Magnetic Properties. Because of the relatively low symmetry (i.e. less than cubic) of the environments in which the Cu^{2+} ion is characteristically found, detailed interpretations of the spectra and magnetic properties are somewhat complicated, even though one is dealing with the equivalent of a one-electron case. Virtually all complexes and compounds are blue or green. Exceptions are generally caused by strong ultraviolet bands—charge-transfer bands—tailing off into the blue end of the visible spectrum and thus causing the substances to appear red or brown. The blue

[24] N. K. Raymond, D. W. Meek and J. A. Ibers, *Inorg. Chem.*, 1968, **7**, 1111; D. J. Hodgson *et al.*, *Chem. Comm.*, **1970**, 786.
[25] Z. Dori, *Chem. Comm.*, **1968**, 714.
[26] B. J. Hathaway *et al.*, *J. Chem. Soc.*, A, **1970**, 806.
[27] A. Vaciago and L. Zambonelli, *J. Chem. Soc.*, A, **1970**, 218.

or green colors are due to the presence of an absorption band in the 600–900 nm region of the spectrum. The envelopes of these bands are generally unsymmetrical, seeming to encompass several overlapping transitions, but definitive resolution into the proper number of sub-bands with correct locations is difficult. Only when polarized spectra of single crystals have been measured has this resolution been achieved unambiguously. The cases of the $Cu(NH_3)_4^{2+}$ ion in $[Cu(NH_3)_4][NH_4](ClO_4)_3 \cdot NH_3$ and $Cu(DPM)_2$ (25-H-XII) are instructive.[28] In both, the single d electron vacancy is expected

(25-H-XII)

to be in the d orbital with lobes directed toward the four ligand atoms. For $Cu(NH_3)_4^{2+}$ this will be the $d_{x^2-y^2}$ orbital if the Cu—N bonds are taken to be along the x- and y-axes. For $Cu(DPM)_2$ this half-empty orbital will be the d_{xy} orbital since the obligatory choice of x- and y-axes is as shown in (25-H-XII). The expected d—d hole transitions in the two cases are then:

$Cu(NH_3)_4^{2+}$	$Cu(DPM)_2$
$x^2 - y^2 \rightarrow xy$	$xy \rightarrow x^2 - y^2$
$x^2 - y^2 \rightarrow z^2$	$xy \rightarrow z^2$
$x^2 - y^2 \rightarrow (xz, yz)$	$xy \rightarrow xz$
	$xy \rightarrow yz$

The polarized spectra show that in $Cu(NH_3)_4^{2+}$ there are three, and in $Cu(DPM)_2$ four, components under the broad, irregular absorption envelopes in the red region of the spectra. These results are compatible with the orbital-energy diagrams shown in Fig. 25-H-4. Thus, in both these square complexes the metal–ligand interactions affect the energies of all the d orbitals about equally except for the xy or $x^2 - y^2$ orbital which becomes strongly antibonding or is destabilized by the formation of the Cu—N or Cu—O σ bonds. This result *cannot* be explained by simple electrostatic crystal-field theory but can be accommodated by MO analysis.

The introduction of ligands along the z-axis would presumably cause the d_{z^2} orbital to rise in energy, approaching the high-lying $d_{x^2-y^2}$ or d_{xy} orbital. For example, in the $Cu(NH_3)_5^{2+}$ ion, which is thought to be square pyramidal, the $d_{x^2-y^2} \rightarrow d_{z^2}$, $d_{x^2-y^2} \rightarrow d_{xy}$, and $d_{x^2-y^2} \rightarrow (d_{xz}, d_{yz})$ transitions are believed to occur at $\sim 10,000$, $\sim 13,000$ and $\sim 18,000 \text{ cm}^{-1}$, respectively.[29]

In the case of $CuCl_5^{3-}$, two d–d transitions are expected and two bands are in fact seen, at $\sim 8,000$ and $\sim 10,000 \text{ cm}^{-1}$.

[28] D. W. Smith, *Inorg. Chem.*, 1966, **5**, 2236; F. A. Cotton and J. J. Wise, *Inorg. Chem.*, 1967, **6**, 917.

[29] A. A. G. Tomlinson and B. J. Hathaway, *J. Chem. Soc., A*, **1968**, 1905.

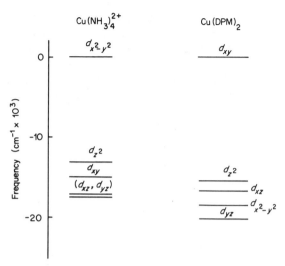

Fig. 25-H-4.   Orbital energy-level diagrams for $Cu(NH_3)_4^{2+}$ and $Cu(DPM)_2$ as deduced from polarized crystal spectra.

The magnetic moments of simple Cu^{II} complexes (those lacking Cu—Cu interactions, for which, see below) are generally in the range 1.75–2.20 B.M. regardless of stereochemistry and independently of temperature except at extremely low temperatures ($<5°$ K).

Binary Copper(II) Compounds.   Black crystalline CuO is obtained by pyrolysis of the nitrate or other oxo salts; above 800° it decomposes to Cu_2O. The hydroxide is obtained as a blue bulky precipitate on addition of alkali hydroxide to cupric solutions; warming an aqueous slurry dehydrates this to the oxide. The hydroxide is readily soluble in strong acids and also in concentrated alkali hydroxides, to give deep blue anions, probably of the type $[Cu_n(OH)_{2n-2}]^{2+}$. In ammoniacal solutions the deep blue tetraammine complex is formed.

The *halides* are the colorless CuF_2, with a distorted rutile structure, the yellow chloride and the almost black bromide, the last two having structures with infinite parallel bands of square CuX_4 units sharing edges. The bands are arranged so that a tetragonally elongated octahedron is completed about each copper atom by bromine atoms of neighboring chains. $CuCl_2$ and $CuBr_2$ are readily soluble in water, from which hydrates may be crystallized, and also in donor solvents such as acetone, alcohol and pyridine.

Salts of Oxo Acids.   The most familiar cupric compound is the blue hydrated sulfate, $CuSO_4 \cdot 5H_2O$, which contains four water molecules in the plane with O atoms of SO_4 groups occupying the axial positions, and the fifth water molecule H-bonded in the lattice. It may be dehydrated to the virtually white anhydrous substance. The hydrated nitrate cannot be fully dehydrated without decomposition. The anhydrous nitrate is prepared by dissolving the metal in a solution of N_2O_4 in ethyl acetate and crystallizing the salt

$Cu(NO_3)_2 \cdot N_2O_4$, which probably has the constitution $[NO^+]$ $[Cu(NO_3)_3^-]$. When heated at 90° this solvate gives the blue $Cu(NO_3)_2$ which can be sublimed without decomposition in a vacuum at 150–200°. There are two forms of the solid, both possessing complex structures in which Cu^{II} ions are linked together by nitrate ions in an infinite array. However, discrete molecules with the kind of structure shown in (25-H-XIII) occur in the vapor phase. It is

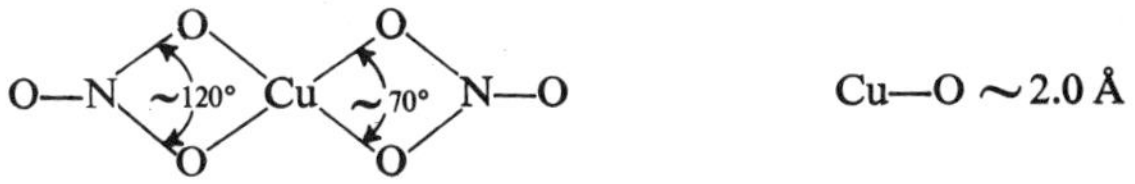

$$Cu-O \sim 2.0\,\text{Å}$$

(25-H-XIII)

not certain whether the molecule is entirely planar, however. Cupric acetate and other carboxylates, which are dimeric, are discussed below, but in solutions the dimers dissociate.

Aqueous Chemistry. Most cupric salts dissolve readily in water and give the aquo ion, which may be written $[Cu(H_2O)_6]^{2+}$, but it must be kept in mind that two of the water molecules are farther from the metal atom than the other four. Addition of ligands to such aqueous solutions leads to the formation of complexes by successive displacement of water molecules. With NH_3, for example, the species $[Cu(NH_3)(H_2O)_5]^{2+}\ldots[Cu(NH_3)_4$-$(H_2O)_2]^{2+}$ are formed in the normal way, but addition of the fifth and sixth molecules of NH_3 is difficult.[30] In fact, the sixth cannot be added to any significant extent in aqueous media but only in liquid ammonia. The reason for this unusual behavior is connected with the Jahn–Teller effect. Because of it, the Cu^{II} ion does not bind the fifth and sixth ligands strongly (even the H_2O). When this intrinsic weak binding of the fifth and sixth ligands is added to the normally expected decrease in the stepwise formation constants (page 649) the formation constants, K_5 and K_6, are very small indeed. Similarly, it is found with ethylenediamine that $[Cu\ en(H_2O)_4]^{2+}$ and $[Cu\ en_2(H_2O)_2]^{2+}$ form readily, but $[Cu\ en_3]^{2+}$ is formed only at extremely high concentrations of en. Many other amine complexes of Cu^{II} are known, and all are much more intensely blue than the aquo ion. This is because the amines produce a stronger ligand field which causes the absorption band to move from the far red to the middle of the red region of the spectrum. For example, in the aquo ion the absorption maximum is at ~ 800 nm, whereas in $[Cu(NH_3)_4$-$(H_2O)_2]^{2+}$ it is at ~ 600 nm, as shown in Fig. 25-H-5. The reversal of the shifts with increasing take-up of ammonia for the fifth ammonia is to be noted, indicating again the weaker bonding of the fifth ammonia molecule.

In halide solutions the equilibrium concentrations of the various possible species depend on the conditions; although $CuCl_5^{3-}$ has only a low formation constant it is precipitated from solutions by large cations of similar charge,

[30] B. J. Hathaway and A. A. G. Tomlinson, *Coordination Chem. Rev.*, 1970, **5**, 1 (an extensive review of Cu^{II}–ammonia complexes).

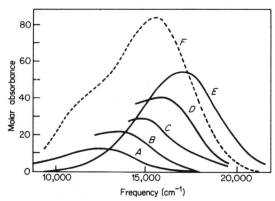

Fig. 25-H-5. Absorption spectra of $[Cu(H_2O)_6]^{2+}$ (A) and of the ammines in $2M$ ammonium nitrate at 25°, $[Cu(NH_3)(H_2O)_5]^{2+}$ (B), $[Cu(NH_3)_2(H_2O)_4]^{2+}$ (C), $[Cu(NH_3)_3(H_2O)_3]^{2+}$ (D), $[Cu(NH_3)_4(H_2O)_2]^{2+}$ (E), and $[Cu(NH_3)_5H_2O]^{2+}$ (F).

e.g., $[Cr(NH_3)_6]^{3+}$; other cations such as Cs^+ or Me_4N^+ may precipitate $CuCl_4^{2-}$.[31]

Many other Cu^{II} complexes may be isolated by treating aqueous solutions with ligands. When the ligands are such as to form neutral, water-insoluble complexes, as in the following equation, the complexes are precipitated and can be purified by recrystallization from organic solvents. The bis(acetylacetonato)copper(II) complex is another example of this type.

$$Cu^{2+}(aq) + 2 \quad \underset{CHO}{\overset{O^-}{\bigcirc}} \quad \longrightarrow \quad Cu\left(\underset{O-C}{\overset{O}{\bigcirc}} \underset{H}{} \right)_2$$

Multidentate ligands which coordinate through oxygen or nitrogen, such as amino acids, form cupric complexes, often of considerable complexity. Thus the well-known blue solutions formed by addition of tartrate to Cu^{2+} solutions (known as Fehling's solution when basic and when *meso*-tartrate is used) may contain monomeric, dimeric or polymeric species at different pH values. One of the dimers, $Na_2[Cu\{(\pm)-C_4O_6H_2\}]\cdot 5H_2O$, has square Cu^{II} coordination, two tartrate bridges and a Cu—Cu distance of 2.99 Å.[32]

Polynuclear Compounds with Magnetic Anomalies.[33] Copper forms many compounds in which Cu—Cu distances are short enough to indicate significant M—M interactions, but in no case are there actual Cu—Cu bonds. An important class is the dinuclear, carboxylato-bridged compounds (25-H-XIV), and the related 1,3-triazinato compound (25-H-XV). In these

[31] See, e.g., I. R. Beattie, T. R. Gibson and G. A. Ozin, *J. Chem. Soc.*, A, **1969**, 534; P. M. Boorman, P. J. Craig and T. W. Swaddle, *J. Chem. Soc.*, A, **1969**, 2970.

[32] R. L. Belford *et al.*, *Chem. Comm.*, **1971**, 508; A. J. Fatiadi, *J. Res. Nat. Bur. Stand.*, *Sect. A*, 1970, **74**, 723; E. Bottari and M. Vicedomini, *J. Inorg. Nuclear Chem.*, 1971, **33**, 1463.

[33] M. Kato, H. B. Jonassen and J. C. Fanning, *Chem. Rev.*, 1964, **64**, 99.

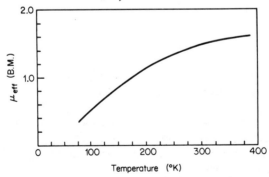

(25-H-XIV)   (25-H-XV)

compounds there is weak coupling of the unpaired electrons, one on each Cu^{II} ion, giving rise to a singlet ground state with a triplet state lying only a few kJ mol^{-1} above it; the latter state is thus appreciably populated at normal temperatures and the compounds are paramagnetic. At 25° μ_{eff} is typically about 1.4 B.M. per Cu atom and the temperature-dependence is very pronounced, as shown in Fig. 25-H-6 for $Cu_2(OCOCH_3)_4 \cdot (H_2O)_2$.

Phenomenologically, the interaction in the dinuclear acetate and many other compounds with similar temperature-dependences of the magnetic moment can be described as antiferromagnetic couplings of the unpaired species on the adjacent Cu^{II} atoms. This is expressed mathematically by introducing a term $-JS_1S_2$ (where $S_1 = S_2 = \frac{1}{2}$ are the spin quantum numbers and J is the coupling constant) into the Hamiltonian operator; the values of J are close to 300 cm^{-1} for the majority of the compounds.

Attempts to specify the detailed nature of this interaction have been plagued by controversy and there are still considerable differences of opinion as to its precise description.[34] It is now generally agreed that the interaction is basically one between orbitals of δ symmetry, but whether this is primarily a direct interaction between $d_{x^2-y^2}$ orbitals of the two metal atoms or one

Fig. 25-H-6.   Temperature-dependence of the effective magnetic moment per Cu atom in $Cu_2(OCOCH_3)_4(H_2O)_2$.

[34] A. E. Hansen and C. J. Ballhausen, *Trans. Faraday Soc.*, 1965, **61**, 631; L. Dubicki and R. L. Martin, *Inorg. Chem.*, 1966, **5**, 2203; D. M. L. Goodgame *et al.*, *Chem. Comm.*, **1969**, 629; R. W. Jotham and S. F. A. Kettle, *Inorg. Chem.*, 1970, **9**, 1390.

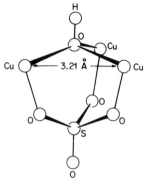

Fig. 25-H-7. Structural unit constituting the core of the compound
$Cu_3(C_6H_5N_2O)_3(OH)(SO_4) \cdot 16.3H_2O$.

which is substantially transmitted through the π orbitals of the bridging carboxyl groups is unsettled. In either of these cases the exact ordering of all energy levels primarily concerned in the spin exchange interation is still in dispute.

Recently an interesting trinuclear Cu^{II} compound

$$Cu_3(C_6H_5N_2O)_3(OH)(SO_4) \cdot 16.3H_2O$$

with appreciable magnetic interaction has been described.[35] Its central part has the structure shown in Fig. 25-H-7 and the magnetic properties can best be explained by assuming that interactions between the initially singly occupied atomic orbitals of the individual Cu^{II} atoms leads to a pattern of molecular orbitals for the set of three in which a doubly degenerate orbital lies lowest. This then is occupied by the three electrons, which results in two of them becoming paired. In view of the large separation of the Cu^{II} atoms, this interaction is considered to be indirect, transmitted through the triply bridging OH group.

25-H-4. Catalytic Properties of Copper Compounds[36]

Copper compounds catalyze an exceedingly varied array of reactions, hetereogeneously, homogeneously, in the vapor phase, in organic solvents and in aqueous solutions. Many of these reactions, particularly if in aqueous solutions, involve oxidation-reduction systems and a Cu^I–Cu^{II} redox cycle. Molecular oxygen can often be utilized as oxidant, e.g., in copper-catalyzed oxidations of ascorbic acid and in the Wacker process (page 798) for conversion of alkenes into aldehydes.

Although the precise mechanisms of the reactions are often not too well known, it seems likely from studies on comparatively simple systems, e.g.,

[35] R. Beckett et al., Austral. J. Chem., 1969, 22, 2527.
[36] O. A. Chaltykyan, Copper Catalytic Reactions, Consultants Bureau, 1966.

the oxidation[37] of $[Cu(MeCN)_4]^+$, that they may involve reactions of the type:

$$Cu^+ + O_2 \rightarrow CuO_2^+$$
$$CuO_2^+ + H^+ \rightarrow Cu^{2+} + HO_2$$
$$Cu^+ + HO_2 \rightarrow Cu^{2+} + HO_2^-$$
$$H^+ + HO_2^- \rightarrow H_2O_2$$

Copper compounds have many uses in organic chemistry for oxidations, coupling reactions, halogenations, etc.[38] (cf. also page 910). Some of these, such as the oxidation of phenols by copper–amine complexes,[39] provide models for phenol-oxidizing enzymes.

25-H-5. Biochemical Activity

Copper is found in both plants and animals[40] and a number of copper proteins, including enzymes, have been isolated. Examples are: (1) *Ascorbic acid oxidase* (M.W. $\approx$ 140,000; 8 gA of Cu) is widely distributed in plants and microorganisms and catalyzes oxidation of ascorbic acid to dehydro-ascorbic acid with O_2 as the electron-acceptor. (2) *Cytochrome oxidase*, the terminal electron-acceptor of the mitochondrial oxidative pathway, contains heme and copper in a 1:1 ratio; a functional role for the copper is indicated. (3) *Tryrosinases*, which catalyze the formation of melanin pigments in a host of plants and animals, were the first enzymes in which copper was shown to be essential to function. (4) Many lower animals, e.g., snails and crabs, contain a cuproprotein as oxygen-carrier, analogous to hemoglobin in mammals. This protein, called *hemocyanin* (though it contains no heme groups!), has a very high molecular weight and there is evidence indicating that it binds one molecule of O_2 per *two* Cu atoms.

Known copper proteins are predominantly oxidases or reversible oxygen-carriers, but little detailed chemical and structural information is yet available.[41]

Copper complexes of amines such as histidine,[42] ethylenediamine and 2,2'-bipyridine have been studied as models for the behavior of copper enzymes which catalyze the decomposition of H_2O_2 to O_2 and H_2O; peroxo species are believed to be intermediates.[42a]

---

[37] R. D. Gray, *J. Amer. Chem. Soc.*, 1969, **91**, 56.

[38] See M. Fieser and L. F. Fieser, *Reagents for Organic Synthesis*, Vols. 1, 2, Wiley, 1967, 1969.

[39] See, e.g., D. G. Hewitt, *Chem. Comm.*, **1970**, 227.

[40] J. Peisach, P. Aisen and W. E. Blumberg, eds., *The Biochemistry of Copper*, Academic Press, 1966; B. L. Vallee and W. Wacker in *The Proteins*, H. Neurath, ed., Academic Press, 1970.

[41] G. Morpurgo and R. J. P. Williams in *Physiology and Biochemistry of Hemocyanins*, Academic Press, 1968.

[42] H. Siegel and D. B. McCormick, *J. Amer. Chem. Soc.*, 1971, **93**, 2041.

[42a] H. Siegel *et al.*, *Angew. Chem. Internat. Edn.*, 1969, **8**, 167; *J. Amer. Chem. Soc.*, 1969, **91**, 1061, 1065, 7758; V. S. Sharma and J. Schubert, *J. Amer. Chem. Soc.*, 1969, **91**, 6291.

25-H-6. Copper(III) Compounds, d^8

Copper(III), which is isoelectronic with Ni^{II}, occurs in several crystalline compounds, but it is unstable in solution. Thus $NaCuO_2$, $KCuO_2$, and other alkali and alkaline-earth cuprates, obtained by heating the appropriate oxide mixture in oxygen, are stable as dry solids but dissolve with decomposition in aqueous base; the Cu^{III} survives no more than a few seconds in solution.[43]

Other Cu^{III} complexes are: (1) $KCu\,bi_2$ (biH_2 = biuret, H_2N-$CONHCONH_2$), obtained by oxidation of K_2Cubi_2 with $K_2S_2O_6$ in $5N$ KOH.[44] (2) $[Cu(IO_6)_2]^{7-}$ and $[Cu(HTeO_6)_2]^{7-}$, obtained by oxidation of alkaline Cu^{II} solutions by ClO^- or other strong oxidizing agents. (3) K_3CuF_6, a pale green crystalline solid, obtained by fluorination of a mixture of KCl and CuCl.[45] (4) $CuBr_2S_2CN(n\text{-}C_4H_9)_2$, dark violet needles, obtained by treatment of $CuS_2CN(n\text{-}C_4H_9)_2$ with Br_2 in CS_2.[46]

All these compounds except K_3CuF_6 are diamagnetic (or show only slight paramagnetism attributable to traces of impurities) and are presumed to contain low-spin, square-coordinated Cu^{III}. This is confirmed by X-ray diffraction for $CuBr_2S_2CN(n\text{-}C_4H_9)_2$. K_3CuF_6 has $\mu_{eff}^{25°} = 2.8$ B.M. and is thus a high-spin octahedral complex.

[43] J. S. Magee and R. H. Wood, *Canad. J. Chem.*, 1965, **43**, 1234; D. Meyerstein, *Inorg. Chem.*, 1971, **10**, 638.
[44] J. J. Bour and J. J. Steggerda, *Chem. Comm.*, 1967, 85.
[45] G. C. Allen and K. D. Warren, *Inorg. Chem.*, 1969, **8**, 1895.
[46] P. T. Beurskens, J. A. Cras, and J. J. Steggerda, *Inorg. Chem.*, 1968, **7**, 810.

26

The Elements of the Second and Third Transition Series

GENERAL COMPARISONS WITH THE FIRST TRANSITION SERIES

In general, the second and the third transition-series elements of a given Group have similar chemical properties but both show pronounced differences from their light congeners. A few examples will illustrate this generalization. Although Co^{II} forms a considerable number of tetrahedral and octahedral complexes and is the characteristic state in ordinary aqueous chemistry, Rh^{II} occurs only in a few complexes and Ir^{II} is unknown. Similarly, the Mn^{2+} ion is very stable, but for Tc and Re the II oxidation state is known only in a few complexes. Cr^{III} forms an enormous number of cationic amine complexes, whereas Mo^{III} and W^{III} form only a few complexes, none of which is especially stable. Again, Cr^{VI} species are powerful oxidizing agents, whereas Mo^{VI} and W^{VI} are quite stable and give rise to an extensive series of polynuclear oxo anions.

This is not to say that there is no valid analogy between the chemistry of the three series of transition elements. For example, the chemistry of Rh^{III} complexes is in general similar to that of Co^{III} complexes, and here, as elsewhere, the ligand field bands in the spectra of complexes in corresponding oxidation states are similar. On the whole, however, there are certain consistent differences of which the above-mentioned comparisons are particularly obvious manifestations.

Some important features of the elements and comparison of these with the corresponding features of the first series are the following:

1. *Radii.* The radii of the heavier transition atoms and ions are known only in a few cases. An important feature is that the filling of the $4f$ orbitals through the lanthanide elements causes a steady contraction, called the *lanthanide contraction* (page 1058), in atomic and ionic sizes. Thus the expected size increases of elements of the third transition series relative to those of the second transition series, due to increased number of electrons and the higher principal quantum numbers of the outer ones, are almost exactly offset, and there is in general little difference in atomic and ionic sizes between the two

heavy atoms of a group whereas the corresponding atom and ions of the first transition series are significantly smaller.

2. *Oxidation states.*   For the heavier transition elements, higher oxidation states are in general much more stable than for the elements of the first series. Thus the elements Mo, W, Tc and Re form oxo anions in high valence states which are not especially easily reduced, whereas the analogous compounds of the first transition series elements, when they exist, are strong oxidizing agents. Indeed, the heavier elements form many compounds such as RuO_4, WCl_6 and PtF_6 that have no analogs among the lighter ones. At the same time, the chemistry of complexes and aquo ions of the lower valence states, especially II and III, which plays such a large part for the lighter elements, is of relatively little importance for most of the heavier ones.

3. *Aqueous chemistry.*   Aquo ions of low and medium valence states are not in general well defined or important for any of the heavier transition elements, and some, such as Zr, Hf and Re, do not seem to form any simple cationic complexes. For most of them anionic oxo and halo complexes play a major role in their aqueous chemistry although some, such as Ru, Rh, Pd and Pt, do form important cationic complexes as well.

4. *Metal-metal bonding.*   In general, although not invariably, the heavier transition elements are much more prone to form strong M—M bonds than are their congeners in the first transition series. The main exceptions to this are the polynuclear metal carbonyl compounds and some related ones, where analogous or similar structures are found for all three elements of a given family. Aside from these, however, it is common to find that the first-series metal will form few or no M—M bonded species whereas the heavier congeners form an extensive series. Examples are the $M_6X_{12}^{n+}$ species formed by Nb and Ta, with no V analogs at all, and the $Tc_2Cl_8^{3-}$ and $Re_2Cl_8^{2-}$ ions, which have no manganese analogs.

General discussions of species containing M—M bonds have been given in previous Chapters (pages 547 and 688).

5. *Magnetic properties.*   Whereas a simple interpretation of magnetic susceptibilities of the compounds of first transition series elements usually gives the number of unpaired electrons and hence the oxidation state and d orbital configuration, more complex behavior is often encountered in compounds of the heavier elements.

One important characteristic of the heavier elements is that they tend to give *low-spin* compounds, which means that in oxidation states where there is an odd number of d electrons there is frequently only one unpaired electron, and ions with an even number of d electrons are very often diamagnetic. There are two main reasons for this intrinsically greater tendency to spin-pairing. First, the $4d$ and $5d$ orbitals are spatially larger than $3d$ orbitals so that double occupation of an orbital produces significantly less interelectronic repulsion. Second, a given set of ligand atoms produces larger splittings of $5d$ than of $4d$ orbitals and in both cases larger splittings than for $3d$ orbitals (see page 577).

When there are unpaired electrons, the susceptibility data are often less easily interpreted. For instance, low-spin octahedral Mn^{III} and Cr^{II} complexes have t_{2g}^4 configurations and hence two unpaired electrons. They have magnetic moments in the neighborhood of 3.6 B.M. which can be correlated with the presence of the two unpaired spins, these alone being responsible for a moment of 2.83 B.M., plus a contribution from unquenched orbital angular momentum. Now, Os^{IV} also forms octahedral complexes with t_{2g}^4 configurations, but these commonly have moments of the order of 1.2 B.M.; such a moment, taken at face value, has little meaning and certainly does not give any simple indication of the presence of two unpaired electrons. Indeed, in older literature it was naïvely taken to imply that there was only one unpaired electron, from which the erroneous conclusion was drawn that the osmium ion was in an odd oxidation state instead of the IV state.

Similar difficulties arise in other cases, and their cause lies in the *high spin–orbit coupling constants* of the heavier ions. Fig. 26-1 shows how the effective magnetic moment of a t_{2g}^4 configuration depends on the ratio of the thermal energy, kT, to the spin–orbit coupling constant, λ. For Mn^{III} and Cr^{II}, λ is sufficiently small that at room temperature ($kT \approx 200$ cm^{-1}) both of these ions fall on the plateau of the curve where their behavior is of the familiar sort. Os^{IV}, however, has a spin–orbit coupling constant that is an order of magnitude higher, and at room temperature kT/λ is still quite small. Thus at ordinary temperatures, octahedral Os^{IV} compounds should (and do) have low, strongly temperature-dependent magnetic moments. Obviously, if measurements on Os^{IV} compounds could be made at sufficiently high temperatures—which is usually impossible—they would have "normal" moments, and, conversely, at very low temperatures Mn^{III} and Cr^{II} compounds would show "abnormally" low moments.

The curve shown in Fig. 26-1 for the t_{2g}^4 case arises because of the following effects of spin–orbit coupling. First, the spin–orbit coupling

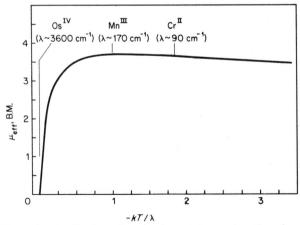

Fig. 26-1. Curve showing the dependence on temperature and on the spin–orbit coupling constant, λ, of the effective magnetic moment of a d^4 ion in octahedral coordination.

splits the lowest triplet state in such a way that in the component of lowest energy the spin and orbital moments cancel one another completely. When λ and hence this splitting are large compared with the available thermal energy, the Boltzmann distribution of systems among the several spin–orbit split components is such that most of the systems are in the lowest one which makes no contribution at all to the average magnetic moment. At $0\,°K$, of course, all systems would be in this non-magnetic state and the substance would become entirely diamagnetic. Second, however, the spin–orbit coupling causes an interaction of this lowest, non-magnetic state with certain high-lying excited states so that the lowest level is not actually entirely non-magnetic at all temperatures, and in the temperature range where kT/λ is much less than unity the effective magnetic moment varies with the square root of the temperature.

Similar difficulties arise for d^1 ions in octahedral fields, when the spin–orbit coupling constant is large. For example, if $\lambda = 500$ (as for Zr^{III}) the non-magnetic ground state, which splits off from the $^2T_{2g}$ term under the influence of spin–orbit coupling, will be so low that a temperature-independent susceptibility corresponding to an effective moment of only ~ 0.8 B.M. at room temperature will be observed. Again, this moment as such has no unique interpretation in terms of the number of unpaired electrons for the ion.

It is beyond the scope of this text to go more deeply into this subject, but it should be borne in mind that, as the examples given demonstrate, the high spin–orbit coupling constants can cause metal ions of the second and third transition series to have magnetic moments at room temperature that cannot be simply interpreted in terms of the number of unpaired electrons present unless measurements are made over a considerable temperature range on magnetically dilute specimens and the results are compared with theoretical calculations such as those represented by the curve in Fig. 26-1 for the low-spin d^4 system. Other systems for which fairly complicated behavior is expected (only octahedral coordination being considered here) are d^1, d^2, d^7, d^8, and d^9. The d^6 systems have no paramagnetism (unless there is some of the temperature-independent type) since they have t_{2g}^6 configurations with no unpaired electrons. The d^3 systems have magnetic moments that are rigorously temperature-independent regardless of the magnitude of λ. The d^5 systems have moments that vary with temperature only for very low values of kT/λ, and even then the temperature-dependence is not severe; nevertheless, these systems can show complicated behavior because of intermolecular magnetic interactions in compounds that are not magnetically dilute.

For general references to compounds and complexes of elements of the second and third series, see Chapter 25, page 804.

26-A. ZIRCONIUM AND HAFNIUM[1]

Because of the effects of the lanthanide contraction, both the atomic radii of Zr and Hf (1.45 and 1.44 Å, respectively) and the radii of the Zr^{4+} and Hf^{4+} ions (0.74 and 0.75 Å, respectively) are virtually identical. This has the effect of making the chemical behavior of the two elements extremely similar, more so than for any other pair of congeneric elements. The chemistry of hafnium has been studied less than that of zirconium but only small differences, e.g., in solubilities and volatilities of compounds, are to be expected or found.

The oxidation states and stereochemistries are summarized in Table 26-A-1.

TABLE 26-A-1

Oxidation States and Stereochemistry of Zirconium and Hafnium

Oxidation state	Coordination number	Geometry	Examples
Zr^0	6	Octahedral(?)	$[Zr(bipy)_3]$?
Zr^I, Hf^I, d^3	?	?	ZrCl, HfCl
Zr^{II}, d^2	?	?	ZrX_2
Zr^{III}, Hf^{III}, d^1	6	Octahedral	$ZrCl_3$, $ZrBr_3$, ZrI_3, HfI_3
Zr^{IV}, Hf^{IV}, d^0	4	Tetrahedral	$ZrCl_4(g)$, $Zr(CH_2C_6H_5)_4$
	6	Octahedral	Li_2ZrF_6, $Zr(acac)_2Cl_2$, $ZrCl_6^{2-}$, $ZrCl_4(s)$
	7	Pentagonal bipyramidal	Na_3ZrF_7, Na_3HfF_7, $K_2CuZr_2F_{12} \cdot 6H_2O$
		Capped trigonal prism	$(NH_4)_3ZrF_7$
		See text, Fig. 26-A-2	ZrO_2, HfO_2(monoclinic)
	8	Square antiprism	$Zr(acac)_4$, $Zr(SO_4)_2 \cdot 4H_2O$
		Dodecahedron	$[Zr(C_2O_4)_4]^{4-}$, $[ZrX_4(diars)_2]$, $[Zr_4(OH)_8(H_2O)_{16}]^{8+}$

The most important difference from titanium is that lower oxidation states are of minor importance. There are few authenticated compounds of these elements except in their tetravalent states. Like titanium, they form interstitial borides, carbides, nitrides, etc., but of course these are not to be regarded as having the metals in definite oxidation states. Increased size also makes the oxides more basic and the aqueous chemistry somewhat more extensive, and permits the attainment of coordination numbers 7 and, commonly, 8 in a number of compounds.

26-A-1. The Elements

Zirconium occurs widely over the earth's crust but not in very concentrated deposits. The major minerals are *baddeleyite*, a form of ZrO_2, and *zircon*, $ZrSiO_4$. The chemical similarity of zirconium and hafnium is well exem-

[1] E. M. Larson, *Adv. Inorg. Chem. Radiochem.*, 1970, **13**, 1 (an extensive review).

plified in their geochemistry, for hafnium is found in Nature in all zirconium minerals in the range of fractions of a percent of the zirconium content. Separation of the two elements is extremely difficult, even more so than for adjacent lanthanides, but it can now be accomplished satisfactorily by ion-exchange or solvent-extraction fractionation methods.[2]

Zirconium metal, m.p. $1855° \pm 15°$, like titanium, is hard and corrosion-resistant, resembling stainless steel in appearance. It is made by the Kroll process (page 809). Hafnium metal, m.p. $2222° \pm 30°$, is similar. Like titanium, these metals are fairly resistant to acids, and they are best dissolved in HF where the formation of anionic fluoro complexes is important in the stabilization of the solutions. Zirconium will burn in air at high temperatures, reacting more rapidly with nitrogen than with oxygen, to give a mixture of nitride, oxide, and oxide nitride, Zr_2ON_2.

26-A-2. Compounds of Zirconium(IV) and Hafnium(IV)

Halides. $ZrCl_4$ is a white solid, subliming at 331°; it is monomeric and tetrahedral in the vapor, but the solid consists of zigzag chains of $ZrCl_6$ octahedra[3] (Fig. 26-A-1), with the Cl atoms forming a distorted *ccp* array. The compounds $ZrBr_4$, $HfCl_4$ and $HfBr_4$ are isotypic. $ZrCl_4$ resembles $TiCl_4$ in its chemical properties. It may be prepared by chlorination of heated zirconium, zirconium carbide or a mixture of ZrO_2 and charcoal; it fumes in moist air, and it is hydrolyzed vigorously by water. Hydrolysis proceeds only part way at room temperature, affording the stable oxide chloride,

$$ZrCl_4 + 9H_2O \rightarrow ZrOCl_2 \cdot 8H_2O + 2HCl$$

$ZrCl_4$ also combines with donors such as ethers, esters, $POCl_3$, and CH_3CN, and with Cl^- ions to form six-coordinate species.[4]

Hexachlorozirconates can be obtained by adding CsCl or RbCl to solutions of $ZrCl_4$ in concentrated HCl. The $ZrCl_6^{2-}$ ion is unstable in solutions and even 15M HCl solutions contain some cationic hydroxo species and the oxide chloride can be crystallized from such solutions.

$ZrCl_4$ also combines with two moles of certain diarsines (as do $TiCl_4$, $HfCl_4$, and several other tetrahalides of these Group IV elements) to form $ZrCl_4(diars)_2$, which has the dodecahedral type of eight-coordinate structure; 1:1 complexes are also formed but these are probably dimeric with octa-

Fig. 26-A-1.   The zigzag $ZrCl_6$ chains in $ZrCl_4$.

[2] See, e.g., W. Fischer *et al.*, *Angew. Chem. Internat. Edn.*, 1966, **5**, 15; I. V. Vinarov, *Russ. Chem. Rev.*, 1967, **36**, 522.
[3] B. Krebs, *Z. anorg. Chem.*, 1970, **378**, 263.
[4] F. M. Chung and A. D. Westland, *Canad. J. Chem.*, 1969, **47**, 195.

hedral Zr. With $CH_3SCH_2CH_2SCH_3$ (DTH), the compound $ZrCl_4(DTH)_2$ is formed and this appears[5a] to be isostructural with $ZrCl_4(diars)_2$. $ZrCl_4$ reacts with carboxylic acids above $100°$ to give $Zr(RCO_2)_4$ compounds which appear to contain 8-coordinate molecules.[5b]

$ZrBr_4$ and ZrI_4 are similar to $ZrCl_4$. ZrF_4 is a white crystalline solid subliming at $903°$ which, unlike the other halides, is insoluble in donor solvents; it has an eight-coordinate structure with square antiprisms joined by sharing fluorines. Hydrated fluorides, $ZrF_4 \cdot 1$ or $3H_2O$, can be crystallized from $HF–HNO_3$ solutions. The trihydrate has an eight-coordinate structure with two bridging fluorines, $(H_2O)_3F_3ZrF_2ZrF_3(H_2O)_3$. The hafnium hydrate has the same stoichiometry but a different structure with chains of $HfF_4(H_2O)$ units linked through four bridging fluorine atoms.[6]

Zirconium Oxide and Mixed Oxides. Addition of hydroxide to zirconium(iv) solutions causes the precipitation of white gelatinous $ZrO_2 \cdot nH_2O$, where the water content is variable; no true hydroxide exists. On strong heating, this hydrous oxide gives hard, white, insoluble ZrO_2. This has an extremely high melting point ($2700°$), exceptional resistance to attack by both acids and alkalis, and good mechanical properties; it is used for crucibles and furnace cores. ZrO_2 in its monoclinic (baddeleyite) form and one form of HfO_2 are isomorphous and have a structure in which the metal atoms are seven coordinate, as shown in Fig. 26-A-2. Three other forms of ZrO_2 have been described, none of which has the rutile structure so often found among MO_2 compounds.

A number of compounds called "zirconates" may be made by combining oxides, hydroxides, nitrates, etc., of other metals with similar zirconium compounds and firing the mixtures at $1000–2500°$. These, like their titanium analogs, are mixed metal oxides; there are no discrete zirconate ions known. $CaZrO_3$ is isomorphous with perovskite, and there are a number of $M^{II}ZrO_4$ compounds having the spinel structure. By dissolving ZrO_2 in molten KOH and evaporating off the excess of solvent at $1050°C$, the crystalline compounds $K_2Zr_2O_5$ and K_2ZrO_3 may be obtained. The former contains ZrO_6 octahedra sharing faces to form chains which, in turn, share

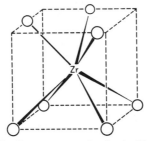

Fig. 26-A-2. Coordination geometry in the baddeleyite form of ZrO_2.

[5] (a) J. B. Hamilton and R. E. McCarley, *Inorg. Chem.*, 1970, **9**, 1339; (b) J. Ludwig and D. Schwartz, *Inorg. Chem.*, 1970, **9**, 607.
[6] D. Hall, C. E. F. Rickard and T. N. Waters, *J. Inorg. Nuclear Chem.*, 1971, **33**, 2395.

edges and corners with other chains. The latter contains infinite chains of ZrO_5 square pyramids (26-A-I).[7]

(26-A-I)

(26-A-II)

Stereochemistry. The Zr^{4+} ion is relatively large, highly charged, and spherical, with no partly filled shell to give it stereochemical preferences. Thus, it is not surprising that zirconium(IV) compounds exhibit high coordination numbers and a great variety of coordination polyhedra. This is well illustrated by the fluoro complexes.[8,9] Li_2ZrF_6 and $CuZrF_6 \cdot 4H_2O$ contain the octahedral ZrF_6^{2-} ion, while ZrF_7^{3-} ions are known with two different structures: the pentagonal bipyramid in Na_3ZrF_7 and a capped trigonal prism (26-A-II) in $(NH_4)_3ZrF_7$. Pentagonal bipyramids sharing an edge form the $Zr_2F_{12}^{4-}$ ion in $K_2CuZr_2F_{12} \cdot 6H_2O$. $N_2H_6ZrF_6$ also contains bicapped trigonal prisms sharing F atoms. ZrF_8 units of several types are found: $Cu_2ZrF_8 \cdot 12H_2O$ has discrete square antiprisms, while $Cu_3Zr_2F_{14} \cdot 16H_2O$ contains $Zr_2F_{14}^{6-}$ ions formed by two square antiprisms sharing an edge.

With oxygen ligands high coordination numbers and varied stereochemistry are also prevalent. Thus, $M_6^I[Zr_2(OH)_2(CO_3)_6] \cdot nH_2O$ ($M^I = K$ or NH_4) contain OH-bridged binuclear units in which each Zr atom is in dodecahedral 8-coordination,[10] and ZrO_7 pentagonal bipyramids are found in $ZrO_2(OH)OMo_2O_4(OH)_5$.[11]

Aqueous Chemistry and Complexes. ZrO_2 is more basic than TiO_2 and is virtually insoluble in an excess of base. There is a more extensive aqueous chemistry of zirconium because of a lower tendency toward complete hydrolysis. Nevertheless hydrolysis does occur and it is very doubtful indeed if Zr^{4+} aquo ions exist even in strongly acid solutions. The hydrolyzed ion is often referred to as the "zirconyl" ion and written ZrO^{2+}. However, there is little, if any, reliable evidence for the existence of such an oxo ion either in solution or in crystalline salts,[12] although the infrared spectrum of one form of ZrOCl is reported to have a band at 877 cm^{-1}, perhaps indicative of a $Zr=O$ group.[13] No such group has ever been directly identified by diffraction methods.

[7] B. M. Gatehouse and D. J. Lloyd, *Chem. Comm.*, **1969**, 606, 727.
[8] J. Fischer *et al.*, *Acta Cryst.*, 1969, A, **25**, 5175; *Chem. Comm.*, **1967**, 328, 329; **1968**, 1137; (b) H. J. Hurst and J. C. Taylor, *Acta Cryst.*, 1970, B, **26**, 2136.
[9] B. Prodic, S. Scavnicar and B. Matovic, *Acta Cryst.*, 1969, A, **25**, 5102.
[10] Yu. E. Gorbunova, G. N. Novitskaya, and V. G. Kuznetsov, *J. Struct. Chem.*, 1970, **11**, 523.
[11] A. Clearfield and R. H. Blessing, *Acta Cryst.*, 1969, A, **25**, S110.
[12] W. P. Griffith and T. D. Wickins, *J. Chem. Soc.*, A, **1967**, 675.
[13] K. Dehnicke and J. Weidlein, *Angew. Chem. Internat. Edn.*, 1966, **5**, 1041.

The most important "zirconyl" salt is $ZrOCl_2 \cdot 8H_2O$ which crystallizes from dilute hydrochloric acid solutions and contains the ion $[Zr_4(OH)_8-(H_2O)_{16}]^{8+}$. Here the Zr atoms lie in a distorted square, linked by pairs of hydroxo bridges, and also bound to four water molecules so that the Zr atom is coordinated by eight oxygen atoms in a distorted dodecahedral arrangement.[14]

In acid solutions, except concentrated HF, where ZrF_6^{2-} and HfF_6^{2-} ions only (MF_7^{3-} species cannot be detected) are present,[15] acid solutions of Zr^{IV} contain polymeric, partially hydrolyzed species. In $1-2M$ perchloric acid $[Zr_3(OH)_4]^{8+}$ and $[Zr_4(OH)_8]^{8+}$ are thought to be the major species. In $2.8M$ HCl the main species appears to be trinuclear, perhaps $[Zr_3(OH)_6Cl_3]^{3+}$, and, as noted above, the stable phase that crystallizes from HCl solutions, $ZrOCl_2 \cdot 8H_2O$, contains tetramers. Studies of solutions are complicated by slowness in the attainment of equilibrium. Chelating agents such as EDTA and NTA form complexes with Zr^{IV}; the $\{Zr[N(CH_2COO)_3]_2\}^{2-}$ ion has been shown to be dodecahedral.[16]

Acetylacetonates of the types $M(acac)_2X_2$ and $M(acac)_3X$ are known. The former have *cis* octahedral configurations but appear to dissociate, whereas the latter are 7-coordinate and, except for $X = I$, do not dissociate.[17]

Some seemingly simple zirconium salts are best regarded as essentially covalent molecules or as complexes; examples are the carboxylates, $Zr(OCOR)_4$, the tetrakis(acetylacetonate), the oxalate and the nitrate. Like its Ti analog, the last of these is made by heating the initial solid adduct of N_2O_5 and N_2O_4 obtained in the reaction

$$ZrCl_4 + 4N_2O_5 \xrightarrow{30°} Zr(NO_3)_4 \cdot xN_2O_5 \cdot yN_2O_4 + 4NO_2Cl.$$

It forms colorless sublimable crystals and IR and Raman spectra suggest[18] that the molecule is isostructural with $Ti(NO_3)_4$ and $Sn(NO_3)_4$. It is soluble in water but insoluble in toluene, whose ring it nitrates. Hafnium gives only $Hf(NO_3)_4 \cdot N_2O_5$. Nitrato complexes, $M(NO_3)_6^{2-}$ are also known but not structurally characterized.

Zirconium forms a borohydride,[19] $Zr(BH_4)_4$, with a highly symmetrical structure (full T_d symmetry) shown in Fig. 26-A-3.

Zirconium solutions with sulfate as anion show considerable difference from Cl^-, NO_3^- or ClO_4^- solutions. Even at low acidities, strong neutral and anionic complexes, some of which may be polymeric, are formed. Crystalline materials with compositions $Zr(SO_4)_2 \cdot nH_2O$ where $n = 4$, 5 or 7 have been structurally characterized. The first has infinite sulfato-bridged sheets with each Zr square antiprismatically coordinated by four H_2O

[14] T. C. W. Mak, *Canad. J. Chem.*, 1968, **46**, 3491.
[15] P. A. W. Dean and D. F. Evans, *J. Chem. Soc., A,* **1967**, 698.
[16] J. L. Hoard, E. W. Silverton and J. V. Silverton, *J. Amer. Chem. Soc.*, 1968, **90**, 2300.
[17] R. B. Von Dreele, J. J. Stezowski and R. C. Fay, *J. Amer. Chem. Soc.*, 1971, **93**, 2887..
[18] J. Weidlein, U. Muller and K. Dehnicke, *Spectrochim. Acta*, 1968, *A*, **24**, 253.
[19] N. Davies, D. Saunders and M. G. H. Wallbridge, *J. Chem. Soc., A,* **1970**, 2915; V. Plato and K. Hedberg, *Inorg. Chem.*, 1971, **10**, 590.

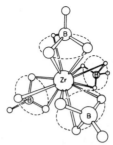

Fig. 26-A-3. The structure of $Zr(BH_4)_4$. [Reproduced by permission from P. H. Bird and M. R. Churchill, *Chem. Comm.*, **1967**, 403; cf. also V. Plato and K. Hedberg, *Inorg. Chem.*, 1971, **10**, 590.]

and four sulfate oxygen atoms; when $n = 5$ or 7, the compound contains the dinuclear ion (26-A-III) in which each Zr is dodecahedrally coordinated.[20]

(26-A-III)

Like Ce^{IV}, Pu^{IV} and other similar ions, Zr^{IV} gives an insoluble iodate which can be crystallized from nitric acid solution; it has an antiprismatic 8-coordinate Zr with bridging IO_3 groups. The oxalate, $Na_4[Zr(C_2O_4)_4] \cdot 3H_2O$ and its Hf analog are isomorphous, with dodecahedral coordination; although isomers are theoretically possible, the anions lose oxalate by hydrolysis and/or aquation so that racemization occurs rapidly.

26-A-3. Lower Oxidation States

The chemistry of lower oxidation states is so far limited to non-aqueous chemistry of lower halides and some complexes thereof.

The three trihalides $ZrCl_3$, $ZrBr_3$ and ZrI_3, are well-established compounds which can be prepared in several ways, e.g., by reduction of the tetrahalides with H_2 or Zr. Their magnetic susceptibilities are low and independent of temperature from 80° to 300 °K. These three compounds and HfI_3 are isotypic; the structure[21] appears to involve a distorted *hcp* array of anions with cations occupying adjacent octahedral interstices so as to form infinite $Zr(X/2)_6$ chains along the c direction. Interatomic pairing of electron spins may then be invoked to account for the magnetic properties.[21]

[20] I. J. Bear and W. G. Mumme, *Acta Cryst.*, 1971, **B27**, 366, 1373.
[21] L. F. Dahl *et al.*, *Inorg. Chem.*, 1964, **3**, 1236.

The ZrX_3 compounds react readily at room temperature with bases such as pyridine, bipyridine, phenanthroline and CH_3CN to form extremely moisture-sensitive compounds such as $ZrX_3 \cdot 2py$, $2ZrX_3 \cdot 5CH_3CN$, of puzzling stoichiometry and, as yet, unknown structures.[22]

The hafnium triiodide phase has a marked tendency to non-stoichiometry, varying in composition from $HfI_{3.0}$ to as high as $HfI_{3.5}$. There is no evidence for any lower iodide phase.[23]

The lower chlorides of Zr and Hf have recently been carefully studied.[24] Non-stoichiometric phases with compositions near $HfCl_{2.5}$ and $ZrCl_{2.5}$ have been observed but as yet not well characterized. A substantially pure $ZrCl_2$ phase also appears to exist. Most interesting, however, are ZrCl and HfCl, which are isotypic, with hexagonal unit cells. They have very metallic properties and HfCl is essentially diamagnetic.[24] $ZrBr_2$ and ZrI_2 have previously been reported but would bear closer study.

The reduction of $ZrCl_4$ with lithium in the presence of bipyridine in THF gives the violet $Zr(dipy)_4$ where, doubtless, there is considerable delocalization of electrons over the ligands. An oxide, ZrO, said to have the NaCl structure, has been mentioned but is not further characterized.

26-A-4. Organometallic Compounds

Most of the organometallic compounds of Zr and Hf are similar to their Ti analogs (page 816). The commonest are the cyclopentadienyl compounds of the type $(h^5\text{-}C_5H_5)_2ZrX_2$. The tetracyclopentadienyls of both Zr and Hf are fluxional and pose a structural problem.[25, 26] Crystallographic results[25] show that, unlike the titanium analog, they are not of the type $(h^5\text{-}C_5H_5)_2(h^1\text{-}C_5H_5)_2M$. The proposal[25] that they are of type $(h^5\text{-}C_5H_5)_3\text{-}(h^1\text{-}C_5H_5)M$ is of dubious validity, since this requires a 20-electron configuration, and it has been suggested[26] that there are two severely tilted rings, of the sort found in $(C_5H_5)_2Mo(NO)X$ compounds, along with one genuine $h^5\text{-}C_5H_5$ and one $h^1\text{-}C_5H_5$ ring. There are also some $(h^5\text{-}C_5H_5)ML_3$ compounds where $L = (acac)^-$ which undergo rapid rearrangements.[27]

Other isolable compounds are the tetrabenzyl[28] and tetraallyl,[29] $(C_6H_5CH_2)_4Zr$ and $(C_3H_4)_4Zr$. No carbonyl compound is known.

[22] G. W. A. Fowles, B. J. Russ and G. R. Willey, *Chem. Comm.*, **1967**, 646.
[23] A. W. Struss and J. D. Corbett, *Inorg. Chem.*, 1969, **8**, 227.
[24] A. W. Struss and J. D. Corbett, *Inorg. Chem.*, 1970, **9**, 1373; E. M. Larson *et al.*, *Chem. Comm.*, **1970**, 281.
[25] V. I. Kulishov, N. G. Bokii and Y. T. Struchkov, *J. Struct. Chem.*, 1971, **11**, 646.
[26] J. L. Calderon *et al.*, *J. Amer. Chem. Soc.*, 1971, **93**, 3592.
[27] J. J. Howe and T. J. Pinnavaia, *J. Amer. Chem. Soc.*, 1970, **92**, 7342.
[28] U. Zucchini, E. Albizzati and U. Giannini, *J. Organometallic Chem.*, 1971, **26**, 357.
[29] G. Wilke *et al.*, *Angew. Chem. Internat. Edn.*, 1966, **5**, 151.

26-B. NIOBIUM AND TANTALUM[1]

Niobium and tantalum, though metallic in many respects, have chemistries in the V oxidation state that are very similar to those of typical non-metals. They have virtually no cationic chemistry but form numerous anionic species. Their halides and oxide halides, which are their most important simple compounds, are mostly volatile and are readily hydrolyzed. In their lower oxidation states they form an extraordinarily large number of metal-atom cluster compounds. Only niobium forms lower states in aqueous solution. The oxidation states and stereochemistries (excluding those in the cluster compounds) are summarized in Table 26-B-1.

TABLE 26-B-1

Oxidation States and Stereochemistries of Niobium and Tantalum

Oxidation state	Coordination number	Geometry	Examples
Nb^{-I}, Ta^{-I}	6	Octahedral (?)	$[M(CO)_6]^-$
Nb^I, Ta^I	7	π-Complex[a]	$(h^5\text{-}C_5H_5)M(CO)_4$
Nb^{II}, Ta^{II}, d^3	6	Octahedral	NbO
Nb^{IV}, Ta^{IV}, d^1	4	Dist. tetrahedral	$Nb(NEt_2)_4$[b]
	6	Octahedral	MX_4, MCl_4py_2, MCl_6^{2-}
	8	Dodecahedral	$NbX_4(diars)_2$
Nb^V, Ta^V, d^0	4	Tetrahedral	$ScNbO_4$
	5	Trig. bipyramid	MCl_5 (vapor)
	5	Distorted tetragonal pyramid	$Nb(NMe_2)_5$[c]
	6	Octahedral	$NaMO_3$ (perovskite), $NbCl_5 \cdot OPCl_3$, $TaCl_5 \cdot S(CH_3)_2$, TaF_6^-, $NbOCl_3$, M_2Cl_{10}, MCl_6^-
	7	?	$TaX_5(diars)$
	7	See text, p. 27	K_2MF_7, K_3NbOF_6
	8	Sq. antiprism	Na_3TaF_8
	8	?	$Nb(troponate)_4^+$
	9	π-Complex[a]	$(h^5\text{-}C_5H_5)_2TaH_3$

[a] If h^5-C_5H_5 is considered as occupying three sites.
[b] D. C. Bradley and M. M. Chisholm, *J. Chem. Soc. A*, **1971**, 1511.
[c] C. Heath and M. B. Hursthouse, *Chem. Comm.*, **1971**, 143.

26-B-1. The Elements

Niobium is 10–12 times more abundant in the earth's crust than tantalum. The main commercial sources of both are the *columbite–tantalite* series of minerals, which have the general composition $(Fe/Mn)(Nb/Ta)_2O_6$, with the ratios Fe/Mn and Nb/Ta continuously variable. Niobium is also obtained from *pyrochlore*, a mixed calcium–sodium niobate. Separation and production of the metals is complex.[1] Both metals are bright, high-melting (Nb, 2468°C; Ta, 2996°C) and very resistent to acids. They can be dissolved, with vigor, in an HNO_3–HF mixture, and, very slowly, in fused alkalis.

[1]   F. Fairbrother, *The Chemistry of Niobium and Tantalum*, Elsevier, Amsterdam, 1967; J. O. Hill, I. G. Worsley and L. G. Hepler, *Chem. Rev.*, 1971, **71**, 127 (oxidation potentials and thermodynamic data).

26-B-2. Pentavalent Niobium and Tantalum, d^0

Oxygen Compounds. The oxides, Nb_2O_5 and Ta_2O_5 are dense white powders that are relatively inert chemically. They are scarcely if at all attacked by acids except concentrated hydrofluoric; they can also be dissolved by fusion with an alkali hydrogen sulfate, an alkali carbonate or hydroxide. They are obtained by dehydrating the hydrous oxides (so-called "niobic" and "tantalic" acids) or by roasting certain other compounds in an excess of oxygen. The hydrous oxides, of variable water content, are gelatinous white precipitates obtained upon neutralizing acid solutions of Nb^V and Ta^V halides. The Ta_2O_5 phase can exist with an excess of Ta atoms as interstitials over the range $TaO_{2.0}$ to $TaO_{2.5}$, and it then shows metallic conductance.[2]

Niobate and tantalate isopolyanions[3] can be obtained by fusing the oxides in an excess of alkali hydroxide or carbonate and dissolving the melts in water. The solutions are stable only at higher pH: precipitation occurs below pH ~ 7 for niobates and ~ 10 for tantalates. The only species that appear to be present in solution are the $[H_xM_6O_{19}]^{(8-x)-}$ ions, $x = 0$, 1 or 2, despite frequent claims for others. The structure of the $M_6O_{19}^{8-}$ ions, found in crystals and believed to persist in solutions, is shown in Fig. 26-B-1.

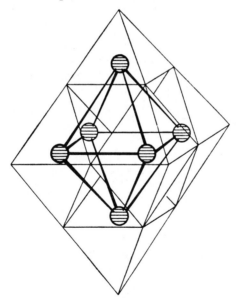

Fig. 26-B-1.   Structure of the $M_6O_{19}^{8-}$ ions (M = Nb or Ta). The central oxygen atom is shared by all six octahedra. [Reproduced by permission from W. M. Nelson and R. S. Tobias, *Inorg. Chem.*, 1963, **2**, 985.]

[2] D. R. Kudrak and M. J. Sienko, *Inorg. Chem.*, 1967, **6**, 880.
[3] M. T. Pope and B. W. Dale, *Quart. Rev.*, 1968, **22**, 527.

Heteropolyniobates and -tantalates are not well known but a few of the former have been prepared and characterized.[4]

With the exception of a few insoluble lanthanide niobates and tantalates, e.g., $ScNbO_4$,[5] which contain discrete, tetrahedral MO_4^{3-} ions, the coordination number of Nb^V and Ta^V with oxygen is essentially always 6. The various "niobates" and "tantalates" are really mixed metal oxides. Thus, for example, the $M^I XO_3$ compounds are perovskites (page 55).

Niobium(v) peroxo species, e.g., $[Nb(O_2)_3phen]^-$, are also known.[6]

Fluorides and Fluoride Complexes. The pentafluorides are made by direct fluorination of the metals or the pentachlorides. Both are volatile white solids (Nb, m.p. 80°, b.p. 235°; Ta, m.p. 95°, b.p. 229°), giving colorless liquids and vapors. They have the tetrameric structure shown in Fig. 26-B-2; they appear to be polymeric also when molten. The oxide fluorides MOF_3 and MO_2F are known.

The metals and the pentoxides dissolve in aqueous HF to give fluoro complexes, whose composition depends markedly on the conditions. Addition of CsF to niobium in 50% HF precipitates $CsNbF_6$, while in weakly acidic solutions hydrolyzed species $[NbO_xF_y \cdot 3H_2O]^{5-2x-y}$, occur and compounds such as $K_2NbOF_5 \cdot H_2O$ may be isolated. Raman and ^{19}F nmr spectra show that $[NbOF_5]^{2-}$ is present in aqueous solution up to about 35% in HF, while $[NbF_6]^-$ becomes detectable beginning at about 25% HF. $[NbF_6]^-$ is normally the highest fluoro complex of Nb^V formed in solution although in 95–100% HF NbF_7^{2-} may possibly be present.[7a] Salts containing the NbF_7^{2-} ion can be crystallized from solutions with very high F^- concentrations. From solutions of low acidity and high F^- concentration, salts of the $[NbOF_6]^{3-}$ ion can be isolated.

There are crystalline tantalum fluoro compounds such as $KTaF_6$, K_2TaF_7 and K_3TaF_8. The TaF_8^{3-} ion, like NbF_7^{2-}, may be stabilized by crystal forces, since in aqueous HF or NH_4F solutions Raman spectra show the

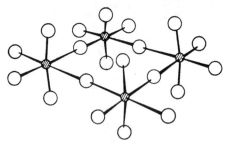

Fig. 26-B-2. The tetrameric structures of NbF_5 and TaF_5 (also MoF_5 and, with slight distortion, RuF_5 and OsF_5). Nb—F bond lengths: 2.06 Å (bridging), 1.77 Å (non-bridging). [Adapted by permission from A. J. Edwards, *J. Chem. Soc.*, **1964**, 3714.]

[4] B. W. Dale, J. M. Buckley and M. T. Pope, *J. Chem. Soc.*, **1969**, 301; C. M. Flynn, Jr., and G. D. Stucky, *Inorg. Chem.*, 1969, **8**, 335.
[5] L. N. Komissarova *et al.*, *Russ. J. Inorg. Chem.*, 1968, **13**, 934.
[6] G. Mathern, R. Weiss and R. Rohmer, *Chem. Comm.*, **1970**, 153.
[7a] J. A. S. Howell and K. C. Moss, *J. Chem. Soc., A*, **1971**, 2481.

presence of TaF_6^- and TaF_7^{2-} ions only. In anhydrous HF solutions of $KTaF_6$ and K_2TaF_7 the only species identified by ^{19}F-nmr spectra is TaF_6^-.[7b] In HF solutions tantalum can be separated from niobium by selective extraction into isobutyl methyl ketone.[8]

The hexafluoro anions can also be made by the dry reaction:

$$M_2O_5 + 2KCl \xrightarrow{\text{BrF}_3} 2KMF_6$$

and the $[MOF_6]^{3-}$ salts can be prepared by bromination of the metals in methanol followed by addition of NH_4F or KF.

The pentafluorides also react as Lewis acids, giving adducts such as $NbF_5 \cdot OEt_2$ and $TaF_5 \cdot SEt_2$; the 2:1 adducts are of the type $[NbF_4L_4]^+$ $[NbF_6]^{-9}$. The clear solutions formed on dissolving the fluorides in water, and which are stable if not boiled, may well contain $[MF_4(H_2O)_4]^+$ and MF_6^-.

Other Halides of Niobium(V) and Tantalum(V). All six of these are yellow to brown or purple-red solids best prepared by direct reaction of the metals with excess of the halogen. The halides melt and boil at 200–300° and are soluble in various organic liquids such as ethers, CCl_4, etc. They are quickly hydrolyzed by water to the hydrous pentoxides and the hydrohalic acid. The chlorides give clear solutions in concentrated hydrochloric acid, forming oxo chloro complexes.

Adducts. The chlorides behave as Lewis acids, forming MCl_6^- and other complexes such as $MCl_5 \cdot L$, where $L = R_2O$, R_2S, $POCl_3$, 2,2'-bipyridine, etc.[10] They also act as Friedel–Crafts catalysts and cause polymerization of acetylenes to arenes. When the chlorides are heated with alcohols or alkali alkoxides, the dimeric alkoxides $[M(OR)_5]_2$ are formed. Lithium dialkylamides also give complexes, for example, $M(NMe_2)_5$.[11] In methanol, interaction with dithiocarbamates gives complexes such as $NbCl(OMe)_2$-$(S_2CNR_2)_2$ which are monomeric and stable in air;[12] the dithiocarbamate $Ta(S_2CNMe_2)_5$ has also been made by a CS_2 insertion reaction (page 777).

The interaction of amines and niobium pentachloride often leads to the formation of reduced species, e.g., pyridine gives $NbCl_4py_2$.

The halides (Cl, Br) can also abstract oxygen from certain oxygen donors, e.g.,

$$NbCl_5 + 3(CH_3)_2SO \rightarrow NbOCl_3 \cdot 2(CH_3)_2SO + (CH_3)_2SCl_2$$
$$(CH_3)_2SCl_2 \rightarrow ClCH_2SCH_3 + HCl$$

VCl_4 and $MoCl_5$ undergo similar oxygen abstraction reactions.

All the pentahalides can be sublimed without decomposition in an atmosphere of the appropriate halogen; in the vapor they are monomeric and probably trigonal bipyramidal. Crystalline $NbCl_5$ has the dimeric

[7b] N. A. Matwiyoff, L. B. Asprey and W. E. Wagerman, *Inorg. Chem.*, 1970, **9**, 2014.
[8] G. B. Alexander, *J. Chem. Educ.*, 1969, **46**, 157.
[9] J. A. S. Howell and K. C. Moss, *J. Chem. Soc., A*, **1971**, 2483.
[10] G. A. Ozin and R. A. Walton, *J. Chem. Soc., A*, **1970**, 2236; C. Djordjevic and V. Katovic, *J. Chem. Soc., A*, **1970**, 3382.
[11] C. Heath and M. B. Hursthouse, *Chem. Comm.*, **1971**, 143.
[12] D. C. Pantaleo and R. C. Johnson, *Inorg. Chem.*, 1970, **9**, 1248.

2.56 A

Fig. 26-B-3.   The dinuclear structure of $NbCl_5$ in the solid. The octahedra are distorted as shown.

structure shown in Fig. 26-B-3; $NbBr_5$, $TaCl_5$ and $TaBr_5$ are isostruc-tural. In CCl_4 and $MeNO_2$, both $NbCl_5$ and $TaCl_5$ are dimeric but in coordinating solvents adducts are formed. It appears probable that NbI_5 has a hexagonal close-packed array of iodine atoms with niobium atoms in octahedral interstices; TaI_5 is not isomorphous and its structure is unknown.

Aside from their probable existence in solutions of the pentachlorides in concentrated hydrochloric acid, the $[MCl_6]^-$ ions appear also to exist in fused mixtures of the pentachlorides with all the alkali metal chlorides except LiCl. There are compounds of the composition M^INbCl_6 or M^ITaCl_6, and it seems reasonable to suppose that they contain $[MCl_6]^-$ ions.

Oxide halides. $NbOCl_3$, $TaOCl_3$, $NbOBr_3$, $TaOBr_3$, $NbOI_3$ and NbO_2I are known. The chlorides are white, and the bromides yellow, volatile solids; they are, however, less volatile than the corresponding pentahalides, and small amounts of the oxide halides which often arise in the preparation of the pentahalides in systems not scrupulously free from oxygen can be rather easily separated by fractional sublimation.

The best methods of preparation are by pyrolysis of the monoetherate of $TaCl_5$ for $TaOCl_3$ and by controlled reaction between the pentahalides and molecular oxygen for the others. They are all hydrolyzed to the hydrous pentoxides by water. From their solutions in concentrated hydrohalic acids and alkali-metal cations, complex oxide halides such as M^INbOCl_4 and M_2NbOCl_5 can be crystallized.

The oxide trihalides are monomeric in the vapor state but the lower volatility, compared to that of the corresponding pentahalides, is under-standable in view of the structure of $NbOCl_3$ (Fig. 26-B-4).

The pentahalides also react with N_2O_4 in ionizing solvents such as CH_3CN to give solvated oxide nitrates, e.g., $NbO_2NO_3 \cdot 0.67CH_3CN$, which

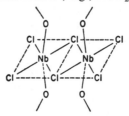

Fig. 26-B-4.   The structure of $NbOCl_3$ in the crystal. The oxygen atoms form bridges between infinite chains of the planar Nb_2Cl_6 groups.

appear to be polymeric with oxo bridges. In the absence of solvent, $NbO(NO_3)_3$ and $TaO(NO_3)_3$ can be obtained.

Other Compounds. These include nitrides, sulfides, silicides, selenides and phosphides, as well as many alloys. Definite hydride phases also appear to exist.[13]

There are no simple salts such as sulfates, nitrates, etc. Sulphates such as $Nb_2O_2(SO_4)_3$ probably have oxo bridges and coordinated sulfato groups. In HNO_3, H_2SO_4 or HCl solutions, Nb^V can exist as cationic, neutral and anionic species, hydrolyzed, polymeric, and colloidal forms in equilibrium, depending on the conditions.

The tropolonate ion gives 8-coordinate complexes with both Nb and Ta, of formula $[(C_7H_5O_2)_4M]^+$.

26-B-3. Lower Oxidation States

Oxides. The only certain lower oxides are NbO and NbO_2. TaO_x compositions from $x = 2$ to 2.5 comprise a Ta_2O_5 phase with interstitial Ta atoms, and not discrete phases or compounds. NbO_2 has a rutile-type structure with pairs of fairly close (2.80 Å) Nb atoms, presumably singly bonded to each other. NbO, which has only a narrow range of homogeneity, has metallic luster and excellent electrical conductivity of the metallic type.[14]

There are complex series of both niobium and tantalum sulfides.

Tetrahalides and Their Derivatives. All possible tetrahalides are known except TaF_4. Niobium(IV) fluoride, is a black involatile, paramagnetic solid in which each Nb atom lies at the center of an octahedron. The other six tetrahalides differ from NbF_4 and resemble each other in their structures and in being diamagnetic.

The tetrachlorides and tetrabromides are all brown-black or black solids, obtainable by reduction of the pentahalides with H_2, Al, Nb or Ta at elevated temperatures. The four compounds are apparently isomorphous. The structure of $NbCl_4$ has been studied in detail; the metal atoms occur in pairs, displaced from the centers of their octahedra towards a common edge so that the Nb—Nb distance is 3.06 Å. The formation of weak Nb—Nb bonds accounts for the diamagnetism.

Niobium(IV) iodide is easily obtained on heating NbI_5 to 300°. It is diamagnetic and trimorphic. One form contains infinite chains of octahedra with the Nb atoms off center so as to form pairs with Nb—Nb distances of 3.31 Å. TaI_4 appears to be similar. The latter can be made most easily by allowing TaI_5 to react with an excess of pyridine to give TaI_4py_2 which, on heating, loses pyridine to give TaI_4.

The tetrahalides form a number of *adducts* with nitrogen, oxygen, and sulfur donors.[15] With unidentate donors these are mostly of the MX_4L_2

[13] J. J. Reilly and R. H. Wiswell, Jr., *Inorg. Chem.*, 1970, **9**, 1678.
[14] G. V. Chandrashekar, J. Mayo and J. M. Honig, *J. Solid State Chem.*, 1970, **2**, 528.
[15] J. B. Hamilton and R. E. McCarley, *Inorg. Chem.*, 1970, **9**, 1333, 1339.

type and are believed to have *cis* octahedral configurations. In some cases two forms, e.g., red and green $NbBr_4py_2$, which may be *cis*- and *trans*-isomers, are known. There are a few 1:1 adducts; these are only slightly paramagnetic, and halogen-bridged bioctahedral structures with M—M bonds have been postulated. With the chelating ligands $MeSCH_2CH_2SMe$ (DTH) and *o*-phenylenebis(dimethylarsine) (diars), the adducts $NbX_4(DTH)_2$ and $MX_4(diars)_2$ are obtained.[16] The structures are probably dodecahedral.

Niobium(IV) chloride reacts with a variety of β-keto-enol molecules to give 8-coordinate, tetrakischelate molecules;[17] these have one unpaired electron and are not isostructural with analogous Zr and Hf compounds. Tantalum(IV) chloride reacts by abstracting oxygen, and analogous complexes are not obtained.

Several Nb^{IV} alkoxy compounds such as salts of $[NbCl_5(OC_2H_5)]^{2-}$, $[NbCl(OC_2H_5)_3py]_2$ and $Nb(OC_2H_5)_4$ are also known. The first, which is a precursor of the others, is obtained by electrolytic reduction of $NbCl_5$ in HCl-saturated alcoholic solution. The compounds are extremely sensitive to air and moisture.[18]

Niobium in aqueous chloride or sulfate solutions can also be reduced by zinc, but the species are poorly characterized.

Other Lower Halides. Besides the Nb^{IV} and Ta^{IV} halides, there is a considerable range of compounds—or phases—in which the formal oxidation numbers range from $+2.0$ to about $+3.1$. Metal–metal bonding, metal atom clusters (page 547) and non-stoichiometry are characteristic features in this field. The anhydrous halides are listed in Table 26-B-2.

TABLE 26-B-2

Anhydrous Lower Halides[a] of Niobium and Tantalum[b]

	Niobium	Tantalum
Fluorides	$[Nb_6F_{12}]F_{6/2}$	—
	"NbF_3"[c]	
Chlorides	$[Nb_6Cl_{12}]Cl_{4/2}$	$[Ta_6Cl_{12}]Cl_{4/2}$
	—	$[Ta_6Cl_{12}]Cl_{6/2}$
	$NbCl_{2.67}$–$NbCl_{3.13}$	$TaCl_{2.9}$–$TaCl_{3.1}$
Bromides	$NbBr_2$	—
	—	$[Ta_6Br_{12}]Br_{4/2}$
	—	$[Ta_6Br_{12}]Br_{6/2}$
	$NbBr_{2.67}$–$NbBr_3$	$TaBr_{2.9}$–$TaBr_{3.1}$
Iodides	NbI_2(?)	—
	$[Nb_6I_8]I_{6/2}$	—
	—	$[Ta_6I_{12}]I_{4/2}$
	$NbI_{2.67}$–NbI_3	TaI_3

[a] $X_{n/2}$ denotes n doubly-bridging halide ions.

[b] Mixed Nb-Ta clusters are known (D. Jura and H. Schäfer, *Z. anorg. Chem.*, 1970, **379**, 122).

[c] "NbF_3" apparently always contains some oxygen. A wide range of $Nb(O, F)_3$ composition exists but it is not certain whether pure NbF_3 is stable.

[16] R. L. Deutscher and D. L. Kepert, *Inorg. Chem.*, 1970, **9**, 2305.

[17] R. L. Deutscher and D. L. Kepert, *Inorg. Chim. Acta*, 1970, **4**, 645.

[18] W. J. Reagan and C. H. Brubaker, Jr., *Inorg. Chem.*, 1970, **9**, 827.

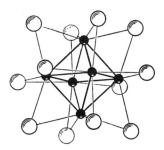

Fig. 26-B-5.   The structure of the $[M_6X_{12}]^{n+}$ units found in many halogen compounds of lower-valent niobium and tantalum. [Reproduced by permission from L. Pauling, *The Nature of the Chemical Bond*, 3rd edn., Cornell Univ. Press, 1960.]

The only apparent dihalides, $NbBr_2$ and NbI_2, though the latter is somewhat uncertain, are incompletely characterized and of unknown structure.

Many of the remaining halides listed in the table are metal-atom cluster compounds and are assigned formulas of the type $[M_6X_{12}]X_{n/2}$. These formulas convey the fact that the compounds are built up of $[M_6X_{12}]^{2+,3+}$ units with bridging halide ions, the latter denoted $X_{n/2}$. For $n = 4$ or 6 the stoichiometries are $MX_{2.33}$ and $MX_{2.5}$, respectively. The structure of the $[M_6X_{12}]$ unit is shown in Fig. 26-B-5; it is an octahedron of metal atoms with a bridging halogen atom along each edge; these $[M_6X_{12}]$ units are characteristic of, though not entirely restricted to, the chemistry of the lower oxidation states of Nb and Ta.[19]

An atypical but very interesting lower halide is Nb_6I_{11}, which, to indicate its structure,[20] is written $[Nb_6I_8]I_{6/2}$. The $[Nb_6I_8]^{3+}$ unit is an octahedron of metal atoms with a bridging iodine atom on each face, an arrangement very prevalent among the lower halides of molybdenum and tungsten. The $[Nb_6I_8]^{3+}$ unit has a single unpaired electron. Above 300° the compound absorbs hydrogen, giving diamagnetic $Nb_6I_{11}H$, where the H atom is believed to occupy the center of the Nb_6 cluster.[21]

The remaining compounds in Table 26-B-2 are the "trihalides" which are characteristically non-stoichiometric. This tendency to non-stoichiometry can be understood as follows. The crystal structure of $NbCl_{2.67}(Nb_3Cl_8)$ consists of a distorted *hcp* array of Cl^- ions, with niobium atoms occupying octahedral interstices in such a way that there are triangular groups of niobium atoms in adjacent octahedra, close enough to be bonded together into metal atom clusters. Thus $NbCl_{2.67}$ (and presumably other niobium halides with this stoichiometry) are structurally well-defined phases in a perfect condition. $NbCl_4$ has a closely related structure, derived by removal of one-third of the niobium atoms from $NbCl_{2.67}$ in such a way that each Nb_3 triangle becomes an Nb_2 pair. It is postulated that solid solutions of

[19] D. Bauer and H. G. von Schnering, *Z. anorg. Chem.*, 1968, **361**, 235.
[20] L. R. Bateman, J. F. Blount and L. F. Dahl, *J. Amer. Chem. Soc.*, 1966, **88**, 1082; A. Simon, H.-G. von Schnering and H. Schäfer, *Z. anorg. Chem.*, 1967, **355**, 295.
[21] A. Simon, *Z. anorg. Chem.*, 1967, **355**, 311.

$NbCl_4$ in $NbCl_{2.67}$ are stable up to about the composition $NbCl_{3.13}$, beyond which the pure $NbCl_4$ phase separates. A similar explanation would evidently be appropriate for the "$NbBr_3$" and "NbI_3" systems, and possibly also for the tantalum "trihalides." However, there is at present a need for further studies before this interesting problem can be regarded as definitively solved.

There is an extensive solution chemistry of lower halide compounds of Nb and Ta, all of it based on the $[M_6X_{12}]^{n+}$ ($n = 2$, 3 or 4) units. Thus, by aqueous chemistry, the compounds $M_6X_{14} \cdot nH_2O$, where M = Nb, Ta, X = Cl or Br and n is usually 7 or 8, can be prepared.[22] The substances[22] contain the central $[M_6X_{12}]^{2+}$ group with the additional halide ions, and some water molecules, coordinated to the vacant external positions on the metal atoms; many other ligands (e.g., DMSO, pyridine-1-oxide, and Ph_3PO) are capable of filling these positions.[23]

The $[M_6X_{12}]^{2+}$ ions, which are diamagnetic, can be oxidized to $[M_6X_{12}]^{3+}$ ions, which have one unpaired electron, and to $[M_6X_{12}]^{4+}$ ions, which are again diamagnetic; salts of many of these, e.g., the compounds $Nb_6Cl_{15} \cdot 7H_2O$, $Ta_6Cl_{15} \cdot 7H_2O$, $Ta_6Br_{15} \cdot 8H_2O$ $Ta_6Br_{16} \cdot 7H_2O$, $Ta_6Cl_{16} \cdot xH_2O$, can be isolated.[24]

There are also anionic complexes of the three $[Nb_6Cl_{12}]^{n+}$ species. When KCl/Nb_6Cl_{14} or $KCl/Nb_3Cl_8/Nb$ mixtures are heated, the compound $K_4Nb_6Cl_{18}$ is obtained,[25] while $(Et_4N)_nNb_6Cl_{18}$ compounds[26] with $n = 4$, 3 and 2, and similar compounds[27] can be prepared from aqueous solutions by oxidation of $[Nb_6Cl_{12}]^{2+}$. In all these compounds there are discrete $\{[Nb_6Cl_{12}]Cl_6\}^{n-}$ complex anions. An esr study of the $\{[Nb_6Cl_{12}]Cl_6\}^{3-}$ species shows[24] that the unpaired electron is delocalized uniformly over six equivalent niobium atoms.

An interesting series of compounds containing the metal-atom cluster shown in Fig. 26-B-6 are obtained[28] as follows:

$$Nb + 5Nb_3X_8 + 4M^IX \rightarrow 4M^INb_4X_{11}$$
$$(M^I = Cs \text{ or } Rb; \qquad X = Cl \text{ or } Br)$$

All four are isotypic; the cesium compounds with Cl and Br have been investigated in detail with the results that the mean lengths of the four perimetral Nb—Nb bonds are 2.95 and 3.05 Å and the lengths of the cross bonds are 2.84 and 2.96 Å in the Cl and Br compound, respectively.

[22] R. A. McCarley et al., Inorg. Chem., 1970, 9, 1343, 1347, 1361.
[23] R. A. Field and D. L. Kepert, J. Less Common Metals, 1967, 13, 378.
[24] B. Spreckelmeyer and H. Schäfer, J. Less Common Metals, 1967, 13, 122, 127; B. Spreckelmeyer, Z. anorg. Chem., 1968, 358, 147; J. H. Espenson et al., J. Amer. Chem. Soc., 1966, 88, 1063; Inorg. Chem., 1971, 10, 1081.
[25] A. Simon, H.-G. von Schnering and H. Schäfer, Z. anorg. Chem., 1968, 361, 235.
[26] R. A. Mackay and R. F. Schneider, Inorg. Chem., 1967, 6, 549.
[27] P. B. Fleming et al., J. Amer. Chem. Soc., 1967, 89, 159.
[28] A. Bröll et al., Z. anorg. Chem., 1969, 367, 1.

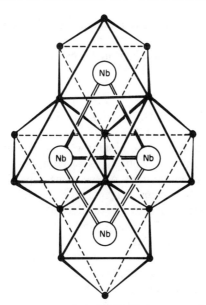

Fig. 26-B-6.   The key structural unit in $M^INb_4X_{11}$ compounds. Small filled circles are halogen atoms and double lines represent M—M bonds. Many of the peripheral halogen atoms are shared with other units.

26-B-4. Organometallic Compounds of Niobium and Tantalum

There is no extensive metal-to-carbon chemistry save in h^5-cyclopentadienyl compounds of the type $(h^5\text{-}C_5H_5)_2MX_3$ and $(h^5\text{-}C_5H_5)M(CO)_4$ although some simple alkyls are known (page 758).

The hydride $(h^5\text{-}C_5H_5)_2TaH_3$ is important since the nmr spectrum, which showed an A_2X pattern for the hydrides, provided proof of the angularity of the metal-to-ring axes (page 744). The niobium analog can be made only under high hydrogen pressure and decomposes rapidly in solution to give an exceedingly reactive $(h^5\text{-}C_5H_5)_2NbH$ species, which has not been isolated although the remarkably stable ethylene adduct $(h^5\text{-}C_5H_5)_2NbH(C_2H_4)$ can be obtained. The latter provides one of the few examples of hydrido alkenes; with an excess of ethylene an ethyl compound is formed. Both the Nb and the Ta compound catalyze hydrogen-atom exchange in aromatic compounds by reactions involving the monohydrido species:

$$(h^5\text{-}C_5H_5)_2TaH + C_6D_6 \rightleftharpoons (h^5\text{-}C_5H_5)_2Ta\!\!\begin{array}{c} C_6D_5 \\ | \\ D \\ | \\ H \end{array} \rightleftharpoons (h^5\text{-}C_5H_5)_2TaD + C_6D_5H$$

Although the oxidative addition of aromatic compounds does not lead to isolable tantalum species, the niobium ones can be characterized [cf. oxidative addition of benzene to $(h^5\text{-}C_5H_5)_2Mo$ (page 774)].[29]

[29] G. W. Parshall et al., J. Amer. Chem. Soc., 1970, 92, 5235; 1971, 93, 3793.

The carbonylate anions $M(CO)_6^-$ are made by reduction of MCl_5 with Na in diglyme at 120° under 200–350 atm of CO and are isolated as $[Na(diglyme)_2]^+$ salts. Various substituted ions such as $[M(CO)_5PPh_3]^-$ can also be prepared.[30]

26-C. MOLYBDENUM AND TUNGSTEN

Molybdenum and tungsten are similar chemically, although there are differences between them in various types of compound that are not easy to explain. Thus some compounds of the same type differ noticeably in their reactivities towards various reagents; for example, $Mo(CO)_6$ but not $W(CO)_6$ reacts with acetic acid to give a diacetate.

Except for compounds with π-acid ligands, there is not a great deal of similarity to chromium. The divalent state, well defined for Cr, is not well known for Mo and W except in strongly M—M bonded compounds; and the high stability of Cr^{III} in its complexes has no counterpart in Mo or W chemistry. For the heavier elements, the higher oxidation states are more common and more stable against reduction.

Both Mo and W have a wide variety of stereochemistries in addition to the variety of oxidation states, and their chemistry is among the most complex of the transition elements. Uranium has sometimes been classed with Mo and W in Group VI, and indeed there are some valid, though often rather superficial, similarities; the three elements form volatile hexafluorides, oxide halides and oxo anions which are similar in certain respects. There is little resemblance to the sulfur group except in regard to stoichiometric similarities, for example, SeF_6, WF_6, SO_4^{2-}, MoO_4^{2-}, and such comparisons are not profitable.

The oxidation states and stereochemistry are summarized in Table 26-C-1.

Molybdenum is one of the biologically active transition elements.[1] It is believed to be a necessary trace element in animal diets, but function and minimum levels have not been established. It is well established that nitrogen-fixing bacteria employ enzymes containing both Mo and Fe. Recently[2] one such enzyme, or a part thereof, has been obtained in pure, crystalline form. One molecular unit appears to have a molecular weight in the range 270,000–300,000 Daltons and to contain two atoms of Mo and forty of Fe. This protein in association with another which contains only iron possesses the ability to catalyze the reduction of N_2 and other substrates. Chemical and structural details are not yet known.

Recently some model systems have been devised to mimic, at a considerably lower level of efficiency, these enzymic activities. The N_2-ase is active

[30] A. Davison and J. E. Ellis, *J. Organometal. Chem.*, 1971, **31**, 239.
[1] J. T. Spence, *Coordination Chem. Rev.*, 1969, **4**, 475 (biochemical aspects of Mo coordination chemistry).
[2] R. C. Burns, R. D. Holsten and R. W. F. Hardy, *Biochem. Biophys. Res. Comm.*, 1970, **39**, 90.

TABLE 26-C-1

Oxidation States and Stereochemistry of Molybdenum and Tungsten

Oxidation state	Coordination number	Geometry	Examples
Mo^{-II}, W^{-II}	5	?	$[Mo(CO)_5]^{2-}$
Mo^0, W^0, d^6	6	Octahedral	$W(CO)_6$, $py_3Mo(CO)_3$, $[Mo(CN)_5NO]^{4-}$, $[Mo(CO)_5I]^-$, $Mo(N_2)_2(diphos)_2$
Mo^I, W^I, d^5	6^a	π-Complex	$(C_6H_6)_2Mo^+$, h^5-$C_5H_5MoC_6H_6$
	7^a		$[h^5$-$C_5H_5Mo(CO)_3]_2$
	6	?	$MoCl(N_2)$ (diphos)$_2$
Mo^{II}, W^{II}, d^4		π-Complex	h^5-$C_5H_5W(CO)_3Cl$
	5	See text	$[Mo(OCOCH_3)_2]_2$, $Mo_2Cl_8^{4-}$
	6	Octahedral	$Mo(diars)_2X_2$, $Mo(CO)_2(diars)I_2$
	7	?	$[Mo(diars)_2(CO)_2X]^+$, $[Mo(CO)_3(diars)Br_2]$
	9	Cluster compounds	Mo_6Cl_{12}, W_6Cl_{12}
Mo^{III}, W^{III}, d^3	6	Octahedral	$[Mo(NCS)_6]^{3-}$, $[MoCl_6]^{3-}$, $[W_2Cl_9]^{3-}$
	7	?	$[W(diars) (CO)_3Br_2]^+$
	8	Dodecahedral (?)	$[Mo(CN)_7(H_2O)]^{4-}$
Mo^{IV}, W^{IV}, d^2	8^a	π-Complex	$(h^5$-$C_5H_5)_2WH_2$, $(h^5$-$C_5H_5)_2MoCl_2$
	9^a	π-Complex	$(h^5$-$C_5H_5)_2WH_3$
	4	Dist. tetrahedral	$Mo(NMe_2)_4$.
	6	Octahedral	$[Mo(NCS)_6]^{2-}$, $[Mo(diars)_2Br_2]^{2+}$, $WBr_4(MeCN)_2$, $MoOCl_2(PR_3)_3$
	6	Trigonal prism	MoS_2
	8	Dodecahedral or square antiprism	$[Mo(CN)_8]^{4-}$, $[W(CN)_8]^{4-}$ $Mo(S_2CNMe_2)_4$, $W(8$-quinolinolate)$_4^b$
Mo^V, W^V, d^1	5	*tbp*	$MoCl_5(g)$
	6	Octahedral	$Mo_2Cl_{10}(s)$, $[MoOCl_5]^{2-}$, WF_6^-
	8	Dodecahedral or square antiprism	$[Mo(CN)_8]^{3-}$, $[W(CN)_8]^{3-}$
Mo^{VI}, W^{VI}, d^0	4	Tetrahedral	MoO_4^{2-}, MoO_2Cl_2, WO_4^{2-}, WO_2Cl_2
	5?	?	$WOCl_4$, $MoOF_4$
	6	Octahedral	MoO_6, WO_6 in poly acids, WCl_6, MoF_6, $[MoO_2F_4]^{2-}$, MoO_3(distorted), WO_3(distorted)
	7	Distorted pentagonal bipyramid	$WOCl_4(diars)$, $K_2[MoO(O_2)ox]$
	8	?	MoF_8^{2-}, WF_8^{2-}
	9	?	WH_6 $(Me_2PhP)_3$

^{a} If C_6H_6 and h^5-C_5H_5 occupy three coordination sites.
^{b} W. D. Bonds, Jr. and R. D. Archer, *Inorg. Chem.*, 1971, **10**, 2057.

only in the presence of an additional iron protein (which is assumed to serve as an electron-transfer system) and stoichiometric amounts of adenosine triphosphate (ATP). Hence the model systems[3a] have included molybdenum, iron, thio compounds, a reducing agent such as BH_4^- or $Na_2S_2O_4$, and ATP. The most effective model system,[3b] which will slowly reduce N_2 to ammonia at atmospheric pressure, is an aqueous solution containing (mmol $\times 10^3$), $Na_2MoO_4(60)$, $FeSO_4(1)$, $NaBH_4$ (390) and 2-aminoethanethiol (60) or

[3a] G. N. Schrautzer *et al.*, *J. Amer. Chem. Soc.*, 1971, **93**, 1608, 1803; W. E. Newton *et al.*, *J. Amer. Chem. Soc.*, 1971, **93**, 268.
[3b] R. E. E. Hill and R. L. Richards, *Nature*, 1971, **233**, 114.

bovine serum albumin (protein with SH groups). In strongly alkaline solutions, mixtures of Mo^{3+}, Mg^{2+} and Ti^{3+} will reduce N_2 to hydrazine. The systems also reduce acetylene, which has been used as a test for nitrogenase-like activity.

26-C-1. The Elements

In respect to occurrence (abundance $\sim 10^{-4}\%$), metallurgy and properties of the metals, molybdenum and tungsten are remarkably similar.

Molybdenum occurs chiefly as *molybdenite*, MoS_2, but also as molybdates such as $PbMoO_4$ (*wulfenite*) and $MgMoO_4$. Tungsten is found almost exclusively in the form of tungstates, the chief ones being *wolframite* (a solid solution and/or mixture of the isomorphous substances $FeWO_4$ and $MnWO_4$), *scheelite*, $CaWO_4$, and *stolzite*, $PbWO_4$.

The small amounts of MoS_2 in ores are concentrated by the foam flotation process; the concentrate is then converted into MoO_3 which, after purification, is reduced with hydrogen to the metal. Reduction with carbon must be avoided because this yields carbides rather than the metal.

Tungsten ores are concentrated by mechanical and magnetic processes and the concentrates attacked by fusion with NaOH. The cooled melts are leached with water, giving solutions of sodium tungstate from which hydrous WO_3 is precipitated on acidification. The hydrous oxide is dried and reduced to metal by hydrogen.

In the powder form in which they are first obtained both metals are dull grey, but when converted into the massive state by fusion are lustrous, silver-white substances of typically metallic appearance and properties. They have electrical conductances approximately 30% that of silver. They are extremely refractory; Mo melts at $2610°$ and W at $3410°$.

Neither metal is readily attacked by acids. Concentrated nitric acid initially attacks molybdenum, but the metal surface is soon passivated. Both metals can be dissolved—tungsten only slowly, however—by a mixture of concentrated nitric and hydrofluoric acid. Oxidizing alkaline melts such as fused KNO_3–NaOH or Na_2O_2 attack them rapidly, but aqueous alkalis are without effect.

Both metals are inert to oxygen at ordinary temperatures, but at red heat they combine with it readily to give the trioxides. They both combine with chlorine when heated, but they are attacked by fluorine, yielding the hexafluorides, at room temperature.

The chief uses of both metals are in the production of alloy steels; even small amounts cause tremendous increases in hardness and strength. "High-speed" steels which are used to make cutting tools that remain hard even at red heat, contain W and Cr. Tungsten is also extensively used for lamp filaments. The elements give hard, refractory and chemically inert interstitial compounds with B, C, N or Si on direct reaction at high temperatures. Tungsten carbide is used for tipping cutting tools, etc.

26-C-2. Oxides, Sulfides and Oxo Acids

Oxides. Many molybdenum and tungsten oxides are known. The simple ones are MoO_3, WO_3; Mo_2O_5; MoO_2, WO_2. Other, non-stoichiometric oxides have been characterized and have complicated structures.[4]

The ultimate products of heating the metals or other compounds such as the sulfides in oxygen are the *trioxides*. They are not attacked by acids but dissolve in bases to form molybdate and tungstate solutions, which are discussed below.

MoO_3 is a white solid at room temperature but becomes yellow when hot and melts at 795° to a deep yellow liquid. It is the anhydride of molybdic acid, but it does not form hydrates directly, although these are known (see below). MoO_3 has a rare type of layer structure in which each molybdenum atom is surrounded by a distorted octahedron of oxygen atoms.

WO_3 is a lemon-yellow solid, m.p. 1200°. Like CrO_3, it has a slightly distorted form of the cubic rhenium trioxide structure (page 976).

Mo_2O_5, a violet solid soluble in warm acids, is prepared by heating the required quantity of finely divided molybdenum with MoO_3 at 750°. When ammonia is added to solutions containing Mo^V, brown $MoO(OH)_3$ is precipitated; this gives Mo_2O_5 on heating.

Molybdenum(IV) oxide, MoO_2, is obtained by reducing MoO_3 with hydrogen or NH_3 below 470° (above this temperature reduction proceeds to the metal) and by reaction of molybdenum with steam at 800°. It is a brown-violet solid with a coppery luster, insoluble in non-oxidizing mineral acids but soluble in concentrated nitric acid with oxidation of the molybdenum to Mo^{VI}. The structure is similar to that of rutile but so distorted that strong Mo—Mo bonds are formed (cf. page 549). WO_2 is similar.

Mixed oxides. On fusion of MoO_3 or WO_3 with alkali or alkaline-earth oxides, mixed oxide systems are obtained that are not related to the molybdates or tungstates made in aqueous solutions (see below). These usually have chain structures with linked MoO_6 polyhedra but the stability of a particular type of structure depends on the cation size. Tungstates may differ from molybdates; thus $K_2Mo_4O_{13}$ has a chain but $K_2W_4O_{13}$ has WO_6 octahedra linked by corners to give six-membered rings with the K^+ ions in the tunnels so formed.[5]

The blue oxides. These are also called *molybdenum blue* and *tungsten blue* and are obtained by mild reduction, for example, by Sn^{II}, SO_2, N_2H_4, H_2S, etc., of acidified solutions of molybdates and tungstates or of suspensions of MoO_3 and WO_3 in water. Moist tungsten(VI) oxide will acquire a blue tint merely on exposure to ultraviolet light.

The "blue oxides" of Mo contain both oxide and hydroxide. There appears to be an entire series of "genotypic" compounds (i.e. having the same basic structure but differing in the charges on cations and anions), with (olive-

---

[4] See, e.g., E. Gebert and R. J. Ackermann, *Inorg. Chem.*, 1966, **5**, 136.
[5] B. M. Gatehouse and P. Leverett, *J. Chem. Soc.*, A, **1971**, 2107.

green) $MoO(OH)_2$ as one limit and MoO_3 as the other. The compounds in which the mean oxidation state of Mo is between 5 and 6 are the blue ones, e.g., $MoO_{2.0}(OH)$ and $MoO_{2.5}(OH)_{0.5}$. A detailed electronic explanation for the blue color has not been found, although the general idea that Mo_3 metal-atom clusters might be responsible has been suggested. In the case of the "blue oxides" of W, similar general results have been obtained. These compounds are comparable to the heteropoly blues (page 957).

Tungsten Bronzes.[6]   The reduction of sodium tungstate with hydrogen at red heat gives a chemically inert substance with a bronze-like appearance. Similar materials are obtained by vapor-phase reaction of alkali metals with WO_3.

The tungsten bronzes are non-stoichiometric substances of general formula $M_n^I WO_3$ ($0 < n \leqslant 1$). The colors vary greatly with composition from golden-yellow for $n \approx 0.9$ to blue-violet for $n \approx 0.3$. Tungsten bronzes with $n > 0.3$ are extremely inert and have semimetallic properties, especially metallic luster and good electrical conductivity in which the charge carriers are electrons. Those with $n < 0.3$ are semiconductors. They are insoluble in water and resistant to all acids except hydrofluoric, and they can be oxidized by oxygen in presence of base to give tungstates(VI):

$$4NaWO_3 + 4NaOH + O_2 \rightarrow 4Na_2WO_4 + 2H_2O$$

Structurally, the sodium tungsten bronzes may be regarded as defective $M^I WO_3$ phases having the perovskite structure. In the defective phase, $M_n^I WO_3$, there are $(1-n)$ W^{VI} atoms, and $(1-n)$ of the Na sites of the pure $NaWO_3$ phase are unoccupied. It appears that completely pure $NaWO_3$ has not been prepared, although phases with sodium enrichment up to perhaps $n \sim 0.95$ are known. The cubic structure collapses to rhombic and then triclinic for $n < \sim 0.3$. In the limit of $n = 0$ we have, of course, WO_3, which, as already noted, has a triclinically distorted ReO_3 structure (page 976). The cubic ReO_3 structure is the same as the perovskite structure with all the large cations removed. Thus the actual range of composition of the tungsten bronzes is approximately $Na_{0.3}WO_3$ to $Na_{0.95}WO_3$.

The semimetallic properties of the tungsten bronzes are associated with the fact that no distinction can be made between W^V and W^{VI} atoms in the lattice, all W atoms appearing equivalent. Thus the n "extra" electrons per mole (over the number for WO_3) are distributed throughout the lattice, delocalized in energy bands somewhat similar to those of metals.

Sulfides.   Of the known sulfides, Mo_2S_3, MoS_4, Mo_2S_5, MoS_3 and MoS_2, only the last three are important. The only tungsten sulfides, WS_2 and WS_3, appear to be similar to their molybdenum analogs.

MoS_2 can be prepared by direct combination of the elements, by heating molybdenum(VI) oxide in hydrogen sulfide, or by fusing molybdenum(VI) oxide with a mixture of sulfur and potassium carbonate. It is the most

[6] P. G. Dickens and M. S. Whittingham, *Quart. Rev.*, 1968, **22**, 30. For a molybdenum bronze see B. M. Gatehouse and D. J. Lloyd, *Chem. Comm.*, **1971**, 13.

stable sulfide at higher temperatures, and the others which are richer in sulfur revert to it when heated in a vacuum. It dissolves only in strongly oxidizing acids such as aqua regia and boiling concentrated sulfuric acid. Chlorine and oxygen attack it at elevated temperatures giving $MoCl_5$ and MoO_3, respectively.

MoS_2 has a structure built of close-packed layers of sulfur atoms stacked to create trigonal prismatic interstices which are occupied by Mo atoms. The stacking is such as to permit easy slippage of alternate layers and thus MoS_2 has mechanical properties (lubricity) similar to those of graphite.

Hydrated, brown MoS_3 is precipitated when hydrogen sulfide is passed into slightly acid solutions of molybdates; the action of H_2S on tungstates in dilute HCl solution gives blue solutions of oxotungsten(v) complexes (page 965). Hydrous MoS_3 dissolves on digestion with alkali sulfide solution to give brown-red thiomolybdates (see below). Hydrated Mo_2S_5 is precipitated from Mo^V solutions. Both hydrous sulfides can be dehydrated.

Simple Molybdates and Tungstates.   The trioxides of molybdenum and tungsten dissolve in aqueous alkali-metal hydroxides, and from these solutions the simple or normal molybdates and tungstates can be crystallized. They have the general formulas $M_2^I MoO_4$ and $M_2^I WO_4$ and contain the discrete tetrahedral ions[7a] MoO_4^{2-} and WO_4^{2-}. These are regular in alkali-metal and a few other salts, but may be distorted in salts of other cations.[7b] It is now certain that the MoO_4^{2-} and WO_4^{2-} ions are also tetrahedral in aqueous solution. Although both molybdates and tungstates can be reduced in solution (see below), they lack the powerful oxidizing property so characteristic of chromates(vi). The normal tungstates and molybdates of many other metals can be prepared by metathetical reactions. The alkali-metal, ammonium, magnesium and thallous salts are soluble in water, whereas those of other metals are nearly all insoluble.

When solutions of molybdates and tungstates are made weakly acid, polymeric anions are formed, but from more strongly acid solutions substances often called molybdic or tungstic acid are obtained. At room temperature the yellow $MoO_3 \cdot 2H_2O$ and the isomorphous $WO_3 \cdot 2H_2O$ crystallize, the former very slowly. From hot solutions, monohydrates are obtained rapidly. These compounds are oxide hydrates. $MoO_3 \cdot 2H_2O$ contains sheets[8] of MoO_6 octahedra sharing corners and is best formulated[8] as $[MoO_{4/2}O(H_2O)] \cdot H_2O$ with one H_2O bound to Mo, the other hydrogen bonded in the lattice.

The oxo anions also give complexes with sulfate and with hydroxo compounds such as glycerol, tartrate ion, and sugars. In base solution, hydrogen peroxide gives peroxo anions believed to be of the type $M_2O_{11}^{2-}$.

Amines generally give ill-defined complexes or salts with molybdates but with diethylenetriamine a unique octahedral complex, $MoO_3 \cdot dien$, is obtained.

[7a] F. X. N. M. Kools, A. S. Koster and G. D. Rieck, *Acta Cryst.*, 1970, **B26**, 1974.
[7b] B. M. Gatehouse and P. Leverett, *J. Chem. Soc.*, A, **1969**, 849.
[8] B. Krebs, *Chem. Comm.*, **1970**, 50.

The *thiomolybdate* and *tungstate* ions are isotypic with SO_4^{2-} in the alkali-metal salts such as K_2MoS_4. The selenomolybdate is known in $(NH_4)_2MoSe_4$.[9] On acidification of the thiomolybdate, H_2S and sulfides are usually formed but the acid H_2WS_4 is known as an unstable red solid.[10] The MoS_4^{2-} and WS_4^{2-} ions can act as bidentate ligands,[11] e.g., in the ion $[Ni(WS_4)_2]^{2-}$, which contains a square NiS_4 coordination group.

26-C-3. Isopoly and Heteropoly Acids and Their Salts[12]

A prominent feature of the chemistry of molybdenum and tungsten is the formation of numerous polymolybdate(VI) and polytungstate(VI) acids and their salts. Only V^V, Nb^V, Ta^V and U^{VI} show comparable behavior but even they do so to a more limited extent.

The poly acids of molybdenum and tungsten are of two types: (*a*) the *isopoly acids* and their related anions, which contain only molybdenum or tungsten along with oxygen and hydrogen, and (*b*) the *heteropoly acids* and anions, which contain one or two atoms of another element in addition to molybdenum or tungsten, oxygen and hydrogen. All the polyanions contain octahedral MoO_6 or WO_6 groups, so that the conversion of MoO_4^{2-} or WO_4^{2-} into poly anions requires an increase in coordination number. It is still not clear why only certain metal oxo ions can polymerize, or why for these metals only certain species, e.g., $Mo_7O_{24}^{6-}$, $HW_6O_{21}^{5-}$ or $Ta_6O_{18}^{6-}$, predominate under a given set of conditions, or why, for chromate, poly-merization stops at $Cr_2O_7^{2-}$. The ability of the metal and oxygen orbitals to overlap to give substantial π bonding, $M=O$, must surely be involved (see page 143), as must the base strength of the oxygen atoms and the ability of the initial protonated species $MO_3(OH)^-$ to expand its coordination sphere by coordination of water molecules. The size of the metal ion is clearly important.

X-ray studies of crystalline compounds of a number of salts of isopoly and heteropoly anions have been made and will be discussed below. By using these structures as a guide, considerable headway has been made in the inter-pretation of solution studies, and, especially in recent work, reliable, inter-nally consistent data have been obtained. It is to be noted, however, that the X-ray studies do not show positions of hydrogen atoms and that, while the basic units determined crystallographically often persist in solutions, hydra-tion and protonation in solution depend on the conditions. Also the fact that a salt with a particular structure crystallizes from solution under certain conditions does not necessarily mean that the same anion is the major species

[9] A. Müller, B. Krebs and E. Diemann, *Angew. Chem. Internat. Edn.*, 1967, **6**, 257.

[10] G. Galtow and A. Franke, *Z. anorg. Chem.*, 1967, **352**, 11, 246.

[11] W. P. Binnie, M. J. Redman and W. J. Mallio, *Inorg. Chem.*, 1970, **9**, 1449; A. Müller, E. Diemann and H.-H. Heinsen, *Chem. Ber.*, 1971, **75**, 1113; *Z. anorg. Chem.*, 1971, **386**, 102.

[12] P. G. Rasmussen, *J. Chem. Educ.*, 1967, **44**, 277; D. L. Kepert, *Inorg. Chem.*, 1969, **8**, 1556; K. H. Tytho and O. Glemser, *Chimia*, 1969, **23**, 494.

in solution—or, in fact, that it even exists in solution. There are clear cases where the ions in solution and the crystals obtained from them are substantially different.

Isopolymolybdates. When a basic solution containing only MoO_4^{2-} and alkali-metal or ammonium ions is acidified, the molybdate ions condense in definite steps to form a series of polymolybdate ions. At the pH where condensation begins for both CrO_4^{2-} and MoO_4^{2-} the species $MO_3(OH)^-$ are formed:

$$MoO_4^{2-} + H^+ \rightarrow MoO_3(OH)^-$$

Presumably coordination of water molecules occurs at this point, perhaps to give $[MoO(OH)_5]^-$. The $M{=}O$ group is known to have a strong *trans*-effect (page 667) and hence OH groups in the *trans*-position could be labile so that the next step could be the formation of an oxo bridge:

$$2[MoO(OH)_5]^- \rightarrow [(HO)_4OMo{-}O{-}MoO(OH)_4]^{2-} + H_2O$$

The subsequent steps must be more complicated and as the pH of the solution is lowered to about pH 6 polymerization is detectable. It was proposed by Lindqvist, primarily because of its existence in crystals, that the first main reaction at this stage is to form the *paramolybdate* ion:

$$7MoO_4^{2-} + 8H^+ \rightarrow Mo_7O_{24}^{6-} + 4H_2O$$

and this has been proved more or less conclusively;[13] in somewhat more acid solutions, the *octamolybdate* ion, $[Mo_8O_{26}]^{4-}$, is formed. These two species in equilibrium seem to be able to account for most of the observed phenomena. There is no reliable evidence that there is any species containing more than one but less than seven Mo atoms in the polymolybdate system. However, there is some evidence that higher polymers may exist. These two main polymeric anions may, of course, exist in hydrated forms or be partially protonated, e.g., $[H_8Mo_7O_{28}]^{6-}$, and there is evidence to suggest that the highly charged ions can also bind counter-ions such as Li^+ or Na^+. In very strong acid solutions, depolymerization occurs and, when hydrochloric acid is used, the species formed above $6M$ HCl appears to be MoO_2Cl_2 or $Cl_2OMo(\mu-O)_2MoOCl_2$.

The so-called dimolybdates, such as $Na_2O \cdot 2MoO_3 \cdot nH_2O$, which sometimes crystallize from solutions in the pH range of 5–6, are probably mixtures of normal molybdates and paramolybdates. The various trimolybdates of general formula $M_2^IO \cdot 3MoO_3 \cdot nH_2O$ appear to contain acid paramolybdate ions or other ions with seven or more Mo atoms.

Tetramolybdates or *metamolybdates*, $M_2^IO \cdot 4MoO_3 \cdot nH_2O$, are formed from concentrated solutions of alkali molybdates treated with 1.5 mols per mol of hydrochloric acid. *Octamolybdates* are obtained similarly from solutions containing 1.75 mols per mol of hydrochloric acid or by adding the calculated amount of MoO_3 to ammonium paramolybdate solution. Both the tetra- and octa-molybdates are derived from the ion $[Mo_8O_{26}]^{4-}$.

[13] J. Aveston, E. W. Anacker, and J. S. Johnson, *Inorg. Chem.*, 1964, **3**, 735; D. S. Honig and K. Kustin, *Inorg. Chem.*, 1972, **11**, 65.

The structures of both the heptamolybdate and octamolybdate ions in crystals are discussed on page 954 in relation to other heteropoly and isopoly anion structures.

Isopolytungstates. The behavior of the tungstate systems is similar to that of the molybdate systems. Again, the degree of aggregation in solution increases as the pH is lowered, and numerous tungstates $M_2^I O \cdot n WO_3 \cdot m H_2 O$, differing in the value of n, have been crystallized from the solutions at different pH's.

In solution, the relationships appear to be those in Fig. 26-C-1. There is some evidence that there may be intermediate stages between paratungstates A and Z and between the former and metatungstate. The species given may be hydrated; e.g., $W_{12}O_{41}^{10-}$ may be also written $H_{10}W_{12}O_{46}^{10-}$. The most important species is paratungstate Z, and crystallization of Na_2WO_4, acidified by addition of 1.167 mols of HCl per mol of W, gives large glassy

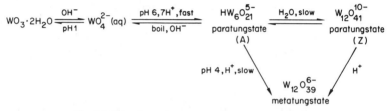

Fig. 26-C-1.   Isopolytungstates in aqueous solution.

crystals. Aqueous solutions of this salt contain mainly $W_{12}O_{41}^{10-}$ which slowly hydrolyzes to a mixture of this species and $HW_6O_{21}^{5-}$ similar to that formed on direct acidification of WO_4^{2-}. The dodecatungstate predominates at high concentrations.

The solution species written above as $W_{12}O_{41}^{10-}$ may, of course, be better represented by a different formula, depending on the true extent of hydration which is not known with certainty; the formula used appears commonly in the literature. In the crystalline substance $(NH_4)_{10}[W_{12}O_{41}] \cdot 11H_2O$ the ion is actually $W_{12}O_{42}H_2^{10-}$ and the structure[14] (omitting the two H atoms) is shown in Fig. 26-C-2d (p. 954).

Heteropoly Acids and Their Salts.   These are formed when molybdate and tungstate solutions containing other oxo anions (e.g., PO_4^{3-}, SiO_4^{4-}) or metal ions are acidified. At least 35 elements are known to be capable of functioning as the hetero atoms.[15] The free acids and most salts of the heteropoly anions are extremely soluble in water and in various oxygenated organic solvents, such as ethers, alcohols, and ketones. When crystallized from water, the heteropoly acids and salts are always obtained in highly hydrated condition. Like the isopoly acids they are decomposed by strong base:

$$34OH^- + [P_2Mo_{18}O_{62}]^{6-} \rightarrow 18MoO_4^{2-} + 2HPO_4^{2-} + 16H_2O$$
$$18OH^- + [Fe_2W_{12}O_{42}]^{6-} \rightarrow 12WO_4^{2-} + Fe_2O_3 \cdot nH_2O + 9H_2O$$

[14] G. Weiss, *Z. anorg. Chem.*, 1969, **368**, 279.
[15] For references see L. C. W. Baker and J. S. Figgis, *J. Amer. Chem. Soc.*, 1970, **92**, 3794.

In contrast to the isopoly acids, many of the heteropoly acids are stable without depolymerization in quite strongly acid solutions; they are often themselves strong acids. In general, heteropolymolybdates and tungstates of small cations, including those of some heavy metals, are water-soluble, but with larger cations insolubility is frequently found. Thus Cs^+, Pb^{2+} and Ba^{2+} salts are usually insoluble, and NH_4^+, K^+ and Rb^+ salts are sometimes insoluble; salts of $[(h^5\text{-}C_5H_5)_2Fe]^+$, R_4N^+, R_4P^+ and alkaloids, are invariably insoluble. Table 26-C-2 lists the principal types of heteropolymolybdates,

TABLE 26-C-2

Principal Types of Heteropolymolybdates

Ratio of hetero atoms to Mo atoms	Principal hetero atoms occurring	Anion formula
1:12	Series A: P^V, As^V, Si^{IV}, Ge^{IV}, $Sn^{IV}(?)$, Ti^{IV}, Zr^{IV}	$[X^{n+}Mo_{12}O_{40}]^{(8-n)-}$
	Series B: Ce^{IV}, Th^{IV}, $Sn^{IV}(?)$	$[X^{n+}Mo_{12}O_{42}]^{(12-n)-}$
1:11	P^V, As^V, Ge^{IV}	$[X^{n+}Mo_{11}O_{39}]^{(12-n)-}$ (possibly dimeric)
1:10	P^V, As^V, Pt^{IV}	$[X^{n+}Mo_{10}O_x]^{(2x-60-n)-}$ (possibly dimeric)
1:9	Mn^{IV}, Ni^{IV}	$[X^{n+}Mo_9O_{32}]^{(10-n)-}$
1:6	Te^{VI}, I^{VII}, Co^{III}, Al^{III}, Cr^{III}, Fe^{III}, Rh^{III}	$[X^{n+}Mo_6O_{24}]^{(12-n)-}$
2:18	P^V, As^V	$[X_2^{n+}Mo_{18}O_{62}]^{(16-2n)-}$
2:17	P^V, As^V	$[X_2^{n+}Mo_{17}O_x]^{(2x-102-2n)-}$
$1m:6m^a$	Ni^{II}, Co^{II}, Mn^{II}, Cu^{II}, Se^{IV}, P^{III}, As^{III}, P^V	$[X^{n+}Mo_6O_x]_m^{m(2x-36-n)-}$

[a] For the tungstate analog of the Co^{II} compound it has been found that $m = 2$.

many of which have exact or similar heteropolytungstate analogs. Table 26-C-3 illustrates the nomenclature according to the IUPAC system.

TABLE 26-C-3

Some Representative Heteropoly Salts and Their Nomenclature[a]

Formula	IUPAC names
$Na_3[P^VMo_{12}O_{40}]$	Sodium 12-molybdophosphate; sodium dodecamolybdophosphate
$H_3[P^VMo_{12}O_{40}]$	12-Molybdophosphoric acid; dodecamolybdophosphoric acid
$K_8[Co_2^{II}W_{12}O_{42}]$	Dimeric potassium 6-tungstocobaltate; dimeric potassium hexatungstocobaltate(II)
$Na_8[Ce^{IV}Mo_{12}O_{42}]$	Sodium 12-molybdocerate(IV); sodium dodecamolybdocerate(IV)

[a] International Union of Pure and Applied Chemistry recommendations.

Structures of Isopoly and Heteropoly Anions. In all cases so far studied definitively by X-ray diffraction, *the tungsten and molybdenum atoms lie at the centers of octahedra of oxygen atoms and the structures are built up of these octahedra by means of shared corners and shared edges (but not shared faces).* In the structural diagrams used here an MoO_6 octahedron is repre-

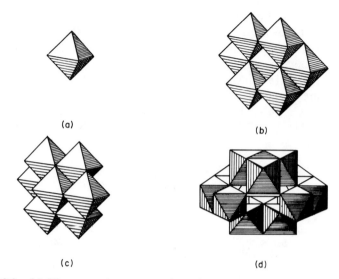

Fig. 26-C-2. (a) Diagrammatic representation of MoO_6 and WO_6 octahedra used in showing structures of some isopoly and heteropoly anions. (b) The structure of the para-molybdate anion, $[Mo_7O_{24}]^{6-}$. (c) The structure of the octamolybdate anion, $[Mo_8O_{26}]^{4-}$ (note that one MoO_6 octahedron is completely hidden by the seven that are shown). (d) The structure of the $[W_{12}O_{42}]^{12-}$ unit in the paratungstate ion.

sented by the sort of sketch shown in Fig. 26-C-2a. It should be stressed that for the polynuclear structures available data indicate that the octahedra are always distorted in such a way that the metal atoms form longer bonds to the shared or inner oxygen atoms than to the unshared or outer ones. For example in the paratungstate ion (Fig. 26-C-2d) there are three types of W—O bond, namely, to unshared, doubly shared and triply shared oxygen atoms. The mean lengths of these are, respectively, 1.76, 1.97, and 2.21 Å.

The isopoly anion structures definitely known from X-ray studies of crystals are the paramolybdate ion, $[Mo_7O_{24}]^{6-}$ in $(NH_4)_6Mo_7O_{24} \cdot 4H_2O$, and the octamolybdate ion, $[Mo_8O_{26}]^{4-}$ in $(NH_4)_4Mo_8O_{26} \cdot 5H_2O$, both of whose structures are shown in Figs. 26-C-2b and 26-C-2c. The metatung-state ion, best formulated in its sodium salt as $Na_6[H_2W_{12}O_{40}] \cdot 3H_2O$ or $Na_6[W_{12}O_{38}(OH)_2]$, has the same structure as the 12-tungsto- and 12-molybdo-hetero anions of type A (see Fig. 26-C-3) to be discussed below. The paratungstate ion in $(NH_4)_{10}[W_{12}O_{41}] \cdot 11H_2O$ is shown in Fig. 26-C-2d.

All the 12-molybdo-hetero anions with P^V, As^V, Ti^{IV} and Zr^{IV} and all the isomorphous 12-tungsto species containing the hetero atoms B^{III}, Ge^{VI}, P^V, As^V and Si^{IV} have the structure shown in Fig. 26-C-3. It may be thought of as consisting of four groups of three MoO_6 or WO_6 octahedra. In each group there is one oxygen atom common to all three octahedra. In the complete structure these groups are so oriented by sharing of oxygen atoms between groups that the four triply shared oxygen atoms are placed at the corners of a central tetrahedron. The hetero atom sits in the center of

Fig. 26-C-3.   The structure of the series A 12-molybdo- and 12-tungs o-heteropoly anions of general formula $[X^{n+}Mo_{12}O_{40}]^{(8-n)-}$.

this tetrahedron in the heteropoly ions, and in the metatungstate ion there is no hetero atom. It will be noted that all the hetero species occurring in the A series (Table 26-C-2) of 12-heteropoly anions are small enough to make a coordination number of four toward oxygen appropriate.

In the 12-hetero acids of type B, the hetero species are larger than those in the ones of type A, and it might be expected that their structures would be such as to have the hetero atoms in central octahedra of oxygen atoms. Their structures are not known, but it is noteworthy that the dodecatungstate structure (Fig. 26-C-2d) has just such a central octahedron.

The 9-molybdo-heteropoly anions are built by packing MoO_6 octahedra so as to produce a central octahedron of oxygen atoms. The structure is shown in Fig. 26-C-4. Closer inspection of this structure reveals that it is dissymmetric and these anions should exist in enantiomorphic forms. No resolution has as yet been reported, however.

Some structural work has been done on 6-molybdo-heteropoly species. In the series $[MMo_6O_{24}H_6]^{3-}$, where $M = Cr$, Al, Fe, Co, Rh or Ga, there is evidence that all are isostructural and the Cr compound has been studied in detail by X-ray diffraction.[16] The structure of the anion is shown in

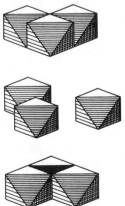

Fig. 26-C-4.   An exploded view of the structure of the 9-molybdoheteropoly anion $[X^{n+}Mo_9O_{32}]^{(10-n)-}$, showing the nine MoO_6 octahedra. When the upper and lower sets of three are moved in so as to share some corners with the equatorial set, a central octahedron of oxygen atoms occupied by the hetero atom is created.

[16] A. Perloff, *Inorg. Chem.*, 1970, 9, 2228.

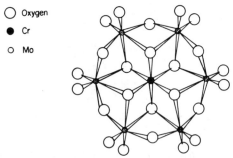

○ Oxygen
● Cr
○ Mo

Fig. 26-C-5. The structure of the $[CrMo_6O_{24}H_6]^{3-}$ ion. [Adapted by permission from A. Perloff, *Inorg. Chem.*, 1970, **9**, 2228.]

Fig. 26-C-5. The most likely sites for the six hydrogen atoms are the oxygen atoms of the central octahedron. There are some MW_6O_{24} species, e.g., $[NiW_6O_{24}H_6]^{4-}$ that evidently have the same structure, and the $[TeMo_6O_{24}]^{6-}$ ion was shown long ago to have this type of structure.

It is probable that the various dimeric 9-molybdo-heteropoly and 9-tungsto-heteropoly anions, of general formula $[X_2Mo_{18}O_{62}]^{6-}$, have the same structure as the $[P_2W_{18}O_{62}]^{6-}$ ion shown in Fig. 26-C-6. This structure can be thought of as consisting of two half-units, each of which is derived from the series A 12-molybdo-heteropoly anion structure (Fig. 26-C-3) by removal of three MoO_6 anions. The 11- and 10-molybdo-heteropoly anions may also be dimeric and consist of appropriate fragments of the 12-molybdo structure, but nothing is definitely known of their structures.

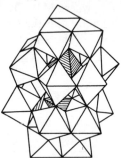

Fig. 26-C-6. The structure of the dimeric anion $(P_2W_{18}O_{62})^{6-}$.

As for the isopoly acids, there have been extensive studies of heteropoly anions in solutions. X-ray studies of solutions of $12WO_3 \cdot SiO_2 \cdot 2H_2O$ have demonstrated the ion $[SiW_{12}O_{40}]^{4-}$ with the structure shown in Fig. 26-C-7, 26-C-7, which is consistent with the crystal structure (Figure 26-C-3).

Finally we note that there is redox chemistry for heteropoly anions. The interconvertibility of the 12-tungstocobaltates(II) and -(III) is not greatly surprising though it involves the novel feature of tetrahedrally coordinated Co^{III}. Of greater novelty and interest is the fact that heteropolymolybdate and -tungstate anions, with a central atom such as P which does

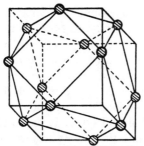

Fig. 26-C-7.   Arrangement of W atoms in the $[SiW_{12}O_{40}]^{4-}$ ion in 0.3 M aqueous solution. [Reprinted by permission from H. A. Levy, P. A. Agron and M. D. Danford, *J. Chem. Phys.*, 1959, **30**, 1486.]

not itself account for the redox properties, can be reduced by addition of 1 to 6 electrons per anion to give the so-called "heteropoly blues." The redox reactions involved occur reversibly and, apparently, without major structural change. Spectroscopic studies[17] indicate that the added electrons reside on individual Mo^V or W^V species with only slow "hopping" from one metal atom to another. The visible spectra may be attributed to intervalence charge-transfer bands. Typical examples of redox series are:

$$PW_{12}O_{40}^{3-} \underset{\longleftarrow}{\overset{e^-}{\longrightarrow}} PW_{12}O_{40}^{4-} \underset{\longleftarrow}{\overset{e^-}{\longrightarrow}} PW_{12}O_{40}^{5-}$$

$$P_2Mo_{18}O_{62}^{6-} \underset{\longleftarrow}{\overset{2e^-,2H^+}{\longrightarrow}} H_2P_2Mo_{18}O_{62}^{6-} \underset{\longleftarrow}{\overset{2e^-,2H^+}{\longrightarrow}} H_4P_2Mo_{18}O_{62}^{6-} \underset{\longleftarrow}{\overset{2e^-,2H^+}{\longrightarrow}} H_6P_2Mo_{18}O_{62}^{6-}$$

26-C-4.   Halides and Halo Complexes

The more important halides are listed in Table 26-C-4. Those containing metal atom clusters will be treated in Section 26-C-8. The others, involving metal oxidation states III to VI, are discussed here.

TABLE 26-C-4

The Fluorides and Chlorides of Molybdenum and Tungsten

II	III	IV	V	VI[a]
	MoF_3 yellow-brown non-volatile	MoF_4 tan non-volatile	$(MoF_5)_4$ yellow m.p. 67°, b.p. 213°	MoF_6 colorless m.p. 17.5°, b.p. 35.0°
		WF_4 red-brown non-volatile	$(WF_5)_4$ yellow disprop. 25°	WF_6 colorless m.p. 2.3°, b.p. 17.0°
α-$MoCl_2$[b] brown	$MoCl_3$ dark red	$MoCl_4$ dark red	$(MoCl_5)_2$ green-black m.p. 194°, b.p. 628°	$MoCl_6$(?) black
		WCl_4 black	$(WCl_5)_x$ green-black	WCl_6 blue-black m.p. 275°, b.p. 346°

[a] Also WBr_6, WF_5Cl, WCl_5F, WCl_4F_2.
[b] The cluster compound Mo_6Cl_{12} is yellow; W_6Cl_{12} is gray.

[17] M. T. Pope *et al.*, *Inorg. Chem*, 1967, **6**, 1147, 1152; 1970, **9**, 662, 667.

Hexahalides. The MF_6 compounds are volatile, colorless, and diamagnetic. Both are readily hydrolyzed and MoF_6 is rather easily reduced and attacks organic matter, while WF_6 is less active in these respects. $MoCl_6$ was claimed[18] in 1967 as a black powder, isotypic to WCl_6, very sensitive to water and prepared by the action of $SOCl_2$ on MoO_3, but further confirmation is desirable. WCl_6 results from direct chlorination of the metal; it is moderately volatile, monomeric in the vapor and soluble in organic liquids such as CS_2, CCl_4, alcohol and ether. It reacts slowly with cold water, rapidly with hot, to give tungstic acid. It acts as an alkylation catalyst for benzene with propene and in presence of aluminum alkyls is a catalyst for alkene metathesis (Chapter 24).[19] WBr_6 can also be obtained by direct halogenation of the metal; it is a dark blue solid which yields WBr_5 on moderate heating. The chlorofluorides are not thermodynamically stable, disproportionating[20] eventually to WF_6 and WCl_6.

Pentahalides. Treatment of molybdenum carbonyl with fluorine diluted in nitrogen at $-75°$ gives a product of composition Mo_2F_9. The nature of this substance has not been investigated, but when it is heated to $150°$ it yields the non-volatile MoF_4 as a residue and the volatile MoF_5 condenses in cooler regions of the apparatus. MoF_5 is also obtained by the reactions:

$$5MoF_6 + Mo(CO)_6 \xrightarrow{25°} 6MoF_5 + 6CO$$
$$Mo + 5MoF_6 \rightarrow 6MoF_5$$
$$Mo + F_2(\text{dilute}) \xrightarrow{400°} MoF_5$$

WF_5 is obtained by quenching the products of reaction of W with WF_6 at $800-1000°K$. It disproportionates above $320°K$ into WF_4 and WF_6.[21] Crystalline MoF_5 and WF_5 (and WOF_4, page 963) have the tetrameric structure common to many pentafluorides.[22]

Mo_2Cl_{10}, which is formed on direct chlorination of the metal, is moderately volatile and monomeric in the vapor, probably having a trigonal-bipyramidal structure. In the crystal, however, chlorine-bridged dimers are formed so that each molybdenum is hexacoordinate. Mo_2Cl_{10} is paramagnetic, the magnetic moment ($\mu_{eff} = 1.64$ B.M. at $293°K$), indicating only negligible coupling of electron spins of the two molybdenum atoms (Mo—Mo = 3.84 Å). Mo_2Cl_{10} is soluble in benzene and also in more polar organic solvents. It is monomeric in solution and is presumably solvated. It readily abstracts oxygen from oxygenated solvents to give oxo species, and it is also reduced by amines to give amido complexes. It is rapidly hydrolyzed by water. Some of its reactions are shown in Figs. 26-C-8 and 26-C-9, which set out the preparative methods for lower chlorides and oxide chlorides.

[18] M. Mercer, *Chem. Comm.*, **1967**, 119.
[19] J. R. Graham and L. H. Slaugh, *Tetrahedron Letters*, **1971**, 787.
[20] G. W. Fraser, C. J. W. Gibbs and R. D. Peacock, *J. Chem. Soc., A*, **1970**, 1708.
[21] J. Schröder and F. J. Grewe, *Chem. Ber.*, 1970, **103**, 1536.
[22] A. J. Edwards, *J. Chem. Soc., A*, **1969**, 909; M. J. Bennett, T. E. Haas and J. T. Purdham, *Inorg. Chem.*, 1972, **11**, 207.

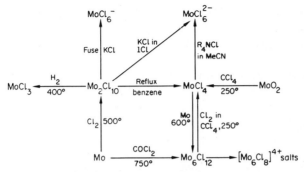

Fig. 26-C-8. Preparation of molybdenum chlorides and chloro complexes.

Green WCl_5 and black WBr_5 are prepared by direct halogenation, the conditions being critical, especially the temperature. Neither structure is known.

The *tetrahalides* include MoF_4 and WF_4, the former arising on disproportionation of Mo_2F_9 as noted above, and both by reduction of the hexahalides with hydrocarbons, e.g., benzene at $\sim 110°$. Both are non-volatile. $MoCl_4$, which is very sensitive to oxidation and hydrolysis, exists in two forms. By the reaction

$$MoCl_5 + C_2Cl_4 \rightarrow \alpha\text{-}MoCl_4$$

a form[23] isomorphous with $NbCl_4$ (page 939) is obtained. The same form, contaminated with carbon, is also obtained when $MoCl_5$ is reduced with hydrocarbons. On being heated to 250°C in the presence of $MoCl_5$, the α-form changes into the high temperature β-form. α-$MoCl_4$ has partial spin-pairing through Mo—Mo interactions whereas the β-form has an *hcp* array of Cl atoms with Mo atoms so distributed in octahedral interstices

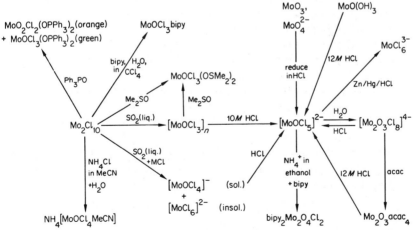

Fig. 26-C-9. Some preparations and reactions of molybdenum pentachloride and of oxomolybdenum compounds.

[23] T. M. Brown and E. L. McCann, III, *Inorg. Chem.*, 1968, **7**, 1227; D. L. Kepert and R. Mandyczewsky, *Inorg. Chem.*, 1968, **7**, 2091.

that no Mo—Mo bond is formed.[24] WCl_4 is best obtained by reducing WCl_6 with Al in a thermal gradient; it disproportionates at 500° to $WCl_2 + 2WCl_5$; it is isotypic with α-$MoCl_4$. $MoBr_4$, WBr_4 and WI_4 all exist but are not well known.

The *trihalides* include MoF_3, $MoCl_3$, $MoBr_3$, MoI_3, WCl_3, WBr_3 and WI_3. MoF_3, a non-volatile brown solid in which Mo atoms are found in octahedra of F atoms, is obtained by reaction of Mo with MoF_6 at $\sim 400°$. $MoCl_3$ has two polymorphs, with *hcp* and *ccp* arrays of Cl atoms and, in each case, Mo atoms in pairs of adjacent octahedral holes at a distance of 2.76 Å across a common edge. The magnetic properties confirm the expected M—M interaction. WCl_3 has an especially interesting cluster structure (page 972).

Molybdenum dichloride. The commonest dichloride is in fact the yellow cluster compound Mo_6Cl_{12} discussed below. The action of dry HCl on $Mo_2(OCOCH_3)_4$ at ca. 300° or better, the reaction sequence

$$Mo(CO)_6 \xrightarrow{Cl_2(l)} Mo(CO)_4Cl_2 \xrightarrow{Heat} MoCl_2$$

gives a brown dichloride[25] which is much more reactive than Mo_6Cl_{12}, dissolving in pyridine, for example, to give red air-sensitive solutions.

Halogeno Complexes. Molydenum(III), but not tungsten(III), forms complexes of the type $[MX_6]^{3-}$. Prolonged electrolytic reduction of a solution of MoO_3 in concentrated hydrochloric acid gives a solution of Mo^{III} in the form of chloro complexes, of which $[MoCl_6]^{3-}$ and $[MoCl_5(H_2O)]^{2-}$ can be precipitated with the larger alkali-metal cations. The salts are red and fairly stable in dry air. In dilute solution, the $MoCl_6^{3-}$ ion is readily aquated.[26a] Thus dissolution of K_3MoCl_6 in aqueous CF_3SO_3H followed by ion exchange purification gives the yellow air-sensitive aquo ion $[Mo(H_2O)_6]^{3+}$. K_3MoCl_6 reacts with molten KHF_2 to produce brown, cubic K_3MoF_6. The magnetic behavior of MoX_6^{3-} salts is usually straightforward, with essentially temperature-independent effective magnetic moments of about 3.8 B.M., corresponding to the expected t_{2g}^3 configuration.

The only pure halo complexes of W^{III} known are $W_2Cl_9^{3-}$ and $W_2Br_9^{3-}$, and there are also analogous molybdenum species. Aqueous HCl solutions of K_2WO_4 can be reduced electrolytically to give solutions of W^{III} from which the $W_2Cl_9^{3-}$ salts can be precipitated with Cs^+, R_4N^+, etc. A similar method[26b] can be used for Mo, by adjusting conditions to favor $Mo_2Cl_9^{3-}$ over $MoCl_6^{3-}$. A procedure leading more smoothly to very pure $Mo_2Cl_9^{3-}$ salts proceeds from $Mo_2(O_2CCH_3)_4$ through $Mo_2Cl_8^{4-}$ (both of which are discussed below):[27]

$$Mo_2(O_2CCH_3)_4 \xrightarrow{HCl(aq)} Mo_2Cl_8^{4-} \xrightarrow[+Cl^-]{-2e} Mo_2Cl_9^{3-}$$

[24] H. Schäfer et al., Z. anorg. Chem., 1967, **353**, 282 (a comprehensive review of molybdenum chlorides other than $MoCl_6$).
[25] G. Holste and H. Schäfer, J. Less Common Metals, 1970, **20**, 164.
[26a] A. R. Bowen and H. Taube, J. Amer. Chem. Soc., 1971, **93**, 3287.
[26b] J. Lewis, R. S. Nyholm and P. W. Smith, J. Chem. Soc., A, **1969**, 57.
[27] M. J. Bennett, J. V. Brencic and F. A. Cotton, Inorg. Chem., 1969, **8**, 1060.

where the second step is carried out electrolytically. $Cs_3Mo_2X_9$ with $X = Cl$ or Br can also be obtained, in the form of relatively large crystals mixed with other products, by treating CsX with MoX_3 at 770–800° in sealed tubes.[28] $W_2Br_9^{3-}$ is best obtained[29] by ligand exchange on $W_2Cl_9^{3-}$ in saturated aqueous HBr at 0°.

The $M_2X_9^{3-}$ species have the confacial bioctahedron structure (1-XLII), which is also possessed by $Cr_2Cl_9^{3-}$. The strength of the interaction between the two d^3 ions increases markedly in the series Cr, Mo, W. Thus, in $Cr_2X_9^{3-}$ there is no M—M bond, and the Cr atoms actually repel each other. The magnetic and spectral properties are those of essentially unperturbed d^3 ions. In $W_2X_9^{3-}$ at the opposite extreme, the interaction is very strong, causing a marked distortion of the structure (W—W, 2.45 Å) and resulting in the absence of unpaired electrons.[28] There must be one σ and two π bonds between the W atoms.[30] In $Mo_2X_9^{3-}$ the situation is intermediate.[28] The three M—M interactions which are very strong in $W_2Cl_9^{3-}$ are of lower strength, giving Mo—Mo distances of 2.67 and 2.78 Å for $Cs_3Mo_2Cl_9$ and $Cs_3Mo_2Br_9$, respectively.[28] In the former, as in both $W_2X_9^{3-}$ species, there is a small temperature-independent paramagnetism. In $Cs_3Mo_2Br_9$, however, the magnetism is temperature-dependent, suggesting perhaps that the Mo—Mo interaction is weak enough to allow some unpairing of electrons at 300°K.

$K_3W_2Cl_9$ reacts with pyridine[31] (and several substituted pyridines) as well as with alcohols[32] to afford compounds in which one W—Cl—W bridge is opened, giving structures with two octahedra sharing an edge;[29] the formulas are $W_2Cl_6py_4$ and $W_2Cl_4(OR)_2(ROH)_4$.

Both Mo and W form octahedral halo complexes in the IV and V oxidation states. Molybdenum(IV) species are postulated as labile intermediates in the reduction of Mo^{VI} by Sn^{II} or other agents and in oxidation of Mo^{III}, and they can be obtained under certain conditions in aqueous solution by mixing Mo^{III} and Mo^{VI} solutions. There are also some isolable anionic complexes of both elements. The dark-green alkali-metal salts, e.g., K_2MoCl_6, are made by the interaction of $MoCl_5$ and MCl in ICl as solvent; the yellow tetraalkylammonium salts can be made in liquid SO_2. Similar red tungsten salts are made by heating WCl_6 and KI at 130°. These compounds evidently contain octahedral anions since they appear to have the K_2PtCl_6 structure. Amine salts are formed by reduction of WCl_6 with aliphatic amines. The compound long formulated as $K_2[W(OH)Cl_5]$ has been shown to be $K_4[Cl_5WOWCl_5]$, with a bent W—O—W group and a small antiferromagnetic interaction between the W atoms.[33]

[28] R. Saillant and R. A. D. Wentworth, *Inorg. Chem.*, 1969, **8**, 1226.
[29] J. L. Hayden and R. A. D. Wentworth, *J. Amer. Chem. Soc.*, 1968, **90**, 5292.
[30] F. A. Cotton, *Accounts Chem. Res.*, 1969, **2**, 240; *Rev. Pure Appl. Chem.*, 1967, **17**, 25.
[31] R. Saillant, J. L. Hayden and R. A. D. Wentworth, *Inorg. Chem.*, 1967, **6**, 1497.
[32] P. W. Clark and R. A. D. Wentworth, *Inorg. Chem.*, 1968, **7**, 1221.
[33] E. König, *Inorg. Chem.*, 1969, **8**, 1278.

The W^{IV} salts and also the orange complex WCl_4py_2, obtained from K_2WCl_6 by action of pyridine or by treating WCl_4 with pyridine, have magnetic moments that are much below the spin-only value for two unpaired electrons. Some compounds are certainly antiferromagnetic and since WCl_6^{2-} salts have crystal structures similar to $IrCl_6^{2-}$ salts where antiferromagnetic interaction occurs through neighboring chlorine atoms, this explanation is probably general.

Hexafluoromolybdates(IV), for example, the dark brown Na_2MoF_6, can be obtained by reduction of MoF_6 with an excess of NaI; the hexafluoro-molybdates(IV) are much more stable with respect to hydrolysis than are the Mo^V species.

The hexafluoromolybdate(V) and hexafluorotungstate(V) anions can be obtained as Na, K, Rb or Cs salts by the reaction:

$$W(Mo)(CO)_6 + M^II + IF_5 \rightarrow M^I[W(Mo)F_6] + 6CO + \text{unidentified products}$$

Here, IF_5 serves both as the fluorinating agent and the solvent. In these complexes there is considerable interionic electron spin coupling which makes them antiferromagnetic (Néel temperatures of the order of 100–150°K). These couplings must take place by overlap of orbitals of the fluoride ions of adjacent $[MF_6]^-$ units in the crystals, that is, by a super-exchange process similar in principle to that in halides and chalconides of some divalent metals of the first transition series. Such an explanation, of course, requires the assumption of significant overlap of metal $d\pi$ (t_{2g}) orbitals with fluoride ion $p\pi$ orbitals.

It is also possible to isolate K_3MoF_8 and K_3WF_8 from the above reaction system under certain conditions. It is not known whether or not these compounds contain octacoordinated metal ions.

Black crystals of the tetraethylammonium salt of $MoCl_6^-$ result from the reaction of $MoCl_5$ with Et_4NCl in CH_2Cl_2. Corresponding green WCl_6^- and WBr_6^- salts can be made similarly or in other ways. The magnetic moments of these complexes are lower than the spin-only value and Et_4NWCl_6 is actually antiferromagnetic. The salts are thermally decomposed to the red W^{IV} complex:

$$2M^IWCl_6 \xrightarrow{280-300°} M_2WCl_6 + WCl_6$$
$$\text{green} \qquad\qquad\qquad \text{red}$$

By reaction of $W(CO)_6$ with IF_5 (as the reaction medium) in presence of KI the compound K_2WF_8 is obtained. The unit cell of this compound and those of $CsWF_7$ and $RbWF_7$ are cubic, but the shape of the coordination polyhedra about the tungsten atom is not yet known. Similar MF_7^- species of uncertain structure are formed by interaction of MF_6 with NOF and NO_2F as the NO^+ and NO_2^+ salts.

A great number of compounds and complexes are known containing a mixture of halogen and other ligand atoms. Those with M=O bonds will be discussed in the next Section.

The pentachlorides react with alcohols and alkoxide ions to yield several types of alkoxy complex. Tungsten forms an extensive series of complexes[34] that includes the paramagnetic $[M(OR)Cl_5]^-$ and $[M(OR)_2Cl_4]^-$ ions as well as the diamagnetic $W_2Cl_2(OR)_8$ and $W_2Cl_4(OR)_6$ molecules which presumably contain octahedrally coordinated metal atoms, bridging Cl atoms and W—W bonds. $MoCl_5$ reacts with alcohols and amines to give products of the types $MoCl_3(OR)_2$ and $MoCl_3(NRR')_2$ which appear generally to be dinuclear with bridging chlorine atoms. The chief products of reaction with phenols are of the type $[MoCl_2(OAr)_3]_2$.

The tungsten hexahalides also react with OH and NH groups, with partial or even complete replacement of the halogen atoms. Thus $W(OPh)_6$ is known and more recently the reaction of WCl_6 with $LiN(CH_3)_2$ at 0° in benzene/petrol has been found to yield small amounts of an orange, crystalline, volatile compound, $W(NMe_2)_6$; this molecule has planar WNC_2 groups so oriented as to give the very rare T_h symmetry.[35] WF_6 reacts with Me_3SiOMe to give a series of compounds $WF_{6-n}(OMe)_n$, $n = 1$–5.[36]

$MoCl_5$ often reacts with ligands to be reduced, giving complexes such as $MoCl_4py_3$, $MoCl_4bipy$ and $MoCl_4(RCN)_2$, and there are tungsten analogs. Little is known in detail about the course of these reactions or the structures of the complexes.

There are also some mixed complexes[37] of Mo^{III}, viz., $MoCl_3py_3$ and $[MoCl_4bipy]^-$.

26-C-5. Oxide Halides

Oxidation State VI. These are of two stoichiometric types: MOX_4 and MO_2X_2. The molybdenum compounds are less stable than those of tungsten and are all fairly rapidly hydrolyzed by water. They are obtained as by-products in the halogenation of the metals unless the metal is first scrupulously reduced and the reaction system vigorously purged of oxygen.

$MoOF_4$ and WOF_4 can both be prepared by the same types of reaction, namely,

$$\left. \begin{array}{l} M + O_2 + F_2 \\ MO_3 + F_2 \\ MOCl_4 + HF \end{array} \right\} \rightarrow MOF_4 \quad (M = Mo \text{ or } W)$$

They are both colorless, volatile solids, not as reactive as the hexafluorides. $MoOF_4$ has octahedral units linked by bringing F's into infinite chains, while WOF_4 has the tetrameric NbF_5 structure[22] (page 936).

WO_2F_2 has been reported but there is some doubt of its actual existence. MoO_2F_2 can be obtained by the action of HF on MoO_2Cl_2; it is a white solid subliming at 270° at 1 atm.

$MoOCl_4$ is best obtained as green crystals (m.p. 101–103°) by evaporation

[34] C. H. Brubaker, Jr., et al., Inorg. Chem., 1969, **8**, 587, 1645; 1970, **9**, 397, 827.
[35] D. C. Bradley et al., Chem. Comm., **1969**, 1261.
[36] L. B. Handy and F. E. Brinckman, Chem. Comm., **1970**, 214.
[37] D. W. Dubois, R. T. Iwamoto and J. Kleinberg, Inorg. Chem., 1969, **8**, 815.

of the red solutions obtained by refluxing MoO_3 with $SOCl_2$; it decomposes to $MoOCl_3$ and Cl_2 even at 25° and is readily reduced by organic solvents to Mo^V species.

MoO_2Cl_2 is best made by the action of chlorine on heated, dry MoO_2. It is fairly volatile and dissolves with hydrolysis in water although in strong HCl an oxide halide species, possibly $Cl_2(O)MoO_2MoOCl_2$, exists.

When MoO_3 is treated with dry hydrogen chloride at 150–200° a pale yellow, very volatile compound soluble in various polar organic solvents is obtained; it has stoichiometry $Mo(OH)_2Cl_2$ but is possibly a dimer.

The two tungsten oxide chlorides are formed together when WO_3 is heated in CCl_4, phosgene or PCl_5 vapor. They are easily separated since $WOCl_4$ is much more volatile than WO_2Cl_2. On strong heating above 200° the following reaction occurs:

$$2WO_2Cl_2 \rightarrow WO_3 + WOCl_4$$

$WOCl_4$ forms scarlet crystals and a red monomeric vapor and is in general highly reactive. It is violently hydrolyzed by water. WO_2Cl_2 occurs as yellow crystals and is not nearly so reactive as $WOCl_4$; it is hydrolyzed only slowly by cold water.

Oxidation State V. The four principal compounds are black $MoOCl_3$ and $MoOBr_3$, olive $WOCl_3$ and brown to black $WOBr_3$. Methods of preparation[38,39] include:

$$WOX_4 + \tfrac{1}{3}Al \rightarrow WOX_3 + \tfrac{1}{3}AlX_3$$
$$2W + WO_3 + \tfrac{9}{2}Br_2 \rightarrow 3WOBr_3$$
$$MoOCl_4 + C_6Cl_6 \rightarrow MoOCl_3$$

All four oxide halides occur in a crystalline form isotypic with $NbOCl_3$, but $MoOCl_3$ has a second (monoclinic) form which has the structure[38,40] shown in Fig. 26-C-10.

Fig. 26-C-10.   A portion of an infinite chain of $MoCl_5O$ octahedra that occurs in monoclinic $MoOCl_3$. Bold and broken circles represent, respectively, upper and lower chlorine atoms; dots are oxygen atoms.

Thio analogs of the oxide halides are also known for Mo^V, W^V and W^{VI}.[41]

[38] M. G. B. Drew and I. B. Tomkins, *J. Chem. Soc., A*, **1970**, 22.
[39] J. Tillack and R. Kaiser, *Angew. Chem. Internat. Edn.*, 1968, 7, 294.
[40] G. Ferguson, M. Mercer, and D. W. A. Sharp, *J. Chem. Soc., A*, **1969**, 2415.
[41] M. G. B. Drew and R. Mandyczewsky, *J. Chem. Soc., A*, **1970**, 2815.

26-C-6. Oxo Complexes[42]

There are many compounds and complexes of molybdenum in the V and VI oxidation states in which there are one or more oxygen atoms bound to the molybdenum by multiple bonds that can be represented Mo=O, since they are evidently approximately double bonds. There are also oxo species with either linear or non-linear M—O—M bridges. While there are certain resemblances in this oxo chemistry to that of Cr, and while tungsten also forms some analogous compounds, by far the most extensive chemistry is known for molybdenum. Some important relationships have been summarized above in Fig. 26-C-9.

Oxomolybdenum(IV) complexes, although as yet not well known, have been prepared,[43] examples being $[MoOCl(diphos)_2]^+$ and $MoOCl_2(PMe_2Ph)_3$.

Molybdenum(V) Complexes.   An extensive range of Mo^V compounds can be obtained by reduction of molybdates or MoO_3 in acid solution either chemically, e.g., by shaking with mercury, or electrolytically. The nature of the resultant species depends critically on the anions present and on conditions of pH and concentration. Probably the most important species, and one which is often used as a source material for preparation of other Mo^V compounds, is the emerald-green ion $[MoOCl_5]^{2-}$ which can be isolated in a variety of salts. The salts can be obtained by reduction of hydrochloric acid solutions of Mo^{VI} or from $MoCl_5$. Solutions of $MoCl_5$ in oxygenated solvents contain oxo species, as we have seen, and green solutions in ethanol or methanol can be used for preparative purposes. The propensity of $MoCl_5$ to abstract oxygen can be utilized for the preparation of oxo complexes such as $MoOCl_3(OSMe_2)_2$, by allowing $MoCl_5$ to react with oxo compounds such as Me_2SO or Ph_3PO.

The pentachlorooxomolybdate salts are most readily obtained by adding, say, KCl to a solution of $MoCl_5$ in concentrated HCl. The salts of $MoOCl_5^{2-}$ are paramagnetic, μ_{eff} ca. 1.67 B.M., indicating a single unpaired electron with slight antiferromagnetic or other effects. The spectra can be assigned on the basis of a strong tetragonal distortion in solution and in crystals.

If sulfur dioxide is used as solvent, salts of the $MoOCl_4^-$ ion can be obtained, e.g.,

$$MoCl_5 + R_4NCl \xrightarrow{SO_2} R_4N^+[MoOCl_4]^- + SOCl_2$$

In solutions the ion is presumably solvated and solvated salts such as $K[MoOCl_4MeCN]$ can be isolated.

The tungsten analogs are also known; e.g., when WO_4^{2-} in $12M$ HCl is reduced, salts of the blue ion $WOCl_5^{2-}$ can be obtained.

When $MoCl_5$ is dissolved in concentrated HCl and solid NaOH is added, the initial green solutions of $MoOCl_5^{2-}$ become darker and almost opaque

[42] (a) P. C. H. Mitchell, *Quart. Rev.*, 1966, **20**, 103; *Coordination Chem. Rev.*, 1966, **1**, 315; these references provide a general review on Mo complexes; (b) W. P. Griffith, *Coordination Chem. Rev.*, 1970, **5**, 459.

[43] A. V. Butcher and J. Chatt, *J. Chem. Soc., A*, **1971**, 2356; S. Midollini and M. Bacci, *J. Chem. Soc., A*, **1970**, 2964; L. Manojlovic-Muir, *J. Chem. Soc., A*, **1971**, 2796.

and finally at less than $4M$ HCl a red brown species is formed. Addition of acid reverses the changes. Two dimers are present, and the equilibria are believed to be the following:[44]

$$\text{MoOCl}_5^{2-} \rightleftharpoons \left[\text{Cl}_4\text{OMo}\underset{\underset{\text{H}}{\text{O}}}{\overset{\overset{\text{H}}{\text{O}}}{\diagup\diagdown}}\text{MoOCl}_4\right]^{4-} \rightleftharpoons [\text{Cl}_4\text{OMoOMoOCl}_4]^{4-}$$

green	dark,	red-brown
diamagnetic	paramagnetic	diamagnetic
>8M HCl	5–6M HCl	1–3M HCl

Essentially all Mo^V oxo complexes contain MoO, Mo_2O_3 or Mo_2O_4 units and a very wide range of anionic and neutral complexes is known. The oxo-bridged species most commonly have either a linear or a bent Mo—O—Mo bridge. The linear bridges involve $d\pi \rightarrow p\pi$ bonding (cf. page 143) and lead to diamagnetism for the complexes; one known structure is the xanthate $\text{Mo}_2\text{O}_3(\text{S}_2\text{COEt})_4$ (26-C-I), which has doubly-bonded

(26-C-I)   (26-C-II)

oxygen atoms *cis* to the bridge. Dioxo bridges are less common[45] but have been proved in the oxalato anion, $[(\text{H}_2\text{O})\text{ox}(\text{O})\text{MoO}_2\text{Mo}(\text{O})\text{ox}(\text{H}_2\text{O})]^{2-}$ and in the histidine complex, $[\text{Mo}_2\text{O}_4(\text{L-histidine})_2]\cdot 3\text{H}_2\text{O}$ (26-C-II).

Other examples are $\text{MoOCl}_3(\text{bipy})$, $\text{Mo}_2\text{O}_3\text{Cl}_4(\text{bipy})_2$, and $\text{Mo}_2\text{O}_4\text{Cl}_2$-$(\text{bipy})_2$, while some quite stable oxomolybdenum(v) species can be obtained with certain anions, notably ethylenediaminetetraacetate and its derivatives. Thus diethylenetriaminepentaacetic acid with Mo^V in $3M$ HCl gives crystalline $\text{H}_3[\text{Mo}_2\text{O}_2(\text{OH})_4\text{DTPA}]$ which is indefinitely stable in air; 8-hydroxyquinoline-5-sulfonic acid also gives stable species.

In view of the interest in models for molybdenum enzyme systems such as xanthine-oxidase, complexes with amino acids, cysteine and organic sulfur compounds have been particularly intensively studied for Mo^V and Mo^{VI}.[46] Complexes with two sulfur bridges similar to the bridged dioxo species are known and can be prepared from solutions of thiomolybdates or by passing H_2S through solutions of oxo-species.

[44] R. Colton and G. G. Rose, *Austral. J. Chem.*, 1968, **21**, 883; M. G. B. Drew and A. Kay, *J. Chem. Soc., A*, **1971**, 1846, 1851.

[45] L. T. J. Delbaere and C. K. Prout, *Chem. Comm.*, **1971**, 162.

[46] See, e.g., A. Kay and P. C. H. Mitchell, *J. Chem. Soc., A*, **1970**, 2421; R. N. Jowitt and P. C. H. Mitchell, *J. Chem. Soc., A*, **1970**, 1702; T. J. Huang and G. P. Haight, *J. Amer. Chem. Soc.*, 1971, **93**, 611; B. Spivac *et al.*, *J. Amer. Chem. Soc.*, 1971, **93**, 5265.

Complexes of Molybdenum(VI) and Tungsten(VI). Molybdenum(VI) oxide dissolves in aqueous HCl to give oxochloride anions. In $12M$ acid $[MoO_2Cl_4]^{2-}$ predominates while in $6M$ acid a species believed to be $[MoO_2Cl_2(H_2O)_2]$ is the main one.[47] Molybdenum(VI) has a pronounced tendency to form oxo complexes, especially of the dioxo type. According to vibrational spectra the overwhelming majority of these have non-linear MoO_2 groups, i.e. the O atoms occupy *cis*-positions in the octahedral ligand sets.[48] Examples are $[MoO_2F_4]^{2-}$, $[MoO_2Cl_4]^{2-}$, $[MoO_2(S_2CNR_2)_2]$ and $[MoO_2(acac)_2]$. In the case of (8-quinolinolato)$_2$MoO$_2$ the *cis*-configuration has been demonstrated by X-ray crystallography.[49]

Like Mo^V, Mo^{VI} forms various binuclear oxo complexes having both Mo=O and Mo—O—Mo groups. The anion in $K_2[Mo_2O_5(C_2O_4)_2(H_2O)_2]$ has the centrosymmetric structure shown in Fig. 26-C-11; the bridge group Mo—O—Mo is linear and symmetrical.

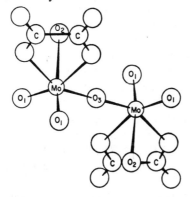

Fig. 26-C-11.   The structure of the anion in $K_2[Mo_2O_5(C_2O_4)_2(H_2O)_2]$.

Tungsten(VI) forms a few oxo complexes, e.g., WOF_5^-, $WO_2F_4^{2-}$, $WO_2Cl_2^{2-}$ and $WO_3F_3^{3-}$; these are not binuclear, but some bent oxo-bridged complex ions are known, e.g., the peroxo species

$$K_2[(O_2)_2(H_2O)(O)W—O—W(O)(H_2O)(O_2)_2].$$

26-C-7. Other Non-cluster Compounds and Complexes of the Lower Oxidation States

The majority of these are obtained by using the hexacarbonyls as starting materials.

Dinuclear Species with Quadruple Bonds. Molybdenum carbonyl reacts readily with carboxylic acids to give yellow crystalline compounds $Mo_2(OOCR)_4$ which are thermally extremely stable. They have the copper

[47] W. P. Griffith and T. D. Wickins, *J. Chem. Soc., A*, **1967**, 675.
[48] W. P. Griffith and T. D. Wickins, *J. Chem. Soc., A*, **1968**, 400; F. W. Moore and R. E. Rice, *Inorg. Chem.*, 1968, 7, 2510.
[49] L. O. Atovmyan and Yu. A. Sokolova, *Chem. Comm.*, **1969**, 649.

acetate structure with an Mo—Mo distance of about 2.12 Å, indicative of a quadruple bond (page 552). The acetate reacts with very concentrated HCl^{50} to yield the $Mo_2Cl_8^{4-}$ ion (page 552) which also contains an Mo—Mo quadruple bond and has the same D_{4h} structure, with Mo—Mo ≈ 2.14 Å, as in $Re_2Cl_8^{2-}$ with which it is isoelectronic (cf. pages 553 and 980). Under slightly oxidizing conditions (page 960) the acetate reacts with HCl to give $Mo_2Cl_8^{3-}$ which has quite a different structure from $Mo_2Cl_8^{4-}$, namely, a $W_2Cl_9^{3-}$-type structure with one bridging Cl absent. There is also a quadruply bonded pair of Mo atoms in tetraallyldimolybdenum (Mo—Mo = 2.18 Å) which is isostructural with its chromium analog.[51] On protonation of $Mo_2(O_2CCH_3)_6$ with strong non-complexing acids,[52] or on conversion of $Mo_2Cl_8^{4-}$ to $Mo_2(SO_4)_4^{2-}$ and removal of sulfate by Ba^{2+} the very air-sensitive cation Mo_2^{4+} is obtained;[26a] this is similar to Rh_2^{4+} (page 1027).

Tungsten(II) shows much less tendency to M—M multiple bonding. The $W_2(O_2CR)_4$ compounds are much less stable than those of Mo and have not, in fact, been unequivocally shown to be isostructural with them.[53] No dinuclear halo complex or allyl derivative of W^{II} is known. However the Mo^{III} and W^{III} alkyls $M_2(CH_2SiMe_3)_6$ are isomorphous and have a short, triple M—M bond (see page 757); the neopentyls are similar.[54]

Carbonyl, Tertiary Phosphine and Related Species. The main chemistry of the carbonyls is discussed in Chapter 22. $Mo(CO)_6$ reacts with liquid chlorine at $-78°$ to give yellow, diamagnetic, apparently dinuclear, 7-coordinate $[Mo(CO)_4Cl_2]_x$. This reacts readily with Ph_3P and Ph_3As to give, for example, $Mo(Ph_3P)_2(CO)_3Cl_2$.[55] The $Mo(CO)_4X_2$ (X = Cl or Br) species react with an excess of an isocyanide to give diamagnetic, 7-coordinate $Mo(CNR)_5X_2$ species.[56]

$Mo(CO)_6$ readily reacts with N, P and As donors with displacement of one to four CO groups. Further reactions with Cl_2, Br_2 or I_2 afford a variety of 6- and 7-coordinate Mo^I, Mo^{II} and Mo^{III} complexes.

There is a wide variety of both Mo and W complexes of tertiary phosphines and other donor ligands that do not contain CO.[57] These are made from the halides or complexes such as $MoCl_4(EtCN)_2$.

For the carbonyl compounds the annexed scheme (Fig. 26-C-12) is representative.

The magnetic properties of most such complexes are complicated. The 7-coordinate Mo^{II} species have temperature-independent magnetic moments of 0.4 to 1.1 B.M., but evidently have no unpaired electrons. The Mo^I and Mo^{III} species are believed to have single unpaired electrons; the Mo^{II}

[50] J. V. Brencic and F. A. Cotton, *Inorg. Chem.*, 1970, **9**, 346, 351.
[51] F. A. Cotton and J. R. Pipal, *J. Amer. Chem. Soc.*, 1971, **93**, 5441.
[52] P Legzdins, G. L. Rempel and J. Wilkinson, *Chem. Comm.*, **1969**, 825.
[53] F. A. Cotton and M. Jeremic, *Synthesis Inorg. Organometal. Chem.*, 1971, **1**, 265.
[54] F. Huq *et al.*, *Chem. Comm.*, **1971**, 1079, 1477.
[55] R. Colton and I. B. Tompkins, *Austral. J. Chem.*, 1966, **19**, 1143.
[56] F. Bonati and G. Minghetti, *Inorg. Chem.*, 1970, **9**, 2642.
[57] J. R. Moss and B. L. Shaw, *J. Chem. Soc., A,* **1970**, 595; A. V. Butcher and J. Chatt, *J. Chem. Soc., A,* **1970**, 2652.

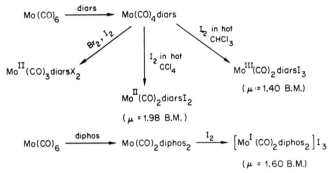

Fig. 26-C-12.   Preparation of some carbonyl-containing molybdenum arsine and phosphine complexes.

species presumably have two, although their magnetic moments are all in the range 1.4–2.0 B.M.

Molybdenum and tungsten form a series of polyhydrides such as $MoH_4(PMePh_2)_4$ and $WH_6(PMe_2Ph)_3$. In some cases the nmr spectra show equivalence of the hydrogen atoms and the molecules are fluxional.[58]

The reaction of $Mo(CO)_6$ with MoF_6 yields $[Mo(CO)_2F_4]_2$; the structure is speculative.[59]

Molybdenum forms a tris(acetylacetonate), an air-sensitive purple–brown solid prepared by heating $Mo(CO)_6$ or K_3MoCl_6 in acetylacetone.

The reaction of ClNO with $Mo(CO)_6$ gives $Mo(NO)_2Cl_2$, a dark green polymeric material that reacts with various additional donors, L, to form $Mo(NO)_2Cl_2L_2$ compounds.

There are a number of $(h^5\text{-}C_5H_5)M(CO)_3X$ (M = Mo or W; X = Cl, H, CH_3, etc.) molecules that have an extensive and important chemistry.

Cyano Complexes.   The best known and most thoroughly studied cyano complexes are the octacyano ions $M(CN)_8^{3-}$ and $M(CN)_8^{4-}$ (M = Mo or W). The interest here has centered on their structures, which appear to be variable with environment. The similar energies of dodecahedral (D_{2d}) and square-antiprismatic (D_{4d}) structures, and the attendent fluxional character for the former, have been mentioned above (page 43). In solid compounds the structures[60,61] found for the $M(CN)_8^{n-}$ ions are as follows (where M indicates that both the Mo and W compounds have the stated structure):

$$
\begin{array}{ll}
\left.\begin{array}{l}
Na_3M(CN)_8\cdot4H_2O \\
[(n\text{-}C_4H_9)_4N]_3[Mo(CN)_8]
\end{array}\right\}
\begin{array}{l}
D_{4d} \\
D_{2d}
\end{array}
& M^V \text{ species}
\end{array}
$$

$$
\begin{array}{ll}
\left.\begin{array}{l}
K_4M(CN)_8\cdot2H_2O \\
H_4M(CN)_8\cdot6H_2O
\end{array}\right\}
\begin{array}{l}
D_{2d} \\
D_{4d}
\end{array}
& M^{IV} \text{ species}
\end{array}
$$

[58] P. Pennella, *Chem. Comm.,* **1971**, 158; P. Jesson *et al., J. Amer. Chem. Soc.,* 1971, **93**, 5261; B. Bell *et al., Chem. Comm.,* **1972**, 34.
[59] T. A. O'Donnell and K. A. Phillips, *Inorg. Chem.,* 1970, **9**, 2611.
[60] L. C. D. Bok, J. G. Leipolt and S. S. Basson, *Acta Cryst.,* 1970, *B*, **26**, 684; J. L. Hoard, T. A. Hamor and M. D. Glick, *J. Amer. Chem. Soc.,* 1968, **90**, 3177.
[61] B. J. Corden, J. A. Cunningham and R. Eisenberg, *Inorg. Chem.,* 1970, **9**, 356.

Thus, the surroundings seem to play a decisive role in stabilizing one or the other geometry. Studies of solutions are, naturally, less conclusive. Most Raman and infrared studies of solutions have been inconclusive, though recently it has been asserted[62] that Raman spectra unequivocally favor the D_{2d} structure for $Mo(CN)_8^{4-}$ in aqueous solution. Esr studies were at first believed to favor D_{4d} symmetry for $Mo(CN)_8^{3-}$ in solution but more recently this has been questioned.[61] The ^{13}C-nmr spectrum of $Mo(CN)_8^{4-}$ in aqueous solution indicates either D_{4d} symmetry or a fluxional D_{2d} structure rearranging through a D_{4d} intermediate (page 43).

The $M(CN)_8^{4-}$ ions in aqueous solution are photochemically converted, through several intermediates, into isolable species long believed to be $[M(CN)_4(OH)_4]^{4-}$ but more recently shown[63] to be six-coordinate, *trans*-dioxo complexes, $[MO_2(CN)_4]^{4-}$.

A cyano complex reported for Mo^{III} is $K_4Mo(CN)_7 \cdot 2H_2O$, which readily oxidizes to $K_4Mo(CN)_8 \cdot 2H_2O$. It has $\mu_{eff} = 1.75$ B.M. at room temperature. The apparent presence of only one unpaired electron has been attributed to d-orbital splitting in the necessarily low symmetry of either an $[Mo(CN)_7]^{4-}$ or an $[Mo(CN)_7(H_2O)]^{4-}$ ion.

Thiocyanate complexes are formed by molybdenum in the III, IV and V oxidation states, the last being of the oxo type, e.g., $[MoO(NCS)_5]^{2-}$. The $[Mo(NCS)_6]^{3-}$ ion has been shown conclusively to have N-bonded thiocyanate ions, and this appears likely to be the case also in all other molybdenum thiocyanato species.

26-C-8. Metal-atom Cluster Compounds

There are a number of compounds of low-valent (mainly oxidation state II) molybdenum and tungsten, which contain metal-atom clusters. The key structural unit is that shown in Fig. 26-C-13, consisting of an octahedron of metal atoms with a bridging atom on each triangular face; the entire unit has full O_h symmetry. Detailed dimensions are available for the $(Mo_6Cl_8)^{4+}$ unit in molybdenum dichloride[64] and for the $(Mo_6Br_8)^{4+}$ unit in

Fig. 26-C-13. The key structural unit, $M_6X_8^{4+}$, found in all metal-atom cluster compounds of Mo^{II} and W^{II}.

[62] T. V. Long, II, and G. A. Vernon, *J. Amer. Chem. Soc.*, 1971, **93**, 1919.
[63] E. G. Arumyunyan, A. S. Antsishkina and E. Y. Balma, *Zhur. Strukt. Khim.*, 1966, **11**, 2400.
[64] H. Schäfer *et al.*, *Z. anorg. Chem.*, 1967, **353**, 281.

$(Mo_6Br_8)Br_4(H_2O)_2$.[65] The $(M_6X_8)^{4+}$ units have the capacity to coordinate six electron-pair donors, one to each metal atom along a four-fold axis of the octahedron, and so far as is known they always do so. Thus, in molybdenum dichloride, "$MoCl_2$," $(Mo_6Cl_8)^{4+}$ units are connected by bridging chlorine atoms (4 per unit) and there are non-bridging Cl atoms in the remaining two coordination positions. In $(Mo_6Br_8)Br_4(H_2O)_2$ the six outer positions are occupied by the four Br atoms and two water molecules. The Mo—Mo distances are 2.62–2.64 Å and consistent with the view[30] that the metal—metal bonds are single.

The usual synthetic routes to these cluster compounds begin with preparation of the anhydrous halides. The preparation of Mo_6Cl_{12}, a yellow, non-volatile solid is shown in Fig. 26-C-8. The tungsten halides, $W_6X_{12}(X = Cl$ or $Br)$, are obtained upon disproportionation of WX_4 at 450–500° or by reduction of WCl_5 or WBr_5 with aluminum in an appropriate temperature gradient.[66] W_6I_{12} is obtained by fusing W_6Cl_{12} with a 10-fold excess of a KI–LiI mixture at 540°.

The bridging groups in the $(M_6X_8)^{4+}$ units undergo replacement reactions only slowly, whereas the six outer ligands are labile; replacement reactions for the latter apparently occur mainly by a dissociative mechanism. It is thus possible to obtain compounds in which a wide variety of ligands occupy the outer positions, such as mixed halides, i.e., $(M_6X_8)Y_4$, a number of salts of the $[(M_6X_8)Y_6]^{2-}$ anions[66,67] and many complexes[68] such as $[(Mo_6Cl_8)Cl_3(Ph_3P)_3]^+$, $[(Mo_6Cl_8)Cl_4(Ph_3P)_2]$, $[(Mo_6Cl_8)L_6]^{4+}$ ($L = Me_2SO$ or Me_2NCHO), and comparable tungsten compounds.[69]

Brief boiling of Mo_6X_{12} ($X = Cl$ or Br) with a methanol solution of NaOMe affords $Na_2[(Mo_6Cl_8)(OMe)_6]$, but prolonged boiling leads to the formation of pyrophoric $Na_2\{[Mo_6(OMe)_8](OMe)_6\}$. Phenoxy groups have also been introduced into both outer and bridging positions.[70] In aqueous solution the $(M_6X_8)^{4+}$ units are unstable to strongly nucleophilic groups such as OH^-, CN^-, and SH^-.

The molybdenum species show little tendency to act as reducing agents, despite the low formal oxidation number, but the tungsten compounds are fairly reactive reductants in aqueous media. It may be recalled that the $(Nb_6X_{12})^{2+}$ and $(Ta_6X_{12})^{2+}$ species (page 942) can sustain reversible oxidation to the $3+$ and $4+$ species. There is no evidence that oxidation (or reduction) is possible for the $(Mo_6X_8)^{4+}$ species, but W_6Cl_{12} and W_6Br_{12} are oxidized by free halogen at elevated temperatures. In the case of W_6Br_{12} the products[71] are W_6Br_{14}, W_6Br_{16} and W_6Br_{18} if the temper-

[65] L. G. Guggenberger and A. W. Sleight, Inorg. Chem., 1969, 8, 2041.
[66] R. D. Hogue and R. E. McCarley, Inorg. Chem., 1970, 9, 1354.
[67] F. A. Cotton, R. M. Wing and R. A. Zimmerman, Inorg. Chem., 1967, 6, 11.
[68] J. E. Fergusson, B. H. Robinson and C. J. Wilkins, J. Chem. Soc., A, 1967, 486.
[69] R. D. Hogue and R. E. McCarley, Abs. A.C.S. National Meeting, April, 1969, No. 178.
[70] P. Nannelli and B. P. Block, Inorg. Chem., 1968, 7, 2423; 1969, 8, 1767.
[71] H. Schäfer and R. Siepmann, Z. anorg. Chem., 1968, 357, 273; R. Siepmann and H.-G. von Schnering, Z. anorg. Chem., 1968, 357, 289.

ature is kept below 150° (above which WBr_6 is obtained). In all of these there has been a 2-electron oxidation of the W_6Br_8 group. W_6Br_{14} may be formulated as $(W_6Br_8)Br_6$, while the others contain bridging Br_4^{2-} units and are formulated as $(W_6Br_8)Br_4(Br_4)_{2/2}$, and $(W_6Br_8)Br_2(Br_4)_{4/2}$.

The reaction of Cl_2 with W_6Cl_{12} at 100° results in an intriguing structural change. The product is of stoichiometry WCl_3 and has been shown to contain the $(W_6Cl_{12})^{6+}$ unit, isostructural with the $(M_6X_{12})^{n+}$ units found characteristically in the cluster compounds of Nb and Ta (page 941), and the complete formulation of WCl_3 is $(W_6Cl_{12})Cl_6$.[72]

26-D. TECHNETIUM AND RHENIUM[1]

Technetium and rhenium are very similar chemically and differ considerably from manganese, despite similarities in the stoichiometries of a few compounds, e.g., the series MnO_4^-, TcO_4^-, ReO_4^- and the metal carbonyls. The most stable and characteristic oxidation state for manganese is the II state, for which the chemistry is mainly that of the high-spin Mn^{2+} cation. Technetium and rhenium have practically no cationic chemistry, form practically no compounds in the II oxidation state and have extensive chemistry in the IV and, especially, the V state. The TcO_4^- and ReO_4^- ions are much less oxidizing than MnO_4^-. A characteristic feature of Re^{III} in its halides is the formation of metal—metal bonds. Technetium also does this to some extent whereas manganese forms no such compounds at all. Indeed rhenium shows a marked tendency to form M—M bonds in oxidation states up to at least IV. The ion $Re_2X_9^-$ has Re—Re = 2.71 Å, and in $La_4Re_6O_{19}$ where the mean oxidation number is +4.33 there are $Re(O)_2Re$ groups with the Re—Re distance 2.42 Å.[2]

While there is very little cationic aqueous chemistry, the oxo anions TcO_4^- and ReO_4^- are well known in aqueous solution. Polarographic studies of the reduction of these ions in aqueous solution containing different anions have been made, but, although standard potential values can be associated with the observed reduction steps, the precise nature of the species present is obscure. The final reduction appears to involve the formation of hydride species (page 989). Some potential data—of limited utility—are as shown in Fig. 26-D-1. The oxidation states and stereochemistry of the elements are summarized in Table 26-D-1.

[72] R. Siepmann, H.-G. von Schnering and H. Schäfer, *Angew. Chem. Internat. Edn*, 1967, **6**, 637.

[1] R. Colton, *The Chemistry of Rhenium and Technetium*, Wiley-Interscience, 1966; R. D. Peacock, *The Chemistry of Technetium and Rhenium*, Elsevier, 1966; K. V. Kotegov, O. N. Pavlov and V. P. Shvedov, *Adv. Inorg. Nuclear Chem.*, 1969, **11**, 1 (technetium); J. E. Fergusson, *Coordination Chem. Rev.*, 1966, **1**, 459 (rhenium complexes).

[2] M. Longo and A. W. Sleight, *Inorg. Chem.*, 1968, **7**, 108.

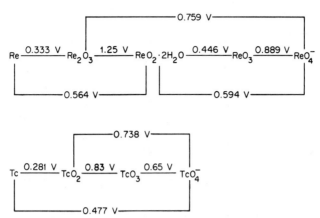

Fig. 26-D-1. Potential data for rhenium in basic solution (R. H. Busey, E. D. Sprague and R. B. Bevan, *J. Phys. Chem.*, 1969, **73**, 1039) and for technetium.

TABLE 26-D-1

Oxidation States and Stereochemistry of Technetium and Rhenium

Oxidation state	Coordin. no.	Geometry	Examples
Tc^{-I}, Re^{-I}	5	?	$[Re(CO)_5]^-$
Tc^0, Re^0, d^7	6	Octahedral	$Tc_2(CO)_{10}$, $Re_2(CO)_{10}$
Tc^I, Re^I, d^6	6^b	π-Complex	h^5-$C_5H_5Re(CO)_2C_5H_8$, h^5-$C_5H_5Re(CO)_3$
	6	Octahedral	$Re(CO)_5Cl$, $K_5[Re(CN)_6]$, $Re(CO)_3py_2Cl$, $[(CH_3C_6H_4NC)_6Re]^+$
Tc^{II}, Re^{II}, d^5	5	?	ReX_2TAS
	6	Octahedral	$Re(diars)_2Cl_2$, $Tc(diars)_2Cl_2$
Tc^{III}, Re^{III}, d^4	$7,8^b$	π-Complex	$(h^5$-$C_5H_5)_2ReH$, $(h^5$-$C_5H_5)_2ReH_2^+$
	5	*tbp*?	$(Ph_3PO)_2ReCl_3$
	6	Octahedral	$[Tc(diars)_2Cl_2]^+$, $ReCl_2acac(PPh_3)_2$ mer-$ReCl_3(PR_3)_3$
		Trigonal prism	$Re(S_2C_2Ph_2)_3$
		Metal atom clusters[a]	$Re_2X_8^{2-}$, $Re_3X_{9+n}^{n-}$, $Re_3X_9L_3$
Tc^{IV}, Re^{IV}, d^3	6^a	Octahedral	K_2TcI_6, K_2ReCl_6, ReI_4py_2, $TcCl_4$, $ReCl_4$, $[Re_2OCl_{10}]^{4-}$, $ReCl_4diars$
	7	?	$[ReCOdiars_2I_2](ClO_4)_2$
Tc^V, Re^V, d^2	5	*tbp*?	$ReCl_5(g)$, ReF_5, $NReCl_2(PPh_3)_2$
		Square pyramid	$[ReOX_4]^-$
	6^a	Octahedral	$ReOCl_3(PPh_3)_2$, $[ReOCl_5]^{2-}$, Re_2Cl_{10}, $Tc(NCS)_6^-$
	7	?	$ReOCl_3TAS$
	8	Dodecahedral(?)	$[Re(diars)_2Cl_4]^+$, $K_3[Re(CN)_8]$
Tc^{VI}, Re^{VI}, d^1	6	Octahedral	ReO_3, ReF_6
	7	?	$ReOCl_6^{2-}$, ReF_7 (?)
	8	Sq. antiprism	ReF_8^{2-}
Tc^{VII}, Re^{VII}, d^0	4^a	Tetrahedral	ReO_4^-, TcO_4^-, ReO_3Cl, Re_2O_7
	6	Octahedral	$ReO_3Cl_3^{2-}$
	7	Pentag. bipyramid	ReF_7
	9	Tricapped trig. prism	ReH_9^{2-}

[a] Most common states.
[b] Assuming h^5-C_5H_5 occupies three coordination sites.

26-D-1. The Elements

Although its existence was predicted much earlier from the Periodic Table, rhenium was first detected, by its X-ray spectrum, only in 1925; a few years later Noddack, Berg and Tacke isolated about a gram of rhenium from molybdenite. Rhenium is now recovered on a fairly substantial scale from the flue dusts in the roasting of molybdenum sulfide ores and from residues in the smelting of some copper ores. The element is usually left in oxidizing solution as perrhenate ion, ReO_4^-. After concentration, the perrhenate is precipitated by addition of potassium chloride as the sparingly soluble salt, $KReO_4$.

All isotopes of technetium are unstable toward β decay or electron capture and traces exist in Nature only as fragments from the spontaneous fission of uranium. The element was named technetium by the discoverers of the first radioisotope—Perrier and Segré. Three isotopes have half-lives greater than 10^5 years, but the only one that has been obtained on a macro scale is ^{99}Tc (β^-, 2.12×10^5 years). Technetium is recovered from waste fission-product solutions after removal of plutonium and uranium. It is an interesting irony that the supply of technetium, which does not exist in Nature, might easily be made to exceed that of Re, which does, because of the increasing number of reactors and the very low ($\sim 10^{-9}\%$) abundance of Re in the earth's crust.

The metals resemble platinum in appearance but are usually obtained as grey powders; Re has a higher melting point (3180°) than any metal except W (3400°). Re and Tc are both obtained by thermal decomposition of NH_4MO_4 or $(NH_4)_2MCl_6$ in H_2. Technetium can also be made by electrolysis of NH_4TcO_4 in $2N$ H_2SO_4, with continuous addition of H_2O_2 to re-oxidize a brown solid also produced. Rhenium can be electrodeposited from H_2SO_4 solutions although special conditions are required to obtain coherent deposits. Both metals crystallize in an *hcp* arrangement. They burn in oxygen above 400° to give the oxides M_2O_7 which sublime away; in moist air the metals are slowly oxidized to the oxo acids. The latter are also obtained by dissolution of the metals in concentrated nitric acid or hot concentrated sulfuric acid. The metals are insoluble in hydrofluoric or hydrochloric acid but are conveniently dissolved by warm bromine water. Rhenium, but not technetium, is soluble in hydrogen peroxide.

At the present time technetium has no uses although the TcO_4^- ion is said to be an excellent corrosion inhibitor for steels.

Rhenium is used in W–Re thermocouples;[3] a Pt–Re alloy catalyst for petroleum re-forming[4] is said to have a longer lifetime than platinum alone.

[3] Y. Tseng, S. Schnatz and E. D. Zysk, *Engelhard Tech. Bull.*, 1970, **11**, 12.
[4] M. J. Sterba and V. Haensel, *World Petroleum*, 1971, **42**, 192.

BINARY COMPOUNDS OF TECHNETIUM AND RHENIUM

26-D-2. Oxides and Sulfides

The known *oxides* are shown in Table 26-D-2. The *heptaoxides*, obtained by burning the metals, are volatile. If acid solutions containing TcO_4^- are evaporated, the oxide is driven off, a fact that can be utilized to isolate and separate technetium; rhenium is not lost from acid solutions on evaporation,

TABLE 26-D-2

Oxides of Rhenium and Technetium[a]

Rhenium		Technetium	
Oxide	Color	Oxide	Color
$Re_2O_3 \cdot xH_2O$	Black		
ReO_2	Brown	TcO_2	Black
ReO_3	Red	$TcO_3(?)$	
Re_2O_5	Blue		
Re_2O_7	Yellow (m.p. 220°)	Tc_2O_7	Yellow (m.p. 119.5°)

[a] Lower hydrated oxides formulated as $ReO \cdot H_2O$ and $Re_2O \cdot 2H_2O$, are obtained by Zn reduction of weakly acid ReO_4^- solutions; they are not fully investigated.

i.e., at 100°, but can be distilled from hot concentrated H_2SO_4. The heptaoxides readily dissolve in water, giving acidic solutions, and Re_2O_7 is deliquescent. The oxides differ structurally and in various physical properties. The structure of Re_2O_7 consists of an infinite array of alternating ReO_4 tetrahedra and ReO_6 octahedra sharing corners,[5] whereas Tc_2O_7 consists of molecules in which TcO_4 tetrahedra share an oxygen atom and the Tc—O—Tc chain is linear.[6] On evaporation of aqueous solutions of Re_2O_7 over P_2O_5 slightly yellow crystals of so-called "perrhenic acid" are obtained. These are actually $Re_2O_7(H_2O)_2$, which is binuclear with both tetrahedral and octahedral rhenium atoms, as in Re_2O_7 itself, i.e. O_3Re—O—$ReO_3(H_2O)_2$.[7] The Re—O—Re bond is essentially linear.

The lower oxides can be obtained either by thermal decomposition of NH_4MO_4 or by heating $M_2O_7 + M$, at 200–300°. The hydrated dioxides, $MO_2 \cdot 2H_2O$, can be obtained by addition of base to M^{IV} solutions, for instance, of $ReCl_6^{2-}$, or, for Tc, by reduction of TcO_4^- in hydrochloric acid with zinc. For rhenium a hydrated sesquioxide, $Re_2O_3 \cdot xH_2O$, readily oxidized by water to $ReO_2 \cdot xH_2O$, has been reported. The pentaoxide has been made by electrolytic reduction of perrhenate in sulfuric acid solution; it decomposes above 200°.

Both TcO_2 and ReO_2 have distorted rutile structures isotypic to that of

[5]   B. Krebs, A. Muller and H. Beyer, *Chem. Comm.*, **1968**, 263.
[6]   B. Krebs, *Angew. Chem. Internat. Edn.*, 1969, **8**, 381; see also H. Selig and S. Fried, *Inorg. Nuclear Chem. Letters*, 1971, **7**, 315.
[7]   J. Beyer, O. Glemser and B. Krebs, *Angew Chem. Internat. Edn.*, 1968, **7**, 295.

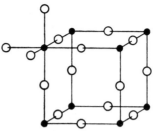

Fig. 26-D-2. The ReO₃ structure. Each metal atom lies at the center of an octahedron of oxygen atoms. This structure is closely related to the perovskite structure (page 55) since the latter is obtained from this one by insertion of a large cation in the center of the cube shown.

MoO_2; thus substantial metal—metal interactions presumably exist, but M—M distances have not been reported.

Rhenium(VI) oxide has a structure that is found for other oxides (e.g., CrO_3, WO_3) and is closely related to the perovskite structure (page 55). It is usually called the ReO_3 structure and is shown in Fig. 26-D-2.

The pairs of *sulfides* TcS_2, ReS_2 and Tc_2S_7, Re_2S_7 are isomorphous. The black heptasulfides are obtained by saturation of 2–6N hydrochloric acid solutions of TcO_4^- or ReO_4^- with hydrogen sulfide. The precipitation is sensitive to conditions and is often incomplete. Treatment of neutral solutions of the oxo anions with thioacetamide or sodium thiosulfate followed by acidification gives a better yield. An excess of sulfur in the precipitates may be extracted with CS_2. ReS_3 is obtained by reduction of Re_2S_7 with hydrogen.

The disulfides are obtained by heating the heptasulfides with sulfur in a vacuum; they have a tendency to be non-stoichiometric.

Rhenium sulfides are effective catalysts for hydrogenation of organic substances and they have the advantage over heterogeneous platinum metal catalysts in that they are not poisoned by sulfur compounds.[8a] An inorganic reduction that they catalyze is that of NO to N_2O at 100°.[8b]

26-D-3. Halides

The known halides of the elements are shown in Table 26-D-3. TcF_6 results from the fluorination of technetium at 400°; it is stable in nickel or dry Pyrex vessels for an extended period of time. On hydrolysis, like ReF_6, it gives a black precipitate, presumably of hydrous TcO_2. $TcCl_4$ is obtained as paramagnetic red crystals by the action of carbon tetrachloride on Tc_2O_7 in a bomb, and is the major product on direct chlorination of the metal. $TcCl_4$ has a structure very similar to that of $ZrCl_4$ (Fig. 26-A-1) in which there are linked $TcCl_6$ octahedra.[9] Both the magnetic behavior (μ_{eff} at

[8a] W. H. Davenport, V. Kollonitsch and C. H. Kline, *Ind. Eng. Chem.*, 1968, **60**, 11 (catalytic uses of Re).
[8b] L. H. Slough, *Inorg. Chem.*, 1964, **3**, 920.
[9] M. Elder and B. R. Penfold, *Inorg. Chem.*, 1966, **5**, 1197.

TABLE 26-D-3

The Halides of Technetium and Rhenium

	$TcCl_4$ red-brown			
	ReF_4 blue, sub.$>300°$	ReF_5 greenish-yellow, m.p. 48°	TcF_6 golden-yellow, m.p. 33°	ReF_7 pale yellow
Re_3Cl_9 dark red	$ReCl_4$ black	$ReCl_5$ dark red-brown m.p. 261	ReF_6 pale yellow, m.p. 18.7°	
Re_3Br_9 red-brown	$ReBr_4$ dark red	$ReBr_5$ dark brown	$ReCl_6$ green-black m.p. 29°	
ReI_2 black	Re_3I_9 black	ReI_4 black		

$300° = 3.48$ B.M.) and the structure (Tc—Tc $= 3.62$ Å) show that there is no Tc—Tc bonding, in contrast to $ReCl_4$ (see below).

ReF_6, is obtained by direct interaction of the elements at 120°. Spectroscopic studies have shown that it is an octahedral molecule and the absorption spectrum of its d^1 system, which is influenced by strong spin–orbit coupling, has been analyzed. Both ReF_6 and TcF_6 have moments much lower than the spin-only value for the same reason. The compound is very sensitive to moisture and on hydrolysis gives ReO_2, $HReO_4$ and HF.

ReF_7 is the only stable heptahalide known, other than IF_7 to which it is crystallographically similar.[10] It can be obtained instead of ReF_6 by direct interaction of F_2 and Re if the reaction is carried out at 400° under pressure. It dissolves in liquid HF, but the ReF_7 molecules appear to remain intact.[11] ReF_5 is obtained along with ReF_4 and oxide fluorides by the reduction of ReF_6 with metal carbonyls. The non-volatile ReF_4 is best made[12] by reduction of ReF_6 with Re at 500°. It is also possible to obtain ReF_7, ReF_6 and some lower fluorides by electrically exploding Re wires in atmospheres of SF_6 or PF_5.[13]

Rhenium hexachloride is prepared[14] by interaction of ReF_6 and BCl_3 at low temperatures; removal of the excess of BCl_3 leaves a green to black, very volatile solid. $ReCl_6$ is hydrolyzed, with disproportionation, to ReO_4^- and hydrated ReO_2 and decomposes to $ReCl_5$ and Cl_2 when heated. If rhenium metal powder (contaminated with KCl) is heated in chlorine at 500°, mainly $ReCl_5$ is obtained, as a dark red-brown vapor that condenses to a dark red-brown solid; chlorination at higher temperatures gives a mixture of $ReCl_5$ and $ReCl_6$. $ReCl_5$ is also obtained by extraction of various salts of $ReCl_6^{2-}$ with boiling CCl_4. Since these salts can be readily obtained from

[10] S. Siegel and D. A. Northrop, *Inorg. Chem.*, 1966, **5**, 2187.
[11] H. Selig and E. L. Garner, *J. Inorg. Nuclear Chem.*, 1968, **30**, 658.
[12] D. E. Lavelle, R. M. Steele and W. T. Smith, Jr., *J. Inorg. Nuclear Chem.*, 1966, **28**, 260.
[13] R. L. Johnson and B. Siegel, *J. Inorg. Nuclear Chem.*, 1969, **31**, 2391.
[14] J. H. Canterford and A. B. Waugh, *Inorg. Nuclear Chem. Letters*, 1971, **7**, 395.

the corresponding perrhenates by reduction with CCl_4 at 400°, or in other ways, this is a useful preparation not needing metal. $ReCl_5$ is rapidly hydrolyzed by water or moist air and is reduced by many ligands,[15] often to Re^{IV} complexes (see below). It has a structure,[16] Re_2Cl_{10}, in which the Re atoms occupy adjacent octahedra, which share an edge, and repel each other (Re—Re = 3.74 Å). The magnetic properties ($\mu = 1.79$ B.M., $\theta = 164°$ for range 79–301°K) imply substantial magnetic exchange interaction even though there is no Re—Re bonding.

Heating $ReCl_5$ in a nitrogen atmosphere causes it to lose chlorine, forming rhenium(III) chloride (see below).

Early reports of the preparation of $ReCl_4$ by action of $SOCl_2$ on $ReO_2 \cdot xH_2O$ merit skepticism,[17] but the substance can be prepared in several ways,[18] viz.:

$$2ReCl_5 + SbCl_3 \rightarrow 2ReCl_4 + SbCl_5$$
$$3ReCl_5 + Re_3Cl_9 \rightarrow 6ReCl_4$$
$$2ReCl_5 + CCl_2{=}CCl_2 \rightarrow 2ReCl_4 + C_2Cl_6$$

$ReCl_4$ is built of a close-packed array of Cl atoms with pairs of adjacent octahedral interstices occupied by Re atoms; the adjacent octahedra have a common face and the Re—Re distance is only 2.73 Å, indicative of a bond.[19] The substance is apparently only metastable and has rather complex reactivity which is not yet well understood.

The pentabromide, obtained by bromination of Re at 650°, decomposes readily to Re_3Br_9 when heated. The tetrabromide and tetraiodide can be made by careful evaporation of solutions of $HReO_4$ in an excess of HBr or HI. The tetraiodide is unstable and when heated at 350° in a sealed tube gives ReI_3. At 110° in nitrogen, ReI_2 is obtained; this is diamagnetic and is believed to be polymeric with Re—Re bonds.

Re^{III} Halides. The chloride, bromide and iodide have been structurally characterized and their true molecular formulas are Re_3X_9. They are not isomorphous, but all[20] consist of Re_3X_9 units connected by sharing of X atoms as shown for Re_3Cl_9 in Fig. 26-D-3. The Re_3X_9 units are metal-atom cluster compounds. The Re—Re distances are ~ 2.48 Å and the M—M bonds are of order 2 as explained elsewhere (page 551). The Re_3X_9 units are so stable as to persist in vapors[21] at temperatures around 600°, and they form the structural basis for much of the chemistry of Re^{III}.

[15] D. A. Edwards and R. T. Ward, *J. Chem. Soc., A*, **1970**, 1617.
[16] K. Mucker, G. S. Smith and Q. Johnson, *Acta Cryst.*, 1968, B, **24**, 874.
[17] I. R. Anderson and J. C. Sheldon, *Inorg. Chem.*, 1968, **7**, 2602.
[18] (a) C. J. L. Lock *et al.*, *Canad. J. Chem.*, 1969, **47**, 1069; J. H. Canterford and R. Colton, *Inorg. Nuclear Chem. Letters*, 1968, **4**, 607; A. Brignole and F. A. Cotton, *Chem. Comm.*, **1971**, 706; (b) F. A. Cotton, W. R. Robinson, and R. A. Walton, *Inorg. Chem.*, 1967, **6**, 223; R. A. Walton, *Inorg. Chem.*, 1971, **10**, 2534.
[19] M. J. Bennett, F. A. Cotton, B. M. Foxman and P. F. Stokely, *J. Amer. Chem. Soc.*, 1967, **89**, 2759.
[20] F. A. Cotton, S. J. Lippard and J. T. Mague, *Inorg. Chem.*, 1965, **4**, 508; M. J. Bennett, F. A. Cotton and B. M. Foxman, *Inorg. Chem.*, 1968, **7**, 1563.
[21] A. Büchler, P. E. Blackburn and J. L. Stauffer, *J. Phys. Chem.*, 1966, **70**, 685; K. Rinke, M. Klein, and H. Schäfer, *J. Less-Common Metals*, 1967, **12**, 497.

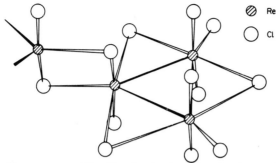

Fig. 26-D-3.   The structure of Re_3Cl_9, showing how the trinuclear molecules are linked by chlorine bridges.

Halo Complexes.   The halides Re_3Cl_9 and Re_3Br_9 can be dissolved in concentrated HCl or HBr, respectively. The exact species present in solution are not known, but by using various large univalent cations, M, substances of the following types have been obtained: $M_3Re_3X_{12}$, $M_2Re_3X_{11}$, MRe_3X_{10} and $M_2Re_4Br_{15}\cdot 3H_2O$. The explanation of the unusual stoichiometries of the first three is simple in terms of the persisting Re_3X_9 unit. The three outer positions (see Fig. 26-D-4) are sites for the attachment of additional ligands, such as X''^- ions, and, depending on the cation, crystal-packing considerations and the equilibria in solution, one, two or all three of these may be filled in the complex anion that is precipitated. The last-mentioned compound consists of $Re_3Br_9(H_2O)_3$ groups where the outer positions are occupied by water molecules[20] and $ReBr_6^{2-}$ ions, the latter having arisen by partial oxidation of the Re^{III}; this compound can be precipitated directly from solutions containing equal molar quantities of Re_3Br_9 and $ReBr_6^{2-}$.

Exchange studies have shown[22] that the order of ligand lability in the $Re_3X_3X_6'X_3''$ species is $X'' \ggg X'$ (non-bridging) $> X$ (bridging); accordingly it possible to get compounds with a mixture of ligands, an example[23] being

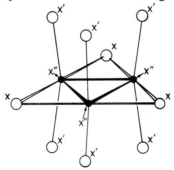

Fig. 26-D-4.   A sketch of an isolated Re_3X_9 unit which has D_{3h} symmetry. The three positions where additional ligands may be attached are indicated by $X'' \rightarrow$.

[22] J. E. Fergusson, *Coordination Chem. Rev.*, 1966, **1**, 459.
[23] M. Elder *et al.*, *Chem. Comm.*, **1969**, 731.

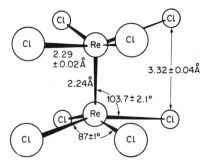

Fig. 26-D-5. The structure of the $Re_2Cl_8^{2-}$ ion.

$CsRe_3Cl_3Br_7(H_2O)_2$, which has an (Re_3Cl_3) core with Br atoms in the X′ positions and with the X″ positions occupied by two H_2O's and a Br^-.

Totally different kinds of Re^{III} halo complexes are obtained by reduction of ReO_4^- in HCl or HBr solution by H_2 or H_3PO_2. These are the $Re_2X_8^{2-}$ ions, which have the sort of structure[24] shown in Fig. 26-D-5. The formulation of the Re—Re quadruple bond whereby the unusual eclipsed structure is explained has been presented elsewhere (page 552). The $Re_2Cl_8^{2-}$ ion can also be conveniently obtained by treatment of Re_3Cl_9 with molten $Et_2NH_2^+Cl^-$ at 127°,[25] and this is perhaps the most convenient method of preparation. The $Re_2X_8^{2-}$ ions can be reduced to $Re_2X_8^{3-}$ and $Re_2X_8^{4-}$ species,[26] which presumably retain the $Re_2X_8^{2-}$ structure. The $Tc_2X_8^{3-}$ ion has, indeed, been shown to have the eclipsed (D_{4h}) structure.[27] On oxidation, with Cl_2 or Br_2, the $Re_2Cl_8^{2-}$ or $Re_2Br_8^{2-}$ ions are converted into the $Re_2X_9^{2-}$ ions[28] which can then be reduced to the $Re_2X_9^{3-}$ ions; the $Re_2Cl_9^{2-}$ ion can also be obtained in other ways.[18b, 29] Further reactions of the $Re_2X_8^{2-}$ ions will be mentioned below (page 987).

It is interesting that reduction of $ReCl_4(CH_3CN)_2$ leads to the ion[30] $[ReCl_4(CH_3CN)_2]^-$, which is more stable than the $Re_2Cl_8^{2-}$ ion.

Rhenium metal reacts with PCl_5 at 500–550 °C to give $[PCl_4]^+[ReCl_6]^-$. Under different conditions $PReCl_8$, which is probably $[PCl_4^+]_2[Re_2Cl_8]^{2-}$ is obtained.[31] Rhenium(v) also forms the ReF_6^- ion by the reaction

$$ReF_6(\text{in excess}) + M^I I \xrightarrow[SO_2]{Hg.} M^I ReF_6 \quad (M^I = Na, \ K, \ Rb \ or \ Cs)$$

There is also an $Re(SCN)_6^-$ ion formed when $ReCl_5$ reacts with molten $KSCN$[32] and a technetium analog.[33] These various ReX_6^- compounds have

[24] F. A. Cotton, B. G. DeBoer and J. Jeremic, *Inorg. Chem*, 1970, **9**, 2143.
[25] R. Bailey and J. McIntyre, *Inorg. Chem.*, 1966, **5**, 1940.
[26] F. A. Cotton, W. R. Robinson and R. A. Walton, *Inorg. Chem.*, 1967, **6**, 1257.
[27] W. K. Bratton and F. A. Cotton, *Inorg. Chem.*, 1970, **9**, 978.
[28] F. Bonati and F. A. Cotton, *Inorg. Chem.*, 1967, **6**, 1353.
[29] E. A. Allen *et al.*, *Inorg. Nuclear Chem. Letters*, 1969, **5**, 239.
[30] G. Rouschias and G. Wilkinson, *J. Chem. Soc., A*, **1968**, 489.
[31] P. W. Frais, A. Guest and C. J. L. Lock, *Chem. Comm.*, **1970**, 1612; and personal communication from C. J. L. Lock.
[32] R. A. Bailey and S. L. Kozak, *Inorg. Chem.*, 1967, **6**, 2155.
[33] T. C. Schwochau and H. H. Pieper, *Inorg. Nuclear. Chem. Letters*, **1968**, 711.

magnetic moments (at 300°) of 1.3–2.7 B.M.; the fluoride salts are strongly antiferromagnetic.

Both Re^{IV} and Tc^{IV} form stable MX_6^{2-} ions. ReF_6^{2-} is quite stable in aqueous solution, even when alkaline; curiously, TcF_6^{2-} has not been reported. The most important and useful complexes are the hexachloro salts, which are obtained by reducing TcO_4^- or ReO_4^- with 8–13M hydrochloric acid, preferably with the addition of KI as reductant. In the case of TcO_4^-, reduction actually gives first[34] a precipitate of $K_2(TcCl_5OH)$ which redissolves in an excess of HCl to give $[TcCl_6]^{2-}$. The salts K_2TcCl_6 (yellow) and K_2ReCl_6 (yellow-green) form large octahedral isomorphous crystals. The solubilities of these and other salts are similar to those of the hexachloroplatinates in that large unipositive ions give insoluble salts. In water, K_2ReCl_6 is hydrolyzed to give $ReO_2 \cdot xH_2O$. In the reduction of TcO_4^- or ReO_4^- by hydrochloric acid, complexes in the intermediate V oxidation state can be isolated (see below). The hexabromo complexes of Tc and Re are made by the action of HBr on the chloro complex, and the hexaiodo complex is prepared by heating the bromide with HI.

Reaction of liquid ReF_6 with alkali fluorides gives relatively unstable salts, M^IReF_7 and $M_2^IReF_8$. In K_2ReF_8 there are discrete ReF_8^{2-} ions with a square antiprismatic configuration.[35]

OXO COMPOUNDS

For rhenium particularly, oxo compounds are predominant and of key importance in the higher oxidation states, especially V and VII. It is convenient to discuss all the oxo compounds and complexes together rather than under the separate oxidation numbers.

26-D-4. Simple Oxo Anions

The MO_4^- Ions. The pertechnetates and perrhenates are among the most important compounds formed by these elements. The aqueous acids or their salts are formed on oxidation of all technetium or rhenium compounds by nitric acid, hydrogen peroxide or other strong oxidizing agents. Pure perrhenic acid has not been isolated, but a red crystalline product, claimed to be $HTcO_4$, has been obtained; both acids are strong acids in aqueous solution. The solubilities of alkali perrhenates generally resemble those of the perchlorates, but pertechnetates are more soluble in water than either (cf. $KReO_4$ 9.8 g/l, $KTcO_4$ 126 g/l, at 20°). Highly insoluble precipitates, suitable for gravimetric determination, are given by tetraphenylarsonium chloride and nitron with both anions.

[34] J. E. Fergusson *et al.*, *J. Chem. Soc.*, A, **1967**, 1423.
[35] P. A. Koz'min, *J. Strukt. Chem.*, 1964, **5**, 60.

The tetrahedral TcO_4^- and ReO_4^- ions are quite stable in alkaline solution, unlike MnO_4^-. They are also much weaker oxidizing agents than MnO_4^-, but they are reduced by HCl, HBr or HI. In acid solutions, the ions can be extracted into various organic solvents such as tributyl phosphate, and cyclic amines extract them from basic solution. Such extraction methods of purification suffer from difficulties due to reduction of the ions by organic material. The anions can be readily absorbed by anion-exchange resins, from which they can be eluted by perchloric acid. The ReO_4^- ion can function as a ligand[36] which coordinates more strongly than ClO_4^- or BF_4^- but less strongly than Cl^- or Br^-.

When $Ba(ReO_4)_2 + BaCO_3$ are heated in stoichiometric proportions and for optimal temperatures and times, $Ba_3(ReO_5)_2$ and $Ba_5(ReO_6)_2$, the so-called meso- and ortho-perrhenates, respectively, are obtained. Their structures are uncertain[37] though spectroscopic data suggest discrete ReO_6^{5-} octahedra in the latter.[38]

26-D-5. Oxohalide Molecules and Anions

Oxohalides. The oxohalides are listed in Table 26-D-4. The majority of them are not extensively characterized; the pertechnetyl and perrhenyl ones,

TABLE 26-D-4

Oxohalides of Technetium and Rhenium

V	VI	VII	
	$TcOF_4^c$ blue	TcO_3F yellow, m.p. 18.3°	
$TcOCl_3$ brown, subl. ~500°	$TcOCl_4$ purple, m.p. ~35°	TcO_3Cl colorless liquid b.p. ~25°	
$TcOBr_3$ brown			
$ReOF_3$ black, non-volatile	$ReOF_4$ blue, m.p. 107.8°	ReO_3F yellow, m.p. 147°	$ReOF_5$ cream, m.p. 34.5°
$ReOCl_3(?)^a$	$ReOCl_4$ Green-brown, m.p. 30°	ReO_3Cl colorless liquid, b.p. 131°	ReO_2F_3 pale yellow, m.p. 90°
$ReOBr_3(?)^b$	$ReOBr_4$ blue, dec. > 80°	ReO_3Br colorless, m.p. 39.5°	

[a] The preparation of what was thought to be $ReOCl_3$ was recently reported [P. W. Frais, C. J. L. Lock and A. Guest, *Chem. Comm.*, **1971**, 75] but subsequent work shows that the substance is not $ReOCl_3$ [C. J. L. Lock, personal communication].
[b] Not adequately characterized.
[c] A green polymorph has discrete trimers $(TcOF_4)_3$ which have asymmetric Tc—F—Tc bridges (A. J. Edwards, G. R. Jones and R. J. C. Sills, *J. Chem. Soc., A*, **1970**, 2521).

[36] H. G. Mayfield, Jr., and W. E. Bull, *Inorg. Chim. Acta*, 1969, **3**, 676.
[37] S. K. Majumdar, R. A. Pacer and C. L. Rulfs, *J. Inorg. Nuclear Chem.*, 1969, **31**, 33.
[38] E. J. Baran and A. Muller, *Z. anorg. Chem.*, 1969, **368**, 168.

MO_3X, are perhaps best known. They are made by various methods including the action of oxygen on halides and of halogens on the oxides. The oxohalides are readily hydrolyzed. Both ReO_3F and ReO_3Cl have been shown to have C_{3v} symmetry, with Re—O distances of 1.69 and 1.76 Å, respectively, indicative of double bonding.

Simple Oxohalogeno Anions.   These can be thought of as derived from oxohalides by addition of halide ions, though this is seldom the actual method of preparation. $Cs[ReOCl_5]$ is actually obtained by addition of CsCl to $ReOCl_4$ in $SOCl_2$.[39] $Cs_2[ReO_3Cl_3]$ is precipitated from a solution of ReO_4^- in concentrated HCl saturated with gaseous HCl; ReO_3Cl may be an intermediate. The ir spectrum suggests a cis-structure for the anion.[40] Besides $ReOCl_5^-$ there is another Re^{VI} oxohalo anion, the blue $ReOF_5^-$, obtained on hydrolysis of K_2ReF_8.

Many of the simple oxohalo anions are formed by the metals in the V oxidation state, the most common being of the types MOX_4^- and MOX_5^{2-}. The $ReOBr_4^-$ ion can be obtained by air oxidation of $Re_3Br_{12}^{3-}$(aq.) or by treatment of a solution of $KReO_4$ in H_2SO_4/CH_3OH with zinc, followed by addition of aqueous HBr. Other X groups may be introduced by metatheses. The $ReOBr_4^-$ ion has C_{4v} symmetry with an Re=O bond length of 1.71–1.73 Å. There is often another ligand (H_2O, CH_3CN) weakly coordinated trans to the oxygen atom.[41]

Salts of $[ReOCl_5]^{2-}$ are best prepared by adding large cations such as Et_4N^+ to $ReCl_5$ in concentrated HCl; however, in concentrated HCl solution Re^V can exist also as $[ReOCl_4(H_2O)]^-$ and $[ReCl_4(OH)_2]^-$.[42] The $[TcOCl_5]^{2-}$ ion also exists.[43]

Rhenium alone is known to form the oxo-bridged species $[Cl_5Re—O—ReCl_5]^{4-}$, in which there is a linear ReORe group. The diamagnetism of this anion is attributable to π interactions through the bridging oxygen atom.

Dioxo Complexes.   Rhenium forms a number of complexes containing the ReO_2 unit. Most contain Re^V and important examples are[44] $[ReO_2(CN)_4]^{3-}$, $[ReO_2en_2]^+$, $[ReO_2py_4]^+$ and, $[ReO_2(NH_3)_4]^+$, all of which appear on spectroscopic evidence to have linear OReO groups confirmed for $[ReO_2(CN)_4]^{3-}$ by X-ray study. The anion $[Re_2O_3(CN)_8]^{6-}$ is postulated on magnetic and spectroscopic evidence to have a curious cyano-bridged structure, $Re(O)_2(CN)_3—CN—Re(O)(CN)_4$. Even some trioxo complexes have been obtained. On passing HCl(g) or HBr(g) into a solution of $KReO_4$ in DMF or DMSO the compounds ReO_3XL_2 (X = Cl or Br; L = DMSO or DMF) are obtained.[45]

[39] R. Colton, *Austral. J. Chem.*, 1965, **18**, 435.
[40] D. E. Grove, N. P. Johnson and G. Wilkinson, *Inorg. Chem.*, 1969, **8**, 1196.
[41] F. A. Cotton and S. J. Lippard, *Inorg. Chem.*, 1965, 4, 1621; 1966, **5**, 9, 416.
[42] J. A. Casey and R. K. Murmann, *J. Amer. Chem. Soc.*, 1970, **92**, 78.
[43] B. Jezowska-Trzebiatowska *et al.*, *J. Strukt. Chem.*, 1967, **8**, 519, 524.
[44] W. P. Griffith, *J. Chem. Soc.*, A, **1969**, 211; R. H. Fenn, A. J. Graham and N. P. Johnson, *J. Chem. Soc.*, A, **1971**, 2880.
[45] H.-A. Lehmann and C. Ringel, *Z. anorg. Chem.*, 1969, **366**, 73.

26-D-6. More Complex Oxo Species

There is a very extensive and important chemistry of oxorhenium(v) compounds containing phosphine ligands; the key molecules in this chemistry are $ReOX_3(PR_3)_2$ (X = Cl or Br; R_3 = Ph_3 or Et_2Ph). The important chloro species, $ReOCl_3(PR_3)_2$, are obtained by interaction of ReO_4^- with PR_3 in ethanol containing HCl. The structure of the green $ReOCl_3(PEt_2Ph)_2$ is distorted octahedral, with Cl opposite to O and the phosphine ligands *trans* to each other. This and other related complexes, including the $ReOX_2(OR)(PR_3)_2$ type, can be obtained in isomeric forms by crystallization from different solvents, but equilibration seems to occur rapidly in solution. Some of the more important reactions of $ReOCl_3(PPh_3)_2$ are shown in Fig. 26-D-6. The halide ion (or other ligand) opposite to the Re=O bond is labile; in ethanol, for example, it is rapidly replaced, giving the $ReOX_2(OEt)(PR_3)_2$ compound.

The compounds $ReO(OEt)I_2L_2$ (L = PPh_3 or py) can be transformed into *trans*-dioxo complexes[46] by the following reactions:

$$ReO(OEt)I_2L_2 \xrightarrow{H_2O} ReO_2L_2I \xrightarrow{CNS^-} \textit{trans-}ReO_2(SCN)_2L_2$$

The reaction[47a] of $ReOCl_3(PR_3)_2$ compounds with carboxylic acids *in the presence of air* leads to formation of two series of binuclear complexes

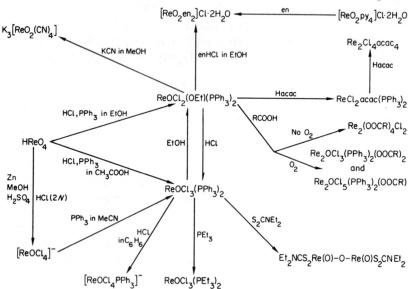

Fig. 26-D-6. Some reactions of $ReOCl_3(PPh_3)_3$. Related compounds, i.e. with various PR_3, AsR_3 or SbR_3 groups in place of PPh_3, are known and have similar though not always identical reactions.

[46] M. Freni *et al.*, *Gazz. Chim. Ital.*, 1969, **99**, 286, 641.

[47] (a) G. Rouschias and G. Wilkinson, *J. Chem. Soc., A*, **1966**, 465; (b) F. A. Cotton and B. M. Foxman, *Inorg. Chem.*, 1968, **7**, 1784; F. A. Cotton, R. Eiss and B. M. Foxman, *Inorg. Chem.*, 1969, **8**, 950.

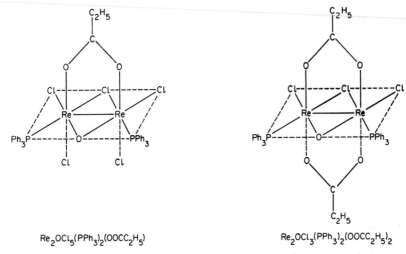

Re$_2$OCl$_5$(PPh$_3$)$_2$(OOCC$_2$H$_5$)   Re$_2$OCl$_3$(PPh$_3$)$_2$(OOCC$_2$H$_5$)$_2$

Fig. 26-D-7.   Structures of two bridged binuclear oxorhenium complexes.

Re$_2$OCl$_5$(PR$_3$)$_2$(OOCR) and Re$_2$OCl$_3$(PR$_3$)$_2$(OOCR)$_2$. Although they were initially formulated slightly differently, these have the structures[47b] shown in Fig. 26-D-7. In both there are Re—Re bonds of length 2.51–2.52 Å, though the formal oxidation numbers are different, namely +4.0 and +3.5. This similarity is presumably due to the constraining effect of the bridging systems, which are nearly identical in the two cases.

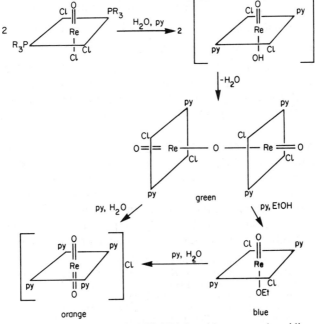

Fig. 26-D-8.   Reaction of ReOCl$_3$(PPh$_3$)$_2$ with water and pyridine.

One of the more unusual reactions of $ReOCl_3(PPh_3)_2$, $ReO(OEt)Cl_2$-$(PPh_3)_2$ or $[ReOX_4]^-$ is that with moist pyridine where the final product is the orange salt $[ReO_2py_4]Cl \cdot 2H_2O$. An intermediate complex $Re_2O_3Cl_4py_4$, can also be isolated. This suggests that the displacement of the labile Cl *trans* to Re=O leads to an unisolated hydroxo species which can condense to the bridged oxo complex. This mechanism is also suggested by the fact that the dioxo species is the final product and by the stability of a corresponding *trans*-oxo-ethoxy compound, as shown in Fig. 26-D-8.

The linear O=Re—O—Re=O group has been confirmed[48] in dithiocarbamate complexes, e.g., $Re_2O_3(S_2NCEt_2)_4$, which are obtained by treating $ReOCl_3(PPh_3)_2$ with $Na_2S_2NCEt_2 \cdot 3H_2O$ in acetone, and in the ion $[(CN)_4OReOReO(CN)_4]^{4-}$.

OTHER COMPLEXES

26-D-7. Complexes with Nitrogen Ligands

Rhenium forms complexes containing Re=NR and Re≡N bonds,[48, 49] representative preparative reactions being

$$ReOCl_3(PPh_3)_2 + PhNH_2 \rightarrow Re(NPh)Cl_3(PPh_3)_2 + H_2O$$

$$ReO(OEt)Cl_2(PPh_3)_2 + N_2H_4 \cdot 2HCl + H_2O \rightarrow$$
$$ReNCl_2(PPh_3)_2 + PPh_3O + EtOH + NH_4Cl$$

$$ReOCl_3(PPh_3)_2 + RNH—NHR \cdot 2HCl \xrightarrow{PPh_3}$$
$$Re(NR)Cl_3(PPh_3)_2 + RNH_3Cl + PPh_3O + HCl$$

The alkylimino complexes react with Cl_2 according to the equation

$$Re(NR)Cl_3(PPh_3)_2 \xrightarrow[CCl_4]{Cl_2 \text{ in}} ReCl_4(PPh_3)_2$$

The Re≡N bond in $ReNCl_2(PEt_2Ph)_3$ exhibits electron-donor properties,[50] like those of a nitrile, in forming adducts with the acceptors BX_3 (X = F, Cl or Br) and $PtCl_2(PEt_3)_2$. $ReNBr_2(PPh_3)_2$ reacts with KCN to give $K_2[ReN(CN)_4] \cdot H_2O$. The Re≡N bonds have the extraordinarily short length of 1.53 Å.[51]

The variations of rhenium-to-nitrogen bond lengths are rather puzzling. Thus, in $ReNCl_2(Ph_3P)_2$, which has a configuration intermediate between *tbp* and *spy*, the Re—N distance is 1.60 Å, whereas in the octahedral $ReNCl_2(PEt_2Ph)_3$ it is 1.79 Å;[52] no satisfactory explanation for this very large difference has been given. An even more curious fact is that the Re—N distances in two $Re(NR)Cl_3(PEt_2Ph)_2$ compounds with R = p-$C_6H_4OCH_3$ or p-$C_6H_4C(O)CH_3$, where the bonds should presumably be of order 2,

[48] R. Shandles, E. O. Schlemper and R. Kent Murinann, *Inorg. Chem.*, 1971, **10**, 2745.
[49] J. Chatt *et al.*, *J. Chem. Soc.*, A, **1970**, 2239, **1971**, 2631, and references therein; V. F. Duckworth *et al.*, *Chem. Comm.*, **1970**, 1082; J. T. Moelwyn-Hughes, A. W. B. Garner and A. S. Howard, *J. Chem. Soc.*, A, **1971**, 2361.
[50] J. Chatt and B. T. Heaton, *J. Chem. Soc.*, A, **1971**, 705.
[51] W. O. Davies *et al.*, *Chem. Comm.*, **1969**, 736.
[52] J. A. Ibers *et al.*, *Inorg. Chem.*, 1967, **6**, 197, 264.

are 1.685–1.710 Å, which is *shorter* than the supposedly triple bond (1.79 Å) in $ReNCl_2(PEt_2Ph)_3$.[53]

Rhenium also forms a series of N_2 complexes (page 710) of the type $Re(N_2)X(PR_3)_4$ where $X = Cl$ or Br and PR_3 is either a monophosphine or half of a chelating diphosphine.[54] The dinitrogen can act as a donor to give bridged species eg., $(PhMe_2P)_4ClReN_2CrCl_3(THF)_2$.

26-D-8. Complexes with Metal—Metal Bonds

A number of complexes can be derived from the trinuclear and dinuclear rhenium(III) halides and halo complexes (page 979) with retention of the Re_3 and Re_2 units.

In the case of the trinuclear systems, the important cases are those in which the X″ positions of Fig. 26-D-4 are occupied by neutral ligands, L, to give molecules of the stoichiometry $Re_3X_9L_3$. In the case where $L = PEt_2Ph$ this type of structure has been verified by X-ray diffraction.

It is also possible to carry out a number of substitution reactions[55] on $Re_2X_8^{2-}$ species. With phosphines, simple 1,2-disubstituted species such as that[56] shown in Fig. 26-D-9a are obtained. Reaction of $Re_2Cl_8^{2-}$ with

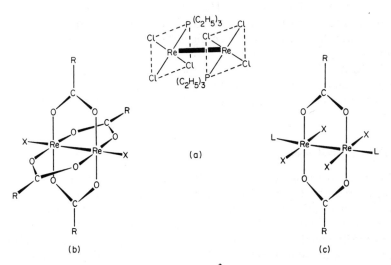

(a)

(b)   (c)

Fig. 26-D-9.   Some complexes from the $Re_2X_8^{2-}$ ions. All contain quadruple Re—Re bonds.

[53] D. Bright and J. A. Ibers, *Inorg. Chem.*, 1968, **7**, 1099; 1969, **8**, 703.
[54] J. Chatt *et al.*, *J. Chem. Soc.*, A, **1971**, 702; B. R. Davies and J. A. Ibers, *Inorg. Chem.*, 1971, **10**, 578.
[55] F. A. Cotton *et al.*, *Inorg. Chem.*, 1967, **6**, 929.
[56] F. A. Cotton and B. M. Foxman, *Inorg. Chem.*, 1968, **7**, 2135.

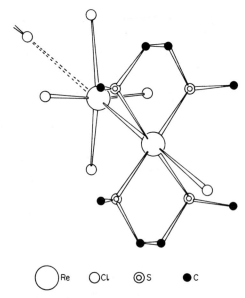

Fig. 26-D-10.   The structure of $Re_2Cl_5(DTH)_2$. The molecules form chains in the crystal and the dashed bond indicates a weaker intermolecular bond.

$CH_3SCH_2CH_2SCH_3$ (DTH) yields $Re_2Cl_5(DTH)_2$, which has the structure[57] shown in Fig. 26-D-10. Here a reduction has occurred and the electron distribution is such that there is only a triple bond between the rhenium atoms. The rotational configuration is then staggered instead of eclipsed.

The $Re_2Cl_8^{2-}$ species react with carboxylic acids[58] to give complexes of the types shown[59] in Fig. 26-D-9b, c. The $Re_2(OOCR)_4Cl_2$ species can also be obtained in excellent yields by the reaction of $ReOCl_3(PPh_3)_2$ with RCOOH.[47a]

26-D-9. Miscellaneous

Rhenium(II) Complexes.   Few of these are well documented. The most likely to be authentic are those containing polydentate phosphine and arsine ligands, namely, $[ReCl(N_2)(diphos)_2]^+$, the complex $ReCl_2(diars)$ obtained by stannite reduction of $[Re(diars)_3]Cl_3$, and also some $ReX_2(TAS)$ and $ReX_2(QAS)$ complexes. In none of these diarsine species is the structure or even the coordination number established since they are too insoluble for molecular weight determinations. Older literature mentions "$ReX_2(PR_3)_2$" complexes, but these are actually the $ReNX_2(PR_3)_2$ species already mentioned.

[57] M. J. Bennett, F. A. Cotton and R. A. Walton, *Proc. Roy. Soc.*, A, 1968, **303**, 175.
[58] F. A. Cotton, C. Oldham and R. A. Walton, *Inorg. Chem.*, 1966, **5**, 1798.
[59] M. J. Bennett *et al.*, *Inorg. Chem.*, 1968, **7**, 1570.; W. K. Bratton and F. A. Cotton, *Inorg. Chem.*, 1969, **8**, 1299.

Rhenium(III) Complexes. Besides the polynuclear complexes treated in earlier Sections, there are a number of mononuclear ones.[30,60] Examples are $ReCl_3(RCN)(PPh_3)_2$,[61] $ReCl_3py_3$, $ReCl_3py_2PPh_3$, $ReCl(acac)_2PPh_3$ and the ion $[ReCl_4(MeCN)_2]^-$ mentioned above. All of these appear to be octahedral and monomeric, with magnetic moments of 1.6–2.0 B.M.

Rhenium(IV) and Technetium(IV) Complexes. Besides the important MX_6^{2-} species mentioned above, there are other Re^{IV} and Tc^{IV} species, though they are not abundant. The addition of halogens to $ReX_3(PR_3)_3$ (X = Cl or Br), as well as reductive interaction of $ReCl_5$ with PPh_3 in acetone,[62] give the $ReX_4(PR_3)_2$ species, which can be obtained also in other ways. *trans*-$[ReCl_4(PEt_2Ph)_2]$ has $\mu_{eff} = 3.64$ B.M. at 20°, in accord with its being an octahedral complex of a d^3 ion. $Re(CO)_5Cl$, on treatment with diars, gives $Re(CO)_3diarsCl$ which can be chlorinated to give $ReCl_4diars$. The tetrahydrofuranate $ReCl_4(THF)_2$ and some similar sulfide complexes are also obtained from $ReCl_5$ by direct interaction.[63]

For technetium, direct reaction of ligands with $TcCl_4$ gives complexes such as $TcCl_4(PPh_3)_2$, $[TcCl_4dipy]$ and $[TcCl_2dipy_2]Cl_2$.[64]

Rhenium is found in trigonal prismatic coordination[65] in $Re(S_2C_2Ph_2)_3$, and it is nine-coordinate in the ReH_9^{2-} ion, which has a tricapped trigonal prismatic (D_{3h}) structure (Fig. 26-D-11).

In addition to the ReH_9^{2-} ion there is a technetium analog. Both of these exhibit a single, sharp pmr signal despite the existence of two different proton environments. This has been attributed to a rapid intramolecular site-exchange process, already described (page 44). The ReH_9^{2-} ion is obtained by reduction of $NaReO_4$ with an excess of sodium in ethanol; the isolation of various pure salts requires careful workup procedures.[66] The ReH_9^{2-} ion

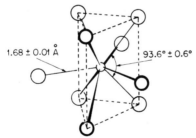

Fig. 26-D-11.   The structure of ReH_9^{2-}; the Re—H distance of 1.68 Å is consistent with the sum of single bond radii (Re = 1.28 Å).

[60] G. Rouschias and G. Wilkinson, *J. Chem. Soc., A,* **1967**, 993; **1968**, 489; P. Gunz and G. J. Leigh, *J. Chem. Soc., A,* **1971**, 2228.
[61] M. G. B. Drew, D. G. Tisley and R. A. Walton, *Chem. Comm.,* **1970**, 600.
[62] H. Gehrke, Jr., and G. Eastland, *Inorg. Chem.,* 1970, **9**, 2722 (see also ref. 14).
[63] E. A. Allen *et al., J. Chem. Soc., A,* **1970**, 788.
[64] J. E. Fergusson and J. H. Hickford, *J. Inorg. Nuclear Chem.,* 1966, **28**, 2293.
[65] R. Eisenberg and J. A. Ibers, *Inorg. Chem.,* 1966, **5**, 411.
[66] A. P. Ginsberg and C. R. Sprinkle, *Inorg. Chem.,* 1969, **8**, 2212; M. Basch and A. P. Ginsberg, *J. Phys. Chem.,* 1969, **73**, 854.

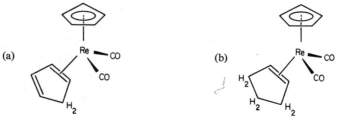

Fig. 26-D-12.   The presumed structures of (a) $C_{10}H_{11}Re(CO)_2$ and (b) its hydrogenation product.

reacts with a variety of phosphines and arsines to give mixed-hydrido species such as $[ReH_8PR_3]^-$ and $ReH_5(AsR_3)_3$; species of the PR_3 type with 7, 5, and uncertain numbers of H atoms can be made from tertiary phosphine complexes such as $ReCl_4(PR_3)_2$ and $ReOCl_3(PR_3)_2$.[67] In all these hydrido species, again, the bound hydrogen atoms are nmr-equivalent, suggesting fluxional character.[68]

   Organo Derivatives.   Few of these are known. The principal ones are $(h^5\text{-}C_5H_5)_2ReH$, $(h^5\text{-}C_5H_5)Re(CO)_3$, several $RRe(CO)_5$ compounds ($R = C_6H_5$, CH_3, C_3F_7, etc.) and the compound $C_{10}H_{11}Re(CO)_2$. The first of these is notable as the first well-characterized, stable transition-metal hydride for which the characteristic high-field nmr chemical shift was observed. Upon treatment with CO at 90° and 250 atm it gives rise to $C_{10}H_{11}Re(CO)_2$, to which structure (a) of Fig. 26-D-12, has been assigned. Catalytic hydrogenation causes the uptake of 1 mol of H_2 per mol, which can be attributed to conversion of (a) into (b).

26-E.   THE PLATINUM METALS

26-E-1.   General Remarks

   Ruthenium, osmium, rhodium, iridium, palladium and platinum are the six heaviest members of Group VIII. They are rare elements; platinum itself is the commonest with an abundance of about $10^{-6}\%$ whereas the others have abundances of the order of $10^{-7}\%$ of the earth's crust. They occur in Nature as metals, often as alloys such as osmiridium, and in arsenide, sulfide and other ores. The elements are usually associated not only with one another but also with the coinage metals copper, silver and gold. The main suppliers are South Africa, Canada and the USSR.

   Since the compositions of the ores differ widely, the extraction methods vary considerably. An important source is the nickel–copper sulfide of South Africa; the ore is physically concentrated by gravitation and flotation, after which it is smelted with lime, coke and sand and bessemerized

[67] J. Chatt and R. S. Coffey, *J. Chem. Soc.*, A, **1969**, 1963.
[68] A. P. Ginsberg, *Chem. Comm.*, 1968, 857; M. Freni *et al.*, *J. Inorg. Nuclear Chem.*, 1969, **31**, 3211.

in a convertor. The resulting Ni–Cu sulfide "matte" is cast into anodes. On electrolysis in sulfuric acid solution, Cu is deposited at the cathode, and Ni remains in solution, from which it is subsequently recovered by electrodeposition, while the platinum metals, silver and gold collect in the anode slimes.[1] The subsequent procedures for separation of the elements are very complicated. Although most of the separations[2] involve classical precipitations or crystallizations, some ion-exchange and solvent-extraction procedures are feasible.

26-E-2. The Metals

Some properties are collected in Table 26-E-1.

TABLE 26-E-1
Some Properties of the Platinum Metals[3]

Element	Melting point (°C)	Form	Best solvent
Ru	~2310	Grey-white, brittle, fairly hard	Alkaline oxidizing fusion
Os	~3050	Grey-white, brittle, fairly hard	Alkaline oxidizing fusion
Rh	1960	Silver-white, soft ductile	Hot conc. H_2SO_4 Conc. $HCl + NaClO_3$ at 125–150°
Ir	2443	Silver-white, hard, brittle	Conc. $HCl + NaClO_3$ at 125–150°
Pd	1552	Grey-white, lustrous, malleable, ductile	Conc. HNO_3, $HCl + Cl_2$
Pt	1769	Grey-white, lustrous, malleable, ductile	Aqua regia

The metals are obtained initially as sponge or powder by ignition of ammonium salts of the hexachloro anions. Almost all complex and binary compounds of the elements give the metal when heated above 200° in air or oxygen; osmium is oxidized to the volatile OsO_4, and at dull red heat ruthenium gives RuO_2, so that reduction in hydrogen is necessary. The finely divided metals are also obtained by reduction of acidic solutions of salts or complexes by magnesium, zinc, hydrogen or other reducing agents such as oxalic acid or formic acid, or by electrolysis under proper conditions.

The metals, as gauze or foil, and especially on supports such as charcoal or alumina on to which the metal salts are absorbed and reduced *in situ* under specified conditions, are widely used as catalysts for an extremely large range of reactions in the gas phase or in solution. One of the biggest uses of platinum is for the reforming, "platforming," of crude oils. Commercial uses in homogeneous reactions are fewer, but Pd salts are used in the

[1] T. Papademetriou and J. R. Grasso, *Engelhard Tech. Bull.*, 1970, **10**, 121.
[2] F. E. Beamish, *Talanta*, 1967, **14**, 991, 1133; 1966, **13**, 773.
[3] For more detailed properties see *Physical Properties of the Precious Metals*, Engelhard Industries, 1965; J. Lopis, *Catalyst Rev.*, 1969, **2**, 161.

Smidt process (page 797) and rhodium chloride in acetic acid synthesis (page 793). Industrially, as well as in the laboratory, catalytic reductions are especially important.[4]

Platinum or its alloys are used for electrical contacts, for printed circuitry, and for plating.

Ru and Os are unaffected by mineral acids below $\sim 100°$ and are best dissolved by alkaline oxidizing fusion, e.g., $NaOH + Na_2O_2$, $KClO_3$, etc. Rh and Ir are extremely resistant to attack by acids, neither metal dissolving even in aqua regia when in the massive state. Finely divided rhodium can be dissolved in aqua regia or hot concentrated H_2SO_4. Both metals also dissolve in concentrated HCl under pressure of oxygen or in presence of sodium chlorate in a sealed tube at 125–150°. At red heat Cl_2 leads to the trichlorides.

Pd and Pt are rather more reactive than the other metals. Pd is dissolved by nitric acid, giving $Pd^{IV}(NO_3)_2(OH)_2$; in the massive state the attack is slow, but it is accelerated by oxygen and oxides of nitrogen. As sponge, Pd also dissolves slowly in HCl in presence of chlorine or oxygen. Platinum is considerably more resistant to acids and is not attacked by any single mineral acid although it readily dissolves in aqua regia and even slowly in HCl in presence of air since

$$PtCl_4^{2-} + 2e = Pt + 4Cl^- \qquad E^0 = 0.75 \text{ V}$$
$$PtCl_6^{2-} + 2e = PtCl_4^{2-} + 2Cl^- \qquad E^0 = 0.77 \text{ V}$$

Platinum is not the inert material that it is often considered to be. There are at least seventy oxidation–reduction and decomposition reactions that are catalyzed by metallic platinum. Examples are the Ce^{IV}–Br^- reaction and the decomposition of N_2H_4 to N_2 and NH_3. It is possible to predict whether catalysis can occur or not from a knowledge of the electrochemical properties of the reacting couples.[5]

Both Pd and Pt are rapidly attacked by fused alkali oxides, and especially by their peroxides, and by F_2 and Cl_2 at red heat. It is of importance in the use of platinum for laboratory equipment that on heating it combines with, e.g., elemental P, Si, Pb, As, Sb, S and Se, so that the metal is attacked when compounds of these elements are heated in contact with platinum under reducing conditions.

Both Pd and Pt are capable of absorbing large volumes of molecular hydrogen, and Pd is used for the purification of H_2 by diffusion (see page 187).

26-E-3. General Remarks on the Chemistry of the Platinum Metals

The chemistries of these elements have some common features but there are nevertheless wide variations depending on differing stabilities of oxidation

[4] See, e.g., P. N. Rylander, *Catalytic Hydrogenation over Platinum Metals*, Academic Press 1968; also the periodicals *Advances in Catalysis, Catalyst Reviews*, etc.
[5] M. Spiro and A. B. Ravnö, *J. Chem. Soc.*, **1965**, 78; M. D. Archer and M. Spiro, *J. Chem. Soc., A*, **1970**, 68, 78.

states, stereochemistries, etc. The principal areas of general similarity are as follows.

1. *Binary compounds*. There are a large number of oxides, sulfides, phosphides, etc., but the most important are the halides.

2. *Aqueous chemistry*. This chemistry is almost exclusively that of complex compounds. Aquo ions of Ru^{II}, Ru^{III}, Rh^{III} and Pd^{II} exist, but complex ions are formed in presence of anions other than ClO_4^-, BF_4^-, or *p*-toluenesulfonate, etc. The precise nature of many supposedly simple solutions, e.g., of rhodium sulfate, is complicated and often unknown.

A vast array of complex ions, predominantly with halide or nitrogen donor ligands, are water-soluble. Exchange and kinetic studies have been made with many of these because of interest in (*a*) *trans*-effects, especially with square Pt^{II}, (*b*) differences in substitution mechanisms between the ions of the three transition-metal series, and (*c*) the unusually rapid electron-transfer processes with heavy-metal complex ions.

Although the species involved may often not be fully identified much potential information has been collected from polarographic and other studies.[6]

3. *Compounds with π-acid ligands*. (*a*) Binary carbonyls are formed by all but Pd and Pt (page 690), the majority of them polynuclear. Substituted polynuclear carbonyls are known for Pd and Pt, and all six elements give carbonyl halides and a wide variety of carbonyl complexes containing other ligands.

(*b*) For Ru, nitric oxide complexes are an essential feature of the chemistry.

(*c*) An especially widely studied area is the formation of complexes with trialkyl- and triaryl-phosphines and related phosphites, and to a lesser extent with R_3As and R_2S. The most important are those with triphenylphosphine and methyl-substituted phosphines, e.g., $PPhMe_2$. The latter are more soluble in organic solvents than PPh_3 complexes, and have also proved particularly useful for the determination of configuration by the nmr virtual coupling method.

Mixed complexes of PR_3 with CO, alkenes, halides and hydride ligands in at least one oxidation state are common for all of the elements.

(*d*) All these elements have a strong tendency to form bonds to carbon, especially with alkenes and alkynes; Pt^{II}, Pt^{IV} and to a lesser extent Pd^{II} have a strong tendency to form σ-bonds, while Pd^{II} very readily forms π-allyl species (page 753).

(*e*) A highly characteristic feature is the formation of hydrido complexes and M—M bonds may be formed when the metal halides in higher oxidation states are reduced, especially in presence of tertiary phosphines or other ligands. Hydrogen abstraction from reaction media such as alcohols or dimethylformamide is common.

---

[6] R. N. Goldberg and L. G. Hepler, *Chem. Rev.*, 1968, **68**, 229 (an authoritative collection of thermodynamic data on compounds of the Pt metals and their oxidation–reduction potentials, and containing also much descriptive chemistry).

(f) For the d^8 ions, Rh^I, Ir^I, Pd^{II}, Pt^{II} the normal coordination is square (though 5-coordinate species are fairly common) and oxidative-addition reactions (page 772) are of great importance.

Finally, it should be noted that platinum metal chemistry is an exceedingly active area of research and that even omitting patents, which are exceedingly numerous, the numbers of research papers are in the many hundreds per year.

BINARY COMPOUNDS

26-E-4. Oxides, Sulfides, Phosphides, etc.

Oxides. The best known anhydrous oxides are listed in Table 26-E-2; the tetraoxides of Ru and Os are discussed later (page 1002). The oxides are generally rather inert to aqueous acids, are reduced to the metal by hydrogen and dissociate on heating. A number of mixed metal oxides, e.g., $BaRuO_3$, $CaIrO_3$, $Tl_2Pt_2O_7$, etc., are known.[7a]

TABLE 26-E-2

Anhydrous Oxides of Platinum Metals[a]

Oxide	Color/form	Structure	Comment
RuO_2[b]	Blue-black	Rutile	From O_2 on Ru at 1250° or $RuCl_3$ at 500–700°; usually O-defective
RuO_4	Orange-yellow cryst., m.p. 25°, b.p. 100°	Tetrahedral molecules	See page 1002
OsO_2	Coppery	Rutile	Heat Os in NO or OsO_4 or dry $OsO_2 \cdot nH_2O$
OsO_4	Colorless cryst. m.p. 40°, b.p. 101°	Tetrahedral molecules	Normal product of heating Os in air; see page 1002
Rh_2O_3	Brown	Corundum	Heat Rh^{III} nitrate or Rh_2O_3(aq)
RhO_2	Black	Rutile	Heat Rh_2O_3(aq) at 700–800° in high-pressure O_2
Ir_2O_3	Brown		Impure by heating $K_2IrCl_6 + Na_2CO_3$
IrO_2	Black	Rutile	Normal product of $Ir + O_2$; dissociates $>1100°$
PdO	Black		From $Pd + O_2$; dissociates 875°. Insol. all acids
PtO_2[c]	Brown		Dehydrate PtO_2(aq); decomp. 650°

[a] In oxygen at 800–1500°, gaseous oxides exist: RuO_3, OsO_3, RhO_2, IrO_3, PtO_2. A number of other solids of uncertain nature exist: RuO_3, OsO_3, Ru_2O_5, Rh_2O_5, Os_2O_3, Ru_2O_3.

[b] For magnetic properties see J. M. Fletcher *et al.*, *J. Chem. Soc.*, A, **1968**, 653.

[c] Two forms and Pt_3O_4 are known: H. R. Hoekstra, S. Siegel and S. Tani, *Inorg. Chem.*, 1968, **7**, 141; *J. Inorg. Nuclear Chem.*, 1969, **31**, 3803; O. Muller and R. Roy, *J. Less Common Metals*, 1968, **16**, 129 (also Rh).

7 (a) R. D. Shannon, D. B. Rogers and C. T. Prewitt, *Inorg. Chem.*, 1971, **10**, 713, 719, 723; (b) J. M. Fletcher *et al.*, *J. Chem. Soc.*, A, **1968**, 653.

Hydrous oxides are commonly precipitated when NaOH is added to aqueous metal solutions, but they are difficult to free from alkali ions and sometimes readily become colloidal. When freshly precipitated, they may be soluble in acids but only with great difficulty or not at all after ageing.

The black precipitate, probably $Ru_2O_3 \cdot nH_2O$, from Ru^{III} chloride solutions is readily oxidized by air, probably to the black $RuO_{2+x} \cdot yH_2O$ which is formed on reduction of RuO_4 or RuO_4^{2-} solutions by alcohol, hydrogen, etc.[7b] Reduction of OsO_4 or addition of OH^- to $OsCl_6^{2-}$ solutions gives $OsO_2 \cdot nH_2O$.

$Rh_2O_3 \cdot nH_2O$ is formed as a yellow precipitate from Rh^{III} solutions. In base solution, powerful oxidants convert it into $RhO_2 \cdot nH_2O$; the latter loses oxygen on dehydration. $Ir_2O_3 \cdot nH_2O$ can be obtained only in moist atmospheres; it is at least partially oxidized by air to $IrO_2 \cdot nH_2O$ which is formed either by action of mild oxidants on $Ir_2O_3 \cdot nH_2O$ or by addition of OH^- to $IrCl_6^{2-}$ in presence of H_2O_2. The precipitation of the oxides of Rh and Ir from buffered $NaHCO_3$ solutions by the action of ClO_2^- or BrO_3^- provides a rather selective separation of these elements.

$PdO \cdot nH_2O$ is a yellow gelatinous precipitate which dries in air to a brown, less hydrated form and at 100° loses more water, eventually becoming black; it cannot be dehydrated completely without loss of oxygen.

When $PtCl_6^{2-}$ is boiled with Na_2CO_3, red-brown $PtO_2 \cdot nH_2O$ is obtained. It dissolves in acids and also in strong alkalis to give what can be regarded as solutions of hexahydroxoplatinate, $[Pt(OH)_6]^{2-}$. The hydrous oxide becomes insoluble on heating to ca. 200°. The brown oxide formed by fusion of $NaNO_3$ and chloroplatinic acid at ca. 550° followed by extraction of soluble salts with water is known as Adams' catalyst and is widely used in organic chemistry for catalytic reductions.

A very unstable Pt^{II} hydrous oxide is obtained by addition of OH^- to $PtCl_4^{2-}$; after drying in CO_2 at 120–150° it approximates to $Pt(OH)_2$, but at higher temperatures gives PtO_2 and Pt.

Sulfides, Phosphides and Similar Compounds. Direct interaction of the metal and other elements such as S, Se, Te, P, As, Bi, Sn or Pb under selected conditions produces dark, often semimetallic, solids that are resistent to acids other than nitric. These products may be stoichiometric compounds and/or non-stoichiometric phases depending upon the conditions of preparation.

The chalcogenides[8] and phosphides are generally rather similar to those of other transition metals; indeed many of the phosphides for example are isostructural with those of the iron group, viz., Ru_2P with Co_2P; RuP with FeP and CoP; RhP_3, PdP_3 with CoP_3 and NiP_3.

Sulfides can also be obtained by passing H_2S into platinum-metal salt solutions. Thus from $PtCl_4^{2-}$ and $PtCl_6^{2-}$ are obtained PtS and PtS_2 respectively; from Pd^{II} solutions PdS, which when heated with S gives PdS_2; the

[8] A. Wold, in *Platinum Group Metals and Compounds*, Adv. Chem. Series, No. 98, *Amer. Chem. Soc.*, **1971**.

Rh^{III} and Ir^{III} sulfides are assumed to be $M_2S_3 \cdot nH_2O$ but exact compositions are uncertain.

26-E-5. Halides of the Platinum Metals[9]

We discuss here primarily the binary halides. All the platinum metals form halogeno complexes in one or more oxidation states and these, as well as the hydrated halides which are closely related to them, are discussed under the respective elements.

Fluorides. These are listed in Table 26-E-3. The most interesting are the hexafluorides, of which only that of Pd is yet unknown. While OsF_6 is the

TABLE 26-E-3

Fluorides of the Platinum Metals

II	III	IV	V	VI
—	RuF_3 brown	RuF_4 sandy-yellow	$[RuF_5]_4$ dark green[b] m.p. 86.5°; b.p. 227°	RuF_6 dark brown m.p. 54°
—	—	OsF_4 yellow	$[OsF_5]_4$ blue-grey[b] m.p. 70°; b.p. 225.9°	OsF_6[a] pale yellow[b] m.p. 33.2°; b.p. 47°
—	RhF_3 red	RhF_4 purple-red	$[RhF_5]_4$ dark red	RhF_6 black
—	IrF_3 black	—	$[IrF_5]_4$ yellow-green m.p. 104°	IrF_6 yellow m.p. 44.8°; b.p. 53.6°
PdF_2 violet	[c]	PdF_4 brick-red	—	—
—	—	PtF_4 yellow-brown	$[PtF_5]_4$ deep red m.p. 80°	PtF_6 dark red m.p. 61.3°; b.p. 69.1°

[a] OsF_7, made under drastic conditions (500°, 400 atm, F_2), dissociates above $-100°$ and is stable only under high-pressure F_2 (O. Glemser *et al.*, *Chem. Ber.*, 1966, **99**, 2652); there is some evidence for OsF_8.
[b] Colorless vapor.
[c] PdF_3 is $Pd^{2+}PdF_6^{2-}$.

normal product of fluorination of Os at 300° and is the most stable compound of the group, the other hexafluorides are the initial products of the fluorinations but, owing to their thermal instability, must be chilled from the gas phase for collection. Platinum wire ignited in fluorine by an electron current continues to react exothermally to give red vapors of PtF_6.

The hexafluorides decrease in stability in the order W > Re > Os > Ir > Pt, and Ru > Rh, dissociating into fluorine and lower fluorides. PtF_6 is one of

[9] J. H. Canterford and R. Colton, *Halides of the Second and Third Row Transition Series*, Interscience-Wiley, 1968 (an exhaustive reference on halides and halide complexes); G. Thiele and K. Broderson, *Fortschr. Chem. Forsch.*, 1968, **10**, 631 (structural chemistry of platinum halides).

the most powerful oxidizing agents known; it reacts with oxygen and xenon to give $O_2^+PtF_6^-$ and $Xe(PtF_6)_n$ (page 414 and page 498). The volatility of the compounds also decreases with increasing mass.

All the hexafluorides are extraordinarily reactive and corrosive substances and normally must be handled in Ni or Monel apparatus although quartz can be used if necessary. Only PtF_6 and RhF_6 actually react with glass (even when rigorously dry) at room temperature. In addition to thermal dissociation, ultraviolet radiation causes decomposition to lower fluorides, even OsF_6 giving OsF_5. The vapors hydrolyze with water vapor, and liquid water reacts violently, e.g., IrF_6 gives HF, O_2, O_3 and $IrO_2(aq)$, while OsF_6 gives OsO_4, HF and OsF_6^-. The hexafluorides are octahedral and their magnetic and spectral properties have been studied in detail.

The *pentafluorides* are obtained by thermal dissociation of MF_6 or by controlled fluorination of the metal. Thus the usual product of fluorination of Ru at 300° is RuF_5 and of $PtCl_2$ at 350° is PtF_5. OsF_5 is usually obtained by uv dissociation of OsF_6 or by reduction with I_2 in IF_5. Fluorination of Rh at 400° gives RhF_5 and of Ir at $\sim360°$ gives IrF_5.

The pentafluorides, which are isomorphous, are also very reactive, hydrolyzable substances. The most unusual feature is the polymerization to give tetramers[10] with non-linear M—F—M bridges similar to the pentafluorides Nb, Ta and Mo (page 936) but somewhat distorted. Some of the color changes on heating, e.g., of green $[OsF_5]_4 \rightarrow$ green liquid $\rightarrow$ blue liquid $\rightarrow$ colorless vapor, are probably due to depolymerization.

The *tetrafluorides* are obtained by reactions such as

$$10RuF_5 + I_2 \rightarrow 10RuF_4 + 2IF_5$$
$$OsF_6 \xrightarrow{\text{W(CO)}_6} OsF_4$$
$$RhCl_3 \xrightarrow{\text{BrF3(1)}} RhF_4 \cdot 2BrF_3 \xrightarrow{\text{Heat}} RhF_4$$
$$Pt \xrightarrow{\text{BrF3(1)}} PtF_4 \cdot 2BrF_3 \xrightarrow{\text{Heat}} PtF_4 \xleftarrow{\text{F}_2} PtBr_4$$
$$PdBr_2 \xrightarrow{\text{BrF3(1)}} Pd^{II}Pd^{IV}F_6 \xrightarrow{\text{F}_2 \cdot 100\text{lb/sq in., } 150°} PdF_4$$

The formation of BrF_3 adducts as above is a fairly common feature in the preparation of heavy-metal fluorides; such adducts may be ionic, i.e., $[BrF_2^+]_2 \cdot MF_6^{2-}$, but it is more likely that they are fluoride-bridged species of the type (26-E-I) or singly bridged polymers.

(26-E-I)

The tetrafluorides are violently hydrolyzed by water. In PdF_4 and PtF_4 the metal is 8-coordinate at the center of two flattened tetrahedra (cf. UCl_4).

Trifluorides. RuF_3 is best obtained by reduction:

$$5RuF_5 + I_2 \xrightarrow{250°} 2IF_5 + 5RuF_3$$

Direct fluorination of Rh or $RhCl_3$ at 500–600° gives RhF_3; the solid is

[10] S. J. Mitchell and J. H. Holloway, *J. Chem. Soc., A,* **1971,** 2789.

unaffected by water or bases. IrF_3, which can be obtained only indirectly by reduction of IrF_6, e.g., by Ir at 50°, is also relatively inert to water. The Rh and Ir trifluorides have a slightly distorted ReO_3 structure.

Difluorides. Palladous fluoride can be obtained by the reaction

$$Pd^{II}Pd^{IV}F_6 + SeF_4 \xrightarrow{\text{Reflux}} 2PdF_2 + SeF_6$$

It is the only simple compound of Pd^{II} that is paramagnetic and the moment is consistent with the observed octahedral coordination. The Pd^{2+} ion also occurs in $Pd^{II}Pd^{IV}F_6$, $Pd^{II}Sn^{IV}F_6$ and $Pd^{II}Ge^{IV}F_6$ which can be obtained by addition of BrF_3 to mixtures of $PdBr_2$ and, e.g., $SnBr_4$.

Chlorides, Bromides and Iodides. The anhydrous halides (other than fluorides—see Table 26-E-3) are listed in Table 26-E-4; they are normally obtained by direct interaction under selected conditions; the higher halides form the lower halides on heating.

TABLE 26-E-4

Anhydrous Chlorides, Bromides and Iodides of Platinum Metals[a]

Oxidation state	Ru	Os	Rh	Ir	Pd	Pt
II	—	—	—	—	$PdCl_2$[b] red	$PtCl_2$[b] black-red
	—	—	—	—	$PdBr_2$ red-black	$PtBr_2$ brown
	—	[e]	—	—	PdI_2 black	PtI_2 black
III	$RuCl_3$[c]	$OsCl_3$ dark-grey	$RhCl_3$ red	$IrCl_3$ brown-red	—	$PtCl_3$ green-black
	$RuBr_3$ dark brown	$OsBr_3$ black	$RhBr_3$ dark red	$IrBr_3$ yellow	—	$PtBr_3$ green-black
	RuI_3 black	OsI_3 black	RhI_3(?) black	IrI_3 black	—	PtI_3(?) black
IV	[d]	$OsCl_4$ black	—	$IrCl_4$(?)	—	$PtCl_4$ red-brown
	—		—		—	
	—	$OsBr_4$ black	—	$IrBr_4$(?)	—	$PtBr_4$ dark red
	—	—	—	IrI_4(?)	—	PtI_4 brown-black

[a] There is some evidence for grey metallic OsI, and lower halides of Rh and Ir.
[b] Two or more polymorphs; $PtCl_2$ yellowish-brown when powdered.
[c] Two forms, α-$RuCl_3$ black, β-$RuCl_3$ brown.
[d] Some evidence for existence of $RuCl_4$ in vapor.
[e] Alleged lower iodides are mixtures of OsI_3 and oxides (H. Schäfer *et al.*, *Z. anorg. Chem.*, 1971, **383**, 49).

With the exception of those of Pd and Pt the halides are generally insoluble in water, rather inert and of little utility for the preparation of complex compounds. Hydrated halides, discussed in a later Section, are normally used for this purpose. We discuss only some of the more important chlorides.

Ruthenium trichloride has two forms. Interaction of the metal with $Cl_2 + CO$ at 370° gives the β-form, which is converted into black leaflets of the α-form at 450° in Cl_2. The latter has a layer lattice and is antiferromagnetic. The so-called *iodide*, which is precipitated from aqueous ruthenium chloride solutions by KI, invariably contains strong OH bands in its infrared spectrum and probably has OH bridges in the lattice.

Osmium tetrachloride is formed when an excess of Cl_2 is used at temperatures above 650°; otherwise a mixture with the *trichloride* is formed. The latter is obtained when $OsCl_4$ is decomposed at 470° in a flow system with a low pressure of chlorine.

Rhodium trichloride has a layer lattice isostructural with $AlCl_3$ and is exceedingly inert. However, when $RhCl_3 \cdot 3H_2O$ (see below) is dehydrated in dry HCl at 180° the red product is much more reactive and dissolves in water or tetrahydrofuran; this property is lost on heating at 300°.

Palladium dichloride and platinum dichloride both exist in two forms, which may be obtained as annexed.

$$Pd \xrightarrow{Cl_2} \begin{cases} \xrightarrow{>550°} \alpha\text{-}PdCl_2 \xrightarrow{\text{slow}} \\ \xrightarrow{<550°} \beta\text{-}PdCl_2 \end{cases}$$

$$H_2PtCl_6 \cdot 6H_2O \xrightarrow{Cl_2 \sim 500°} PtCl_4 \xrightarrow{>350°} \beta\text{-}PtCl_2 \xrightarrow[\text{1-2 days}]{500°} \alpha\text{-}PtCl_2$$

$$Pt \xrightarrow[\text{gradient}]{Cl_2, 650° \rightarrow 500°} \alpha\text{-}PtCl_2$$

Unlike the nickel halides or PdF_2 which are ionic and paramagnetic, these chlorides are molecular or polymeric and diamagnetic and Born–Haber calculations indicate that ionic lattices would be endothermic.

The β-forms are isomorphous and have the molecular structure (26-E-II) with M_6Cl_{12} units; the structures appear to be stabilized mainly by the halogen bridges rather than by metal—metal bonds.[11] The molecular behavior is shown by the fact that Pt_6Cl_{12} is soluble in benzene.[12]

Although the structure of β-$PtCl_2$ is uncertain, it differs from that of α-$PdCl_2$ which has the flat chain (26-E-III).

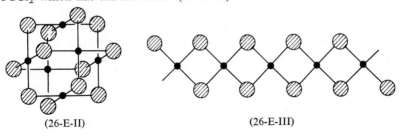

(26-E-II)   (26-E-III)

[11] R. Mattes, *Z. anorg. Chem.*, 1969, **364**, 290.
[12] U. Wiese *et al.*, *Angew. Chem. Internat Edn.*, 1970, **9**, 158.

Palladous chloride is soluble in hydrochloric acid, forming the ion $[PdCl_4]^{2-}$ and will also react with many ligands, L, such as amines, benzonitrile and phosphines to give complexes of the types L_2PdCl_2 and $[LPdCl_2]_2$; $PtCl_2$ is similar.

The greenish-black *trichloride*, $PtCl_3$, contains both Pt^{II} and Pt^{IV} with units of $[Pt_6Cl_{12}]$ and an infinite chain, $(1/\infty)$ $[PtCl_2{}^aCl_{4/2}{}^b]$ containing distorted $PtCl_6$ octahedra linked by common edges, similar to the chain of PtI_4. The chloride and the similar $PtBr_3$ and PtI_3 are made by thermal gradient reactions of Pt and halogen.[12]

Platinic chloride is commonly made by heating chloroplatinic acid to 300° in Cl_2. The reddish-brown crystals are readily soluble in water, presumably giving ions such as $[PtCl_4(OH)_2]^{2-}$, and in polar solvents. The structure is not yet known but PtI_4 has PtI_6 octahedra linked by iodide bridges with two *cis*-non-bridging I atoms.[13]

Oxohalides and Halogeno Complexes. A number of fluoro and other oxohalides of the platinum metals have been characterized: $RuOF_4$, OsO_3F_2, $OsOF_5$, $PtO_xF_{3-x}(?)$, $OsOCl_4$ and Os_2OCl_6. These, and fluoro complex anions, are of small importance or utility, but some of the chloro complex anions will be discussed in later Sections.

26-F. RUTHENIUM AND OSMIUM[1]

26-F-1. General Remarks: Stereochemistry

The chemistry of ruthenium and osmium bears little resemblance to that of iron except in compounds such as sulfides or phosphides, and in complexes with ligands such as CO, PR_3, or h^5-C_5H_5. The higher oxidation states, VI and VIII, are much more readily obtained than for iron and there is an extensive and important chemistry of the tetraoxides, MO_4, oxohalides and oxo anions. There are analogies between the chemistries of Ru, Os and Re especially in oxo, nitrogen and nitrido compounds.

For ruthenium, the principal lower oxidation states are 0, II and III, whereas for osmium they are 0, III and IV. For neither element is there good evidence for the I oxidation state (other than formally in compounds with metal—metal bonds) although some brown air-sensitive solutions obtained by reduction of $RuCl_3$ in dimethylacetamide by hydrogen may contain a Ru^I species.[2] The oxidation states and stereochemistries are summarized in Table 26-F-1.

The 0 state, d^8. The chemistry in this state is primarily one of the metal carbonyls; mononuclear and polynuclear carbonyls are known for both

[13] K. Broderson, G. Thiele and B. Holle, *Z. anorg. Chem.*, 1969, **369**, 154.
[1] W. P. Griffith, *The Chemistry of the Rarer Platinum Metals; Os, Ru, Ir and Rh*, Interscience-Wiley, 1969; B. R. James, *Inorg. Chem. Acta Rev.*, 1970, **4**, 73 (homogeneous catalysis by ruthenium complexes).
[2] B. Hui and B. R. James, *Chem. Comm.*, **1969**, 198.

TABLE 26-F-1

Oxidation States and Stereochemistry of Ruthenium and Osmium

Oxidation state	Coordination number	Geometry	Examples
Ru^{-II}	4	Tetrahedral(?)	$Ru(CO)_4^{2-}(?)$, $[Ru(diphos)_2]^{2-}$
Ru^0, Os^0	5	tbp?	$Ru(CO)_5$, $Os(CO)_5$, $Ru(CO)_3(PPh_3)_2$
Ru^I, d^7	6^d		$[h^5\text{-}C_5H_5Ru(CO)_2]_2$, $[Os(CO)_4X]_2$
Ru^{II}, Os^{II}, d^6	5	See text	$RuCl_2(PPh_3)_3$
	5	tbp	$RuHCl(PPh_3)_3$
	6^a	Octahedral	$[RuNOCl_5]^{2-}$, $[Ru(bipy)_3]^{2+}$, $[Ru(NH_3)_6]^{2+}$, $[Os(CN)_6]^{4-}$, $RuCl_2CO(PEtPh_2)_3$, $OsHCl(diphos)_2$
Ru^{III}, Os^{III}, d^5	$6^{a,b}$	Octahedral	$[Ru(NH_3)_5Cl]^{2+}$, $[RuCl_5H_2O]^{2-}$ $[Os(dipy)_3]^{3+}$, K_3RuF_6, $[OsCl_6]^{3-}$
Ru^{IV}, Os^{IV}, d^4	$6^{a,b}$	Octahedral	K_2OsCl_6, K_2RuCl_6, $[Os(diars)_2X_2]^{2+}$, RuO_2^c
	7	?	$OsH_4(PMePh)_3$
Ru^V, Os^V, d^3	5 in vapor(?)		RuF_5
	6	Octahedral	$KRuF_6$, $NaOsF_6$, $(RuF_5)_4$
Ru^{VI}, Os^{VI}, d^2	4	Tetrahedral	RuO_4^{2-}
	5	?	$OsOCl_4$
	6^b	Octahedral	RuF_6, OsF_6, $[OsO_2Cl_4]^{2-}$, $[OsO_2(OH)_4]^{2-}$, $[OsNCl_5]^{2-}$
Ru^{VII}, Os^{VII}, d^1	4	Tetrahedral	RuO_4^-
	6	Octahedral	$OsOF_5$
Ru^{VIII}, Os^{VIII}, d^0	4	Tetrahedral	RuO_4, OsO_4, $[OsO_3N]^-$
	5	?	OsO_3F_2
	6	Octahedral	$[OsO_3F_3]^-$, $[OsO_4(OH)_2]^{2-}$

[a] Most common states for Ru.

[b] Most common states for Os.

[c] Metal—metal bond present.

[d] If $h^5\text{-}C_5H_5$ assumed to occupy three coordination sites.

elements. Both types undergo substitution reactions and in the polynuclear species the clusters are often retained. They also undergo protonation reactions, and a variety of hydrido species are known. We do not deal explicitly with these compounds but some of the chemistry has been noted elsewhere (Chapters 22, 23 and 24) and certain aspects are described later in this Section.

The II state, d^6. An enormous number of Ru and Os complexes with CO, PR_3 and similar π-acid ligands are known. For other ligands, the main chemistry is that of chloro, ammonia and other amine ligands, and again large numbers of complexes exist. The aquo ion, $[Ru(H_2O)_6]^{2+}$ has been prepared, but it is readily oxidized to $[Ru(H_2O)_6]^{3+}$. For osmium the best characterized complexes are those with aromatic amines.

All Ru^{II} and Os^{II} complexes are octahedral and diamagnetic as expected for the t_{2g}^6 configuration. Although these compounds are fairly labile, the reactions often proceed with retention of configuration, suggesting an associative mechanism.

The III state, d^5. There is an extensive chemistry with both π-acid and σ-donor ligands. Ruthenium(III) species are more common than those of

Os^{III}. All the complexes are of low-spin type with one unpaired electron and are octahedral.

It is to be noted that there are a number of complexes of Ru that have fractional oxidation numbers, e.g., 2.5 as in $[Ru_2(OOCCH_3)_4]^+$ and more commonly 3.5 as in the "ruthenium reds" (see below).

The IV state, d^4. In this state most complexes are neutral or anionic, although a few cationic species such as $[Os(diars)_2X_2]^+$ are known; but compared to the II and III states relatively few complexes have been prepared. However, it is the main state for Os, where the ion $OsCl_6^{2-}$ is very stable.

Again, compounds in mixed oxidation states are common, and in perchlorate or other non-complexing media polynuclear complexes are likely to exist.

Ru^{IV} and Os^{IV} complexes all have octahedral or distorted octahedral structures and should thus have t_{2g}^4 electron configurations. This configuration is especially subject to anomalous magnetic behavior when the spin–orbit coupling constant of the metal ion becomes high as it is in Os^{IV} (page 925). The chief effect in this case is that the effective magnetic moment is brought far below the spin-only value (2.84 B.M.), typical values for Os^{IV} complexes at room temperature being in the range 1.2–1.7 B.M. As the temperature is lowered, μ_{eff} decreases as the square root of the absolute temperature. Ru^{IV} complexes have practically normal moments at room temperature (2.7–2.9 B.M.), but these also decrease with $T^{1/2}$ as the temperature is lowered. Little is known of the *d–d* transitions in Ru^{IV} and Os^{IV} complexes since the relevant absorption bands are severely masked by strong charge transfer bands

Ruthenium and *osmium* are readily recovered and separated by utilizing the high volatility of their tetraoxides, which can be distilled from aqueous solution. Nitric acid is sufficient to oxidize osmium compounds, but for ruthenium more powerful oxidants are required. Since OsO_4 is the commercial source and since the usual starting material for the preparation of ruthenium compounds is "$RuCl_3 \cdot 3H_2O$" obtained by reducing RuO_4 with concentrated HCl, we discuss the high oxidation states first.

26-F-2. Oxo compounds of Ruthenium and Osmium (VI), (VII) and (VIII)

Ruthenium and osmium tetraoxides and oxo anions provide some of the more unusual and useful features of the chemistry. The major compounds or ions are shown in Table 26-F-2.

The Tetraoxides, RuO_4 and OsO_4. These volatile, crystalline solids (Table 26-E-2, page 994) are both toxic substances with characteristic, penetrating, ozone-like odors. OsO_4 is a particular hazard to the eyes on account of its ready reduction by organic matter to a black oxide, a fact utilized in its employment in dilute aqueous solution as a biological stain.

Ruthenium tetraoxide is obtained when acid ruthenium solutions are treated with oxidizing agents such as MnO_4^-, $AuCl_4^-$, BrO_3^- or Cl_2 ; the oxide can be distilled from the solutions or swept out by a gas stream. It may also be obtained by distillation from concentrated perchloric acid solutions or by passing Cl_2 into a melt of ruthenate in NaOH.

TABLE 26-F-2

Some Oxo Compounds and Ions of Ru and Os

VIII	VII	VI
RuO_4	RuO_4^-	RuO_4^{2-}
		$RuO_2Cl_4^{2-}$
OsO_4		
OsO_3N^-		$[OsO_2X_4]^{2-}$ [a]
$OsO_4X_2^{2-}$ [a]	$OsOF_5$	$[OsO_2(OH)_2X_2]^{2-}$ [b]

[a] X = F or OH.
[b] X = Cl, CN or 0.5ox, etc.

Osmium tetraoxide can be obtained by burning osmium or by oxidation of osmium solutions with nitric acid, peroxodisulfate in sulfuric acid or similar agents.

Both compounds have a tetrahedral structure.[3] They are extremely soluble in CCl_4 and can be extracted from aqueous solutions by it. RuO_4 is quite soluble in dilute sulfuric acid giving golden-yellow solutions; OsO_4 is sparingly soluble. The tetraoxides are powerful oxidizing agents. Above $\sim 180°$, RuO_4 can explode, giving RuO_2 and O_2, and it is decomposed slowly by light; OsO_4 is more stable in both respects. OsO_4 finds specialized use in organic chemistry[4] since it can add to olefinic double bonds to give a cis-ester that can be reduced to the cis-dihydroxo compound by Na_2SO_3 :

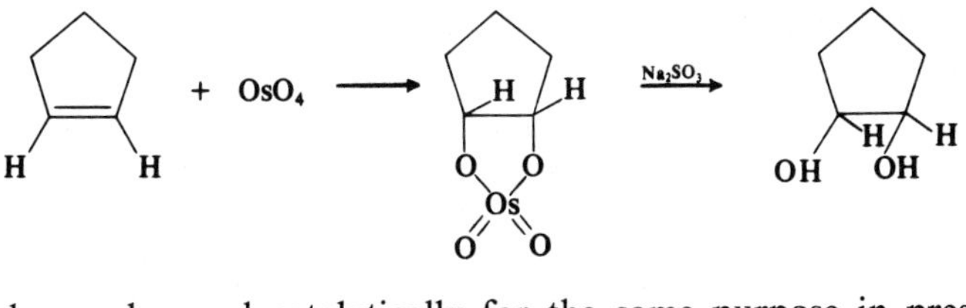

The oxide can be used catalytically for the same purpose in presence of H_2O_2 or ClO_3^-. Ruthenium tetraoxide reacts more vigorously with organic substances but also has some uses as an oxidant;[5] a convenient catalytic method uses $RuCl_3$ in sodium hypochlorite solutions.[6]

Both RuO_4 and OsO_4 are soluble in alkali hydroxide solutions, but the

[3] R. S. McDowell and M. Goldblatt, Inorg. Chem., 1971, 10, 625.
[4] L. F. Fieser and M. Fieser, Reagents for Organic Synthesis, Vols. 1 and 2, Wiley, 1967, 1969; P. N. Rylander, Engelhard Tech. Bull., 1969, 9, 90.
[5] P. N. Rylander, Engelhard Tech. Bull., 1969, 9, 135.
[6] S. Wolfe, S. K. Hasan and J. R. Campbell, Chem. Comm., 1970, 1420.

behavior is quite different. RuO_4 is reduced by hydroxide first to per-ruthenate(VII), which in turn is further reduced to ruthenate(VI):

$$4RuO_4 + 4OH^- \rightarrow 4RuO_4^- + 2H_2O + O_2$$
$$4RuO_4^- + 4OH^- \rightarrow 4RuO_4^{2-} + 2H_2O + O_2$$

On the other hand, OsO_4 gives the ion $[OsO_4(OH)_2]^{2-}$, discussed below. This difference between Ru and Os appears to be due to the ability of the $5d$ metal oxo anion to increase the coordination shell. Similar behavior occurs for ReO_4^-, which in concentrated alkali gives yellow *meso*-perrhenate:

$$ReO_4^- + 2OH^- = ReO_4(OH)_2^{3-} = ReO_5^{3-} + H_2O$$

Ruthenates(VI) and -(VII). There is a close similarity between Ru and Mn in the oxo anions, both MO_4^- and MO_4^{2-} being known.

The fusion of Ru or its compounds with alkali in presence of an oxidizing agents gives a green melt containing the *perruthenate* ion, RuO_4^-. Because of the high alkali concentration, on dissolution in water a deep orange solution of the stable ruthenate(VI) ion, RuO_4^{2-}, is obtained. However, if RuO_4 is collected in ice-cold $1M$ KOH, black crystals of $KRuO_4$ can be obtained, which are stable when dry. Perruthenate solutions, which are a yellowish–green, are reduced by hydroxyl ion, and kinetic studies suggest that unstable intermediates with coordinated OH^- are involved—this contrasts with the case of $3d$ metal oxo anions where there is no evidence for addition of OH^-. Since H_2O_2 is also formed in the reduction and RuO_4^- is incompletely reduced to RuO_4^{2-} by H_2O_2, a step such as

$$[RuO_4(OH)_2]^{2-} \rightarrow RuO_4^{2-} + H_2O_2$$

is plausible.

The tetrahedral RuO_4^{2-} ion is moderately stable in alkaline solution. It is paramagnetic with two unpaired electrons in contrast to osmate(VI). It may be noted that most ruthenium species in alkaline solution are specifically oxidized to RuO_4^{2-} by $KMnO_4$; hypochlorite gives a mixture of the RuO_4^- and RuO_4^{2-} ions, while Br_2 gives RuO_4^-. The RuO_4^- ion can be conveniently reduced to RuO_4^{2-} by iodide ion, although further reduction can occur with an excess of I^-.

Osmates. OsO_4 is moderately soluble in water and its absorption spectrum in the solution is the same as in hexane, indicating that it is still tetrahedral. However, in strong alkaline solution coordination of OH^- ion occurs and a deep red solution is formed:

$$OsO_4 + 2OH^- \rightarrow [OsO_4(OH)_2]^{2-}$$

from which red salts such as $K_2[OsO_4(OH)_2]$ can be isolated. These "perosmates" or "osmenates" have *trans*-hydroxo groups. The reduction of such perosmate solutions by alcohol or other agents gives the osmate(VI) ion, which is pink in aqueous solutions but blue in methanol; its salts are also obtained in the alkaline oxidative fusion of the metal; in solution and in salts the ion is octahedral, $[OsO_2(OH)_4]^{2-}$. Unlike the corresponding RuO_4^{2-} ion, it is diamagnetic. The diamagnetism of the ion, its substituted derivatives such as $[OsO_2Cl_4]^{2-}$, and $[RuO_2Cl_4]^{2-}$, all of which have *trans*-dioxo groups, can be explained in terms of ligand-field theory. If the z axis passes through the two oxide ligands and the x and y axes through OH, there will be a tetragonal splitting of the e_g level into two singlets, $d_{x^2-y^2}$ and d_{z^2} whereas the t_{2g} level gives a singlet, d_{xy} and a doublet, $d_{xz} d_{yz}$. The oxide ligands will form Os$=$O bonds by π overlap mainly with d_{xz} and d_{yz} and will thus destabilize those orbitals, leaving a low-lying d_{xy} orbital which will be occupied by the two electrons, leading to diamagnetism.

Other Oxo Species. When RuO_4 is treated with gaseous HCl and Cl_2, hygroscopic crystals of $(H_3O)_2[RuO_2Cl_4]$ are produced from which Rb and Cs salts can be obtained. The ion is hydrolyzed by water:

$$2Cs_2RuO_2Cl_4 + 2H_2O \rightarrow RuO_4 + RuO_2 + 4CsCl + 4HCl$$

but there is evidence for other Ru^{VI} species in solution. If RuO_4 in dilute H_2SO_4 is reduced with Na_2SO_3 or $FeSO_4$, green solutions are obtained. These contain Ru^{VI}; although the precise nature of the species is not known, it is probably $[RuO_2(SO_4)_2]^{2-}$. The green ion can be formed by mixing freshly prepared Ru^{IV} solutions which are an intense brown with RuO_4 in dilute H_2SO_4:

$$Ru^{IV} + Ru^{VIII} \rightarrow 2Ru^{VI}$$

The green solutions decompose within a few hours to Ru^{IV} and like all ruthenium species in the VI, VII and VIII oxidation states, are reduced to Ru^{III} by an excess of iodide ion.

The osmate ion $[OsO_2(OH)_4]^{2-}$, can undergo substitution reactions with various ions such as Cl^-, Br^-, CN^-, $C_2O_4^{2-}$ and NO_2^-, to give orange or red crystalline salts, sometimes referred to as osmyl derivatives. They can also be obtained directly from OsO_4 with which, for example, aqueous KCN gives the salt $K_2[OsO_2(CN)_4]$. This particular ion is unaffected by hydrochloric or sulfuric acid, but the other oxo anions are not very stable in aqueous solutions, although they are considerably more stable than the ruthenyl salts mentioned above.

There are also derivatives of the type $[OsO_2(OH)_2X_2]^{2-}$, and the action of alkali fluorides on OsO_4 at low temperature gives red or brown salts of the ion $[OsO_4F_2]^{2-}$ which are soluble, but unstable, in water.

Fluorination of OsO_2 at 250° gives green $OsOF_5$ (m.p., 59°).[7] Oxo nitrido compounds are described below.

26-F-3. Ruthenium and Osmium Chloro Complexes

In view of the importance of chloride complexes in aqueous solution, we discuss those in different oxidation states together.

Ruthenium Chloro Complexes. Fig. 26-F-1 gives some common reactions of these chloro species.

Ruthenium(IV) Chloro Complexes. The reduction of RuO_4 by HCl in presence of KCl gives red crystals of $K_4[Ru_2OCl_{10}]$, which was the first example of the ions $M_2OX_{10}^{4-}$ (X = Cl or Br, and M = Ru or Os). This complex has the structure shown in Fig. 26-F-2. The diamagnetism of the

[7] N. Bartlett, N. K. Jha and J. Trotter, *J. Chem. Soc., A*, **1968**, 536, 543.

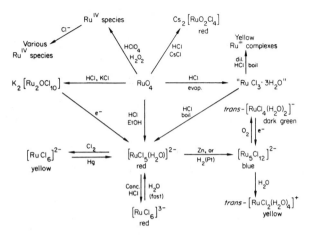

Fig. 26-F-1. Some reactions of ruthenium chloro complexes.

ion can be understood by using a simple MO treatment for the Ru—O—Ru group. If we assume that these atoms lie along the z axis of a coordinate system, and further that the ligand field around each Ru^{IV} ion is essentially octahedral, the ruthenium(IV) ions will then each have a $d_{xy}^2 d_{xz} d_{yz}$ configuration before reaction with the oxygen. By the action of the d_{xz} orbitals on each Ru^{IV} and the p_x orbital of oxygen, three three-center MO's, one bonding, one approximately non-bonding and one antibonding, will be formed. The four electrons (one from each Ru^{IV} and two from oxygen) will occupy the lower two of these MO's. The same kind of interaction will occur in the Rud_{yz}—Op_y—Rud_{yz} set of orbitals, and thus all electron spins become paired.

The reduction of $[Ru_2OCl_{10}]^{2-}$ usually proceeds to Ru^{III} species, so that the simple *hexachlororuthenate*(IV) ion or its salts are best prepared by the oxidation of Ru^{III} chloro species by chlorine. Although the ion is yellow in solution the salts are dark brown or purple; they are isomorphous with other octahedral Os, Ir, Pd and Pt salts.

In aqueous solution, $[RuCl_6]^{2-}$ is rather unstable. Perchlorate solutions containing Ru^{IV}, which may be obtained by reducing RuO_4 in $HClO_4$ by H_2O_2, contain a variety of poorly characterized species,[8] probably in part polynuclear and possibly with mixed oxidation states; addition of Cl^- to such solutions produces a series of color changes whose sequence and persistence times depend upon pH and the Cl^- concentration. Polynuclear species, and chloro, aquo and hydroxo species, are doubtless involved.

Ru—O, 1.8 Å; Ru—Cl, 2.34 Å
∠ RuORu, 180° ; ∠ ClRuO, 90°

Fig. 26-F-2. The structure of the $[Ru_2Cl_{10}O]^{4-}$ ion.

[8] R. M. Wallace and R. C. Propst, *J. Amer. Chem. Soc.*, 1969, **91**, 3779.

Ruthenium(III) Chloro Complexes. These are among the best characterized complexes.

When RuO_4 is collected in concentrated HCl and the solution is evaporated, a dark red, deliquescent, crystalline material is obtained. This commercial product, usually called "$RuCl_3 \cdot 3H_2O$," is the starting point for the preparation of most ruthenium compounds. Whilst containing Ru^{III} species, such as $RuCl_3 \cdot (H_2O)_3$, it appears also to contain some polynuclear Ru^{IV} complexes. It is readily soluble in water, ethanol, acetone and similar solvents, so that reactions with organic-soluble compounds such as phosphines and alkenes can be readily induced. Although the deep red aqueous solutions initially do not precipitate AgCl on treatment with Ag^+, they darken rapidly owing to hydrolysis. When dilute HCl solutions are heated or are shaken with Hg (as a reductant for Ru^{IV}) the resulting yellow solutions contain not only the aquo ion $[Ru(H_2O)_6]^{3+}$ but chloro species such as *cis*-$RuCl_3(H_2O)_3$, *mer*-$RuCl_3(H_2O)_3$, $[RuCl(H_2O)_5]^{2+}$, and *cis*- and *trans*-$[RuCl_2(H_2O)_4]^+$. These can be separated by ion-exchange and identified by their electronic spectra. Only with very high concentrations of Cl^- is $[RuCl_6]^{3-}$ formed. The rate of replacement of Cl^- by H_2O increases with the number of Cl^- ions present, so that while the aquation of $[RuCl_6]^{3-}$ to $[RuCl_5H_2O]^{2-}$ is of the order of seconds in water, the half-reaction time for conversion of $[RuCl(H_2O)_5]^{2+}$ into $[Ru(H_2O)_6]^{3+}$ is about a year. The yellow *trans*-$[RuCl_2(H_2O)_4]^+$ and green *trans*-$[RuCl_4(H_2O)_2]^-$ are best obtained by oxidation of $[Ru_5Cl_{12}]^{2-}$ (see below).

Ruthenium(III) chloro species catalyze the reduction of Fe^{3+} by H_2O and also the hydration of acetylenes. With CO or formic acid, Ru^{III} chloro species are formed, as discussed below (cf. also Figure 26-F-3).

Ruthenium(II) Chloro Complexes. It has long been known that deep inky-blue solutions are obtained when solutions of Ru^{III} chloro complexes in HCl solution are reduced electrolytically or chemically, e.g., by Ti^{3+}, or by H_2 (2 atm) in presence of platinum black. Blue solutions are also obtained on treating the acetate, $Ru_2(CO_2CH_3)_4Cl$, (oxidation state of Ru, +2.5), and the hexammine,[9] $[Ru^{II}(NH_3)_6]Cl_2$, with HCl. The constitution of the various blue species is not settled but they may all contain ruthenium +2.5. Some[10] are cationic, of the type $Ru_2Cl_{3+n}^{(2-n)+}$, or anionic, depending on the chloride concentration; a salt of ion $Ru_5Cl_{12}^{2-}$ is known. The blue salt from the ammine appears to be $[Ru_2(NH_3)_6Cl_4(H_2O)]Cl$.

The chloride solutions are air-sensitive and form *trans*-$[RuCl_4(H_2O)_2]^-$; in absence of air, oxidation by water gives hydrogen and yellow *trans*-$[RuCl_2(H_2O)_4]^+$. The blue solutions prepared by reduction of $RuCl_3 \cdot 3H_2O$ in methanol provide a useful source for the preparation of other Ru^{II} and Ru^{III} complexes.[11] Electrolytic reduction of solutions of $RuCl_3 \cdot 3H_2O$,

[9] F. Bottomley and S. B. Tong, *Can. J. Chem.*, 1971, **49**, 3739.
[10] E. E. Mercer and P. E. Dumas, *Inorg. Chem.*, 1971, **10**, 2755.
[11] J. D. Gilbert, D. Rose and G. Wilkinson, *J. Chem. Soc., A*, **1970**, 2765.

from which much Cl^- has been removed by addition of $AgBF_4$, followed by ion-exchange separation gives solutions containing pink $[Ru(H_2O)_6]^{2+}$; this is oxidized by air or ClO_4^-:[12]

$$[Ru(H_2O)_6]^{3+} + e = [Ru(H_2O)_6]^{2+}   E^0 = 0.23 \text{ V}$$

Osmium Chloro Complexes. The chemistry of Os–Cl complexes is similar to that of Ru but is less extensive.

The reduction of OsO_4 by HCl in presence of KCl leads to $K_2[OsO_2Cl_4]$ and $K_4[Os_2OCl_{10}]$, which are similar to the Ru species. If alcohol or Fe^{2+} is used as reductant in HCl, the yellow orange ion $[OsCl_6]^{2-}$ is formed and can be isolated as various salts whose colors, orange to brown, depend on the nature of the cation. The salts are isomorphous with $PtCl_6^{2-}$ salts. The ion can be reduced to $OsCl_6^{3-}$ but, unlike the Ru analog, there is no evidence for further reduction to Os^{II} species. The Os^{III} complex is rather unstable and in solution is hydrolyzed to the hydrated oxide.

26-F-4. The Chemistry of Nitrogen Donor Ligands

It is again most convenient to consider such compounds together.

Ruthenium(II) Ammines.[13]   There is an exceedingly wide range of ruthenium ammine complexes and they are of special interest because of the ready formation of nitrogen (N_2) complexes.

The most common starting materials are $[Ru(NH_3)_5N_2]Cl_2$ and $[Ru(NH_3)_6]Cl_2$. The former, which was the first N_2 complex to be prepared,[14] is readily produced by action of hydrazine on aqueous ruthenium chloride solutions; there is an osmium analog. The hexammine,[15] made as orange crystals by the Zn dust reduction of strongly ammoniacal, NH_4Cl^+ containing solutions of ruthenium halides may be contaminated with the nitrogen complex, but this can be detected by its infrared band at ca. 2100 cm^{-1}. The hexamine is strongly reducing,

$$[Ru(NH_3)_6]^{3+} + e = [Ru[NH_3)_6]^{2+},   E_0 = 0.24 \text{ V}$$

but it is sufficiently substitution-inert for the electron-transfer to proceed by an outer-sphere mechanism.[16] The aquation of $[Ru(NH_3)_6]^{2+}$ is, unusually, dependent on the H^+ ion concentration and unlike aquation reactions of, e.g., $[Cr(NH_3)_6]^{3+}$. This led to the suggestion of a protonated intermediate, $[Ru(NH_3)_6H]^{3+}$ although no direct evidence was obtained; there is, of course, considerable precedent for protonation of the metal in electron-rich systems (cf. page 772).

Some of the more important reactions of Ru^{II} and Ru^{III} ammines are given in Fig. 26-F-3.

[12] R. R. Buckley and E. E. Mercer, *J. Phys. Chem.*, 1966, **70**, 3103; T. W. Kallen and J. E. Earley, *Inorg. Chem.*, 1971, **10**, 1149; C. Creutz and H. Taube, *Inorg. Chem.*, 1971, **10**, 2664.
[13] P. C. Ford, *Coordination Chem. Rev.*, 1970, **5**, 75; J. Chatt et al., *J. Chem. Soc., A*, **1971**, 3168; A. M. Zwickel and C. Creutz, *Inorg. Chem.*, 1971, **10**, 2395.
[14] A. D. Allen et al., *J. Amer. Chem. Soc.*, 1967, **89**, 5595.
[15] F. M. Lever and A. R. Powell, *J. Chem. Soc., A*, **1969**, 1477.
[16] H. Taube et al., *Inorg. Chem.*, 1968, **7**, 1976, 2369; 1970, **9**, 2627.

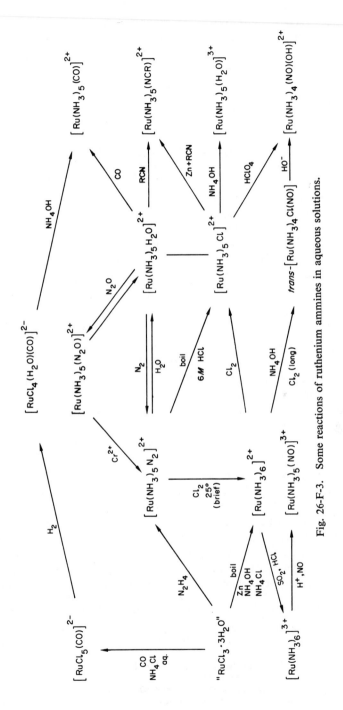

Fig. 26-F-3.   Some reactions of ruthenium ammines in aqueous solutions.

The $Ru(NH_3)_5^{2+}$ group has remarkable π-bonding properties, which are shown, not only by the formation of N_2, CO and similar complexes, but also by the fact that in nitrile complexes, $[Ru(NH_3)_5NCR]^{2+}$, the CN stretching frequency is substantially lower than in the free nitrile, whereas in other metal nitrile complexes, even in $[Ru^{III}(NH_3)_5NCR]^{3+}$, the frequency is invariably higher.[17]

The aquopentammine reacts with both N_2[18] and N_2O,[19] according to the equations

$$Ru(NH_3)_5(H_2O)^{2+} + N_2O \rightleftharpoons Ru(NH_3)_5N_2O^{2+} + H_2O \qquad K \approx 7$$
$$Ru(NH_3)_5(H_2O)^{2+} + N_2 \rightleftharpoons Ru(NH_3)_5N_2^{2+} \qquad K \sim 3 \times 10^4$$
$$Ru(NH_3)_5N_2^{2+} + Ru(NH_3)_5H_2O^{2+} \rightleftharpoons [(NH_3)_5Ru-N-N-Ru(NH_3)_5]^{4+} \qquad K \sim 7 \times 10^3$$

The last reaction, forming the bridged ion, exemplifies the donor properties of coordinated N_2 (cf. page 986). The Ru—N—N—Ru group is nearly linear and the N—N distance, 1.124 Å, is only slightly longer than in N_2 itself (1.0976 Å). The bonding can be described by MO theory[20] similar to that for Ru—O—Ru discussed above. The N_2O complex is very rapidly reduced by Cr^{2+} to form the N_2 complex, the overall reaction being

$$[Ru(NH_3)_5N_2O]^{2+} + 2Cr^{2+} + 2H^+ \rightarrow [Ru(NH_3)_5N_2]^{2+} + 2Cr^{3+} + H_2O$$

It is evident that coordination of N_2O lowers the N—O bond strength, but the precise course of the reduction is still uncertain. Finally, azido Ru^{III} ammines are unstable:

$$[Ru(NH_3)_5N_3]^{2+} \rightarrow [Ru(NH_3)_5N_2]^{2+} + \tfrac{1}{2}N_2$$

but, in presence of acid, bridged dimers are also formed and a nitrene complex, Ru=NH, may be involved.[21]

There are many other ammine derivatives. One series, formed by the action of $NaHSO_3$ on Ru^{III} ammines, contain SO_2, HSO_3^- or SO_3^{2-}; in $[Ru(NH_3)_4SO_2Cl]Cl$, the $RuSO_2$ group is planar and bound through sulfur probably by π-interaction.[22]

Other Complexes of N-Ligands with Ru^{II} and Os^{II}. Ethylenediamine and other amines form similar complexes with Ru^{II}, but few amine complexes are known for Os^{II}. However, with aromatic amines such as 2,2'-bipyridine extremely stable cations, e.g., $[Os(bipy)_3]^{2+}$, are formed. They are obtained by reducing $K_2[RuCl_5H_2O]$ or $(NH_4)_2OsBr_6$ in presence of the ligand. The *levo*-isomers are formed when sodium $(+)$-tartrate is used as reductant and this unusual asymmetric synthesis implies an intermediate with coordinated reductant whereby the stereochemistry is directed. The ion $[Ru(bipy)_3]^{2+}$

[17] R. E. Clarke and P. C. Ford, *Inorg. Chem.*, 1970, **9**, 227, 495; P. C. Ford, *Chem. Comm.*, **1971**, 7.
[18] J. N. Armor and H. Taube, *J. Amer. Chem. Soc.*, 1970, **92**, 6170; E. L. Farquhar, L. Rosnock and S. E. Gill, *J. Amer. Chem. Soc.*, 1970, **92**, 417; C. M. Elson *et al.*, *Chem. Comm.*, **1970**, 875.
[19] F. Bottomley and J. R. Crawford, *Chem. Comm.*, **1971**, 177; J. N. Armor and H. Taube, *J. Amer. Chem. Soc.*, 1971, **93**, 6476.
[20] I. M. Treitel *et al.*, *J. Amer. Chem. Soc.*, 1969, **91**, 6512.
[21] L. A. P. Kane-Maguire *et al.*, *J. Amer. Chem. Soc.*, 1970, **92**, 5865.
[22] L. H. Vogt, J. L. Kotz and S. E. Wiberley, *Inorg. Chem.*, 1965, **4**, 1157.

has an intense red color and like other platinum-metal bipyridine complexes is luminescent and can act as a photosensitizing agent[23].

Ruthenium(III) and Osmium(III) Amine and Ammine Complexes.   Only a few amine complexes of Os^{III} are known, but for Ru^{III} there are several types and some of their reactions have been given in Fig. 26-F-3. Both elements form 2,2′-bipyridine and 1,10-phenanthroline complexes.

The reduction of ammines of the type $[Ru(NH_3)_5L]^{3+}$ with Cr^{2+} and other reducing agents has been studied in detail.[24] The reactions are similar to those of Co^{III} complexes except that the electron enters the t_{2g} rather than the e_g level and the resulting Ru^{II} complexes are diamagnetic. Bridged species are probably intermediates.

In some reactions, the rate-determining step appears to be attack on the d electron density of the metal atom. Thus the hexaammine, $[Ru(NH_3)_6]^{3+}$, undergoes aquation only very slowly at room temperature but reacts rapidly with NO:

$$[Ru(NH_3)_6]^{3+} + H^+ + NO \rightarrow [Ru(NH_3)_5NO]^{3+} + NH_4^+.$$

By binding with the unpaired d electron of Ru^{III} a 7-coordinate intermediate or activated complex could be obtained.[25]

"Ruthenium Red."   A characteristic of ruthenium complex ammine chemistry is the formation of highly colored red or brown species usually referred to as ruthenium reds. Thus, if commercial ruthenium chloride is treated with ammonia in air for several days, a red solution is obtained. Alternatively, if Ru^{III} chloro complexes are reduced by refluxing ethanol and the resulting solution is treated with ammonia and exposed to air at 90° with addition of more ammonia at intervals, again a red solution is obtained. Crystallization of the solutions gives ruthenium red.

The structure of the species appears to be that of a linear trinuclear ion with oxygen bridges[26a] between the metal atoms,

$$(NH_3)_5Ru^{III}—O—Ru^{IV}(NH_3)_4—O—Ru^{III}(NH_3)_5]^{6+};$$

this type of structure has been confirmed by X-ray diffraction[26b] for the ethylenediamine derivative, $[(NH_3)_5RuORu(en)_2ORu(NH_3)_5]Cl_6$.

Since the average oxidation state of Ru is $3\frac{1}{3}$, the metal atoms must be in different formal oxidation states. The diamagnetism can be ascribed to Ru—O—Ru π-bonding as in the $[M_2OX_{10}]^{4-}$ ions. The above ion can be oxidized in acid solution by air, Fe^{3+} or Ce^{4+} to a brown paramagnetic ion of the same constitution but with charge $+7$. It is likely that there are corresponding trinuclear chloro complexes, such as $[Ru_3O_2Cl_6(H_2O)_6]$, in the violet aqueous solutions of $RuCl_3 \cdot 3H_2O$.

A different type of oxo-bridged species is $[Ru_3O(OOCCH_3)_6L_3]^+$,

[23] J. N. Demas and A. W. Adamson, *J. Amer. Chem. Soc.*, 1971, **93**, 1801.

[24] W. G. Movius and R. G. Linck, *J. Amer. Chem. Soc.*, 1970, **92**, 2677.

[25] J. N. Armor, H. A. Scheidegger and H. Taube, *J. Amer. Chem. Soc.*, 1968, **90**, 5928.

[26] (a) J. E. Early and T. Fealey, *Chem. Comm.*, **1971**, 331; W. P. Griffith, *J. Chem. Soc.*, A, **1969**, 2270, and references therein; C. A. Clausen, III, R. A. Prados and, M. L. Good, *Inorg. Nuclear Chem. Letters*, 1971, **7**, 485; (b) P. M. Smith *et al.*, *Inorg. Chem.*, 1971, **10**, 1943; (c) F. A. Cotton *et al.*, *Chem. Comm.*, **1971**, 967.

$L = H_2O$; py. These have a central 3-coordinate O and are similar to the Cr_3O and Fe_3O acetate complexes (page 837) but differ in that they undergo reversible reduction reactions.[26c]

26-F-5. Nitric Oxide Complexes of Ruthenium and Osmium

The formation of nitric oxide complexes is a marked feature of ruthenium chemistry; those of Os have been less well studied, but where known they are even more stable than the Ru analogs.

The group RuNO can occur in both anionic and cationic octahedral complexes in which it is remarkably stable, being able to persist through a variety of substitution and oxidation–reduction reactions. Ruthenium solutions or compounds that have at any time been treated with nitric acid can be suspected of containing nitric oxide bound to the metal. The presence of NO may be detected by infrared absorption ca. $1930–1845 \text{ cm}^{-1}$.

Almost any ligand can be present along with the RuNO group; those of phosphines are considered in the next Section but conventional complexes are $[Ru(NO)Cl_5]^{2-}$, $[Ru(NO)(NH_3)_4Cl]^{2+}$ and $Ru(NO)[S_2CNMe_2]_3$. The complexes can be obtained in a variety of ways and the source of NO can be HNO_3, NO, NO_2 or NO_2^-. A few examples will illustrate the preparative methods. If RuO_4 in $\sim 8M$ HCl is evaporated with HNO_3, a purple solution is obtained from which the addition of ammonium chloride precipitates the salt $(NH_4)_2[RuNOCl_5]$. If this salt is boiled with ammonia, it is converted into the golden-yellow salt $[RuNO(NH_3)_4Cl]Cl_2$. When commercial "ruthenium chloride" in HCl solution is heated with NO and NO_2, a plum-colored solution is obtained from which brick-red $RuNOCl_3 \cdot 5H_2O$ can be obtained. The addition of base to the solution gives a dark brown gelatinous precipitate of $RuNO(OH)_3 \cdot H_2O$. When this hydroxide is boiled with $8M$ HNO_3 and the solution evaporated, red solutions are obtained from which ion-exchange separation has allowed identification of species such as $[Ru(NO)(NO_3)_4H_2O]^-$, $[Ru(NO)(NO_3)_2(H_2O)_3]^+$, $[Ru(NO)(NO_3)(H_2O)_4]^{2+}$ and $[Ru(NO)(H_2O)_5]^{3+}$. Other complex anions are present as well as neutral species, of which the main one is $[Ru(NO)(NO_3)_3(H_2O)_2]$, which can be extracted into tributyl phosphate.

There are only a few cases known where the RuNO group is attacked; one is similar to the well known attack of OH^- on FeNO to give $FeNO_2$ (page 781), namely,

$$[RuX(bipy)_2(NO)]^{2+} + 2OH^- \rightleftarrows RuX(bipy)_2NO_2 + H_2O$$

and is reversed by acid.[27]

The vast majority of RuNO complexes are of the general type $Ru(NO)L_5$, in which the metal atom is *formally* in the divalent state, if we postulate electron transfer from NO to the metal as Ru^{III} followed by donation from NO^+. For iron, very few such octahedral complexes are known except with

[27] T. J. Meyer, J. B. Godwin and N. Winterton, *Inorg. Chem.*, 1971, **10**, 471, 2150.

cyanide as an associated ligand, and the different behavior of the two elements could be attributed in part to the relatively low stabilization energy of the $t_{2g}^3 e_g^2$ ion for ruthenium and the consequent readiness of Ru^{III} to accept an electron from NO giving Ru^{II} (t_{2g}^6); the larger size of Ru^{3+} (~ 0.72) than of Fe^{3+} (~ 0.64) would also favor better $d\pi-p\pi$ overlap for NO π bonding.

X-ray studies have suggested that the Ru—N—O bond may be non-linear in some complexes, but in $Na_2[Ru(NO_2)_4(NO)(OH)]$ the group is linear with a short Ru—N bond consistent with the configuration $R=N^+=\ddot{O}:$; the NO and OH groups are *trans* to each other. The high *trans*-position of NO has been shown by the ready replacement of ligands such as NH_3 or Cl^- *trans* to it.

26-F-6. Tertiary Phosphine and Related Complexes of Ru and Os

In common with other platinum metals, an intensively studied area is the chemistry involving trialkyl- and triaryl-phosphines, the corresponding phosphites and, to a lesser extent, the arsines. An extremely wide range of complexes is known, mainly of the II state, although compounds in the 0, III, and less commonly, IV state are known;[28] other ligands commonly associated with the PR_3 group are halogens, H, alkyl and aryl groups, CO, NO, alkenes, etc.

The main preparative routes[28,29] are as follows.

(a) Interaction of "$RuCl_3\cdot 3H_2O$", K_2OsCl_6 or other halide species with PR_3 in an alcohol or other solvent. In many of these reactions, either hydride or CO may be abstracted from the solvent molecule, leading to hydrido or carbonyl species. Sodium borohydride is also used as reducing agent.[29]

(b) Complexes in the 0 oxidation state may be obtained either by reduction of halides such as $RuCl_2(PPh_3)_3$ with Zn in presence of CO or by reaction of metal carbonyls with phosphines. Reactions of polynuclear carbonyls such as $Ru_3(CO)_{12}$ with phosphines tend to preserve the cluster structure.[30]

(c) For the M^0 and M^{II} species, oxidative-addition reactions with halogens or other molecules give oxidized species.

(d) Carbonyl-containing complex ions such as $M(CO)Cl_5^{2-}$, *cis*-$M(CO)_2Cl_4^{2-}$ and $M(CO)_3Cl_3^-$ are formed by action of CO or formic acid on Ru and Os chloro complexes,[31] and addition of phosphines to such solutions gives replacement products.

Some typical reactions for ruthenium, which are generally representative also for osmium, are given in Fig. 26-F-4. It is to be noted that phosphines

[28] See, e.g., J. Chatt *et al.*, *J. Chem. Soc., A,* **1969**, 854; **1971**, 895, 1169; M. S. Lupin and B. L. Shaw, *J. Chem. Soc., A,* **1968**, 740; T. A. Stephenson, *J. Chem. Soc., A,* **1970**, 889 (contains many references); F. G. Moers, *Chem. Comm.,* **1971**, 79.

[29] S. D. Robinson *et al.*, *J. Chem. Soc., A,* **1970**, 639, 2947; J. C. S. Dalton, **1972**, 1.

[30] See, e.g., F. L'Eplattenier and F. Calderazzo, *Inorg. Chem.,* **1968**, 7, 1290; M. I. Bruce and F. G. A. Stone, *Angew. Chem. Internat. Edn.,* **1968**, 7, 427.

[31] M. J. Cleare and W. P. Griffith, *J. Chem. Soc.,* **1969**, 372; J. A. Stanko and S. Chaipayungpundhu, *J. Amer. Chem. Soc.,* **1970**, **92**, 5580.

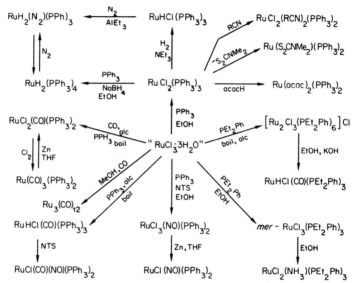

Fig. 26-F-4. Some reactions of tertiary phosphine complexes of ruthenium. (alc = 2-methoxyethanol; NTS = N-methyl-N-nitroso-p-toluenesulfonamide.)

with differing steric and basic properties often give different types of product. The reaction conditions are also critical; for example, for trialkylphosphines and $RuCl_3$ in ethanol, short reaction times give $RuCl_3(PR_3)_3$, whereas prolonged reactions give $[Ru_2Cl_3(PR_3)_6]Cl$.

We can deal only with a few selected compounds. The red-brown crystalline $RuCl_2(PPh_3)_3$ is of interest for two reasons. First, it is an unusual example of 5-coordinate Ru^{II}. The sixth position is actually blocked by a hydrogen atom of one of the phenyl rings so that a quasi-octahedral structure results. The occupancy of a coordination site by a non-bonding ligand atom so as to achieve a preferred coordination is known elsewhere (cf. page 1036). This situation is clearly closely related to the situation in reactions where the α-H atom of a phenyl ring is transferred to the metal or is lost (Chapter 24 and below). Secondly, $RuCl_2(PPh_3)_3$ is readily converted into $RuHCl(PPh_3)_3$ by H_2 in presence of a base. The product is one of the most active catalysts yet known for the homogeneous hydrogenation of alkenes and is remarkably specific towards 1-alkenes (page 789); it also catalyzes the oxidation of cyclohexene by oxygen[32a] and the addition of HCN to olefins to give nitriles.[32b] The hydridochloride differs from the dichloride in having a trigonal-bipyramidal structure; evidently steric factors involving all the ligands are important. The hydridochloride reacts with aldehydes[32c] losing hydrogen to form a complex which has a π-bonded keto group.

[32] (a) S. Cenini A. Fusi and G. Capparella, Inorg. Nucl. Chem. Letts., 1972, 8, 127. (b) Du Pont, French Patent 1603, 513 (1971). (c) R. R. Hitch, S. K. Gondal and C. T. Sears, Chem. Comm., 1971, 777.

The dihydrides, $RuH_2(PPh_3)_n$, ($n = 3$ or 4), may be obtained by boro-hydride reduction. They undergo reversible additions of H_2, CO, N_2, etc. and also polymerize acrylonitrile.[29,33] Some of the alkylarylphosphines and their phosphite analogs, e.g., $H_2Ru[P(OMe)_3]_4$, provide examples of the unusual non-rigid octahedral molecules (see Chapter 1, page 43).[34]

A number of the complexes, especially of Ru^{III}, have halide-bridged structures, some with three bridges,[35] as in $[Ru_2Cl_3(PEt_2Ph)_6]^+$. The Ru^{III} and Os^{III} complexes have octahedrally coordinated metal atoms.

The compounds in the IV state are mainly those of osmium,[36] one interesting species being $OsH_4(PMe_2Ph)_3$ which is one of a series:

$$WH_6(PR_3)_3, \quad ReH_5(PR_3)_3, \quad OsH_4(PR_3)_3, \quad IrH_3(PR_3)_3,$$

where the coordination number decreases from 9 to 6.

Osmium also forms a variety of nitrogen complexes,[37] e.g., $OsCl_2(N_2)$-$(PEt_3)_2$, by the interaction of $OsCl_3(PEt_3)_3$ with zinc in tetrahydrofuran under N_2. The analogous Ru complexes cannot be isolated but by using chelating phosphines nitrogen complexes for Fe, Ru and Os of the type trans-$M(H)(N_2)(diphos)_2^+$ can be isolated.[38] Chelate phosphine complexes are of interest also in that they provided the first example of hydrogen transfer from a ligand to a metal (page 783).

Finally, there are nitrosyl complexes in several oxidation states (cf. Fig. 26-F-4), examples being $RuCl(CO)(NO)(PPh_3)_2$ and $OsCl(NO)_2(PPh_3)_2]^+$. The latter and its analog $RuCl(NO)_2(PPh_3)_2^+$, which is made by action of $NO^+PF_6^-$ on $RuCl(NO)(PPh_3)_2$ (a compound similar to $IrCl(CO)(PPh_3)_2$ in its oxidative-addition chemistry) are of particular interest as they have both a linear and a bent Ru—N—O group.[40]

26-F-7. Other Ruthenium(II) and (III) Complexes

In addition to the complexes discussed above, there is a variety with oxygen ligands of which the oxalates, e.g., $[Ru(ox)_3]^{3-}$, and acetylacetonate, $Ru(acac)_3$, are well characterized. Many complexes of sulfur donors are also known.[41]

The complex, $[Ru_2(OOCCH_3)_4]Cl$ is obtained along with the oxo-centered trimer noted above on refluxing $RuCl_3 \cdot 3H_2O$ with acetic (or other carboxy-

[33] T. I. Eliades, R. O. Harris and M. C. Zia, Chem. Comm., 1970, 1709.
[34] P. Meakin et al., J. Amer. Chem. Soc., 1970, 92, 3484.
[35] K. A. Raspin, J. Chem. Soc., A, 1969, 461.
[36] B. E. Mann, C. Masters and B. L. Shaw, Chem. Comm., 1970, 1041; B. Bell, J. Chatt, and G. J. Leigh, Chem. Comm., 1970, 576; F. L'Eplattenier, Inorg. Chem., 1969, 8, 965; H. P. Ganz and G. J. Leigh, J. Chem. Soc., A, 1971, 2229.
[37] J. Chatt, G. J. Leigh, and R. L. Richards, J. Chem. Soc., A, 1970, 2243; P. K. Maples, F. Basolo and R. G. Pearson, Inorg. Chem., 1971, 10, 765.
[38] G. M. Bancroft et al., J. Chem. Soc., A, 1970, 2146.
[39] K. R. Laing and W. R. Roper, J. Chem. Soc., A, 1970, 2149.
[40] C. G. Pierpont et al., J. Amer. Chem. Soc., 1970, 92, 4761; J. M. Waters and K. R. Whittle, Chem. Comm., 1971, 518.
[41] J. Chatt, G.J. Leigh and A.P. Storace, J. Chem. Soc. A, 1971, 1380.

lic) acid. The Cl atoms act as bridges between $Ru_2(OOCR)_4^+$ units in which the average oxidation state is 2.5.[42]

26-F-8. Complexes of Ruthenium(V) and Osmium(V), d^3

This oxidation state is unfavorable and there is no simple compound, save the fluorides and a few complexes. The octahedral hexafluoro complexes can be prepared by various non-aqueous reactions of which the following are representative:

$$RuCl_3 + M^ICl + F_2 \xrightarrow{\ 300°\ } M^I[RuF_6]$$
$$Ru + M^{II}Cl_2 + BrF_3 \longrightarrow M^{II}[RuF_6]_2$$
$$OsCl_4 + M^ICl + BrF_3 \longrightarrow M^I[OsF_6]$$

The colors vary with the mode of preparation, probably owing to presence of traces of impurities. For example, $KRuF_6$ samples prepared by high-temperature fluorination are pale blue, whereas those from bromine trifluoride solution may be pale pink or cream.

The fluororuthenates(v) dissolve in water with evolution of oxygen, undergoing reduction to $[RuF_6]^{2-}$ and also producing traces of RuO_4. The osmium salts dissolve in water without reaction, but when base is added oxygen is evolved and $[OsF_6]^{2-}$ is formed.

The $[MF_6]^-$ ions have t_{2g}^3 configurations with three unpaired electrons. Their magnetic moments are independent of temperature, averaging ~ 3.7 B.M. for the $[RuF_6]^-$ salts and ~ 3.2 B.M. for the $[OsF_6]^-$ salts. The differences from the spin-only moment (3.87 B.M.) may be due in part to certain second-order spin–orbit coupling effects but, since observed moments are perhaps lower than can be explained by this process alone, probably also to antiferromagnetic interactions.

26-F-9. Nitrido Complexes of Ru and Os

The osmiamate ion, $[OsO_3N]^-$, was the first example to be prepared of complex ions in which nitrogen is bound to a transition metal by a multiple bond.

When OsO_4 in KOH solution is treated with strong ammonia, the yellowish-brown color of $[OsO_4(OH)_2]^{2-}$ changes to yellow and from the solution orange-yellow crystals of $K[OsO_3N]$ can be obtained. This ion has been shown to be distorted tetrahedral (C_{3v}). The infrared spectrum shows three main bands, at 1023, 858 and 890 cm^{-1}, the first these being displaced on isotopic substitution with ^{15}N, which confirms the assignment as the Os—N stretching frequency; the high value suggests considerable Os—N multiple-bond character, and we can formally write this Os≡N.

Although the osmiamate ion is stable in alkaline solution, it is readily reduced by HCl or HBr, and from the resulting red solutions red crystals of salts such as $K_2[Os^{VI}NCl_5]$ can be obtained.[43] This chloronitrido anion is

[42] M. J. Bennett, K. G. Caulton and F. A. Cotton, *Inorg. Chem.*, 1969, **8**, 1.
[43] D. Bright and J. A. Ibers, *Inorg. Chem.*, 1969, **8**, 709.

diamagnetic and presumably has two electrons in a low-lying d level (cf. the osmyl derivatives above). On further reduction with acidified stannous chloride, salts of the ion $[Os^{III}NH_3Cl_5]^{2-}$ can be obtained.

The action of *tert*-butylamine on a petroleum solution of OsO_4 gives yellow crystals of $OsO_3NC(CH_3)_3$ which are soluble in organic solvents. Some other amines act similarly.

Binuclear nitrido complexes of both Os and Ru are known which may be anionic or cationic depending on the ligands present.[44] The ruthenium complex $K_3[Ru_2NCl_8(H_2O)_2]$ is obtained by reduction of $K_2[RuNOCl_5]$ with HCl and $SnCl_2$; ammonia converts it into $[Ru_2N(NH_3)_8Cl_2]^{3+}$, and other similar conversions in which the Ru—N—Ru group is retained can be made.

The structure and bonding are analogous to those of the $[Ru_2OCl_{10}]^{4-}$ ion but with a linear Ru^{IV}—N—Ru^{IV} group. Aqueous solutions of the salt are useful for the electrolytic deposition of ruthenium.[45]

26-G. RHODIUM AND IRIDIUM[1]

26-G-1. General Remarks: Stereochemistry

Rhodium and iridium differ from ruthenium and osmium in not forming oxo anions or volatile oxides. Their chemistry centers mainly around the oxidation states $-$I, 0, I and III for rhodium and I, III and IV for iridium. Oxidation states exceeding IV are limited to hexafluorides and to salts of the IrF_6^- ion.

The $-$*I and* 0 *oxidation states.* These are mainly concerned with the carbonylate anions, polynuclear carbonyls, and substituted carbonyls with ligands such as PPh_3 (see Chapter 23).

The I *oxidation state,* d^8. Both square and 5-coordinate diamagnetic species exist mainly with CO, tertiary phosphines and alkenes as ligands and commonly with halide or H^- ions. Oxidative-addition reaction leading to Rh^{III} and Ir^{III} species are an important feature of the chemistry (cf. page 772).

The II *oxidation state,* d^7. There is no evidence for the existence of complexes comparable to those of Co^{2+}, such as $Co(NH_3)_6^{2+}$ or $CoCl_4^{2-}$, although such Rh^{II} species may be intermediates in reductions.

The best defined species are certain phosphine-stabilized ones with metal-to-carbon bonds, the bridged carboxylates $Rh_2(OOCR)_4$ and the ion Rh_2^{4+}.

The III *oxidation state,* d^6. Both elements form a wide range of "normal,"

44 M. J. Cleare and W. P. Griffith, *J. Chem. Soc., A,* **1970,** 1117; M. Ciechanowicz and A. C. Skapski, *J. Chem. Soc. A,* **1971,** 1792.
45 L. Greenspan, *Engelhard Tech. Bull.,* 1970, **11,** 76; C. W. Bradford, M. J. Cleare and H. Middleton, *Platinum Metal Rev.,* 1969, **13,** 90.
1 W. P. Griffith, *The Chemistry of the Rarer Platinum Metals,* Interscience–Wiley, 1968; B. R. James, *Coordination Chem. Rev.,* 1966, **1,** 505 (reactions and catalytic properties of Rh^I and Rh^{III} in solution).

octahedral and diamagnetic complexes with nitrogen and oxo ligands. In addition, extensive series of complexes with CO, PR_3 and similar ligands are known, many of which are obtained by oxidative addition from the square M^I species.

The IV oxidation state, d^7. This is of little importance for Rh and only a few complexes have been characterized. However, the IV state for Ir is well defined and a number of stable paramagnetic complex ions exist.

The oxidation states and stereochemistries are summarized in Table 26-G-1.

TABLE 26-G-1

Oxidation States and Stereochemistries of Rhodium and Iridium

Oxidation state	Coordination number	Geometry	Examples
Rh^{-I}, Ir^{-I}	4	Tetrahedral	$Rh(CO)_4^-$, $[Ir(CO)_3PPh_3]^-$, $IrNO(PPh_3)_3^c$
Rh^0, Ir^0	4	Cluster	$Rh_6(CO)_{16}$, $Ir_4(CO)_{12}$
Rh^I, Ir^I, d^8	$4^{a,b}$	Planar	$[Rh(CO)_2Cl]_2$, $C_8H_{12}RhCl(AsPh_3)$, $IrClCO(PEt_3)_2$
	5	*tbp*	$HRh(diphos)_2$, $HIrCO(PPh_3)_3$, $HRh(PF_3)_4$, $(C_8H_{10})_2RhSnCl_3$
Rh^{II}, d^7	?	?	$[Rh_2I_2(CNPh)_8]^{2+}$
	4	Square	$[Rh\{S_2C_2(CN)_2\}_2]^{2-}$, $RhCl_2[P(o\text{-}MeC_6H_4)_3]_2$
	5	?	$[Rh(bipy)_2Cl]^+$
	5	Cu^{II} acetate struct.	$[Rh(OCOR)_2]_2$
	6	Cu^{II} acetate struct.	$[Ph_3PRh(OCOCH_3)_2]_2$
Rh^{III}, Ir^{III}, d^6	5	*tbp*	$IrH_3(PR_3)_2$
	5	Square pyramidal	$RhI_2(CH_3)(PPh_3)_2$
	$6^{a,b}$	Octahedral	$[Rh(H_2O)_6]^{3+}$, $RhCl_6^{3-}$, $IrH_3(PPh_3)_3$, $RhCl_3(PEt_3)_3$, $IrCl_6^{3-}$, $[Rh(diars)_2Cl_2]^+$, RhF_3, $IrF_3(ReO_3$ type$)$
Rh^{IV}, Ir^{IV}, d^5	6^b	Octahedral	K_2RhF_6, $[Ir(C_2O_4)_3]^{2-}$, $IrCl_6^{2-}$ IrO_2 (rutile type)
Ir^V, d^4	6	Octahedral	$CsIrF_6$
	7	?	$IrH_5(PPhEt_2)_2$
Rh^{VI}, Ir^{VI}, d^3	6	Octahedral	RhF_6, IrF_6

^{a} Most common states for Rh.
^{b} Most common states for Ir.
^{c} V. G. Albano, P. Belloni and M. Sansoni, *J. Chem. Soc., A.,* **1971**, 2420.

26-G-2. Complexes of Rhodium(I) and Iridium(I), d^8

There is a very extensive chemistry for both elements in the I state, but it is exclusively one involving π-acid ligands such as CO, PR_3, alkenes, etc. Some of this chemistry has already been discussed in Chapters 23 and 24.

Both square and 5-coordinate species are formed, the latter often being produced by addition of neutral ligands to the former, e.g.,

$$trans\text{-}IrCl(CO)(PPh_3)_2 + CO \rightleftarrows IrCl(CO)_2(PPh_3)_2$$

The criteria for relative stability of 5- and 4-coordinate species are by no means fully established. Substitution reactions of square species, which are

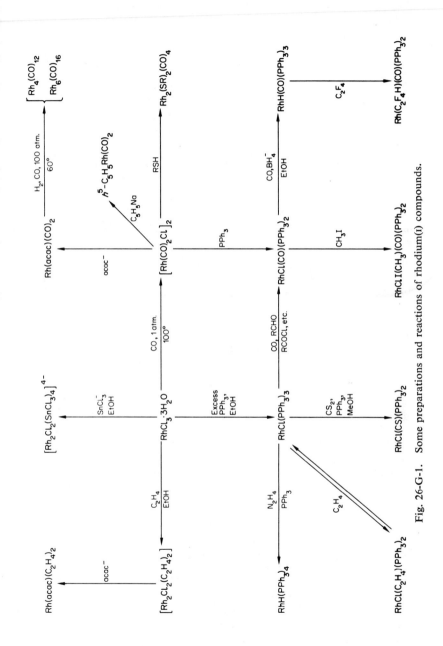

Fig. 26-G-1. Some preparations and reactions of rhodium(I) compounds.

often rapid, proceed by an associative pathway involving 5-coordinate intermediates, e.g.,[2]

$$RhCl(C_8H_{12})SbR_3 + amine \rightleftharpoons RhCl(C_8H_{12})amine + SbR_3$$

The Rh^I and Ir^I complexes are invariably prepared by some form of reduction, either of similar M^{III} complexes or of halide complexes such as RhCl$_3 \cdot 3H_2O$ or K$_2$IrCl$_6$ in presence of the complexing ligand. As noted under ruthenium, alcohols, aldehydes or formic acid may furnish CO and/or H under certain circumstances; the ligand itself may also act as a reducing agent.

Rhodium. Some preparations and reactions of Rh^I complexes are shown in Figs. 26-G-1 and 26-G-4. We can discuss only a few of the more important compounds.

Bis(dicarbonyl)dichlorodirhodium, [Rh(CO)$_2$Cl]$_2$, is most easily obtained by passing CO saturated with ethanol over RhCl$_3 \cdot 3H_2O$ at ca. 100°, when it sublimes as red needles. It has the structure shown in Fig. 26-G-2, where the coordination around each Rh atom is planar, and there are bridging chlorides with a marked dihedral angle, along the Cl–Cl line. There appears to be some direct interaction between electrons in rhodium orbitals perpendicular to the planes of coordination. In rhodium and iridium β-diketonates such as Rh(CO)$_2$(acac) and in Ir(CO)$_2$(8-quinolinolate), there is distinct metal—metal interaction in the lattice, leading to semi-conductor behaviour.[3]

This carbonyl chloride is a useful source of other rhodium(I) species and it is readily cleaved by donor ligands to give *cis*-dicarbonyl complexes, e.g.,

$$[Rh(CO)_2Cl]_2 + 2L \rightarrow 2RhCl(CO)_2L$$
$$[Rh(CO)_2Cl]_2 + 2Cl^- \rightarrow 2[Rh(CO)_2Cl_2]^-$$
$$[Rh(CO)_2Cl]_2 + acac^- \rightarrow 2Rh(CO)_2(acac) + 2Cl^-$$

trans-*Chlorocarbonylbis(triphenylphosphine)rhodium*, trans-RhCl(CO)-(PPh$_3$)$_2$. Although not as widely studied as its iridium analog, this is an important compound and is an intermediate in the preparation of the more important complex RhH(CO)(PPh$_3$)$_3$ discussed below.

The yellow crystalline complex is best obtained by the reduction of RhCl$_3 \cdot 3H_2O$ in ethanol, with formaldehyde as both a source of CO and as reductant. It is also formed by PPh$_3$ bridge-cleavage from [Rh(CO)$_2$Cl]$_2$ and by action of CO on RhCl(PPh$_3$)$_3$ (see below).

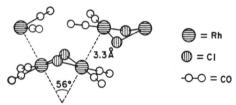

⊜ = Rh

⬛ = Cl

-O-O = CO

Fig. 26-G-2. The structure of crystalline Rh(CO)$_2$Cl]$_2$.

[2] L. Cattalini, R. Ugo and A. Orio, *J. Amer. Chem. Soc.*, 1968, **90**, 4800.
[3] R. Ugo *et al.*, *J. Organometallic Chem.*, 1968, **11**, 159, 341; L. L. Interrante and F. P. Bundy, *Chem. Comm.*, **1970**, 584.

Although it is readily oxidized by Cl_2 to $RhCl_2(CO)(PPh_3)_2$, the oxidative-adducts are generally less stable than those of trans-$IrCl(CO)(PPh_3)_2$ and the equilibria such as

$$trans\text{-}RhCl(CO)(PPh_3)_2 + HCl \rightleftharpoons Rh^{III}HCl_2(CO)(PPh_3)_2$$

generally lie well to the left-hand side.

Other rhodium carbonyl species.   In addition to $Rh(CO)_2Cl_2^-$ noted above, other carbonyl anions are known and are best made by the action of CO^{4a} or formic acid[4b] on $RhCl_3$ solutions. In the CO reduction there is an intermediate Rh^{III} complex $[RhCl_5CO]^{2-}$ and reduction probably occurs by transfer of H_2O to CO, giving a formato intermediate which then loses CO_2 (cf. page 781).[4] The overall reaction is hence

$$Rh^{III} + H_2O + 3CO \rightarrow Rh^I(CO)_2 + CO_2 + 2H^+$$

The final product, $Rh(CO)_2Cl_2^-$, can be reoxidized in HCl solution by O_2, so that there is a catalytic cycle for oxidation of CO to CO_2:

$$[Rh(CO)_2Cl_2]^- + O_2 + 2H^+ + 3Cl^- \rightarrow [Rh(CO)Cl_5]^{2-} + CO_2 + H_2O$$

Hydridocarbonyltris(triphenylphosphine)rhodium.   This yellow crystalline solid has a *tbp* structure with equatorial phosphine groups. It is best prepared from $RhCl(CO)(PPh_3)_2$ by the reaction

$$trans\text{-}RhCl(CO)(PPh_3)_2 + PPh_3 \xrightarrow[\text{EtOH}]{\text{NaBH}_4} RhH(CO)(PPh_3)_3$$

but it is also formed by action of $CO + H_2$ under pressure with virtually any rhodium compound in presence of an excess of PPh_3.

Whilst the complex undergoes a range of reactions[5] its main importance is as a hydroformylation catalyst for alkenes (see page 790).

Chlorotris(triphenylphosphine)rhodium.   This remarkable red-violet crystalline solid is formed by reduction of ethanolic solutions of $RhCl_3 \cdot 3H_2O$ with an excess of triphenylphosphine. It is widely used as a homogeneous hydrogenation catalyst (page 788), but it also undergoes a wide range of oxidative-addition and other reactions (Fig. 26-G-1). It catalyzes the oxidation of cyclohexene and other molecules by oxygen, but these reactions appear to be free-radical in nature.[6] It also abstracts CO readily from metal carbonyl complexes and from organic compounds[7] such as acyl chlorides and aldehydes, often at room temperature. With aldehydes, the reaction is a type of reversal of the hydroformylation reaction, involving first oxidative addition of RCHO to the metal.[8] The metal product is always trans-$RhCl(CO)(PPh_3)_2$, and this complex itself may act as a decarbonylation catalyst but at high temperatures.

[4]   (a) B. R. James, G. L. Rempel and F. T. T. Ng, *J. Chem. Soc., A*, 1969, 2454; C. K. Thomas and G. Petrov, *Inorg. Chem.*, 1971, **10**, 566; (b) M. J. Cleare and W. P. Griffith, *J. Chem. Soc., A*, **1970**, 2738; D. Forster, *Inorg. Chem.*, 1969, **8**, 2556.
[5]   See e.g., C. K. Brown *et al.*, *J. Chem. Soc., A*, **1971**, 850, 3120.
[6]   J. E. Blum *et al.*, *J. Chem. Soc., B*, **1969**, 1000; J. E. Baldwin and J. C. Swallow, *Angew. Chem. Internat. Edn.*, 1969, **8**, 601; L. W. Fine, M. Grayson and V. H. Suggs, *J. Organometallic Chem.*, 1970, **22**, 219.
[7]   J. Tsuji and K. Ohno, *Synthesis*, **1969**, 157 (a review).
[8]   M. C. Baird, C. J. Nyman and G. Wilkinson, *J. Chem. Soc., A*, **1968**, 232.

The structure of crystalline $RhCl(PPh_3)_3$ is not planar, like that of most other Rh^I species, but is distorted towards tetrahedral.[9]

Iridium. The most important of Ir^I complexes are $trans\text{-}IrCl(CO)(PPh_3)_2$ and its analogs with other phosphines. These compounds have been much studied as they provide some of the clearest examples of oxidative-addition reactions (Chapter 23, page 772), since the equilibria

$$trans\text{-}IrX(CO)(PR_3)_2 + AB \rightleftharpoons Ir^{III}XAB(CO)(PR_3)_2$$

lie well to the oxidized side and the oxidized compounds are usually stable octahedral species, unlike many of their rhodium analogs.

The complexes are usually made by refluxing sodium chloroiridate and phosphine in 2-methoxyethanol or diethylene glycol under an atmosphere of CO.

For both rhodium and iridium, a variety of related cationic[10] and some anionic species (which may be either 4- or 5-coordinate) are known, e.g.,

$$trans\text{-}IrCl(CO)(PR_3)_2 + CO \xrightarrow{\quad}
\begin{cases}
\xrightarrow{Na/Hg} Na[Ir(CO)_3(PR_3)] \\
\xrightarrow{NaClO_4} [Ir(CO)_3(PR_3)_2]ClO_4
\end{cases}$$

$$trans\text{-}IrCl(CO)(PR_3)_2 \xrightarrow{diphos} [Ir(diphos)_2]Cl$$

$$Rh(diene)acac \xrightarrow{HClO_4, PPh_3} [Rh(diene)(PPh_3)_2]^+ (ClO_4) + acac H$$

The cationic nitrosyl, $[IrClNO(PPh_3)_2]^+$, which is isoelectronic with the above carbonyl, also undergoes oxidative-addition reactions; for iridium there is quite an extensive NO chemistry.[11] Although a nitrogen analog of the carbonyl can be obtained[12] by the reaction

$$trans\text{-}IrCl(CO)(PPh_3)_2 + PhCON_3 \xrightarrow[0°]{CHCl_3} IrCl(N_2)(PPh_3)_2 + PhCONCO$$

it is very unstable.

The carbonyl can be readily converted into the 5-coordinate hydride:

$$trans\text{-}IrCl(CO)(PPh_3)_2 \underset{}{\overset{CO}{\rightleftharpoons}} IrCl(CO)_2(PPh_3)_2 \xrightarrow[EtOH]{NaBH_4} IrH(CO)_2(PPh_3)_2$$

and this is of interest in that it is much more stable than its rhodium analog and hence allows many prototypes for intermediates in the hydroformylation sequence to be isolated (page 792).

Again like rhodium, iridium forms alkene complexes. Examples are the cyclooctene or 1,5-cyclooctadiene (COD) compounds, e.g., $[IrCl(COD)_2]_2$, formed by boiling $(NH_4)_2IrCl_6$ with the olefin in alcohols; this product can be converted into $IrCH_3(COD)(PMe_2Ph)_2$, which shows unusual fluxional behavior.[13] Ethylene forms the unusual 5-coordinate $IrCl(C_2H_4)_4$.[14]

[9] P. B. Hitchcock, M. McPartlin and R. Mason, *Chem. Comm.*, **1969**, 1367.

[10] A. J. Deeming and B. L. Shaw, *J. Chem. Soc.*, A, **1971**, 376; M. C. Hall, B. T. Kilbourn and K. A. Taylor, *J. Chem. Soc.*, A, **1970**, 2538; L. M. Haines, *Inorg. Chem.*, 1971, **10**, 1684, 1693; R. R. Schrock and J. A. Osborn, *J. Amer. Chem. Soc.*, 1971, **93**, 2397; M. Green et al, *J. Chem. Soc.*, A, **1971**, 2334.

[11] C. A. Reed and W. R. Roper, *J. Chem. Soc.*, A, **1970**, 3054; D. P. Mingos and J. A. Ibers, *Inorg. Chem.*, 1970, **9**, 1105.

[12] J. Chatt, D. P. Melville and R. L. Richards, *J. Chem. Soc.*, A, **1969**, 2841.

[13] J. R. Shapley and J. A. Osborne, *J. Amer. Chem. Soc.*, 1970, **92**, 6976.

[14] A. van der Ent and T. C. van Soest, *Chem. Comm.*, **1970**, 225.

Finally, by action of PPh_3 on $[IrCl(COD)]_2$, $IrCl(PPh_3)_3$ can be obtained.[15] The latter differs from $RhCl(PPh_3)_3$ in reacting irreversibly with H_2 to give $IrClH_2(PPh_3)_3$. Since this octahedral species does not dissociate in solution it does not act as a hydrogenation catalyst for olefins (page 789) at 25° although it will do so under ultraviolet irradiation. By contrast, the bis species, $IrCl(PPh_3)_2$, which is made *in situ* by action of PPh_3 on the cyclooctene complex $[IrCl(C_8H_{14})_2]_2$, is an active catalyst.[16] These observations clearly show the necessity for having a vacant site for coordination of olefin on the hydrido complex.

26-G-3. Complexes of Rhodium(III) and Iridium(III), d^6

Both elements form a large number of octahedral complexes, cationic, neutral, and anionic; in contrast to Co^{III} complexes, reduction of Rh^{III} or Ir^{III} does not give rise to divalent complexes (except in a few special cases). Thus, depending on the nature of the ligands and on the conditions, reduction may lead to the metal—usually with halogens, water or amine ligands present—or to hydridic species of M^{III} or to M^I when π-bonding ligands are involved.

While being similar to Co^{III} in giving complex anions with CN^- and NO_2^-, Rh and Ir differ in readily giving octahedral complexes with halides, e.g., $[RhCl_5H_2O]^{2-}$ and $[IrCl_6]^{3-}$, and with oxygen ligands such as oxalate, EDTA, etc.

The cationic and neutral complexes of all three elements are generally kinetically inert, but the anionic complexes of Rh^{III} are usually labile. By contrast, anionic Ir^{III} complexes are inert and the preparation of such complexes is significantly harder than for the corresponding Rh species.

Rhodium complex cations have proved particularly suitable for studying *trans*-effects in octahedral complexes.[17]

In their magnetic and spectral properties the Rh^{III} complexes are fairly simple. All the complexes, and indeed all compounds of rhodium(III), are diamagnetic. This includes even the $[RhF_6]^{3-}$ ion, of which the cobalt analog constitutes the only example of a high-spin Co^{III}, Rh^{III} or Ir^{III} ion in octahedral coordination. Thus the inherent tendency of the octahedral d^6 configuration to adopt the low-spin t_{2g}^6 arrangement (see page 556), together with the relatively high ligand field strengths prevailing in these complexes of tripositive higher-transition-series ions, as well as the fact that all $4d^n$ and $5d^n$ configurations are more prone to spin pairing than their $3d^n$ analogs, provide a combination of factors that evidently leaves no possibility of there being any high-spin octahedral complex of Rh^{III} or Ir^{III}.

The visible spectra of Rh^{III} complexes have the same explanation as do those

[15] M. A. Bennett and D. L. Milner, *J. Amer. Chem. Soc.*, 1969, **91**, 6983.
[16] H. van Gaal, H. G. A. M. Cuppers and A. van der Ent, *Chem. Comm.*, **1970**, 1694.
[17] See A. J. Poë and K. Shaw, *J. Chem. Soc.*, A, **1970**, 393.

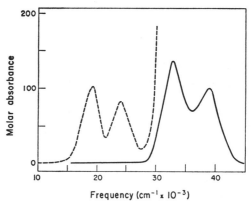

Fig. 26-G-3. The visible spectra of the $[RhCl_6]^{3-}$ ($---$) and the $[Rh(NH_3)_6]^{3+}$ (———) ions.

of Co^{III} complexes. As illustrated in Fig. 26-G-3 for the $[Rh(NH_3)_6]^{3+}$ and $[RhCl_6]^{3-}$ ions, there are in general two bands towards the blue end of the visible region, which, together with any additional absorption in the blue due to charge-transfer transitions (see page 616), are responsible for the characteristic orange, red, yellow or brown colors of rhodium(III) compounds. These bands are assigned as transitions from the $^1A_{1g}$ ground state to the $^1T_{1g}$ and $^1T_{2g}$ upper states just as shown in the energy level diagram for Fe^{II} and Co^{III} (Appendix). The spectra of Ir^{III} complexes have a similar interpretation.

The Rhodium Aquo Ion.   Unlike cobalt, rhodium gives a stable yellow aquo ion, $[Rh(H_2O)_6]^{3+}$. It is obtained by dissolution of $Rh_2O_3(aq)$ in cold mineral acids, or, as the perchlorate, by repeated evaporation of $HClO_4$ solutions of $RhCl_3(aq)$. Exchange studies with $H_2^{18}O$ confirm the hydration number as 5.9 ± 0.4. The ion is acidic, $pK_a\sim3.3$, giving $[Rh(H_2O)_5OH]^{2+}$ in solutions less than about $0.1M$ in acid. The crystalline deliquescent perchlorate is isomorphous with other salts containing octahedral cations, e.g., $[Co(NH_3)_6](ClO_4)_3$. The aquo ion also occurs in alums, $M^IRh(SO_4)_2\cdot12H_2O$, and in the yellow sulfate, $Rh_2(SO_4)_3\cdot14H_2O$, obtained by vacuum-evaporation at $0°$ of solutions of $Rh_2O_3(aq)$ in H_2SO_4. A red sulfate, $Rh_2(SO_4)_3\cdot6H_2O$, obtained by evaporation of the yellow solutions at $100°$, gives no precipitate with Ba^{2+} ion and is presumably a sulfato complex.

An aquoiridium(III) ion is not established. A sulfite, $Ir_2(SO_3)_3\cdot6H_2O$, crystallizes from solutions of $Ir_2O_3(aq)$ in water saturated with SO_2, and a sulfate may be isolated from sulfuric acid solutions of the hydrous oxide with exclusion of air, but the structures of these compounds are unknown.

The Rh^{III}–Cl System.   The species formed when $[Rh(H_2O)_6]^{3+}$ is heated with dilute hydrochloric acid have been studied by ion-exchange and, the yellow cations, $[RhCl(H_2O)_5]^{2+}$ and $[RhCl_2(H_2O)_4]^+$, and their formation constants and spectra, have been characterized. Additional acid gives

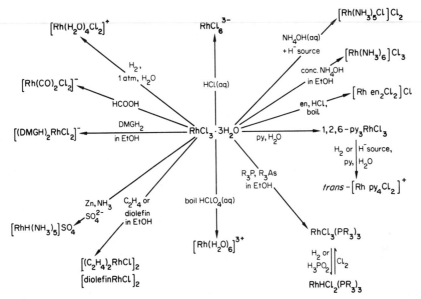

Fig. 26-G-4.   Some reactions of rhodium trichloride.

cis- and *trans-*isomers of red $[RhCl_3(H_2O)_3]$, the red anions, $[RhCl_4(H_2O)_2]^-$ and $[RhCl_5(H_2O)]^{2-}$, and finally, the rose-pink $[RhCl_6]^{3-}$.

One of the most important of Rh^{III} compounds and the usual starting material for the preparation of rhodium complexes (see Figs. 26-G-1 and 26-G-4) is the dark red, crystalline, deliquescent trichloride, $RhCl_3 \cdot nH_2O$; n is usually 3 or 4. This is obtained by dissolving hydrous Rh_2O_3 in aqueous hydrochloric acid and evaporating the hot solutions. It is very soluble in water and alcohols, giving red-brown solutions. The precise natures of the hydrate and of its aqueous solutions are uncertain. The fresh solutions do not give precipitates with Ag^+ ion and presumably contain one or more neutral, probably polynuclear, chloro complexes. On boiling, the solution gives $[Rh(H_2O)_6]^{3+}$, and heating it with an excess of HCl gives $[RhCl_6]^{3-}$. *Hexachlororhodates* are usually obtained by heating Rh metal and an alkali chloride (plus a little carbon) in chlorine, extracting the melt with water and crystallizing.

Rhodium trichloride has been used to catalyze a number of organic reactions. Thus in aqueous emulsions it induces the stereoregular polymerization of butadiene to *trans-*polybutadiene and it catalyzes the isomerization of various alkenes in ethanolic solutions.

Several dark green, hydrated Ir^{III} halides are obtained by dissolution of $Ir_2O_3(aq)$ in the appropriate acid. The nature of several species formed by aquation of $[IrCl_6]^{3-}$, e.g., $[Ir(H_2O)Cl_5]^{2-}$, $[Ir(H_2O)_2Cl_4]^-$, and $[Ir(H_2O)_3Cl_3]$, have been studied in great detail.[18]

[18] See, e.g., A. J. P. Domingos, A. M. T. S. Domingos and J. M. Peixoto Cabral, *J. Inorg. Nuclear Chem.*, 1969, **31**, 2563.

Cationic Complexes. Both Rh and Ir give cobalt-like ammines of the types $[ML_6]^{3+}$, $[ML_5X]^{2+}$, and $[ML_4X_2]^+$, of which $[Rh(NH_3)_5Cl]Cl_2$ is a typical example. The salts are made in various ways, but usually by the interaction of aqueous solutions of $RhCl_3(aq)$ with the ligand.

The formation of complex ions from $RhCl_3(aq)$, $[Rh(H_2O)Cl_5]^{2-}$ or $[RhCl_6]^{3-}$ is often catalyzed by the addition of reducing agents that can furnish hydride ions; ligands such as ethylenediamine may also themselves act in this way. The effect of ethanol was discovered by Delépine long before the general nature of such catalysis was recognized. It now appears that many rhodium complexes have been made only because ethanol was used as a solvent. One example of the catalysis is the action of pyridine, which with $RhCl_3(aq)$ gives mainly $Rhpy_3Cl_3$ and with aqueous $[Rh(H_2O)Cl_5]^{2-}$ gives $[Rhpy_2Cl_4]^-$. On addition of alcohol, hydrazine, BH_4^- or other reducing substances—even molecular hydrogen at 25° and $\leqslant 1$ atm—conversion into *trans*-$[Rhpy_4Cl_2]^+$ rapidly occurs. Kinetic studies of this reaction suggest that Rh^I complexes, rather than hydridic ones, are involved in the catalysis since $[Rh(CO)_2Cl]_2$ and $Rh(acac)(CO)_2$ are more effective than hydride-producing substances.[19] Further evidence comes from the reaction of $[Rh(H_2O)_6]^{3+}$ in ethanolic ClO_4^- solution with 2,2'-bipyridine, when an air-sensitive brown complex, $[Rh^I bipy_2]ClO_4 \cdot 3H_2O$ can be isolated.[20]

Similar catalysis of the formation of Ir^{III} complexes occurs, but the rates are still slow compared to those of rhodium systems. Thus to convert $Na_2Ir^{IV}Cl_6$ into py_3IrCl_3 and *trans*-$[Irpy_4Cl_2]Cl$ a bomb reaction is normally used. Quite rapid conversions are obtained as follows:

$$Na_2IrCl_6 \xrightarrow[\text{boil 30 min.}]{NaH_2PO_2(aq) + py} cis\text{-}py_3IrCl_3 \xrightarrow{\text{6 hr}} [Ir\,py_4Cl_2]Cl$$

$$H_2IrCl_6 \xrightarrow[\substack{2\text{-methoxyethanol,} \\ \text{boil 10 min.}}]{HCl\ in} IrCl_6^{3-} \xrightarrow[\text{1 hr}]{+py,\ boil} mer\text{-}py_3IrCl_3$$

trans-$[Iren_2Cl_2]^+$ can be similarly obtained by using hypophosphorous acid as catalyst.[21]

Hydrido complexes. With NH_3 or amines, quite stable octahedral hydrido complexes can be obtained for rhodium. Thus the reduction of $RhCl_3 \cdot 3H_2O$ in NH_4OH by Zn in presence of SO_4^{2-} leads to the white, air-stable, crystalline salt $[RhH(NH_3)_5]SO_4$. In aqueous solution the ion dissociates:

$$[RhH(NH_3)_5]^{2+} + H_2O = [RhH(NH_3)_4H_2O]^{2+} + NH_3 \qquad K \sim 2 \times 10^{-4}$$

Various substitution reactions with other amines can be carried out, and with alkenes remarkably stable alkyl derivatives, e.g., $[RhC_2H_5(NH_3)_5]SO_4$, can be obtained.[22a] Action of NaCN in dry methanol on $[RhCl(CO)_2]_2$ gives[22b] $Na_3[RhH(CN)_5]$. Air-oxidation of hydrido complexes forms first hydroperoxo and then blue superoxo species.[23]

[19] J. V. Rund, *Inorg. Chem.*, 1968, **7**, 25; R. D. Gillard, B. T. Heaton and D. H. Vaughan, *J. Chem. Soc., A*, **1971**, 1840.

[20] G. C. Kulasingam, W. R. McWhinnie and J. D. Miller, *J. Chem. Soc., A*, **1969**, 521.

[21] R. A. Bauer and F. Basolo, *Inorg. Chem.*, 1969, **8**. 2231, 2237.

[22] (a) K. Thomas and G. Wilkinson, *J. Chem. Soc., A*, **1970**, 356; (b) R. A. Jewsbury and J. P. Maher, *J. Chem. Soc., A*, **1971**, 1840.

[23] A. W. Addison and R. D. Gillard, *J. Chem. Soc., A*, **1970**, 2523.

Neutral Complexes. Interaction of acetylacetone and hydrous Rh_2O_3 gives the trisacetylacetonate, which has been resolved into enantiomeric forms. It undergoes a variety of electrophilic substitution reactions of the coordinated ligand, such as chlorination. The stereochemistry and racemization of the *cis*- and *trans*-isomers of the unsymmetrical trifluoroacetylacetonate have been studied by nmr spectroscopy; the compound is extremely stable to isomerization.

Neutral complexes with CO, PR_3, pyridines, etc., as ligands may be made directly from $RhCl_3 \cdot 3H_2O$ or Na_3IrCl_6 but, as noted above, they are also commonly and often readily obtained by oxidative addition to the M^I complexes. Typical formulas are MCl_3L_3, $MHCl_2I_3$, $MCl_3(CO)L_2$, etc. The dimethyl sulfoxide complex of iridium $IrCl_3(OSMe_2)_3$, catalyzes, *via* a hydrido intermediate, the reduction of cyclohexanones to axial alcohols and the hydrogenation of $\alpha\beta$-unsaturated ketones.[24]

26-G-4. Complexes of Rhodium(II), d^7

Heating methanol solutions of $RhCl_3 \cdot 3H_2O$ with sodium salts of carboxylic acids leads to dark green solutions from which methanol solvates of the binuclear carboxylates, $[Rh(OOCR)_2MeOH]_2$, can be crystallized.[25] These and the unsolvated compounds have the tetra-bridged structure (page 549).[26] The end positions can be occupied by ligands other than methanol; with oxygen donors the adducts are green or blue but with π-acids such as PPh_3 they are orange to red.[27]

By action of strong non-complexing acids on the acetate, the green diamagnetic ion Rh_2^{4+} can be obtained in solution;[25] it is also formed[28] by Cr^{2+} reduction of $[Rh(H_2O)_5Cl]^{2+}$. Although exceedingly deliquescent solids can be obtained by evaporation of solutions of the acetate in CF_3SO_3H, so far no salts of the Rh_2^{4+} ion have been crystallized.

Finally, a dimethylglyoxime complex, $Rh_2(DMGH)_4(PPh_3)_2 \cdot H_2O \cdot C_3H_7OH$, has a metal—metal bond but, in contrast to the carboxylates, no bridging group.[29] Reduced rhodium dimethylglyoximates behave similarly to the cobaloxime, vitamin B_{12} analogs (page 890);[30] reduction[31] of *trans*-$[Rh\,en_2Cl_2]^+$ electrochemically gives a species alleged to be $[en_2(H_2O)\,RhRh-en_2(H_2O)]^{4+}$.

Paramagnetic Rh^{II} complexes. Interaction of the acetate, sodium maleonitrile dithiolate and a quaternary ammonium hydroxide in methanol gives a red solution and a green salt containing the anion (26-G-I)

[24] H. B. Henbest and R. B. Mitchell, *J. Chem. Soc.*, B, **1970**, 785; M. McPartlin and R. Mason, *J. Chem. Soc.*, A, **1970**, 2206; M. Gullotti *et al.*, *J. Chem. Soc.*, C, **1971**, 2652.
[25] P. Legzdins *et al.*, *J. Chem. Soc.*, A, **1970**, 3322.
[26] F. A. Cotton *et al.*, *J. Amer. Chem. Soc.*, 1970, **92**, 2926.
[27] L. Dubicki and R. L. Martin, *Inorg. Chem.*, 1970, **9**, 673.
[28] H. Maspero and H. Taube, *J. Amer. Chem. Soc.*, 1968, **90**, 7361.
[29] K. G. Caulton and F. A. Cotton, *J. Amer. Chem. Soc.*, 1971, **93**, 1914.
[30] J. M. Weber and G. N. Schrauzer, *J. Amer. Chem. Soc.*, 1970, **92**, 726.
[31] R. D. Gillard, B. T. Heaton and D. H. Vaughan, *J. Chem. Soc.*, A, **1971**, 734.

(26-G-I)

This salt is paramagnetic, but the significance of the oxidation state of the metal is uncertain (see page 724). Although traces of a paramagnetic species, in addition to $RhCl(PPh_3)_3$, are formed when PPh_3 reacts with $RhCl_3 \cdot 3H_2O$ in ethanol, when tris-o-tolylphosphine is used a blue-green paramagnetic complex $RhCl_2(o\text{-tol}_3P)_2$ is obtained.[32] It is believed that the square Rh^{II} is stabilized because a CH_3 group of each phosphine ligand lies above and below the plane to give a pseudooctahedral complex [cf. $RuCl_2(PPh_3)_3$, page 1014].

26-G-4. Complexes of Rhodium(IV) and Iridium(IV), d^5

Rhodium. Oxidation of Rh^{III} sulfate solutions with O_3 or with sodium bismuthate gives red solutions that may contain Rh^{IV}. Higher states, even V and VI, have been postulated in reactions of Rh^{III} with hypobromite, although this seems unlikely.[33]

The only well-defined species are the halides RhF_6^{2-} and $RhCl_6^{2-}$. The yellow, readily hydrolyzed salts of the former are obtained when $RhCl_3$ and an alkali chloride are treated with F_2 or BrF_3. The magnetic moments of ca. 1.8 B.M. are consistent with a t_{2g}^5 configuration.

The dark green compound Cs_2RhCl_6 is made by oxidation of ice-cold solutions of $RhCl_6^{3-}$ by Cl_2 in presence of CsCl. It is isomorphous with Cs_2PtCl_6. The salt decomposes in water.

Iridium. By contrast, the IV state is comparatively stable for iridium. The existence of crystalline hexafluoroiridates is to be expected, but also hexachloroiridates and hexabromoiridates and a variety of aquated complex ions such as $[IrCl_3(H_2O)_3]^+$, $[IrCl_5(H_2O)]^-$ and $[IrCl_4(H_2O)_2]$ have been characterized. A number of trimeric oxo- and nitrido-bridged complexes,[34] e.g., the green $K_4[Ir_3N(SO_4)_6(H_2O)_3]$ made by boiling Na_3IrCl_6 and $(NH_4)_2SO_4$ in concentrated H_2SO_4, contain iridium in a mean oxidation state $3\frac{2}{3}$ (cf. page 637).

Hexachloroiridates(IV) can be made by chlorinating a mixture of iridium powder and an alkali-metal chloride, or, in solution, by adding the alkali-metal chloride to a suspension of hydrous IrO_2 in aqueous HCl. The black

[32] M. A. Bennett and P. A. Longstaff, *J. Amer. Chem. Soc.*, 1969, **91**, 6267; see also C. Masters and B. L. Shaw, *J. Chem. Soc., A,* **1971**, 3678.
[33] F. Pantani, *Talanta*, 1962, **9**, 15.
[34] D. B. Brown *et al.*, *Inorg. Chem.*, 1970, **9**, 2315; M. Ciechanowicz *et al.*, *Chem. Comm.*, **1971**, 876.

crystalline sodium salt, Na_2IrCl_6 which is very soluble in water, is the usual starting material for the preparation of other Ir^{IV} complexes.

The so-called "chloroiridic acid," is made by treating the ammonium salt with aqua regia; it is soluble in ether and hydroxylated solvents and is probably $(H_3O)_2IrCl_6 \cdot 4H_2O$.

In basic solution the dark red-brown $IrCl_6^{2-}$ is rather unstable, undergoing spontaneous reduction within minutes[35] to pale yellow-green $IrCl_6^{3-}$:

$$2IrCl_6^{2-} + 2OH^- \rightleftharpoons 2IrCl_6^{3-} + \tfrac{1}{2}O_2 + H_2O$$

From known potentials the *acid* reaction can be written:

$$2IrCl_6^{2-} + H_2O = 2IrCl_6^{3-} + \tfrac{1}{2}O_2 + 2H^+ \qquad K = 7\times10^{-8}\ \text{atm}^{\frac{1}{2}}\ \text{mol}^2\ \text{l}^{-2}\ (25°)$$

Thus in strong acid, say $12M$ HCl, $IrCl_6^{3-}$ is partially oxidized to $IrCl_6^{2-}$ in the cold and completely on heating, while in strong base (pH > 11) $IrCl_6^{2-}$ is rapidly and quantitatively reduced to $IrCl_6^{3-}$.

$IrCl_6^{2-}$ is readily and quantitatively reduced to $IrCl_6^{3-}$ by KI or sodium oxalate. In neutral solutions slow reduction of $IrCl_6^{2-}$ occurs spontaneously. A variety of organic compounds can be oxidized by $IrCl_6^{2-}$.[36]

Octahedral Ir^{IV}, t_{2g}^5, has one unpaired electron. For pure $IrCl_6^{2-}$ salts the μ_{eff} values are low, 1.6–1.7 B.M., owing to antiferromagnetic interactions; on dilution with isomorphous $PtCl_6^{2-}$ salts, normal values are found.

26-G-5. Complexes of Rhodium(v) and Iridium(v), d^4

Only the hexafluoro ions MF_6^- of Rh^V and Ir^V are known. The salts are made by reactions such as

$$RhF_5 + CsF \xrightarrow{IF_5} CsRhF_6$$
$$IrBr_3 + CsCl \xrightarrow{BrF_3} CsIrF_6$$

The red-brown Rh salt is isomorphous with $CsPtF_6$. The iridium salts are pink, with magnetic moments ca. 1.25 B.M. at 273°, which are temperature-dependent, suggesting strong spin–orbit coupling and possibly antiferromagnetic interaction. They dissolve in water, evolving O_2 and being reduced to IrF_6^{2-}. The only other known complexes are the multihydrides $IrH_5(PR_3)_2$.[37]

26-H. PALLADIUM AND PLATINUM[1]

26-H-1. General Remarks: Stereochemistry

The principal oxidation states of Pd and Pt are II and IV. There is a limited chemistry in the 0 state with tertiary phosphine and carbonyl ligands. The higher states of V and VI occur only in fluoro compounds (page 997). By

[35] D. W. Fine, *Inorg. Chem.*, 1969, **8**, 1014.
[36] R. Cecil, A. J. Fear and J. S. Littler, *J. Chem. Soc., B*, **1970**, 632.
[37] B. E. Mann, C. Masters and B. L. Shaw, *J. Inorg. Nuclear Chem.*, 1971, **33**, 2195.
[1] J. R. Miller, *Adv. Inorg. Chem. Radiochem.*, 1962, **4**, 133 (a comparison of Ni, Pd and Pt chemistry); see also references under Organometallic Compounds, page 1038.

contrast with Ni, there is no evidence for the III state in compounds, although unstable Pt^{III} species may be formed as intermediates in oxidation of Pt^{II} complexes or on light or electron irradiation,[2] and a dicarbollyl complex,[3] formally at any rate, contains Pd^{3+}.

The 0 state, d^{10}. These compounds are generally similar to those of Ni^0 except that no binary carbonyl exists. The main complexes are those of tertiary phosphines, notably those with triphenylphosphine; the PF_3 compounds $M(PF_3)_4$ resemble $Ni(PF_3)_4$ in being volatile. Many of the compounds with CO and phosphines are cluster compounds.

The II state, d^8. The Pd^{2+} ion occurs in PdF_2 (page 998) and is paramagnetic. In aqueous solution, however, the $[Pd(H_2O)_4]^{2+}$ ion is diamagnetic and is presumably square. In general, however, Pd^{II} and Pt^{II} complexes are square or 5-coordinate and are diamagnetic. They can be of all possible types, e.g., ML_4^{2+}, ML_3X^+, *cis-* and *trans-*ML_2X_2, MLX_3^- and MX_4^- where X is uninegative and L is a neutral ligand. Similar types with chelate acido or other chelate ligands are common.

As a rule, Pd^{II} and Pt^{II} show a preference for nitrogen (in aliphatic amines and in NO_2), halogens, cyanide. and heavy donor atoms, such as P, As, S, and Se and relatively small affinity for oxygen and fluorine. The strong binding of the heavy-atom donors is due in great measure to the formation of metal—ligand π bonds by overlap of filled $d\pi$ orbitals (d_{xz}, d_{xy}, and d_{yz}) on the metal with empty $d\pi$ orbitals in the valence shells of the heavy atoms. This π bonding has been discussed in connection with the *trans-*effect (page 669). Cyanide ions, nitro groups and carbon monoxide are also bound in a manner involving π bonding which results in these cases from overlap of filled metal $d\pi$ orbitals with empty $p\pi$ antibonding molecular orbitals of these ligands. In such complexes, there is usually considerable similarity of Ni to Pd and Pt.

The formation of cationic species even with non-π-bonding ligands and anionic species with halide ions contrasts with the chemistry of the isoelectronic Rh^I and Ir^I where most of the complexes involve π-bonding. The difference is presumably a reflection of the higher charge. Further, although Pd^{II} and Pt^{II} species add neutral molecules to give 5 and 6 coordinate species, they do so with much less ease; also the oxidative-addition reactions characteristic of square d^8 complexes tend to be reversible except with strong oxidants, presumably owing to the greater promotional energy for M^{II}—M^{IV} than for M^I—M^{III}.

Palladium(II) complexes are somewhat less stable in both the thermodynamic and the kinetic sense than their Pt^{II} analogs, but otherwise the two series of complexes are usually similar. The kinetic inertness of the Pt^{II} (and also Pt^{IV}) complexes has allowed them to play a very important role in the development of coordination chemistry. Many studies of geometrical iso-

[2] R. C. Wright and G. S. Laurence, *Chem. Comm.*, **1972**, 132; J. Halpern and M. Pribanič, *J. Amer. Chem. Soc.*, 1968, **90**, 5942.
[3] L. F. Warren and M. F. Hawthorne, *J. Amer. Chem. Soc.*, 1970, **92**, 1157.

merism and reaction mechanisms[4] have had a profound influence on our understanding of complexes. Both elements readily give allylic species, whereas platinum more commonly forms σ-bonded and alkene and alkyne complexes (Chapters 23 and 24).

The IV state, d^6. Although Pd^{IV} compounds exist they are generally less stable than those of Pt^{IV}. The coordination number is invariably 6. The substitution reactions of platinum(IV) complexes are greatly accelerated by presence of Pt^{II} species.[4] Solutions also readily undergo photochemical reactions in light.

Certain platinum compounds such as *cis*-$Pt^{IV}(NH_3)_2Cl_4$ and *cis*-$Pt^{II}(NH_3)_2Cl_2$ have remarkable physiological properties and may also be potential anticancer agents.[5a] Some rhodium complexes, e.g., *trans*-$[Rh\ py_4Cl_2]Cl$, also appear to have antibacterial activity.[5b]

The oxidation states and stereochemistries are summarized in Table 26-H-1.

TABLE 26-H-1

Oxidation States and Stereochemistry of Palladium and Platinum

Oxidation state	Coordination number	Geometry	Examples
Pd^0, Pt^0, d^{10}	3	Planar	$Pd(PPh_3)_3$
	4	Distorted	$Pt(CO)\ (PPh_3)_3$
	4	Tetrahedral	$Pt(Ph_2PCH_2CH_2PPh_2)_2$, $Pd(PF_3)_4$
Pd^{II}, Pt^{II}, d^8	$4^{a,b}$	Planar	$[PdCl_2]_n$, $[Pd(NH_3)_4]Cl_2$, PdO, PtO, $PtCl_4^{2-}$, $PtHBr(PEt_3)_2$, $[Pd(CN)_4]^{2-}$, PtS, $[Pd\ py_2Cl]_2$, PdS, $Pt(PEt_3)_2(C_6F_5)_2$
	5	*tbp*	$[Pd(diars)_2Cl]^+$, $[Pt(SnCl_3)_5]^{3-}$
	6	Octahedral	PdF_2(rutile type), $[PtNOCl_5]^{2-}$, $Pd(diars)_2I_2$ $Pd(DMGH)_2$,^{c}
Pd^{IV}, Pt^{IV}, d^6	6^b	Octahedral	$[Pt(en)_2Cl_2]^{2+}$, $PdCl_6^{2-}$, $[Pt(NH_3)_6]^{4+}$, $[Me_3PtCl]_4$
Pt^V, d^5	6	Octahedral	$[PtF_5]_4$
		Octahedral	PtF_6^-
Pt^{VI}, d^4	6	Octahedral	PtF_6

^{a} Most common states for Pd.
^{b} Most common states for Pt.
^{c} Has planar set of N atoms with weak Pd—Pd bonds completing a distorted octahedron.

26-H-2. Complexes of Palladium(II) and Platinum(II), d^8

The presence of non-bonding electron density and two vacant coordination sites on square d^8 complexes leads, as noted in earlier Sections, to a susceptibility to both nucleophilic and electrophilic attack. In solutions of square complexes, the additional sites may be occupied by solvent molecules

[4] D. S. Martin, Jr., *Inorg. Chim. Acta Rev.*, 1967, **1**, 87; W. R. Mason, *Coord. Chem. Rev.*, 1972, **7**, 241.
[5] (a) B. J. Leonard *et al.*, *Nature*, 1971, **234**, 43; (b) R. D. Gillard, *Platinum Metals Rev.*, 1970, **14**, 50.

and many of the reactions of square complexes proceed by an associative mechanism (cf. page 665).

Halogeno Anions. The ions MCl_4^{2-} are among the most important species as their salts are commonly used as source materials for the preparation of other complexes in the II and the 0 oxidation states. The yellowish $PdCl_4^{2-}$ ion is formed when $PdCl_2$ is dissolved in aqueous HCl or when $PdCl_6^{2-}$ is reduced with Pd sponge. The red $PtCl_4^{2-}$ ion is normally made by reduction of $PtCl_6^{2-}$ with a stoichiometric amount of hydrazine hydrochloride, oxalic acid or other reducing agent. The sodium salt cannot be obtained pure. In water, solvolysis of $PtCl_4^{2-}$ is extensive but the rate is slow:[6]

$$PtCl_4^{2-} + H_2O = PtCl_3(H_2O)^- + Cl^- \qquad K = 1.34 \times 10^{-2} \; M \; (25°, \mu \, 0.5)$$
$$PtCl_3(H_2O)^- + H_2O = PtCl_2(H_2O)_2 + Cl^- \qquad K = 1.1 \times 10^{-3} \; M$$

so that a $10^{-3} M$ solution of K_2PtCl_4 at equilibrium contains only 5% of $PtCl_4^{2-}$ with 53% of mono- and 42% of bis-aquo species.

For both metals, bromo and iodo complex anions occur and if large cations such as Et_4N^+ are used, salts of halogeno-bridged ions $M_2X_6^{2-}$ may be obtained.[7] Both MX_4^{2-} and $M_2X_6^{2-}$ are square, but in crystals the ions in K_2MCl_4 are stacked one above the other. However, unlike other stacks containing MCl_4^{2-} ions discussed below, the M—M distances (Pd 4.10 Å and Pt 4.13 Å) are too large for any chemical bonding; similarly in the dimeric ions there is no evidence for metal—metal interaction.[8]

The catalytic and other reactions of $PdCl_4^{2-}$ salts have been discussed elsewhere (page 525; page 797).

The Aquopalladium(II) Ion; Compounds of Oxo Acids. Palladium, but not platinum, forms the aquo ion, and brown deliquescent salts such as $[Pd(H_2O)_4](ClO_4)_2$ can be crystallized from solutions of PdO in dilute non-complexing acids. In 3.94M $HClO_4$ the formal potential Pd/Pd^{2+} is -0.979 V at 25°.[9]

A sublimable anhydrous *nitrate* is obtained by treating the hydrated nitrate with liquid N_2O_4; it appears to have both bridging and non-bridging NO_3 groups.

A more important compound is the *acetate*, which is obtained as brown crystals. In the crystal it is a trimer $[Pd(OOCMe)_2]_3$, where the metal atoms are in a triangle with bridging acetate groups; the Pd atoms are about 0.25 Å out of the plane of the square of oxygen atoms.[10] Heating in benzene causes dissociation to a monomer, and similar cleavage of the trimer can be achieved by the action of donor ligands on the acetate and other carboxylates to give yellow *trans*-complexes, $Pd(OOCR)_2L_2$.

Platinous acetate is not isomorphous with the Pd compound and is not cleaved by ligands; there is some hazard in its preparation from the action

[6] L. I. Elding, *Acta Chem. Scand.*, 1970, **24**, 1331, 1341, 1527.
[7] P. M. Henry, *Inorg. Chem.*, 1971, **10**, 373.
[8] P. Day, M. J. Smith and R. J. P. Williams, *J. Chem. Soc.*, A, **1968**, 668.
[9] R. M. Izatt *et al.*, *J. Chem. Soc.*, A, **1970**, 2514.
[10] A. C. Skapski and M. L. Smart, *Chem. Comm.*, **1970**, 658.

of acetic acid on HNO_3 solutions of $Na_2Pt(OH)_6$ and the solutions have been known to explode.

Palladous acetate has been widely studied in connection with vinyl acetate synthesis (page 798). It acts to some extent like Hg^{II} and Pb^{IV} acetates in attacking benzene and other aromatic hydrocarbons in acid media.[11] Thus, in acetic acid, it specifically attacks the side chain of toluene.

Neutral Complexes.[12]   There are an enormous number of Pd and Pt complexes of the general formula $MXYL_1L_2$, where X and Y are anionic groups and L are neutral donor ligands such as NR_3, PR_3, SR_2, CO, alkenes, etc. In addition to ions such as Cl^-, SCN^-, etc., X or Y may also be H or an alkyl or aryl group. A common palladium compound often used as a source material is bis(benzonitrile)dichloropalladium, $PdCl_2(C_6H_5CN)_2$, made by dissolving $PdCl_2$ in the ligand and crystallizing.

Besides the mononuclear species, there are a considerable number of *bridged binuclear complexes* of the type (26-H-I) of which (26-H-II) is a

(26-H-I)   (26-H-II)   (26-H-III)

specific example. For the triphenylarsinepalladium analog of (26-H-II), linkage isomers are known and the mode of bonding is solvent-dependent.[13]

For Pt^{II}, the bridging tendencies of the anions are in the order $SnCl_3^-$ $<RSO_2^- <Cl^- <Br^- <I^- <R_2PO^- <SR^- <PR_2^-$.[14] The strong tendency of SR and PR_2 groups to form the four-membered rings (26-H-III) may be accounted for by delocalized bonding arising from overlap of filled metal d_{xz} and d_{yz} orbitals with empty d orbitals on sulfur or phosphorus.

Bridged halide complexes are the commonest encountered, and bridged species are quite generally subject to cleavage by donor ligands to give mononuclear species, e.g.,

When the bridges are Cl^- or Br^-, the equilibria generally lie towards the mononuclear complexes. It might be supposed that such bridge-splitting

[11] R. G. Brown and J. M. Davidson, *J. Chem. Soc.*, A, **1971**, 1321; M. O. Unger and R. A. Fouty, *J. Org. Chem.*, 1969, **34**, 18; P. M. Henry, *J. Org. Chem.*, 1971, **36**, 1886.

[12] M. Orchin and P. J. Schmidt, *Coordination Chem. Rev.*, 1968, **3**, 123, 345 (pyridine and pyridine N-oxide complexes of Pt^{II}).

[13] D. W. Meek, P. E. Nicpon and I. Meek, *J. Amer. Chem. Soc.*, 1970, **92**, 5351; J. L. Burmeister, R. L. Hassel and R. D. Phelan, *Chem. Comm.*, 1970, 679; G. Beran and G. J. Palenik, *Chem. Comm.*, **1970**, 1354.

[14] J. Chatt and D. M. P. Mingos, *J. Chem. Soc.*, A, **1969**, 1770.

reactions should give the *trans*-mononuclear complexes, and indeed *trans*-isomers are probably the initial products of cleavage reactions. However, since the relative stabilities of *cis*- and *trans*-isomers depend on the particular nature of the ligands involved and on the solvent (cf., discussion page 775) mixtures may result. In some cases, as in the cleavage of phosphine complexes with CO or C_2H_4, *cis*-products, are obtained presumably because they are thermodynamically more stable than the *trans*-isomers owing to the high *trans*-effect of π-bonding ligands.

Cationic Species.   There are a number of these, perhaps, the most important being those with amines, e.g., $[Pd(NH_3)_4]^{2+}$, $[Pt\ en_2]^{2+}$, etc. They are usually readily formed by direct interaction of halide solutions with the ligand, and a variety of salts can be obtained.

Axial Interactions in Square Complexes. Metal–metal interactions.   In solvents, the vacant sites on a square species are doubtless occupied by solvent molecules, but in crystals there is the possibility of interaction of the metal with the metal of another square or with an atom or atoms of the ligands on another square.

In certain crystalline compounds of Pd^{II}, Pt^{II} and Ni^{II} there is evidence of interaction between the metal atoms of square units lined up in stacks, even though the metal—metal distance is too long to be considered as representing bonding. Thus in $PtenCl_2$,[15] which has the structure shown in Fig. 26-H-1a, a weak bond can be ascribed to interactions between d orbitals on adjacent metal atoms. The well known stacked nickel and palladium dimethylglyoxime complexes provide another example,[16] but the platinum dimethylglyoxime complex has a different structure where there is weak intermolecular interaction between each Pt atom and the oxygen of adjacent $Pt(DMGH)_2$ units (cf. below).

Metal—metal interaction is of considerable importance in a number of complex salts of Pd and Pt anionic and cationic species.

One of the best known examples is Magnus' green salt, $[Pt(NH_3)_4][PtCl_4]$. Others include $[Pd(NH_3)_4][Pd(SCN)_4]$, $[Pt(CH_3NH_2)_4][PtBr_4]$, and $[Cu(NH_3)_4][PtCl_4]$. They generally have the same structure as Magnus' green salt in which anions and cations are stacked with parallel planes creating chains of metal atoms. It appears that whenever *both* M and M' are Pt^{II} the substance shows a green color (although the constituent cations are colorless or pale yellow and the anions red). Marked dichroism with high absorption of light polarized in the direction of the metal chains, and also much increased electrical conductivity, have been observed.[17] If steric hindrance is too

[15] D. S. Martin *et al.*, *Inorg. Chem.*, 1970, **9**, 1276.
[16] Y. Ohashi, I. Hanazaki and S. Nagakura, *Inorg. Chem.*, 1970, **9**, 2551.
[17] L. V. Interrante *et al.*, *Inorg. Chem.*, 1971, **10**, 1169, 1174; P. S. Gomm *et al.*, *J. Chem. Soc., A*, **1971**, 2154.

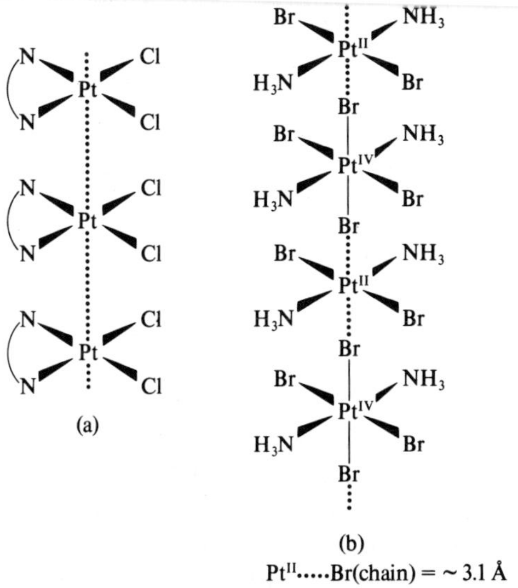

Fig. 26-H-1.   (a) Linear stacks of planar $PtenCl_2$ molecules. (b) Chains of alternating Pt^{II} and Pt^{IV} atoms with bridging bromide ions in $Pt(NH_3)_2Br_3$.

great, as in $[Pt(EtNH_2)_4][PtCl_4]$, a different structure is adopted and the compound has a pink color, which is merely the sum of the colors of its constituent ions.

There is a related class of compound with chain-like structures that contain both M^{II} and M^{IV} but differ from the above in that the metal units are linked by halide bridges.[18] There is similar behavior in that there is high electrical conductivity[17] along the direction of the $-Cl-M^{II}-Cl-M^{IV}-$ chains, e.g., in $[Pd^{II}(NH_3)_2Cl_2][Pd^{IV}(NH_3)_2Cl_4]$. Thus Wolfram's red salt has octahedral $[Pt(EtNH_2)_4Cl_2]^+$ and planar $[Pt(EtNH_2)_4]^{2+}$ ions linked in chains, the other four Cl^- ions being within the lattice. A typical structure is shown in Fig. 26-H-1.

Metal—ligand interactions. There is also good evidence that there can be interaction between a bound ligand and the metal atom either inter- or intra-molecularly.

In the complex *trans*-$PdI_2(PMe_2Ph)_2$ the α-H atoms of the phenyl group of the coordinated phosphine occupy an axial position (cf. pages 783 and 1014) and the *trans*-axial position is occupied by an iodine of an adjacent molecule, so that a quasi-7-coordinate complex (26-H-IV) results.[19]

In ammine complexes such as $PtCl_2(NH_3)_2$, infrared evidence suggests that anomalous N—H stretching frequencies can be attributed to a type of hydrogen-bonding interaction between H and the filled d_{xy} or d_{xz} orbital of

[18] M. B. Robin and P. Day, *Adv. Inorg. Chem. Radiochem.*, 1967, **10**, 351; H. Krogman, *Angew. Chem. Internat. Edn.*, 1969, **8**, 35; P. Day, *Inorg. Chim. Acta Revs.*, 1969, **3**, 81.
[19] N. A. Bailey and R. Mason, *J. Chem. Soc., A*, **1968**, 2594.

(26-H-IV)

the metal. Palladium complexes do not show this effect, probably owing to the smaller spatial extension of the $4d$ orbitals.

Finally, another case is the complex ion $[Pd(Et_4dien)Cl]^+$, where the ethyl groups block off the axial positions so that kinetically, in substitution reactions, the ion behaves as an octahedral rather than a square complex.[20]

Five-coordinate Complexes of Pd^{II} and Pt^{II}. It is generally agreed that many substitution and isomerization reactions of square complexes normally proceed by an associative path involving distorted 5-coordinate intermediates[21] and there is good evidence for solvation, for example, of $PtCl_2(n\text{-}Bu_3P)_2$ by CH_3CN. The *cis–trans*-isomerization of $PtCl_2(n\text{-}Bu_3P)_2$ and similar Pd complexes, where the isomerization is immeasurably slow in the absence of an excess of phosphine, is very fast when phosphine is present. The isomerization doubtless proceeds by pseudorotation (page 40) of the 5-coordinate state. In this case an ionic mechanism is unlikely since polar solvents actually slow the reaction. Photochemical isomerizations, on the other hand, appear to proceed through tetrahedral intermediates, while thermal isomerizations involve an ionic mechanism.

There are a number of quite stable, isolable 5-coordinate complexes of Pd^{II} and Pt^{II}. First, multifunctional ligands such as tris-[o-(diphenylarsino)-phenylarsine] (QAS) give salts of the type $[Pd(QAS)X]^+X$, and other similar species can be made from certain tertiary phosphines.

Although $PtCl_4^{2-}$ does not appear to coordinate with an excess of Cl^- ion, PdX_4^{2-}, and $[Pd\ phen_2]^{2+}$ appear to do so in solution.[22] However, the best characterized cases for monodentate ligands are those with trichloro-stannate(II) as ligand (cf. page 330). The ion $[Pt(SnCl_3)_5]^{3-}$ can be isolated as an R_4N^+ or R_4P^+ salt from the red solutions obtained by adding an excess of $SnCl_3^-$ to $PtCl_4^{2-}$ in $3M$ HCl or $SnCl_2$ to ethanolic solutions of Na_2PtCl_4. The $[Pt(SnCl_3)_5]^{3-}$ ion may exist only in the solid state; the nature of the $SnCl_3^- \text{–} PtCl_4^{2-}$ solutions is exceedingly complex and depends on the concen-

[20] J. B. Goddard and F. Basolo, *Inorg. Chem.*, 1968, **7**, 936, and references therein.
[21] P. Haake and R. M. Pfeiffer, *J. Amer. Chem. Soc.*, 1970, **92**, 4986, 5242.
[22] See, e.g., C. M. Harris, S. E. Livingstone and I. H. Rees, *J. Chem. Soc.*, **1959**, 1505.

trations, acidity, time and temperature. There are several displacement reactions involved, the first of which is

$$[PtCl_4]^{2-} + SnCl_3^- \rightleftharpoons [PtCl_3(SnCl_3)]^{2-} + Cl^-$$

The $SnCl_3^-$ ion has a strong *trans*-effect (cf. page 667), so that the *trans*-Cl^- would be expected to be labile, giving *trans*-$[PtCl_2(SnCl_3)_2]^{2-}$ readily. Both the *trans*-isomer and the more thermodynamically stable yellow *cis*-isomer can be present in solutions. At higher tin concentrations other species up to the maximum must exist; the equilibria appear also to be very labile. In acetone solution, there is obtained an anion $[Pt_3Sn_2Cl_{20}]^{4-}$. Several complexes with such Pt_3Sn_2 clusters are known; they are *tbp* with axial Sn atoms.[23]

The Pt—Sn complexes catalyze the hydrogenation of ethylene and some other olefinic compounds; this action is doubtless connected with the ready dissociation of the complexes in solution, promoted by the *trans*-effect of $SnCl_3^-$, which leaves vacant sites for coordination of olefin and of hydrogen (cf. also page 771).

Finally, a number of dithio complexes such as dithiocarbamates and dithiophosphates form adducts with phosphine complexes, e.g., $Pt(S_2CNEt_2)$-$(PMePh_2)$. Some of these show nmr behavior in solution suggesting equilibrium between planar and 5-coordinate species.[24]

Octahedral Complexes of Pd^{II} and Pt^{II}. These are very few and, while they may be octahedral in the solid state, e.g., *trans*-$MI_2(diars)_2$, dissociation probably occurs in solution.

Phosphine Complexes. The complexes of tertiary phosphines with Pd^{II} and Pt^{II} have been extensively studied; many complexes are known and these have often been studied in detail, especially from the nmr viewpoint. Thus ideas of π-bonding and *trans*-effects can be obtained through $^{31}P-^{195}Pt$ coupling constants.[25] These coupling constants are a sensitive function of the *trans*-ligand and relatively insensitive to the *cis*-ligand and are thus useful in making structural assignments. For example, *trans*-$Pt(C_6H_5)Cl(PEt_3)_2$ has only a single resonance with J_{Pt-P} 2800 Hz, whereas the *cis*-isomer has two ^{31}P resonances with J_{Pt-P} values of 1580 and 4140 Hz. Orders of *trans*-effects of ligands obtained in this way are similar to those obtained by other methods. The hydrido and alkylphosphine complexes have been particularly well studied and are discussed below.

Because of their kinetic inertness, Pt^{II} complexes of the type $PtCl_2L_2$ allow separation of optical isomers of olefins, allenes, tertiary phosphines or arsines. The complexes are optically stable at 25° and the enantiomorphic forms can be separated by suitable manipulations.[26]

[23] A. Terzis, T. C. Strekas and T. G. Spiro, *Inorg. Chem.*, 1971, **10**, 2617.
[24] J. P. Fackler, Jr., J. A. Fetchin and W. C. Seidel, *J. Amer. Chem. Soc.*, 1969, **91**, 1217.
[25] J. F. Nixon and A. Pidcock, *Ann. Rev. NMR Spectroscopy*, 1969, **2**, 346; F. H. Allen, A. Pidcock and C. R. Waterhouse, *J. Chem. Soc.*, A, **1970**, 2080, 2087.
[26] B. Bosnich and S. W. Wild, *J. Amer. Chem. Soc.*, 1970, **92**, 459; P. H. Boyle, *Quart. Rev.*, 1971, **25**, 323.

Organometallic and Hydrido Complexes of Palladium(II) and Platinum(II).[27] The alkene, allyl, and alkyne complexes of Pd and Pt have been discussed to a large extent in earlier Sections, as have the use of palladium species in catalysis (Chapters 23 and 24). Here we deal mainly with complexes of the type $MXHL_2$ and $MXRL_2$ where X is halogen, R is an alkyl or aryl group and L is usually a tertiary phosphine.

Hydrides. The hydrides have the *trans*-structure, and the *cis*-species appear not to exist.[28] Most of the studies have been on platinum compounds since the comparable palladium (and also nickel) hydrido complexes are usually less stable thermally; some compounds such as *trans*-$PdClH(PR_3)_2$ with R = cyclohexyl or Ph have been made.[29]

The phosphine and arsine hydrides are obtained from the corresponding halides (the *cis*-isomer is usually most reactive) by the action of a variety of hydrogen-transfer agents such as KOH in ethanol, H_2 at 50 atm/95°, $LiAlH_4$ in THF or, most conveniently, 90% aqueous hydrazine, e.g.,

$$cis\text{-}PtBr_2(Et_3P)_2 \rightarrow trans\text{-}PtHBr(Et_3P)_2$$

The KOH-alcohol reduction is a general one for the preparation of hydrido or hydridocarbonyl complexes, but detailed mechanistic studies are lacking.

The hydrido compounds of Pt^{II} are usually air-stable, colorless, crystalline solids, soluble in organic solvents and sublimable. Their chemical reactions resemble those of other hydrido species. They will also add HCl or CH_3I to give octahedral Pt^{IV} complexes, but these usually readily lose the added molecules.

Cationic hydrido species can also be obtained by reactions of the type[30]

$$trans\text{-}PtHCl(AsPh_3)_2 + NaClO_4 + CO \rightarrow [PtH(CO)(AsPh_3)_2]ClO_4 + NaCl$$

and a rare example of a complex with an alkene and hydride bound to platinum has been noted on page 786.

The nmr and ir spectra of hydrido species have been of interest because of the information that can be obtained concerning *trans*-effects (cf. ^{31}P-nmr above). There is a strong dependence of Pt—H stretching frequencies, proton chemical shifts and [195]Pt—H and [31]P—H coupling constants, depending on the ligand *trans* to hydride.[30,31] For example, the Pt—H

[27] F. R. Hartley, *Chem. Rev.*, 1969, **69**, 799 (an extensive review of all types of organo compounds and their use in catalysis); F. R. Hartley, *Organometallic Chem. Rev., A*, 1970, **6**, 119 (starting materials for the preparation of organo compounds); R. J. Cross, *Organometallic Chem. Rev., A*, 1967, **2**, 97 (σ-complexes of Pt^{II} with C, H and Group IV, elements); J. H. Nelson and H. B. Jonassen, *Coordination Chem. Rev.*, 1971, **6**, 27 (monoolefin and acetylene complexes of Ni, Pd and Pt); U. Belluco *et al., Inorg. Chim. Acta Rev.*, 1969, **3**, 19 (unsaturated hydrocarbon complexes of Pt^{II}); P. M. Maitlis, *The Organic Chemistry of Palladium*, Vols. 1, 2, Academic Press, 1971 (a comprehensive account including catalytic reactions); U. Belluco *et al., Inorg. Chim. Acta Rev.*, 1970, **4**, 7 (complexes with Group IV donor ligands); see also references for Chapters 23 and 24.

[28] I. Collamati, A. Furlani and G. Attioli, *J. Chem. Soc., A*, **1970**, 1694.

[29] M. L. H. Green, H. Munakata and T. Saito, *J. Chem. Soc., A*, **1971**, 469; K. Kudo *et al., Chem. Comm.*, **1970**, 1701.

[30] M. J. Church and M. J. Mays, *J. Chem. Soc., A*, **1970**, 1938.

[31] Cf. P. R. Dean and J. C. Green, *J. Chem. Soc., A*, **1968**, 3047; E. R. Birnbaum, *Inorg. Nuclear Chem. Letters*, **1971**, **7**, 233.

stretching frequencies and chemical shifts increase for the *trans*-ligand in the order $Cl < Br < I < NCS < SnCl_3 < CN$.

Alkyls and aryls.   These substances are usually white air-stable crystalline solids and are obtained by action of lithium alkyls or Grignard reagents on the halides.[32] A typical example is *trans*-$PtBrCH_3(PEt_3)_2$. The alkyls can also be made by addition of alkenes at Pt—H bonds (page 779) or by insertion reactions into other Pt—C bonds.[33] Many closely related compounds with Pt or Pd bound to Si, Ge or other metallic elements have been made, some typical ones being those with Pt—SiR_3 groups.[34] They can be made by the action of sodium salts on the platinum halides or by addition of halides to Pd^0 or Pt^0 compounds (see below).

As for the hydrido compounds, the nmr spectra of alkyl complexes have been studied in detail.[35] It may be noted that platinum acetylacetonate complexes often have Pt—C bonds to the γ-carbon atom rather than the usual metal—oxygen bonding, and the acetylacetonate is thus unidentate as in PtCl(acac) (diphos).[36]

Three-membered ring complexes.   Addition of a variety of unsaturated molecules, especially to the zerovalent triphenylphosphine complexes of Pd and Pt, gives rise to complexes that have the metal atom as part of a three-membered ring and formally at any rate in the divalent state. Examples have been given in Chapter 24 (page 773), and some are given in Fig. 26-H-3 (page 1043) but others are the CS_2 complex (26-H-V)[37] and the fumaronitrile complex (26-H-VI) (see also pages 729 and 751).[38]

(26-H-V)   (26-H-VI)

The question of bonding in such complexes and in related olefin and acetylene complexes,[39] which can be considered as Pt^0 complexes (see below), and whether it is best to use an MO or VB treatment, has been discussed elsewhere (page 729).

[32] See, e.g., R. G. Cross and R. Wardle, *J. Chem. Soc.*, A, **1970**, 841.
[33] H. C. Clark and R. J. Puddephat, *Inorg. Chem.*, 1970, **9**, 2670.
[34] J. Chatt, C. Eaborn and P. N. Kapoor, *J. Chem. Soc.*, A, **1970**, 881.
[35] See, e.g., M. J. Church and M. J. Mays, *J. Chem. Soc.*, A, **1968**, 3074; F. H. Allen and A. Pidcock, *J. Chem. Soc.*, A, **1968**, 2700; C. D. Cook and K. Y. Wan, *J. Amer. Chem. Soc.*, 1970, **92**, 2595.
[36] R. Mason, G. B. Robertson and P. L. Pauling, *J. Chem. Soc.*, A, **1969**, 485; G. Hulley, B. F. G. Johnson and J. Lewis, *J. Chem. Soc.*, A, **1970**, 1732.
[37] R. Mason and A. I. M. Rae, *J. Chem. Soc.*, A, **1970**, 1767.
[38] C. Pannatoni *et al.*, *J. Chem. Soc.*, B, **1970**, 371.
[39] G. Bombieri *et al.*, *J. Chem. Soc.*, A, **1970**, 1313; C. D. Cook *et al.*, *J. Amer. Chem. Soc.*, 1971, **93**, 1904.

26-H-3. Complexes of Palladium(IV) and Platinum(IV), d^6

Palladium. The Pd^{IV} complexes are more stable than simple Pd^{IV} compounds, but only a few are known, and apart from the complex formed on dissolution of Pd in concentrated nitric acid (page 992) they are mainly the octahedral halide anions. The fluoro complexes of Ni, Pd and Pt are all very similar and are rapidly hydrolyzed by water.[40] The chloro and bromo ions are stable to hydrolysis but are decomposed by hot water to give the Pd^{II} complex and halogen. The red $PdCl_6^{2-}$ ion is formed when Pd is dissolved in aqua regia or when $PdCl_4^{2-}$ solutions are treated with chlorine.

Oxidation of K_2PdCl_4 by persulfate in presence of KCN gives the yellow salt, $K_2[Pd(CN)_6]$.[41]

The diamine complexes, such as $Pdpy_2Cl_4$, which is obtained as a deep orange crystalline powder when $Pdpy_2Cl_2$ suspended in chloroform is treated with chlorine, are of marginal stability. They lose chlorine or bromine rapidly in moist air. Other complexes such as $Pd(NH_3)_2(NO_2)_2Cl_2$ are stable.

Platinum. In marked contrast to Pd^{IV}, platinum(IV) forms many thermally stable and kinetically inert complexes. So far as is known, Pt^{IV} complexes are invariably octahedral and, in fact, Pt^{IV} has such a pronounced tendency to be six-coordinated that in some of its compounds quite unusual structures are adopted. An apparent exception to the rule is $h^5\text{-}C_5H_5Pt(CH_3)_3$ but, as with other $h^5\text{-}C_5H_5$ complexes, the ring can be considered as occupying three positions of an octahedron. Several interesting examples of this tendency of Pt^{IV} to be 6-coordinate exist where novel bonding is required for this to be achieved (see below).

The most extensive and typical series of Pt^{IV} complexes are those which span the entire range from the hexammines, $[PtAm_6]X_4$, including all intermediates such as $[PtAm_4X_2]X_2$ and $M^I[PtAmX_5]$, to $M_2^I[PtX_6]$. Some of these are particularly notable as examples of the classical evidence that led Werner to assign the coordination number 6 to Pt^{IV}. The amines which occur in these complexes include ammonia, hydrazine, hydroxylamine and ethylenediamine, and the acido groups include the halogens, thiocyanate, hydroxide and nitro group. Although not all these groups are known to occur in all possible combinations in all types of compounds, it can be said that, with a few exceptions, they are generally interchangeable.

The most important Pt^{IV} compounds are salts of the red *hexachloroplatinate* ion, $PtCl_6^{2-}$. The "acid", commonly referred to as chloroplatinic acid, is an oxonium salt (page 167); this or its Na or K salt is the normal starting material for the preparation of many Pt compounds. The ion is formed on dissolving Pt in aqua regia or HCl saturated with Cl_2. The oxidation of $PtCl_4^{2-}$ and other Pt^{II} complexes by Cl_2 and Br_2 has been studied in detail.[41a]

[40] N. A. Matwiyoff *et al.*, *Inorg. Chem.*, 1969, **8**, 750.
[41] H. Siebert and A. Siebert, *Z. anorg. Chem.*, 1970, **378**, 160.
[41a] M. M. Jones and K. A. Morgan, *J. Inorg. Nuclear Chem.*, 1972, **34**, 259, 275.

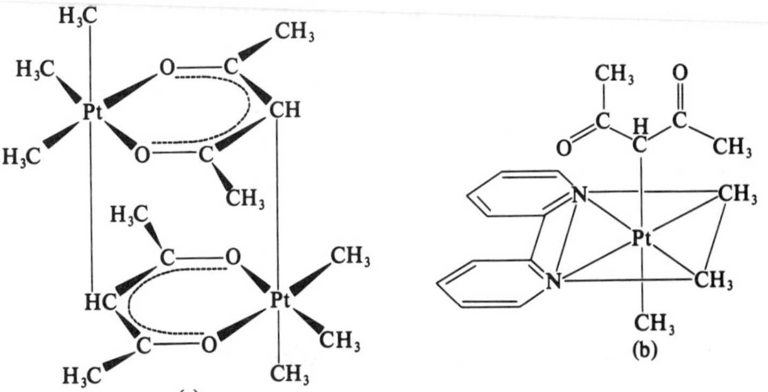

Fig. 26-H-2. (a) The molecular structure of trimethylplatinum acetylacetonate dimer, showing how the Pt^{IV} attains octahedral coordination. (b) Schematic representation of the molecular structure of the bipyridine adduct of trimethylplatinum(IV) acetylacetonate.

Platinum(IV)*-to-carbon bonds.*[27,42] A characteristic feature of Pt^{IV} is the stability of compounds with a $(CH_3)_3Pt$ group. In all of these, Pt^{IV} is octahedrally coordinated.[43] Thus the halides are tetrameric, e.g., $[Me_3PtCl]_4$ with three-way halogen bridges.[43b] In aqueous solutions the very stable octahedral ion[44] $[Me_3Pt(H_2O)_3]^+$ is formed with non-coordinating anions such as BF_4^- or ClO_4^-. Tetramethylplatinum does not exist and what was thought to be this compound was actually $[Me_3Pt(OH)]_4$.[45]

Trimethylplatinum acetylacetonate, $[(CH_3)_3Pt(O_2C_5H_7)]_2$, long known to be a dimer in non-coordinating solvents, has the structure shown in Fig. 26-H-2a.[46] The acetylacetone functions as a tridentate ligand, the third donor atom, besides the two oxygen atoms which are normally the only donor atoms, being the middle carbon atom of the ring. The donor ability of this carbon atom can perhaps best be understood in terms of resonance structures of the chelate ring as shown in (26-H-VII). Ordinarily only (26-H-VIIa) and (26-H-VIIb) are considered, but (26-H-VIIc) is also quite valid and must predominate in this platinum(IV) dimer. Other Me_3Pt derivatives also have octahedrally coordinated Pt.

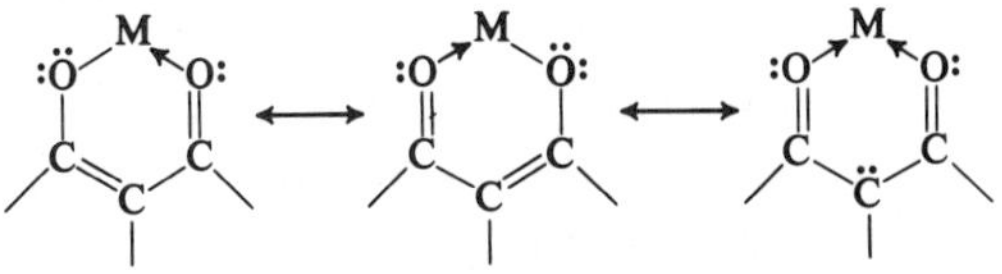

(26-H-VIIa)　　　(26-H-VIIb)　　　(26-H-VIIc)

[42] J. S. Thayer, *Organometallic Chem. Rev.*, A, 1970, **5**, 53.
[43] (a) See, e.g., J. D. Ruddick, and B. L. Shaw, *J. Chem. Soc., A*, **1969**, 2801; (b) R. N. Hargreaves and M. R. Truter, *J. Chem. Soc., A*, **1971**, 90.
[44] D. E. Clegg, J. R. Hall and N. S. Hain, *Austral. J. Chem.*, 1970, **23**, 1981; K. Kite and D. R. Rosseinsky, *Chem. Comm.*, **1971**, 205.
[45] P. A. Bulliner, V. A. Maroni and T. G. Spiro, *Inorg. Chem.*, 1970, **9**, 1887.
[46] R. N. Hargreaves and M. R. Truter, *J. Chem. Soc., A*, **1969**, 2282; for review of C-bonded β-diketonates of Pt and other metals see D. Gibson, *Coordination Chem. Rev.*, 1969, **4**, 225.

In the monomeric compound, $(CH_3)_3Pt(dipy)(O_2C_5H_7)$, six rather than seven coordination is achieved by the formation of only one bond to the acetylacetonate ion (Fig. 26-H-2b) as in the Pt^{II} complexes noted earlier. The great strength of the Pt—C bond is shown by the fact that in the preparation of this compound from $[(CH_3)_3Pt(O_2C_5H_7)]_2$ by the action of bipyridine, it is the Pt—O rather than Pt—C bonds that are broken.

Finally, the compound made by the action of cyclopropane on chloroplatinic acid is similar to the tetrameric alkyls but with a platino cyclobutane ring.[47]

26-H-4. Complexes of Palladium(0) and Platinum(0), d^{10} [48]

There is a reasonably extensive chemistry of the 0 state, which is of importance especially for triphenylphosphine complexes; the latter undergo a wide variety of oxidative-addition reactions.

In general, the compounds are similar to those of Ni^0 except that no analog of $Ni(CO)_4$ is known. This is believed to be due to a poorer tendency to π-bonding, possibly associated with the greater ionization potentials of Pd and Pt. It is significant that when triphenylphosphine, which is a better σ-donor but poorer π-acceptor, is present the carbonyl compounds are relatively stable, e.g., $Pd(CO)(PPh_3)_3$ and $Pt(CO)_2(PPh_3)_2$.[49]

The most extensive series of compounds is that with tertiary phosphines, examples being $Pt(PF_3)_4$, $Pt[P(CF_3)_2F]_4$, $Pt(diphos)_2$, etc. The fluorophosphine compounds can be readily obtained by interaction of, e.g., PF_3 or $PF(CF_3)_2$, at high pressure with the chlorides.[50, 51] They are volatile liquids but $Pd(PF_3)_4$ decomposes above $-20°$. They can be rapidly substituted by other phosphines by an S_N1 mechanism[51] although, presumably because of removal of electron density from the metal by inductive effects, they do not undergo oxidative-addition reactions. They are tetrahedral.

Although binary carbonyls are not known, what may be a polymeric carbonyl $[Pt(CO)_2]_n$ is obtained as an air-sensitive substance by action of CO on ethanolic Na_2PtCl_4; it reacts with PPh_3 to give $Pt_3(CO)_3(PPh_3)_4$.[52] Phosphine-substituted carbonyls, many of them clusters, are known, viz., $Pt(CO)L_3$, $Pd_3(CO)_3L$, $Pt_4(CO)_5L_4$.[53]

Triphenylphosphine Complexes.[54]   These are the most important of the zerovalent species. They are made by action of hydrazine on ethanolic

[47] S. E. Binns et al., J. Chem. Soc., A, 1969, 1227.
[48] R. Ugo, Coordination Chem. Rev., 1968, 3, 319.
[49] V. G. Albano, G. M. B. Ricci and P. L. Belloni, Inorg. Chem., 1969, 8, 2109; Chem. Comm., 1969, 899; M. Kudo, M. Hidai and Y. Uchida, J. Organ. Chem., 1971, 33, 393.
[50] J. F. Nixon and M. D. Sexton, J. Chem. Soc., A, 1970, 321.
[51] R. D. Johnson, F. Basolo and R. G. Pearson, Inorg. Chem., 1971, 10, 247.
[52] G. Booth and J. Chatt, J. Chem. Soc., A, 1969, 2131.
[53] P. Chini et al., J. Chem. Soc., A, 1970, 1538, 1542; R. G. Vranka et al., J. Amer. Chem. Soc., 1969, 91, 1574.
[54] For references see D. M. Blake and C. J. Nyman. J. Amer. Chem. Soc., 1970, 92, 5359; R. Ugo et al., Inorg. Chim. Acta, 1970, 4, 390; H. C. Clark and K. Itoh, Inorg. Chem., 1971, 10, 1707.

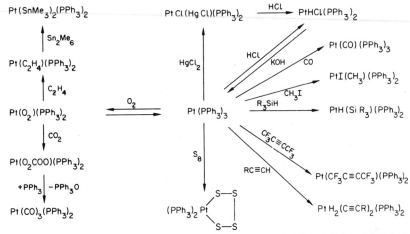

Fig. 26-H-3. Oxidative-addition and related reactions of tris(triphenylphosphine)platinum. Reactions with Pt(PPh₃)₄ and the corresponding palladium complexes are similar.

solutions of K_2PdCl_4 or K_2PtCl_4. Hydrido intermediates occur and these can be dehydrogenated by strong bases such as KOH. Depending on the reaction conditions, either $M(PPh_3)_3$ or $M(PPh_3)_4$ may be obtained as white crystalline solids. Their main characteristic is dissociation:[55]

$$Pt(PPh_3)_4 \underset{}{\overset{-PPh_3}{\rightleftharpoons}} Pt(PPh_3)_3$$

but a proposed second dissociation to $Pt(PPh_3)_2$ probably does not occur in absence of traces of oxygen. The main driving force for this may be steric. The compounds are notable for the wide range of oxidative-addition reactions[54, 56] that they undergo to give either normal square M^{II} complexes or complexes with three-membered rings. Examples are given in Fig. 26-H-3.

A variety of related but formally zerovalent compounds with olefins[57] and acetylenes[58] are also known. These are of the types $(R_3P)_2Pt$-$(R^1R^2C=CR^3R^4)$ and $(R_3P)_2Pt(R^1C≡CR^2)$. These complexes also have three-membered rings but in $(Ph_3P)_2Pt(CF_2=CCl_2)$, for example, where the two P atoms, Pt and the two C atoms are coplanar, the metal could be regarded as three-coordinate with the olefin occupying only one coordination position.

The oxygen complexes, e.g., $(Ph_3P)_2Pt(O_2)$ can act as oxidants for CO, CO_2, SO_2, PPh_3, etc., and peroxo intermediates have been characterized (Chapter 24, page 785). Finally, unusual Pd and Pt compounds of diben-

[55] J. Halpern and A. L. Pickard, Inorg. Chem., 1970, 9, 2798.
[56] See, e.g., S. Cenini et al., J. Chem. Soc., A, 1971, 113, 409, 416, 522; D. M. Roundhill, Inorg. Chem., 1970, 9, 254; J. Chatt and D. P. Mingos, J. Chem. Soc., A, 1970, 1243; R. D. W. Kemmitt, R. D. Peacock and J. Stocks, J. Chem. Soc., A, 1971, 846.
[57] For structures and references see J. N. Francis, A. McAdam, and J. A. Ibers, J. Organometallic Chem., 1971, 29, 131, 149; R. van der Linde and R. O. de Jongh, Chem. Comm., 1971, 563.
[58] J. H. Nelson, J. J. R. Reed and H. B. Jonassen, J. Organometallic Chem., 1971, 29, 163; M. A. Bennett et al., J. Amer. Chem. Soc., 1971, 93, 3797.

zylidineacetone, $M(PhCH{=}CHCOCH{=}CHPh)_n$, $n = 2, 3$, are obtained by interaction of the ligand with Na_2PdCl_4 or K_2PtCl_4 in hot methanol. The ligand appears to be bound *via* a π-carbonyl group rather than by the olefinic groups.[59]

26-I. SILVER AND GOLD

Like copper, silver and gold have a single s electron outside a completed d shell, but in spite of the similarity in electronic structures and ionization potentials there are few resemblances between Ag, Au and Cu, and there are no simple explanations for many of the differences.

Apart from obviously similar stoichiometries of compounds in the same oxidation state (which do not always have the same structure) there are some similarities within the group—or at least between two of the three elements:

1. The metals all crystallize with the same face-centered cubic (*ccp*) lattice.

2. Cu_2O and Ag_2O have the same body-centered cubic structure where the metal atom has two close oxygen neighbors and every oxygen is tetrahedrally surrounded by four metal atoms.

3. Although the stability constant sequence for halo complexes of many metals is $F > Cl > Br > I$, Cu^I and Ag^I belong to the group of ions of the more noble metals for which it is the reverse.

4. Cu^I and Ag^I (and to a lesser extent Au^I) form very much the same types of ion and compound, such as $[MCl_2]^-$, $[Et_3AsMI]_4$ and K_2MCl_3.

5. Certain complexes of Cu^{II} and Ag^{II} are isomorphous, and Ag^{III}, Au^{III} and Cu^{III} also give similar complexes.

The only stable cationic species, apart from complex ions, is Ag^+ and by contrast the Au^+ ion is exceedingly unstable with respect to the disproportionation

$$3Au^+(aq) = Au^{3+}(aq) + 2Au(s) \qquad K \approx 10^{10}$$

Gold(III) is invariably complexed in all solutions, usually as anionic species such as $[AuCl_3OH]^-$. The other oxidation states, Ag^{II} Ag^{III} and Au^I, are either unstable to water or exist only in insoluble compounds or complexed species. Intercomparisons of the standard potentials are of limited utility, particularly since these strongly depend on the nature of the anion; some useful ones are:

$$Ag^{2+} \xrightarrow{1.98} Ag^+ \xrightarrow{0.799} Ag$$
$$Ag(CN)_2^- \xrightarrow{-0.31} Ag + 2CN^-$$
$$AuCl_4^- \xrightarrow{1.00} Au + 4Cl^-$$
$$Au(CN)_2^- \xrightarrow{-0.6} Au + 2CN^-$$

Gold(II) occurs *formally* in dithiolene compounds (page 724) and in the dicarbollyl $[Au(B_9C_2H_{11})_2]^{2-}$ but otherwise possibly exists only as a transient intermediate in reactions.

The oxidation states and stereochemistry are summarized in Table 26-I-1.

[59] K. Moseley and P. M. Maitlis, *Chem. Comm.*, **1971**, 982.

TABLE 26-I-1

Oxidation States and Stereochemistry of Silver and Gold

Oxidation state	Coordination number	Geometry	Examples
Ag^I, d^{10}	2^a	Linear	$[Ag(CN)_2]^-$, $[Ag(NH_3)_2]^+$, AgSCN
	3	Trigonal	$(Me_2NC_6H_4PEt_2)_2AgI$
	4^a	Tetrahedral	$[Ag(SCN)_4]^{3-}$, $[AgIPR_3]_4$, $[AgSCNPPr_3]_n$, $[Ag(PPh_3)_4]^+ClO_4^-$
	6	Octahedral	AgF, AgCl, AgBr(NaCl structure)
Ag^{II}, d^9	4	Planar	$[Ag\ py_4]^{2+}$
	6	Distorted octahedral	Ag (2,6-pyridinedicarboxylate)$_2 \cdot H_2O$
Ag^{III}, d^8	4	Planar	AgF_4^-, $\frac{1}{2}$ of Ag atoms in AgO, $[Ag(ebg)_2]^{3+b}$
	6	Octahedral	$[Ag(IO_6)_2]^{7-}$, Cs_2KAgF_6
Au^I, d^{10}	2^a	Linear	$[Au(CN)_2]^-$, $Et_3P \cdot AuC \equiv C \cdot C_6H_5$; $(AuI)_n$
	4	Tetrahedral	$[Au(diars)_2]^+I^-$
Au^{III}, d^8	4^a	Planar	$AuBr_4^-$, Au_2Cl_6, $[(C_2H_5)_2AuBr]_2$, R_3PAuX_3, $K[Au(NO_3)_4]$
	5	Trigonal bipyramidal	$[Au(diars)_2I]^{2+}$
	6	Octahedral	$AuBr_6^{3-}$, trans-$[Au(diars)_2I_2]^+$

[a] Most common states.
[b] See diagram 26-I-IV, page 1050.

26-I-1. The Elements

The elements are widely distributed in Nature in the free state and in sulfides and arsenides; silver also occurs as the chloride, AgCl. Silver is often recovered from the work-up of copper and lead ores. The elements are usually extracted by treatment with cyanide solutions in presence of air and are recovered from them by addition of zinc. They are purified by electro-deposition.

Silver is a white, lustrous, soft, and malleable metal (m.p. 961°) with the highest known electrical and thermal conductivity. It is chemically less reactive than copper, except toward sulfur and hydrogen sulfide, which rapidly blacken silver surfaces. The metal dissolves in oxidizing acids and in cyanide solutions in presence of oxygen or peroxide.

Gold[1] is a soft, yellow metal (m.p. 1063°) with the highest ductility and malleability of any element. It is chemically unreactive and is not attacked by oxygen or sulfur, but reacts readily with halogens or with solutions containing or generating chlorine such as aqua regia; and it dissolves in cyanide solutions in presence of air or hydrogen peroxide to form $[Au(CN)_2]^-$. The reduction of solutions of $AuCl_4^-$ by various reducing agents such as $SnCl_3^-$ may, under suitable conditions, give highly colored solutions containing colloidal gold.

Both silver and gold form many alloys and, in the case of gold, some sub-

[1] E. M. Wise, ed., *Gold: Recovery, Properties and Applications*, Van Nostrand, 1964.

stances can be regarded as compounds, e.g., Cs^+Au^- or $AuTe_2$;[2a] gold also forms many very stable gaseous molecules, e.g., AlAu or NiAu.[2b]

Since the chemistries of Ag and Au compounds differ so much, we treat them separately.

26-I-2. Silver Compounds[3]

Silver(I), d^{10}, Compounds. The argentous state (Ag^I) is the normal and dominant oxidation state. The well known argentous ion, Ag^+, is evidently solvated in aqueous solution but an aquo ion does not occur in salts, practically all of which are anhydrous. $AgNO_3$, $AgClO_3$ and $AgClO_4$ are water-soluble but Ag_2SO_4 and $AgOOCCH_3$ are sparingly so. The salts of oxo anions are primarily ionic in nature, but although the water-insoluble halides AgCl and AgBr have the NaCl structure there appears to be appreciable covalent character in the Ag···X interactions, while in compounds such as AgCN and AgSCN, which have chain structures (26-I-I and 26-I-II), the bonds are considered to be predominantly covalent.

$$-Ag-C-N-Ag-C-N-$$

(26-I-I)

(26-I-II)

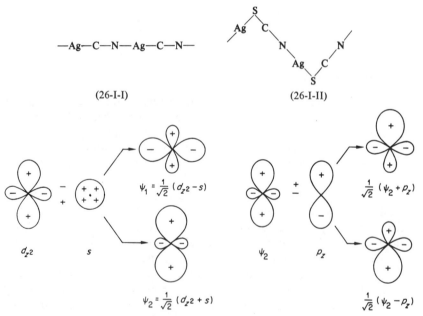

Fig. 26-I-1. Sketches showing the hybrid orbitals formed from a d_{z^2} and an s orbital, ψ_1 and ψ_2, and the hybrids that can be formed from ψ_2 and a p_z orbital. In each sketch the z axis is vertical and the actual orbital is the figure generated by rotating the sketch about the z axis.

2 (a) R. F. L. Andon, J. F. Martin and K. C. Mills, *J. Chem. Soc., A*, **1971**, 1788. (b) K. A. Gingerich and H. C. Finkelbein, *Chem. Comm.*, **1969**, 901.
3 L. M. Gedansky and L. G. Hepler, *Engelhard Tech. Bull.*, 1969, **9**, 117 (collected thermodynamic and potential data for silver and its compounds, in solid and aqueous solutions; 100 references).

Ag^I and Au^I, along with Cu^I and Hg^{II}, show a pronounced tendency to exhibit linear, 2-fold coordination. This may be due to a relatively small energy difference between the filled d orbitals and the unfilled s orbital ($4d$, $5s$ for Ag^I), which permits extensive hybridization of the d_{z^2} and s orbitals, as shown in Fig. 26-I-1. The electron pair initially in the d_{z^2} orbital occupies ψ_1, giving a circular region of relatively high electron density from which ligands are somewhat repelled, and regions above and below this ring in which the electron density is relatively low. Ligands are attracted to the latter regions. By further mixing of ψ_2 with the p_z orbital, two hybrid orbitals suitable for forming a pair of linear covalent bonds can be formed.

Binary Compounds. *Argentous oxide.* The addition of alkali hydroxide to Ag^+ solutions produces a dark brown precipitate that is difficult to free from alkali ions. It is strongly basic and its aqueous suspensions are alkaline:

$$\tfrac{1}{2}Ag_2O(s) + \tfrac{1}{2}H_2O = Ag^+ + OH^- \qquad \log K = -7.42 \;(25°, 3\,M\;NaClO_4)$$
$$\tfrac{1}{2}Ag_2O(s) + \tfrac{1}{2}H_2O = AgOH \qquad \log K = -5.75$$

they absorb carbon dioxide from the air to give Ag_2CO_3. The oxide decomposes above $\sim 160°$ and is readily reduced to the metal by hydrogen. Silver oxide is more soluble in strongly alkaline solution than in water, and $AgOH$ and $Ag(OH)_2^-$ are formed. The treatment of water-soluble halides with a suspension of silver oxide is a useful way of preparing hydroxides, since the silver halides are insoluble. Analogously to copper (page 906) and gold, alkali-metal silver oxides contain $Ag_4O_4^{4-}$ units.

Argentous sulfide. The action of hydrogen sulfide on argentous solutions gives black Ag_2S, which is the least soluble in water of all silver compounds ($\log K_{SP} \approx -50$). The black coating often found on silver articles is the sulfide; this can be readily reduced by contact with aluminum in dilute sodium carbonate solution.

Argentous halides. The *fluoride* is unique in forming hydrates such as $AgF \cdot 4H_2O$, which are obtained by crystallizing solutions of Ag_2O in aqueous HF. The other well known halides are precipitated by the addition of X^- to Ag^+ solutions;[4] the color and insolubility in water increase $Cl < Br < I$. Silver chloride can be obtained as rather tough sheets that are transparent over much of the infrared region and have been used for cell materials. Silver chloride and bromide are light-sensitive and have been intensively studied because of their importance in photography.

Argentous Complexes. There is a great variety of silver complexes, which exist in solution or in the solid state. Since the most stable Ag^+ complexes have the linear structure, $L—Ag—L^+$, chelating ligands cannot form such simple ions and hence tend to give polynuclear complex ions. For monodentate ligands, the species AgL^+, AgL_2^+, AgL_3^+ and AgL_4^+ can exist, but the constants K_1 and K_2 are usually high whereas K_3 and K_4 are relatively small; hence the main species are usually of the linear AgL_2 type. The coordination number, however, is sensitive to the nature of the ligand and anions

---

[4] R. Ramette, *J. Chem. Educ.*, 1960 **37**, 348 (solubility and equilibria of AgCl).

and a variety of types can occur because of the possibilities of sp^2 and sp^3 bonding of Ag^+, in addition to the linear hybridization discussed earlier.[5]

Complexes are not well known for oxygen ligands. Ligands with the donor atoms N, P, As, S, Se, etc., give many complex compounds, but it is doubtful whether π-bonding of the type important in the earlier transition groups makes any substantial contribution to the bonding where P, S, etc. are the donor atoms. Thus, from studies on the complex formation by donor atoms in acids such as $RC_6H_4As(CH_2COOH)_2$, which gives only a 1:1 complex, and similar P, S and Se compounds, the greater stability of these complexes than of their N analogs has been attributed to the polarizing power of Ag^+ and the high polarizability of the "soft" donor atoms.[6]

Halogeno Complexes. The insoluble halides such as AgCl are appreciably soluble in concentrated HNO_3 and HCl. They also dissolve in solutions of high halide ion concentration[7] and also in solutions of $AgNO_3$ or $AgClO_4$ in CH_3CN, $(CH_3)_2SO$, acetone and other solvents.[7, 8] Complex anions or cations are thus formed:

$$AgX + nX^- \rightleftarrows AgX_{n+1}^{n-}$$
$$AgX + nAg^+ \rightleftarrows Ag_{n+1}X^{n+}.$$

The stability of the ions is generally $I > Br > Cl$. In Cs_2AgI_3 there are chains of AgI_4 sharing corners, while in $[Me_4N]Ag_2I_3$ the AgI_4 tetrahedra share edges.[9] Salts of $Ag_4I_5^-$ and $Ag_{13}I_{15}^{2-}$ have channel structures and behave as solid electrolytes.[10]

The halides also dissolve, giving complexes, in solutions of NH_3, CN^- and $S_2O_3^{2-}$, although AgI is but sparingly soluble in ammonia. Silver cyanide, which is also insoluble in water, dissolves in CN^- and also in liquid HF where an HCN donor complex, $AgNCH^+$, is formed by protonation.[11]

Other Complexes. Salts of the complex ions mentioned above can usually be readily isolated and, in addition, tertiary phosphine or phosphite complexes of the types $[AgL_4]ClO_4$, AgL_2X, AgL_3X are known. For ethylenediamine, the complex obtained depends on the conditions, but 2-coordination of Ag^I is preserved either by bridging as in $ClAgNH_2CH_2CH_2NH_2$-AgCl or by polymerization.

Organo compounds of Ag^I are discussed below.

Silver(II) Compounds, d^9.[12] The Ag^{2+} ion is well defined and numerous complexes are known but only one binary compound, AgF_2; the oxide, AgO, is actually $Ag^IAg^{III}O_2$ and is accordingly discussed under Ag^{III}.

Silver(II) *fluoride* is obtained as a dark brown solid by heating AgF or other silver compounds in F_2. It appears to be an authentic Ag^{II} compound

[5] See, e.g., phosphine complexes; E. L. Mutterties and C. W. Allegranti, *J. Amer. Chem. Soc.*, 1970, **92**, 4114.
[6] L. D. Pettit and A. Royston, *J. Chem. Soc.*, A, **1969**, 1970.
[7] D. C. Luehrs, R. Iwamoto and J. Kleinberg, *Inorg. Chem.*, 1966, **5**, 201.
[8] D. C. Luehrs and K. Abote, *J. Inorg. Nuclear Chem.*, 1969, **31**, 549.
[9] P. A. Tucker and P. Woodward, *J. Chem. Soc.*, A, **1971**, 1337.
[10] S. Geller and M. D. Lind, *J. Chem. Phys.*, 1970, **52**, 5854.
[11] M. F. A. Dove and J. G. Hallett, *J. Chem. Soc.*, A, **1969**, 278.

but is antiferromagnetic, having μ_{eff} (298° K) well below the spin-only value for one unpaired electron. It is a useful fluorinating agent (cf. page 489). It is hydrolyzed at once by moisture.

The argentic ion, Ag^{2+}, which is paramagnetic with an unpaired electron, is obtained in $HClO_4$ or HNO_3 solution by oxidation of Ag^+ with ozone or by dissolution of AgO in acid.

The potentials for the Ag^{2+}/Ag^+ couple, $+2.00$ V, in $4M$ $HClO_4$ and $+1.93$ V in $4M$ HNO_3, show that Ag^{2+} is a powerful oxidizing agent. There is evidence for complexing by NO_3^-, SO_4^{2-} and ClO_4^- in solution, and the electronic spectra in $HClO_4$ solutions are dependent on acid concentration, for example. The ion is reduced by water, even in strongly acid solution, but the mechanism is complicated.[13]

Many oxidations, e.g., of oxalate, by the peroxodisulfate ion are catalyzed by Ag^+ ion and the kinetics are best interpreted by assuming initial oxidation to Ag^{2+}, which is then reduced by the substrate.[14a] Decarboxylation[14b] of carboxylic acids is also promoted by Ag^{II} complexes such as (26-I-III).

(26-I-III)

Complexes. Numerous complexes of Ag^{II} are known and they are normally prepared by persulfate oxidation of Ag^+ solutions containing the complexing ligand.

With neutral ligands, cationic species such as $[Ag\,py_4]^{2+}$, $[Ag(dipy)_2]^{2+}$ and $[Ag(phen)_2]^{2+}$ form crystalline salts, while with uninegative chelating ligands such as 2-pyridinecarboxylate, neutral species such as (26-I-III) are obtained.[15]

The Ag^{II} complexes have $\mu_{eff} = 1.75$–2.2 B.M., consistent with the d^9 configuration, and their electronic spectra accord with square coordination. The salt $[Ag\,py_4]S_2O_8$ and the bispicolinate (26-I-III) are isomorphous with the planar copper(II) analogs. Exceptions to the rule are the unusual 2,3- and 2,6-pyridinedicarboxylates, $Ag(C_7H_4NO_4)_2 \cdot H_2O$, which have a very distorted octahedral structure.[16]

Silver(III), d^8, Compounds.[12] **Binary Compounds.** A black oxide, that is not readily purified but is probably Ag_2O_3, is obtained by anodic oxidation of Ag^+ in alkaline solution. The black so-called "silver(II) oxide," which is

[12] J. A. McMillan, *Chem. Rev.*, 1962, **62**, 65 (an extensive review of higher oxidation states of Ag).
[13] A. Viste et al., *Inorg. Chem.*, 1971, **10**, 631.
[14] J. M. Anderson and J. K. Kochi, (a) *J. Amer. Chem. Soc.*, 1970, **92**, 1651; (b) *J. Org. Chem.*, 1970, **35**, 986.
[15] See, e.g., M. G. B. Drew, R. W. Matthews and R. A. Walton, *J. Chem. Soc., A*, **1970**, 1405; W. H. Thorpe and J. K. Kochi, *J. Inorg. Nuclear Chem.*, 1971, **33**, 3958.
[16] M. G. B. Drew et al., *J. Amer. Chem. Soc.*, 1969, **91**, 7769; *J. Chem. Soc, A*, **1571**, 2559.

better characterized, is made by oxidation of Ag_2O in NaOH solution with $S_2O_8^{2-}$ or, as single crystals, by controlled electrolysis of $2M$ $AgNO_3$ solutions.[17]

The oxide is a semi-conductor, is stable to $\sim 100°$, and dissolves in acids evolving oxygen but giving some Ag^{2+} in solution. It is a powerful oxidizing agent. Since AgO is diamagnetic, it cannot in fact be Ag^{II} oxide. Neutron diffraction shows that it is $Ag^{I}Ag^{III}O_2$ with two types of silver atoms in the lattice, one with linear coordination to two oxygen atoms (Ag^I) and the other square with respect to oxygen (Ag^{III}). When AgO is dissolved in acid Ag^{II} is formed, probably according to the equation

$$AgO^+ + Ag^+ + 2H^+ \rightarrow 2Ag^{2+} + H_2O$$

but, in presence of complexing agents in alkaline solution, Ag^{III} complexes are obtained (see below). The separation of Ag^I and Ag^{III} can be made by the reaction[18]

$$4AgO + 6KOH + 4KIO_4 \rightarrow 2K_5H_2[Ag^{III}(IO_6)_2] + Ag_2O + H_2O$$

We can mention here the unusual salts of stoichiometry $Ag_7O_8^+$ HF_2^- that are obtained as black needles by electrolysis of aqueous solutions of AgF. These contain a polyhedral ion $Ag_6O_8^+$ that acts as a clathrate to enclose Ag^+ and HF_2^-. The salts thus contain Ag^I and Ag^{III} with an average oxidation state of $2\frac{3}{7}$.[19]

Complexes. The most readily obtained complexes of Ag^{III} are made by oxidation of Ag_2O in strongly alkaline solution by $S_2O_8^{2-}$ in presence of periodate or tellurate ions. Representative compounds are: $K_6H[Ag(IO_6)_2] \cdot 10H_2O$, and $Na_6H_3[Ag(TeO_6)_2] \cdot 18H_2O$. These species are analogous to those of Cu^{III} and, by analogy, the structure of the periodate anion is probably of the type (26-I-IV), where one or more of the oxygen atoms bound to iodine can be protonated and such OH groups H-bonded to water. An Ag^{III} complex of remarkable stability is the one with ethylenedibiguanide (26-I-V), which is obtained as the red sulfate when Ag_2SO_4 is treated with aqueous potassium peroxodisulfate in the presence of ethylenedibiguanid-

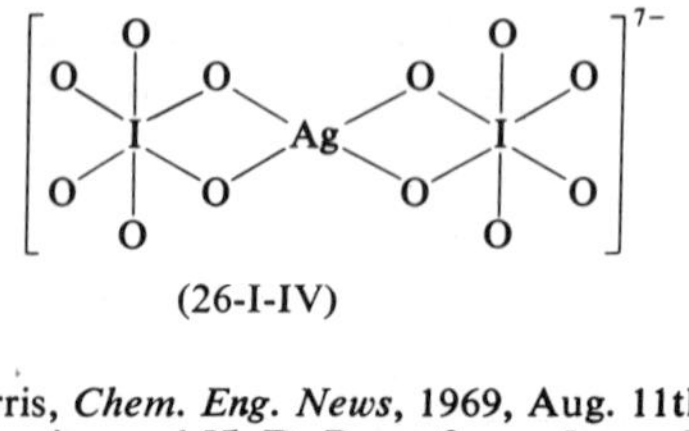

(26-I-IV) 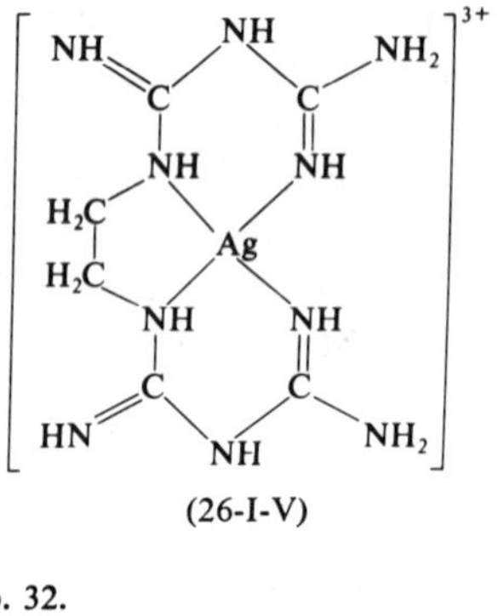   (26-I-V)

[17] L. Norris, *Chem. Eng. News*, 1969, Aug. 11th, p. 32.
[18] J. L. Servian and H. D. Buenafama, *Inorg. Nuclear Chem. Letters*, **1969**, 337.
[19] A. C. Gossard *et al.*, *J. Amer. Chem. Soc.*, 1967, **89**, 7121.

inium sulfate. The hydroxide, nitrate and perchlorate have been prepared metathetically. These salts are diamagnetic and oxidize two equivalents of iodide ion per gram-atom of silver.

Less accessible are the yellow fluoro complexes, e.g., $KAgF_4$ or Cs_2KAgF_6, obtained by action of F_2 at 300° on a stoichiometric mixture of the alkali chlorides and silver nitrate. The AgF_6^{3-} ion appears to be octahedral according to its magnetism and electronic absorption spectrum[20] ($\mu_{eff} = 2.6$ B.M. at 298°; $\Delta_0 = 18,400$ cm^{-1}).

Organosilver Compounds.[21] The organo compounds are all of Ag^I. The σ-bonded alkyls and aryls, which can be prepared by usual alkylating agents, are very unstable; thus CH_3Ag, a yellow solid, decomposes rapidly above $-30°$. However, comparatively stable fluoroalkyls can be obtained[22] by reactions such as

$$AgF + CF_3CF{=}CF_2 \xrightarrow{\text{MeCN}} (CF_3)_2CFAg(MeCN)$$

These compounds are probably clusters when non-solvated.

The only important classes are compounds derived from acetylenes, olefins, and aromatic compounds (Chapter 23).

Unsaturated hydrocarbon compounds. Virtually all olefins and many aromatic compounds give complexes when the hydrocarbon is shaken with aqueous solutions of $AgNO_3$ or $AgBF_4$. In many cases, crystalline solids may be obtained, but even where solids are not formed the equilibrium constants can be determined by studying the distribution of Ag$^+$ between aqueous and organic solvent phases. The stoichiometry of the solid complexes, depending on the conditions, may be 3:1, 2:1 or 1:1 and the structures of many of these have been determined; Ag$^+$ can be bound to 1, 2 or 3 double bonds. In some cases, this leads to the formation of polymers as in the rather similar Cu^I–alkene complexes (page 911).

The formation of crystalline Ag$^+$ complexes may provide a simple way of purifying particular alkenes, or for separation of mixtures, e.g., of 1,3-, 1,4- and 1,5-cyclooctadiene, or of the optical isomers of α- and β-pinene, because of the differing stabilities of the complexes.

Complexes are also formed by other unsaturated compounds such as alcohols or esters, and if suitable functional groups are present it is possible for chelate compounds to be formed; thus 2-allylpyridine, NC_5H_4—$CH_2CH{=}CH_2$, is bound to Ag$^+$ by both the double bond and the N atom.

Alkynyl compounds are also similar to those formed by Cu^I (page 910). Interaction of acetylene with Ag$^+$ solutions gives a yellow precipitate:

$$C_2H_2 + 2Ag^+ \rightarrow AgC{\equiv}CAg + 2H^+$$

while substituted alkynes give white precipitates $[RC{\equiv}CAg]_n$ which are insoluble and probably polymeric owing to π-bonding between Ag$^+$

[20] G. C. Allen and K. D. Warren, *Inorg. Chem.*, 1969, **8**, 1895.
[21] C. D. M. Beverwijk *et al.*, *Organometallic Chem. Rev.*, A, 1970, **5**, 215 (an extensive review).
[22] W. T. Miller, R. H. Snider and R. J. Hummel, *J. Amer. Chem. Soc.*, 1969, **91**, 6532.

and the triple bond, as discussed for copper compounds. The compounds may be broken down by more strongly coordinating ligands such as phosphines, but the products are also polymeric for the same reason; e.g., $[Et_3PAgC{\equiv}CC_6H_5]_n$ is dimeric in nitrobenzene.

26-I-3. Gold Compounds[23]

The chemistry of gold in either oxidation state is mainly one of complex compounds. Mössbauer studies can be made with ^{197}Au.[24]

Binary Compounds. There is no good evidence for aurous oxide but *auric oxide*, Au_2O_3, can be made as a brown powder by dehydration of the brown precipitate of $Au(OH)_3$ formed on addition of OH^- to $AuCl_4^-$ solutions. It decomposes to gold and oxygen at $\sim 150°$, but at high oxygen pressure other oxide phases and also alkali gold oxides, e.g., CsAuO, and which contain a square $Au_4O_4^{4-}$ unit,[25] have been observed. Gold(III) hydroxide is weakly acidic and dissolves in OH^- solutions to give species possibly of the type $[Au(OH)_4]^-$.

Gold(III) fluoride is best made by fluorination of Au_2Cl_6 at $300°$ and forms orange crystals, which decompose to the metal at $500°$. It has a unique structure[26] with square AuF_4 units linked into a chain by *cis*-fluoride bridges; the F atoms of adjacent chains interact weakly with the axial sites.

The *chloride* and *bromide*, which form red crystals, are made by direct interaction at $200°$; both are dimers, Au_2X_6, in the solid and in the vapor. They dissolve in water to give hydrolyzed species, but in HX the ions AuX_4^- are obtained (see below).

Aurous chloride, formed by heating Au_2Cl_6 at $160°$, is a yellow powder which dissociates at higher temperatures and is decomposed by water; the iodide has chains I—Au—I—Au— with linear I—Au—I bonds.

Gold(I) Complexes. There are several aurous complexes stable in aqueous solution, the most important ones being $Au(CN)_2^-$, $AuCl_2^-$ and the thiosulfate species. The cyanide complex is very stable ($K = 4 \times 10^{28}$) and is formed when AuCN is treated with an excess of cyanide or, more usually, when gold is treated with an alkali cyanide in presence of air or hydrogen peroxide. Crystalline compounds such as $K[Au(CN)_2]$ can be obtained, and the free acid, $HAu(CN)_2$, is isolable by evaporation of its solutions; as in other free cyano acids, a hydrogen-bonded lattice with —CN—H—NC— bonds is formed.

Numerous complexes of Au^I with substituted phosphine, arsine and sulfide ligands as well as carbon monoxide can be obtained. Unlike the copper(I) and silver(I) complexes, which are polymeric with four-coordinated metal

[23] L. M. Gedansky and L. G. Hepler, *Engelhard Tech. Bull.*, 1969, **10**, 5 (thermochemical and potential data for compounds and aqueous solutions).
[24] J. S. Charlton and D. I. Nichols, *J. Chem. Soc., A*, **1970**, 1484.
[25] O. Muller, R. E. Newnham, and R. Roy, *J. Inorg. Nuclear Chem.*, 1969, **31**, 2966; K. Hesterman and R. Hoppe, *Z. anorg. Chem.*, 1968, **360**, 113.
[26] F. W. B. Einstein *et al.*, *J. Chem. Soc., A*, **1967**, 478.

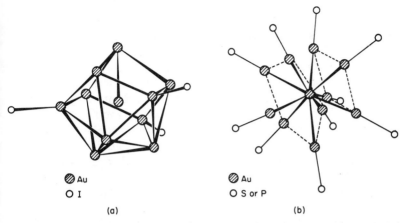

Fig. 26-I-2. The structures of Au_{11} cluster compounds: (a) $Au_{11}I_3[P(p\text{-}ClC_6H_4)_3]_7$; (b) $Au_{11}(SCN)_3(PPh_3)_7$. [Reproduced with permission from V. G. Albano et al., Chem. Comm., 1970, 1210; and M. McPartlin, R. Mason and L. Malatesta, Chem. Comm. 1969, 334.]

atoms, the gold(I) complexes are usually, though not invariably, mono-meric.[27] Thus the action of phosphorus trichloride or trialkylphosphines on auric chloride in ether reduces the Au^{III} to give crystals of R_3PAuCl. Halides of this type react with sodium salts, e.g., with $NaCo(CO)_4$ to form $Ph_3PAuCo(CO)_4$. The reduction of the triarylphosphine complex, R_3PAuX with $NaBH_4$ forms polynuclear gold cluster compounds.[28a] Two of these have been shown to have an Au_{11} cluster, as shown in Fig. 26-I-2, consisting, formally at any rate, of three Au^I and nine Au^0 atoms. The structure can be roughly described as an icosahedral one with a central gold atom; nine apices are each occupied by gold atoms while the remaining three are occu-pied by only a single gold atom. It seems quite certain that gold alkyl sulfides, $(AuSR)_n$, and similar gold compounds that are obtained from sulfurized terpenes, are highly soluble in organic solvents ("liquid gold"), and are used for decorating china and glass articles, also contain gold clusters since the gold content of these materials is exceptionally high. Much still remains to be learned about these complex polygold compounds. The dithiocarbamates, $[Au(S_2CNR_2)]_2$, are dimeric with both a metal—metal bond and a $—CS_2—$ bridge.[28b]

Gold(III) Complexes. Gold(III) is isoelectronic with Pt(II), and its com-plexes therefore show many structural similarities to those of Pt(II).

Halogeno complexes. Salts of the tetrafluoroaurate ion are obtained by the action of BrF_3 on a mixture of gold and an alkali chloride; $KAuF_4$ is isomorphous with $KBrF_4$ and has a square AuF_4^- ion.[29]

[27] See, e.g., D. I. Nichols and A. S. Charleston, J. Chem. Soc., A, 1969, 2581; A. D. Westland, Canad. J. Chem., 1969, 47, 4135.
[28] (a) F. Cariati et al., Inorg. Chim. Acta, 1971, 5, 172; Chem. Comm., 1971, 1423; (b) F. J. Farrell and T. G. Spiro, Inorg. Chem., 1971, 10, 1606.
[29] A. J. Edwards and G. R. Jones, J. Chem. Soc., A, 1969, 1936.

When gold is dissolved in aqua regia or Au_2Cl_6 is dissolved in HCl and the solution of $AuCl_4^-$ is evaporated, chloroauric acid can be obtained as yellow crystals $[H_3O]^+[AuCl_4]^- \cdot 3H_2O$. Other water-soluble salts such as $KAuCl_4$ and $NaAuCl_4 \cdot 2H_2O$ are readily obtained. In water, hydrolysis to $AuCl_3OH^-$ occurs.[30] From the dilute hydrochloric acid solutions the gold can be solvent-extracted with a very high partition coefficient into ethyl acetate or diethyl ether; the species in the organic solvent appears to be $[AuCl_3OH]^-$, which is presumably associated in an ion pair with an oxonium ion. Gold is readily recovered from such solutions, e.g., by precipitation with SO_2.

Other anionic complexes. There are several other anions such as $[Au(CN)_4]^-$, $[Au(NO_3)_4]^-$ in which NO_3 is unidentate,[31] and sulfato complexes. Au(III) differs from Pt(II) in its greater affinity for oxygen donor ligands.

Cationic complexes. There are numerous four-coordinate complexes such as $[AuCl_2py_2]Cl$, $[AuphenCl_2]Cl$. Chloroauric acid reacts with diethylenetriamine to give the ammonium tetrachloroaurate, $[Au \; dienCl]Cl_2$ or $[Au(dienH)Cl]Cl$, depending on the concentration and pH.[32] The kinetics of substitution of various anions in $[AudienCl]^{2+}$ have been compared with those for planar Pt^{II}; there is evidence that axial interactions occur here also in solution, e.g.,

Electronic spectra have been interpreted on an M.O. basis.

Gold(III) complexes have been obtained by using a chelating diarsine ligand from the reaction with sodium tetrachloroaurate(III) in presence of sodium iodide. The iodide $[Au(diars)_2I_2]I$ and other cations, $[Au(diars)_2I]^{2+}$ and $[Au(diars)_2]^{3+}$, can be obtained. It is held that these are species with six, five and four coordination for Au^{III} with octahedral,[33] trigonal bipyramidal and planar structures, respectively. Chelating phosphine complexes also exist.

Organogold compounds.[34] Alkyl derivatives of gold were among the first organometallic compounds of transition metals to be prepared. Both gold(I) and gold(III) compounds with σ-bonds to carbon, as well as olefin complexes, are known.

Gold(I) complexes are mainly of the type RAuL, where L is a stabilizing ligand such as R_2S, R_3P or RNC. They are made from the corresponding

[30] F. H. Fry, G. A. Hamilton and J. Turkevitch, *Inorg. Chem.*, 1966, **5**, 1943.
[31] C. D. Garner and S. C. Wallwork, *J. Chem. Soc.*, A, **1970**, 3092.
[32] C. F. Wieck and F. Basolo, *Inorg. Chem.*, 1966, **5**, 576.
[33] V. F. Duckworth and N. C. Stephenson, *Inorg. Chem.*, 1969, **8**, 1661.
[34] B. Armer and H. Schmidbauer, *Angew. Chem. Internat. Edn*, 1970, **9**, 101 (an extensive review).

halides by action of lithium alkyls or Grignard reagents such as FC_6H_4MgBr.[35] Acetylides such as $(R_3PAuC\equiv CR')_n$ are also known.

Gold(III) alkyls also are usually stable only when other ligands such as triphenylphosphine are present, as in $(CH_3)_3AuPPh_3$,[36a] but $Au(C_6F_5)_3$ is stable in ether;[36b] cis-$AuCl(C_6H_5)_2(PPh_3)$ is planar.[37] The main chemistry here is of the dialkyl species, R_2AuX, and especially of the methyl compounds which have been extensively studied[38] and form complex ions such as $Me_2Au(OH)_2^-$, $Me_2AuBr_2^-$, $Me_2Au(H_2O)_2^+$. The halides are dimeric with halogen bridges, but the cyanide and hydroxide form tetramers with linear bridges Au—CN—Au in the former and angular hydroxo bridges in the latter.

[35] L. G. Vaughan and W. A. Sheppard, J. Amer. Chem. Soc., 1969, 91, 6151, and references therein.

[36] (a) A. Shiotani, H.-F. Klein and H. Schmidbauer, J. Amer. Chem. Soc., 1971, 93, 1557; (b) L. G. Vaughan and W. A. Sheppard, J. Organometallic Chem., 1970, 22, 739.

[37] R. W. Baker and P. Pauling, Chem. Comm., 1969, 745.

[38] R. S. Tobias et al., Inorg. Chem., 1971, 10, 2639.

27

The Lanthanides;
also Scandium and Yttrium

***

GENERAL

27-1.

The lanthanides—or lanthanons, as they are sometimes called—are, strictly, the fourteen elements that follow lanthanum in the Periodic Table and in which the fourteen $4f$ electrons are successively added to the lanthanum configuration. Since the term lanthanide is used to indicate that these elements form a closely allied group, for the chemistry of which lanthanum is the prototype, the term is often taken as including lanthanum itself. Table 27-1 gives some properties of the atoms and ions. The electronic configurations are not all known with complete certainty owing to the great complexity of the electronic spectra of the atoms and ions and the attendant difficulty of analysis.

The chemistry of these elements is predominantly ionic and is determined primarily by the size of the M^{3+} ion. Since *yttrium*, which lies above La in Transition Group III and has a similar $+3$ ion with a noble-gas core, has both atomic and ionic radii lying close to the corresponding values for Tb and Dy (a fact resulting from the lanthanide "contraction" to be discussed below), this element is also considered here. It is generally found in Nature along with the lanthanides and resembles Tb^{III} and Dy^{III} in its compounds. The lighter element in Group III, *scandium*, has a smaller ionic radius, so that its chemical behavior is intermediate between that of Al and of the lanthanides, but it is also considered in this Chapter.

27-2. Oxidation States

The first three ionization potentials, the sum of which is given in Table 27-1, are comparatively low; so the elements are highly electropositive and their compounds essentially all ionic in nature.

All the lanthanides, as well as Sc and Y, form M^{3+}. Certain lanthanides show $+2$ or $+4$ states but these are always less stable than the $+3$ state.

TABLE 27-1

Some Properties of Lanthanide Atoms and Ions

Atomic number	Name	Symbol	Electronic configuration[a,b]				ΣI (eV)[c]	E_0 (V)[d]	Radii $M^{3+e,f}$
			Atom	M^{2+}	M^{3+}	M^{4+}			
57	Lanthanum	La	$5d6s^2$	$5d$	[Xe]	—	36.2	−2.52	1.061
58	Cerium	Ce	$4f^1 5d^1 6s^2$	$4f^2$	$4f$	[Xe]	36.4	−2.48	1.034
59	Praseodymium	Pr	$4f^3 6s^2$	$4f^3$	$4f^2$	$4f$	37.55	−2.47	1.013
60	Neodymium	Nd	$4f^4 6s^2$	$4f^4$	$4f^3$	$4f^2$	38.4	−2.44	0.995
61	Promethium	Pm	$4f^5 6s^2$	—	$4f^4$	—	—	−2.42	0.979
62	Samarium	Sm	$4f^6 6s^2$	$4f^6$	$4f^5$	—	40.4	−2.41	0.964
63	Europium	Eu	$4f^7 6s^2$	$4f^7$	$4f^6$	—	41.8	−2.41	0.950
64	Gadolinium	Gd	$4f^7 5d6s^2$	$4f^7 5d$	$4f^7$	—	38.8	−2.40	0.938
65	Terbium	Tb	$4f^9 6s^2$	$4f^9$	$4f^8$	$4f^7$	39.3	−2.39	0.923
66	Dysprosium	Dy	$4f^{10} 6s^2$	$4f^{10}$	$4f^9$	$4f^8$	40.4	−2.35	0.908
67	Holmium	Ho	$4f^{11} 6s^2$	$4d^{11}$	$4f^{10}$	—	40.8	−2.32	0.894
68	Erbium	Er	$4f^{12} 6s^2$	$4f^{12}$	$4f^{11}$	—	40.5	−2.30	0.881
69	Thulium	Tm	$4f^{13} 6s^2$	$4f^{13}$	$4f^{12}$	—	41.85	−2.28	0.869
70	Ytterbium	Yb	$4f^{14} 6s^2$	$4f^{14}$	$4f^{13}$	—	43.5	−2.27	0.858
71	Lutetium	Lu	$4f^{14} 5d6s^2$	—	$4f^{14}$	—	40.4	−2.25	0.848

[a] Only the valence-shell electrons, that is, those outside the [Xe] shell, are given.

[b] A dash indicates that this oxidation state is not known in any isolable compound. Properties of M^{2+} ions stabilized by substitution for Ca^{2+} in a CaF_2 lattice have been studied: see D.S. McClure and Z. Kiss, *J. Chem. Phys.*, 1963, **39**, 3251, and K.E. Johnson and J.N. Sandoe, *J. Chem. Soc.*, A, **1969**, 1644.

[c] 1st to 3rd, derived from Born–Haber cycles of oxides and arsenides; see M.M. Faktor and R. Hanks, *J. Inorg. Nuclear Chem.*, 1969, **31**, 1649, and ref. 1.

[d] For $M = M^{3+} + 3e$; $Y = -2.37$ V; $Sc = -1.88$ V.

[e] D.H. Templeton and C.H. Dauben, *J. Amer. Chem. Soc.*, 1954, **76**, 5237.

[f] Radius of $Y^{3+} = 0.88$ Å, and of $Sc^{3+} = 0.68$ Å; W.H. Zachariasen in G.T. Seaborg and J.J. Katz, *The Actinide Elements*, McGraw-Hill, New York, 1954.

To a certain extent, the occurrence of M^{2+} and M^{4+} can be correlated with the electronic structures and ionization potentials.[1] There appears to be a special stability associated with an empty, half-filled, or filled f shell just as, to a lesser degree, the same phenomenon is seen in the regular transition series (notably, Mn^{2+}) and in the ionization potentials of the first short period. This stability is a reflection of the fact that the exchange-energy loss on ionization falls to zero after the half-filled shell and leads to a large drop in the third ionization potential between Eu and Gd. These exchange-energy differences increase from zero for $f^1 \rightarrow f^0$ until $f^7 \rightarrow f^6$ and again from zero for $f^8 \rightarrow f^7$ until $f^{14} \rightarrow f^{13}$. There is some evidence for a break in I_3 around the $\frac{3}{4}$-filled shell, which would suggest that Dy^{2+} might be stable with respect to Dy^{3+}, but no isolable Dy^{2+} compound is known.

Scandium, yttrium and lanthanum form only the M^{3+} ions, since removal of three electrons leaves the noble-gas configuration. Similarly Lu and Gd form only M^{3+} since they then have the stable $4f^{14}$ and $4f^7$ configurations, respectively. In all these five cases, never *less* than three electrons are

[1] For discussion see D. A. Johnson, *J. Chem. Soc.*, A, **1969**, 1525, 1528, 2578; also D. A. Johnson, *Thermodynamic Aspects of Inorganic Chemistry*, Cambridge University Press, 1968; and O. Johnson, *J. Chem. Educ.*, 1970, **47**, 431.

removed under chemical conditions because the M^{2+} or M^+ ions would be much larger than the M^{3+} ions. Thus the energy saved in the ionization step would be less than the additional lattice or hydration energies of the salts of the small M^{3+} ions compared with the lattice or hydration energies of the M^{2+} or M^+ ions.

The most stable M^{2+} and M^{4+} ions are formed by elements that can attain the f^0, f^7, or f^{14} configuration by so doing. Thus Ce^{4+} and Tb^{4+} attain the f^0 and f^7 configurations, respectively, whereas Eu^{2+} and Yb^{2+} have the f^7 and f^{14} configurations, respectively. These facts seem to support the view that "special stability" of the f^0, f^7 and f^{14} configurations is important in determining the existence of oxidation states other than III. This argument becomes less convincing, however, when we note that Sm and Tm give M^{2+} species having f^6 and f^{13} configurations but no M^+ ion, whereas Pr and Nd give M^{4+} ions with configurations f^1 and f^2 but no penta- or hexa-valent species. Admittedly, the Sm^{II} and especially Tm^{II}, Pr^{IV}, and Nd^{IV} states are very unstable, but the idea that stability is already favored by the mere *approach* to an f^0, f^7 or f^{14} configuration, even though such a configuration is not actually attained, is of dubious validity. The existence of Nd^{2+}, f^4, and evidence even for Pr^{2+} and Ce^{2+} in lattices, provide particularly cogent reasons for believing that, although the special stability of f^0, f^7, f^{14} may be one factor, there are other thermodynamic and kinetic factors that are of equal or greater importance in determining the stability of oxidation states.

27-3. The Lanthanide Contraction

This term has been used above in discussing the elements of the third transition series, since it has certain important effects on their properties. It consists of a significant and steady decrease in the size of the atoms and ions with increasing atomic number; that is, La has the greatest, and Lu the smallest radius (Table 27-1; see also Fig. 28-2). Note that the radius of La^{3+} is about 0.18 Å larger than that of Y^{3+}, so that if the fourteen lanthanide elements did not intervene we might have expected Hf^{4+} to have a radius ~ 0.2 Å greater than that of Zr^{4+}. Instead, the shrinkage, amounting to 0.21 Å, almost exactly wipes out this expected increase and results in almost identical radii for Hf^{4+} and Zr^{4+}, as noted previously.

The cause of this contraction is the same as the cause of the less spectacular ones that occur in the d-block transition series, namely, the imperfect shielding of one electron by another in the same subshell. As we proceed from La to Lu, the nuclear charge and the number of $4f$ electrons increase by one at each step. The shielding of one $4f$ electron by another is very imperfect (much more so than with d electrons) owing to the shapes of the orbitals, so that at each increase the effective nuclear charge experienced by each $4f$ electron increases, thus causing a reduction in size of the entire $4f^n$ shell. The accumulation of these successive contractions is the total lanthanide contraction.

It should be noted also that the decrease, though steady, is not quite regular, the biggest decreases occurring with the first f electrons added; there also appears to be a larger decrease after f^7, that is, between Tb and Gd. Certain chemical properties of lanthanide compounds show corresponding divergences from regularity as a consequence of the ionic size (see below).

27-4. Magnetic and Spectral Properties

In several aspects, the magnetic and spectral behavior of the lanthanides is fundamentally different from that of the d-block transition elements. The basic reason for the differences is that the electrons responsible for the properties of lanthanide ions are $4f$ electrons, and the $4f$ orbitals are very effectively shielded from the influence of external forces by the overlying $5s^2$ and $5p^6$ shells. Hence the states arising from the various $4f^n$ configurations are only slightly affected by the surroundings of the ions and remain practically invariant for a given ion in all of its compounds.

The states of the $4f^n$ configurations are all given, to a useful approximation, by the Russell–Saunders coupling scheme. In addition, the spin–orbit coupling constants are quite large (order of 1000 cm^{-1}). The result of all this is that, with only a few exceptions, the lanthanide ions have ground states with a single well-defined value of the total angular momentum, J, with the next lowest J state at energies many times kT (at ordinary temperatures equal to $\sim 200 \text{ cm}^{-1}$) above and hence virtually unpopulated (Fig. 27-1 a).

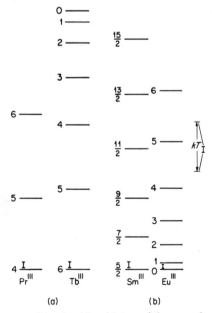

(a)   (b)

Fig. 27-1. J states for several lanthanides: (a) two of the several cases where the J states are widely separated compared to kT and (b) the two cases where the separations are of the same order as, or less than, kT.

Thus the susceptibilities and magnetic moments should be given straight-forwardly by formulas considering only this one well-defined J state, and indeed such calculations give results that are, with only two exceptions, in excellent agreement with experimental values (Fig. 27-2). For Sm^{3+} and Eu^{3+}, it turns out that the first excited J state is sufficiently close to the ground state (Fig. 27-1 b) for this state (and in the case of Eu^{3+} even the second and third excited states) to be appreciably populated at ordinary temperatures. Since these excited states have higher J values than the ground state, the actual magnetic moments are higher than those calculated by considering the ground states only. Calculations taking into account the population of excited states afford results in excellent agreement with experiment (Fig. 27-2).

It should be emphasized that magnetic behavior depending on J values is qualitatively different from that depending on S values—that is the "spin-only" behavior—which gives a fair approximation for many of the d-block transition elements. Only for the f^0, f^7 and f^{14} cases, where there is no orbital angular momentum ($J = S$), do the two treatments give the same answer. For the lanthanides the external fields do not either appreciably split the free ion terms or quench the orbital angular momentum.

Because the f orbitals are so well shielded from the surroundings of the ions, the various states arising from the f^n configurations are split by external fields only to the extent of $\sim 100\ cm^{-1}$. Thus when electronic transitions, called f–f transitions, occur from one J state of an f^n configuration to another J state of this configuration, the absorption bands are *extremely sharp*. They are similar to those for free atoms and are quite unlike the broad bands observed for the d–d transitions. Virtually all the absorption bands found in the visible and near-ultraviolet spectra of the lanthanide $+3$ ions have this

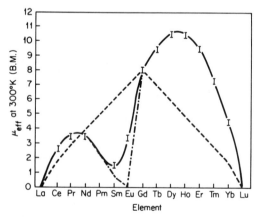

Fig. 27-2. Measured and calculated effective magnetic moments (BM) of lanthanide M^{3+} ions at 300° K. I's are ranges of experimental values; ———— gives values calculated for appropriate J ground states with allowance for the Sm and Eu anomalies; . . gives values calculated without allowance for the Sm and Eu anomalies; ----- gives calculated spin-only values.

line-like character. Although the absorption bands normally show little dependence on the nature of the ligand, certain of the lanthanide M^{3+} ions —those with 3 or more f electrons or equivalent holes—do have one or more transitions that show an increase in intensity when H_2O is replaced by other ligands.[2] There are also bands[3] due to ligand–metal charge transfer with reducing ligands.

For the M^{2+} ions trapped in host lattices such as CaF_2 or BaF_2 the absorption bands are generally broad; a detailed interpretation of the bands has been given.[4]

The colors and electronic ground states of the M^{3+} ions are given in Table 27-2; the color sequence in the La–Gd series is accidentally repeated in the series Lu–Gd. As implied by the earlier discussion, insofar as the colors are due to f–f transitions they are virtually independent of the environment of the ions.

TABLE 27-2

Colors and Electronic Ground States of the M^{3+} Ions

Ion	Ground State	Color	Ion	Ground state
La	$^1S_0$	Colorless	Lu	$^1S_0$
Ce	$^2F_{5/2}$	Colorless	Yb	$^2F_{7/2}$
Pr	$^3H_4$	Green	Tm	$^3H_6$
Nd	$^4I_{9/2}$	Lilac	Er	$^4I_{15/2}$
Pm	$^5I_4$	Pink; yellow	Ho	$^5I_8$
Sm	$^6H_{5/2}$	Yellow	Dy	$^6H_{15/2}$
Eu	$^7F_0$	Pale pink	Tb	$^7F_6$
Gd	$^8S_{7/2}$	Colorless	Gd	$^8S_{7/2}$

An important feature in the spectroscopic behavior is that of fluorescence or luminescence of certain lanthanide ions, notably Y and Eu, when used as activators in lanthanide oxide, silicate, or transition-metal oxide lattices. Oxide phosphors are used in color television tubes. Certain of the +2 ions trapped in CaF_2 lattices, as well as organic cation salts of complex anions such as $[Eu\ \beta\text{-diket}_4]^-$, show laser activity.

27-5. Coordination Numbers and Stereochemistry

The coordination numbers and stereochemistry of the ions are given in Table 27-3. In both ionic crystals and in complexes, coordination numbers exceeding 6 are the general rule rather than the exception. Indeed the number of lanthanide compounds in which the coordination number of 6 has been unequivocally established is small; many complexes that could have been so formulated are known to have solvent molecules bound to the metal, leading

[2] For discussion and references, see L. I. Katzin, *Inorg. Chem.*, 1969, **8**, 1649.
[3] J. C. Barns, *J. Chem. Soc.*, **1964**, 3880.
[4] K. E. Johnson and J. N. Sandoe, *J. Chem. Soc.*, A, **1969**, 1694.

TABLE 27-3

Oxidation States, Coordination Numbers and Stereochemistry of Lanthanide Ions

Oxidation state	Coordination number	Geometry	Examples
+2	6	NaCl type	EuTe, SmO, YbSe
	6	CdI_2 type	YbI_2
	8	CaF_2 type	SmF_2
+3	6	Octahedral	$[Er(NCS)_6]^{3-}$, $[Sc(NCS)_2 \text{ bipy}_2]^+$
	6	$AlCl_3$ type	MCl_3 (Tb–Lu)
	7	Monocapped trig. prism	Gd_2S_3, $Y(acac)_3 \cdot H_2O$
	7	Face-capped dist. octahedron	$Y(PhCOC_6H_4COMe)_3 \cdot H_2O$
	7	ZrO_2 type	ScOF
	8	Distorted square antiprism	$Y(acac)_3 \cdot 3H_2O$; $La(acac)_3(H_2O)_2$
	8	Dodecahedral	$Cs[Y(CF_3COCHCOCF_3)_4]$; $NH_4[Pr(TTA)_4]H_2O$
	8	Bicapped trig. prism	Gd_2S_3; $MX_3(PuBr_3$ type)
	9	Tricapped dist. trig. prism	$[Nd(H_2O)_9]^{3+}$; $Y(OH)_3$; $K[La \text{ edta}] \cdot 8H_2O$; $La_2(SO_4)_3 \cdot 9H_2O$
	9	Complex	LaF_3, MCl_3 (La–Gd)
	10	Complex	$La_2(CO_3)_3 \cdot 8H_2O$
	10	Bicapped dodecahedron	$Ce(NO_3)_5^{2-}$
	12	Distorted icosahedron	$Ce(NO_3)_6^{3-}$
+4	6	Octahedral	Cs_2CeCl_6
	8	Archim. antiprism	$Ce(acac)_4$
	8	Dist. square antiprism (chains)	$(NH_4)_2CeF_6$
	8	CaF_2 type	CeO_2
	10	Complex[a]	$Ce(NO_3)_4(OPPh_3)_2$
	12	Distorted icosahedron	$(NH_4)_2[Ce(NO_3)_6]$

[a] Contains four bidentate NO_3 groups (Mazhar-ul-Haque et al., Inorg. Chem., 1971, 10, 115.

to the common coordination numbers 7, 8 and 9. Compared to the d-block transition elements the lanthanide ions form comparatively few complexes with ligands other than oxygen, and even those with nitrogen ligands are often readily hydrolyzed. There is no complex of this series in which π-bonding plays any significant role. These differences can be attributed in part to the unavailability of the f orbitals for formation of hybrid orbitals, which might lead to covalent bond strength, and in part to the fact that the lanthanide ions are rather large (radii 0.85–1.06 Å) compared with those of the transition elements (e.g., Cr^{3+} and Fe^{3+}, radii 0.60–0.65 Å), which lowers electrostatic forces of attraction.

As a result of the lanthanide contraction, changes in coordination number through the series La–Lu are to be expected.

Solid State. In crystals there may be a pronounced change in structural type at one or more points in the series; at such change-over points a compound may have two or more crystal forms. Examples are the following:

(a) The anhydrous halides, MCl_3, of La–Gd are all 9-coordinate with a

UCl_3-type lattice; $TbCl_3$ and one form of $DyCl_3$ are of the $PuBr_3$ type, while finally the Tb–Lu chlorides have the octahedral $AlCl_3$-type structure.[5]

(b) In the M_2S_3 series[6] there are three main structure types, La–Dy (orthorhombic, 8- or 7-coordinate); Dy–Tm (monoclinic, 7- or 6-coordinate); Ho_2S_3 actually has half its atoms 6- and half 7-coordinate; Yb and Lu (corundum type, 6-coordinate).

(c) The heats of vaporization of certain volatile chelates show irregular decreases from Pr to Lu, but with a plateau at Gd.[7]

Solutions. In solution similar adjustments can occur, but here, in view of the possible complications, the differences are more difficult to characterize unequivocally.[8] Various physical properties of solutions suggest generally that the lighter members La^{3+}–Nd^{3+} have 9-coordination, whereas the series Gd^{3+}–Lu^{3+} are generally 8-coordinate. It must be borne in mind, however, that while a smaller ion can be expected to have a small number of near neighbors on account of its size, for example in complex ions and even in solutions, nevertheless for solvated ions the greater polarizing power of the smaller ion will lead to a greater solvated radius as determined by some physical parameter such as ionic-exchange behavior. Examples are the following:

(a) Certain stability[8] and rate data[9] for complex formation suggest a break at about Gd^{3+}.

(b) Certain bands in the absorption spectra of Nd^{3+} solutions and complexes depend on the concentration, solvent, and presence of an excess of anions. The changes have been ascribed to changes in coordination and the symmetry of the molecule or ion.[10]

(c) Rate data for substitution processes in MSO_4^+ ions determined by ultrasonic methods suggest a probable change for the coordination sphere[11] from 8 in the La^{3+} end to 9 in the Lu^{3+} end.

However, it must be noted that, although earlier entropy data for M^{3+} ions suggested a change in hydration number, recent studies indicate this not to be so.[12]

27-6. Occurrence and Isolation

The lanthanide elements were originally known as the Rare Earths from their occurrence in oxide (or, in old usage, earth) mixtures. They are not rare elements, and the absolute abundances in the lithosphere are relatively high.

[5] B. Morosin, *J. Chem. Phys.*, 1968, **49**, 3007.
[6] C. T. Prewitt and A. W. Sleight, *Inorg. Chem.*, 1968, **7**, 1090, 2283.
[7] R. E. Sievers *et al.*, *Inorg. Chem.*, 1969, **8**, 1649.
[8] S. Sierkievski, *J. Inorg. Nuclear Chem.*, 1970, **32**, 419; L. J. Nugent, *J. Inorg. Nuclear Chem.*, 1970, **32**, 3485 (discussions of irregularities in the formation of lanthanide and actinide complexes); A. Fratiello *et. al.*, *Inorg. Chem.*, 1971, **10**, 2552.
[9] H. B. Silver, R. D. Farina and J. H. Swinehart, *Inorg. Chem.*, 1969, **8**, 89.
[10] D. G. Karraker, *Inorg. Chem.*, 1968, **7**, 473; 1967, **6**, 1863.
[11] N. Purché and C. A. Vincent, *Trans. Faraday Soc.*, 1967, **63**, 2745.
[12] R. J. Hinckey and J. W. Cobble, *Inorg. Chem.*, 1970, **9**, 917.

Thus even the scarcest, thulium, is as common as bismuth ($\sim 2 \times 10^{-5}$ wt. %) and more common than As, Cd, Hg or Se, which are not usually considered rare. Substantial deposits are located in Scandinavia, India, the Soviet Union, and the United States, with a wide occurrence in smaller deposits in many other places. Many minerals make up these deposits, one of the most important being *monazite*, which usually occurs as a heavy dark sand of variable composition. Monazite is essentially a lanthanide orthophosphate, but significant amounts of thorium (up to 30%) occur in most monazite sands. The distribution of the individual lanthanides in minerals is usually such that La, Ce, Pr and Nd make up about 90% with Y and the heavier elements together constituting the remainder. Monazite and other minerals carrying lanthanides in the +3 oxidation state are usually poor in Eu which, because of its relatively strong tendency to give the +2 state, is often more concentrated in the calcium group minerals.

Promethium[13] occurs in Nature only in traces in uranium ores (4×10^{-15} g per kilo) as a spontaneous fission fragment of ^{238}U.[14] Milligram quantities of pink ^{147}Pm^{3+} salts can be isolated by ion-exchange methods[15] from fission products from nuclear reactors where ^{147}Pm (β^-, 2.64 y) is formed.

The lanthanides are separated from most other elements by precipitation of oxalates or fluorides from nitric acid solution. The elements are separated from each other by ion-exchange, which is carried out commercially on a large scale. Cerium and europium are normally first removed, the former by oxidation to Ce^{IV} and removal by precipitation of the iodate which is insoluble in $6M$ HNO$_3$ or by solvent extraction, and the latter by reduction to Eu^{2+} and removal by precipitation as insoluble EuSO$_4$.

The ion-exchange behavior depends primarily on the hydrated ionic radius, and La should be most tightly bound and Lu least; hence the elution order is Lu $\rightarrow$ La. This trend is accentuated by use of appropriate complexing agents at an appropriate pH; the ion of smallest radius also forms the strongest complexes and hence the preference for the aqueous phase is enhanced. One of the best ligands is α-hydroxyisobutyric acid, (CH$_3$)$_2$CH(OH)COOH,[16] at 25° and pH 3.35, but EDTAH$_4$ and other hydroxo or amino carboxylic acids can also be used. The eluates are treated with oxalate ion, and the insoluble oxalates are ignited to the oxides.

Although Ce^{IV} (also Zr^{IV}, Th^{IV} and Pu^{IV}) is readily extracted from nitric acid solutions by tributyl phosphate dissolved in kerosene or other inert solvent and can thus be readily separated from the +3 lanthanide ions, the trivalent lanthanide nitrates can also be extracted under suitable conditions with various phosphate esters or acids. Extractability under given conditions increases with increasing atomic number; it is higher in strong acid or high nitrate concentrations.

[13] P. Weigel, *Fortschr. Chem. Forsch.*, 1969, **12**, 539 (review).
[14] M. Attrep and P. K. Koroda, *J. Inorg. Nuclear Chem.*, 1968, **30**, 699.
[15] E. J. Wheelwright, *J. Inorg. Nuclear Chem.*, 1969, **31**, 3287.
[16] R. L. Ritzman *et al.*, *J. Inorg. Nuclear Chem.*, 1966, **28**, 2758.

27-7. The Metals

The lighter metals (La–Gd) are obtained by reduction of the trichlorides with Ca at 1000° or more, whereas for others (Tb, Dy, Ho, Er, Tm and also Y) the trifluorides are used because the chlorides are too volatile. Pm is made by reduction of PmF_3 with Li. Eu, Sm and Yb trichlorides are reduced only to the dihalides by Ca, but the metals can be prepared by reduction of the oxides, M_2O_3, with La at high temperatures.

The metals are silvery-white and very reactive. They all react directly with water, slowly in the cold, rapidly on heating, to liberate hydrogen. Their high potentials (Table 27-1) are indicative of their electropositive character. They tarnish readily in air and all burn easily to give the sesquioxides, except cerium which gives CeO_2. Yttrium is remarkably resistant to air even up to 1000° owing to formation of a protective oxide coating. The metals react exothermically with hydrogen, though heating to 300–400° is often required to initiate the reaction. The resulting hydride phases MH_2 and MH_3, which are usually in a defect state, have remarkable thermal stability, in some cases up to 900° (see page 186). The metals also react readily with C, N_2, Si, P, S, halogens and other non-metals at elevated temperatures.

The atomic volumes, densities, and some other physical properties of the metals change smoothly along the series, except for Eu and Yb and occasionally Sm and Tm. For example, Figure 27-3 shows a plot of atomic volumes and heats of vaporization. The deviations occur with just those lanthanides that have the greatest tendency to exist in the divalent state; presumably these elements tend to donate only two electrons to the conduction bands of the metal, thus leaving larger cores and affording lower binding forces.

Metallic Eu and Yb dissolve in liquid ammonia at −78° to give blue solutions, golden when concentrated. The spectra of the blue solutions, which decolorize slowly, are those expected for M^{2+} and solvated electrons.[17]

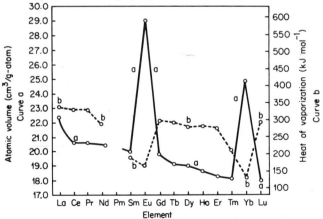

Fig. 27-3. The atomic volumes (curve a) and heats of vaporization (curve b) of the lanthanide metals.

[17] S. Salot and J. C. Warf, *J. Amer. Chem. Soc.*, 1968, **90**, 1932.

THE TRIVALENT STATE

27-8. Binary Compounds

The trivalent state is the characteristic one for all the lanthanides. They form *oxides*, M_2O_3, which resemble the Ca–Ba group oxides and absorb carbon dioxide and water from the air to form carbonates and hydroxides, respectively. The *hydroxides*, $M(OH)_3$, are definite compounds, having hexagonal structures, and not merely hydrous oxides. The basicities of the hydroxides decrease with increasing atomic number, as would be expected from the decrease in ionic radius. The hydroxides are precipitated from aqueous solutions by ammonia or dilute alkalis as gelatinous precipitates. They are not amphoteric.

Among the *halides*, the fluorides are of particular importance because of their insolubility. Addition of hydrofluoric acid or fluoride ions precipitates the fluorides from M^{3+} solutions even $3M$ in nitric acid and is a characteristic test for lanthanide ions. The fluorides, particularly of the heavier lanthanides, are slightly soluble in an excess of HF owing to complex formation. They may be redissolved in $3N$ nitric acid saturated with boric acid which removes F^- as BF_4^-. The chlorides are soluble in water, from which they crystallize as hydrates, the La–Nd group often with $7H_2O$ and the Nd–Lu group (including Y) with $6H_2O$; other hydrates may also be obtained.

The anhydrous chlorides cannot easily be obtained from the hydrates because, when heated, these lose hydrochloric acid (to give the oxo chlorides, MOCl) more readily than they lose water. (However, scandium and cerium give Sc_2O_3 and CeO_2, respectively). The chlorides[18] are made by heating oxides with ammonium chloride:

$$M_2O_3 + 6NH_4Cl \xrightarrow{\sim 300°} 2MCl_3 + 3H_2O + 6NH_3$$

or as methanolates, $MCl_3 \cdot 4CH_3OH$, by treating the hydrates with 2,2-dimethoxypropane (cf. page 465).[19]

The bromides and iodides are rather similar to the chlorides.

Numerous other binary compounds are obtained by direct interaction at elevated temperatures; examples are the semi-conducting sulfides,[20] M_2S_3, which can also be made by reaction of MCl_3 with H_2S at 1100°. Group V compounds, MX, where X = N, P, As, Sb or Bi, which have the NaCl structure; borides, MB_4, MB_6; and carbides, MC_2 and M_2C_3 (pages 290 and 1075).

[18] K. Johnson and J. R. Mackenzie, *J. Inorg. Nuclear Chem.*, 1970, **32**, 43.
[19] L. F. Quil and G. W. Clink, *Inorg. Chem.*, 1967, **6**, 1433.
[20] See, e.g., N. L. Eatough, A. W. Webb and H. T. Hall, *Inorg. Chem.*, 1969, **8**, 2069; A. W. Nebb and H. T. Hall, *Inorg. Chem.*, 1970, **9**, 1084.

27-9. Oxo Salts

Hydrated salts of common acids, which contain the ions $[M(H_2O)_n]^{3+}$, are readily obtained by dissolving the oxide in acid and crystallizing.

Double salts are very common, the most important being the double nitrates and double sulfates, such as $2M(NO_3)_3 \cdot 3Mg(NO_3)_2 \cdot 24H_2O$, $M(NO_3)_3 \cdot 2NH_4NO_3 \cdot 4H_2O$, and $M_2(SO_4)_3 \cdot 3Na_2SO_4 \cdot 12H_2O$. The solubilities of double sulfates of this type fall roughly into two classes: the cerium group, La–Eu, and the yttrium group, Gd–Lu and Y. Those of the Ce group are only sparingly soluble in sodium sulfate, whereas those of the Y group are appreciably soluble. Thus, a fairly rapid separation of the entire group of lanthanides into two sub-groups is possible. Various of the double nitrates were used in the past for further separations by fractional crystallization procedures.

The precipitation of the *oxalates* from dilute nitric acid solution is a quantitative and fairly specific separation procedure for the lanthanides, which can be determined gravimetrically in this way, with subsequent ignition to the oxides. The actual nature of the oxalate precipitate depends on conditions. In nitric acid solutions, where the main ion is Hox^-, ammonium ion gives double salts $NH_4M ox_2 \cdot yH_2O$ ($y = 1$ or 3). In neutral solution, ammonium oxalate gives the normal oxalate with lighter, but mixtures with heavier, lanthanides. Washing the double salts with $0.1N$ HNO_3 gives, with some ions, the normal oxalates. The phosphates are sparingly soluble in dilute acid solution. Although carbonates exist,[21] many are basic; the normal carbonates are best made by hydrolysis of the chloroacetates:

$$2M(C_2Cl_3O_2)_3 + (x+3)H_2O \rightarrow M_2(CO_3)_3 \cdot xH_2O + 3CO_2 + 6CHCl_3$$

$La_2(CO_3)_3 \cdot 8H_2O$ has a complex structure with 10-coordinate La, and both uni- and bi-dentate carbonate ions.[22]

27-10. Complexes[23]

The *aquo ions*, M^{3+}, are hydrolyzed in water:

$$[M(H_2O)_n]^{3+} + H_2O \rightleftharpoons [M(OH)(H_2O)_{n-1}]^{2+} + H_3O^+$$

and the tendency to hydrolysis increases with increasing atomic number, as would be expected from the contraction in radii. Yttrium also gives predominantly MOH^{2+} but also $M_2(OH)_2^{4+}$ ions; for Ce^{3+}, however, only about 1% of the metal ion is hydrolyzed without forming a precipitate and in this case the main equilibrium appears to be

$$3Ce^{3+} + 5H_2O \rightleftharpoons [Ce_3(OH)_5]^{4+} + 5H^+$$

Halogeno Complexes. In aqueous solutions, rather weak complexes, $MF^{2+}(aq)$, are formed with fluoride ion, but there is little evidence for

[21] R. P. Turcotte, J. O. Sawyer, and L. Eyring, *Inorg. Chem.*, 1969, **8**, 238; F. Fromage and A. Morgant, *Bull. Soc. Chim. France*, **1967**, 2611.
[22] D. B. Shim and H. A. Eick, *Inorg. Chem.*, 1968, **7**, 1340.
[23] T. Moeller *et al.*, *Chem. Rev.*, 1965, **65**, 1; *Progr. Sci. Tech. Rare Earths*, 1968, **3**, 61.

complex anion formation; this is a distinction as a group from the actinide elements which do form complexes in strong HCl solutions. However, by the use of non-aqueous media such as ethanol or acetonitrile, salts of the weak complexes MX_6^{3-} can be prepared;[24] the iodo complexes are exceedingly weak, dissociating in non-aqueous solvents even in presence of an excess of I^-, and they are attacked by moisture and oxygen.

In presence of $AlCl_3$, the anhydrous chlorides are surprisingly volatile and this has been attributed to the formation of complexes,[25] e.g.,

$$NdCl_3(s) + 2Al_2Cl_6(g) \rightarrow NdAl_4Cl_{15}(g)$$

Oxygen Ligands.   By far the most stable and common of lanthanide complexes are those with chelating oxygen ligands. The use of EDTA-type anions and hydroxo acids such as tartaric or citric, for the formation of water-soluble complexes is of great importance in ion-exchange separations, as noted above. All of these can be assumed to have coordination numbers exceeding 6, as in $[La(OH_2)_4 \, EDTAH] \cdot 3H_2O$.

The complexes of β-diketones such as acetylacetone have been extensively studied, particularly since some of the fluorinated derivatives give complexes that are volatile[26] and suitable for gas-chromatographic separation.

The preparation of β-diketonates[26a] by conventional methods *invariably* gives hydrated or solvated species such as $[M(acac)_3] \cdot C_2H_5OH \cdot 3H_2O$. It is often difficult to remove solvent, especially water, without decomposition, although prolonged drying over $MgClO_4$ leads to the very hygroscopic $M(\beta\text{-dik})_3$. The solvated species whose structures are known have invariably coordination numbers >6 (see Table 27-3). A further characteristic of the tris-β-diketonates is complexing to give anionic species $[M(\beta\text{-dik})_4]^-$, the alkali-metal salts of which are sometimes appreciably volatile.

An important use of β-diketonate complexes that are soluble in organic solvents, such as those derived from 1,1,2,2,3,3-heptafluoro-7,7-dimethyl-4,6-octanedione, especially of Eu and Pr, is for shift reagents in nmr spectrometry. The paramagnetic complex deshields the protons of complicated molecules, and vastly improved separation of the resonance lines may be obtained.[26b]

The complexes of monodentate oxygen ligands are less stable than those of chelates and tend to dissociate in aqueous solution; the ions associate in aqueous solution with both nitrate and sulfate to give species such as MNO_3^{2+}, MSO_4^+, etc., and anionic carbonate complexes are known. Many crystalline compounds or salts have been obtained from the lanthanide salts in ethanolic solutions (*a*) with hexamethylphosphoramide,[27] which appears to give 6-coordinate species, $[M(HMPA)_6](ClO_4)_3$, (*b*) with triphenyl-

[24] J. L. Ryan, *Inorg. Chem.*, 1969, **8**, 2053.
[25] H. A. Øye and D. M. Gruen, *J. Amer. Chem. Soc.*, 1969, **91**, 2229.
[26] See, e.g., H. A. Swain and D. G. Karraker, *J. Inorg. Nuclear Chem.*, 1971, **33**, 2851.
[26a] M. F. Richardson and R. E. Sievers, *Inorg. Chem.*, 1971, **10**, 498; J. Selbin, N. Ahmad and N. Bhacca, *Inorg. Chem.*, 1971, **10**, 1383.
[26b] See W. DeW. Horrocks, Jr., and J. P. Snipe, III, *J. Amer. Chem. Soc.*, 1971, **93**, 6800.
[27] J. T. Donoghue *et al.*, *J. Inorg. Nuclear Chem.*, 1969, **31**, 1431.

phosphine oxide or -arsine oxide,[28] and pyridine N-oxides,[29] e.g., $M(NO_3)_3$-$(OAsPh_3)_4$, which have coordinated nitrate, and $[M(PyO)_8](ClO_4)_3$, and (c) with $DMSO$,[30] e.g., $(DMSO)_nM(NO_3)_3$.

Nitrogen Ligands. Until comparatively recently few N complexes were known. Under anhydrous conditions, in CH_3CN as solvent, polyamine complexes such as Men_3Cl_3 and $[M\ dien_4(NO_3)](NO_3)_2$ have been obtained.[31] These are decomposed by water and are insoluble in organic solvents. There are similar complexes of aromatic amines, such as $[M\ (terpyridine)_3](ClO_4)_3$ and $[M\ phen_3](SCN)_3$. These are usually hydrated and doubtless have coordination numbers exceeding six, as is found for $[Pr(terpy)(H_2O)_5Cl]Cl_2 \cdot 3H_2O$, which is 9-coordinate with a monocapped square antiprism structure.[31a] The pyridine complexes are very weak and exist only in solution.[32] A variety of thiocyanato complexes,[33] shown by ir spectroscopy to be N-bonded, are also known, e.g., $M(NCS)_3(OPPh_3)_4$ and $[M(NCS)_6]^{3-}$. In aqueous solution thiocyanato complexes have appreciable formation constants, and SCN^- can be used as an eluant for ion-exchange separations.

Sulphur Ligands. The only complexes are the readily hydrolyzed dithiocarbamates $M(dtc)_3$ made from MBr_3 and NaS_2CNR_2 in ethanol, and the anionic species $[M(dtc)_4]^-$ obtained by using an excess of the reagent.[33a]

Organo Compounds.[34] The best characterized organo compounds of M^{3+} are the thermally very stable, but air- and water-sensitive, cyclopentadienides, $M(C_5H_5)_3$, $M(C_5H_5)Cl_2 \cdot 3THF$, and $M(C_5H_5)_3CNC_6H_{11}$. These are ionic and have magnetic moments similar to those of the M^{3+} ions; they sublime in a vacuum at $\sim 200°$, presumably as molecules. The structure of $(C_5H_5)_3Sm$ in the crystal is complicated and one of the rings appears to act as a bridge group.[35]

For the M^{2+} ions, thermally stable cyclopentadienides are also known[36] and, by reaction of metallic Eu and Yb with cyclooctatetraene in liquid ammonia, thermally stable materials of stoichiometry C_8H_8M are produced[36a] that presumably contain M^{2+} and $C_8H_8^{2-}$. Interaction of KC_8H_8 in THF and the halides of Ce, Pr, Nd, Sm or Eu gives salts of stoichiometry

[28] D. R. Cousins and F. A. Hart, *J. Inorg. Nuclear Chem.*, 1967, **29**, 1754, 2965.

[29] V. N. Krisnamurthy and S. Soundararajan, *Can. J. Chem.*, 1967, **45**, 189.

[30] S. K. Ramalingam and S. Soundararajan, *J. Inorg. Nuclear Chem.*, 1967, **29**, 1763.

[31] J. H. Forsberg and C. A. Wathen, *Inorg. Chem.*, 1971, **10**, 1379; L. J. Charpentier and T. Moeller, *J. Inorg. Nuclear Chem.*, 1970, **32**, 3574.

[31a] L. J. Radonovich and M. D. Glick, *Inorg. Chem.*, 1971, **10**, 1463.

[32] T. Moeller *et al.*, *J. Amer. Chem. Soc.*, 1969, **91**, 7274; *J. Inorg. Nuclear Chem.*, 1970, **32**, 333.

[33] D. R. Cousins and F. A. Hart, *J. Inorg. Nuclear Chem.*, 1968, **30**, 3009; J. L. Martin *et al.*, *J. Amer. Chem. Soc.* 1968, **90**, 4493.

[33a] T. H. Siddall and W. E. Stewart, *J. Inorg. Nuclear Chem.*, 1970, **32**, 1147.

[34] H. Gysling and M. Tsutsui, *Adv. Organometallic Chem.*, 1970, **9**, 361 (a review of organo-lanthanides and -actinides).

[35] C.-H. Wong, T.-Y. Lee and Y.-T. Lee, *Acta Cryst.*, 1970, *B*, **25**, 2580.

[36] G. W. Watt and E. W. Gillow, *J. Amer. Chem. Soc.*, 1969, **91**, 775.

[36a] R. G. Hayes and J. L. Thomas, *J. Amer. Chem. Soc.*, 1969, **91**, 6876.

$K[M(COT)_2]$ which appear to have the cyclooctatetraenide (COT) ion and the metal ion in a ferrocene-like sandwich structure.[36b]

Scandium and yttrium form phenyl compounds, $M(C_6H_5)_3$, but these are very sensitive to air and water; lanthanum and praseodymium give complexes that appear to be $Li[M(C_6H_5)_4]$.

Poorly characterized and pyrophoric materials that may be alkyl complexes are obtained from the action of metal halides with methyllithium.[37]

Europium, samarium and ytterbium also give Grignard-like reagents, RMgI, when the metal reacts with alkyl halides in THF.[37a]

SCANDIUM[38]

27-11.

Scandium, $[Ar]3d4s^2$, is the congener of Al in Group III and the first member of the Sc, Y, La, Ac group. Its ionic radius, ~ 0.7 Å, is considerably smaller than the radii of Y and the lanthanides and hence its chemistry more closely resembles that of its congener, aluminum. Indeed, the fact that in its compounds coordination numbers exceeding 6 are not known emphasizes this resemblance. There is no evidence for a lower oxidation state, although the interaction of Sc and ScX_3 leads to phases with unusual stoichiometries,[39] e.g., $ScI_{2.17}$ and $ScBr_{2.3}$.

Scandium is quite common, its abundance being comparable to that of As and twice that of B. It is not readily available, owing partly to a lack of rich sources and partly to the difficulty of separation. The separation from the lanthanides can be achieved by using a cation-exchange method and oxalic acid as eluant.[39a]

Oxides. The oxide Sc_2O_3 is less basic than those of the lanthanides and is similar to Al_2O_3. Addition of base to Sc^{3+} solutions gives the hydrous oxide $Sc_2O_3 \cdot nH_2O$, but a hydroxide, $ScO(OH)$, with the same structure as $AlO(OH)$ is known. The hydrous oxide is amphoteric, dissolving in concentrated NaOH. The salt, $Na_3[Sc(OH)_6] \cdot 2H_2O$, can be crystallized, but the anion is stable only at hydroxide ion concentrations $> 8M$. "Scandates" such as $LiScO_2$ can also be made by fusing Sc_2O_3 and alkali oxides.

The Sc^{3+} ion is more readily hydrolyzed than the lanthanide ions; in perchlorate solutions the main species[40] are $ScOH^{2+}$, $Sc_2(OH)_2^{4+}$, $Sc_3(OH)_4^{5+}$ and $Sc_3(OH)_5^{4+}$, but in chloride media species such as $ScCl^{2+}$ also occur.

[36b] F. Mares, K. Hodgson and A. Streitwieser, Jr., *J. Organometallic Chem.*, 1970, **24**, C 68; see also, K. O. Hodgson and K. N. Raymond, *Inorg. Chem.*, 1972, **11**, 171.

[37] F. A. Hart, A. G. Massey and M. Saran, *J. Organometallic Chem.*, 1970, **21**, 147; R. S. P. Coutts and P. C. Wailes, *J. Organometallic Chem.*, 1970, **25**, 117.

[37a] D. F. Evans, G. V. Fazakerley and R. F. Phillips, *J. Chem. Soc., A*, **1971**, 1931.

[38] G. A. Melson and R. W. Stotz, *Coordination Chem. Rev.*, 1971, **7**, 133 (a review of coordination chemistry).

[39] B. C. McCollum and J. D. Corbett, *Chem. Comm.*, **1968**, 1666.

[39a] K. A. Orlandini, *Inorg. Nuclear Chem. Lett.*, 1969, **5**, 325.

[40] D. L. Cole, L. D. Rich, J. Owen and E. M. Eyring, *Inorg. Chem.*, 1969, **8**, 682.

Halides. The fluoride is insoluble in water but dissolves readily in an excess of HF or in NH_4F to give fluoro complexes such as ScF_6^{3-}, and the similarity to Al is confirmed by the existence of a cryolite phase, Na_3ScF_6 as well as $NaScF_4$ in the $NaF-ScF_3$ system.[41] The chloride, $ScCl_3$, is readily hydrolyzed by water; it sublimes at a much lower temperature than the lanthanide halides, and this can be associated with the different structure of the solid,[42] which is isomorphous with $FeCl_3$ (page 859). Unlike $AlCl_3$ it does not act as a Friedel–Crafts catalyst.

Oxo Salts and Complexes. Simple hydrated oxo salts are known, as well as some double salts such as $K_2SO_4 \cdot Sc_2(SO_4)_3 \cdot nH_2O$ which is very insoluble in K_2SO_4 solution. Ammonium double salts such as the tartrate, phosphate and oxalate are also insoluble in water.

The β-diketonates resemble those of Al rather than of the lanthanides; thus the acetylacetonate is normally anhydrous and may be sublimed around 200°. The TTA complex can be extracted from aqueous solutions at pH 1.5–2 by benzene, while the 8-quinolinolate (cf. Al) can be quantitatively extracted by $CHCl_3$; Sc^{3+} can also be extracted from aqueous sulfate solutions by a quaternary ammonium salt.[43]

Although the yellow color formed on addition of thiocyanate to acid Sc^{3+} solutions was originally thought to be due to a complex, it arises from the organic sulfur compound $NH_2\overline{C}{=}\overline{NC(S)SS}$, and no definite complex can be obtained from the aqueous solution.

A number of octahedral complexes such as $[Sc(DMSO)_6](ClO_4)_3$ and $[Sc\text{ bipy}_3](SCN)_3$ can be made.[44]

THE IV OXIDATION STATE

27-12. Cerium(IV).

This is the only tetrapositive lanthanide species sufficiently stable to exist in aqueous solution as well as in solid compounds. The terms ceric and cerous are commonly used to designate the IV and III valence states.

The only binary solid compounds of Ce^{IV} are the oxide, CeO_2, the hydrous oxide, $CeO_2 \cdot nH_2O$, and the fluoride, CeF_4. The dioxide, CeO_2, white when pure, is obtained by heating cerium metal, $Ce(OH)_3$, or any of several Ce^{III} salts of oxo acids such as the oxalate, carbonate or nitrate, in air or oxygen; it is a rather inert substance, not attacked by either strong acids or alkalis; it can, however, be dissolved by acids in the presence of reducing agents (H_2O_2, Sn^{II}, etc.), giving then Ce^{III} solutions. Hydrous ceric oxide, $CeO_2 \cdot nH_2O$, is a yellow, gelatinous precipitate obtained on treating Ce^{IV} solutions with bases; it redissolves fairly easily in acids. CeF_4 is prepared by treating

[41] R. E. Thoma and R. H. Karraker, *Inorg. Chem.*, 1966, **5**, 1933.
[42] N. N. Greenwood and R. L. Tranter, *J. Chem. Soc.*, A, **1969**, 2878.
[43] R. W. Cattrall and J. E. Slater, *Inorg. Chem.*, 1970, **9**, 598.
[44] N. P. Crawford and G. A. Melson, *J. Chem. Soc.*, A, **1969**, 427, 1049; **1970**, 141.

anhydrous $CeCl_3$ or CeF_3 with fluorine at room temperature; it is relatively inert to cold water and in reduced to CeF_3 by hydrogen at 200–300°.

Ce^{IV} in solution is obtained by treatment of Ce^{III} solutions with very powerful oxidizing agents, for example, peroxodisulfate or bismuthate in nitric acid. The aqueous chemistry of Ce^{IV} is similar to that of Zr, Hf and, particularly, tetravalent actinides. Thus Ce^{IV} gives phosphates insoluble in $4N$ HNO_3 and iodates insoluble in $6N$ HNO_3, as well as an insoluble oxalate. The phosphate and iodate precipitations can be used to separate Ce^{IV} from the trivalent lanthanides. Ce^{IV} is also much more readily extracted into organic solvents by tributyl phosphate and similar extractants than are the M^{III} lanthanide ions.

The hydrated ion, $[Ce(H_2O)_n]^{4+}$, is a fairly strong acid and, except at very low pH, hydrolysis and polymerization occur. It is probable that the $[Ce(H_2O)_n]^{4+}$ ion exists only in concentrated perchloric acid solution. In other acid media there is coordination of anions, which accounts for the dependence of the potential of the Ce^{IV}/Ce^{III} couple on the nature of the acid medium:

$$Ce^{IV} + e = Ce^{III} \qquad E^0 = +1.28\,(2M\ HCl),\ +1.44\,(1M\ H_2SO_4),$$
$$+1.61\,(1M\ HNO_3),\ +1.70\,(1M\ HClO_4)$$

Comparison of the potential in sulphuric acid, where at high SO_4^{2-} concentrations the major species is $[Ce(SO_4)_3]^{2-}$, with that for the oxidation of water:

$$O_2 + 4H^+ + 4e = 2H_2O \qquad E^0 = +1.229$$

shows that the acid Ce^{IV} solutions commonly used in analysis are metastable. The oxidation of water is kinetically controlled but can be temporarily catalyzed by fresh glass surfaces.

Cerium(IV) is used as an oxidant, not only in analysis, but also in organic chemistry where it is commonly used in acetic acid. The solid acetate, which is bright yellow, can be made by ozone oxidation of Ce^{III} acetate.[45] The solutions oxidize aldehydes and ketones at the α-carbon atom, and benzaldehyde, for example, gives benzoin while ammonium hexanitratocerate will oxidize toluenes to aldehydes. These oxidations appear to proceed by the initial formation of 1:1 complexes; the complexes of alcohols are red.[46]

Complex anions are formed quite readily and some of the salts previously considered to be double must be re-formulated, notably the analytical standard "ceric ammonium nitrate" which can be crystallized from HNO_3. This is actually $(NH_4)_2[Ce(NO_3)_6]$ with bidentate NO_3 groups both in the crystal and in solution.[47] In NH_4F solutions of CeF_4 the solid phase in equilibrium is $(NH_4)_4[CeF_8]$ although $(NH_3)CeF_7 \cdot H_2O$ can be grown from 28% NH_4F solutions; when heated, the octafluorocerate gives $(NH_4)_2CeF_6$.[48]

[45] N. E. May and J. K. Kochi, *J. Inorg. Nuclear Chem.*, 1968, **30**, 884.
[46] W. S. Trahanovsky *et al.*, *J. Amer Chem. Soc.*, 1969, **91**, 5060, 5068.
[47] T. A. Beineke and J. Delgaudio, *Inorg. Chem.*, 1968, **7**, 715.
[48] R. A. Penneman and A. Rosenzweig, *Inorg. Chem.*, 1969, **8**, 627.

In aqueous solution, Ce^{IV} oxidizes concentrated HCl to Cl_2, but the reaction of CeO_2 with HCl in dioxane gives orange needles of the oxonium salt of the $[CeCl_6]^{2-}$ ion; the corresponding pyridinium salt is thermally stable to 120° and is used to prepare ceric alkoxides:

$$(C_5H_5NH)_2CeCl_6 + 4ROH + 6NH_3 \rightarrow Ce(OR)_4 + 2C_5H_5N + 6NH_4Cl$$

The isopropoxide is crystalline, subliming in a vacuum at 170°, but other alkoxides, prepared from the isopropyl compound by alcohol exchange, are non-volatile and presumably polymerized by Ce—O(R)—Ce bridges.

27-13. Other Tetravalent Compounds

Praseodymium(IV). Only a few solid compounds are known, the commonest being the black non-stoichiometric oxide formed on heating Pr^{III} salts or oxide in air. The oxide system which is often formulated as Pr_6O_{11} is actually very complicated,[49] with five stable phases each containing Pr^{3+} and Pr^{4+} between Pr_2O_3 and the true dioxide PrO_2.

When alkali fluorides mixed in the correct stoichiometric ratio with Pr salts are heated in F_2 at 300–500°, compounds such as $NaPrF_5$ or Na_2PrF_6 are obtained. The action of dry HF on the latter gives PrF_4, although this cannot be obtained by direct fluorination of PrF_3.[50]

The tetravalence of Pr in these compounds has been established by magnetic, spectral, and X-ray data.

Pr^{IV} is a very powerful oxidizing agent, the Pr^{IV}/Pr^{III} couple being estimated as +2.9 V. This potential is such that Pr^{IV} would oxidize water itself, so that its non-existence in solution is not surprising. Pr_6O_{11} dissolves in acids to give aqueous Pr^{III} and liberate oxygen, chlorine, etc., depending on the acid used.

There is some evidence that $Pr(NO_3)_4$ is partially formed by action of N_2O_5 and O_3 on PrO_2.

Terbium(IV). The chemistry resembles that of Pr^{IV}. The Tb–O system is complex and non-stoichiometric and, when oxo salts are ignited under ordinary conditions, an oxide of approximately the composition Tb_4O_7 is obtained. This formula ($TbO_{1.75}$) is the nearest approach, using small whole numbers, to the true formula of the stable phase obtained, which varies from $TbO_{1.71}$ to $TbO_{1.81}$, depending on the preparative details. For the average formula Tb_4O_7, Tb^{III} and Tb^{IV} are present in equal amounts. TbO_2, with a fluorite structure, can be obtained by oxidation of Tb_2O_3 with atomic oxygen at 450°. Colorless TbF_4, isostructural with CeF_4 and ThF_4, is obtained by treating TbF_3 with gaseous fluorine at 300–400°, and compounds of the type M_nTbF_{n+4} (M = K, Rb, or Cs; $n \geqslant 2$) are known.

No numerical estimate has been given for the Tb^{IV}/Tb^{III} potential, but it

[49] J. M. Warmkessel, S. H. Lin and L. Eyring, *Inorg. Chem.*, 1969, **8**, 875; S. Randers, K. C. Patel, and C. N. R. Rao, *J. Chem. Soc., A*, **1970**, 64.
[50] L. B. Asprey, J. S. Coleman and M. J. Reisfeld, Advances in Chemistry Series, No. 71, p. 122, Amer. Chem. Soc., Washington D.C., 1967.

must certainly be more positive than $+1.23$ V since dissolution of any oxide containing Tb^{IV} gives only Tb^{III} in solution and oxygen is evolved. TbF_4 is even less reactive than CeF_4 and does not react rapidly even with hot water.

Neodymium(IV) and Dysprosium(IV). Claims of the preparation of higher oxides of these elements, supposedly containing Nd^{IV} and Dy^{IV}, are erroneous. Even treatment of Nd_2O_3 with atomic oxygen gives no Nd^{IV}-containing product. Only in the products of fluorination of mixtures of RbCl and CsCl with $NdCl_3$ and $DyCl_3$ is there fair evidence for the existence of Nd^{IV} and Dy^{IV}. Apparently such compounds as Cs_3NdF_7 and Cs_3DyF_7 can be formed, at least partially, in this way.

THE DIVALENT STATE

27-14.

The divalent state, as M^{2+}, is well-established for both solutions and solid compounds of Sm, Eu and Yb (Table 27-4). Less well-established are Tm^{2+} and Nd^{2+}, but the $+2$ ions of all the lanthanides can be prepared and stabilized in CaF_2, SrF_2 or BaF_2 lattices by reduction of, e.g., MF_3 in CaF_2 with Ca.

TABLE 27-4
Properties of the Lanthanide M^{2+} Ions

Ion	Color	E^0 (V)^a	Crystal radius (Å)
Sm^{2+}	Blood-red	-1.55	1.11
Eu^{2+}	Colorless	-0.43	1.10
Yb^{2+}	Yellow	-1.15	0.93

^{a} For $M^{3+} + e = M^{2+}$. The potentials for these and other lanthanide and actinide elements have been estimated by using a relation between the potential and the wave number of the lowest-energy electron-transfer band (see L. J. Nugent, R. D. Baybarz and J. L. Burnett, *J. Phys. Chem.*, 1969, **73**, 1178).

Aqueous Solutions. The most stable of the ions, Eu^{2+}, can be readily made by reducing aqueous Eu^{3+} solutions with Zn or Mg. The other ions require the use of sodium amalgam, but all three can be prepared by electrolytic reduction in aqueous solution or in halide melts.[51]

The ions Sm^{2+} and Yb^{2+} are quite rapidly oxidized by water as well as by air, but Eu^{2+} solutions are readily handled. Solutions of Eu^{2+} have been much studied in order to compare electron-transfer mechanisms with those of other one-electron reducing agents.[52] The ion is rapidly oxidized by O_2 but the rates of reaction with other ions often differ from expectations based on

[51] See, e.g., K. E. Johnson, J. R. McKenzie and J. N. Sandoe, *J. Chem. Soc.*, A, **1968**, 2644.
[52] See, e.g., J. H. Espenson *et al.*, *J. Amer. Chem. Soc.*, 1969, **91**, 599, 7311.

the standard potential. Thus Eu^{2+} reduces V^{3+} more slowly than does Cr^{2+} ($E^0 = -0.41$ V), but Eu^{2+} and Cr^{2+} will not reduce ClO_4^- although the weaker reductant, V^{2+} ($E^0 = -0.25$ V), will do so. The reasons for such differences are not entirely clear but are probably connected with the electronic configuration and ability to form transition states with bridging groups (see page 675).

The lanthanide M^{2+} ions resemble the Group II ions, especially Ba^{2+}. Thus the sulfates are insoluble whereas the hydroxides are soluble, and the Eu^{2+} ion can be readily separated from the other lanthanides by Zn reduction followed by the precipitation of other hydroxides by carbonate-free ammonia. The resemblance is also shown by the fact[53] that the complexity constant of Eu^{2+} towards EDTA lies between those of Ca^{2+} and Sr^{2+}.

Solid Europium Compounds. Solid europium oxide, chalcogenides, halides, carbonate, phosphate, etc., may be obtained by reduction of the corresponding Eu^{3+} compounds or, metathetically, from $EuCl_2$. The metal reacts with liquid ammonia at 50° to give the orange amide, $Eu(NH_2)_2$, which gives EuN when heated.[54] The compounds are usually isostructural with the Sr^{2+} or Ba^{2+} analogs. However, definite lower fluorides and chlorides appear to exist except for La, Ce, Pr, Gd, Tb and Er.[57]

Other Compounds. For some of the other elements, claims of lower oxides are spurious.[55] Thus an "oxide" Sm_2O appears to be the hydride SmH_2 or $SmH_{1-x}O_x$, while that of Yb is an oxide carbide $Yb_2^{II}OC$.[55a]

By reaction of the metal with triiodides, or for Nd and Gd the chlorides,[56] metal-like solids are obtained. These may be formulated as $M^{3+}(X^-)_2e$ with the odd valence electron located in a metallic conduction band. The compound $GdCl_{1.58}$ gives crystals with a unique chain of metal atoms arranged in octahedra.

Other formally divalent compounds are the hydrides, MH_2 (page 186), and the sulfides, MS, e.g., the golden-yellow LaS. These sulfides have the NaCl-type structure and are metallic conductors; they are best formulated as $M^{3+}(S^{2-})e$. Similarly the carbides, MC_2, which give acetylene on hydrolysis, are $M^{3+}(C_2^{2-})e$. The sulfides of Sm, Eu and Yb are, however, in the M^{2+} state according to their magnetic susceptibilities.

Finally, we note that the blue solutions of Sm, Eu and Yb in liquid ammonia react with 2,2'-bipyridine or 1,10-phenanthroline (L) to give compounds of stoichiometry ML_4 whose magnetic properties are best described

[53] J. C. Barnes and P. A. Bristol, *Inorg. Nuclear Chem. Lett.*, 1969, **5**, 565.
[54] R. Juza and C. Hadenfeldt, *Naturwiss.*, 1968, **55**, 228; cf. also L. M. Slaugh, *Inorg. Chem.*, 1967, **6**, 851.
[55] See G. J. McCarthy and W. B. White, *J. Less Common Metals*, 1970, **22**, 409.
[55a] G. J. McCarthy, W. B. White and R. Roy, *Inorg. Chem.*, 1969, **8**, 1236; J. M. Haschke and H. A. Eick, *Inorg. Chem.*, 1970, **9**, 851.
[56] D. A. Lokken and J. D. Corbett, *J. Amer. Chem. Soc.*, 1970, **92**, 1799 and references therein.
[57] J. J. Stezowski and H. A. Eick, *Inorg. Chem.*, 1970, **9**, 1102; D. E. Johnson, *J. Chem. Soc., A*, **1969**, 2578; P. E. Caro and J. D. Corbett, *J. Less Common Metals*, 1969, **18**, 1.

by a radical anion formulation,[58] e.g., $M^{2+}(bipy^-)_2(bipy)_2$. Whether compounds of similar stoichiometry, e.g., the black and sensitive Nd bipy$_4$, made[59] by interaction of the tribromide and Li bipyridine in THF, are to be similarly formulated is not clear. The diamide, $Yb(NH_2)_2$,[60] is obtained by interaction of Yb with NH_3 under pressure.

[58] G. R. Feistel and T. P. Mathai, *J. Amer. Chem. Soc.*, 1968, **90**, 2988.
[59] S. Herzog and R. Schuster, *Z. Chem.*, 1967, **7**, 26, 281.
[60] J. C. Warf and V. Gutmann, *J. Inorg. Nuclear Chem.*, 1971, **33**, 1583.

Further Reading

Asprey, L. B., and B. B. Cunningham, *Progr. Inorg. Chem.*, 1960, **2**, 267 (unusual oxidation states of some actinide and lanthanide elements).

Bagnall, K. W., ed., *Lanthanides and Actinides*, Butterworth, 1972.

Brown, D., *Halides of the Lanthanides and Actinides*, Interscience–Wiley, 1968 (an authoritative monograph).

Callow, R. J., *The Industrial Chemistry of the Lanthanons, Yttrium, Thorium and Uranium*, Pergamon Press, 1967.

———, *The Rare Earth Industry*, Pergamon Press, 1966 (sources, recovery and uses).

Hirschhorn, I. S., *Chem. Tech.*, **1971**, 314 (uses of lanthanide metals).

Lanthanide–Actinide Chemistry, F. R. Fields and T. Moeller, eds., Advances in Chemistry Series No. 71, Amer. Chem. Soc., Washington D.C., 1967 (conference reports).

Misumi, S., S. Kida and M. Aihari, *Coordination Chem. Rev.*, 1968, **3**, 189 (spectra and solution properties).

Moeller, T., *The Chemistry of the Lanthanides*, Reinhold, 1963 (an introductory treatment, but authoritative and thorough at its level).

Progress in the Science and Technology of the Rare Earths, Vol. 1, 1964 (series with reviews on extraction, solution chemistry, magnetic properties, analysis, halides, oxides, etc.).

Sinha, S. P., *Complexes of the Rare Earths*, Pergamon Press, 1966 (emphasizes spectra).

———, *Europium*, Springer Verlag, 1968.

The Rare Earths, Spedding F. H., and A. M. Daane, eds., Wiley, 1951 (contains detailed discussions of occurrence, extraction procedures, preparation, and properties of metals and alloys; also describes applications).

Topp, N. E., *The Chemistry of the Rare Earth Elements*, Elsevier, 1965.

Vickery, R. C., *Analytical Chemistry of the Rare Earths*, Pergamon Press, 1961.

———, *Chemistry of Yttrium and Scandium*, Pergamon Press, 1961.

Wybourne, B. G., *Spectroscopic Properties of Rare Earths*, Wiley, 1965 (comprehensive discussion of atomic spectra and especially spectra of salts).

Yost, D. M., H. Russell and C. S. Garner, *The Rare Earth Elements and Their Compounds*, Wiley, 1947 (a classical book containing early references; still of value).

28

The Actinide Elements

GENERAL REMARKS

28-1. Occurrence

The actinide elements, all of whose isotopes are radioactive, are listed in Table 28-1, together with some of their properties. The principal isotopes obtained in macroscopic amounts are given in Table 28-2.

TABLE 28-1

The Actinide Elements and Some of Their Properties

z	Sym-bol	Name	Electronic structure	z	Sym-bol	Name	Electronic structure
89	Ac	Actinium	$6d7s^2$	100	Fm	Fermium	$5f^{12}7s^2$
90	Th	Thorium	$6d^27s^2$	101	Md	Mendelevium	$5f^{13}7s^2$
91	Pa	Protactinium	$5f^26d7s^2$	102	No	Nobelium[a]	$5f^{14}7s^2$
			or $5f^16d^27s^2$	103	Lr	Lawrencium[b]	$5f^{14}6d7s^2$
92	U	Uranium	$5f^36d7s^2$	104	Rf	Rutherfordium[b]	
						(eka-hafnium)	
93	Np	Neptunium	$5f^57s^2$	105		"Hahnium"	
						(eka-tantalum)	
94	Pu	Plutonium	$5f^67s^2$	106		eka-tungsten	
95	Am	Americium	$5f^77s^2$	107		eka-rhenium	
96	Cm	Curium	$5f^76d7s^2$	108		eka-osmium	
97	Bk	Berkelium	$5f^86d7s^2$	109		eka-iridium	
			or $5f^97s^2$	110		eka-platinum	
98	Cf	Californium	$5f^{10}7s^2$	111		eka-gold	
99	Es	Einsteinium	$5f^{11}7s^2$	112		eka-mercury	

[a] The name "nobelium" was given to an unconfirmed isotope but has now been adopted for authentic isotopes.

[b] The names "joliotium" (103) and "kurchatovium" (104) are used by Russian writers (cf. I. Zvara *et al.*, *J. Inorg. Nuclear Chem.*, 1970, **32**, 1885.

TABLE 28-1 (continued)

Element	Metal		E^0 (V)^c	Ionic radius (Å)^d	
	m.p. (°C)	Radius (Å)		M^{3+}	M^{4+}
Ac	1050	1.88	~ -2.6	1.11	—
Th	$\sim$1750	1.795	e	—	0.90
Pa	1560	1.64		—	0.96
U	1132	1.57	-1.80	1.03	0.93
Np	639	1.56	-1.83	1.01	0.92
Pu	640	1.60	-2.03	1.00	0.90
Am	944	1.74	-2.36	0.99	0.89
Cm	1350	1.75		0.985	0.88
Bk	986^f	1.76		0.98	
Cf				0.977	

^{c} For $M^{3+} + 3e = M$ and $1M$ HClO$_4$ solutions.
^{d} In the octahedral fluorides, MF$_6$, the M—F bond distance also decreases with increasing Z, namely, U—F, 1.994 Å; Np—F, 1.981 Å; Pu—F, 1.969 Å.
^{e} For Th^{4+} + 4e = Th, $E^0 = -1.90$ V.
^{f} J. A. Fahey et al., Inorg. Nucl. Chem. Letts., 1972, 101.

TABLE 28-2

Principal Actinide Isotopes Available in Macroscopic Amounts^a

Isotope	Half-life	Source
^{227}Ac	21.7 yr	Natural; ^{226}Ra$(n\gamma)^{227}$Ra $\xrightarrow[41.2\ min]{\beta^-}$ ^{227}Ac
^{232}Th	1.39×10^{10} yr	Natural; 100% abundance
^{231}Pa	3.28×10^5 yr	Natural; 0.34 p.p.m. of U in uranium ores
^{235}U	7.13×10^8 yr	Natural 0.7204% abundance
^{238}U	4.50×10^9 yr	Natural; 99.2739% abundance
^{237}Np	2.20×10^6 yr	^{235}U$(n\gamma)^{236}$U$(n\gamma)^{237}$U $\xrightarrow[6.75\ d]{\beta^-}$ ^{237}Np
		(and ^{238}U$(n, 2n)^{237}$U)
^{238}Pu	86.4 yr	^{237}Np$(n\gamma)^{238}$Np $\xrightarrow[2.1\ d]{\beta^-}$ ^{238}Pu
^{239}Pu	24,360 yr	^{238}U$(n\gamma)^{239}$U $\xrightarrow[23.5\ min]{\beta^-}$ ^{239}Np $\xrightarrow[2.35\ d]{\beta^-}$ ^{239}Pu
^{242}Pu	3.79×10^5 yr	Successive $n\gamma$ in ^{239}Pu
^{244}Pu	8.28×10^7 yr	Successive $n\gamma$ in ^{239}Pu
^{241}Am	433 yr	^{239}Pu$(n\gamma)^{240}$Pu$(n\gamma)^{241}$Pu $\xrightarrow[13.2\ yr]{\beta^-}$ ^{241}Am
^{243}Am	7650 yr	Successive $n\gamma$ on ^{230}Pu
^{242}Cm	162.5 d	^{241}Am$(n\gamma)^{242m}$Am $\xrightarrow[16.0\ hr]{\beta^-}$ ^{242}Cm
^{244}Cm	17.6 yr	^{239}Pu$(4n\gamma)^{243}$Pu $\xrightarrow[5.0\ hr]{\beta^-}$ ^{243}Am$(n\gamma)^{244}$Am $\xrightarrow[26\ min]{\beta^-}$ ^{244}Cm
^{249}Bk	314 d	Successive $n\gamma$ on ^{239}Pu
^{252}Cf	2.57 yr	Successive $n\gamma$ on ^{242}Pu

^{a} ^{237}Np and ^{239}Pu are available in multi-kilogram quantities; ^{238}Pu, ^{242}Pu, ^{241}Am, ^{243}Am and ^{244}Cm in amounts of 100 g or above; ^{244}Pu, ^{252}Cf, ^{249}Bk, ^{248}Cm (4.7×10^5 yr), ^{253}Es (20 d), ^{254}Es (1.52 yr for α) in milligram and ^{257}Fm (94 d) in μg quantities. Other long-lived isotopes are known but can be obtained in traces by use of accelerators, e.g., ^{247}Bk ($\sim 10^4$ yr).

The terrestrial occurrence of Ac, Pa, U, and Th is due to the half-lives of the isotopes ^{235}U, ^{238}U and ^{232}Th which are sufficiently long to have enabled the species to persist since genesis. They are the sources of actinium and protactinium formed in the decay series and found in uranium and thorium ores. The half-lives of the most stable isotopes of the trans-uranium elements are such that any primordial amounts of these elements appear to have disappeared long ago. However, neptunium and plutonium have been isolated in traces from uranium[1a] minerals in which they are formed continuously by neutron reactions such as

$$^{238}U \xrightarrow{n\gamma} {}^{239}U \xrightarrow{\beta^-} {}^{239}Np \xrightarrow{\beta^-} {}^{239}Pu$$

The neutrons arise from spontaneous fission of ^{235}U or from α,n reactions of light elements present in uranium minerals.

Traces of ^{244}Pu have recently been found in a very old cerium-containing mineral. This plutonium is believed to be primordial in origin but its formation by cosmic ray interactions is not yet rigorously excluded.[1b]

The first transuranium elements, neptunium and plutonium, were obtained in tracer amounts from bombardments of uranium by McMillan and Abelson and by Seaborg, McMillan, Kennedy, and Wahl, respectively, in 1940. Both elements are obtained in substantial quantities from the uranium fuel elements of nuclear reactors. Only plutonium is normally recovered and is used as a nuclear fuel since, like ^{235}U, it undergoes fission; its nuclear properties apparently preclude its use in hydrogen bombs. Certain isotopes of the heavier elements are made by successive neutron capture in ^{239}Pu in high-flux nuclear reactors ($> 10^{15}$ neutrons cm^{-2} sec^{-1}). Others are made by the action of accelerated heavy ions of B, C, N, O or Ne on Pu, Am or Cm.

The elements above fermium ($Z = 100$) exist only as short-lived species, the most stable being ^{258}Md (53 d), ^{255}No (185 sec), ^{256}Lr (45 sec), and ^{261}Rf (ca 70 sec). The nuclear properties of isotopes, such as their decay modes and half-lives, can be predicted with some accuracy from nuclear systematics.[1c] For elements beyond Pa ($Z = 91$), spontaneous nuclear fission becomes increasingly important and is indeed the limiting factor for the life of a given isotope. The element $Z = 104$ appears to be hafnium-like. As the $6d$ shell is filled the subsequent elements up to $Z = 114$ should have chemical properties analogous to those of the elements Hf up to Pb.

28-2. Electronic Structures; Comparison with Lanthanides

The atomic spectra of the heavy elements are very complex and it is difficult to identify levels in terms of quantum numbers and configurations.[2] For

[1] (a) See W. A. Myers and M. Lindner, *J. Inorg. Nucl. Chem.*, 1971, **33**, 3233; (b) D. C. Hoffman *et al.*, *Nature*, 1971, **234**, 123, 132; (c) G. T. Seaborg, *Ann. Rev. Nuclear Sci.*, 1968, **18**, 53.
[2] For discussion and data see M. Fred in *Lanthanide and Actinide Chemistry*, Advances in Chemistry Monograph No. 71, p. 188, Amer. Chem. Soc., Washington D.C., 1967.

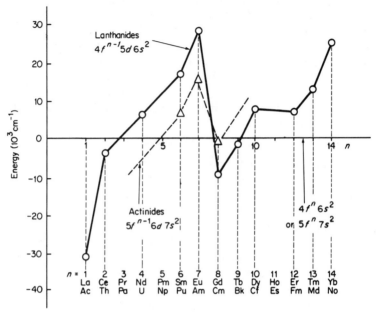

Fig. 28-1. Approximate relative positions of $f^n s^2$ and $f^{n-1} ds^2$ conformations for neutral lanthanide and actinide atoms. [Reproduced by permission from ref. 2 with additional data kindly provided by Dr. M. Fred.]

chemical behavior the lowest configuration is of greatest importance and the competition between $5f^n 7s^2$ and $5f^{n-1} 6d 7s^2$ is of interest. Fig. 28-1 shows the approximate relative positions for both the lanthanides and the actinides. For the elements in the first half of the f shell it appears that less energy is required for the promotion of $5f \rightarrow 6d$ than for the $4f \rightarrow 5d$ promotion in the lanthanides; there is thus a greater tendency to supply more bonding electrons with the corollary of higher valences in the actinides. The second half of the actinides resemble the lanthanides more closely.

Another difference is that the $5f$ orbitals have a greater spatial extension relative to the $6s$ and $6p$ orbitals than the $4f$ orbitals have relative to the $5s$ and $5p$ orbitals. The greater spatial extension of the $5f$ orbitals has been shown experimentally; the electron-spin resonance spectrum of UF_3 in a CaF_2 lattice shows structure attributable to the interaction of fluorine nuclei and the electron spin of the U^{3+} ion. This implies a small overlap of $5f$ orbitals with fluorine and constitutes an f covalent contribution to the ionic bonding. With the neodymium ion a similar effect is *not* observed. Because they occupy inner orbitals the $4f$ electrons in the lanthanides are not accessible for bonding purposes and virtually no compound in which $4f$ orbitals are used can be said to exist.

In the actinide series, therefore, we have a situation in which the energies of the $5f$, $6d$, $7s$ and $7p$ orbitals are about comparable over a range of atomic numbers (especially U—Am), and, since the orbitals also overlap spatially, bonding can involve any or all of them. In the chemistries, this situation is indicated by the fact that the actinides are much more prone to complex formation than are the lanthanides, where the bonding is almost exclusively ionic. Indeed the actinides can even form complexes with certain π-bonding ligands as well as forming complexes with halide, sulfate and other ions. The difference from lanthanide chemistry is usually attributed to the contribution of covalent-hybrid bonding involving $5f$ electrons.

A further point is that, since the energies of the $5f$, $6d$, $7s$ and $7p$ levels are comparable, the energies involved in an electron shifting from one to another, say $5f$ to $6d$, may lie *within* the range of chemical binding energies. Thus the electronic structure of the element in a given oxidation state may vary between compounds and in solution be dependent on the nature of the ligands. It is accordingly also often impossible to say *which* orbitals are being utilized in bonding or to decide meaningfully whether the bonding is covalent or ionic.

28-3. Ionic Radii

The ionic radii for the commonest oxidation states are given in Table 28-1 and are compared with those of the lanthanides in Fig. 28-2. There is clearly an "actinide" contraction, and the similarities in radii of both series correspond to similarities in their chemical behavior for properties that depend on the ionic radius, such as thermodynamic results for hydrolysis of halides. It is also generally the case that similar compounds in the same oxidation state have similar crystal structures that differ only in the parameters.

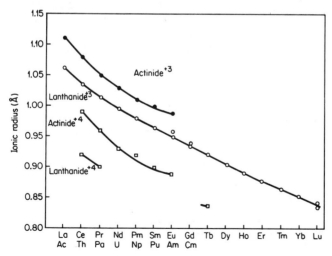

Fig. 28-2. Radii of actinide and lanthanide ions. [Reproduced by permission from D. Brown, *Halides of the Lanthanides and Actinides*, Interscience-Wiley, 1968.]

28-4. Absorption Spectra and Magnetic Properties

The electronic absorption spectra of the actinide ions, like those of the lanthanides, are due to transitions within the $5f^n$ levels and consist of narrow bands, relatively uninfluenced by ligand fields; the intensities are generally about ten times those of the lanthanide bands.

Spectra involving only one f electron are simple, consisting of only a single transition $^2F_{5/2}-^2F_{7/2}$. For the f^7 configuration (Cm^{3+}; cf. Gd^{3+}) the lowest excited state lies about 4 eV above the ground level, so that these ions show only charge-transfer absorption in the ultraviolet.

The magnetic properties of the actinide ions are considerably harder to interpret than those of the lanthanide ions. The experimental magnetic moments are usually lower than the values calculated by using Russell–Saunders coupling, and this appears to be due both to ligand field effects similar to those operating in the d transition series and to inadequacy of this coupling scheme. Since $5f$ orbitals can participate to some extent in covalent bonding, ligand effects are to be expected.

For the ions Pu^{3+} and Am^{3+}, the phenomenon noted for Sm^{3+} and Eu^{3+} is found; since the multiplet levels are comparable to kT, anomalous temperature-dependence of the susceptibilities is found.

28-5. Oxidation States; Stereochemistry

The known oxidation states of the actinides are given in Table 28-3. With the exception of Th and Pa the common, and for trans-americium elements the dominant, oxidation state is $+3$ and the behavior is similar to the $+3$ lanthanides. Thorium and the other elements in the $+4$ state show resemblances to Hf or Ce^{IV}, while Pa and the elements in the $+5$ state show some resemblances to, say, Ta^V. Exceptions to the latter statement are the $+5$ state of U, Np, Pu and Am in the *dioxo ions* MO_2^+; these ions and the MO_2^{2+} ions of the $+6$ states show unusual and exceptional behavior, as discussed below.

TABLE 28-3

Oxidation States of the Actinide Elements

Ac	Th	Pa	U	Np	Pu	Am	Cm	Bk	Cf	Es	Fm	Md	No
						2[c]			2[d]	2[d]	2[d]	2[d]	2[d]
3[a]	3[b]	3	3	3	3	3	3	3	3	3	3	3	3
	4	4	4	4	4	4	4	4					
		5	5	5	5	5			5 ?[e]				
			6	6	6	6							
				7	7								

[a] Bold number signifies most stable state.
[b] Solid state only.
[c] In CaF_2 lattice.
[d] So far only in solution.
[e] Not characterized properly.

Although there is certainly some doubt concerning the extent of covalent bonding in actinide compounds, the angular distributions and relative strengths of various orbital combinations using f orbitals have been worked out theoretically in a manner similar to that for the schemes for light elements. Examples are: sf, linear; sf^3, tetrahedral; sf^2d, square; and d^2sf^3, octahedral. These hybridizations *could* be considered to hold in PuO_2^{2+}, $NpCl_4$ and UCl_6, for example. However, in view of the closeness in the energy levels of electrons in the valence shells, and the mutual overlap of orbitals of comparable size in these heavy atoms, several equally valid descriptions can be chosen in a particular case. In such circumstances, the orbitals actually used must be some mixture of all the possible limiting sets and it is not justified to treat the bonding in terms of any single set, except as a convenient first approximation.

Examples of the stereochemistry of actinide compounds and complexes are given in Table 28-4. For the $+3$ oxidation state, where the resemblance

TABLE 28-4

Stereochemistry of Actinides

Oxidation state	Coordination number	Geometry[a]	Examples
$+3$	5	*tbp*	AcF_3, $BaUF_6$ (LaF$_3$-type)
	6	Octahedral	Macac$_3$, $[M(H_2O)_6]^{3+}$
	8	Bicapped trig. prism	$PuBr_3$, $[AmCl_2(H_2O)_6]^+$
	9		UCl_3, $AmCl_3$ (also La–GdCl$_3$)
$+4$	6	Octahedral	UCl_6^{2-}, $UCl_4(PEt_3)_2$
	8	Cubic(O_h)	$(Et_4N)_4[U(NCS)_8]$
		Dodecahedral	$[Th\,ox_4]^{4-}$, $Th(S_2CNEt_2)_4$
		Fluorite str.	ThO_2, UO_2
		Square antiprism	$ThI_4(s)$, $Uacac_4$, $(NH_4)_2UF_6$
	9	Distorted square antiprism	$(NH_4)_2ThF_8$
	10	Bicapped square antiprism	$K_4Th\,ox_4 \cdot 4H_2O$
	10	?	M(tropolonate)$_5^-$, M = Th or U
	12	Irreg. icosahedral	$[Th(NO_3)_6]^{2-}$
$+5$	6	Octahedral	UF_6^-, α–UF_5 (infinite chain)
	7		β–UF_5
	8	Cubic(O_h)	Na_3MF_8 (M = U, Np)
	9	Complex	PaF_7^{2-} in K_2PaF_7
$+6$	6	Octahedral	UF_6, Li_4UO_5 (distorted), UCl_6
	6–8	See text	MO_2^{2+} complexes
	8	?	$M_2^IUF_8$

[a] For detailed discussion of crystal structures, many of which are most complicated, see A. F. Wells, *Structural Inorganic Chemistry*, 3rd edn., Oxford University Press, 1962, p. 959.

to the lanthanides is distinct, octahedral coordination is often found, but higher coordination numbers (e.g., 9 in UCl_3) are also common. Eight-coordination is especially a characteristic of the $+4$ oxidation state. An example here is Th acac$_4$, which is isomorphous with the uranium analog and has a structure based on a slightly distorted square antiprism. This structure is that predicted on purely electrostatic grounds, and the volatility of the compound is no criterion of covalent bonding but only a reflection of the almost spherical nature of the molecules and valence saturation of the outer atoms. Well-defined 10-coordinate anions, $M(\text{tropolonate})_5^-$ have been prepared for Th^{IV} and U^{IV}.[3]

GENERAL CHEMISTRY OF THE ACTINIDES

In view of the close similarities in preparations and properties of actinide compounds in a given oxidation state it is convenient to discuss some general features and to follow this by additional descriptions for the separate elements. Methods of chemical separations of the elements are also discussed subsequently.

28-6. The Metals

The metals may all be prepared by a method applicable on either a 10^{-6} g scale, as in the first preparation of Cm, or on a multikilogram scale. This is the reduction of one of the anhydrous fluorides, MF_3 or MF_4, with the vapor of Li, Mg, Ca or Ba at 1100–1400°; chlorides or oxides can be reduced similarly. On large scales, e.g., for uranium, Mg or Ca is normally used.

There are other procedures: thus very pure Th is made from ThI_4 by thermal decomposition (de Boer process). Electrolytic methods are not commonly used, but Th can be obtained from a melt of ThF_4, KCN and NaCl. Americium has been obtained by a method depending upon the volatility, which is greater than that of the other actinides:

$$2La + Am_2O_3 \xrightarrow{1200°} 2Am\uparrow + La_2O_3$$

Curium has also been made on a gram scale by extraction from a melt of $MgCl_2$, MgF_2 and CmO_2 with molten Zn—Mg alloy, the excess of which is then distilled off;[4] uranium can also be obtained as an amalgam, from which it can be recovered, by action of Na/Hg on uranyl acetate.[5]

The melting points of the metals are given in Table 28-1.

Actinium. This is silvery-white and glows in the dark owing to its radioactivity, which also contributes to its disintegration and high reactivity.

Thorium. The metal is white but tarnishes in air. It can be readily machined and forged. It is highly electropositive, resembling the lanthanide metals, and is pyrophoric when finely divided. It is attacked by boiling water, by oxygen at 250°, and by N_2 at 800°. Dilute HF, HNO_3, and H_2SO_4, and concentrated HCl or H_3PO_4, attack thorium only slowly, and concentrated

[3] E. L. Muetterties, *J. Amer. Chem. Soc.*, 1969, **91**, 4420; J. Selbin and D. Ortego, *J. Inorg. Nuclear Chem.*, 1968, **30**, 315.
[4] I. D. Eubanks and M. C. Thompson, *J. Inorg. Nuclear Chem.*, 1969, **31**, 187.
[5] Y. Kobayashi and T. Ishimori, *J. Inorg. Nuclear Chem.*, 1969, **31**, 981.

nitric acid makes it passive. The attack of hot 12M hydrochloric acid on thorium gives a black residue that appears to be a complex hydride approximating to $ThO_{1.3}Cl_{0.7}H_{1.3}$.

Protactinium. This is a relatively unreactive, shiny and malleable metal; it tarnishes in air.

Uranium. For its use in nuclear reactors, uranium must be exceedingly pure and free from elements such as B or Cd which have high absorption capacities for thermal neutrons.

Uranium is one of the densest metals (19.07 g cm^{-3} at 25°) and has three crystalline modifications. It forms a wide range of intermetallic compounds —U_6Mn, U_6Ni, USn_3, etc.—but, owing to the unique nature of its crystal structures, it cannot form extensive ranges of solid solutions. Uranium is chemically reactive and combines directly with most elements. In air, the surface is rapidly converted into a yellow and subsequently a black non-protective film. Powdered uranium is frequently pyrophoric. The reaction with water is complex; boiling water forms UO_2 and hydrogen, the latter reacting with the metal to form a hydride, which causes disintegration. Uranium dissolves rapidly in hydrochloric acid (a black residue often remains; cf. Th) and nitric acid, but slowly in sulfuric, phosphoric or hydrofluoric acid. It is unaffected by alkalis. An important reaction of uranium is that with hydrogen, forming the hydride (see page 186) which is a useful starting material for the synthesis of uranium compounds.

Neptunium. This silvery metal resembles U but is denser (20.45 g cm^{-3}) and has three modifications.

Plutonium. The metal is again similar to U chemically; it is pyrophoric and must be handled with extreme care owing to the health hazard. Also, above a certain critical size the pure metal can initiate a nuclear explosion. The metal is unique in having at least six allotropic forms below its melting point, each with a different density, coefficient of expansion, and resistivity; and curiously, if the phase expands on heating, the resistance decreases. Plutonium forms numerous alloys.

Americium. This is the first actinide metal to resemble a lanthanide. It melts higher and has a much lower density (13.7 g cm^{-3}) than its predecessors. It is more electropositive than Pu, being comparable to a light lanthanide.

Curium. The metal resembles its analog, Gd, in having a relatively high melting point and in its magnetic properties.

28-7. The +2 Oxidation State

This is an unusual state for the actinides and is at present confined to Am (the $5f$ analog of Eu) and the Cf–No group. This can be associated with the greater energy of promotion $5f \rightarrow 6d$ than of promotion $4f \rightarrow 5d$ in the lanthanides, and the +2 state is thus more stable at the end of the series.

The Am^{2+} ion is known only in CaF_2 lattices where it has been characterized by optical and esr spectra. For Cf, Es, Fm and Md, there is evidence for M^{2+} in solution; the potentials are given in Table 28-5.

TABLE 28-5

Actinide Potentials, $M^{3+} + e = M^{2+}$, E^0 in volts

Element	Measured	Calculated[a]	Element	Measured	Calculated[a]
Am	—	−2.6	Es	−1.6	−1.6
Cm	—	−5.0	Fm	—	−1.3
Bk	—	−3.4	Md	−0.15	
Cf	−1.9	−2.0	No	+1.45	

[a] From the relation between the wave number of the lowest-energy electron-transfer band of M^{3+} and E^0 (L. J. Nugent, R. D. Baybarz and J. L. Burnett, *J. Phys. Chem.*, 1969, 73, 1178).

The Md^{2+} ion is especially stable,[6] more so than even Eu^{2+} or Yb^{2+}. Md^{3+} can be reduced by Zn/Hg and can be coprecipitated with sulfates. As expected, the properties of the +2 ions, where known, are similar to those of the +2 lanthanides or Ba^{2+}.

28-8. The +3 Oxidation State

This is the common state for all the actinides except Th and Pa, and it is the normal state for Ac, Am and trans-Am elements.

The general chemistry closely resembles that of the lanthanides. The halides, MX_3, may be readily prepared and are easily hydrolyzed to MOX. The oxides, M_2O_3, are known only for Ac, Pu and heavier elements. In aqueous solution there are M^{3+} ions, and insoluble hydrated fluorides and oxalates can be precipitated. Isomorphism of crystalline solids is common.

Of the +3 ions, U^{3+} is the most readily oxidized, even by air or more slowly by water.

Since the ionic sizes are comparable for both series, the formation of complex ions and their stability constants are similar, so that it is difficult to separate actinide from lanthanide elements, though it can be done, as described below (page 1111), by ion-exchange or solvent-extraction procedures.

28-9. The +4 Oxidation State

This state is more common than in the lanthanides where only Ce^{IV} is known in solution. It is the principal state for Th; Pa^{IV}, U^{IV} and Pu^{IV} are reasonably stable in solution; Am^{IV} and Cm^{IV} are much more easily reduced and exist only as complex ions in concentrated fluoride solution of low acidity or, for Am, also in phosphate solutions.[7] The elements after Bk cannot be oxidized. Again the general chemistry is lanthanide-like, with

[6] J. Malý, *J. Inorg. Nuclear Chem.*, 1969, 31, 741.
[7] E. Yaniv, M. Givon and Y. Marcus, *J. Inorg. Nuclear Chem.*, 1969, 31, 369.

sparingly soluble hydroxides and hydrated fluorides and phosphates. Other points of importance are: (*a*) The dioxides, MO_2 from Th to Bk, all have the fluorite lattice. (*b*) The tetrafluorides, MF_4, are isostructural with lanthanide tetrafluorides. (*c*) The chlorides and bromides are known only for Th, Pa, U, and Np, presumably owing to the inability of the halogen to oxidize the heavier metals; and for iodides only those of Th, Pa and U exist. (*d*) Oxo halides, MOX_2, can be made for Th—Np, e.g., by the reaction:[8]

$$3MX_4 + Sb_2O_3 \xrightarrow{\text{Heat}} 3MOX_2 + 2SbX_3 \uparrow_{450^\circ}$$

(*e*) Hydrolysis, complexation and disproportionation are important in aqueous solution, as discussed below.

28-10. The +5 Oxidation State[9]

The +5 state is the normal oxidation state for Pa, and there is quite a close resemblance to the chemistry of Nb and Ta.

For the other elements, comparatively few solid compounds have been made. The halides are known only for Pa and U. Salts of fluoro anions such as MF_6^-, MF_7^{2-} and MF_8^{2-} are known for Pa–Pu, although the Np and Pu compounds can be made only by solid-state reactions. Oxo chlorides, $MOCl_3$, are known for Pa, U and Np.

An important difference from the Nb and Ta Group is the formation of the dioxo ion, MO_2^+, which is of great importance for U, Np, Pu and Am chemistry. These ions are discussed further below, but we note here that their stability in aqueous solution is determined by the ease of disproportionation, e.g.:

$$2UO_2^+ + 4H^+ \rightleftharpoons U^{4+} + UO_2^{2+} + 2H_2O \qquad K = 1.7 \times 10^6$$

and the stability order is $U < Pu < Np \sim Am$.

28-11. The +6 Oxidation State

In simple compounds this state occurs only in the hexafluorides, MF_6, of U, Np and Pu. There is no evidence for AmF_6.

The principal chemistry of the +6 state in solids and in solutions is that of the dioxo ions, MO_2^{2+}, of U, Np, Pu and Am. These unique ions are discussed below.

28-12. The +7 Oxidation State

The existence of this state has been only recently shown, by Russian workers, and so far is known only for Np and Pu.[10]

The action of ozone on suspensions of "neptunates" (cf. uranates,

[8] K. W. Bagnell, D. Brown and J. F. Easy, *J. Chem. Soc., A,* **1968**, 288.
[9] J. Selbin and J. D. Ortego, *Chem. Rev.,* 1969, **69**, 657 (an extensive review on uranium(v) chemistry).
[10] R. C. Thompson and J. C. Sullivan, *J. Amer. Chem. Soc.,* 1970, **92**, 3028; S. K. Awasthi *et al., Inorg. Nucl. Chem. Letters,* 1971, **7**, 145.

page 1100), $Na_2NP_2O_7 \cdot nH_2O$, followed by addition of $[Co(NH_3)_6]^{3+}$, gives a salt of the ion NpO_5^{3-} or more likely,[11] $[NpO_4(OH)_2]^{3-}$. Some salts are isomorphous with the Tc and Re analogs. The oxidation state $+7$ was confirmed for Np by way of the characteristic Mössbauer spectra where the isomer shifts may be correlated with the number of $5f$ electrons present.[11] Electrolytic or ozone oxidation of Np^V or Np^{VI} in NaOH gives a green solution of NpO_5^{3-}, which is only slowly reduced at $25°$:

$$NpO_5^{3-} + H_2O + e = NpO_4^{2-} + 2OH^- \qquad E = 0.58 \text{ V } (1M \text{ NaOH})$$

For plutonium,[12] the oxides PuO_2 and Li_2O are exposed to oxygen at $430°$; the resulting Li_5PuO_6 gives a green unstable aqueous solution. This behavior is similar to that of Re^{VII} and Tc^{VII}.

28-13. The Dioxo Ions MO_2^+ and MO_2^{2+}

These ions are remarkably stable with respect to the strength of the M—O bond (see below). Unlike some other oxo ions they can persist through a variety of chemical changes, and they behave like cations whose properties are intermediate between those of M^+ or M^{2+} ions of similar size but greater charge. The MO_2 group even appears more or less as an "yl" group in certain oxide and oxo ion structures, and further, whereas MoO_2F_2 or WO_2F_2 are molecular halides, in UO_2F_2 there is a linear O—U—O group with F bridges. The stability of UO_2^{2+} and PuO_2^{2+} ions in aqueous solution is shown by the very long half-life for exchange with $H_2^{18}O$ of $> 10^4$ hours; the exchange can be catalyzed by the presence of reduced states or, for PuO_2^{2+}, by self-reduction due to radiation effects.

In both crystalline compounds and in solutions the oxo ions are evidently linear. The ions form a great variety of complexes with negative ions and neutral molecules. Crystallographic data show that four, five, or six ligand atoms can lie in the equatorial plane of the O—M—O group; the ligand atoms may or may not be entirely coplanar depending on the circumstances. Planar 5- and 6-coordination in the equatorial plane is commonest and appears to give geometry more stable than the puckered hexagonal configurations. Planar 5-coordination best allows rationalization of a number of hydroxide and other structures, as well as the behavior of polynuclear uranyl ions in hydrolyzed solutions. An example is the structure of the anion in the complex salt, sodium uranyl acetate shown in Fig. 28-3; the carboxylate groups are bidentate and equivalent. Similar structures have been found in other species[13a] such as $UO_2(NO_3)_2(H_2O)_2$, $UO_2(NO_3)_2[OP(OEt)_3]_2$ and $Rb[UO_2(NO_3)_3]$.

For such heavy atoms there are difficulties in accurately locating oxygen atoms and assessing M—O distances, but it is certain that the M—O distances are *not constant*. Thus for UO_2^{2+} the range appears to be from ca. 1.6 to

[11] K. Fröhlich, P. Gütlich and C. Keller, *Angew. Chem. Internat. Edn.*, 1972, **11**, 57.
[12] C. Keller and H. Seiffert, *Angew. Chem. Internat. Edn.*, 1969, **8**, 279.
[13a] N. Kent Dalley, M. H. Mueller and S. H. Simonsen, *Inorg. Chem.*, 1971, **10**, 323.

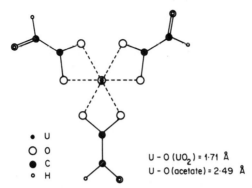

- • U
- ○ O
- ● C
- ○ H

U - O (UO₂) = 1·71 Å
U - O (acetate) = 2·49 Å

Fig. 28-3. Structure of the anion in Na[UO₂(OCOCH₃)₃.]

ca. 2.0 Å. Accordingly there has been extensive use of infrared data for correlating bond lengths, r (Å), and force constants, k (millidynes $Å^{-1}$), using the rule

$$r_{U-O} = 1.08 \, k^{-1/3} + 1.17$$

Where comparison is possible, the rule generally agrees with X-ray data.[13b]

For MO_2^{2+} the bond strengths, as well as chemical stabilities towards reduction, decrease in the order $U > Np > Pu > Am$. It also appears that the force constants for U—O bonds are high, indicating a multiplicity greater than two. Appropriate d and f orbitals can be combined into molecular orbitals to give one σ plus two π bonds. The MO's are filled at UO_2^{2+}, and succeeding electrons are fed into non-bonding orbitals.[14] The MO scheme allows detailed interpretation of spectroscopic and magnetic data for the oxo ions. It also provides an explanation of the U–Am stability sequence and the non-existence of PaO_2^{2+}. The latter is connected with the fact that, for Pa, the $6d$ is higher than the $5f$ level, whereas for $U(5f^3 6d^1 7s^2)$ it is the reverse, so that, for Pa, the $5f\sigma-2p\sigma$ metal–oxygen overlap is poor. The instability of UO_2^+ is probably also connected with the sensitivity of the energy of the $5f$ electrons to total charge, thus critically affecting the U–O overlap.

28-14. Actinide Ions in Aqueous Solution[15]

The formal reduction potentials of actinide ions in aqueous solution are given in Table 28-6, from which it is clear that the electropositive character of the metals increases with increasing Z and that the stability of the higher oxidation states decreases. A comparison of various actinide ions is given in Table 28-7. It must be noted also that, for comparatively short-lived isotopes decaying by α-emission or spontaneous fission, heating and chemical effects due to the high level of radioactivity occur in both solids and aqueous solu-

[13b] J. I. Bullock, *J. Chem. Soc., A*, **1969**, 781 (data for many UO_2^{2+} compounds).
[14] J. T. Bell, *J. Inorg. Nuclear Chem.*, **1969**, **31**, 703; see also S. P. McGlynn and J. K. Smith, *J. Mol. Spectroscopy*, **1961**, **6**, 164, 188.
[15] A. D. Jones and G. R. Choppin, *Actinide Chem. Rev.*, **1969**, **1**, 311 (an extensive review of complex ions).

TABLE 28-6

Formal Reduction Potentials of the Actinides for $1M$ Perchloric Acid Solutions at 25°
(in volts; brackets [ ] indicate estimate)

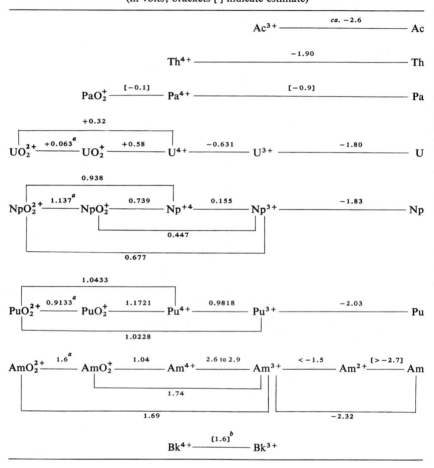

Notes:

1. $PaO_2^+ + 4H^+ = Pa + 2H_2O$, $E = [-1.0]$.

2. Couples involving oxygen-transfer, for example, $UO_2^{2+} + 4H^+ + 2e = U^{4+} + 2H_2O$ are *irreversible* and are of course hydrogen ion dependent. Couples such as PuO_2^{2+}/PuO_2^+ *are* reversible.

[a] The true E^0 values for U, Np, Pu and Am are 0.163, 1.236, 1.013 and 1.7 V, respectively (J. R. Brand and J. W. Cobble, *Inorg. Chem.*, 1970, **9**, 912).

[b] By direct measurement in $0.1 M H_2SO_4$, $E = 1.43$ V (R. C. Propst and M. L. Hyder, *J. Inorg. Nuclear Chem.*, 1970, **32**, 2205).

tions, e.g., for ^{238}Pu, ^{241}Am and ^{242}Cm the heat output is calculated as 0.5, 0.1 and 122 watts g^{-1}, respectively. Radiation-induced decomposition of water leads to H and OH radicals, H_2O_2 production, etc., and in solution higher oxidation states such as Pu^V, Pu^{VI} and Am^{IV-VI} are reduced. Chemi-

TABLE 28-7

The Principal Actinide Ions[a] in Aqueous Solution

Ion	Color[b]	Preparation	Stability
U^{3+}	Red-brown	Na or Zn/Hg on UO_2^{2+}	Slowly oxidized by H_2O, rapidly by air to U^{4+}
Np^{3+}	Purplish	$H_2(Pt)$ or electrolytic	Stable in water; oxidized by air to Np^{4+}
Pu^{3+}	Blue-violet	SO_2, NH_2OH on higher states	Stable to water and air; easily oxidized to Pu^{4+}
Am^{3+}	Pink	I^-, SO_2, etc., on higher states	Stable; difficult to oxidize
Cm^{3+}	Pale yellow		Stable; not oxidized chemically
U^{4+}	Green	Air or O_2 on U^{3+}	Stable; slowly oxidized by air to UO_2^{2+}
Np^{4+}	Yellow-green	SO_2 on NpO_2^+ in H_2SO_4	Stable; slowly oxidized by air to NpO_2^{2+}
Pu^{4+}	Tan	SO_2 or NO_2^- on PuO_2^{2+}	Stable in $6M$ acid; disproportionates in low acid $\rightarrow Pu^3 + PuO_2^{2+}$
Am^{4+c}	Pink-red	$Am(OH)_4$ in $15M$ NH_4F	Stable in $15M$ NH_4F; reduced by I^-
Cm^{4+c}	Pale yellow	CmF_4 in $15M$ CsF	Stable only 1 hour at $25°$
UO_2^+	?	Transient species	Stability greatest pH 2–4; disproportionates to U^{4+} and UO_2
NpO_2^+	Green	Np^{4+} and hot HNO_3	Stable; disproportionates only in strong acid
PuO_2^+	?	Hydroxylamine on PuO_2^{2+}	Always disproportionates; most stable at low acidity
AmO_2^+	Pale yellow	Am^{3+} with OCl^-, cold $S_2O_8^{2-}$	Disproportionates in strong acid; reduced (2% per hour) by products of own α-radiation
UO_2^{2+}	Yellow	Oxidize U^{4+} with HNO_3, etc.	Very stable; difficult to reduce
NpO_2^{2+}	Pink	Oxodize lower states with Ce^{4+},	Stable; easily reduced
PuO_2^{2+}	Yellow-pink	MnO_4^-, O_3, BrO_3^-, etc.	Stable; fairly easy to reduce
AmO_2^{2+}	Rum		Reduced (4% per hour) by products from own α-radiation

[a] Ac^{3+}, Th^{4+}, Cm^{3+}, and ions of Pa are colorless.

[b] Depends on concentration and nature of ions.

[c] As fluoro complex, MF_6^{2-}.

cal reactions observable with a short-lived isotope, e.g., ^{242}Cm (163 d), may differ when a longer-lived isotope is used; thus Cm^{IV} can be observed only when ^{244}Cm (17.6 yr) is employed.

The possibility of several cationic species introduces complexity into the aqueous chemistries, particularly of U, Np, Pu and Am. Thus all four oxidation states of Pu can coexist in appreciable concentrations in a solution. The solution chemistries and the oxidation–reduction potentials are further complicated by the formation in the presence of ions other than perchlorate, of cationic, neutral or anionic species. Further, even in solutions of low pH, hydrolysis and the formation of polymeric ions occurs. Thirdly, there is the additional complication of disproportionation of certain ions, which is particularly dependent on the pH.

Since extrapolation to infinite dilution is impossible for most of the actinide ions, owing to hydrolysis—for example, Pu^{4+} cannot exist in solution below $0.05M$ in acid—only approximate oxidation potentials can sometimes be given. The potentials are sensitive to the anions and other conditions.

The actinides have a far greater tendency to complex formation than the lanthanides. Thus there are extensive series of halogeno complexes, and complex ions are given with most oxo anions such as NO_3^-, SO_4^{2-}, ox^{2-}, CO_3^{2-}, phosphate, etc. A vast amount of data exists on complex ion formation in solution since this has been of primary importance in connection with solvent-extraction, ion-exchange behavior, and precipitation reactions involved in the technology of actinide elements. The general tendency to complex ion formation decreases in the direction controlled by factors such as ionic size and charge, so that the order is generally $M^{4+} > MO_2^{2+} > M^{3+} > MO_2^+$. For anions, the order of complexing ability is generally: uninegative ions, $F^- > NO_3^- > Cl^- > ClO_4^-$; binegative ions, $CO_3^{2-} > ox^{2-} > SO_4^{2-}$.

28-15. Complexes

Most complex chemistry of the actinides is in aqueous solution. However, a number of neutral complexes and complexes between halides and neutral donor ligands such as Ph$_3$PO or Me$_2$SO are known. There are very few complexes formed by π acid ligands, providing a notable contrast to the d-block elements; for example, there is no evidence for the bonding of NO or alkenes. However, uranium carbonyls have been trapped in matrixes at 4°K.[16a]

The only stable organo compounds[16b] are those such as (C$_5$H$_5$)$_3$UCl and (C$_5$H$_5$)$_3$Am or (C$_8$H$_8$)$_2$Th and (C$_8$H$_8$)$_2$Pu,[17] for which the "sandwich-type" structure and bonding (cf. page 739) can be explained by appropriate hybridization schemes involving f orbitals. These air-sensitive substances are

[16a] J. L. Slater et al., J. Chem. Phys., 1971, 55, 5129.
[16b] H. Gysling and M. Tsutsui, Adv. Organometal. Chem., 1970, 9, 361.
[17] P. G. Laubereau and J. H. Burns, Inorg. Chem., 1970, 9, 1091; B. Kanellakopulos et al., J. Organometal. Chem., 1970, 25, 123, 507; D. Karraker et al., J. Amer. Chem. Soc., 1971, 93, 7343; K. O. Hodgson et al., Chem. Comm., 1971, 1592.

made by the interaction in THF of, e. g., UCl_4 and C_5H_5Na or the potassium salt of the cyclooctatetraene anion, $C_8H_8^{2-}$. An allyl, $(C_3M_5)_4U$, is unstable above $-20°$.[18]

ACTINIUM

28-16. The Element and its Compounds

Actinium was originally isolated from uranium minerals in which it occurs in traces, but it is now made on a milligram scale by neutron capture in radium (Table 28-2). The actinium $+3$ ion is separated from the excess of radium and isotopes of Th, Po, Bi and Pb formed simultaneously by ion-exchange elution or by solvent-extraction with thenoyltrifluoroacetone.

The general chemistry of Ac^{3+} in both solid compounds and solution, where known, is very similar to that of lanthanum, as would be expected from the similarity in position in the Periodic Table and in radii (Ac^{3+}, 1.10; La^{3+}, 1.06 Å) together with the noble-gas structure of the ion. Thus actinium is a true member of Group III, the only difference from lanthanum being in the expected increased basicity. The increased basic character is shown by the stronger absorption of the hydrated ion on cation-exchange resins, the poorer extraction of the ion from concentrated nitric acid solutions by tributyl phosphate, and the hydrolysis of the trihalides with water vapor at $\sim 1000°$ to the oxo halides AcOX; the lanthanum halides are hydrolyzed to oxide by water vapor at $1000°$.

The crystal structures of actinium compounds, where they have been studied, for example, in AcH_3, AcF_3, Ac_2S_3 and AcOCl, are the same as those of the analogous lanthanum compounds.

The study of even milligram amounts of actinium is difficult owing to the intense γ radiation of its decay products which rapidly build up in the initially pure material.

THORIUM

Thorium is widely distributed in Nature and there are large deposits of the principal mineral, *monazite*, a complex phosphate containing uranium, cerium, and other lanthanides. The extraction of thorium from monazite is complicated, the main problems being the destruction of the resistant sand and the separation of thorium from cerium and phosphate. One method involves a digestion with sodium hydroxide; the insoluble hydroxides are removed and dissolved in hydrochloric acid. When the pH of the solution is adjusted to 5.8, all the thorium and uranium, together with about 3% of the lanthanides, are precipitated as hydroxides. The thorium is recovered by tributyl phosphate extraction from $>6M$ hydrochloric acid solution or by

[18] G. Lugli *et al.*, *Inorg. Chim. Acta*, 1969, **3**, 252.

extraction with isobutyl methyl or other ketone from nitric acid solutions in presence of an excess of a salt such as aluminum nitrate as "salting-out" agent.

28-17. Binary Compounds of Thorium

Some typical thorium compounds are listed in Table 28-8.

TABLE 28-8

Some Thorium Compounds

Compound	Form	Melting point (°C)	Properties
ThO_2	White, crystalline; fluorite structure	3220	Stable, refractory, soluble in $HF + HNO_3$
ThN	Refractory solid	2500	Slowly hydrolyzed by water
ThS_2	Purple solid	1905	Metal-like; soluble in acids
$ThCl_4$	Tetragonal white crystals	770	Soluble in and hydrolyzed by H_2O; Lewis acid
$Th(NO_3)_4 \cdot 5H_2O$	White crystals, orthorhombic		Very soluble in H_2O, alcohols, ketones, ethers
$Th(IO_3)_4$	White crystals		Precipitated from 50% HNO_3; very insoluble
$Th(C_5H_7O_2)_4$	White crystals	171	Sublimes in a vacuum 160°
$Th(BH_4)_4$	White crystals	204	Sublimes in a vacuum about 40°
$Th(C_2O_4)_2 \cdot 6H_2O$	White crystals		Precipitated from up to $2M$ HNO_3

Oxide and Hydroxide. The only oxide, ThO_2, is obtained by ignition of oxo acid salts or of the hydroxide. The latter is insoluble in an excess of alkali hydroxides, although it is readily peptized by heating it with Th^{4+} or Fe^{3+} ions or dilute acids; the colloid exists as fibers that are coiled into spheres in concentrated sols but uncoil on dilution. Addition of hydrogen peroxide to Th^{4+} salts gives a highly insoluble white precipitate of variable composition which contains an excess of anions in addition to peroxide; the composition is approximately $Th(O_2)_{3.2}X_{0.5}^-O_{0.15}^{2-}$ but it is usually referred to as thorium peroxide.

Halides. The anhydrous halides may be prepared by dry reactions such as

$$ThO_2 + 4HF(g) \xrightarrow{600°} ThF_4 + 2H_2O$$
$$ThO_2 + CCl_4 \xrightarrow{600°} ThCl_4 + CO_2$$

They are all white crystalline solids which, with the exception of ThF_4, can be sublimed in a vacuum at 500–600°. The hydrated tetrafluoride is precipitated by aqueous hydrofluoric acid from Th^{4+} solutions; it can be dehydrated by heat in an atmosphere of hydrogen fluoride. The other halides are soluble in acid and are partially hydrolyzed by water. They behave as Lewis acids and form complexes with ammonia, amines, ketones, alcohols and donor molecules generally.

The *oxo halides*, $ThOX_2$, can be obtained by interactions of ThO_2 and ThX_4 at 600°; they appear to have —Th—O—Th—O chains.

Other Binary Compounds. Various borides, sulfides, carbides, nitrides, etc., have been obtained by direct interaction of the elements at elevated temperatures. Like other actinide and lanthanide metals, thorium also reacts at elevated temperatures with hydrogen. Products with a range of compositions can be obtained, but two definite phases, ThH_2 and Th_4H_{15}, have been characterized.

28-18. Oxo Salts, Aqueous Solutions and Complexes of Thorium

Thorium salts of strong mineral acids usually have varying amounts of water of crystallization. The most common salt and the usual starting material for preparation of other thorium compounds is the nitrate, $Th(NO_3)_4 \cdot 5H_2O$. This salt is very soluble in water as well as in alcohols, ketones, ethers and esters. Various reagents give insoluble precipitates with thorium solutions, the most important being hydroxide, peroxide, fluoride, iodate, oxalate and phosphate; the last four give precipitates even from strongly acid ($6M$) solutions and provide useful separations of thorium from elements other than those having $+3$ or $+4$ cations with similar properties.

The thorium ion, Th^{4+}, is more resistant to hydrolysis than other $4+$ ions but undergoes extensive hydrolysis in aqueous solution at pH higher than ~ 3; the species formed are complex and dependent on the conditions of pH, nature of anions, concentration, etc. In perchlorate solutions the main ions appear to be $Th(OH)^{3+}$, $Th(OH)_2^{2+}$, $Th_2(OH)_2^{6+}$, $Th_4(OH)_8^{8+}$, while the final product is the hexamer $Th_6(OH)_{15}^{9+}$; of course, all these species carry additional water.[19] Hexameric ions exist also for Nb^V and for Ce^{IV} and U^{IV}; $[M_6O_4(OH)_4]^{12+}$ ions are found in crystals of the sulfates. The metal atoms are linked by hydroxo or oxo bridges. In crystals of the hydroxide, $Th(OH)_4$, or the compound $Th(OH)_2CrO_4 \cdot H_2O$, chain-like structures have been identified, the repeating unit being $Th(OH)_2^{2+}$; in solution, the polymers may have similar form (28-I) or may additionally be cross-linked.

$$(28\text{-Ia}) \qquad\qquad (28\text{-Ib})$$

The high charge on Th^{4+} makes it susceptible to complex formation, and in solutions with anions other than perchlorate, complexed species, which may additionally be partially hydrolyzed and polymeric, are formed. Equilibrium constants for reactions such as the following have been measured:

$$Th^{4+} + nCl^- = ThCl_n^{4-n}$$
$$Th^{4+} + NO_3^- = Th(NO_3)^{3+}$$
$$Th^{4+} + 2HSO_4^- = Th(HSO_4, SO_4)^+ + H^+$$

[19] W. E. Bacon and G. H. Brown, *J. Phys. Chem.*, 1969, **73**, 4163.

A number of salts of *complex anions* have been isolated; some of the more important are K_4Th $ox_4 \cdot 4H_2O$,[20a] $M^{II}[Th(NO_3)_6] \cdot 8H_2O$, where the NO_3 groups are bidentate, $(NH_4)_4ThF_8$,[20b] and complexes of EDTA and related acids.

Neutral complexes are formed by 8-quinolinol[21] and β-diketones; an example of the latter type is the tetrakis-(1,3-diphenylacetonate), which has a square-antiprismatic structure.[22]

The dithiocarbamate,[23] obtained from $ThCl_4$ and NaS_2CNEt_2 is air-sensitive and isomorphous with U, Np and Pu analogs. Other complexes include *adducts*[24] such as $ThCl_4$ $phen_2$ and $Th(NO_3)_4 \cdot 2Ph_3PO$.

28-19. Lower Oxidation States

There is no evidence for the existence of any low oxidation state in solution, and little, if any, for its existence in the solid state.

A diamagnetic, gold-colored solid that is air-sensitive and has low electrical resistance is obtained on heating ThI_4 and Th at 800°. Despite the stoichiometry ThI_2, this compound has a complex layer structure[25] corresponding to the formulation $Th^{4+}(I^-)_2(e^-)_2$ and is thus similar to the lanthanide "lower" iodides (page 1075). Sulfides of stoichiometry ThS and Th_2S_3 are doubtless similar, with Th^{4+} and S^{2-} ions and electrons in conduction bands.

PROTACTINIUM

Protactinium as ^{231}Pa occurs in pitchblende, but even the richest ores contain only about 1 part of Pa in 10^7. The isolation of protactinium from residues in the extraction of uranium from its minerals is difficult, as indeed is the study of protactinium chemistry generally, owing to the extreme tendency of the compounds to hydrolyze. In aqueous solution, polymeric ionic species and colloidal particles are formed, and these are carried on precipitates and adsorbed on vessels; in solutions other than those containing appreciable amounts of mineral acids or complexing agents or ions such as F^-, the difficulties are almost insuperable.

Protactinium can be recovered from solutions $2-8M$ in nitric or hydrochloric acids by extraction with tributyl phosphate, isobutyl methyl ketone or other organic solvents. The protactinium can be stripped from the solvent by aqueous acid fluoride solutions; the addition to these solutions of Al^{3+} ion or boric acid, which form stronger complexes with fluoride ion than

[20] (a) M. N. Akhtar and A. J. Smith, *Chem. Comm.*, **1969**, 705. (b) R. A. Penneman *et al.*, *Inorg. Chem.*, 1969, **8**, 1379; *Acta Cryst.*, 1969, *B*, **25**, 1958.
[21] B. C. Baker and D. T. Sawyer, *Inorg. Chem.*, 1969, **8**, 1160.
[22] C. Wiedenheft, *Inorg. Chem.*, 1969, **8**, 1174.
[23] D. Brown, G. Holah and C. E. F. Rickard, *J. Chem. Soc.*, A, **1970**, 423.
[24] B. C. Smith and M. A. Wassef, *J. Chem. Soc.*, A, **1969**, 1817.
[25] L. J. Guggenberger and R. A. Jacobsen, *Inorg. Chem.*, 1968, **7**, 2257.

protactinium, then allows re-extraction and further purification of protactinium. Anion-exchange procedures involving mixtures of hydrofluoric and hydrochloric acid as elutants can also be used, since in these solutions protactinium fluoro or chloro anions are formed. About 125 g of protactinium were isolated in a twelve-stage process from 60 tons of accumulated sludges of uranium extraction from Belgian Congo ore by the United Kingdom Atomic Energy Authority; previously only about 1 g had ever been isolated. The method involved leaching of Pa from the residues with $4M$ HNO_3–$0.5M$ HF, followed by extraction of Pa^V from these solutions by 20% tributyl phosphate in kerosene. After collection of the Pa on a hydrous alumina precipitate it was purified further by extraction from HCl–HF solution with dibutyl ketone, anion-exchange separation from HCl solution, and finally precipitation from dilute H_2SO_4 by H_2O_2.

28-20. Protactinium(V) Compounds[26]

Comparatively few compounds have been characterized and some of these and their preparations are given in Fig. 28-4.

The *pentoxide*, Pa_2O_5, obtained by ignition of other compounds in air has a cubic lattice; heating it *in vacuo* affords a black sub-oxide phase, $PaO_{2.3}$, and finally PaO_2, but the real situation is more complex.

The hydrous pentoxide is similar to that of niobium and spectra suggest that M—O—M bonds are present.

The *pentafluoride* is obtained as a white hygroscopic solid by fluorination of PaF_4; it is less volatile than the pentafluorides of V, Nb and Ta but does sublime *in vacuo* above 500°. It is very soluble in $1M$ or stronger HF but evaporation of aqueous solutions gives only mixtures. Action of HF gas on

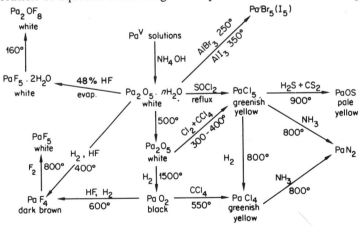

Fig. 28-4.   Preparation of protactinium compounds.

[26] D. Brown, *Adv. Inorg. Chem. Radiochem.*, 1969, **12**, 1 (general review of Pa); R. Muxart, R. Guillaumont and G. Bouissières, *Actinides Rev.*, 1969, **1**, 233 (solid Pa^V and Pa^{IV} compounds).

hydrous Pa_2O_5 at $60°$ gives $PaF_5 \cdot H_2O$; when heated this gives oxo-fluorides.[27]

The *pentachloride*, $PaCl_5$, sublimes at $160°$ in a vacuum; it is readily hydrolyzed to oxo halides and is soluble in THF. It has a structure quite unlike that of UCl_5 or $TaCl_5$ (page 938) with infinite chains of irregular pentagonal bipyramidal $PaCl_7$ groups sharing edges.

Aqueous Chemistry.[28] The chemistry of Pa in solution is somewhat like that of Nb and Ta, but Pa is even less tractable because of hydrolysis and the formation of colloidal hydroxo species.

In perchloric acid, cationic species, probably $PaO(OH)_2^+$ and $PaO(OH)^{2+}$, exist; but, when Cl^-, NO_3^- or other complexing anions are present, a whole range of species from cationic to anionic may exist depending upon the conditions. A number of anionic complexes are well-established. The fluoro complexes, which resist hydrolysis, have been well studied and salts of the ions PaF_6^-, PaF_7^{2-} and PaF_8^{3-} have been isolated. In K_2PaF_7 there are PaF_9 groups linked by double, unsymmetrical bridges into infinite chains, while Na_3PaF_8, which is isostructural[29] with the U and Np analogs has the Pa at the center of a slightly distorted cube of F atoms; in $RbPaF_6$ the Pa atom is also 8-coordinate.

Salts of chloro and bromo anions of Pa^V can be made by interaction of PaX_5 with MX in solvents such as $SOCl_2$ or CH_3CN.[30]

Other stable complex anions are those formed by $C_2O_4^{2-}$, SO_4^{2-}, citrate and tartrate. Neutral complexes with β-diketones such as TTA and with alkyl phosphates may be extracted from aqueous solutions by benzene or CCl_4.

28-21. Protactinium(IV) Compounds

The fluoride, PaF_4, is high-melting and insoluble in HNO_3–HF solution; and this salt and $PaCl_4$ are isomorphous with corresponding Th and U halides. PaO_2 is isomorphous with the dioxides of Th–Am inclusive.

The lower oxidation state can also be obtained in aqueous solution by reduction of Pa^V solutions with Cr^{2+} or Zn amalgam, but the solutions are rapidly oxidized by air. The solutions of $PaCl_4$ in HCl, H_2SO_4 and $HClO_4$ have very similar absorption spectra, being similar to $Ce^{III}(4f^1)$. Further, the absorption and esr spectra of Pa^{IV} incorporated in Cs_2ZrCl_6 are more compatible with a $5f^1$ configuration for Pa^{IV} than with a $6d^1$ configuration, and similar studies where U^{IV} is incorporated again indicate $5f^2$ configuration for the ion. Chloro, bromo[30] and other complex ions of Pa^{IV} have been characterized. An organo compound, $(C_5H_5)_4Pa$, has also been made.[31]

[27] D. Brown and J. F. Easy, *J. Chem. Soc., A*, **1970**, 3378.
[28] R. Guillaumont, G. Bouissières and R. Muxart, *Actinides Rev.*, 1968, **1**, 135 (extensive review of Pa^V and Pa^{IV} in solutions).
[29] D. Brown, J. F. Easy and C. E. F. Rickard, *J. Chem. Soc., A*, **1969**, 1161.
[30] D. Brown and P. J. Jones, *J. Chem. Soc.*, **1967**, 243, 247, 719.
[31] F. Baumgarten *et al.*, *Angew. Chem. Internat. Edn.*, 1969, **8**, 202.

URANIUM[32]

Uranium was discovered by Klaproth in 1789. Until the discovery of uranium fission by Hahn and Strassman in 1939, uranium had little commercial importance; its ores were sources of radium and small quantities were used for coloring glass and ceramics, but the bulk of the uranium was discarded. Uranium is important as a nuclear fuel; its chemical importance lies in its being the prototype for the succeeding three elements.

While it is more abundant than Ag, Hg, Cd or Bi, uranium is widely disseminated with relatively few economically workable deposits. The most important ores are the oxide, *uraninite* (one form is *pitchblende*), which has variable composition approximating to UO_2, and uranium vanadates.

The methods of extraction of uranium are numerous and complex, but the final stages usually employ the extraction of uranyl nitrate from aqueous solutions by ether or some other organic solvent.

28-22. Uranium Compounds

The chemistry of uranium compounds has been studied in great detail and only the more important aspects can be described here. Generally, the stoichiometries, structures, and properties of Np, Pu and Am compounds are similar to those of uranium; in the III and the IV state the properties are similar to those of lanthanide compounds.

Uranium Oxides. The U–O system is one of the most complex oxide systems known, owing in part to the multiplicity of oxidation states of comparable stability; deviations from stoichiometry are the rule rather than the exception, and stoichiometric formulas must be considered as ideal compositions. In the dioxide, UO_2, for example, about 10% excess oxygen atoms can be added before any notable structural change is observable, and the UO_2 phase extends from UO_2 to $\sim UO_{2.25}$. The main oxides are: UO_2, brown-black; U_3O_8, greenish-black; and UO_3, orange-yellow. Each of these oxides has several crystalline modifications of different thermal and thermodynamic stabilities and colors. The *trioxide*, UO_3, is obtained by decomposition at 350° of uranyl nitrate or, better, of "ammonium diuranate" (see below). One polymorph has a structure that can be considered as uranyl ion linked by U—O—U bonds through the equatorial oxygens to give layers. The same type of structure occurs also in UO_2F_2 (F bridges) and certain uranates. The other oxides can be obtained by the reactions

$$3UO_3 \xrightarrow{700°} U_3O_8 + \tfrac{1}{2}O_2$$
$$UO_3 + CO \xrightarrow{350°} UO_2 + CO_2$$

All the oxides readily dissolve in nitric acid to give UO_2^{2+} salts. The addition of hydrogen peroxide to uranyl solution at pH 2.5–3.5 gives a pale yellow

[32] E. H. P. Cordfunke, *The Chemistry of Uranium*, Elsevier, 1969 (a monograph, including nuclear applications); *The Recovery of Uranium*, Internat. Atomic Energy Agency, Proceedings Series, STI-PUB-262, Vienna, 1971.

precipitate, of formula approximately $UO_4 \cdot 2H_2O$. The U^{VI} peroxo system is exceedingly complex; this particular peroxide is best formulated as UO_2^{2+} $(O_2^{2-}) \cdot 2H_2O$; on treatment with NaOH and H_2O_2 it gives the very stable salt, $Na_4[UO_2(O_2)_3] \cdot 9H_2O$, whose anion consists of linear UO_2 with three peroxo groups in the equatorial plane.[33]

Uranates. The fusion of uranium oxides with alkali or alkaline-earth carbonates, or thermal decomposition of salts of the uranyl acetate anion, gives orange or yellow materials generally referred to as uranates,[34] e.g.,

$$2UO_3 + Li_2CO_3 \rightarrow Li_2U_2O_7 + CO_2$$
$$Li_2U_2O_7 + Li_2CO_3 \rightarrow 2Li_2UO_4 + CO_2$$
$$Li_2UO_4 + Li_2CO_3 \rightarrow Li_4UO_5 + CO_2$$

Other metal oxides can also be incorporated and such ternary substances are best regarded as mixed oxides. The uranates are generally of stoichiometry $M_2^I U_x O_{3x+2}$, but $M_4^I UO_5$, $M_3^{II} UO_6$, etc., are known. In contrast to Mo or W, there appear to be no iso- or hetero-poly anions for U in solution. A useful material obtained by addition of aqueous NH_3 to $UO_2(NO_3)_2$ solutions is the so-called "ammonium diuranate." This is mainly the hydrate $UO_2(OH)_2 \cdot H_2O$.

Alkaline-earth uranates do not contain discrete ions such as UO_4^{2-}; they have unsymmetrical oxygen coordination such that two U—O bonds are short, ca. 1.92 Å, constituting a sort of uranyl group, with other longer U—O bonds in the plane normal to this UO_2 axis linked into chains or layers. However, Na_4UO_5 and $M_3^{II}UO_6$ do not have such uranyl groups; the former has strings of UO_6 octahedra sharing opposite corners so as to give infinite —U—O—U—O— chains with a planar UO_4 group normal to the chain; the U—O bonds in the chain are longer than those in the UO_2 group.

Other Binary Compounds. Direct reaction of uranium with B, C, Si, N, P, As, Sb, Se, S, Te, etc., leads to semimetallic compounds that are often non-stoichiometric, resembling the oxides. Some of them, for example, the silicides, are chemically inert, and the sulfides,[35] notably US, can be used as refractories.

Uranium Halides. The principal halides are listed in Table 28-9; they have been studied in great detail and chemical, structural and thermodynamic properties are well-known.

Fluorides. The *trifluoride*, UF_3, is a high melting, non-volatile, crystalline solid resembling the lanthanum fluorides, and is insoluble in water or dilute acids; the preparation is by the aluminum reduction of UF_4:

$$UF_4 + Al \xrightarrow{900°} UF_3 + AlF$$

The hydrated tetrafluoride can be obtained by precipitation from U^{4+} solution, and the anhydrous fluoride by reactions such as:

$$UO_2 \xrightarrow[500-600°]{C_2Cl_4F_2} UF_4$$

[33] N. W. Alcock, *J. Chem. Soc. A*, **1968**, 1588.
[34] E. H. P. Cordfunke and B. O. Loopstra, *J. Inorg. Nuclear Chem.*, 1971, **33**, 2427.
[35] F. Grønvald *et al.*, *J. Inorg. Nuclear Chem.*, 1968, **30**, 2117, 2127.

TABLE 28-9

Uranium Halides[a]

+3	+4	+5	+6
UF_3 green	UF_4 green	UF_5 white-blue	UF_6 white
UCl_3 red	UCl_4 green	U_2Cl_{10} red-brown	UCl_6 black
UBr_3 red	UBr_4 brown	—	—
UI_3 black	UI_4 black	—	—

[a] Other fluorides in addition to UF_4 and UF_6 are known: U_2F_9, U_4F_{14} and U_5F_{22} are black; Pa also forms a fluoride of uncertain stoichiometry, and Pu gives a red solid Pu_4F_{17}.

The non-volatile solid *tetrafluoride* is insoluble in water but is readily soluble in solutions of oxidizing agents. The *hexafluoride*, UF_6, is obtained by the action of fluorine at ca. 400° on the lower fluorides; it forms colorless crystals, m.p. 64.1°, with a vapor pressure of 115 mm at 25°. This is the only readily accessible volatile uranium compound, and its physical properties have been intensively studied, primarily because it is used in the separation of uranium isotopes by gaseous diffusion in order to produce pure ^{235}U nuclear fuel. The structure is octahedral in the gas and has a small tetragonal distortion in the molecular crystals. UF_6 is a powerful fluorinating agent, converting many substances into fluoro compounds, e.g., CS_2 into SF_4, $(CF_3)_2S_3$, etc., and it is also hydrolyzed rapidly by water. The intermediate fluorides, UF_5, U_2F_9 and U_4F_{14}, are made by interaction of UF_6 and UF_4; they disproportionate quite readily, e.g.,

$$3UF_5 \rightleftarrows U_2F_9(s) + UF_6(g)$$

UF_5 is made by treating UF_4 with fluorine at 240° or UF_6 with HBr at 65°; it has a polymeric chain structure. U_2F_9 has crystallographically identical U atoms, each 9-coordinate; the black color evidently results from charge transfer transitions giving formally +4 and +5 atoms in the excited state.

Chlorides. UCl_3 can be made only in anhydrous conditions, for example, by the action of hydrogen chloride on UH_3; the aqueous solutions obtained by reduction of acid solutions of UO_2^{2+} by zinc amalgam are readily reoxidized to U^{4+} by air. The most important chloride is UCl_4, which is best made by liquid-phase chlorination of UO_3 by refluxing with hexachloropropene. The primary product is believed to be UCl_6 which decomposes thermally. UCl_4 is soluble in polar organic solvents and in water. The penta- and hexachloride are both soluble in carbon tetrachloride; they are violently hydrolyzed by water. UCl_5 disproportionates when heated but can be isolated by chilling the gaseous products in the reaction:

$$2UCl_4 + Cl_2 \underset{<250°}{\overset{500°}{\rightleftarrows}} U_2Cl_{10} \xrightarrow{100-180°} UCl_4 + UCl_6$$

The vapor[36] (and probably the CCl_4 solution) contains U_2Cl_{10} molecules that probably have octahedral U^V with a double halogen bridge. In the crystal

[36] D. M. Green and R. L. McBeth, *Inorg. Chem.*, 1969, **8**, 2625.

the U_2Cl_{10} unit is well-defined and is similar to other M_2Cl_{10} halides of Ta, Mo and Re (cf. pages 938 and 978).

Halogeno Complexes. All the halides can form halogeno complexes, those with F^- and Cl^- being the best known. They can be obtained by interaction of the halide and alkali halides in melts or in solvents such as $SOCl_2$, or in the case of fluorides sometimes in aqueous solution.

Fluoride complexes such as green UF_5^-, UF_6^{2-}, UF_7^{3-} and UF_8^{4-} can be made by sealed-tube reactions, or by dissolution of UF_4 in RbF, but perhaps the most interesting is the stabilization of U^V in aqueous solutions (see below) as a fluoro complex ion. Thus the deep blue solutions of UF_5 in 48% HF are only slowly oxidized by air and on cooling give large blue crystals of $HUF_6 \cdot 2.5H_2O$. On dilution with water, hydrolysis to UO_2^{2+} and insoluble UF_4 occurs, but addition of Rb or Cs fluorides gives stable blue salts that are isostructural with $CsNb(Ta)F_6$. These salts are best made by interaction of ClF_5 and MF in liquid HF as solvent. Absorption spectra indicate nearly octahedral symmetry.

For U^{VI} the ions UF_7^- and UF_8^{2-} are established in sodium and potassium salts.

For U^{IV} and U^V, yellow salts such as K_2UCl_6, $(Me_4N)UCl_6$, and $(Me_4N)_3UCl_8$ are known; UCl_5 also reacts with PCl_5 to give $[PCl_4]^+[UCl_6]^-$ and when U_3O_8 is boiled with hexachloropropene a dark red complex of trichloroacryloyl chloride (L) of stoichiometry UCl_5L is formed. Other adducts can be made from this by displacement reactions with, e.g., $SOCl_2$.[37]

Oxo Halides. The stable uranyl compounds, UO_2X_2, are soluble in water. They are made by reactions such as

$$UCl_4 + O_2 \xrightarrow{350°} UO_2Cl_2 + Cl_2$$
$$UO_3 + 2HF \xrightarrow{400°} UO_2F_2 + H_2O$$

Other Compounds and Complexes. One of the few U^V compounds stable to disproportionation is the *alkoxide*, $[U(OEt)_5]_2$, which contains U^V with octahedral coordination achieved by a double alkoxide bridge.[38] Finally, there are well-established U^{IV} complexes with β-diketones,[39] 8-quinolinol,[40] and diethyl dithiocarbamate,[41] which are similar to those of Th, Np and Pu^{IV}.

28-23. Aqueous Chemistry of Uranium

Uranium ions in aqueous solution can give very complex species because, in addition to the four oxidation states, complexing reactions with all ions other than ClO_4^- as well as hydrolytic reactions leading to polymeric ions

[37] J. Selbin et al., *Inorg. Chem.*, 1968, **7**, 976; *Chem. Comm.*, **1969**, 759.
[38] D. G. Karraker, T. H. Siddall and W. E. Stewart, *J. Inorg. Nuclear Chem.*, 1969, **31**, 711.
[39] C. Wiedenheft, *Inorg. Chem.*, 1969, **8**, 1174; T. H. Siddell and W. E. Stewart, *Chem. Comm.*, **1969**, 922; H. Titze, *Acta. Chem. Scand.*, 1970, **24**, 405.
[40] B. C. Baker and D. J. Sawyer, *Inorg. Chem.*, 1969, **8** 1160.
[41] D. Brown, D. G. Holah and C. E. F. Rickard, *J. Chem. Soc.*, A, **1970**, 786.

occur under appropriate conditions. The formal potentials for $1M$ $HClO_4$ have been given in Table 28-6; in presence of other anions the values differ: thus for the U^{4+}/U^{3+} couple in $1M$ $HClO_4$ the potential is -0.631 V, but in $1M$ HCl it is -0.640 V. The simple ions and their properties are also listed in Table 28-7. Aqueous solutions of uranium salts have an acid reaction due to hydrolysis, which increases in the order $U^{3+} < UO_2^{2+} < U^{4+}$. The uranyl and U^{4+} solutions have been particularly well studied. The main hydrolyzed species of UO_2^{2+} at $25°$ are UO_2OH^+ $(UO_2)_2(OH)_2^{2+}$ and $(UO_2)_3(OH)_5^+$, but the system is a complex one and the species present depend on the medium; at higher temperatures the monomer is most stable but the rate of hydrolysis to UO_3 of course increases. The solubility of large amounts of UO_3 in UO_2^{2+} solutions is also attributable to formation of UO_2OH^+ and polymerized hydroxo bridged species.

The U^{4+} ion is only slightly hydrolyzed in molar acid solutions:

$$U^{4+} + H_2O \rightleftharpoons U(OH)^{3+} + H^+ \qquad K_{25°} = 0.027 \; (1M \; HClO_4, NaClO_4)$$

but it can also give polynuclear species in less acid solutions.

The U^{4+} ion gives insoluble precipitates with F^-, PO_4^{3-} and IO_3^- from acid solutions (cf. Th^{4+}).

The uranium(V) ion, UO_2^+, is extraordinarily unstable towards disproportionation and has a transitory existence under most conditions, although evidence for its occurrence can be obtained polarographically. It is also an intermediate in photochemical reductions of uranyl ions in presence of sucrose and similar substances. The ion is most stable in the pH range 2.0–4.0 where the disproportionation reaction to give U^{4+} and UO_2^{2+} is negligibly slow. By contrast, reduction of UO_2^{2+} in dimethyl sulfoxide gives UO_2^+ in concentrations sufficiently high to allow the spectrum to be obtained and disproportionation occurs with a half-life of about an hour.[42] As noted above, U^V can be stabilized in HF solutions as UF_6^-, as well as in concentrated Cl^- and CO_3^{2-} solutions.[42]

Spectroscopic and other studies have shown that in aqueous solutions of UO_2^{2+} and U^{4+}, complex ions are often readily formed, for example,

$$U^{4+} + Cl^- \rightleftharpoons UCl^{3+} \qquad K = 1.21 \; (\mu = 2.0; 25°)$$
$$U^{4+} + 2HSO_4^- \rightleftharpoons U(SO_4)_2 + 2H^+ \qquad K = 7.4 \times 10^3 \; (\mu = 2.0; 25°)$$
$$UO_2^{2+} + Cl^- \rightleftharpoons UO_2Cl^+ \qquad K = 0.88 \; (\mu = 2.0; 25°)$$
$$UO_2^{2+} + 2SO_4^{2-} \rightleftharpoons UO_2(SO_4)_2^{2-} \qquad K = 7.1 \times 10^2 \; (\mu = 2.0; 25°)$$

Nitrate complexes also exist, and nitrate solutions of U^{IV} contain $[UNO_3(H_2O)_4]^{3+}$ and similar species; in concentrated nitric acid it appears that $[U(NO_3)_6]^{2-}$ can be formed and the cesium salt can be precipitated.

The nature of the reduction of UO_2^{2+}, especially by Cr^{2+}, has been studied and it appears that there is a bright green intermediate complex ion, probably $[(H_2O)_5Cr^{III}—O—U^VO(H_2O)_n]^{4+}$, which reacts further to give Cr^{III} and U^{IV}. A similar intermediate occurs in the reduction of PuO_2^{2+}, and for Np an intermediate has been separated by ion-exchange. It is pertinent to note here

[42] D. Cohen, *J. Inorg. Nuclear Chem.*, 1970, **32**, 3525.

that the reverse process, oxidation of U^{4+} by various agents, has been studied in detail; this is possible only because of the slow exchange of UO_2^{2+} with water. Using ^{18}O tracer it was found that PbO_2, H_2O_2 or MnO_2 gave UO_2^{2+} where virtually all the O came from the solid oxidant whereas for O_2 and O_3 only one oxygen atom is transferred from the oxidant to U^{IV}.

Complex ions are also formed with citrate and anions of other organic acids, thiocyanate and phosphates. The phosphates are important in view of the occurrence of uranium in phosphate minerals, and species such as $UO_2H_2PO_4^+$ and $UO_2H_3PO_3^{2+}$, and at high concentrations anionic complexes are known.

28-24. Uranyl Salts

These are the only common uranium salts and the most important one is the nitrate which crystallizes with six, three or two molecules of water depending on whether it is obtained from dilute, concentrated or fuming nitric acid. The most unusual and significant property of the nitrate is its solubility in numerous ethers, alcohols, ketones, and esters—it distributes itself between the organic and an aqueous phase. The nitrate is also readily extracted from aqueous solutions, and this operation has become classical for the separation and purification of uranium since, with the exception of the other actinide MO_2^{2+} ions, few other metal nitrates have any extractability. A great deal of information is available, and phase diagrams for the $UO_2(NO_3)_2$–H_2O–solvent systems have been determined. The effect of added salts, for example, $Ca(NO_3)_2$ or NH_4NO_3, as "salting-out" agents is to increase substantially the extraction ratio to technically usable values. Studies of the organic phase have shown that $UO_2(NO_3)_2$ is accompanied into the solvents by $4H_2O$ molecules, but there is little or no ionization and the nitrate is undoubtedly coordinated in the equatorial plane of the UO_2 system. An important extractant for uranyl nitrate, that does not require a salting-out agent for useful ratios, is tributyl phosphate. Anhydrous uranyl nitrate is obtained by the reactions

$$U + N_2O_4(l) \xrightarrow{\text{MeCN}} UO_2(NO_3)_2 \cdot N_2O_4 \cdot 2MeCN \xrightarrow{163°} UO_2(NO_3)_2$$

Other uranyl salts are given by organic acids, sulfate, halides, etc.; the water-soluble acetate in presence of an excess of sodium acetate in dilute acetic acid gives a crystalline precipitate of $NaUO_2(OCOCH_3)_3$.

NEPTUNIUM, PLUTONIUM[43] AND AMERICIUM

28-25. Isolation of the Elements

Although several isotopes of these elements are known, the ones that can be obtained in macroscopic amounts are given in Table 28-2. Both ^{237}Np

[43] J. M. Cleveland, *The Chemistry of Plutonium*, Gordon and Breach, 1970.

and ^{239}Pu are found in the uranium fuel elements of nuclear reactors, from which plutonium is isolated on a kilogram scale. ^{237}Np also occurs in substantial amounts and is recovered primarily for conversion by neutron irradiation of NpO_2 into ^{238}Pu, which is used as a power source for satellites. Americium is produced from intense neutron irradiations of pure plutonium. The problems involved in the extraction of these elements include the recovery of the expensive starting material and the removal of hazardous fission products that are formed simultaneously in amounts comparable to the amounts of the synthetic elements themselves. Not only are the chemical problems themselves quite formidable, but the handling of highly radioactive solutions or solids (in the case of plutonium the exceedingly high toxicity is an additional hazard since even a μg is potentially a lethal dose) has necessitated the development of remote control operations. For the large-scale extractions from fuel elements, detailed studies of the effects of radiation on structural and process materials have also been required. There are numerous procedures for the separation of Np, Pu and Am variously involving precipitation, solvent-extraction, differential volatility of compounds and so on, and we can give only the briefest outline. The most important separation methods are based on the following chemistry.

1. *Stabilities of oxidation states.* The stabilities of the major ions involved are: $UO_2^{2+} > NpO_2^{2+} > PuO_2^{2+} > AmO_2^{2+}$; $Am^{3+} > Pu^{3+} \gg Np^{3+}$, U^{4+}. It is thus possible (see also Table 28-7) by choice of suitable oxidizing or reducing agents to obtain a solution containing the elements in different oxidation states; they can then be separated by precipitation or solvent-extraction. For example, Pu can be oxidized to PuO_2^{2+} while Am remains as Am^{3+}—the former could be removed by solvent-extraction or the latter by precipitation of AmF_3.

2. *Extractability into organic solvents.* As noted previously, the MO_2^{2+} ions can be extracted from nitrate solutions into organic solvents. The M^{4+} ions can be extracted into tributyl phosphate in kerosene from $6M$ nitric acid solutions; the M^{3+} ions can be similarly extracted from $10-16M$ nitric acid; and neighboring actinides can be separated by a choice of conditions.

3. *Precipitation reactions.* Only M^{3+} and M^{4+} give insoluble fluorides or phosphates from acid solutions; the higher oxidation states give either no precipitate or can be prevented from precipitation by complex formation with sulfate or other ions.

4. *Ion-exchange procedures.* Although ion-exchange procedures, both cationic and anionic, can be used to separate the actinide ions, they are best suited for small amounts of material. Since they have found most use in the separation of the trans-americium elements, these procedures are discussed below.

The following examples are for the separation of plutonium from uranium; similar procedures involving the same basic principles have been devised to separate Np and Am. The initial starting material in plutonium extraction is a solution of the uranium fuel element (plus its aluminum or other protective jacket) in nitric acid. The combination of oxidation–reduction cycles coupled with solvent extraction and/or precipitation methods removes the bulk of fission products (FP's); however, certain elements—notably ruthenium, which forms cationic, neutral, and anionic nitrosyl complexes—may require special elimination steps. The initial uranyl nitrate solution contains Pu^{4+} since nitric acid cannot oxidize this to Pu^V or Pu^{VI}.

1. *Isobutyl methyl ketone (hexone) method.* This is shown in Scheme 28-1.

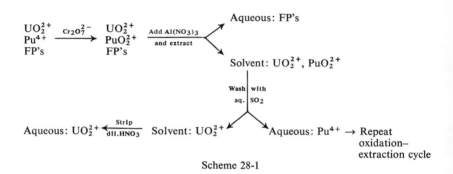

Scheme 28-1

2. *Tributyl phosphate method.* The extraction coefficients from $6N$ nitric acid solutions into 30% tributylphosphate in kerosine are $Pu^{4+} > PuO_2^{2+}$; $Np^{4+} \sim NpO_2^+ \gg Pu^{3+}$; $UO_2^{2+} > NpO_2^+ > PuO_2^{2+}$; the M^{3+} ions have very low extraction coefficients in $6M$ acid, but from $12M$ hydrochloric acid or $16M$ nitric acid the extraction increases and the order is $Np < Pu < Am < Cm < Bk$.

Thus in the U–Pu separation, after addition of NO_2^- to adjust all of the plutonium to Pu^{4+}, we have Scheme 28-2.

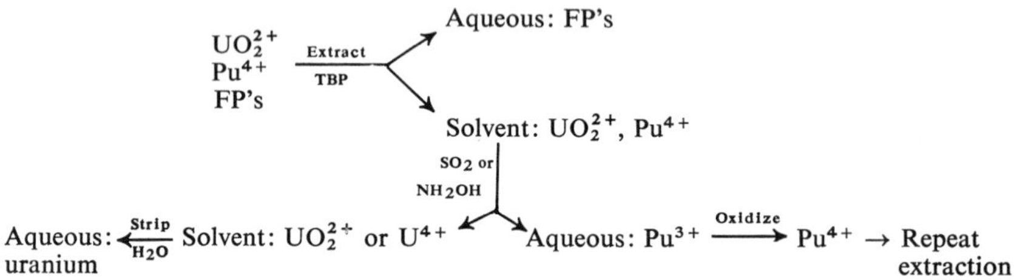

$$UO_2^{2+}$$
$$Pu^{4+} \xrightarrow[\text{TBP}]{\text{Extract}}$$
$$FP's$$

Aqueous: FP's

Solvent: UO_2^{2+}, Pu^{4+}

$$\downarrow \begin{array}{c} SO_2 \text{ or} \\ NH_2OH \end{array}$$

Aqueous: $\xleftarrow[H_2O]{\text{Strip}}$ Solvent: UO_2^{2+} or U^{4+} Aqueous: $Pu^{3+} \xrightarrow{\text{Oxidize}} Pu^{4+} \rightarrow$ Repeat
uranium   extraction

Scheme 28-2

The extraction of ^{237}Np involves similar principles of adjustment of oxidation state and solvent extraction; Pu is reduced by ferrous sulfamate plus hydrazine to unextractable Pu^{III}, while Np^{IV} remains in the solvent from which it is differentially stripped by water to separate it from U.

3. *Lanthanum fluoride cycle.* This classical procedure was first developed by McMillan and Abelson for the isolation of neptunium, but it is applicable elsewhere and is of great utility. For the U–Pu separation again, we have Scheme 28-3. The cycle shown is repeated with progressively smaller amounts of lanthanum carrier and smaller volumes of solution until plutonium

becomes the bulk phase. This fluoride cycle has also been used in combination with an initial precipitation step for Pu^{4+} with bismuth phosphate as a carrier.

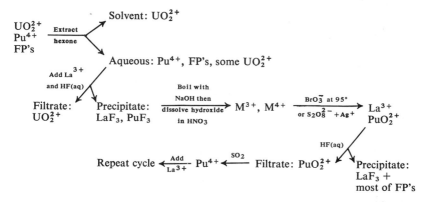

Scheme 28-3

28-26. Binary Compounds

Oxides. All three oxide systems have various solid solutions and other non-stoichiometric complications. The monoxides are interstitial compounds. The important oxides of Np, Pu and Am are the *dioxides*, which are obtained on heating the nitrates or hydroxides of any oxidation state in air; they are isostructural with UO_2. Ordinarily, PuO_2 is non-stoichiometric and may have different colors, but ignition at 1200° gives the stoichiometric oxide. The oxide Np_3O_8, isomorphous with U_3O_8, can be obtained under specific conditions. The action of ozone on suspensions of the M^{IV} hydroxides gives rise to the hydrated *trioxides*, brown $NpO_3 \cdot 2H_2O$ and $NpO_3 \cdot H_2O$, and red-gold $PuO_3 \cdot 0.8H_2O$, but, in contrast to U, which also gives $UO_3 \cdot 0.8H_2O$, no anhydrous trioxide is known. Above 300° black Np_2O_5 is obtained. $NpO_3 \cdot 2H_2O$ and Np_2O_5 can also be made by oxidation in $LiClO_4$ melts. Reduction of black AmO_2 with hydrogen at 600° gives the pink dimorphic Am_2O_3, which is the first lanthanide-like sesquioxide in the actinide series.

Halides. The halides are listed in Table 28-10.

The Np, Pu and Am halides, which are isostructural with and chemically similar to those of uranium, clearly show the decrease in stability of compounds in the higher oxidation states, and this trend continues in the succeeding elements. The preparative methods used are also similar to those for uranium, for example,

$$NpO_2 + \tfrac{1}{2}H_2 + 3HF(g) \xrightarrow{500°} NpF_3 + 2H_2O$$
$$PuF_4 + F_2 \xrightarrow{500°} PuF_6$$
$$AmO_2 + 2CCl_4 \xrightarrow{800°} AmCl_3 + 2COCl_2 + \tfrac{1}{2}Cl_2$$

The fluorides, MF_3 and MF_4, can be precipitated from aqueous solutions

TABLE 28-10

Halides of Np, Pu and Am[a]

+3	+4	+6
NpF$_3$, purple-black	NpF$_4$, green	NpF$_6$, orange, m. p. 55.1°
PuF$_3$, purple	PuF$_4$, brown	PuF$_6$, red-brown, m. p. 51.6°[b]
AmF$_3$, pink	AmF$_4$, tan	—
NpCl$_3$, white	NpCl$_4$, red-brown	
PuCl$_3$, emerald	—	
AmCl$_3$, pink	—	
NpBr$_3$, green	NpBr$_4$, red-brown	
PuBr$_3$, green	—	
AmBr$_3$, white	—	
NpI$_3$, brown	—	
PuI$_3$, green	—	
AmI$_3$, yellow		

[a] Certain oxo halides M^{III}OX, M^{V}OF$_3$ and M^{VI}O$_2$F$_2$ are known.

[b] Unlike the situation for U, intermediate fluorides are not formed in conversion of MF$_4$ into MF$_6$ (cf. L. E. Trevarrow, T. J. Gerding and M. J. Steindler, *J. Inorg. Nuclear Chem.*, 1968, **30**, 2671).

in hydrated form. The hexafluorides have been much studied since they are volatile; the melting points and stabilities decrease in the order U > Np > Pu. PuF$_6$ is so very much less stable than UF$_6$ that, at equilibrium, the partial pressure of PuF$_6$ is only 0.004% of the fluorine pressure. Hence PuF$_6$ formed by fluorination of PuF$_4$ at 750° must be quenched immediately by a liquid-nitrogen probe. The compound also undergoes self-destruction by α-radiation damage, especially in the solid; it must also be handled with extreme care owing to the toxicity of Pu. PuF$_6$ contains two non-bonding 5f electrons and should be paramagnetic; however, like UF$_6$, where all the valence electrons are involved in bonding, it shows only a small temperature-independent paramagnetism. This observation has been explained by ligand field splitting of the f levels to give a lower-lying orbital that is doubly occupied.

NpF$_6$ has a 5f^1 configuration according to esr and absorption spectra, the octahedral field splitting the seven-fold orbital degeneracy of the 5f electron and leaving a ground state that has only spin degeneracy. This quenching of the orbital angular momentum is similar to that in the first-row d-transition group. It provides further evidence for the closeness of the energy levels of the 5f and the valence electrons in actinides, in contrast to the much lower energies of the 4f electrons in lanthanides. NpF$_6$ is slightly distorted in the solid, and its magnetic behavior depends on its environment when diluted with UF$_6$.

Other Compounds. A substantial number of compounds, particularly of plutonium, are known, and most of them closely resemble their uranium analogs. The hydride systems of Np, Pu and Am are more like that of thorium than that of uranium and are complex. Thus non-stoichiometry up to MH$_{2.7}$ is found in addition to stoichiometric hydrides such as PuH$_2$ and AmH$_2$.

As with uranium, many complex salts are known, e.g., Cs$_2$PuCl$_6$, NaPuF$_5$,

$KPuO_2F_3$, $NaPu(SO_4)_2 \cdot 7H_2O$, and $CsNp(NO_3)_6$. A simple solid hydrated nitrate, $Pu(NO_3)_4$, is obtained by evaporation of Pu^{IV} nitrate solution; at 150–180° in air this gives $PuO_2(NO_3)_2$. Pu^V also occurs in salts of the PuF_7^{2-} and PuF_6^- ions.

28-27. Aqueous Chemistry of Neptunium, Plutonium[44] and Americium

The formal reduction potentials have been given in Table 28-6 and the general stabilities of the ions in Table 28-7.

Aqueous solutions of Pu, Am^V, Am^{VI} and especially Am^{IV} undergo rapid self-reduction due to their α-radiation.

For Np, the potentials of the four oxidation states are separated, like those of uranium, but in this case the NpO_2^+ state is comparatively stable. Earlier evidence that NpO_2^{2+} was reduced by Cl^- has been shown to be due to catalysis by platinum and the rate is very slow. With Pu, however, the potentials are not well separated and in $1 M$ $HClO_4$ all four species can coexist in appreciable concentrations; PuO_2^+ becomes increasingly stable with decreasing acidity since the couples are strongly hydrogen ion dependent. The Am ions stable enough to exist in finite concentrations are Am^{3+}, AmO_2^+ and AmO_2^{2+}; the Am^{3+} ion is the usual state since powerful oxidation is required to achieve the higher oxidation states. Alkaline solutions are more favorable for the stabilization of Am^{IV}, and for $1 M$ basic solution the $Am(OH)_4$–$Am(OH)_3$ couple has a value of $+0.5$ V, nearly 2 V less than for the Am^{4+}/Am^{3+} couple in acid solution. Thus pink $Am(OH)_3$ can be readily converted into black $Am(OH)_4$ (or $AmO_2(aq)$) by the action of hypochlorite. This black hydroxide is also soluble in $13 M$ ammonium fluoride solutions to give stable solutions from which $(NH_4)_4AmF_8$ can be precipitated; the anion in this salt probably has the square-antiprism structure, as does AmF_4.

As with uranium, the solution chemistry is complicated owing to hydrolysis and polynuclear ion formation, complex formation with anions other than perchlorate, and disproportionation reactions of some oxidation states. The tendency of ions to displace a proton from water increases with increasing charge and decreasing ion radius, so that the tendency to hydrolysis increases in the same order for each oxidation state, that is, $Am > Pu > Np > U$ and $M^{4+} > MO_2^{2+} > M^{3+} > MO_2^+$; simple ions such as NpO_2OH^+ or $PuOH^{3+}$ are known, in addition to polymeric species that in the case of plutonium can have molecular weights up to 10^{10}.

The complexing tendencies decrease, on the whole, in the same orders as the hydrolytic tendencies. The formation of complexes shifts the oxidation potentials, sometimes influencing the relative stabilities of oxidation states; thus the formation of sulfate complexes of Np^{4+} and NpO_2^{2+} is strong enough to cause disproportionation of NpO_2^+. The disproportionation reac-

---

[44] J. M. Cleveland, *Co-ordination Chem. Rev.*, 1970, **5**, 101 (an extensive review of aqueous complexes of plutonium).

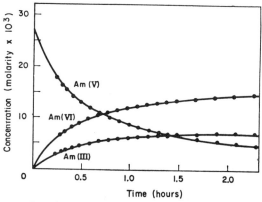

Fig. 28-5.   Disproportionation of AmO_2^+ in $6M$ perchloric acid at 25°. Net reaction $3AmO_2^+ + 4H^+ = 2AmO_2^{2+} + Am^{3+} + 2H_2O$. [Reproduced by permission from J.S. Coleman, *Inorg. Chem.*, 1963, **2**, 53.]

tions have been studied in some detail; Figs. 28-5 and 28-6 illustrate some of the complexities involved.

Typical of these disproportionations are the following at low acidity:

$$3Pu^{4+} + 2H_2O \rightleftharpoons PuO_2^{2+} + 2Pu^{3+} + 4H^+$$
$$2Pu^{4+} + 2H_2O \rightleftharpoons PuO_2^+ + Pu^{3+} + 4H^+$$
$$PuO_2^+ + Pu^{4+} \rightleftharpoons PuO_2^{2+} + Pu^{3+}$$

At 25° and $1M\ HClO_4$, we have

$$K = \frac{[Pu^{VI}][Pu^{III}]}{[Pu^V][Pu^{IV}]} = 10.7$$

which indicates that measurable amounts of all four states can be present.

An example of complex formation is provided by carbonate; for Am this provides a useful separation from Cm since $Cm(OH)_3$ is insoluble in $NaHCO_3$ and cannot be oxidized to soluble complexes. However, treatment of Am^{3+} in $2M\ Na_2CO_3$ with O_3 at 25° gives a red-brown Am^{VI} carbonate complex anion of uncertain composition; however, at 90°, reduction to the ion $[Am^VO_2CO_3]^-$ occurs unless $S_2O_8^{2-}$ is present. The Am^{VI}/Am^V couple in

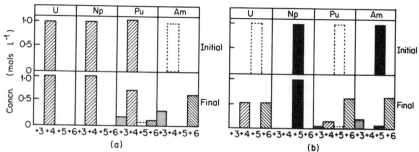

Fig. 28-6.   Disproportionation reactions of (a) tetra- and (b) penta-positive ions in $1M$ acid at 25°. [Reproduced by permission from J. J. Katz and G. T. Seaborg, *The Chemistry of the Actinide Elements*, Methuen, London, 1957.]

$0.1M$ NaHCO$_3$ is estimated to be ca. 1 V. Np^V, Pu^V and Am^V carbonate complexes can also be obtained by oxidation of dilute HNO$_3$ solutions with O$_3$, reduction of MO$_2^{2+}$ with KI, and addition of KHCO$_3$. The KM^{V}O$_2$CO$_3$ salts are isostructural, with layers held together by K$^+$ ions.

Hexacoordinate AmF$_6^{2-}$ salts also exist. The precipitation reactions of Np, Pu and Am are generally similar to those of uranium in the corresponding oxidation states, for example, of NaM^{VI}O$_2$(OCOCH$_3$)$_3$ or MF$_3$.

THE TRANS-AMERICIUM ELEMENTS

The isotope ^{242}Cm was first isolated among the products of α-bombardment of ^{239}Pu, and its discovery actually preceded that of americium. Isotopes of the other elements were first identified in products from the first hydrogen bomb explosion (1952) or in cyclotron bombardments. Although Cm, Bk and Cf have been obtained in macro amounts (Table 28-2), much of the chemical information has been obtained on the tracer scale. The remaining elements have been characterized only by their chemical behavior on the tracer scale in conjunction with their specific nuclear decay characteristics.

For these elements, the correspondence of the actinide and the lanthanide series becomes most clearly revealed. The position of curium corresponds to that of gadolinium where the f shell is half-filled. For curium, the +3 oxidation state is the normal state in solution, although, unlike gadolinium, a solid tetrafluoride, CmF$_4$, has been obtained. Berkelium has +3 and +4 oxidation states, as would be expected from its position relative to terbium, but the +4 state of terbium does not exist in solution whereas for Bk it does.

The remaining elements, from Cf onward, have only the +3 state. The great similarity between the +3 ions of Am and the trans-americium elements has meant that the more conventional chemical operations successful for the separation of the previous actinide elements are inadequate and most of the separations require the highly selective procedures of ion-exchange discussed below; solvent-extraction of the M^{3+} ions from 10–16M nitric acid by tributyl phosphate also gives reasonable separations.

28-28. Ion-exchange Separations

Ion-exchange has been indispensable in the characterization of the trans-americium elements and is also important for some of the preceding elements, particularly for tracer quantities of material. We have seen in the case of the lanthanides (Chapter 27) that the +3 ions can be eluted from a cation-exchange column by various complexing agents, such as buffered citrate, lactate or 2-hydroxybutyrate solutions, and that the elution order follows the order of the hydrated ionic radii so that the lutetium is eluted first and lanthanum last.

By detailed comparison with the elution of lanthanide ions and by extrapolating data for the lighter actinides such as Np^{3+} or Pu^{3+}, the order of elution of the heavier actinides can be accurately forecast. Even a few *atoms* of the element can be identified because of the characteristic nuclear radiation.

The main problems in the separations are (*a*) separation of the actinides as a group from the lanthanide ions (which are formed as fission fragments in the bombardments which produce the actinides) and (*b*) separation of the actinide elements from one another.

The former problem can be solved by the use of concentrated hydrochloric acid as eluting agent; since the actinide ions form chloride complexes more easily, they are desorbed first from a cation-exchange resin, thus effecting a *group* separation; conversely, the actinides are more strongly adsorbed on anion-exchange resins. Although some of the actinide ions are themselves separated in the concentrated hydrochloric acid elutions on cationic columns, the resolution is not too satisfactory, particularly for Cf and Es. A more effective group separation employs $10M$ LiCl as eluant for a moderately cross-linked, strongly basic anion-exchange column operating at elevated temperatures (up to $\sim 90°$). In addition to affording a lanthanide–actinide separation, fractionation of the actinide elements into groups Pu, Am–Cm, Bk, and Cf–Es can be obtained. Except for unexplained reversals observed in the elution order of Gd and Ho, and of Cm and Es, the elution sequences proceed in the order of increasing Z, with La the least strongly absorbed.

The actinide ions are effectively separated from each other by elution with citrate or similar elutant; some typical elution curves in which the relative positions of the corresponding lanthanides are also given are displayed in Fig. 28-7. It will be noted that a very striking similarity occurs in the spacings of corresponding elements in the two series. There is a distinct break between Gd and Tb and between Cm and Bk, which can be attributed to the small change in ionic radius occasioned by the half-filling of the $4f$ and $5f$ shells, respectively. The elution order is not always as regular as that in Fig. 28-7.

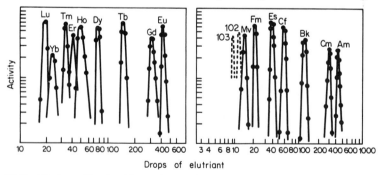

Fig. 28-7.   Elution of lanthanide $+3$ ions (left) and actinide $+3$ ions (right) from Dowex 50 cation-exchange resin. Buffered ammonium 2-hydroxybutyrate was the eluant. The predicted positions of elements 102 and 103 (unobserved here) are shown by broken lines. [Reproduced by permission from J. J. Katz and G. T. Seaborg, *The Chemistry of the Actinide Elements*, Methuen, London, 1957.]

With some complexing agents, e.g., thiocyanate, more complicated elution orders are found but nevertheless useful purifications, e.g., of Am, can be developed.[45]

After separation by ion-exchange, the actinides may be precipitated by fluoride or oxalate in macroscopic amounts or collected on an insoluble fluoride precipitate for trace quantities.

28-29. Compounds of the Elements

Curium. Solid curium compounds are known, for example, CmF_3, CmF_4, $CmCl_3$, $CmBr_3$, white Cm_2O_3 (m.p. 2265°)[46] and black CmO_2. Where X-ray structural studies have been made—and these are difficult since amounts of the order of 0.5×10^{-6} g must be used in order to avoid fogging of the film by radioactivity and because of destruction of the lattice by emitted particles—the compounds are isomorphous with other actinide compounds.

In view of the position of Cm in the actinide series, numerous experiments have been made to ascertain if Cm has only the $+3$ state in solution; no evidence for a lower state has been found. Concerning the $+4$ state, the potential of the Cm^{4+}/Cm^{3+} couple must be greater than that of Am^{4+}/Am^{3+}, which is 2.6 to 2.9 V, so that solutions of Cm^{4+} must be unstable. When CmF_4, prepared by dry fluorination of CmF_3, is treated with $15M$ CsF at 0°, a pale yellow solution is obtained which appears to contain Cm^{4+} as a fluoro complex. The solution exists for only an hour or so at 10° owing to reduction by the effects of α-radiation; its spectrum resembles that of the isoelectronic Am^{3+} ion.

The solution reactions of Cm^{3+} closely resemble the lanthanide and actinide $+3$ ions, and the fluoride, oxalate, phosphate, iodate and hydroxide are insoluble. There is some evidence for complexing in solution, although the complexes appear to be weaker than those of preceding elements.

Magnetic measurements on CmF_3 diluted in LaF_3 and also the close resemblance of the absorption spectra of CmF_3 and GdF_3 support the hypothesis that the ion has the $5f^7$ configuration.

Berkelium. As the analog of Tb, berkelium could be expected to show the $+4$ state and does so, not only in solid compounds but also in solution.[47a] Thus Bk^{3+} solutions can be oxidized by BrO_3^-, and the Bk^{4+} ion can be co-precipitated with Ce^{4+} or Zr^{4+} as phosphate or iodate. The ion can also be extracted (cf. Ce^{4+}) by hexane solutions of bis-(2-ethylhexyl) hydrogen phosphate or similar complexing agents. Bk^{IV} is a somewhat weaker oxidant than Ce^{IV}.[47b]

[45] See, e.g., J. S. Coleman, L. B. Asprey and R. C. Chisholm, *J. Inorg. Nuclear Chem.*, 1969, **31**, 1166.
[46] P. K. Smith, *J. Inorg. Nuclear Chem.*, 1969, **31**, 241; for the Cm–O₂ system, see T. Chikalla and L. Eyring, *J. Inorg. Nuclear Chem.*, 1969, **31**, 85.
[47] (a) R. C. Propst and M. L. Hyder, *J. Inorg. Nuclear Chem.*, 1970, **32**, 2205; (b) B. Weaver and J. N. Stevenson, *J. Inorg. Nuclear Chem.*, 1971, **33**, 1877.

Comparatively few solid compounds are known, viz., BkF_3 (LaF_3 type), $BkCl_3$ (UCl_3 type), $BkOCl$,[48a] BkF_4 (UF_4 type),[48b] Bk_2O_3 and BkO_2 (fluorite type)[49a] and the hydrate $[BkCl_2(H_2O)_6]Cl$.[49b] The resemblance to Ce^{IV} is also shown by the isolation of orange Cs_2BkCl_6; this is obtained by action of $CsCl$ and Cl_2 on a concentrated HCl solution of the green hydrous oxide, which is in turn obtained by oxidizing Bk^{III} in $2M$ H_2SO_4 with BrO_3^- at 90° and adding ammonia.[50]

Californium. The only evidence for the existence of an oxidation state higher than $+3$, possibly Cf^V in view of the fact that the $5f$ shell would be half-filled in this state, comes from electrolytic oxidation.[51]

A number of solid compounds have been made, e.g., Cf_2O_3, $CfCl_3$, $CfOCl$, the corresponding bromides and iodides, and Cf_2S_3.[52]

Einsteinium. The crystal structures of $EsCl_3$ and $EsOCl$ have been determined[53] and the absorption spectrum of Es^{3+} recorded.

Elements 100–103 and the Trans-actinides. The elements 100–103 appear to have the $+3$ state, but the $+2$ state seems to be more stable than the $+2$ state at the end of the lanthanide series. Thus Md^{3+} is readily reduced to Md^{2+}, and for No the $+2$ state is the most stable. The last actinide, lawrencium, appears to be most stable in the $+3$ state (cf. lutetium).[54]

The elements from 104 onwards should show characteristic group behavior and indeed ion-exchange studies[55] with ^{261}Rf indicate hafnium-like behavior.

There is still dispute regarding the first discovery of an isotope of element 104. The Russian claims appear to be inconsistent and are not supported.[56]

Although at the present time there is no confirmed report of heavier elements, a claim for the isolation of an isotope of element 112 having been withdrawn, predictions of the nuclear properties suggest that various isotopes would have long lives provided they can be made.

[48] (a) J. R. Peterson and B. B. Cunningham, *J. Inorg. Nuclear. Chem.*, 1968, **30**, 823, 1775; (b) T. K. Keenan and L. B. Asprey, *Inorg. Chem.*, 1969, **8**, 235.
[49] (a) R. D. Baybarz, *J. Inorg. Nuclear Chem.*, 1968, **30**, 1769; (b) J. H. Burns and J. R. Peterson, *Inorg. Chem.*, 1971, **10**, 147.
[50] L. R. Morss and J. Fuger, *Inorg. Chem.*, 1969, **8**, 1433.
[51] R. L. Propst and M. L. Hyder, *Nature*, 1969, **221**, 1142.
[52] S. Fried, D. Cohen, S. Siegel and B. Tani, *Inorg. Nuclear Chem. Lett.*, 1968, **4**, 495.
[53] D. K. Fujita, B. B. Cunningham, T. C. Parsons and J. R. Peterson, *J. Inorg. Nuclear Chem.*, 1969, **31**, 245, 307.
[54] R. Silva, T. Sikkeland, N. Nurmia and A. Ghiorso, *Inorg. Nuclear Chem. Lett.*, 1970, **6**, 733.
[55] A. Ghiorso et al., *Inorg. Nuclear Chem. Lett.*, 1970, **6**, 871; *Nature*, 1971, **229**, 603.
[56] (a) I. Zvara et al., (b) A. Ghiorso et al., *Inorg. Nucl. Chem. Lett.*, 1971, **7**, 1109, 1117.

Further Reading

Actinide Reviews, A. H. W. Aten, Jr., ed., Elsevier, Vol. 1, 1968 (various topics including nuclear properties, separations, chemistry, etc.).
Asprey, L. B., and R. A. Penneman, *Chem. Eng. News*, 1967, July 31, p. 75 (a good general review).
Bagnall, K. W., *Coordination Chem. Rev.*, 1967, **2**, 145 (coordination chemistry of actinide halides).

Bagnall, K. W., ed., *Lanthanides and Actinides*, Butterworth, 1972.

Brown, D., *Halides of Lanthanides and Actinides*, Wiley-Interscience, 1968 (an exhaustive monograph).

Comyns, A. E., *Chem. Rev.*, 1960, **60**, 115 (a comprehensive review of the coordination chemistry of the actinides).

Friedlander, G., J. W. Kennedy and J. M. Miller, *Nuclear and Radiochemistry*, 2nd edn., Wiley, New York, 1964 (discussion of radioactivity and nuclear stability).

Hyde, E. K., I. Perlman and G. T. Seaborg, *The Nuclear Properties of the Heavy Elements*, Vols. I–III, Prentice-Hall, New Jersey, 1964 (comprehensive reference treatise on nuclear structure, radioactive properties, and fission).

Katz, J. J., and G. T. Seaborg, *The Chemistry of the Actinide Elements*, Methuen, London, 1957 (a lucidly written reference text).

Keller, C., *Angew. Chem. Internat. Edn.*, 1965, **4**, 903 (synthesis of trans-curium elements by heavy-ion bombardments).

Keller, C., *The Chemistry of the Transuranium Elements*, Verlag Chemie, 1971 (comprehensive monograph).

Lanthanide/Actinide Chemistry, Advances in Chemistry Series No. 71, Amer. Chem. Soc., Washington D.C., 1967 (symposium reports).

Makarov, E. S., *Crystal Chemistry of Simple Compounds of* U, Th, Pu, *and* Np, (transl.) Consultants Bureau, New York, 1959.

Martin, F. S., and G. L. Miles, *Chemical Processing of Nuclear Fuels*, Butterworths, London, 1958 (procedures for isolating actinides from reactor fuels).

Oetting, F., *Chem. Rev.*, 1967, **67**, 261 (thermodynamic properties of plutonium compounds).

Rabinowitch, E., and R. Linn Belford, *Spectroscopy and Photochemistry of Uranyl Compounds*, Macmillan, New York, 1964 (extensive review).

Rand, M. H., and O. Kubaschewski, *The Thermodynamic Properties of Uranium Compounds*, Wiley, New York, 1963.

Roberts, L. E. J., *Quart. Rev.*, 1961, **15**, 442 (the actinide oxides).

Seaborg, G. T., *Man-Made Transuranium Elements*, Prentice-Hall, 1963 (lucid and well illustrated introduction enriched with historical development).

Yaffé, L., ed., *Nuclear Chemistry*, 2 vols., Academic Press, 1968.

Appendix

ENERGY LEVEL DIAGRAMS

Here and on the following two pages is a complete set of semiquantitative energy level diagrams for the d^2–d^8 configurations in octahedral symmetry. (After Tanabe and Sugano, *J. Phys. Soc. Japan*, **9**, 753 (1954).)

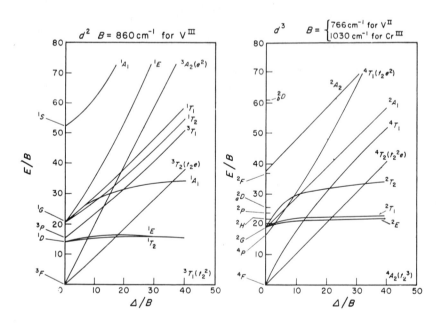

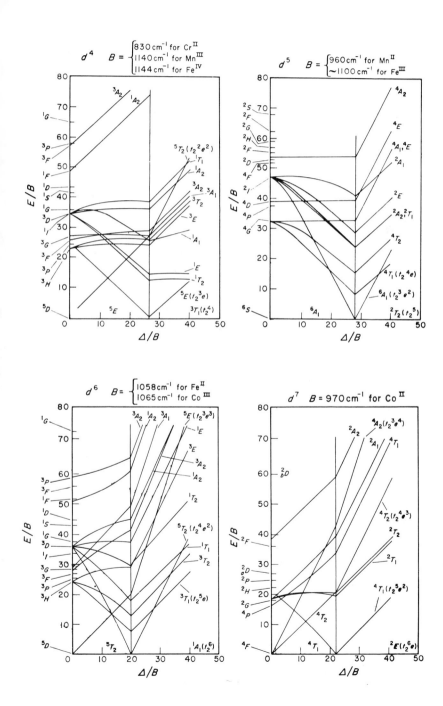

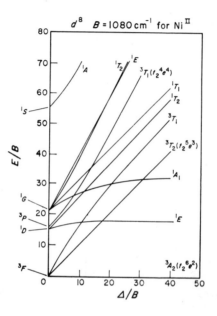

General References

The references and reading lists within each of the chapters are intended to cover the secondary literature, that is reviews and monographs relating specifically to the topics in the chapter. Here we mention some more general sources of information in inorganic chemistry.

Comprehensive Compendia

Gmelin's Handbuch der anorganischen Chemie, Verlag Chemie, Weinheim. This continuing series contains exhaustive treatments of the chemistry of all the elements with the exception of that part of carbon chemistry conventionally designated as organic chemistry. Some volumes are old but supplementary volumes (which run about a decade behind current literature) are issued regularly. Newer volumes have subheadings in English.

Mellor's Comprehensive Treatise on Inorganic and Theoretical Chemistry, Longmans, Green, London and Wiley, New York. Less extensive or up to date than Gmelin, but recently supplements have again begun to appear.

Pascal's Nouveau Traité de Chimie Minérale, Masson, Paris. Similar to Gmelin though less comprehensive. Recent volumes are more up to date and critical.

Shorter Compendia and Reference Texts

Brauer, G., ed., *Handbook of Preparative Inorganic Chemistry*, 2nd ed., Academic Press, New York, 1963. Contains descriptions of techniques and tested methods of preparation for many compounds.

Ferraro, J. R., *Low Frequency Vibrations of Inorganic and Coordination Compounds*, Heyden, 1971.

Hampel, E. F., ed., *The Encyclopedia of the Chemical Elements*, Reinhold, New York. Especially good on occurrence, refinement, uses and physical properties of the individual elements.

Moody, G. J., and J. D. R. Thomas, *Dipole Moments in Inorganic Chemistry*, Edward Arnold, 1971.

Nakamoto, K., *Infrared Spectra of Inorganic and Coordination Compounds*, 2nd ed., Wiley–Interscience, 1970.

Rao, C. N. R., and J. R. Ferraro, *Spectroscopy in Inorganic Chemistry*, Vol. 1, 1970, Academic Press.

Remy, H., *Treatise on Inorganic Chemistry* (translated by J. S. Anderson), Vols. I, II, Elsevier, New York, 1956. This book contains much detailed information on the elements and their compounds.

Shriver, D. F., *The Manipulation of Air Sensitive Compounds*, McGraw-Hill, 1969.

Sidgwick, N. V., *The Chemical Elements and Their Compounds* (2 Volumes), Oxford University Press, 1950. A unique and personal book worth perusal by serious students. Much of the interpretative discussion is now quite dated.

Sneed, M. C., J. L. Maynard and R. C. Brasted (with other contributors), *Comprehensive Inorganic Chemistry*, Van Nostrand, New York. A series of volumes on the elements, their compounds and some other selected topics which are useful sources of information but less exhaustive than others listed.

Wells, A. F., *Structural Inorganic Chemistry*, 3rd ed., Clarendon Press, Oxford, 1962. The third edition of this well-known book is an exceedingly comprehensive source book for experimental methods of structural chemistry and detailed solid-state structures of oxides, sulfides, silicates, metals and alloys, etc., as well as of compounds of a number of the elements. It can be strongly recommended as general reading for the student.

Winnacker, K., and L. Küchler, eds., *Chemische Technologie*, 3rd ed., Vols. 1, 2, Carl Hanser Verlag, Münich, 1969. Technology of inorganic compounds.

Nomenclature of Inorganic Chemistry, 2nd ed., IUPAC, Butterworth, 1971.

Annual Reviews and Series

Carlin, R. L., ed., *Progress in Transition Metal Chemistry*, Dekker, New York, Vol. 1, 1965.

Eméleus, H. J., and A. G. Sharpe, eds., *Advances in Inorganic Chemistry and Radiochemistry*, Academic Press, Vol. 13, 1970. Annual volumes.

Jolly, W. L., ed., *Preparative Inorganic Reactions*, Interscience–Wiley, Vol. 7, 1971. A continuing series of volumes describing and illustrating preparative methods. Compounds are treated by classes.

Jonassen, H. G., and A. Weissberger, eds., *Technique of Inorganic Chemistry*, Interscience–Wiley, New York. A series of volumes containing articles on particular experimental methods.

Lippard, S. J., ed., *Progress in Inorganic Chemistry*, Interscience–Wiley, Vol. 14, 1971. Annual volumes.

Annual Reports, Quarterly Reviews and Specialist Periodical Reports of the Chemical Society, London, provide frequent up to date, though not necessarily exhaustive reviews. Of particular interest are the Specialist Reports: *Spectroscopic Properties of Inorganic and Organometallic Compounds*, Vol. 1, 1968, which includes n.m.r., n.q.r. and other spectroscopy; *Electronic Structure and Magnetism of Inorganic Compounds*, Vol. 1, 1972; *Inorganic Reaction Mechanisms*, Vol. 1, 1972; *Inorganic Chemistry of Transition Elements*, Vol. 1, 1972; *Organometallic Chemistry*, Vol. 1, 1972; *Chemical Society Reviews*, which is the new Chem. Soc. publication combining (in 1972) the Quart. Rev., and the R.I.C. reviews.

Inorganic Syntheses, McGraw-Hill, Vol. 13, 1972. An annual series giving checked, detailed synthetic procedures of important inorganic compounds and brief descriptions of their properties.

MTP International Review of Science, Inorganic Chemistry Series One; University Park Press—Butterworths. Reference Series 1971-1972.

Index

M

Periodic Table of the Elements

Period	Group Ia	Group IIa	Group IIIa	Group IVa	Group Va	Group VIa	Group VIIa	Group VIII			Group Ib	Group IIb	Group IIIb	Group IVb	Group Vb	Group VIb	Group VIIb	Group O
1 1s	1 H																1 H	2 He
2 2s2p	3 Li	4 Be											5 B	6 C	7 N	8 O	9 F	10 Ne
3 3s3p	11 Na	12 Mg											13 Al	14 Si	15 P	16 S	17 Cl	18 Ar
4 4s3d 4p	19 K	20 Ca	21 Sc	22 Ti	23 V	24 Cr	25 Mn	26 Fe	27 Co	28 Ni	29 Cu	30 Zn	31 Ga	32 Ge	33 As	34 Se	35 Br	36 Kr
5 5s4d 5p	37 Rb	38 Sr	39 Y	40 Zr	41 Nb	42 Mo	43 Tc	44 Ru	45 Rh	46 Pd	47 Ag	48 Cd	49 In	50 Sn	51 Sb	52 Te	53 I	54 Xe
6 6s (4f) 5d 6p	55 Cs	56 Ba	57* La	72 Hf	73 Ta	74 W	75 Re	76 Os	77 Ir	78 Pt	79 Au	80 Hg	81 Tl	82 Pb	83 Bi	84 Po	85 At	86 Rn
7 7s (5f) 6d	87 Fr	88 Ra	89** Ac															

*Lanthanide series 4f	58 Ce	59 Pr	60 Nd	61 Pm	62 Sm	63 Eu	64 Gd	65 Tb	66 Dy	67 Ho	68 Er	69 Tm	70 Yb	71 Lu
**Actinide series 5f	90 Th	91 Pa	92 U	93 Np	94 Pu	95 Am	96 Cm	97 Bk	98 Cf	99 Es	100 Fm	101 Md	102 No	103 Lr